D1376776

The chemistry of the
cyano group

THE CHEMISTRY OF FUNCTIONAL GROUPS

A series of advanced treatises under the general editorship of Professor Saul Patai

The chemistry of alkenes (published in 2 volumes)
The chemistry of the carbonyl group (published)
The chemistry of the ether linkage (published)
The chemistry of the amino group (published)
The chemistry of the nitro and nitroso groups (published in 2 parts)
The chemistry of carboxylic acids and esters (published)
The chemistry of the carbon–nitrogen double bond (published)
The chemistry of the cyano group (published)

$-C\equiv N$

The chemistry of the cyano group

Edited by

ZVI RAPPOPORT

The Hebrew University of Jerusalem

1970

INTERSCIENCE PUBLISHERS

a division of John Wiley & Sons

LONDON—NEW YORK—SYDNEY—TORONTO

Library of Congress Catalogue Card No. 70–116165
ISBN 0 471 70913 1

Printed in Northern Ireland at The Universities Press Belfast

This volume is dedicated to the
memory of my father

Foreword

Over twenty years have passed since the appearance of Migrdichian's treatise on cyano compounds. During these decades many new developments have emerged such as the new area of 'cyanocarbon chemistry' or the complex formation by cyano and polycyano compounds, and extensive mechanistic knowledge regarding the reactions and behaviour of cyano compounds accumulated. While some of the new topics have been briefly reviewed elsewhere it was felt that the publication of a comprehensive volume covering most of the new aspects of the chemistry of the cyano group would now be justified, and I hope that this volume fulfills such expectations.

As in the other books of the series 'The Chemistry of Functional Groups' the authors were asked not to present an encyclopaedic coverage of all known reactions and compounds, but to emphasize mechanistic aspects and recent advances in the field. Somewhat exceptional is the detailed chapter on 'Cyanocarbon and polycyano compounds' which gives almost all the information available on this fascinating class of compounds.

The editor is glad that all the invited chapters materialized and believes that the slight postponement of the publishing date caused by the late arrival of several chapters was well worth the delay.

I acknowledge with great pleasure the advice and the cooperation of Professor Saul Patai of the Hebrew University. Thanks are also due to Professor Lennart Eberson of the University of Lund who read and commented on several of the chapters, as well as to the editorial and production staff of the publisher for their help in the production of the book.

Jerusalem, January 1970. ZVI RAPPOPORT

The Chemistry of the Functional Groups
Preface to the series

The series 'The Chemistry of the Functional Groups' is planned to cover in each volume all aspects of the chemistry of one of the important functional groups in organic chemistry. The emphasis is laid on the functional group treated and on the effects which it exerts on the chemical and physical properties, primarily in the immediate vicinity of the group in question, and secondarily on the behaviour of the whole molecule. For instance, the volume *The Chemistry of the Ether Linkage* deals with reactions in which the C–O–C group is involved, as well as with the effects of the C–O–C group on the reactions of alkyl or aryl groups connected to the ether oxygen. It is the purpose of the volume to give a complete coverage of all properties and reactions of ethers in as far as these depend on the presence of the ether group, but the primary subject matter is not the whole molecule, but the C–O–C functional group.

A further restriction in the treatment of the various functional groups in these volumes is that material included in easily and generally available secondary or tertiary sources, such as Chemical Reviews, Quarterly Reviews, Organic Reactions, various 'Advances' and 'Progress' series as well as textbooks (i.e. in books which are usually found in the chemical libraries of universities and research institutes) should not, as a rule, be repeated in detail, unless it is necessary for the balanced treatment of the subject. Therefore each of the authors is asked *not* to give an encyclopaedic coverage of his subject, but to concentrate on the most important recent developments and mainly on material that has not been adequately covered by reviews or other secondary sources by the time of writing of the chapter, and to address himself to a reader who is assumed to be at a fairly advanced post-graduate level.

With these restrictions, it is realized that no plan can be devised for a volume that would give a *complete* coverage of the subject with *no* overlap between chapters, while at the same time preserving the readability of the text. The Editor set himself the goal of attaining *reasonable* coverage with *moderate* overlap, with a minimum of cross-references between the chapters of each volume. In this manner, sufficient freedom is given to each author to produce readable quasi-monographic chapters.

The general plan of each volume includes the following main sections:

(a) An introductory chapter dealing with the general and theoretical aspects of the group.

(b) One or more chapters dealing with the formation of the functional group in question, either from groups present in the molecule, or by introducing the new group directly or indirectly.

(c) Chapters describing the characterization and characteristics of the functional groups, i.e. a chapter dealing with qualitative and quantitative methods of determination including chemical and physical methods, ultraviolet, infrared, nuclear magnetic resonance, and mass spectra; a chapter dealing with activating and directive effects exerted by the group and/or a chapter on the basicity, acidity or complex-forming ability of the group (if applicable).

(d) Chapters on the reactions, transformations and rearrangements which the functional group can undergo, either alone or in conjunction with other reagents.

(e) Special topics which do not fit any of the above sections, such as photochemistry, radiation chemistry, biochemical formations and reactions. Depending on the nature of each functional group treated, these special topics may include short monographs on related functional groups on which no separate volume is planned (e.g. a chapter on 'Thioketones' is included in the volume *The Chemistry of the Carbonyl Group*, and a chapter on 'Ketenes' is included in the volume *The Chemistry of Alkenes*). In other cases, certain compounds, though containing only the functional group of the title, may have special features so as to be best treated in a separate chapter as e.g. 'Polyethers' in *The Chemistry of The Ether Linkage*, or 'Tetraaminoethylenes' in *The Chemistry of the Amino Group*.

This plan entails that the breadth, depth and thought-provoking nature of each chapter will differ with the views and inclinations of the author and the presentation will necessarily be somewhat uneven. Moreover, a serious problem is caused by authors who deliver their manuscript late or not at all. In order to overcome this problem at least to some extent, it was decided to publish certain volumes in several parts, without giving consideration to the originally planned logical order of the chapters. If after the appearance of the originally planned parts of a volume, it is found that either owing to non-delivery of chapters, or to new developments in the subject, sufficient material has accumulated for publication of an additional part, this will be done as soon as possible.

The overall plan of the volumes in the series 'The Chemistry of the Functional Groups' includes the titles listed below:

The Chemistry of the Alkenes (published in two volumes)
The Chemistry of the Carbonyl Group (Volume 1 *published, Volume* 2 *in preparation)*
The Chemistry of the Ether Linkage (published)
The Chemistry of the Amino Group (published)
The Chemistry of the Nitro and Nitroso Group (published in two parts)
The Chemistry of Carboxylic Acids and Esters (published)
The Chemistry of the Carbon–Nitrogen Double Bond (published)
The Chemistry of the Cyano Group (published)
The Chemistry of the Amides (in press)
The Chemistry of the Carbon–Halogen Bond
The Chemistry of the Hydroxyl Group (published)
The Chemistry of the Carbon–Carbon Triple Bond
The Chemistry of the Azido Group (in preparation)
The Chemistry of Imidoates and Amidines
The Chemistry of the Thiol Group
The Chemistry of the Hydrazo, Azo and Azoxy Groups
The Chemistry of Carbonyl Halides (in preparation)
The Chemistry of the SO, SO_2, $—SO_2H$ *and* $—SO_3H$ *Groups*
The Chemistry of the $—OCN$, $—NCO$ *and* $—SCN$ *Groups*
The Chemistry of the $—PO_3H_2$ *and Related Groups*

Advice or criticism regarding the plan and execution of this series will be welcomed by the Editor.

The publication of this series would never have started, let alone continued, without the support of many persons. First and foremost among these is Dr. Arnold Weissberger, whose reassurance and trust encouraged me to tackle this task, and who continues to help and advise me. The efficient and patient cooperation of several staff-members of the Publisher also rendered me invaluable aid (but unfortunately their code of ethics does not allow me to thank them by name). Many of my friends and colleagues in Jerusalem helped me in the solution of various major and minor matters and my thanks are due especially to Prof. Y. Liwschitz, Dr. Z. Rappoport and Dr. J. Zabicky. Carrying out such a long-range project would be quite impossible without the non-professional but none the less essential participation and partnership of my wife.

SAUL PATAI

The Hebrew University,
Jerusalem, ISRAEL

Contents

1. Electronic structure in molecules, especially in cyano compounds 1
E. Clementi

2. Introduction of the cyano group into the molecule 67
Klaus Friedrich and Kurt Wallenfels

3. Basicity, hydrogen bonding and complex formation 123
Just Grundnes and Peter Klaboe

4. Detection and determination of nitriles 167
David J. Curran and Sidney Siggia

5. Directing and activating effects of the cyano group 209
William A. Sheppard

6. Nitrile reactivity 239
Fred C. Schaefer

7. Reduction of the cyano group 307
Mordecai Rabinovitz

8. Additions to the cyano group to form heterocycles 341
A. I. Meyers and J. C. Sircar

9. Cyanocarbon and polycyano compounds 423
E. Ciganek, W. J. Linn and O. W. Webster

10. Molecular complexes of polycyano compounds 639
L. Russell Melby

11. Radicals with cyano groups 671
H. D. Hartzler

12. The biological function and formation of the cyano group 717
J. P. Ferris

13. Syntheses and uses of isotopically labelled cyanides 743
Louis Pichat

14. Nitrile oxides 791
Ch. Grundmann

15. Isonitriles 853
Peter Hoffmann, Dieter Marquarding, Helmut Kliimann and Ivar Ugi

16. Rearrangement reactions involving the cyano group 885
Joseph Casanova, Jr.

Author index 947

Subject index 1013

Contributing Authors

Joseph Casanova Jr.	California State College, Los Angeles, California, U.S.A.
E. Ciganek	E. I. du Pont de Nemours, Wilmington, Delaware, U.S.A.
E. Clementi	I.B.M. Research Laboratory, San Jose, California, U.S.A.
David J. Curran	University of Massachusetts, Amherst, Massachusetts, U.S.A.
James P. Ferris	Rensselaer Polytechnic Institute, Troy, New York, U.S.A.
Klaus Friedrich	University of Freiburg, Freiburg, Germany.
Ch. Grundmann	Mellon Institute, Pittsburgh, Pennsylvania, U.S.A.
Just Grundnes	University of Oslo, Norway.
H. D. Hartzler	E.I. du Pont de Nemours, Wilmington, Delaware, U.S.A.
Peter Hoffmann	E.I. du Pont de Nemours, Wilmington, Delaware, U.S.A.
Peter Klaboe	University of Oslo, Norway.
Helmut Kliimann	Bayer AG, Leverkusen, Germany.
W. J. Linn	E.I. du Pont de Nemours, Wilmington, Delaware, U.S.A.
Dieter Marquarding	Bayer AG, Leverkusen, Germany.
L. Russell Melby	E.I. du Pont de Nemours, Wilmington, Delaware, U.S.A.
A. I. Meyers	Louisiana State University, New Orleans, Louisiana, U.S.A.
Louis Pichat	Centre d'Etudes Nucléaires de Saclay, Gif-sur-Yvette, France.
Mordecai Rabinovitz	The Hebrew University, Jerusalem, Israel.
Fred C. Schaefer	American Cyanamid Co., Stamford, Connecticut, U.S.A.
William A. Sheppard	E.I. du Pont de Nemours, Wilmington, Delaware, U.S.A.
S. Siggia	University of Massachusetts, Amherst, Massachusetts, U.S.A.
J. C. Sircar	Louisiana State University, New Orleans, Louisiana, U.S.A.
Kurt Wallenfels	University of Freiburg, Freiburg, Germany.
O. W. Webster	E.I. du Pont de Nemours, Wilmington, Delaware, U.S.A.
Ivar Ugi	Bayer AG, Leverkusen, Germany.

CHAPTER I

Electronic structure in molecules, especially in cyano compounds

E. CLEMENTI

IBM Research Laboratory, San Jose, California, U.S.A.

I. INTRODUCTION 1
II. MULTICONFIGURATION SELF-CONSISTENT-FIELD THEORY . . 10
III. ELECTRON POPULATION ANALYSIS 20
IV. BOND ENERGY ANALYSIS 23
V. CORRELATION ENERGY AND STATISTICAL METHODS . . . 27
VI. ANALYSIS OF THE 'EXPERIMENTAL' CORRELATION ENERGY . . 35
VII. MOLECULAR EXTRA CORRELATION ENERGY 43
VIII. REMARKS ON THE ELECTRONIC STRUCTURE OF THE —CN GROUP . 46
IX. REFERENCES 63

I. INTRODUCTION

It is clearly impossible to review the field of theoretical chemistry in one chapter, even if it is a long one. Therefore, it has been decided to concentrate on a few points. An introduction is given in which a sketch is presented of two dominant theories of the last 40 years and a step by step development of the molecular-orbital theory is briefly exposed. A second section is devoted to the most promising avenue available today, namely, the multiconfiguration self-consistent-field theory. Perhaps the impatient reader should only concentrate on the chemical implications following equation (22). The third and fourth sections are added to give the basis of a popularized version of electronic density and energy in molecules. The fifth, sixth and seventh sections are devoted to the correlation energy problem. After the reader will have suffered through such sections, it is hoped he will be in a position to feel confident in 'guessing' the correlation energy of any molecule in which he is interested.

A final section is added with direct reference to the —CN group. With this in mind, we can start, after offering an apology for planning to confront the organic chemist with many equations. This is done not because of any sadistic feeling, but because the author is convinced that: (*a*) theoretical chemistry has made finally those steps which allows it to be a predictive tool, (*b*) that there is a great need of more exchange between those of us who use theory and those of us who work mostly in a laboratory. The chemistry of tomorrow will certainly be less and less the empirical science that we know today, and to achieve this goal, theory and experiment should intermix more and more.

Whereas it is pointless to remind ourselves that molecules are built up of atoms, there are still divergent opinions among us chemists concerning the extent of the usefulness in formulating theory whereby a molecule is assumed from the start to be built from a set of atoms or ions (of course somewhat deformed in their electronic structure).

Pauling[1] in the early thirties did very much with the basic assumption that it is worthwhile to retain the electronic structure of the atoms as the starting point in the electronic structure of molecules. Pauling's theory is referred to as the valence-bond approximation. An x-ray analysis of a chemical compound seems to be the clearest proof that indeed this is the best conceptual approach.

On the opposite side, Mulliken[2] and his school immediately reject the necessity of the above starting point by introducing from the onset the concept of 'molecular orbitals', i.e. of electrons which share the space around not a single nucleus, but as many as energetically convenient.

Attempts to put together the two points of view have been made frequently. For example, one can go back to Moffitt's 'atoms in molecules' technique[3].

Others attempted to define the range in which one approach is better than the other. Here we refer, for example, to Kotani's analysis[4].

Clearly, an appropriate extension of Pauling's viewpoint (not existing today) and presently available extensions of the 'molecular orbital' approximation should give identical results. We refer to Coulson's book[5] on this subject (but it must be pointed out that

Coulson's work was written prior to much work in the field of theoretical chemistry in the last decade and, therefore, does not represent an up-to-date analysis).

In order not to give an incorrect representation of the 'molecular orbital' approximation, it must be emphasized that this model does not claim that one should disregard completely the parent atoms in a given molecule, but rather that one does not have to do so; that the molecular orbitals will go naturally to atomic orbitals as the molecule dissociates. The main point is that most of the electronic structure of the separated atoms is so drastically changed by the time we are at the equilibrium configuration geometry, that it is more of a hindrance than a help to consider the separated atoms and their concomitant deformation when we study a molecule.

The concept of heavy distortion in the electronic cloud when one proceeds from the separated atoms to the molecule is very well within chemical tradition. Indeed, much before quantum theory chemists assumed (and correctly) that, for example, the carbon atom has several ways to bind corresponding to different arrangements of the surrounding atoms; thus, the concept of trigonal and tetragonal structures. Today we would rather speak of 'valence states', but this is no more than remembering that, in a given external field, the valence electrons (in particular) and the inner-shell electrons (to a smaller degree) of a particular atom experience, not only a central field of force alone, but the external field as well.

The valence-bond approximation is progressively losing its impact, particularly in theoretical chemistry and somewhat more slowly in chemistry in general. It is becoming more and more apparent that with it, any 'all-electron' treatment of a moderately complex molecule is unfeasible, even with modern high-speed computers. In addition, a full valence-bond treatment, which considers all the electrons of a molecule, introduces unreasonably highly positive and negative ions of dubious physical meaning. (For example, a valence-bond treatment of benzene will use a large number of structures, including C^{+}, C^{2+}, C^{3+}, C^{-}, C^{2-}, C^{3-}, etc.) Its appeal remains in its basic simplicity and in having brought about the concept of resonating structures, which remains a basic concept in theoretical interpretations of chemistry (however, this concept is *not* a necessary one).

The molecular-orbital theory has the advantage of being conceptually based on atomic theory, with techniques that can be tested for atoms. It is important to note that an electronic theory of molecules should in principle and in practice be applicable to the limiting case of a single atom. Electrons do not change nature going from atoms to molecules, and the same should hold for any model which describes the electronic structure either of atoms or of molecules.

We shall not elaborate further on the above differences between the two starting points. We simply wish to remind the reader of some historical evolution in theoretical chemistry. Nor shall we claim that one technique is correct and the other is incorrect, since, in principle, one can have the same results from both. Pragmatically, we shall devote our attention to the molecular-orbital approach and its improved assumptions, since via such a technique we can today perform computations of increasing sophistication, to the point that we can quantitatively predict some of the chemistry and understand some more of it.

'Æsthetical' considerations, namely economy in the number of basic assumptions, very much influence the author's choice and in addition it is frankly admitted to the reader that the author has a biased preference toward the molecular-orbital point of view.

Rather dogmatically, and in part for sake of brevity, a start is made by stating that the Schrödinger equation, foundation of today's quantum mechanics, is good enough for chemistry. It can also be stated, but in a somewhat more guarded way, that today's understanding of relativistic quantum electron dynamics is good enough for chemistry. From this it follows that it is worthwhile to attempt a solution of the Schrödinger equation. As known, this is the problem; a problem not too much of principle, but very much of practice, of substantial numerical difficulty for the parts and of extreme numerical difficulties for the whole. Therefore, let us make simplifying assumptions in a step-like fashion and let us first solve the larger part of the problem and then let us go to smaller and smaller details. The good sense and logic of the approach seem to be commendable. However, as we shall see later, this approach required a large amount of work, indeed several decades of hard work, to reach a stage with which the chemist should be most dissatisfied since, in general, no quantitative agreement with experimental data was obtained. Therefore, since the chemist is an eminently empirical scientist, continuous effort took place toward finding shortcuts, regardless of the basic theoretical validity of such attempts. Lately, another factor appeared, namely the electronic computer (and the cost of computations!). Here again the shortcuts were and are introduced on the ground that theory 'must be cheap to be good'*.

Other techniques have been proposed in addition to the valence-bond and the molecular-orbital approximation. Of special interest are those computational techniques which can be applied to molecular systems with few electrons. For example, Kolos and coworkers[6] have obtained very accurate wave functions for a number of electronic states in the H_2 molecule at many internuclear distances. Kolos' computer program can definitely not only supplement but compete with most sophisticated spectrographs. The resolution of his program is within fractions of a wave number and the use of the program presents so little difficulty as to be correctly handled by any 'technician'. The results of Kolos and coworkers provide a concrete, although partial, example of what one wishes to obtain from theoretical chemistry. However, most molecules have more than two electrons and exact wave functions are at present not easily obtainable. Nevertheless, a number of important steps have been made and there is good reason for optimism in the future.

Some of the steps which have been taken will be outlined, together with their accomplishments and limitations.

A pleasant characteristic of the molecular-orbital theory is that each progressive improvement, or step, has a natural physical explanation. Rather arbitrarily we shall present the molecular-orbital theory as a five-step evolution.

The *first step* is the LCAO–MO approximation. There are actually two approximations in the above step: the first is the MO approximation, the second is the LCAO approximation of an MO. As known, the short notation LCAO–MO stands for 'linear combination of atomic orbitals–molecular orbitals'.

The MO is a one-electron function which is factored into a spatial component and a spin component. The expression 'one-electron function' means that only the coordinates of one electron are explicitly used in a given MO. This factorization into spatial and spin

* Whereas there is no doubt that technology should be related to economy, nevertheless, one should be careful in accepting the rule which attempts to appraise the value of knowledge only or predominantly in terms of cost. The justification for such comments in this theoretical exposition of the electronic structure in molecules is prompted by the existence of a belief current among chemists which is very insensitive to the last two generations efforts in theoretical chemistry and which would not mind eliminating its support, paradoxically, now, when preliminary and solid evidences are finally available that computational chemistry is developing into a practical tool of enormous consequence for future chemistry.

components is permissible, since generally one uses a Hamiltonian which does not explicitly contain spin-dependent terms. The MO's are the exact analogue of the atomic orbitals which describe the electrons in an atom to a first approximation. Indeed, one can read several chapters of the classical works of Condon and Shortely[7], replace the word 'AO' with the word 'MO' and one will then read a book on molecular physics instead of atomic physics.

This situation has some important consequences: namely, a large amount of testing and development for molecular wave function techniques can be done with atoms. For this reason atomic and molecular examples are freely mixed in the following sections of this review.

If the molecule contains $2n$ electrons (let us consider a closed-shell case for simplicity), the MO approximation will distribute the electrons in $2n$ molecular orbitals $\varphi_1, \varphi_2 \ldots \varphi_{2n}$. Since there are two possible spin orientations (α and β spins), a space distribution function has either spin α or β and, therefore, the $2n$-electron system is described by n space functions and $2n$ spin orbitals. Thus φ_1 and φ_2 will have the same space distribution (will depend on the coordinates of one electron alone), but, in accordance with the Pauli exclusion principle, will have different spin functions. It is stressed that the one-electron model is justified only because it simplifies the treatment. Indeed, in the very beginning of quantum theory, Hylleraas introduced a wave function for the He atom in which one orbital is described in terms of the coordinates of *both* electrons.

The total wave function Ψ of the $2n$-electron system is then

$$\Psi = \frac{1}{\sqrt{(2n)!}} \begin{vmatrix} \varphi_1(1) & \cdots & \varphi_1(2n) \\ \varphi_2(1) & \cdots & \varphi_2(2n) \\ \cdot & & \cdot \\ \cdot & & \cdot \\ \cdot & & \cdot \\ \varphi_{2n}(1) & \cdots & \varphi_{2n}(2n) \end{vmatrix}$$

where the number between parentheses indicates a given electron. This determinant wave function guarantees that any interchange of two electrons (i and j) brings about a sign change in the wave function. This is the Pauli constraint for fermions. The energy for such a system is given by the relation

$$E = \langle \Psi^* | H | \Psi \rangle$$

where the Hamiltonian H is

$$H = -\sum_{ia} \tfrac{1}{2}\Delta_i^2 - \sum_{ia} \frac{Z_a}{r_i} + \sum_{ij} \frac{1}{r_{ij}} - \sum_{ab} \frac{Z_a Z_b}{R_{ab}}$$

The first term is the kinetic operator for the ith electron, the second term is the potential operator between the ith electron and the ath nucleus (with charge Z_a), the third term is the electron–electron potential between the ith and the jth electrons at a distance $r_{ij} = r_i - r_j$ and finally the last term is the nucleus–nucleus potential with R_{ab} the distance between the ath and bth nucleus of respective charges Z_a and Z_b.

The first and second terms are subsequently referred to as the one-electron Hamiltonian and will be indicated as h_0. The total energy for such a determinant was given by Slater, and it is

$$E = 2 \sum_i h_i + \sum_{ij} (2J_{ij} - K_{ij}) + E_{NN}$$

where

$$\begin{aligned} h_i &= \langle \varphi_i^* |\, h_0 \,| \varphi_i \rangle \\ J_{ij} &= \langle \varphi_i(1)^* \varphi_j(2)^* |\, r_{12}^{-1} \,| \varphi_i(1) \varphi_j(2) \rangle \\ K_{ij} &= \langle \varphi_i(1)^* \varphi_j(2)^* |\, r_{12}^{-1} \,| \varphi_i(2) \varphi_j(1) \rangle \\ E_{NN} &= \sum_{ab} (Z_a Z_b / R_{ab}) \end{aligned}$$

As known, J and K are usually referred to as Coulomb and exchange terms, respectively. The above equation can be rewritten as

$$\begin{aligned} E &= \sum_i (2h_i + J_{ii}) + \sum_{i>j} (2J_{ij} - K_{ij}) + E_{NN} \\ &\equiv \sum_i \tau_i + \sum_{i>j} \rho_{ij} + E_{NN} \end{aligned}$$

where $\tau_i \equiv 2h_i + J_{ii}$ and $\rho_{ij} = (2J_{ij} - K_{ij})$.

What form should the MO have? Clearly, the molecular orbitals are subjected to symmetry constraints (as in the case of atomic orbitals) and any molecular orbital will transform as an irreducible representation of the molecular symmetry group. This statement, however, is not a sufficient one; indeed it tells us mainly how the molecular orbital should *not* be. In principle we could insist on the analogy between atomic one-electron functions and molecular one-electron functions and 'tabulate' the MO in a way analogous to the method of Hartree and Fock in the 1930's. This would ensure that we have the best possible molecular orbitals. It is noted that numerical Hartree–Fock functions for diatomic molecules are a somewhat tempting possibility; this, however, has not seriously been explored at the present.

Nevertheless, chemistry is concerned with more than only diatomic molecules. An answer is provided by the LCAO approximation, in which the MO's are built up as linear combinations of atomic functions. We refer to Mulliken's classical series of papers for the early development and application of the LCAO–MO approximation[2].

The *second step* in the evolution of quantum theory is the introduction of self-consistency. Again, the physical model is provided by atomic physics, namely by the Hartree–Fock model. The LCAO approximation to the MO requires the best possible linear combination: this is what one intends for self-consistency. A good review paper on this subject is the one by Roothaan[8]. There the self-consistent-field technique in the LCAO–MO approximation (SCF LCAO–MO) is systematically exposed for the closed-shell case.

Up to now we are strictly in the one-electron approximation. The electrons interact among themselves only via the average field and the MO has no explicit electron–electron parameter. Fortunately, the Pauli principle keeps electrons with parallel spin (in different MO's) away from each other, but it has nothing to offer to electrons with antiparallel spin in the same MO. The full catastrophe might be appreciated by recalling that in the SCF LCAO–MO approximation, two fluorine atoms are incapable of giving molecular bonding when brought together; i.e. the SCF LCAO–MO does not recognize the existence of the F_2 molecule[9]. Of course, it does not require a computation of F_2 to realize this point. For example, when Roothaan's work appeared (1950), another less familiar paper was written by Fock[10], to a large degree solving the problem and introducing the concept of two-electron molecular functions or 'geminals', as they are called today. At the same time Lennard-Jones and collaborators[11] put forward a classical series of papers in which part of the correlation problem was tentatively solved, but at the expense of drastic orthogonality restrictions. For a variety of reasons, neither of the two avenues was numerically explored and in the meantime a third possibility slowly emerged.

Hylleraas[12], and later Boys[13], proposed the possibility of using not only one determinant, but as many as needed. This technique is known as the configuration interaction or superposition of configuration technique; since the first designation is more common, it will be adopted hereafter (CI for short).

Let us consider, for example, the case of the beryllium atom in its ground state. Considering atoms instead of molecules is appropriate, since we are really considering electrons. The electronic configuration is $1s^2 2s^2$ and therefore the Hartree–Fock function Ψ_0 is

$$\Psi_0 = \frac{1}{\sqrt{4!}} \begin{vmatrix} 1s(1) & 1s(2) & 1s(3) & 1s(4) \\ \overline{1s}(1) & \overline{1s}(2) & \overline{1s}(3) & \overline{1s}(4) \\ 2s(1) & 2s(2) & 2s(3) & 2s(4) \\ \overline{2s}(1) & \overline{2s}(2) & \overline{2s}(3) & \overline{2s}(4) \end{vmatrix} \tag{1}$$

(where the bar designates β spin).

Let us consider the following functions

$$\Psi_1 = \{1s(1)\overline{1s}(2)2p(3)\overline{2s}(4)\}$$
$$\Psi_2 = \{2s(1)\overline{2s}(2)2p(3)\overline{2p}(4)\}$$
$$\Psi_3 = \{2p(1)\overline{2p}(2)2p(3)\overline{2p}(4)\} \cdots$$

and having ensured that each Ψ has ^1S symmetry, let us build the following functions

$$\Psi = a_0\Psi_0 + a_1\Psi_1 + a_2\Psi_2 + \cdots \tag{2}$$

By optimizing the orbitals in each function and by variationally selecting the CI coefficients a_0, a_1, a_2, . . . we shall have a solution necessarily as good as, or better than, Ψ_0, and if the series of the above equations is sufficiently long, then we shall reach an exact solution. The only trouble is that the necessary series is *too* long. The slow convergence of the series is due to the fact that in most cases one insists on using a $1s$ orthogonal to the $2s$ and to the $3s$, a $2p$ orthogonal to the $3p$, etc., with $2p$, $3p$, $3s$ functions overlapping very little the $1s$ and $2s$ functions. If the added functions overlap very little, they will interact very little and correlate equivalently.

However, let us assume that when we construct Ψ_0 we construct Ψ_1, Ψ_2, etc., *at the same time*, and we do not insist on the best possible Ψ_0, but on the best possible Ψ; then the variational principle, used simultaneously on both the a's (the CI coefficients) and the φ's (the atomic orbitals), will ensure that the Ψ_i will overlap as much as possible. This is accomplished in *the multiconfiguration* SCF LCAO–MO technique (MC SCF LCAO–MO), *the third step*. Before entering into the details of the MC SCF LCAO–MO theory, let us briefly mention the fourth and fifth steps.

A *fourth step* in the molecular-orbital theory is the inclusion of relativistic effects. There is little work done in this area at present (and this is not only true for molecular functions, but for atomic functions as well). Recent advances in organometallic chemistry, with heavy metals as constituents, demand a relativistic interpretation of the electronic structure. Even in molecules containing low Z atoms, the importance of spin–orbit effects in transition intensities is demanding more studies and computations in this area. It is gratifying to note that a simple perturbation treatment on atoms (Hartmann and Clementi[14], Clementi[15]) gives energies as good as a full relativistic Hartree–Fock treatment[16].

Finally, the 'electronic structure' of molecules should always be considered a limiting case of the vibronic structure of molecules. Real molecules vibrate (and rotate and translate, too) and, therefore, the question of how much we can rely on the Born–Oppenheimer approximation should not be ignored. This is a *fifth step* and the reader will be referred to the work of Kolos and coworkers for more details[17].

II. MULTICONFIGURATION SELF-CONSISTENT-FIELD THEORY

The MC SCF theory seems to have been first proposed by Frankel[18] (1934) and by D. R. Hartree, W. Hartree and Swviles[19] (1939) and Yutsis[20] (1952). Recently, it has been reanalysed and applied by Yutsis, Vizbaraite, Strotkite and Bauzaitis[21] (1962), Veillard[22] (1966), Veillard and Clementi[23] (1966), Clementi[24] (1967), and Das and Wahl[25] (1966).

We shall first consider the simpler case of two configurations[22] and expand it later to many configurations[23].

Let us consider a configuration of the type $1s^2 2s^2 2p^n$ (called configuration A) and a configuration of the type $1s^2 2s^0 2p^{n+2}$ (called configuration B). States of like symmetry from A and B will interact and the resultant function will be (we are interested in its lowest eigenvalue)

$$\Psi = A\Psi_A + B\Psi_B \tag{3}$$

with energy

$$E = (\Psi^* | H | \Psi) = A^2 E_A + B^2 E_B + ABE_{AB}$$

The SCF theory can be used in solving first Ψ_A and then Ψ_B, and a secular equation can be solved for $\Psi = A\Psi_A + B\Psi_B$. This is standard configuration interaction. However, the problem can be solved in one step, i.e. an optimal Ψ_A and Ψ_B can be found so that when the two interact, an optimal Ψ is given. In other words, for a given basis set in Ψ_A and Ψ_B an optimal two-determinant combination can be obtained. The standard SCF guarantees an optimal Ψ_A or an optimal Ψ_B, the MC SCF guarantees an optimal Ψ, *but not an optimal* Ψ_A *or* Ψ_B. For the specific case in consideration, the E_A, E_B and E_{AB} are standard energy expressions, namely

$$E_A = \tau_{1s} + \tau_{2s} + 2\rho_{1s,2s} + f_1 2 \sum_m h_m + f_1 \sum_{mn} (2a_1 J_{mn} - b_1 K_{mn}) + 2 \sum_m \rho_{1s,m} + \sum_m \rho_{2s,m} \tag{4}$$

$$E_B = \tau_{1s} + f_2 \sum_m h_m + f_2 \sum_{mn} (2a_2 J_{mn} - b_2 K_{mn}) + 2 \sum_m \rho_{1s,m} \tag{5}$$

$$E_{AB} = c \sum_m K_{2s,m} \tag{6}$$

where τ, ρ, h, J, K have been previously defined (see section I), f_1 and f_2 are occupation numbers ($f_1 = n/6$, $f_2 = (n + 2)/6$ where n is the number of electrons in the $2p$ shell), a_1, b_1 and a_2, b_2 and c are numerical constants which ensure proper book-keeping in the energy expression and are called vector coupling coefficients. The indices m and n refer to the $2p$ orbitals.

The SCF technique is then applied, that is an infinitesimal variation is applied on each orbital in Ψ_A and Ψ_B, which brings about a variation δE in the energy. The optimal solutions are those for which $\delta E = 0$. The orbitals are constrained to be orthonormal, and the mixing coefficients A and B are subjected to the relation $A^2 + B^2 = 1$. The constraints are sufficient in number to ensure that a unique solution for the problem is found. The MC SCF technique in this respect parallels the traditional SCF technique.

Let us analyse the results for Be(^1S), B(^2P) and C(^3P) with electronic configurations $1s^2 2s^2$, $1s^2 2s^2 2p$ and $1s^2 2s^2 2p^2$. The MC SCF functions are found to be

$$\text{Be} \quad \Psi(^1S) = 1s^2[0{\cdot}9484(2s)^2 - 0{\cdot}317(2p)^2] \tag{7}$$

$$\text{B} \quad \Psi(^2P) = 1s^2[0{\cdot}9728(2s)^2 - 0{\cdot}2316(2p)^3] \tag{8}$$

$$\text{C} \quad \Psi(^3P) = 1s^2[0{\cdot}9888(2s)^2 - 0{\cdot}1490(2p)^4] \tag{9}$$

The experimental energies are −14·6685, −24·6580, −37·8557 a.u., respectively. The relativistic corrections[14,15] are computed to be −0·0022, −0·0061, −0·0138 a.u. The single-determinant energy

(Hartree–Fock) is −14·5730, −24·5290, −37·6886 a.u., respectively. When this energy is mass corrected, we have −14·5721, −24·5278, −37·6869 a.u., respectively.

The correlation energy is defined as the difference between the experimental energy and the sum of the Hartree–Fock and relativistic energies. In the following table we shall give the correlation energy (E_{Corr}) for the three atoms under consideration. Introduction

	E_{Exp}	E_{HF}	E_{Rel}	E_{Corr}
Be(^1S)	−14·6685	− (−14·5721	− 0·0022)	= −0·0942 a.u. = −2.563 eV
B(^2P)	−24·6580	− (−24·5278	− 0·0061)	= −0·1241 a.u. = −3·378 eV
C(^3P)	−37·8557	− (−37·6869	− 0·0138)	= −0·1550 a.u. = −4·217 eV

of the second configuration lowers the correlation energy error by 0·0424, 0·0311, 0·0173 a.u., respectively. The remaining error is partly due to the $1s^2$ electrons and this is about 0·0443, 0·0447, 0·0451 a.u., respectively (these values are taken from the two-electron isoelectronic series). In Table 1 we shall give the percent error of the Hartree–Fock (HF) energy, the Hartree–Fock plus relativistic correction, of the two-configuration SCF calculation plus

TABLE 1. Energy contributions to the total energy.

Atom	HF (%)	HF + R (%)	MC SCF + R (%)	MC SCF + R + E^{Corr} % (1s)	Remainder (%)
Be(^1S)	99·3428	99·3578	99·6474	99·9495	0·0505
B(^2P)	99·4720	99·4967	99·6228	99·8041	0·1959
C(^3P)	99·5541	99·5906	99·6362	99·7554	0·2446

relativistic correction, and of the two-configuration calculation plus relativistic correction and the $1s^2$ correlation energy contribution. The remaining error is 0·0074 a.u. for Be(^1S), 0·0483 a.u. for B(^2P) and 0·1126 a.u. for C(^3P). It is noted that there are two electrons with parallel spin in addition to the $1s^22s^2$ electrons in C(^3P), one unpaired electron in B(^2P). The following algebra is quite tempting: $2 \times 0{\cdot}0483 + 0{\cdot}0074 = 0{\cdot}1040$ a.u. to be compared with 0·1126 a.u. above reported. More accurately we should not use the value of 0·0074 which was obtained for the beryllium atom, but rather the value 0·0093 which is derived for $C^{2+}(^1S)$. Therefore, we have not explained the correlation effect by the amount of

$$0{\cdot}1126 - 0{\cdot}1059 = 0{\cdot}0067 \text{ a.u.}$$

This error is a cumulation of small contributions like the neglect of the relativistic energy difference between the $1s^2(a2s^22p^2 + b2p^4)$ and the $1s^22s^22p^2$ configurations, oversimplifications in the estimate of the p–p correlation correction, the use of the correlation energy for the $1s^2(C^{4+})$ in $C(^3P)$ and other small errors.

How many configurations should be added in order to obtain an accurate $Be(^1S)$ ground-state energy is not certain without a numerical check. One could expect, however, that two configurations are sufficient to give 90% of the $1s^2$ correlation (one for radial correlation, ns^22s^2, and one for angular correction, np^22s^2). If we are correct, the configurations $1s^22s^2$, ns^22s^2, np^22s^2, $1s^22p^2$ should improve the Hartree–Fock energy by about 0·08 or 0·085 a.u. (to be compared with 0·0942 a.u., the total correlation correction). A standard configuration interaction treatment would require over 20 configurations to reach this energy.

We shall now extend the MC SCF theory to the case of n configurations for a closed shell ground state.

We assume that the $2n$ electrons of a given closed-shell system are distributed in n doubly occupied orbitals $\varphi_1 \cdots \varphi_n$ and we shall refer to this set as the '(n)' set. A second set of orbitals $\varphi_{(n+1)} \cdots \varphi$ is used and this will be referred to as the '$(\omega - n)$' set. We consider *all* the possible excitations from the (n) set to the $(\omega - n)$ set, i.e. we consider $n(\omega - n)$ configurations. A given excitation from the (n) set to the $(\omega - n)$ set will be indicated as $t \rightarrow u$ where t is a number from 1 to n and u is a number from $n + 1$ to ω.

We shall designate as the complete multiconfiguration–self-consistent-field (CMC SCF) technique the one where a given orbital of the (n) set is excited to all orbitals of the $(\omega - n)$ set; if an orbital of the (n) set is excited to one or more, *but not all* orbitals of the $(\omega - n)$ set, then we shall describe the technique as *incomplete* MC SCF (IMC SCF).

In the following, the CMC formalism is described following the analysis of Veillard and Clementi[23]. (A computer program for molecules of general geometry has been coded for the CMC formalism.) The wave function of the system is

$$\Psi = a_{00}\psi_{00} + \sum_{t=1}^{n}\sum_{u=1}^{\omega-n} a_{tu}\psi_{tu} \tag{10}$$

If one wishes to exclude a number of occupied orbitals from the excitations, i.e. if a number of orbitals are left uncorrelated, then this requires simply starting the summation over the index t at some value of t larger than one. We shall use t or t' as indices for the (n) set and u or u' as indices for the $(\omega - n)$ set.

The energy corresponding to Ψ is

$$E = \langle\psi_{00}^*| H |\psi_0\rangle a_{00}^2 + \sum_{t=1}^{n}\sum_{u=1}^{\omega-n} a_{tu}^2 \langle\psi_{tu}^*| H |\psi_{tu}\rangle \tag{11}$$

$$+ \sum_{t=1}\sum_{u=1} a_{tu} \sum_{t'=1}^{n}\sum_{u'=1}^{\omega-n} a_{t'u} \langle\psi_{tu}^*| H |\psi_{t'u}\rangle$$

$$+ 2a_{00} \sum_{t=1}^{n}\sum_{u=1}^{\omega-n} a_{tu} \langle\psi_{00}^*| H |\psi_{tu}\rangle$$

$$E = a_{00}^2 E_{00} + \sum_{t}\sum_{u} a_{tu}^2 E_{tu} + \sum_{t}\sum_{u} a_{tu} \sum_{t'} a_{t'u} I_{tt'}(1 - \delta_{tt'})$$

$$+ \sum_{t}\sum_{u} a_{tu} \sum_{u'} a_{tu'} I_{uu'}(1 - \delta_{uu'}) + 2a_{00} \sum_{t}\sum a_{tu} I_{tu} \tag{12}$$

where

$$E_{00} = \sum_{t=1}^{n} 2h_t + \sum_{t=1}^{n}\sum_{t'=1}^{n} P_{tt'} \tag{13}$$

$$E_{tu} = E_{00} - 2h_t + 2h_u - \sum_{t'=1}^{n} 2P_{tt'}$$

$$+ \sum_{t=1}^{n} 2P_{tu} - 2P_{tu} + P_{uu} + P_{tt} \tag{14}$$

$$I_{ij} = K_{ij} \tag{15}$$

$$P_{ij} = J_{ij} - \tfrac{1}{2}K_{ij} \tag{16}$$

By simple algebraic manipulations the energy expression can be rewritten as

$$E = 2\sum_{t=1}^{n} h_t + \sum_{t'=1}^{n} P_{tt'} - A_t h_t + 2\sum_{t'=1}^{n} P_{tt'} - P_{tt} \tag{17}$$

$$+ \sum_{u=1}^{\omega-n} 2B_u h_u + P_{uu} + 2\sum_{t=1}^{n} P_{tu}$$

$$+ 2\sum_{t=1}^{n}\sum_{u=1}^{\omega-n} a_{tu}(a_{00}K_{tu} - 2a_{tu}P_{tu})$$

$$+ \sum_{t=1}^{n}\sum_{t'=1}^{n} A_{tt'}K_{tt'}(1 - \delta_{tt'}) + \sum_{u=1}^{\omega-n}\sum_{u'=1}^{\omega-n} B_{uu'}K_{uu'}(1 - \delta_{uu'})$$

where

$$1 = a_{00}^2 + \sum_{t=1}^{n} \sum_{u=1}^{\omega-n} a_{tu}^2 \tag{18}$$

$$A_{tt'} = \sum_{u=1}^{\omega-n} a_{tu} a_{t'u} \tag{19a}$$

$$A_t \equiv A_{tt} = \sum_{u=1}^{\omega-n} a_{tu}^2 \tag{19b}$$

$$B_{uu'} = \sum_{t=1}^{n} a_{tu} a_{tu'} \tag{20a}$$

$$B_u \equiv B_{uu} = \sum_{t=1}^{n} a_{tu}^2 \tag{20b}$$

The coefficients a_{tu}, A_u and B_u are related by the following equations

$$1 = a_{00}^2 + \sum_{t=1}^{n} A_t = a_{00}^2 + \sum_{u=1}^{\omega-n} B_u \tag{21}$$

or

$$\sum_{t=1}^{n} A_t = \sum_{u=1}^{\omega-n} B_u \tag{22}$$

The coefficient A_t represents the 'fraction of an electron' which is excited from the φ_t orbital of the (n) set to the φ_u orbitals of the entire $(\omega - n)$ set. The coefficient B_u represents the 'fraction of an electron' in the φ_u orbital of the $(\omega - n)$ set as a result of the excitation from the entire (n) set. It is, therefore, tempting to reexamine the configuration structure of a $2n$-electron system. The standard electronic configuration for the $2n$ electrons is a product of n orbitals. For example φ_{00} has configuration

$$\varphi_1^2 \varphi_2^2 \cdots \varphi_n^2$$

Let us call such a configuration a 'zero-order electronic configuration'. The MC SCF LCAO–MO function will be a set of $(\omega n - n^2)$ zero-order configurations with appropriate coefficients, a_{tu}. It is rather difficult to visualize in a simple way the effect of such a somewhat long expansion. However, we can make use of the A_t and B_u coefficients and write the following configuration

$$\underbrace{\varphi_1^{2(1-A_1)} \varphi_2^{2(1-A_2)} \cdots \varphi_n^{2(1-A_n)}}_{(n)\ \text{set}} \quad \underbrace{\varphi_{n+1}^{2B_1} \varphi_{n+2}^{2B_2} \cdots \varphi_\omega^{2B(\omega-n)}}_{(\omega-n)\ \text{set}}$$

which we shall refer to as the 'complete electronic configuration'. The set of (n) orbitals has a fractional occupation equal to $(1 - A_t)$

for the orbital φ_t, whereas the remaining orbitals (the φ_u's of the $(\omega - n)$ set) will have, in general, relatively small fractional occupation values, B_u. Clearly, the sum of the fractions of electrons annihilated from the (n) set is equal to the sum of the fractions created in the $(\omega - n)$ set, since $\sum_t A_t = \sum_u B_u$ (equation 22).

The energy E_{00} defined in equation (13) is formally the SCF MO closed-shell energy expansion; however, the φ_t in the CMC SCF formalism are not equal to the φ_t of the Hartree–Fock formalism. If we indicate with E_{HF} the usual Hartree–Fock energy, we can state that E_{HF} is somewhat lower than E_{00}; by an amount which is almost proportional to the correlation error of E_{HF}, as can be seen by analysis of Clementi and Veillard's ICM results for first-row atoms. We now define a quantity $E_c = E - E_{00}$ which is larger than the correlation energy by the amount that E_{00} is larger than E_{HF}. It is noted that the correlation energy is commonly defined as $E - E_{\mathrm{HF}}$. *Therefore the* CMC SCF *formalism differs from most many-body techniques presented to date insofar as we do not assume the Hartree–Fock energy to be the zero-order energy.*

We shall briefly analyse the energy expression (14) in terms of E, E_c and E_{00}. For this purpose we introduce the following definitions.

$$E_c(t) = -2h_t + 2\sum_{t'=1} 2P_{tt'} - 2P_{tt} \tag{23}$$

$$E_c(u) = 2h_u + 2P_{uu} + 2\sum_t 2P_{tu} \tag{24}$$

$$E_c(tu) = 2a_{00}K_{tu} - 4a_{tu}P_{tu} \tag{25}$$

$$E_c(tt') = K_{tt'}(1 - \delta_{tt'}) \tag{26}$$

$$E_c(uu') = K_{uu'}(1 - \delta_{uu'}) \tag{27}$$

We can now write

$$\begin{aligned} E = E_{00} + \sum_t A_t E_c(t) &= \sum_u B_u E_c(u) + \sum_{tu} a_{tu} E_c(tu) \\ &+ \sum_{tt'} A_{tt'} E_c(tt') + \sum_{uu'} B_{uu'} E_c(uu') \end{aligned} \tag{28}$$

The first term is the contribution to E given by the one-electron model. The second term is a correction to E_{00} obtained by annihilation of electrons in the (n) set. The third term is the energy of the electrons created in the $(\omega - n)$ set. The fourth term is interaction of created and annihilated electrons. The fifth term is the interaction energy resulting from any pair of electrons in a φ_t orbital interacting with any pair of electrons in a $\varphi_{t'}$ orbital. Therefore, it is the pair–pair interaction in the (n) set. The last term is the pair–pair interaction in the $(\omega - n)$ set.

Inspection of the energy expression (equation 28) reveals the reason for the often-found poor agreement between computed orbital energies ε_t and ionization potentials or excitation potentials in the standard SCF computations, where $E_{00} = \sum_t (\varepsilon_t + h_t)$. As known, one reason is that the orbitals in the excited configuration or in the ionized molecule often differ sufficiently from the ground-state orbitals even in the SCF LCAO–MO approximation. The second reason is clearly obvious by inspection of equation (28), namely that the numerical values of the A_t and B_u coefficients will, in general, vary from the ground state to the excited states of a neutral molecule or from the ground state of the neutral molecule to the ground state of the ionized molecule.

It is tempting to consider the possibility of a semiempirical scheme, whereby the correct ionization potential or the correct excitation energies are obtained by empirically determining the A_t and B_u fractional occupation values. It is noted that the justification of the use of empirical parameters in the Pariser–Parr technique lies exactly in the fact that the one-electron approximation assumes $A_t = B_u = 0$, whereas in an exact theory A_t and B_u are different from zero.

Let us now continue with the development of the complete MC SCF LCAO–MO theory. We wish to obtain the best φ_t's and φ_u's, making use of the variational principle, i.e. by requiring that $(\partial E/\partial \varphi_t) = 0$ and $(\partial E/\partial \varphi_u) = 0$. In addition we have to satisfy the equation $(\partial E/\partial a_{00}) = 0$ and $(\partial E/\partial a_{tu}) = 0$ in order to obtain the best multiconfiguration expansion. We shall make use of the Lagrangian multiplier technique for determining φ_t and φ_u, and of the solution of the secular equation for determining the a_{tu} coefficients.

Let us define the following operators:

$$F_t = (1 - A_t)h + 2(1 - A_t - A_{t'})P_t + 2A_tP_t + 2\sum_u B_uP_u + \sum_u (a_{00}a_{tu}K_u - 2a_{tu}{}^2P_u) + \sum_{ut'} A_{tt'}K_{t'}(1 - \delta_{tt'}) \quad \text{(29a)}$$

and

$$F_u = B_u(h + 2P_u + \sum_t 2P_t) + \sum_t (a_{tu}a_{00}K_t - 2a_{tu}{}^2P_{tu}) + \sum_{tu'} B_{uu'}K_{u'}(1 - \delta_{uu'}) \quad \text{(29b)}$$

where $P_{ij} = \langle \varphi_i^* | P_j | \varphi_i \rangle$ and $K_{ij} = \langle \varphi_j | K_i | \varphi_j \rangle$. Differentiation of E with respect to the variational parameters φ_t, φ_u, a_{00}, A_u, B_u brings about the following relation:

$$\begin{aligned}\delta E = {} & 2\langle \delta\varphi_t | F | \varphi_t \rangle + 2\langle \varphi_t | F_t | \delta\varphi_t \rangle + 2\langle \delta\varphi_u | F_u | \varphi_u \rangle \\ & + 2\langle \varphi_u | F_u | \delta\varphi_u \rangle + \sum_t \delta A_t(-2h_t - 4 \sum_{t'} P_{tt'}) \\ & + \sum_u \delta B_u (2h_u + P_{uu} + \sum_t 4P_{tu}) \\ & + 2\delta a_{00} \sum_t \sum_u a_{tu} K_{tu} + 2 \sum_t \sum_u \delta a_{tu} [2a_{00} K_{tu} - 8a_{tu} P_{tu} \\ & + 2 \sum_{t'} a_{t'u} K_{tt'} (1 - \delta_{tt'} + 2 \sum_{u'} a_{tu'} K_{uu'} (1 - \delta_{uu'}))] \end{aligned} \quad (30)$$

The variational principle is satisfied for φ_t and φ_u if $(\partial E / \partial \varphi_t) = 0$ and $(\partial E / \partial \varphi_u) = 0$. However, the variation in the φ's is constrained by imposition of the orthogonality relations

$$\langle \varphi_i | \varphi_j \rangle = \delta_{ij} \quad (31)$$

where the indices i and j run over the full (n) and $(\omega - n)$ sets. By setting equation (30) to zero, then by differentiation of the above equation and finally by joining the resulting equations, we obtain the relation which defines φ_t and φ_u

$$|F_t - \sum_{t'}^{(t)} | \varphi_{t'} \rangle \langle \varphi_{t'} | F_t \varphi_t \rangle - \sum_u | \varphi_u \rangle \langle \varphi_u | F_u | \varphi_t \rangle | \varphi_t \rangle = | \varphi_t \rangle \theta_{tt} \quad (32a)$$

$$|F_u - \sum_{u'}^{(u)} | \varphi_{u'} \rangle \langle \varphi_{u'} | F_u | \varphi_u \rangle - \sum_t | \varphi_t \rangle \langle \varphi_t | F_t | \varphi_u \rangle | \varphi_u \rangle = | \varphi_u \rangle \theta_{uu} \quad (32b)$$

which can be rewritten as

$$F_t - T_t - T_u | \varphi_u \rangle = | \varphi_t \rangle \theta_{tt} \quad (33a)$$

$$F_u - U_u - U_t | \varphi_u \rangle = | \varphi_u \rangle \theta_{uu} \quad (33b)$$

where T_t and T_u are the second and third operators in equation (33a) and U_u and U_t are the second and third operators in equation (33b).

In the past, use has been made of the 'virtual orbitals' in the configuration interaction technique. It is noted in this regard, that the φ_u of the $(\omega - n)$ set are quite different from the virtual orbitals of a standard SCF LCAO–MO computation. The reason is that virtual orbitals have very little physical meaning: they are obtained from diagonalization of the Fock equation and are orthogonal to the occupied orbitals. The variational principle cannot act on them, however, since they do not contribute to the total energy. In general, the virtual orbitals have very little overlap with the occupied orbitals and therefore are of little use in correlating the electrons of the occupied orbitals. A discussion on this point can be found in Yutzis' review paper[21] as well as in Weiss' work[25].

It is noted that a given φ_u will mainly be used to correlate one, or at most two or three, φ_t and therefore the remaining $(n - 1)$ or $(n - 2)$ or $(n - 3)$ φ_t's which are promoted to that given φ_u will add little to the correlation correction. However, by including the $(n - 1)$ or $(n - 2)$ or $(n - 3)$ remaining set of φ_u's we will include part of the pair–pair correlation in the total energy at no extra cost. In addition, the inclusion of the additional excitation allows us to make use of the equality $\sum B_u = \sum A_t$ with the simple physical meaning for each B_u and A_t as previously explained. Therefore, for a given A_t and a given B_u there are one or at most very few leading terms in the $\sum_u a_{tu}{}^2$ or in the $\sum_t a_{tu}{}^2$ summation, respectively. The IMC SCF treatments consider only the leading terms in A_t or B_u, and this requires a more accurate optimization of the *basis set* for the φ_t and φ_u which is very time consuming in the computation.

Recently, the ICM technique was applied to the first-[26] and second-[27] row atoms for two configurations. (The theory was applied to open- and closed-shell atoms.) For closed shells the ICM SCF LCAO–MO theory was developed by Das and Wahl for the H_2, Li_2 and F_2 molecules. Since only one φ_t orbital was excited, due to program limitations, the results are quite good for H_2, good for Li_2 (from a molecular viewpoint) and rather poor for F_2 (as expected).

It is finally noted that Nesbet[28] has applied to atoms the Bethe–Goldstone formalism which is to some extent analogous to the CMC SCF LCAO–MO formalism (but which does not fully employ self consistency and, therefore, has to work with larger numbers of configurations).

It seems that most of the earlier literature on the subject was not noticed by those workers who were mainly involved in machine computations. This is somewhat unfortunate, because the CMC SCF technique does not require any large increase in computational effort. The likely reason for the retarded explosion of CMC SCF computations is that an undue amount of expectation was placed on the Hartree–Fock technique, despite quite extended theoretical proof to the contrary available in the last 15 to 20 years. In this respect the work of Nesbet indicates full awareness of the problem[28]. The same can be stated for the work of Löwdin[29,30] where many of the pitfalls of the Hartree–Fock technique have been predicted, and where much of the CMC SCF theory had been developed along different lines, in Löwdin's 'natural orbitals'.

III. ELECTRON POPULATION ANALYSIS

The SCF wave functions can be analysed indirectly via a study of the physical properties of the molecule under consideration (such as moments, polarizabilities, vibrational analysis, etc.,) or directly by what is known as 'electron population analysis'. In the following, we shall briefly expose the method of Mulliken[31], here somewhat modified and extended.

From the previous exposition of the SCF approximation, a molecular orbital is written as

$$\varphi_{\lambda i} = \sum_{p} c_{\lambda i p} \chi_{\lambda p} \tag{34}$$

where λ, i, p are indices which refer to symmetry representation, a specific orbital and a specific basis set, respectively. The basis set is in general a symmetry-adapted function (SAF), i.e. it transforms as λ. In the LCAO approximation the $\chi_{\lambda p}$ is a linear combination of functions, designated by γ_q, centred on the atoms. The linear combination coefficients of the SAF are determined on the basis of symmetry alone, and we can write

$$\varphi_{\lambda i} = \sum_{p} c_{\lambda i p} \sum_{q} d_{\lambda p q} \gamma_q \tag{35}$$

By combining the c's and the d's into a new coefficient w we have

$$\varphi_{\lambda i} = \sum_{ms} w_{\lambda i m s} \gamma_{ms}$$

where for each λ and i the index m refers to a given atom and the index s refers to a given γ_q on the m atom.

For real functions, the electronic density of $\varphi_{\lambda i}$ is

$$(\varphi_{\lambda i})^2 = \sum_{ms} \sum_{m's'} w_{\gamma i m s} w_{\lambda i m' s'} \langle \gamma_{ms} \mid \gamma_{m's'} \rangle \tag{36}$$

This relation is the base of Mulliken's analysis. The above sum can be written as

$$\begin{aligned} (\varphi_{\lambda i})^2 &= \sum_{mm'} \sum_{ss'} w_{\lambda i m s} w_{\lambda i m' s'} \langle \gamma_s \mid \gamma_{s'} \rangle_{mm'} \\ &= \sum_{mm'} \sum_{ss'} [mm'ss']_{\lambda i} \end{aligned} \tag{37}$$

where for $m = m'$ and $s = s'$ the overlap $\langle \gamma_s \mid \gamma_{s'} \rangle$ is unity since the SAF as well as the γ are normalized. For each MO, Table 2 can be constructed:

TABLE 2. Graphical representation of the population analysis.

		$m = 1$			$m = 2$		
m'	s'	$s = 1$	$s = 2$		$s = (i + 1)$	$s = (i + 2)$	...
1	1	[1111]	[1121]	...	$[21(i + 1)1]$	$[21(i + 2)1]$	......
1	2	[1112]	[1122]	...	$[21(i + 1)2]$	$[21(i + 2)2]$	......
1	.	.	.	...	.	.	......
	.	.	.	...	.	.	......
	.	.	.	...	.	.	......
2	$i + 1$	$[121(i + 1]$	$[122(i + 1)]$	...	[ ]	[ ]	......
2	$i + 2$	$[121(i + 2]$	$[122(i + 2)]$	...	[ ]	[ ]	......
2	.	.	.	...	.	.	......
	.	.	.	...	.	.	......
	.	.	.	...	.	.	......
.	.	.	.	...	.	.	......
.	.	.	.	...	.	.	......
.	.	.	.	...	.	.	......

There is one such table for each MO, $\varphi_{\lambda i}$, and each table is by construction symmetrical. We shall call 'quadrant' the matrix of numbers with a given m and m' and indicate this by $\{mm'\}_{\lambda i}$. The sum of its terms is indicated as $S\{mm'\}_{\lambda i}$. The diagonal elements of a quadrant are indicated as $\{mm'\}_{d\lambda i}$ and its sum as $S\{mm'\}_{d\lambda i}$.

For $m = m'$, the quadrant $\{mm\}_{\lambda i}$ contains quantities which are specific to the atomic set for the atom m; for $m \neq m'$ the quadrant $\{mm'\}_{\lambda i}$ contains quantities which are specific to atomic sets of the atoms m and m'.

For the atom m the following definitions, borrowed from Mulliken, are given[31]:

Net atomic population
$$P_m = \sum_{\lambda} \sum_{i} S\{mm\}_{\lambda i} \tag{38}$$

Overlap population with atom m'
$$P_{mm'} = \sum_{\lambda} \sum_{i} S\{mm'\}_{\lambda i} \tag{39}$$

Gross atomic population
$$G_m = P_m + \sum_{m'} P_{mm'} \tag{40}$$

Let us now focus our attention on the quadrant $\{mm\}$. The atomic set (γ's) of any such quadrant will be of s, p, d, etc., type; therefore, within the $\{mm\}$ quadrant, we can have subquadrants of the type $\{m_s m_s\}$, $\{m_s m_p\}$, $\{m_p m_p\}$, etc., designated in general as $\{m_l m_{l'}\}$ where l and l' are the angular quantum numbers for the γ's. In full analogy to the previous definitions for the quadrants $\{mm'\}$ we can define $S\{m_l m_{l'}\}_{a\lambda i}$ for the subquadrants $\{m_l m_{l'}\}$. With this in mind we can introduce the following definitions:

Non-hybrid net atomic $$P_{m_l} = \sum_{\lambda}\sum_{\lambda} S\{m_l m_l\}_{\lambda i} \tag{41}$$

Hybrid net atomic $$P_{m_{ll'}} = \sum_{\lambda}\sum_{i} S\{m_l m_{l'}\}_{\lambda i} \tag{42}$$

Non-hybrid overlap $$P_{m_l m'_l} = \sum\sum_{i} S\{m_l m_{l'}\}_{\lambda i} \tag{43}$$

Hybrid overlap $$P_{m_l m'_l} = \sum_{i} S\{m_l m_{l'}\}_{\lambda i} \tag{44}$$

Non-hybrid gross atomic $$G_{m_l} = P_{m_l} + \sum_{\lambda}\sum_{i}\sum_{m'} P_{m_l m'_l} \tag{45}$$

Hybrid gross atomic $$G_{m_{ll'}} = P_{m_{ll'}} + \sum_{\lambda}\sum_{i}\sum_{m'} P_{m_l m'_{l'}} \tag{46}$$

Hybridization is a very familiar concept in theoretical chemistry. However, its meaning is often used in an exceedingly restrictive sense, usually when we have more than one atom. However, hybridization is no more than polarization, and therefore we can talk of hybridization between two atoms or between two electrons on the same atom. As a consequence we have *internal hybridization* (within a given atom and due to the electrons of that atom) as well as external hybridization (within a given atom and due to a field originated outside the atom). External hybridization is the familiar one. An example of internal hybridization is the beryllium ground-state atom previously discussed. As we know the 2s orbital is strongly hybridized (internally) with the 2p orbitals. Therefore, the correlation problem in atoms can be viewed as a problem of describing in the best possible way the internal hybridization, and the correlation problem in molecules can be viewed as a problem of describing in the best possible way the internal polarization of the component atoms *plus* the external hybridization. Personally, the present author would not mind the absence of the word 'hybridization' in theoretical chemistry, since the term 'polarization' seems to be more accurate. However, it shall continue to be used in deference to previous workers in the field.

It is stressed that the above definitions have meaning only for a given basis set. Therefore, they provide quantitative data of qualitative character. However, it is exactly this *type* of data which we like to analyse in order to obtain some correlation between molecular structures. An exact wave function for a molecule provides a tool for obtaining exact expectation values. These can be obtained as an alternative from experimental data. However, taken alone, neither an exact list of expectation values nor an accurate list of experimental data constitutes understanding of the electronic structure of molecules.

Let us consider the population analysis in the CMC SCF LCAO–MO formalism. Since the φ_t's and the φ_u's are orthogonal, we shall have that $\langle \psi_{tu} \mid \psi_{tu'} \rangle = \langle \psi_{t'u} \mid \psi_{tu} \rangle = 0$, and recalling equation (10), we shall have

$$\langle \Psi \mid \Psi \rangle = a_{00}{}^2 \langle \psi_{00} \mid \psi_{00} \rangle + \sum_t \sum_u a_{tu}{}^2 \langle \psi_{tu} \mid \psi_{tu} \rangle$$

For each of the determinants ψ_{00} or ψ_{tu} the previous definitions for the electronic population are valid. Therefore, we shall have

Net atomic population $$\bar{P}_m = a_{00}^2 P_m^{00} + \sum_t \sum_u a_{tu}{}^2 P_m^{tu} \qquad (47)$$

Overlap population $$\bar{P}_{m'} = a_{00}^2 P_{mm'}^{00} + \sum_t \sum_u a_{tu}{}^2 P_{mm'}^{tu} \qquad (48)$$

Gross atomic population $$\bar{G}_m = \bar{P}_m + \sum_{m'} \bar{P}_{mm'} \qquad (49)$$

and equivalent expressions for the hybrid and non-hybrid populations.

IV. BOND ENERGY ANALYSIS

It is customary in the literature to report wave functions, expectation values, orbital energies and total energies. The molecular total energy is then compared with the total energy of the separated atoms and deductions on bond energies are made.

In theoretical chemistry the orbital energies are used mainly in connection with the Koopman theorem[32], but cannot be identified with bond energies.

In this work we shall introduce the definitions of the 'bond energy analysis'[24]. We shall make use of a number of bond energy classifications which are derived at first from the usual SCF LCAO–MO energy expression for the total energy (see equation 13):

$$E = 2\sum_i h_i + \sum_{ij} P_{ij} + E_{NN}$$

The above energy expression can be written as

$$\begin{aligned} E = 2\sum_i \sum_A h_{iA} + \sum_{AB} h_{iAB} + \sum_{ABC} h_{iABC} \\ + \sum_i \sum_j \sum_A P_{ijA} + \sum_{AB} P_{ijAB} + \sum_{ABC} P_{ijABC} \\ + \sum_{ABCD} P_{ijABCD} + \sum_{AB} E_{NA.NB} \end{aligned} \tag{50}$$

where A, B, C, D are indices running over the atoms and where h_{iA}, h_{iAB}, h_{iABC} are the one-, two- and three-centre components of h_i, $E_{NA.NB}$ is the two-centre component of E_{NN} and P_{ijA}, P_{ijAB}, P_{ijABC}, P_{ijABCD} are the one-, two-, three- and four-centre components of the electron–electron interaction energy.

We shall then denote as the zero-order energy diagram

$$E_0 = \sum_A E_A = 2\sum_A \sum_i j_{iA} + 2\sum_{ij} P_{ijA} \tag{51}$$

as the first-order energy diagram ($A \neq B$)

$$E_1 = E_{AB} = \sum_{AB} \sum_i h_{iAB} - E_{NA.NB} - P_{ijAB} \tag{52}$$

as the second-order diagram ($A \neq B \neq C$)

$$E_2 = E_{ABC} = \sum_{ABC} \sum_i h_{iABC} - \sum_{ij} P_{ijABC} \tag{53}$$

and finally as the third-order diagram ($A \neq B \neq C \neq D$)

$$E_3 = \sum_{ABCD} E_{ABCD} = \sum_{ABCD} \sum_{ij} P_{ijABCD} \tag{54}$$

The E_0 should be compared with the sum of the energy of the separated atoms. The correlation correction within E_0 can be estimated directly from atomic energy computation and can be taken as equal to the electronic population within 'a given atom', not summing up the full contribution to a given m column, but only the first quadrant.

E_1, the first-order diagram energy, is the first 'bond' energy, and links any two atoms in a molecule, two at a time. Clearly, the *classical chemical formulae are a representation of the first-order diagram*—an incomplete one, however, since in them only some of the nearest-neighbour atoms are connected. So we usually do not write a bond between neighbouring hydrogen atoms in benzene, although there is clearly an interaction between them. The first-order energy differs from the other bond energies in that it includes the nuclear–nuclear repulsion, clearly present only in E_1. This quantity is numerically quite significant; it is part of the potential energy and, therefore, in view of the virial theorem, one can expect that E_1 will be the dominant part of the binding energy. A large number of other evidences, both theoretical and experimental, are known to support this point. As is the case for E_0, E_1 is composed of a number of terms which will satisfy molecular symmetry considerations. In other words, the equivalent atoms in E_0 will have equal energy, so the equivalent bonds in E_1 will be equal in energy. The correlation energy correction associated with E_1 for the atoms A and B will clearly depend on the electronic density between atoms A and B. Here we should be careful in the use of atomic correlation energy computations, since no bond analogy can be uniquely drawn between the electrons in an atom and those in a molecule (see, however, Wigner's work for a possible analogy).

E_2, the second-order diagram energy, is the three-atom interaction less the direct atom–atom pair interaction. This term does not contain a nuclear–nuclear repulsion term, and the one-electron energy (kinetic and potential) is relatively small.

E_3, the third-order diagram energy, is the only term which does not include one-electron terms; therefore, it is totally a potential energy (Coulomb and exchange) term. This term includes, by definition, mainly long-range interaction. E_3 as well as E_2 are in effect neglected in many semiquantitative computations (like the Pariser–Parr approximation).

The population analysis, as previously described, has an energy analogue in E_0 and E_1, not in E_2 and E_3. Classical chemistry formulae have a partial analogue in E_0 and E_1.

The present breakdown of the total energy should represent a natural frame for transferability of bond energy for which there is a large body of thermodynamic evidence within families of compounds. It provides a framework which will not change by introducing more and more configurations in the MC SCF LCAO–MO formalism. It seems to offer advantages to *ab initio* vibrational analysis and to offer easy interpretation of the overall constancy of group frequency in different molecules. Finally, it maintains the basic ideas of the molecular-orbital theory but associates it intimately with our basic intuitive approach in chemistry, i.e. that there are atoms in molecules.

This breakdown does not follow traditional ideas on the number of bonds one would like to see associated with an atom. So we could have in a molecule of N atoms a hydrogen atom with $(N - 1)$ first-order bonds. However, this number will be reduced: (*a*) by consideration of the nearest neighbours and (*b*) by quantitative consideration of the energy associated with each bond.

It is noted that there are at most four atomic bonds in our analysis. This is an effect of having chosen a basis set centred at the atoms. Alternatively, one could use localized orbitals and this would alter the number of atoms involved in bonds.

The above discussion brings about the following conclusions: (*a*) a chemical bond is an arbitrary concept and can be defined in as many ways as one wishes, (*b*) of the many possible representations, some are more useful than others and (*c*) the 'chemical bond' concept is to a certain extent simply a book-keeping device, of great importance, however.

One might prefer to have an exact correspondence between electron population analysis and bond energy analysis. However, the starting point of the electron population is $(\varphi_{\lambda i})^2$ and this can lead only to a subdivision involving one and two centres, whereas the bond energy analysis, in view of our definitions, involves one, two, three and four centres.

Let us briefly digress to CMC SCF LCAO–MO bond energy diagrams and compare these with SCF LCAO–MO bond energy diagrams. Let us start with the first-order diagrams and the first-order energy $E_0 = \sum_A E_A{}^0$. The main correction to these diagrams can be obtained from atomic computations, more explicitly from associating to the $E_A{}^0$ a correction proportional to the $\{mm\}$ and the atomic correlation less degeneracy effects (see Veillard and Clementi[26,27] for numerical values). This is a very simple correction to introduce. Alternatively, the correction to $E_A{}^0$ can be computed using the device of correcting the P_{ij} integrals with a pseudopotential (for example, see Clementi's work[33]). This alternative is not too different from the previous one, since the pseudopotential uses the overlap between orbitals as a parameter. A formal equation for E_0 can be most easily obtained from equation (17).

The first-order diagrams and their energy $E_1 = \sum_{A>B} E_{AB}$ are the most important in general for molecular correlation corrections. The correction to these diagrams should be proportional for a given atom to $\sum_{m' \neq m} \{mm'\}$. Alternatively, one could use a pseudopotential computed correction in a manner exactly equivalent to our work on atoms. Again the formal expression of E_1 can be obtained from equation (17).

The second- and third-order diagrams cannot be corrected by the use of population analysis parameters. However, it can be done by the use of the pseudopotential technique or by the CMC SCF LCAO–MO method. The total energy will, therefore, be subdivided as

$$E = \sum_{A} (E_A + \eta_A) + \sum_{AB} (E_{AB} + \eta_{AB}) + \sum_{ABC} (E_{ABC} + \eta_{ABC}) + \sum_{ABCD} (E_{ABCD} + \eta_{ABCD}) \tag{55}$$

where the E's are the energies computed in the SCF LCAO–MO approximation and the η are the energies obtained either from the CMC SCF LCAO–MO theory or from empirical correction to the E, or a proper mixture of both.

One of the expectations in proposing the bond energy diagram partitioning of E is that by a systematic comparison of a number of molecules, a simple correlation will emerge which will put 'transferability of bond energies' on a sound basis. It is noted that in the early attempts at studying molecular kinetics, the approach of using Morse-type potentials between any pair of atoms was often adopted. The present analysis in some respects does exactly that, if one considers only zero- and first-order diagrams. In addition it does much more. Therefore it is expected that this type of analysis will be of help in the formulation of a theory for reaction mechanism.

Other problems, like vibrational analysis, the study of the barrier to internal rotation and charge transfer can be explained quite naturally in this framework of analysis. We shall return later to these points.

V. CORRELATION ENERGY AND STATISTICAL METHODS

What is wrong with the wave function Ψ_0 as given in equation (1)? The answer is available, for example, in the work by Wigner in 1934[33]. If we ask what the statistical relation is between the position of two electrons, say electrons 1 and 2, from the wave function (1), we have to solve the integral

$$\int \Psi_0{}^*\delta(x - x_1)\delta(x - x_2)\Psi_0 \prod_i^{(1,2)} \mathrm{d}\tau_i \tag{56}$$

where x stands for the cartesian and spin coordinates, and the integral is carried over all electron coordinates except for those of electrons 1 and 2 under consideration. The above integrals yield (by indicating with σ the spin variables)

$$\text{normalization factor } x \sum_{i=1}^{2n}\sum_{j=1}^{2n} \times \left[1 - \int \varphi_i(1)\varphi_i{}^*(2)\varphi_j{}^*(1)\varphi_j(2)\,\mathrm{d}\sigma_i(1)\,\mathrm{d}\sigma_i(2)\,\mathrm{d}\sigma_j(1)\,\mathrm{d}\sigma_j(2)\right] \tag{57}$$

For the case of two electrons with parallel spin, the second term above is, in general, different from zero. For the case where the two electrons have opposite spin the second term of the equation above is zero. This means that the wave function (1) allows a pair of electrons with parallel spin to 'feel' each other's relative position, to be correlated, whereas for the case of electrons with antiparallel spin the wave function allows any position with equal probability. Thus, the Hartree–Fock determinant introduces correlation in pairs of electrons with parallel spin, but does not correlate those of antiparallel spin. Since the second term of equation (57) comes about because of the antisymmetry of the wave functions required for fermions, the second term's effect is referred to as the 'Fermi hole'. Thus, in the Hartree–Fock wave function we have Fermi potential holes for the system of parallel spin, but no hole between two electrons with antiparallel spin. Since clearly two electrons with opposite spin should never occupy the same position simultaneously, the Hartree–Fock function lacks a mechanism to provide for a potential hole experienced by an electron with spin α, when in the neighbourhood of an electron with spin β. The hole not present in the wave function (1), the Hartree–Fock function, is referred to as the 'Coulomb hole'.

The energy gained by the system in introducing the Coulomb hole in the wave function which represents such a system is the correlation energy. This name was introduced by Wigner in the work previously quoted[33]. This energy gain can be redefined therefore as the energy difference between the exact non-relativistic energy and the Hartree–Fock energy[30].

The energy gained by the system in introducing the Fermi hole in the wave function which represents such a system is called precorrelation energy. This energy can be redefined, therefore, as the energy difference between the Hartree and the Hartree–Fock energy[34].

To introduce the Fermi hole, the form of the wave function was changed from a simple product (Hartree) to a determinant (Slater). To introduce the Coulomb hole, we again have to change the form of the wave function. A large number of papers have been devoted to this subject since the late 1920's[35] and we have no intention to review the extremely abundant literature. We note that one possibility was put forward as early as 1930, namely to add corrections to the Hartree–Fock energy using wave functions of the form[33]

$$\Psi = \Psi_0 + \lambda\phi \tag{58}$$

where Ψ_0 gives the Hartree–Fock energy and $\lambda\phi$ the correlation correction*.

It is also noted that the total density distribution given by Hartree's functions is not far different from the density distribution given by Hartree–Fock functions.

It is, therefore, reasonable to assume that the density distribution variation needed for the introduction of a Coulomb hole (relative to the Hartree–Fock density) is also small. Therefore, we can use the Hartree–Fock density as the correct density and attempt to extract relations which should give the correlation energy from knowledge of the Hartree–Fock density. This task was solved by Wigner[33] who did propose a relation between density and correlation energy. We shall refer to this work as the 'statistical estimate of the correlation energy'.

We shall now consider in detail the quantitative aspect of the correlation energy to atoms and then we shall discuss molecules. In this analysis, the emphasis is not on historical developments, but rather on the possibilities of current methods.

* In equation (58) we have combined equation (2) and equation (14) of Wigner[33]. It should be noted that Wigner's starting point was not Ψ_0, but an incorrect form of Ψ_0, which, however, has the same energy as Ψ_0. More specifically, Wigner decomposed Ψ_0 from a determinant of dimension $2n$ into a product of two determinants with dimension n; this was, however, done because of the fact that the two-determinant product gives an energy equal to the one given by Hartree–Fock. In presenting equation (58) as Wigner's equation, we do some injustice to the formalism, but not to the spirit of Wigner's equation (4).

TABLE 3. Correlation energy computed by adapting statistical models.

Atom		Correct values[a]	Gombas First term[b]	Gombas Second term[c]	Gombas Total	Modified relation Best a_1[d]	Modified relation Corr. energy[e]
He	^1S	−0·0421	−0·0316	−0·0263	−0·0579	0·0237	−0·0421
Li	^2S	−0·0453	−0·0448	−0·0393	−0·0841	0·0180	−0·0476
Be	^1S	−0·0944	−0·0610	−0·0551	−0·1161	0·0276	−0·0947
B	^2P	−0·1240	−0·0782	−0·0738	−0·1520	0·0285	−0·1263
C	^3P	−0·1551	−0·0958	−0·0958	−0·1917	0·0289	−0·1580
C	^1D	−0·1659	−0·0956	−0·0954	−0·1910	0·0309	−0·175
C	^1S	−0·1956	−0·0953	−0·0948	−0·1901	0·0366	−0·175
N	^4S	−0·1861	−0·1137	−0·1208	−0·2345	0·0292	−0·1882
N	^2D	−0·2032	−0·1135	−0·1202	−0·2337	0·032	−0·209
N	^2P	−0·2274	−0·1134	−0·1199	−0·2333	0·035	−0·209
O	^3P	−0·2339	−0·1314	−0·1475	−0·2789	0.0318	−0·2408
O	^1D	−0·2617	−0·1313	−0·1472	−0·2785	0·0355	−0·265
O	^1S	−0·2985	−0·1312	−0·1467	−0·2779	0·0406	−0·2649
F	^2P	−0·3160	−0·1493	−0·1768	−0·3261	0·0378	−0·2980
Ne	^1S	−0·381	−0·1672	−0·2084	−0·3756	0·0406	−0·3594
Na	^2S	−0·386	−0·1807	−0·2302	−0·4109	0·0381	−0·347
Mg	^1S	−0·428	−0·1967	−0·2537	−0·4504	0·0388	−0·4417
Al	^2P	−0·459	−0·2130	−0·2774	−0·4904	0·0384	−0·4667
Si	^3P	−0·494	−0·2303	−0·3030	−0·5333	0·0383	−0·4910
Si	^1D	−0·505	−0·23	−0·3026	−0·5326	0·0392	−0·533
Si	^1S	−0·520	−0·2298	−0·3022	−0·5320	0·0404	−0·532
P	^4S	−0·521	−0·2479	−0·3300	−0·5779	0·0375	−0·5132
P	^2D	−0·539	−0·2477	−0·3297	−0·5774	0·0388	−0·559
P	^2P	−0·555	−0·2475	−0·3295	−0·5770	0·04	−0·558
S	^3P	−0·595	−0·2655	−0·3582	−0·6237	0·0400	−0·5818
S	^1D	−0·606	−0·2654	−0·3580	−0·6234	0·0407	−0·631
S	^1S	−0·624	−0·2653	−0·3576	−0·6229	0·042	−0·630
Cl	^2P	−0·667	−0·2833	−0·3879	−0·6712	0·0420	−0·6545
A	^1S	−0·732	−0·3013	−0·419	−0·7203	0·0433	−0·7310
Li^+	^1S	−0·0435	−0·033	−0·0353	−0·0683	0·0237	−0·0438
Be^{2+}	^1S	−0·0443	−0·0337	−0·0421	−0·0758	0·0237	−0·0447
B^{3+}	^1S	−0·0448	−0·0341	−0·0478	−0·0719	0·0237	−0·0453
C^{4+}	^1S	−0·0451	−0·0344	−0·0525	−0·0869	0·0237	−0·0456
N^{5+}	^1S	−0·0453	−0·0346	−0·0566	−0·0912	0·0237	−0·0459
O^{6+}	^1S	−0·0455	−0·0347	−0·0602	−0·0949	0·0237	−0·0460
F_7^+	^1S	−0·0456	−0·0348	−0·0635	−0·0983	0·0237	−0·0462
Ne^{8+}	^1S	−0·0457	−0·0349	−0·0664	−0·1013	0·0237	−0·0463
Kr^{36+}	^1S	−0·0470	−0·0355	−0·1038	−0·1393	0·0237	−0·0471

[a] From reference 27.

[b] First term is the contribution to the correlation energy from $-0{\cdot}0357 \int \rho^{4/3}(0{\cdot}0562 + \rho^{1/3})^{-1}\, d\rho$.

[c] Second term is the contribution to the correlation energy from $-0{\cdot}0311 \int \rho ln(1 + 2{\cdot}39\rho^{1/3})\, d\rho$.

[d] Best value of the constant a_1 (see equation) using only the first term and fitting the correct correlation energy (first column).

[e] Best empirical value of the correlation energy using equation (59).

The statistical model of Wigner has been revised a number of times in attempts to extend its validity from the cases where the density is high to cases of smaller density. Gombas[36] attempted to give an additional expression which covers both the high- and low-density region. His expression is for the correlation energy E_c

$$E_c = \int \rho^2 \varepsilon_c(\rho)\, d\tau$$

$$= \int \rho^2 a_1 \rho^{\frac{1}{3}} (a_2 + \rho^{\frac{1}{3}})^{-1}\, d\tau + \int \rho^2 b_1 \ln (1 + b_2 \rho^{\frac{1}{3}})\, d\tau \quad (59)$$

where $a_1 = 0{\cdot}0357$, $a_2 = 0{\cdot}0562$, $b_1 = -0{\cdot}0311$, $b_2 = 2{\cdot}39$ are Gombas' constants.

If in the first term we put $a_2 = 0$, then clearly it contributes to the correlation energy E_c by an amount a_2N where N is the number of electrons. In Table 3 the value of the correlation correction using Hartree–Fock functions is given for the first- and second- row atoms.

The overall agreement with experimental correlation energies is not bad as seen by comparing column 1 (the experimental correlation energy) with the sum of the first and second term (columns 4, 2 and 3, respectively). If we consider the statistical model as a useful fitting formula then we can improve the situation. This was done by Clementi and Salez[37]. First, we put $b_1 = 0$, then we obtained the best value of a_1 for $a_2 = 0{\cdot}0562$ (second last column of Table 3). This done, we expressed the value of a_1 analytically, giving rise to the following modified relation

$$E_c = \alpha \int (\rho^{\frac{4}{3}}(0{\cdot}0562) + \rho^{\frac{1}{3}})^{-1}\, d\tau \quad (60)$$

where α is a numerical constant obtained from the relation (for an n-electron system):

$$\alpha = 0{\cdot}0237 + (0{\cdot}0279n - 0{\cdot}08176)(n + 3{\cdot}45)^{-1}$$

The computed correlation energies using equation (50) are given in the last column of Table 3. In Figure 1 we report the experimental correlation energy and the 'statistical' correlation energy computed either with equation (59) or with the empirical set of constants (equation 60).

It could be noted that the equation (60) is very near to the original formulation of Wigner, a part of the structure included by Clementi

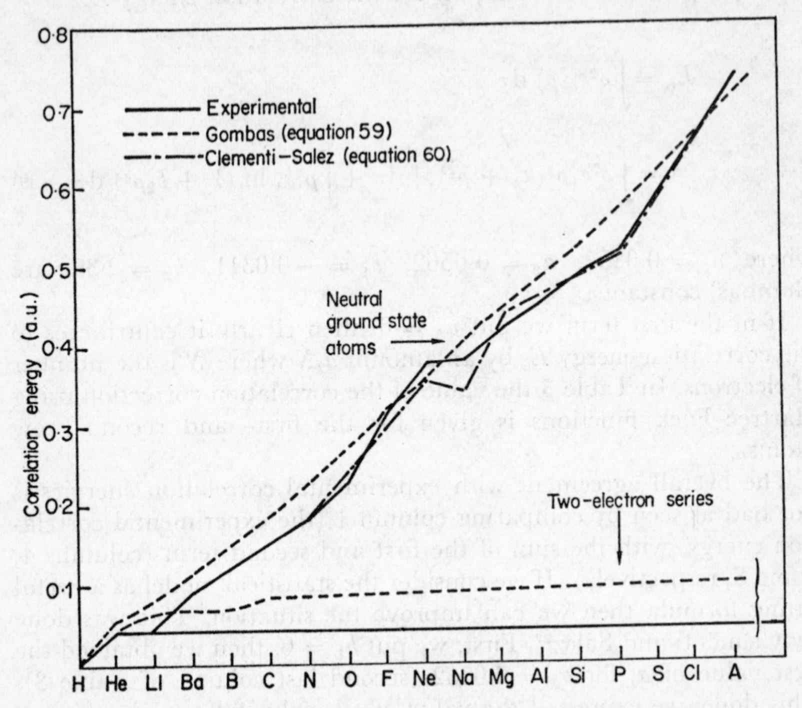

FIGURE 1. Correlation energy from statistical models. The solid line gives the experimental correlation energy for neutral atoms from helium to argon and for the isoelectronic series He, Li^+, Be^{2+}, ... A^{16+}. The dash–dash and the dash–line report the correlation energy using Gombas or Clementi and Salez relations. For the two-electron isoelectronic series, the Clementi–Salez data are in exact agreement with the experimental one.

and Salez in the constant a_1. Therefore in principle good estimates of the correlation energy *would have been easily available for the last three decades if one had had Hartree–Fock functions.*

A systematic attempt to obtain Hartree–Fock functions for atoms and ions, and to extract from it the correlation energy was made by the author starting in the early 1960's. By now we have Hartree–Fock functions for the first period neutral atoms and positive ions[38], and for the second period neutral atoms and positive ions[39], for the third period neutral atoms and positive ions[40], for the negative ions of the first[41], second[42] and third[43] periods. We have, in addition, simpler functions for the neutral atoms from Kr to Rn[44].

From such wave functions we have computed by perturbation theory the relativistic correction[45,46]. This done, we were finally in a position to present the first systematic body of 'experimental' correlation energies[47], by subtracting the Hartree–Fock and the relativistic energy from the experimental energy[47].

Rather than solving for equation (60), we can obtain the correlation energy by a different technique introduced by the author[45]. This technique uses the Hartree–Fock formalism, but adds to the Hartree–Fock operator, F (i.e. the average field effect of all the electrons), an operator of the form $\sum J_{ij}'$ where i and j refer to particles. The operator $\sum J_{ij}'$ does not allow the electron i to go nearer to the electron j than a prescribed amount. This prescription is used for electrons with antiparallel spin. The radius of the sphere which limits the access of electron i to electron j is proportional to an inverse power of the density, following Wigner. It is noted that in order to determine the radius of the excluded volume, we have not made use of the Wigner or Gombas relations; we have the case of the He and the Ne atoms set up in such a way as to give the 'experimental' correlation energy, which we had previously computed. The results of this technique are presented in Figure 2. It is clear that the overall agreement with experimental data is not too bad. This technique again uses the Hartree–Fock density as a starting point, and does not provide correlated functions, only the correlation energy. The method we have introduced gives the correct correlation energy for the two-electron series He, Li^+, etc., up to Kr^{34+} (Kr^{34+} being the highest two-electron ion that we have tested). The statistical methods previously discussed (with the exception of the semiempirical modification by Clementi and Salez) tend to give a much larger correlation for the highest ions in the two-electron series.

It can, therefore, be concluded that there are sufficient indications to prove that knowledge of the Hartree–Fock density is sufficient for obtaining a good estimate of the correlation energy[45].

However, present methods without semiempirical parameters fail to give the detailed structure of the correlation energy[48]. This is not surprising, since the statistical methods stress the total density of a system but not the angular momentum dependency, since the method has been designed primarily for closed-shell systems. For

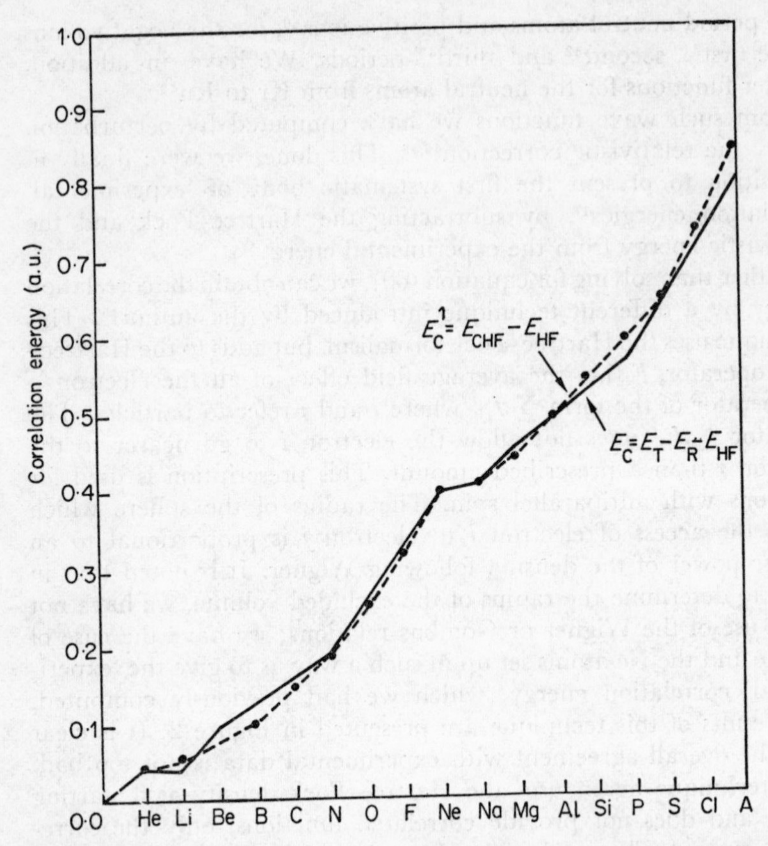

FIGURE 2. Experimental correlation energy and Coulomb–Hartree–Fock method. The solid line gives the correlation energy for neutral atoms from helium to argon. The dashed line gives the correlation energy as computed by correcting the electron–electron interaction (Coulomb–Hartree–Fock technique) E_c, E_T, E_R, E_{HF} stand for correlation energy, total energy, relativistic energy and Hartree–Fock energy. E_c^1 is the correlation energy obtained by subtracting the E_{HF} from the Coulomb–Hartree–Fock energy.

example, in the carbon atom the ^3P state ($1s^22s^22p^2$) and the ^1D state ($1s^22s^22p^2$) have about the same density, but quite different correlation energy. As long as the statistical methods do not depend explicitly on the angular momentum of a given state, it is hard to see how one can obtain agreement with the 'experimental' correlation energy. The Coulomb hole technique we have introduced could, however, give such agreement since there the correlation energy does not depend on the total density, but on the Coulomb integrals.

Finally, it should be pointed out that one can consider the Wigner or Gombas expression as the first term of a convergent series (indeed, this is the case). Other terms of the series will contain not only a density-dependent term but derivatives of the density-dependent term. A theoretical study on this point has been recently presented by Brueckner[49].

Physically one can understand the need of such terms by remembering that: (*a*) the total energy is related to the kinetic energy via the virial theorem and (*b*) that the kinetic energy function at a given point varies quite differently from the total density function. (That is, it has maxima in the neighbourhood of density minima, and is constant in the neighbourhood of the density maxima.) Therefore, terms containing derivatives of the density (and the kinetic energy is a derivative of the density) should be included. In a different language one can restate the same physical concept by noticing that the volume excluded from collision of electron j and electron i will be dependent on the kinetic energy of the colliding particles. Figure 3 gives the plot of the kinetic energy as a function of the radial distance, the density and the $(a_2 + \rho^{\frac{1}{3}})^{-1}$ plot for the $1s$, $2s$ and $2p$ orbitals of the neon atom. It is noted that for $\rho = 0$ (as at the nodes, where the kinetic energy has a finite value different from zero) the existence of the constant a_2 in Gombas' relation (or in Wigner's relation) gives a correlation contribution for the region of zero density (as one would have by introducing derivatives of the density-dependent terms).

VI. ANALYSIS OF THE 'EXPERIMENTAL' CORRELATION ENERGY

The 'experimental' correlation energies for the first two periods of the periodic table (He to Ar) are computed in Table 4. The computation requires knowledge of the experimental total energy (column 1), of the Hartree–Fock energy[50] (column 2) and of the relativistic correction (columns 3, 4 and 5). The corrections for the Lamb shift are obtained from the semiempirical relations below and seem to be somewhat large. However, no sufficiently accurate theoretical formulation of the Lamb shift correction is available for many-electron systems. Subtracting from the experimental total energy the Hartree–Fock energy (mass corrected) and the relativistic corrections, one obtains the experimental correlation energies given in the first column of Table 3.

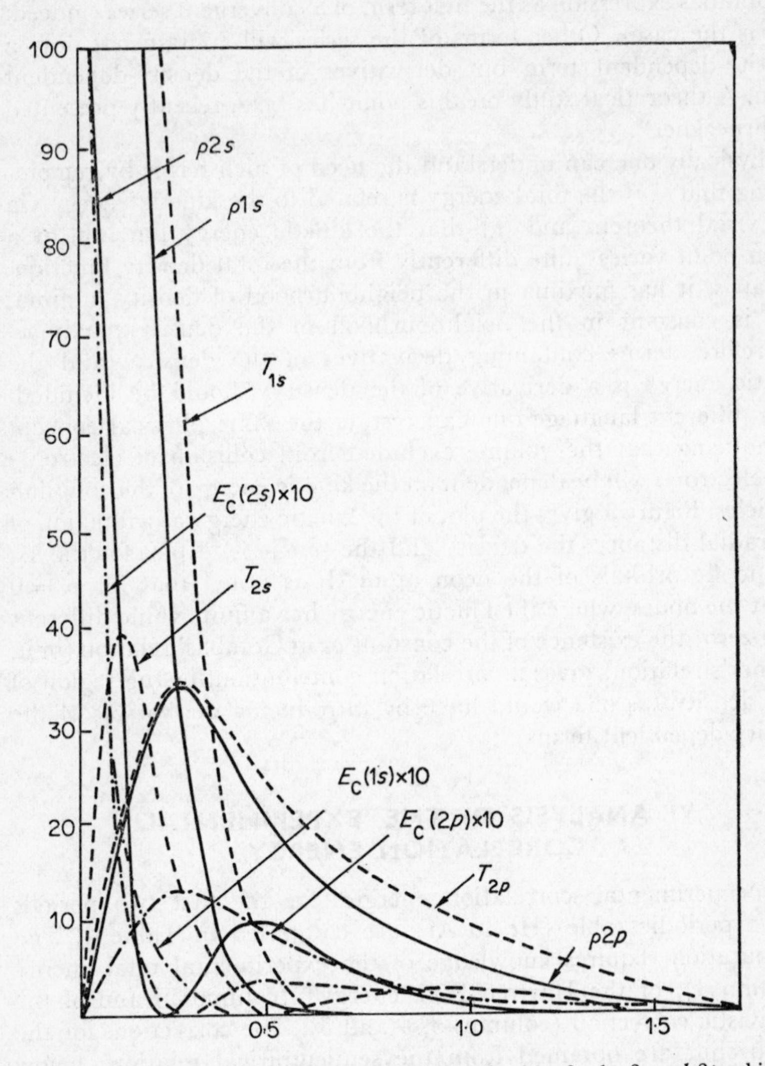

FIGURE 3. Kinetic energy, correlation energy and densities for 1*s*, 2*s* and 2*p* orbitals of neon atom (^1S). The densities (denoted by ρ), the kinetic energies (T) and the correlation energy (E_c) are plotted as a function of R, the distance from the nucleus (in a.u.).

TABLE 4. Experimental, Hartree–Fock and relativistic data for first and second row (in a.u.)[a].

Case		Experimental	Hartree–Fock[47]	HF mass corr.	Relativistic[48,49]	Corr.[48]
He	^1S	−2·9038	−2·8616799	−2·86129	−0·00007	0·000022
Li	^2S	−7·4780	−7·4327257	−7·43214	−0·00055	0·000106
Be	^1S	−14·6685	−14·573020	−14·57229	−0·00220	0·000323
B	^2P	−24·6579	−24·529052	−24·52782	−0·00603	0·000740
C	^3P	−37·8558	−37·688611	−37·68690	−0·01381	0·001439
C	^1D	−37·8093	−37·631317	−37·62961	−0·01377	0·001439
C	^1S	−37·7572	−37·549535	−37·54783	−0·01379	0·001439
N	^4S	−54·6122	−54·400911	−54·39879	−0·02732	0·002500
N	^2D	−54·5246	−54·296152	−54·29404	−0·02736	0·002500
N	^2P	−54·4808	−54·228087	−54·22598	−0·02739	0·002500
O	^3P	−75·1101	−74·809369	−74·80683	−0·04940	0·004000
O	^1D	−75·0378	−74·729213	−74·72667	−0·04940	0·004000
O	^1S	−74·9562	−74·610955	−74·60842	−0·04935	0·004000
F	^2P	−99·8053	−99·40928	−99·40644	−0·08289	0·006015
Ne	^1S	−129·056	−128·54701	−128·54355	−0·13121	0·008614
Na	^2S	−162·441	−161·85889	−161·85506	−0·20021	0·011856
Mg	^1S	−200·333	−199·61458	−199·61011	−0·29505	0·015791
Al	^2P	−242·752	−241·87665	−241·87177	−0·42062	0·020460
Si	^3P	−289·927	−288·85426	−288·84867	−0·58351	0·025887
Si	^1D	−289·898	−288·81500	−288·80941	−0·58382	0·025887
Si	^1S	−289·857	−288·75845	−288·75286	−0·58414	0·025887
P	^4S	−342·025	−340·71866	−340·71268	−0·79111	0·032085
P	^2D	−341·973	−340·64872	−340·64274	−0·79110	0·032085
P	^2P	−341·940	−340·60316	−340·59718	−0·79126	0·032085
S	^3P	−399·144	−397·50475	−397·49801	−1·05076	0·039051
S	^1D	−399·102	−397·45210	−397·44536	−1·05084	0·039051
S	^1S	−399·043	−397·37444	−397·36770	−1·05090	0·039051
Cl	^2P	−461·514	−459·48187	−459·47482	−1·37168	0·046765
Ar	^1S	−529·303	−526·81734	−526·81017	−1·76094	0·055190

[a] The first and second columns are case identifications, the following columns are the experimental total energy, the computed Hartree–Fock energy, the same energy mass corrected, the computed relativistic energy (with the spin–spin and spin–orbit energy estimated) and the Lamb shift correction.

The correlation energies for the $3d$ and $4s$ configurations are available elsewhere[47]. In Figure 4, we give the correlation energy for the atoms from He to Ne and in this figure we have included the isoelectronic series for the ground-state atoms from 2 to 9 electrons.

By simple inspection of Figure 4 we note the following points. First, the correlation energy for the two-electron isoelectronic series

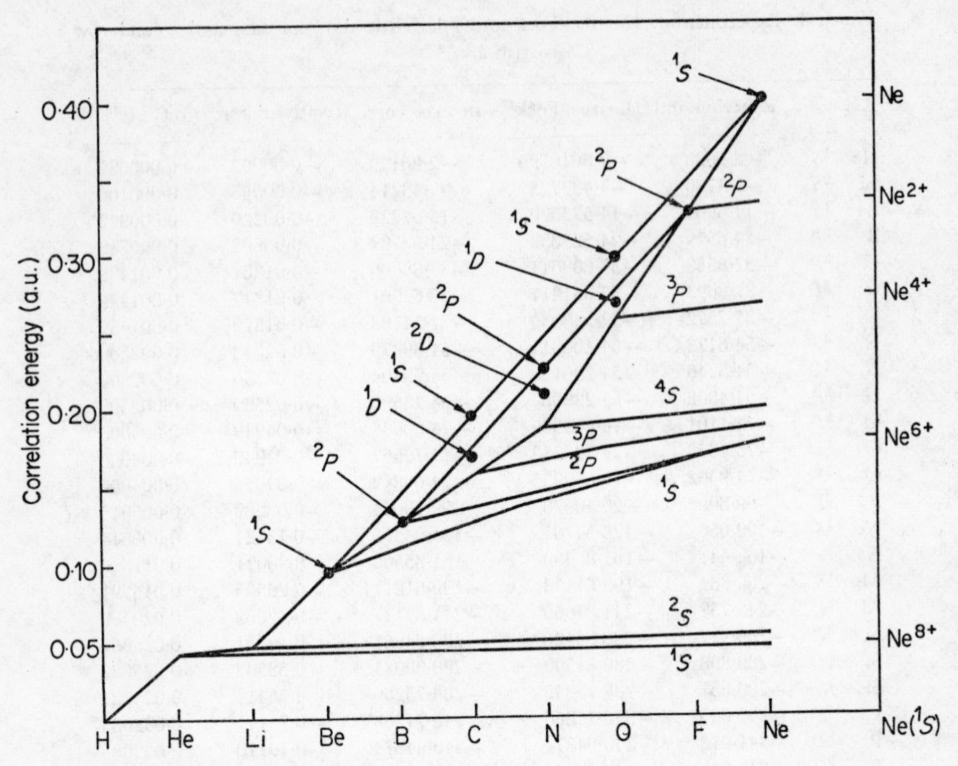

FIGURE 4. Correlation energy for first-row atoms. This figure reports the correlation energy for the neutral atoms and the corresponding isoelectronic series (up to $Z = 10$) and for the excited states of the ground state electronic configuration for the neutral atoms. The state designations followed by pointing arrows refer to neutral atoms. The state designations without pointing arrows designate the isoelectronic series. On the right side of the figure, the increase in correlation energy for the neon case is repeated in order to show the correlation energy increase which accompanies an addition of two electrons starting with $Ne^{8+}(^1S)$.

is about constant from He to Ne^{8+}. The correlation energy for the three-electron isoelectronic series is only slightly larger than the corresponding values for the two-electron series. This is no surprise because there is only a main pair in the He and Li atoms (the $1s^2$). In Li ($1s^2 2s^1$, ^2S) there is an increase of correlation because the $1s$ electron finds a $2s$ electron with opposite spin. However, the $1s$ and the $2s$ orbitals do not overlap too much. Since only the fraction of the electron which overlaps requires correlation, the Hartree–Fock functions for Li introduce an error nearly as large as in He.

In the beryllium atom we have an additional pair of electrons relative to He. This time the added pair introduces a large increase in correlation. The added correlation in Be should be about twice as much as the correlation in He as indicated from the statistical model. We shall talk of intrasubshell and intersubshell correlation energy and of intrashell and intershell correlation energy. The electrons in a subshell have the same n and l, but differ in m and spin quantum number. The electrons in a shell have identical n but different l, m and spin quantum numbers. Therefore, the correlation of the $1s^2$ pair is intrasubshell correlation energy and also intrashell since for $n = 1$ only $l = 0$ is allowed.

The correlation of the $2s^2$ pair is intrasubshell. The correlation of the $2s^12p^1$ pair is intersubshell. The correlation of the $1s^12s^1$ or $1s^12p^1$ is an intershell correlation. In general, we can say that intra terms are more important than inter terms and subshell cases are more important than shell cases for the correlation energy effects.

The electronic configuration for the ground state of Be is $1s^22s^2(^1S)$. However, this configuration is only approximately correct. The reason is as follows. To a crude approximation the Be atom has an energy corresponding to the sum of the energies of the four electrons in the nuclear field alone. That is to say that the main contribution to the energy is the electron–nucleus attraction and the kinetic energy of the electron in the field of the bare nucleus. This approximation is more and more correct as the nuclear charge gets higher. Therefore, four electrons of configuration $1s^22s^2$ in Ne^{6+} are better represented by the above approximation than the four electrons of configuration $1s^22s^2$ in Be. The above approximation is the hydrogen-like approximation. For hydrogen, the $2s$ and $2p$, or the $3s$, $3p$, $3d$ energy levels are degenerate in energy. Therefore, insofar as the hydrogenic approximation holds for Be (or Ne^{6+}) we should expect a degeneracy between the $1s^22s^2$ and the $1s^22p^2$ configurations. This degeneracy effect has been quantitatively considered by Hartree and collaborators in 1934[19–21]. Linderberg and Shull[51] and Froman[52] reexamined the problem and referred to it as the 'Z effect', for obvious reasons.

The problem was then reanalysed by McKoy and Sinanoglu[53] using some data of empirical nature and some computation. Clementi and Veillard[22,26,27] continued Hartree's work by: (*a*) reintroducing the problem into the self-consistent multiconfiguration framework and (*b*) by making no use of empirical data. Our data[26,27] give a total correlation of $-0{\cdot}094$ a.u. for Be in the $1s^22s^2$

configuration and a total correlation of $-0{\cdot}052$ a.u. for the configuration $= 0{\cdot}94839(1s^22s^2) - 0{\cdot}31710(1s^22p^2)$, where the values $0{\cdot}94839$ and $-0{\cdot}31710$ are the expansion coefficients obtained from the self-consistent-field multiconfiguration technique. McKoy and Sinanoglu obtained for the total correlation energy in the two

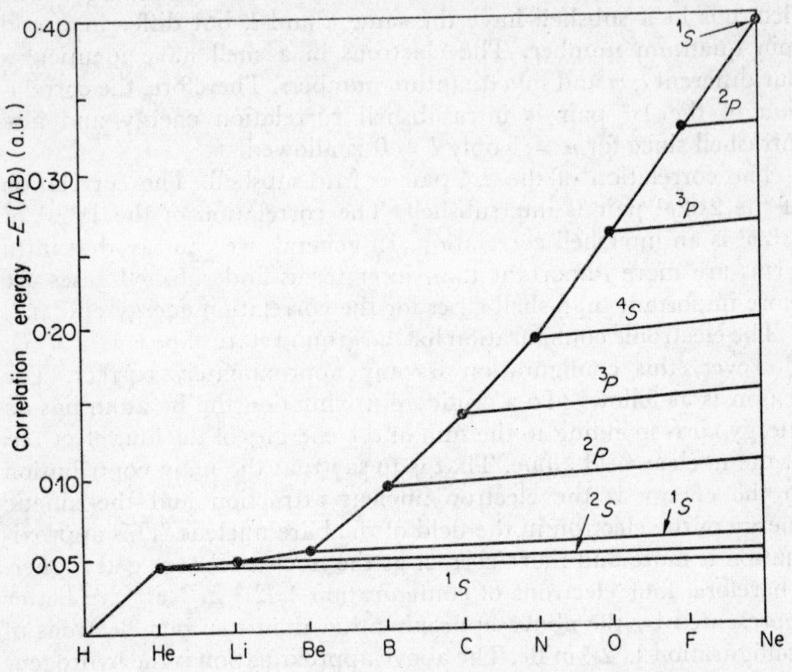

FIGURE 5. Correlation energy for first-row atoms after subtraction of the 'Z effect' energy. The data reported are for the ground state of the first-row atoms and the corresponding isoelectronic series (up to $Z = 10$).

configurations $1s^22s^2$ and $1s^22p^2$ a value in error by over 20%. In Figure 5 the 'degeneracy' effect has been subtracted from the total correlation energy by Clementi and Veillard[26]. Since the large fraction of the $2s$ electrons in Be and its isoelectronic series of positive ions have been substituted by $2p$ electrons, the correlation energy in Figure 4 is not much larger than the correlation energy of Li.

Nesbet[54] has recently offered the most comprehensive tabulation of computed (rather than 'experimental') correlated functions for the first period. The breakdown in energy contribution for the Be atoms obtained by Nesbet is as follows: $-0{\cdot}0418$ a.u. for the $1s^2$

intrashell correlation energy, $-0{\cdot}0453$ a.u. for the second intrashell correlation energy (the one usually referred to as $2s^2$ electrons), $-0{\cdot}006$ a.u. for the intershell correlation energy, namely, $1s$ and $2s$ (again $2s^2$ is here to be taken as designation of the second pair, rather than specification of quantum numbers). The values $-0{\cdot}0418$ a.u. and $-0{\cdot}0453$ a.u. are quite similar and this breakdown corresponds to what was assumed by simple inspection of Figure 4[50].

Let us continue in analysing the 'experimental' correlation energies and consider boron ($1s^22s^22p^1$, ^2P), carbon ($1s^22s^22p^2$, ^3P) and nitrogen ($1s^22s^22p^3$, ^4S) and their isoelectronic series for the positive ions. No new intrapair correlation is added from what is present in the Be atom, since the spins in $2p^1$, $2p^2$ and $2p^3$ are parallel for the ground states. However, the density of the $2s$ follows quite closely the density of $2p$ orbitals (neglecting the $2s$ orthogonality). Since the statistical model is density dependent, we should expect an increase in the correlation energy. From a different point of view we could say that in the same region of space of the $2s$ pair, new electrons, the $2p$, are added (again here $2s$ refers both to $2s$ proper and its promotion to $2p$) and, therefore, we expect intersubshell correlation.

Clearly, it does not matter too much whether the $2s$ is a proper $2s$ or partly $2s$ and partly $2p$. The net effect is that there are sets of three, four and five electrons in the L shell for boron, carbon and nitrogen and only one has different spin from the rest of the set. The 'degeneracy effect' is still present for boron ($1s^22s^22p^1$ and $1s^22s^02p^3$ both have a ^2P state) or carbon ($1s^22s^22p^2$ and $1s^22s^02p^4$ both have a ^3P state), but not for nitrogen (the $1s^22s^22p^3$ has a ^4S state, but $1s^22s^02p^5$ has no ^4S state). By comparing Figures 4 and 5 this can be clearly noted, especially if one compares the isoelectronic series. This much is obtained from analysis of the 'experimental' data.

On returning to Nesbet's work we find that the $1s^2$ correlation energy is nearly the same as in He or Li, or Be. The $(2s^m2p^n)$ correlation energies are $-0{\cdot}071$ a.u., $-0{\cdot}098$ a.u. and $-0{\cdot}127$ a.u. and the interaction between $(1s^2)$ and $(2s^m2p^n)$, i.e. the intershell correlations, are $-0{\cdot}009$ a.u., $-0{\cdot}012$ a.u. and $-0{\cdot}014$ a.u., respectively. By inspection of Figure 5, the constant increase in the $2s^m2p^n$ correlation is expected. Since in Figure 5 the $1s^2-(2s^m2p^n)$ intershell correlation is not distinguishable from the $(2s^m2p^n)$ intrashell correlation, all one can conclude for the latter is that it should increase by about $0{\cdot}04$ a.u., which compares very well with Nesbet's data.

Let us consider now the oxygen, fluorine and neon atoms and the corresponding isoelectronic series of positive ions (note that additional data for the isoelectronic series are given in the original publications). The configurations for these atoms are $1s^2 2s^2 2p^4$, $1s^2 2s^2 2p^5$, $1s^2$, $2s^2$, $2p^6$ and the corresponding ground states are ^3P, ^2P and ^1S. We have sets of six, seven and eight electrons sharing nearly the same space in the L shell; these sets consist of two electrons with α spin and four with β spin, three with α spin and four with β spin, and four with α spin and four with β spin, respectively. Therefore, we have a larger number of pair interactions, and a higher increment of correlation than for B, C and N. This is clear from Figures 4 and 5.

The decomposition given by Nesbet in terms of shells is $-0{\cdot}019$ a.u., $-0{\cdot}022$ a.u. and $-0{\cdot}025$ a.u. for the K–L intershell correlation, and $-0{\cdot}187$ a.u., $-0{\cdot}251$ a.u. and $-0{\cdot}317$ a.u. for the L intrashell correlation energy. (A more detailed breakdown is available from the Nesbet work in terms of subshells.)

Let us now comment on the excited states of the lowest electronic configurations, reported in Figure 4 only for the neutral atoms.

First of all, we should notice that for a state of some multiplicity and angular momentum (like the ^1S for He, Be, C, O, Ne or the ^2P for B, N, F) the correlation energy is a linear function of the number of the electrons.

Second, we should notice that for a given atom the correlation energy decreases with increasing value of the angular momentum; for example, the *singlet states* with angular momentum S have more correlation than the *singlets* with angular momentum D. The same holds for D and P.

Therefore, we have a correlation energy which works in the opposite way to Hund's rules. This is of interest because if the correlation energy difference between states is larger than the term separation then Hund's rules will not hold.

The explanation of this behaviour for the excited states is to be found in the fact that the larger (from S to P, from P to D, from D to F, etc.) the angular momenta, the more structured is the electronic density.

For S states, the distribution varies only along the r coordinates (in the r, θ, φ, space of an atom). For higher angular momentum the density varies not only along r but along θ and φ. This point will be reanalysed later in this work in connection with the decrease of the correlation energy in the ten-electron systems Ne, HF, H_2O, NH_3 and CH_4. All these systems are singlets and have a ground state of totally symmetric representation. However, the electronic density is structured, localized, and the more so, the lower is the correlation energy. The decrease in correlation energy for an increment in angular momentum is not small and of the order of about −0·015 a.u. This is approximately the decrease along the Ne, HF, H_2O, NH_3 and CH_4 series.

The reason for the inability of the statistical method to obtain such variations (see Table 3) is that in equation (59) or in equation (60) the angular part does not appear in the integration since it is derived only for closed shells. It is clear, therefore, that the statistical method could, in principle, give such variations directly, by appropriate expansion of equations (59) or (60).

Additional examples of variation in correlation energy with total angular momentum can be found in the *d*-electron series (Sc to Zn) and are discussed elsewhere[50].

VII. MOLECULAR EXTRA CORRELATION ENERGY

From our previous discussion on the atomic correlation energy it is clear that the correlation energy in a molecule should be related to the correlation energy in the component atoms. First of all, from the statistical analysis of the correlation energy, we know that the correlation energy is proportional to the electronic density, since in no way is the statistical analysis restricted to an atom rather than to a molecule.

From the statistical model we know of the additivity of the correlation, to a first-order approximation. Indeed, the total density is the sum of the orbital densities. Whenever in a molecule an orbital has mainly atomic character (like those of the inner shells), its correlation is that of the corresponding atom. Transferability of correlation energy data from atoms to molecules or solids[33] has been obvious since 1934.

However, there are differences in the density of a molecule when compared with the density one would obtain by simply superimposing atomic densities. Therefore, we would expect differences in the molecular correlation energies of the corresponding atoms.

The energy difference obtained by subtracting from the correlation energy of a molecule the correlation energy of the component atoms (in the correct dissociation state) has been called molecular extra correlation energy (Clementi[55], 1962). The name implicitly contains the connotation of an increase in correlation energy from the sum of those belonging to the corresponding atoms. The main reason for such an increase is that the formation of the molecule always increases the number of electron pairs. This increase is two-fold: the most important one is due to the fact that electrons previously unpaired in the separated atoms pair during molecule formation. For example, the hydrogen atom has one unpaired spin, the fluorine atom has nine electrons, eight paired and one unpaired, but in the hydrogen fluoride molecule the two unpaired spins (the one of H and the one of F) pair up ($H(^2S) + F(^2P) \rightarrow HF(^1\Sigma^+)$). Extreme cases of pairing in diatomic molecules clearly correspond to extreme unpairing (higher spin multiplicity) of separated atoms like N_2 from $N(^4S)$, P_2 from $P(^4S)$, Mn_2 from $Mn(^6S)$.

The second factor in increasing the number of pairs is that more interpair correlation energy contributions will occur, because of the increase in the number of electrons.

The molecular extra correlation energy can be a substantial fraction of the total dissociation energy.

Let us indicate with *hf*, *c*, *r* quantities obtained from the Hartree–Fock methods, or due to correlation or relativistic effects, respectively. Let us indicate the component atoms of an N-atom molecule, m, by the subscript a, and finally let us use the symbols D for dissociation energy and E for total energy.

The exact dissociation energy of a molecule is then given by the relation

$$D = D_{hf} + D_c + D_r$$

or

$$D = D_{hf} + \text{MECE} + D_r \tag{61}$$

where

$$D_{hf} = \left(\sum_{a=1}^{N} E_{a,hf}\right) - E_{m,hf}$$

$$D_c = \left(\sum_{a=1}^{N} \mathrm{E}_{a,c}\right) - E_{m,c} \equiv \text{molecular extra correlation energy (MECE)}$$

$$D_r = \left(\sum_{a=1}^{N} E_{a,r}\right) - E_{m,r}$$

In other words, we partition the dissociation energy into a Hartree–Fock component, a correlation component and a relativistic energy component. The quantity D_{hf} is obtained directly from the molecular computations of Hartree–Fock type (now standardly available) and from the atomic Hartree–Fock energies[56].

The quantity D_c is the definition of molecular extra correlation energy (MECE) previously discussed.

The quantity D_r is the relativistic part of the dissociation energy. It is customary to assume that $\sum_{a=1}^{N} E_{a,r} = E_{m,r}$ and, therefore, to discount such terms. We do not agree with such an optimistic approach except for molecules containing first- and possibly second-row atoms. For molecules containing third-row atoms, we expect D_r to have a value of from a few tenths of a calorie up to a kilocalorie in extreme cases. However, such contributions can be computed safely from perturbation theory. For molecules containing fourth-row (and higher) atoms, the relativistic contribution to the dissociation energy is not expected to increase nearly as much as the atomic relativistic energy, but to increase somewhat more than merely in the 0·1 to few kilocalories range. The above comments apply in particular to compounds where the heavy atom is highly ionic, in the sense given by population analysis.

Let us take a few examples of dissociation energy. For the N_2 molecule in its ground state ($^1\Sigma^+$) the experimental dissociation energy is 9·90 eV, the computed dissociation energy in the Hartree–Fock approximation is $D_{hf} = 5{\cdot}18$. The molecular extra correlation energy is therefore 4·72 eV, taking $D_r = 0{\cdot}000$ eV.

Let us now consider HCN in its ground state ($^1\Sigma^+$). The computed dissociation energy is $D_{hf} = 8{\cdot}85$ eV, the experimental dissociation energy $D = 13{\cdot}55 \pm 0{\cdot}05$ eV and MECE $= 4{\cdot}7 \pm 0{\cdot}05$ eV. It is noted that HCN and N_2 molecules are isoelectronic and that for HCN also we form three new pairs in the reaction

$$H(^2S) + C(^3P) + N(^4S) \longrightarrow HCN(^1\Sigma^+)$$

The agreement of the extramolecular correlation of 4·72 eV for N_2 and 4·70 ± 0·05 eV indicates how transferable are data within molecular systems. It is noted that MECE for N_2 is somewhat larger than MECE for N_2: the reason is that in HCN the electronic charges are more localized than in HCN.

Let us now consider N_2^+ in the $^2\Pi_u$ and $^2\Sigma_g$ states. The MECE are 4·14 eV and 5·73 eV. This is a transposition to molecular cases of the variation of correlation energy with angular momentum as previously discussed in atoms.

In the past few years an increasingly large body of data is available so that it is expected that we shall be very shortly in a position to estimate the molecular extra correlation energy from atomic data and the Hartree–Fock dissociation energy, to within an error of a few percent[56].

VIII. REMARKS ON THE ELECTRONIC STRUCTURE OF THE —CN GROUP

The —CN group is a 14-electron group; 13 electrons are supplied by the C and N atoms (6 + 7) and one is used to bind the CN group. The latter one is supplied by R in R—CN where R is either an atom or a radical (aliphatic or aromatic). In this way we might consider the CN^- negative ion as the prototype of the —CN group. The —CN group can bind via σ-electrons (these are symmetric around the C—N axis, generally chosen to be the z axis and are primarily built with the $1s$, $2s$ and $2p_\sigma = 2p_z$ electrons on C and N) or via the π-electrons (which are built up with the $2p_x$ and $2p_y$ electrons) or both. In addition, the CN group can give or receive charges from R and in that case the binding has an ionic contribution. Finally, the 14th electron supplied by R creates a new pair and this brings about a contribution to MECE of 1·5 ± 5 eV.

Let us briefly analyse H—CN in the ground-state configuration $1\sigma^2 2\sigma^2 3\sigma^2 4\sigma^2 5\sigma^2 1\pi^4$, which yields a $^1\Sigma^+$ state. The $1\sigma^2$ and the 2σ are about 100% of the $1s^2$ on C and N. These four electrons are the inner shell of the —CN group. Of the ten remaining electrons, six are of the σ-type and four are of π-type: these are the valency electrons. The $3\sigma^2$ is mainly contributed by $2s$ on C and $2s$ on nitrogen. The $4\sigma^2$ and $5\sigma^2$ contribute both to the nitrogen lone pair and to the H—C bond. Note that HCN is isoelectronic with N_2 where the last two σ-orbitals are the lone pairs on the two nitrogens. In HCN one of the lone pair of N_2 is essentially transformed into the H—C bond.

It can be said that three of the four π-electrons are contributed by the nitrogen and one by the carbon if we take as 'reference field' the isolated atoms. The three from the nitrogen spill over the carbon and a bit further, indeed over the hydrogen, too. Therefore, we have π-charge-transfer from CN to R = H and CN is a π-donor, H being a π-acceptor. As a result of the electropositivity of the hydrogen atom, it will donate σ charges to the CN group; i.e. the R = H is a σ-donor and CN is a σ-acceptor. A simple description of the charge transfer in HCN is given, for example, by Clementi and Clementi[57].

A much better wave function for HCN has been computed by McLean and Yoshimine[58]. Let us analyse (Table 5) the orbital energies obtained by McLean and Yoshimine. For comparison let

TABLE 5. Orbital energies for selected RCN molecules (in a.u.).

Case	H—CN[a]	F—CN[b]	(O—CN⁻)[c]	(S—CN⁻)[d]	(NC—CN)[e]	(HCC—CN)[f]	(Cl—CN)[g]	CN($^2\Pi$)[h]
$\varepsilon(1\sigma)$	−15·5970	−26·4375	−20·3177	−91·7031	−15·6598	−15·6191	−104·9336	−15·6173
$\varepsilon(2\sigma)$	−11·2888	−15·6115	−15·2657	−15·3120	−15·6598	−11·3281	−15·6063	−11·3236
$\varepsilon(3\sigma)$	−1·2348	−11·4018	−11·0716	−11·0670	−11·3604	−11·3160	−11·3479	−1·2199
$\varepsilon(4\sigma)$	−0·8178	−1·7673	−1·1851	−8·7079	−11·3593	−11·3122	−10·6579	−0·5828
$\varepsilon(5\sigma)$	−0·5817	−1·2523	−0·9063	−6·3903	−1·3057	−1·2612	−8·1244	
$\varepsilon(6\sigma)$		−0·9308	−0·4694	−0·9502	−1·2804	−1·1100	−1·2717	
$\varepsilon(7\sigma)$		−0·6012	−0·2928	−0·7823	−0·9091	−0·9193	−1·1949	
$\varepsilon(8\sigma)$				−0·3903	−0·6446	−0·7715	−0·7570	
$\varepsilon(9\sigma)$				−0·3059	−0·6293	−0·6004	−0·5884	
$\varepsilon(1\pi)$	−0·4952	−0·8141	−0·3675	−6·3872	−0·6027	−0·5527	−8·1218	−0·5194
$\varepsilon(2\pi)$		−0·4984	−0·1349	−0·2785	−0·5004	−0·4376	−0·5983	−0·2429
$\varepsilon(3\pi)$				−0·1124			−0·4646	
$\varepsilon(1s)$	−11·3255	−15·6289	−20·6686	−26·3829	−92·0362	−0·5000	−104·8846	
$\varepsilon(2s)$	−0·7056	−0·9452	−1·2443	−1·5725	−9·0330		−10·6077	
$\varepsilon(3s)$					−0·8957		−1·0731	
$\varepsilon(2p)$	−0·4333	−0·5675	−0·6319	−0·7300	−6·7110		−8·0725	
$\varepsilon(3p)$					−0·3835		−0·5065	
Case	C(^3P)	N(^4S)	O(^3P)	F(^2P)	S(^3P)	H(^2S)	Cl(^2P)	

[a] CN distance is 2·1791 a.u.; R—CN distance is 2·0143 a.u.
[b] CN distance is 2·2016 a.u.; R—CN distance is 2·3811 a.u.
[c] CN distance is 2·2810 a.u.; R—CN distance is 2·2130 a.u.
[d] CN distance is 2·300 a.u.; R—CN distance is 2·9500 a.u.
[e] CN distance is 2·1860 a.u.; R—CN distance is 2·6080 a.u.
[f] CN distance is 2·1864 a.u.; R—CN distance is 2·6116 a.u.
[g] CN distance is 2·1978 a.u.; R—CN distance is 3·0784 a.u.
[h] CN in the $^2\Pi$ state (CN distance is 2·214 a.u.).

us at the same time consider the orbital energies of the molecules[58]

$$FCN(^1\Sigma^+);\ 1\sigma^2 2\sigma^2 \cdots 7\sigma^2 1\pi^4 2\pi^4$$

$$OCN^-(^1\Sigma^+);\ 1\sigma^2 2\sigma^2 \cdots 7\sigma^2 1\pi^4 2\pi^4$$

$$SCN^-(^1\Sigma^+);\ 1\sigma^2 2\sigma^2 \cdots 9\sigma^2 1\pi^4 2\pi^4 3\pi^4$$

$$NC\text{—}CN(^1\Sigma_g^+);\ 1\sigma^2 2\sigma^2 \cdots 9\sigma^2 1\pi^4 2\pi^4$$

$$HCC\text{—}CN(^1\Sigma_g^+);\ 1\sigma^2 2\sigma^2 \cdots 9\sigma^2 1\pi^4 2\pi^4$$

$$Cl\text{—}CN(^1\Sigma^+);\ 1\sigma^2 2\sigma^2 \cdots 9\sigma^2 1\pi^4 2\pi^4 3\pi^4$$

and the orbital energies of the $C(^3P)$, $N(^4S)$, $O(^3P)$, $F(^2P)$, $S(^3P)$ and $Cl(^2P)$ atoms.

Of interest are, for example, the last π-orbital of the neutral molecules: for all the molecules of Table 6, the orbital energy is about $-0{\cdot}49 \pm 0{\cdot}05$ a.u. For the negative species OCN^- and SCN^- the highest filled π-orbital has an energy much lower (less bound) and is about $-0{\cdot}13 \pm 0{\cdot}02$ a.u. This constancy of the π-orbital energy is expected to hold for most molecules. It is noted that in the example given, R varies from strongly electropositive (R = H) to neutral (R = CN) to electronegative (R = F) and, therefore, the sample molecules in Table 6 should cover most RCN cases. The orbital energy of the R—CN bond is somewhat more specific than the π-orbital energy (for the highest orbital) and this is expected since the ionicity and covalency are quite different and reflect themselves mainly in the R—CN σ-bond.

TABLE 6. Orbital energies and total energies (in a.u.) due to perturbing point charges in the cyano group.

Case	$-\varepsilon(1\sigma)$	$-\varepsilon(2\sigma)$	$-\varepsilon(3\sigma)$	$-\varepsilon(4\sigma)$	$-\varepsilon(5\sigma)$	$-\varepsilon(1\pi)$	Total energy
$Z^{0\cdot2}(5)CN^-$	15·3297	11·0197	0·9678	0·3745	0·2310	0·2283	−92·3551
$Z^{0\cdot2}(3)CN^-$	15·3439	11·0341	0·9811	0·3901	0·2562	0·2416	−92·3765
$Z^{0\cdot2}(2)CN^-$	15·3498	11·0337	0·9862	0·4017	0·2794	−0·2473	−92·3792
$Z^{0\cdot4}(5)CN^-$	15·3659	11·0573	1·0025	0·4096	0·2713	0·2628	−92·3932
$Z^{0\cdot4}(3)CN^-$	15·3952	11·0876	1·0302	0·4426	−0·3216	0·2900	−92·4413
$Z^{0\cdot4}(2)CN^-$	15·4097	11·0909	1·0426	1·4764	0·3641	0·3034	−92·4562
$Z^{-0\cdot2}(5)CN^-$	15·2595	10·9472	0·9000	0·3064	0·1515	0·1612	−92·2823
$Z^{-0\cdot2}(3)CN^-$	15·2464	10·9346	0·8876	0·2927	0·1270	0·1488	−92·2658
$Z^{-0\cdot2}(2)CN^-$	15·2437	10·9404	0·8854	0·2875	0·1044	0·1458	−92·2735
$Z^{-0\cdot4}(5)CN^-$	15·2253	10·9122	0·8670	0·2734	0·1123	0·1285	−92·2476
$Z^{-0\cdot4}(3)CN^-$	15·2005	10·8888	0·8433	0·2476	0·0641	0·1046	−92·2193
$Z^{-0\cdot4}(2)CN^-$	15·1987	10·9069	0·8424	0·2415	0·0220	0·1018	−92·2456
CN^-	15·2942	10·9830	0·9336	0·3401	0·1911	0·1945	−92·3182

The computed dipole moments are[59] 3·29 D for HCN (exp. 2·95), 7·22 D for OCN^-, 2·28 D for FCN (exp. 2·17), 6·93 for SCN^-, 3·05 D for ClCN (exp. 2·80) and 4·13 for HCC—CN. In general, one can state that the agreement between the computed dipole moment and the experimental dipole moment is rather good. It is noted that the dipole moment given here had been computed in the Hartree–Fock approximation. No equivalent work with better wave functions has been done up to date.

Let us analyse in more detail some recent results[60] from an SCF computation of the CN^- radical. The electronic configuration is $1\sigma^2 2\sigma^2 3\sigma^2 4\sigma^2 5\sigma^2 1\pi^4$ and the corresponding orbital energies are: $\varepsilon(1\sigma) = -15{\cdot}2942$ a.u.; $\varepsilon(2\sigma) = -10{\cdot}9830$ a.u.; $\varepsilon(3\sigma) = -0{\cdot}9336$ a.u.; $\varepsilon(4\sigma) = -0{\cdot}3401$ a.u.; $\varepsilon(5\sigma) = -0{\cdot}1911$ a.u.; $\varepsilon(1\pi) = -0{\cdot}1945$ a.u., respectively. The total Hartree–Fock electronic energy is $-92{\cdot}3182$ a.u.; the sum of the Hartree–Fock energies of the separated atoms $C(^3P)$ and $N(^4S)$ is

$$-37{\cdot}6886 \text{ a.u.} - 54{\cdot}4009 \text{ a.u.} = -92{\cdot}0895 \text{ a.u.}$$

Therefore, the Hartree–Fock binding energy is $92{\cdot}3182 - 92{\cdot}0895 = 0{\cdot}2287$ a.u. (or 6·223 eV). The total Hartree–Fock energy for the CN molecule ($^2\Pi$ state) is $-91{\cdot}8510$ a.u. and, therefore, in the Hartree–Fock approximation $CN^-(^1\Sigma^+)$ is more stable than $CN(^2\Pi)$ by 0·4672 a.u. The molecular extra correlation energies (MECE) in CN^- should not be too different from the molecular extra correlation energy in HCN, previously discussed (about $4{\cdot}7 \pm 0{\cdot}05$ eV).

Another estimate of the molecular extra correlation energy in CN^- is obtained by considering the N_2 molecule (isoelectronic with CN^-) and in this case MECE for N_2 is 4·72 eV. On this basis we shall estimate the binding energy of CN^- as $6{\cdot}223 + 4{\cdot}7 \sim 10{\cdot}9$ eV.

A final comparison might be in order, namely the experimental binding energy of CO 11·242 eV; the binding of CN^- is therefore not dissimilar from the binding of CO^-. (Our computations for CN^- have been performed at a CN distance of 1·7774 Å.)

Since we have taken the CN^- radical as prototype of the —CN group, we report some euristic computations which should elucidate the —CN role in RCN compounds. The R group will appear to the CN group either as positively or negatively charged. For example, we can think that HCN is really $H^{\delta+}$ and $CN^{\delta-}$, where δ is some value between 1 and 0. Since the —CN group has 14 electrons as the CN^- groups, we shall simulate RCN by a study of the electronic charge rearrangement of CN^- in the presence of point charges.

For example, let us recompute the SCF wave function of CN^- in the presence of a positive point charge 0·2 located 5 a.u. from the carbon atom on the CN axis. For simplicity we shall write this 'compound' $Z^{0\cdot 2}(5)(CN)^-$ where the notation indicates that a point charge Z, of value +0·2, has been added 5 a.u. from the C atom along the CN axis in CN^-.

In Table 6 we compare CN^- with a variety of $Z^x(Y)CN^-$ and we report the energy variation in CN^- when point charges +0·2, +0·4, −0·2 and −0·4 are placed 5, 3 and 2 a.u. from the C atom on the CN axis. All the orbital energies are affected by the point charges and not only the valency electrons. This is important, since it tells us that a probe in the inner shell is as important as a probe in the valency shell in understanding the substitution (i.e. fields variation) effect between different R's in R—CN. The positive charges lower the orbital energies, i.e. the electrons are more tightly bound, and as a consequence the total energy of the system is increased. The lowering increases either by increasing the point charge (from +0·2 to +0·4, for example) or by setting the same point charge nearer and nearer to the carbon atom. The negative charges decrease the stability of the system, i.e. the orbital energy and the total energy decrease. The extreme case, i.e. ±0·4 charge at 2·0 a.u. from the CN group, should correspond to the case of a highly positively or negatively charged R. It should be noted that we have taken as reference CN^- so as to have 14 electrons, since the point charge does not introduce any additional electron to the system. Therefore, this analysis is relevant only in describing the classical electrostatic effect and relative to the CN^- molecule. However, it gives us, by analogy, some of the gross features to be expected in —CN, namely, that if the atom which adjoins the —CN group has an overall positive or negative charge, then one should expect the —CN group to interact as indicated in Table 6.

Let us now turn to the isocyano group, and let us repeat the point-charge simulation, this time by placing positive (+0·2 or +0·4) or negative (−0·2 or −0·4) charges at various distances (2, 3 and 5 a.u.) from the nitrogen atom, along the CN axis in the CN^- field. In Table 7 we have used the notation $CN^-(X)Z^Y$ to indicate that the point charge Z of value Y has been placed at a distance X from the nitrogen atom.

The orbital energies for the 1σ, 2σ are by and large equal to those of Table 6. The 4σ and 5σ in the $CN^-(X)Z^Y$ are respectively less stable and more stable than the corresponding $Z^Y(X)CN^-$ case, upon point-charge perturbation. Note that the 4σ and 5σ correspond to the lone pair on nitrogen and the lone pair on carbon (the bond between R and CN, in our model). The π-orbital energies are not too different in Tables 5 and 6; however, one should notice a trend towards a decrease in stability by comparing Table 6 with Table 7.

These differences are explained by referring to the net atomic charges on the carbon and nitrogen atoms (Table 7) computed for

TABLE 7. Orbital energies and total energies (in a.u.) due to perturbing point charges in the isocyano group.

Case	$-\varepsilon(1\sigma)$	$-\varepsilon(2\sigma)$	$-\varepsilon(3\sigma)$	$-\varepsilon(4\sigma)$	$-\varepsilon(5\sigma)$	$-\varepsilon(1\pi)$	Total energy
$(CN)^-(5)Z^{0.2}$	15·3275	11·0167	0·9670	0·3766	0·2230	0·2283	−92·3538
$(CN)^-(3)Z^{0.2}$	15·3427	11·0326	0·9827	0·4004	0·2384	0·2440	−92·3743
$(CN)^-(2)Z^{0.2}$	15·3469	11·0413	0·9921	0·4270	0·2486	0·2527	−92·3807
$(CN)^-(5)Z^{0.4}$	15·3619	11·0510	1·0012	0·4137	0·2554	0·2628	−92·3906
$(CN)^-(3)Z^{0.4}$	15·3938	11·0841	1·0337	0·4633	0·2862	0·2952	−92·4354
$(CN)^-(2)Z^{0.4}$	15·4030	11·1019	1·0533	0·5198	0·3051	0·3128	−92·4558
$(CN)^-(5)Z^{-0.2}$	15·2619	10·9501	0·9008	0·3043	0·1595	0·1613	−92·2836
$(CN)^-(3)Z^{-0.2}$	15·2484	10·9356	0·8864	0·2826	0·1444	0·1468	−92·2668
$(CN)^-(2)Z^{-0.2}$	15·2454	10·9277	0·8781	0·2609	0·1318	0·1388	−92·2678
$(CN)^-(5)Z^{-0.4}$	15·2302	10·9179	0·8686	0·2691	0·1283	0·1287	−92·2500
$(CN)^-(3)Z^{-0.4}$	15·2053	10·8902	0·8412	0·2280	0·0980	0·1008	−92·2199
$(CN)^-(2)Z^{-0.4}$	15·2012	10·8757	0·8261	0·1907	0·0697	0·0858	−92·2296
CN^-	15·2942	10·9830	0·9336	0·3401	0·1311	0·1345	−92·3182

the $Z^Y(X)CN^-$ and $CN^-(X)Z^Y$ cases. The positive point charge pulls electrons towards itself in any case; however, the pull is much more effective when the positive charge is on the N side than when on the C side. The negative point charge pushes electrons away; however, the repulsion is more effective when the negative charge is on the nitrogen side than when on the carbon side. This is, however, only a very crude description of the electrostatic effects due to point charges. There are complications, which can not be explained by simple reference to Table 8, and even for simple point charges, the repulsion or attraction for the electrons varies orbital by orbital. Of course, we would expect that the inner-shell electrons will be effected less than the valency electrons. After all, the inner-shell electrons are much more energetic and will be therefore less perturbed; in addition, the inner-shell electrons do not experience directly the perturbing point charge, but only through the screening of the valency electrons. However, we do not refer to such obvious

TABLE 8. Gross atomic population analysis variations due to perturbing point charges[a].

Case	Total charge on C	N	Case	Total charge on C	N
$Z^{0.2}(5)CN^-$	6·1456	7·8546	$CN^-(5)Z^{0.2}$	6·0733	7·9267
$Z^{0.2}(3)CN^-$	6·1673	7·8327	$CN^-(3)Z^{0.2}$	6·0400	7·9599
$Z^{0.2}(2)CN^-$	6·1403	7·8598	$CN^-(2)Z^{0.2}$	6·0210	7·9790
$Z^{0.4}(5)CN^-$	6·1908	7·8092	$CN^-(3)Z^{0.4}$	6·0505	7·9494
$Z^{0.4}(3)CN^-$	6·2366	7·7632	$CN^-(3)Z^{0.4}$	5·9850	8·0150
$Z^{0.4}(2)CN^-$	6·1940	7·8060	$CN^-(2)Z^{0.4}$	5·9476	8·0524
$Z^{-0.2}(5)CN^-$	6·0596	7·9403	$CN^-(3)Z^{-0.2}$	6·1348	7·8651
$Z^{-0.2}(3)CN^-$	6·0410	7·9589	$CN^-(3)Z^{-0.2}$	6·1700	7·8300
$Z^{-0.2}(2)CN^-$	6·0826	7·9175	$CN^-(2)Z^{-0.2}$	6·1918	7·8082
$Z^{-0.4}(5)CN^-$	6·0195	7·9804	$CN^-(5)Z^{-0.4}$	6·1723	7·8277
$Z^{-0.4}(3)CN^-$	5·9871	8·0129	$CN^-(2)Z^{-0.4}$	6·2936	7·7063
$Z^{-0.4}(2)CN^-$	6·0868	7·9131	CN^-	6·1018	7·8983

[a] The four decimal figures here reported are excessive in a physical sense but are given only for numerical accuracy.

orbital deformation, but to a variation in the direction of the migration: the point charge pulls some of the orbital toward itself (or repels it) but this pull in one orbital is partially compensated in a repulsion (or attraction) in another orbital. In other words, some orbitals experience charge transfer in one direction and some different orbitals experience charge transfer in the opposite direction upon perturbation by the same point charge.

This is indicated in Table 9, where we report the gross charge on the carbon and nitrogen atoms for CN^-, $Z^{0.2}(2)CN^-$, $Z^{-0.2}(2)CN^-$, $(CN)^-(2)Z^{0.2}$ and $(CN)^-(2)Z^{-0.2}$. For example, in $Z^{0.2}(2)CN^-$ the pull towards the positive point charge has an overall effect of about 0·04 electrons increase on the carbon atom at the expense of the nitrogen atom. We have an increase of 0·02 electrons in the 3σ, of 0·24 electrons in the 4σ, a decrease of 0·26 electrons in the 5σ and an increase of 0·05 electrons in the 1π-orbital.

It is likely that a good characterization of the —CN group is its extreme versatility in exchanging charges from C to N in a complex (two-way) fashion. It is important to stress that this two-way charge-transfer effect is not due to the presence of outside electrons, but can be found by a simple point charge, i.e. it is a 'classical electrostatic effect'. The real picture will be by necessity even more complex when we replace the point charge with electropositive or electronegative atoms. Indeed, in such a case not only is there an electrostatic effect

TABLE 9. Orbital by orbital charge migration due to point-charge perturbation.

Case	CN^-	$Z^{0.2}(2)CN^-$	$Z^{-0.2}(2)CN^-$	$(CN)^-(2)Z^{0.2}$	$(CN)^-(2)Z^{-0.2}$
$1\sigma(C)$	−0·0029	−0·0028	−0·0030	−0·0030	−0·0027
$1\sigma(N)$	2·0029	2·0028	2·0030	2·0030	2·0027
$2\sigma(C)$	1·9999	1·9999	2·0000	2·0000	1·9999
$2\sigma(N)$	0·0001	0·0001	0·0000	0·0000	0·0001
$3\sigma(C)$	0·3306	0·3495	0·3148	0·3322	0·3290
$3\sigma(N)$	1·6694	1·6505	1·6852	1·6677	1·6709
$4\sigma(C)$	0·5287	0·7621	0·4098	0·3993	0·7385
$4\sigma(N)$	1·4713	1·2379	1·5902	1·6007	1·2615
$5\sigma(C)$	1·7520	1·4916	1·9071	1·8758	1·5533
$5\sigma(N)$	0·2480	0·5084	0·0929	0·1242	0·4467
$1\pi(C)$	1·4933	1·5398	1·4539	1·4167	1·5737
$1\pi(N)$	1·5067	2·4602	1·5461	2·5833	2·4262
Total (C)	6·1018	6·1402	6·0825	6·0210	6·1918
Total (N)	7·8982	7·8597	7·9175	7·9790	7·8082

but the R group of RCN can either donate or accept electrons, and these are either of σ- or π-type or both. The charge transfer between R and —CN can be related to one of the 16 possible cases represented in Table 10. In Table 10 we consider the cases where either R or CN can be a π-donor or acceptor or a σ-donor or acceptor. One could state that most of the chemistry of the —CN group can be explained by the 16 cases in Table 10.

TABLE 10. Charge transfer between R and CN groups.

	CN, σ-donor $\leftarrow\sigma$	CN, π-donor $\leftarrow\pi$	CN, σ-acceptor $\sigma\rightarrow$	CN, π-acceptor $\pi\rightarrow$
R; σ-donor; $\sigma\rightarrow$	1 $\left(\begin{matrix}\leftarrow\sigma\\ \sigma\rightarrow\end{matrix}\right)$	2 $\left(\begin{matrix}\leftarrow\pi\\ \sigma\rightarrow\end{matrix}\right)$	3 $\left(\begin{matrix}\sigma\rightarrow\\ \sigma\rightarrow\end{matrix}\right)$	4 $\left(\begin{matrix}\pi\rightarrow\\ \sigma\rightarrow\end{matrix}\right)$
R; π-donor; $\pi\rightarrow$	5 $\left(\begin{matrix}\leftarrow\sigma\\ \pi\rightarrow\end{matrix}\right)$	6 $\left(\begin{matrix}\leftarrow\pi\\ \pi\rightarrow\end{matrix}\right)$	7 $\left(\begin{matrix}\sigma\rightarrow\\ \pi\rightarrow\end{matrix}\right)$	8 $\left(\begin{matrix}\pi\rightarrow\\ \pi\rightarrow\end{matrix}\right)$
R; σ-acceptor; $\leftarrow\sigma$	9 $\left(\begin{matrix}\leftarrow\sigma\\ \leftarrow\sigma\end{matrix}\right)$	10 $\left(\begin{matrix}\leftarrow\pi\\ \leftarrow\sigma\end{matrix}\right)$	11 $\left(\begin{matrix}\sigma\rightarrow\\ \leftarrow\sigma\end{matrix}\right)$	12 $\left(\begin{matrix}\pi\rightarrow\\ \leftarrow\sigma\end{matrix}\right)$
R; π-acceptor; $\leftarrow\pi$	13 $\left(\begin{matrix}\leftarrow\sigma\\ \leftarrow\pi\end{matrix}\right)$	14 $\left(\begin{matrix}\leftarrow\pi\\ \leftarrow\pi\end{matrix}\right)$	15 $\left(\begin{matrix}\sigma\rightarrow\\ \leftarrow\pi\end{matrix}\right)$	16 $\left(\begin{matrix}\pi\rightarrow\\ \leftarrow\pi\end{matrix}\right)$

Let us analyse in some detail this table. Clearly, cases 1, 6, 11 and 16 (i.e. the cases along the diagonal of the table) cannot exist, since the charge-transfer characteristics of the R and CN group are mutually exclusive. It is noted, however, that cases 1, 6, 11 and 16 could occur whenever the R and CN groups interact with a third group (be it an atom, molecule or set of molecules).

The cases 2, 5, 12 and 15 are two-way charge-transfer cases (note that the table is redundant insofar as case 2 and case 15 as well as case 5 and case 12 are identical).

The cases 3, 4, 7, 8, 9, 10, 13 and 14 are one-way charge-transfer cases. The arrows point out this fact, and we have used the convention of having an arrow pointing to the right whenever charges flow from R to CN, and pointing to the left whenever charges flow from CN to R. The charge-transfer table can be extended easily from the RCN case to any molecule. The 16 cases originated by our choice of analysing two groups (R and CN) and two types of charges (σ and π). Clearly, this is an arbitrary choice. For example, we could have chosen rather than σ- or π-electrons, inner-shell and valency-shell electrons, or molecular orbitals or localized orbitals. In addition, a more adequate generalization of Table 10 would have repeated on the vertical column the four CN possibilities. In this way we could analyse charge transfer between groups (R with CN) and within groups (R within itself and CN within itself). Then we could talk of one-way and two-way charge transfer between two different groups and of carriers (one-way) and transformers (two-way) charge transfer within one group[61]. We could ask . . . what is the meaning of a charge transfer *within* a group (or atom)? Equally well, however, we could ask ourselves what is the meaning of a charge transfer *between* groups. The answer to both questions is that electrons *do not* belong to atoms or molecules. Electrons simply feel fields and induce fields. We chemists think of electrons as belonging to atoms or molecules and in order to both understand electrons and to express ourselves we use 'reference fields'. By definition, reference fields are arbitrary; however, these are sometimes useful concepts for communication reasons. Classical examples of charge transfer within a group which makes use of 'reference states' are the valency states. To a large degree the same comment holds for the concept of hybridization, electron affinity, electronegativity and polarizability.

Let us briefly survey some population analysis results for the vinylisocyanide molecule $CH_2{=}CHNC$. The molecule has been computed in the xy plane with the x axis passing through the two carbons of the ethylene group ($CH_2=CHR$). This plane has reflection symmetry which is the only symmetry for $CH_2=CHNC$. The three carbon atoms are hereafter designated as $C_{(1)}$, $C_{(2)}$ and $C_{(3)}$ where $C_{(1)}$ is the carbon atom of CH_2, $C_{(2)}$ is the carbon atom of CH and $C_{(3)}$ is the carbon of the —NC groups. The H_2 of CH_2 are designated as $H_{(5)}$ and $H_{(6)}$ and $H_{(7)}$ is the hydrogen of CHNC. The following net charges have been computed[60].

$$\begin{array}{ll} N & 1s^{1\cdot 997}2s^{1\cdot 277}2p_x^{1\cdot 379}2p_y^{1\cdot 289}2p_z^{1\cdot 423} \\ C_{(1)} & 1s^{1\cdot 999}2s^{1\cdot 150}2p_x^{1\cdot 059}2p_y^{1\cdot 188}2p_z^{0\cdot 986} \\ C_{(2)} & 1s^{1\cdot 999}2s^{1\cdot 104}2p_x^{1\cdot 038}2p_y^{0\cdot 985}2p_z^{0\cdot 037} \\ C_{(3)} & 1s^{1\cdot 999}2s^{1\cdot 597}2p_x^{0\cdot 070}2p_y^{0\cdot 955}2p_z^{0\cdot 554} \end{array}$$

$$H_{(5)} \quad 1s^{0\cdot 783} \qquad H_{(6)} \quad 1s^{0\cdot 779} \qquad H_{(7)} \quad 1s^{0\cdot 762}$$

Again, the hydrogens are σ-donors, having released about 0·2 electrons per hydrogen atom. The nitrogen σ-charge is the sum of the $1s$, $2s$, $2p_x$ and $2p_y$ charges, namely 5·942 electrons.

We shall now analyse the computed electronic population of methyl cyanide CH_3CN and methyl isocyanide CH_3NC. For CH_3CN the gross charges are

$$\begin{array}{ll} C_{(1)} & 1s^{1\cdot 999}2s^{0\cdot 918}2p_\sigma^{1\cdot 005}2p_\pi^{1\cdot 993} \\ C_{(2)} & 1s^{1\cdot 999}2s^{1\cdot 226}2p_\sigma^{0\cdot 997}2p_\pi^{2\cdot 432} \\ N & 1s^{1\cdot 998}2s^{1\cdot 672}2p_\sigma^{1\cdot 470}2p_\pi^{2\cdot 034} \\ H & 1s_\sigma^{0\cdot 239}1s_\pi^{0\cdot 514} \end{array}$$

where $C_{(1)}$ is the carbon atom of the CN group. Only one hydrogen is reported since the three hydrogens are symmetrical. The equivalent data for CH_3—NC are:

$$\begin{array}{ll} C_{(1)} & 1s^{1\cdot 999}2s^{1\cdot 592}2p_\sigma^{1\cdot 121}2p_\pi^{1\cdot 147} \\ C_{(2)} & 1s^{1\cdot 999}2s^{1\cdot 209}2p_\sigma^{0\cdot 868}2p_\pi^{2\cdot 417} \\ N & 1s^{1\cdot 999}2s^{1\cdot 285}2p_\sigma^{1\cdot 223}2p_\pi^{2\cdot 855} \\ H & 1s_\sigma^{0\cdot 235}1s_\pi^{0\cdot 528} \end{array}$$

The hydrogen differs very little in the cyano compound relative to the isocyano compound. The carbon atom of the CH_3 group (designated as $C_{(2)}$) is not too different in the two compounds, however, somewhat more in the CH_3NC than in CH_3CN (the excess charges on $C_{(2)}$ in CH_3CN is 0·654 whereas it is 0·493 in CH_3NC). The main difference, as expected, is in the isocyanide or cyanide group. In CH_3CN the nitrogen has an excess charge of 0·17 electrons versus 0·36 electrons in CH_3NC. The CN carbon has a charge deficiency of 0·14 electrons in CH_3NC and a charge deficiency of 0·08 electrons in CH_3CN. The main overall difference is that in the cyano compound there is charge sign alternation, namely $(H_3)^+$—C^-—C^+—N^- whereas in the isocyano compound there is charge reenforcement $(H_3)^+$—C^-—N^-—C^+.

After this preliminary comparison between CH_3CN and CH_3NH, let us analyse the results in more detail. First, we shall analyse the overall σ- and π- (correctly *a* and *e*) charges in CH_3CN. From the above data we can conclude that there is σ- and π-donation from CH_3 to CN and the net result is $[CH_3]^{+0{\cdot}088}[CN]^{-0{\cdot}088}$ and we should note that the small charge transfer is about twice as much of σ-electrons as of π. The CH_3 group has four π-electrons and five σ-electrons. The H_3 group can be explained by assuming that each hydrogen atom has 0·3333 σ-electrons and 0·6666 π-electrons. The carbon atom in CH_3 has 'reference' configuration $1s^22s^12p_\sigma{}^12p_\pi{}^2$. Hence, from the computed data reported above we find that the H_3 group donates 0·283 σ-electrons and 0·458 π-electrons. The excess of 0·060 σ-electrons and 0·026 π-electrons, left from the donation by H_3 and acceptance from C is transferred to the CN group.

In conclusion, the H_3 group donates charges to the carbon atom and to the CN group. Therefore, H_3 is a σ- and π-donor, the carbon atom is a σ- and π-acceptor. In addition, the carbon passes over to the CN group the excess donation from the H_3 group, and, we shall therefore say that the C atom is not only a σ-, π-acceptor, but also a σ-, π-'carrier'. By carrier we mean that the atom transmits charges taken from a group (or atom) to a second group (or atom). The CN group is an overall acceptor, as noted above. The detailed charge-transfer mechanism is as follows. The carbon atom of CN has configuration $1s^22s^12p^12p_\pi^2$ or four σ-electrons and two π-electrons. The computation indicates a donation of 0·078 σ-charges and 0·007 π-charges. The nitrogen atom has five σ-electrons and two π-electrons and is an acceptor by 0·140 σ-charges and 0·034 π-charges. These charges are 0·078 σ from the C of CN and 0·060 from the σ of H_3; from the π-charges 0·007 are donated to the N atom from the C of CN and 0·026 from the H_3. Therefore, the C atom of CN is both a σ-, π-donor and σ-, π-carrier; the nitrogen atom is an acceptor both of σ- and π-charges.

The charge transfer goes through the following stages:

donor → acceptor, carrier → donor and carrier → acceptor

Let us now analyse the CH_3—NC molecule. The H_3 group (one σ-electron and two π) donates 0·295 σ-charges and 0·416 π-charges; the carbon atom of CH_3 (four σ-electrons and two π-electrons) accepts 0·076 σ-charges and 0·416 π-charges from the H_3 group; the nitrogen atom (five σ-electrons and two π-electrons) donates 0·493 σ-charges and accepts 0·853 π-charges and, finally, the carbon atom of the CN group (four σ-electrons and two π-electrons) accepts 0·712 σ-charges and donates 0·853 π-charges. The H_3 group is a σ- and π-donor; the C atom of CH_3 is a σ- and π-acceptor, the nitrogen atom is a σ-donor and π-acceptor, and finally, the last carbon atom is a σ-acceptor and π-donor. Since σ-charges are transferred from H_3 to the C of CN, then the C of CH_3 and N are σ-carriers. Whereas in the CH_3CN molecule we had only one-way charge transfer, in CH_3NC we have two-way charge transfer between the CN atoms of the CN group. The process is, therefore, donor → acceptor and carrier → donor, acceptor, and carrier → acceptor or donor. To be exact, there is a negligible π-transfer between the carbon atoms (0·002) but this is within the rounding error. A schematic account of the entire energy-transfer process is given in Table 11.

The charge-transfer ability of the CN group has important consequences in the dipole moment of RCN and CNR compounds. The dipole moment μ of CN^- is very small, 0·14922 a.u. (or 0·3792 D) from our computation. This is of no surprise since the dipole moment of CN should not be too different from the dipole moment of N_2, namely zero, or of CO(0·112 D).

Before analysing the dipole moment variation in —CN and CN— compounds, let us discuss in some detail the dipole moment of CN^-. There are, of course, two components in the dipole moment, one due to the nuclear charges, the second due to the electronic charges. The nuclear charge component, μ_1, is the sum of the products of the charges and the distance from a given origin. The electronic component μ_2 is the sum of the expectation value of R for the orbitals of the wave function. In CN (or in any linear molecule) μ_2 is the sum over all electrons of the expectation value of z, if z is the molecular axis, since the expectation values for x and y are zero.

Whereas for neutral molecules the choice of the dipole moment origin is of no importance (i.e. the dipole moment is invariant with respect to the origin), this is not the case for non-neutral species.

TABLE 11. Energy-transfer process for the CH_3NC molecule.

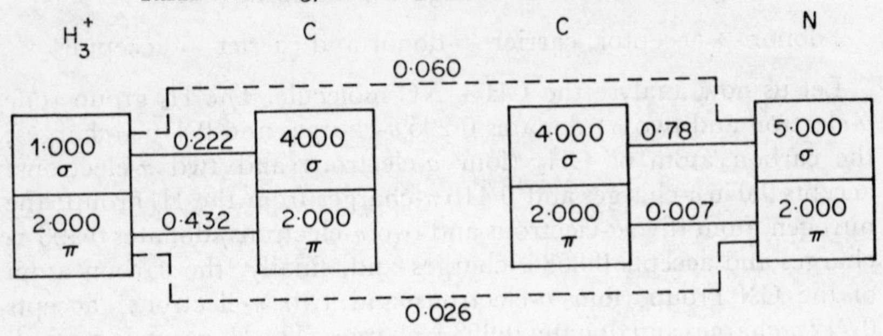

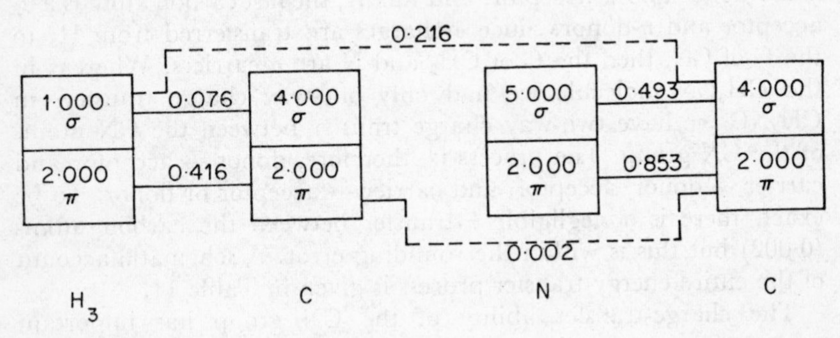

We can prove this most simply. Let us select an origin O and a second origin O′, for example, along the z axis so that we shall have $O = O' - z_0$; thus we shall have the following two expressions for the dipole moment with respect either to O or to O′ (assuming N nuclei of charge z and n electrons)

$$\mu_{(O)} = \mu_{1(O)} + \mu_{2(O)} = \sum_{i=1}^{N} Z_i z_{i(O)} - \sum_{j=1}^{n} \langle \varphi_j | \, z \, | \varphi \rangle_O \tag{62}$$

$$\mu_{(O')} = \mu_{2(O')} + \mu_{2(O')} = \sum_{i=1}^{N} Z_i z_{i(O')} - \sum_{j=1}^{n} \langle \varphi_j | \, z \, | \varphi_j \rangle_O \tag{63}$$

We can express $\mu_{(O')}$ in terms of $\mu_{(O)}$ and we have

$$\mu_{(O')} = \mu_{1(O')} + \mu_{2(O')} = \sum_{i=1}^{N} Z_i (z_i - z_O) - \sum_{j=1}^{n} \langle \varphi_j | \, z - z_O \, | \varphi_j \rangle_O \tag{64}$$

The term $\langle \varphi_j | \, z_O \, | \varphi_j \rangle = z_O \langle \varphi_i \mid \varphi_i \rangle = z_O n$, is the number of electrons multiplied by the constant z_O. Therefore, the term within parentheses is equal to zero for neutral molecules, and $\mu_{(O')} = \mu_{(O)}$. For non-neutral molecules, or for molecules in an external point charge field, the term within parentheses is not zero. For example, in CN^-, the term within parentheses is $(13 - 14) = -1$ since there are 13 positive charges (six from C, seven from N) but 14 electrons.

In the following we shall analyse the dipole moment variation of CN^- due to positive or negative charges either in the vicinity of C or of N. We shall find that the very small dipole moment of CN^-, becomes a very substantial dipole moment; part of this increase is due to μ_1, but a large part is due to μ_2. Since our point charges simulate the addition of an R group to —CN, this analysis indicates that we should expect a large dipole moment in RCN or CNR compounds. The sensitivity of μ_2 to outside fields is a manifestation of the CN ability to have internal two-way charge transfer.

It might be instructive to report the details of the value of μ for CN^-.

$$\begin{aligned}\mu_2 &= (-2{\cdot}2222 \times 2) - (0{\cdot}0007 \times 2) - (1{\cdot}3985 \times 2) \\ &\quad - (2{\cdot}0123 \times 2) + (0{\cdot}1197 \times 2) - (1{\cdot}3939 \times 4) \\ &= -16{\cdot}6036 \text{ a.u.}\end{aligned}$$

The above values are the computed electronic dipole moment components for the electrons in $1\sigma^2$, $2\sigma^2$, $3\sigma^2$, $4\sigma^2$, $5\sigma^2$ and $1\pi^4$ molecular orbitals. By taking as origin the carbon atoms, the value of μ_1 is 15·55616 a.u. Therefore, if the origin is at the carbon atom, the dipole moment for CN^- is $\mu = -16{\cdot}6036 + 15{\cdot}55616 = -1{\cdot}04744$ a.u. Since CN^- is not neutral, we should compute μ with respect to a more sensible origin, namely the centre of positive charges. The distance between C and N is 2·222309 a.u. and using a mass of 12 for carbon and a mass of 14 for nitrogen, the centre of positive charges is 1·1966 a.u. from the carbon atom. Translation of the origin from C to the centre of positive charges, gives a new dipole moment of 0·14922 a.u.

In Table 12 we analyse the dipole moment variation due to positive or negative point charges either on the carbon or on the nitrogen side. In the first column we give the case identification, in the second column, the nuclear charge contribution to μ_O (origin with reference to the carbon atom), in the third column the electronic contribution to μ_O, in the fourth column $\mu_{O'}$, (origin at the centre of charges). The fifth, sixth and seventh columns give the electronic contribution from a single electron in the 4σ, 5σ and 1π molecular orbitals, respectively. A casual analysis reveals rather substantial dipole moments. One would, of course, expect a variation

TABLE 12. Dipole moment in CN^- with and without perturbation[a].

Case	μ_1	μ_2	$\mu_{(O)}$	$-Rc(\sum Z_N - 14)$	$\mu_{(O')}$	$-\mu(4\sigma)^1$	$-\mu(5\sigma)^2$	$-\mu(1\pi)^1$
$Z^{+0\cdot4}(5)CN^-$	14·7562	−16·2718	−1·5156	0·7180	−0·7977	1·8523	−0·0429	1·3566
$Z^{+0\cdot4}(3)CN^-$	14·3562	−16·0584	−1·7022	0·7180	−0·9843	1·4747	0·2810	1·3340
$Z^{+0\cdot2}(3)CN^-$	14·9562	−16·3342	−1·3780	0·9593	−0·4207	1·7883	0·0364	1·3644
$Z^{+0\cdot2}(2)CN^-$	15·1562	−16·3218	−1·1656	0·9593	−0·2083	1·5102	0·3191	1·3601
$Z^{-0\cdot4}(5)CN^-$	17·5562	−16·9140	+0·6422	−1·6753	−1·0331	2·1391	−0·1704	1·4298
$Z^{-0\cdot4}(3)CN^-$	16·7562	−17·1292	−0·3730	−1·6753	−2·0483	2·2855	−0·2569	1·4495
$Z^{-0\cdot4}(2)CN^-$	16·3562	−17·0418	−0·6856	−1·6753	−2·3609	2·4178	−0·4264	1·4470
$Z^{-0\cdot2}(5)CN^-$	16·5562	−16·7608	−0·2046	−1·4360	−1·6406	2·0793	−0·1479	1·4120
$Z^{-0\cdot2}(3)CN^-$	16·1562	−16·8660	−0·7098	−1·4360	−2·1458	2·1711	−0·2111	1·4223
$Z^{-0\cdot2}(2)CN^-$	15·9562	−16·8458	−0·8896	−1·4360	−2·3254	2·2749	−0·3275	1·4231
CN^-	15·5562	−16·6036	−1·0474	1·1967	0·1492	2·0123	0·1197	1·3939
$CN^-(2)Z^{-0\cdot2}$	14·7117	−16·2652	−1·5535	−1·4360	−2·9894	1·5826	+0·2571	1·3472
$CN^-(3)Z^{-0\cdot2}$	14·5117	−16·3170	−1·8053	−1·4360	−3·2412	1·8129	+0·0246	1·3561
$CN^-(5)Z^{-0\cdot2}$	14·1117	−16·4510	−2·3393	−1·4360	−3·7752	1·9415	−0·0789	1·3733
$CN^-(2)Z^{-0\cdot4}$	13·8672	−15·9192	−2·0519	−1·6753	−3·7273	1·0259	0·7561	1·2996
$CN^-(5)Z^{-0\cdot4}$	12·6672	−16·3018	−3·6346	−1·6753	−5·3098	1·8670	−0·0324	1·3525
$CN^-(2)Z^{+0\cdot2}$	16·4006	−16·9348	−0·5342	0·9593	+0·4231	2·2949	−0·3547	1·4396
$CN^-(3)Z^{+0\cdot2}$	16·6006	−16·8948	−0·2942	0·9593	+0·6631	2·1833	−0·2321	1·4309
$CN^-(5)Z^{+0\cdot2}$	17·0006	−16·7606	+0·2400	0·9593	+1·1973	2·0801	−0·1549	1·4143
$CN^-(2)Z^{+0\cdot4}$	17·2451	−17·2608	−0·0157	0·7180	+0·7022	2·4710	−0·4896	1·4864
$CN^-(3)Z^{+0\cdot4}$	17·6451	−17·1930	+0·4521	0·7180	+1·1701	2·3305	−0·3164	1·4671
$CN^-(5)Z^{+0\cdot4}$	18·4451	−16·9236	+1·5248	0·7180	+2·2394	2·1457	−0·1842	1·4345

[a] All quantities given in atomic units (1 a.u. = 2·54154 D).

in the nuclear contribution (since we add a point charge). What is of interest is the strong variation in the electronic contribution (column 3), or the strong variation in $\mu(4\sigma)$, $\mu(5\sigma)$ and $\mu(1\pi)$.

We shall now consider the variation in the contribution to the dipole moment due to one of the π-electrons ($\mu(1\pi)$ column). The extrema of the variation are $-1{\cdot}2996$ and $-1{\cdot}4844$ a.u. to be compared with $-1{\cdot}3939$ a.u. for CN^-. The extrema in the variation of the electronic part (μ_2 column) are $-17{\cdot}2608$ and $-15{\cdot}9192$ a.u. to be compared with $-16{\cdot}6036$ a.u. for CN^-. Thus, we notice a maximum variation of about 0·4 due to the four π-electrons, to be compared with a maximum variation of about 0·6 to 0·7 a.u. for the total of μ_2. Therefore, the π-electrons alone, despite their 'mobility', cannot account for more than about 50% of the variation in dipole moment. The remainder is due to the σ-electrons. (In Table 12 we have not reported the values of $\mu_2(1\sigma)$, $\mu_2(2\sigma)$ and $\mu_2(3\sigma)$ since these remain essentially constant upon perturbation of either positive or negative charges.) The 4σ- and 5σ-orbitals represent the lone pairs on the carbon and the nitrogen atoms. By variation in the *sp* hybridization, the lone pair can extend itself toward the outside of the molecule or centre itself more uniformly around the carbon or nitrogen atoms. Therefore, with little (or no) variation in the gross charge, the lone pairs can profoundly vary their contribution to the dipole moment. Indeed this is what is reported in Table 12. The $\mu_2(4\sigma)$ is 2·0123 a.u. for CN^-, but upon perturbation this value changes to the extrema of 1·0259 and 2·4710 a.u. The $\mu_2(5\sigma)$ is 0·1197 a.u. in CN^-, but upon perturbation of the negative or positive charges, this value changes to the extrema of $-0{\cdot}4896$ a.u. and $+0{\cdot}7561$ a.u. In addition, the 4σ- and 5σ-orbitals collaborate in reducing the overall variation. Thus, for example, when $\mu_2(4\sigma) = 2{\cdot}4710$ a.u., the $\mu_2(5\sigma) = -0{\cdot}4896$ and when $\mu_2(5\sigma) = 0{\cdot}7561$ a.u., the $\mu_2(4\sigma) = 1{\cdot}0259$ a.u.

In conclusion, the overall result in the dipole moment indicates that the π- and 4σ- and 5σ-electrons very strongly react to perturbing charges and, by shifting their electronic cloud, can substantially increase the dipole moments in RCN and CNR compounds over the dipole moment value for CN^-.

We can be more specific: a positive charge in the vicinity of the carbon atom in CN^- is less effective in increasing the dipole moment of CN^- than a negative charge. A positive charge in the vicinity of the nitrogen atom in CN^- is less effective in increasing the dipole moment of CN^- than a negative charge.

A positive charge in the vicinity of CN^- simulates an atom (or atoms) which has lost part of its electrons in RCN or CNR. Similarly, a negative charge in the vicinity of CN^- simulates an atom (or atoms) which has gained some fraction of electrons in RCN or CNR. However, we now have to consider the fact that we are using CN^- in simulating —CN or CN—.

Let us consider, for example, LiCN or CNLi[62]. The molecule will be highly polar and to a first approximation it will be $Li^+(CN)^-$ or $(CN^-)Li^+$; the analysis of LiCCH[63] is rather conclusive in this regard. In this extreme case we can expect to use Table 12 to obtain an estimate of the dipole moment by considering the point charge of about $+1{\cdot}0$ at a distance of about 3·5 a.u. (which seems to be a reasonable distance for LiC bound in LiCN). From Table 12 we can interpolate a value of $\mu_2 = 16{\cdot}34$ a.u. for $Z^{+0{\cdot}2}(3{\cdot}5)CN^-$ and a value of $\mu_2 = 16{\cdot}11$ a.u. for $Z^{+0{\cdot}4}(3{\cdot}5)CN^-$; from these two values we can extrapolate $\mu_2 = 15{\cdot}44$ a.u. for $Z^{+1{\cdot}0}(3{\cdot}5)CN^-$. The nuclear component, μ_1, can be easily computed as $\mu_1 = 12{\cdot}0562$ a.u., yielding a total dipole moment estimate of $-15{\cdot}44 + 12{\cdot}0562 = -3{\cdot}38$ a.u. for LiCN. Let us now estimate from Table 12 the dipole moment for CNLi; again we use Table 12 to obtain $\mu_2 = 16{\cdot}875$ a.u. for $CN^-(3{\cdot}5)Z^{+0{\cdot}2}$ and $\mu_2 = 17{\cdot}150$ a.u. for $CN^-(3{\cdot}5)Z^{+0{\cdot}4}$. By extrapolation we obtain $\mu_2 = 17{\cdot}835$ a.u. for $CN^-(3{\cdot}5)Z^{+1{\cdot}0}$; since $\mu_1 = 21{\cdot}2788$ the dipole moment estimated for CNLi, assuming $(CN)^-Li^+$ and assuming a N—Li distance of 3·5 a.u., is 3·44 a.u. or nearly equal to the dipole moment of LiCN.

Work in progress in our laboratory indicates that the above estimates are substantially (within 15–20%) correct.

It seems, therefore, that Table 12 could be of help in understanding the general characteristics of RCN and CNR compounds.

This brings to an end our discussion on the electronic structure for the —CN group. There are several very important aspects in the chemistry of the CN group that we have not even mentioned. For a rather extended survey of dipole moment data in cyano compounds, the reader is referred to a paper of Sundararian[64]. For some semiempirical study on RCN compounds the reader is referred, for example, to the work of Moffat[65]. In general, one can obtain from the above semiempirical work, a good set of 'rules' for correlating experimental data, with sufficient reliability as to be valuable even for prediction.

The preliminary data reported here should make it clear that theoretical computation which accounts for all the electrons in a molecule has become feasible in the last few years. At present, the main hope of many theoretical chemists is in obtaining better wave functions than the Hartree–Fock functions. However, in the meantime a larger number of self-consistent-field all-electron wave functions are being computed and analysed. For historical prospective it is worthwhile to note that the first crude self-consistent-field wave functions for the first row diatomic molecules were computed in early 1960. The systematic work on Hartree–Fock wave functions for diatomic and linear molecules took place between 1964 and 1967. Medium polyatomic molecules with up to ten second-row atoms and a few hydrogen atoms have been computed in the last two to three years. There is, therefore, very much reason to be optimistic even if somewhat disappointed by the lack of a systematic analysis on the cyano group.

Several computations here reported were hurriedly performed quite recently in order to present some examples in this chapter. A number of additional computations are in progress at our laboratory.

IX. REFERENCES

1. L. C. Pauling, *The Nature of the Chemical Bond*, Cornell University Press, Ithaca, New York, 1960.
2. R. S. Mulliken, *J. Chem. Phys.*, **3**, 375 (1935).
3. W. Moffit, *Proc. Roy. Soc.* (*London*), **A210**, 245 (1951).
4. M. Kotany, *Rev. Mod. Phys.*, **32**, 266 (1960).
5. C. A. Coulson, *Valency*, Oxford University Press, Oxford, 1959.
6. S. Kolos and L. Wolniewicz, *J. Chem. Phys.*, **41**, 3663 (1964); *J. Chem. Phys.*, **43**, 2429 (1965).
7. E. V. Condon and G. H. Shortley, *The Theory of Atomic Spectra*, Cambridge University Press, Cambridge, 1957.
8. C. C. J. Roothaan, *Rev. Mod. Phys.*, **23**, 69 (1951).
9. A. C. Wahl, *J. Chem. Phys.*, **41**, 2600 (1964).
10. V. Fock, *Izv. Akad. Nauk SSSR, Ser. Fiz. Mat. Nauk*, **18**, 161 (1954).
11. A. C. Hurley, J. E. Lennard-Jones and J. A. Pople, *Proc. Roy. Soc.* (*London*), **A220**, 446 (1953).
12. E. Hylleraas, *Z. Physik*, **54**, 347 (1929); *Z. Physik*, **65**, 759 (1930).
13. E. S. Boys and G. B. Cook, *Rev. Mod. Phys.*, **32**, 285 (1960).
14. H. Hartmann and E. Clementi, *Phys. Rev.*, **133**, A1295 (1964).
15. E. Clementi, *J. Mol. Spec.*, **12**, 18 (1964).
16. Young-Ki Kim, *Phys. Rev.*, **154**, 17 (1967).
17. W. Kolos and L. Wolneiwicz, *J. Chem. Phys.*, **41**, 3674 (1964).
18. J. Frenkel, *Wave Mechanics, Advanced General Theory*, Clarendon Press, Oxford, 1934.
19. D. R. Hartree, W. Hartree and B. Swirles, *Phil. Trans. Roy. Soc.* (*London*), **A238**, 223 (1939).

20. A. P. Yutsis, *Zh. Eksperim. i. Teor. Fiz.*, **23**, 129 (1952); *Zh. Eksperim. i. Teor. Fiz.* **24**, 425 (1954); A. P. Yutsis, *Soviet Phys.*, **2**, 481 (1956). See also T. L. Gilbert, *J. Chem. Phys.*, **43**, S248 (1956).
21. A. P. Yutsis, Ya. I. Vizbaraire, T. D. 'Strotskire and A. A. Bandzaitis, *Op. Spectr. USSR Eng. Transl.*, **12**, 83 (1962).
22. A. Veillard, *Theoret. Chim. Acta* **4**, 22 (1966).
23. A. Veillard and E. Clementi, *Theoret. Chim. Acta*, **7**, 133 (1967).
24. E. Clementi, *J. Chem. Phys.*, **46** (3842), 1967.
25. G. Das and A. C. Wahl, *J. Chem. Phys.*, **44**, 87 (1966).
26. E. Clementi and A. Veillard, *J. Chem. Phys.*, **44**, 3050 (1965).
27. E. Clementi and A. Veillard, *J. Chem. Phys.*, September 1, 1968.
28. R. K. Nesbet, *Phys. Rev.*, **155**, 51, 56 (1967).
29. See, for example: P.-O. Löwdin, *Technical Note No. 2 (1957)*, *Technical Note No. 48 (1960)*, Quantum Chemistry Group, Uppsala University, Uppsala (Sweden); P.-O. Löwdin, *Preprint No. 53 (1964)*, *Preprint No. 65 (1964)*, *Quantum Theory Project for Research in Atomic, Molecular and Solid State Chemistry and Physics*, University of Florida, Gainesville, Florida.
30. P.-O. Löwdin, *Advances in Chemical Physics*, Vol. 2 (Ed. I. Prigogine), Interscience, New York, 1959.
31. R. S. Mulliken, *J. Chem. Phys.*, **23**, 1833, 1841, 2338, 2343 (1955).
32. See, for example: J. C. Lorquett. *Rev. Mod. Phys.*, **32**, 312 (1960).
33. E. Wigner, *Phys. Rev.*, **46**, 1002 (1934) and E. Wigner and F. Seitz, *Phys. Rev.*, **43**, 804 (1933).
34. E. Clementi, *J. Chem. Phys.*, **38**, 2248 (1963).
35. e.g. E. A. Hylleraas, *Z. Physik*, **48**, 469 (1928); H. Bethe, *Z. Physik*, **47**, 815 (1929).
36. P. Gombas, *Pseudopotentiale*, Springer-Verlag, New York, 1967.
37. E. Clementi and C. Salez, 'Correlation Energy in Atomic Systems. VI', unpublished results.
38. E. Clementi, *J. Chem. Phys.*, **38**, 1001 (1963).
39. E. Clementi, C. C. J. Roothaan and M. Yoshimine, *Phys. Rev.*, **127**, 1618 (1962).
40. E. Clementi, *J. Chem. Phys.*, **41**, 295 (1964).
41. E. Clementi and A. D. McLean, *Phys. Rev.*, **133**, A419 (1964).
42. E. Clementi, A. D. McLean, D. L. Raimondi and M. Yoshimine, *Phys. Rev.*, **133**, A1274 (1964).
43. E. Clementi, *Phys. Rev.*, **135**, A980 (1964).
44. E. Clementi, D. L. Raimondi and W. P. Reinhardt, *J. Chem. Phys.*, **47**, 1300 (1967).
45. H. Hartman and E. Clementi, *Phys. Rev.*, **133**, A1295 (1964).
46. E. Clementi, *J. Mol. Spectr.*, **12**, 18 (1964).
47. E. Clementi, 'First Row Neutral Atoms and Positive Ions,' *J. Chem. Phys.*, **38**, 2248 (1963); E. Clementi, 'Second Row Neutral Atoms and Positive Ions,' *J. Chem. Phys.*, **39**, 175 (1963); E. Clementi, 'Third Row Neutral Atoms and Positive Ions,' *J. Chem. Phys.*, **42**, 2783 (1965). For a partial survey of correlation energy data, see E. Clementi, *IBM J. Res. Develop.*, **9**, 1 (1965). In this work the Coulomb Hartree–Fock technique is described. See also references 26 and 27 for a critical review of correlation energy data.
48. J. C. Slater, 'Proc. Int. Symp. Atoms, Molecules Solid State Theory,' *Intern. J. Quantum Chem.*, **1967**, 783.
49. K. A. Brueckner and S. K. Ma, *Document IRPA 67-150 (June 1967)*, University of California, La Jolla, California.

50. E. Clementi, 'Tables of Atomic Functions,' *IBM J. Res. Develop. Suppl.*, **9**, 2 (1965).
51. J. Linderberg and H. Shull, *J. Mol. Spectr.*, **5**, 1 (1960).
52. A. Fröman, *Phys. Rev.*, **112**, 870 (1960); *Rev. Mod. Phys.*, **32**, 317 (1960).
53. V. McKoy and O. Sinanoğlu, *J. Chem. Phys.*, **41**, 268 (1964).
54. R. K. Nesbet, *IBM Tech. Report RJ 497* (1968). This work solves equation (58) following the work of H. A. Bethe and J. Goldstone, *Proc. Roy. Soc.* (*London*), **A238**, 551 (1957), and therefore is referred to as 'Bethe–Goldstone' model.
55. E. Clementi, *J. Chem. Phys.*, **35**, 33 (1962).
56. E. Clementi, *J. Chem. Phys.*, **38**, 2780 (1963); *J. Chem. Phys.*, **32**, 487 (1963). K. D. Carlson and P. N. Shancke, *J. Chem. Phys.*, **40**, 613 (1964). See also R. K. Nesbet in *Advances in Quantum Chemistry*, Vol. 3 (Ed. P.-O. Löwdin), Academic Press, New York, 1967 for complete list of references.
57. E. Clementi and H. Clementi, *J. Chem. Phys.*, **11**, 2824 (1962).
58. A. D. McLean and M. Yoshimine, 'Tables of Molecular Wave Functions', a supplement to a paper by McLean and Yoshimine, *IBM J. Res. Develop.*, **12**, 206 (1967).
59. A. D. McLean and M. Yoshimine, *Intern. J. Quantum Chem.*, **1**, 313 (1967).
60. E. Clementi and D. Klint, unpublished results.
61. See R. S. Mulliken, *J. Am. Chem. Soc.*, **72**, 600 (1350); *J. Am. Chem. Soc.*, **24**, 811 (1952) for charge transfer *between two groups* and E. Clementi, *J. Chem. Phys.*, **47**, 2323 (1967) for charge transfer *within one group*.
62. B. Bak and E. Clementi, unpublished results.
63. A. Veillard, *J. Chem. Phys.*, **48**, 2012 (1968).
64. R. Sundararjan, *Indian J. Chem.*, **1**, 503 (1963).
65. J. B. Moffat, *Can. J. Chem.*, **42**, 1323 (1964); see also H. E. Popkie and J. B. Moffat, *Can. J. Chem.*, **43**, 624 (1965).

CHAPTER 2

Introduction of the cyano group into the molecule

Klaus Friedrich and Kurt Wallenfels

University of Freiburg, Freiburg/Br., Germany

I. Introduction 67
II. Preparation of Nitriles by Addition of Hydrogen Cyanide . 68
A. Addition to Carbon–Carbon Multiple Bonds 68
B. Addition to Carbon–Oxygen Double Bonds 72
C. Addition to Carbon–Nitrogen Multiple Bonds 75
III. Preparation of Nitriles by Substitution 77
A. Reaction of Hydrogen Cyanide or Its Salts with Organic Compounds 77
1. Halides 77
2. Aryl sulphonates 84
3. Alcohols, esters and ethers 84
4. Nitro or amino compounds 86
5. Diazonium salts 87
B. Reaction of Cyanogen and Similar Compounds with Aliphatic and Aromatic Compounds 88
IV. Preparation of Nitriles by Elimination 92
A. Starting from Aldehydes, Ketones and Their Derivatives . 92
B. Starting from Carboxylic Acids and Their Derivatives . . 96
V. Miscellaneous Methods 103
VI. Preparation of Nitriles Using Molecules or Molecular Fragments with Cyano Groups 105
VII. References 110

I. INTRODUCTION

The synthetic methods for the preparation of nitriles can be related mainly to four reaction types: addition, substitution, elimination and conversion of other nitriles.

The addition reactions which are important in nitrile synthesis imply the nucleophilic addition of hydrogen cyanide or its salts to multiple bonds between carbon, oxygen and nitrogen atoms.

The second reaction type is the substitution of a suitable leaving group in an organic compound by a cyano group. Other reactions leading to nitriles, such as those between organometallic compounds and various cyano compounds like cyanogen or its halides, are also treated here.

While the two above-mentioned methods introduce the cyano group as a whole, the third method consists of an elimination reaction which converts a suitable carbon–nitrogen system already present in the molecule to a nitrile function.

The conversion of nitriles to other nitriles without affecting the cyano group is a widely used method for the preparation of nitriles. Generally, halogenation, alkylation and acylation reactions of nitriles do not possess features very much different from the corresponding reactions of esters, ketones or nitro compounds. Therefore these methods will not be treated in this chapter. On the other hand, it seemed important to emphasize the nitrile syntheses which use condensation and cycloaddition reactions of molecules and molecular fragments already containing the nitrile function.

In the following text the preparative aspects of nitrile chemistry will be emphasized. The literature references were chosen on the basis of their applicability to problems usually encountered in the research laboratory. Therefore, patents will only be mentioned if their methods can be applied within the possibilities and the scale of a laboratory.

Two excellent reviews have been published some time ago which are of interest for the preparative chemist in the nitrile field and which provide much information concerning synthetic methods[1–3]. The second review contains numerous procedural details.

II. PREPARATION OF NITRILES BY ADDITION OF HYDROGEN CYANIDE

A. Addition to Carbon–Carbon Multiple Bonds

The addition of hydrogen cyanide to olefinic hydrocarbons does not proceed very satisfactorily. Even in the presence of catalysts and at high temperatures the yields are low[1,2].

Contrary to this, acetylenes readily add hydrogen cyanide. The reaction requires a Nieuwland-type catalyst[3], an aqueous solution of cuprous chloride, ammonium chloride and hydrogen chloride. Thus a yield of 80% acrylonitrile is obtained at 80° from acetylene and hydrogen cyanide[4]:

$$CH\equiv CH + HCN \longrightarrow CH_2=CHCN$$

Under the same conditions vinylacetylene gives 1-cyano-1,3-butadiene.

The rather inert carbon–carbon double bond becomes susceptible to base-catalysed addition of hydrogen cyanide when polarized by an alkoxy substituent. Vinyl ethers react with hydrogen cyanide to give 1-alkoxy-1-cyanoethanes:

$$ROCH=CH_2 + HCN \longrightarrow ROCH(CN)CH_3$$

The catalysts used are alkali cyanides or pyridine[5]. The addition of hydrogen cyanide to vinyl esters of organic acids proceeds equally well, catalysts being again alkali cyanides or potassium acetate[6]. Acetaldehyde enhances the rate of the reaction considerably and the following mechanism has been suggested:

$$CH_3CHO + HCN \longrightarrow CH_3CH(CN)OH \xrightarrow{CH_2=CHOAc} CH_3CH(CN)OAc + CH_3CHO$$

Under the influence of the basic catalyst, hydrogen cyanide adds to the acetaldehyde. The resulting cyanohydrin is subsequently acylated by the vinyl ester, thus producing the 1-acyloxy-1-cyano-ethane, and reforms the acetaldehyde which continues the reaction[7].

An example of the addition of hydrogen cyanide to a vinyl amine is the reaction with 1,1-bis(dimethylamino)ethylene, which yields an *N*,*N*-acetal of an acyl cyanide[8]:

$$CH_2=C(NMe_2)_2 + HCN \longrightarrow CH_3C(NMe_2)_2CN$$

The reaction requires no additional catalyst and is complete after 12 hours at room temperature.

The base-catalysed addition of hydrogen cyanide is favoured by electron-withdrawing substituents at the olefinic double bond[16]. Consequently, a wide variety of α,β-unsaturated nitriles, esters, ketones, nitro and sulphonyl compounds undergo addition of hydrogen cyanide in the presence of basic catalysts[1,2]. As might be expected from the direction of the polarization of the carbon–carbon double bond in these compounds, the nitrile group appears in the β position to the activating substituent in the reaction product. Thus the potassium cyanide-catalysed addition of hydrogen cyanide to acrylonitrile gives succinodinitrile in nearly quantitative yield[9]:

$$CH_2{=}CHCN + HCN \longrightarrow NCCH_2CH_2CN$$

In this example water-free hydrogen cyanide is used, yet the most efficient method for obtaining addition products of hydrogen cyanide to an activated olefin is the reaction of potassium cyanide in ethanol–water[10,11]:

$$PhCH{=}C(COOEt)_2 \xrightarrow{KCN} \underset{\displaystyle CN}{PhCH}CH_2COOK$$

The example also shows the shortcomings of this procedure. The reaction conditions facilitate partial hydrolysis of the intermediate addition product and decarboxylation. In order to avoid subsequent reactions, the addition may be effected at room temperature in a slightly acidic solution. In the following example hydrochloric acid is added to a mixture of potassium cyanide and diethyl benzalcyanoacetate[12]:

$$PhCH{=}C(CN)COOEt \longrightarrow \underset{\displaystyle CN}{PhCH}{-}\underset{\displaystyle CN}{CH}COOEt$$

The last two examples showed olefins bearing a second electron-withdrawing substituent at the α-carbon atom. This naturally enhances the electron deficiency of the olefinic double bond[13] and helps to overcome the counter-balancing effect of the phenyl group at the β-carbon atom, which for instance in the case of cinnamic acid derivatives prevents the addition of hydrogen cyanide[14]. The influence of substituents in a β-phenyl group has been investigated in the reaction of potassium cyanide with benzalmalononitriles[15], and except for the *p*-hydroxy derivative, a correlation has been found between the logarithms of the rate constants for the cyanide addition and the Hammett σ values. For a general review on nucleophilic attacks on olefins see reference 16.

The anion resulting from the addition of the cyanide anion to the olefinic bond may undergo a Michael-type addition to a second molecule of the starting compound. This side reaction may become important in cases where the olefin carries electron-withdrawing substituents at both carbon atoms of the double bond, such as dimethylfumarate[13]. Similar complications are found with α-cyanocinnamic acid[17] and benzalacetophenone[18]:

$$\text{PhCH=CHCOPh} \xrightarrow{\text{KCN, AcOH}} \underset{\text{CN}}{\text{PhCHCH}_2\text{COPh}} \longrightarrow \text{PhCHCH}_2\text{COPh} \atop \underset{\text{CN}}{\text{PhCCH}_2\text{COPh}}$$

The structure of the final product needs confirmation, since an attack by the initially formed carbanion will give an isomeric compound. With α,β-unsaturated ketones as in the example shown above, the addition to the carbon–carbon double bond may be followed by a cyanohydrin formation of the β-ketonitrile[19]. A method which avoids the undesirable side reactions encountered in the normal hydrocyanation, uses absolute hydrogen cyanide in the presence of trialkylaluminium compounds, preferably triethylaluminium, in tetrahydrofuran[20]. In the example given one mole of 4-cholesten-3-one reacts with two moles of hydrogen cyanide and three moles of triethylaluminium at 25° to give a mixture of diastereomeric 5-cyano-3-cholestanones in 85% yield.

Owing to the greater tendency to form cyanohydrins, α,β-unsaturated aldehydes add hydrogen cyanide exclusively to the carbonyl group[13].

Quinones represent a special case of α,β-unsaturated ketones. Thus, *p*-benzoquinone reacts with hydrogen cyanide to yield 2,3-dicyanohydroquinone together with hydroquinone[21]:

$$2\ (p\text{-benzoquinone}) + 2\ \text{HCN} \longrightarrow \text{2,3-dicyanohydroquinone} + \text{hydroquinone}$$

The normal course of the reaction between a *p*-benzoquinone and a reagent of the type HX is thought to proceed via a 1,4-addition of HX to the C=C—C=O system. The intermediate then enolizes to the mono-X-hydroquinone, which subsequently is reoxidized by a second molecule of benzoquinone to mono-X-benzoquinone. This again adds HX to give the 2,5-di-X-hydroquinone:

$$p\text{-benzoquinone} \xrightarrow{\text{HX}} \text{X-hydroquinone} \xrightarrow{\text{oxidation}} \text{X-benzoquinone} \xrightarrow{\text{HX}} \text{2,5-di-X-hydroquinone}$$

This mechanism has also been suggested for the reaction between hydrogen cyanide and *p*-benzoquinone[22], with the modification that in the second addition the cyano group already present in the molecule directs the new cyanide anion to the *ortho* position. It has recently been pointed out that the reaction cannot imply the formation of monocyanobenzoquinone because it should have a higher redox potential than the unsubstituted benzoquinone. Furthermore, monocyanobenzoquinone has been synthesized and is an extremely water-sensitive compound which would not have any chance of surviving in the aqueous reaction medium. An alternative mechanism for the formation of 2,3-dicyanohydroquinone which needs further confirmation is suggested[23]. It is assumed that the 1,4-addition product of hydrogen cyanide to *p*-benzoquinone adds a second hydrogen cyanide molecule in 1,6-fashion, thus giving a 2,3-dicyano-2,3-dihydrohydroquinone. The latter is then oxidized by benzoquinone to the 2,3-dicyanohydroquinone.

B. Addition to Carbon–Oxygen Double Bonds

The reaction of the carbonyl group with hydrogen cyanide yields α-hydroxynitriles, commonly called cyanohydrins. The first synthesis of an aromatic cyanohydrin was reported in 1832[24], that of an aliphatic one in 1867[25]. A great improvement was the use of bases to catalyse the reaction[26]. The cyanohydrin formation is an equilibrium reaction comprising two steps:

$$R^1R^2C{=}O + {}^{-}CN \rightleftharpoons R^1R^2C(O^-)(CN) \overset{H^+}{\rightleftharpoons} R^1R^2C(OH)(CN)$$

of which the first nucleophilic addition step is rate-determining[27]. The dissociation constant of a cyanohydrin to its components is governed by several factors. Generally, aldehydes yield the most stable cyanohydrins, followed by aliphatic ketones, whereas aromatic-aliphatic ketones give the least stable cyanohydrins. Purely aromatic ketones do not add hydrogen cyanide. The ring size has a pronounced influence on the stability of the cyanohydrins derived from cyclic ketones. Here, cyclohexanones and cyclopentanones give the most stable cyanohydrins, while with larger rings the stability decreases[28].

Several methods have been developed for the synthesis of cyanohydrins[29]. The use of water-free hydrogen cyanide and a catalyst such as alkali cyanide, hydroxide or carbonate mostly gives good yields[30]:

$$\begin{array}{l} R^1C{=}O \\ | \\ R^2CHCN \end{array} + HCN \xrightarrow{KOH} \begin{array}{l} R^1C(CN)(OH) \\ | \\ R^2CHCN \end{array}$$

Another example is the conversion of 3,5-diethoxycarbonyl-1,2-cyclopentandione into the corresponding bis(cyano)hydrin[31]:

EtOOC–(cyclopentane-1,2-dione)–COOEt + 2 HCN ⟶ EtOOC–(1,2-dicyano-1,2-dihydroxycyclopentane)–COOEt

Some highly enolized ketones, however, such as oxaloacetic esters and benzoylacetic esters, fail to add hydrogen cyanide[29].

The addition may be carried out in aqueous solution by adding mineral acid to a solution of the carbonyl compound and alkali cyanide[32]. An example is the synthesis of mandelonitrile[33] or the preparation of trifluoroacetoacetic ester cyanohydrin[34]:

$$PhCHO + KCN + HCl \longrightarrow PhCH(OH)(CN) + KCl$$

$$CF_3COCH_2COOEt \xrightarrow{KCN,\ HCl} CF_3C(OH)(CN)CH_2COOEt$$

If hexafluoroacetone is treated with an alkali cyanide in tetrahydrofuran, the alkali salts of the cyanohydrin are obtained, and these, after addition of mineral acid, yield the free hydroxynitrile. Chlorofluoroacetones react similarly[35]. Instead of a mineral acid, an organic acid such as acetic acid may be used to generate the hydrogen cyanide. For instance, it can be formed by hydrolysing acetic anhydride, as in the following example[36]:

cyclohexanone $\xrightarrow{KCN, Ac_2O, H_2O}$ 1-hydroxycyclohexanecarbonitrile (HO, CN)

The occurrence of free hydrogen cyanide is avoided if the bisulphite adduct of the carbonyl compound is treated with an aqueous solution of an alkali cyanide[37,38]:

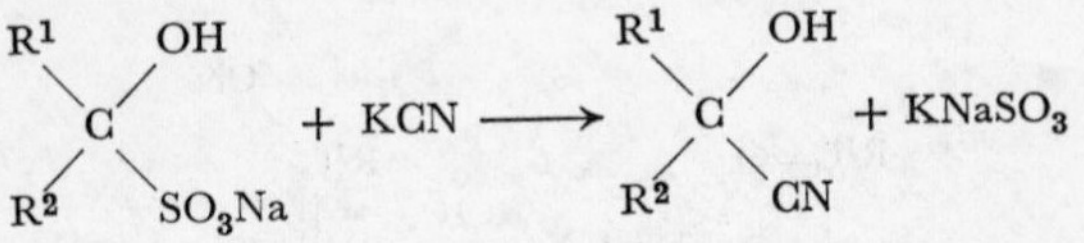

$$R^1R^2C(OH)(SO_3Na) + KCN \longrightarrow R^1R^2C(OH)(CN) + KNaSO_3$$

Since the cyanohydrin formation is an equilibrium reaction, a cyanohydrin with suitable dissociation constant, preferably acetone cyanohydrin, may serve as a hydrogen cyanide donor for another carbonyl compound in the presence of a basic catalyst[39]. In the following example a second molecule of hydrogen cyanide adds to the carbon–carbon double bond[40]:

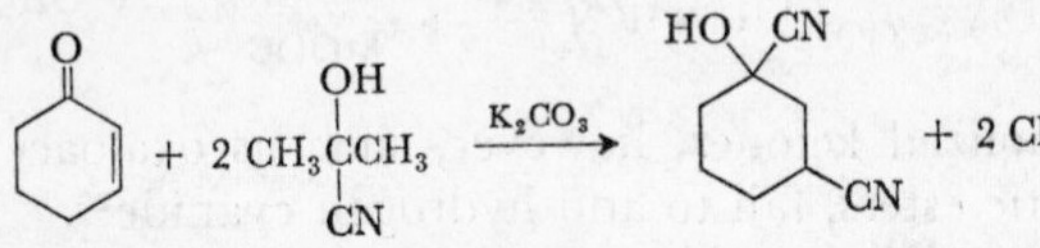

$$\text{cyclohexenone} + 2\,CH_3C(OH)(CN)CH_3 \xrightarrow{K_2CO_3} \text{1-hydroxy-1,3-dicyanocyclohexane} + 2\,CH_3COCH_3$$

Optically active cyanohydrins have been synthesized by addition of hydrogen cyanide to carbonyl compounds in the presence of optically active amines[41]. As mentioned above, with α,β-unsaturated aldehydes no difficulties are encountered, but α,β-unsaturated ketones may add hydrogen cyanide to the olefinic double bond. Another complication may arise with α-halogeno aldehydes and ketones. With these it is preferable to use the method employing water-free hydrogen cyanide and a trace of basic catalyst as shown below[42]:

$$EtCOCH_2Cl + HCN \xrightarrow{K_2CO_3} EtC(OH)(CN)CH_2Cl$$

Use of alkali cyanides in water or water–ethanol mixtures usually leads to glycido nitriles[43] as, for instance, in the case of desyl chloride[44]:

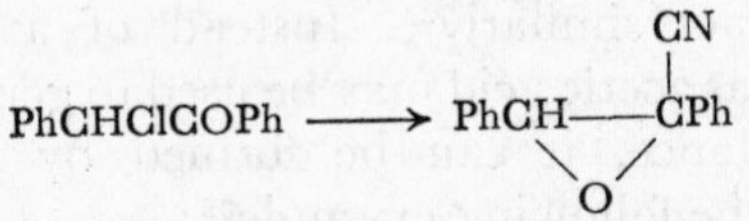

$$PhCHClCOPh \longrightarrow PhCH\overset{O}{—}C(CN)Ph$$

Closely related to the cyanohydrin synthesis is the preparation of α-amino nitriles[45]. In 1850, Strecker obtained alanine after the hydrolysis of the reaction product from acetaldehyde, ammonia and aqueous hydrogen cyanide[46]. Later the intermediate in this reaction was shown to be α-aminopropionitrile[47]. The method was subsequently applied to ketones as well, and it was found that primary and secondary amines could be used instead of ammonia. The initial procedure, adding hydrogen cyanide to the carbonyl–amine adduct, was extended by the finding that the same α-amino nitrile could be synthesized by reacting the cyanohydrin with the amine[48]. The free hydrogen cyanide may be replaced by ammonium cyanide[49] or equimolar amounts of potassium cyanide and ammonium chloride[50]. A number of monoalkylamino nitriles has been synthesized by different methods starting from aldehydes or ketones. No additional catalysts were needed, since the amines proved to be sufficiently basic[51]. The Strecker synthesis has also been applied to aliphatic diamines. Here the carbonyl compound was added to a mixture of the diamine hydrochloride and potassium cyanide in ethanol–water[52]. Substituted hydrazines, needed as precursors for azodinitriles, have been synthesized from the corresponding ketones, hydrazine salt and alkali cyanide in water[53]:

$$2\ R^1R^2C{=}O + N_2H_4{\cdot}HX + 2\ KCN \longrightarrow R^1R^2C(CN){-}NH{-}NH{-}C(CN)R^1R^2$$

The application of ammonium chloride is shown in the reaction leading to methyleneaminoacetonitrile[54]:

$$2\ HCHO + NaCN + NH_4Cl \longrightarrow CH_2{=}N{-}CH_2CN + NaCl + 2\ H_2O$$

Sometimes it may prove advantageous to prepare first the bisulphite adduct of the carbonyl compound, then to add the amine, followed by the cyanide[55] (Knoevenagel–Bucherer method[56]):

$$n\text{-}PrCHO \xrightarrow[3.\ KCN]{1.\ NaHSO_3;\ 2.\ CH_3NH_2} n\text{-}PrCH(NHCH_3)(CN)$$

C. Addition to Carbon–Nitrogen Multiple Bonds

The azomethine system $>C{=}N{-}$ adds hydrogen cyanide to give α-amino nitriles. As in the case of the carbonyl group, the carbon atom represents the positive end of the dipole and is attacked by the cyanide anion. Thus aldimines and ketimines react with absolute hydrogen cyanide[57,58]:

$$R^1CH{=}N{-}R^2 + HCN \longrightarrow R^1CH(CN){-}NHR^2$$

The addition of hydrogen cyanide to a variety of Schiff bases has been reported[59]. Further examples are hydrazones[60], ketazines[61] and oximes[62]:

$$RCH{=}NOH + HCN \longrightarrow RCH(CN){-}NHOH$$

The hydrogen cyanide may be used as the water-free liquid or replaced by an aqueous solution of sodium bisulphite and sodium cyanide[63]. The resulting α-hydroxylaminonitriles may be oxidized by *p*-benzoquinone to α-cyano oximes or by nitric acid–ammonium nitrate to α-dinitronitriles[64].

With potassium cyanide as catalyst, even the C≡N triple bond of nitriles adds absolute hydrogen cyanide in an exothermic reaction[65]:

$$Cl_3C{-}CN \xrightarrow{HCN} Cl_3C{-}C(CN){=}NH \xrightarrow{HCN} Cl_3C{-}C(CN)_2{-}NH_2$$

The resulting ketimine yields α-dicyano amines with a second molecule of reagent.

The acylated products of the addition of hydrogen cyanide to quinolines and isoquinolines are called 'Reissert compounds'[66]. They are synthesized, for example, by reacting quinoline with benzoyl chloride in aqueous potassium cyanide solution:

$$\text{quinoline} + KCN + PhCOCl \longrightarrow \text{1-benzoyl-1,2-dihydroquinoline-2-carbonitrile (N–COPh, C(H)CN)} + KCl$$

Treatment of the product with phosphorus pentachloride gives 2-cyanoquinoline. Other heterocyclic bases such as pyridine do not undergo the Reissert reaction. On the other hand, it is possible to add cyanide anion to pyridinium salts. Thus 4-cyanopyridine may be synthesized via the *N*-oxide[67]:

$$\text{pyridine } N\text{-oxide} \xrightarrow{Me_2SO_4} \text{N-methoxypyridinium (N–OMe, +)} \xrightarrow{KCN} \text{4-cyanopyridine}$$

Electron-withdrawing substituents enhance the electrophilicity of the pyridinium cation considerably. *N*-methyl-3,5-dicyanopyridinium tosylate adds cyanide to give the 1,2-adduct, which above 120° rearranges to the 1,4-isomer[68]:

$$\text{N-methyl-3,5-dicyanopyridinium } TsO^- \xrightarrow{HCN} \text{1,2-adduct (2-H, 2-CN)} \rightleftharpoons \text{1,4-isomer (4-H, 4-CN)}$$

Rate and equilibrium constants for the 1,4-addition of cyanide ion to various *N*-substituted 3-carbamoylpyridinium ions have been measured. With increasing electron-withdrawing power of the nitrogen substituent the rate constant increases and the dissociation constant to the components decreases[69].

III. PREPARATION OF NITRILES BY SUBSTITUTION

A. Reaction of Hydrogen Cyanide or Its Salts with Organic Compounds

I. Halides

The reaction between organic halogen compounds and metal cyanides is a frequently used nitrile synthesis[1,2]. As in other nucleophilic displacement reactions, the reactivity of the organic halogen compound increases from chlorine to iodine; fluorine, however, being rather inert. Often a certain amount of isonitrile is formed along with the nitrile. Alkali cyanides predominantly yield nitriles, whereas heavy metal cyanides such as copper, silver and mercury cyanide give increasing amounts of isonitrile. At reaction temperatures above 150° the yield of nitrile increases, since at about that temperature, considerable isomerization of isonitriles to nitriles already takes place. The tendency of heavy metal cyanides to yield isonitriles may be explained by the ability of heavy metal ions to promote S_N1 reactions of halogen compounds. The ambient cyanide anion then will react via its centre of highest electron density, i.e. the nitrogen atom, and thus yield an isonitrile. On the other hand, the reaction of alkali cyanides, especially with primary halides, follows the S_N2 mechanism, and here the cyanide anion reacts with its more nucleophilic carbon atom to give a nitrile[70]. Primary halides usually give good yields of nitriles, but with secondary and tertiary halides, an increased tendency for dehydrohalogenation, effected by the basic alkali cyanides, is observed. Another complication, originating from the basicity of alkali cyanides, is the conversion of halides into alcohols or ethers by solvent attack or the solvolysis of the nitrile.

In the synthesis of nitriles from saturated aliphatic halides, potassium or sodium cyanide in various solvents are usually employed, mainly in alcohols such as ethanol[71], methanol or their mixtures with water as in the example below[72]:

$$\mathrm{EtOCH_2CH_2Br + NaCN \xrightarrow{EtOH,\ H_2O} EtOCH_2CH_2CN + NaBr}$$

Water[73], acetone[74] and their mixtures[75] are also common solvents:

$$\mathrm{HC{\equiv}C(CH_2)_5Br + KCN \xrightarrow{acetone,\ H_2O} HC{\equiv}C(CH_2)_5CN + KBr}$$

The addition of sodium iodide catalyses the exchange reaction considerably. The effect is due to a preceding conversion of the chloro or bromo compounds into the corresponding iodo compounds, which show a higher reactivity.

Other suitable solvents are ethylene glycol, di- and polyethylene glycols[76] and their ethers[77]. By the use of glycols and their derivatives, higher reflux temperatures are attainable and a sufficient amount of the alkali cyanide is dissolved. These solvents must be anhydrous to avoid hydrolytic side reactions.

During the last few years, several dipolar aprotic solvents have come into use for nucleophilic substitution reactions. These solvents are often superior to the above-mentioned ones in that they allow considerably shorter reaction times and lower temperatures. Such solvents are formamide[78], especially dimethylformamide (DMF), and dimethylsulphoxide (DMSO). Thus α-chloro ethers in DMF or DMSO react vigorously at 20° with sodium cyanide to give the corresponding α-cyano ethers[79]. In other solvents alkali cyanides do not react with α-chloro ethers and in water or alcohols where a reaction takes place, the products are not stable. Another route to α-cyano ethers is the reaction of the chloro compounds with cuprous cyanide without a solvent[80]. High yields of aliphatic dinitriles are obtained from the corresponding chloro compounds with sodium cyanide in DMSO[81]:

$$\mathrm{Cl(CH_2)_4Cl + 2\ NaCN \xrightarrow[100^\circ]{DMSO} NC(CH_2)_4CN + 2\ NaCl}$$

The yields attainable with chloro compounds in this solvent are usually higher than those from the corresponding iodides in ethanol–water[82]. A review of DMSO as solvent has been published[83].

Similar advantages for nitrile syntheses as shown by DMF and DMSO are also shown by the hexamethylamide of phosphoric acid (HMPT)[84], a most versatile solvent which recently has been reviewed[85].

Because of the different reactivity of the halogens it is possible to exchange a bromine or chlorine atom, whereas a fluoro substituent remains unaffected[86]:

$$\mp(CH_2)_nBr + NaCN \xrightarrow{EtOH,\ H_2O} \mp(CH_2)_nCN + NaBr$$

α,ω-Dichloro paraffins react with sodium cyanide in stages which can be separated, so that ω-chloro nitriles may be produced[87]:

$$Cl(CH_2)_nCl + NaCN \xrightarrow[n = 4\text{–}8,\ 10]{EtOH,\ H_2O} Cl(CH_2)_nCN + NaCl$$

In the reaction of a dihalide, the nitrile group introduced first has an accelerating effect on the exchange reaction of the second halogen if the reaction centres are connected by an even number of carbon atoms. An odd number of carbon atoms has the reverse effect. With increasing length of the carbon chain this influence diminishes. The benzene ring has the same effect as an aliphatic chain[88].

Alkene halohydrins easily react with alkali cyanides[89]. Because of the intermediacy of epoxides the nitrile group may appear at the former position of the hydroxy group[90]:

$$CH_3CHClCH_2OH \longrightarrow \left[\begin{array}{c} CH_3CH\text{—}CH_2 \\ \diagdown O \diagup \end{array}\right] \longrightarrow CH_3CH(OH)CH_2CN$$

As described above, α-halogenated ketones with alkali cyanides yield epoxynitriles[43,44].

Halogens attached to an unactivated olefinic carbon atom are difficult to substitute. Contrary to this, a halogen in an allylic position will react quite readily with cyanide anions. The following example shows the different reactivity of chlorine in vinylic and allylic positions[91]:

$$ClCH_2CH{=}CHCl + NaCN \longrightarrow NCCH_2CH{=}CHCl + NaCl$$

The reaction of allylic halides with cyanides is often accompanied by an allylic rearrangement. Thus allyl bromide with sodium cyanide yields eventually crotononitrile[92]:

$$CH_2{=}CHCH_2Br \longrightarrow CH_2{=}CHCH_2CN \longrightarrow CH_3CH{=}CHCN$$

The above-mentioned complications encountered by the use of alkali cyanides, such as dehydrohalogenation or solvent attack, which are consequences of the basic properties of the alkali cyanides, may be avoided if heavy metal cyanides are applied. Most frequently cuprous cyanide is used. The formation of crotononitrile in the last example is completely suppressed if the exchange is made with cuprous cyanide, the reaction product then being pure allyl cyanide[93]. In other cases, however, especially if instead of bromo compounds[94] the chloro derivatives are used[95], allylic rearrangements have been observed[96].

At higher temperatures the conversion of vinyl halides to the corresponding nitriles with cuprous cyanide can be achieved. *Trans-* and *cis*-diiodoethylenes give at 150° fumaronitrile and maleonitrile, respectively[97]:

$$\text{ICH{=}CHI} \xrightarrow{\text{CuCN}} \text{NCCH{=}CHCN}$$

Another example is the synthesis of triphenylacrylonitrile from triphenylvinyl bromide and cuprous cyanide at 250° [98].

Vinyl halides with electron-withdrawing groups at the carbon in the β position are susceptible to nucleophilic attacks. The reactions of halogeno quinones fall within this scheme. The halogeno quinones may be regarded as vinylogues of acid halides and this explains their normal substitution reactions with amines or with alkoxide ions to give the 2,5-disubstituted derivatives. The acidic compound obtained by the reaction of tetrachlorobenzoquinone with potassium cyanide in ethanol–water was therefore assumed to be 2,5-dicyano-3,6-dihydroxybenzoquinone (cyanoanilic acid), formed by hydrolysis of an intermediate 2,5-dicyano-3,6-dichlorobenzoquinone[99]. However, it has recently been shown to be 2,3-dichloro-5,6-dicyanohydroquinone. The corresponding bromo- and iodobenzoquinones react similarly and the same is found with 2,3-dichloronaphthoquinones[100]:

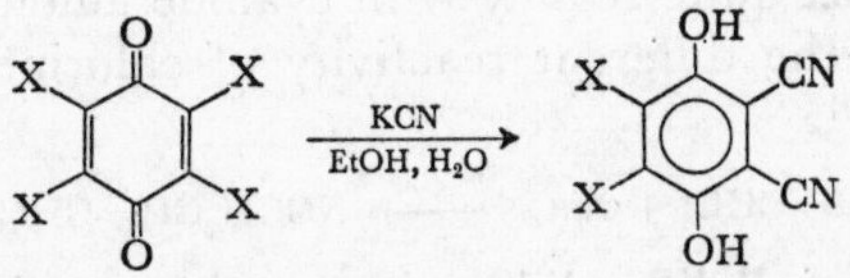

$$\xrightarrow[\text{EtOH, H}_2\text{O}]{\text{KCN}}$$

As in the reaction of *p*-benzoquinone with hydrogen cyanide, the second nitrile group enters the position *ortho* to the first one. Two halogen atoms in the *para* position are substituted by cyanide if there are already two *p*-nitrile groups in the quinone molecule[101]:

$$\text{2,5-X}_2\text{-3,6-(CN)}_2\text{-1,4-benzoquinone} \xrightarrow[\text{MeOH}]{\text{KCN}} \text{2,3,5,6-(CN)}_4\text{-hydroquinone}$$

The solvent in this reaction serves as reducing agent for the intermediate tetracyanobenzoquinone. Ion-exchange resins can be used instead of metal cyanides in reactions with halides. Thus the cyanide form of amberlite IR 400 reacts in 95% ethanol at 65° with benzyl bromide to give a 53% yield of benzyl cyanide. The reaction may also be conducted in aprotic solvents such as ether, tetrahydrofuran, benzene or dimethylformamide, although yields will then drop to about 20%[102]. The main advantage of the method is the avoidance of undesirable side reactions such as solvolytic cleavage of the products.

Acetone cyanohydrin may be used for the preparation of nitriles from halides. A variety of bromo compounds have been reacted with this reagent in the presence of a basic catalyst. The exchange has been effected in methanol at 100° in an autoclave. The best yields were obtained with benzyl bromides[103].

Instead of using its salts, hydrogen cyanide itself can be reacted with aliphatic halides in water at a controlled pH value (3·0–4·5) in the presence of cuprous salts. By this method a mixture of propargyl cyanides and cyanoallenes is obtained from propargylic chlorides[104]:

$$ClCH_2C{\equiv}CR \xrightarrow{HCN} NCCH_2C{\equiv}CR + NCCH{=}C{=}CHR$$

Allyl chlorides give the corresponding nitriles without rearrangement and by the same method dinitriles may be prepared, thus 1,4-dichloro-2-butene gives 1,4-dicyano-2-butene[105]. Tertiary acetylenic bromides yield cyanoallenes[106]:

$$R^1R^2CBrC{\equiv}CH \xrightarrow{HCN} R^1R^2C{=}C{=}CHCN$$

Acyl cyanides may be synthesized from acyl halides with heavy metal cyanides[107]. The best yields are obtained with acyl bromides and cuprous cyanide[108]:

$$RCOBr + CuCN \longrightarrow RCOCN + CuBr$$

Catalytic amounts of phosphorus compounds exert a rate-increasing effect.

With acyl chlorides longer reaction times are required. The conversion of trifluoroacetyl chloride into the nitrile with silver cyanide requires three days at 80–95° in a sealed vessel[109]:

$$CF_3COCl + AgCN \longrightarrow CF_3COCN + AgCl$$

Anhydrous alkali cyanides are unreactive towards acyl halides, while solvents like water or alcohols would at once solvolyse acyl cyanides in most cases. If the high temperatures necessary for the reaction between acyl halides and metal cyanides have to be avoided, another method may be adopted. It involves the reaction of acyl halides with anhydrous hydrogen cyanide and an organic base, usually pyridine[110]:

$$RCOCl + HCN + C_5H_5N \longrightarrow RCOCN + C_5H_5N{\cdot}HCl$$

The pyridine is added to the solution of the acyl halide and hydrogen cyanide in an inert solvent.

For the preparation of aromatic nitriles from halides the reagent of choice is cuprous cyanide. In a few cases other cyanides such as silver or zinc cyanide or complex cyanides like potassium ferrocyanide have been used[111]. The reaction of aryl halides with cuprous cyanide without a solvent was originally investigated by von Braun[112], applying temperatures up to 250°. An example for this method is the synthesis of 9-cyanophenanthrene[113]:

9-Bromophenanthrene (Br) $\xrightarrow{CuCN}$ 9-cyanophenanthrene (CN)

The catalytic effect of cuprous cyanide in the reaction of aqueous or alcoholic potassium cyanide with aryl halides at 200° was reported by Rosenmund[114]. The method used by von Braun is now called the Rosenmund–von Braun synthesis. A detailed study of the reaction showed that it is autocatalytic; small amounts of a nitrile reduce the induction period and cupric salt accelerates the conversion[115].

It is advantageous to use a solvent which is capable of complexing copper salts. Aromatic amines such as pyridine or quinoline are mostly employed[111]. An example is the synthesis of α-naphthonitrile[116]:

1-Bromonaphthalene (Br) $\xrightarrow{CuCN,\ pyridine}$ 1-naphthonitrile (CN)

Although aryl chlorides undergo the reaction, the corresponding bromides require shorter reaction times. Further examples of the pyridine method are the preparation of 2,4,6-triethylbenzonitrile[117] and 2,4,6-tricyanomesitylene[118]. Usually the yield is improved by longer reaction times and higher temperatures, but there seems to exist an optimum time for each reaction. A 70% yield of nitrile is obtained from 2-bromoacetophenone after 1·5 hours at 210°, but it is substantially diminished if the refluxing is carried on for an additional 5 hours[119].

The dipolar aprotic solvents, dimethylformamide, *N*-methylpyrrolidone and dimethylsulphoxide have also found application in the synthesis of aromatic nitriles. The reaction conditions and methods of isolation of the products have been investigated for the reaction of aryl halides with cuprous cyanide in DMF[120] and in *N*-methylpyrrolidone[121]. Reaction times are 2–3 hours for bromo compounds and 10–25 hours for chloro compounds. Tricyanomesitylene is obtained in good yield from the bromo compound in DMF with cuprous cyanide[122] and the use of DMSO or DMF as solvent for the preparation of nitriles from chloronitrobenzenes has been described[123]:

NO_2, Cl, Cl (benzene ring) + CuCN ⟶ NO_2, CN, Cl (benzene ring) + CuCl

Since copper(II) salts exert a rate-accelerating influence, one may conclude that cupric ions are participating in the substitution reaction and a mechanism has been proposed[115]. Still more work has to be done, however, before the reaction will be thoroughly understood, especially the influence of the substituents[124,125].

There are possible side reactions in those cases where *ortho* substituents may interfere by ring closure with the nitrile function. Thus *o*-bromocarboxylic acids tend to give dicarboxylic acid imides instead of the desired *o*-cyano acids[126], a reaction which corresponds

HOOC, Br, Br, COOH (benzene ring) —CuCN→ HN, NH, O, O, O, O (bis-imide)

to the formation of phthalimide from *o*-cyanobenzoic acid at 180–190°[127] and the ring closure of *o*-cyanobenzamide to give 'phthalimidine'[128]. Aromatic *o*-dihalides generally yield phthalocyanines when heated with cuprous cyanide[129], although it is possible to synthesize the mononitriles from *o*-dibromo compounds in moderate yields by carefully controlling the reaction time[130]:

$$\text{2,3,5,6-Br}_4\text{-1,4-(OMe)}_2\text{C}_6 \xrightarrow{\text{CuCN, DMF}} \text{2,5-Br}_2\text{-3,6-(CN)}_2\text{-1,4-(OMe)}_2\text{C}_6$$

2. Aryl sulphonates

The alkali salts of aromatic sulphonic acids may be converted to the corresponding nitriles by melting them with alkali cyanides[131]. Temperatures up to 400° are necessary. The modest-to-fair yields

$$ArSO_3Na + NaCN \longrightarrow ArCN + Na_2SO_3$$

are somewhat increased by the use of potassium ferrocyanide[132]. By this method the ten different dicyanonaphthalenes have been synthesized starting from the corresponding sodium cyanonaphthalenesulphonates, the yields ranging from 8 to 76%[133]. The reaction may be conducted in diluents such as mineral oil or sand[134]. The phthalocyanine formation does not seem to be so predominant with *o*-haloarylsulphonic acids as in the case of *o*-dihalides[135].

Two examples with heteroaromatic compounds are the synthesis of 3-cyanopyridine from the sodium sulphonate with sodium cyanide at 340–400° [136] in 46% yield, and the reaction of sodium 2,4-dimethylpyrimidinesulphonate with potassium cyanide at 260° [137].

3. Alcohols, esters and ethers

Saturated aliphatic alcohols and ethers undergo reaction with hydrogen cyanide only at elevated temperatures and in the presence of catalysts[138], but allylic alcohols may be converted into nitriles by hydrogen cyanide in aqueous solution, using a Nieuwland-type catalyst[139]:

$$CH_2{=}CHCH_2OH + HCN \xrightarrow[100^\circ]{CuCl,\ NH_4Cl,\ H_2O} CH_2{=}CH_2CH_2CN$$

Under similar conditions 2-butene-1,4-diol gives 1,4-dicyano-2-butene. It is very likely that the reactions proceed through the intermediate formation of the halides, since the esterification of allyl alcohol with hydrogen chloride is known to be strongly promoted by cuprous chloride[140]. The reaction of acetylenic alcohols with a mixture of cuprous cyanide, potassium cyanide and hydrobromic acid may take a similar way[141].

Esters of sulphuric and phosphoric acids have found limited application for the synthesis of simple aliphatic nitriles[142], for example acetonitrile[143]:

$$(CH_3O)_2SO_2 + KCN \longrightarrow CH_3CN + CH_3OSO_3K$$

The conversion of more complicated alcohols into nitriles is best accomplished by the reaction of their sulphonic esters with metal cyanides:

$$R^1OSO_2R^2 + MCN \longrightarrow R^1CN + R^2SO_3M$$
$$M = \text{metal}$$

Generally, the esters of *p*-toluenesulphonic acid and methylsulphonic acid are employed. They are usually treated with alkali cyanides in solvents such as ethanol[144] or dimethylformamide[145]. An example is the reaction of cholestan-3β-yl tosylate with calcium cyanide in *N*-methylpyrrolidone-*t*-butyl alcohol, which yields the 3α-nitrile in 81% yield[146]:

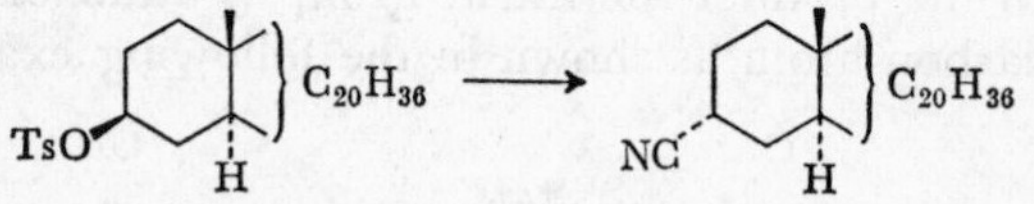

TsO, H, $C_{20}H_{36}$ ⟶ NC, H, $C_{20}H_{36}$

The formation of cyclopropyl cyanides is reported from the reaction of 2,2-dialkyl-1,3-propanediol tosylates with potassium cyanide in ethylene glycol[147]:

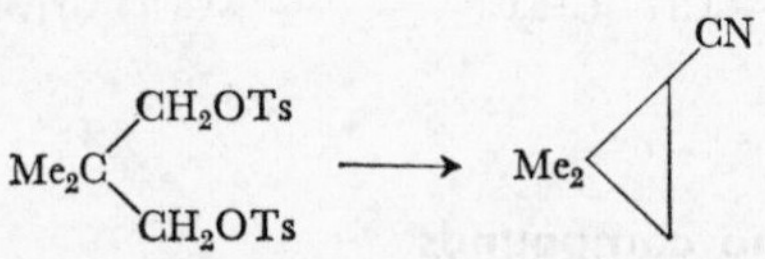

$Me_2C(CH_2OTs)_2$ ⟶ Me_2(cyclopropyl)CN

Open-chain aliphatic esters generally do not react with hydrogen cyanide or its salts to give nitriles, but small ring lactones such as γ-butyrolactone[148] or phthalide[149] yield open-chain cyano compounds:

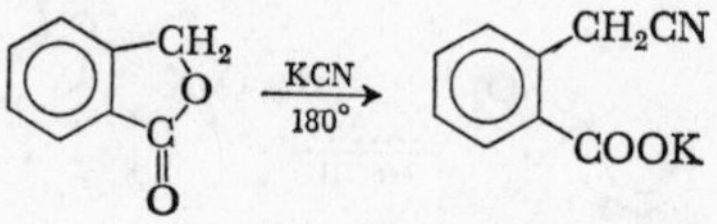

phthalide (CH_2, O, C=O) $\xrightarrow[180°]{KCN}$ $C_6H_4(CH_2CN)(COOK)$

Ortho esters exchange one alkoxyl group when treated with hydrogen cyanide and zinc chloride at elevated temperatures[150]:

$$R^1C(OR^2)_3 + HCN \rightleftharpoons R^1C(CN)(OR^2)_2 + R^2OH$$

Similarly, diethyl acetals of carboxylic acid amides can be reacted with hydrogen cyanide, although under milder conditions[7]:

$$RC(OEt)_2NMe_2 + HCN \longrightarrow RC(CN)(OEt)NMe_2 + EtOH$$

The methoxyl groups of 2,5-dicyano-3,6-dimethoxybenzoquinone are very reactive and are easily substituted by cyanide anion[101]:

$$\text{2,5-(NC)}_2\text{-3,6-(MeO)}_2C_6O_2 \xrightarrow{KCN,\ MeOH} \text{1,4-(HO)}_2C_6(CN)_4$$

While the normal ether linkage requires elevated temperatures and catalysts to react with hydrogen cyanide, the epoxides readily give hydroxynitriles when treated with hydrogen cyanide and a basic catalyst[151]. In the product the nitrile group is attached to the least substituted carbon atom as shown in the following examples[152,153]:

$$CH_3C{\equiv}C{-}C(CH_3)(O)CH_2 \xrightarrow[EtOH,\ H_2O]{KCN} CH_3C{\equiv}C{-}C(OH)(CH_3){-}CH_2CN$$

$$CH_2(O)CH{-}CH_2Cl \xrightarrow[H_2O]{KCN} NCCH_2CH(OH)CH_2CN$$

4. Nitro or amino compounds

Because of the complexity of the products[154], the reaction of alkali cyanides with polynitrobenzenes and -phenols has found only limited application. A relatively simple example is the substitution of a hydrogen atom by cyanide in *m*-dinitrobenzene[155]:

$$\text{1,3-}(O_2N)_2C_6H_4 \xrightarrow[MeOH]{KCN} \text{2-CN-1-}O_2N\text{-3-MeO-}C_6H_3$$

While in this reaction a nitro group is also substituted by solvent attack, in other cases, like that of picric acid, reduction may occur[156]:

$$\text{2,4,6-}(O_2N)_3C_6H_2OH \xrightarrow[\text{EtOH}]{\text{KCN}} \text{(OK, } O_2N\text{, NHOH, NC, CN, } NO_2 \text{ substituted benzene)}$$

The conversion of *m*- and *p*-nitrohalobenzenes into halobenzoic acids (von Richter rearrangement)[157] very likely proceeds through an intermediate nitrile[158].

A few cases of the substitution of amino groups by cyanide are known. Thus a diethylamino group is replaced by a cyano substituent during an apparent cyanomethylation of indole[159]:

$$\text{Indole} + NCCH_2NEt_2 \longrightarrow \text{3-}(CH_2NEt_2)\text{indole} + HCN$$

$$\longrightarrow \text{3-}(CH_2CN)\text{indole} + HNEt_2$$

Ammonium salts have been found to undergo a similar reaction. Compared to the amines, where the leaving group is a substituted amide ion, the reaction of ammonium salts proceeds with ease because the leaving group is neutral. An example is the synthesis of ferrocenylacetonitrile from a quaternary ammonium salt[160]:

$$C_5H_5FeC_5H_4CH_2\overset{+}{N}Me_3\ I^- \xrightarrow{\text{KCN}} C_5H_5FeC_5H_4CH_2CN + KI + NMe_3$$

or the formation of β-benzoylpropionitrile[161]:

$$PhCOCH_2CH_2NMe_2{\cdot}HCl \xrightarrow{\text{HCN}} [\text{cyanohydrin}] \longrightarrow PhCOCH_2CH_2CN$$

5. Diazonium salts

In 1884 Sandmeyer discovered a nitrile synthesis which consists of the reaction of an aromatic diazonium salt with an aqueous solution of cuprous cyanide and potassium cyanide[162]. With some modifications this is still a very important method for the preparation of aromatic nitriles[163]:

$$ArN_2Cl + KCu(CN)_2 \longrightarrow ArCN + N_2 + KCl + CuCN$$

The modifications largely consist in the preparation and the nature of the complex cyanide. The original procedure uses cupric sulphate and an excess of potassium cyanide to prepare the double salt which is formed with concomitant loss of cyanogen. This can be avoided by starting from cuprous chloride as reported in the synthesis of tolunitriles[164], or by adding dry cuprous cyanide to a potassium cyanide solution[165]. Another way to reduce the losses connected with the cyanogen evolution is the use of a potassium cuprammonium cyanide which exists only in solution[166]:

$$CuSO_4 + NH_3 + KCN \longrightarrow K_2[CuNH_3(CN)_4] + K_2SO_4$$

The cuprous cyanide may be replaced in the Sandmeyer synthesis by catalytic amounts of freshly precipitated copper powder[167]. Although this method is more effective in some cases, it has apparently not found much application. The double salt from potassium cyanide and nickel cyanide has been used instead of the copper salt[168,169]. The loss of hydrogen cyanide is minimized by neutralization of the diazonium salt solution prior to its addition to the cyanide[170,171].

The generally accepted mechanism of the Sandmeyer reaction is that of a radical reaction[172]. The rate-determining step is the reduction of the diazonium cation by the univalent copper, which is subsequently regenerated during the combination of the aryl radical with the cyanide anion[173]:

$$ArN_2^+ + Cu^+ \longrightarrow Ar^\bullet + N_2 + Cu^{2+}$$
$$Cu^{2+} + Ar^\bullet + CN^- \longrightarrow ArCN + Cu^+$$

B. Reaction of Cyanogen and Similar Compounds with Aliphatic and Aromatic Compounds

The preparation of nitriles by reaction of halides with metal cyanides finds its counterpart in the reaction of cyanogen halides with organometallic compounds[174]. A typical example is the synthesis of diethyl cyanomalonate from sodium diethyl malonate and cyanogen chloride[175]:

$$EtOOCCH_2COOEt \xrightarrow{ClCN,\ NaOEt} EtOOC\underset{\displaystyle CN}{\underset{|}{C}}HCOOEt$$

Cyanogen chloride appears to be the reagent best suited for this type of reaction, since the other cyanogen halides in most cases yield halogenated compounds which subsequently may react with the starting material[176].

The nitrile group can be introduced in a similar way into acetoacetic ester[177], malononitrile[178] or cyanosulphones[179]. The silver salt of diethyl cyanomalonate, upon treatment with cyanogen chloride, gives diethyl dicyanomalonate[176]. Depending on the amounts of reagents used, cyclopentadiene reacts with cyanogen chloride and sodium hydride to give successively the mono-, di- and tricyanocyclopentadienide anion[180]:

$$C_5H_6 \xrightarrow{ClCN,\ NaH} [C_5H_{5-n}(CN)_n]^- \ Na^+ \qquad n = 1\text{–}3$$

Soon after their discovery, organomagnesium compounds were applied to the synthesis of nitriles. The reaction may take two different courses[181]:

$$RMgX^1 + X^2CN \longrightarrow RCN + MgX^1X^2 \qquad (1)$$
$$RMgX^1 + X^2CN \longrightarrow RX^2 + MgX^1CN \qquad (2)$$

With X^2 = chlorine the main reaction is represented by equation (1), if R is a primary aliphatic, acetylenic or aromatic group. If R is a secondary, alicyclic or tertiary group or if X^2 is bromine or iodine, reaction (2) predominates.

β-Ketonitriles are obtained by the treatment of enamines with cyanogen chloride[182]:

$$\text{pyrrolidino-cycloalkene }(CH_2)_n \xrightarrow{1.\ ClCN;\ 2.\ H^+} O{=}C\text{-cycloalkane}(CN)\ (CH_2)_n$$

Instead of cyanogen chloride, cyanogen may be employed for the preparation of nitriles from the carbanions of active methylene compounds. Thus diethyl malonate, acetylacetone or acetoacetic ester add cyanogen in the presence of catalytic amounts of sodium ethoxide to give iminonitriles which, upon subsequent treatment with base, lose hydrogen cyanide and yield *β*-ketonitriles[183]:

$$CH_3COCH_2COCH_3 + (CN)_2 \xrightarrow{EtONa} CH_3COCH(C(=NH)CN)COCH_3 \xrightarrow{OH^-} CH_3COCH(CN)COCH_3$$

Fair yields of nitriles are obtained from Grignard compounds if during their reaction with cyanogen the latter is kept in excess[181]. This method also works with alicyclic compounds[184]:

$$\text{3-}CH_3\text{-}C_6H_{10}\text{-}MgCl \xrightarrow{(CN)_2} \text{3-}CH_3\text{-}C_6H_{10}\text{-}CN$$

Furthermore, organomagnesium or organolithium compounds yield nitriles when treated with substituted cyanamides as shown in the following example[185]:

$$\text{2,4,6-trimethylpyridine} \xrightarrow{\text{PhLi, PhN(CN)Me}} \text{4,6-dimethyl-2-pyridyl-}CH(CN)_2$$

A similar method uses aryl cyanates which add to the carbanions of active methylene compounds to give imido ethers as intermediates. The latter lose the corresponding phenol and yield the nitrile[186]:

$$\text{enamine (piperidino-C=CH, }(CH_2)_n\text{ ring)} + ArOCN \longrightarrow \left[\text{piperidino-C=C-C(=NH)-OAr, }(CH_2)_n\text{ ring}\right] \xrightarrow{H_3O^+}$$

$Ar = C_6H_5$; p-C_6H_4Cl; p-$C_6H_4NO_2$

$$\text{C(=O)-CH-CN, }(CH_2)_n\text{ ring} + ArOH + \text{piperidinium } (N^+H_2)$$

Instead of the magnesium compounds, the alkali salts or enamines can be used. So far mainly aliphatic compounds have been dealt with.

Aromatic nitriles may be synthesized by using cyanogen and its derivatives under Friedel–Crafts conditions[187]. Although cyanogen[188] and cyanogen chloride[189] were applied together with aluminium chloride for the cyanogenation of aromatic compounds, the yields of nitriles were only moderate. A closer examination of the reaction revealed that the use of freshly prepared cyanogen chloride or bromide and finely ground aluminium chloride improves the yields considerably[190]. This latter modification is called the Friedel–Crafts–Karrer method. For example, a further cyanation of the above-mentioned tricyanocyclopentadienide anions is achieved with cyanogen chloride and aluminium chloride[180]:

$$C_5H_2(CN)_3^- \longrightarrow C_5H(CN)_4^- \longrightarrow C_5(CN)_5^-$$

Another possibility is the application of aryl cyanates under Friedel–Crafts conditions to synthesize aromatic nitriles[186]:

$$R-C_6H_4 + ArOCN \longrightarrow R-C_6H_4-CN + ArOH$$

With ferric chloride anhydrous hydrogen cyanide may be used for the preparation of cyanoferrocenes[191]:

$$(C_5H_5)_2Fe^+FeCl_4^- \xrightarrow{HCN} C_5H_5FeC_5H_4CN$$

If one of the cyclopentadienyl rings carries an alkyl substituent, the cyano group enters the same ring. In the case of a cyano- or chloroferrocene the cyanation occurs at the unsubstituted ring.

Aromatic compounds, especially those having electron-releasing substituents such as hydroxy, alkoxy or alkyl groups, are amenable to the Houben–Fischer nitrile synthesis, which uses trichloroacetonitrile for the introduction of the cyano group[192]:

$$ArH + Cl_3CCN \xrightarrow{AlCl_3,HCl} Ar-\overset{\overset{\displaystyle NH\cdot HCl}{\|}}{C}-CCl_3 \xrightarrow{OH^-} ArCN + CHCl_3$$

Although formally classified as substitution reactions, it is very likely that all the reactions described in this section proceed by an addition–elimination mechanism:

$$R-Y + X-CN \longrightarrow R-\overset{\overset{\displaystyle N-Y}{\|}}{C}-X \longrightarrow R-CN + Y-X$$

where X may be halogen, CN, *o*-aryl, $N(CH_3)C_6H_5$ or CCl_3 and Y = H, Li, Na or Mg halides.

Cyanogen halides may be used in the photochemical cyanation of organic compounds. Thus, open-chain and cyclic aliphatic hydrocarbons yield nitriles when irradiated together with cyanogen chloride at wavelengths of 250–500 mμ in the presence of carbonyl compounds such as acetyl chloride. Good yields are obtained in the photocyanation of aliphatic ethers[193a]. Similarly benzene and other aromatic compounds are cyanated when irradiated with cyanogen iodide[193b].

While the photochemical cyanation as a radical process does not show much selectivity, the anodic cyanation may be compared with the aromatic electrophilic substitution. Electrolysis of aromatic compounds in a methanolic solution of sodium cyanide yields nitriles. The yields and the position of the introduced cyano group is greatly influenced by the other substituents[194].

IV. PREPARATION OF NITRILES BY ELIMINATION

A. Starting from Aldehydes, Ketones and Their Derivatives

The most frequently used method to convert aldehydes to nitriles is the dehydration of the corresponding oximes[1]:

$$\text{RCHO} \xrightarrow{H_2NOH} \text{RCH=NOH} \xrightarrow{-H_2O} \text{RCN}$$

This may be effected by a number of reagents, of which one of the most important is acetic anhydride. An example is given in the synthesis of 1,3,5-trichloro-2,4,6-tricyanobenzene[195]:

$$C_6Cl_3(CH{=}NOH)_3 \xrightarrow{Ac_2O} C_6Cl_3(CN)_3$$

Further applications of this reagent are illustrated in the preparations of cinnamonitrile[196], veratronitrile[197], 2,6-dichlorobenzonitrile[198] and 4-cyanocyclohexene[199]. Sometimes it is advantageous to effect the dehydration with acetic anhydride in the presence of bases such as sodium acetate, as in the synthesis of pentaacetylgluconpremonitrile[200]:

$$\text{HOCH}_2(\text{CHOH})_4\text{CHO} \longrightarrow \text{HOCH}_2(\text{CHOH})_4\text{CH=NOH} \xrightarrow{Ac_2O,\ NaOAc} \text{AcOCH}_2(\text{CHOAc})_4\text{CN}$$

or in the following example, where the preparation of the oxime and its subsequent dehydration is performed in one step in excess pyridine[201]:

$$\text{RCHO} + H_2NOH{\cdot}HCl + Ac_2O \xrightarrow[100°]{\text{pyridine}} \text{RCN}$$

The oxime acetate is an intermediate in the dehydration by acetic anhydride. Its isolation and subsequent decomposition may be advantageous as shown in the preparation of carbonyl cyanide[202]:

$$\text{NCCOCH=NOAc} \xrightarrow{160\text{–}180°} \text{NCCOCN} + \text{AcOH}$$

Acetyl chloride[203] and thionyl chloride[204] are more vigorous reagents, and are therefore applied in solvents such as ether[205,206] or DMF[206]:

$$C_6(CN)_5CH{=}NOH \xrightarrow[20\text{–}30°]{SOCl_2,\ DMF} C_6(CN)_6$$

Other reagents well suited for the dehydration of oximes are benzoyl chloride[207], arylsulphonyl chlorides[208], ethyl chloroformate[209], phosphorus pentoxide[210], triethyl phosphate[211] and phenyl isocyanate[212].

O,N-bis(trifluoroacetyl)hydroxylamine is used in a one-step method, together with pyridine, to convert aldehydes to nitriles[213]:

$$\text{RCHO} + \text{CF}_3\text{CONHOCOCF}_3 \xrightarrow{\text{pyridine}} \text{RCN}$$

Since the nitrile formation from the acyl derivatives of oximes proceeds through a *trans* elimination, the *anti*-oximes and their derivatives are the precursors of the nitriles[214].

Apart from the familiar dehydrating agents mentioned above, treatment with aqueous alkali[215] or acids will convert oximes to nitriles. Thus, one-step methods have been developed by which a mixture of hydroxylammonium salt and formic acid–sodium formate[216] or acetic acid–sodium acetate[217] are reacted with the aldehyde to give the nitrile. Similarly a facile one-step preparation transforms aliphatic aldehydes, aldehyde–bisulphite adducts, aldehyde trimers or oximes to nitriles by treatment with hydroxylammonium chloride and a small amount of hydrochloric acid in ethanol[218]. Compared with the above-mentioned methods, the dehydrations of oximes at elevated temperatures in the presence of alumina or thoria catalysts appear less important[219]. A number of syntheses has to be reported which do not proceed via oximes.

Treatment of an aldehyde with hydrazoic acid in sulphuric acid (Schmidt reaction) gives rise to a nitrile and minor amounts of formylamine[220].

In a quite unexpected reaction, aromatic aldehydes are converted to nitriles when heated in acetic acid together with 1-nitropropane and diammonium hydrogen phosphate[221]:

$$\text{ArCHO} + n\text{-PrNO}_2 + (\text{NH}_4)_2\text{HPO}_4 \longrightarrow \text{ArCN}$$

By condensation of chloramine with aldehydes[222] and treatment of the resulting chloroimines with base, nitriles have been synthesized[223]:

$$\text{4-pyridyl-CHO} \xrightarrow[0°]{\text{H}_2\text{NCl}} \text{4-pyridyl-CH{=}N{-}Cl} \xrightarrow{\text{NEt}_3} \text{4-pyridyl-CN}$$

Nitriles are also obtained by passing aldehydes and ammonia over thoria at 220–240° [224] or by generating the imines in solution and oxidizing them by iodine[225], lead tetraacetate[226] or oxygen[227]:

$$\text{RCHO} + \text{NH}_3 + \text{O}_2 \xrightarrow[\text{MeOH}]{\text{CuCl}_2,\ \text{NaOMe}} \text{RCN}$$

Since the oxidation is carried out in a basic medium, aromatic aldehydes give better yields than base-sensitive aliphatic aldehydes.

A nitrile synthesis corresponding to the Hofmann elimination is the degradation of aldehyde trialkylhydrazonium salts by base[228]:

$$\text{RCH}{=}\text{N}{-}\overset{+}{\text{N}}\text{Me}_3\text{X}^- \xrightarrow{\text{NaOMe}} \text{RCN} + \text{NMe}_3 + \text{NaX}$$

Similarly, the reaction between 1,1-dimethylhydrazine and acrolein in disodium hydrogen phosphate solution and subsequent addition of alkali produces β-dimethylaminopropionitrile[229]. Here and in the following example the intermediate is a pyrazolinium salt[230]:

$$\text{CH}_3\text{NHNH}_2 + \text{CH}_2{=}\text{CHCHO} \longrightarrow \text{(1-methyl-2-pyrazoline)} \xrightarrow{\text{CH}_3\text{I}} \text{(1,1-dimethylpyrazolinium)}\ \text{I}^-$$

$$\xrightarrow{\text{OH}^-} (\text{CH}_3)_2\text{NCH}_2\text{CH}_2\text{CN}$$

Another elimination reaction analogous to the Cope reaction of amine oxides is the oxidation of aldehyde dialkylhydrazones with hydrogen peroxide[231]:

$$\text{R(H)C}{=}\text{N}{-}\text{N(CH}_3)_2{\rightarrow}\text{O} \longrightarrow \text{RCN} + (\text{CH}_3)_2\text{NOH}$$

$$\text{R} = \text{Ar}^-, \text{PhCH}{=}\text{CH}^-$$

Azomethines resulting from the condensation of aromatic or heterocyclic aldehydes with 4-amino-1,2,4-triazole or its derivatives yield nitriles by pyrolysis or base-catalysed elimination[232]:

$$\text{RCH}{=}\text{N}{-}\text{N(1,2,4-triazol-4-yl)} \longrightarrow \text{RCN} + \text{HN(1,2,4-triazole)}$$

There are some reports on formation of nitriles from aromatic aldazines upon heating[233] or chlorination at elevated temperatures[234].

Although normally ketoximes cannot be converted to nitriles, α-oximino ketones or acids undergo a so-called 'abnormal' or 'second-order' Beckmann rearrangement, giving nitriles. For example, benzil monoxime yields benzonitrile when treated with polyphosphoric acid[235]:

$$\mathrm{PhCO\overset{\overset{NOH}{\|}}{C}Ph \longrightarrow PhCOOH + PhCN}$$

α-Oximino ketones possessing *anti* configuration always yield a nitrile and a carboxylic acid by a second-order Beckmann rearrangement, whether this is brought about by strong acids, acid chlorides or by an acylating agent and a base[236].

This rearrangement may also be effected by heat, as in the example of isatin monoxime[237]. Similarly the sodium salt of 1-nitroso-2-naphthol yields 2-cyanocinnamic acid[238].

$$\text{R-isatin 3-oxime (C=NOH, C=O, NH)} \longrightarrow \text{R-}C_6H_3(CN)(NH_2) + CO_2$$

The α-oximino carboxylic acids are correspondingly converted to nitriles, water and carbon dioxide by treatment with acids[239], acetic anhydride[240] or heat alone[241]:

$$\mathrm{R\overset{\overset{NOH}{\|}}{C}COOH \longrightarrow RCN + CO_2 + H_2O}$$

Aldehydes can be transformed to nitriles with chain extension by condensing them with rhodamine, cleaving the product by alkali and converting the resulting α-mercaptoacrylic acid to the α-oximino acid. The final step yielding the nitrile is accomplished by acetic anhydride[242]:

$$\text{5-(RCH=)rhodanine (S, NH, C=S, C=O)} \xrightarrow{OH^-} \text{RCH=C(5-ring: C(=O), NH, C(=O), S)} \xrightarrow{OH^-} \mathrm{RCH{=}\underset{\underset{SH}{|}}{C}COOH} \xrightarrow{NH_2OH}$$

$$\longrightarrow \mathrm{RCH_2\underset{\underset{NOH}{\|}}{C}COOH \xrightarrow{Ac_2O} RCH_2CN}$$

The same result is attained on a shorter route by preparing the aldehyde cyanohydrin, which then is converted to the α-chloronitrile by thionyl chloride and dehalogenated with zinc[38].

The transformation of α-keto acids can be effected by a one-step procedure with hydroxylamine in water or in ethanol–pyridine[243].

So far, rearrangements of ketoximes with an adjacent carbonyl or carboxyl group have been discussed. Ketoximes bearing an amino or ether substituent at the β position show a similar behaviour. Thus the fragmentation of β-keto ether oximes occurs under Beckmann conditions with reagents such as phosphorus pentachloride or thionyl chloride[244], and the corresponding reaction of β-oximino

$$R^4(R^2R^3C(OR^1))C{=}N{-}OH \longrightarrow R^4CN + R^1OH + R^2COR^3$$

amines is facilitated if the oxime is converted to its ethers or esters. While the benzyl ethers of the *anti*-oximes can be prepared, the corresponding esters of *p*-toluenesulphonic or picric acid undergo cleavage during their synthesis[245]:

$$(R^1)_2N{-}\overset{|}{\underset{|}{C}}{-}\underset{R^2}{\underset{|}{C}}{=}N{-}X \longrightarrow (R^1)_2\overset{+}{N}{=}C\langle \;\; X^- + R^2CN$$

$$X = p\text{-}OSO_2C_6H_4CH_3,\; p\text{-}OC_6H_3(NO_2)_2,\; p\text{-}OCOPh$$

By reaction of β-keto aldehydes with hydroxylamine, isoxazoles rather than the oximes are obtained. Owing to the manner of their preparation, these isoxazoles possess a hydrogen in the 3-position and are therefore cleaved by bases to yield β-ketonitriles[246]:

$$R^1COCHR^2CHO \xrightarrow{H_2NOH} \text{(5-}R^1\text{-4-}R^2\text{-isoxazole)} \xrightarrow{\text{base}} R^1COCHR^2CN$$

Since isoxazoles may be synthesized by various methods[246,247] they represent a useful starting material for the preparation of β-ketonitriles[248]. The method is especially important in the case of compounds like the rather unstable cyanoacetone which has been prepared by basic cleavage of 5-methylisoxazole and used *in situ* for the Michael addition to α,β-unsaturated ketones[249]. Similarly, cyanoacetaldehyde can be generated by the reaction of isoxazole with alkali, and subsequently trapped by ethylation with diethyl sulphate[250]:

$$\text{isoxazole} \xrightarrow{(EtO)_2SO_2,\,NaOH} EtOCH{=}CHCN$$

B. Starting from Carboxylic Acids and Their Derivatives

The most important method for the conversion of carboxylic acids into the corresponding nitriles consists in the dehydration of the amides[251]:

$$RCONH_2 \xrightarrow{-H_2O} RCN$$

In cases where the desired nitrile is sufficiently volatile to allow distillation, phosphorus pentoxide may be used as dehydrating agent, usually at temperatures between 100–250° [252]:

$$RCONH_2 + P_2O_5 \longrightarrow RCN + 2\,HPO_3$$

Many applications of this method for saturated[253], unsaturated[254] and aromatic nitriles[255] are known. The main limitation is the presence of other groups capable of reacting with phosphorus pentoxide, such as hydroxy, or primary and secondary amines. Tertiary amines do not interfere in the dehydration, since the preparation of acid-sensitive nitriles can be accomplished with phosphorus pentoxide in benzene in the presence of triethylamine[256]:

$$(EtO)_2CHCH_2CONH_2 \xrightarrow{P_2O_5,\ Et_3N} (EtO)_2CHCH_2CN$$

To facilitate the heat transfer and prevent excessive frothing of the reaction mixture, diluents such as sea sand or mineral oil have been recommended[251]. A very effective reagent for the conversion of amides to nitriles is phosphorus pentachloride[257], the preparation of malononitrile being an example[258]:

$$NCCH_2CONH_2 + PCl_5 \longrightarrow NCCH_2CN + POCl_3 + 2\,HCl$$

According to recent studies, the mechanism of the dehydration of amides with phosphorus pentachloride involves the formation of an acyl phosphorimidic trichloride, which then is cleaved to give the nitrile and phosphoryl chloride[259]:

$$RCONH_2 + PCl_5 \xrightarrow{-2HCl} RCON{=}PCl_3 \longrightarrow RCN + POCl_3$$

Mono- and dialkylamides, which are normally inert towards dehydrating agents, react with phosphorus pentachloride or pentabromide to give nitriles and the corresponding alkyl halides (von Braun reaction)[260]:

$$ArCONR_2 + PCl_5 \longrightarrow ArCN + 2\,RCl + POCl_3$$

The observation that $\frac{1}{3}$ or $\frac{1}{4}$ of a mole of phosphorus pentachloride is sufficient to transform one mole of acid amide to the nitrile led to the application of phosphoryl chloride as a dehydrating agent[261]. Apart from its cheapness, it has the advantage of not attacking carbonyl groups. Hydroxy substituents may be converted to chloro substituents, especially in heterocycles (see below). Phosphoryl chloride is applied alone, as in the example shown below, for it acts as a solvent at the same time[262]:

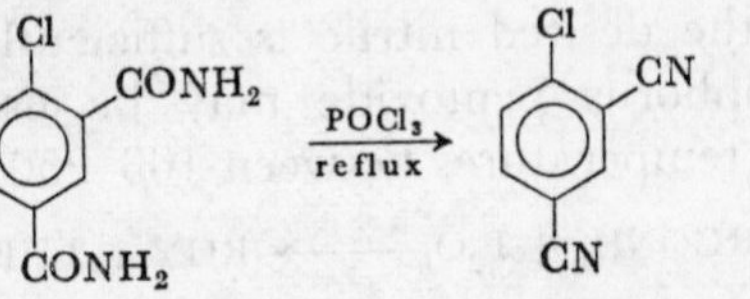

$$\text{2-Cl-C}_6\text{H}_3\text{-1,4-(CONH}_2)_2 \xrightarrow[\text{reflux}]{POCl_3} \text{2-Cl-C}_6\text{H}_3\text{-1,4-(CN)}_2$$

A number of aliphatic[263] and aromatic nitriles[264] has been prepared in this way. Solvents like acetonitrile[31] or 1,2-dichloroethane[265] may be employed in order to moderate vigorous reactions. The addition of sodium chloride has a yield-increasing effect and facilitates the working-up of the reaction mixture by converting the resulting metaphosphoric acid into its sodium salt[31,266].

Very often phosphorus oxychloride is applied together with a base such as pyridine[267]. An example of the sometimes rather special conditions required for a dehydration reaction is the synthesis of tetracyanofuran. Dehydration of the tetramide of furan-tetracarboxylic acid can only be effected by pure phosphoryl chloride. On the other hand, the only reagent which converts 3,4-dicyanofuran-2,5-dicarbonamide into tetracyanofuran is a mixture of phosphoryl chloride and pyridine[268]:

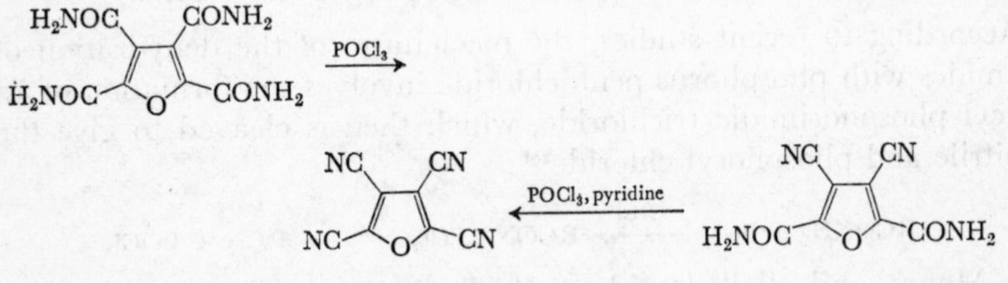

$$\text{furan-2,3,4,5-(CONH}_2)_4 \xrightarrow{POCl_3} \text{3,4-(NC)}_2\text{-furan-2,5-(CONH}_2)_2 \xrightarrow{POCl_3,\ pyridine} \text{furan-2,3,4,5-(CN)}_4$$

Treatment of the amides of *N*-acylamino acids or peptides with phosphoryl chloride–pyridine yields the corresponding nitriles with retention of the optical activity[269]. Instead of pyridine, other bases have found application. In the following example, 5-cyano-4,6-dichloro-2-methylpyrimidine is obtained by refluxing the dihydroxycarbonamide with phosphoryl chloride and *N*,*N*-dimethylaniline[270]:

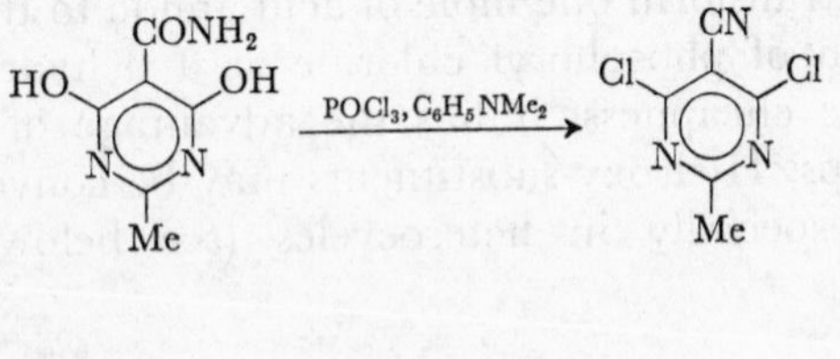

$$\text{4,6-(HO)}_2\text{-2-Me-pyrimidine-5-CONH}_2 \xrightarrow{POCl_3,\ C_6H_5NMe_2} \text{4,6-Cl}_2\text{-2-Me-pyrimidine-5-CN}$$

Thionyl chloride may be used for the dehydration of amides either in pure form as shown below[271]:

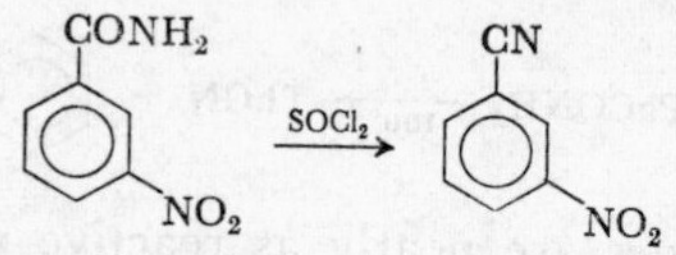

$$m\text{-}NO_2C_6H_4CONH_2 \xrightarrow{SOCl_2} m\text{-}NO_2C_6H_4CN$$

or together with organic bases or *N*-acylated amines such as dimethylformamide[126]:

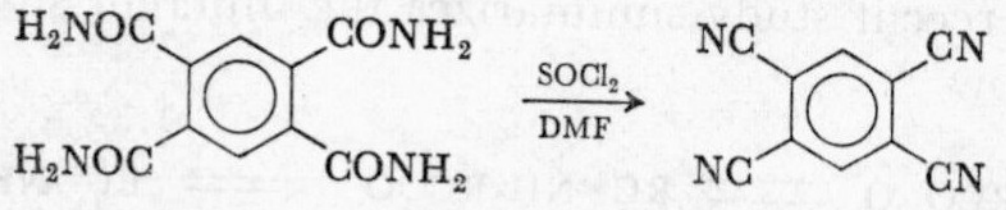

$$C_6H_2(CONH_2)_4 \xrightarrow[DMF]{SOCl_2} C_6H_2(CN)_4$$

Under careful temperature control, mixtures of thionyl chloride and dimethylformamide are a general and convenient dehydrating agent, especially for the preparation of aromatic nitriles[100,272]. The formation of dimethylformamide chloride has been reported to occur in the reaction of thionyl chloride with dimethylformamide[273]. It seems likely that this compound is the actually dehydrating agent in the above-mentioned reactions.

Other acid chlorides that have been applied to the nitrile synthesis from amides are phosgene in the presence of pyridine or acylated secondary amines[274], and arylsulphonyl chlorides in

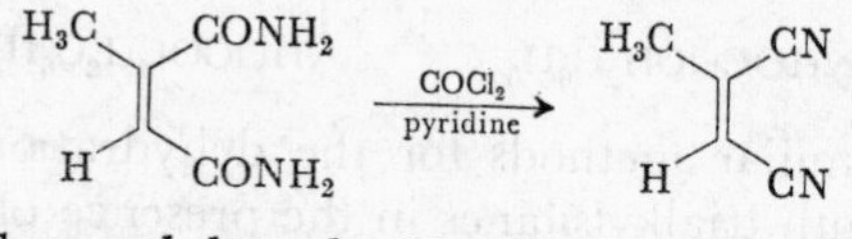

$$H_3C(CONH_2)C{=}CH(CONH_2) \xrightarrow[pyridine]{COCl_2} H_3C(CN)C{=}CH(CN)$$

pyridine[275]. Methanesulphonyl chloride in pyridine is used in a one-step synthesis of aromatic *o*-cyano esters from the corresponding *o*-dicarboxylic acid monoamides[276]:

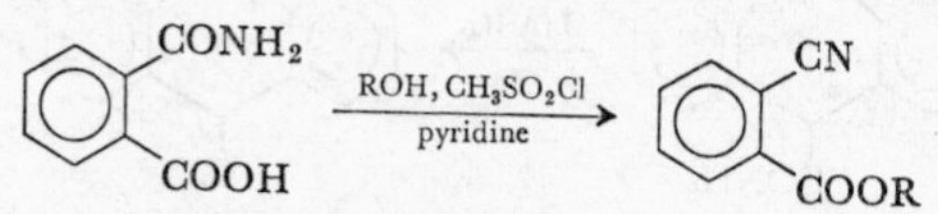

$$o\text{-}C_6H_4(CONH_2)(COOH) \xrightarrow[pyridine]{ROH,\ CH_3SO_2Cl} o\text{-}C_6H_4(CN)(COOR)$$

In a similar way treatment of the amide of a maleic acid with ethyl chloroformate and triethylamine gives the β-cyanocarboxylic acid ester[277]:

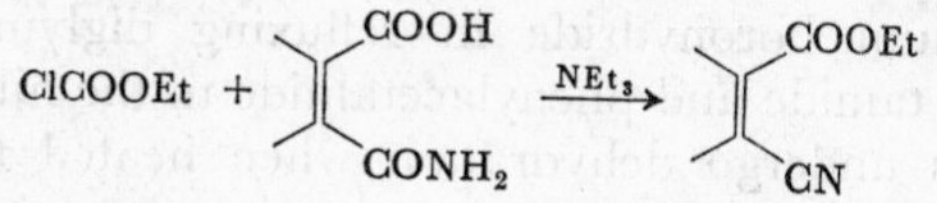

$$ClCOOEt + {>}C(COOH){-}C(CONH_2){<} \xrightarrow{NEt_3} {>}C(COOEt){-}C(CN){<}$$

Pyrocatechyl phosphorus trichloride easily converts benzamide into benzonitrile in nearly quantitative yield[278]:

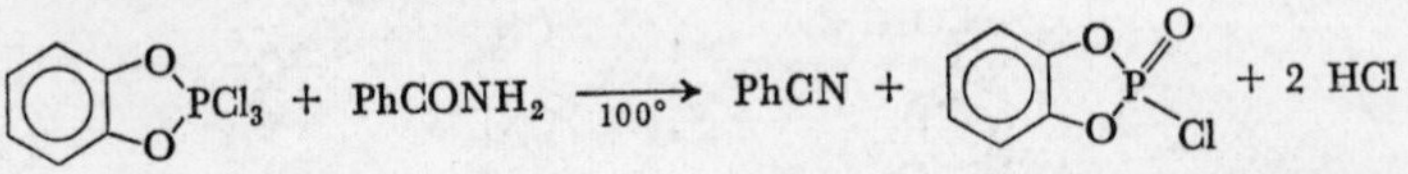

$$C_6H_4O_2PCl_3 + PhCONH_2 \xrightarrow{100°} PhCN + C_6H_4O_2P(O)Cl + 2\ HCl$$

Since acid anhydrides are nearly as reactive as acid chlorides it seems obvious to use them in amide dehydration. Yet the reaction between an anhydride and an amide only proceeds to an equilibrium[279]. A recent study summarizes the different steps involved, as shown below[280]:

$$RCONH_2 + (RCO)_2O \rightleftharpoons RC(OCOR){=}\overset{+}{N}H_2\,RCOO^- \rightleftharpoons RC(OCOR){=}NH + RCOOH$$

$$RC(OCOR){=}\overset{+}{N}H_2\,RCOO^- \rightleftharpoons RCN + 2\ RCOOH \rightleftharpoons RC(OCOR){=}NH \rightleftharpoons (RCO)_2NH$$

Dicyclohexylcarbodiimide (DCC), commonly used as a reagent in peptide synthesis, has been shown to act as a dehydrating agent for amides in pyridine solution[281]:

$$CH_2CONH_2{-}CHCOOH{-}NHCOCH_2C_6H_5 \xrightarrow{DCC} CH_2CN{-}CHCOOH{-}NHCOCH_2C_6H_5$$

Other less familiar methods for the dehydration of amides are the treatment with trialkylsilanes in the presence of zinc chloride[282] or with complex hydrides. In the latter case, certain amides instead of yielding the corresponding amines give mainly nitriles[283]:

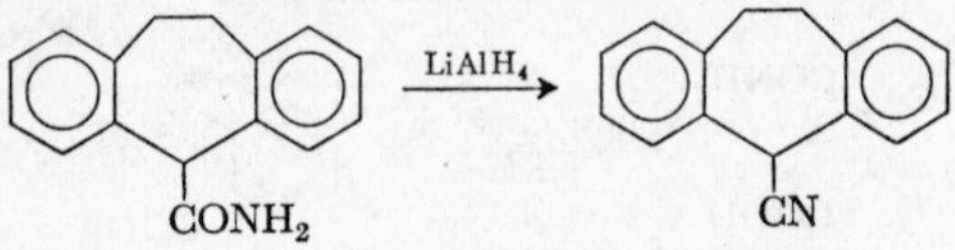

$$\text{(dibenzocycloheptene)}{-}CONH_2 \xrightarrow{LiAlH_4} \text{(dibenzocycloheptene)}{-}CN \quad 92\%$$

$$Ph_2CHCONH_2 \xrightarrow{LiAlH_4,\ HgCl_2} Ph_2CHCN \quad 81{\cdot}5\%$$

Similarly, sodium borohydride in refluxing diglyme transforms benzamide, acetamide and phenylacetamide to the nitriles[284].

Acid amides undergo dehydration when heated together with

sulphamic acid[285] or ammonium sulphamate[286]:

$$RCONHPh \xrightarrow[150\text{–}200°]{H_2NSO_3NH_4} RCN$$

This method is of general applicability and gives good yields.

Thioamides may be converted to nitriles by treatment with mercuric chloride and methylamine in methanol[287]:

$$RC(=S)NH_2 \xrightarrow[MeOH]{HgCl_2,\ CH_3NH_2} RCN$$

The same reaction takes place without reagents at elevated temperatures. It is the final step in a nitrile synthesis in which methyl groups activated by an aromatic ring are transformed to nitriles by heating the compound with elemental sulphur and anhydrous ammonia[288]:

$$ArCH_3 + 3\,S + NH_3 \rightleftharpoons ArC(=S)NH_2 + 2\,H_2S$$

$$ArC(=S)NH_2 \rightleftharpoons ArCN + H_2S$$

N-Substituted thioamides also undergo thermal conversion to nitriles. Thus *N*-benzylthiobenzamide at 400° yields benzonitrile, hydrogen sulphide and stilbene[289].

Several methods are known for the direct conversion of carboxylic acids to nitriles[290]. A very simple way to obtain nitriles is to treat the acid at sufficiently high temperatures with dry ammonia. By this method higher fatty acid nitriles such as stearonitrile or lauronitrile have been prepared[291]. Another possibility is to subject the ammonium salt of an acid to dehydration, either by chemical means such as phophorus pentoxide, or thermally[290]. Reaction of carboxylic acids with appropriate amides at elevated temperatures affords a route to nitriles as shown in the following example, where sebaconitrile is obtained by reaction of the acid with urea[292]:

$$HOOC(CH_2)_8COOH \xrightarrow{(H_2N)_2CO} \text{diamide} \longrightarrow NC(CH_2)_8CN + HOOC(CH_2)_8CN$$

Aryl sulphonamides have also found use in a similar conversion[293]:

$$ArCOOH \xrightarrow[225°]{PhSO_2NH_2} ArCN$$

A modification of this method is the reaction of carboxylic acids with aryl sulphonamides and phosphorus pentachloride[294]:

$$p\text{-}O_2N\text{-}C_6H_4\text{-}COOH + ArSO_2NH_2 + 2\,PCl_5 \longrightarrow p\text{-}O_2N\text{-}C_6H_4\text{-}CN + ArSO_2Cl + 2\,POCl_3 + 3\,HCl$$

A more detailed investigation revealed that the phosphorus pentachloride reacts with the aryl sulphonamide to give a sulphonyl phosphorimidic trichloride, which in turn reacts with the carboxylic acid chloride present in the mixture. The resulting acyl phosphorimidic trichloride is then cleaved thermally to give the nitrile and phosphoryl chloride[295]:

$$Ar^1COCl + Ar^2SO_2{-}N{=}PCl_3 \longrightarrow Ar^2SO_2Cl + Ar^1CO{-}N{=}PCl_3 \xrightarrow{\sim 200^\circ} Ar^1CN + POCl_3$$

The arylsulphonyl phosphorimidic trichloride may be prepared separately. In this case two moles of it are required for the reaction with one mole of carboxylic acid.

Another reagent which shows similar structural features is the trimeric phosphonitrilic chloride. It can be used in the conversion of sodium salts of carboxylic acids to nitriles[296]:

$$RCOONa \xrightarrow[110^\circ\text{–}240^\circ]{(NPCl_2)_3} RCN$$

A simple conversion of carboxylic acids to nitriles is possible with chlorosulphonyl isocyanate. Its reaction with carboxylic acids yields the mixed anhydrides of the acid and *N*-chlorosulphonyl carbamic acid which in turn lose carbon dioxide during their preparation. The resulting *N*-chlorosulphonyl carbonamides, upon treatment with dimethylformamide, eliminate chlorosulphonic acid and give the nitrile[297]:

$$RCOOH + O{=}C{=}NSO_2Cl \longrightarrow [RCOOCONHSO_2Cl] \xrightarrow{-CO_2} RCONHSO_2Cl \xrightarrow{DMF} RCN + ClSO_3H{\cdot}DMF$$

Chlorosulphonyl isocyanate is also used to introduce a cyano group into an olefin[298]:

$$O{=}C{=}N{-}SO_2Cl + PhCH{=}CH_2 \longrightarrow \underset{\substack{| \qquad\quad | \\ N \qquad\; C \\ ClSO_2 \qquad\quad O}}{PhCH{-}CH_2} \xrightarrow{DMF} PhCH{=}CHCN + ClSO_3H{\cdot}DMF$$

V. MISCELLANEOUS METHODS

Various nitrile syntheses do not fit into the scheme of the preceding sections and will therefore be mentioned here.

Isonitriles start to rearrange to nitriles at temperatures around 150°. Correspondingly, reactions that give isonitriles in the first step will produce nitriles if performed at a sufficiently high temperature. Thus pyrolysis of *N*-formyl amines at temperatures around 500° in the presence of a silica-gel catalyst yields nitriles[299]:

$$o\text{-}CH_3C_6H_4NHCHO \longrightarrow o\text{-}CH_3C_6H_4CN$$

Another route is to reflux formanilides with zinc dust[300]. Instead of the formyl group the dichloro- or trichloroacetyl group may be used. Trichloroacetanilides are converted to nitriles in fair yields when passed over quartz chips in a nitrogen stream at 570° [301]:

$$PhNHCOCCl_3 \longrightarrow PhCN + COCl_2 + HCl$$

Phenyl isocyanate is obtained as a by-product in low yield.

Aryl isothiocyanates can be desulphurized by treatment with triphenyl phosphite, the resulting isonitriles subsequently rearranging to nitriles[302]:

$$Ar{-}N{=}C{=}S + (PhO)_3P \longrightarrow ArCN + (PhO)_3P{=}S$$

The method is useful for the preparation of alkyl and cycloalkyl cyanides and especially sterically hindered aromatic cyanides such as 2,6-diethylbenzonitrile.

As with aldimines, which can be oxidized to give nitriles, primary amines also undergo this reaction. Unbranched aliphatic amines are treated with lead tetraacetate in refluxing benzene to give nitriles[303]:

$$RCH_2NH_2 \xrightarrow{Pb(AcO)_4} RCN$$

Other oxidants for this reaction are iodine pentafluoride[304] and nickel peroxide[305].

The oxidation of amino acids provides a simple route to nitriles. Usually they are oxidized by hypobromite[306] or hypochlorite in aqueous solution as shown in the following example[307]:

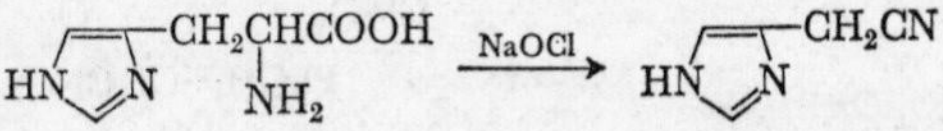

$$\text{(imidazol-4-yl)}-CH_2CH(NH_2)COOH \xrightarrow{NaOCl} \text{(imidazol-4-yl)}-CH_2CN$$

Another oxidant which has been used in this context is *N*-bromosuccinimide[308].

1,2-Diaminobenzenes undergo ring cleavage to muconic acid dinitriles when treated with lead tetraacetate[309] or nickel peroxide[310]. These reactions presumably involve the formation of nitrenes as in the decomposition of 1,2-diazidobenzenes[311]:

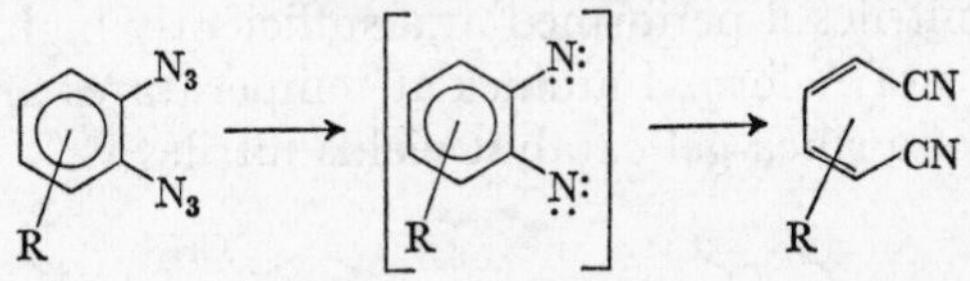

$$R\text{-}C_6H_3(N_3)_2 \longrightarrow \left[R\text{-}C_6H_3(\ddot{N}:)_2 \right] \longrightarrow R\text{-muconic dinitrile (CN, CN)}$$

The decomposition of 2-benzoylvinylazide by hydrogen chloride in acetic acid is reported to give ω-cyanoacetophenone[312]:

$$PhCOCH{=}CHN_3 \longrightarrow N_2 + PhCOCH_2CN$$

A few methods have been reported by which nitriles are synthesized from aliphatic nitro compounds. The alkali salts of primary nitroparaffins react with diethyl phosphorochloridate to give nitriles as the main product[313]:

$$2\,[PhCH{=}NO_2]K + (EtO)_2PCl \longrightarrow PhCN + PhCH_2NH_2 + (EtO)_2P(=O)Cl + KCl$$

Another route from nitro compounds to nitriles consists of the treatment of 1-bromo-1-nitroalkanes with triphenyl phosphine[314]:

$$C_7H_{15}CHBr(NO_2) \xrightarrow{Ph_3P} C_{17}H_{15}CN$$

The same reagent also transforms *N*-bromoamides to nitriles.

Orthocarboxylic esters, orthocarbonates and acetals are cleaved by acetyl cyanide in moderate yields[315]. An example is shown below:

$$(EtO)_3CH + CH_3COCN \xrightarrow[80°]{} (EtO)_2CHCN + AcOEt$$

Another nitrile synthesis using acyl cyanides is their addition to carbon–nitrogen double bonds, which proceeds smoothly at room temperature[316,317]:

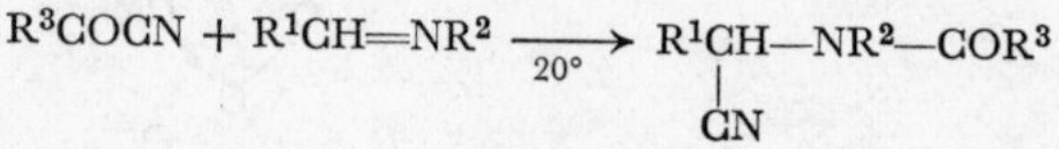

$$R^3COCN + R^1CH{=}NR^2 \xrightarrow[20^\circ]{} R^1CH(CN){-}NR^2{-}COR^3$$

Esters of cyanoformic acid may be pyrolysed at 700–800° to give nitriles[318]:

$$ROCOCN \longrightarrow RCN + CO_2$$

By this method acetonitrile, phenylacetonitrile and malononitrile have been synthesized.

VI. PREPARATION OF NITRILES USING MOLECULES OR MOLECULAR FRAGMENTS WITH CYANO GROUPS

In order to give a more complete picture of the methods of nitrile synthesis, it is necessary to mention a number of reactions which introduce the nitrile group together with a part of the final molecule. Some of these reactions, such as the alkylation of nitriles[319] or cyanoethylation[320] are described elsewhere.

The condensation of active methylene compounds with activating nitrile groups with carbonyl or related compounds provides a simple route to cyanoethylenes:

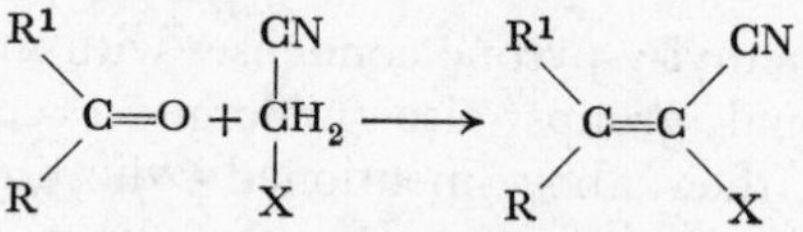

$$R^1(R)C{=}O + CH_2(CN)X \longrightarrow R^1(R)C{=}C(CN)X$$

$$X = COOR^2, COOH, CN, Ar$$

The reaction is base catalysed[321]. Common catalysts are sodium ethoxide[322], sodium amide[323], pyridine[324], piperidine[325] or its acetate[326] and ammonium acetate[326,327]. In some cases the primary condensation products may undergo a further reaction with a second mole of the starting nitrile, as shown below[328]:

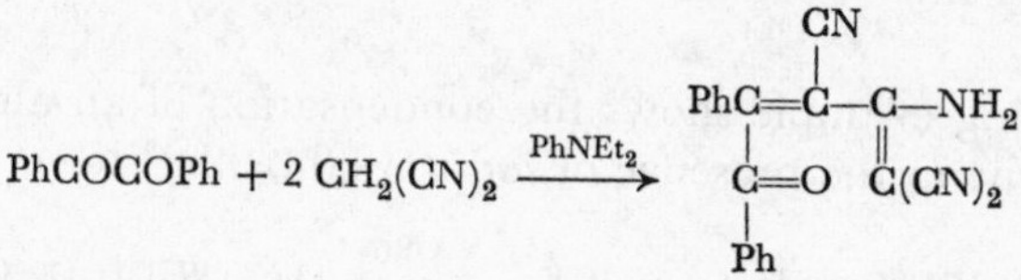

$$PhCOCOPh + 2\ CH_2(CN)_2 \xrightarrow{PhNEt_2} PhC{=}C(CN){-}C(NH_2){=}C(CN)_2 \text{ (with } C{=}O \text{ bearing Ph on the first carbon)}$$

If pyridine is used instead of diethylaniline, benzil and malononitrile give the mono condensation product[329]:

$$PhCOCOPh + CH_2(CN)_2 \xrightarrow{\text{pyridine}} PhC(=C(CN)_2)-C(=O)Ph$$

Another complication which may arise is the Michael addition of a second mole of the active methylene compound, e.g. malononitrile, to the product. In the following example the condition of the equilibrium depends on R[330]:

$$RCH(CH(CN)_2)_2 \rightleftharpoons RCH{=}C(CN)_2 + CH_2(CN)_2$$

The cleavage and exchange reactions of alkylidene malononitriles have been studied[331].

Carbonyl compounds can also be condensed with active methylene compounds with an activating nitrile group by refluxing in acetic anhydride[332]:

$$C_7H_6{=}O + CH_2(CN)_2 \longrightarrow C_7H_6{=}C(CN)_2$$

Similarly 2,6-dimethyl-γ-pyrone condenses with malononitrile[333].

Other functional groups also undergo a condensation with nitriles. Thus the above-mentioned dicyanomethylenecycloheptatriene is obtained from ethoxytropylium fluoroborate and malononitrile[334]. The cyclic acetal of fluorenone with tetrachloropyrocatechol reacts with malononitrile in refluxing butanol to give fluorenylidenemalononitrile[335]. Coumarin diethyl acetal and malononitrile in refluxing ethanol yield the corresponding condensation product[336]:

$$\text{2,2-(OEt)}_2\text{-chromene} + CH_2(CN)_2 \longrightarrow \text{chromen-2-ylidene}{=}C(CN)_2$$

The following example shows the condensation of an amidine with malononitrile in the presence of sodium ethoxide[337]:

$$PhCONHCH_2C(NH_2){=}NH + CH_2(CN)_2 \xrightarrow{\text{NaOEt}} PhCONHCH_2C(NH_2){=}C(CN)_2$$

The Wittig reaction can be applied to the synthesis of monocyanoethylenes from aldehydes[338], as in the example below[339]:

$$Ph_3P + ClCH_2CN \longrightarrow Ph_3P\begin{matrix} Cl \\ CH_2CN \end{matrix} \xrightarrow{NaOH}$$

$$Ph_3P{=}CHCN \xrightarrow{ArCHO} ArCH{=}CHCN$$

While monocyanophosphoranes can still be reacted with aldehydes, the introduction of a second cyano group stabilizes the ylid to such an extent that it does not undergo the Wittig reaction[340].

The use of halogenoacetonitriles has been reported for the Reformatsky reaction[341].

The formation of a dicyanocyclopropane in the reaction of bromomalononitrile with tetramethylethylene was thought to proceed via dicyanocarbene[342]. A more detailed investigation showed that the bromomalononitrile adds in the first step to the olefinic bond, and a subsequent dehydrobromination yields the cyclopropane[343]:

$$Me_2C{=}CMe_2 + BrCH(CN)_2 \longrightarrow Me_2CBr{-}C(Me_2)CH(CN)_2 \xrightarrow{-HBr} \underset{C(CN)_2}{Me_2C{-}CMe_2}$$

Dibromomalononitrile undergoes a similar reaction[344].

The base-catalysed condensation of activated olefins with halomethanes bearing electron-withdrawing substituents is a general synthesis of polyfunctional cyclopropanes. No base is required in the reaction between alkylidene malononitriles and bromomalononitrile[345]:

$$R^1R^2C{=}C(CN)_2 + BrCH(CN)_2 \longrightarrow R^1R^2C\begin{matrix} C(CN)_2 \\ | \\ C(CN)_2 \end{matrix}$$

The thermal decomposition of dicyanodiazomethane affords dicyanocarbene which attacks aromatic compounds with the formation of norcaradiene derivatives[346]:

$$N_2C(CN)_2 + C_6H_6 \xrightarrow[\Delta]{-N_2} \text{7,7-dicyanonorcaradiene}$$

While in the above example the addition product is sufficiently stable to permit isolation, the reaction of cyanonitrene, generated by the decomposition of cyanogen azide, yields *N*-cyanoazepine[347]:

N_3CN + (R-substituted benzene) $\xrightarrow[45-60°]{-N_2}$ (R-substituted azepine N—CN)

R = H, Me, CO_2Me, Cl, F, CF_3, CCl_3

The reaction of thermally generated cyanonitrene with cyclooctatetraene gives a stable 1,4-adduct and a labile 1,2-adduct[348]:

(cyclooctatetraene) $\xrightarrow[\Delta]{N_3CN}$ (NCN-bridged 1,4-adduct) + (NCN 1,2-adduct)

The formation of four-membered rings containing cyano groups may be achieved by 2 + 2 cycloadditions of *α,β*-unsaturated nitriles to electron-rich olefins. Thus tetracyanoethylene (TCNE) adds to vinyl ethers and similar compounds[349]:

$CHX{=}CH_2$ + $C(CN)_2{=}C(CN)_2$ ⟶ (X-substituted cyclobutane with $(CN)_2$, $(CN)_2$)

X = RO, RS, $R^1(R^2CO)N$, $PhSO_2NR$

Similarly tetramethoxyethylene adds TCNE to give the corresponding four-membered ring[350].

By photodimerization of solid fumarodinitrile, *cis,trans,cis*-1,2,3,4-tetracyanocyclobutane is obtained[351].

If a normal Diels–Alder 2 + 4 cycloaddition is not possible, TCNE will react with a diene to give the 2 + 2 cycloaddition product[352]:

(methylenecyclobutene) + TCNE ⟶ (spiro product with $(NC)_2$, $(NC)_2$)

The carbonyl group in acetyl cyanide undergoes photocycloaddition to olefins[353]:

(olefin) + CH_3COCN $\xrightarrow{h\nu}$ (oxetane with O, CH_3, CN)

Cycloadditions leading to five-membered cyano-substituted rings are the reactions of the highly reactive tetracyanoethylene oxide (TCNEO) with aromatic compounds. Benzene is attacked at

130–150°:

$$C_6H_6 + (NC)_2C\text{—}O\text{—}C(CN)_2 \longrightarrow C_6H_6[C(CN)_2]_2O$$

Olefins add TCNEO to give the corresponding tetracyanotetrahydrofurans. Pyridine does not undergo 3 + 2-cycloaddition but instead gives the nitrogen ylid[354]:

$$C_5H_5N + TCNEO \xrightarrow{0°} C_5H_5N^{+}\text{—}C^{-}(CN)_2 + O{=}C(CN)_2$$

The addition of dicyanoacetylene to the 'bent' σ-bond of bicyclo[2.1.0]pentane may also be considered as a 3 + 2 addition[355]:

$$C_5H_8 + NC\text{—}C{\equiv}C\text{—}CN \xrightarrow{20°} \text{(bicyclo[2.2.1]heptene)}(CN)_2 + \text{cyclopentenylidene}{=}C(CN)CH_2CN$$

Quite recently a new convenient synthesis of dicyanoacetylene has been reported. Here also many references concerning the cycloaddition reactions of dicyanoacetylene may be found[356]. The Diels–Alder reaction of α,β-unsaturated nitriles has found many applications[357]. Tetracyanoethylene, with its highly electron-deficient double bond, exhibits a remarkable reactivity[358]. The addition reactions of fumaronitrile, maleonitrile, acrylonitrile and TCNE to a variety of cyclic dienes have been investigated[359].

By addition of acrylonitrile to cyclopentadiene the *endo* product is obtained[360]:

$$C_5H_6 + CH_2{=}CHCN \longrightarrow \text{endo-norbornenecarbonitrile}$$

The corresponding reaction with methacrylonitrile, however, is reported to give mainly the *exo*-nitrile[361]. The stereochemistry of the addition of 1,2-bis(trifluoromethyl)fumaronitrile to electron-rich allenes has been studied[362].

Several methods may be used to link two nitrile-containing moieties. The electrolytic hydrodimerization of acrylonitrile and related compounds has become an industrially important process[363]:

$$2\,CH_2{=}CHCN + 2\,e^- + 2\,H_2O \longrightarrow NC(CH_2)_4CN + 2\,OH^-$$

Cyanoalkyl radicals formed by the decomposition of azo-bis-(alkyl)nitriles combine to give the corresponding *vicinal* dinitriles[53]:

$$R^1(CN)(R)C{-}N{=}N{-}C(CN)(R^1)R \xrightarrow{-N_2} R^1(CN)(R)C{-}C(CN)(R^1)R$$

Carbon-to-carbon coupling has been found to occur to a certain extent during the electrolysis of cyanoacetic acids[364].

VII. REFERENCES

1. D. T. Mowry, *Chem. Rev.*, **42**, 189 (1948).
2. P. Kurtz, 'Methoden zur Herstellung und Umwandlung von Nitrilen und Isonitrilen', *Methoden der organischen Chemie* (Houben–Weyl), Band VIII, Georg Thieme Verlag, Stuttgart, 1952, p. 247.
3. J. A. Nieuwland, W. S. Calcott, F. B. Downing and A. S. Carter, *J. Am. Chem. Soc.*, **53**, 4197 (1931).
4. P. Kurtz, *Ann. Chem.*, **572**, 36 (1951).
5. E. Bauer, *French Pat.*, 892,870; *Chem. Zentr.*, **1948**(1), 170.
6. *Brit. Pat.*, 573,627; *Chem. Abstr.*, **43**, 3027 (1949).
7. P. Kurtz, *Ann. Chem.*, **572**, 36 (1951).
8. H. Bredereck, G. Simchen and P. Horn, *Angew. Chem.*, **77**, 508 (1965); *Int. Ed. Engl.*, **4**, 951 (1965).
9. P. Kurtz, *Ann. Chem.*, **572**, 26 (1951).
10. J. Bredt and J. Kallen, *Ann. Chem.*, **293**, 344 (1896);
11. L. Higginbotham and A. Lapworth, *J. Chem. Soc.*, **1922**, 49.
12. A. Lapworth and W. Baker in *Organic Syntheses*, Coll. Vol. 1, (Ed. H. Gilman), John Wiley and Sons, New York, 1941, p. 752.
13. A. Michael and N. Weiner, *J. Am. Chem. Soc.*, **59**, 744 (1937); P. Kurtz, *Ann. Chem.*, **572**, 34 (1951).
14. J. Bredt and J. Kallen, *Ann. Chem.*, **293**, 338 (1896).
15. R. B. Pritchard, C. E. Lough, J. B. Reesor, H. L. Holmes and D. J. Currie, *Can. J. Chem.*, **45**, 775 (1967).
16. S. Patai and Z. Rappoport in *The Chemistry of Alkenes* (Ed. S. Patai), Interscience, London, 1964, pp. 469–584.
17. J. Thiele and J. Meisenheimer, *Ann. Chem.*, **306**, 247 (1899).
18. W. Davey and D. J. Tivey, *J. Chem. Soc.*, **1958**, 1230; C. F. Allen and R. K. Kimball in *Organic Syntheses*, Coll. Vol. 2 (Ed. A. H. Blatt), John Wiley and Sons, New York, (1943), p. 498.
19. D. T. Mowry, *Chem. Rev.*, **42**, 230 (1948); P. Kurtz, 'Methoden zur Herstellung und Umwandlung von Nitrilen und Isonitrilen', *Methoden der organischen Chemie* (Houben–Weyl), Band VIII, Georg Thieme Verlag, Stuttgart 1952, p. 272.
20. W. Nagata, M. Yoshioka and S. Hirai, *Tetrahedron Letters*, **1962**, 461.

21. J. Thiele and J. Meisenheimer, *Ber.*, **33**, 675 (1900).
22. C. F. H. Allen and C. V. Wilson, *J. Am. Chem. Soc.*, **63**, 1756 (1941).
23. K. Wallenfels, D. Hofmann and R. Kern, *Tetrahedron*, **21**, 2239 (1965).
24. F. Winkler, *Ann. Chem.*, **4**, 246 (1832).
25. A. Gautier and M. Simpson, *Compt. Rend.*, **65**, 414 (1867).
26. F. Urech, *Ann. Chem.*, **164**, 255 (1872); H. Kiliani, *Ber.*, **21**, 916 (1888); A. Lapworth, *J. Chem. Soc.*, **1903**, 85.
27. V. Franzen, *Chem. Z.*, **80**, 379 (1956); H.-H. Hustedt and E. Pfeil, *Ann. Chem.*, **640**, 15 (1961).
28. A. M. El-Arbady, *J. Org. Chem.*, **21**, 828 (1956); L. Ruzicka, P. Plattner and H. Wild, *Helv. Chim. Acta*, **28**, 613 (1945); V. Prelog and M. Kobelt, *Helv. Chim. Acta*, **32**, 1187 (1949); O. H. Wheeler and J. Z. Zabicky, *Can. J. Chem.*, **36**, 656 (1958).
29. D. T. Mowry, *Chem. Rev.*, **42**, 231 (1948); P. Kurtz, 'Methoden zur Herstellung und Umwandlung von Nitrilen und Isonitrilen', *Methoden der organischen Chemie* (Houben–Weyl), Band VIII, Georg Thieme Verlag, Stuttgart, 1952, p. 275.
30. J. A. Ultee, *Ber.*, **39**, 1856 (1906); *Rec. Trav. Chim.*, **28**, 10 (1909); W. F. Beech and H. A. Piggott, *J. Chem. Soc.*, **1955**, 425.
31. R. C. Cookson and K. R. Friedrich, *J. Chem. Soc.*, **1966**, 1641; K. R. Friedrich, *Angew. Chem.*, **78**, 449 (1966); *Int. Ed. Engl.*, **5**, 420 (1966).
32. F. Urech, *Ann. Chem.*, **164**, 255 (1872); R. F. B. Cox and R. T. Stormont in *Organic Syntheses*, Coll. Vol. 2, (Ed. A. H. Blatt), John Wiley and Sons, 1943, p. 7.
33. A. Spiegel, *Ber.*, **14**, 239 (1881).
34. J. Burdon, T. J. Smith and J. C. Tatlow, *J. Chem. Soc.*, **1961**, 4519.
35. T. Mill, J. O. Rodin, R. M. Silverstein and C. Woolf, *J. Org. Chem.*, **29**, 3715 (1964).
36. R. L. Frank, R. E. Berry and O. L. Shotwell, *J. Am. Chem. Soc.*, **71**, 3889 (1949); W. D. Emmons and J. P. Freeman, *J. Am. Chem. Soc.*, **77**, 4387 (1955).
37. H. Bucherer and A. Grolée, *Ber.*, **39**, 1224 (1906).
38. W. H. Davies, A. W. Johnson and H. A. Piggott, *J. Chem. Soc.*, **1945**, 352.
39. J. N. Nasarov, A. A. Akhrem and A. V. Kamernitskii, *Zh. Obshch. Khim.*, **25**, 1345 (1955); *Chem. Abstr.*, **50**, 4950 (1956); P. Kurtz, 'Methoden zur Herstellung und Umwandlung von Nitrilen und Isonitrilen', *Methoden der organischen Chemie* (Houben–Weyl), Band VIII, Georg Thieme Verlag, Stuttgart, 1952, p. 276.
40. A. V. Kamernitskii and A. A. Akhrem, *Zh. Obshch. Khim.*, **30**, 754 (1960); *Chem. Abstr.*, **55**, 413 (1961); B. E. Betts and W. Davey, *J. Chem. Soc.*, **1958**, 4193.
41. V. Prelog and M. Wilhelm, *Helv. Chim. Acta.*, **37**, 1634 (1934); H. Albers and E. Albers, *Z. Naturforsch.*, **98**, 122 (1954).
42. R. Justoni and M. Terruzzi, *Gazz. Chim. Ital.*, **78**, 166 (1948).
43. R. Justoni and M. Terruzzi, *Gazz. Chim. Ital.*, **78**, 155 (1948).
44. E. P. Kohler and F. W. Brown, *J. Am. Chem. Soc.*, **55**, 4299 (1933).
45. D. T. Mowry, *Chem. Rev.*, **42**, 236 (1948); P. Kurtz, 'Methoden zur Herstellung und Umwandlung von Nitrilen und Isonitrilen', *Methoden der organischen Chemie* (Houben–Weyl), Band VIII, Georg Thieme Verlag, Stuttgart, 1952, p. 279.
46. A. Strecker, *Ann. Chem.*, **75**, 27 (1850); *Ann. Chem.*, **91**, 349 (1954).
47. E. Erlenmeyer and S. C. Passavant, *Ann. Chem.*, **200**, 120 (1880).
48. F. Tiemann, *Ber.*, **14**, 1957 (1881); F. Tiemann and L. Friedländer, *Ber.*, **14**, 1967 (1881); F. Tiemann and R. Piest, *Ber.*, **14**, 1982 (1881); F. Tiemann, *Ber.*, **13**, 381 (1880).

49. N. Ljubavin, *Ber.*, **14**, 2686 (1881); W. Gulewitsch, *Ber.*, **33**, 1900 (1900); H. Bucherer, *Ber.*, **39**, 2033 (1906).
50. N. Zelinsky and G. Stadnikoff, *Ber.*, **39**, 1722 (1906).
51. L. J. Exner, L. S. Luskin and P. L. de Benneville, *J. Am. Chem. Soc.*, **75**, 4841 (1953).
52. H. Zahn and H. Wilhelm, *Ann. Chem.*, **579**, 1 (1953).
53. C. G. Overberger, M. T. O'Shaughnessy and H. Shalit, *J. Am. Chem. Soc.*, **71**, 2661 (1949); C. G. Overberger, T. B. Gibb Jr., S. Chibnik, P. T. Huang and J. J. Monagle, *J. Am. Chem. Soc.*, **74**, 3290 (1952); C. G. Overberger, H. Biletch, A. B. Finestone, J. Lilker and J. Henbest, *J. Am. Chem. Soc.*, **75**, 2078 (1953); C. G. Overberger and A. Lebovits, *J. Am. Chem. Soc.*, **76**, 2722 (1954); C. G. Overberger, P. T. Huang and M. B. Berenbaum in *Organic Syntheses*, Coll. Vol. 4, (Ed. N. Rabjohn), John Wiley and Sons, New York, 1963, p. 274.
54. R. Adams and W. D. Langley in *Organic Syntheses*, Coll. Vol. 1 (Ed. H. Gilman), John Wiley and Sons, New York, 1941, p. 355.
55. A. H. Cook and S. F. Cox, *J. Chem. Soc.*, **1949**, 2334; C. F. H. Allen and J. A. van Allen in *Organic Syntheses*, Coll. Vol. 3 (Ed. E. C. Horning), John Wiley and Sons, New York, 1955, p. 275; H. M. Taylor and C. R. Hauser in *Organic Syntheses*, Vol. 43 (Ed. B. C. McKusick), John Wiley and Sons, New York, 1963, p. 25.
56. D. T. Mowry, *Chem. Rev.*, **42**, 238 (1948); P. Kurtz, 'Methoden zur Herstellung und Umwandlung von Nitrilen und Isonitrilen', *Methoden der organischen Chemie* (Houben–Weyl), Band VIII, Georg Thieme Verlag, Stuttgart, 1952, p. 280.
57. D. T. Mowry, *Chem. Rev.*, **42**, 241 (1948); P. Kurtz, 'Methoden zur Herstellung und Umwandlung von Nitrilen und Isonitrilen', *Methoden der organischen Chemie* (Houben–Weyl), Band VIII, Georg Thieme Verlag, Stuttgart, 1952, p. 284.
58. M. R. Triollais, *Bull. Soc. Chim. France*, **1947**, 966.
59. W. v. Miller and J. Plöchl, *Ber.*, **25**, 2020 (1892).
60. *U.S. Pat.*, 2,580,919, *Chem. Abstr.*, **46**, 7119 (1952).
61. C. G. Overberger and M. B. Berenbaum, *J. Am. Chem. Soc.*, **73**, 2618 (1951); C. G. Overberger, W. F. Hale, M. B. Berenbaum and A. B. Finestone, *J. Am. Chem. Soc.*, **76**, 6185 (1954).
62. C. D. Hurd and J. M. Longfellow, *J. Org. Chem.*, **16**, 761 (1951).
63. L. Neelakantan and W. H. Hartung, *J. Org. Chem.*, **23**, 964 (1958).
64. L. W. Kissinger and H. E. Ungnade, *J. Org. Chem.*, **25**, 1471 (1960).
65. *Ger. Pat.*, 1,053,500; *Chem. Abstr.*, **55**, 9286 (1961).
66. A. Reissert, *Ber.*, **38**, 1603, 3415 (1905); for a review see R. L. Cobb, *Chem. Rev.*, **55**, 511 (1955).
67. W. E. Feely, *U.S. Pat.*, 2,991,285; *Chem. Abstr.*, **56**, 7282 (1962).
68. K. Wallenfels and H. Diekmann, *Ann. Chem.*, **621**, 166 (1959); K. Wallenfels and W. Hanstein, *Angew. Chem.*, **77**, 861 (1965), *Int. Ed. Engl.*, **4**, 869 (1965); K. Wallenfels and W. Hanstein, *Ann. Chem.*, **709**, 151 (1967).
69. R. N. Lindquist and E. H. Cordes, *J. Am. Chem. Soc.*, **90**, 1269 (1968).
70. E. S. Gould, *Mechanism and Structure in Organic Chemistry*, Holt, Rinehart and Winston, New York, 1959, p. 297; N. Kornblum, R. A. Smiley, R. K. Blackwood and D. C. Iffiand, *J. Am. Chem. Soc.*, **77**, 6269 (1955).
71. M. Mousseron and J. Jullien, *Compt. Rend.*, **231**, 410 (1950).
72. G. C. Harrison and H. Diehl in *Organic Syntheses*, Coll. Vol. 3 (Ed. E. C. Horning), John Wiley and Sons, New York, 1955, p. 372; C. S. Marvel and E. M. McColm in *Organic Syntheses*, Coll. Vol. 1 (Ed. H. Gilman), John Wiley and Sons, New York, 1941, p. 536.

73. A. Lapworth and W. Baker in *Organic Syntheses*, Coll. Vol. 1 (Ed. H. Gilman), John Wiley and Sons, New York, 1941, p. 181; E. C. Kendall and B. McKenzie in *Organic Syntheses*, Coll. Vol. 1 (Ed. H. Gilman), John Wiley and Sons, New York, 1941, p. 256; F. Johnson and J. P. Panella in *Organic Syntheses*, Vol. 46 (Ed. E. J. Corey), John Wiley and Sons, New York, 1966, p. 48.
74. *U.S. Pat.*, 2,734,908; *Chem. Abstr.*, **50**, 15587 (1956); K. Rorig, K. Johnston, R. W. Hamilton and T. J. Telinsky in *Organic Syntheses*, Coll. Vol. 4, (Ed. N. Rabjohn), John Wiley and Sons, New York, 1963, p. 576.
75. M. S. Newman and J. H. Wotiz, *J. Am. Chem. Soc.*, **71**, 1292 (1949).
76. R. N. Lewis and P. V. Susi, *J. Am. Chem. Soc.*, **74**, 840 (1952); A. Brandström, *Acta Chim. Scand.*, **10**, 1197 (1956).
77. *U.S. Pat.*, 2,415,261; *Chem. Abstr.*, **41**, 3119 (1947); J. W. Ferguson, *Proc. Indiana Acad. Sci.*, **63**, 131 (1953); *Chem. Abstr.*, **49**, 8094 (1955).
78. *Japan. Pat.*, 12418 (1961); *Chem. Abstr.*, **56**, 8536 (1962).
79. P. A. Argabright and D. W. Hall, *Chem. Ind.* (*London*), **1964**, 1365.
80. N. E. Rigler and H. R. Henze, *J. Am. Chem. Soc.*, **58**, 474 (1936); H. R. Henze, V. B. Duff, W. H. Matthews, J. W. Melton and E. O. Forman, *J. Am. Chem. Soc.*, **64**, 1222 (1942).
81. *Ger. Pat.*, 1,177,136; *Chem. Abstr.*, **61**, 14539 (1964).
82. R. A. Smiley and C. Arnold, *J. Org. Chem.*, **25**, 257 (1960); L. Friedman and H. Shechter, *J. Org. Chem.*, **25**, 877 (1960).
83. D. Martin, A. Weise and H.-J. Niclas, *Angew. Chem.*, **79**, 340; *Int. Ed. Engl.*, **6**, 318 (1967).
84. B. T. Freure and H. J. Decker, *U.S. Pat.*, 3,026,346; *Chem. Abstr.*, **57**, 11032 (1962).
85. H. Normant, *Angew. Chem.*, **79**, 1029 (1967); *Int. Ed. Engl.*, **6**, 1046 (1967).
86. F. L. M. Pattison, W. J. Cott, W. C. Howell and R. W. White, *J. Am. Chem. Soc.*, **78**, 3484 (1956).
87. *Brit. Pat.*, 768,303; *Chem. Abstr.*, **51**, 16515 (1957).
88. D. T. Mowry, *Chem. Rev.*, **42**, 197, 199 (1948).
89. E. C. Kendall and B. McKenzie in *Organic Syntheses*, Coll. Vol. 1 (Ed. H. Gilman), John Wiley and Sons, New York, 1941, p. 256.
90. A. Dewael, *Bull. Soc. Chim. Belges*, **33**, 504 (1924).
91. P. van der Straeten and A. Bruylants, *Bull. Soc. Chim. Belges*, **66**, 345 (1957); P. Kurtz and H. Schwarz, *U.S. Pat.*, 2,665,297; *Chem. Abstr.*, **49**, 1780 (1955).
92. R. Lespieau, *Bull. Soc. Chim. France*, [3] **33**, 55 (1905).
93. P. Bruylants, *Bull. Soc. Chim. Belges*, **31**, 176 (1922); J. V. Supniewski and P. L. Salzberg in *Organic Syntheses*, Coll. Vol. 1 (Ed. H. Gilman), John Wiley and Sons, New York, 1941, p. 46.
94. R. Delaby, *Compt. Rend.*, **203**, 1521 (1936).
95. R. Breckpot, *Bull. Soc. Chim. Belges*, **39**, 465, 468 (1930).
96. J. F. Lane, J. Fentress and L. T. Sherwood Jr., *J. Am. Chem. Soc.*, **66**, 545 (1944); D. T. Mowry, *Chem. Rev.*, **42**, 199 (1948); P. Kurtz, 'Methoden zur Herstellung und Umwandlung von Nitrilen und Isonitrilen', *Methoden der organischen Chemie* (Houben–Weyl), Band VIII, Georg Thieme Verlag, Stuttgart, 1952, p. 298.
97. J. Jennen, *Bull. Acad. Belg.*, **12**, 1169 (1936); *Bull. Soc. Chim. Belges*, **46**, 199 (1937).
98. C. F. Koelsch, *J. Am. Chem. Soc.*, **58**, 1330 (1936).
99. M. M. v. Richter, *Ber.*, **44**, 3469 (1911).
100. K. Wallenfels, D. Hofmann, G. Bachmann and R. Kern, *Tetrahedron*, **21**, 2239 (1965).
101. K. Wallenfels and G. Bachmann, *Angew. Chem.*, **73**, 142 (1961).

102. M. Gordon and C. E. Griffin, *Chem. Ind.* (*London*), **1962**, 1019; Y. Urata, *Nippon Kagaku Zasshi*, **83**(10), 1105 (1962); *Chem. Abstr.*, **59**, 11240 (1963).
103. J. N. Nazarov, A. V. Semenovskii and A. V. Kamernitskii, *Izv. Akad. Nauk. SSSR. Otdel. Khim. Nauk*, **1957**, 976; *Chem. Abstr.*, **52**, 4549 (1958).
104. P. Kurtz, H. Schwarz and H. Disselnkötter, *Ann. Chem.*, **624**, 1 (1959); *Ger. Pat.*, 1,064,504; *Chem. Abstr.*, **55** 11307 (1961).
105. P. Kurtz, H. Schwarz and H. Disselnkötter, *Ann. Chem.*, **631**, 21 (1960).
106. Y. Pasternak and G. Peiffer, *Compt. Rend.*, **259**, 1142 (1964).
107. D. T. Mowry, *Chem. Rev.*, **42**, 209 (1948); P. Kurtz, 'Methoden zur Herstellung und Umwandlung von Nitrilen und Isonitrilen', *Methoden der organischen Chemie* (Houben–Weyl), Band VIII, Georg Thieme Verlag, Stuttgart, 1952, p. 304; T. S. Oakwood and C. S. Weisgerber in *Organic Syntheses*, Coll. Vol. 3 (Ed. E. C. Hornung), John Wiley and Sons, New York, 1955, p. 112.
108. *Ger. Pat.*, 835,141, *Chem. Abstr.*, **52**, 10160 (1958); *East Ger. Pat.*, 9569, *Chem. Abstr.*, **52** 17113 (1958).
109. R. H. Patton and J. H. Simmons, *J. Am. Chem. Soc.*, **77**, 2016 (1955).
110. L. Claisen, *Ber.*, **41**, 1023 (1898).
111. D. T. Mowry, *Chem. Rev.*, **42**, 207 (1948); P. Kurtz, 'Methoden zur Herstellung und Umwandlung von Nitrilen und Isonitrilen', *Methoden der organischen Chemie* (Houben–Weyl), Band VIII, Georg Thieme Verlag, Stuttgart, 1952, p. 302.
112. J. v. Braun and G. Manz, *Ann. Chem.*, **488**, 116 (1931).
113. J. E. Callen, C. A. Dornfeld and G. H. Coleman in *Organic Syntheses*, Coll. Vol. 3 (Ed. E. C. Horning), John Wiley and Sons, New York, 1955, p. 212.
114. K. W. Rosenmund and E. Struck, *Ber.*, **52**, 1749 (1919).
115. C. F. Koelsch and A. G. Whitney, *J. Org. Chem.*, **6**, 795 1941).
116. M. S. Newman, *J. Am. Chem. Soc.*, **59**, 2473 (1937); *Organic Syntheses*, Coll. Vol. 3 (Ed. E. C. Horning), John Wiley and Sons, New York, 1955, p. 631.
117. R. C. Fuson, J. W. Kreisley, N. Rabjohn and M. L. Ward, *J. Am. Chem. Soc.*, **68**, 533 (1946).
118. C. D. Weis, *J. Org. Chem.*, **27**, 2964 (1962).
119. K. Akanuma, H. Amamiya, T. Hayashi, K. Watanabe and K. Hata, *Nippon Kagaku Zasshi*, **81**, 333 (1960); *Chem. Abstr.*, **56**, 406 (1962); see also J. H. Helberger, *Ann. Chem.*, **529**, 211 (1937).
120. L. Friedman and H. Shechter, *J. Org. Chem.*, **26**, 2522 (1961).
121. M. S. Newman and H. Boden, *J. Org. Chem.*, **26**, 2525 (1961)
122 K. Walenfels and K. Friedrich, *Tetrahedron Letters*, **1963**, 1223.
123. P. ten Haken, *Brit. Pat.*, 861,898; *Chem. Abstr.*, **55**, 17581 (1961).
124. C. F. Koelsch, *J. Am. Chem. Soc.*, **58**, 1328 (1936).
125. D. T. Mowry, *Chem. Rev.*, **42**, 208 (1948).
126. E. A. Lawton and D. D. McRitchie, *J. Org. Chem.*, **24**, 26 (1959).
127. O. Allendorff, *Ber.*, **24**, 2347 (1891).
128. A. Braun and J. Tscherniac, *Ber.*, **40**, 2709 (1907).
129. H. De Diesbach and E. von der Weid, *Helv. Chim. Acta*, **10**, 886 (1927); R. P. Linstead, *Ber.*, **72A**, 93 (1939).
130. K. Friedrich, unpublished results.
131. V. Merz, *Z. Chem.*, **4**, 33, 396 (1868).
132. O. Witt, *Ber.*, **6**, 448 (1873).
133. E. F. Bradbrook and R. P. Linstead, *J. Chem. Soc.*, **1936**, 1739.
134. D. T. Mowry, *Chem. Rev.*, **42**, 193 (1948); P. Kurtz, 'Methoden zur Herstellung

und Umwandlung von Nitrilen und Isonitrilen', *Methoden der organischen Chemie* (Houben–Weyl), Band VIII, Georg Thieme Verlag, Stuttgart, 1952, p. 308.
135. L. Darmstaedter and H. Wichelhaus, *Ann. Chem.*, **152**, 298 (1869).
136. S. M. McElvain and M. A. Goese, *J. Am. Chem. Soc.*, **65**, 2233 (1943).
137. E. Ochiai and H. Yamanaka, *Pharm. Bull.* (*Tokyo*), **3**, 173; *Chem. Abstr.*, **50**, 7810 (1956).
138. D. T. Mowry, *Chem. Rev.*, **42**, 244 (1948); P. Kurtz, 'Methoden zur Herstellung und Umwandlung von Nitrilen und Isonitrilen', *Methoden der organischen Chemie* (Houben–Weyl), Band VIII, Georg Thieme Verlag, Stuttgart, 1952, p. 286.
139. P. Kurtz, *Ann. Chem.*, **572**, 49 (1951).
140. A. Dewael, *Bull. Soc. Chim. Belges*, **39**, 40 (1930).
141. P. M. Greaves, S. R. Landor and D. R. J. Laws, *Chem. Commun.*, **1965**, 321.
142. D. T. Mowry, *Chem. Rev.*, **42**, 192 (1948); P. Kurtz, 'Methoden zur Herstellung und Umwandlung von Nitrilen und Isonitrilen', *Methoden der organischen Chemie* (Houben–Weyl), Band VIII, Georg Thieme Verlag, Stuttgart, 1952, p. 306.
143. P. Walden, *Ber.*, **40**, 3215 (1907).
144. R. Grewe and E. Nolte, *Ann. Chem.*, **575**, 1 (1952).
145. M. S. Newman and T. Otsuka, *J. Org. Chem.*, **23**, 797 (1958); M. S. Newman and R. M. Wise, *J. Am. Chem. Soc.*, **78**, 450 (1956).
146. H. B. Henbest and W. R. Jackson, *J. Chem. Soc.*, **1962**, 954.
147. E. R. Nelson, M. Maienthal, L. A. Lane and A. A. Benderly, *J. Am. Chem. Soc.*, **79**, 3467 (1957).
148. D. T. Mowry, *Chem. Rev.*, **42**, 288 (1948).
149. C. C. Price and R. G. Rogers in *Organic Syntheses*, Coll. Vol. 3 (Ed. E. C. Horning), John Wiley and Sons, New York, 1955, p. 174.
150. J. G. Erickson, *J. Am. Chem. Soc.*, **73**, 1338 (1951); *U.S. Pat.*, 2,519,957, *Chem. Abstr.*, **44**, 10728 (1950); D. T. Mowry, *Chem. Rev.*, **42**, 290 (1948).
151. E. Erlenmeyer, *Ann. Chem.*, **191**, 270, 273 (1878); R. Rambaud, *Bull. Soc. Chim. France*, **3**(5), 138 (1936).
152. F. Y. Perveev and K. Golodova, *Zh. Obshch. Khim.*, **32**, 2092 (1962); *Chem. Abstr.*, **58**, 5511.
153. F. Johnson and J. P. Panella in *Organic Syntheses*, Vol. 46 (Ed. E. J. Corey), John Wiley and Sons, New York, 1966, p. 48.
154. D. T. Mowry, *Chem. Rev.*, **42**, 211 (1948); P. Kurtz, 'Methoden zur Herstellung und Umwandlung von Nitrilen und Isonitrilen', *Methoden der organischen Chemie* (Houben–Weyl), Band VIII, Georg Thieme Verlag, Stuttgart, 1952, p. 314.
155. A. Russell and W. G. Tebbens in *Organic Syntheses*, Coll. Vol. 3 (Ed. E. C. Horning), John Wiley and Sons, New York, 1955, p. 293.
156. W. Borsche and E. Böcker, *Ber.*, **37**, 4388 (1904).
157. V. v. Richter, *Ber.*, **4**, 21, 459, 553 (1873); J. F. Bunnett, J. F. Cormack and F. C. McKay, *J. Org. Chem.*, **15**, 481 (1950).
158. J. F. Bunnett and M. M. Rauhut, *J. Org. Chem.*, **21**, 944 (1956).
159. H. Hellman and E. Lingens, *Chem. Ber.*, **87**, 940 (1954).
160. D. Lednicer and C. R. Hauser in *Organic Synthesis*, Vol. 40 (Ed. M. S. Newman), John Wiley and Sons, New York, 1960, p. 45.
161. E. B. Knott, *J. Chem. Soc.*, **1947**, 1190; E. Haggett and S. Archer, *J. Am. Chem. Soc.*, **71**, 2255 (1949).
162. T. Sandmeyer, *Ber.*, **17**, 2653 (1884); D. T. Mowry, *Chem. Rev.*, **42**, 213 (1948).
163. P. Kurtz, 'Methoden zur Herstellung und Umwandlung von Nitrilen und

Isonitrilen', *Methoden der organischen Chemie* (Houben–Weyl), Band VIII, Georg Thieme Verlag, Stuttgart, 1952, p. 311; H. H. Hodgson, *Chem. Rev.*, **40**, 251 (1947); E. Pfeil, *Angew. Chem.*, **65**, 155 (1953).

164. H. T. Clarke and R. R. Read in *Organic Syntheses*, Coll. Vol. 1 (Ed. H. Gilman), John Wiley and Sons, New York, 1941, p. 514.
165. H. J. Barber, *J. Chem. Soc.*, **1943**, 79.
166. H. Hagenest and F. Stauf, *U.S. Pat.*, 1,879,209; *Chem. Abstr.*, **27**, 997 (1933); *U.S. Pat.*, 1,962,559; *Chem. Abstr.*, **28**, 4848 (1934).
167. L. Gattermann, *Ber.*, **23**, 1218, 1223 (1890); see also O. Stephenson and W. A. Waters, *J. Chem. Soc.*, **1939**, 1796.
168. A. Korczynski, W. Mrozinski and W. Vielau, *Compt. Rend.*, **171**, 182 (1920).
169. H. H. Hodgson and F. Heyworth, *J. Chem. Soc.*, **1949**, 1131.
170. H. T. Clarke and R. R. Reade, *J. Am. Chem. Soc.*, **46**, 1001 (1924).
171. R. Ikan and E. Rapoport, *Tetrahedron*, **23**, 3823 (1967).
172. D. T. Mowry, *Chem. Rev.*, **42**, 214 (1948); J. R. Kochi, *J. Am. Chem. Soc.*, **79**, 2942 (1957); S. C. Dickerman, K. Weiss and A. K. Ingberman, *J. Am. Chem. Soc.*, **80**, 1904 (1958); W. A. Cowdrey and D. S. Davies, *Quart. Rev.*, **1952**, 358.
173. D. T. Mowry, *Chem. Rev.*, **42**, 214 (1948); E. S. Gould, *Mechanism and Structure in Organic Chemistry*, Holt, Rinehart and Winston, New York, 1959, p. 729.
174. D. T. Mowry, *Chem. Rev.*, **42**, 215 (1948); P. Kurtz, 'Methoden zur Herstellung und Umwandlung von Nitrilen und Isonitrilen', *Methoden der organischen Chemie* (Houben–Weyl), Band VIII, Georg Thieme Verlag, Stuttgart, 1952, p. 316.
175. A. Haller, *Compt. Rend.*, **95**, 142 (1882).
176. G. Mignonac and O. W. Rambeck, *Compt. Rend.*, **188**, 1298 (1929); *Bull. Soc. Chim. France*, **45**, 337 (1929).
177. A. Haller and A. Held, *Ann. Chim.*, [6] **17**, 222 (1889).
178. H. Schmidtmann, *Ber.*, **29**, 1171 (1896); A. Hantzsch and G. Osswald, *Ber.*, **32**, 643 (1899).
179. F. Arndt, H. Scholz and E. Frobel, *Ann. Chem.*, **521**, 95 (1935).
180. O. W. Webster, *J. Am. Chem. Soc.*, **88**, 3046 (1966).
181. V. Grignard, *Compt. Rend.*, **152**, 388 (1911); V. Grignard and E. Belliet, *Compt. Rend.*, **155**, 44 (1912); **158**, 457 (1914); V. Grignard and C. Courtot, *Compt. Rend.*, **154**, 361 (1912); *Bull. Soc. Chim. France*, **17**, 228 (1915).
182. M. E. Kuehne, *J. Am. Chem. Soc.*, **81**, 5400 (1959).
183. W. Traube, *Ber.*, **31**, 191, 2938 (1898); *Ann. Chem.*, **332**, 104 (1904).
184. M. Mousseron and F. Winternitz, *Bull. Soc. Chim. France*, **1948**, 79.
185. H. Lettré, P. Jungmann and J.-Ch. Saalfeld, *Chem. Ber.*, **85**, 397 (1952).
186. D. Martin, S. Rackow and A. Weise, *Chem. Ber.*, **98**, 3662 (1965).
187. D. T. Mowry, *Chem. Rev.*, **42**, 220 (1948); P. Kurtz, 'Methoden zur Herstellung und Umwandlung von Nitrilen und Isonitrilen', *Methoden der organischen Chemie* (Houben–Weyl), Band VIII, Georg Thieme Verlag, Stuttgart, 1952, p. 318.
189. C. Friedel and J. M. Crafts, *Bull. Soc. Chim. France*, **29**, 2 (1878); *Ann. Chim.* **1**(6), 528 (1884).
188. M. A. Degrez, *Bull. Soc. Chim. France*, **13**(3), 738 (1895); D. Vorländer, *Ber.*, **44**, 2455 (1911).
190. P. Karrer and E. Zeller, *Helv. Chim. Acta*, **2**, 482 (1919); P. Karrer, A. Rebmann and E. Zeller, *Helv. Chim. Acta*, **3**, 261 (1920).
191. A. N. Nesmeyanov, E. G. Perewalowa and L. P. Jurjewa, *Chem. Ber.*, **93**, 2729 (1960).

192. D. T. Mowry, *Chem. Rev.*, **42**, 221 (1948); P. Kurtz, 'Methoden zur Herstellung und Umwandlung von Nitrilen und Isonitrilen', *Methoden der organischen Chemie* (Houben–Weyl), Band VIII, Georg Thieme Verlag, Stuttgart, 1952, p. 319; J. Houben and W. Fischer, *J. Prakt. Chem*, **123**(2), 313 (1929); *Ber.*, **63**, 2464 (1930); **66**, 339 (1933).
193. (a) E. Müller and H. Huber, *Chem. Ber.*, **96**, 670, 2319 (1963).
(b) N. Karasch and L. Göthlich, *Angew Chem.*, **74**, 651 (1962); *Int. Ed. Engl.*, **1**, 459 (1962); L. Eberson and S. Nilson, *Disc. Faraday Soc.*, in press (1968).
194. K. Koyama, T. Susuki and S. Tsutsumi, *Bull. Chem. Soc. Japan*, **23**, 2675 (1966); V. D. Parker and B. E. Burgert, *Tetrahedron Letters*, **1965**, 4065; S. Tsutsumi and K. Koyama, *Disc. Faraday Soc.*, in press (1968); L. Eberson and S. Nilson, *Disc. Faraday Soc.*, in press (1968).
195. K. Wallenfels, F. Witzler and K. Friedrich, *Tetrahedron*, **23**, 1353, 1845 (1967).
196. D. T. Mowry, *J. Am. Chem. Soc.*, **69**, 573 (1947); H. Plaut and J. J. Ritter, *J. Am. Chem. Soc.*, **73**, 4076 (1951).
197. J. S. Buck and W. S. Ide in *Organic Syntheses*, Coll. Vol. 2 (Ed. A. H. Blatt), John Wiley and Sons, New York, 1942, p. 622.
198. H. Koopmann, *Rec. Trav. Chim.*, **80**, 1075 (1961).
199. H. Fiesselmann, *Ber.*, **75**, 881 (1942).
200. H. T. Clarke and S. M. Nagy in *Organic Syntheses*, Coll. Vol. 3 (Ed. E. C. Horning), John Wiley and Sons, New York, 1955, p. 690; see also M. F. Browne and R. L. Shiner, *J. Org. Chem.*, **22**, 1320 (1957).
201. C. Trabert, *Arch. Pharm.*, **294**, 246 (1961); *Chem. Abstr.*, **55**, 19746 (1961).
202. R. Malachowski, L. Jurkiewicz and J. Wojtowicz, *Ber.*, **70**, 1012 (1937); **71**, 2239 (1938); O. Glemser and V. Häusser, *Z. Naturforsch.*, **3B**, 159 (1948); see also T. L. Cairns, R. A. Carboni, D. D. Coffman, V. A. Engelhardt, R. E. Heckert, E. L. Little, E. G. McGeer, B. C. McKusick and W. J. Middleton, *J. Am. Chem. Soc.*, **80**, 2775 (1958).
203. B. Lach, *Ber.*, **17**, 1571 (1884).
204. C. Moureu, *Bull. Soc. Chim. France*, **11**(3), 1067 (1894).
205. F. P. Doyle, W. Ferrier, D. O. Holland, M. D. Mehta and J. H. C. Nayler, *J. Chem. Soc.*, **1956**, 2856.
206. K. Wallenfels and K. Friedrich, *Tetrahedron Letters*, **1963**, 1223.
207. E. Restelli de Labriola and V. Deulofeu, *J. Org. Chem.*, **12**, 726 (1947).
208. M. O. Forster and H. M. Judd, *J. Chem. Soc.*, **1910**, 254; E. Müller and B. Narr, *Z. Naturforsch.*, **16B**, 845 (1961); T. J. Bentley, J. F. McGhie and D. H. R. Barton, *Tetrahedron Letters*, **1965**, 2497.
209. O. L. Brady and G. P. McHugh, *J. Chem. Soc.*, **1923**, 1190.
210. R. Scholl and J. Adler, *Monatsh. Chem.*, **39**, 240 (1918).
211. T. Mukaiyama and T. Hata, *Bull. Soc. Chem. Japan*, **34**, 99 (1961).
212. T. Mukaiyama and H. Nohira, *J. Org. Chem.*, **26**, 782 (1961).
213. J. H. Pomeroy and C. A. Craig, *J. Am. Chem. Soc.*, **81**, 6340 (1959).
214. D. T. Mowry, *Chem. Rev.*, **42**, 251 (1948); T. Mukaiyama, K. Tonouka and K. Inoue, *J. Org. Chem.*, **26**, 2202 (1961).
215. L. Claisen and O. Manasse, *Ber.*, **20**, 2194 (1887); M. Passerini, *Gazz. Chim. Ital.*, **56**, 122 (1926).
216. T. van Es, *J. Chem. Soc.*, **1965**, 1564.
217. J. H. Hunt, *Chem. Ind.* (*London*), **1961**, 1873.
218. J. A. Findlay and C. S. Teng, *Can. J. Chem.*, **45**, 1014 (1967).

219. A. Mailhe and F. de Godon, *Bull. Soc. Chim. France*, **23**(4), 18 (1918); A. Mailhe, *Ann. Chim.*, **13**(9), 206 (1920).
220. K. F. Schmidt, *Ber.*, **57**, 704 (1924); C. Schuerch Jr., *J. Am. Chem. Soc.*, **70**, 2293 (1948).
221. H. M. Blatter, H. Lukaszewski and G. de Stevens, *J. Am. Chem. Soc.*, **83**, 2203 (1961); *Organic Syntheses*, Vol. 43 (Ed. B. C. McKusick), John Wiley and Sons, New York, 1963, p. 58. J. E. Hodgkins and J. A. King, *J. Am. Chem. Soc.*, **85**, 2679 (1963).
222. C. R. Hauser and A. G. Gillaspie, *J. Am. Chem. Soc.*, **52**, 4517 (1930).
223. E. J. Poziomek, D. N. Kramer and W. A. Mosher, *J. Org. Chem.*, **25**, 2135 (1960).
224. A. Mailhe and F. de Godon, *Compt. Rend.*, **166**, 215 (1918).
225. A. Misono, T. Osa and A. Koda, *Bull. Soc. Chim. Japan*, **39**, 854 (1966), *Chem. Abstr.*, **65**, 2169.
226. K. N. Parameswaram and O. M. Friedman, *Chem. Ind. (London)*, **1965**, 988.
227. W. Brackman and P. J. Smit, *Rec. Trav. Chim.*, **82**, 757 (1963).
228. R. F. Smith and L. E. Walker, *J. Org. Chem.*, **27**, 4372 (1962).
229. B. V. Ioffe and K. N. Zelenin, *Dokl. Akad. Nauk. SSSR.*, **134**, 1094 (1960); *Chem. Abstr.*, **55**, 8284 (1961).
230. B. V. Ioffe and K. N. Zelenin, *Dokl. Akad. Nauk. SSSR.*, **144**, 1303 (1962); **154**, 864 (1964); *Chem. Abstr.*, **57**, 13602j; B. V. Ioffe and K. N. Zelenin, *Tetrahedron Letters*, **1962**, 481; *Zh. Obshch. Khim.*, **33**, 3231 (1963), *Chem. Abstr.*, **60**, 5331.
231. R. F. Smith, J. A. Albright and A. M. Waring, *J. Org. Chem.*, **31**, 4100 (1966).
232. H. G. O. Becker and H. J. Timpe, *Z. Chem. (Leipzig)*, **4**, 304 (1964); *Chem. Abstr.*, **61**, 13232 (1964).
233. N. P. Buu-Hoi and G. Saint-Ruf, *Bull. Soc. Chim. France*, **1967**, 955.
234. E. Klingsberg, *J. Org. Chem.*, **25**, 572 (1960).
235. R. T. Conley and F. A. Mikulski, *J. Org. Chem.*, **24**, 97 (1959).
236. A. F. Ferris, *J. Org. Chem.*, **25**, 12 (1960).
237. G. R. Bedford and M. W. Partridge, *J. Chem. Soc.*, **1959**, 1633; see also A. Etienne and A. Staehelin, *Bull. Soc. Chim. France*, **1954**, 743.
238. W. Davies and H. G. Poole, *J. Chem. Soc.*, **1927**, 2661.
239. K. L. Waters, *Chem. Rev.*, **41**, 587 (1947).
240. C. E. Kaslow and D. J. Cook, *J. Am. Chem. Soc.*, **67**, 1969 (1945); D. Barnard and L. Bateman, *J. Chem. Soc.*, **1950**, 926.
241. R. Adams and A. W. Schrecker, *J. Am. Chem. Soc.*, **71**, 1191 (1949); H. R. Snyder and J. K. Williams, *J. Am. Chem. Soc.*, **76**, 1298 (1954).
242. C. Gränacher, M. Gerö, A. Ofner, A. Klopfenstein and E. Schlatter, *Helv. Chim. Acta*, **6**, 458 (1923); P. L. Julien and B. M. Sturgis, *J. Am. Chem. Soc.*, **57**, 1126 (1935); H. E. Fischer and H. Hibbert, *J. Am. Chem. Soc.*, **69**, 1208 (1947).
243. A. Ahmed and I. D. Spenser, *Can. J. Chem.*, **39**, 1340 (1961); K. N. F. Shaw, A. McMillan, A. G. Gudmundson and M. D. Armstrong, *J. Org. Chem.*, **23**, 1171 (1958).
244. R. K. Hill, *J. Org. Chem.*, **27**, 29 (1962).
245. H. Fischer, C. A. Grob and E. Renk, *Helv. Chim. Acta*, **42**, 872 (1959); H. Fischer and C. A. Grob, *Tetrahedron Letters*, **1960**(26), 22.
246. A. Quilico, 'Isoxazoles and Related Compounds' in *The Chemistry of Heterocyclic Compounds* (Ed. R. H. Wiley), Interscience, New York, 1962, p. 44 1962.
247. K. v. Auwers and H. Wunderling, *Ber.*, **67**, 1062 (1934); L. Claisen, *Ber.*, **36**, 3665 (1903).

248. C. Musante, *Gazz. Chim. Ital.*, **69**, 523 (1939); I. S. Tagaki and H. Yasuda, *Yakugaku Zasshi*, **79**, 467 (1959); *Chem. Abstr.*, **53**, 18003 (1959); W. E. Shelbey and W. J. Johnson, *J. Am. Chem. Soc.*, **67**, 1745 (1945); S. Tagaki, H. Yasuda and A. Yokoyama, *Yakugaku Zasshi*, **81**, 1639 (1961); *Chem. Abstr.*, **56**, 8584 (1962).
249. C. H. Eugster, L. Leichner and E. Jenny, *Helv. Chim. Acta*, **46**, 543 (1963).
250. R. J. Tarsio and L. Nicholl, *J. Org. Chem.*, **22**, 192 (1957).
251. D. T. Mowry, *Chem. Rev.*, **42**, 257 (1948); P. Kurtz, 'Methoden zur Herstellung und Umwandlung von Nitrilen und Isonitrilen', *Methoden der organischen Chemie* (Houben-Weyl), Band VIII, Georg Thieme Verlag, Stuttgart, 1952, p. 330; J. C. Thurman, *Chem. Ind.* (London), **1964**, 752.
252. J. Dumas, *Compt. Rend.*, **25**, 383 (1847).
253. F. Kraft and B. Stauffer, *Ber.*, **15**, 1728 (1882); *Organic Syntheses*, Coll. Vol. 3, (Ed. E. C. Horning), John Wiley and Sons, New York, 1955, p. 493.
254. C. Moreu and J. C. Bongrand, *Ann. Chim.*, **14** (9), 13 (1920); E. Gryszkiewicz-Trochimowski, W. Schmidt and O. Gryszkiewicz-Trochimowski, *Bull. Soc. Chim. France*, **1948**, 593; A. J. Saggiomo, *J. Org. Chem.*, **22**, 1171 (1957).
255. G. M. Bennett and R. L. Wain, *J. Chem. Soc.*, **1936**, 1108; P. T. Teague and W. A. Short in *Organic Syntheses*, Coll. Vol. 4, (Ed. N. Rabjohn), John Wiley and Sons, New York, 1963, p. 706.
256. S. M. McElvain and R. L. Clarke, *J. Am. Chem. Soc.*, **69**, 2657 (1947).
257. C. Gerhardt, *Ann. Chim.*, **53**(3), 302 (1858).
258. B. B. Corson, R. W. Scott and C. E. Vose in *Organic Syntheses*, Coll. Vol. 2 (Ed. A. H. Blatt), John Wiley and Sons, New York, 1943, p. 379.
259. A. V. Kirsanov, *Izv. Akad. Nauk. SSSR. Otdel. Khim. Nauk.*, **1954**, 646; *Chem. Abstr.*, **49**, 13161 (1955).
260. J. v. Braun, *Ber.*, **37**, 2812 (1904); W. R. Vaughan and R. D. Carlson, *J. Am. Chem. Soc.*, **84**, 769 (1962).
261. J. v. Braun and W. Rudolph, *Ber.*, **67**, 275 (1934).
262. K. Wallenfels, F. Witzler and K. Friedrich, *Tetrahedron*, **23**, 1353 (1967).
263. A. R. Surrey in *Organic Syntheses*, Coll. Vol. 3 (Ed. E. C. Horning), John Wiley and Sons, New York, 1955, p. 535.
264. M. Robba, *Ann. Chim. (Paris)*, **5**, 351 (1960); R. D. Battershell and H. Bluestone, *French Pat.*, 1,397,521; *Chem. Abstr.*, **63**, 4212 1(965); A. S. Bailey, B. R. Henn and J. M. Langdon, *Tetrahedron*, **19**, 161 (1963).
265. J. H. Clark, R. G. Shepherd, H. W. Marson, J. Krapcho and R. O. Roblin Jr., *J. Am. Chem. Soc.*, **68**, 1046 (1946).
266. C. H. Kao, J. Y. Yen and S. L. Chien, *J. Chinese Chem. Soc.*, **2**, 240 (1934); *Chem. Abstr.*, **29**, 725 (1935).
267. R. Delaby, G. Tsatsas, X. Lusinchi and M. C. Jendrot, *Bull. Soc. Chim. France*, **1956**, 1294; **1958**, 409; B. Liberek, *Chem. Ind.* (*London*), **1961**, 987.
268. C. D. Weis, *J. Org. Chem.*, **27**, 3514 (1962).
269. D. W. Walley, J. W. B. Hershey and H. A. Jodlowsky, *J. Org. Chem.*, **28**, 2012 (1963).
270. Z. Budesinsky and J. Kopecky, *Coll. Czech. Chem. Commun.*, **20**, 52 (1955).
271. A. Michaelis and H. Siebert, *Ann. Chem.*, **274**, 312 (1893); see also H. Goldstein and R. Voegeli, *Helv. Chim. Acta*, **26**, 1125 (1943).
272. J. C. Thurman, *Chem. Ind.* (*London*), **1964**, 752.
273. H. H. Bosshard, R. Mori, M. Schmid and H. Zollinger, *Helv. Chim. Acta*, **42**, 1653 (1959).

274. P. M. Brown, D. B. Spiers and M. Whalley, *J. Chem. Soc.*, **1957**, 2882; see also A. Einhorn and C. Mettler, *Chem. Ber.*, **35**, 3647 (1902).
275. C. R. Stephens, E. J. Bianco and F. J. Pilgrim, *J. Am. Chem. Soc.*, **77**, 1701 (1955).
276. L. A. Carpino, *J. Am. Chem. Soc.*, **84**, 2197 (1962).
277. C. K. Sauers and R. J. Cotter, *J. Org. Chem.*, **26**, 6 (1961).
278. H. Gross and J. Gloede, *Chem. Ber.*, **96**, 1387 (1963).
279. A. Braun and J. Tscherniac, *Ber.*, **40**, 2709 (1907).
280. D. Davidson and H. Skovronek, *J. Am. Chem. Soc.*, **80**, 376 (1958).
281. C. Ressler and H. Ratzkin, *J. Org. Chem.*, **26**, 3356 (1961).
282. R. Calas, E. Frainnet and A. Bazouin, *Compt. Rend.*, **254**, 2357 (1962).
283. L. G. Humber and M. A. Davis, *Can. J. Chem.*, **44**, 2113 (1966); M. S. Newman and T. Fukunaga, *J. Am. Chem. Soc.*, **82**, 693 (1960).
284. S. E. Ellzey Jr., C. H. Mack and W. J. Comnick Jr., *J. Org. Chem.*, **32**, 846 (1967).
285. A. V. Kirsanov and Y. M. Zolotov, *Zh. Obshch. Khim.*, **20**, 284 (1950), *Chem. Abstr.*, **44**, 6385 (1950).
286. J. L. Boivin, *Can. J. Res.*, **28B**, 671 (1950), *Chem. Abstr.*, **45**, 4643 (1951); P. E. Gagnon, J. L. Boivin and C. Haggart, *Can. J. Chem.*, **34**, 1662 (1956).
287. G. Shaw and D. W. Butler, *J. Chem. Soc.*, **1959**, 4040.
288. W. G. Toland, *J. Org. Chem.*, **27**, 869 (1962).
289. R. Boudet, *Compt. Rend.*, **228**, 756; *Bull. Soc. Chim. France*, **1951**, 846.
290. D. T. Mowry, *Chem. Rev.*, **42**, 334 (1948).
291. G. Reutenauer and C. Paquot, *Compt. Rend.*, **223**, 578 (1946).
292. B. S. Biggs and W. S. Bishop in *Organic Syntheses*, Coll. Vol. 3 (Ed. E. C. Horning), John Wiley and Sons, New York, 1955, p. 768; B. S. Biggs and W. S. Bishop, *J. Am. Chem. Soc.*, **63**, 944 (1941).
293. P. Oxley, T. D. Robson, A. Koebner and W. F. Short, *Brit. Pat.*, 583,586; *Chem. Abstr.*, **41**, 2750 (1947); P. Oxley, M. W. Partridge, T. D. Robson and W. F. Short, *J. Chem. Soc.*, **1946**, 763; A. D. Kemp, *Nature*, **159**, 509 (1947).
294. C. S. Miller in *Organic Syntheses*, Coll. Vol. 3 (Ed. E. C. Horning), John Wiley and Sons, New York, 1955, p. 646.
295. A. V. Kirsanov, *Izv. Akad. Nauk. SSSR. Otdel. Khim. Nauk.*, **1950**, 426; *Chem. Abstr.*, **45**, 1503 (1951); A. V. Kirsanov, *Zh. Obshch. Khim.*, **22**, 274 (1952), *Chem. Abstr.*, **46**, 11135 (1952); A. V. Kirsanov and E. A. Abrazhanova, *Sb. Statei Obshch. Khim.*, **2**, 865 (1953); *Chem. Abstr.*, **49**, 6820 (1955).
296. I. I. Bezman and W. R. Reed, *J. Am. Chem. Soc.*, **82**, 2167 (1960).
297. G. Lohaus, *Chem. Ber.*, **100**, 2719 (1967); R. Graf, *Angew. Chem.*, **80**, 183 (1968); J. Bretschneider and K. Wallenfels, *Tetrahedron*, **24**, 1078 (1968).
298. R. Graf, *Angew. Chem.*, **80**, 186 (1968).
299. F. Becke, J. Gnad, H. Hager, G. Mutz and O. Swoboda, *Chemiker Z.*, **89**, 807 (1965).
300. P. J. C. Fierens and J. van Rysselberge, *Bull. Soc. Chim. Belges*, **61**, 215 (1952).
301. T. Mukeiyama, M. Tokizawa and H. Takei, *J. Org. Chem.*, **27**, 803 (1962).
302. J. Moffat, H. R. Havens, W. H. Burton and S. M. Katzman, *Chem. Eng. News*, **42**(48), 31 (1964).
303. M. L. Mihailovic, A. Stojiljkovic and V. Andrejevic, *Tetrahedron Letters*, **1965**, 461; see also ref. 226.
304. T. E. Stevens, *J. Org. Chem.*, **26**, 2531 (1961).
305. K. Nakagawa and T. Tsuji, *Chem. Pharm. Bull.*, **11**, 296 (1963).
306. W. H. McGregor and F. H. Carpenter, *Biochemistry*, **1**, 53 (1962).

307. *Biochem. Prep.*, **5**, 97 (1957).
308. G. W. Stevenson and J. M. Luck, *J. Biol. Chem.*, **236**, 715 (1961).
309. K. Nakagawa and H. Onoue, *Chem. Commun.*, **1965**, 396.
310. K. Nakagawa and H. Onoue, *Tetrahedron Letters*, **1965**, 1433.
311. J. H. Hall and E. Patterson, *J. Am. Chem. Soc.*, **89**, 5856 (1967); J. A. VanAllan, W. J. Priest, A. S. Marshall and G. A. Reynolds, *J. Org. Chem.*, **33**, 1100 (1968).
312. R. A. Abramowitch and B. A. Davis, *Chem. Rev.*, **64**, 172 (1964); A. N. Nesmeyanov and M. I. Rybinskaya, *Izv. Akad. Nauk. SSSR. Otdel. Khim. Nauk*, **1962**, 816.
313. T. Mukaiyama and H. Nambu, *J. Org. Chem.*, **27**, 2201 (1962).
314. S. Tripett and D. M. Welker, *J. Chem. Soc.*, **1960**, 2976.
315. H. Böhme and R. Neidlein, *Chem. Ber.*, **95**, 1859 (1962).
316. A. Dornow and S. Lüpfert, *Chem. Ber.*, **90**, 1780 (1957).
317. A. Dornow and S. Lüpfert, *Chem. Ber.*, **89**, 2718 (1956).
318. W. S. Sheppard, *J. Org. Chem.*, **27**, 3756 (1962).
319. A. C. Cope, H. L. Holmes and H. O. House, *Org. Reactions*, **9**, 107 (1957).
320. H. A. Bruson, *Org. Reactions*, **5**, 79 (1949).
321. S. Patai and Y. Israeli, *J. Chem. Soc.*, **1960**, 2025; see also F. S. Prout, *J. Org. Chem.*, **18**, 928 (1953).
322. S. Wawzonek and E. M. Smolin in *Organic Syntheses*, Coll. Vol. 3 (Ed. E. C. Horning), John Wiley and Sons, New York, 1963, p. 715.
323. K. J. Rorig, *U.S. Pat.*, 2,745,866; *Chem. Abstr.*, **51**, 1272 (1957); S. Wawzonek and E. M. Smolin in *Organic Syntheses*, Coll. Vol. 3 (Ed. E. C. Horning), John Wiley and Sons, New York, 1963, p. 387; C. Runti and L. Sindellari, *Boll. Chim. Farm.*, **99**, 499 (1960), *Chem. Abstr.*, **55**, 10468 (1961).
324. E. A. La Lancette, *J. Am. Chem. Soc.*, **83**, 4867 (1961).
325. A. C. Cope and K. E. Hoyle, *J. Am. Chem. Soc.*, **63**, 733 (1941). M. Cordier and J. Moreau, *Compt. Rend.*, **214**, 621 (1942).
326. A. C. Cope and K. E. Hoyle, *J. Am. Chem. Soc.*, **63**, 733 (1941).
327. J. M. Patterson in *Organic Syntheses*, Vol. 40 (Ed. M. S. Newman), John Wiley and Sons, New York, 1960, p. 46.
328. D. M. W. Anderson, F. Bell and J. L. Duncan, *J. Chem. Soc.*, **1961**, 4705.
329. K. Friedrich, unpublished results.
330. H. Hart and F. Freeman, *Chem. Ind. (London)*, **1963**, 332.
331. S. Patai and Z. Rappoport, *J. Chem. Soc.*, **1962**, 377.
332. *Japan Pat.*, 13071; *Chem. Abstr.*, **59**, 9914 (1963).
333. L. L. Woods, *J. Am. Chem. Soc.*, **80**, 1440 (1958).
334. K. Hafner, H. W. Riedel and M. Danielisz, *Angew. Chem.*, **75**, 344 (1963); *Int. Ed. Engl.*, **2**, 215 (1963).
335. N. Latif and N. Mishriky, *Can. J. Chem.*, **44**, 1271 (1966).
336. H. Meerwein, W. Florian, N. Schön and G. Stopp, *Ann. Chem.*, **641**, 20 (1961).
337. A. A. Goldberg and W. Kelly, *J. Chem. Soc.*, **1947**, 1375.
338. A. Maercker, *Org. Reactions*, **14**, 270 (1965).
339. S. S. Novikov and G. A. Sherkhgeimer, *Izv. Akad. Nauk. SSSR., Otdel. Khim. Nauk.*, **1960**, 2061; *Chem. Abstr.*, **55**, 13353 (1961); see also G. P. Schiemenz and H. Engelhard, *Chem. Ber.*, **94**, 578 (1961); W. Stilz and H. Pommer, *Ger. Pat.*, 1,108,208; *Chem. Abstr.*, **56**, 11422 (1962); H. Pommer and W. Stilz, *Ger. Pat.*, 1,116,652; *Chem. Abstr.*, **57**, 2267 (1962).
340. L. Horner and H. Oediger, *Chem. Ber.*, **91**, 437 (1958).

341. L. K. Vinograd and N. S. Vulf'son, *Zh. Obshch. Khim.*, **29**, 245 (1959); *Chem. Abstr.* **53**, 21786 (1959); *Chem. Abstr.*, **54**, 10940 (1960).
342. J. S. Swanson and D. J. Renaud, *J. Am. Chem. Soc.*, **87**, 1394 (1965).
343. P. Boldt, L. Schulz and J. Etzemüller, *Chem. Ber.*, **100**, 1281 (1967).
344. K. Torssell and K. Dahlqvist, *Acta. Chem. Scand.*, **16**, 346 (1962).
345. H. Hart and Y. C. Kim, *J. Org. Chem.*, **31**, 2784 (1966).
346. E. Ciganek, *J. Am. Chem. Soc.*, **89**, 1454, 1458 (1967); **87**, 652 (1965).
347. F. D. Marsh and H. E. Simmons, *J. Am. Chem. Soc.*, **87**, 3529 (1965).
348. A. G. Anastassiou, *J. Am. Chem. Soc.*, **90**, 1527 (1968).
349. J. K. Williams, D. W. Wiley and B. C. McKusick, *J. Am. Chem. Soc.*, **84**, 2210 (1962).
350. R. W. Hoffmann and H. Häuser, *Angew. Chem.*, **76**, 346 (1964); *Int. Ed. Engl.*, **3**, 380 (1964).
351. G. W. Griffin, J. E. Basinski and L. I. Peterson, *J. Am. Chem. Soc.*, **84**, 1012 (1964).
352. A. T. Blomquist and Y. C. Meinwald, *J. Am. Chem. Soc.*, **79**, 5316 (1957); J. K. Williams, *J. Amer. Chem. Soc.*, **81**, 4013 (1959).
353. Y. Shigemitsu, Y. Odaina and S. Tsutsumi, *Tetrahedron Letters*, **1967**, 55.
354. W. J. Linn, O. W. Webster and R. E. Benson, *J. Am. Chem. Soc.*, **85**, 2032 (1963); **87**, 3651 (1965); W. J. Linn and R. E. Benson, *J. Am. Chem. Soc.*, **87**, 3657 (1965); see also R. Hoffmann, *J. Am. Chem. Soc.*, **90**, 1475 (1968).
355. P. G. Gassman and K. T. Mansfield, *J. Am. Chem. Soc.*, **90**, 1517, 1524 (1968).
356. E. Ciganek and C. G. Krespan, *J. Org. Chem.*, **33**, 541 (1968).
357. H. L. Holmes, 'The Diels–Alder Reaction of Ethylenic and Acetylenic Dienophiles', *Org. Reactions*, **4**, 60 (1948).
358. W. J. Middleton, R. E. Heckert, E. L. Little and C. G. Krespan, *J. Am. Chem. Soc.*, **80**, 2783 (1958).
359. P. Scheiner and W. R. Vaughan, *J. Org. Chem.*, **26**, 1923 (1961); P. Scheiner, K. U. Schmiegel, G. Smith and W. R. Vaughan, *J. Org. Chem.*, **28**, 2960 (1963).
360. W. R. Boehme, E. Schipper, W. G. Scharpf and J. Nichols, *J. Am. Chem. Soc.*, **80**, 5488 (1958); see also M. Schwarz and M. Maienthal, *J. Org. Chem.*, **25**, 449 (1960).
361. J. Gillois-Doucet, *Ann. Chim. Paris*, **10**, 497 (1955).
362. S. Proskow, H. E. Simmons and T. L. Cairns, *J. Am. Chem. Soc.*, **88**, 5254 (1966).
363. M. M. Baizer, *Tetrahedron Letters*, **1963**, 973; *J. Org. Chem.*, **29**, 1670 (1964).
364. L. Eberson, *J. Org. Chem.*, **27**, 2329 (1962); L. Eberson and S. Nilson, *Acta Chem. Scand.*, **22**, in press (1968).

CHAPTER 3

Basicity, hydrogen bonding and complex formation

JUST GRUNDNES and PETER KLABOE

University of Oslo, Oslo, Norway

I. INTRODUCTION . . . 124
II. BASICITY OF THE CYANO GROUP . . . 125
A. Basicity Constants . . . 125
B. Nitrilium Salts . . . 127
1. The 1:2 nitrile–hydrogen halide compounds . . . 128
2. *N*-substituted nitrilium salts . . . 129
3. Unsubstituted nitrilium salts . . . 130
4. Solvation of metal ions by nitriles . . . 131
a. Alkali metal ions . . . 132
b. Heavy metal ions . . . 132
III. WEAK COMPLEXES . . . 134
A. Infrared and Raman Shifts . . . 134
B. Hydrogen Bonding . . . 136
1. Intermolecular hydrogen bonding . . . 136
a. Infrared and ultraviolet studies . . . 136
b. Nuclear magnetic resonance studies . . . 139
c. Calorimetric studies . . . 140
d. Miscellaneous studies . . . 141
2. Intramolecular hydrogen bonding . . . 142
3. Adsorbed species . . . 143
C. Charge-Transfer Complexes . . . 143
1. Ultraviolet and visible studies . . . 144
2. Infrared and Raman studies . . . 146
3. x-Ray crystallographic studies . . . 146
D. Dipole–Dipole Association . . . 147
1. Heteroassociation . . . 147
2. Self-association . . . 148
E. Various Interactions . . . 149
IV. NON-IONIZED COORDINATION COMPLEXES . . . 150
A. Coordination Complexes with Group III Elements . . . 150
1. With boron compounds . . . 150

a. Boron halides and organoboron compounds 150
b. Boron hydrides 151
2. With aluminium and gallium compounds 152
B. Coordination Complexes of Heavy Metal Salts 153
1. Far infrared studies 154
2. Chelates 154
3. Conformations of dinitriles in coordination complexes . . 156
V. Donor Strength of the Cyano Group 156
A. Inductive Effects 157
B. Conjugation Effects 157
C. Steric Effects 158
VI. References 159

I. INTRODUCTION

The present chapter will be concerned with the basicity and complex formation of the cyano group in organic compounds. Primarily we shall discuss mononitriles RC≡N, although a few references will be made to dinitriles. The complexes of polynitriles like tetracyanoethylene are discussed in Chapter 10.

The relatively few studies of the hydrogen bonding to isonitriles RN≡C have been included. However, the broad field of isonitrile complexes of metal salts has been covered by Malatesta[1] and will not be treated in this chapter. Moreover, the coordination chemistry of the cyanide ion $(C{\equiv}N)^-$ is outside the scope of this review.

The coordination ability of the nitriles is fundamentally related to the electronic structure of the cyano group (Chapter 1). In the cyano group the nitrogen and carbon atoms are approximately diagonally (*sp*) hybridized. The bonding therefore consists of a σ-bond and further of two π-bonds at right angles to each other, giving a linear arrangement R—C≡N. The large bond moment[2] (~3·5 D) is mainly caused by the lone-pair orbital centred on the nitrogen atom and directed along the CN axis. Furthermore, the π-orbitals are displaced towards the nitrogen, and the charge distribution may therefore be represented as in **1**, in which the dashed line represents the direction of the lone-pair orbital.

$$\mathrm{R{-}\overset{\delta+}{C}{\equiv}\overset{\delta-}{N}\,- - -}$$

(1)

The lone-pair electrons are mainly responsible for the coordination of the cyano group. Thus, protonation occurs at the nitrogen atom, and hydrogen bonding and complex formation to Lewis acids generally take place through the lone pair. However, weak complex formation involving the C≡N π-electrons can occur. The large dipole moment of the cyano group can lead to dipole–dipole interactions in the pure nitriles (self-association) or from nitriles to other molecules with polar groups. Furthermore, association may take place as a result of interaction between the partial positive charge on the carbon atom and lone-pair electrons on other molecules. Therefore, the coordination chemistry of the nitriles is a very broad field and it is essential for understanding the reactions of the cyano group. Many chemical reactions proceed through complexes as intermediates and the solvent properties of the nitriles are certainly connected with their coordination ability.

Most of the studies concerned with the basicity and complexes of the cyano group include only acetonitrile, and often the conclusions drawn from various types of measurements are in confusing disagreement. Very few systematic studies of large series of nitriles have been presented. Therefore the influence of structural factors on the basicity of various nitriles is at present uncertain. The fundamental concepts regarding complexes of ethers, also valid for the nitriles, have been excellently discussed by Searles and Tamres in an earlier volume of this series[3]. All of this background material will not be repeated in this chapter, and we will aim at a description of the experimental material relevant to the understanding of the basicity and complexing ability of the cyano group. The literature has been surveyed to the end of 1967. Emphasis has been placed on recent developments which have clarified controversial problems in this field, and no extensive compilation of data has been attempted.

II. BASICITY OF THE CYANO GROUP

A. Basicity Constants

The position of many weak organic bases on the Hammett H_0 scale[4] has been a controversial subject since the introduction of this useful extension of the familiar pH concept[3,5]. The problems involved in the various experimental approaches (kinetic, cryoscopic, titrimetric, conductometric, distribution and indicator methods) have been discussed by Arnett[6].

From titration studies of acetonitrile with perchloric acid in glacial acetic acid and the use of indicators, a pK_a value around −4·2 was reported[7]. The value of −4·32 was obtained for propionitrile titrated with sulphuric acid in formic acid[8], when converted to the present H_0 scale[9]. However, a pK_a of −4 is at the limit of the working range for acid–base titrations in acetic acid[6]. Furthermore, the apparent change in acidity with added nitrile may not be caused by protonation, but by a medium effect[10].

Much lower pK_a values were deducted from the old cryoscopic data of Hantzsch[11], since acetonitrile was only half protonated in ca. 100% sulphuric acid. These results were confirmed by the conductometric studies[12] of acetonitrile and benzonitrile in 100% sulphuric acid. From these data it can be inferred that acetonitrile and benzonitrile are half protonated in 99·6 and 99·8% sulphuric acid, respectively. From the fundamental equation (1) and the H_0 values for sulphuric acid[9], the pK_a value −10·1 was calculated for acetonitrile and −10·4 for benzonitrile[6].

$$pK_a = H_0 + \log \frac{[BH^+]}{[B]} \tag{1}$$

The Raman[13] and proton magnetic resonance[10] studies of Deno and coworkers definitely confirm the cryoscopic and conductometric results. The slow increase in the C≡N stretching frequency in solutions from 0 to 80% sulphuric acid was interpreted as due to hydrogen-bonded species

$$(MeC{\equiv}N \cdots H_3O^+ \text{ and } MeC{\equiv}N \cdots HOSO_3H)$$

only (section III.A). The proton magnetic resonance studies[10] reveal that acetonitrile and propionitrile are half protonated in 100% and 98% sulphuric acid, respectively, while chloroacetonitrile requires 30% oleum for half protonation (p$K_a \sim -12{\cdot}8$). These results agree with the expected inductive effects of the alkyl groups on the basicity. By this method the direct ratio between the protonated and unprotonated species can be estimated. It is therefore an advantage using these spectroscopic methods, since distinction can be made between 'real' protonated and hydrogen-bonded species. However, the high nitrile concentrations required and the inaccurate estimation of band areas make a precise pK_a determination impossible from such spectroscopic results.

In spite of the experimental uncertainty and the fact that the pK_a values might not be true thermodynamic values[10], the work of Deno and coworkers settles the basicity of simple, aliphatic nitriles towards the proton and demonstrates lower basicity than, for example, for ketones[3], ethers[5] and alcohols[13].

The dissociation of a Brønsted–Lowry acid in different solvents is determined by the basicity of the solvent, but the dielectric constant and the solvation ability for the anionic species present are also important. Experimental and theoretical approaches to the study of ionic reactions in acetonitrile have recently been discussed by Coetzee[14].

It has been known for a long time that organic acids dissociate to a smaller extent in acetonitrile than in water, indicating a smaller basicity of the former solvent[15]. Among the common mineral acids, which are strong in water, only perchloric acid is extensively dissociated in acetonitrile[16] or benzonitrile[17]. The other acids are weak in acetonitrile as inferred from the dissociation constants[18] (pK_a equal to 5·5 for hydrobromic, 7·25 for sulphuric, 8·9 for nitric and hydrochloric acids). Benzonitrile was also found to behave as a differential solvent towards the hydrogen halides, the acid strength increasing in the order $HCl < HBr < HI$[19].

Coetzee and McGuire have proposed to use the difference in the dissociation constants of the protonated form of Hammett bases in an organic solvent and in water as a semiquantitative measure of the solvent basicity relative to water[20]. They found that the dissociation constant for the conjugated acid of 4-chloro-2-nitro-*N*-methylaniline was about 10^5 times lower in acetonitrile than in water, reflecting the lower proton affinity of the nitrile. The weaker basicity of the nitriles relative to acetone, also found by other methods, was demonstrated.

B. Nitrilium Salts

In analogy with other Lewis bases like the ethers, the nitriles should react with protonic acids H—Y or with Lewis acids of the type R—Y, with the transfer of a cation to the nitrogen of the nitrile, and with formation of a nitrilium salt. Such salts have in fact been isolated, and evidence for coordination of metal atoms is also available. Due to the electronic structure of the cyano group, however, this part of the nitrile chemistry is quite complicated compared with, for example, the oxonium salt chemistry of the ether linkage[3]. This is particularly demonstrated by the reactions between nitriles and the hydrogen halides. These systems give complicated conductance–concentration relationships in solution[21,22] and reaction products of varying composition (RCN·*n*HX or 2RCN·*n*HX) as well as cyclic products may be formed[23] (see Chapter 6). Hydrogen chloride is a very important catalyst in many organic reactions involving the cyano group, and these and other aspects of the nitrile–hydrogen halide chemistry have been extensively covered in two reviews by Zil'berman[23,24].

I. The 1:2 nitrile–hydrogen halide compounds

The simplest nitrilium salt should have the structure

$$[RC\equiv NH]^+X^-$$

where X is a halogen. Salts of this composition have not been isolated and are evidently very unstable. Instead, products containing two hydrogen halide molecules are obtained. These were formerly believed to have structures **2**[25] and **3**[26], but later work was interpreted in terms of structure **4**[27].

$$R{-}C(X)_2{-}NH_2 \quad \textbf{(2)} \qquad [R{-}C\equiv NH]^+\,HX_2^- \quad \textbf{(3)} \qquad \left[R{-}C(=NH_2)X\right]^+ X^- \quad \textbf{(4)}$$

The infrared spectra have definitely shown that the solid 1:2 compounds between acetonitrile and hydrogen bromide and hydrogen iodide have structure **4**[28]. The same structure was also found for the corresponding methyl thiocyanate–hydrogen bromide reaction product[29]. This conclusion has recently been verified by x-ray[30] and neutron diffraction studies[31] of MeC≡N·2HCl and MeC≡N·2HBr.

Infrared spectra of hydrogen chloride solutions in acetonitrile have some evidence for HCl_2^- ions, corresponding to the nitrilium salt **3**[32]. The importance of **4** as an intermediate in acid-catalysed reactions has been stressed by Zil'berman[33].

The first step in the reaction between nitriles and hydrogen halides is most probably the formation of a hydrogen bonded 'outer' complex RC≡N: · · · HX. From this starting point, Janz and Danyluk[19,21] have interpreted the formation of the compounds **3** and **4** in terms of the Mulliken charge-transfer theory. These compounds also account for the increase in conductivity with time observed for these systems[19,22,32,34]. However, a quantitative rationalization of the conductance–concentration plot is difficult because of the variety of ionic species formed, at least in more concentrated hydrogen halide solutions[24]. The hydrohalogenation products described here are fundamentally different from the low melting compounds reported by Murray and Schneider[35] (see section III.B.1.d).

2. *N*-substituted nitrilium salts

Although nitrilium salts certainly were intermediates in the reactions of nitriles with diazonium salts[36,37], it was left to Meerwein[38] and Klages[27] and their collaborators to isolate and characterize these interesting compounds. The *N*-substituted nitrilium salts have the formula **5** where R^1 is alkyl or aryl, R^2 is alkyl, aryl or acyl and

$$[R^1-C\overset{...}{\equiv}N-R^2]^+X^-,$$

(5)

X^- is an anion, such as tetrafluoroborate or hexachloroantimonate. The formula indicates that the positive charge is not localized on either the carbon or the nitrogen atom.

Reactions of preparative importance leading to *N*-substituted nitrilium salts may be classified as follows:

(a) The reaction between an aliphatic or aromatic nitrile and a trialkyloxonium salt of complex halogen acids[38], exemplified by the reaction (2). The reaction proceeds quickly by gentle heating and evaporation of the ether.

$$PhC\equiv N + [Et_3O]^+AlCl_4^- \longrightarrow [PhC\overset{...}{\equiv}NEt]^+AlCl_4^- + Et_2O \qquad (2)$$

Tetrafluoroborate and hexachloroantimonate salts can also be prepared by this method.

(b) *N*-alkyl nitrilium salts can be prepared from alkyl halides and nitrile–metal halide complexes[38]. The simplest member of the alkylated series (*N*-methylacetonitrilium hexachloroantimonate) was synthesized by this method (reaction 3)[39].

$$MeC\equiv N:SbCl_5 + MeCl \longrightarrow [MeC\overset{...}{\equiv}NMe]^+SbCl_6^- \qquad (3)$$

With primary alkyl halides the reaction is slow, the ionization of the alkyl halide being the rate-determining step[38]. The rate of the reaction therefore increases in the order primary < secondary < tertiary alkyl chlorides. In the last case a diluting solvent may be necessary. The alkyl halides react less readily with the nitrile adducts of tin tetrachloride, iron trichloride and aluminium trichloride than with the antimony pentachloride compounds[38].

(c) The reaction between iminochlorides $R^1C(Cl){=}NR^2$ and electrophilic metal salts like antimony pentachloride or aluminium trichloride is very useful. *N*-Aryl nitrilium salts which cannot be prepared by reactions (2) and (3) are obtained by this procedure (reaction 4)[27,38].

$$R^1{-}\underset{\displaystyle Cl}{\underset{|}{C}}{=}NR^2 + MCl_n \longrightarrow [R^1{-}C{\equiv}N{-}R^2]^+\,MCl_{n+1}^- \qquad (4)$$

$$(M = \text{metal})$$

The compounds are easily synthesized by mixing the components in a suitable solvent, avoiding excess of the metal halide. The yield may amount to 80%.

(d) A fourth method of great interest is the reaction between aliphatic and aromatic nitriles and aryldiazonium fluoroborates (reaction 5)[38]. This procedure is limited to the diazonium salts

$$R^1C{\equiv}N + [R^2N{=}N]^+BF_4^- \longrightarrow [R^1C{\equiv}NR^2]^+BF_4^- + N_2 \qquad (5)$$

which are decomposed at relatively low temperatures. At higher temperatures (>50–60°) the nitrilium salts react with another nitrile molecule and more complicated products are formed.

The *N*-acylation of nitriles in the presence of an electrophilic metal halide has been demonstrated. For example, *N*-benzoyl benzonitrilium trichlorozincate was obtained in good yield (reaction 6)[38].

$$PhC{\equiv}N{:}ZnCl_2 + PhCOCl \longrightarrow [PhC{\equiv}NCOPh]^+ZnCl_3^- \qquad (6)$$

In most cases, the *N*-acyl nitrilium salts react with another nitrile molecule, and Meerwein used this reaction in an elegant route leading to quinazolines[40].

The electronic structure of the nitrilium ion may be represented by the resonance forms **6** and **7**. It reacts in general according to

$$\underset{(6)}{R^1{-}\overset{+}{C}{=}N{-}R^2} \longleftrightarrow \underset{(7)}{R^1{-}C{\equiv}\overset{+}{N}{-}R^2}$$

the carbonium ion structure **6**, and therefore readily adds nucleophilic agents like OH^-, water, alcohols, etc. These reactions have been used to establish the constitution of the salts[27,38] this being confirmed by infrared spectroscopy[41].

3. Unsubstituted nitrilium salts

The simple nitrilium salts $[RC{\equiv}NH]^+X^-$ ($X^- = Cl^-, Br^-, I^-$) have not been isolated, but products containing two molecules of hydrogen halides are formed. Since the synthesis of *N*-substituted nitrilium salts with complex anions was straightforward, Klages and collaborators[17] tried to prepare an N—H-containing salt from the reaction between hydrogen chloride and the benzonitrile–antimony pentachloride adduct. However, two molecules of hydrogen chloride reacted according to equation (7). The structure of this salt is

$$PhC{\equiv}N{:}SbCl_5 + 2\,HCl \longrightarrow \left[\begin{array}{c}PhC{=}NH_2\\ |\\ Cl\end{array}\right]^+ SbCl_6^- \qquad (7)$$

analogous to **4**, the 1:2 nitrile–hydrogen halide adducts. However, when the benzonitrile–tin tetrachloride adduct was used, only one molecule of hydrogen chloride reacted, giving the benzonitrilium hexachlorostannate (8)[17]. With *p*-nitrobenzonitrile, the hexachloroantimonate and not the hexachlorostannate could be prepared, while both salts were obtained with acetonitrile.

$$[PhC{\equiv}NH]_2^{\,+}\ SnCl_6^{\,2-}$$

(8)

The reaction temperature determines which of the salts is formed. At very low temperature there is no reaction between the nitrile–metal halide adduct and the hydrogen chloride, at intermediate temperatures the nitrilium salt may be formed, and at higher temperatures two molecules of hydrogen chloride are added directly. When the intermediate temperature range is narrow, no nitrilium salt is obtained[17]. Since the unsubstituted nitrilium salt has a higher hydrogen chloride vapour pressure than **4** a disproportionation occurs easily, especially at the higher temperatures[17] (equation 8).

$$2\,[RC{\equiv}NH]^+\,SbCl_6^- \longrightarrow \left[RC\begin{array}{l}{/\!\!/}NH_2\\ \backslash Cl\end{array}\right]^+ SbCl_6^- + RC{\equiv}N{:}SbCl_5 \qquad (8)$$

4. Solvation of metal ions by nitriles

The high solubility of many metal salts in nitriles is partly caused by the ability of the metal cation to coordinate nitrile molecules in the first coordination sphere. A striking example is silver nitrate, which is soluble in acetonitrile to an extent of 0·4 mole fraction at room temperature[42]. It was reported that the solubility of silver halides in different solvents, including acetonitrile, parallels the relative solvating ability for silver ions, as determined by electrochemical methods[43].

For some metal salts the interactions lead to non-ionic complexes, as in the adducts between acetonitrile and tin tetrachloride[44] or zinc dichloride[45]. In these compounds, there are covalent bonds between the chlorine atoms and the central metal atom, and they are classified as molecular complexes. The coordination ability of the metal may also be utilized by nitrile molecules only, leading to ionic structures which may be considered as metallonitrilium salts. The type of complexes formed depends both on the cation and the anion of the salt. The ionic compounds being favoured for typical metals of Groups I and II, and for complex anions with low coordinating ability like the tetrafluoroborate and perchlorate.

a. Alkali metal ions. It was found that when alkali or alkaline earth salts were dissolved in acetonitrile, the infrared[46] and Raman[47] spectra of the nitrile changed in the same way as when it is engaged in hydrogen bonding or halogen charge-transfer interactions (see section III.A). In acetonitrile solutions of, for example, lithium perchlorate, the C≡N and C—C stretching frequencies were split into two components. One peak remained at the same position as in the free solvent, the other was shifted to higher wave numbers. The shift was independent of the salt concentration, and was identical for two salts with different anions[46]. These observations strongly suggest that the new bands were due to acetonitrile molecules bonded to the cations through the nitrogen atom, while the unshifted bands correspond to acetonitrile molecules not engaged in solvation. It was reported that in 0·99M lithium perchlorate solutions, one lithium ion coordinated approximately two molecules of the donor[47].

The magnitude of the C≡N shift increased in the order

$$Na^+ < Li^+ < Mg^{2+}$$

indicating stronger interaction with increasing charge density of the cation[47]. However, silver(I) and copper(I) ions gave a larger shift than sodium. This probably means that covalent as well as ion-dipole forces are important in the interaction between the nitrile and these ions. The heat of formation for the lithium and magnesium solvates with acetonitrile and acrylonitrile have been reported[48].

b. Heavy metal ions. Hathaway and coworkers[49,50] have demonstrated that transition metal ions can be completely coordinated by acetonitrile molecules only. Wickenden and Krause prepared $Ni(MeCN)_2(ClO_4)_2$, $Ni(MeCN)_4(ClO_4)_2$ and $Ni(MeCN)_6(ClO_4)_2$[51]. In the first and second complexes the perchlorate groups enter the coordination sphere of nickel(II) as bidentate and monodentate ligands, respectively. The metal atom is octahedrally coordinated in each case. Proton magnetic resonance studies of $Co(ClO_4)_2$ and $Ni(ClO_4)_2$ in acetonitrile, however, could be interpreted in terms of a single species $M(MeCN)_6^{2+}$ over a wide range of concentrations and temperatures[52]. The electronic spectrum of $Mn(ClO_4)_2$ in acetonitrile was consistent with the formation of the octahedral solvated species $Mn(MeCN)_6^{2+}$ [53].

Janz and collaborators[42,54] made a detailed study of the infrared and Raman spectra of silver nitrate in acetonitrile, covering the complete range of silver nitrate solubility. The spectra of the solid adducts $AgNO_3{\cdot}MeCN$ and $AgNO_3{\cdot}2MeCN$ [42] facilitated the interpretation in terms of the formation of 1:1 and 1:2 complexes in equilibrium (equations 9 and 10).

$$Ag^+ + MeC{\equiv}N \rightleftharpoons (MeC{\equiv}N)Ag^+ \quad (9)$$

$$(MeC{\equiv}N)Ag^+ + MeC{\equiv}N \rightleftharpoons (MeC{\equiv}N)_2Ag^+ \quad (10)$$

At a mole fraction 0·33 of silver nitrate all three species were detected, and cooling displaced both equilibria to the right. Decreasing silver ion concentration favoured the formation of the 1:2 complex. It was pointed out[42] that the chemical shift observed in the proton magnetic resonance spectra of these solutions[55] might be also rationalized in terms of the two equilibria (9 and 10). The formation constants for these complexes in water solution have been determined by polarographic methods[56]. In other solvents, an additional 1:3 complex $(MeC{\equiv}N)_3Ag^+$ was suggested and corresponding data for copper(I) reported[56]. It should be emphasized that the infrared measurements as well as viscosity studies[57] give the number of donor molecules directly attached to the metal ion, while certain types of electrochemical methods[58] give the total solvation number as distinct from the number in the primary solvation sphere.

A cation is reduced at a potential which among other factors depends upon the solvation energy of the ion. Increased solvation leads to a more 'difficult' reduction of the ion corresponding to a higher negative potential. Polarographic methods should therefore be well suited to study the relative solvation ability of various nitriles and to compare the nitriles with other solvents.

It appears that most cations give more negative half-wave potentials in water than in acetonitrile, in agreement with the relative proton affinities of these solvents[59–62]. Exceptions are silver (I) and copper(I) which are firmly bonded to the nitriles by specific interactions[62].

Coetzee and coworkers[63] studied the reduction of various cations relative to the rubidium ion in several nitriles[63]. Variations in the half-wave potentials for a series of nitriles were quite small. However, it was clearly demonstrated that benzonitrile has a lower solvation ability towards metal ions than propionitrile, indicating a lower electron availability in the former compared to the latter case.

III. WEAK COMPLEXES

The nitriles form a variety of weak complexes for which the heat of formation is less than 10 kcal/mole with other organic and inorganic molecules. Such complexes cannot generally be isolated except at low temperatures[64], but their formation can be studied by a variety of optical and other physical methods when the complex C is in equilibrium with the components A and B (equation 11).

$$\mathrm{A} + \mathrm{B} \rightleftharpoons \mathrm{C} \tag{11}$$

These complexes have almost invariably been studied in solution, but since solvation effects influence the equilibrium (11), vapour-phase studies should be very valuable. A large number of these weak complexes are reported to have 1:1 stoichiometry. In this case the formation constant (K_c) can be calculated from equation (12)

$$K_c = \frac{[\mathrm{C}]}{[\mathrm{A} - \mathrm{C}][\mathrm{B} - \mathrm{C}]} \tag{12}$$

if the equilibrium concentration of the complex [C] can be determined, as often done by spectrophotometric methods. The heat and entropy of formation can be calculated from the temperature dependence of the K_c values, although the recent development in calorimetric measurements promise a higher accuracy obtained by this more direct method. For many weak nitrile complexes neither the formation constant nor the heat of formation have been determined, but the strength of interaction has been estimated from spectral perturbations in the ultraviolet, visible or infrared regions, proton magnetic resonance chemical shifts, etc.

A. Infrared and Raman Shifts

Among the physical methods applied to the studies of hydrogen bonding, charge-transfer complexes and strong coordination complexes with the nitriles, infrared spectroscopy plays a dominating role. Therefore, it seems appropriate, to describe briefly the infrared spectra of the nitriles and the origin of the frequency shifts.

The vibrational spectra of acetonitrile has been studied in great detail[65,66] and normal coordinate calculations[67,68] carried out. Propionitrile[69,70], benzonitrile[71], succinonitrile[72,73] and malononitrile[74] and higher homologues[75] have bands around 2250 cm^{-1}, which to a very good approximation can be considered a 'true' C≡N stretching frequency. It was first observed by Russian workers that this absorption was shifted to higher frequencies when acetonitrile formed complexes with some metal halides[76] and upon hydrogen bonding to phenol and trichloroacetic acid[77]. Complex formation from electronegative atoms linked by a double bond X=Y[78] (e.g. C=O, C=S, P=O, As=O) invariably leads to a red shift of the X=Y stretching frequency and correspondingly to a weaker X=Y bond. The 'blue shifts' observed for the nitriles (and partly in isonitriles[79]) have been a controversial topic, and several explanations have been proposed, including kinematic coupling[80], changing hybridization[81] and bond electron repulsion[82]. Normal coordinate analysis of acetonitrile adducts reveals that the C≡N force constant increases significantly upon coordination to boron trifluoride and metal halides, whereas the kinematic coupling is less significant[83]. A recent molecular-orbital calculation by Purcell[84] indicates that coordination by the lone-pair electrons results in stronger σ-bonding and slightly weakened π-bonding in acetonitrile. For methyl isonitrile the π-bonding is considerably more destabilized in agreement with the fact that the N≡C stretching frequency can be blue-shifted or red-shifted on coordination[79].

For acetonitrile the bands at 919 cm^{-1} and 380 cm^{-1} (liquid), assigned mainly to the C—C stretching and the C—C≡N bending modes, respectively, rise in frequency on coordination from the nitrogen. This observation has been extensively used for diagnostic purposes. The combination band[65] at 2293 cm^{-1} in Fermi resonance with the C≡N stretching band is also blue-shifted upon interaction[45,85]. The threefold axis of acetonitrile is retained on coordination[45], and a linear[86] or near linear[64] arrangement C—C≡N: · · · A appears from x-ray data[87].

In some instances the C≡N stretching frequency of the nitriles decreased on coordination[88–91], and in these cases it was suggested that the π-electrons of the C≡N triple bond interacted with the Lewis acid. Enhanced C≡N stretching band intensity in the infrared[92] and Raman[93] is observed when nitriles form hydrogen bonds, or on complex formation to the halogens[85] and to metal halides[44].

B. Hydrogen Bonding

Various attempts have been made to describe hydrogen bonding in terms of the charge-transfer theory[94] (see section III.C). However, so far it has not been possible to interpret the various experimental data related to hydrogen bonding coherently from the charge-transfer viewpoint. We shall therefore describe empirically the hydrogen bonding of the nitriles as based upon the methods of observation.

1. Intermolecular hydrogen bonding

Due to the strongly directional character of the lone-pair electrons on the *sp* hybridized nitrogen atom, a nitrile should be a good hydrogen-bonding base[95]. However, the nitriles were not included among the 'well-recognized hydrogen-bonding compounds' in Pimentel and McClellan's book[96]. The cyano group is certainly a poor π-electron donor, resulting in weak intermolecular hydrogen bonding in, for example, *o*-cyanophenol (see section III.B.2). Systematic studies of intermolecular hydrogen bonding with nitriles in later years have shown that the heat of formation for the acetonitrile complex with a 'standard' acid like phenol is only slightly lower than for acetone or tetrahydrofuran[97,98].

The interaction between a nitrile RC≡N and a Brønsted–Lowry acid H—Y leading to a hydrogen-bonded complex (**9**) represented by equation (13) is the simplest type of interaction in these systems.

$$\text{R—C}\equiv\text{N: + H—Y} \rightleftharpoons \underset{\textbf{(9)}}{\text{R—C}\equiv\text{N:}\cdots\text{H—Y}} \qquad (13)$$

Spectroscopic methods are well suited for the study of such equilibria. When the hydrogen-bonded complex is formed, characteristic shifts in the spectra of both the electron donor (the base) and the electron acceptor (the acid) take place. These shifts can be utilized for an estimation of the strength of interaction between the donor and the acceptor, and may be correlated with other properties of the complex.

a. Infrared and ultraviolet studies. The most conspicuous changes in the infrared spectrum of the complex relative to those of the components are the shift to lower frequencies and the broadening of the H—Y stretching band. Furthermore, an increase of the C≡N stretching frequency is observed, indicating that the complex is formed via the nitrogen lone-pair electrons, and not via the

π-electrons of the triple bond. The existence of an isosbestic point is a strong indication that a simple equilibrium is involved[97].

The results of many hydrogen-bonding studies with nitriles are reported in terms of the difference $\Delta\nu_{Y-H}$ between the Y—H stretching frequency in the free acceptor and in the complex ($\Delta\nu_{Y-H} = \nu_{Y-H}(\text{free}) - \nu_{Y-H}(\text{bonded})$). These shifts are valuable as one criterion of the hydrogen-bonding ability of the base. In some cases the ν_{Y-H}(bonded) has been determined in a two-component system, with the nitrile as a solvent. However, this technique is not reliable, as first pointed out by Tsubomura[99] and later demonstrated by several other investigators[100-102]. In the acetonitrile–phenol system in carbon tetrachloride, ν_{O-H}(bonded) decreases from 3460 cm^{-1} for a very dilute solution to 3405 cm^{-1} in pure acetonitrile[101]. It was shown that *t*-butyl cyanide and cyclohexyl cyanide gave smaller shifts $\Delta\nu_{O-H}$ than acetonitrile when measured in binary systems, contrary to what was expected from the inductive effects. At lower nitrile concentrations in carbon tetrachloride, the relative shifts change and at infinite dilution the sequence of the $\Delta\nu_{O-H}$ values is as expected[101].

Mitra studied frequency shifts $\Delta\nu_{O-H}$ for various nitriles with phenol in carbon tetrachloride[100]. In a systematic investigation Allerhand and Schleyer[103] measured the frequency shifts $\Delta\nu_{O-H}$ of phenol and methanol with nitriles RC≡N in order to clarify the variation with the substituent R. For 46 saturated and unsaturated nitriles the shift with phenol varied between 174 cm^{-1} (cyclohexyl cyanide) to 62 cm^{-1} (trichloroacetonitrile), and between 85 (cyclohexyl cyanide) and 51 cm^{-1} (chloroacetonitrile) with methanol, at nitrile concentrations of 0·04M. White and Thompson[104] also determined the frequency shift and the formation constants for several phenol–nitrile complexes in carbon tetrachloride. A linear correlation between $\Delta\nu_{O-H}$ and the logarithm of the formation constant was found.

Other molecules than phenol and methanol which have been studied as acceptors are pyrrole, indole[97] and n-amyl alcohol[105]. The interactions between nitriles and sterically hindered phenols have been studied by various authors and will be described in section V.C.

The interactions between acetylenes and nitriles in two-component systems (equation 14) (R^2 = H[106], Ph[107] and n-C_5H_{11}[108]), have been

$$R^1\text{—C}\equiv\text{N:} + \text{H—C}\equiv\text{C—}R^2 \rightleftharpoons R^1\text{—C}\equiv\text{N:}\cdots\text{H—C}\equiv\text{C—}R^2 \qquad (14)$$

demonstrated by the decreased C—H stretching and increased C—H bending frequencies. Several authors have reported lower H—Y stretching frequencies for molecules dissolved in acetonitrile compared to inert solvents or to vapour phase[109–114]. The $\Delta\nu_{O-D}$ shifts of methanol-d[115] and D_2O[116] in various nitriles, reported by Gordy and Stanford, are tabulated in Arnett's review[6]. A double minimum potential for the O—H vibration of the hydrogen-bonded proton was verified by studies in the overtone region[117,118].

In some studies the heat of formation ($\Delta H°$) for hydrogen-bonded nitrile complexes in carbon tetrachloride solution has been determined by infrared technique. The heats of formation for complexes between acrylonitrile and some hydroxylic compounds were reported[119,120]. From the formation constants at room temperature and at 60°, $\Delta H° = -4{\cdot}2 \pm 0{\cdot}8$ kcal/mole was reported for the acetonitrile–phenol system[121]. The value $\Delta H° = -3{\cdot}9$ kcal/mole was obtained by Mitra[100] for the same complex and he also reported a linear correlation between $\Delta\nu_{O-H}$ and $-\Delta H°$ for the phenol complexes with a few nitriles. In the acrylonitrile–phenol system $\Delta H° = -3{\cdot}6$ kcal/mole was reported. The $\Delta H°$ values for methanol and pyrrole with acetonitrile were $-2{\cdot}3$ and $-1{\cdot}9$ kcal/mole, respectively[100]. Fritzsche[97] made an extensive study of several donors, including acetonitrile, with phenol and indole dissolved in carbon tetrachloride. This study allows a comparison of the hydrogen-bonding ability for different groups of donors, obtained under the same experimental conditions. For acetonitrile–phenol $\Delta H°$ was $-4{\cdot}30 \pm 0{\cdot}35$ kcal/mole, as compared to $-4{\cdot}70 \pm 0{\cdot}30$ kcal/mole for acetone and $-5{\cdot}00 \pm 0{\cdot}20$ kcal/mole for *p*-dioxan. The heat of formation for the acetonitrile–phenol complex determined by infrared spectroscopy agrees reasonably well with the calorimetric value[98].

The first step in the reaction between nitriles and hydrogen chloride is probably the formation of the hydrogen-bonded complex $RC{\equiv}N{:}\cdots HCl$ (section II.B.1). Dilute solutions of acetonitrile (0·1–0·4M) and of hydrogen chloride (0·02–0·1M) in carbon tetrachloride revealed the infrared characteristics of an equilibrium between the free molecules and the complex[122a]. The heat of formation was estimated to be $-7{\cdot}8$ kcal/mole from Sokolov's method[122b].

At low concentrations of hydrogen fluoride in acetonitrile, the hydrogen-bonded complex was the predominant species even without an inert solvent. The ion HF_2^- was formed, however, in more concentrated solutions[123]. Infrared and Raman studies of acetic acid in acetonitrile were interpreted in terms of a 1:1 complex[124].

The spectra of water in acetonitrile recorded in the O—H stretching region[111] and from 14·000–700 cm^{-1} [109] have been reported. Later the interaction between water and the nitrile in carbon tetrachloride was investigated in more detail[125]. At low nitrile concentration a 1:1 complex (**10**) was formed. The spectrum consisted of a sharp band at 3690 cm^{-1}, assigned to the free O—H stretching group, and a broad band at 3615 cm^{-1}, assigned to the bonded O—H group. In more concentrated solutions evidence was found for a 2:1 complex (**11**). Here the water symmetry is retained and the symmetric and asymmetric O—H stretching modes give two equally broad bands at 3640 and 3545 cm^{-1} [125].

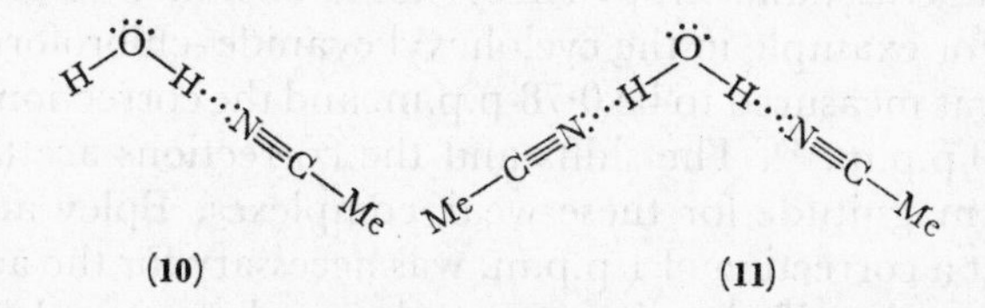

The aliphatic nitriles display their n–π^*-transition in the high frequency ultraviolet region and perturbations of this band on hydrogen bonding has therefore not been studied. The ultraviolet bands of phenol are red-shifted when hydrogen bonds are formed to bases. These ultraviolet shifts have been employed to determine the formation constant and heat of formation for hydrogen-bonded acetonitrile[126,127].

Isonitriles RN≡C form relatively strong hydrogen bonds and these molecules represent the first known example of hydrogen bonding to a carbon atom[103,128–130]. The infrared shifts $\Delta\nu_{O-H}$ of phenol and methanol are larger for isonitriles than for the corresponding nitriles[103]. The formation constant for the 1:1 complex between benzyl isocyanide and n-amyl alcohol appears larger than that for benzyl cyanide[129]. When aniline interacts with benzyl isocyanide both the hydrogens of the amino group may be involved, resulting in 1:2 stoichiometry[130]. With phenylacetylene, the interesting arrangement C—H · · · :C≡N was demonstrated.

b. Nuclear magnetic resonance studies. Several proton magnetic resonance studies involving hydrogen bonding to nitriles have been made. These investigations include particularly acetonitrile[131–140], but propionitrile[141], chloroacetonitrile, isobutyronitrile[132], cyclohexyl cyanide[142,143], benzyl cyanide[130] and methyl thiocyanate[144] have also been studied. Among the isonitriles, methyl isocyanide[136], cyclohexyl isocyanide[142] and benzyl isocyanide[130] were compared with the corresponding nitriles. The acceptors included chloroform[134,135,141–143], phenylacetylene[130], dimethylamine[137], water[131,133], methanol[136], trifluoroethanol[140], *t*-butyl alcohol[132] and phenol[138,139]. Much of this work has recently been reviewed by Laszlo[145].

It was shown by ^{15}N and ^{13}C resonances that the hydrogen bonding occurs to the nitrogen in nitriles and to the carbon in isonitriles[136]. The hydrogen-bonded shift of the acceptor may be correlated with other criteria for the strength of hydrogen bonding. These shifts, however, reported for identical systems, deviate considerably because of different experimental procedures. Moreover, a correction for the C≡N anisotropy effect should be added to the measured shifts[135]. For example in the cyclohexyl cyanide–chloroform complex the shift was measured to be 0·78 p.p.m. and the correction calculated to be 0·60 p.p.m.[143]. The shifts and the corrections are therefore of the same magnitude for these weak complexes. Epley and Drago[98] found that a correction of 1 p.p.m. was necessary for the acetonitrile–phenol complex if the linear correlation between the hydrogen bonded shift and the heat of formation observed for a variety of donors is valid also for the nitrile. Various workers have found linear relationships between the hydrogen-bonded shifts and the infrared shift $\Delta\nu_{O-H}$[98,131,133,140].

Unfortunately, hydrogen-bonded shifts for large series of nitriles have not been reported, but for acetonitrile, isobutyronitrile and chloroacetonitrile with *t*-butyl alcohol[132] the shifts (1·7, 1·7 and 1·9 p.p.m.) do not agree with the sequence of the infrared shifts $\Delta\nu_{O-H}$ for the corresponding phenol complexes (159, 166 and 117 cm^{-1})[103].

The proton magnetic resonance dilution shift technique was used to determine the formation constants and heats of formation for a few complexes. With the present technique the experimental uncertainties in the formation constants seem quite large and only very different bases could be distinguished[142]. From the heats of formation, cyclohexyl cyanide is a stronger base than cyclohexyl isocyanide towards chloroform[142], but methyl isocyanide is stronger than acetonitrile towards methanol[136]. However, the heat of formation for the acetonitrile–methanol system (−0·8 kcal/mole)[136] does not agree with the value obtained from infrared work (−2·3 kcal/mole)[100].

c. Calorimetric studies. From the difference in the heat of mixing between n-butyl alcohol–acetonitrile and acetonitrile–hexane, both extrapolated to infinite dilution, a heat of formation for the former complex equal to $-2 \cdot 5$ kcal/mole was calculated[146]. This value was lower than for acetone and n-butyl ether.

Recently, Drago[98] and Arnett[147] with collaborators have studied hydrogen bonding calorimetrically in a three-component system, claiming an accuracy superior to the spectrophotometric methods. The value[98] $\Delta H^\circ = -4 \cdot 65 \pm 0 \cdot 06$ kcal/mole for the acetonitrile–phenol complex should be considered the most accurate reported for this system, and further studies of the hydrogen-bonding properties of the nitriles by this method would be very desirable.

d. Miscellaneous studies. The solid reaction products between nitriles and hydrogen halides of 1:2 composition are salts (see section II.B.1). However, freezing-point diagrams of binary mixtures of hydrogen chloride with acetonitrile, propionitrile and butyronitrile reveal low-melting 1:1 molecular compounds, (e.g. $MeC{\equiv}N{\cdot}HCl$, m.p. $-97 \cdot 2°$)[35]. The structures are interpreted as binary hydrogen-bonded adducts in which the covalent bond in hydrogen chloride is retained. These nitriles also form complexes $RC{\equiv}N{\cdot}nHCl$ ($n = 5$ and 7) and $2RC{\equiv}N{\cdot}3HCl$. Murray and Schneider suggested a structure for the 1:5 compounds in which one HCl molecule is bonded to the nitrogen lone-pair electrons and the other four molecules are situated in a plane vertical to the molecular axis and bonded by the π-electrons of the C≡N triple bond (**12**)[35].

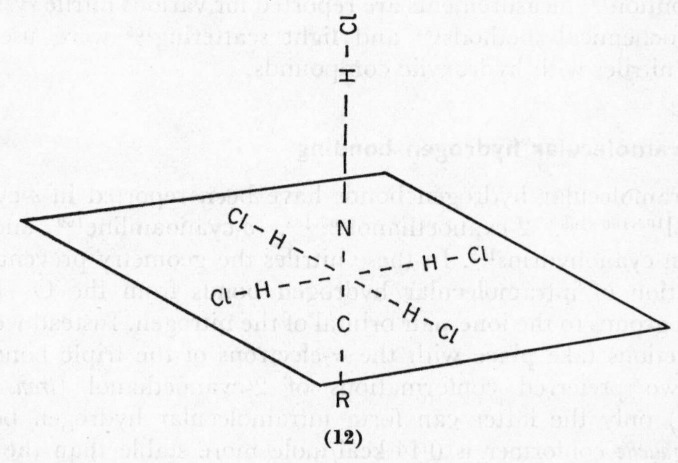

(12)

When the viscosity, density and dielectric constant for acetonitrile mixtures with water and methanol are plotted versus the mole fraction of the nitrile, large deviations from linearity are observed[148].

Surprisingly, the molar polarization plots[148,149] for these systems are linear, possibly because the partial molar volume of both components vary only slightly with the concentration. In an early work, dielectric data for various binary mixtures of nitriles with phenols indicated 1:1 interactions[150], later interpreted as hydrogen-bonded complexes[151]. The dipole moment for the acetonitrile–chloroform complex in carbon tetrachloride was larger than the sum of the dipole moments for the interacting molecules[152] as a result of the mutual polarization.

In liquid hydrogen cyanide the strong hydrogen bonds result in a polymer chain $(HCN)_n$ [153] persisting in the vapour phase[154]. From dielectric measurements the intermolecular hydrogen bond energy has been estimated to be 4·6 kcal/mole in the liquid, for the reaction (15)[155].

$$(HCN)_n + HCN \rightleftharpoons (HCN)_{n+1} \qquad (15)$$

An azeotropic mixture has been reported for acetonitrile and water[156,157], but not for the acetonitrile–chloroform system[158]. Several thermodynamic and physical parameters like heat of mixing, partial pressure, density, viscocity, surface tension and index of refraction are reported for acetonitrile–water[148,159], acetonitrile–ethanol[159] and acetonitrile–carboxylic acids[160]. Solubility[161,162] and distribution[163] measurements are reported for various nitrile systems. Electrochemical methods[164] and light scattering[165] were used to study nitriles with hydroxylic compounds.

2. Intramolecular hydrogen bonding

Intramolecular hydrogen bonds have been reported in *o*-cyanophenol[103,166–168], 2-cyanoethanol[103,168], *o*-cyanoaniline[169] and in certain cyanohydrins[168]. In these nitriles the geometry prevents the formation of intramolecular hydrogen bonds from the O—H or N—H groups to the lone-pair orbital of the nitrogen. Instead weaker interactions take place with the π-electrons of the triple bond. In the two preferred conformations of 2-cyanoethanol (*trans* and *gauche*) only the latter can form intramolecular hydrogen bonds. The *gauche* conformer is 0·14 kcal/mole more stable than the *trans* form, hydrogen-bond interaction contributing to most of this difference[170]. In 2-cyanopropanol and 2-cyanobutanol no intramolecular hydrogen bonding was observed[103].

3. Adsorbed species

The interaction between a solid surface and adsorbed gases or liquids has been extensively studied by infrared[171] and vapour pressure[172] measurements. Acetonitrile adsorbed on hydroxylated silica surface shifts the O—H stretching frequency by 305 cm^{-1} indicating hydrogen bonding[173]. Significant shifts to higher frequencies of the C≡N stretching band for hydrogen cyanide, acetonitrile and acrylonitrile adsorbed on silica and zeolite surfaces, revealed interaction from the nitrogen atom[174]. Infrared spectra of chloroacetonitrile and methyl thiocyanate adsorbed on aluminium chloride gave similar results[175]. Isomerization of methyl thiocyanate to methylisothiocyanate was observed.

C. *Charge-Transfer Complexes*

The description of donor–acceptor complexes which has been generally adopted was presented by Mulliken[176] and is applicable to a variety of chemical interactions. This theory is a generalization of the familiar Brønsted–Lowry and Lewis acid–base definitions to include the weak intermolecular forces arising from a small displacement of electronic charge from the donor (base) to the acceptor (acid). The quantum-mechanical description of the donor–acceptor interaction is covered by Briegleb[177]. In this description the ground state wave function (ψ_N) of the complex between an electron donor D and an acceptor A can be written as in equation (16); $\psi_0(D, A)$ and $\psi_1(D^+A^-)$ represent the two extreme forms, called the 'no-bond'

$$\psi_N = a\psi_0(D, A) + b\psi_1(D^+A^-) \tag{16}$$

and the 'dative' wave functions, respectively. In the former the electrons remain in the donor molecule, in the latter an electron from an occupied orbital of the donor is transferred to an unoccupied orbital of the acceptor. In a weak complex a^2 is much larger than b^2. The wave function for the excited state is given in equation (17) in

$$\psi_E = a^*\psi_1(D^+A^-) - b^*\psi_0(D, A) \tag{17}$$

which $a^* \sim a$, $b^* \sim b$ and $a^{*2} \gg b^{*2}$. Thus in the ground state the no-bond structure (D, A), and in the excited state the dative structure (D^+A^-) make the dominating contribution to the respective wave functions. The complex can be brought from the ground state (ψ_N) to the excited state (ψ_E) through the absorption of a proper light quantum $h\nu$, and this transition gives rise to a spectral band characteristic of the complex, called the 'charge-transfer' band. The addition compounds are called donor–acceptor or charge-transfer complexes, and they have been extensively studied in the last two decades[177–179].

Ordinary aliphatic and aromatic nitriles act as so-called *n*-donors through the lone-pair electrons on the nitrogen. In addition, the nitriles have the inherent ability to act as so-called π-donors through the π-electrons of the C≡N triple bond, although such interactions have not been reported except in the special case of hydrogen bonding (see section III.B.2).

The acceptors are usually classified[177] as σ-acceptors (the halogens, interhalogens and the pseudohalogen iodine cyanide) and π-acceptors which are molecules with a π-electron system with high electron affinity (trinitrobenzene, chloranil and quinones). Therefore, polynitriles like tetracyanoethylene do not act as donors, but as acceptors and this is discussed in Chapter 10.

I. Ultraviolet and visible studies

Only charge-transfer complexes between nitriles and the halogens have been studied. Undoubtedly because of the weak forces involved, this field has not been extensively investigated. The nitrile–halogen interactions have mostly been inferred from spectral perturbations in the visible region. Iodine, bromine and the interhalogens iodine monochloride and iodine monobromide, when dissolved in inert solvents, have adsorption bands in the region 520–400 mμ. Upon complex formation these bands are displaced to shorter wavelengths[180–182]. The degree of the blue shifts may serve as a measure for the strength of the interaction[183], and the small shifts[184–186] observed with the nitriles indicate that they form quite weak complexes with the halogens. The formation constants have been calculated from the visible absorbance using the Benesi–Hildebrand equation[187] or one of its many modifications[177], and the available results[188–192] are listed in Table 1. Compared to other well-known iodine complexes such as those with trimethylamine[193] ($\Delta H° = -12{\cdot}1$), diethyl sulphide[194] ($\Delta H° = -7{\cdot}82$) or diethyl ether[195] ($\Delta H° = 4{\cdot}2$ kcal/mole) the nitriles form very weak complexes as apparent from the values in Table 1. According to the Mulliken theory[176] the position and intensity of the charge-transfer band can be correlated with

6

TABLE 1. Formation constants K_c and enthalpies of formation ΔH° for various halogen complexes calculated from spectroscopic data for the visible region.

	ICl		IBr		I_2		Br_2			
	K_c (l/mole)	$-\Delta H^\circ$ (kcal/mole)	K_c (l/mole)	$-\Delta H^\circ$ (kcal/mole)	K_c (l/mole)	$-\Delta H^\circ$ (kcal/mole)	K_c (l/mole)	Solvent	Temperature (°c)	Reference
Acetonitrile	8·4	4·9	1·5	4·1	0·4	1·9		CCl_4	25	189
Chloroacetonitrile	2·4	5·3	0·6	3·1	0·2	1·5		CCl_4	25	189
Dichloroacetonitrile	0·7	3·7	0·2	2·0				CCl_4	25	189
Trichloroacetonitrile	0·2	5·4						CCl_4	25	189
Propionitrile	9·7	5·2	2·3		0·4		0·1	CCl_4	20	185
Valeronitrile					1·0	3·0		C_7H_{16}	25	190
Benzonitrile	8·1		2·1		0·8		0·2	CCl_4	20	184
Acrylonitrile	5·3		1·2		0·5			CCl_4	20	186
p-Toluonitrile	13·8							CCl_4	20	191
Cyclohexyl cyanide	19·9							CCl_4	20	191
Dimethylcyanamide	120[a]	7·3	18·8	5·6	1·8[b]	2·8		CCl_4	20	192

[a] at 26°.
[b] at 23°.

intrinsic properties of the donor and the acceptor. For the nitrile–halogen complexes, however, no charge-transfer band has been reported, except an uncertain tail below 230 mμ for the acrylonitrile–iodine complex[186]. The nitriles have very high ionization potentials[196]; 12·39, 11·85, 9·95 and 10·75 ev for acetonitrile, propionitrile, benzonitrile and acrylonitrile, respectively, and the position of the charge-transfer band should be at higher frequencies than for donors with lower ionization potentials[183]. Various empirical[197,198] and theoretical[199] relationships between the ionization potential and the energy of the charge-transfer transition have been proposed, and from these correlations the charge-transfer band should be situated in the accessible ultraviolet region. However, the low formation constants make the detection difficult.

Popov and coworkers[188,200] reported that for iodine monochloride, iodine monobromide or iodine solutions in pure acetonitrile, the electrical conductance increased with time, and the visible absorption spectra changed[184,185,201]. The effect was most pronounced for the interhalogen complexes and a passage from an 'outer' to an 'inner' complex (equation 18) was proposed[176].

$$\text{MeC}{\equiv}\text{N}:\cdots\text{I—X} \rightleftharpoons [\text{MeC}{\equiv}\text{NI}]^+\text{X}^- \rightleftharpoons \text{MeC}{\equiv}\text{NI}^+ + \text{X}^- \qquad (18)$$

2. Infrared and Raman studies

When nitriles interacted with iodine monochloride in carbon tetrachloride solution, the C≡N stretching frequencies were shifted 8 to 14 cm^{-1} [85]. The formation constants were calculated from the infrared spectra. Fumaronitrile, having two cyano groups, can form 1:1 as well as 1:2 complexes with iodine monochloride[202].

Person and coworkers studied perturbations in the infrared spectra of the acceptors on complex formation. The stretching frequencies in iodine monochloride[203,204], iodine monobromide[205] and iodine cyanide[206] were shifted to lower frequencies. The infrared-inactive I—I frequency becomes active on complex formation and was recently observed at 205 cm^{-1} in iodine complexes with acetonitrile[207]. In spite of the yellow-brown colours of these complexes, Klaboe[208] obtained Raman spectra of the solutions by helium–neon laser excitation.

3. x-Ray crystallographic studies

At low temperatures a 2:1 acetonitrile–bromine adduct was prepared and the crystal structure determined[64]. The bromine forms a bridge between two acetonitrile molecules in an approximately linear arrangement as found in other weak donor–halogen complexes[209]. The interatomic distances reveal[64] a stretching of the Br—Br bond and a shortening of the C≡N bond relative to the free molecules, in agreement with the spectroscopic observations.

The cyanogen halides X—C≡N (X = Cl[210], Br[211], I[212]) crystallize in infinite chains, with weak intermolecular C≡N:···X bonding. Similar linear arrangements have been observed for the halogeno cyanoacetylenes X—C≡C—C≡N (X = Cl, Br[213], I[214]) and a Raman band[215] was observed at 79 cm^{-1} in the solid iodo compound, interpreted as an intermolecular vibrational mode.

D. Dipole–Dipole Association

Because of the high bond moments of the cyano group[2] the nitriles are self-associated and they interact with other molecules having polar groups by dipole–dipole forces. Various methods have been used to study these phenomena.

I. Heteroassociation

Thompson and coworkers[93,216–219] found that the position, and particularly the intensity, of the C≡N stretching vibration of nitriles varied with solvent polarity, which in part might be attributed to dipole–dipole interaction. Ritchie and his colleagues studied the areas of the infrared C≡N stretching band for acetonitrile[220] and benzonitriles[221] with increasing amounts of polar molecules like dimethyl sulphoxide, dimethyl formamide and acetone dissolved in carbon tetrachloride. They interpreted the data in terms of complex formation. The formation constants were independent of the substituent on the benzene ring for various benzonitrile–dimethyl sulphoxide systems (Table 2). Therefore,

TABLE 2. Dissociation constants of dipole–dipole complexes[220].

Constituents of complex		K_{diss} (M)
Benzonitrile	Dimethyl sulphoxide	1·1
p-Chlorobenzonitrile	Dimethyl sulphoxide	1·1
p-Nitrobenzonitrile	Dimethyl sulphoxide	1·1
m-Nitrobenzonitrile	Dimethyl sulphoxide	1·1
p-Anisonitrile	Dimethyl sulphoxide	0·85
Acetonitrile	Dimethyl sulphoxide	0·85
Benzonitrile	Dimethyl formamide	1·6

interactions involving the π-electrons of the benzene ring or dipolar effects involving the entire molecules could be eliminated and the complex formation was ascribed to specific interactions between dipoles of the cyano and the sulphoxide groups.

This interaction has been further studied from variations in the frequency and intensity of the sulphoxide S=O stretching frequency[222]. A 1:1 *p*-chlorobenzonitrile–dimethyl sulphoxide complex was verified cryoscopically[223] and a geometry with antiparallel alignment of the C≡N and the S=O dipoles was suggested[220]. The equilibrium constants for the conversion between *cis*- and *trans*-4-*t*-butylcyclohexyl cyanide in dimethyl sulphoxide as compared to tetrahydrofuran were studied[224,225]. It was suggested[220] that complex formation with the sulphoxide group stabilizes the *trans* relative to the *cis* isomer. Taft and his students[226] observed an increased electron shielding in the fluorine atom when *p*-fluorobenzonitrile was dissolved in dimethyl sulphoxide or dioxan, as compared to an inert solvent. Evidently the electron-withdrawing effect of the cyano group decreased in the polar solvents, probably as a result of electron donation from the lone-pair electrons of the solvent molecules to the positively charged carbon atom.

2. Self-association

The nitriles have considerably higher boiling points, viscosities and heats of vaporization than the corresponding hydrocarbons, strongly suggesting association. Since *t*-butyl cyanide with no hydrogens on the α-carbon atom has nearly the same boiling point elevation compared to the hydrocarbon as the n-butyl and *s*-butyl cyanides, this association cannot be caused by hydrogen bonding[227]. However, in cyano compounds with acidic hydrogens, such as succinonitrile and glutaronitrile, hydrogen bonding may contribute to the association[161,162].

Saum[227] interpreted the viscosity data for several aliphatic nitriles in terms of 96% dimers at 30° for acetonitrile, gradually reduced to 56% for n-heptyl cyanide, and suggested the presence of dipole–dipole interaction. Later work on the liquid nitriles by dielectric methods[228,229] indicates a lower proportion of dimers. The second virial coefficient for acetonitrile vapour[230] was measured and the heat of dimerization calculated (5·2 kcal/mole), but dielectric polarization data[231] gave a lower value (3·8 kcal/mole).

Antiparallel dipole pairs were proposed from measurements of the second virial coefficients[230] and from infrared intensity data[77]. The dielectric work[229] was best interpreted in terms of this model (**13**). Murray and Schneider[35] favoured a skewed configuration (**14**)

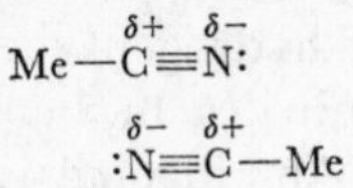

$$\begin{array}{c} \overset{\delta+}{\text{Me}-\text{C}}\equiv\overset{\delta-}{\text{N}}: \\ :\overset{\delta-}{\text{N}}\equiv\overset{\delta+}{\text{C}}-\text{Me} \end{array} \qquad (\mathbf{13})$$

$$\text{Me}-\overset{\delta+}{\text{C}}\equiv\overset{\delta-}{\text{N}}: \quad \text{(skewed: Me}-\overset{\delta+}{\text{C}}\equiv\overset{\delta-}{\text{N}}:) \qquad (\mathbf{14})$$

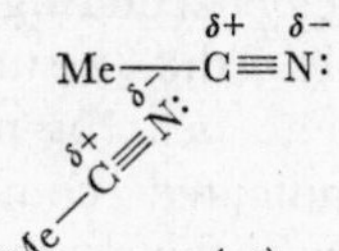

since the mutual repulsions of the high charge densities at the C≡N triple bond should make the antiparallel configuration unfavourable. This assumption is supported by nuclear magnetic resonance studies[226].

E. Various Interactions

Dinitrogen tetroxide forms molecular complexes with various organic bases and it is reported[232] that molecules with lone-pair electrons (n-donors) generally form 2:1 complexes, but the aromatic molecules (π-donors) form 1:1 complexes. Melting-point curves for acetonitrile and benzyl cyanide with dinitrogen tetroxide reveal 2:1 complexes, but benzonitrile gives a 1:1 complex, indicating a π-complex from the aromatic ring in this case[232]. The vapour pressure curves show relatively small negative deviations from Raoult's law and it was concluded[233] that dipole–dipole interactions play a larger role in these complexes than the donor–acceptor contribution.

The so-called clathrates are non-stoichiometric compounds in which small molecules can be encaged in certain compounds[234]. Powell and coworkers have studied the hydroquinone–acetonitrile compound by an x-ray crystallographic technique[235]. The dielectric relaxation time[236,237] of this compound was 3×10^7 times larger than for hydroquinone–methanol. Although the type of interactions are uncertain, the infrared spectrum of the acetonitrile–hydroquinone clathrate indicates that the former molecules are encaged without any specific bonding to the cyano group[238].

Various very weak interactions of uncertain origin have been reported between nitriles and aromatic molecules. The presence of an added ultraviolet absorption around 290 mμ in mixtures of p-xylene with propionitrile was interpreted as a result of complexation in which the nitrile was tentatively classified as an electron acceptor[239]. From the highly temperature-dependent chemical shifts a complex formation between acetonitrile and toluene was postulated[240], ascribed to hydrogen bonding from the methyl hydrogens to the benzene ring[241]. Since the same effect was observed for propionitrile, the assumption of hydrogen bonding was questioned[242]. The heat of mixing for nitriles in benzene[243] and toluene[244,245] indicated complexation. An incongruent melting point was observed for acetonitrile–benzene, revealing a 1:2 complex, but the freezing point curves for substituted benzenes gave no indication of complexes[246].

IV. NON-IONIZED COORDINATION COMPLEXES

The nitriles react with many compounds which can act as Lewis acids to form non-ionic addition compounds through the lone-pair electrons of the cyano group. Particularly, the chemistry of the coordination compounds between nitriles and metal salts is a very broad field. A considerable number of complexes between compounds of Group III elements and acetonitrile are reported and have been treated among other coordination compounds in several reviews[247–251]. However, with organometallic compounds of alkali metals (e.g. butyllithium) chemical reactions occur[252,253]. The organometallic compounds of the alkaline earth elements, exemplified by the Grignard reagents also react with the nitriles, but molecular complexes have been proposed[254,255] as intermediates. A few complexes between acetonitrile and alkaline earth halides have been reported[256,257].

A. Coordination Complexes with Group III Elements

I. With boron compounds

a. Boron halides and organoboron compounds. The coordination compounds of nitriles and boron halides were reviewed[258] in 1965 and only some recent work will be commented on.

The boron trihalides are planar molecules and act as strong Lewis acids whereby they assume tetrahedral structure[259]. It is well established that the acid strengths of boron halides decrease in the order $BI_3 > BBr_3 > BCl_3 > BF_3$, and this sequence is also verified for the BX_3 complexes with acetonitrile from calorimetric studies[260,261], proton[262] and ^{19}F [263] chemical shifts, infrared studies[264] and normal coordinate analysis[82] (Table 3). The relative stabilities of the addition compounds between boron trifluoride and benzonitrile, *o*-, *m*- and *p*-toluonitrile and 2,4,6-trimethylbenzonitrile have been determined by vapour pressure measurements[265]. Boron isothiocyanate $B(NCS)_3$ forms an addition compound with acetonitrile, characterized by infrared spectroscopy[266].

TABLE 3. Thermodynamic and spectral data for some BX_3 acetonitrile complexes.

	$-\Delta H^\circ$ (kcal/mole)	Reference	δ^a (Hz)	Reference	$\nu_{C\equiv N}$ (cm^{-1})	$\Delta\nu^b$ (cm^{-1})	Reference
BF_3	26·5	260	161·8	262	2359	111	44
BCl_3	33·8	260	182·8	262	2357	109	44
BBr_3	39·4	261	189·9	262	2320	72	264
BH_3	25	270					

[a] The proton chemical shifts of $CH_3CN{\cdot}BX_3$ relative to tetramethylsilane in nitrobenzene solution (0·01M).
[b] $\Delta\nu = \nu_{C\equiv N}$ (complex) $- \nu_{C\equiv N}$ (acetonitrile).

Triphenylboron forms molecular addition compounds[267] with isonitriles to the carbon atom (**15**).

$$R{-}N{\equiv}C{:}BPh_3$$
(15)

Infrared bands in the region 2225–2275 cm^{-1} were assigned to the N≡C stretching frequency in the complexes. With trialkylboron, however, a cyclization reaction to piperazine derivatives takes place[268].

b. Boron hydrides. The boron hydrides are electron-deficient molecules and interact with Lewis bases like nitriles and isonitriles[269]. The simplest boron hydride (diborane) reacts reversibly with acetonitrile and propionitrile to form the complex $RCN{:}BH_3$ which decomposes above 20° to trialkylborazine[270]. A large number of reaction products between the higher boron hydrides and nitriles have been described; those formed from decaborane have been reviewed by Hawthorne[271] and the crystal structures in general by Lipscomb[272]. Decaborane ($B_{10}H_{14}$) reacts with acetonitrile to form a solid crystalline compound $2MeCN{\cdot}B_{10}H_{12}$ in which two hydrogens are substituted with acetonitrile molecules[273]. The ^{11}B nuclear magnetic resonance spectra of this compound and the corresponding propionitrile and ethyl isonitrile derivatives have been studied[274] and the various peaks assigned to non-equivalent substituent positions in the molecule.

The complex $MeCN:B_9H_{13}$ has been prepared[275] and the crystal structure reveals a linear arrangement BNCC[276]. In the compound $3MeCN \cdot B_{20}H_{16}$ only two acetonitrile molecules are coordinated to boron, the third being a molecule of crystallization[277]. In this molecule a B_{12} icosahedron shares three boron atoms with a B_{11} icosahedral fragment and is the first known example of two icosahedra sharing a face.

A compound $2EtNC \cdot B_{10}H_{12}$ has been prepared in which the N≡C stretching frequency is 113 cm^{-1} higher than for ethyl isonitrile itself[278]. In this complex the ethyl isonitrile molecule is more tightly bonded to the boron through the lone pair than the propionitrile molecule in the corresponding nitrile complex.

2. With aluminium and gallium compounds

Complexes between aluminium chloride and acetonitrile have been inferred from phase studies[279]. These include $AlCl_3 \cdot 2MeCN$, $2AlCl_3 \cdot 3MeCN$ and $AlCl_3 \cdot MeCN$.

Unlike the organoboron molecules R_3B, which apparently do not interact with nitriles[280], the organoaluminium compounds R_3Al form 1:1 adducts. Thus an equimolar mixture of triphenylaluminium and benzonitrile in benzene solution gave a complex $PhC{\equiv}N:AlPh_3$ studied by proton magnetic resonance[281]. The adduct had an infrared band at 2260 cm^{-1} characteristic for the C≡N group[282], but by heating the solid adduct, a further reaction gave a crystalline product $Ph_2C = NAlPh_2$ with a band at 1660 cm^{-1} attributed to the C=N linkage[282,283].

Wade and coworkers studied complexes formed between trimethylaluminium, triethylaluminium, triphenylaluminium and benzonitrile[283], acetonitrile and *t*-butyl cyanide[284]. In each case 1:1 adducts were formed and the C≡N stretching bands had higher frequencies than in the free nitriles. When heated (110–240°) addition to the triple bond and dimerization occurred.

Alkylaluminium chlorides R_mAlCl_{3-m} ($R = Et$ or Me and $m = 1, 2$) also form stable 1:1 complexes with nitriles, and several of these were investigated by Pasynkiewicz and coworkers[285,286]. The heat of formation for benzonitrile adducts were studied calorimetrically and correlated with the increase of the C≡N stretching frequency.

No adducts between gallium trihalides and nitriles have been characterized. A complex between trimethylgallium and acetonitrile was reported[287]. Recently, the organogallium adducts with nitriles R—C≡N:GaR$_3$ were studied in more detail, and a blue-shifted C≡N stretching frequency was observed[288]. Unlike the trialkylaluminium complexes[283,284], no migration of alkyl groups from gallium to the nitrile carbon was observed. Under reduced pressure these adducts dissociate into the components.

Indirect evidence from conductometric measurements on methyl chlorosilanes in acetonitrile was interpreted in terms of complex formation[289].

B. Coordination Complexes of Heavy Metal Salts

The non-ionized molecular complexes formed between nitriles and the metal compounds, which have a tendency to additional covalent bond formation, have been studied extensively in later years, particularly by inorganic chemists. An excellent condensed survey of the properties, stoichiometry and structure of alkyl cyanides with metal halides has been given by Walton[258] in 1965, and some aryl cyanide complexes were also discussed. The complexing ability of silicon, germanium and tin tetrahalides towards acetonitrile has been summarized[251], and very recently the complexes of Group V pentahalides with this nitrile[290,291] have been included in monographs. Since Walton's article gives an adequate survey of this field, we will limit this chapter to some new developments and we will not try to cover the specialized, though interesting, work which has been reported lately.

The 1:1 compound, prepared as a white powder on adding dry acetonitrile to a chloroform solution of antimony pentachloride in molecular ratios at 0°, was found to be monomeric and undissociated in benzene solution[292]. The dipole moment of 7·8 D was consistent with an octahedral model with a nitrogen–antimony coordinate bond. However, when antimony pentachloride was dissolved in acetonitrile, a very high electrical conductivity was observed, increasing with time[293,294]. The solid compound obtained by evaporating the solvent was formulated as the ionic salt $[SbCl_4(MeCN)_2]^+SbCl_6^-$. This structure was consistent with spectroscopic measurements, although a molecular 1:1 complex in benzene solution was not excluded[295]. Recently, the crystal structure has been determined, revealing a molecular 1:1 complex with the five chlorine atoms and the nitrogen end of the nitrile bonded to the antimony in a somewhat distorted octahedron[296].

1. Far infrared studies

Recently there has been a considerable interest in the low frequency infrared spectra of molecular complexes, because of the potentialities of this technique to study metal–ligand vibrations, stereochemistry of the complexes, etc.[297,298]. A disagreement concerning the assignments of the bands below 500 cm^{-1} in the spectrum of the 1:2 adducts between tin tetrahalides and nitriles[299–301] has been solved. Farano and Grasselli[302] showed that the bands around 400 cm^{-1} were independent of the halogens, and therefore ligand vibrations (C—C≡N bending modes), shifted to higher wave numbers on complexation. Bands in the 300–370 cm^{-1} region, present in the tin tetrachloride adducts but missing in the corresponding tin tetrabromide and tin tetraiodide complexes, were assigned as Sn—Cl stretching modes. The complex $SnCl_4 \cdot 2RC{\equiv}N$ showed two bands in the 140–265 cm^{-1} region which could not be ligand vibrations, and they were assigned to the asymmetric and symmetric Sn—N stretching frequencies, as proposed by Beattie and Rule[300].

Far infrared spectra of other nitrile–metal halide complexes are reported[303,304], as well as the spectra of acetonitrile and acrylonitrile adducts of chromium, molybdenum and tungsten hexacarbonyls[305].

2. Chelates

The complexes between the Group IV halides and dinitriles $N{\equiv}C{-}(CH_2)_n{-}C{\equiv}N$ have been extensively studied in later years. These metals prefer the usual coordination number six[258]. Therefore coordination to the same metal atom from both ends of the dinitrile molecule might take place, giving monomeric 1:1 chelate compounds. Because of the steric requirements connected with the linear C—C≡N → M arrangement these bonds should only be expected for $n > 2$. However, chelates are not formed from such dinitriles and Group IV tetrahalides[88,306], nor in the ionic compounds such as copper(I) perchlorate[307].

The short-chained dinitriles might also give chelates, employing the π-electron systems of the cyano groups. This possibility was ruled out, since an increase in the C≡N stretching frequency was invariably found. However, Jain and Rivest[89,308] recently prepared

1:1 complexes between aminonitriles $R_2N(CH_2)_nC{\equiv}N$ and titanium, tin and zirconium tetrahalides. From molecular weight and conductivity data these complexes were shown to be monomeric, non-ionic compounds. Moreover, for diethylaminoacetonitrile ($n = 1$) a decrease in the C≡N frequency from 40 to 100 cm^{-1} was observed, indicating that chelation to the triple bond had taken place. For $n = 2$ both types of chelates were formed, one with normal coordination from the lone pair of the cyano group, and one in which chelation takes place through the triple bond. When $n = 3$ the normal chelates were formed.

The first example of a cyano group involved in π-bonding to transition metals was a compound prepared from dialkylcyanamides (R_2N—C≡N) and nickel tetracarbonyl[90]. It has also been shown that succinonitrile may coordinate to manganese through its C≡N triple bonds. For the compounds $Mn(CO)_3(NCCH_2CH_2CN)X$ (X = Cl, Br, I) the structure **16** was inferred from infrared and proton magnetic resonance studies[91].

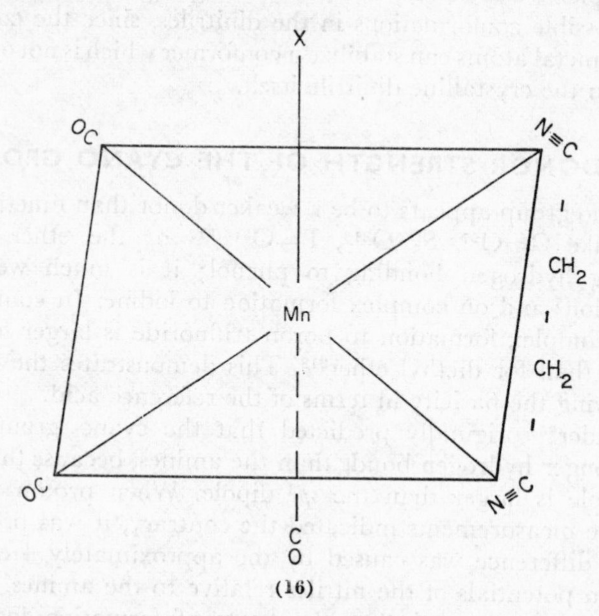

(16)

The complexes between titanium and tin tetrahalides with alkyl thiocyanates (RSCN and $NCS(CH_2)_2SCN$) are interesting because in these adducts the sulphur instead of the nitrogen seems to be preferred for coordination[309].

3. Conformations of dinitriles in coordination complexes

It is well known that succinonitrile[72,73] and 2-halogeno propionitriles[70,310] exist as mixtures of *trans* and *gauche* conformations in the vapour and the liquid, but crystallize in the *gauche* form. Kubota and coworkers[306] demonstrated that polymeric structures arise when tin tetrachloride or titanium tetrachloride coordinate with dinitriles, the bidentate ligands forming bridges. Succinonitrile assumes *trans* conformation in these coordination compounds as well as with silver[311] and copper[312] salts. A series of complexes with the general formula $Cu[NC(CH_2)_nCN]_2NO_3$ have been studied by x-ray crystallographic and infrared spectroscopic technique, and found to exist in *gauche*[313] ($n = 2$) and *gauche–gauche*[314] ($n = 3$) conformations. The *gauche–gauche* conformation was further established for silver(I) salts, but in tin(IV) and titanium(IV) the *trans–trans* conformation prevailed[315]. Thus infrared spectral studies of coordination complexes can be very valuable in assigning infrared frequencies to the possible conformations in the dinitriles, since the coordinate bonds to metal atoms can stabilize a conformer which is not ordinarily present in the crystalline dinitrile itself.

V. DONOR STRENGTH OF THE CYANO GROUP

The cyano group appears to be a weaker donor than functional oxo groups like C=O[97], S=O[316], P=O[316,317] or the ether linkage[3], regarding hydrogen bonding to phenol; it is much weaker on protonation[5] and on complex formation to iodine. In contrast, the heat of complex formation to boron trifluoride is larger for acetonitrile[260] than for diethyl ether[318]. This demonstrates the necessity of specifying the basicity in terms of the reference acid.

Schneider[95] originally predicted that the cyano group should form stronger hydrogen bonds than the amines, because the *sp* lone-pair dipole is larger than the sp^3 dipole. When proton magnetic resonance measurements indicated the contrary, it was proposed[141] that the difference was caused by the approximately 4 ev higher ionization potentials of the nitriles relative to the amines. Berkeley and Hanna[142] reported that the heats of formation for the *N*-methylpyrrolidine–chloroform and the cyclohexyl cyanide–chloroform systems are $-3{\cdot}9$ and $-2{\cdot}6$ kcal/mole respectively. They interpreted the hydrogen bond in terms of an electrostatic attractive energy and a quantum-mechanical overlap repulsion energy. Their calculations resulted in a longer hydrogen-bond length for the nitriles compared to the amines, closely corresponding to the observed differences in the hydrogen-bond energies.

A. Inductive Effects

The relative basicities of the nitriles towards Lewis acids appear to be quite similar. However, some significant variations are found, particularly when electron-withdrawing substituents are present. A meaningful correlation between the inductive effects of the substituents R and the basicity of the nitriles RC≡N is difficult, due to the lack of basicity constants and heats of formation for hydrogen-bonded or charge-transfer complexes. There are many investigations in which acetonitrile and other organic bases have been studied, but very few systematic studies of the various nitriles are available. Since small differences in basicity are involved, the data should be obtained by the same experimental method and preferably in the same laboratory.

A few comprehensive studies have been made of variations in the O—H stretching frequency of phenol hydrogen bonded to nitriles. Allerhand and Schleyer[103] obtained linear correlations between the Taft σ^*-constants of the substituents R for a large number of aliphatic nitriles RC≡N and the shifts $\Delta\nu_{O-H}$. The same conclusion was reached by White and Thompson[104]. A linear relationship between the heat of formation and the frequency shifts $\Delta\nu_{O-H}$ may be valid[98] at least for a series of related donors[319]. Therefore, the inductive effects seem to influence the hydrogen-bonding ability of the nitriles as expected.

For the charge-transfer complexes between iodine, iodine monochloride and iodine monobromide with some nitriles, a linear relationship[189] was proposed between σ^* and the logarithms of the formation constants for each acceptor, respectively. A linear correlation between σ^* and the heat of formation for the halogen complexes should therefore be valid[320]. Qualitatively, the effects of electronegative groups in reducing the complex formation constants to iodine monochloride[85] and to tin tetrachloride[321] have been demonstrated.

B. Conjugation Effects

The effects of conjugation on the basicity of the cyano group is not clear, and both positive and negative resonance effects may be present. Very few thermodynamic functions for the hydrogen bond or charge-transfer complexes with aromatic nitriles are known. Therefore, the donor properties of the aromatic and aliphatic nitriles cannot be directly compared. Spectral parameters may give some indications, however. Smaller shifts in the O—H frequencies of phenol and methanol hydrogen bonded to benzonitrile and acrylonitrile compared to those with acetonitrile and propionitrile are reported[103], indicating stronger interaction with the saturated, aliphatic compounds. The spectral shift for benzonitrile–phenol, however, is larger than predicted from the Taft σ^*-plot valid for aliphatic nitriles. This indicates a positive resonance effect, although not large enough to compensate for the negative inductive effect of the phenyl group. For acrylonitrile the shift was smaller than predicted from the σ^*-plot, which points to a negative resonance effect in this molecule. The results are consistent with the data for charge-transfer studies between these donors and the halogens[184–186,189]. In dimethyl cyanamide[192] the strong interaction with the halogens definitely establishes conjugation between the lone-pair electrons on the amine nitrogen and the nitrile group.

Substituents with a positive resonance effect such as the methoxy group increase the donor strength of benzonitrile[321]. The formation constants for the reaction between n-amyl alcohol and *p*-alkylbenzonitriles $RC_6H_4C{\equiv}N$ did not increase monotonically in the following sequence when R = H, Me, Et, i-Pr, *t*-Bu, and a Baker–Nathan order for the substituents was indicated[105].

C. Steric Effects

Due to their linear C—C≡N arrangement, the nitriles will as a rule not be subject to steric hindrance regarding complexation. When the acceptor atom is strongly sterically screened, steric factors may prevent the close proximity of the donor and the acceptor necessary for the formation of a stable complex. Several workers have studied the hydrogen bonding between sterically hindered phenols and acetonitrile.

With two methyl or isopropyl groups as *ortho* substituents in phenol, the formation constants are slightly smaller than for the phenol complex[104]. However, the infrared spectra of 2,6-di-*t*-butylphenol[104,322,323] and 2,6-di-*t*-butyl-4-methylphenol (ionol)[324,325] in solutions of acetonitrile revealed the free O—H stretching band around 3630 cm^{-1}. Only in the pure nitrile solutions did a very weak, broad band around 3500 cm^{-1} indicate the existence of a complex. Therefore in these systems only a small fraction of the donor molecules penetrate the potential barrier created by the bulky groups. Large substituents on the nitrile (e.g. in *t*-butyl cyanide) have small effects[104] on the hydrogen bonding to phenol.

Although steric hindrance can prevent the formation of oxonium salts by the addition of tertiary alkyl chloride (*t*-butyl chloride) to metal halide etherates, the corresponding nitrilium salt can easily be prepared[38].

VI. REFERENCES

1. L. Malatesta in *Progress in Inorganic Chemistry*, Vol. 1, Interscience, New York, 1959.
2. C. P. Smyth, *Dielectric Behavior and Structure*, McGraw-Hill, New York, 1955, p. 245.
3. S. Searles and M. Tamres in *The Chemistry of the Ether Linkage* (Ed. S. Patai), Interscience, London, 1967, Chapter 6.
4. L. P. Hammett and A. J. Deyrup, *J. Am. Chem. Soc.*, **54**, 2721 (1932).
5. V. A. Palm, Ü. L. Haldna and A. J. Talvik in *The Chemistry of the Carbonyl Group* (Ed. S. Patai), Interscience, London, 1967.
6. E. M. Arnett in *Progress in Physical Organic Chemistry*, Vol. 1, Interscience, New York, 1963.
7. H. Lemaire and H. J. Lucas, *J. Am. Chem. Soc.*, **73**, 5198 (1951).
8. L. P. Hammett and A. J. Deyrup, *J. Am. Chem. Soc.*, **54**, 4239 (1932).
9. M. A. Paul and F. A. Long, *Chem. Rev.*, 1 (1957).
10. N. C. Deno, R. W. Gaugler and M. J. Wisotsky, *J. Org. Chem.*, **31**, 1967 (1966).
11. A. Hantzsch, *Z. Physik. Chem. (Leipzig)*, **65**, 41 (1909).
12. M. Liler and D. Kosanovic, *J. Chem. Soc.*, **1958**, 1084.
13. N. C. Deno and M. J. Wisotsky, *J. Am. Chem. Soc.*, **85**, 1735 (1963).
14. J. F. Coetzee in *Progress in Physical Organic Chemistry*, Vol. 4, Interscience, New York, 1967.
15. M. Kilpatric and M. L. Kilpatric, *Chem. Rev.*, **13**, 131 (1933).
16. J. F. Coetzee and I. M. Kolthoff, *J. Am. Chem. Soc.*, **79**, 6110 (1957).
17. F. Klages, R. Ruhnau and W. Hauser, *Ann. Chem.*, **626**, 60 (1959).
18. I. M. Kolthoff, S. Bruckenstein and M. K. Chantooni, *J. Am. Chem. Soc.*, **83**, 3927 (1961).
19. G. J. Janz and I. Ahmad, *Electrochim. Acta*, **9**, 1539 (1964).
20. J. F. Coetzee and D. K. McGuire, *J. Phys. Chem.*, **67**, 1810 (1963).
21. G. J. Janz and S. S. Danyluk, *Chem. Rev.*, **60**, 209 (1960).
22. E. N. Zil'berman, T. S. Ivcher and E. M. Perepletchikova, *Zh. Obshch. Khim.*, **31**, 2037 (1961).
23. E. N. Zil'berman, *Usp. Khim.*, **31**, 1309 (1962).
24. E. N. Zil'berman, *Usp. Khim.*, **29**, 709 (1960).
25. H. Biltz, *Ber.*, **25**, 2533 (1892).
26. A. Hantzsch, *Ber.*, **69**, 667 (1931).
27. F. Klages and W. Grill, *Ann. Chem.*, **594**, 21 (1955).
28. E. Allenstein and A. Schmidt, *Spectrochim. Acta*, **20**, 1451 (1964).

29. E. Allenstein and P. Quis, *Chem. Ber.*, **97**, 3162 (1964).
30. B. Matkovic, S. W. Peterson and J. M. Williams, *Croat. Chem. Acta*, **39**, 139 (1967).
31. S. W. Peterson and J. M. Williams, *J. Am. Chem. Soc.*, **88**, 2866 (1966).
32. G. J. Janz and S. S. Danyluk, *J. Am. Chem. Soc.*, **81**, 3850 (1959).
33. (a) E. N. Zil'berman and A. E. Kulikova, *Zh. Obshch. Khim.*, **29**, 1694 (1959).
    (b) E. N. Zil'berman, *Zh. Obshch. Khim.*, **30**, 1277 (1960).
34. G. J. Janz and S. S. Danyluk, *J. Am. Chem. Soc.*, **81**, 3846 (1959); **81**, 3854 (1959).
35. F. E. Murray and W. G. Schneider, *Can. J. Chem.*, **33**, 797 (1955).
36. W. E. Hanley and W. A. Waters, *J. Chem. Soc.*, **1939**, 1792.
37. G. G. Makarova and A. N. Nesmeyanov, *Izv. Akad. Nauk SSSR, Otdel. Khim. Nauk*, **1954**, 1019; *Chem. Abstr.*, **50**, 240 (1956).
38. H. Meerwein, P. Laasch, R. Mersch and J. Spille, *Chem. Ber.*, **89**, 209 (1956).
39. J. E. Gordon and G. C. Turrel, *J. Org. Chem.*, **24**, 269 (1959).
40. H. Meerwein, P. Laasch, R. Mersch and J. Nentwig, *Chem. Ber.*, **89**, 224 (1956).
41. G. C. Turrel and J. E. Gordon, *J. Chem. Phys.*, **30**, 895 (1965).
42. G. J. Janz, M. J. Tait and J. Meier, *J. Phys. Chem.*, **71**, 963 (1967).
43. D. C. Luehrs, R. T. Iwamoto and J. Kleinberg, *Inorg. Chem.*, **5**, 201 (1966).
44. H. J. Coerver and C. Curran, *J. Am. Chem. Soc.*, **80**, 3522 (1958).
45. J. C. Evans and L. Y.-S. Lo, *Spectrochim. Acta*, **21**, 1033 (1965).
46. I. S. Perelygin, *Opt. i Spectroskopiya*, **13**, 360 (1962).
47. Z. Kęcki and J. Witanowski, *Roczniki Chem.*, **38**, 691 (1964).
48. I. S. Pominov and A. Z. Gadzhiev, *Izv. Vysshikh Uchebn. Zavedenii Fiz.*, **8**, 19 (1965); *Chem. Abstr.*, **64**, 2802 (1966).
49. B. J. Hathaway, D. G. Holah and J. D. Postlethwaite, *J. Chem. Soc.*, **1961**, 3215.
50. B. J. Hathaway, D. G. Holah and A. E. Underhill, *J. Chem. Soc.*, **1962**, 2444.
51. A. E. Wickenden and R. A. Krause, *Inorg. Chem.*, **4**, 404 (1965).
52. N. A. Matwiyoff and S. W. Hooker, *Inorg. Chem.*, **6**, 1127 (1967).
53. S. I. Chan, B. M. Fung and H. Lütje, *J. Chem. Phys.*, **47**, 2121 (1967).
54. C. B. Baddiel, M. J. Tait and G. J. Janz, *J. Phys. Chem.*, **69**, 3634 (1965).
55. H. Schneider and H. Strehlow, *Z. Physik. Chem. (Frankfurt)*, **49**, 44 (1966).
56. S. E. Manahan and R. T. Iwamoto, *J. Electroanal. Chem.*, **14**, 213 (1967).
57. G. J. Janz, A. E. Marcinkowsky and I. Ahmad, *J. Electrochem. Soc.*, **112**, 104 (1965).
58. H. Strehlow and H.-M. Koepp, *Z. Electrochem.*, **62**, 373 (1957).
59. S. Wawzonek and M. E. Runner, *J. Electrochem. Soc.*, **99**, 457 (1952).
60. A. I. Popov and D. H. Geske, *J. Am. Chem. Soc.*, **79**, 2074 (1957).
61. I. M. Kolthoff and J. F. Coetzee, *J. Am. Chem. Soc.*, **79**, 870 (1957).
62. I. M. Kolthoff and J. F. Coetzee, *J. Am. Chem. Soc.*, **79**, 1852 (1957).
63. J. F. Coetzee, D. K. McGuire and J. L. Hedrick, *J. Phys. Chem.*, **67**, 1814 (1963).
64. K. M. Marstokk and K. O. Strømme, *Acta Cryst.*, **24B**, 713 (1968).
65. P. Venkateswarlu, *J. Chem. Phys.*, **19**, 293 (1951).
66. J. C. Evans and H. J. Bernstein, *Can. J. Chem.*, **33**, 1476 (1955).
67. I. Nakagawa and T. Shimanouchi, *Spectrochim. Acta*, **18**, 513 (1962).
68. J. L. Duncan, *Spectrochim. Acta*, **20**, 1197 (1964).
69. N. E. Duncan and G. J. Janz, *J. Chem. Phys.*, **23**, 434 (1955).
70. P. Klaboe and J. Grundnes, *Spectrochim. Acta*, **24A**, 1905 (1968).
71. J. H. S. Green, *Spectrochim. Acta*, **17**, 607 (1961).
72. W. E. Fitzgerald and G. J. Janz, *J. Mol. Spectry*, **1**, 49 (1957).
73. T. Fujiyama, K. Tokumaru and T. Shimanouchi, *Spectrochim. Acta*, **20**, 415 (1964).
74. T. Fujiyama and T. Shimanouchi, *Spectrochim. Acta*, **20**, 829 (1964).

75. R. E. Kitson and N. E. Griffith, *Anal. Chem.*, **24**, 334 (1952).
76. A. Terenin, W. Filimonov and D. Bystrov, *Z. Electrochem.*, **62**, 180 (1958).
77. E. L. Zhukova, *Opt. i Spektroskopiya*, **5**, 270 (1958).
78. I. Lindqvist, *Inorganic Adduct Molecules of Oxo Compounds*, Springer-Verlag, Berlin, 1963.
79. F. A. Cotton and F. Zingales, *J. Am. Chem. Soc.*, **83**, 351 (1961).
80. T. L. Brown and M. Kubota, *J. Am. Chem. Soc.*, **83**, 4175 (1961).
81. H. A. Brune and W. Zeil, *Z. Naturforsch.*, **169**, 1251 (1961).
82. I. R. Beattie and T. Gilson, *J. Chem. Soc.*, **1964**, 2293.
83. K. F. Purcell and R. S. Drago, *J. Am. Chem. Soc.*, **88**, 919 (1966).
84. K. F. Purcell, *J. Am. Chem. Soc.*, **89**, 247 (1967).
85. E. Augdahl and P. Klaboe, *Spectrochim. Acta*, **19**, 1665 (1963).
86. J. L. Hoard, T. B. Owen, A. Buzzell and O. N. Salmon, *Acta Cryst.*, **3**, 130 (1950).
87. D. Britton in *Perspectives in Structural Chemistry*, Vol. 1, John Wiley and Sons, New York, 1967, p. 109.
88. S. C. Jain and R. Rivest, *Can. J. Chem.*, **41**, 2130 (1963).
89. S. C. Jain and R. Rivest, *Inorg. Chem.*, **6**, 467 (1967).
90. H. Bock and H. tom Dieck, *Chem. Ber.*, **99**, 213 (1966).
91. M. F. Farona and N. J. Bremer, *J. Am. Chem. Soc.*, **88**, 3735 (1966).
92. G. L. Caldow, D. Cunliffe-Jones and H. W. Thompson, *Proc. Roy. Soc.* (*London*) *Ser. A*, **254**, 17 (1960).
93. J. P. Jesson and H. W. Thompson, *Proc. Roy. Soc.* (*London*) *Ser. A.*, **268**, 68 (1962).
94. (a) R. S. Mulliken, *J. Chim. Phys.*, **61**, 20 (1964).
(b) H. Tsubomura, *J. Chem. Phys.*, **24**, 927 (1956).
(c) S. Nagakura, *J. Chim. Phys.*, **61**, 217 (1964).
(d) K. Szczepaniak and A. Tramer, *J. Phys. Chem.*, **71**, 3035 (1967).
95. W. G. Schneider, *J. Chem. Phys.*, **23**, 26 (1955).
96. G. C. Pimentel and A. L. McClellan, *The Hydrogen Bond*, W. H. Freeman, San Francisco, 1960.
97. H. Fritzche, *Ber. Bunsenges. Physik. Chem.*, **68**, 459 (1964).
98. T. D. Epley and R. S. Drago, *J. Am. Chem. Soc.*, **89**, 5770 (1967).
99. H. Tsubomura, *J. Chem. Phys.*, **23**, 2130 (1955).
100. S. S. Mitra, *J. Chem. Phys.*, **36**, 3286 (1962).
101. A. A. Allerhand and P. R. Schleyer, *J. Am. Chem. Soc.*, **85**, 371 (1963).
102. M. Horák, P. Poláková, M. Jakoubková, J. Movravec and J. Plíva, *Collection Czech. Chem. Commun.*, **31**, 622 (1966).
103. A. Allerhand and P. R. Schleyer, *J. Am. Chem. Soc.*, **85**, 866 (1963).
104. S. C. White and H. W. Thompson, *Proc. Roy. Soc.* (*London*) *Ser. A*, **291**, 460 (1966).
105. L. L. Ferstandig, *Tetrahedron*, **20**, 1367 (1963).
106. A. V. Iogansen and G. A. Kurkchi, *Optika i Spektroskopiya*, **13**, 480 (1962).
107. L. L. Caldow and H. W. Thompson, *Proc. Roy. Soc.* (*London*) *Ser. A*, **254**, 1 (1960).
108. P. V. Huong and G. Turrel, *J. Mol. Spectry*, **25**, 185 (1968).
109. E. Greinacher, W. Lüttke and R. Mecke, *Z. Electrochem.*, **59**, 23 (1955).
110. M.-L. Josien and P. Saumagne, *Bull. Soc. Chim. France*, **1956**, 937.
111. P. Saumagne and M.-L. Josien, *Bull. Soc. Chim. France*, **1958**, 813.
112. J. Lascombe, M. Haurie and M.-L. Josien, *J. Chim. Phys.*, **59**, 1233 (1962).
113. L. J. Bellamy, H. E. Hallam and R. L. Williams, *Trans. Faraday Soc.*, **54**, 1120 (1958).
114. L. J. Bellamy and H. E. Hallam, *Trans. Faraday Soc.*, **55**, 220 (1959).

115. W. Gordy and C. S. Stanford, *J. Chem. Phys.*, **9**, 204 (1941).
116. W. Gordy, *J. Chem. Phys.*, **9**, 215 (1941).
117. Y. Sato and S. Nagakura, *J. Chem. Soc. Japan, Pure Chem. Sect.*, **76**, 1007 (1955); *Chem. Abstr.*, **50**, 5406 (1956).
118. C. L. Bell and G. M. Barrow, *J. Chem. Phys.*, **31**, 300 (1959).
119. S. Wada, *Bull. Chem. Soc. Japan*, **35**, 707 (1962).
120. S. Wada, *Bull. Chem. Soc. Japan*, **35**, 710 (1962).
121. M. St. C. Flett, *J. Soc. Dyers Colourists*, **68**, 59 (1952).
122. (a) I. S. Perelygin, N. R. Safiullina and V. G. Romanov, *Zh. Fiz. Khim.*, **39**, 394 (1965). (b) N. D. Sokolov, *Usp. Fiz. Nauk.*, **57**, 205 (1955).
123. R. M. Adams and J. J. Katz, *J. Mol. Spectry*, **1**, 306 (1957).
124. M. Haurie and A. Novak, *J. Chim. Phys.*, **64**, 679 (1967).
125. C. C. Mohr, W. D. Wilk and G. M. Barrow, *J. Am. Chem. Soc.*, **87**, 3048 (1965).
126. M. D. Joesten and R. S. Drago, *J. Am. Chem. Soc.*, **84**, 3817 (1962).
127. C. P. Nash, D. K. Fujita and E. D. Retherford, *J. Phys. Chem.*, **71**, 3187 (1967).
128. P. R. Schleyer and A. Allerhand, *J. Am. Chem. Soc.*, **84**, 1322 (1962).
129 L. L. Ferstandig, *J. Am. Chem. Soc.*, **84**, 1323 (1962).
130. L. L. Ferstandig, *J. Am. Chem. Soc.*, **84**, 3533 (1962).
131. G. Mavel, *J. Phys. Radium*, **20**, 834 (1959).
132. B. Lemanceau, C. Lussan and N. Souty, *J. Chim. Phys.*, **59**, 148 (1962).
133. J. R. Holmes, D. Kivelson and W. C. Drinkard, *J. Am. Chem. Soc.*, **84**, 4677 (1962).
134. B. B. Howard, C. F. Jumper and M. T. Emerson, *J. Mol. Spectry*, **10**, 117 (1963).
135. P. J. Berkeley and M. W. Hanna, *J. Phys. Chem.*, **67**, 846 (1963).
136. D. Loewenstein and A. Margalit, *J. Phys. Chem.*, **69**, 4152 (1965).
137. C. S. Springer and D. W. Meek, *J. Phys. Chem.*, **70**, 481 (1966).
138. D. P. Eyman and R. S. Drago, *J. Am. Chem. Soc.*, **88**, 1617 (1966).
139. G. Socrates, *Trans. Faraday Soc.*, **63**, 1083 (1967).
140. K. F. Purcell and S. T. Wilson, *J. Mol. Spectry*, **24**, 468 (1967).
141. G. J. Korinek and W. G. Schneider, *Can. J. Chem.*, **35**, 1157 (1957).
142. P. J. Berkeley and M. W. Hanna, *J. Chem. Phys.*, **41**, 2530 (1964).
143. P. J. Berkeley and M. W. Hanna, *J. Am. Chem. Soc.*, **86**, 2990 (1964).
144. I. Gränacher, *Helv. Phys. Acta*, **34**, 272 (1961).
145. P. Laszlo in *Progress in Nuclear Magnetic Resonance Spectroscopy*, Vol. 3, Pergamon Press, Oxford, 1967, Chapter 6.
146. S. Murakami and R. Furishiro, *Bull. Chem. Soc. Japan*, **39**, 720 (1966).
147. E. M. Arnett, T. S. S. R. Murty, P. R. Schleyer and L. Joris, *J. Am. Chem. Soc.*, 89, 5955 (1967).
148. G. P. Cunningham, G. A. Vidulich and R. L. Kay, *Chem. Eng. Data*, **12**, 336 (1967).
149. E. A. S. Cavell, H. G. Jerrard, B. A. W. Simmonds and J. A. Speed, *J. Phys. Chem.*, **69**, 3657 (1965).
150. P. Laurent, *Compt. Rend.*, **199**, 582 (1934).
151. C. H. Giles, T. J. Rose and D. G. M. Vallance, *J. Chem. Soc.*, **1952**, 3799.
152. C. F. Jumper and B. B. Howard, *J. Phys. Chem.*, **70**, 588 (1966).
153. G. E. Coates and J. E. Coates, *J. Chem. Soc.*, **1944**, 77.
154. W. F. Giauque and R. A. Ruehrwein, *J. Am. Chem. Soc.*, **61**, 2626 (1939).
155. R. H. Cole, *J. Am. Chem. Soc.*, **77**, 2012 (1955).
156. F. D. Mashlan and E. A. Stoddard, *J. Phys. Chem.*, **60**, 1146 (1956).
157. Z. P. Chirikova, L. A. Galata, Z. N. Kotova and L. S. Kofman, *Zh. Fiz. Khim.*, **40**, 918 (1966).

158. F. Mato and M. Sanchez, *Ann. Real Soc. Espan. Fiz. Quim., Ser. B*, **63**, 1 (1967); *Chem. Abstr.*, **67**, 26265 y (1967).
159. A.-L. Vierk, *Z. Anorg. Chem.*, **261**, 283 (1950).
160. M. Usanovich and V. Dulova, *Zh. Obshch. Khim.*, **16**, 1978 (1946); *Chem. Abstr.*, **41**, 6795 (1947).
161. M. J. Copley, G. F. Zellhoefer and C. S. Marvel, *J. Am. Chem. Soc.*, **61**, 3550 (1939).
162. M. J. Copley, G. F. Zellhoefer and G. S. Marvel, *J. Am. Chem. Soc.*, **62**, 227 (1940).
163. R. Collander, *Acta Chem. Scand.*, **3**, 717 (1949).
164. B. V. Tronov and N. D. Strel'nikova, *Izv. Tomsk Politekh. Inst.*, **83**, 98 (1956); *Chem. Abstr.*, **53**, 12804 (1959).
165. M. F. Vyks and L. I. Lisnyanskii, *Vodorodnaya Svyaz. Akad. Nauk SSSR, Inst. Khim. Fiz., Sb. Statei*, **1964**, 108; *Chem. Abstr.*, **62**, 7243 (1965).
166. S. B. Hendricks, O. R. Wulf, G. E. Hilbert and U. Liddel, *J. Am. Chem. Soc.*, **58**, 1991 (1936).
167. V. Prey and H. Berbalk, *Monatsh. Chem.*, **82**, 990 (1951).
168. M. St. C. Flett, *Spectrochim. Acta*, **10**, 21 (1957).
169. P. J. Krueger, *Can. J. Chem.*, **40**, 2300 (1962).
170. P. J. Krueger and H. D. Mettee, *Can. J. Chem.*, **43**, 2888 (1965).
171. L. H. Little, *Infrared Spectra of Adsorbed Species*, Academic Press, London, 1966.
172. A. V. Kiselev, *Zh. Fiz. Khim.*, **38**, 2753 (1964).
173. V. Y. Davydov, A. V. Kiselev and B. G. Kuznetsov, *Zh. Fiz. Khim.*, **39**, 2058 (1965).
174. K. T. Geodakyan, A. V. Kiselev and V. I. Lygin, *Zh. Fiz. Khim.*, **40**, 1584 (1966).
175. A. Bertoluzza and G. B. Bonino, *Atti Accad. Naz. Lincei Rend. Classe Sci. Fis. Mat. Nat.*, **39**, 232 (1965); **40**, 179 (1966).
176. R. S. Mulliken, *J. Phys. Chem.*, **56**, 801 (1952).
177. G. Briegleb, *Elektronen-Donator-Acceptor-Komplexe*, Springer-Verlag, Berlin, 1961.
178. L. J. Andrews and R. M. Keefer, *Molecular Complexes in Organic Chemistry*, Holden-Day, San Francisco, 1964.
179. J. Rose, *Molecular Complexes*, Pergamon Press, Oxford, 1967.
180. N. S. Bayliss and A. L. G. Rees, *J. Chem. Phys.*, **8**, 377 (1940).
181. R. S. Mulliken, *Proc. Int. Conf. Coord. Comp., Amsterdam, 1955*, Butterworths, London, p. 336.
182. S. Nakagura, *J. Am. Chem. Soc.*, **80**, 520 (1958).
183. J. Walkley, D. N. Glew and J. H. Hildebrand, *J. Chem. Phys.*, **33**, 621 (1960).
184. P. Klaboe, *J. Am. Chem. Soc.*, **84**, 3458 (1962).
185. P. Klaboe, *J. Am. Chem. Soc.*, **85**, 871 (1963).
186. P. Klaboe, *Acta Chem. Scand.*, **17**, 1179 (1963).
187. H. A. Benesi and J. H. Hildebrand, *J. Am. Chem. Soc.*, **71**, 2703 (1949).
188. A. I. Popov and W. A. Deskin, *J. Am. Chem. Soc.*, **80**, 2976 (1958).
189. W. B. Person, W. C. Golton and A. I. Popov, *J. Am. Chem. Soc.*, **85**, 891 (1963).
190. J. A. Maguire, A. Bramley and J. J. Banewicz, *Inorg. Chem.*, **6**, 1752 (1967).
191. F. Shah-Malak and J. H. P. Utley, *Chem. Commun.*, **1967**, 69.
192. E. Augdahl and P. Klaboe, *Acta Chem. Scand.*, **19**, 807 (1965).
193. H. Yada, J. Tanaka and S. Nagakura, *Bull. Chem. Soc. Japan*, **33**, 1660 (1960).
194. H. Tsubomura and R. P. Lang, *J. Am. Chem. Soc.*, **83**, 2085 (1961).
195. M. Brandon, M. Tamres and S. Searles, *J. Am. Chem. Soc.*, **82**, 2129 (1960).
196. J. D. Morrison and A. J. C. Nicholson, *J. Chem. Phys.*, **20**, 1021 (1952).
197. S. H. Hastings, J. L. Franklin, J. C. Schiller and F. A. Matsen, *J. Am. Chem. Soc.*, **75**, 2900 (1953).

198. H. McConnel, J. S. Ham and J. R. Platt, *J. Chem. Phys.*, **21**, 66 (1953).
199. G. Briegleb and J. Czekalla, *Z. Elektrochem.*, **63**, 6 (1959).
200. A. I. Popov and N. E. Skelly, *J. Am. Chem. Soc.*, **77**, 3722 (1955).
201. R. E. Buckles and J. F. Mills, *J. Am. Chem. Soc.*, **76**, 4845 (1954).
202. C. H. J. Wells, *Spectrochim. Acta*, **22**, 2125 (1966).
203. W. B. Person, R. E. Erickson and R. E. Buckles, *J. Am. Chem. Soc.*, **82**, 29 (1960).
204. W. B. Person, R. E. Humphrey, W. A. Deskin and A. I. Popov, *J. Am. Chem. Soc.*, **80**, 2049 (1958).
205. Y. Yagi, A. I. Popov and W. B. Person, *J. Phys. Chem.*, **71**, 2439 (1967).
206. W. B. Person, R. E. Humphrey and A. I. Popov, *J. Am. Chem. Soc.*, **81**, 273 (1959).
207. J. Yarwood and W. B. Person, *J. Am. Chem. Soc.*, **90**, 594 (1968).
208. P. Klaboe, *J. Am. Chem. Soc.*, **89**, 3667 (1967).
209. O. Hassel and C. Rømming, *Quart. Rev. (London)*, **16**, 1 (1962).
210. R. B. Heiart and G. B. Carpenter, *Acta Cryst.*, **9**, 889 (1956).
211. S. Geller and A. L. Schawlow, *J. Chem. Phys.*, **23**, 779 (1955).
212. J. A. A. Ketelaar and J. W. Zwartsenberg, *Rec. Trav. Chim.*, **58**, 448 (1939).
213. T. Bjorvatten, *Acta Chem. Scand.*, **22**, 410 (1968).
214. B. Borgen, O. Hassel and C. Rømming, *Acta Chem. Scand.*, **16**, 2469 (1962).
215. P. Klaboe and E. Kloster-Jensen, *Spectrochim. Acta*, **23A**, 1981 (1967).
216. M. R. Mander and H. W. Thompson, *Trans. Faraday Soc.*, **53**, 1402 (1957).
217. J. P. Jesson and H. W. Thompson, *Spectrochim. Acta*, **13**, 217 (1958).
218. H. W. Thompson and D. J. Jewell, *Spectrochim. Acta*, **13**, 254 (1959).
219. G. H. J. Facer and H. W. Thompson, *Proc. Roy. Soc. (London) Ser. A.*, **268**, 79 (1962).
220. C. D. Ritchie and A. L. Pratt, *J. Am. Chem. Soc.*, **86**, 1571 (1964).
221. C. D. Ritchie, B. A. Bierl and R. J. Honour, *J. Am. Chem. Soc.*, **84**, 4687 (1962).
222. T. Cairns, G. Eglinton and D. T. Gibson, *Spectrochim. Acta*, **20**, 31 (1964).
223. C. D. Ritchie and A. L. Pratt, *J. Phys. Chem.*, **67**, 2498 (1963).
224. N. L. Allinger and W. Szkrybalo, *J. Org. Chem.*, **27**, 4601 (1962).
225. R. Rickborn and F. R. Jensen, *J. Org. Chem.*, **27**, 4606 (1962).
226. R. W. Taft, E. Price, I. R. Fox, I. C. Lewis, K. K. Andersen and G. T. Davis, *J. Am. Chem. Soc.*, **85**, 709 (1963).
227. A. M. Saum, *J. Polymer Sci.*, **42**, 57 (1960).
228. W. Dannhauser and A. F. Flueckinger, *J. Chem. Phys.*, **38**, 69 (1963).
229. W. Dannhauser and A. F. Flueckinger, *J. Phys. Chem.*, **68**, 1814 (1964).
230. J. D. Lambert, G. A. H. Roberts, J. S. Rowlinson and V. J. Wilkinson, *Proc. Roy. Soc. (London) Ser. A.*, **196**, 113 (1949).
231. A. D. Buckingham and R. E. Raab, *J. Chem. Soc.*, **1961**, 5511.
232. C. C. Addison and J. C. Sheldon, *J. Chem. Soc.*, **1956**, 1941.
233. C. C. Addison and J. C. Sheldon, *J. Chem. Soc.*, **1956**, 1941.
234. H. M. Powell in *Non-Stoichiometric Compounds* (Ed. L. Mandelcorn), Academic Press, New York, 1963, Chapter 7.
235. S. C. Wallwork and H. M. Powell, *J. Chem. Soc.*, **1956**, 4855.
236. J. S. Dryden and R. J. Meakins, *Nature*, **169**, 324 (1952).
237. J. S. Dryden, *Trans. Faraday Soc.*, **49**, 1333 (1953).
238. M. Davies and W. C. Child, *Spectrochim. Acta*, **21**, 1195 (1965).
239. R. F. Weimer and J. M. Prausnitz, *Spectrochim. Acta*, **22**, 77 (1966).
240. J. V. Hatton and W. G. Schneider, *Can. J. Chem.*, **40**, 1285 (1962).
241. R. J. Abraham, *J. Chem. Phys.*, **34**, 1062 (1961).

242. A. D. Buckingham, T. Schaefer and W. G. Schneider, *J. Chem. Phys.*, **34**, 1064 (1961).
243. I. Brown and W. Fock, *Australian J. Chem.*, **8**, 361 (1955); **9**, 180 (1956); **10**, 417 (1957).
244. K. Amaya, *Bull. Chem. Soc. Japan*, **34**, 1271, 1278, 1349 (1961).
245. R. V. Orye and J. M. Prausnitz, *Trans. Faraday Soc.*, **61**, 1338 (1965).
246. J. R. Goates, J. B. Ott and A. H. Budge, *J. Phys. Chem.*, **65**, 2162 (1961).
247. D. R. Martin and J. M. Canon in *Friedel–Crafts and Related Reactions*, Vol. 1 (Ed. G. A. Olah) Interscience, London, 1963, Chapter 6.
248. N. N. Greenwood and R. L. Martin, *Quart. Rev. (London)*, **8**, 1 (1954).
249. N. N. Greenwood and K. Wade in *Friedel–Crafts and Related Reactions*, Vol. 1 (Ed. G. A. Olah) Interscience, London, 1963, Chapter 7.
250. F. G. A. Stone, *Chem. Rev.*, **58**, 101 (1958).
251. I. R. Beattie, *Quart. Rev. (London)*, **17**, 382 (1963).
252. G. Sumrell, *J. Org. Chem.*, **19**, 817 (1954).
253. E. M. Kaiser and C. R. Hauser, *J. Am. Chem. Soc.*, **88**, 2348 (1966).
254. C. G. Swain, *J. Am. Chem. Soc.*, **69**, 2306 (1947).
255. S. J. Storfer and E. I. Becker, *J. Org. Chem.*, **27**, 1868 (1962).
256. R. Fricke and F. Ruschhaupt, *Z. Anorg. Chem.*, **146**, 103 (1925).
257. E. T. McBee, O. R. Pierce and D. D. Meyer, *J. Am. Chem. Soc.*, **77**, 83 (1955).
258. R. A. Walton, *Quart. Rev. (London)*, **19**, 126 (1965).
259. C. M. Bax, A. R. Katritzky and L. E. Sutton, *J. Chem. Soc.*, **1958**, 1254.
260. A. W. Laubengayer and D. S. Sears, *J. Am. Chem. Soc.*, **67**, 164 (1945).
261. J. M. Miller and M. Onyszchuk, *Can. J. Chem.*, **43**, 1877 (1965).
262. J. M. Miller and M. Onyszchuk, *Can. J. Chem.*, **44**, 899 (1966).
263. R. W. Taft and J. W. Carten, *J. Am. Chem. Soc.*, **86**, 4199 (1964).
264. W. Gerrard, M. F. Lappert, H. Pyszora and J. W. Wallis, *J. Chem. Soc.*, **1960**, 2182.
265. H. C. Brown and R. B. Johannesen, *J. Am. Chem. Soc.*, **72**, 2934 (1950).
266. D. B. Sowerby, *J. Am. Chem. Soc.*, **84**, 1831 (1962).
267. G. Hesse, H. Witte and G. Bittner, *Ann. Chem.*, **687**, 9 (1965).
268. G. Hesse and H. Witte, *Ann. Chem.*, **687**, 1 (1965).
269. R. T. Holzmann, *Production of the Boranes and Related Research*, Academic Press, New York, 1967, Chapter 13.
270. H. J. Emeléus and K. Wade, *J. Chem. Soc.*, **1960**, 2614.
271. M. F. Hawthorne in *Advances in Inorg. Chem. and Radiochem.*, Vol. 5, Academic Press, New York, 1963, p. 307.
272. W. N. Lipscomb, *Boron Hydrides*, W. A. Benjamin, New York, 1963.
273. R. Schaeffer, *J. Am. Chem. Soc.*, **79**, 1006 (1957).
274. D. E. Hyatt, F. R. Scholer and L. J. Todd, *Inorg. Chem.*, **6**, 630 (1967).
275. B. M. Graybill, A. R. Pitocelli and M. F. Hawthorne, *Inorg. Chem.*, **1**, 626 (1962).
276. F. E. Wang, P. G. Simpson and W. N. Lipscomb, *J. Am. Chem. Soc.*, **83**, 492 (1961).
277. J. H. Enemark, L. B. Friedman and W. N. Lipscomb, *Inorg. Chem.*, **5**, 2165 (1966).
278. D. E. Hyatt, D. A. Owen and L. J. Todd, *Inorg. Chem.*, **5**, 1749 (1966).
279. C. D. Schmulbach, *J. Inorg. Nucl. Chem.*, **26**, 745 (1964).
280. J. E. Lloyd and K. Wade, *J. Chem. Soc.*, **1964**, 1649.
281. T. Mole, *Australian J. Chem.*, **16**, 801 (1963).
282. G. K. J. Gibson and D. W. Hughes, *Chem. Ind. (London)*, **13**, 544 (1964).
283. J. E. Lloyd and K. Wade, *J. Chem. Soc.*, **1965**, 2662.

284. R. Jennings, J. E. Lloyd and K. Wade, *J. Chem. Soc.*, **1965**, 5083.
285. K. Starowieyski and S. Pasynkiewicz, *Roczniki Chem.*, **40**, 47 (1966).
286. K. Starowieyski, S. Pasynkiewicz and M. Boleslawski, *J. Organometal. Chem.*, **10**, 393 (1967).
287. G. E. Coates and R. G. Hayter, *J. Chem. Soc.*, **1953**, 2519.
288. J. R. Jennings and K. Wade, *J. Chem. Soc.* (*A*), **1967**, 1222.
289. A. P. Kreshov, V. A. Drozdov and S. Kubiak, *Zh. Obshch. Khim.*, **31**, 3099 (1961).
290. M. Webster, *Chem. Rev.*, **66**, 87 (1966).
291. L. Kolditz in *Halogen Chemistry* Vol. 2 (Ed. V. Gutmann), Academic Press, London, 1967, pp. 115–168.
292. S. R. Jain and S. Soundararajan, *Z. Anorg. Allgem. Chem.*, **337**, 214 (1965).
293. L. Kolditz and H. Preiss, *Z. Anorg. Allgem. Chem.*, **310**, 242 (1961).
294. L. Kolditz, Chr. Kürschner and U. Calov, *Z. Anorg. Allgem. Chem.*, **329**, 172 (1964).
295. I. R. Beattie and M. Webster, *J. Chem. Soc.*, **1963**, 38.
296. H. Binas, *Z. Anorg. Allgem. Chem.*, **352**, 271 (1967).
297. R. J. H. Clark, *Spectrochim. Acta*, **21**, 955 (1965).
298. D. M. Adams, *Metal–Ligand and Related Vibrations*, Edward Arnold, London 1967.
299. I. R. Beattie, G. P. McQuillan, L. Rule and M. Webster, *J. Chem. Soc.*, **1963**, 1514.
300. I. R. Beattie and L. Rule, *J. Chem. Soc.*, **1964**, 3267.
301. R. C. Aggarwal and P. P. Singh, *J. Inorg. Nucl. Chem.*, **28**, 1651 (1966).
302. M. F. Farona and J. G. Grasselli, *Inorg. Chem.*, **6**, 1675 (1967).
303. R. A. Walton, *Spectrochim. Acta*, **21**, 1795 (1965).
304. R. A. Walton, *Can. J. Chem.*, **44**, 1480 (1966).
305. M. F. Farona, J. G. Grasselli and B. L. Ross, *Spectrochim. Acta*, **23A**, 1875 (1967).
306. M. Kubota and S. R. Schulze, *Inorg. Chem.*, **3**, 853 (1964).
307. M. Kubota and D. L. Johnston, *J. Inorg. Nucl. Chem.*, **29**, 769 (1967).
308. S. C. Jain and R. Rivest, *Can. J. Chem.*, **45**, 139 (1967).
309. S. C. Jain and R. Rivest, *Can. J. Chem.*, **43**, 787 (1965).
310. W. S. Orwille-Thomas and E. Wyn-Jones, *J. Chem. Soc.* (*A*), **1966**, 101.
311. M. Kubota, D. L. Johnston and I. Matsubara, *Inorg. Chem.*, **5**, 386 (1966).
312. T. Nomura and Y. Saito, *Bull. Chem. Soc. Japan*, **39**, 1468 (1966).
313. Y. Kinoshita, I. Matsubara and Y. Saito, *Bull. Chem. Soc. Japan*, **32**, 741 (1959).
314. Y. Kinoshita, I. Matsubara and Y. Saito, *Bull. Chem. Soc. Japan*, **32**, 1216 (1959).
315. M. Kubota and D. L. Johnston, *J. Am. Chem. Soc.*, **88**, 2451 (1966).
316. D. Hadzi, C. Klofutar and S. Oblak, *J. Chem. Soc.* (*A*), **1968**, 905.
317. T. Gramstad and S. I. Snaprud, *Acta Chem. Scand.*, **16**, 999 (1962).
318. D. E. McLaughlin, M. Tamres and S. Searles, *J. Am. Chem. Soc.*, **82**, 5621 (1960).
319. E. Grunwald and W. C. Coburn, *J. Am. Chem. Soc.*, **80**, 1322 (1958).
320. W. B. Person, *J. Am. Chem. Soc.*, **84**, 536 (1962).
321. T. L. Brown and M. Kubota, *J. Am. Chem. Soc.*, **83**, 331 (1961).
322. J. J. Wren and P. M. Lenthen, *J. Chem. Soc.*, **1961**, 2557.
323. I. Brown, G. Eglinton and M. Martin-Smith, *Spectrochim. Acta*, **19**, 463 (1963).
324. L. J. Bellamy and R. L. Williams, *Proc. Roy. Soc.* (*London*) *Ser. A*, **254**, 119 (1960).
325. M. Horák, J. Moravec and J. Plíva, *Spectrochim. Acta*, **21**, 919 (1965).

CHAPTER 4

Detection and determination of nitriles

DAVID J. CURRAN and SIDNEY SIGGIA

University of Massachusetts, Amherst, Massachusetts, U.S.A.

I. QUALITATIVE CHEMICAL METHODS 168
A. Hydrolysis 168
B. Reduction 170
C. Formation of Iminoalkylmercaptoacetic Acid Hydrochloride . 171
D. Reaction with Grignard Reagent 171
E. Reaction with Phloroglucinol 172
F. Test with Hydroxylamine Hydrochloride 172
G. Spot Tests 172
H. *N*-(Diphenylmethyl)amide Derivatives 173
II. QUANTITATIVE CHEMICAL METHODS 173
A. Hydrolytic Methods 173
B. Reduction Methods 175
C. Colorimetric Methods 177
III. SEPARATION TECHNIQUES 177
A. Distillation 177
B. Extraction 178
C. Liquid Chromatography 178
D. Gas Chromatography 178
IV. ELECTROCHEMICAL APPROACHES 179
A. Aromatic Nitriles 179
B. Aliphatic Nitriles 186
C. Polycyanocarbons 190
D. Miscellaneous 190
V. ULTRAVIOLET ABSORPTION 190
VI. INFRARED ABSORPTION 194
VII. MASS SPECTROMETRY 197
VIII. NUCLEAR MAGNETIC RESONANCE 201
IX. REFERENCES 202

The approaches for detection and determination of the nitrile group which are discussed in this chapter are as follows: chemical methods, separation techniques, electrochemical approaches, optical absorption, mass spectrometry and nuclear magnetic resonance. There is no preferred approach; one must adapt the sample and situation to the method which best suits the case.

I. QUALITATIVE CHEMICAL METHODS

A. Hydrolysis

Nitriles hydrolyse readily in both alkaline and acid media[1,2]. The

$$\mathrm{RC{\equiv}N} \xrightarrow[\mathrm{H^+}]{\mathrm{H_2O}} \mathrm{R\overset{O}{\overset{\|}{C}}NH_2} \xrightarrow[\mathrm{H^+}]{\mathrm{H_2O}} \mathrm{RC}\begin{smallmatrix}\diagup\!\!\!\!O\\ \\ \diagdown OH\end{smallmatrix} + \mathrm{NH_4^+}$$

$$\mathrm{RC{\equiv}N} \xrightarrow[\mathrm{OH^-}]{\mathrm{H_2O}} \mathrm{R\overset{O}{\overset{\|}{C}}NH_2} \xrightarrow[\mathrm{OH^-}]{\mathrm{H_2O}} \mathrm{RC}\begin{smallmatrix}\diagup\!\!\!\!O\\ \\ \diagdown O^-\end{smallmatrix} + \mathrm{NH_3}$$

detection of ammonia by its odour on treatment of the nitrile with sodium hydroxide is one method of detection[3]. However, it is not specific for nitriles since amides also yield ammonia, as seen in the above equations. Both amides and nitriles also yield carboxylic acids, so the drawbacks of the hydrolytic approach are obvious, especially since amides and nitriles often occur together, each being derivable from the other.

In some cases the hydrolysis can be stopped at the amide stage and the amide isolated and identified[1,4,5]. The procedure does not have too general a utility but can be applied when the amide precipitates.

Method[4]*:* 0·5 g sample is dissolved in a mixture of 2N sodium hydroxide and 5 ml of ethanol. 1 ml of 30% hydrogen peroxide is added, and the mixture is gently heated and well shaken. A vigorous evolution of oxygen takes place, and the amide, if insoluble, begins to precipitate in 10 to 30 min.

$$\mathrm{RCN} + 2\,\mathrm{H_2O_2} \longrightarrow \mathrm{R\overset{O}{\overset{\|}{C}}NH_2} + \mathrm{O_2} + \mathrm{H_2O}$$

The more general hydrolytic detection method is acid or base hydrolysis all the way to the acid. The acid is isolated and identified by its own chemical and physical properties or through its *p*-bromophenacyl ester[1,2,5]. Infrared identification of the acid is faster, but it is often not too discriminating between close homologues. x-Ray diffraction patterns on the sodium, or other salts, are usually more definitive.

Acid Hydrolysis Method[2]*:* About 25 ml of 75% sulphuric acid and 1 g of sodium chloride are heated to 150–160 °C under reflux and 5 g of the nitrile are added in 0·5 ml portions, with vigorous shaking after each addition. The mixture is heated at 160 °C with stirring for 30 minutes, and at 190° for another 30 minutes. It is then cooled and poured on 100 g of crushed ice and the precipitate, if any, is collected, treated with a slight excess of 10% sodium hydroxide and the insoluble amide filtered. The filtrate is acidified and the acid is recrystallized.

The authors of this chapter suggest that if the carboxylic acid does not precipitate, the acidic hydrolysis medium should be extracted with benzene, diethyl ether or petroleum ether. The free acid which is partially extracted can then be separated from the solvent by distillation, and addition of alcoholic sodium hydroxide will sometimes result in the precipitation of the sodium salt of the acid. Care should be taken to wash the organic layer fairly free of sulphuric acid before precipitation in order to avoid contamination of the product with sodium sulphate. However, if infrared absorption or x-ray diffraction are used to identify the sodium carboxylate, any sodium sulphate which is present can be readily distinguished. It is more important to avoid a large excess of NaOH since it may be precipitated in the organic phase. The excess sodium hydroxide may be removed by washing the precipitated sodium salts with a 50–50 alcohol–benzene mixture, if the solubility of the carboxylate in this mixture is low.

Alkaline Hydrolysis Method[1]*:* 1 g of the sample is treated with 5 ml 30–40% NaOH and heated under reflux. The odour of ammonia suggests a nitrile or primary amide. After sufficient time, usually $\frac{1}{2}$–2 h, the mixture is cooled, acidified and the free acid is isolated as above.

Revira and Palfray[6] used potassium hydroxide hydrolysis in a glycerine or diethylene glycol solvent. The higher boiling point of these solvents improve the results compared to those in water.

B. Reduction

Nitriles can be reduced to the corresponding amines, which can be isolated and identified directly or by making a solid derivative.

$$RC{\equiv}N \xrightarrow{H_2} RCH_2NH_2$$

Catalytic reductive hydrogenation has the advantage of avoiding other reagents in the system, but it can be unwieldy for analytical purposes. Sodium and alcohol as a reducing system is commonly used[7,8]. Lithium aluminium hydride has also been used[9]. The reduction was found to be quantitative for a few nitriles. Most nitriles did reduce to some degree but did not proceed to completion. The qualitative application was not followed up.

Method: Reduction of Nitriles to Amines and Conversion to Substituted Phenylthioureas[10]. In a dry flask, fitted with a reflux condenser, 20 ml of absolute ethanol and 1 g of an aliphatic nitrile or 2 g of an aromatic nitrile are placed. 1·5 g of finely cut sodium slices are added as rapidly as possible without causing the reaction to become too vigorous. When the reduction is complete (10–15 min), the mixture is cooled to 20 °c, and 10 ml of concentrated hydrochloric acid is added dropwise, with vigorous shaking of the mixture. The acidity of the reaction mixture is checked with litmus paper. About 20 ml of ethanol and water are distilled off, the mixture is cooled and 15 ml of 40 % sodium hydroxide solution is added drop by drop, with shaking. The reaction is vigorous, and care must be exercised to avoid adding the alkali too fast. At the end of the addition, the mixture is heated until the distillation of the amine is complete. Distillation is stopped when the remaining mixture becomes very viscous.

To the distillate 0·5 ml of phenyl isothiocyanate is added, and the mixture is shaken vigorously for 3–5 min. The precipitated phenyl thiourea derivative is collected, washed with a little cold 50 % ethanol and recrystallized.

If the phenyl thiourea is not a suitable derivative for particular amines or if more data are needed, the amine may be extracted or distilled from the reduction medium and identified. Neutral equivalent, infrared, mass spectral or nuclear magnetic resonance spectral determination of the amine can be used. x-Ray diffraction patterns of its chloroplatinate derivative[11] can also be used.

C. Formation of Iminoalkylmercaptoacetic Acid Hydrochloride

Nitriles react with mercaptoacetic (thioglycolic) acid to form the corresponding iminoalkylmercaptoacetic acid hydrochlorides[7,12].

$$RC{\equiv}N + HSCH_2COOH + HCl \longrightarrow RC\begin{matrix} /\!\!/ NH{\cdot}HCl \\ \backslash SCH_2COOH \end{matrix}$$

Method[7]*:* 0·5 g of nitrile is mixed with 1·0 g of thioglycolic acid and 7 ml of absolute ether in a test tube. The mixture is saturated with dry hydrogen chloride at an ice–salt temperature, the tube is stoppered and kept in the ice bath until crystals separate out. They may form after one hour for aliphatic nitriles, or may take as long as 24 h with aromatic nitriles. The crystals are collected, washed with absolute ether and dried in a vacuum desiccator.

These derivatives often do not give real melting points but decomposition points. Since this is not desirable, a neutral equivalent can be determined. In addition, infrared or x-ray (patterns) comparison with derivatives of known nitriles provide excellent identification information. If the derivative does not compare with any of the known derivatives, its nuclear magnetic resonance spectra may be of help in deducing its identity.

D. Reaction with Grignard Reagent

Nitriles add Grignard reagent; on hydrolysis the addition compound forms a ketone which can be identified by any of the standard means[7,12,13]. The use of the reaction as an analytic method is not

$$RC{\equiv}N + R^1MgX \longrightarrow \underset{\textstyle R^1}{\underset{|}{RC}}{=}NMgX \xrightarrow[HX]{H_2O} \begin{matrix} R \\ R^1 \end{matrix}\!\!>\!C{=}O + NH_4X + MgX_2$$

recommended, since so many compounds react with Grignard reagents. While the nitrile is one of the few to give a carbonyl compound, the reaction with the other reactive components in the sample generally consumes so much reagent that little, if any, is left for reaction with the nitrile. This test is excellent on an isolated, rather pure nitrile.

The ketone is identified by formation of its 2,4-dinitrophenylhydrazone, oxime or semicarbazone.

E. Reaction with Phloroglucinol

The Hoesch reaction is also used to identify nitriles. The nitrile reacts with phloroglucinol[12,14] to obtain the alkyl trihydroxyphenyl ketone which is usually a solid.

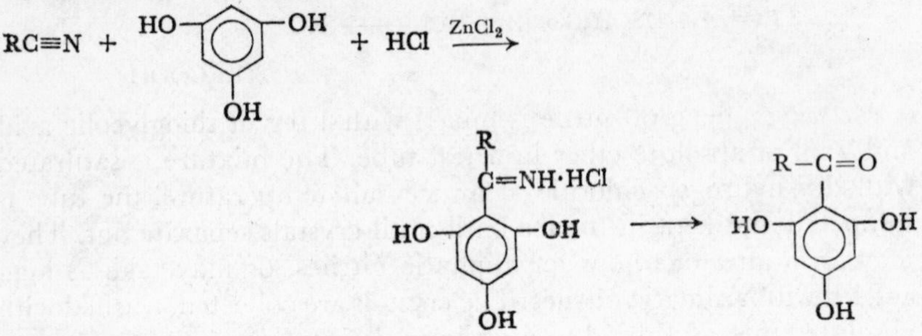

$$RC{\equiv}N + C_6H_3(OH)_3 + HCl \xrightarrow{ZnCl_2} (HO)_3C_6H_2C(R){=}NH{\cdot}HCl \longrightarrow (HO)_3C_6H_2C(R){=}O$$

F. Test with Hydroxylamine Hydrochloride

Nitriles normally do not give the hydroxamic acid test. However, under the following conditions, they do.

Method[15,16]: 30 mg of sample is dissolved in 2 ml of 1N hydroxylamine hydrochloride in propylene glycol. 1 ml of 1N potassium hydroxide in propylene glycol is also added, and the mixture is boiled for two minutes, and then cooled. 0·5–1·0 ml of 5% ferric chloride solution is added. A red to violet colour indicates nitriles.

Anilides can also give positive tests by the above method and hence can interfere. Interference from many other carboxylic acid derivatives such as esters, anhydrides and acid halides should also be suspected.

G. Spot Tests

When nitriles are fused with fluorescein chloride[17], a symmetrical dialkyl rhodamine which exhibits a yellow to yellow-green fluorescence is formed. Amides and primary amines generally interfere. During the test the nitrile is probably hydrolysed to the amide.

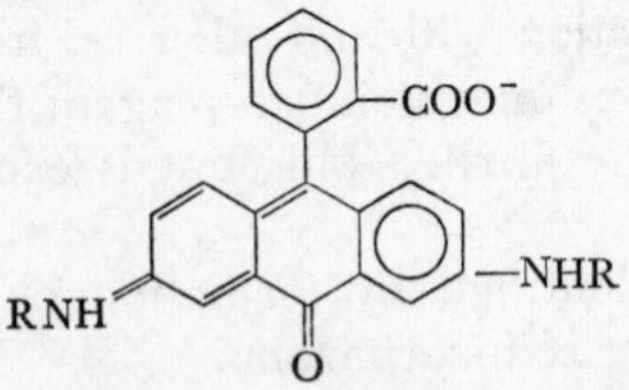

COO⁻

RNH —NHR

O

Another spot test approach involves pyrolysis of nitriles in the presence of sulphur, resulting in the production of thiocyanic acid[18]. This acid produces a red stain on a filter paper moistened with 1% acidified ferric chloride. Both aromatic and aliphatic nitriles give positive tests. However, false positive tests can be given by other nitrogen-containing organic materials.

H. N-(Diphenylmethyl)amide Derivatives[19]

Aliphatic and aromatic nitriles react with benzhydrol (Ph_2CHOH) in acetic acid to form the corresponding *N*-(diphenylmethyl)amide derivatives. The authors[19] used the melting point of the derivatives

$$RCN + Ph_2CHOH \longrightarrow RC(=O)NHCHPh_2$$

to identify the nitriles, but x-ray diffraction can also be used.

Method[19]: 1·0 to 2·0 ml of nitrile is mixed with 1 g of benzhydrol and 4 ml of acetic acid. 0·5 ml of concentrated sulphuric acid is added, the mixture is heated for 30 min at 60 °c and then allowed to stand at room temperature for several hours. The mixture is poured into 15 ml of ice water and the precipitate formed is filtered off, washed with water and recrystallized from ethanol.

II. QUANTITATIVE CHEMICAL METHODS

The quantitative chemical methods for determining nitriles revolve mainly around the hydrolysis reaction. Attempts have been made also to use the reduction reaction to the amine, but only limited success has been obtained.

A. Hydrolytic Methods

As discussed in section I.A, nitriles can be hydrolysed by strong acid or alkalis. However, the concentration of acids or base must usually be so high that the analysis cannot be performed by titration of the excess acid or base. The hydrolysis can be carried out with strong caustic, and the ammonia formed can be then measured quantitatively by Kjeldahl digestion. Siggia, Hanna and Serencha[20] used this approach in some of their kinetic work. Amides will interfere, since they will react quantitatively by the same method. The direct hydrolysis approach should not be considered general since some nitriles do not hydrolyse quantitatively to acid and ammonia, probably due to side reactions.

Method[20]*:* A sample containing approximately 1·6 mequiv. is placed in a Kjeldahl distillation flask along with 200 ml of water and 10 ml of tetrahydrofuran, and 10 ml of 50% aqueous sodium hydroxide solution is added. The mixture is heated electrically just enough to obtain smooth boiling, and nitrogen is passed through the system at a rate of about 2 bubbles per second. The distillate is received in 100 ml of 4% boric acid solution which has been adjusted to pH 4·00. After the distillation is complete, the boric acid is titrated back to pH 4·00 using 0·04N hydrochloric acid.

By measuring the rate of evolution of ammonia, Siggia, Hanna and Serencha were able to determine mixtures of nitriles. Successful results were obtained with acetonitrile, butyronitrile, valeronitrile, benzonitrile, *p*-anisonitrile and binary mixtures of these nitriles.

Whitehurst and Johnson[21] broadened the hydrolytic approach by first hydrolysing the nitrile to the amide using hydrogen peroxide,

$$RC{\equiv}N + H_2O_2 \longrightarrow RC\begin{matrix}{/\!\!/}O \\ \diagdown NH_2\end{matrix} + H_2O + O_2$$

and then hydrolysing the amide with potassium hydroxide. However, the hydrogen peroxide also oxidizes aldehydes to the acids, and since this approach measures base consumed, acid amides and esters will interfere. A separate analysis can be applied to correct for these groups. Benzonitrile, acrylonitrile, ethylene cyanohydrin and 3-methylpropionitrile gave high results. Acetonitrile, propionitrile, butyronitrile and succinonitrile were analysed successfully.

Method for Relatively Pure Nitriles[21]*:* To two alkali resistant flasks 50 ml of 1N potassium hydroxide and 100 ml of 3% hydrogen peroxide are added. One solution is used as a blank. A sample containing 6 to 10 mequiv. of nitrile is added to the other solution and the samples are allowed to stand at room temperature for 5 min. A few glass beads are then added (boiling chips cause caustic consumption). The flasks are then heated and the solution distilled until a volume of approximately 10 ml remains. The flasks are cooled and the distillation systems are washed with about 100 ml of distilled water. 50 ml of 0·5N sulfuric acid is added to each flask. The contents are titrated with 0·5N sulphuric acid to the disappearance of the pink colour of phenolphthalein indicator. The blank is subtracted from the sample titration.

Method for Low Concentration of Nitriles in Water[21]*:* This method is the same as above except for the use of 25 ml of 0·2N potassium hydroxide, 20 ml of peroxide and 200 ml of sample. Another difference is that the end titration uses 0·1N sulphuric acid.

This method was used down to 5 p.p.m. concentration levels of nitrile.

Mitchell and Hawkins[22] described a hydrolytic method whereby they introduce a known quantity of water, then hydrolyse the nitrile and determine the excess water with Karl Fischer reagent. Successfully used were the following nitriles: acetonitrile, propionitrile, butyronitrile, valeronitrile, adiponitrile, sebaconitrile, phenylacetonitrile, *m*-toluonitrile, *p*-toluonitrile, *p*-chlorobenzonitrile and *β*-naphthonitrile. *o*-Toluonitrile, α-naphthonitrile, methyleneaminoacetonitrile and cyanogen gave low results. Amides appear to interfere only slightly. Apparently this hydrolysis of nitriles proceeds only to the amide stage and no further.

Method[22]*:* To a sample containing up to 10 mequiv. of nitrile, 20 ml of hydrolysis reagent is added. The hydrolysis reagent consists of 300 g of dry boron trifluoride gas and 6·5 ml of water dissolved in 500 ml of glacial acetic acid. The mixture is kept at 80 ± 2 °C for 2 h, cooled spontaneously to room temperature and then cooled further with chopped ice while 15 ml of dry pyridine are added. The solution is then titrated with Karl Fischer reagent.

B. Reduction Methods

Reduction of nitriles to the corresponding amines has not led to very general quantitative chemical methods. Only two approaches have reported any success, and these are limited. Siggia and Stahl[23] used lithium aluminium hydride and successfully determined benzonitrile, butyronitrile, *N*-capronitrile and *p*-chlorobenzonitrile. However, low results were obtained with acetonitrile, acrylonitrile, succinonitrile, adiponitrile, phenylacetonitrile, 3-butenonitrile, γ-phenoxybutyronitrile, lactonitrile, *m*-nitrobenzonitrile and l-naphthonitrile.

Huber[24] used catalytic hydrogenation of the nitrile to the amine (in a non-aqueous solvent) followed by titration of the amine. Glacial acetic acid is used as solvent and platinum oxide as catalyst. Good results were obtained with aceto-, benzo-, adipo- and succinonitriles. However, Siggia and D'Andrea[25], using a similar approach, had success with only benzonitrile using palladium on charcoal as catalyst. Evidently, the catalyst is critical in the feasibility of this method. It is advisable to check the method and available catalyst for each nitrile on which it is attempted. Amides do not generally interfere.

Hydride Method[23]*:* An exactly weighed sample containing approximately 0·0006 moles of nitriles is placed in a Kjeldahl flask and 5 ml of lithium aluminium hydride reagent are added. The lithium aluminum hydride reagent consists of 10 g of fresh, dry hydride, refluxed with 500 ml of anhydrous ethyl ether for several hours. The insoluble portions of the reagent settle rapidly on cooling and clear solution can be pipetted from the supernatant liquid. The solution, if protected from atmospheric moisture, is usable for about one month.

The sample is allowed to react with the reagent for 15 min at room temperature, and the flask is attached to a Kjeldahl distillation apparatus. Water is added dropwise to the reaction mixture until the excess hydride is decomposed. 10 ml of 6N sodium hydroxide are then added and the amine is steam distilled into a receiver containing 50 ml of 0·02N sulphuric acid. About 50 ml of distillate are collected and the excess acid is titrated with 0·02N sodium hydroxide to the green end-point of methyl purple indicator.

In the cases where high-boiling amines, which do not steam distil very well in the above aqueous system, are obtained, ethylene glycol is added. After the excess hydride reagent is decomposed, the flask is washed with water followed by 6N sodium hydroxide and 25 ml of ethylene glycol. In this case, the regular Kjeldahl distillation apparatus is not used, but instead an arrangement which permits constant addition of ethylene glycol to prevent sodium hydroxide from splashing into the receiver[23].

The solution is evaporated rapidly almost to dryness, and 25 ml portions of ethylene glycol are added slowly so that boiling does not stop. The addition of ethylene glycol is continued until 100 ml have been distilled. The distillate is titrated potentiometrically with 0·02N sulphuric acid.

Hydrogenation Method[24]*:* 10 ml of glacial acetic acid and 100 mg of catalyst are placed in an hydrogenation apparatus and the catalyst is saturated with hydrogen. A sample of about 0·5 to 1·0 mmoles of nitrile is introduced as a solution in glacial acetic acid. The hydrogenation is carried out until hydrogen is no longer consumed. The solution obtained is titrated with 0·1N perchloric acid in glacial acetic acid either potentiometrically or with Napthol benzein or crystal violet indicators.

Huber[23] used a specific hydrogenation apparatus, which consists merely of a reaction vessel equipped with a dropping funnel arrangement for the introduction of solutions. This reaction vessel is attached to a gas burette for monitoring the hydrogen absorption. Other hydrogenation apparatus can be used equally well and modifications to the method can be applied.

C. Colorimetric Methods

There appear to be no general colorimetric methods for the nitriles as a class. However, there are colorimetric tests for specific types of nitriles or for specific, individual compounds. Szewczvk[26] determined α-aminonitriles by a complex reaction involving bromine and a benzidine–pyridine reagent. Ashworth and Schupp[27] used the Janovsky reaction to determine phenylacetonitrile colorimetrically.

III. SEPARATION TECHNIQUES

A. Distillation

Since many nitriles have sufficient vapour pressure, conventional distillation provides one method of separation. However, there are also some rather unconventional distillation approaches which have been used. Meissner[28] described a method for separating nitriles from water effluents or from complex mixtures using a 'cascade distillation'. The method involves successive distillation from solutions at different pH to remove specific fractions.

Valentine[29] describes an 'extractive' distillation to separate *m*-dicyanobenzene from *m*-diaminomethylbenzene. In this case, glycerol or 1,2,6-hexanetriol is added for a codistillation. *m*-Cyanobenzylamine was also separated from *m*-diaminomethylbenzene by a similar process.

Maslan and Robertson[30] used a combination of extraction and extractive distillation to separate acrylonitrile, acetonitrile and propionitrile from each other when present in an aqueous solution.

Concentration of components, though not complete separation, is achieved by extraction. The extracts are then subjected to distillation in a column into which water is introduced at the top. Resolution of the acetonitrile occurs at this stage while the acrylonitrile and propionitrile are concentrated. The latter two are then separated by fractional distillation.

B. Extraction

A method was described[31] by which low molecular weight nitriles, obtained by cracking of high molecular weight nitriles, were separated by using extraction with aqueous solutions of alcohols, phenols, amines or carboxylic acids. Specific alcohols used were methanol, pentanol, cyclohexanol and allyl alcohol. Specific amines were aniline, ethylamine, diethylamine, allylamine, diallylamine, benzylamine, pyridine and cyclohexylamine. Acids used were acetic, acrylic, valeric and caproic.

Ralston and Pool[32] used similar extractions on cracked high molecular weight nitriles, except that they extracted the resulting hydrocarbon–nitrile system with aqueous aliphatic alcohol solutions.

C. Liquid Chromatography

Pool[33] cracked stearonitrile and obtained a mixture of aliphatic nitriles and aliphatic hydrocarbons which he resolved on a silica gel adsorbent.

Acetonitrile was separated from C_4 hydrocarbons using synthetic zeolites. At 20 °c and 4–5 atmospheres pressure, using the sodium form of the zeolite, 80–90 % recovery was achieved[34].

D. Gas Chromatography

Since many nitriles are volatile, gas chromatography provides a convenient way for their detection and determination. Caution should be exercised in the frequent cases where nitriles are accompanied by amides. Primary amides can dehydrate in the injection ports or on the columns, to form the corresponding nitriles and more nitrile than was present originally in the sample will be detected. In addition, nitriles can hydrolyse in the heated equipment if water is present in the sample; this would yield low results.

The gas chromatographic literature consists almost exclusively of operating parameters. This is summarized in Table 1.

IV. ELECTROCHEMICAL APPROACHES

The electrochemistry of cyano compounds is varied and interesting. One review on the electrochemical reduction of nitriles has been published[49]. It appears that all types of nitriles can be reduced electrochemically under the proper conditions. Among the factors which determine the products of electrolysis are: the type of cyano compound involved, the nature of the solvent employed and the nature of the electrode used. By way of illustration, acetonitrile is a well-known solvent for polarographic and related electrochemical studies. Very negative potentials may be attained using tetraalkylammonium salts as supporting electrolytes. In general, saturated aliphatic nitriles are regarded as being electroinactive under polarographic conditions, but several instances will be cited later where acetonitrile has been electrochemically reduced to ethylamine in aqueous media using electrodes other than mercury.

A. Aromatic Nitriles

The subject of the reduction of aromatic nitriles is in a state of some confusion but Manousek and Zuman[50] have proved by wave height studies, controlled potential coulometry and u.v. spectra that *p*-acylbenzonitriles, *p*-$RCOC_6H_4CN$, where R is H or CH_3, undergo reduction at mercury cathodes to the corresponding benzylamines in 0·1M H_2SO_4. Polarographic reduction occurred at −0·70 V versus SCE and one well-defined wave was observed. The reduction scheme was written as in equation (1), i.e. it was proposed

$$R{-}\overset{O}{\overset{\|}{C}}{-}C_6H_4{-}CN + H^+ \rightleftharpoons R{-}\overset{O}{\overset{\|}{C}}{-}C_6H_4{-}C\overset{+}{N}H \xrightarrow[4H^+]{4e^-} R{-}\overset{O}{\overset{\|}{C}}{-}C_6H_4{-}CH_2NH_2 \quad (1)$$

that protonation occurred at the nitrogen of the cyano group. The wave height was limited by the rate of protonation and a decrease in wave height was found at pH values greater than 1·5. Compounds of the type XC_6H_4CN, where X = *p*-COO^-, *p*-SO_2NH_2 and *m*-SO_2NH_2, were studied using the same techniques but in neutral solution. The existence of free cyanide ion in solution after controlled potential electrolysis was established by anodic polarography

TABLE 1. Gas chromatography conditions for analysis of nitriles.

Column packing	Column parameters	Inject port temp. (°c)	Column temp. (°c)	Carrier gas	Flow rate	Detector	Nitriles tested	Reference
Carbowax 4000 monostearate on Firebrick C22 (30–60 mesh)	—	—	220–235	He	52 ml/min	—	C_8–C_{20}	35
Sterchamol grains with Apiezon L	80 cm × 4 mm	—	265	N_2	174 ml/min	—	Higher aliphatic	36
Kieselguhr 0·2–0·3 mm with Apiezon L	3·3 m × 4 mm	350–400	275	H_2	50–100 ml/min	—	C_{12}–C_{18}	37
Polyethylene glycol or glycerol	—	—	—	—	—	—	—	38
Diethylene glycol polyester adipate with 2% H_3PO_4 or Chromosorb W (30–60 mesh)	2 m × 6 mm	—	70–220 (programmed and un-programmed)	—	—	Hot wire thermal conductor	Acrylo-, propio-, butyro-, succino-, adipo-.	39
20% Polyester and H_3PO_4 (85%) 3% on Chromosorb W (60–80 mesh)	—	—	185–220	—	—	—	C_{10}–C_{22}	40

LAC (2) 25% and H_3PO_4 3%	1·2 m	—	207	He	50 ml/min	—	C_{14}–C_{16}	41
Na caproate and Apiezon L on Celite 22a	—	—	210	—	—	—	C_4–C_{10} dinitriles	42
—	1·5 and 3·3 mm	—	'high temp.'	H_2	—	—	—	43
15% Silicone wax on diatomite	2·8 m × 0·6 mm	—	—	H_2	—	Thermal conductor	C_7–C_{20}	44
10% Paraffin wax on Haloport F	5 m × 4 mm	—	70	H_2	60 ml/min	—	Aceto-, acrylo-, proprio-, crotono-, metha-crylo-	45
10% liquid phase on INZ Brick 0·25–0·50 mm	4 m × 6 mm	—	—	He	80 ml/min	—	Aceto and others	46
15% Pentaerythritol stearate on Chromosorb W (30–60 mesh)	300 × 0·4 mm	—	170	He	—	Hot wire	Ethyl cyano-acetate	47
2% Versamid (900)	3 m	308	100–280 programmed	He	129·2 ml/min	Thermal con-duction	Nitrilotris-acetonitrile, iminodi-acetonitrile	48

and chemical tests. This, together with the polarographic and controlled potential coulometric determination of the number of electrons involved in the reduction, leads to equation (2) for the electrochemical reaction:

$$XC_6H_4CN + 2e^- + H^+ \longrightarrow XC_6H_5 + CN^- \quad (2)$$

The work of Manousek and Zuman is in accord with earlier observations on the electrochemical reduction of aromatic nitriles which were carried out using the classical technique of constant current electrolysis. In this case, the current is held more or less constant by manual adjustment of the voltage applied to the cell. Since there is no external control of the potential of the working electrode, this technique is neither very selective nor very efficient. It seems likely that in most cases where this technique is applied in aqueous solutions, hydrogen is also evolved at the electrode and may have an influence on the products of electrolysis. The earliest report is that of Ahrens[51], who isolated benzylamine from the electrochemical reduction of benzonitrile in acid solution at a lead electrode. Similar results were obtained by Ogura[52], not only in sulphuric acid, but in solutions of sodium hydroxide and of ammonium sulphate. The yield, based on product, varied from 24 % in 2·5 % H_2SO_4 to 49·5 % in 7·5 % $(NH_4)_2SO_4$. *p*-Methylbenzonitrile produced a 57 % yield of the corresponding amine in the latter electrolyte. Lead amalgam, zinc amalgam, mercury and platinum electrodes were also employed with no increase in yield. Ohta[53] and Sawa[54] reported the reduction of benzonitrile in hydrochloric acid solution at palladium-plated platinum, and Ni–Pd alloy (5 % Ni) electrodes, respectively. The product yield in the latter case was 96 %. A plausible explanation for the high yield was provided by Ohta, who suggested that intermediate aldimines may be formed in the course of the electrochemical reduction, and reaction of these with the primary amine to produce secondary amine (a side-product) would be suppressed by the presence of HCl, which would bind the primary amine as the amine hydrochloride.

Further evidence for the generality of the reduction of nitriles with aromatic character to amines is obtained from the work of Matukawa[55], who isolated 2-methyl-4-amino-5-aminomethyl-pyrimidine from the electrochemical reduction of the corresponding 5-cyano compound at a palladium-coated platinum electrode in hydrochloric acid solution. The results of Manousek and Zuman are also analogous to the results of a number of investigations for the polarographic reduction of 2- and 4-cyanopyridine. Laviron[56] and Volke, Kubicek and Santavy[57] found a four-electron irreversible polarographic reduction in acid solution and a two-electron irreversible reduction in basic solution. Volke and Holubek[58] later reinvestigated the reduction of 4-cyanopyridine in alkaline solution and confirmed the production of cyanide ion and a two-electron process in solutions buffered at pH 10–11. 2-Cyanopyridine undergoes alkaline hydrolysis at very high pH and Laviron[56] determined a rate constant of $0{\cdot}70 \times 10^{-3}$ l/mole s by following the disappearance of the polarographic reduction wave of 2-cyano-pyridine. Jarvie, Osteryoung and Janz[59] developed a polarographic method of analysis for 2-cyanopyridine. Both the half-wave potential and the diffusion current were functions of pH. For best results, it was recommended that the pH be controlled in the region of 5–6. Very good accuracy and precision were reported. Unfortunately, very acidic supporting electrolytes were not investigated, but it is possible that greater sensitivity could be attained in more acidic solutions.

The polarographic reduction behaviour of aromatic nitriles in non-aqueous solvents is very different from that observed in aqueous solutions. Sevast'yanova and Tomilov stated earlier that aromatic nitriles are not reduced at the dropping mercury electrode[60] (DME) but later reported[61] that benzonitrile in *N,N*-dimethylformamide (DMF) with 0·05M tetrabutylammonium iodide as the supporting electrolyte reduced at the DME in a one-electron step which was independent of any added proton donor (H_2O). No dimeric or polymeric products were found and 2,3-dihydrobenzonitrile was identified. The process is described in equations (3)–(4). An extensive

$$C_6H_5\text{—CN} + 1e^- \rightleftharpoons [C_6H_5\text{—CN}]^{\overline{\cdot}} \qquad (3)$$

$$2\,[C_6H_5\text{—CN}]^{\overline{\cdot}} + 2\,H^+ \longrightarrow C_6H_5\text{—CN} + C_6H_7\text{—CN} \qquad (4)$$

study on the polarography of aromatic nitriles in DMF is found in the work of Rieger and coworkers[62], who used d.c. polarography, triangular wave polarography (cyclic voltammetry) and electron spin resonance to examine the electrolytic generation of nitrile radicals. Polarographic data are shown in Table 2. Benzonitrile was reduced reversibly to the radical anion. Cyclic voltammetry

TABLE 2. Polarographic data for reduction of nitriles[62].

Compound	First wave $E_{\frac{1}{2}}{}^{a}$ (v)	n^{b}	Second wave $E_{\frac{1}{2}}{}^{a}$ (v)	n^{b}
Benzonitrile	−2·74	1	—	—
Phthalonitrile	−2·12	1	−2·76	2
Isophthalonitrile	−2·17	1	—	—
Terephthalonitrile	−1·97	1	−2·64 −2·74	1[c] 1[c]
Pyromellitonitrile	−1·02	1	—	—
4-Toluonitrile	−2·75	1	—	—
4-Aminobenzonitrile	−3·12	$\frac{3}{2}$	—	—
4-Fluorobenzonitrile	−2·69	$\frac{3}{2}$	—	—
4-Cyanopyridine	−2·03	1	−2·87	1
4-Nitrobenzonitrile	−1·25	1	−1·9[e]	2
3,5-Dinitrobenzonitrile	−0·96	1	−1·5[e,f]	1
4-Cyanobenzoic acid	−1·91	1	−2·53	1
Tetracyanoquinodimethane	−0·19[d,g]	—	−0·75[d,g]	—
Tetracyanoethylene	−0·17[d,g]	—	−1·17[d,g]	—
Tetramethylammonium 1,1,2,3,3-pentacyanopropenide	−1·76	1	−2·30[h]	1
Sodium 1,1,3,3-tetracyano-2-dimethylaminopropenide	−2·75[d]	—	?	—
4-Chlorobenzonitrile	−2·4[d,e]	—	—	—
4-Anisonitrile	−2·95[d]	—	—	—
Benzoylacetonitrile	−2·09[d]	—	—	—
Benzoyl cyanide	−1·45[d,e]	—	—	—
Sodium 1,1,3,3-tetracyano-2-ethoxypropenide	−2·25[d,e]	—	−2·40[d,i]	—

[a] Potentials measured in volts versus Ag–$AgClO_4$ electrode.

[b] Estimated number of electrons in the reduction as obtained from qualitative observations rather than careful measurements.

[c] Two waves poorly separated.

[d] Because of drift in the potential of the Ag–$AgClO_4$ reference electrode, these potentials may be inconsistent with the other polarographic data.

[e] Wave obscured by maximum.

[f] Three-electron wave at −2·1 v.

[g] Polarogram run in acetonitrile.

[h] Third one-electron wave at −2·75 v.

[i] Third wave at −2·75 v obscured by maximum.

did not indicate that the reduction was complicated by chemical kinetic factors. Evidence for reaction (4) might be obtained by this technique by addition of water to the electrolyte. A second polarographic wave observed in a number of cases was attributed most frequently to further reduction of the radical anion to the dianion.

This was the case for phthalonitrile and the experimental evidence supported the following reaction scheme (equations 5–8):

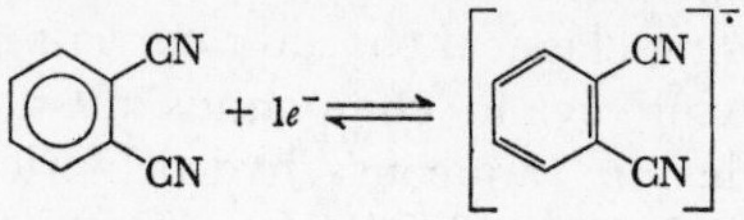

$$o\text{-}C_6H_4(CN)_2 + 1e^- \rightleftharpoons [o\text{-}C_6H_4(CN)_2]^{\cdot -} \qquad E_{\frac{1}{2}} \quad -2{\cdot}12\ \text{v} \tag{5}$$

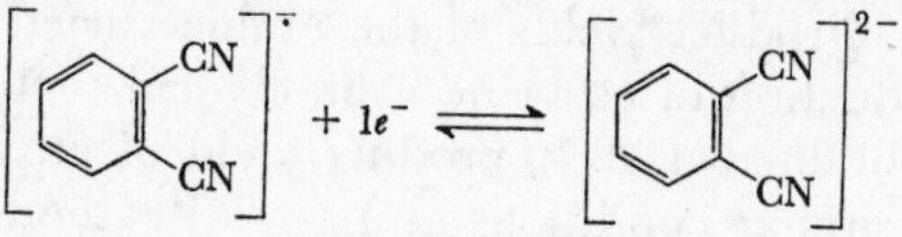

$$[o\text{-}C_6H_4(CN)_2]^{\cdot -} + 1e^- \rightleftharpoons [o\text{-}C_6H_4(CN)_2]^{2-} \qquad E_{\frac{1}{2}} \quad -2{\cdot}76\ \text{v} \tag{6}$$

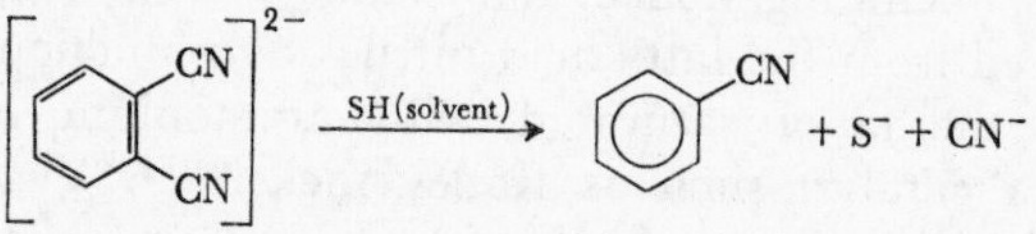

$$[o\text{-}C_6H_4(CN)_2]^{2-} \xrightarrow{\text{SH (solvent)}} C_6H_5CN + S^- + CN^- \tag{7}$$

$$C_6H_5CN + 1e^- \rightleftharpoons [C_6H_5CN]^{\cdot -} \qquad E_{\frac{1}{2}} \quad -2{\cdot}74\ \text{v} \tag{8}$$

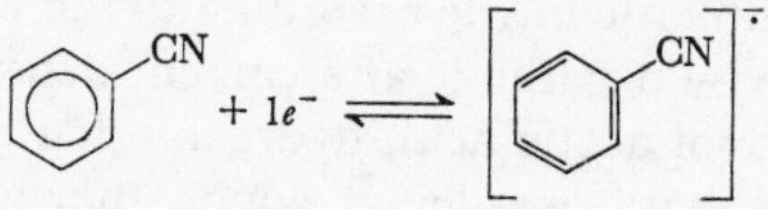

In the case of 4-aminobenzonitrile, the following sequence of reactions was given (equations 9–11):

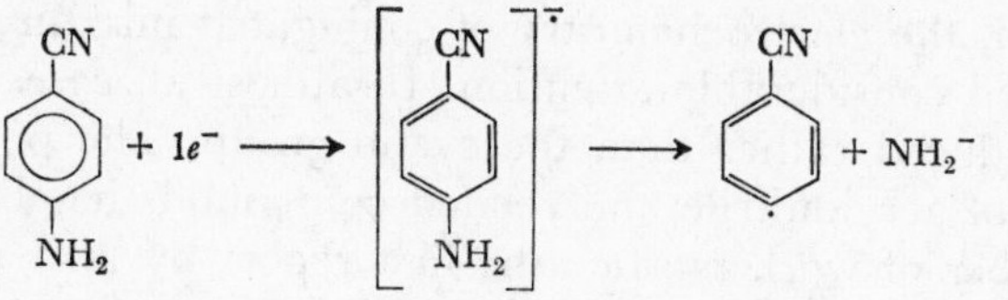

$$p\text{-}NH_2C_6H_4CN + 1e^- \longrightarrow [p\text{-}NH_2C_6H_4CN]^{\cdot -} \longrightarrow p\text{-}{}^{\cdot}C_6H_4CN + NH_2^- \tag{9}$$

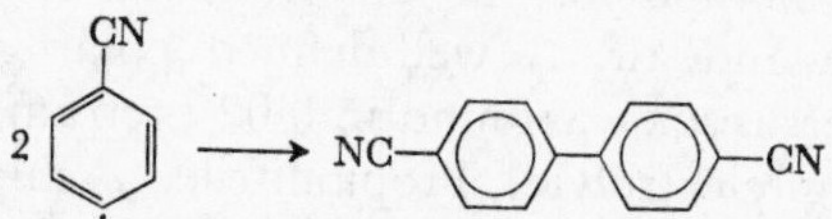

$$2\ p\text{-}{}^{\cdot}C_6H_4CN \longrightarrow NC\text{-}C_6H_4\text{-}C_6H_4\text{-}CN \tag{10}$$

$$NC\text{-}C_6H_4\text{-}C_6H_4\text{-}CN + 1e^- \rightleftharpoons [NC\text{-}C_6H_4\text{-}C_6H_4\text{-}CN]^{\cdot -} \tag{11}$$

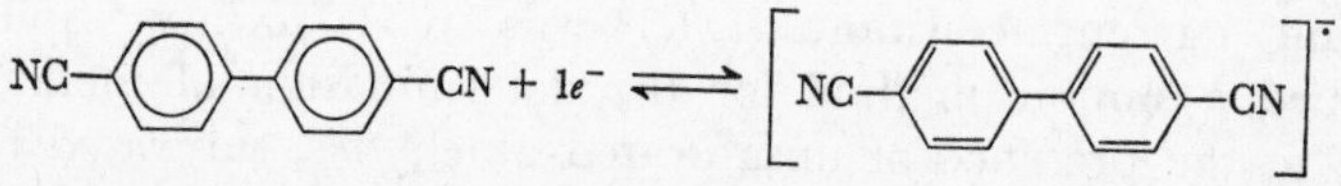

Evidence for the formation of the radical anion was found for most of the compounds studied. As shown above, those compounds having a second reduction wave display very complicated behaviour, depending on the fate of the dianion.

B. Aliphatic Nitriles

Saturated aliphatic nitriles have been reduced to the corresponding amines using classical constant current electrolysis. Ohta[53] reduced the following cyano compounds at a palladium-plated platinum electrode in aqueous hydrochloric acid: CH_3CN, $NCCH_2CN$, $NC(CH_2)_2CN$, $NC(CH_2)_4CN$, $NCCH_2CH_2OH$ and $NCCH_2COOH$. Product yields of the amine varied from a low of 30 % for the reduction of ethylene cyanohydrin to 100 % for acetonitrile. Sawa obtained a 96 % product yield of ethylamine hydrochloride under similar conditions at a Ni–Pd cathode[54]. Some fatty acid amines, including dodecyl-, hexadecyl- and octadecyl-, have been prepared from the fatty acid nitriles by reduction at a copper cathode in methanol saturated with ammonium chloride and containing a catalyst such as Raney nickel[63,64]. Yields of up to 85 % were found at a current density of 5·0 A/dm^2. Benzyl cyanide is reduced as an aliphatic nitrile according to the work of Kawamura and Suzuki[65]. A Ni–Pd cathode at a current density of 2 A/dm^2 was used in a catholyte of acetic acid, hydrochloride acid and water.

Unsaturated aliphatic nitriles in which the double bond is not in conjugation with the cyano group have been reported to be inactive under polarographic conditions[66]. They do not seem to have been studied under other electrochemical conditions. On the other hand, the electrochemistry of conjugated unsaturated nitriles has received considerable attention. In almost all cases the double bond is reduced rather than the cyano group. The polarographic reduction of acrylonitrile and related compounds has been studied by a number of workers since the first report by Bird and Hale[67]. They devised a method for the determination of acrylonitrile in water, butadiene and air. A well-defined polarographic wave at $E_{\frac{1}{2}} = -2{\cdot}05$ V versus SCE was found in 0·02M tetramethylammonium iodide supporting electrolyte. Propionitrile, β-chloropropionitrile, ethylene cyanohydrin and cyanoacetamide did not interfere. The diffusion current constant (i_d/C) was 8·54 μA/mM. Spillane[68] described a similar method for the determination of methacrylonitrile in the presence of unsaturated aldehydes, saturated nitriles and water. In 0·1N $(CH_3)_4NBr$ as the supporting electrolyte, the half-wave potential was $-2{\cdot}07$ V versus SCE. α,β-Unsaturated aldehydes did not interfere since their reduction is at more negative potentials than that of the nitrile wave. A precision of about 3 %

was reported. Oxygen was not removed from the solution prior to reduction and the diffusion current was proportional to concentration, but did show some time dependence. Clearly, acrylonitrile cannot be distinguished from methacrylonitrile polarographically. Strause and Dyer[69] extended the work of Bird and Hale to the analysis of acrylonitrile in the presence of potassium persulphate, hydrogen cyanide, formaldehyde and other products resulting from the persulphate-initiated attack of oxygen on acrylonitrile. A calibration technique was used because of interferences. Solutions of acrylonitrile from 5×10^{-4} to 1×10^{-3} mM were analysed. Daves and Hamner[70] introduced a separation step for the determination of acrylonitrile by the method of Bird and Hale. Azeotropic distillation of acrylonitrile with methanol was effective for a large variety of water samples containing large numbers of interferences. The reduction behaviour is little affected by adding ethyl alcohol to the solvent. Platonova[71] reported a half-wave potential of −2·05 v in alcohol–water solvents using 0·01M tetraethylammonium chloride supporting electrolyte. The reduction was described as a two-electron irreversible process. The diffusion current was proportional to concentration over the range 0·3 to 12·04mM.

Claver and Murphy[72] determined residual acrylonitrile in styrene–acrylonitrile copolymer by polarography in DMF containing 5% water and 0·1M tetrabutylammonium iodide. The half-wave potential was −1·63 v versus a mercury pool reference electrode. 30 to 200 p.p.m. of acrylonitrile were determined with an accuracy of 3·6%. Crompton and Buckley[73] extended the limit of detection in this solvent system to two p.p.m. Murphy and coworkers[74] studied the reduction of acrylonitrile in anhydrous DMF and concluded that condensation reactions were taking place. A more extensive study[60,61,75] included a comparison of the polarographic and linear sweep voltammetric (oscillopolarographic) behaviour of a number of α,β-unsaturated nitriles at mercury electrodes in aqueous solutions, dry DMF and DMF with added water. In aqueous solutions and in DMF when the concentration of acrylonitrile was less than about 0·7mM, the reduction yielded a single two-electron wave. However, in DMF, as the concentration of acrylonitrile is increased, the limiting current begins to decrease. Further, the limiting current in DMF may be increased upon addition of water. Apparently, the product of electrochemical reduction depends to a large extent on the proton donor properties of the solvent. Sevast'yanova and Tomilov proposed the following scheme for the reaction (equations 12–13):

$$RCH{=}CHCN + 2\,e^- \longrightarrow RCHCHCN^{2-} \qquad (12)$$

$$RCHCHCN^{2-} + 2\,H^+ \longrightarrow RCH_2CH_2CN \qquad (13)$$

where reduction to the dianion is the rate-determining step. This conclusion is in agreement with the work of Lasarov, Trifonov and Vitanov[76] on acrylonitrile in DMF using polarography, microcoulometry and chronopotentiometry. These workers appear to have extended their current–voltage curves to more negative potentials beyond the reduction wave for acrylonitrile than others, and observed that in more concentrated solutions, the acrylonitrile wave began to split into two waves, both of which were irreversible. The limiting current for the first wave was diffusion controlled and proportional to the nitrile concentration, while that for the second wave was kinetically controlled and not linearly dependent on the concentration. The height of the first wave was nearly independent of any added water but the second wave showed a non-linear rise which seemed to approach a limiting value in the region of 1 to 10% water added. On the basis of i.r. spectra of the electrolysis solutions, it was proposed that, in the absence of water, polymerization took place through reaction of the dianion with a molecule of acrylonitrile. In the presence of water, the product evidently depends on the relative amounts of unreacted nitrile and water. Lasarov and coworkers suggested that protonation occurred according to equation (13), but at high concentrations of acrylonitrile, hydrodimerization, i.e. coupling at the β position and acquisition of hydrogen at the α position (equation 14), was proposed by

$$RCHCHCN^{2-} + RCH{=}CHCN + 2\,H^+ \longrightarrow \begin{array}{l} RCHCH_2CN \\ \;| \\ RCHCH_2CN \end{array} \qquad (14)$$

Sevast'yanova and Tomilov[60]. Baizer[77,78] and Baizer and Anderson[79,80] have studied the electrolysis of concentrated solutions of acrylonitrile in aqueous tetramethylammonium *p*-toluenesulphonate at lead and mercury cathodes. Adiponitrile was produced in nearly 100% yield and at nearly 100% current efficiency if the solution pH was controlled. If the concentration of acrylonitrile was below 10%, or the supporting electrolyte contained alkali metal cations, increasing amounts of propionitrile appeared as product. Using the same supporting electrolyte in aqueous DMF, Wiemann and Bouguerra[81] reported hydrodimerization of crotononitrile to β,β'-dimethyladiponitrile.

The electrochemical reduction mechanism of α,β-unsaturated nitriles is changed if an electron-acceptor group is present in conjugation with the double bond. In aqueous solution, cinnamonitrile and vinylacrylonitrile[60,61,75,82] show a single polarographic wave, but in anhydrous DMF, two one-electron steps are observed[60,61,75]. The reduction proceeds through a stable anion radical. The product formed in the presence of water depends on the stability and rate of protonation of the intermediate ion radical formed. Electrolysis of cinnamonitrile at a mercury cathode in anhydrous DMF and i.r. spectra of the resulting solutions produced evidence for the dimeric 2,3-diphenyladiponitrile, while vinylacrylonitrile showed polymerization under the same conditions. As water was added, the latter compound showed a decreasing tendency to form higher polymers and a dimer, probably 2,2′-dicyanobicyclobutane, formed. Kaabak, Tomilov and Varshavskii[83] used classical constant current electrolysis to reduce vinylacrylonitrile at lead-, zinc-, tin- and mercury-plated graphite electrodes. The ratio of the products obtained, i.e. 3- and 4-pentenonitriles and the above dimer, did not depend to any significant extent on the nature of the cathode or the current density. A complex mechanism is presented to explain the products produced, and it is interesting that adsorption of vinylacrylonitrile is proposed as the first step in the process. This is the first suggestion that surface adsorption may play a role in the mechanism of the electrochemical reduction of nitriles, despite the clear evidence obtained by Franklin and Sothern[84] for the adsorption of propio-, aceto-, butyro-, capro-, isocapro- and phenylacetonitriles on platinized platinum electrodes in aqueous acid solutions. The Freundlich adsorption isotherm was obeyed. Further evidence for the presence of free radical ions in the reduction of unsaturated nitriles is found in the work of Bargain on some α-ethylenic nitriles[85].

Tomilov, Kaabak and Varshavskii[86] also examined the constant current reduction of vinylaceto-, methacrylo- and crotononitriles in 0·7N NaOH at Cu, Sn, Zn and graphite cathodes. At the latter three electrodes, reduction of the double bond occurred with the formation of monomeric and dimeric products. At the copper cathode, some primary and secondary amine was formed in addition to reduction of the double bond. Similar electrolyses of neutral potassium phosphate solutions of 1,4-dicyano-1-butene[87] yielded principally adiponitrile, but as the rate of product formation depended on the concentration of starting material, the process was not practical as a preparative method. A dimer was also obtained in low yield. The same technique applied to α,β-unsaturated nitriles in alkaline solution at a tin cathode[88] produced organotin compounds. Tetra(β-cyanoethyl)tin and tetra(β-cyanopropyl)tin were obtained from acrylonitrile and methacrylonitrile, respectively.

C. Polycyanocarbons

The electrochemical reduction of polycyanocarbons has been little studied. Peover[89] reported a one-electron reversible reduction of tetracyanoethylene, tetracyanobenzene and tetracyanoquinodimethane in acetonitrile using d.c. and a.c. polarography. Comment was made about second waves but these were not studied. Table 2 shows results obtained in DMF for a number of cyanohydrocarbons. Unfortunately, some of these compounds are reduced to the anion radical when dissolved in DMF[62] and further work is needed.

D. Miscellaneous

Although not involving the direct electroreduction of a cyano compound, the work of Zuman and Santavy should be mentioned[90,91]. Equilibrium constants for the formation of cyanohydrins in alkaline solution (equation 15) were determined.

$$\mathrm{ArCHO} + \mathrm{CN^-} \rightleftharpoons \mathrm{ArC(O^-)(CN)H} + \mathrm{H^+} \rightleftharpoons \mathrm{ArCH(OH)(CN)} \tag{15}$$

Reaction (15) was studied for 24 aldehydes by following the decrease in height of the polarographic wave due to reduction of the aldehyde.

V. ULTRAVIOLET ABSORPTION

Saturated aliphatic nitriles have been little studied. They show only a weak characteristic absorption in the near ultraviolet region[92–94]. Schurz and coworkers[94] reported the wavelength of maximum absorption at approximately 270 mμ. Cutler[95] examined the vacuum ultraviolet spectra of aceto- and propionitriles from 1000 to 1800 Å. A Rydberg series was found for MeCN having a series limit corresponding to an ionization potential of 11·96 ev. The first member of the series occurred at 1292 Å. The electronic transition was assigned to excitation of a π electron in the triple bond.

An ultraviolet study has been made which supports the possibility of weak donor–acceptor complexes between aromatic hydrocarbons and polar organic solvents including propionitrile[96]. Using *p*-xylene as the donor and n-heptane as the solvent, the formation constants for the complex were $0{\cdot}071 \pm 0{\cdot}002$, $0{\cdot}063 \pm 0{\cdot}002$ and $0{\cdot}061 \pm 0{\cdot}002$ l/mole at 25 °c, for wavelengths of 229, 229·5 and 230 mμ, respectively. The charge-transfer absorption bands of aromatic hydrocarbons with polycyano compounds occur at longer wavelengths. Wallenfels and Friedrich[97] studied complexes of pyrene with several polycyano compounds and provided the following wavelengths (mμ) of maximum absorption of the charge-transfer complexes in CH_2Cl_2, unless otherwise stated: 1,2,4,5-tetracyanobenzene (495 mμ in $CHCl_3$ and 460 mμ in CH_3CN), tetracyano-*m*-xylene (439 mμ), pentacyanotoluene (533, 411 mμ), hexacyanobenzene (637, 450 mμ), tetracyanoethylene (724, 495 mμ), 2,3-dicyanobenzoquinone (718, 478 mμ) and tetracyanobenzoquinone (1129, 617, 485 mμ). Peover[98] found that the energy of the longest wavelength of maximum absorption of the complexes of a number of polycyano compounds with hexamethylbenzene or pyrene was linear with the polarographic half-wave potentials of the acceptor molecules. An interesting feature in the region between 220 and 270 mμ was observed by Almange[99] for compounds of the type $(R^1)(R^2)C(CN)CH(CN)_2$ where R^1 = Ph and R^2 = H or Ph. The molar extinction coefficient increased with the polarity of the solvent and with dilution for a given solvent. The possibility of this being due to a tautomeric equilibrium was discounted when further study involving i.r. spectrometry and some acid–base chemistry revealed that the compounds ionized according to equation (16). The

$$R^1R^2C(CN)CH(CN)_2 \rightleftharpoons [R^1R^2C(CN)C(CN)_2]^- + H^+ \qquad (16)$$

extinction coefficient of the anion is much larger than that of the unionized molecule in the spectral region of interest, and accounts for the observed behaviour. Boyd[100] has studied the reversible protonation of cyanocarbon acids in solutions of strong acids ($HClO_4$ and H_2SO_4), and devised an acidity scale, H_-, for the protonation of anions (equation 17) similar to the Hammett acidity function for uncharged bases:

$$H_- \equiv pK_a + \log (C_{A^-}/C_{HA}) \qquad (17)$$

Since either the free anion or the protonated form or both absorb

TABLE 3. Ionization constants of cyanocarbon acids at 25 °c.

Cyanocarbon acid	pK_a $HClO_4$	pK_a H_2SO_4	λ_{max}(mμ) for longest wavelength band: Free anion	λ_{max}(mμ) for longest wavelength band: Protonated form
p-(Tricyanovinyl)phenyldicyanomethane	0·60	—	607	332
Methyl dicyanoacetate	−2·78	−2·93	235	<200
Hexacyanoheptatriene	−3·55	−3·90	645	347
Cyanoform	−5·13	−5·00	210	<200
Tricyanovinyl alcohol	−5·3	−5·0	295	275
Bis(tricyanovinyl)amine	−6·07	−5·98	467	366
Tetracyanopropene	<−8		344	?
Pentacyanopropene	<−8·5	—	395	?
Hexacyanoisobutylene	pK_1 <−8·5	—	370	?
	pK_2 (2·5)	—	336	370

in the ultraviolet–visible region, these acids serve as their own indicators in the determination of their acid strengths. Values for pK_a determined in this fashion are shown in Table 3, where the stronger acids were determined relative to methyl dicyanoacetate, which in turn was relative to *p*-(tricyanovinyl)phenyldicyanomethane.

Heilman and coworkers[101,102] examined a number of unsaturated aliphatic nitriles. The effect of substitution on wavelength in 96% ethanol can be seen from the data shown in Table 4 for compounds

TABLE 4. Ultraviolet spectra of unsaturated nitriles of the type $R^2R^3C{=}CR^1CN$.

R^1	R^2	R^3	λ_{max}(mμ)	ε
H	H	H	203	6100
Me	H	H	203	8900
H	H	C_6H_{13}	206	5100
Et	H	Me	208	10,600–12,500
Me	H	Me	208	10,400–9950
H	Me	Et	210–212	11,500–11,800 (β,γ isomer, no maximum)
H	Me	Pr	213	6300–9700 (β,γ isomer, no maximum)
t-Bu	H	H	203	8500
H	H	*t*-Bu	206	9300
Me	H	Am	209	13,800
Me	Me	Me	215	12,500
Me	Et	Et	218	15,400
H	Me	C_6H_{13}	213	11,600–14,000 (β,γ isomer, no maximum)

TABLE 5. Rules for substitutent effects on the ultraviolet absorption of unsaturated nitriles.

Type of substitution	$\Delta\lambda^a$ (mμ)	Intensity
Mono, R_α	1–2	$\Delta\varepsilon_{R\beta} < \Delta\varepsilon_{R\alpha}$
Mono R_β	2–4	
Di, $R_{\alpha,\beta}$	5–7	$\Delta\varepsilon_{R\alpha,\beta} > \Delta\varepsilon_{R\beta,\beta}$
Di, $R_{\beta,\beta}$	9–12	
Tri, $R_{\alpha,\beta,\beta}$	12	10,000–12,000

^{a} With reference to CH_2=CHCN.

of the type $R^2R^3C{=}CR^1CN$. Table 5 shows the empirical rules deduced[102] for the effect of substitution. These results were interpreted in terms of hyperconjugation and inductive effects. The work included also one unsaturated cyclic nitrile not shown in Table 4: 1-cyanocyclohexene (λ_{max} = 208 mμ, ε = 12,500). Wheeler[103] extended the work to include a number of unsaturated cyclic nitriles either by direct laboratory examination of the compounds or from literature data. His findings are summarized in Table 6. These wavelength shifts, as measured from the wavelength of maximum absorption for the unsubstituted compound (203 mμ) are somewhat larger than would be predicted by the substitution rules given in Table 5. This has been interpreted by Wheeler as being due to ring strain in the cyclic compounds. The u.v. absorption spectra of a series of nitriles RCH=CHCN, (R = $CH_3(CH_2)_n$, n = 1–8) were examined by Castille and Ruppol[104]. As expected, the *trans* isomers produced more intense bands than the *cis* isomers except for 1-pentene.

A number of studies of K band shifts of aromatic compounds have included the cyano group[105–108]. Bloor[108] has interpreted the K band spectrum as due to intramolecular charge transfer, which in the cyano case involves charge transfer from the hydrocarbon to

TABLE 6. Substitution effects on the ultraviolet spectra of unsaturated cyclic nitriles.

Type of substitution	λ_{max}(mμ)
α,β Six-membered ring	210–212
β,β Six-membered ring	216
α,β Five-membered ring	216–225
β,β Five-membered ring	210
α,β,β Five-membered ring	222–224

the substituent. Good agreement between calculated (MO) and observed K band transition energies was achieved. In further work[109], a linear relationship between the polarographic half-wave potential and the energy of the lowest unoccupied molecular orbital was found.

VI. INFRARED ABSORPTION

The infrared absorption of the cyano group has been known for many years. Bell[110], who reported that aryl and alkyl cyanides absorbed at 4·5 and 4·4 μ, respectively, refers to the fact that Coblentz obtained the spectra of aceto- and benzonitrile as early as 1905. However, Kitson and Griffith[111] performed the first extensive study which included saturated and unsaturated aliphatic nitriles and aromatic nitriles. Bellamy[112] summarized the frequency ranges for the C≡N stretching vibration as follows: saturated alkyl nitriles 2260–2240 cm^{-1}, aryl nitriles 2240–2220 cm^{-1}, α,β-unsaturated alkyl nitriles 2235–2215 cm^{-1}, isocyanates 2275–2240 cm^{-1}. Later, Besnainou, Thomas and Bratoz[113], also summarizing experimental work in the literature, agreed with Bellamy's frequency range for saturated nitriles, but pointed out that alkyl substitution on the β carbon lowered the frequency as much as 16 cm^{-1} from 2250 cm^{-1}. Their range for aromatic nitriles was 2240–2221 cm^{-1}. Their findings also showed that unsaturated non-aromatic groups shifted the frequency to lower wavenumbers. There are several situations, however, where the nitrile band has been shifted to frequencies lower than those indicated by any of the above ranges. Baldwin[114] has studied the i.r., u.v. and n.m.r. spectra of 4-aminospiro[Δ^3-cyclohexene-1,9′-fluorene]-3-carbonitrile (**1**) and 4-amino-1′-oxo-3′,4′-dihydrospiro[Δ^3-cyclohexene-1,2′(1′H)-naphthalene]-3-carbonitrile (**2**) and summarized the literature information on other

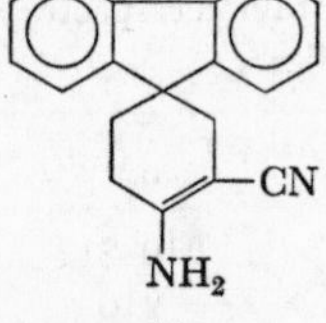

-CN
NH_2
(**1**)

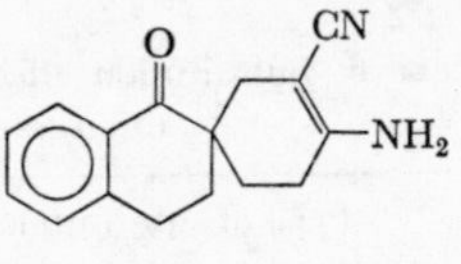

O CN
-NH_2
(**2**)

β-amino-α,β-unsaturated nitriles. The cyano group stretching frequency for seventeen compounds is shifted 17–50 cm^{-1} below the lowest frequency listed earlier for α,β-unsaturated nitriles (2215 cm^{-1}). Earlier workers, referenced by Baldwin, had considered the enamine form as the true structure for a number of the compounds listed. This author considered the abnormally large frequency shifts to be due to the existence of compounds **1** and **2** in the enamine tautomeric form. This permits a number of charge-separated resonance forms to be written (equation 18):

$$H_2N{-}C{=}C{-}C{\equiv}N \longleftrightarrow H_2N{-}\overset{+}{C}{-}C{=}C{=}\overset{-}{N} \longleftrightarrow$$
$$H_2\overset{+}{N}{=}C{-}C{=}C{=}\overset{-}{N} \longleftrightarrow H_2\overset{+}{N}{=}C{-}\overset{-}{\ddot{C}}{-}C{\equiv}N \quad (18)$$

Baldwin also cites similar frequency shifts (5–46 cm^{-1}) for β-hydroxy- and β-alkoxy-α,β-unsaturated nitriles and similar but somewhat smaller shifts (5–15 cm^{-1}) for nitroacetonitrile anions for which analogous resonance forms can be written. Supporting evidence from other i.r. bands and the u.v. and n.m.r. spectra are presented to reinforce the arguments for the enamine structure. α-Aminonitrile hydrochlorides also show absorption bands at frequencies lower than 2215 cm^{-1}. Kazitsyna, Lokshin and Glushkova[115] examined the spectra of the hydrochlorides of compounds

$$R^1{-}C(R^2)(C{\equiv}N){-}NH{-}(CH_2)_n{-}NH{-}C(R^2)(C{\equiv}N){-}R^1 \quad \textbf{(3)}$$

Compound number	n	R^1	R^2
3a	2	Ph	Me
3b	2	Cyclohexylidene	Cyclohexylidene
3c	6	Me	Me
3d	6	H	Ph
3e	6	Me	Ph
3f	6	Cyclohexylidene	Cyclohexylidene
3g	6	Me	Et
3h	6	H	i-Pr

of the type **3**. The cyano group stretching frequencies were respectively: **3a,** 2210–2215; **3b,** 2207; **3c,** 2203; **3d,** 2215; **3e,** 2211; **3f,** 2206; **3g,** 2207; and **3h,** 2213 cm^{-1}. The stretching frequency of the cyano group is also shifted to lower wavenumbers for carbanions. Yukhnovski[116] obtained the spectra of α-naphthylacetonitrile, β-naphthylacetonitrile and phenylacetonitrile in hexamethylphosphortriamide. The respective bands were at 2113, 2106 and 2102 cm^{-1}. The order of frequencies was explained on the basis of increasing tendency of the aryl group to conjugate with the bridge carbon atom and therefore produce less conjugation with the cyano group.

The nitrile band intensity is described by Besnainou and coworkers[113] as relatively strong for saturated nitriles, enhanced for nitriles containing unsaturated non-aromatic groups and very strong for aromatic nitriles. The intensity is quenched when oxygen containing groups are introduced[111,112]. Cross and Rolfe[117] measured molar extinction coefficients for a few aliphatic nitriles and benzonitrile in $CHCl_3$. Skinner and Thompson[118] measured integrated band intensities for a number of alkyl cyanides and substituted benzonitriles. Flett[119] compared peak extinction coefficients and band area intensities for a number of functional groups, including twenty nitriles, and recommended use of the total area under the band. A number of factors act to affect the band intensity and/or frequency of the carbon–nitrogen stretching frequency in nitriles. In many cases, the band intensity is more subject to change than is the band frequency. This is particularly true for substitution effects in both aliphatic and aromatic nitriles. A detailed discussion of the large amount of work available is beyond the scope of this presentation but Table 7 provides a summary of some of the work

TABLE 7. Factors affecting the frequency and/or intensity of nitrile band absorption.

Factor	Comments	References
1. Hydrogen bonding	A number of proton donors with aceto-, propio- and other nitriles	125, 126
	Adsorption on surfaces containing hydroxyl groups	127
2. Salt effects	Principally perchlorate salts. RCN bonded to the cation. ν_{CN} shifts to higher frequencies	128–130
3. Lewis acid–base interactions and charge-transfer complexes	Band shifts to higher frequencies. Typical Lewis acids: $SnCl_4$, BF_3, ICl	131–137
4. Solvent effects	Frequency shifts to lower frequencies but only slightly. Intensity is very solvent dependent	138–146
5. Substitution effects	Substituted benzonitriles—Hammett $\sigma\rho$ correlations.	147–156
	Aliphatic nitriles	157–160
	Other nitriles and miscellaneous	161–164
6. Steric effects	Polycyclic aromatic nitriles	165, 166
	Other nitriles	167

that has been reported and includes hydrogen bonding, solvent effects, substitution effects, salt effects and others.

A few specific analyses based on the nitrile band absorption have been presented. Dinsmore and Smith[120] determined the nitrile content of buna-N rubber using the band at 2237 cm^{-1}. Nelson and coworkers determined nitriles in high molecular weight fatty amines[121]. The band at 2247 cm^{-1} was used and results were interpreted in terms of the total fatty acid nitrile present. The reaction mixture obtained in amide synthesis was analysed for nitrile content spectrophotometrically at 2250 cm^{-1} [122]. Potassium bromide pellet techniques were used to determine the nitrile groups in polyacrylonitrile[123] and in cyanoethyl cellulose[124].

VII. MASS SPECTROMETRY

Aliphatic nitriles have received most of the attention in mass spectrometric studies of cyano compounds. Interpretation of the spectra is complicated by the presence of isobaric carbon–hydrogen and carbon–nitrogen ions and by complex fragmentation processes. Nevertheless, the technique is a useful tool for the identification of cyano compounds and for molecular weight determination, but structure determinations are considerably more difficult. McLafferty[168] reported the first extensive study of the mass spectrometry of saturated aliphatic nitriles. Spectral correlations for eighteen compounds were presented. Ninety-degree sector type instruments were used with inlet temperatures usually at 200 °C and ionizing voltages of 75 ev. The molecular ion is the base peak for acetonitrile and is strong for propionitrile, but in all aliphatic nitriles higher than propionitrile this peak was weak or missing. Some tendency for the molecular ion intensity to increase with molecular weight for nitriles higher than propionitrile was observed but the peaks were still weak. Molecular weight determination is aided by the presence of pressure-dependent $M + 1$ peaks. In most cases, significant peaks also exist for $M - 1$ ions, although the intensity falls off with increasing molecular weight. Although it might be reasoned that the $M - 1$ ion arises from loss of an α hydrogen, it was shown by high resolution mass spectrometry and deuterium labelling that all positions in the alkyl chain of n-hexyl cyanide contribute to the peak for this compound[169]. Evidence was also obtained that the formation of the $M - 1$ ion in the spectrum of isohexyl cyanide cannot be accounted for solely by the loss of a $C_{(2)}$ hydrogen atom from the molecular ion.

In contrast to other electronegative groups, aliphatic nitriles do not show significant peaks due to alpha bond cleavage, as the peaks due to the ions, C_nH_{2n+1}, and CN^+ are very weak. The peak at *m/e* 41 is intense for all nitriles and is the most abundant peak in the spectra of straight-chain nitriles from n-butyro- to n-undecano-nitrile. McLafferty[168] suggested that the contribution to this peak from $C_3H_5^+$ was not significant and proposed that the $C_2H_3N^+$ ion arose from beta cleavage and a hydrogen atom rearrangement according to equation (19):

$$H_2C(H)\text{–}CH_2\text{–}CH_2\text{–}C{\equiv}\overset{\circ}{N}{}^+ \longrightarrow H_2C{=}CH_2 + CH_2{=}C{=}NH^+ \qquad (19)$$

There is now considerable evidence that the hydrocarbon ion does contribute significantly to the peak intensity at *m/e* 41. Using high resolution mass spectrometry, Rol[170] found that the $C_3H_5^+$ ion contributed 6% of the intensity of the peak for the nitrile of n-butane. Carpenter and coworkers[169] examined the intensity of the same peak in the spectrum of n-hexyl cyanide and found a 40% contribution from $C_2H_3N^+$ and a 60% one from $C_3H_5^+$. The formation of the former ion via a McLafferty rearrangement (equation 19) is still in some doubt, since studies of the deuterated analogues of n-hexyl cyanide were inconclusive in providing proof[169]. Further, experiments at 12 ev failed to show enhancement of the *m/e* 41 peak which would be expected if indeed the fragmentation proceeded by this rearrangement.

Two other prominent peaks in the mass spectra of both branched and straight chain nitriles occur at *m/e* 55 and 54. Beugelmans and coworkers[171] reported the intensity ratio of $C_4H_7^+/C_3H_5N^+$ as 2·94/1 for isohexyl cyanide, while the ratio is 5·65/1 for n-hexyl cyanide[169]. Rol[170] has suggested that the mechanism for the production of the *m/e* 55 ion from the molecular ion may involve seven- or eight-membered rings in the case of branched-chain nitriles. It was not possible to confirm a mechanism from studies of deuterated n-hexyl cyanide, but a weak metastable peak at *m/e* 53 showed that the peak at *m/e* 54 was due in part to the loss of hydrogen from *m/e* 55 species[169].

Loss of ethylene and ethyl radical is characteristic of aliphatic nitriles and is a complex process, as shown by the data in Table 8 for the M − 28 peak of deuterium substituted n-hexyl cyanide[169].

TABLE 8. Origin of the M − 28 peak in the mass spectrum of n-hexyl cyanide[a].

Compound (7-6-5-4-3-2-CN, C-1)	Percentage of carbon atoms eliminated as ethylene
2,2-d_2	20%
3,3-d_2	21%
4,4-d_2	9%
5,5-d_2	10%
6,6-d_2	76%
7,7,7-d_3	75%

[a] Isotropic purities were calculated from the spectra of the corresponding alkyl ethers which were synthesized from the deuterated alkyl bromides used in the preparation of the deuterated n-hexyl cyanides.

It is clear from Table 8 that $C_{(2)}$ and $C_{(3)}$, $C_{(4)}$ and $C_{(5)}$, and $C_{(6)}$ and $C_{(7)}$ all contribute to the loss of ethylene. This can be best explained by postulating cyclic intermediates in the fragmentation process. The preferred scheme by these authors is depicted in equation (20):

$-CH_2{=}CH_2$ (6, 7) → m/e 83; $-CH_2{=}CH_2$ (2, 3) → m/e 83; 5 ⦚ 6 → $(M - C_2H_5)$ ≡ m/e 82 (20)

While it was not possible to ascertain quantitatively the origin of the m/e 82 peak, it was concluded that loss of hydrogen from the precursor of mass 83 was the most important route. Other cyclic intermediates were proposed to explain peaks at m/e 96, 68 and 29, and it was concluded that such intermediates are characteristic of the fragmentation processes of nitriles. Obviously, high resolution instrumentation is necessary for structure studies on aliphatic nitriles, and it is discouraging to note the observation of Carpenter and coworkers that the mass spectra are dependent upon the particular instrument used[169].

The mass spectra of a few other types of nitriles have been reported. Griffin and Peterson[172] obtained the spectra of the *cis* and *trans* isomers of tricyanocyclopropane. Both isomers produced a parent peak at *m/e* 117 and the major fragments were interpreted according to cyclopropyl structures. Beynon and coworkers[173] studied a series of oxygen- and nitrogen-containing compounds in order to examine ion–molecule reactions occurring in the mass spectrometer. Adiponitrile was among the compounds included. As might be expected, $M + 1$ peaks were observed, but in addition a peak at *m/e* 149 was seen for adiponitrile. This was attributed to the addition of CH_3CN to the parent ion. The mass spectrum of dicyanoheptafulvene has been reported by Kitahara and Kato[174]. Mass spectrometry was used for the identification of the *syn* and *anti* forms of *N*-fluoroiminonitriles by Logothetis and Sausen[175]. Dibeler, Reese and Franklin[176] have made probable assignments for the principal peaks in the mass spectrum of dicyanodiacetylene. The base peak corresponds to the parent ion. The mass spectra of methyl, ethyl, ethyl isopropyl, ethyl isopropyl-d_3, ethyl *s*-butyl, ethyl isobutyl, ethyl n-butyl and methyl n-butyl cyanoacetates were studied by Bowie and coworkers[177] using high resolution instrumentation. The spectra showed low abundance molecular ions and lacked the $M - 1$ and $M + 1$ peaks characteristic of the alkyl nitriles. In contrast, significant $M + 2$ peaks were observed. Important fragmentation paths involved loss of CO_2 or HCO_2 from the ester group and corresponding alkyl migrations of a methyl or ethyl group. Williams and coworkers[178] have reported the only work on α,β-unsaturated nitriles except for the work on dicyanoheptafulvene. For compounds of the type

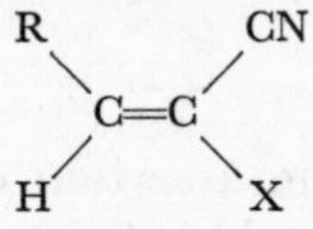

$$\begin{matrix} R & & CN \\ & C{=}C & \\ H & & X \end{matrix}$$

ions corresponding to RCN^+ were observed in the spectra for

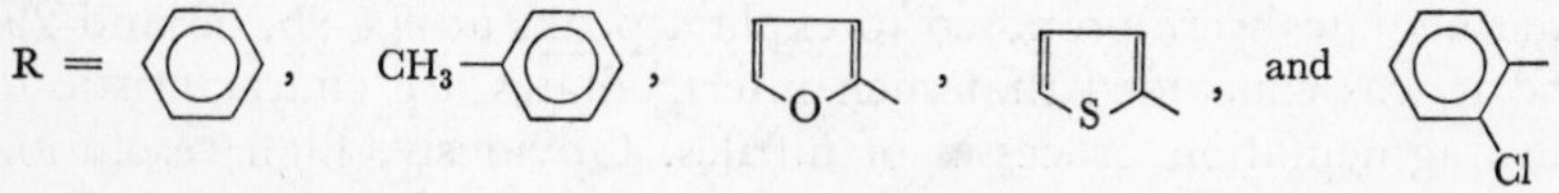

R = phenyl, CH_3–phenyl, 2-furyl, 2-thienyl, and 2-chlorophenyl (Cl)

and X = CO_2Et, CO_2H, CO_2Bu-*t* and CN. This is reputed to be the first report of ion formation involving bond formation between 1,2-groups attached to double bonded carbon atoms. Another important fragmentation process for compounds containing the ester group involved an alkoxyl or hydroxyl group migration according to equation (21).

$$\left[(R^2)(H)C{=}C(CN)C{=}O(OR^1)\right]^{+\cdot} \longrightarrow \left[H{-}C(R^2)(R^1O){+}C(CN){=}C{=}O\right]^{+\cdot} \longrightarrow H{-}C^{+}(R^2)(R^1O) + NC\dot{C}{=}C{=}O \qquad (21)$$

There seem to be no published reports on the mass spectrometry of aromatic nitriles, although Beynon[179] states that the —CN group is always found in the spectra of such compounds and parent peaks can usually be observed.

VIII. NUCLEAR MAGNETIC RESONANCE

Negita and Bray[180] determined the ^{14}N resonance frequencies, coupling constants and asymmetry parameters for trichloroacetonitrile, 4-cyanopyridine, 2-cyanopyridine and benzonitrile. These are given in Table 9.

Schmidt, Brown and Williams[181a] carried out ^{14}N n.m.r. studies on compounds containing the —CN group. The only organonitrile studied was CH_3CN; the other compounds were inorganic cyanides, thiocyanates and cyanates. The chemical shift for the ^{14}N resonance in CH_3CN when run as a liquid was 1500 ± 40 milligauss with a

TABLE 9. N.m.r. parameters for nitriles.

Compound	Resonance frequency (Hz)	Coupling constant (Hz)	Asymmetry parameters (%)
Trichloroacetonitrile	3·0337 and 3·0444	4·0521	0·53
4-Cyanopyridine	2·9073 and 2·9353	3·8951	1·44
2-Cyanopyridine	2·8978 and 3·0396	3·9583	7·16
Benzonitrile	2·8098 and 3·0183	3·8854	10·73

line width of 240 ± 30 milligauss. The relative shift* ($\Delta H/H \times 10^6$) was 140. ^{14}N n.m.r. studies of nitriles was also carried out by Witanowski[181b] who reports the ^{14}N chemical shifts for a series of RCN compounds. He reports ^{14}N shifts to lower fields as the electronegativity of R increases. The effect is reversed for RNC.

Cross and Harrison[182] studied the long-range shielding of steroidal angular protons by cyano groups. This was accomplished by determining the methyl proton resonance shifts. This study is more oriented to the effect of the cyano group on the molecular configuration rather than on any properties of the cyano group itself.

Jacquesy and coworkers[183] determined the effect of the substituents on the 19-methyl group of 5α-cholestane. Nitrile substituents are considered.

Lippmaa, Olivson and Past[184] studied the ^{13}C chemical shifts of nitriles among other compounds. These chemical shift data include all data recorded prior to 1965.

Cyanoacrylic acids were investigated by Hamelin[185]. The compounds were of the type $R^1R^2C{=}C(CN)COOEt$, where R^1 was *p*-methoxyphenyl, methyl, hydrogen, chloro or nitro and R^2 was methyl or ethyl.

Cyanophenanthrenes were investigated as part of a study by Bartle and Smith[186] using proton n.m.r. Coupling constants and chemical shifts are given.

^{19}F n.m.r. was used to determine the electronic properties of β,β-dicyano- and α,β,β-tricyanovinyl groups on fluorobenzenes. The tricyanovinyl group is more strongly electron-withdrawing than a nitro group and the dicyanovinyl group is comparable to a nitro group. ^{19}F n.m.r. chemical shifts were used to determine the extent of complexing and interaction.

The conformation of cyanoethylated α-methylcyclohexanones[187] was investigated with n.m.r. The chemical shifts for the different conformations of the methyl group are recorded.

Chemical shifts and coupling constants were obtained for the three stereoisomers of 2,4,6-tricyanoheptane[188]. The purpose of this study was to shed light on the tacticity of polyacrylonitrile by looking at these two-unit model compounds.

IX. REFERENCES

1. A. I. Vogel, *A Textbook of Practical Organic Chemistry*, 3rd ed., John Wiley and Sons, New York, 1957, p. 1077.

* Relative shift refers to the shift compared with the ^{14}N resonance of the nitrogen atom in the nitrate ions contained in a saturated aqueous solution of ammonium nitrate.

2. R. L. Shriner, R. C. Fuson and D. Y. Curtin, *The Systematic Identification of Organic Compounds*, 4th ed., John Wiley and Sons, New York, 1956, pp. 256, 258.
3. S. Trofimenko and J. W. Sease, *Anal. Chem.*, **30,** 1432 (1958).
4. S. Veibel, *The Identification of Organic Compounds*, 6th ed., G.E.C. Gad, 1966, p. 329.
5. N. D. Cheronis, J. B. Entrikin and E. M. Hodnett, *Semimicro Qualitative Organic Analysis*, 3rd ed., Interscience, New York, 1965, pp. 620–621.
6. S. Revira and L. Palfray, *Compt. Rend.*, **211,** 396 (1940).
7. N. D. Cheronis, J. B. Entrikin and E. M. Hodnett, *Semimicro Qualitative Organic Analysis*, 3rd ed., Interscience, New York, 1965, pp. 621–623.
8. H. B. Cutter and M. Taras, *Ind. Eng. Chem.*, **13,** 830 (1941).
9. S. Siggia and C. R. Stahl, *Anal. Chem.*, **27,** 550 (1955).
10. R. L. Shriner, R. C. Fuson and D. Y. Curtin, *The Systematic Identification of Organic Compounds*, 4th ed., John Wiley and Sons, New York, 1956, p. 259.
11. C. W. Gould and S. T. Gross, *Anal. Chem.*, 25, 749 (1953).
12. R. L. Shriner, R. C. Fuson and D. Y. Curtin, *The Systematic Identification of Organic Compounds*, 4th ed., John Wiley and Sons, New York, 1956, pp. 257–260.
13. R. L. Shriner and T. A. Turner, *J. Am. Chem. Soc.*, **52,** 1267 (1930).
14. H. P. Howells and J. G. Little, *J. Am. Chem. Soc.*, **54,** 2451 (1932).
15. N. D. Cheronis, J. B. Entrikin and E. M. Hodnett, *Semimicro Qualitative Organic Analysis*, 3rd ed., Interscience, New York, 1965, p. 402.
16. S. Soloway and A. Lipschitz, *Anal. Chem.*, **24,** 898 (1952).
17. F. Feigl, *Spot Tests*, Vol. 2, Elsevier, New York, 1954, p. 197.
18. F. Feigl, V. Gentil and E. Jungreis, *Mikrochim. Acta*, **3,** 47 (1959).
19. B. Prajsnar, A. Maslankiewicz and Z. Najzarek, *Chem. Anal.* (*Warsaw*), **10**(6), 1221 (1965).
20. S. Siggia, J. G. Hanna and N. M. Serencha, *Anal. Chem.*, **36,** 227 (1964).
21. D. H. Whitehurst and J. B. Johnson, *Anal. Chem.*, **30,** 1332 (1958).
22. J. Mitchell Jr. and W. Hawkins, *J. Am. Chem. Soc.*, **67,** 777 (1945).
23. S. Siggia and C. R. Stahl, *Anal. Chem.*, **27,** 550 (1955).
24. W. Z. Huber, *Anal. Chem.*, **197,** 236 (1963).
25. S. Siggia and R. D'Andrea, unpublished work.
26. A. Szewczvk, *Chem. Anal.* (*Warsaw*), **4,** 971 (1959).
27. M. R. F. Ashworth and R. Schupp, *Mikrochim. Acta*, **1967**(2), 366.
28. B. Meissner, *Wasserwirtsch.-Wassertech.*, **8,** 417 (1958).
29. R. S. Valentine, *U.S. Pat.* 2,900,309 (1959).
30. F. D. Maslan and N. C. Robertson, *Brit. Pat.* 802,114 (1958).
31. Armour and Co., *Brit. Pat.* 479,023 (1938).
32. A. W. Ralston and W. O. Pool, *U.S. Pat.* 2,107,904 (1937).
33. W. O. Pool, *U.S. Pat.* 2,107,904 (1937).
34. T. A. Markova, V. S. Vinogradova and L. S. Kofman, *Materialy Vses. Soveshch. po Tseolitam*, 2nd ed., Leningrad, *1964*, 283 (1965).
35. W. E. Line, H. M. Hickman and R. A. Morrissette, *J. Am. Oil Chemists Soc.*, **36,** 20 (1959).
36. J. Pokorny and I. Zeman, *Sb. Vysoke Skoly Chem.-Technol. v. Praze Oddil Fak. Potravinareske Technol.*, **4**(2), 255 (1960).
37. V. Vasilescu, *Fette Seifen Anstrichmittel*, **63,** 132 (1961).
38. Y. Murata and T. Takenishi, *Kogyo Kagaku Zasshi*, **64,** 787 (1961).
39. Y. Arad-Talmi, M. Levy and D. Vofsi, *J. Chromatog.*, **10,** 417 (1963).
40. L. D. Metcalfe, *J. Gas Chromatog.*, **1,** 7 (1963).

41. L. S. Gray, L. D. Metcalfe and S. W. Leslie, *Gas Chromatog. Int. Symp.*, **4,** 203 (1963).
42. E. Mugnaini and G. Gambelli, *Chim. Ind.*, **45,** 44 (1963).
43. V. Vasilescu, *Abhandl. Deut. Akad. Wiss. Berlin, Kl. Chem. Geol. Biol.*, **1963,** 131.
44. A. I. Parimskii and I. K. Shelomov, *Maslob.-Zhir Prom.*, **30,** 28 (1964).
45. M. Taramasso and A. Guerra, *J. Gas. Chromatog.*, **3,** 189 (1965).
46. N. M. Nicholaeva and B. R. Serebryakov, *Gaz. Khromatogr., Moscow,* (3) 28 (1965).
47. M. N. Moskvitin, L. I. Anan'eva and P. I. Petrovich, *Zavod. Lab.*, **33,** 136 (1967).
48. N. H. Weissert and R. A. Coelho, *J. Gas Chromatog.*, **5,** 160 (1967).
49. A. P. Tomilov, L. V. Kaabak and S. L. Varshavskii, *Khim. Prom.*, **1962,** 562; *Chem. Abstr.*, **59,** 1282a (1963).
50. O. Manousek and P. Zuman, *Chem. Commun.*, **1965,** 158.
51. F. B. Ahrens, *Z. Elektrochem.*, **5,** 99 (1896).
52. K. Ogura, *Mem. Coll. Sci. Kyoto Imp. Univ.*, **12A,** 339 (1929); *Chem. Abstr.*, **24,** 2060 (1930).
53. M. Ohta, *Bull. Chem. Soc. Japan,* **17,** 485 (1942).
54. S. Sawa (Shionigi Drug Mfg. Co.), *Japan. Pat.* 5019 (1951).
55. T. Matukawa, *Japan. Pat.* 133,464 (1939).
56. E. Laviron, *Compt. Rend.*, **250,** 3671 (1960).
57. J. Volke, R. Kubicek and F. Santavy, *Collection Czech. Chem. Commun.*, **25,** 1510 (1960).
58. J. Volke and J. Holubek, *Collection Czech. Chem. Commun.*, **28,** 1597 (1963).
59. J. M. S. Jarvie, R. A. Osteryoung and G. J. Janz, *Anal. Chem.*, **28,** 264 (1956).
60. I. G. Sevast'yanova and A. P. Tomilov, *J. Gen. Chem. USSR Eng. Transl.*, **33,** 2741 (1963).
61. I. G. Sevast'yanova and A. P. Tomilov, *Soviet Electrochem.*, **3,** 563 (1967).
62. P. H. Rieger, I. Bernal, W. H. Reinmuth and G. K. Fraenkel, *J. Am. Chem. Soc.*, **85,** 683 (1963).
63. P. B. Janardhan, *Indian Pat.* 44,230 (1952).
64. P. B. Janardhan, *J. Sci. Ind. Res.* (*India*), **12B,** 183 (1953); *Chem. Abstr.*, **47,** 10378c (1953).
65. F. Kawamura and S. Suzuki, *J. Chem. Soc. Japan Ind. Chem. Sect.*, **55,** 476 (1952); *Chem. Abstr.*, **48,** 3167e (1954).
66. M. I. Bobrova and A. N. Matveeva, *J. Gen. Chem. USSR Eng. Transl.*, **27,** 1219 (1957).
67. W. L. Bird and C. H. Hale, *Anal. Chem.*, **24,** 586 (1952).
68. L. Spillane, *Anal. Chem.*, **24,** 587 (1952).
69. S. F. Strauss and E. Dyer, *Anal. Chem.*, **27,** 1906 (1955).
70. G. W. Daues and W. F. Hamner, *Anal. Chem.*, **29,** 1035 (1957).
71. M. N. Platnova, *Zh. Anal. Khim.*, **3,** 310 (1956).
72. G. C. Claver and M. E. Murphy, *Anal. Chem.*, **31,** 1682 (1959).
73. T. R. Crompton and D. Buckley, *Analyst*, **90,** 76 (1965).
74. M. Murphy, M. G. Carangelo, M. B. Ginaine and M. C. Markham, *J. Polymer Sci.*, **54,** 107 (1961).
75. I. G. Sevast'yanova and A. P. Tomilov, *Soviet Electrochem.*, **2,** 1026 (1966).
76. S. Lasarov, A. Trifonov and T. Vitanov, *Z. Phys. Chem.* (*Leipzig*), **226,** 221 (1964).
77. M. M. Baizer, *Tetrahedron Letters*, **15,** 973 (1963).
78. M. M. Baizer, *J. Electrochem. Soc.*, **111,** 215 (1964).
79. M. M. Baizer and J. D. Anderson, *J. Electrochem. Soc.*, **111,** 223 (1964).
80. M. M. Baizer and J. D. Anderson, *J. Electrochem. Soc.*, **111,** 226 (1964).

81. J. Wiemann and M. L. Bouguerra, *Compt. Rend.*, **265**, 751 (1967); *Chem. Abstr.*, **68**, 68447k (1968).
82. M. I. Bobrova and A. N. Matveeva-Kudasheeva, *Zh. Obshch. Khim.*, **28**, 2929 (1958).
83. L. V. Kaabak, A. P. Tomilov and S. L. Varshavskii, *J. Gen. Chem. USSR Eng. Transl.*, **34**, 2121 (1964).
84. T. C. Franklin and R. D. Sothern, *J. Phys. Chem.*, **58**, 951 (1954).
85. M. Bargain, *Compt. Rend. Congr. Nat. Soc. Savantes Sect. Sci.*, **88**(1), 139 (1963); *Chem. Abstr.*, **64**, 17041g (1966).
86. A. P. Tomilov, L. V. Kaabak and S. L. Varshavskii, *J. Gen. Chem. USSR*, **33**, 2737 (1963).
87. Yu. D. Smirnov, S. K. Smirnov and A. P. Tomilov, *Zh. Org. Khim.*, **4**, 216 (1968).
88. L. V. Kaabak and A. P. Tomilov, *Zh. Obshch. Khim.*, **33**, 2808 (1963).
89. M. E. Peover, *Trans. Faraday Soc.*, **58**, 2370 (1962).
90. P. Zuman and F. Santavy, *Chem. Listy*, **47**, 267 (1953).
91. P. Zuman and F. Santavy, *Collection Czech. Chem. Commun.*, **19**, 174 (1954).
92. J. Bielecke and V. Henri, *Compt. Rend.*, **156**, 1860 (1913).
93. M. Grunfeld, *Ann. Chim.*, **20**, 304 (1933).
94. J. Schurz, H. Sah and A. Ullrich, *Z. Physik. Chem. (Frankfurt)*, **21**, 185 (1959).
95. J. A. Cutler, *J. Chem. Phys.*, **16**, 136 (1948).
96. R. F. Weimer and J. M. Prausnitz, *Spectrochim. Acta*, **22**, 77 (1966).
97. K. Wallenfels and D. Friedrich, *Tetrahedron Letters*, **1963**, 1223.
98. M. E. Peover, *Trans. Faraday Soc.*, **58**, 2370 (1962).
99. J. P. Almange, *Compt. Rend.*, **261**, 2676 (1965).
100. R. H. Boyd, *J. Phys. Chem.*, **67**, 737 (1963).
101. R. Heilmann, J. M. Bonnier and M. de Gaudemaris, *Compt. Rend.*, **244**, 1787 (1957).
102. R. Heilmann and J. M. Bonnier, *Compt. Rend.*, **248**, 2495 (1959).
103. O. H. Wheeler, *J. Org. Chem.*, **26**, 4755 (1961).
104. A. Castille and E. Ruppol, *Bull. Soc. Chim. Belg.*, **44**, 351 (1935).
105. E. C. Lim, *Spectrochim. Acta*, **19**, 1967 (1963).
106. G. Leandri and D. Spinelli, *Boll. Sci. Fac. Chim. Ind. Bologna*, **15**(3), 90 (1957); *Chem. Abstr.*, **52**, 2533d (1958).
107. A. Burawoy, P. Brockhurst, R. S. Berry and A. R. Thompson, *Tetrahedron*, **10**, 102 (1960).
108. J. E. Bloor, *Can. J. Chem.*, **39**, 2256 (1961).
109. J. E. Bloor, G. R. Gilson and D. D. Shillady, *J. Phys. Chem.*, **71**, 1238 (1967).
110. F. K. Bell, *J. Am. Chem. Soc.*, **57**, 1023 (1935).
111. R. E. Kitson and N. E. Griffith, *Anal. Chem.*, **24**, 334 (1952).
112. L. J. Bellamy, *The Infrared Spectra of Complex Molecules*, 2nd ed., Methuen, 1958, p. 263.
113. S. Besnainou, B. Thomas and S. Bratoz., *J. Mol. Spectry*, **21**, 113 (1966).
114. S. Baldwin, *J. Org. Chem.*, **26**, 3288 (1961).
115. L. A. Kazitsyna, B. V. Lokshin and O. A. Glushkova, *Zh. Obshch. Khim.*, **32**, 1391 (1962).
116. I. N. Yukhnovski, *Teor. Eksp. Khim.*, **3**, 410 (1967).
117. L. H. Cross and A. C. Rolfe, *Trans. Faraday Soc.*, **47**, 354 (1951).
118. M. W. Skinner and H. W. Thompson, *J. Chem. Soc.*, **1955**, 487.
119. M. St. C. Flett, *Spectrochim. Acta*, **18**, 1537 (1962).
120. H. L. Dinsmore and D. C. Smith, *Anal. Chem.*, **20**, 11 (1948).
121. J. P. Nelson, L. R. Peterson and A. J. Milun, *Anal. Chem.*, **33**, 1882 (1961).

122. Sh. A. Zelenaya, T. I. Pantelei and O. F. Koutenko, *Zavodsk Lab.*, **30**, 1077 (1964).
123. K. Doerffel, *Wiss. Z. Tech. Hochsch. Chem. Leuna-Merseberg*, **7**, 71 (1965); *Chem. Abstr.*, **64**, 814a (1966).
124. Yu. P. Putiev and Yu. T. Tashpulatov, *Uzbebsk. Khim. Zh.*, **10**, 41 (1966); *Chem. Abstr.*, **66**, 20069v (1967).
125. K. T. Geodakyan, A. V. Kiselev and V. I. Lygin, *Zh. Fiz. Khim.*, **40**(7), 1589 (1966).
126. A. Allerhand and P. v. R. Schleyer, *J. Am. Chem. Soc.*, **85**, 866 (1963).
127. S. S. Mitra, *J. Chem. Phys.*, **36**, 3286 (1962).
128. I. S. Perelygin, *Optika i Spektroskopiya*, **13**, 360 (1962).
129. I. S. Pominov and A. Z. Gadzhiev, *Izv. Vysshikh Uchebn Zavedenii, Fiz*, **8**(5), 19 (1965); *Chem. Abstr.*, **64**, 2802d (1966).
130. M. Kubota and D. L. Johnston, *J. Inorg. Nuclear Chem.*, **29**(3), 769 (1967).
131. H. J. Coever and C. Curran, *J. Am. Chem. Soc.*, **80**, 3522 (1958).
132. H. A. Brune and W. Zeil, *Z. Naturforsch*, **166**, 1251 (1961).
133. K. F. Purcell and R. S. Drago, *J. Am. Chem. Soc.*, **88**, 919 (1966).
134. K. F. Purcell, *J. Am. Chem. Soc.*, **89**, 247 (1967).
135. K. F. Purcell, *J. Am. Chem. Soc.*, **89**, 6139 (1967).
136. K. F. Purcell, *J. Am. Chem. Soc.*, **88**, 919 (1966).
137. E. Augdahl and P. Klaboe, *Spectrochim. Acta*, **19**(10), 1665 (1963).
138. N. S. Bayliss, A. R. H. Cole and L. H. Little, *Spectrochim. Acta*, **1959**, 12.
139. A. Foffani, C. Pecile and F. Pietra, *Nuovo Cimento*, **13**, 213 (1959); *Chem. Abstr.*, **53**, 21159 (1960).
140. G. L. Caldow, D. Cunliffe-Jones and H. W. Thompson, *Proc. Roy. Soc.* (*London*), **A254**, 17 (1960).
141. Ya. A. Kushnikov and L. P. Krasnomolova, *Izv. Akad. Nauk. Kaz. SSR, Ser. Khim.*, **1962**(1), 54; *Chem. Abstr.*, **58**, 2027f (1963).
142. R. Heess, *Abhandl. Deut. Akad. Wiss. Berlin, Kl. Math., Physik. Tech.*, **1964**(6), 97; *Chem. Abstr.*, **63**, 5117g (1965).
143. S. Tanaka, K. Tanabe and H. Kamada, *Spectrochim Acta.*, **23A**(2), 209 (1967).
144. K. Venkateswarlu and C. Balasubramanian, *Proc. Indian Acad. Sci. Sect. A*, **51**, 151 (1960).
145. C. D. Ritchie, B. A. Bierl and R. J. Honour, *J. Am. Chem. Soc.*, **84**, 4687 (1962).
146. J. P. Jesson and H. W. Thompson, *Spectrochim. Acta.*, **13**, 217 (1958).
147. M. St. C. Flett, *Trans. Farday Soc.*, **44**, 767 (1948).
148. G. G. Gualberto and P. Sensi, *Ann. Chim* (*Rome*), **46**, 816 (1956).
149. H. W. Thompson and G. Steel, *Trans. Faraday Soc.*, **52**, 1451 (1956).
150. M. R. Mander and H. W. Thompson, *Trans. Faraday Soc.*, **53**, 1402 (1957).
151. M. F. Amr El Sayed, *Dissertation Abstr.*, **20**, 2593 (1960).
152. T. L. Brown, *J. Am. Chem. Soc.*, **80**, 794 (1958).
153. A. Cabana and C. Sandorfy, *Proc. Int. Meeting Mol. Spectry, 4th Bologna, 1959*, **2**, 750 (1962); *Chem. Abstr.*, **59**, 4679c (1963).
154. J. K. Wilmshurst, *Can. J. Chem.*, **37**, 1896 (1959).
155. M. F. Amr El Sayed, *J. Inorg. Nucl. Chem.*, **10**, 168 (1959).
156. P. P. Shorygin, M. A. Geiderikh and T. I. Ambrush, *Zh. Fiz. Khim.*, **34**(2), 335 (1960).
157. V. P. Roshchupkin and E. M. Popov, *Tr. Komis. po Spektroskopii, Akad. Nauk. SSSR*, **1964**(1), 193; *Chem. Abstr.*, **63**, 1723h (1965).
158. P. Sensi and G. Gallo, *Gazz. Chim. Ital.*, **85**, 224 (1955).

159. L. P. Krasnomolova and Yu. A. Kushnikov, *Izv. Akad. Nauk. Kaz. SSR, Ser. Khim.*, **17**(4), 31 (1967); *Chem. Abstr.*, **68**, 64330u (1968).
160. P. P. Shorygin, V. P. Roshchupkin and Kh. Khomenko, *Dokl. Akad. Nauk. SSSR*, **159**(2), 391 (1964).
161. M. F. Amr El Sayed and R. K. Sheline, *J. Inorg. Nucl. Chem.*, **6**, 187 (1958).
162. I. N. Yukhnovski, *Dokl. Akad. Nauk. SSSR*, **168**(5), 117 (1966).
163. I. N. Yukhnovski, *Compt. Rend. Acad. Bulg. Sci.*, **20**(2), 97 (1967); *Chem. Abstr.*, **67**, 69019p (1967).
164. G. Pala and T. Bruzzese, *Ann. Chim. (Rome)*, **54**(5), 349 (1964).
165. H. P. Figeys and J. Nasielski, *Spectrochim. Acta*, **22**, 2055 (1966).
166. M. Figeys-Fauconnier, H. P. Figeys, G. Geuskens and J. Nasielski, *Spectrochim. Acta*, **18**, 689 (1962).
167. D. G. I. Felton and S. F. D. Orr, *J. Chem. Soc.*, **1955**, 2170.
168. F. W. McLafferty, *Anal. Chem.*, **34**, 26 (1962).
169. W. Carpenter, Y. M. Sheikh, A. M. Duffield and C. Djerassi, *Org. Mass Spectry*, **1**, 3 (1968).
170. N. C. Rol, *Rec. Trav. Chim.*, **84**, 413 (1965).
171. R. Beugelmans, D. H. Williams, H. Budzikiewicz and C. Djerassi, *J. Am. Chem. Soc.*, **86**, 1386 (1964).
172. G. W. Griffin and L. I. Peterson, *J. Org. Chem.*, **28**, 3219 (1963).
173. J. H. Beynon, G. R. Lester, R. A. Saunders and A. E. Williams, *Trans. Faraday Soc.*, **57**, 1259 (1961).
174. Y. Kitahara and T. Kato, *Chim. Pharm. Bull. (Tokyo)*, **12**, 916 (1964).
175. A. L. Logothetis and G. N. Sausen, *J. Org. Chem.*, **31**, 3689 (1966).
176. V. H. Dibeler, R. M. Reese and J. L. Franklin, *J. Am. Chem. Soc.*, **83**, 1813 (1961).
177. J. H. Bowie, R. Grigg, S.-O. Lawesson, P. Madsen, G. Schroll and D. H. Williams, *J. Am. Chem. Soc.*, **88**, 1699 (1966).
178. D. H. Williams, R. G. Cooks, J. H. Bowie, P. Madsen, G. Schroll and S.-O. Lawesson, *Tetrahedron*, **23**, 3173 (1967).
179. J. H. Beynon, *Mass Spectrometry and Its Applications to Organic Chemistry*, p. 404, Elsevier, New York, 1960.
180. H. Negita and P. J. Bray, *J. Chem. Phys.*, **33**, 1876 (1960).
181. (a) B. M. Schmidt, L. C. Brown and D. Williams, *J. Mol. Spectry*, **3**, 30 (1959); (b) M. Witanowski, *Tetrahedron*, **23**(11), 4299 (1967).
182. A. D. Cross and I. T. Harrison, *J. Am. Chem. Soc.*, **85**, 3223 (1963).
183. J. C. Jacquesy, R. Jacquesy, J. Levisalles, J. P. Peter and H. Rudler, *Bull. Soc. Chim. France*, **1964**(9), 2224.
184. E. Lippmaa, A. Olivson and J. Past, *Eesti NSV Teaduste Akad. Toimetised, Fuusikilas Mat. Teaduste Seeria*, **14**(3), 473 (1965).
185. J. Hamelin, *Compt. Rend.*, **C283**(7), 553 (1966).
186. K. D. Bartle and J. A. S. Smith, *Spectrochim. Acta*, **A23**(6), 1715 (1967).
187. P. Dufey, J. Delmau and J. C. Duplan, *Bull. Soc. Chim. France*, **1967**(4), 1336.
188. M. Murano and R. Yamadera, *J. Polymer Sci.*, **A16**(4) 843 (1968).

CHAPTER 5

Directing and activating effects of the cyano group

WILLIAM A. SHEPPARD

E. I. du Pont de Nemours and Co.
Wilmington, Delaware, U.S.A.

I. SIZE AND SHAPE . . . 209
II. ELECTRICAL CHARACTER . . . 211
A. Dipole Moments . . . 214
B. Complexes . . . 216
C. Theoretical Calculations . . . 216
III. SUBSTITUENT EFFECTS . . . 217
A. The Cyano Group . . . 217
B. Cyanocarbon Groups . . . 221
IV. REACTIVITY . . . 226
A. Hydrocarbons . . . 226
B. Olefins . . . 228
C. Aromatic Derivatives . . . 230
V. SPECTRAL PROPERTIES . . . 231
A. Effect of Cyano Substitution on Infrared and Ultraviolet Spectra 231
B. N.m.r. Shielding Effects of the Cyano Group . . . 233
VI. SUMMARY . . . 235
VII. REFERENCES . . . 235

The cyano group is recognized as a powerful electron-withdrawing substituent. The purpose of this chapter is to quantitatively compare the cyano group to other well-known, strong electron-withdrawing groups (such as a nitro or trifluoromethyl) and to review and critically evaluate the changes in physical and chemical properties of a molecule that result from substitution of one or more cyano groups.

I. SIZE AND SHAPE

The size and shape of the cyano group make it unique relative to other substituents. The cyano group is effectively a rod surrounded by a cylindrical cloud of π-electrons which can interact with an adjacent

π-system regardless of rotatory orientation (see Figure 1, data for bond lengths and Van der Waal radii obtained from standard reference works[1]). It has a strong dipole, oriented with the negative end towards the nitrogen. In contrast, the nitro group is triangular with negative charge on the two oxygens which, if revolving, spreads the effective charge over a hemispherical area. It can interact with an adjacent π-system only when it is properly oriented; if forced to rotate, resonance interaction is lost. A trifluoromethyl group closely approximates a hemisphere with high charge density spread over the surface; it cannot interact effectively with a π-system although it does show some apparent resonance effects.

In substitution on an organic residue, the width (or diameter) of the group is an important factor; in this regard, the cyano group is much smaller than nitro or trifluoromethyl and about the same size as chloro or bromo groups.

An obvious consequence of the rod-like shape of a cyano group is that most organic molecules can be polysubstituted by cyano groups without steric interference, so that interaction with an adjacent π-system is not disturbed. One piece of experimental evidence in support of this argument is the u.v. spectra of some nitro and cyano *ortho*-substituted benzenes (Table 1). The *ortho*-methyl and *ortho*-cyano substituents on benzonitrile show a normal auxochromic effect with

TABLE 1. U.v. spectra of cyano- and nitro-substituted benzenes[a].

Compound	λ_{max} (mμ)	log ε	Reference
C_6H_5CN	221	4·08	4
	269	2·92	
	276	2·90	
o-$NCC_6H_4CH_3$	228	4·02	5
	276	3·15	
	284	3·16	
o-$C_6H_4(CN)_2$	236	4·01	4
	280	3·23	
	290	3·31	
$C_6H_5NO_2$	261	3·92	6
o-$O_2NC_6H_4CH_3$	259	3·72	6
o-$C_6H_4(NO_2)_2$	No maximum[b] over 210		7

[a] All spectra measured in methanol or ethanol.

[b] The broad absorption with no maximum observed for *o*-dinitrobenzene was interpreted as evidence that both nitro groups are forced out of the plane of the ring.

an increase in intensity of absorption. In contrast, an *ortho*-methyl substituent in nitrobenzene shows a slight hypsochromic shift with decreased absorption, and an *ortho*-nitro group so drastically distorts the interaction of the nitro groups with the ring that no well-defined absorption is observed.

The existence of stable percyano compounds such as hexacyanobenzene and tetracyanoethylene with no significant distortion of the basic carbon skeleton[2] is further testimony to the small steric requirements of the cyano group. In contrast, pernitro analogues are unknown or highly unstable. Even in hexabromobenzene the bromines are alternately bent 12° out of the plane of the ring[3], apparently to relieve steric interactions. However, this comparison is dangerous until more experimental data are available. The benzene ring is assumed planar in hexabromobenzene, whereas Littke and Wallenfels[2] do not give any information on the spatial position of the nitrile groups in hexacyanobenzene.

The small radius of the cyano group was also considered as a possible explanation for the absence of an appreciable enthalpy change in the axial–equatorial interconversion of the cyano group in 4-*t*-butylcyclohexanecarbonitrile[8]. An apparent anomalous slow rate of platinum-catalysed exchange reaction between *meta* aromatic ring protons in benzonitriles and deuterium oxide was attributed to unusual geometry of a nitrile group compared to other substituents[9]. In this case, the small radius was not critical, but the long length of the cyano group was thought to cause a steric problem in the bonding of the aromatic system to the catalyst surface, essential for *meta* exchange.

II. ELECTRICAL CHARACTER

Nitriles have high dipole moments which lead to strong dipole–dipole interaction and strong molecular associations. The electron distribution for a nitrile group, which leads to a strongly polar force field, is two π-orbitals directed at right angles to each other and centred mainly in the region between the C and N atoms, and a lone pair centred on the nitrogen atom. Thus the cyano group outwardly appears as a cylinder of π-electrons capped on the end by an unshared pair of electrons. The lone pair is probably best represented by a digonally hybridized *sp* orbital directed 180° along the axis of the C—N bond; it is no doubt largely responsible for the high dipole moment of nitriles, and is an excellent donor site for complexing with Lewis acids.

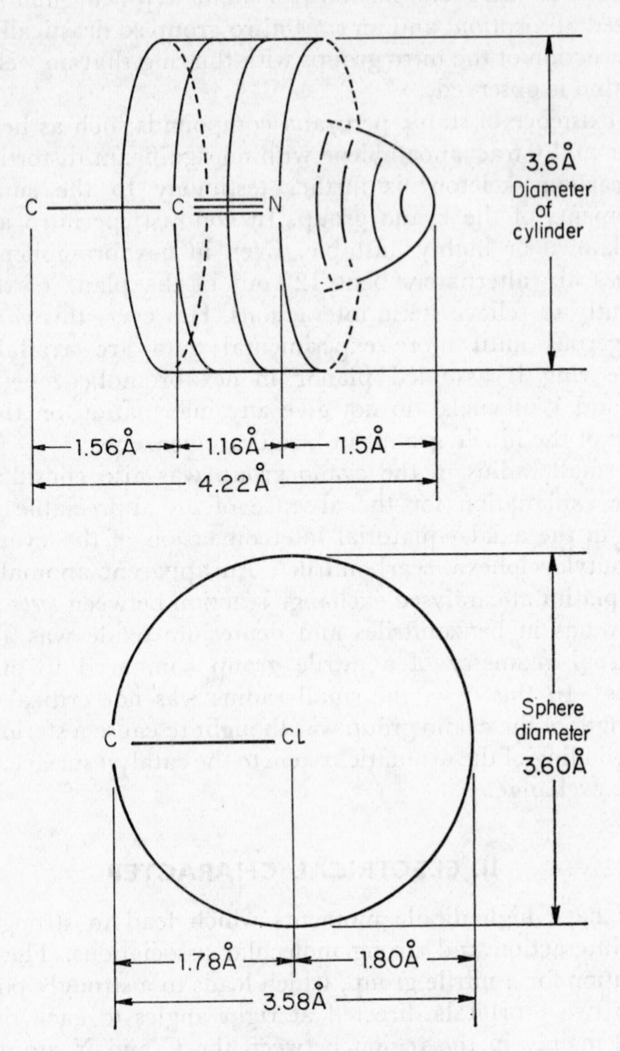

FIGURE 1. Size and shape of substituent groups.
Top: nitrile group, bottom: chloro group.

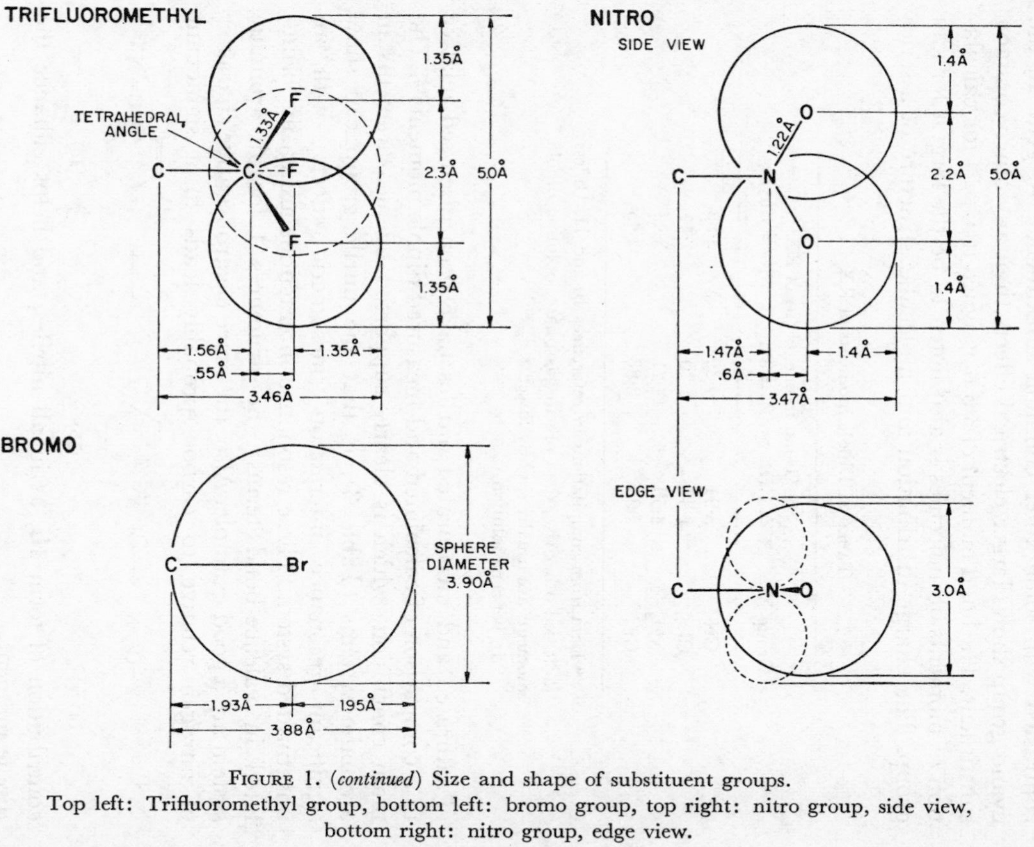

FIGURE 1. (*continued*) Size and shape of substituent groups.
Top left: Trifluoromethyl group, bottom left: bromo group, top right: nitro group, side view, bottom right: nitro group, edge view.

A. Dipole Moments

Dipole moments of a series of nitriles are given in Table 2 and contrasted with some other common substituents. In general, the cyano group shows larger electrical effects than any other common substituent. The bond moment of 3·6 D assigned to C≡N for calculation of moments of molecules is also noted to be the largest for any group. The charge distribution and molecular structure of a series

TABLE 2. Dipole moments of RX.

Group X	Group moment[a] (D) of RX R = CH_3	R = Ph	R = vinyl
CN	3·94	4·39	3·88
Cl	1·87	1·70	1·44
NO_2	3·50	4·21	3·44[b]
CF_3	2·35	2·86	2·45

[a] Data from compilations in references 10 and 11. Unless indicated otherwise, data are in gas phase. Solution dipole moments are usually slightly lower.
[b] In benzene solution.

of saturated and unsaturated nitriles have been discussed relative to a comparison of calculated and measured dipole moments[12]. The main conclusion, which is clearly supported by measurements in aromatic systems (Table 3), is that the nitrile group can show significant resonance interaction (mesomeric effects) with an unsaturated system. These resonance interactions are easily formulated in valence-bond terms; the structures **1** for an aromatic system are a good example. Also, the more electronegative character of nitrogen relative to carbon probably leads to a significant

$$\ddot{X}\text{—}C_6H_4\text{—}C{\equiv}N \longleftrightarrow \ddot{X}\text{—}C_6H_4\text{—}\overset{+}{C}{=}\overset{-}{N} \longleftrightarrow \overset{+}{X}{=}C_6H_4{=}C{=}N^-$$

(**1a**) (**1b**) (**1c**)

contribution of form **1b**, as in all nitriles, and helps enhance the dipole moment.

The data in Table 3 show the strong resonance interaction of both cyano and nitro groups with an electron-rich π-system, since the dipole moments are considerably enhanced over the predicted value. (Note that the resonance interaction for a trifluoromethyl

TABLE 3. Dipole moments and resonance interactions.

Compound	Dipole moment (D)[a] Measured	Calculated	Δ
$H_2N-C_6H_4-CN$	6·46	5·4	1·06
$H_2N-C_6H_4-NO_2$	6·35	5·2	1·15
$H_2N-C_6H_4-CF_3$	4·28	3·86	0·44
$H_2N-C_6(CH_3)_4-NO_2$ (with H_3C, CH_3, H_3C, CH_3 ring substituents)	4·98	~5	~0

[a] Data from references 13 and 14.

group is surprisingly large although no classical resonance interactions are possible[15].) The nitro group appears to be slightly better than the cyano group at resonance electron withdrawal. However, *ortho*-methyl substituents prevent the nitro group from achieving coplanarity with the ring, and resonance interaction is lost in aminonitrodurene. Because of its size and shape, the cyano group is predicted to still show a resonance-enhanced dipole moment in a compound such as aminocyanodurene. Data in Table 4 provide experimental support for this prediction. The difference in dipole moments of the fluorobenzene derivatives for pairs 1 and 2 relative

TABLE 4. Dipole moments of fluorobenzene derivatives[16].

Compound	Dipole moment (D)
1. $C_6F_5NO_2$	1·94
C_6F_5CN	2·34
2. *p*-$HC_6F_4NO_2$	3·28
p-HC_6F_4CN	3·62
3. *p*-$FC_6H_4NO_2$	2·69
p-FC_6H_4CN	2·67

to pair 3 is best explained as steric inhibition of resonance of the nitro group by *ortho*-fluorines, whereas the cyano group shows no such effect.

B. Complexes

As pointed out above, the unshared pair of electrons and the electronegative character of nitrogen make nitriles excellent donor compounds. Stable 1:1 molecular addition complexes are formed between aliphatic nitriles and a variety of good acceptor molecules such as boron halides[17–19] (see Chapter 3). x-Ray analysis[18] has shown that the acetonitrile–boron trifluoride complex has the linear structure expected from complexing on the unshared electron pair protruding from the nitrogen. The bond lengths in the complexed acetonitrile are not changed significantly. With stronger acceptor

$$H_3C{-}C{\equiv}N \cdots BF_3$$

1·44 Å 1·13 Å 1·64 Å

molecules, the π-electrons on the nitrile can also donate; thus complexes such as $CH_3CN{\cdot}5HCl$ can be detected by freezing-point studies[19]. A consequence of the high complexing ability of cyano groups is that nitriles often strongly associate in solution[19], and solvent interactions may be large. Such interactions can change the electronic and steric character of the organic molecule to which the CN is attached and alter chemical reactivity. For example, the ^{19}F nuclear magnetic resonance chemical shift in *p*-fluorobenzonitrile changed up to 13 p.p.m. (downfield) when the compound was complexed with a series of boron reagents[20a]. Again the existence of 1:1 complexes was demonstrated, and the relative shift was shown to provide a quantitative measure of acceptor strength of the boron complexing agent. A similar type complexing (called specific solvation) of nitriles with dimethyl sulphoxide was shown to be of importance in the equilibrium of *cis*- and *trans*-4-*t*-butylcyclohexylcarbonitrile and on the relative rates of borohydride reduction of cyclohexanone and dihydroisophorone[20b].

C. Theoretical Calculations

The simple orbital picture given above for a cyano group is adequate to explain qualitatively the chemical and physical properties of cyanocarbons. A more quantitative picture should be derived by molecular-orbital calculations, but no detailed calculation on cyanocarbons has been published. σ- and π-bonding effects in the coordination of the cyano group has been discussed, and a series of LCAO–MO calculations on both free and coordinated cyano groups, including CN^-, NCN^{-2}, OCN^-, ClCN, CH_3CN, HCN and CH_3NC, were correlated with Lewis-base properties and nuclear quadrupole resonance data[21].

Qualitative theoretical calculations were used to investigate the factors that stabilize the bis(thiocarbonyl) or 1,2-dithiete nuclei of the substituted $[C_2S_2]$ system **2**[22]. The results qualitatively agreed

(**2**)

X = CN, CF_3, $N(CH_3)_2$

with experimental observations on the stability and chemical reactivity of the system.

III. SUBSTITUENT EFFECTS

The most common method of quantitatively evaluating the electronic effects of a group is by substituent constants. These are determined experimentally through pK_a measurements or kinetic rate studies on a probe group, such as carboxyl, ammonium or hydroxyl, that is sufficiently remote from the substituent to have no significant steric effect[23]. Recently, ^{19}F nuclear magnetic resonance of fluorobenzenes has provided a useful technique for measuring substituent parameters, particularly for examining the effects on the molecule in an undisturbed ground state[24] and for studying solvent or complexing interactions[20a,25].

A copious amount of work has been reported on substituent effects, and a variety of schemes have been proposed to classify the effects and assign substituent parameters[23]. For simplicity, we restrict our discussion to the standard σ parameters as first proposed by Hammett[23d] and the inductive and resonance parameters, σ_I and σ_R, as defined by Taft[23].

A. The Cyano Group

The inductive effect of a group is determined directly by pK_a measurements on saturated aliphatic systems or indirectly by analysis

TABLE 5. pK and rate data used to determine substituent effects.

System		pK or rate parameters for substituent X				Comments	Reference
		H	CN	NO_2	Cl		
Ionization of acetic acids, XCH_2CO_2H, in H_2O, 25°, pK_a		4·76	2·43	1·68	2·86	Calculated σ_I values; CN 0·56, NO_2 0·63	26
Ionization of [2,2,2]-bicyclooctanecarboxylic acids (4-X-bicyclo[2.2.2]octane-1-CO_2H)	pK_a in water, 25°	6·75	5·90			Study made to determine relative importance of through-space versus through-molecule transmission of inductive effects	27
	log K/K_H (in 50% ethanol–water, 25°)	0·00	0·93	1·058	0·739		
Ionization of benzoic acids, $XC_6H_4CO_2H$, water, 25°, pK_a	*meta*	4·20	3·60	3·45	3·83	Classical system for establishment of Hammett σ-parameter scale	26, 28
	para		3·55	3·44	3·99		
Ionization of phenols, XC_6H_4OH, water, 25°, pK_a	*meta*	9·99	8·61	8·38	9·13	*Para* values used to calculate σ parameters	29a, b
	para		7·97	7·15	9·42		
Ionization of anilinium ions, $XC_6H_4\overset{+}{N}H_3$, water, 25°, pK_a	*meta*	4·60	2·75	2·46	3·52	*N,N*-dimethylanilines give comparable results; largest resonance effects for electron-withdrawing groups shown by this system	29a, b
	para		1·74	1·00	3·98		
Solvolysis of *t*-cumyl chlorides, $XC_6H_4CCl(CH_3)_2$ in 90% aqueous acetone at 25°, rate constant, $10^5\, k_1 (s^{-1})$	*meta*	12·4	0·0347	0·0108	0·194	System for defining the σ^+ parameters	29c, d
	para		0·0126	0·00319	3·78		

of the substituent parameters in aromatic systems[23a]. The extent of resonance interaction of a substituent varies with the electron supply or demand in the remainder of the molecule, and is most easily determined from the difference between the σ_m and σ_p values measured in aromatic systems. More extensive analysis can be used to calculate σ_R, the resonance parameter. Pertinent pK, rate data, and ^{19}F n.m.r. chemical shifts used to measure the electronic effect of the cyano group are given in Tables 5 and 6. These data clearly show the strong electron-withdrawing effect of the cyano group as compared with other common substituents. Substituent parameters calculated from the pK_a of anilinium ions and ^{19}F measurements on fluorobenzenes are given in Table 7, together with comparison data for other substituents. The cyano group is clearly almost as powerful an

TABLE 6. Electronic effects of cyanocarbon substituents: ^{19}F n.m.r. chemical shift measurements on substituted fluorobenzenes and substituent parameters[24,30,32].

	δ in solvent in p.p.m.[a]					
Substituent	CCl_4 or $CFCl_3$	Benzene	CH_3CN	Methanol	σ_I	σ_R
CN *meta*	−2·69	−2·75	−2·68	−3·10	0·47	
para	−9·07	−9·05	−9·90	−10·25		0·22
NO_2 *meta*	−3·45	−3·40	−3·33	−3·53	0·56	
para	−9·45	−9·20	−10·35	−10·35		0·20
CH_3 *meta*	1·18	1·10	1·13	1·23	−0·08	
para	5·40	5·40		5·55		−0·14
CH_2CN *meta*	−1·1	—	—		0·24	
para	1·20		—	1·35		−0·08
$CH(CN)_2$ *meta*[32]	—	−3·29	−2·85	−3·13	0·55	
para	—	−2·30	−2·27	−2·41		−0·03
$C(CN)_3$ *meta*	−6·46	−6·33	−5·04	−5·38	0·98	
para	−6·71	−5·80	−5·80	−5·84		−0·02
$CH{=}CH_2$ *meta*	0·65	0·60	0·30	0·58	0·01	
para	1·40		—	1·30		−0·03
CH=CHCN *meta*	−1·31	—	—	—	0·27	
cis *para*	−4·98	—	—	—		0·10
CH=CHCN *meta*	−1·28	—	—	—	0·27	
trans *para*	−4·58	—	—	—		0·09
$CH{=}C(CN)_2$ *meta*	—	−2·27	−2·37	−2·56	0·41	
para	—	−10·93	−10·95	−11·25		0·29
$C(CN){=}C(CN)_2$ *meta*	—	−4·12	−3·62	−3·80	0·67	
para	—	−13·19	−12·71	−12·48		0·31

[a] At infinite dilution relative to fluorobenzene.

TABLE 7. Substituent parameters[30].

Group	Method[a]	σ_m	σ_p	σ_I	σ_R
CN	A	0·65	1·02	0·53	0·41
	B	0·56	0·66	0·52	0·08
	F	0·58	0·69	0·47	0·22
$CH{=}C(CN)_2$	A	0·55	1·20	0·43	0·75
	F	0·45	0·70	0·41	0·29
$C{=}C(CN)_2$ (with CN on C)	A	0·77	1·70	0·60	1·08
	F	0·83	0·98	0·67	0·31
$C(CN)_3$	F	0·97	0·96	0·98	−0·02
NO_2	A	0·70	1·17	0·55	0·61
	B	0·71	0·78	0·68	0·14
	F	0·67	0·77	0·56	0·20
CF_3	A[31]	0·49	0·62	0·44	0·18
	F[31]	0·44	0·49	0·39	0·10
SO_2CF_3	A	1·00	1·65	0·84	0·73
	F	0·90	1·06	0·75	0·31
Cl	A	0·37	0·23	0·45	−0·20
	F	0·36	0·26	0·44	−0·18

[a] Method F, from ^{19}F chemical shift measurements in benzene; Method A, from pK_a of anilinium or *N,N*-dimethylanilinium ions; Method B, from pK_a of benzoic acids.

electron-withdrawing substituent as the nitro group, both by inductive and resonance mechanisms as judged by the high positive value of σ. In fact, it appears to be the most powerful electron-withdrawing substituent of the groups composed of one or two atoms*. This result is in accord with the qualitative molecular-orbital picture of electron density distribution as discussed in section II. The large, highly directional dipole of the cyano group has particular consequence in transmission of substituent effects.

This question is discussed in the following section on cyanocarbons. However, the relative changes in σ_I and σ_R values for cyano compared with nitro, as determined by different methods, must

* The nitroso group (NO) appears to withdraw electrons more strongly by resonance interactions but is inductively weaker than a cyano group[24]. The halogens, which are highly electronegative and inductively strong electron-withdrawing groups (but electron-donating by resonance because of the unshared electron pair), are not as effective as cyano groups.

originate from the different type of charge distribution in the two groups.

The reduced steric requirements of cyano relative to nitro is also shown by pK_a data of phenols given in Table 8. Methyl groups *ortho* to the interacting substituent prevent resonance interaction

TABLE 8. Steric effects in pK_a's.

Compound	pK_a Measured	pK_a Calculated	Reference
HO–$C_6H_2(CH_3)_2$–CN (3,5-dimethyl-4-cyanophenol)	8·21	8·21	35
HO–$C_6H_2(CH_3)_2$–NO_2 (3,5-dimethyl-4-nitrophenol)	8·25	7·40	36
$H_3\overset{+}{N}$–$C_6H_2(CH_3)_2$–NO_2	2·49	1·16	37

of nitro with a *p*-hydroxy or a *p*-ammonium group, while the cyano group gives the expected resonance enhancement of pK_a (inductive effect of the methyl groups is taken into account).

B. Cyanocarbon Groups

An accumulation of cyano groups leads to molecules with unusual properties, both physical and chemical (see Chapter 9). A qualitative estimation of the electronic effects of cyano substitution on carbon has been made[33] by the comparison of $>C{=}$C–CN to =N–, =C(CN)$_2$ to =O (carbonyl oxygen) and –C(CN)$_3$ to F.

However, quantitative determination of changes in electronic properties of various groups when highly substituted by cyano groups has been measured by substituent effect studies[30].

For a methyl group, the electronic changes that occur when the hydrogens are systematically replaced by cyano groups have been measured by ^{19}F n.m.r. chemical shifts in fluorotoluenes (see

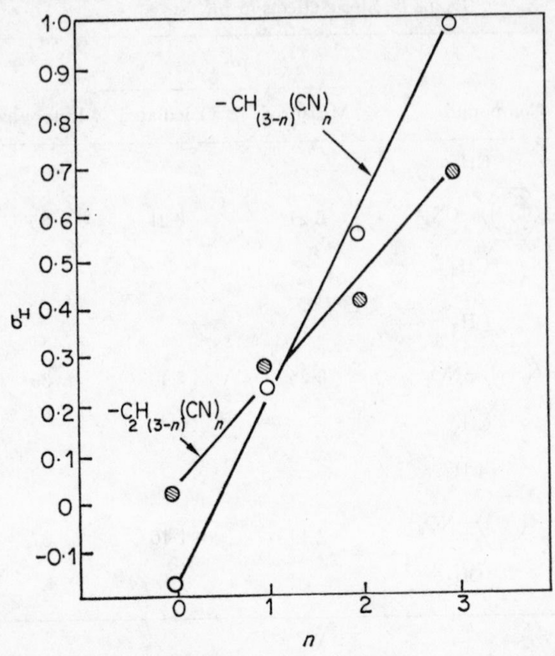

FIGURE 2. Inductive effect of methyl or vinyl groups versus cyano substitution.

Table 6). As expected, the strong inductive withdrawal by three cyano substituents drastically changes the electronic properties of the methyl group. Inductively, the tricyanomethyl group, $C(CN)_3$, is one of the strongest electron-withdrawing groups that has been reported[30,34]. The inductive withdrawal by a methyl group increases linearly with cyano substitution (Figure 2). Overall, the inductive effect appears larger than predicted. This fact is most clearly shown for the tricyanomethyl group, which is predicted to have a σ_I value of 0·5–0·7 (calculated by using σ_I of 0·47 for the cyano group and a transmission factor of 0·3–0·5 per carbon) instead of approximately 1·0. This enhanced inductive effect is suggested to result from a through-space interaction; the dipoles of the cyano groups on the tetrahedral carbon of the methyl group are oriented in a different direction and should have more through-space effect than a cyano group directly on the aromatic ring. This conclusion is supported by abnormal solvent effects for the percyano substituted groups. Normally, more polar solvents such as acetonitrile or hydrogen-bonding solvents like methanol enhance the downfield shifts induced by strong neutral electron-withdrawing groups (note nitro and cyano in Table 6), but for tricyanomethyl and tricyanovinyl groups the largest downfield shifts are for the less polar solvents. In methanol or acetonitrile[30b], the shift is upfield relative to that in benzene or $CFCl_3$. A dampening of the transmittance of through-space effect by hydrogen bonding or dipole–dipole interactions must occur. If the transmittance of charge effect were mainly through the molecule, then interaction with polar solvents, particularly on the negative end of the dipole on nitrogen, should enhance the electron-withdrawing effects.

The resonance effect for the tricyanomethyl group varies from small negative to positive values depending on the solvent used for the ^{19}F chemical shift measurement. The dicyanomethyl and cyanomethyl groups show increasing −R effects, approaching that for the methyl group*. Variations in the resonance effects, particularly for the tricyanomethyl group, must result from specific solvent interactions. No mutual interactions of the nitro or cyano type are possible. Small +R effects could arise by hyperconjugation interactions involving cyanide ion such as in **3** (in analogy to fluoride

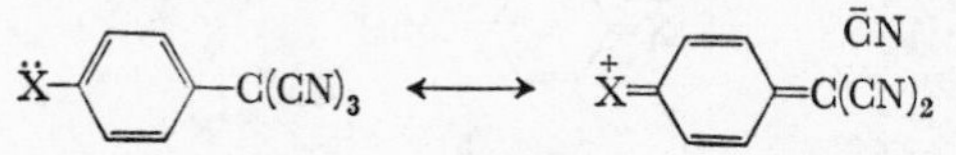

$$\ddot{X}\text{—}C_6H_4\text{—}C(CN)_3 \longleftrightarrow \overset{+}{X}\text{=}C_6H_4\text{=}C(CN)_2 \;\; \bar{C}N$$

(3)

ion hyperconjugation proposed for the CF_3 group[31]), by a π-inductive mechanism[38], or by some sort of p–π interaction[31] (using p-electrons of CN). In any case, the resonance effects are minor. In fact, the negligible resonance effects for $C(CN)_3$ are argued as strong evidence against the proposal that the +R effect of a CF_3 group results by a π-inductive mechanism. However, the differences in resonance effects of CF_3 and $C(CN)_3$ may result from grossly different charge distribution in F and CN; in other words, the differences may be a simple consequence of diffuse versus highly

* The American convention as defined by Taft is used throughout this discussion. A negative value means electron-donating and positive means electron-withdrawing, in contrast to the English school which employs the reverse convention.

oriented dipoles. This problem merits more study. A final point is that a gradual regular change in $-R$ for dicyanomethyl and monocyanomethyl groups can be used to argue against any significant contribution from hyperconjugation interaction of the acidic proton as in **4**. However, this effect may only be observed by using an electron-poor probe group X.

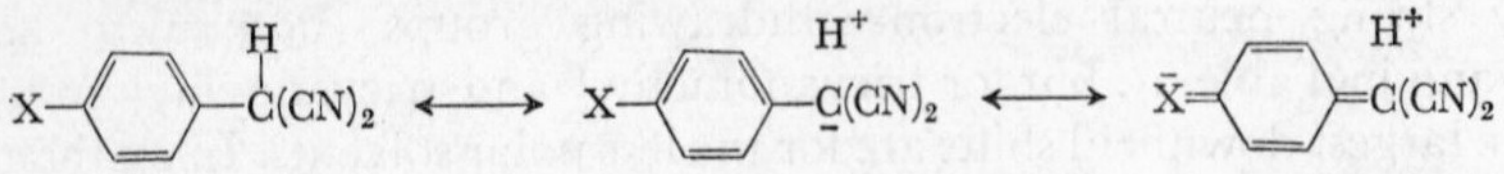

$$X-C_6H_4-C(H)(CN)_2 \longleftrightarrow X-C_6H_4-\bar{C}(CN)_2\;H^+ \longleftrightarrow \bar{X}=C_6H_4=C(CN)_2\;H^+$$

(**4**)

A large change in the electronic properties of a vinyl group also results from cyano substitutions. The tricyanovinyl group withdraws electrons much more strongly than nitro from both the *meta* and *para* positions (σ_m and σ_p values given in Table 7 are larger); its overall effect is almost as strong as that of the SO_2CF_3 group (which is reported to be the strongest neutral electron-withdrawing group known[39]). The relative extent to which resonance interactions and inductive effects contribute to the overall electron withdrawal is estimated from the inductive and resonance parameters σ_I and σ_R. The inductive effect is enhanced slightly, but most of the increase in electron-withdrawing power of the tricyanovinyl group over cyano or nitro groups appears to come from very large resonance interactions. A major contribution by a charge-separated resonance form **5**

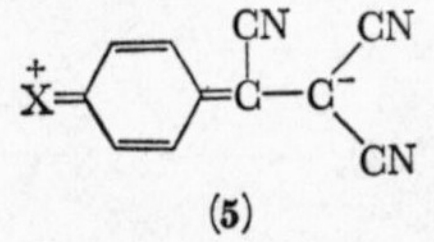

$$\overset{+}{X}=C_6H_4=C(CN)-\bar{C}(CN)_2$$

(**5**)

is expected since the negative charge is strongly stabilized by the two cyano groups α to it. Indeed, hydrocyanocarbons are highly acidic and sometimes can only be isolated as an anion[40]. The β,β-dicyanovinyl group is also strongly electron-withdrawing; it does not appear to be as strong inductively as a nitro group but again shows strong resonance interaction. The tricyanovinyl group shows an enhanced resonance effect over the dicyanovinyl group which is difficult to explain by a classical resonance picture. Apparently the α-cyano group interacts with the π-ethylenic system by an inductive mechanism or through orbital overlap to provide an additional stabilization for form **5**. The stabilizing effect of cyano groups on quinomethanes (e.g. tetracyanoquinodimethane[41]) may be related to this enhancement of the resonance effect.

A similar question has arisen over the relatively large intensity of absorption for 2,6-dimethyl-4-(1,1,2,2-tetracyanomethyl)aniline (λ_{max} 264 mμ, ε 11,700)[42]. In this case, no-bond resonance structures can be drawn (direct interactions with the NH_2 group analogous to those in **3** for a $C(CN)_3$ group), but some better explanation is needed.

Even one β-cyano group significantly changes the electron-withdrawing power of the vinyl group. Again an almost linear correlation is found for the σ_I value of the vinyl group with the extent of cyano substitution (see Figure 2). However, the effect is not as marked as with the methyl group, probably because of the greater distance of the β-cyano substituents. The β,β-dicyanovinyl group does appear to be significantly weaker inductively.

The negligible difference in electronic properties between the *cis* and *trans* forms of the β-cyanovinyl group (Table 6) was not expected. The size of cyano is such that even in the *cis* conformation, the vinyl group can obtain coplanarity with the ring (the coplanar configuration is shown possible on models and substantiated by study of ultraviolet spectra[43]). Consequently, the resonance interaction is expected to remain essentially unchanged, as is found when judged by the σ_R values. However, the field effect through space of the cyano group in the *cis* form (**6**) is expected to be drastically different from that of the cyano group in the *trans* configuration (**7**), but instead the inductive effect is found to be the same (σ_I of 0·27 for both). A possible explanation is that the negative end of the nitrile dipole is oriented towards the ring in the *cis* form and fortuitously cancels out enhanced field effect from the close proximity. An interesting

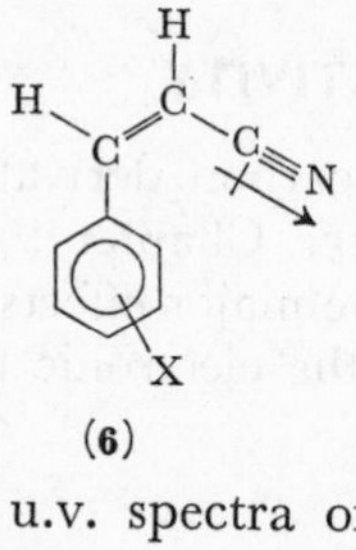

(**6**) 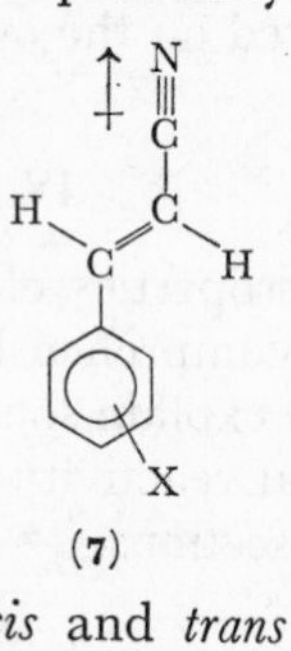 (**7**)

point is that the u.v. spectra of the *cis* and *trans* forms are also almost identical.

The effect of cyano groups on stabilizing the cyclopentadienide system has been clearly shown by Webster[44]. The available electron density of tetracyanocyclopentadienide as a substituent can be

TABLE 9. Acidities of tetracyanocyclopentadienide derivatives relative to benzene analogues[44].

	pK_a (H_2O)		pK_a (H_2O)
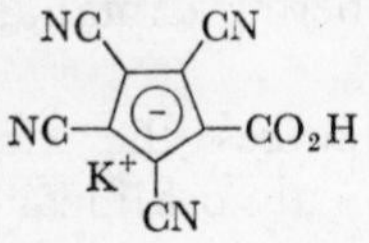 Potassium tetracyanocyclopentadienide carboxylic acid (K^+; NC, CN, NC, CN, CO_2H)	4·1	$PhCO_2H$	4·2
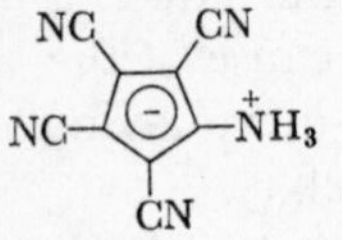 Tetracyanocyclopentadienide with $\overset{+}{N}H_3$ (NC, CN, NC, CN)	2·0	$Ph\overset{+}{N}H_3$	4·6
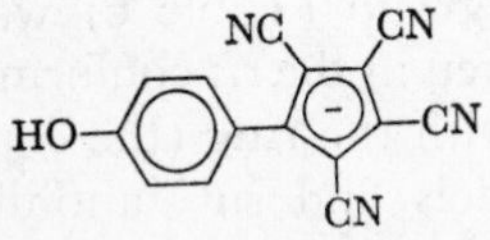 HO–C₆H₄–tetracyanocyclopentadienide (NC, CN, CN, CN)	9·66 $\sigma_p = 0·12$	PhOH	9·92

estimated from pK_a measurements given in Table 9. Inductively, the tetracyanocyclopentadienide system is slightly more electronegative than a benzene ring (note pK_a of the acid corresponding to benzoic acid and σ_p value of 0·12 from pK_a of the phenol derivative). This means that the negative charge on the cyclopentadienide system is completely distributed on the cyano groups; additional charge density, if available, such as from an NH_2 substituent, can also be distributed on the cyano groups.

IV. REACTIVITY

The chemical properties of cyanocarbon derivatives have already been reviewed comprehensively (see Chapter 9). Our purpose in this section is to explain some of the major effects of cyano substitution on chemical reactivity using the electronic pictures developed in the previous sections.

A. Hydrocarbons

The most impressive chemical property of polycyanocarbons is that they are often strong acids and form very stable salts. A negative charge is stabilized by a cyano group both inductively and, when alpha to the charge, by resonance interactions:

$$\mathrm{N{\equiv}C{-}\overset{|}{\underset{|}{C}}H \xrightarrow{-H^+} N{\equiv}C{-}\overset{|}{\underset{|}{C}}^- \longleftrightarrow {}^-N{=}C{=}C\!\!<}$$

Several cyano groups adjacent to the negative charge provide a cumulative effect that leads to highly stable anions and even dianions. These anions are useful synthetic intermediates (see Chapter 9, section XI). However, from a physical point of view they provide an interesting new class of anions in which the negative charge is highly delocalized over the molecular surface.

Boyd[40,45] carried out an extensive study of cyanocarbon acids and salts, and developed quantitative correlations of structure with acidity and the Hammett-type acidity scale H_- for cyanocarbon anions. The effect of cyano groups on acidity is clearly illustrated by the data in Table 10. The acidity of hydrogen increases by large increments (several orders of magnitude) with each additional α-cyano group. The effect on the acidity of the cyanohydrocarbon

TABLE 10. Effect of cyano groups on acidity.

Compound		Acidity (pK_a)	Reference
CH_4		~58; chemically very inert;	49
CH_3CN		~25; slightly acidic; reacts only with very strong bases in aprotic systems	
$CH_2(CN)_2$		11·2	50
$CHBr(CN)_2$		~5	50
$CH(CN)_3$		−5·1	40, 45
CH_3NO_2		10·2	50
$CH_2(NO_2)_2$		3·37	51
$CHCl(NO_2)_2$		3·80	51
$CH_3CO_2CH_3$		Hydrogens chemically inert	
$CH_2(CN)CO_2C_2H_5$		>9	50
$CHBr(CN)CO_2C_2H_5$		~6	40, 45
$CH(CN)_2CO_2CH_3$		−2·9	40, 45
	n		
H H cyclopentadiene–$(CN)_n$	0	15	46
	1	9·78	
	2	0·81 to 2·40 (4 isomers)	
	3	−6·1 to −7·2 (2 isomers)	
	4	−9	

is much more striking than in substituent effect studies because the cyano groups are at the site of reaction instead of at a remote location in the molecule. The stronger electron-withdrawing effect of the nitro over the cyano group is clearly shown by the more acidic character of the nitromethanes. Again the small cylindrical cyano group has a distinct advantage in that steric interactions are at a minimum even when the cyano is attached to a highly substituted reaction site (compare pK_a of halomethanes in Table 10).

Webster has studied the acidity of a series of polycyanocyclopentadienes[46]. Again the pK_a decreases by increments of 5 to 6 (10^6 in terms of ionization constants!) with substitution of each additional cyano; in this series the cyano group is not alpha to the ionizable hydrogen but in the unsaturated system. In the extreme cases, tetracyanocyclopentadienide salt is only protonated by extremely strong acids such as perchloric in acetonitrile, and the conjugate acid of pentacyanocyclopentadienide salt (where a cyano group is alpha to the proton) is too strong to be prepared with available reagents. As pointed out in section III.B, the four cyano groups in tetracyanocyclopentadienide completely delocalize the negative charge, so that the cyclopentadienide ring is about equivalent in reactivity to a normal benzene ring.

B. Olefins

Polycyanoolefins are highly electrophilic because the electron density of the π-bond is delocalized into the cyano substituents. As a result, these olefins are abnormal compared to hydrocarbon olefins in that they

(a) are highly reactive to attack by nucleophiles,
(b) cycloadd to electron-rich olefins and dienes,
(c) interact with π-bases to form stable charge-transfer or π-complexes,
(d) add an electron to form stable ion radicals.

The first two topics are adequately covered in Chapter 9. The subject of charge-transfer complexes has recently been reviewed[47], and molecular complexes of polycyano compounds are discussed in Chapter 10. Charge-transfer complexes are proposed as intermediates in some chemical reactions of cyanocarbons. The strength of the complex is dependent on both the structure of the cyanocarbon and the π-base; the relative acceptor strength is readily determined from the energy (λ_{max}) of the charge-transfer absorption in the ultraviolet spectra. In similar structural types, the acceptor strength is proportional to the number of cyano substituents on the π-system. Another question of interest is the extent of charge transfer or ionization in the complex. The current opinions, based on both experimental data and theoretical calculations, are that ionization is negligible[48], particularly of aromatic π-bases with cyanocarbons, and that stabilization is the result of frontier orbital interactions.

The ^{19}F n.m.r. chemical shifts in fluoro(tricyanovinyl)benzenes

TABLE 11. Effect of π-complexing with *N,N*-dimethyl-*p*-toluidine (DMT) on ^{19}F chemical shifts[30].

System (%)	Estimated percentage of π-complexes	$-\delta$ of $FC_6H_4C(CN){=}C(CN)_2$ relative to PhF (p.p.m.) *meta*	*para*
DMT/CCl_3F (98:2)	80–100	3·17	11·96
$DMT/CH_3CN/CCl_3F$ (50:48:2)	50–100	3·63	12·81
$DMT/CH_3CN/CCl_3F$ (25:73:2)	30–80	3·67	12·87
$Dioxane/CCl_3F$ (98:2)	0	3·65	11·86

are a good probe to determine changes in electron density in the π-acid as a result of π-complexing, particularly since the ^{19}F shift is very sensitive to changes in the π-system[24]. Any ionization in complexing should cause a large shift upfield relative to uncomplexed form; however, the shifts were in the range noted for solvent effects (Table 11 and compare also with data in Table 6). This result supports the other evidence that no significant ionization occurs in the complex.

The high electron affinity that makes tetracyanoethylene (TCNE) among the strongest of π-acids can be rationalized by simple molecular-orbital calculations that suggest that the lowest vacant π-orbital of the molecule has bonding characteristics (see Chapter 10 and reference 52). A direct consequence is that TCNE is easily reduced to a stable ion radical,[53] $TCNE^{\overline{\cdot}}$, and a variety of other stable polycyanocarbon ion radicals are easily prepared[54,55]. The complete delocalization of the odd electron into the cyano groups is clearly shown by e.p.r. studies[52,56,57].

C. Aromatic Derivatives

In electrophilic aromatic substitution reactions, +R substituents are *meta*-directing, whereas −R substituents direct *ortho–para*. Cyano, like nitro, is a typical +R substituent and gives chiefly *meta* orientation in typical electrophilic reactions such as nitration and chlorination. Of course, both groups strongly deactivate the aromatic ring towards electrophilic substitution because the combination of +I and +R effects destabilize intermediates or transition states formed

TABLE 12. Isomer distributions in the chlorination and nitration of monosubstituted benzenes[58].

Substituent group	Chlorination				Nitration			
	o-	*m*-	*p*-	o_f/p_f	*o*-	*m*-	*p*-	o_f/p_f
Nitro	17·6	80·9	1·5	5·9	6·4	93·2	0·3	11·0
Cyano	23·2	73·9	2·9	4·0	17·1	80·7	2·0	4·3
Trifluoro-methyl	15·7	80·2	4·1	1·9	—	—	—	—
Bromo	39·7	3·4	56·9	0·35	36·5	1·2	62·4	0·3
Chloro	36·4	1·3	62·3	0·29	29·6	0·9	69·5	0·21
Fluoro					8·7		91·3	0·05

by attack of the X^+ reagent. Polycyanoaromatic derivatives are unreactive to electrophilic substitution.

Some isomer distribution data are given in Table 12 for several +R groups and, for comparison, some −R groups (halogens). For +R groups, cyano appears to give the largest amount of *ortho* substitution. However, the *ortho–para* ratio is higher for a nitro than a cyano group. These results can be rationalized in terms of the smaller steric size of the cyano group and more effective dipole interaction of the nitro group (due to geometry) which positions the X^+ reagent for *ortho* substitution. In the halogens, the largest group surprisingly gives the largest amount of *ortho* substitution.

Another important reaction of cyano-substituted aromatic derivatives is nucleophilic substitution[59]. Strong electron-withdrawing substituents such as cyano, nitro, carboxy, or sulphonyl groups activate substituents such as halogens, alkoxy, or thioalkyl to replacement by nucleophilic reagents such as alkoxides, amines, azide or thiocyanate. Because a large negative charge develops in the transition state (note that the rate of substitution is proportional to the σ^- parameter of the group), nitro is a slightly better activating group than cyano. Both of these groups can accept electrons by resonance and activate the *ortho* and *para* positions much more strongly than the *meta*. Many important activating groups, nitro in particular, must be coplanar with the ring to exert their maximum activating effect. *Ortho* substituents sterically prevent nitro groups from becoming coplanar and can decrease the rate of reaction as much as 10^3. However, cyano groups do not show appreciable steric effect on rate[60] because the resonance interaction is not disturbed by *ortho* substituents.

Polycyanobenzene derivatives are particularly reactive towards nucleophilic displacement[61]. As examples, the halides in 1-halo-2,4,6-tricyanobenzenes are displaced extremely easily[61]. Note, however, that more drastic conditions are required for the dicyano derivative compared with the tricyano compound[61a].

Br, NC, CN, CN —(morpholine, 25°)→ N(morpholino), NC, CN, CN

Br, CN, CN —(morpholine, 80°)→ N(morpholino), CN, CN

F, NC, CN, F, F, CN —(NH_3, benzene, 25°)→ NH_2, NC, CN, H_2N, NH_2, CN

V. SPECTRAL PROPERTIES

A. Effect of Cyano Substitution on Infrared and Ultraviolet Spectra

The effect of cyano groups on infrared absorptions is that normally resulting from substitution by strong electron-withdrawing groups.

In an unsaturated system, the nitrile can conjugate to give the expected shifts (frequency decrease of about 30 cm^{-1}). However, direct comparisons are difficult because of practical problems, such as loss of absorption due to symmetry, and complexity in spectral absorption leading to overlap of absorption, uncertainty in assignments and splitting. Comparison of Raman absorptions should be much better. Of course, the nitrile absorption in the region of 2,215 to 2,275 cm^{-1} responds to structural changes in the rest of the molecule and is useful for characterization purposes[62].

Ultraviolet spectral absorptions of cyano compounds also only show the expected effects from an electron-withdrawing substituent

TABLE 13. Ultraviolet spectra comparison.

Compound	X	U.v. spectra[a] λ_{max}	U.v. spectra[a] ε
PhX	CN	276	800
		269	830
		229	10,400
		221	12,000
	$CO_2C_2H_5$	280	750
		273	924
		228	12,500
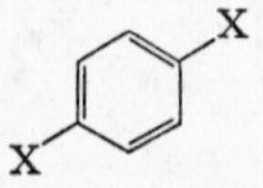 p-XC_6H_4X	CN	290	1600
		280	1600
		245	24,500
		236	21,500
	CO_2CH_3	294	1400
		285	1670
		240	19,600
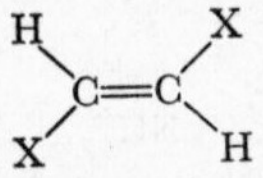 trans-HXC=CXH	CN	220	16,000
	CO_2CH_3	211	16,500
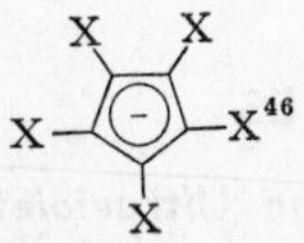 $C_5X_5^-$ (cyclopentadienide)[46]	CN	291	10,250
		255	95,500
	CO_2CH_3	295	15,000
		265	50,000

[a] Spectra in ethanol or methanol. Unless indicated otherwise, data from references 4 and 67.

and from conjugation with an unsaturated system. Empirically, the ultraviolet spectra of nitrile derivatives closely resemble those of the corresponding carboalkoxy analogues[63] (see Table 13 for comparative data). This type of comparison can be used as evidence in structure determinations[44]. The u.v. absorption of cyanocarbon anions are usually very intense, with lower energy shifts than the corresponding unsaturated system. For example, pentacyanopropenide ion[64] has λ_{max} 412 and 393 mμ (ε 22,100 and 22,600) compared to hexacyanobutadiene[65] with λ_{max} 302 mμ (ε 15,300). This behaviour is typical in other systems but is much more striking in the cyanocarbon series than in unsubstituted hydrocarbons because many of the cyanocarbon anions are intensely coloured.

B. N.m.r. Shielding Effects of the Cyano Group

Triple bonds such as —C≡C— and —C≡N have axial symmetry; a circulation is induced in the cylindrical shell of π-electrons by an external magnetic field so that a resultant field is developed as shown in Figure 3. Nuclei along the axis of the triple bond will be shielded whereas nuclei in the area above and below the bond will be

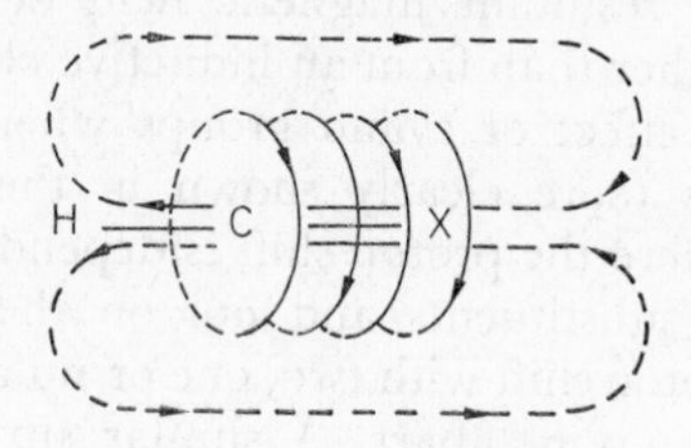

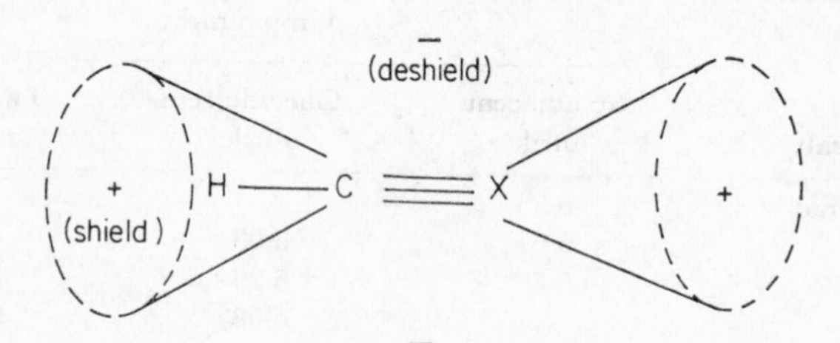

FIGURE 3. Ring currents in nitriles and acetylenes.

TABLE 14. N.m.r. chemical shift data for nitriles and acetylenes.

Compound	δ(p.p.m.)[68,a]	Compound	δ(p.p.m.)[66]
HC≡N	3·70	HC≡CR	2·35[b]
CH_3R	0·9		
$CH_3C{\equiv}N$	2·0	$HC{\equiv}CCH_3$	1·80[c]
$CH_2(C{\equiv}N)_2$	3·5		
$Na^+\bar{C}H(C{\equiv}N)_2$	2·8		

[a] At infinite dilution in $CDCl_3$ for HCN and in acetonitrile for nitriles.
[b] Average value for series of unconjugated acetylenes in solution.
[c] Gas phase, single resonance for both acetylenic and methyl hydrogens.

deshielded relative to their normal chemical shift. Thus the hydrogen in both hydrogen cyanide and acetylene has a chemical shift at much higher field than expected for a hydrogen bonded to an electronegative *sp*-hybridized carbon (see Table 14).

The shift for the methyl hydrogens in acetonitrile is at δ of 2·0 p.p.m. A second nitrile group deshields the hydrogen by 1·5 p.p.m. This shift is suggested to result from interaction of the hydrogen with the resultant magnetic field of the nitrile in the deshielding area rather than from an inductive electron withdrawal.

The deshielding effect of cyano groups when oriented parallel to the hydrogen is more clearly shown in the cyanocyclopentadienide system[68] where the proton shift is dependent on the number of adjacent cyano substituents and not on their overall number (Table 15). The proton shift with two, one or no adjacent hydrogens occurs roughly 0·5 p.p.m. apart. A similar situation is noted for benzenoid systems.

TABLE 15. N.m.r. chemical shifts for cyclopentadienide protons[68,69].

	δ in p.p.m.[a]		
R_4N^+ salt	No adjacent nitrile	One adjacent nitrile	Two adjacent nitriles
Cyclopentadienide	5·55		
Cyano-	5·67	6·08	
1,2-Dicyano-	5·76	6·29	
1,3-Dicyano-		6·08	6·59
1,2,3-Tricyano-		6·27	
1,2,4-Tricyano-			6·67
1,2,3,4-Tetracyano-			6·77

[a] δ at infinite dilution in acetonitrile relative to internal tetramethylsilane.

Another example of this deshielding effect when the proton is oriented in the deshielding zone of the nitrile is shown by the *cis–trans* pair of isomers **8** and **9**[66a]. The deshielding effect is again approximately 0·5 p.p.m.

H_A H_B H_C H_D H_E C≡N **(8)**

H_A H_B H_C H_D N≡C H_E **(9)**

	(8)		(9)	
δ (p.p.m.) for protons	A to D	7·8–8·0	A to C	ca. 8·0
			D	8·4
	E	5·77	E	5·72

VI. SUMMARY

The cyano group is a strong electron-withdrawing substituent; its inductive effect can be considerably enhanced by resonance interactions. Quantitative comparisons show that cyano is slightly weaker than a nitro group (both inductively and by resonance interactions) but stronger than most other common substituents.

The cyano group is a unique substituent because it is a rod-shaped molecule surrounded by a cylinder of π-electrons. It has relatively small steric requirements (between chlorine and bromine) and, unlike the nitro group, does not show steric interference to resonance interaction from nearby substituents.

The physical and chemical properties of molecules change as expected from substitution by the electronegative cyano group, but the highly directional dipole of the latter can produce some unusual effects. Because of the size and shape of the cyano group, polycyano substitution is easily achieved. The resulting cyanocarbons have unique chemical and physical properties that can be understood in terms of the cumulative effects derived from the electronic character of the cyano group.

VII. REFERENCES

1. L. Pauling, *The Nature of the Chemical Bond*, 3rd ed., Cornell University Press, Ithaca, New York, 1960; *Tables of Interatomic Distances and Configuration in Molecules and Ions*, Special Publication No. 11, The Chemical Society, London, 1958.
2. W. Littke and K. Wallenfels, *Tetrahedron Letters*, **1965**, 3365.
3. O. Bastiansen and O. Hassel, *Acta Chem. Scand.*, **1**, 489 (1947).
4. DMS, *U.v. Atlas of Organic Compounds*, Plenum Press, New York, 1966.
5. G. Leandri and D. Spinelli, *Boll. Sci. Chim. Ind. Bologna*, **15**, 90 (1957).

6. V. Baliah and V. Ramakrishnan, *J. Indian Chem. Soc.*, **35**, 151 (1958).
7. C. P. Conduit, *J. Chem. Soc.*, **1959**, 3273.
8. (a) B. Rickborn and F. R. Jensen, *J. Org. Chem.*, **27**, 4606 (1962);
   (b) N. L. Allinger and W. Szkrybalo, *J. Org. Chem.*, **27**, 4601, (1962).
9. R. R. Fraser and R. N. Renaud, *J. Am. Chem. Soc.*, **88**, 4365 (1966).
10. C. P. Smyth, *Dielectric Behavior and Structure*, McGraw-Hill, New York, 1955.
11. A. L. McClellan, *Tables of Experimental Dipole Moments*, W. H. Freeman, San Francisco, 1963.
12. S. Soundararajan, *Indian J. Chem.*, **1**(12), 503 (1963).
13. G. W. Wheland, *Resonance in Organic Chemistry*, John Wiley and Sons, New York, 1955.
14. J. D. Roberts, R. L. Webb and E. A. McElhill, *J. Am. Chem. Soc.*, **72**, 408 (1950).
15. W. A. Sheppard, *J. Am. Chem. Soc.*, **87**, 2410 (1965).
16. R. D. Chambers, personal communication.
17. W. Gerrard, M. F. Lappert and J. W. Wallis, *J. Chem. Soc.*, **1960**, 2178.
18. J. L. Hoard, T. B. Owen, A. Buzzell and O. N. Salmon, *Acta. Cryst.*, **3**, 130 (1950).
19. F. E. Murray and W. G. Schneider, *Can. J. Chem.*, **33**, 797 (1955).
20. (a) R. W. Taft and J. W. Carten, *J. Am. Chem. Soc.*, **86**, 4199 (1964).
    (b) C. D. Ritchie and A. L. Pratt, *J. Am. Chem. Soc.*, **86**, 1571 (1964).
21. K. F. Purcell, *J. Am. Chem. Soc.*, **89**, 6139 (1967).
22. H.E. Simmons, D. C. Blomstrom and R. D. Vest, *J. Am. Chem. Soc.*, **84**, 4782 (1962).
23. For a review of methods of measurements and interpretation of substituent constants see:
    (a) R. W. Taft Jr., 'Separation of Polar, Steric and Resonance Effects in Reactivity,' in *Steric Effects in Organic Chemistry* (Ed. M. S. Newman), John Wiley and Sons, New York, 1956, Chap. 13;
    (b) C. D. Ritchie and W. F. Sager, *Progr. Phys. Org. Chem.*, **2**, 323 (1964);
    (c) R. W. Taft, *J. Phys. Chem.*, **64**, 1805 (1960);
    (d) L. P. Hammett, *Physical Organic Chemistry*, McGraw-Hill, New York, 1940, Chap. 7.
24. R. W. Taft, E. Price, I. R. Fox, I. C. Lewis, K. K. Andersen and G. T. Davis, *J. Am. Chem. Soc.*, **85**, 709, 3146 (1963).
25. R. W. Taft, G. B. Klingensmith and S. Ehrenson, *J. Am. Chem. Soc.*, **87**, 3620 (1965).
26. H. C. Brown, D. H. McDaniel and O. Häfliger in *Determination of Organic Structures by Physical Methods* (Ed. E. A. Braude and F. C. Nachod), Academic Press, New York, 1955, Chap. 14.
27. (a) J. D. Roberts and W. T. Moreland, Jr., *J. Am. Chem. Soc.*, **75**, 2167 (1953);
    (b) F. W. Baker, R. C. Parish and L. M. Stock, *J. Am. Chem. Soc.*, **89**, 5677 (1967).
28. D. H. McDaniel and H. C. Brown, *J. Org. Chem.*, **23**, 420 (1958).
29. (a) M. M. Fickling, A. Fischer, B. R. Mann, J. Packer and J. Vaughn, *J. Am. Chem. Soc.*, **81**, 4226 (1959);
    (b) A. I. Biggs and R. A. Robinson, *J. Chem. Soc.*, **1961**, 388;
    (c) H. C. Brown and Y. Okamoto, *J. Am. Chem. Soc.*, **80**, 4979 (1958);
    (d) Y. Okamoto, T. Inukai and H. C. Brown, *J. Am. Chem. Soc.*, **80**, 4969 (1958).
30. (a) W. A. Sheppard, *Trans. N.Y. Acad. Sci.*, [2] **29**, 700 (1967);
    (b) W. A. Sheppard and R. M. Henderson, *J. Am. Chem. Soc.*, **89**, 4446 (1967).
31. W. A. Sheppard, *J. Am. Chem. Soc.*, **87**, 2410 (1965).
32. W. A. Sheppard, unpublished results.
33. K. Wallenfels, *Chimia (Aarau)*, **20**, 303 (1966).
34. J. K. Williams, E. L. Martin and W. A. Sheppard, *J. Org. Chem.*, **31**, 919 (1966).

35. G. W. Wheland, R. M. Brownell and E. C. Mayo, *J. Am. Chem. Soc.*, **70,** 2492 (1948).
36. H. Kloosterziel and H. J. Backer, *Rec. Trav. Chim.*, **72,** 185 (1953).
37. F. Kieffer and P. Rumpf, *Compt. Rend.*, **230,** 1874, 2302 (1950); M. Gillois and P. Rumpf, *Bull. Soc. Chim. France*, **1954,** 112.
38. (a) M. J. S. Dewar, *Hyperconjugation*, The Ronald Press, New York, 1962, p. 154; (b) M. J. S. Dewar and A. P. Marchand, *J. Amer. Chem. Soc.*, **88,** 354 (1966).
39. W. A. Sheppard, *J. Am. Chem. Soc.*, **85,** 1314 (1963).
40. R. H. Boyd, *J. Phys. Chem.*, **67,** 737 (1963).
41. D. S. Acker and W. R. Hertler, *J. Am. Chem. Soc.*, **84,** 3370 (1962).
42. Z. Rappoport and E. Shohamy, *J. Chem. Soc. (B)*, **1969,** 77.
43. E. Lippert and W. Lüder, *Z. Physik. Chem.*, **33,** 60 (1962).
44. O. W. Webster, *J. Am. Chem. Soc.*, **88,** 4055 (1966).
45. R. H. Boyd, *J. Phys. Chem.*, **65,** 1834 (1961); *J. Am. Chem. Soc.*, **83,** 4288 (1961).
46. O. W. Webster, *J. Am. Chem. Soc.*, **88,** 3046 (1966).
47. E. M. Kosower, *Progr. Phys. Org. Chem.*, **3,** 81 (1965).
48. (a) M. J. S. Dewar and C. C. Thompson Jr., *Tetrahedron Suppl.*, **7,** 97 (1966); (b) H. H. Greenwood and R. McWeeny, *Advan. Phys. Org. Chem.*, **4,** 73 (1966) (see particularly pp. 112–118).
49. R. P. Bell, *The Proton in Chemistry*, Cornell University Press, Ithaca, New York, 1959, Chap. 6.
50. R. G. Pearson and R. L. Dillon, *J. Am. Chem. Soc.*, **75,** 2439 (1953).
51. H. G. Adolph and M. L. Kamlet, *J. Am. Chem. Soc.*, **88,** 4761 (1966).
52. W. D. Phillips, J. C. Rowell and S. I. Weissman, *J. Chem. Phys.*, **33,** 626 (1960).
53. O. W. Webster, W. Mahler and R. E. Benson, *J. Am. Chem. Soc.*, **84,** 3678 (1962).
54. O. W. Webster, *J. Am. Chem. Soc.*, **86,** 2898 (1964);
55. L. R. Melby, R. J. Harder, W. R. Hertler, W. Mahler, R. E. Benson and W. E. Mochel, *J. Am. Chem. Soc.*, **84,** 3374 (1962).
56. P. H. Rieger, I. Bernal, W. H. Reinmuth and G. K. Fraenkel, *J. Am. Chem. Soc.*, **85,** 683 (1963).
57. P. H. Rieger and G. K. Fraenkel, *J. Chem. Phys.*, **37,** 2795 (1962).
58. L. M. Stock and H. C. Brown, *Advan. Phys. Org. Chem.*, **1,** 35 (1963).
59. J. F. Bunnett, *Quart. Rev.*, **12,** 1 (1958).
60. W. C. Spitzer and G. W. Wheland, *J. Am. Chem. Soc.*, **62,** 2995 (1940).
61. (a) K. Wallenfels, F. Witzler and K. Friedrich, *Tetrahedron*, **23,** 1353 (1967); (b) K. Wallenfels, F. Witzler and K. Friedrich, *Tetrahedron*, **23,** 1845 (1967).
62. L. J. Bellamy, *The Infrared Spectra of Complex Molecules*, 2nd ed., John Wiley and Sons, New York, 1958.
63. E. Wallace, private communication.
64. W. J. Middleton, E. L. Little, D. D. Coffman and V. A. Engelhardt, *J. Am. Chem. Soc.*, **80,** 2795 (1958).
65. O. W. Webster, *J. Am. Chem. Soc.*, **86,** 2898 (1964).
66. (a) D. W. Mathieson, *Nuclear Magnetic Resonance*, Academic Press, New York, 1967, p. 32;
(b) J. W. Emsley, J. Feeney and L. H. Sutcliffe, *High Resolution Nuclear Magnetic Resonance Spectroscopy*, Vols. 1 and 2, Pergamon Press, Oxford, 1965.
67. *Organic Electronic Spectral Data*, Vols. 1–4, Interscience, New York, 1966.
68. O. W. Webster, private communication.
69. O. W. Webster, *J. Am. Chem. Soc.*, **88,** 3046 (1966).

CHAPTER 6

Nitrile reactivity

FRED C. SCHAEFER

American Cyanamid Co., Stamford, Conn., U.S.A.

I. General Considerations . . . 240
II. Interaction of Nitriles with Acids . . . 243
A. Halogen Acids; Formation of Imidyl Halides . . . 243
B. Oxy Acids . . . 245
C. Carboxylic Acids . . . 246
III. Nitrile–Metal Complexes . . . 249
A. Complexes with Electrophilic Metal Halides . . . 250
B. The Gattermann and Houben–Hoesch Syntheses . . . 254
C. Metal Carbonyl Complexes . . . 255
IV. Hydration of Nitriles . . . 256
A. Base Catalysis . . . 257
B. Acid Catalysis . . . 258
C. Addition of Hydrogen Peroxide . . . 262
V. Nucleophilic Addition of Hydroxy, Amino and Sulphydryl Compounds . . . 263
A. Addition of Alcohols and Phenols . . . 263
1. The Pinner synthesis of imidates . . . 264
2. Base-catalysed addition of alcohols . . . 267
3. Imidates as activated nitriles . . . 268
B. Addition of Ammonia and Amines . . . 269
1. Synthesis of amidinium salts . . . 270
2. Reaction of nitriles with sodamide . . . 273
C. Addition of Sulphydryl Compounds . . . 274
VI. Nucleophilic Addition of Carbanions . . . 276
A. The Grignard Reaction . . . 276
B. Base-Catalysed Dimerization and Trimerization of Nitriles . . . 282
VII. Linear Polymerization of Nitriles . . . 287
VIII. Electrophilic Addition of Carbonium Ions . . . 290
A. *N*-Alkylation with Olefins and Alcohols . . . 291
B. Acid-Catalysed Reactions of Nitriles with Aldehydes . . . 294
IX. Addition of Free Radicals to the Cyano Group . . . 296
A. Fluorination of the Cyano Group . . . 296
B. Miscellaneous Reactions of Free-Radical Character . . . 297
X. References . . . 298

I. GENERAL CONSIDERATIONS

This chapter will examine the fundamental factors of nitrile reactivity as they are demonstrated in the typical nitrile reactions. Variation of reactivity with structure and the role of catalysts will be major aspects considered. Hopefully, the treatment will provide a rational basis for discussion of future developments in nitrile chemistry. Descriptive chemistry has been drawn upon for illustrative purposes, but a review of the reactions of nitriles is not intended. Broad coverage is provided by Migrdichian's *The Chemistry of Organic Cyanogen Compounds* (W. A. Benjamin, New York, 1947) and P. A. S. Smith's *The Chemistry of Open-Chain Organic Nitrogen Compounds* (Reinhold, New York, 1965).

The reactivity of nitriles is fundamentally due to the polarization of the C≡N triple bond, which arises from the greater electronegativity of nitrogen compared to carbon, $RC{\equiv}N \longleftrightarrow RC^{+}{=}N^{-}$. Nucleophilic reagents will attack at the electrophilic carbon atom while the nitrogen atom is a weakly basic site. Interaction with an acid A^{+} enhances the polarization and gives a species having increased susceptibility to nucleophilic attack,

$$RC{\equiv}N^{+}A \longleftrightarrow RC^{+}{=}NA$$

Although the preponderance of nitrile reactions are acid- or base-catalysed additions of nucleophiles at the nitrile carbon atom, examples of alkylation at the nitrogen atom by cationic species are also important. Reactions of radical character are rare, although important.

Although it is common to think in terms of propionitrile or benzonitrile and to describe nitrile reactions as they are observed with these rather ordinary members of the class, it is practically and theoretically important to appreciate that the nitriles offer a very broad spectrum of reactivity, with the group R in RC≡N ranging from dialkylamino to perfluoroalkyl. The range of substituted benzonitriles which has been considered in some reactivity studies represents only a small part of this spectrum. Often it is instructive to examine the reactions of an extreme member of the class, where intermediates may have longer lifetimes or where reactions may be much more rapid than with the conventional members. Similarly, the reactions observed for particular nitriles may suggest new versions to try with the less familiar relatives. An important aspect of the study of nitrile reactions is an appreciation of the influence of substituents on nitrile reactivity.

The combined inductive and mesomeric effects of the substituent R have a major influence on the C≡N polarization. Propionitrile shows slightly lower reactivity than acetonitrile in various nucleophilic addition reactions, as would be expected from the slightly greater electron-releasing inductive effect of the ethyl compared to the methyl group. Electronegative substituents (as in **1**), however, can markedly enhance the nucleophilicity of the nitrile carbon atom and the reactivity. Usually such substituents will also reduce the electron density at the nitrogen atom[1] and somewhat reduce the effectiveness of acid catalysts. In the cyanates, cyanamides (**2**) and cyanogen halides, the R group is both electron-attracting inductively and electron-donating by a mesomeric effect. Experience shows that the inductive effect is dominant in the reactivity of these nitriles toward nucleophiles.

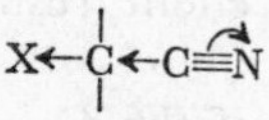

X←C←C≡N (1) R_2N←C≡N (2)

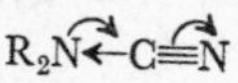

For the most part, our knowledge of nitrile reactivity has been developed with the aid of analogies to carbonyl compounds, a relatively small number of kinetic studies on limited classes of nitriles, and a liberal amount of extrapolation, all occasionally checked against a century of more or less reliable experience. Experimental work correlating reactivity with structure over a wide range is rare. Although a generally satisfactory qualitative understanding has emerged, it would be gratifying to have a correlation between nitrile reactivity and some fundamental physical property. The following interesting approaches are under consideration.

Promise of new insight into the electronic state of the C≡N bond in RC≡N as it is influenced by the nature of R is offered through ^{14}N nuclear quadrupole resonance spectroscopy. An inductively electron-releasing or -attracting substituent influences the polarity of the π- and σ-components of the C≡N bond in the same sense but not necessarily to the same extent. The difference is reflected in the quadrupole coupling constant. Conjugative and hyperconjugative effects which disturb the cylindrical symmetry of the triple bond can also be discerned by this technique and appear to be substantial. Although quadrupole resonance spectroscopy examines the fundamental nitrile structure, the electronic interactions involved are poorly understood at present, and it is not yet possible to correlate the spectroscopic data with reactivity[2].

The ^{14}N n.m.r. chemical shifts for a substantial series of nitriles have been reported[3]. Values ranging from 149 p.p.m. (relative to nitrate ion) for Et_2NCN to 83 p.p.m. for 3-cyanopyridine oxide were observed. The general trend in the observed shifts is clearly in the direction of lower field with increasingly electronegative substituents, although the effects of the phenyl and vinyl groups are anomalously large for correlation with any reactivity series. Although deshielding of the nitrogen nucleus may not be easily related to the electron density at the carbon atom, which is of primary interest, this work holds considerable promise.

Work on the ^{13}C resonance in nitrile groups may prove to give direct evidence of the ground state electron distribution, but very little has been done in this area. Extensive work on carbonyl compounds, however, has given excellent results and appears to point the way[4].

The infrared stretching frequency of the C≡N bond is sensitive to electronic perturbations which alter the bond order. Electronegative substituents inductively reduce the polarization and so tend to raise the frequency of the absorption band. However, mesomeric electron donation from an adjacent unsaturated centre has an opposite effect which is considerably greater[5,6]. The latter effect is much less a factor in chemical reactivity toward nucleophiles, and consequently no broad correlation of reactivity with frequency has been possible. The position of the absorption band is also affected by the force constant of the adjacent bond to the group R. Consequently when R is hydrogen or a heteroatom, a substantial additional shift in the resonance frequency may be superimposed on any effect due to the C≡N polarity[7].

A further proposal has been to correlate reactivity with the intensity of the nitrile absorption band[8]. This is a measure of the change in dipole moment during vibration and thus of the polar properties of the band[7]. It was in fact found that for a series of benzonitriles the absorption of the C≡N band becomes weaker as electronegative substituents are introduced, and a fairly satisfactory linear correlation between the logarithm of the absorbance and the respective Hammett σ constants was established. Alternatively, correlation of σ^* constants with the square root of the absorption intensity has been recommended[9]. This approach does not appear promising for extension to other nitrile classes, unless special treatment can be given to those nitriles in which mesomeric and electromeric effects lead to exceptionally high absorbance. The extremely weak absorption of α-oxygenated aliphatic nitriles also seems to be out of line compared with the intensity of α-halogenated nitriles, and other puzzling relationships have been noted.

II. INTERACTION OF NITRILES WITH ACIDS

A. Halogen Acids; Formation of Imidyl Halides

Acid catalysis is of major importance in reactions of nitriles, and accordingly the interactions of nitriles with strong acids have received much attention. The compounds formed with the hydrogen halides have been of particular interest because of their probable involvement as intermediates in many widely used reactions. The nature of the nitrile–hydrogen halide compounds and their role in various reactions of nitriles with nucleophiles were discussed in 1962 in an excellent review[10] by Zil'berman, who has been a major contributor to this subject.

The special nature of the nitrile–hydrogen halide compounds stems from the substantial nucleophilicity of the halide ions compared to the anions of the oxy acids. Addition of a proton to the weakly basic nitrogen atom is followed or perhaps accompanied by coordination of a halide ion at the electrophilic carbon. The imidyl halide formed (**3**) is usually sufficiently basic to accept a second proton to form a salt (equation 1). This representation is to some

$$\mathrm{RCN} \underset{}{\overset{\mathrm{H^+}}{\rightleftharpoons}} \left[\begin{array}{c} \mathrm{RC{\equiv}\overset{+}{N}H} \\ \updownarrow \\ \mathrm{R\overset{+}{C}{=}NH} \end{array}\right] \overset{\mathrm{Cl^-}}{\rightleftharpoons} \underset{(\mathbf{3})}{\mathrm{RC(=NH)Cl}} \overset{\mathrm{H^+}}{\rightleftharpoons} \mathrm{RC(=\overset{+}{N}H_2)Cl} \qquad (1)$$

extent an oversimplification of matters which perhaps are not fully agreed upon in detail[10]. While the existence of imidyl halide salts is now firmly established[11,12] it is important to note that in the reversible reaction (1) the formation of **3** is frequently slow and is favoured by high hydrogen halide concentration. Several authors have described intermediate compositions which are undoubtedly mixtures of the compounds in this sequence.

The weak basicity of the simple nitriles is *a priori* attributable to either the lone pair of electrons on nitrogen or to the π-electrons of the triple bond. The fact that complexes of nitriles with Lewis acids invariably are found to be bound through the lone pair[13] is good evidence against the π-base concept. Nevertheless, electromeric enhancement of the π-electron availability in dialkylcyanamides and presumably in 4-methoxybenzonitrile, for example, markedly accelerates the formation of the imidyl chloride salt. Strongly electronegative substituents reduce the nitrile basicity while favouring bonding to the nucleophilic halide ion. This can lead to imidyl halides of low basicity.

In addition to the simple imidyl halide salts, some nitriles are known to form dimeric products with hydrogen halides. A particularly clear case is that of dimethylcyanamide which reacts very rapidly with added hydrogen chloride at room temperature to give the crystalline compound **5**. When this is heated above 100°, hydrogen chloride is rapidly evolved, and **6** is produced in essentially quantitative yield. This dimerization probably proceeds by dis-

$$Me_2NCN \xrightarrow{HCl} \underset{(\mathbf{4})}{Me_2NC(=NH)Cl} \underset{<50^\circ}{\overset{HCl}{\rightleftharpoons}} \underset{(\mathbf{5})}{Me_2NC(=\overset{+}{N}H_2)Cl\ \overset{-}{Cl}} \xrightarrow[100^\circ]{-HC} \underset{(\mathbf{6})}{\left[Me_2NC(=\overset{+}{N}H_2)N{=}C(NMe_2)Cl\right]Cl^-}$$

placement of chloride ion from **5** by the nucleophilic chloroformamidine base (**4**). An important difference between the monomeric and dimeric structures **5** and **6** is the stability of the conjugate base of the latter against elimination of hydrogen chloride. Whereas **5** merely reverts to dimethylcyanamide when treated with basic reagents, **6** undergoes displacement of chloride to give a variety of products[14].

The simple aliphatic imidyl chloride hydrochlorides show analogous behaviour[15]. However, the α-halogenated monomeric compounds are stable only at low temperature, addition of hydrogen halide to the nitrile at room temperature leading directly to the dimer (**7**)[16]. The *N*-(α-haloalkylidene)amidines (**7**) in general are weakly basic, and their salts tend to release hydrogen halide *in vacuo*, which accounts for much of the lack of agreement regarding their composition[15].

Hydrogen cyanide forms similar complex hydrohalides which deserve special mention. The only definite compounds which have been obtained under the usual reaction conditions are the sesquihydrohalides, 2HCN·3HX (X = Cl or Br). These compounds have

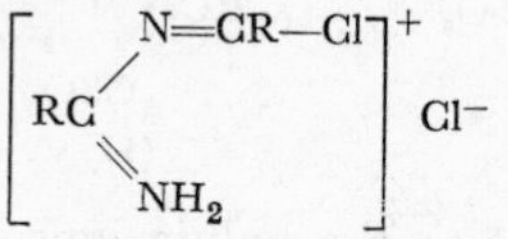

$$\left[RC \begin{matrix} \diagup N{=}CR{-}Cl \\ \diagdown\!\!\diagdown NH_2 \end{matrix} \right]^+ Cl^- \qquad [NH_2{=}CH{-}NH{-}CHX_2]^+X^-$$

(7) (8)

been shown to release hydrogen chloride *in vacuo* to give 2HCN·HX. They are now firmly established as members of the imidyl halide dimer class[11]. The sesquihydrohalides are modified, however, to the extent of incorporating a third hydrohalide molecule by addition to the C=N bond to form a dihalomethyl group (**8**). Monomeric formimidyl chloride has recently been prepared as the hexachloroantimonate salt by addition of hydrogen chloride to the $HCN{\cdot}SbCl_5$ complex[17] (section III.A).

Intramolecular dimerization of imidyl halides has become a useful route to a variety of heterocyclic systems. Usually a 1,2- or 1,3-dicyano compound is treated with hydrogen halide in an inert medium[18], but in some cases the reaction takes place rapidly in concentrated aqueous acid[19] (reaction 2).

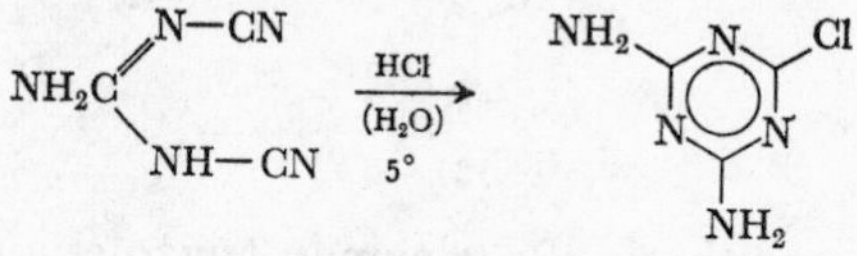

$$NH_2C\begin{matrix} \diagup\!\!\diagup N{-}CN \\ \diagdown NH{-}CN \end{matrix} \xrightarrow[(H_2O)\ 5^\circ]{HCl} \text{2,4-diamino-6-chloro-1,3,5-triazine } (NH_2,\ Cl,\ NH_2 \text{ on ring } C_3N_3) \qquad (2)$$

It is probable that nitrile–hydrohalide interaction will intervene in many reactions where both components are present. In some such cases it has been demonstrated that allowance of time for the formation of an imidyl chloride intermediate permits much improved yields[20,21].

B. Oxy Acids

Anhydrous sulphuric and phosphoric acids form crystalline 1:1 complexes with hydrogen cyanide[22a], and presumably other nitriles react analogously. The suggestion[22a] that these compounds are of the type **9** is unlikely because sulphuric acid catalyses the trimerization of some aromatic nitriles. This fact is more in line with the nitrilium salt formulation **10**, but an imidyl sulphate or phosphate structure **11** is also possible[22b]. The sulphate and the phosphate anions are

$$\mathrm{HC(=O)NHSO_3H}\quad(\mathbf{9})\qquad [\mathrm{RC{\equiv}NH}]^+\mathrm{HSO_4}^-\quad(\mathbf{10})\qquad \mathrm{RC(=\overset{+}{N}H_2)OSO_3^-}\quad(\mathbf{11})$$

relatively poor nucleophiles compared to the halide ions, but the covalent imidyl structure provides an easy explanation of the ability of nitriles to catalyse phosphorylation reactions[23] (for example reaction 3). Probably **10** and **11** are in equilibrium in solution.

$$2\,\mathrm{PhCH_2OP(O)(OH)_2} + \mathrm{Me_2NC{\equiv}N} \longrightarrow \left[\mathrm{PhCH_2OP(=O)(OH)OC(=NH)NMe_2}\right] \longrightarrow (\mathrm{PhCH_2OP(=O)(OH)-})_2\mathrm{O} + \mathrm{Me_2NCONH_2} \qquad (3)$$

Aromatic nitriles react with cold sulphur trioxide to give diaryloxathiadiazine dioxides (**12**)[24]. These compounds have been found

$$2\,\mathrm{ArC{\equiv}N} + \mathrm{SO_3} \longrightarrow (\mathbf{12}) \xrightarrow[-\,\mathrm{SO_3}]{\mathrm{NH_2C(NH)NH_2}} \text{2,4-diaryl-6-amino-}s\text{-triazine}\ \ 70\text{–}80\%$$

very useful in the synthesis of *s*-triazines, triazoles and oxadiazoles[25]. Acetonitrile reacts differently giving a product of unknown structure, $3(CH_3CN)\cdot 2SO_3$[26].

C. Carboxylic Acids

The reaction of nitriles with carboxylic acids[27] is frequently cited as a synthesis for imides. In special cases it is effective, but the generality is greatly exaggerated. With the simple nitriles and acids little reaction can be observed below 200°, and an equilibrium is then reached in which very little imide is present[28]. Equilibrium is reached fairly rapidly, however, and this has been shown to permit nitrile–carboxylic acid interchange to occur[29,30] (reaction 4). On the other hand, dimethylcyanamide and cyanoacetic acid react exothermically at room temperature to give *N*-cyanoacetyl-*N'N'*-dimethylurea in good yield[31] (reaction 5).

$$PhCH_2CN + MeCO_2H \rightleftharpoons PhCH_2CONHCOMe \rightleftharpoons PhCH_2CO_2H + MeCN \quad (4)$$

$$Me_2NCN + NCCH_2CO_2H \xrightarrow{25°} Me_2NCONHCOCH_2CN \quad (5)$$

The relatively few examples of this reaction which have been investigated show that: (*a*) the reaction is strongly promoted if it occurs intramolecularly to form a 5-ring imide, (*b*) it is favoured by increased acid strength and (*c*) electrophilic nitriles have greater reactivity. It has also been shown that mineral acids catalyse the reaction so that the weak carboxylic acids can be used at much lower temperatures[28,30,32]. The reaction must proceed by coordination of a carboxylate anion at the electrophilic carbon of the protonated nitrile to give an isoimide (**13**) which rearranges to the imide. Analogous isoimides which are obtained by addition of acids to carbodiimides, isomerize rapidly to imides[33]. At 200° the reverse reaction must be comparable in rate.

$$R^1C{\equiv}N + R^2CO_2^- \overset{H^+}{\rightleftharpoons} R^1C({=}NH)OCOR^2 \rightleftharpoons R^1CONHCOR^2$$

(**13**)

An example of the combined effect of acid catalysis and favourable geometry[34a] is shown by the facile hydrolysis of 3-cyano-3,3-diphenylpropionic acid (**14**). Rather than impede the reaction as expected, the two phenyl groups in this molecule bring the functional groups into juxtaposition; hydrolysis to the diacid is complete within minutes. The monophenyl compound hydrolyses much more slowly to the carboxamide, while 4-cyano-4,4-diphenylglutaric acid is very resistant[34b].

$$Ph_2C(CN)CH_2CO_2H \xrightarrow[90°]{\text{conc. HCl}} \left[\begin{array}{l} Ph_2C{-}C{=}NH \\ \;|\qquad\; \backslash \\ \;|\qquad\;\; O \\ \;|\qquad\; / \\ CH_2{-}C{=}O \end{array} \right] \xrightarrow{H_2O} Ph_2C(CO_2H)CH_2CO_2H$$

(**14**)

Of much greater interest than the imide formation itself, is the intermediate occurrence of the isoimide **13**, a highly reactive species, rarely represented in a stable compound[35]. Such a molecule is potentially useful as an acylating agent, and considerable interest in this possibility has been evident. The reactivity of cyanamides towards carboxylic acids has been studied recently, and has given a clear picture of the pattern for other nitriles.

Just as cyanamide is able to bring about phosphorylation through the intermediacy of an imidyl phosphate (section II.B), so it can be used successfully to activate an amino acid for peptide synthesis[36] (reaction 6). Undoubtedly a 'high energy carboxylate' analogous to

$$NH_2CH_2CO_2Et + BzOCONHCPhMeCO_2H + NH_2CN \xrightarrow[THF]{85°} BzOCONHCPhMeCONHCH_2CO_2Et + NH_2CONH_2 \quad (6)$$

13 is an intermediate. When cyanamide is heated with various carboxylic acids at 100–150°, mixtures of the ureide **16** and the amide **17** are obtained, the ratio of **17** to **16** increasing as the molecular weight increases in the aliphatic series. The product distribution under different conditions and with differing acid strength indicates the reaction sequence shown[37] (reaction 7).

$$RCO_2H + NH_2CN \longrightarrow \underset{(\mathbf{15})}{RCO_2\text{—}C(=NH)NH_2} \xrightarrow{RCO_2H} (RCO)_2O + NH_2CONH_2$$

$$(RCO)_2O + NH_2CONH_2 \longrightarrow \underset{(\mathbf{16})}{RCONHCONH_2} \longrightarrow \underset{(\mathbf{17})}{RCONH_2} + [HOCN] \quad (7)$$

More recently[38] the question of the intermediate occurrence of an anhydride in this process has been examined, since direct rearrangement of the *O*-acylpseudourea **15** to **16** would not be surprising. A 2:1 mixture of chloroacetic acid and dimethylcyanamide in CD_3CN was held at 48° and examined at intervals by n.m.r. It was shown that the gradual loss of dimethylcyanamide was accompanied by corresponding appearance of dimethylurea and chloroacetic anhydride (**18**), while no intermediate compound reached a detectable concentration. Slower changes followed (reaction 8), and corresponding n.m.r. resonances have been observed. At room

$$Me_2NC{\equiv}N + 2\,ClCH_2CO_2H \longrightarrow Me_2NCONH_2 + \underset{(\mathbf{18})}{(ClCH_2CO)_2O} \longrightarrow Me_2NCONH\text{—}COCH_2Cl \longrightarrow \underset{\text{ring: }H_2C\text{—}O\text{—}C(NMe_2){=}N\text{—}C(=O)}{\text{oxazolinone}} \quad (8)$$

temperature the yield of **18** was 72% in three days. With acetic acid, a maximum yield of about 10% acetic anhydride was obtained by heating at 60° for two hours.

When cyanamide and a carboxylic acid are heated in an alcoholic solvent, ester formation is preferred to acylation of the urea. Yields are 3–15 times as high as those obtained under the same conditions in the absence of cyanamide[31].

In these cases, the carboxylic acid itself acts as a catalyst, the stronger acids being more reactive. It is not surprising then that hydrogen chloride can catalyse the reactions of weak acids. Imidyl chloride hydrochlorides react slowly at 0° with carboxylic acids to give both the anhydride and the chloride of the original acid[10,39,40] (reaction 9). The acyl chloride may arise by reorganization of **19** or

$$R^1C(=\overset{+}{N}H_2\bar{C}l)Cl + R^2CO_2H \longrightarrow \underset{(\mathbf{19})}{R^1C(=\overset{+}{N}H_2\bar{C}l)OCOR^2} \begin{cases} \longrightarrow R^1C(=NH)OH + R^2COCl \\ \xrightarrow{R^2CO_2H} R^1C(=\overset{+}{N}H_2\bar{C}l)OH + (R^2CO)_2O \xrightarrow{HCl} R^1C(=NH)OH + R^2COCl \end{cases} \quad (9)$$

by reaction of hydrogen chloride with the anhydride. The amide recovered is derived solely from the original nitrile. Acid-catalysed acetolysis of hydrogen cyanide must take the same course[41]. It is noteworthy that in the presence of 0·1 M hydrochloric or sulphuric acid the latter reaction is about 10^6 times as fast as hydration in an aqueous system. The difference must be primarily due to greater effective acidity of the mineral acid in acetic acid compared to water.

III. NITRILE–METAL COMPLEXES

The nucleophilic reactivity of nitriles demonstrated by their formation of salts with proton acids is also reflected in a very wide variety of metal complexes. These are formed by donation of the nonbonding electrons of the nitrogen atom, and their stability is influenced by the electronegativity of substituents in the nitrile. In essentially all complexes investigated, the R—C—N—metal system is linear. Formation of the donor–acceptor bond increases the frequency of the C≡N stretching band substantially, a feature which is considered diagnostic for such structures[13,42]. Distortion of the C≡N bond by the positive charge on the nitrogen atom causes enhanced electrophilicity at the carbon atom and underlies the wide use of electrophilic metal compounds, particularly metal halides, as catalysts in nitrile reactions. Interest in these complexes has been increasing also because of the use of the simple nitriles as aprotic solvents in synthetic inorganic chemistry. The inorganic and physicochemical aspects of the metal halide complexes have been reviewed by Walton[13].

A. Complexes with Electrophilic Metal Halides

Nitrile complexes with halides of boron, aluminium, antimony, tin and the transition metals have been studied extensively. In descriptions of the organic chemistry of these species, they are usually represented as simple donor–acceptor pairs, for example $RCN \rightarrow AlCl_3$. However, in most cases they have probably more complicated structures, frequently dimeric and often with more than one nitrile ligand on the metal atom. The reactivity of the nitrile may well be influenced by steric factors within the coordination sphere, and interaction with other ligands may play an unexpected role. While impedance to reaction from such causes could be important, facilitation of reaction might also arise from beneficial orientation of a cocomplexed reactant.

Nitrile complexes with the strongly electrophilic metal halides react with alkyl chlorides to give nitrilium salts (**20**)[22b,43,44]. There

$$R^1C{\equiv}\overset{+}{N}{\rightarrow}SbCl_5 + R^2C{-}Cl \longrightarrow \left[R^1C{\equiv}\overset{+}{N}CR^2 \longleftrightarrow R^1\overset{+}{C}{=}NCR^2 \right] SbCl_6^- \qquad (10)$$

(**20**)

appears to be no appreciable steric factor in these reactions, and tertiary chlorides react extremely rapidly. Secondary chlorides are much slower and primary are very sluggish, although all give excellent yields with $CH_3CN{\cdot}SbCl_5$. The reaction rate also depends on the metal used, the complex reactivity decreasing in the order: $SbCl_5 > FeCl_3 > AlCl_3 > SnCl_4$. The detailed mechanism of these reactions has not been studied. Evidently high S_N1-type halide reactivity is accelerating; possibly a four-centre concerted reorganization occurs as indicated in equation (10).

In rare cases, it is also possible to add hydrogen chloride to the nitrile complex to give the unsubstituted nitrilium salt (e.g. **21**).

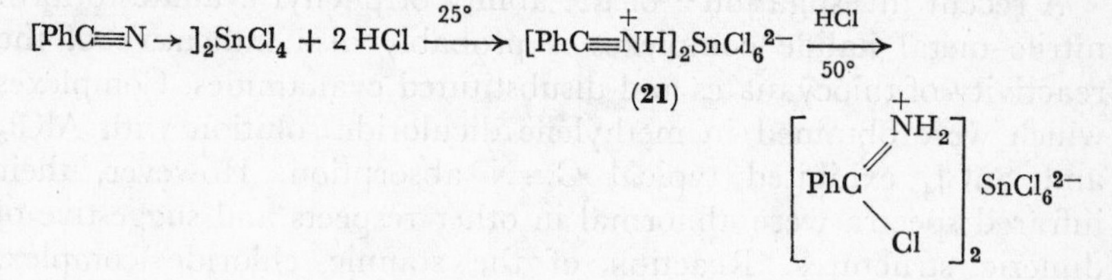

$$[PhC{\equiv}N \rightarrow]_2SnCl_4 + 2\,HCl \xrightarrow{25^\circ} [PhC{\equiv}\overset{+}{N}H]_2SnCl_6^{2-} \xrightarrow[50^\circ]{HCl} \left[PhC(=\overset{+}{N}H_2)Cl \right]_2 SnCl_6^{2-}$$

(**21**) (**22**)

At 50° the salt **21** reacts with a second equivalent of hydrogen chloride to give the imidyl chloride hexachlorostannate (**22**). More commonly hydrogen chloride does not react at all or gives the imidyl chloride directly[45,58].

Acyl chlorides usually react with nitrile–metal halide complexes at about 150° to give salts containing two nitrile residues per acyl group, regardless of the reactant ratio used[43,46–48]. The product oxonium salts (**23**) owe their stability to the substantial resonance energy of this heterocyclic system. This reaction involves nucleo-

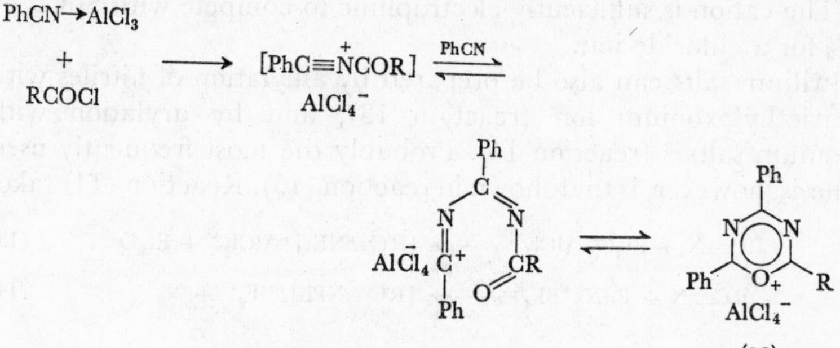

$$PhCN{\rightarrow}AlCl_3 + RCOCl \longrightarrow [PhC{\equiv}\overset{+}{N}COR]\,AlCl_4^- \xrightleftharpoons{PhCN}$$

$$AlCl_4^-\ PhC^+{=}N{-}C(Ph){=}N{-}C(R){=}O \rightleftharpoons \text{2,4-diphenyl-6-R-1,3,5-oxadiazinium } AlCl_4^- \quad (11)$$

(**23**)

$$\mathbf{23} + NH_2OH \longrightarrow \text{3,5-diphenyl-1,2,4-oxadiazole} \quad (12)$$

philic attack of a nitrile at the electrophilic carbon atom in the nitrilium salt, a step frequently observed in the reactions of such nitrile complexes. One may speculate that such interaction is facilitated by the geometry around the complexed metal atom.

The oxonium salts (**23**) have proven to be useful intermediates for the synthesis of several classes of heterocycles[46]. One such case is conversion to an oxadiazole by reaction with hydroxylamine (reaction 12). Scrambling of the original nitrile group and acyl chloride radicals takes place in reaction (11) through reversible ring opening of unsymmetrically substituted oxonium salts[43].

A recent investigation[49] of the ability of phenyl cyanate to form nitrile–metal halide complexes is probably also indicative of the reactivity of thiocyanates and disubstituted cyanamides. Complexes which were obtained in methylene dichloride solution with $AlCl_3$ and $SnCl_4$ exhibited typical C≡N absorption. However, their infrared spectra were abnormal in other respects and suggestive of dimeric structures. Reaction of the stannic chloride complex, possibly **24**, with *t*-butyl chloride and subsequent hydrolysis gave the pseudourea **26** in 70% yield. The imidyl chloride complex **25** is a logical intermediate. It appears that the potential electro-

$$\underset{(\mathbf{24})}{\begin{matrix} PhOC{\equiv}N \\ \vdots \quad\; \searrow \\ N \quad\;\; SnCl_3 \\ \| \quad\; \nearrow \\ PhOC{\leftarrow}Cl \end{matrix}} \xrightarrow{t\text{-}BuCl} \underset{(\mathbf{25})}{\begin{matrix} \quad\quad\; Bu\text{-}t \\ PhOC{=}N \\ / \quad \searrow SnCl_4 \\ N \\ \| \\ PhOC{-}Cl \end{matrix}} \longrightarrow \underset{(\mathbf{26})}{\begin{matrix} PhOC{-}NHBu\text{-}t \\ \| \\ NCO_2Ph \end{matrix}}$$

meric effect of the aryloxy group does not stabilize the nitrilium ion. The cation is sufficiently electrophilic to compete with $SbCl_5$ or $AlCl_3$ for a chloride ion.

Nitrilium salts can also be prepared by alkylation of nitriles with the triethyloxonium ion (reaction 13), and by arylation with diazonium salts[43] (reaction 14). Probably the most frequently used synthesis, however, is that shown in reaction (15). Reaction (14) takes

$$RC{\equiv}N + Et_3O^+AlCl_4^- \longrightarrow [RC{\equiv}NEt]^+AlCl_4^- + Et_2O \qquad (13)$$

$$RC{\equiv}N + PhN_2^+BF_4^- \longrightarrow [RC{\equiv}NPh]^+BF_4^- + N_2 \qquad (14)$$

$$R^1C\begin{matrix} \nearrow Cl \\ \searrow\!\!\searrow NR^2 \end{matrix} + SbCl_5 \longrightarrow [R^1C{\equiv}NR^2]^+SbCl_6^- \qquad (15)$$

place satisfactorily at low temperature with simple aliphatic nitriles and phenylacetonitrile. However, with aromatic nitriles the *N*-aryl nitrilium salt undergoes nucleophilic attack by a second nitrile molecule to give a dimeric intermediate which cyclizes to a quinazoline[50] (reaction 16) by way of an intramolecular Houben–Hoesch reaction (see section III.B). If the *N*-aryl group is blocked in both *o*-positions, the *N*-arylated dimer salt (**27**) can be isolated*. The

$$[PhC{\equiv}NPh]^+AlCl_4^- \xrightarrow{PhCN} \left[\begin{matrix} PhC{=}NPh \\ | \\ N{=}CPh \end{matrix}\right]^+ AlCl_4^- \longrightarrow \text{(2,4-diphenylquinazolinium)}\ AlCl_4^- \quad (16)$$

(**27**)

nitrilium salt may of course be generated by any of the routes above for use in this quinazoline synthesis. Its formation is favoured by electron-releasing substituents in the nitrile reactant, as would be expected. Alkyl thiocyanates and disubstituted cyanamides gave excellent yields, while mono- and trichloroacetonitrile and ethyl cyanoacetate reacted very sluggishly[50]. Aryl cyanates, which by several tests are less nucleophilic than aromatic nitriles, gave lower yields[51].

Nitrilium salts have been used in many interesting ways in recent years. Reactions can occur at either the nitrile carbon, as in the quinazoline synthesis above, or at the nitrogen atom, although the former is much more common. An example[43] of the latter type is illustrated in reaction (17).

$$[PhC{\equiv}NCH_2OCH_3]^+SbCl_6^- \xrightarrow[\text{pyridine}]{} PhCN + [C_5H_5NCH_2OCH_3]^+SbCl_6^- \quad (17)$$

Several unusual complexes are formed by $PtCl_2$ with aliphatic nitriles in the presence of amines or ammonia. The compound $Pt(MeCN)_2(NH_3)_4Cl_2{\cdot}H_2O$ has been shown to have structure **28** containing acetamidine ligands, evidently synthesized within the coordination sphere[52].

$$\left[\begin{matrix} & NH_2 & & NH_3 & & & \\ & | & & \downarrow & & & \\ CH_3{-}&C{=}NH & \rightarrow & Pt & \leftarrow & HN{=}C & {-}CH_3 \\ & & & \uparrow & & | & \\ & & & NH_3 & & NH_2 & \end{matrix}\right] Cl_2$$

(**28**)

A variety of aliphatic nitriles react with phosphorus pentachloride at room temperature to give excellent yields of *N*-chlorovinylphosphoroimidic trichlorides of varying complexity[53]. The reaction is believed to proceed from a complex of the nitrile with the

* The order of events in this case casts doubt on the need to postulate the dimeric complex **24** as a precursor of **26** in the discussion above.

electrophilic species PCl_4^+. Some typical reactions of the product from chloroacetonitrile (**29**) are indicated below:

$$ClCH_2CN \xrightarrow{PCl_5} ClCH_2C{\equiv}\overset{+}{N}PCl_4{\cdot}PCl_6^- \xrightarrow[-HCl]{PCl_5} \left[\begin{array}{c} Cl_3\overset{+}{P} \qquad Cl \\ C{=}C \\ Cl \qquad N{=}PCl_3 \end{array} \right] PCl_6^-$$

(**29**)

$$\mathbf{29} \xrightarrow{200^\circ} CCl_3CCl_2N{=}PCl_3$$

$$\mathbf{29} \xrightarrow{MeOPCl_2} Cl_2P{-}C(Cl){=}C(Cl){-}N{=}PCl_3$$

$$\mathbf{29} \xrightarrow{SO_2} Cl_2P(=O){-}C(Cl){=}C(Cl){-}N{=}PCl_3$$

B. The Gattermann and Houben–Hoesch Syntheses

In these fundamentally identical reactions (reaction 18), hydrogen cyanide or a nitrile is activated by interaction with hydrogen chloride and (usually) an electrophilic metal halide to give an electrophilic species which will attack an electron-rich aromatic system. The process is comparable to the Friedel–Crafts acylation reaction, and superficially the reagent appears to be an imidyl chloride. The aldimine or ketimine salt produced is normally hydrolysed to a carbonyl compound without isolation. The history and practice of both the aldehyde synthesis (Gattermann reaction)[54] and the ketone synthesis (Houben–Hoesch reaction)[10,55] have been reviewed recently.

$$C_6H_4(OH)_2 + RCN \xrightarrow[0^\circ \text{ in } Et_2O]{ZnCl_2,\ HCl} C_6H_3(OH)_2(RC{=}\overset{+}{N}H_2) \xrightarrow{H_2O} C_6H_3(OH)_2(RC{=}O) \qquad (18)$$

Despite high interest in these reactions and much critical study, their detailed mechanisms remain speculative. Nevertheless, it is undoubtedly true that the active electrophile is a nitrilium ion, probably in equilibrium with the imidyl chloride. Either 1:2 or 2:1 nitrile–metal complexes would be admissible possibilities. The most effective species appears to vary with the nitrile, the metal halide and the reactant ratio used.

Zil'berman[20,56] has made a major advance in this field by finding that maximum yields are obtained in the Houben–Hoesch reaction if the $RCN \cdot ZnCl_2 \cdot HCl$ reagent is mixed in ether and allowed to age at 0° for several hours before the aromatic reactant is added. Furthermore, there is an optimum duration for the ageing process, before or after which the yields are greatly reduced. Obviously the complex-forming and reorganization stages are slow, and it is an unstable complex which is most reactive. It is clear from this work that past reported yields in many cases probably can be greatly improved. By first preparing the imidyl chloride salt it was even possible in some cases to obtain improved yields of ketimine without the use of the metal halide catalyst.

Experiments with $SnCl_4$ and $TiCl_4$ as catalysts gave very poor yields in the reaction of benzonitrile with resorcinol; $AlCl_3$ gave a moderate yield, but $ZnCl_2$ was far better[57]. This variation is presumably related to the complicated effect of the metal on the nitrilium ion–imidyl chloride equilibrium[58] (section III.A).

Correlations of the yields in the Houben–Hoesch reaction are subject to some doubt, but there are sufficient examples to show that the effectiveness of various types of nitriles, $RC{\equiv}N$, increases with the electronegativity of the group R.

C. Metal Carbonyl Complexes

Nitriles can displace carbon monoxide from various metal carbonyls to give several types of complexes. Chromium, tungsten and molybdenum hexacarbonyls react with excess acetonitrile at 80° to give trisacetonitriletricarbonyls which have substantial stability and can be used in the preparation of other complex compounds[59]. Complexes with fewer nitrile ligands are obtained at lower temperature by photochemical reaction[60]. The complex **30** reacts with acrylonitrile at room temperature to give **31**, which cannot be obtained directly; di- and triacrylonitrile complexes can be prepared similarly. The infrared and n.m.r. spectra of **31** and the dinitrile complex unequivocally show the linear $—C{\equiv}N \rightarrow W$ structure[60,61]. The C=C bond is not involved, and **31** will react with cyclopentadiene to give a Diels–Alder adduct (reaction 19)[61].

$$\underset{(\mathbf{30})}{(MeCN)W(CO)_5} \xrightarrow{CH_2{=}CHCN} \underset{(\mathbf{31})}{(CH_2{=}CHCN)W(CO)_5} \xrightarrow{C_5H_6} [\text{bicyclic } CH_2\text{-bridged adduct}]{-}CN \cdot W(CO)_5 \qquad (19)$$

The triacrylonitrile complex, however, is bonded through the π-electrons of the C=C function in more common fashion. A complex $(CH_2{=}CHCN)_2Mo(CO)_2$ has also been reported, which is considered to be a polymer containing bidentate acrylonitrile ligands as bridges[62].

Dialkylcyanamides which are not sterically impeded react with boiling nickel carbonyl to give air-sensitive complexes of the composition $[(R_2NCN)Ni(CO)]_2$. Thermal decomposition in high vacuum regenerates the starting dialkylcyanamide, showing that this structure is not chemically altered. Spectroscopic and other physical properties of these complexes support the structure **32** in which zero-valent nickel is bonded to the amino nitrogen through its lone-pair electrons, and to the nitrile group through its π-electrons. Addition of a halocarbon to a solution of the complex causes a spectral shift to the high C≡N absorption frequency of a linear nitrile–metal complex[63].

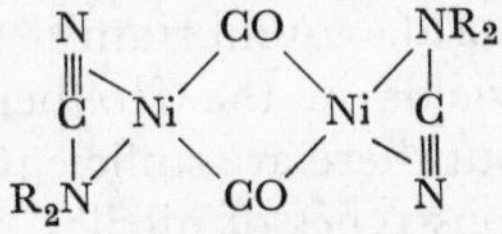

N CO NR₂ C Ni Ni C R₂N CO N

(32)

Acetonitrile reacts with decaborane on heating to give a bis-nitrilium salt[64] (equation 20). This reaction has also been used to

$$2\ MeCN + B_{10}H_{14} \longrightarrow (MeC{\equiv}\overset{+}{N})_2B_{10}H_{12}^{2-} + H_2 \qquad (20)$$

prepare polymeric products from dinitriles[65]. The ease of reaction is dependent on the basicity of the nitrile; cyanogen does not react, malononitrile is sluggish and diethylcyanamide will displace acetonitrile. In satisfying analogy to strong proton acids, iron carbonyls have been shown to catalyse trimerization of PhCN[66].

IV. HYDRATION OF NITRILES

Hydrolysis of nitriles in either alkaline or acidic media has long been an important synthetic route to carboxylic acids. Under the conditions commonly employed, hydrolysis of the intermediate amide is much faster than the initial hydration of the nitrile (equation 21).

$$RCN + H_2O \longrightarrow RCONH_2 \xrightarrow{H_2O} RCO_2H + NH_3 \qquad (21)$$

However, several effective procedures have been developed for interrupting the process at the amide stage. Examples of amide and acid syntheses and pertinent bibliographies have been compiled[67]. This discussion will be primarily concerned with the hydration process.

A. Base Catalysis

Base-catalysed hydration of nitriles has proven to be mechanistically straightforward. The sodium hydroxide-catalysed reaction obeys the bimolecular rate law very closely over a wide range of alkali concentration[68]. This is in accord with the mechanism suggested by analogy with ester hydrolysis (Scheme 1).

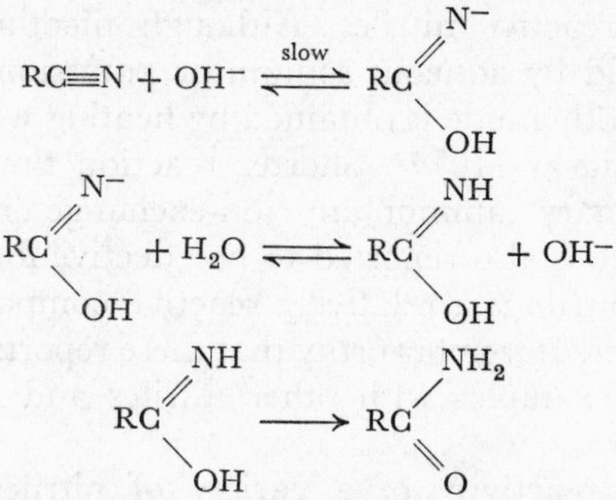

$$RC{\equiv}N + OH^- \xrightleftharpoons{\text{slow}} RC(=N^-)OH$$

$$RC(=N^-)OH + H_2O \rightleftharpoons RC(=NH)OH + OH^-$$

$$RC(=NH)OH \longrightarrow RC(=O)NH_2$$

SCHEME 1.

Wiberg has studied the hydroxide-catalysed hydration of a limited series of benzonitriles[69] and shown that the substituent effects are in line with the expected accelerating influence of electron-attracting R groups. His data fit the Hammett correlation satisfactorily, with a ρ-value quite comparable to that for alkaline hydrolysis of benzoate esters[70]. A particular point of interest was the high negative entropy of activation in these reactions (−13 to −17 e.u.). A comparable value was derived from the data for propionitrile[68]. Rates of alkaline hydrolysis, and hence of hydration to the amide, have also been reported for the following aliphatic nitriles which gave the reactivity order: succinonitrile (relative rate constant, 2·62), acrylonitrile (1·39), adiponitrile (1·18), acetonitrile (0·945), methacrylonitrile (0·362). Energies of activation were found to increase in the inverse order sufficiently (33·6 to 47·5 kcal/mole) to explain the relative rates[71].

The relative rates of propionitrile and propionamide hydrolysis have been shown to be about 1 : 10 over a wide range of alkalinity[68]. This relationship is not general, however, as has sometimes been assumed. For example, the Hammett ρ-value is 2·31 for alkali-catalysed hydration of benzonitriles and 1·06 for hydrolysis of benzamides[70]. Consequently these correlation lines must intersect at some high σ-value. The equivalent of the necessary high σ-values are readily attained in electronegatively substituted aliphatic nitriles, cyanopyridines, etc. An illustration is found in the observation that hydration of 2-cyano-1,10-phenanthroline is 15 times as fast as hydrolysis of 2-carbamido-1,10-phenanthroline in 1·0M sodium hydroxide[72].

As these considerations suggest, synthesis of amides by hydroxide-catalysed hydration of nitriles is most successful with mild bases and with relatively reactive nitriles. Although nicotinonitrile is converted to the acid by aqueous sodium or potassium hydroxide, an 80 % yield of nicotinamide is obtained by heating with concentrated aqueous ammonia at 110° [73]. Shorter reaction time is sufficient if a basic quaternary ammonium ion-exchange resin is used[74]. Magnesium oxide is also reported to be effective for the same reaction[75]. Nicotinonitrile is a relatively reactive compound from other evidence, however. It is noteworthy that these reports of its hydration do not contain examples with other nitriles and no generality is claimed.

Study of the reactivity of a variety of nitriles with aqueous amines has yielded pertinent information[76,77]. Simple aliphatic or aromatic nitriles do not react at a significant rate with aqueous primary or secondary amines at 100°, and only very slow reaction (to give *N*-substituted amides and the carboxylic acids) takes place at 150°. If the nitrile is strongly activated by electronegative substituents, however, as in malononitrile, cyanoacetic esters or cyanoacetamides, reactions take place readily at 100° [77,78]. Since the products of these reactions are usually *N*-substituted amides, it is not certain that base-catalysed hydration is in fact the initial step. If this is the case, tertiary amines would be expected to serve as hydration catalysts. Succinonitrile and 3-methoxycarbonylpropionitrile react slowly with aqueous amines at 100°, the small activating effects of the β-substituents perhaps being reinforced by formation of cyclic intermediates[77,78]. (Compare section II.C.)

B. Acid Catalysis

The mechanism of acid-catalysed hydration of nitriles has been studied intensively. The kinetic data of Krieble[41,79] and Winkler[80] and their associates have provided the basic description of the variation of rates of hydration of simple nitriles as a function of acid concentration and reaction temperature. Specific effects of the acid anion were disclosed as well as some substituent effects on nitrile reactivity. Nevertheless, a comprehensive analysis of the reaction mechanism did not emerge until the work of Kilpatrick[81] on the acid-catalysed conversion of cyanamide to urea. It is instructive to consider first the complexities of the hydration of cyanamide and then to relate the conclusions to other nitriles.

All observations on nitrile hydration in dilute acid are in accord with the steps of Scheme 2. Although inspection suggests that

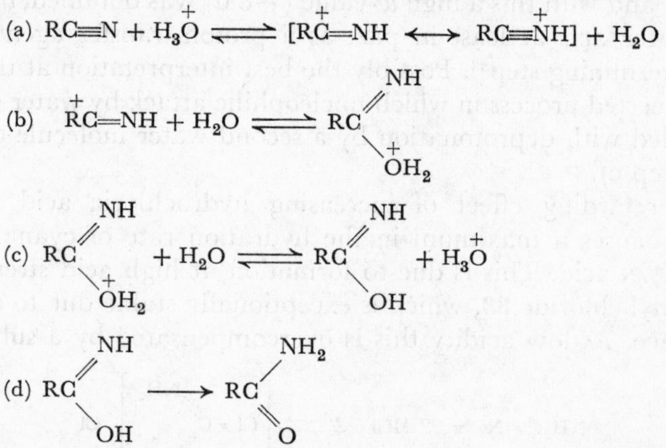

$$\text{(a)}\quad RC{\equiv}N + H_3\overset{+}{O} \rightleftharpoons [R\overset{+}{C}{=}NH \longleftrightarrow RC{\equiv}\overset{+}{N}H] + H_2O$$

$$\text{(b)}\quad R\overset{+}{C}{=}NH + H_2O \rightleftharpoons RC(=NH)\overset{+}{O}H_2$$

$$\text{(c)}\quad RC(=NH)\overset{+}{O}H_2 + H_2O \rightleftharpoons RC(=NH)OH + H_3\overset{+}{O}$$

$$\text{(d)}\quad RC(=NH)OH \longrightarrow RC(NH_2)=O$$

SCHEME 2.

step (b) would in all probability be rate-determining, Hammett found that the rate of hydration of cyanamide in 0·5–5M nitric acid is approximately proportional to the H_0 acidity function. By the Zucker–Hammett hypothesis, this implied a mechanism which did not include water in the rate-determining step. The slow step was instead conceived as isomerization of the conjugate acid $NH_2CN{\cdot}H^+$ to a more active species which reacted rapidly with water[82a]. Kilpatrick's subsequent work, however, showed that hydrohalic acids cause strong retardation of hydration at high concentration, while in trichloroacetic acid and in acetate or dichloroacetate buffers hydration is more rapid than anticipated from the hydrogen ion concentration. Clearly the unimolecular postulate was untenable, and the complex data also were not satisfied by considering step (b) alone as rate-determining. In particular, the apparent base catalysis by carboxylate anions required explanation. An unconventional proposal that the rate-determining step was (c), in which deprotonation of the hydrated cation competes with more rapid ejection of the water molecule[81], has not been entirely accepted because proton exchange reactions at oxygen or nitrogen are normally very fast[82b,c]. Bunnett has discussed this question in the light of his empirical *w*-criterion of mechanism[82b]. Values of *w* for HNO_3-catalysed hydration (+0·8 to +1·5) were indicative of a reaction in which the rate-determing step is nucleophilic attack by water on the conjugate acid of the substrate, thus step (b). Recently, however, it has been shown that the nitrate ion causes a large salt effect which could lead to unusually low *w*-values. Perchloric acid gives a more accurate picture, and with this a high *w*-value (+8·0) was obtained, implying that water acts at least in part as a proton-transfer agent in the rate-determining step[83]. Possibly the best interpretation at this time is a concerted process in which nucleophilic attack by water (step b) is coupled with deprotonation by a second water molecule or other base (step c).

The retarding effect of increasing hydrochloric acid concentration causes a maximum in the hydration rate of cyanamide at about 1·7N acid. This is due to formation at high acid strengths of the imidyl chloride **33**, which is exceptionally stable due to amidine resonance. At low acidity this is overcompensated by a substantial

$$NH_2C{\equiv}N \ + \ 2\ HCl \ \rightleftharpoons \ \left[Cl{-}C\!\begin{smallmatrix} \cdot\cdot\cdot NH_2 \\ \cdot\cdot\cdot NH_2 \end{smallmatrix} \right]^+ Cl^-$$

(33)

positive electrolyte effect. The retardation is very marked; in 3N hydrochloric acid the hydration rate at 30° is one-tenth that in 3N nitric acid[81].

Other nitriles have been studied in less depth, but counterparts of several features of the cyanamide hydration process have been observed[81]. At hydrochloric or hydrobromic acid concentrations above about 4N, several common nitriles give higher reaction rates than are observed with sulphuric acid. In these concentrated solutions, the nitrile is to some extent converted to the protonated imidyl halide (equation 1, section II.A). Unlike **33**, these species are not stabilized by resonance and hydrolyse at higher rates than the nitrilium salts (Scheme 2) so that the composite rate is abnormally high. This change in mechanism is in agreement with a progressive decrease in the apparent activation energy of the reaction as the acid concentration is increased. A smaller but similar effect occurs in sulphuric acid, but other complications in such systems make the cause obscure. The hydration of propionitrile in

10M hydrochloric acid is about twenty times as fast as in 10M sulphuric acid; hydrogen cyanide and lactonitrile show still greater effects. Carboxylate anions could also act as nucleophiles in this sense[82c] and in so doing would give isoimide structures (**15**, section II.C). As discussed earlier, these would be high-energy carboxylates and could be responsible for part of the specific catalytic activity of carboxylic acids.

Acceleration of the rate of hydrolysis of a nitrile by conversion to an imidyl chloride deserves further comment. In dilute acid, the reactive species is clearly the mesomeric cation **34**. If this were well represented by **34b**, coordination with chloride ion would compete with hydrolysis and be decelerating. If, as is much more reasonable, the cation is close to **34a**, conversion to **35** and **36** produces an increase in the effective concentration of species susceptible to hydrolysis. The carbon atom of the functional group must be more electrophilic in **36** than in the hybrid cation **34**.

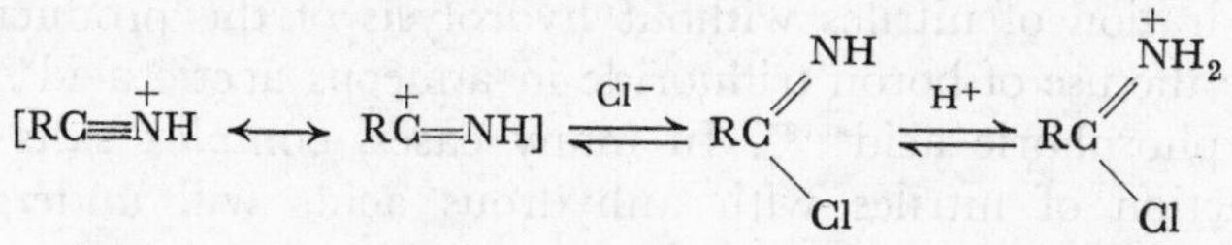

$$[\mathrm{RC{\equiv}\overset{+}{N}H} \longleftrightarrow \mathrm{R\overset{+}{C}{=}NH}] \underset{}{\overset{\mathrm{Cl^-}}{\rightleftharpoons}} \mathrm{RC({=}NH)Cl} \overset{\mathrm{H^+}}{\rightleftharpoons} \mathrm{RC({=}\overset{+}{N}H_2)Cl}$$

(**34a**) (**34b**) (**35**) (**36**)

An interesting case has been reported of concerted general acid catalysis by metal ions in nitrile hydration by hydroxide ion[72]. The second order rate constant for the reaction of the Ni^{2+} complex of the cyanophenanthroline **37** with hydroxide ion to give the amide is 10^7 times greater than that for **37** itself. Activation enthalpies for the two cases are essentially equal, the entire rate difference being attributable to the exceptionally large positive entropy of activation in the metal ion-catalysed reaction. It is proposed that the Ni^{2+} cation is appropriately situated in the complex to bond to the developing imino anion in the transition state (**38**). This interaction could result in reduced solvation and the very favourable entropy change. The presence of the metal also facilitates attack by ethoxide ion to give the imidate.

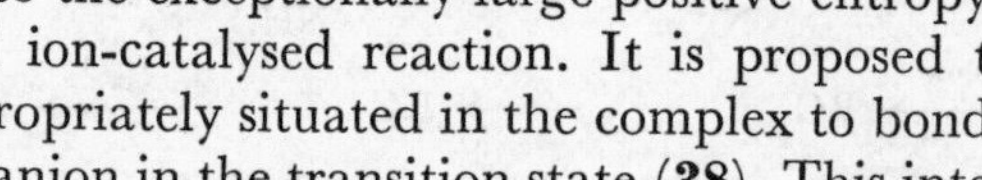

N N C≡N

(**37**)

N N Ni^{2+} C—OH $\bar{N}$

(**38**)

As the discussion above suggests, the rate of nitrile hydration in concentrated acids can substantially exceed the rate of further hydrolysis of the product amide. Practice has anticipated theory to some extent, and concentrated sulphuric, hydrochloric and phosphoric acids have long been used in practical amide syntheses[67]. A particularly important illustration is the conversion of acrylonitrile to acrylamide, which is carried out industrially by heating the nitrile briefly with an equimolar amount of sulphuric acid monohydrate at about 100°[84] (reaction 22). The reaction product, acrylamide sulphate (**39**), can be processed to yield acrylamide

$$CH_2{=}CHCN + H_2SO_4 + H_2O \longrightarrow \underset{\mathbf{(39)}}{CH_2{=}CHCONH_2{\cdot}H_2SO_4} \qquad (22)$$

itself or may be reacted with an alcohol or with additional water to an give an acrylate or acrylic acid.

Additional special techniques which have been recommended for the hydration of nitriles without hydrolysis of the product amide include the use of boron trifluoride in aqueous acetic acid[85,86], and of polyphosphoric acid[86,87]. In many cases, complex salts formed by reaction of nitriles with anhydrous acids will undergo easy hydrolysis when water is added.

C. Addition of Hydrogen Peroxide

The Radziszewski reaction, in which a nitrile is converted to an amide by reaction with hydrogen peroxide in weakly basic solution (reaction 23), is very closely related to hydroxide-catalysed hydration. It has been shown by Wiberg[69,88] that the reaction rates for several

$$RCN + HO_2^- \xrightarrow{\text{slow}} RC\begin{matrix}{/\!\!/ N^-}\\{\backslash O{-}OH}\end{matrix} \longrightarrow \underset{\mathbf{(40)}}{RC\begin{matrix}{/\!\!/ NH}\\{\backslash O{-}OH}\end{matrix}} \xrightarrow{H_2O_2} RC\begin{matrix}{/ NH_2}\\{\backslash\!\!\backslash O}\end{matrix} + O_2 \qquad (23)$$

substituted benzonitriles obey the Hammett correlation as required by the mechanism shown ($\rho = 1{\cdot}66$). The reaction is over 10^4 times as fast as that with hydroxyl ion, the difference being related to much more favourable entropy of activation in the hydroperoxide addition step; enthalpies of activation were found to be approximately the same for the two cases. The exceptional nucleophilicity of the hydroperoxide anion in this reaction has been attributed to participation of the proton on oxygen in a cyclic transition state

(**41**), in which the developing negative charge on nitrogen is partially 'solvated' intramolecularly. This suggestion[72] is closely related to

$$\mathrm{R{-}C{\equiv}N \cdots H}, \quad {}^{-}\mathrm{O{-}O}$$

(**41**)

the explanation for facilitation of the hydroxide-catalysed hydration of 2-cyano-1,10-phenanthroline by complexed metal ions (see section IV.B).

Various modifications of the Radziszewski reaction have been recommended[67]. Yields of amides are frequently excellent.

Utilization of the transient peroxyimidic acid intermediate (**40**) of the Radziszewski reaction as an oxidizing reagent is an important concept introduced by Payne[89]. This species proves to be a highly reactive electrophilic oxidant toward any available reducing agent. It will efficiently epoxidize an olefin, convert pyridine to its *N*-oxide, and oxidize aniline to azoxybenzene. In the absence of such substrates, oxidation of hydrogen peroxide itself produces oxygen and the carboxamide. Optimum conditions for the epoxidation process have been published[90]. As a general procedure, 50–90% aqueous hydrogen peroxide is added gradually to an excess of both the olefin and acetonitrile in methanol, while the pH is held at 7·5–8·0 by concurrent addition of alkali. Acetonitrile is the most practical reagent, although reactions with benzonitrile are faster. Alternatively phosphate buffer, which inhibits reaction of **40** with hydrogen peroxide, can be used to obviate the need for close control of the alkalinity of the reaction mixture[91]. Trichloroacetonitrile and isopropylidenemalononitrile[92], highly activated reactants in many instances, were found to react rapidly even at pH 5–6. The peroxyimidic acids should find many synthetic applications. Their advantage over peroxides lies in their applicability under alkaline conditions[90, 93].

V. NUCLEOPHILIC ADDITION OF HYDROXY, AMINO AND SULPHYDRYL COMPOUNDS

A. Addition of Alcohols and Phenols

Addition of an alcohol or phenol to the C≡N bond gives an alkyl or aryl ester of the related imidic acid (an 'imidate', formerly called an 'imino ether'). The corresponding thioimidates are also known, but are of much less interest because of the poor availability of the sulphydryl compounds. Following rather intense activity around 1900, interest in the imidates has been low until recent years. A review of their chemistry by Roger and Neilson appeared in 1961[94] and has been supplemented by Zil'berman[10].

I. The Pinner synthesis of imidates

The classic procedure for the preparation of alkyl imidate salts is the Pinner synthesis[95] (reaction 24). Although numerous modifications in reactant ratio, choice of solvents and mode of operation

$$R^1C{\equiv}N + R^2OH + HCl \xrightarrow[\text{ether, etc.}]{\sim 0^\circ} \left[R^1C\begin{matrix} {/\!\!/} \overset{+}{N}H_2 \\ \diagdown OR^2 \end{matrix} \right] Cl^- \quad (24)$$

have been advanced, very little significant improvement has been made on Pinner's basic procedure. Reaction of equivalent amounts of the three components commonly gives yields in excess of 90% within 24 hours.

The reaction is almost completely general with respect to the nitrile reactant, limited only by steric factors and by the great ease with which the alkyl imidate hydrochlorides from strongly electronegatively substituted nitriles undergo the 'Pinner cleavage' (reaction 25), in some cases under the synthesis conditions. In practice, the alcohol component is usually methanol or ethanol

$$R^1C\begin{matrix} {/\!\!/} \overset{+}{N}H_2Cl^- \\ \diagdown O{-}R^2 \end{matrix} \longrightarrow R^1C\begin{matrix} / NH_2 \\ {\diagdown\!\!\diagdown} O \end{matrix} + R^2Cl \quad (25)$$

because in the subsequent reactions of greatest interest this group is usually lost. However, benzyl alcohol and secondary alcohols can be used successfully. Recently it has been shown that strongly acidic alcohols, such as 2,2,2-trichloroethanol[96] and the 2-nitroalkanols[97] give stable imidate hydrochlorides with trichloroacetonitrile. In these compounds, the electron-attracting substituents oppose displacement of the protonated trichloroacetimidyl group, which is an exceptionally good leaving group[96]. The Pinner synthesis is usually also satisfactory with simple phenols. It commonly competes with the Houben–Hoesch reaction despite the use of a metal halide catalyst in the latter process, and in some cases the imidate has been formed exclusively. Aryl imidate hydrochlorides do not undergo the Pinner cleavage on heating and may be prepared normally from the highly activated nitriles.

Hydrogen chloride is used almost exclusively in the Pinner synthesis. Although hydrogen bromide and iodide have been recommended[98] to increase the speed of the reaction, this is rarely a significant problem. On the other hand, the imidate hydrobromides and hydroiodides have lower thermal stabilities[97].

Although there has been almost no study of the mechanism of the Pinner reaction, it seems quite safe to assume that it proceeds analogously to the solvolysis of nitriles in water or carboxylic acids. As in those cases, perhaps the clearest data come from a study of the alcoholysis of cyanamide[81]. Solvolysis of cyanamide in methanol containing 0·1N hydrogen chloride (reaction 26) is 150 times as fast as in water (reaction 27), probably largely because of the greater nucleophilic reactivity of the alcohol compared to water. Surprisingly the bimolecular rate constant drops off with increasing

$$NH_2CN + MeOH + HCl \longrightarrow \left[NH_2C\begin{matrix} \overset{+}{N}H_2 \\ OMe \end{matrix} \right] Cl^- \qquad (26)$$

$$NH_2CN + H_2O + HCl \longrightarrow NH_2C\begin{matrix} \overset{+}{N}H_2 \\ OH \end{matrix} \underset{}{\overset{-H^+}{\rightleftharpoons}} NH_2CONH_2 \qquad (27)$$

acid concentration even at this low level. Evidently, as in the hydration reaction (section IV.B), formation of chloroformamidine hydrochloride impedes solvolysis. A maximum rate must occur at some still lower hydrogen chloride concentration. In the practical synthesis of *O*-alkylpseudoureas from ethanol and higher alcohols by reaction (26), separation of crystalline chloroformamidine hydrochloride usually can be observed in the early stage of the reaction.

If the parallel with hydration holds for alcoholysis of nitriles in general, it may be expected that intermediate imidyl chloride formation will normally enhance the rate of conversion to imidate. Thus hydrogen halides should be better catalysts than the oxy acids. There is no experimental basis, however, upon which to judge the relative importance in the reaction of the imidyl halide versus the nitrilium cation.

Alkyl thiocyanates[99] give dialkyl iminothiocarbonates in the

Pinner synthesis (reaction 28), but alkyl cyanates are decomposed

$$\mathrm{RSCN + EtOH + HCl \rightleftharpoons (RS)(EtO)C{=}\overset{+}{N}H_2Cl^-} \qquad (28)$$

to the alkyl chloride and cyanic acid which reacts with the alcohol to form a carbamate[100]. Aryl cyanates give unstable alkyl aryl iminocarbonates which decompose easily to the aryl carbamate (reaction 29) and will react with additional alcohol to give methyl orthocarbonate[101] (reaction 30). This reactivity is impressive in

$$\mathrm{PhOCN + MeOH + HCl \longrightarrow (PhO)(MeO)C{=}\overset{+}{N}H_2Cl^- \longrightarrow PhOCONH_2 + MeCl} \qquad (29)$$

$$\mathrm{(PhO)(MeO)C{=}\overset{+}{N}H_2Cl^- \xrightarrow{MeOH} C(OMe)_4 + PhOH} \qquad (30)$$

view of the stability of *O*-methylpseudourea hydrochloride[102]. Although the cyanates and cyanamides have somewhat comparable reactivity, reactions of the latter lead to amidine structures which are strongly stabilized by resonance.

A variation on the Pinner synthesis consists of initial preparation of the imidyl chloride followed by reaction of this with an alcohol[14,15,21]. Frequently excellent yields are obtained in this way, although normally there is no advantage over the conventional procedure. However, as pointed out earlier (section II.A), the 'imidyl halides' are somewhat unpredictable in composition, and unexpected results may occur. For example, the dimeric imidyl chloride salt from chloroacetonitrile reacts with ethanol to give a high yield of chloroacetamidine[103] (reaction 31). This kind of

$$\mathrm{ClCH_2C(Cl){=}N{-}C({=}\overset{+}{N}H_2Cl^-)CH_2Cl + 3\,EtOH \longrightarrow ClCH_2C({=}\overset{+}{N}H_2Cl^-)NH_2 + ClCH_2CO_2Et + Et_2O + HCl} \qquad (31)$$

anomaly is particularly likely to be encountered with the more reactive nitriles.

2. Base-catalysed addition of alcohols

Base-catalysed alcoholysis of nitriles (reaction 32), originally discovered by Nef[104], is a very simple and efficient method for the conversion of many nitriles to imidates. It has only rarely been applied until recently, however, because it is unsuited for use with

$$R^1C{\equiv}N + R^2OH \underset{}{\overset{R^2O^-}{\rightleftharpoons}} R^1C(=NH)OR^2 \qquad (32)$$

the simple nitriles. The reaction was reintroduced in 1960 with a study of its scope, suitable procedures and new illustrations of its utility[105].

The equilibrium conversion of nitrile to imidate in methanol at 25° in the presence of a catalytic amount of sodium methoxide is low with the simple nitriles, but as shown in Table 1, the reaction

TABLE 1. Base-catalysed conversion of nitriles to methyl imidates (1·0M in methanol at 25°).

Nitrile	Equilibrium conversion (%)
MeCN	2
$PhCH_2CN$	12
Et_2NCH_2CN	22
Me_3COCH_2CN	91
$ClCH_2CN$	93
$(EtO)_2CHCN$	95
PhCN	20
$4\text{-}ClC_6H_4CN$	45
$3\text{-}NO_2C_6H_4CN$	86
$3\text{-}C_5H_4NCN$	71
Me_2NCN	100

is strongly promoted by electronegative substituents and gives high conversions with many interesting nitriles. In most cases, the equilibrium is reached in 15–100 minutes, and the solution of the imidate base can often be used directly for a subsequent reaction without separation of residual unreacted nitrile. The fact that reaction (32) is reversible accounts for the instability of some imidate bases toward alkali. Autocatalysis of dissociation during storage can be appreciable in those cases where the equilibrium is unfavourable.

The mechanism of reaction (32) appears to be straightforward. Those aspects of the kinetics which have been examined[105,106] are in general accord with other base-catalysed addition reactions of nitriles. It is worth noting that the data on base-catalysed imidate formation probably offers the broadest comparison available of nitrile reactivity versus structure.

This base-catalysed imidate synthesis neatly compliments the Pinner reaction, each being well suited to those cases for which the other is least useful. For the large intermediate group of nitriles which react well by either method, the less familiar basic procedure frequently offers advantage.

In recent years, there has been particular interest in this technique for use with cyanogen[107,108], trichloroacetonitrile[96,109] and perfluoroaliphatic nitriles[110]. In these cases, alkali carbonates or cyanides or tertiary amines are satisfactory catalysts. Trichloroacetonitrile has been used successfully with a variety of complex primary and secondary alcohols, and even with tertiary alcohols. Trifluoroacetonitrile does not add tertiary alcohols because of competing alkoxide-catalysed trimerization, which could well involve a related mechanism (reaction 33).

$$CF_3CN + RO^- \longrightarrow CF_3C(=N^-)OR \underset{}{\overset{ROH}{\rightleftharpoons}} CF_3C(=NH)OR$$

$$CF_3C(=N^-)OR \underset{}{\overset{CF_3CN}{\rightleftharpoons}} CF_3C(OR)=N-C(CF_3)=N^- \xrightarrow{CF_3CN} \text{2,4,6-tris(trifluoromethyl)-1,3,5-triazine} \quad (33)$$

The imidates obtained from these very active nitriles react extremely rapidly with acids, including carboxylic acids and hydrogen fluoride, in the sense of the Pinner cleavage (reaction 25), and their study has provided an insight into the detailed mechanism of this reaction[96].

3. Imidates as activated nitriles

Imidates can be expected to be more reactive than nitriles in acid-catalysed reactions (the cyanamide derivatives are probably a unique exception) for the same reason that imidyl chlorides are much more reactive than the corresponding nitriles (section IV.B). The carbon atom of the functional group in a protonated imidate cation is highly electrophilic, and an alkoxide is a sufficiently good leaving group to permit nucleophilic displacement reactions. With nitrogen nucleophiles in particular, amidine resonance in the reaction product provides much of the driving force for displacement of the alkoxide (reaction 34). This type of imidate reaction gives

$$\mathrm{{>}NH + R^1C(=\overset{+}{N}H_2)OR^2 \rightleftharpoons {>}\overset{+}{H}N{-}C(R^1)(OR^2){-}NH_2 \longrightarrow \left[{>}N\cdots C(R^1)\cdots NH_2\right]^+ + R^2OH} \qquad (34)$$

the same product as would be expected from the nitrile. In practice it is frequently found that substantial advantage can be gained in this way. A few examples are given below.

[1] The two-step conversion of a nitrile to an ester by reaction (35) takes place within minutes at room temperature[105].

$$\mathrm{ClCH_2CN + MeOH \xrightarrow{NaOMe} ClCH_2C(=NH)OMe \xrightarrow[H_2O]{HCl} ClCH_2CO_2Me + NH_4Cl} \qquad (35)$$

[2] Amidines are obtained very readily by reaction of imidate hydrochlorides with alcoholic ammonia[95] or by reaction of imidate bases with ammonium salts[105], whereas acid-catalysed addition of ammonia to a nitrile is a slow, high-temperature process (section V.B).

[3] A limited number of *s*-triazines can be obtained by strong acid-catalysed trimerization of nitriles; imidates, in general, trimerize readily at ordinary temperature in the presence of acid catalysts[111] (reaction 36).

$$\mathrm{3\ MeC(=NH)OEt \xrightarrow{HOAc} 2,4,6\text{-}Me_3\text{-}s\text{-}triazine + 3\ EtOH} \qquad (36)$$

B. Addition of Ammonia and Amines

Unsubstituted amidines are usually prepared by reaction of an imidate salt with ammonia (equation 37) or by the equivalent reaction of an imidate base with an ammonium salt. For large scale usage, however, a single step synthesis from the nitrile is desirable. Two broadly useful solutions to this need are discussed below.

$$R^1C(=\overset{+}{N}H_2)OR^2 + NH_3 \longrightarrow \left[R^1C(\cdots NH_2)(\cdots NH_2)\right]^+ + R^2OH \qquad (37)$$

I. Synthesis of amidinium salts

Nitriles react with ammonia or amines as with other nucleophiles to establish an equilibrium (equation 38) in which the conversion to amidine is favoured by electron-attracting substituents in the nitrile. Very strongly activated nitriles such as trichloroacetonitrile[112] or trifluoroacetonitrile[113] will add ammonia and amines at low temperature rapidly and essentially quantitatively. Such amidines will themselves react with additional nitrile to give unstable imidylamidines (reaction 39). Hydrogen cyanide, with its

$$RC\equiv N + NH_3 \rightleftharpoons RC(=NH)NH_2 \qquad (38)$$

$$2\ CCl_3CN + NH_3 \xrightarrow[H_2O]{10^\circ} CCl_3C(=NH){-}N{=}C(NH_2){-}CCl_3 \qquad (39)$$

uniquely unhindered functional group, reacts readily with primary amines to give substituted formamidines[114–116] (reaction 40) and in a somewhat more complicated way with secondary alicyclic amines

$$HCN + RNH_2 \longrightarrow HC(=NH)NHR \longrightarrow HC(=NR)NHR + NH_3 \qquad (40)$$

$$2\ HCN + 2\ C_5H_{10}NH \xrightarrow{-NH_3} \left[HC(\cdots NC_5H_{10})(\cdots NC_5H_{10})\right]^+ CN^- \rightleftharpoons (C_5H_{10}N)_2CHCN \qquad (41)$$

(reaction 41). In this case the products exhibit salt-like properties in strongly solvating media such as phenol but are covalent under usual conditions[116]. These reactions also take place very rapidly in aqueous amine solutions, in which case hydrolysis to the *N*-substituted amide immediately ensues[117]. (Compare section IV.A.)

With most nitriles, ammonia and amines cause very little conversion to amidine under any conditions investigated[118]. This failure probably has two causes: (*a*) the equilibrium is itself unfavourable, particularly at the high temperatures which have sometimes been used, and (*b*) the activation energy must be very high and the rate of reaction very slow with the simple nitriles. When liberated from their salts, the free amidines are moderately stable at ordinary temperatures; acetamidine, for example, can be distilled at 90° at low pressure without appreciable decomposition[31]. If, however, the reaction can be carried out in such a way that the amidine is stabilized as an amidinium salt, higher temperatures can be employed to accelerate the addition reaction. Such a technique has long been known for the conversion of cyanamides to guanidine salts[119] and for the synthesis of *N*-arylbenzamidines[120] and consists simply of the fusion of an ammonium salt with the nitrile (reaction 42).

$$R^1CN + R^2R^3NH\cdot HX \longrightarrow \left[R^1C\begin{matrix} NH_2 \\ NR^2R^3 \end{matrix} \right]^+ X^- \qquad (42)$$

Oxley and Short and their collaborators[118] studied nearly all the variables in this general scheme and found it to be particularly useful for the synthesis of *N*-substituted amidines from aromatic nitriles. Optimum conditions include the use of sufficiently high temperatures (180–250°) to obtain a melt and a large molar excess of the salt. The anion present is quite critical, and best results are obtained with benzenesulphonates or thiocyanates. Ammonium thiocyanate gives moderately good yields of the unsubstituted amidines, but recovery of the product in useful form is usually awkward. An exception is the preparation of *p*-chlorobenzamidine thiocyanate in 90% isolated yield[121]. Thiourea is an equivalent reagent for this purpose due to its isomerization to ammonium thiocyanate at the reaction temperature.

A major breakthrough was achieved in 1961 with the discovery that unsubstituted amidine salts can be prepared in high yield from a wide variety of aliphatic and aromatic nitriles by reaction with ammonium chloride or bromide in the presence of ammonia under pressure at 125–150° [121]. There are no significant side reactions in this amidine synthesis, provided the nitrile is not degraded by ammonia at 150°. Consideration of the factors involved in maximizing the amidine yield is instructive. The problem is to achieve an adequate reaction rate in the slow step (reaction 43a) at a temperature where the equilibrium concentration of amidine salt is high. This prohibits use of temperatures much above 150°. A high ammonia concentration is needed to promote the forward reaction

$$\mathrm{RCN} + \mathrm{NH_3} \underset{(a)}{\rightleftharpoons} \mathrm{RC}\begin{matrix}\nearrow \mathrm{NH} \\ \searrow \mathrm{NH_2}\end{matrix} \underset{(b)}{\overset{\mathrm{NH_4X}}{\rightleftharpoons}} \left[\mathrm{RC}\begin{matrix}\nearrow \mathrm{NH_2} \\ \searrow \mathrm{NH_2}\end{matrix}\right]^+ \mathrm{X^-} + \mathrm{NH_3} \quad (43)$$

and to act as a solvent when no other is used. Too much is disadvantageous, however, because it necessitates an increase in the salt content to ensure protonation of the amidine product (reaction 43b). Practical reactant ratios are: 1·0 mole nitrile/1·2–20 moles NH_3/1–4 moles NH_4X. To reduce the working pressure at the reaction temperature it is advantageous to use methanol as a solvent. This has essentially no effect on the yield or reaction rate. Equilibrium conversions with ammonium chloride and ammonium bromide are essentially the same, but reactions with the latter are substantially faster. The expected activating effect of electronegative substituents is observed in faster reactions at lower temperatures.

Acid catalysis cannot be a significant factor in these reactions in which free ammonia is present. However, some contribution to the success of the salt fusion process may come from the slightly acidic character of the ammonium salts. Something approaching a concerted reaction may be involved (equation 44).

$$\begin{matrix}\mathrm{R^1C{\equiv}N} \\ \mathrm{R_2{}^2H\overset{+}{N}{-}H}\end{matrix} \rightleftharpoons \begin{matrix}\mathrm{R^1C{=}N} \\ | \quad\quad \searrow \mathrm{H} \\ \mathrm{R_2{}^2\overset{+}{N}H}\end{matrix} \rightleftharpoons \left[\mathrm{R^1C}\begin{matrix}\nearrow \mathrm{NH_2} \\ \searrow \mathrm{NR_2{}^2}\end{matrix}\right]^+ \quad (44)$$

Hydroxylamine is an exceptionally nucleophilic amino compound and reacts readily with all types of nitriles to give amidoximes (equation 45). The reaction can be carried out in an aqueous system, but better yields are usually obtained with solutions of free

$$\mathrm{RC{\equiv}N} + \mathrm{NH_2OH} \longrightarrow \mathrm{RC}\begin{matrix}\nearrow\!\!\!\nearrow \mathrm{NOH} \\ \searrow \mathrm{NH_2}\end{matrix} \quad (45)$$

hydroxylamine in ethanol or butanol[122]. The reaction rate reflects the activating influence of electronegative substituents in the nitrile[123].

Monomeric imidyl halides do not react with ammonia or amines to give amidines. Instead the original nitrile is regenerated (reaction 46).

$$RC(=\overset{+}{N}H_2Cl^-)Cl + 2\,NH_3 \longrightarrow RC{\equiv}N + 2\,NH_4Cl \quad (46)$$

2. Reaction of nitriles with sodamide

For many applications, amidines are used in the free base form, frequently with alkoxide present. For such purposes a particularly attractive amidine synthesis is available in the reaction of a nitrile with sodium or potassium amide in liquid ammonia, ether or benzene (reaction 47)[124–126]. The suitability of this reaction for use with both aliphatic and aromatic nitriles is well established[126]

$$RC{\equiv}N + NaNH_2 \rightleftharpoons \left[RC\begin{smallmatrix}NH\\NH\end{smallmatrix}\right]^- Na^+ \quad (47)$$

despite earlier doubts, yet it has seen remarkably little use. It is in fact met most frequently as a side reaction in alkylation or acylation of nitriles by the sodamide procedure[125,127].

The sodium salt of the amidine can be converted to the free base by addition of water, but isolation in this form can be discouraging. Thus low yields of recovered long-chain aliphatic amidines have been reported[126], although yields of derived pyrimidines (**42**) in the 45–55% range have been obtained by direct reaction of the crude potassium salts with diethyl malonate[128].

$$\left[RC\begin{smallmatrix}NH\\NH\end{smallmatrix}\right]^- K^+ + CH_2(CO_2Et)_2 \longrightarrow$$

(**42**)

In a variation of this amidine synthesis, magnesium derivatives of secondary amines have been found to give good yields in reactions with several aliphatic and aromatic nitriles and with diethylcyanamide[129]. Very recently sodium hydrazide has also been used successfully[130] to prepare amidrazones (reaction 48). These compounds in turn form bis(diethylalumino) derivatives (**43**) which

$$RC{\equiv}N + NaNHNH_2 \xrightarrow[Et_2O]{0^\circ} \left[RC \begin{matrix} \cdot\cdot\cdot NH \\ \cdot\cdot\cdot N{-}NH_2 \end{matrix} \right]^- Na^+ \quad (48)$$

also add to nitrile groups (reaction 49). Related reactions of industrial importance include the dimerization of cyanamide to cyano-

$$R^1C \begin{matrix} \nearrow NH \\ \searrow NHNH_2 \end{matrix} + 2\ Et_3Al \xrightarrow[benzene]{20^\circ} R^1C \begin{matrix} \cdot\cdot\cdot NH \\ \cdot\cdot\cdot N{-}HNAlEt_2 \end{matrix} \cdot AlEt_2 + 2\ C_2H_6$$

(**43**)

$$\mathbf{43} + R^2C{\equiv}N \xrightarrow[2.\ H_2O]{1.\ 70^\circ \text{ in benzene}} R^1C(NH_2){=}N{-}N{=}C(NH_2)R^2 \quad (49)$$

guanidine (reaction 50) and the reaction of the latter with nitriles to form diamino-*s*-triazines (reaction 51).

$$NH_2CN + \bar{N}HCN \xrightarrow[H_2O]{pH\ 9} NH_2C(NH_2){=}NCN \quad (50)$$

$$RCN + NH_2C(NH_2){=}NCN \xrightarrow{KOH} \text{2-R-4,6-(NH}_2)_2\text{-}s\text{-triazine} \quad (51)$$

C. Addition of Sulphydryl Compounds

Addition of hydrogen sulphide to the C≡N bond provides a dependable route to thioamides (reaction 52). The reaction has commonly been carried out by heating the nitrile with excess

$$RCN + H_2S \xrightarrow{HS^-} \left[RC \begin{matrix} \nearrow NH \\ \searrow SH \end{matrix} \right] \longrightarrow RC \begin{matrix} \nearrow NH_2 \\ \searrow\!\!\searrow S \end{matrix} \quad (52)$$

hydrogen sulphide in alcoholic solution under pressure with an alkali hydrosulphide serving as catalyst[131]. Amines also catalyse the addition, and improved techniques using triethylamine in pyridine at room temperature[132] or diethylamine in excess nitrile or in dimethylformamide at 60° [133] have been recommended. Under such conditions good yields are often obtained in a few hours. This very clean reaction, for which Kindler proposed the now self-evident mechanism (reaction 52), was used by Kindler[133] with a series of substituted benzonitriles in a very early investigation of the effects of substituents on reaction rates. Kindler's data were used by Hammett to compute a ρ-value of 2·14 for the addition of hydrogen sulphide to benzonitriles in alkaline solution at 60°.

The very reactive ethyl cyanate[100] reacts slowly with hydrogen sulphide in the absence of a catalyst, but in all other cases reported, including aryl cyanates[135], a base has been needed. Addition of hydrogen sulphide to cyanamide in water at pH 8–9 is an industrially important method for the synthesis of thiourea (reaction 53)[137].

$$NH_2CN + H_2S \longrightarrow NH_2\overset{\overset{S}{\|}}{C}NH_2 \qquad (53)$$

Hydrogen sulphide is occasionally used as a catalyst in reactions of nitriles with amines. It may be presumed that thioamide intermediates are involved which undergo displacement of HS^- by the amine[138] (reaction 54). Simple nitriles do not react appreciably with aqueous amines at 100–150°, but if hydrogen sulphide is present

$$RCH_2CN + \begin{array}{c} NHCH_2CH_2NH_2 \\ | \\ CH_2 \\ | \\ CH_2 \\ | \\ NHCH_2CH_2NH_2 \end{array} \xrightarrow[120°]{H_2S} \begin{array}{c} RCH_2C{\overset{N\text{—}}{\underset{N\text{—}}{\lessgtr}}} \\ | \\ (CH_2)_2 \\ | \\ RCH_2C{\overset{N\text{—}}{\underset{N\text{—}}{\lessgtr}}} \end{array} \qquad (54)$$

reaction occurs readily to give *N*-substituted amides. The steps of reaction (55) are presumably involved[76]. Similarly, in an anhydrous system, hydrogen cyanide gives *N*-substituted thioamides under mild

$$R^1CN + H_2S \xrightarrow[H_2O]{R^2NH_2} \left[R^1C{\overset{SH}{\underset{NH}{\lessgtr}}} \right] \longrightarrow R^1C{\overset{NHR^2}{\underset{NH}{\lessgtr}}} \longrightarrow R^1C{\overset{NHR^2}{\underset{O}{\lessgtr}}} \qquad (55)$$

conditions[139]. Catalysis of the trimerization of benzonitrile by hydrogen sulphide in the presence of ammonia at 250° undoubtedly has a related explanation[140].

Thiophenols and alkanethiols have been successfully added to cyanates[135] without catalysis, although basic catalysts are required with other nitriles[136]. The reaction is reversible, as is the base-catalysed addition of alcohols, but the equilibrium lies well on the thioimidate side. The *O,S*-disubstituted iminothiocarbonates (**44**) obtained from aryl cyanates decompose very easily under alkaline conditions to give thiocyanates[135] (reaction 56). In most such cases

$$\mathrm{PhOCN} + 4\text{-}t\text{-}\mathrm{BuC_6H_4SH} \longrightarrow \underset{(\mathbf{44})}{(\mathrm{PhO})(4\text{-}t\text{-}\mathrm{BuC_6H_4S})\mathrm{C{=}NH}} \longrightarrow \mathrm{PhOH} + 4\text{-}t\text{-}\mathrm{BuC_6H_4SCN} \quad (56)$$

the intermediate adduct **44** is not isolable, and the thiocyanate is obtained directly.

Under conditions resembling the Pinner synthesis, nitriles will react with thioamides to give diimidyl sulphide salts (e.g. **45**) of widely varying stability. Thioamides can be prepared conveniently by carrying out the reaction with excess thioacetamide in hot dimethylformamide and evaporating the acetonitrile formed in the equilibration[141] (reaction 57). In many cases, isomerization to an *N*-thioacylamidine is observed[142] (reaction 58). On the other hand, cyanamide reacts with thiourea, or abstracts hydrogen sulphide

$$\mathrm{RCN} + \mathrm{MeCSNH_2} \underset{\mathrm{DMF}}{\overset{\mathrm{HCl}}{\rightleftharpoons}} \underset{(\mathbf{45})}{\left[\mathrm{R\overset{\overset{\displaystyle NH_2}{\|}}{C}{-}S{-}\overset{\overset{\displaystyle NH_2}{\|}}{C}Me}\right]^{2+}} 2\,\mathrm{Cl^-} \rightleftharpoons \mathrm{RCSNH_2} + \mathrm{MeCN} \quad (57)$$

$$\left[\mathrm{R^1\overset{\overset{\displaystyle NH_2}{\|}}{C}{-}S{-}\overset{\overset{\displaystyle NH_2}{\|}}{C}R^2}\right]^{2+} 2\,\mathrm{Cl^-} \longrightarrow \mathrm{R^1C(NH_2){=}N{-}CSR^2} + 2\,\mathrm{HCl} \quad (58)$$

from thioamides to form **46**, a very stable compound[143]. A related

$$\mathrm{NH_2CN} + \mathrm{RCSNH_2} \xrightarrow{\mathrm{HCl}} \mathrm{RCN} + \mathrm{NH_2CSNH_2}$$

$$\mathrm{NH_2CN} + \mathrm{NH_2CSNH_2} \xrightarrow{\mathrm{HCl}} \underset{(\mathbf{46})}{\left[\mathrm{NH_2\overset{\overset{\displaystyle NH_2}{\|}}{C}{-}S{-}\overset{\overset{\displaystyle NH_2}{\|}}{C}NH_2}\right]^{2+}} 2\,\mathrm{Cl^-}$$

reaction of aryl cyanates[144] is indicated in reaction (59).

$$\mathrm{ROCN} + \mathrm{H_2S_2O_3} \xrightarrow{\mathrm{H_2O}} \mathrm{RO\overset{\overset{\displaystyle S}{\|}}{C}NH_2} + \mathrm{H_2SO_4} \quad (59)$$

VI. NUCLEOPHILIC ADDITION OF CARBANIONS

A. The Grignard Reaction

Reaction of a nitrile with a Grignard reagent is a widely applicable synthetic route to ketimines[145,146] which by easy hydrolysis give ketones (reaction 60). A detailed review by Kharasch and Reinmuth[146] (1954) summarizes extensive earlier studies of this modifica-

$$R^1CN + R^2MgX \longrightarrow R^1C(=NMgX)R^2 \xrightarrow{CH_3OH} R^1C(=NH)R^2 \xrightarrow{H_2O} R^1C(=O)R^2 \qquad (60)$$

tion of the Grignard reaction and provides a valuable assessment of its scope and limitations as well as illustrative procedures.

Investigations of the reactivities of various substrates toward PhMgBr have shown nitriles to be relatively unreactive[147]. Consequently long reaction times and relatively high temperatures are frequently required, although excellent yields are common in uncomplicated cases. The familiar favourable effect of an electronegative substituent is evident in the exothermic addition of MeMgI to 2-cyanoquinoline.

In the last decade, new insight into the nature of the Grignard reagent has led to refined kinetic studies and intensive investigation of the detailed mechanisms of both the nitrile and the carbonyl addition reactions. The emerging theory is substantially identical for these two cases, and the reader is referred to Eicher's discussion of the reactions of organometallic reagents with carbonyl compounds[148] for a review of current concepts regarding the reagent solution.

Attempts to extend the results of study of a particular Grignard reaction to other cases have frequently been disappointing. The solvent, substrate, reactant concentrations and the temperature all appear to influence the nature of the reactive organomagnesium species and the detailed mechanism. Appreciation for the influence of these factors has been due in large part to the recent work of Becker and his associates[149–153]. For the present purposes, it is sufficient to note that the complexity of even the normal addition reaction is such that small changes in conditions may cause large effects.

Insofar as the nitrile reactant is concerned, however, there has been little change in the mechanistic picture since the proposal of Swain in 1947[147]. Subsequent developments have been consistent with his concept of initial equilibration of the basic nitrile with the organomagnesium reagent, a Lewis acid (reaction 61). Formation of the complex **47** is expected to result in enhanced electrophilic character at the nitrile carbon atom and in weakening of the R—mg bond* in the reagent (an *ate-complex* effect, in Wittig's

* The symbol R—mg denotes the organometallic bond under discussion, while avoiding definition of the remaining unknown and probably complex structure of the reagent[154].

terminology[155]). Intramolecular bond reorganization in **47** gives **48** in what may be termed an assisted four-centre reaction[154]. In some

$$R^1CN + R^2\text{—mg} \rightleftharpoons \underset{\substack{R^2\text{—mg} \\ (47)}}{R^1C{\equiv}N} \longrightarrow \underset{(48)}{(R^1)(R^2)C{=}Nmg} \qquad (61)$$

cases good evidence indicates that a dimeric magnesium complex is involved in a six-centre transition state[149].

The dependence of the rate of the Grignard addition reaction on the structures of both the nitrile and the organic ligand transferred was investigated under carefully controlled conditions. Data for the initial second order rates of a series of *p*-substituted benzonitriles with Et_2Mg gave an excellent Hammett correlation ($\rho = +1{\cdot}57$). This result is in agreement with qualitative comparisons reported in earlier studies with conventional Grignard reagents[146], and confirms the expected rate enhancement by substituents which tend to increase the electrophilic character of the nitrile carbon atom.

In a complimentary study[152], the initial reaction rates were measured for the addition of *p*-substituted phenylmagnesium bromides to benzonitrile. A satisfactory linear correlation was found between these rates and the Taft σ° inductive effect constants of the substituents. The ρ value ($-2{\cdot}85$) obtained is consistent with the concept of an attacking carbanion activated by inductively electron-releasing groups.

The choice of solvent is a matter of practical significance[151]. Strongly basic solvents compete with the nitrile in formation of an organomagnesium complex. In a less predictable way, the solvent may also influence the nature of the reagent and its reactivity through precipitation of MgX_2 or modification of the ionic character of the organometallic bond. The substrate itself can be a similar source of complications through substituents which may be present.

When the Grignard addition reaction is used with aliphatic nitriles a serious complication analogous to the enolization of ketones in these conditions arises due to the acidity of α-hydrogen atoms. Acid–base interaction causes liberation of hydrocarbon and formation of the nitrile anion[146] (reaction 62). This in turn leads to dimerization and trimerization of the nitrile[156,157] (see section VI.B). Under comparable conditions the yield of ketone with

$$R^1CH_2CN + R^2\text{—mg} \rightarrow [R\bar{C}^1H\text{—}C{\equiv}N \leftrightarrow R^1CH{=}C{=}\bar{N}]mg^+ + R^2H \qquad (62)$$

PhMgBr increases from 35% for $PhCH_2CN$ to 90% for n-$BuCH_2CN$ as the acidity of the nitrile decreases. With MeMgBr the ketone yields are much lower, less than 40% with PrCN. This difference is due to the increased basicity of the reagent and is still more serious with ethyl, isopropyl and *t*-butyl reagents which give virtually no ketone with MeCN by the normal addition procedure. If, however, the Grignard reagent is added to excess MeCN (inverse addition), n-alkyl methyl ketones are obtained in 20–40% yields[156]. This result may have its explanation in the marked difference in solvent polarity prevailing in the two cases, and in the effect that this would have on the nature of the complex reagent. The ionization reaction is plausibly represented as occurring by an assisted six-centre transition state (**49**) competitive with the four-centre addition process.

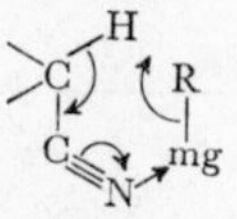

H C R C mg N

(**49**)

In cases where the cyano group is substituted for hydrogen in a strongly acidic hydrocarbon, interaction with a Grignard reagent can result in 'reductive displacement' of the cyano group. The mechanism is indicated in reaction (63). Typical examples with Ph_3CCN and disubstituted malononitriles in the older literature

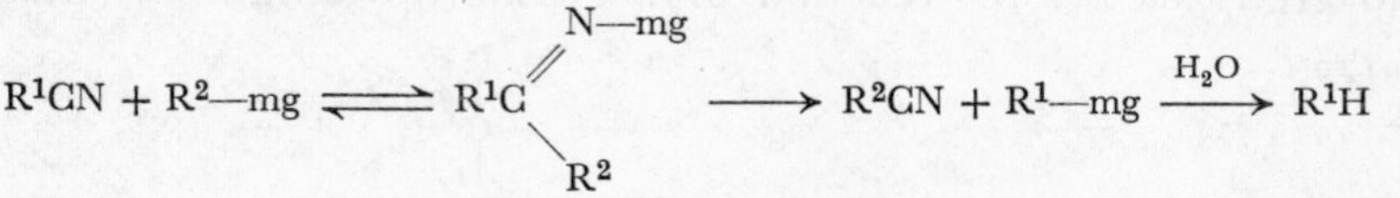

$$R^1CN + R^2\text{—mg} \rightleftharpoons R^1C(=N\text{—mg})R^2 \longrightarrow R^2CN + R^1\text{—mg} \xrightarrow{H_2O} R^1H \qquad (63)$$

were explained in substantially these terms by Kharasch[156]. Equilibria of this nature are probably common, although rarely identified. A later example is the reaction of Ph_2CHCN with EtMgBr[158] (reaction 64). Carbonation of the reaction mixture produces **50** in 55% yield.

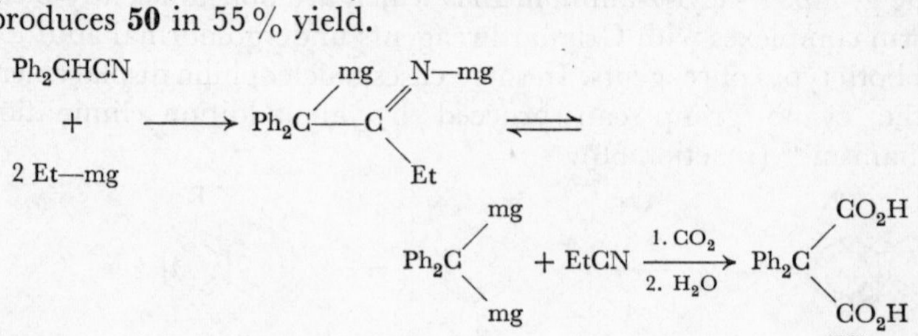

$$Ph_2CHCN + 2\ Et\text{—mg} \longrightarrow Ph_2C(\text{mg})\text{—}C(=N\text{—mg})Et \rightleftharpoons$$

$$Ph_2C(\text{mg})_2 + EtCN \xrightarrow[2.\ H_2O]{1.\ CO_2} Ph_2C(CO_2H)_2 \qquad (64)$$

(**50**)

No instance has been reported of reduction of the cyano group to an aldimine by the Grignard reagent, as further analogy to ketone reactivity[157] might predict.

Nitriles in which the cyano group is weakly bonded to the α-carbon react with Grignard reagents in a coupling process (reaction 65). These include many α-aminonitriles, ketone cyanohydrins and acyl cyanides.

$$R^1C{\equiv}N + R^2{-}mg \longrightarrow R^1R^2 + mg^+(CN)^- \quad (65)$$

The mechanism of this reaction of α-aminonitriles has been studied recently[159]. If the amino group bears a hydrogen atom, it is metallated by the reagent. Dissociation of the cyanide ion, assisted by the neighbouring metal, then leads to an imine which adds additional reagent to give the 'coupling' product (reaction 66). Yields are 40–75 % with PhMgBr and nearly quantitative with PhLi. If

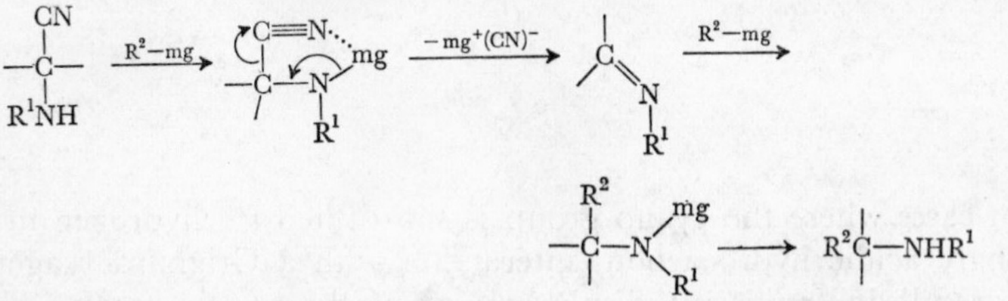

(66)

the amino group is disubstituted, the Grignard reagent can still coordinate with this basic centre and promote dissociation of the cyano group as before (reaction 67). Triisopropylamine has thus

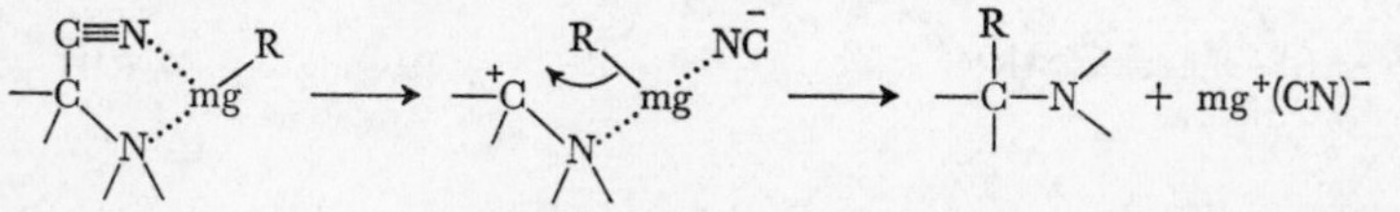

(67)

been obtained in 54 % yield by reaction of α-diisopropylaminopropionitrile with MeMgCl[160]. Lithium reagents have much less tendency to form such complexes and instead add normally to the nitrile group. *N*-acyl-α-aminonitriles which are not sufficiently basic to form complexes with Grignard reagents undergo normal addition with both types of reagents. In some cases, nucleophilic displacement of the cyano group can proceed by an addition–elimination mechanism[146] (reaction 68).

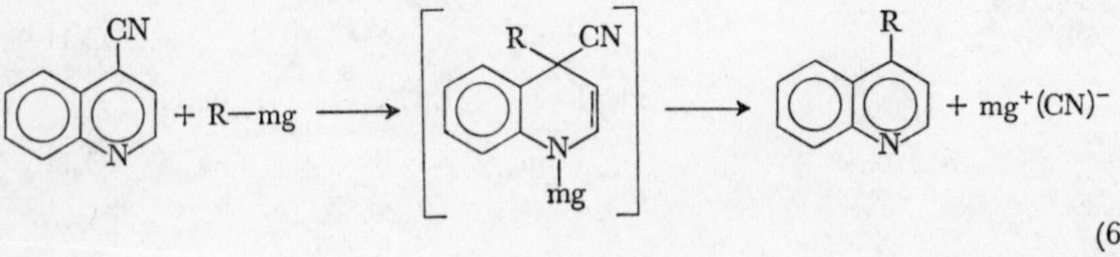

(68)

The ketimine complexes formed in the normal Grignard reaction rarely react further to give carbinamine derivatives. However, alkoxyacetonitriles, highly activated substrates, have given excellent yields of tertiary amines[161] (reaction 69). This may be an example of

$$EtOCH_2CN + 2\,CH_2{=}CHCH_2MgBr \longrightarrow EtOCH_2C(CH_2{-}CH{=}CH_2)_2{-}N(MgBr)_2 \xrightarrow{(H_2O)} EtOCH_2C(CH_2{-}CH{=}CH_2)_2NH_2 \quad (69)$$

the abnormal reactivity of the allyl reagent since these nitriles[162] and dimethoxyacetonitrile[163] react normally with a variety of other Grignard reagents.

Similar but less efficient double addition has also been observed with cyanogen, the reactions of which can be controlled to give a variety of products[108]. Addition of cyanogen to excess reagent solution at about room temperature results in an unsymmetrical 1:2 adduct (**52**) which can be hydrolysed to give a glycinonitrile (**53**). Acid hydrolysis gives the ketone **54.** If the reaction is conducted at

$$\underset{(\mathbf{54})}{RCOR} \xleftarrow[H_2O]{HCl} \underset{(\mathbf{52})}{R_2C(CN)N(mg)_2} \xrightarrow[H_2O]{NH_4Cl} \underset{\substack{(\mathbf{53})\\6\text{–}33\%}}{R_2C(CN)NH_2}$$

$$\mathbf{52} \xleftarrow{25^\circ} \mathbf{51}$$

$$(CN)_2 + R{-}mg \longrightarrow \underset{(\mathbf{51})}{RC(CN){=}N{-}mg} \xrightarrow{H_2O} \underset{\substack{(\mathbf{57})\\35\text{–}70\%}}{RCN} + mg^{+}(CN)^{-}$$

$$\mathbf{51} \xrightarrow{-70^\circ} \underset{(\mathbf{55})}{\begin{matrix} RC{=}N{-}mg \\ | \\ RC{=}N{-}mg \end{matrix}} \xrightarrow{H_2O} \underset{(\mathbf{56})}{RCOCOR}$$

−70°, the symmetrical 1:2 adduct **55** is obtained, and hydrolysis gives the diketone **56.** With excess cyanogen, reaction stops at the 1:1 adduct (**51**), which decomposes on hydrolysis to give the nitrile **57**. Diethylzinc also reacts in this manner, giving propionitrile. Comparable behaviour with aryl cyanates has been reported recently[102].

Alkyl- and aryllithium reagents[164] closely resemble Grignard reagents in their reactions but exhibit subtle differences. Addition reactions with C≡N and C=O bonds are frequently comparable to Grignard additions, but the lithium reagents are generally more reactive by virtue both of their greater ionic character and the small size of the lithium atom. Benzonitriles containing *p*-methoxy or *p*-dimethylamino groups rapidly give good yields of ketimine with MeLi, although Grignard reagents do not react readily[165], and PhLi reacts at least 100 times faster than PhMgBr with PhCN in ether at 25° [147]. However, the greater basicity of both the lithium reagents and the product ketimine salts can cause more complex reactions (see section VII).

Although the Reformatsky reaction[166] was first applied to nitriles in 1901 concurrently[167] with the application of the Grignard reagents, only very recently has it received further consideration[168,169]. It seems probable that Kagan's work[169] will lead to greater interest in view of the excellent yields he has achieved (reaction 70). Ethyl

$$\begin{matrix} \text{RCN} \\ + \\ \text{Me}_2\text{CBrCO}_2\text{Et} \end{matrix} \xrightarrow[\text{2. H}^+]{\text{1. Zn, benzene, 78°}} \text{R}\overset{\overset{\text{NH}}{\|}}{\text{C}}\text{CMe}_2\text{CO}_2\text{Et} \longrightarrow \underset{75\text{–}83\%}{\text{RCOCMe}_2\text{CO}_2\text{Et}} \quad (70)$$

α-bromopropionate gave nearly as good results as those indicated, and substantial yields (22–67 %) were also obtained with ethyl bromoacetate.

Recently reactions of organoaluminium reagents with nitriles have been investigated[170]. Benzonitrile forms 1:1 complexes with several trialkyl- and triarylaluminium compounds, and these isomerize at 130–200° to ketimine derivatives (reaction 71). Similar

$$\text{PhCN} + \text{Et}_3\text{Al} \longrightarrow [\text{PhC}{\equiv}\text{N} \rightarrow \text{AlEt}_3]_n \xrightarrow{\Delta} \begin{matrix} \text{Ph} \\ \quad\diagdown \\ \quad\quad \text{C}{=}\text{N}{-}\text{AlEt}_2 \\ \quad\diagup \\ \text{Et} \end{matrix} \quad (71)$$

complexes with aliphatic nitriles tend to eliminate hydrocarbon on heating and form polymeric products.

B. Base-Catalysed Dimerization and Trimerization of Nitriles

Those nitriles having an α-methylene group are sufficiently acidic to give carbanions which will react in the manner of the organometallic reagents discussed above. Thus β-aminoacrylonitriles[171] are obtained in accordance with reaction (72), and from them

β-ketonitriles are obtained by acid hydrolysis. Such reactions can usually be carried to high conversion if steric factors or competitive

$$[\mathrm{R\bar{C}HCN} \longleftrightarrow \mathrm{RCH{=}C{=}\bar{N}}] \overset{\mathrm{RCH_2CN}}{\underset{+}{\rightleftharpoons}} \mathrm{RCH_2C(=N^-)CH(R)CN} \xrightarrow{\mathrm{H_2O}} \mathrm{RCH_2C(NH_2){=}C(R)CN} \qquad (72)$$

reactions do not interfere. Conversely, nitrile dimerization is a more or less serious side reaction in Grignard reactions[157], in base-catalysed amidine syntheses and in alkylation and acylation of nitriles. Depending on the conditions used, the dimer may also react further to give trimeric pyridine or pyrimidine derivatives.

The concurrent, competitive reactions of addition of the reagent at the cyano group versus proton abstraction can be controlled to some extent. As the metal ion is varied for any organic ligand, the ionic character, the basicity of the anion and the tendency toward proton abstraction increase in the order: Mg < Li < Na < K[172]. Less polar reaction media inhibit ionization of the reagents and favour the four-centre addition mechanism[154].

Reynolds (1951) has summarized the earlier work in this field and has reported improved procedures and preferred basic catalysts[173]. In general, good yields of nitrile dimers can be obtained by reactions in ethyl ether, and trimers are obtained in boiling dibutyl ether. Only pyrimidine trimers ('cyanalkines'[174]) are obtained from the simple aliphatic nitriles, but phenylacetonitrile and probably other relatively acidic nitriles can be converted to pyridine derivatives by a proper choice of catalyst. Reaction (73) occurs when the

$$3\ \mathrm{PhCH_2CN} + \mathrm{NaNH_2} \xrightarrow[\text{3 h reflux}]{\mathrm{Et_2O}} \text{pyrimidine (2-PhCH}_2\text{, 4-NNa, 5-Ph, 6-CH}_2\text{Ph)} \quad 89\% \qquad (73)$$

theoretical amount of base is used. Reaction (74) occurs with a 1:1 ratio of nitrile to diisopropylaminomagnesium bromide catalyst (DIPAM). There is no obvious explanation for selective formation of different products in these two cases. Possibly equilibration produces the more acidic diaminopyridine when excess base is present.

Malononitrile trimerizes exothermically at room temperature in the presence of ammonia[175,176]. The product has been assigned the

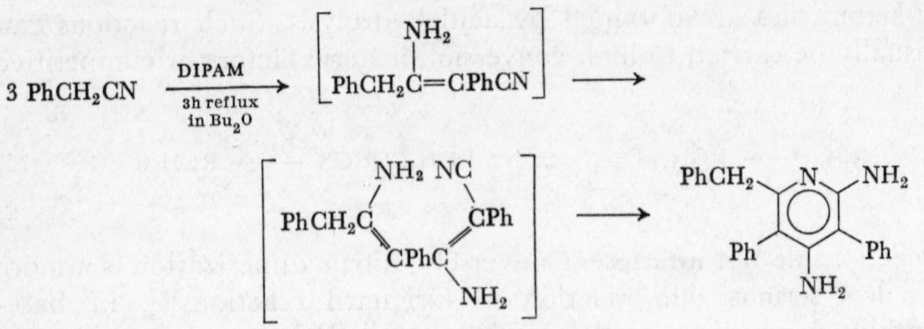

$$3\ PhCH_2CN \xrightarrow[\text{3h reflux in } Bu_2O]{\text{DIPAM}} \left[PhCH_2C(NH_2){=}CPhCN \right] \longrightarrow$$

$$\left[PhCH_2C(NH_2){=}CPhC(NH_2){=}CPh{-}CN \right] \longrightarrow \text{2-amino-3,5-diphenyl-4-amino-6-benzylpyridine} \qquad (74)$$

structure **58** on the basis of considerable physical evidence[177]. A different trimer, suggested to be **59**, is obtained under more vigorous

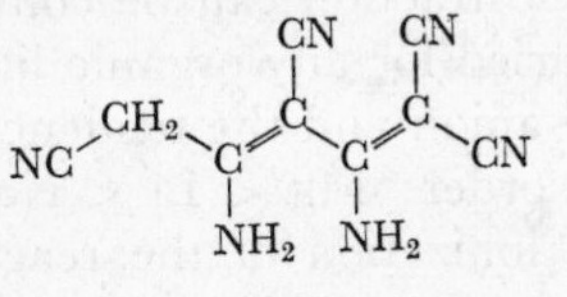

$$NC{-}CH_2{-}C(NH_2){=}C(CN){-}C(CN){=}C(NH_2){-}CN$$

(58)

(59): H₂N–, –CN, CN, NH₂, NH, H, N substituted bicyclic structure

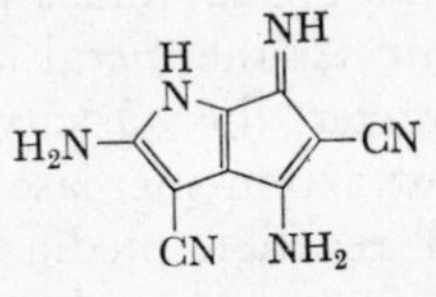

(59)

basic conditions. Presumably dimerization is an intermediate step in these reactions and takes place exceptionally easily. The dimer itself (**60**) was not prepared until recently[178].

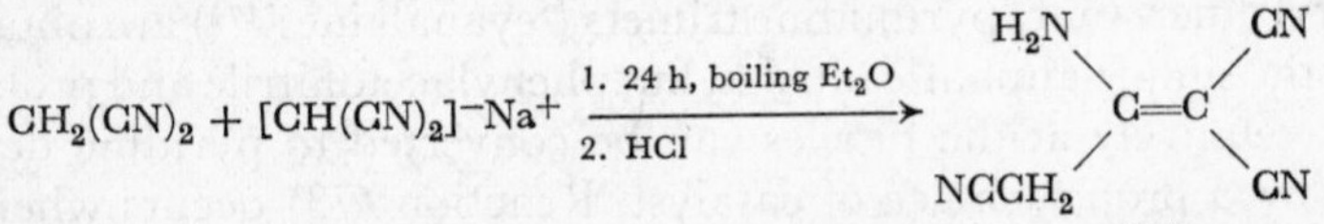

$$CH_2(CN)_2 + [CH(CN)_2]^-Na^+ \xrightarrow[\text{2. HCl}]{\text{1. 24 h, boiling } Et_2O} (H_2N)(NCCH_2)C{=}C(CN)_2$$

(60)

Thorpe[179] first demonstrated that base-catalysed nitrile dimerization could be applied intramolecularly to give cyclic β-ketonitriles. Subsequently Ziegler[171,180] refined the procedure and adapted it to the synthesis of large rings. Sodium and lithium *N*-alkylanilides were found to be exceptionally good reagents for such reactions.

If benzonitrile is heated with metallic sodium, the salt **61** can be obtained in 85 % yield[181]. The mechanism is indicated in reaction (75)[182]. Anker and Cook[182] reasoned that the high temperature was required only for the initial formation of PhNa and found that nearly quantitative yields of analogous alkyltriphenyldihydrotriazines could be obtained by using a variety of alkyllithium reagents at room temperature. Phenyllithium, less basic than the sodium compound, reacted similarly but also gave a low yield of ketimine,

$$PhCN \xrightarrow[80°]{Na,\ benzene} NaCN + PhNa \xrightarrow{PhCN} PhC(=NNa)Ph \xrightarrow{PhCN}$$

$$Ph_2C{=}N{-}C(=NNa)Ph \xrightarrow{PhCN} Ph_2C{=}N{-}CPh{=}N{-}CPh{=}NNa \longrightarrow$$

$$[\text{triazine anion, Ph, Ph, Ph, Ph}]^{-}\ Na^{+} \quad \textbf{(61)} \qquad (75)$$

while $PhCH_2Li$ or Ph_2CHNa produced pyrazolines instead (reaction 76). Analogous reactions are involved in the trimerization of

$$PhCN + Ph_2CHNa \longrightarrow PhC(=NNa){-}CHPh_2 \rightleftharpoons PhC(=NH){-}C(Na)Ph_2 \xrightarrow{PhCN}$$

$$PhC(=NH){-}CPh_2{-}C(=NNa)Ph \longrightarrow \text{pyrazoline (N—NH, H, Ph, Ph, Ph, Ph)} \qquad (76)$$

benzonitrile by sodium hydride[183], metal amides[183,184], Grignard reagents[150,182] and diethylzinc[185].

An example of α-methylene reactivity in the *ketimine anion* has recently been reported by Becker's group[153]. Reaction of Et_2Mg with excess PhCN in tetrahydrofuran gave a mixture of **63** and **64**. Although **63** might be formed by addition of two molecules of

$$PhC{\equiv}N + Et_2Mg \xrightarrow{THF} PhC(Et){=}N{-}mg\ \textbf{(62a)} \rightleftharpoons PhC(=NH){-}CH(CH_3){-}mg\ \textbf{(62b)} \longrightarrow PhC(=NH){-}CH(CH_3){-}CPh(=Nmg)$$

$$\xrightarrow{16\%\ H_2O} PhCOCH(CH_3)COPh\ \textbf{(64)}$$

$$\longrightarrow PhC(=NH){-}CH(CH_3){-}C(Ph){=}N{-}C(Ph){=}Nmg \xrightarrow{54\%} \text{2,4,6-triphenyl-5-methylpyrimidine}\ \textbf{(63)}$$

PhCN to **62a** followed by cyclization[153], this path would be expected to give an *s*-triazine derivative. The common pathway via (**62b**) to both **63** and **64** seems more plausible. When the reaction was run in ethyl ether, however, the amidine **65** was obtained in addition to some **63**, showing the occurrence of the alternative process. Amidines of this type are also obtained when alkali metal phosphides

$$\mathbf{62a} + \text{PhCN} \longrightarrow \text{PhC}(\text{Et}){=}\text{N}{-}\text{C}(\text{Ph}){=}\text{Nmg} \xrightarrow{H_2O} \text{PhC}(={\text{NH}}){-}\text{N}{=}\text{CPhEt}$$

18%

(65)

are used as basic catalysts[185].

There are numerous examples of base-catalysed codimerization of unlike nitriles. In those cases where only one reactant has an α-methylene group and the other has a more electrophilic C≡N group, the reaction is highly specific[187]. Good control is also achieved when one nitrile is appreciably more acidic[188] (reaction 77).

$$\text{PhCH}_2\text{CN} + \text{CH}_2(\text{CN})\text{CO}_2\text{Et} \xrightarrow{\text{NaOEt}} \text{PhCH}_2\text{C}(\text{NH}_2){=}\text{C}(\text{CN})\text{CO}_2\text{Et} \qquad (77)$$

Another example has been recently illustrated with aryl cyanates[101] (reaction 78). Adducts such as **66** usually dissociate in the basic

$$\text{PhOCN} + \text{CH}_2(\text{CO}_2\text{Et})_2 \xrightarrow[<30^\circ]{\text{NaOEt}} \underset{(\mathbf{66})}{\text{PhOC}(={\text{NH}}){-}\text{CH}(\text{CO}_2\text{Et})_2} \longrightarrow \text{PhOH} + (\text{EtO}_2\text{C})_2\text{CHCN} \qquad (78)$$

system to give a new nitrile. If, however, a tertiary amine catalyst is used, **66** can be isolated or converted at higher temperature to a pyrimidine. Further unusual examples are the nucleophilic addition of a benzylidenephosphorane (**67**) to benzonitrile[189] (reaction 79) and the analogous reaction with dimethylsulphonium methylide[190].

$$\text{PhC}{\equiv}\text{N} + \underset{(\mathbf{67})}{\text{MeEtPhP}{=}\text{CHPh}} \xrightarrow[12\text{ h}]{\text{ether}} \text{PhC}(={\text{N}^-}){-}\text{CH}(\text{Ph})\overset{+}{\text{P}}\text{MeEtPh} \xrightarrow[\text{H}_2\text{O,MeOH}]{\text{KOH}}$$

$$\underset{78\%}{\text{PhCOCH}_2\text{Ph}} + \text{MeEtPhP}{\rightarrow}\text{O} \qquad (79)$$

VII. LINEAR POLYMERIZATION OF NITRILES

A high polymer having the backbone

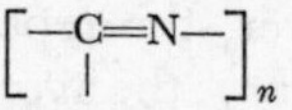

$$\left[\begin{array}{c}-\mathrm{C}{=}\mathrm{N}- \\ | \end{array}\right]_n$$

might be expected to display novel properties as an organic semiconductor by virtue of its extended conjugation and the lone-pair electrons on the nitrogen atoms. During the past decade some progress toward such materials has been reported. Although the accomplishment is impressive when measured against the inherent difficulties, it must be admitted that the literature in this field tends toward extravagent claims.

The most attractive proposition for the synthesis of the desired polymers is linear polymerization of nitriles, $n\,\mathrm{RC{\equiv}N} \rightarrow (\mathrm{—RC{=}N—})_n$. It is clear that strong acids or bases catalyse dimerization and trimerization reactions in this sense; the difficulty is to prevent cyclization of the linear trimer to a resonance-stabilized *s*-triazine or pyrimidine ring system. Moreover, if higher polymers can be produced in an indirect way, a question remains concerning their possible reversion to the aromatic trimers.

In 1961 Kargin and his coworkers[191] reported experiments in which they attempted to take advantage of the greater basicity of the linear polymeric nitrile structure (*a polyamidine*) compared to the monomer or trimer to provide a driving force for high polymer formation. Nitrile complexes with metal halides were found to trimerize on heating at 200–250°, and the trimers were more slowly converted, in turn, to complexes of higher polymers. The organic products were recovered by decomposing the metal complexes in water. They were soluble in concentrated sulphuric acid, in which viscosity measurements indicated polymers of 20–30 units. Infrared spectra supported the linear polymer structure. Comparable results were obtained with benzonitrile, capronitrile and propionitrile, and substantially faster reactions occurred when 1–6 % HPO_3 was present to provide a proton source. It seems probable that the aliphatic nitrile polymers would contain the

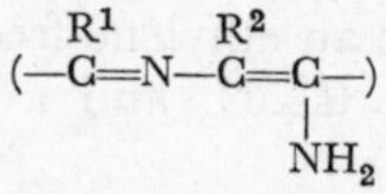

$$(\mathrm{—}\overset{\mathrm{R^1}}{\mathrm{C}}\mathrm{{=}N{—}}\overset{\mathrm{R^2}}{\mathrm{C}}\mathrm{{=}}\underset{\mathrm{NH_2}}{\underset{|}{\mathrm{C}}}\mathrm{—})$$

group to some extent (the trimers observed as intermediates were aminopyrimidines). A triphenyl-*s*-triazine–$TiCl_4$ complex polymerized similarly.

Subsequently, polymerization of acetonitrile under similar conditions by Oikawa[192] has been considered to give comparable linear polymers, and study of their physical properties has shown similarities to polyacetylene[193]. However, the reported analytical data for the products are very unsatisfactory; the average composition, $C_{10}H_{10}N_3O_2$ (theory $C_{10}H_{10}N_5$), indicates drastic loss of nitrogen by elimination of ammonia (which was indeed observed) and hydrolysis or hydration in the metal removal procedure. Either type of degradation would be compatible with the enamine structure above.

Related work[194] is claimed to have given linear polycyanamide. Urea was heated with zinc chloride at 350–500°, presumably to produce cyanamide as a metal complex which polymerized in the manner discussed above (reaction 80). However, the reported

$$NH_2CONH_2 \xrightarrow{ZnCl_2} NH_2C{\equiv}N{\cdot}ZnCl_2 \longrightarrow \left[-\underset{}{\overset{NH_2}{\overset{|}{C}}}{=}N-\overset{NH_2}{\overset{|}{C}}{=}N- \right]_n \qquad (80)$$

properties of the product are remarkably like those of melamine, or a hydrolysis or deamination product, which would reasonably be expected from such a reaction.

In acrylonitrile and syndiotactic methacrylonitrile polymers, the cyano groups can be so ordered that linear polymerization appears easy, while cyclic trimerization would be sterically excluded. Tabata[195] has been able to polymerize acrylonitrile and methacrylonitrile in the solid state (−196°) with ionizing radiation to give products believed to have the ladder structure **68**. It has been

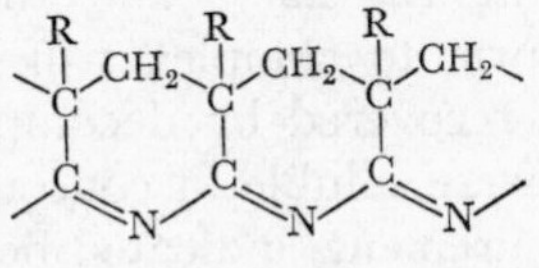

(68)

suggested that the vinyl polymer backbone is formed initially, followed by ionic polymerization of the C≡N groups, which is probably facilitated by favourable orientation in the crystal. If acrylonitrile dissolved in ethylene is irradiated with γ-rays at −78°, ionic polymerization gives an ethylene-free polymer in which C=N groups predominate over C≡N, and a substantially open linear structure is probable[196].

When polyacrylonitrile is treated with alkaline reagents or is heated, yellow to red colouration results. This is probably caused by development to a greater or lesser extent of the structure **68**[197].

Polymerization of cyanogen under various conditions produces a black solid known as paracyanogen. Its physical properties and stability have led to the suggestion that it has the planar conjugated structure **69**[198]. Recently Peska[199] has described the preparation of

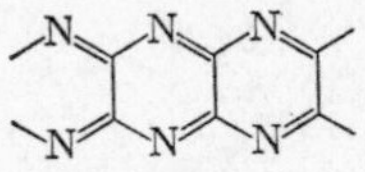

(**69**)

cyanogen oligomers in low-temperature reactions initiated by butyllithium, lithium naphthalene and lithium benzophenone. The black solids obtained were lithium salts which dissolved in polar organic solvents. The butyllithium product approximated to a trimer and was suggested to have the structure **70**, consistent with the

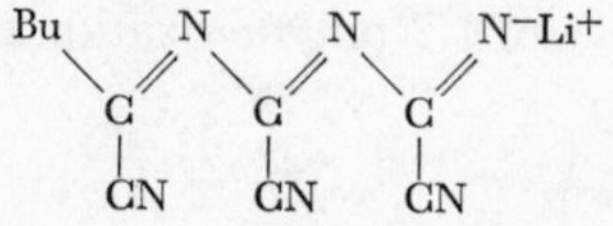

(**70**) 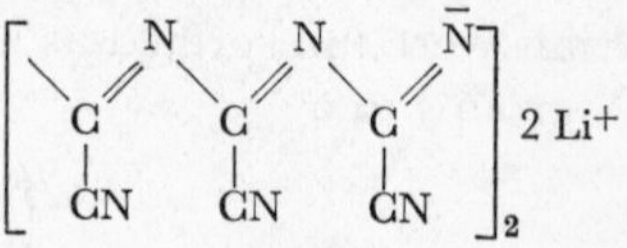 (**71**)

infrared spectrum. A dihydro-*s*-triazine structure is excluded by the colour. The other initiators apparently reacted as lithium atoms; the naphthalene or benzophenone was substantially recovered, and a dibasic salt (**71**) was obtained. Some question may well be raised concerning the non-involvement of the second cyano group.

Polymerization of hydrogen cyanide in the presence of a base gives a brown-black solid which has been known as azulmic acid. Its composition is somewhat variable but tends to approach that of the 'double ladder' structure **76**. The complex polymerization process is believed to proceed through the stages shown in reaction (81)[200,201]. The backbone polymer **74** is formed by base-catalysed polymerization of the dimer **72**. 'Lacing-up' of the favourably juxtaposed cyano groups of the *trans* conformation of **74** to **75** and **76** is comparable to the examples already discussed and leads to a very compact rod structure. Side reactions in the earlier stages give diaminomaleonitrile[202], purines and pyrimidines[201]. There is currently much interest in the possibility of prebiotic synthesis of such biologically important molecules from hydrogen cyanide and ammonia[203,204].

Deichert and Tobin subjected several nitriles to a 20,000-volt, 60-cycle electric discharge and obtained dark, acetone-soluble

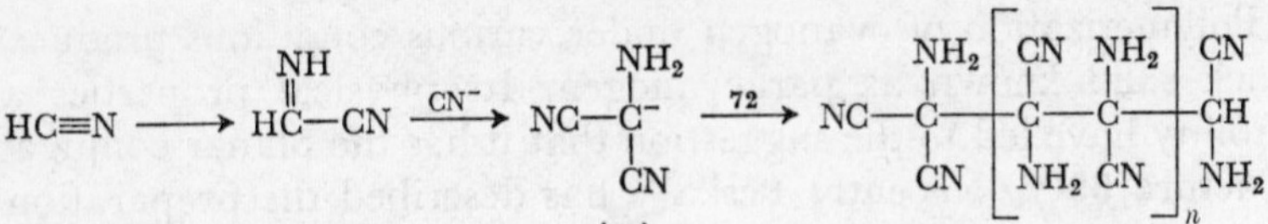

$$HC{\equiv}N \longrightarrow \underset{(\mathbf{72})}{HC({=}NH){-}CN} \xrightarrow{CN^-} \underset{(\mathbf{73})}{NC{-}C^-(NH_2)(CN)} \xrightarrow{\mathbf{72}} \underset{(\mathbf{74})}{NC{-}C(NH_2)(CN){-}[C(CN)(NH_2){-}C(NH_2)(CN)]_n{-}CH(CN)(NH_2)}$$

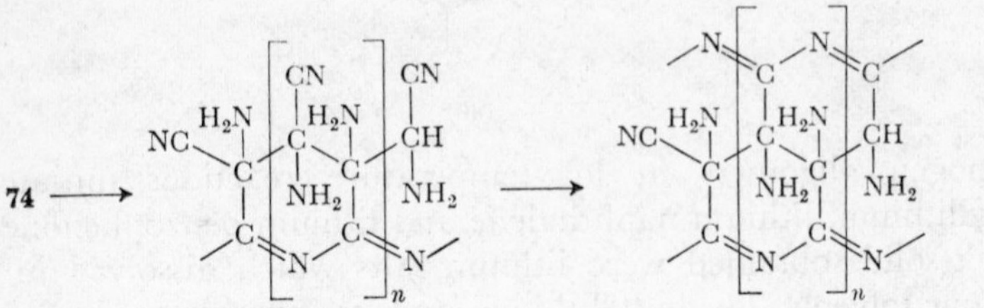

$$\mathbf{74} \longrightarrow (\mathbf{75}) \longrightarrow (\mathbf{76}) \qquad (81)$$

polymers of low molecular weight[205]. The infrared spectra of these materials were interpreted as indicative of 1,2-polymerization of the C≡N groups to a

$$\left({-}\underset{\displaystyle R}{\underset{|}{C}}{=}N{-} \right)_n$$

structure. With acrylonitrile there was also evidence for the (—C=C=N—) unit in the polymer. Involvement of the C≡N group in this manner might be initiated by exceptional stretching of the polar bond in the intense electric field.

VIII. ELECTROPHILIC ADDITION OF CARBONIUM IONS

Alkylation of nitriles by $Et_3O{\cdot}BF_4$ was discussed in section III.A as one route to nitrilium salts, and some other examples of nitrile nucleophilicity were noted there. Among them *N*-acylation of nitrile–metal halide complexes was considered. However, Bredereck has recently shown that dimethylcyanamide, which is relatively basic, can be *N*-acylated without such activation[206]. Depending on the acyl chloride used, imidyl chloride adducts (**77**) or pyrylium salts (**78**) can be obtained. Other reactions in which the weakly

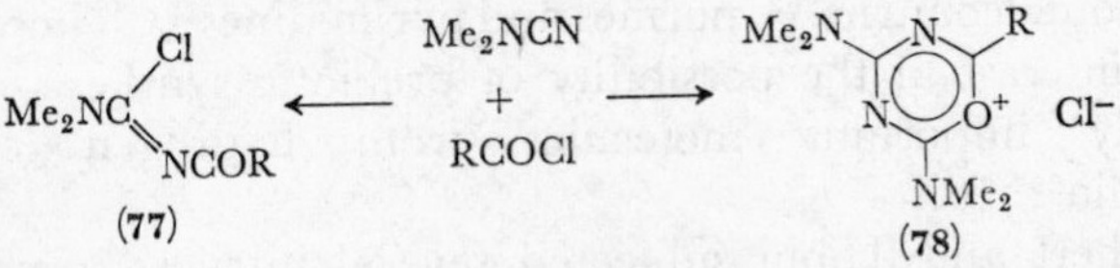

$$\underset{(\mathbf{77})}{Me_2NC(Cl){=}NCOR} \longleftarrow \begin{matrix} Me_2NCN \\ + \\ RCOCl \end{matrix} \longrightarrow \underset{(\mathbf{78})}{\text{Me}_2\text{N},\ \text{R},\ \text{NMe}_2\text{-substituted } C_3N_2O^+ \text{ ring} \quad Cl^-}$$

basic nitrogen atom is attacked by electrophilic species have been longer known, but occur under conditions which lead directly to amides. Two basic reaction types which can be distinguished are discussed here. Nitrile reactivity of this kind has previously been surveyed by Zil'berman (1960)[207].

A. N-Alkylation with Olefins and Alcohols

Probably the earliest report of *N*-alkylation by an olefin is that of Wieland[208] who obtained *N*-cyclohexylformamide after reaction of hydrogen cyanide with cyclohexene in the presence of hydrogen chloride and $AlCl_3$, in an attempt to accomplish a reaction analogous to the Gattermann aldehyde synthesis (reaction 82). Subsequently,

$$\text{HCN} + \text{C}_6\text{H}_{10} \xrightarrow[\text{HCl}]{\text{AlCl}_3} \text{C}_6\text{H}_{11}\text{N{=}CHCl} \xrightarrow{\text{H}_2\text{O}} \text{C}_6\text{H}_{11}\text{NHCHO} \qquad (82)$$

this was extended in Germany to other nitriles and olefins, and a variety of strong acids were used[209]. Concurrent development is disclosed in a duPont patent[210] covering substantially the same ground and including the use of alcohols as substitutes for olefins. Independent discovery of the general reaction by Ritter and his coworkers, however, led to an extensive series of papers[211] which have become the primary references on this subject.

Olefins are protonated in concentrated sulphuric acid to give carbonium ions which will react if a nitrile is present to give an *N*-alkylnitrilium ion. Subsequent hydrolysis gives the amide (reaction 83). Covalent structures are probably in equilibrium with

$$\text{>C{=}C<} \xrightarrow[\text{H}_2\text{SO}_4]{96\%} \left[\text{>C}^+\text{—CH<}\right]\text{HSO}_4^- \xrightarrow[25\text{–}50°]{\text{RCN}}$$
$$\textbf{(79)}$$

$$\left[\text{RC}{\equiv}\overset{+}{\text{N}}\text{—}\overset{|}{\underset{|}{\text{C}}}\text{—CH<}\right]\text{HSO}_4^- \xrightarrow{\text{H}_2\text{O}} \text{RCONH}\overset{|}{\underset{|}{\text{C}}}\text{—CH<} \qquad (83)$$
$$\textbf{(80)}$$

the ionic species **79** and **80**[212]. This synthesis of *N*-substituted amides has proven to be widely useful[213]. Although highest yields are obtained with 1,1-disubstituted olefins, very successful reactions have been carried out with trialkylethylenes[211], terminal linear alkenes[212] and haloalkenes[211,214]. Moreover, benzylic alcohols and glycols[215] and a wide variety of secondary and tertiary alcohols and diols which also give carbonium ions under the reaction conditions, can be substituted for the olefin reactant. This has permitted such syntheses[211,216] as shown in reaction (84).

$$Me_2C(OH)CH_2CH_2CMe_2OH + MeCN \xrightarrow{H_2SO_4} \left[\begin{array}{c} CMe_2 \\ CH_2 \quad N \\ Me_2C{=}CH \quad \overset{+}{C}{-}Me \end{array} \right] \longrightarrow$$

$$\left[\begin{array}{c} CMe_2 \\ CH_2 \quad N \\ Me_2\overset{+}{C}{-}CH{-}C{-}Me \end{array} \right] \longrightarrow \begin{array}{c} Me \quad Me \\ N \\ {=}\!{-}Me \\ Me_2C \\ 80\% \end{array} \qquad (84)$$

It is difficult to discern from the published work how the structure of the nitrile influences the success of these reactions. Good results have been achieved with a rather wide variety of nitriles[211,212,217], however, and probably in many instances low yields could be improved by closer study. With HCN, $CHCl_2CN$ and CCl_3CN, amidation of camphene can be carried out without Wagner rearrangement by operating at low temperatures; with acetonitrile and benzonitrile the isobornyl derivative cannot be avoided. This distinction has been attributed to slower reactions with the latter nitriles[218a]. However, these would *not* normally be expected to be less reactive than the less basic halogenated nitriles toward a carbonium ion. Possibly the more basic nitriles are made less reactive by protonation in the reaction mixture.

Mesityl oxide and diacetone alcohol both react with MeCN or PhCN in 96 % sulphuric acid to give **81**[218b]. Acetone gives the same product, initially undergoing acid-catalysed dimerization (reaction 85). This reaction is broadly applicable to ketones which readily

$$2\,CH_3COCH_3 \longrightarrow (CH_3)_2\overset{OH}{\overset{|}{C}}{-}CH_2COCH_3 \xrightarrow{RCN} \underset{(\mathbf{81})}{RCONHC(CH_3)_2CH_2COCH_3} \qquad (85)$$

self-condense in acid media and has also been demonstrated with diethyl benzalmalonate prepared *in situ*.

The applicability of the reaction to simple, commercially important raw materials has made it attractive for exploitation industrially. It provides a route to *N-t*-butylacrylamide and *N*-isopropylacrylamide, for example, which are of interest in polymers. Hydrolysis of the *N*-alkylamides offers a route to secondary and tertiary carbinamines based on readily available olefins. Although many such amides resist hydrolysis, those derived from HCN[212] or ClCN[219] readily give the free amine.

Nitriles do not react with trityl perchlorate or with dimethyl sulphoxide at ordinary temperatures. However, a solution of all three reagents slowly produces **82** in good yield[220]. This appears to be a general reaction for nitriles and is attributed to participation by the solvent (reaction 86). Aryl cyanates react more rapidly than simple

$$\mathrm{R{-}C{\equiv}N\cdots O{=}S(CH_3)_2} + \mathrm{Ph_3C^+} \xrightarrow{\text{DMSO}} \mathrm{R{-}C(O{-}\overset{+}{S}(CH_3){-}CH_2{-}H){=}N{-}CPh_3} \longrightarrow \underset{(\mathbf{82})}{\mathrm{RC(=O){-}NHCPh_3}} + \mathrm{CH_3{-}\overset{+}{S}{=}CH_2} \qquad (86)$$

aliphatic and aromatic nitriles; curiously, disubstituted cyanamides react only if one substituent is aryl. Attempts to extend the reaction to other carbonium ions were unsuccessful, and it is suggested that these are too tightly solvated by dimethyl sulphoxide.

Wieland's original experiments have been amplified by Cannon[221] who showed that cyclohexyl chloride gave the same results as cyclohexene with a variety of nitriles. Diphenylmethyl bromide is readily converted by silver sulphate to a carbonium ion which reacts with nitriles to give a nitrilium salt from which the *N*-alkylated amide is obtained[222].

A further interesting and practical way of generating a reactive carbonium ion in the presence of a nitrile is by the reaction of a halogen with an olefin[223a] (reaction 87). In cases where the nitrile

$$\mathrm{R^1CH{=}CH_2} + \mathrm{Cl_2} \longrightarrow \mathrm{Cl^-} + \mathrm{R^1\overset{+}{C}H{-}CH_2Cl} \xrightarrow{\mathrm{R^2CN}} \underset{(\mathbf{83})}{\mathrm{R^2C(Cl){=}N{-}CHR^1CH_2Cl}} \xrightarrow{\mathrm{H_2O}} \mathrm{R^2CONHCHR^1CH_2Cl} \qquad (87)$$

has no α-hydrogen, the imidyl halide **83** can be isolated. Several variations of both the nitrile and the olefin have been successfully used; either bromine or chlorine is suitable.

An interesting recent development is *N*-alkylation of the solvent acetonitrile by carbonium ions generated by electrochemical oxidation of aliphatic carboxylic acids[223b], polymethylbenzenes[223c] or alkyl iodides[223d]. Such processes may have practical advantages over the usual Ritter reaction conditions when acid-sensitive groups are involved[223b].

A somewhat related reaction is *N*-alkylation of the *anions* of certain nitriles in which the α-carbon is highly shielded[224] (reaction 88). Trimethylchlorosilane reacts similarly with acetonitrile in the

$$\left[\begin{matrix}Me_3C \\ \quad\diagdown \\ \quad C{-}CN \\ \quad\diagup \\ Me_2CH\end{matrix} \longleftrightarrow \begin{matrix}Me_3C \\ \quad\diagdown \\ \quad C{=}C{=}\bar{N} \\ \quad\diagup \\ Me_2CH\end{matrix}\right] + Me_2CHI \longrightarrow \begin{matrix}Me_3C \\ \quad\diagdown \\ \quad C{=}C{=}NCHMe_2 \\ \quad\diagup \\ Me_2CH\end{matrix} \quad (88)$$

56%

presence of sodium to give a 25% yield of $Me_3SiCH{=}C{=}NSiMe_3$[225].

B. Acid-Catalysed Reactions of Nitriles with Aldehydes

It has long been known[226] that aliphatic aldehydes react with nitriles in concentrated sulphuric acid to give *N,N'*-alkylidenebis-amides (equation 89). However, it was not until interest arose in the

$$2\ R^1C{\equiv}N + R^2CHO + H_2O \xrightarrow{H_2SO_4} R^1CONHCHR^2NHCOR^1 \quad (89)$$

application of the reaction to the synthesis of polyamides that its details were investigated[227,228]. The reaction is particularly effective with formaldehyde (as trioxan) and proceeds exothermically to high yields at room temperature in 85–98% sulphuric acid. Excess sulphuric acid is often used as a solvent and diluent, but formic acid, phosphoric acid, acetic anhydride, chloroform and other media can be used. Hydrogen chloride, boron trifluoride and other acids are effective but are less practical than sulphuric acid. Benzonitrile reacts considerably more rapidly than MeCN or aliphatic dinitriles, but the latter give over 90% yields in one to four hours.

There has been very little study of the mechanism *per se*, but it is accepted on good grounds that *N*-alkylation of the nitrile proceeds through nucleophilic attack on a cationic species derived from the aldehyde. The reactivity of the aldehyde toward such attack is enhanced by protonation, in a manner fully analogous to acid catalysis of nitrile reactions as discussed in section I (reaction 90). The cation **84** may be immediately hydrated by the water normally present in the reaction mixture, or may remain in equilibrium with a covalent imidyl sulphate (**86**). In either case, this initial step will be followed by acid-catalysed dissociation of the hydroxyl group, leaving a new carbonium ion (**87**) to react with a second nitrile

$$R^1CH{=}O \xrightarrow{H_2SO_4} \left[\begin{array}{c} R^1CH{=}\overset{+}{O}H \\ \updownarrow \\ R^1\overset{+}{C}H{-}OH \end{array}\right] \xrightarrow{R^2CN} \underset{(\mathbf{84})}{\left[\begin{array}{c} R^2C{\equiv}\overset{+}{N}{-}CHR^1{-}OH \\ \updownarrow \\ R^2\overset{+}{C}{=}N{-}CHR^1{-}OH \end{array}\right]} \quad (90)$$

$$\mathbf{84} \xrightarrow{OH^-} \underset{(\mathbf{85})}{R^2CONHCHR^1OH} \xrightarrow[-H_2O]{H^+} \underset{(\mathbf{87})}{R^2CONH\overset{+}{C}HR^1} \xrightarrow{R^2CN}$$

$$\mathbf{84} \xrightarrow{HSO_4^-} \underset{(\mathbf{86})}{R^2C(=NCHR^1{-}OH)(OSO_3H)} \longrightarrow \mathbf{87}$$

$$R^2CONHCHR^1{-}\overset{+}{N}{\equiv}CR^2 \xrightarrow{H_2O} \underset{(\mathbf{88})}{R^2CONHCHR^1{-}NHCOR^2} \quad (91)$$

molecule (reaction 91). Final hydration to the diamide (**88**) may await dilution of the reaction mixture with water. It is clear, from the reaction rates and from the fact that the condensation also occurs in strictly anhydrous systems, that initial hydrolysis to the amide is not a step in the main sequence. Amides do, however, react with aldehydes to give *N*-hydroxyalkylamides (**85**) which condense to **88** under acidic conditions[229] as well as react with nitriles as indicated in reaction (91). Reactions analogous to (90) and (91) also occur when a mixture of nitrile and aldehyde is saturated with hydrogen chloride at low temperature. With acetonitrile and acetaldehyde an imidyl chloride salt analogous to **86** can be isolated[10].

If the ratio of formaldehyde to nitrile is increased over the 1:2 requirement for methylenebis(amide) formation, the reaction tends to stop after the first stage, and yields are reduced. At higher temperatures, however, 1,3,5-triacylhexahydro-*s*-triazines (**89**) are produced. These compounds are most efficiently prepared[230] by reaction of at least 1 : 1 ratios of nitrile to formaldehyde in the presence of catalytic amounts of anhydrous sulphuric acid at 80–100° (reaction 92). It is evident that this reaction involves the same basic mechanism as indicated in reactions (90) and (91). Here, however,

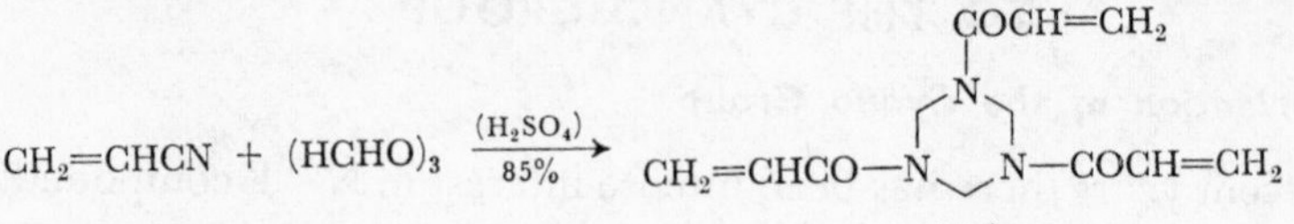

$$CH_2{=}CHCN + (HCHO)_3 \xrightarrow[85\%]{(H_2SO_4)} \underset{(\mathbf{89})}{\text{1,3,5-tris}(CH_2{=}CHCO)\text{-hexahydro-}s\text{-triazine}} \quad (92)$$

(Structure **89**: six-membered ring N–CH₂–N–CH₂–N–CH₂ bearing COCH=CH₂ on each N: $CH_2{=}CHCO{-}N$, $N{-}COCH{=}CH_2$, and $COCH{=}CH_2$ on the top N.)

the final product crystallizes directly from the reaction mixture, no hydration step being required.

A variation of this principle has been utilized recently in the amidomethylation of aromatic compounds[231] (reaction 93). In this work an intermediate imidyl sulphate (**90**) could be isolated.

$$\text{1,3-}Me_2C_6H_4 + 2\,MeCN + 2\,HCHO \xrightarrow[85-90^\circ]{HOAc,\ H_2SO_4} \text{2,4-}Me_2C_6H_2\text{-1,5-}(CH_2N{=}C(Me)OSO_3H)_2\ \textbf{(90)} \xrightarrow{H_2O} \text{2,4-}Me_2C_6H_2\text{-1,5-}(CH_2NHCOMe)_2\ \ 75\% \tag{93}$$

In the preparation of aldehydes by the Stephen reaction, a nitrile is reduced by stannous chloride in the presence of hydrogen chloride. It is recognized that if the product aldimine chlorostannate does not separate readily from solution, *N*,*N'*-alkylidenebis(amides) may be obtained instead of the expected aldehyde[232,233]. Depending on the procedure used, this result might be due to either: (*a*) *N*-alkylation of unchanged nitrile by the aldimine cation in analogy with reaction (90), followed by the analogue of reaction (91)[232], or (*b*) reduction of a predimerized species followed by *N*-alkylation of the nitrile (reaction 94)[234,235].

$$2\,RCN \xrightarrow{HCl} \left[RC({=}NH_2)\text{—}N{=}CRCl\right]^+ Cl^- \xrightarrow[HCl]{SnCl_2} \left[RC({=}NH_2)\text{—}N{=}CHR\right]^+ \xrightarrow{RCN}$$

$$\left[RC({=}NH)\text{—}NH\text{—}CHR\text{—}\overset{+}{N}{\equiv}CR\right] \xrightarrow{H_2O} RCONHCHRNHCOR \tag{94}$$

IX. ADDITION OF FREE RADICALS TO THE CYANO GROUP

A. Fluorination of the Cyano Group

In recent years there has been intense interest in N—F compounds as high energy oxidants, and fluorination of many nitrogen compounds has been thoroughly studied[236]. The products obtained from nitriles depend very much on the reagents and reaction conditions employed. Addition to the C≡N group and cleavage of the molecular skeleton are observed, and usually a large number of products form in any reaction. In every case these may be adequately explained by postulating free-radical intermediates which couple, rearrange or react further with fluorine. The best-defined reactions are observed with simple hydrogen-free substrates, but considerable study has been devoted to acetonitrile as well[237].

Vigorous fluorinating agents attack the cyano group, as indicated in reaction (95)[238]. The radical **91** generally reacts immediately to

$$CF_3C{\equiv}N \xrightarrow[100^\circ]{AgF_2} \underset{(\mathbf{91})}{CF_3CF{=}N\cdot} \longrightarrow \underset{(\mathbf{92})}{CF_3CF_2{-}\ddot{N}:} \xrightarrow{90\%} \underset{(\mathbf{93})}{C_2F_5N{=}NC_2F_5} \quad (95)$$

give the nitrene **92** which dimerizes to **93**, but with a more controlled flow process using fluorine, $CF_3CF{=}NF$ can be a major product[239]. Azo compounds such as **93** can be further fluorinated to $C_2F_5NF_2$ [236] (90% yield). Chlorine monofluoride, a relatively gentle reagent, undergoes reaction (96)[240]. Fluorocarbons and perfluoroalkylamines,

$$R_FCN + ClF \xrightarrow[65-75\%]{-78^\circ} R_FCF_2NCl_2 \quad (96)$$

e.g. $(CF_3)_2NF$, are common products arising from cleavage fragments.

B. Miscellaneous Reactions of Free-Radical Character

The most important reaction of free-radical nature which nitriles undergo is catalytic hydrogenation to amines (Chapter 7). In this case, the normally prohibitively endothermic addition of a radical species to the C≡N bond is circumvented by concerted processes in the adsorbed layer on the metal surface. Aliphatic nitriles are not reduced polarographically, and acetonitrile is an excellent solvent for use in this technique. On the other hand, aromatic nitriles and certain polycyanoolefins can be reduced electrolytically or by alkali metals to give anion radicals in which a small part of the spin density is probably in the cyano group, principally on the nitrogen atom[241].

Nitriles are unreactive toward oxygen, and acetonitrile is a convenient solvent for ozonization reactions. The cyano group does, however, tend to stabilize an α-radical and may thus influence reactions in the group to which it is attached. The relatively stable radical **94** produced thermally from azoisobutyronitrile has some *N*-radical character and in part dimerizes to the ketenimine **95**[242].

$$2\ Me_2\dot{C}{-}C{\equiv}N \longleftrightarrow Me_2C{=}C{=}\dot{N} \longrightarrow Me_2C{=}C{=}N{-}CMe_2CN$$

(**94**) (**95**)

Cycloaddition reactions of nitriles with dienes take place at elevated temperatures[243] with considerable analogy to the Diels–Alder reaction (see Chapter 8). It is notable, however, that only those nitriles having strongly electronegative substituents give appreciable yields, indicating a major polar factor in the reaction. Similarly, nitriles can be trimerized at high pressure and temperature, but these reactions are very slow in the absence of an electrophilic or nucleophilic catalyst[244].

X. REFERENCES

1. S. Soundararajan, *Indian J. Chem.*, **1**, 503 (1963).
2. A. Colligiani, L. Guibé, P. J. Haigh and E. A. C. Lucken, *Mol. Phys.*, **14**, 89 (1968).
3. M. Witanowski, *Tetrahedron*, **23**, 4299 (1967); D. Herbison-Evans and R. E. Richards, *Mol. Phys.*, **8**, 19 (1964).
4. J. B. Stothers, *Quart. Rev. (London)*, **19**, 144 (1965).
5. L. J. Bellamy, *The Infrared Spectra of Complex Molecules*, 2nd ed., Methuen, London, 1958, pp. 263–266.
6. M. R. Yagudaev and A. D. Grebenyuk, *J. Org. Chem. USSR*, **2**, 1794 (1966).
7. N. B. Colthup, L. H. Daly and S. E. Wiberley, *Introduction to Infrared and Raman Spectroscopy*, Academic Press, New York, 1964, Chap. 4.
8. H. W. Thompson and G. Steel, *Trans. Faraday Soc.*, **52**, 1451 (1956); D. Hadzi in *Infrared Spectroscopy and Molecular Structure* (Ed. M. Davies). Elsevier, Amsterdam, 1963, p. 256.
9. T. L. Brown, *J. Phys. Chem.*, **64**, 1798 (1960).
10. E. N. Zil'berman, *Russian Chem. Rev.*, **31**, 615 (1962).
11. E. Allenstein, A. Schmidt and V. Beyl, *Chem. Ber.*, **99**, 431 (1966).
12. E. Allenstein and A. Schmidt, *Spectrochim. Acta*, **20**, 1451 (1964); S. W. Peterson and J. M. Williams, *J. Am. Chem. Soc.*, **88**, 2866 (1966).
13. R. A. Walton, *Quart. Rev. (London)*, **19**, 126 (1965).
14. I. Hechenbleikner, *U.S. Pat.* 2,704,297 (1955); *Chem. Abstr.*, **50**, 1902f (1956); *U.S. Pat.* 2,719,174 (1955), *Chem. Abstr.*, **50**, 8718i (1956); *U.S. Pat.* 2,768,204 (1956); *Chem. Abstr.*, **51**, 16544d (1957).
15. L. E. Hinkel and G. J. Treharne, *J. Chem. Soc.*, **1945**, 866.
16. A. Ya. Lazaris, E. N. Zil'berman and O. D. Strizhakov, *J. Gen. Chem. USSR*, **32**, 890 (1962).
17. E. Allenstein and A. Schmidt, *Chem. Ber.*, **97**, 1863 (1964).
18. F. Johnson and R. Madroñero in *Advances in Heterocyclic Chemistry*, Vol. 6 (Ed. A. R. Katritzky), Academic Press, New York, 1966, pp. 95–146.
19. J. J. Roemer and D. W. Kaiser, *U.S. Pat.* 2,658,893 (1953); *Chem. Abstr.*, **48**, 12813f (1954).

20. E. N. Zil'berman and N. A. Rybakova, *J. Gen. Chem. USSR*, **30**, 1972 (1960).
21. E. N. Zil'berman and A. E. Kulikova, *J. Gen. Chem. USSR*, **29**, 1671 (1959).
22. (a) A. W. Cobb and J. H. Walton, *J. Phys. Chem.*, **41**, 351 (1937); G. Berger and S. C. J. Olivier, *Rec. Trav. Chim.*, **46**, 600 (1927);
(b) For recent work see G. Olah and T. E. Kiovsky, *J. Am. Chem. Soc.*, **90**, 4666 (1968).
23. G. W. Kenner, C. B. Reese and A. Todd, *J. Chem. Soc.*, **1958**, 546; A. Schimple, R. M. Lemmon and M. Calvin, *Science*, **147**, 149 (1965); G. Steinman, R. M. Lemmon and M. Calvin, *Proc. Nat. Acad. Sci. U.S.*, **52**, 27 (1964).
24. P. Eitner, *Ber.*, **25**, 464 (1892).
25. H. Weidinger and J. Kranz, *Chem. Ber.*, **96**, 2070 (1963).
26. P. Eitner, *Chem. Ber.*, **26**, 2833 (1893); J. I. Jones, *Chem. Commun.*, **1967**, 128.
27. A. Gautier, *Ann. Chem.*, **150**, 187 (1869).
28. D. Davidson and H. Skovronek, *J. Am. Chem. Soc.*, **80**, 376 (1958).
29. C. E. Colby and F. D. Dodge, *Am. Chem. J.*, **13**, 1 (1891); W. G. Toland and L. L. Ferstandig, *J. Org. Chem.*, **23**, 1350 (1958).
30. E. N. Zil'berman, *J. Gen. Chem. USSR*, **29**, 3312 (1959).
31. F. C. Schaefer, unpublished work.
32. C. M. Hendry, *J. Am. Chem. Soc.*, **80**, 973 (1958).
33. H. G. Khorana, *Chem. Rev.*, **53**, 145 (1953); G. Doleschall and K. Lembert, *Tetrahedron Letters*, 781, 1195 (1963).
34. (a) S. Wideqvist, *Arkiv Kemi*, **3**, 289 (1951);
(b) F. Salmon-Legagneur, *Bull. Soc. Chim. France*, **25**, 1052 (1958).
35. M. L. Ernst and G. L. Schmir, *J. Am. Chem. Soc.*, **88**, 5001 (1966), and references cited therein.
36. G. Losse and H. Weddige, *Ann. Chem.*, **636**, 144 (1960).
37. G. E. Kretov and A. P. Momsenko, *J. Gen. Chem. USSR*, **31**, 3651 (1961).
38. F. C. Schaefer and J. E. Lancaster, unpublished work.
39. A. Colson, *Bull. Soc. Chim. France*, [3] **17**, 55 (1897); *Ann. Chim. Phys.*, [7] **12**, 250 (1897).
40. V. K. Krieble and R. H. Smellie, *U.S. Pat.* 2,390,106 (1945); *Chem. Abstr.*, **40**, 1868 (1946).
41. V. K. Krieble, F. C. Duennebier and E. Colton, *J. Am. Chem. Soc.*, **65**, 1479 (1943).
42. T. L. Brown and M. Kubota, *J. Am. Chem. Soc.*, **83**, 4175 (1961).
43. H. Meerwein, P. Laasch, R. Mersch and J. Spille, *Chem. Ber.*, **89**, 209 (1956).
44. J. E. Gordon and G. C. Turrell, *J. Org. Chem.*, **24**, 269 (1959).
45. F. Klages and W. Grill, *Ann. Chem.*, **626**, 60 (1959).
46. R. R. Schmidt, *Chem. Ber.*, **98**, 334 (1965).
47. P. Eitner and F. Krafft, *Ber.*, **25**, 2263 (1892).
48. F. Krafft and A. v. Hansen, *Ber.*, **22**, 803 (1889).
49. D. Martin and A. Weisse, *Chem. Ber.*, **100**, 3747 (1967).
50. H. Meerwein, P. Laasch, R. Mersch and J. Nentwig, *Chem. Ber.*, **89**, 224 (1956).
51. D. Martin and A. Weisse, *Chem. Ber.*, **100**, 3736 (1967).
52. N. C. Stephenson, *J. Inorg. Nucl. Chem.*, **24**, 801 (1962); see also G. Rouschias and G. Wilkinson, *J. Chem. Soc. (A)*, **1968**, 489.
53. V. I. Shevchenko, P. P. Kornuta, N. D. Bodnarchuk and A. V. Kirsanov, *J. Gen. Chem. USSR*, **36**, 743 (1966); V. I. Shevchenko and V. P. Kukhar, *J. Gen. Chem. USSR*, **36**, 747 (1966); V. I. Shevchenko and P. P. Kornuta, *J. Gen. Chem. USSR*, **36**, 1270, 1642 (1966); V. I. Shevchenko and N. D. Bodnarchuk, *J. Gen. Chem.*

USSR, **36**, 1645 (1966); V. I. Shevchenko, P. P. Kornuta and A. V. Kirsanov, *J. Gen. Chem. USSR*, **35**, 1602, 1962 (1965); V. I. Shevchenko, N. D. Bodnarchuk and A. V. Kirsanov, *J. Gen. Chem. USSR*, **33**, 1312, 1553 (1963).

54. G. A. Olah and S. J. Kuhn in *Friedel–Crafts and Related Reactions*, (Ed. G. A. Olah), Vol. 3, Part 2, Interscience, New York, 1964, p. 1191 onwards; W. E. Truce, *Org. React.*, **9**, 37 (1957).
55. W. A. Ruske in *Friedel–Crafts and Related Reactions*, (Ed. G. A. Olah), Vol. 3, Part 1, Interscience, New York, 1964, p. 383 onwards.
56. E. N. Zil'berman and N. A. Rybakova, *J. Gen. Chem. USSR*, **32**, 581 (1962).
57. E. N. Zil'berman and N. A. Rybakova, *Kinetika i Kataliz*, **5**, 538 (1964).
58. F. Klages, R. Ruhnau and W. Hauser, *Ann. Chem.*, **626**, 60 (1959).
59. D. P. Tate, W. R. Knipple and J. M. Augl, *Inorg. Chem.*, **1**, 433 (1962).
60. B. L. Ross, J. G. Grasselli, W. M. Ritchey and H. D. Kaesz, *Inorg. Chem.*, **2**, 1023 (1963).
61. D. P. Tate, J. M. Augl and A. Buss, *Inorg. Chem.*, **2**, 427 (1963).
62. D. P. Tate, A. A. Buss, J. M. Augl, B. L. Ross, J. G. Grasselli, W. M. Ritchey and F. J. Knoll, *Inorg. Chem.*, **4**, 1323 (1965); see also M. L. Ziegler, *Angew. Chem.*, **80**, 239 (1968).
63. H. Bock and H. tom Dieck, *Chem. Ber.*, **99**, 213 (1966).
64. R. Schaeffer, *J. Am. Chem. Soc.*, **79**, 1006 (1957); J. van der Maas Reddy and W. M. Lipscomb, *J. Chem. Phys.*, **31**, 610 (1959).
65. J. Green, M. M. Fein, N. Mayes, G. Donovan, M. Israel and M. S. Cohen, *J. Polymer Sci.*, **B2**, 987 (1964).
66. S. F. A. Kettle and L. E. Orgel, *Proc. Chem. Soc.* (*London*), **1959**, 307.
67. H. Henecka and P. Kurz in *Methoden der Organischen Chemie* (Houben-Weyl), (Ed. E. Müller), Vol. 8, Georg Thieme Verlag, Stuttgart, 1952, pp. 427–432, 661–663; R. B. Wagner and H. D. Zook, *Synthetic Organic Chemistry*, John Wiley and Sons, New York, 1953, pp. 412–415, 570.
68. B. S. Rabinovitch and C. A. Winkler, *Can. J. Research*, **20B**, 185 (1942).
69. K. B. Wiberg, *J. Am. Chem. Soc.*, **77**, 2519 (1955).
70. L. P. Hammett, *Physical Organic Chemistry*, McGraw-Hill, New York, 1940, p. 189.
71. V. A. Linetskii and B. R. Serebryakov, *Izv. Vysshikh Uchebn. Zavedenii, Neft i Gaz*, **8**(1), 72 (1965); **8**(2), 62 (1965); *Chem. Abstr.*, **62**, 11521c, 15745b (1965).
72. R. Breslow, R. Fairweather and J. Keana, *J. Am. Chem. Soc.*, **89**, 2135 (1967).
73. C. F. Krewson and J. F. Couch, *J. Am. Chem. Soc.*, **65**, 2256 (1943).
74. A. Galat, *J. Am. Chem. Soc.*, **70**, 3945 (1948).
75. E. J. Gasson and D. J. Hadley, *Brit. Pat.* 777,517 (1956); *Chem. Abstr.*, **52**, 7361c (1958).
76. P. L. de Benneville, C. L. Levesque, L. J. Exner and E. Hertz, *J. Org. Chem.*, **21**, 1072 (1956).
77. L. J. Exner, M. J. Hurowitz and P. L. de Benneville, *J. Am. Chem. Soc.*, **77**, 1103 (1955).
78. M. J. Hurowitz, L. J. Exner and P. L. de Benneville, *J. Am. Chem. Soc.*, **77**, 3251 (1955).
79. V. K. Krieble and A. L. Peiker, *J. Am. Chem. Soc.*, **51**, 3368 (1929); V. K. Krieble and C. I. Noll, *J. Am. Chem. Soc.*, **61**, 560 (1939).
80. B. S. Rabinovitch, C. A. Winkler and A. R. P. Stewart, *Can. J. Research*, **20B**, 121 (1942); J. D. McLean, B. S. Rabinovitch and C. A. Winkler, *Can. J. Research*, **20B**, 168 (1942); B. S. Rabinovitch and C. A. Winkler, *Can. J. Research*, **20B**, 221 (1942).

81. M. J. Sullivan and M. L. Kilpatrick, *J. Am. Chem. Soc.*, **67**, 1815 (1945); M. L. Kilpatrick, *J. Am. Chem. Soc.*, **69**, 40 (1947).
82. (a) F. A. Long and M. A. Paul, *Chem. Revs.*, **57**, 935 (1957);
(b) J. F. Bunnett, *J. Am. Chem. Soc.*, **83**, 4956, 4968, 4973, 4978 (1961);
(c) R. B. Martin, *J. Am. Chem. Soc.*, **84**, 4130 (1962).
83. B. R. Mole, J. P. Murray and J. G. Tillett, *J. Chem. Soc.*, **1965**, 802.
84. W. M. Thomas in *Encyclopedia of Polymer Science and Technology* (Eds. H. F. Mark, N. G. Gaylord and N. M. Bikales), Vol. 1, Interscience, New York, 1964, p. 197.
85. C. R. Hauser and D. S. Hoffenberg, *J. Org. Chem.*, **20**, 1448 (1955).
86. C. R. Hauser and C. J. Eby, *J. Am. Chem. Soc.*, **79**, 725 (1957).
87. H. R. Snyder and C. T. Elston, *J. Am. Chem. Soc.*, **76**, 3039 (1954).
88. K. B. Wiberg, *J. Am. Chem. Soc.*, **75**, 39611 (1953).
89. G. B. Payne and P. H. Williams, *J. Org. Chem.*, **26**, 651 (1961).
90. G. B. Payne, P. H. Deming and P. H. Williams, *J. Org. Chem.*, **26**, 659 (1961).
91. Y. Ogata and Y. Sawaki, *Tetrahedron*, **20**, 2065 (1964).
92. G. B. Payne, *J. Org. Chem.*, **26**, 663 (1961).
93. G. B. Payne, *Tetrahedron*, **18**, 763 (1962).
94. R. Roger and D. G. Neilson, *Chem. Revs.*, **61**, 179 (1961).
95. A. Pinner, *Die Imidoäther und ihre Derivate*, Oppenheim, Berlin, 1892.
96. F. Cramer and H-J. Baldauf, *Chem. Ber.*, **92**, 370 (1959).
97. M. L. Shul'man, G. A. Shvekhgeimer and R. A. Miftakhova, *J. Org. Chem. USSR.*, **3**, 840 (1967).
98. D. J. Morgan, *Chem. Ind.* (*London*), **1959**, 854.
99. A. Knorr, *Ber.* **49**, 1735 (1916).
100. K. A. Jensen, M. Due, A. Holm and C. Wentrup, *Acta Chem. Scand.*, **20**, 2091 (1966).
101. E. Grigat and R. Pütter, *Angew. Chem. Int. Ed. Engl.*, **6**, 206 (1967).
102. R. H. McKee, *Am. Chem. J.*, **26**, 209 (1901).
103. E. N. Zil'berman and A. Ya. Lazaris, *J. Gen. Chem. USSR*, **31**, 1224 (1961).
104. J. U. Nef, *Ann. Chem.*, **287**, 265 (1895).
105. F. C. Schaefer and G. A. Peters, *J. Org. Chem.*, **26**, 412 (1961).
106. M-C. Chiang and T-C. Tai, *Hua Hsueh Pao*, **30**, 312 (1964); *Chem. Abstr.*, **61**, 6885a (1965).
107. H. M. Woodburn, A. B. Whitehouse and B. G. Pautler, *J. Org. Chem.*, **24**, 210 (1959).
108. T. K. Brotherton and J. W. Lynn, *Chem. Rev.*, **59**, 868 (1959).
109. F. Cramer, K. Pawelzik and H-J. Baldauf, *Chem. Ber.*, **91**, 1049 (1958); F. Cramer, K. Pawelzik and F. W. Lichtenthaler, *Chem. Ber.*, **91**, 1555 (1958).
110. H. C. Brown and C. R. Wetzel, *J. Org. Chem.*, **30**, 3724, 3729 (1965).
111. F. C. Schaefer and G. A. Peters, *J. Org. Chem.*, **26**, 2778 (1961).
112. K. Dachlauer, *Reactions of Trichloroacetonitrile*, PB58825, Office of the Publication Board, Department of Commerce, Washington, D.C.; *Ger. Pat.* 671,785 (1939); J. C. Grivas and A. Taurins, *Can. J. Chem.*, **36**, 771 (1958); **39**, 761 (1961); H. J. Backer and W. L. Wanmaker, *Rec. Trav. Chim.*, **70**, 638 (1951); F. C. Schaefer, I. Hechenbleikner, G. A. Peters and V. P. Wystrach, *J. Am. Chem. Soc.*, **81**, 1466 (1959).
113. W. L. Reilly and H. C. Brown, *J. Org. Chem.*, **22**, 698 (1957); H. C. Brown and P. D. Schuman, *J. Org. Chem.*, **28**, 1122 (1963); H. C. Brown and C. R. Wetzel, *J. Org. Chem.*, **30**, 3734 (1965); H. C. Brown, P. D. Schuman and J. Turnbull, *J. Org. Chem.*, **32**, 231 (1967); R. N. Johnson and H. M. Woodburn, *J. Org. Chem.*, **27**, 3958 (1962).

114. J. G. Erickson, *J. Org. Chem.*, **20**, 1569 (1955).
115. W. Jentsch and M. Seefelder, *Chem. Ber.*, **98**, 1342 (1965).
116. M. Seefelder, *Chem. Ber.*, **99**, 2678 (1966).
117. P. L. de Benneville, J. S. Strong and V. T. Elkind, *J. Org. Chem.*, **21**, 772 (1956)
118. P. Oxley and W. F. Short, *J. Chem. Soc.*, **1949**, 449 and earlier papers cited therein; J. Cymerman, J. W. Minnis, P. Oxley and W. F. Short, *J. Chem. Soc.*, **1949**, 2097.
119. E. Erlenmeyer, *Ann. Chem.*, **146**, 258 (1868).
120. A. Bernthsen, *Ber.*, **10**, 1235, 1238 (1877).
121. F. C. Schaefer and A. P. Krapcho, *J. Org. Chem.*, **27**, 1255 (1962).
122. F. Eloy and R. Lenaers, *Chem. Rev.*, **62**, 155 (1962).
123. C. S. Hollander, R. A. Yoncoskie and P. L. de Benneville, *J. Org. Chem.*, **23**, 1112 (1958).
124. E. F. Cornell, *J. Am. Chem. Soc.*, **50**, 3311 (1928).
125. K. Ziegler and H. Ohlinger, *Ann. Chem.*, **495**, 84 (1932); K. Ziegler, *U.S. Pat.* 2,049,582 (1936).
126. G. Newberry and W. Webster, *J. Chem. Soc.*, **1947**, 738.
127. C. J. Eby and C. R. Hauser, *J. Am. Chem. Soc.*, **79**, 723 (1957).
128. J. H. Clark and H. Marson, private communication.
129. R. P. Hullin, J. Miller and W. F. Short, *J. Chem. Soc.*, **1947**, 394.
130. Th. Kauffmann, L. Ban, W. Burkhardt, E. Rauch and J. Sobel, *Angew. Chem. Int. Ed. Engl.*, **4**, 1090 (1965); Th. Kauffmann, L. Ban and D. Kuhlmann, *Angew. Chem. Int. Ed. Engl.*, **6**, 256 (1967).
131. K. Kindler, *Ann. Chem.*, **431**, 187 (1923).
132. A. E. S. Fairfull, J. L. Lowe and D. A. Peak, *J. Chem. Soc.*, **1952**, 742.
133. E. E. Gilbert, E. J. Rumanowski and P. E. Newallis, *J. Chem. Eng. Data*, **13**, 130 (1968).
134. K. Kindler, *Ann. Chem.*, **450**, 1 (1926).
135. E. Grigat and R. Pütter, *Chem. Ber.*, **97**, 3022 (1964).
136. H. C. Brown and R. Pater, *J. Org. Chem.*, **27**, 2858 (1962).
137. S. A. Miller and B. Bann, *J. Appl. Chem. (London)*, **6**, 89 (1956).
138. A. Marxer, *J. Am. Chem. Soc.*, **79**, 467 (1957); see also C. Zervos and E. H. Cordes, *J. Am. Chem. Soc.*, **90**, 6892 (1968).
139. P. L. de Benneville, J. S. Strong and V. T. Elkind, *J. Org. Chem.*, **21**, 772 (1956).
140. W. G. Toland, *J. Org. Chem.*, **27**, 869 (1962).
141. E. C. Taylor and J. A. Zoltewicz, *J. Am. Chem. Soc.*, **82**, 2656 (1960).
142. J. Goerdeler and H. Porrmann, *Chem. Ber.*, **94**, 2856 (1961).
143. P. Chabrier, S. Renard and E. Renier, *Compt. Rend.*, **235**, 64 (1952).
144. E. Grigat, R. Pütter and E. Mühlbauer, *Chem. Ber.*, **98**, 3777 (1965).
145. P. L. Pickard and T. L. Tolbert, *J. Org. Chem.*, **26**, 4887 (1961).
146. M. S. Kharasch and O. Reinmuth, *Grignard Reactions of Non-Metallic Substances*, Prentice-Hall, New York, 1954, Chap. 10.
147. C. G. Swain, *J. Am. Chem. Soc.*, **69**, 2306 (1947).
148. T. Eicher in *The Chemistry of the Carbonyl Group* (Ed. S. Patai), Interscience, New York, 1966, Chapter 13.
149. S. J. Storfer and E. I. Becker, *J. Org. Chem.*, **27**, 1868 (1962).
150. J. D. Citron and E. I. Becker, *Can. J. Chem.*, **41**, 1260 (1963).
151. A. A. Scala and E. I. Becker, *J. Org. Chem.*, **30**, 3491 (1965).
152. H. Edelstein and E. I. Becker, *J. Org. Chem.*, **31**, 3375 (1966).
153. A. A. Scala, N. M. Bikales and E. I. Becker, *J. Org. Chem.*, **30**, 303 (1965).
154. R. E. Dessy and F. Paulik, *J. Chem. Educ.*, **40**, 185 (1963).

155. G. Wittig, *Angew. Chem.*, **70**, 65 (1958).
156. P. Bruylants, *Bull. Acad. Roy. Belg.*, [5] **8**, 7 (1922); A. Bruylants, *Bull. Soc. Chim. France*, **1958**, 1291.
157. C. R. Hauser and W. J. Humphlett, *J. Org. Chem.*, **15**, 359 (1950).
158. F. F. Blicke and E-P. Tsao, *J. Am. Chem. Soc.*, **75**, 5587 (1953).
159. J. Yoshimura, Y. Ohgo and T. Sato, *Bull. Chem. Soc. Japan*, **38**, 1809 (1965).
160. F. Kuffner and W. Koechlin, *Monatsh. Chem.*, **93**, 476 (1962).
161. B. B. Allen and H. R. Henze, *J. Am. Chem. Soc.*, **61**, 1790 (1939).
162. V. Migrdichian, *The Chemistry of Organic Cyanogen Compounds*, Reinhold, New York, 1947, Chap. 14.
163. J. O. Bledsoe, *Dissertation Abstr.*, **26**, 1913 (1966).
164. E. A. Braude, *Prog. Org. Chem.*, **3**, 185 (1955).
165. H. Gilman and M. Lichtenwalter, *Rec. Trav. Chim.*, **55**, 561 (1936); H. Gilman and R. H. Kirby, *J. Am. Chem. Soc.*, **55**, 1265 (1933).
166. R. L. Shriner, *Org. React.*, **1**, 1 (1942).
167. E. E. Blaise, *Compt. Rend.*, **132**, 38, 478 (1901).
168. J. Cason, K. L. Rinehart and J. D. Thornton, *J. Org. Chem.*, **18**, 1594 (1953).
169. H. B. Kagan and Y.-H. Suen, *Bull. Soc. Chim. France*, **1966**, 1819; A. Horean, J. Jacques, H. B. Kagan and Y.-H. Suen, *Bull. Soc. Chim. France*, **1966**, 1823.
170. J. E. Lloyd and K. Wade, *J. Chem. Soc.*, **1965**, 2662; J. R. Jennings, J. E. Lloyd and K. Wade, *J. Chem. Soc.*, **1965**, 5083.
171. J. P. Schaefer and J. J. Bloomfield, *Org. React.*, **15**, 1 (1967).
172. W. I. Sullivan, F. W. Swamer, W. J. Humphlett and C. R. Hauser, *J. Org. Chem.*, **26**, 2306 (1961).
173. G. A. Reynolds, W. J. Humphlett, F. W. Swamer and C. R. Hauser, *J. Org. Chem.*, **16**, 165 (1951).
174. Frankland and Kolbe, *Ann. Chem.*, **65**, 269 (1848); E. von Meyer, *J. Prakt. Chem.*, [2] **39**, 262 (1880).
175. R. Schenk and H. Finken, *Ann. Chem.*, **462**, 267 (1928).
176. D. M. W. Anderson, F. Bell and J. L. Duncan, *J. Chem. Soc.*, **1961**, 4705.
177. H. Junek and H. Sterk, *Z. Naturforsch.* **22b**, 732 (1967).
178. R. A. Carboni, D. D. Coffman and E. G. Howard, *J. Am. Chem. Soc.*, **80**, 2838 (1958).
179. C. W. Moore and J. F. Thorpe, *J. Chem. Soc.*, **93**, 165 (1908).
180. K. Ziegler in *Methoden der Organischen Chemie* (Houben-Weyl), Vol. 4, Part 2, Georg Thieme Verlag, Stuttgart, 1955, pp. 758–764.
181. J. J. Ritter and R. D. Anderson, *J. Org. Chem.*, **24**, 208 (1959); A. Lottermoser, *J. Prakt. Chem.*, **54**, 132 (1896).
182. R. M. Anker and A. H. Cook, *J. Chem. Soc.*, **1941**, 323.
183. F. H. Case, *J. Org. Chem.*, **31**, 2398 (1966); F. W. Swamer, G. A. Reynolds and C. R. Hauser, *J. Org. Chem.*, **16**, 43 (1951).
184. T. K. Brotherton and J. F. Bunnett, *Chem. Ind.* (*London*), **1957**, 80.
185. E. Frankland and J. C. Evans, *J. Chem. Soc.*, **37**, 563 (1880).
186. K. Issleib and R. D. Black, *Z. Anorg. Allgem. Chem.*, **336**, 234 (1965); *Chem. Ber.*, **98**, 1093 (1965).
187. A. Dornow, I. Kühlcke and F. Baxmann, *Chem. Ber.*, **82**, 254 (1949).
188. E. F. J. Atkinson, H. Ingham and J. F. Thorpe, *J. Chem. Soc.*, **91**, 578 (1907).
189. A. Bladé-Font, W. E. McEwen and C. A. VanderWerf, *J. Am. Chem. Soc.*, **82**, 2646 (1960).
190. H. König, H. Metzger and C. Seelert, *Chem. Ber.*, **98**, 3724 (1965).

191. V. A. Kargin, V. A. Kabanov, V. P. Zubov and A. B. Zazin, *Dokl. Akad. Nauk SSSR*, **139**, 605 (1961); V. A. Kabanov, V. P. Zubov, V. P. Kovaleva and V. A. Kargin, *J. Polymer Sci. (C)*, **1964**(4), 1009; V. P. Zubov, I. P. Terekhina, V. P. Kabanov and V. A. Kargin, *Vysokomolekul. Soedin.*, *Geterotsepnye Vysokomolekul. Soedin.*, **1964**, 147; V. P. Zubov, E. T. Zakharenko, V. P. Kabanov and V. A. Kargin, *Vysokomlekul. Soedin.*, *Geterotsepnye Vysokomolekul. Soedin.*, **1964**, 186; see also Y. E. Doroshenko, *Russian Chem. Rev.*, **36**, 563 (1967).
192. E. Oikawa and S. Kambara, *Polymer Letters*, **2**, 649 (1964); *Bull. Chem. Soc. Japan*, **37**, 1849 (1964).
193. E. Oikawa, K. Mari and G. Saito, *Bull. Chem. Soc. Japan*, **39**, 1182 (1966).
194. A. F. Lurin and Ya. M. Paushkin, *Polymer Sci. USSR*, **6**, 1626 (1964).
195. Y. Tabata, E. Oda and H. Sobue, *J. Polymer Sci.*, **45**, 469 (1960).
196. Y. Tabata, K. Hara and H. Sobue, *J. Polymer Sci.*, **2A**, 4077 (1964).
197. Y. Tabata, I. Hiroi and M. Taniyama, *J. Polymer Sci.*, **2A**, 1567 (1964), and references cited therein; see also J. N. Hay, *J. Polymer Sci.*, **6A1**, 2127 (1968).
198. L. L. Bircumshaw, F. M. Taylor and D. H. Wiffin, *J. Chem. Soc.*, **1954**, 931.
199. J. Peska, M. J. Benes and O. Wichterle, *Collection Czech. Chem. Commun.*, **31**, 243 (1966).
200. T. Völker, *Angew. Chem.*, **69**, 728 (1957); **72**, 379 (1960).
201. W. Ruske, *Rev. Chim. Acad. Rep. Populaire Roumaine*, **7**, 1245 (1962).
202. J. P. Ferris and L. E. Orgel, *J. Am. Chem. Soc.*, **88**, 3829 (1966); *J. Org. Chem.*, **30**, 2365 (1965).
203. J. Oro and S. S. Kamat, *Nature*, **190**, 442 (1961); J. Oro, *Nature*, **191**, 1193 (1961); J. Oro and A. P. Kimball, *Arch. Biochem. Biophys.*, **94**, 217 (1961); **96**, 293 (1962).
204. R. Sanchez, J. Ferris and L. E. Orgel, *Science*, **153**, 72 (1966).
205. W. G. Deichert and M. C. Tobin, *J. Polymer Sci.*, **45**, S39 (1961); M. C. Tobin and W. G. Deichert, *U.S. Pat.* 3,155,629 (1964); *Chem. Abstr.*, **62**, 5356g (1965).
206. K. Bredereck and R. Richter, *Chem. Ber.*, **99**, 2461 (1966).
207. E. N. Zil'berman, *Russ. Chem. Rev.*, **29**, 331 (1960).
208. H. Wieland and E. Dorrer, *Chem. Ber.*, **63**, 404 (1930).
209. PB Report 70344, Office of the Publication Board, Department of Commerce, Washington, D.C., frames 16376–88.
210. W. Gresham and W. Grigsby, *U.S. Pat.* 2,457,660 (1948); *Chem. Abstr.*, **43**, 3028g (1949).
211. J. J. Ritter and P. P. Minieri, *J. Am. Chem. Soc.*, **70**, 4045 (1948); J. J. Ritter and J. Kalish, *J. Am. Chem. Soc.*, **70**, 4048 (1948); F. R. Benson and J. J. Ritter, *J. Am. Chem. Soc.*, **71**, 4128 (1949); L. W. Hartzel and J. J. Ritter, *J. Am. Chem. Soc.*, **71**, 4130 (1949); R. M. Lusskin and J. J. Ritter, *J. Am. Chem. Soc.*, **72**, 5577 (1950); H. Plaut and J. J. Ritter, *J. Am. Chem. Soc.*, **73**, 4076 (1951); E. Tillmans and J. J. Ritter, *J. Org. Chem.*, **22**, 839 (1957); A. I. Meyers and J. J. Ritter, *J. Org. Chem.*, **23**, 1918 (1958).
212. T. Clark, J. Devine and D. W. Dicker, *J. Am. Oil Chemists Soc.*, **41**, 78 (1964).
213. F. Möller in *Methoden der Organischen Chemie* (Houben-Weyl), (Ed. E. Müller), Vol. 2, Georg Thieme Verlag, Stuttgart, 1957, pp. 994–9.
214. S. Julia and C. Papantoniou, *Compt. Rend.*, **260**, 1440 (1965).
215. C. L. Parris and R. M. Christenson, *J. Org. Chem.*, **25**, 331 (1960).
216. A. I. Meyers, *J. Org. Chem.*, **25**, 145, 1147 (1960).
217. D. Giraud-Clenet and J. Anatol, *Compt. Rend., Ser. C.*, **262**, 224 (1966).
218. (a) N. K. Kochetkov, A. Ya. Khorlin and K. I. Lopatina, *J. Gen. Chem. USSR*, **29**, 77 (1959);

(b) A. Ya. Khorlin, O. S. Chizkov and N. K. Kochetkov, *J. Gen. Chem. USSR*, **29**, 3373 (1959).
219. E. M. Smolin, *J. Org. Chem.*, **20**, 295 (1955).
220. D. Martin and A. Weise, *Ann. Chem.*, **702**, 86 (1967).
221. G. W. Cannon, K. K. Grebber and Y.-K. Hsu, *J. Org. Chem.*, **18**, 516 (1953).
222. J. Cast and T. S. Stevens, *J. Chem. Soc.*, **1953**, 4180.
223. (a) T. L. Cairns, P. J. Graham, P. L. Barrick and R. S. Schreiber, *J. Org. Chem.*, **17**, 751 (1952); A. Hassner, L. A. Levy and R. Gault, *Tetrahedron Letters*, **1966**, 3119;
(b) L. Eberson and K. Nyberg, *Acta Chem. Scand.*, **18**, 1567 (1964);
(c) L. Eberson and K. Nyberg, *Tetrahedron Letters*, **1966**, 2389;
(d) L. L. Miller and A. K. Hoffmann, *J. Amer. Chem. Soc.*, **89**, 593 (1967).
224. M. S. Newman, T. Fukunaga and T. Miwa, *J. Am. Chem. Soc.*, **82**, 873 (1960).
225. M. Prober, *J. Am. Chem. Soc.*, **78**, 2274 (1956).
226. E. Hepp and G. Spiess, *Ber.*, **9**, 1424 (1876); E. Hepp, *Ber.*, **10**, 1649 (1877).
227. D. T. Mowry and E. L. Ringwald, *J. Am. Chem. Soc.*, **72**, 4439 (1950).
228. E. E. Magat, B. F. Faris J. E. Reith and L. F. Salisbury, *J. Am. Chem. Soc.*, **73**, 1028 (1951); E. E. Magat, L. B. Chandler, B. F. Faris, J. E. Reith and L. F. Salisbury, *J. Am. Chem. Soc.* **73**, 1031 (1951); E. E. Magat and L. F. Salisbury, *J. Am. Chem. Soc.* **73**, 1035 (1951).
229. A. Einhorn, *Ann. Chem.*, 343, 207 (1905).
230. M. A. Gradsten and M. W. Pollack, *J. Am. Chem. Soc.*, **70**, 3079 (1948); R. Wegler and A. Ballauf, *Chem. Ber.*, **81**, 527 (1948).
231. C. L. Parris and R. M. Christenson, *J. Org. Chem.*, **25**, 1888 (1960).
232. J. A. Knight and H. D. Zook, *J. Am. Chem. Soc.*, **74**, 4560 (1952).
233. L. Turner, *J. Chem. Soc.*, **1956**, 1686.
234. O. Bayer in *Methoden der Organischen Chemie* (Houben-Weyl) (Ed. E. Müller), Vol. 7, Part 1, Georg Thieme Verlag, Stuttgart, 1954, pp. 299–301.
235. E. N. Zil'berman and P. S. Pyryalova, *J. Gen. Chem. USSR*, **33**, 3348 (1963).
236. J. K. Ruff, *Chem. Rev.*, **67**, 665 (1967).
237. S. P. Makarov, I. V. Ermakova and V. A. Skysanskii, *J. Gen. Chem. USSR*, **36**, 1426 (1966); M. Lustig, *Inorg. Chem.*, **6**, 1064 (1967).
238. J. A. Young, W. S. Durrell and R. D. Dresdner, *J. Amer. Chem. Soc.*, **82**, 4553 (1960); J. K. Ruff, *J. Org. Chem.*, **32**, 1675 (1967); W. J. Chambers, C. W. Tullock and D. D. Coffman, *J. Am. Chem. Soc.*, **84**, 2337 (1962); J. B. Hynes, B. C. Bishop and L. A. Bigelow, *J. Org. Chem.*, **28**, 2811 (1963); O. Glemser, H. Schröder and H. Haeseler, *Z. Anorg. Allgem. Chem.*, **282**, 80 (1955).
239. B. C. Bishop, J. B. Hynes and L. A. Bigelow, *J. Am. Chem. Soc.*, **86**, 1827 (1964).
240. J. B. Hynes and T. E. Austin, *Inorg. Chem.*, **5**, 488 (1966); J. B. Hynes, B. C. Bishop and L. A. Bigelow, *Inorg. Chem.*, **6**, 417 (1967).
241. P. H. Rieger, I. Bernal, W. H. Reinmuth and G. K. Fraenkel, *J. Am. Chem. Soc.*, 683 (1963); P. H. Rieger and G. K. Fraenkel, *J. Chem. Phys.*, **37**, 2795 (1962); M. T. Jones, *J. Am. Chem. Soc.*, **88**, 5060 (1966).
242. M. Talat-Erben and S. Bywater, *J. Am. Chem. Soc.*, **77**, 3712 (1955); G. S. Hammond, C. S. Wu, O. D. Trapp, J. Warkentin and R. T. Keys, *J. Am. Chem. Soc.*, **82**, 5394 (1960).
243. G. J. Janz in *1,4-Cycloaddition Reactions* (Ed. J. Hamer), Academic Press, New York, 1967, pp. 97–125.
244. T. L. Cairns, A. W. Larchar and B. C. McKusick, *J. Am. Chem. Soc.*, **74**, 5633 (1952); W. L. Reilly and H. C. Brown, *J. Org. Chem.*, **22**, 698 (1957); I. S. Bengelsdorf, *J. Am. Chem. Soc.*, **80**, 1442 (1958).

CHAPTER 7

Reduction of the cyano group

MORDECAI RABINOVITZ

The Hebrew University of Jerusalem, Jerusalem, Israel

I. INTRODUCTION 307
II. REDUCTION TO ALDEHYDES 308
A. Introduction 308
B. The Stephen Synthesis of Aldehydes 308
C. Selective Hydrogenation 309
1. Catalytic hydrogenation 309
2. Chemical reduction 313
D. Hydride Reduction 314
1. Lithium aluminium hydride 314
2. Indirect aldehyde synthesis. 315
3. Variation of the hydride reagent 315
III. REDUCTION TO ALDIMINES 319
IV. REDUCTION TO AMINES 319
A. Introduction. 319
B. Catalytic Hydrogenation 320
C. Chemical Reduction 325
1. Metal–hydrogen donor reductions 325
2. Hydride reduction 326
V. REDUCTION TO HYDROCARBONS 331
A. Reduction of Nitrile to Methyl Group. 331
B. Reductive Decyanation and Ring Closure 333
VI. MISCELLANEOUS 336
VII. REFERENCES 337

I. INTRODUCTION

The cyano group of the nitriles is unsaturated. As a consequence it enters into reaction with a large variety of reagents, resulting in partial or complete loss of unsaturation. Important among these reactions is reduction, in which hydrogen is added to the triple bond, and similarly to other carboxylic acid derivatives this group yields products of several degrees of reduction (equations 1–3):

$$RC{\equiv}N \xrightarrow{H_2} RCH{=}NH \longrightarrow RCHO \tag{1}$$

$$RC{\equiv}N \xrightarrow{2\,H_2} RCH_2NH_2 + RCH_2NHCH_2R \tag{2}$$

$$RC{\equiv}N \xrightarrow{3\,H_2} RCH_3 \tag{3}$$

The most important reduction products are aldehydes and amines[1–4] which can be utilized in many organic reactions.

II. REDUCTION TO ALDEHYDES

A. Introduction

The conversion of nitriles to aldehydes (equation 1) is a synthetic route of considerable importance. An intermediate in this conversion is the aldimine which is further hydrolysed to the aldehyde (equation 4):

$$RC{\equiv}N + H_2 \longrightarrow RCH{=}NH \xrightarrow{H_2O} RCHO + NH_3 \tag{4}$$

The availability of the nitriles, coupled with their reduction to aldimines, led to seeking of methods in which aldimines are used as precursors to aldehydes via their hydrolysis. The reduction of nitriles to aldehydes has been reviewed in the literature[1–4]. Two principal methods were used: (*a*) the well-known Stephen method[5] and (*b*) selective hydrogenation[3].

B. The Stephen Synthesis of Aldehydes

In the Stephen method, dry hydrogen chloride in ether reacts with a nitrile to form the corresponding imino chloride salt, and the latter intermediate is then reduced. The reduction is performed by anhydrous stannous chloride, and on hydrolysis the resulting aldiminium chloride yields an aldehyde (equation 5):

$$RCN \xrightarrow{HCl} \underset{\substack{|\\ Cl}}{RC}{=}NH{\cdot}HCl \xrightarrow{SnCl_2} (\underset{\substack{|\\ H}}{RC}{=}NH)_2SnCl_4 \xrightarrow{H_2O} \underset{\substack{|\\ H}}{RC}{=}O \tag{5}$$

A similar reduction and a similar intermediate are involved in the Sonn–Müller synthesis of aldehydes from amides[6] (equation 6):

$$\underset{\substack{\|\\ O}}{ArCNHPh} \xrightarrow{PCl_5} \underset{\substack{|\\ Cl}}{ArC}{=}NPh{\cdot}HCl \xrightarrow[HCl]{SnCl_2} \underset{\substack{|\\ H}}{ArC}{=}NPh \xrightarrow{H_2O} \underset{\substack{|\\ H}}{ArC}{=}O \tag{6}$$

The Stephen method is satisfactory for the preparation of aromatic aldehydes where the yields are generally good[7,8]. Some modifications of this method have been proposed[9,10]. The Stephen method has also been applied to the reduction of some heterocyclic nitriles[3] and it can also be used for the preparation of a few aliphatic aldehydes. However, the yields of aliphatic aldehydes are generally low[2,4]. Recently, this reaction was investigated for determination of optimum reaction conditions[11]. Generally, electron-donating substituents increased the nucleophilic properties of the nitrile group and facilitated aldimine formation, while electron-attracting ones enhanced the rate of the first step. In contrast to straight-chain aliphatic nitriles, which give amides, nitriles with more than seven carbon atoms branched at the vicinity of the nitrile group are readily converted, albeit in low yields, to the corresponding aldehydes[11].

A modified form of the Stephen reagent has been used in comparing the reduction rates of a number of unsubstituted aliphatic and substituted aromatic nitriles[12]. The usually prepared reagent is not suitable for kinetic studies since it consists of two immiscible layers. A homogeneous reagent was prepared by adding acetyl chloride to stannous chloride dihydrate in ether. The reaction is neither first nor second order, and the substituent effects are similar to those described above[11].

C. Selective Hydrogenation

1. Catalytic hydrogenation

Excellent yields of aromatic aldehydes were obtained, according to Backeberg and Staskun[13], from the reduction of nitriles with Raney nickel in the presence of sodium hypophosphite[13,14] in aqueous acetic acid or aqueous acetic acid–pyridine (equation 7):

$$C_6H_5{-}C{\equiv}N \xrightarrow[NaH_2PO_2]{Raney\ Ni/H_2} C_6H_5{-}CHO + NaH_2PO_3 \qquad (7)$$

This was found as a result of the observation that Raney nickel liberates hydrogen from water in the presence of sodium hypophosphite (which becomes oxidized to the phosphite), while at the same time retaining its catalytic activity. This method does not give good yields with hindered nitriles such as *o*-toluonitrile and α-naphthonitrile. The latter compounds also give low yields in the

Stephen synthesis and these were ascribed to steric and solubility factors[12]. However, by stirring the hindered nitriles with moist Raney nickel in formic acid at 75–80°, the aldehydes were formed in 60–75 % yield[15]. The formic acid serves as a hydrogen source[16], and the Raney nickel catalyses both its decomposition and the reduction of the nitrile. Thus the formic acid method supplements the sodium hypophosphite reduction. The two methods are compared in Table 1.

TABLE 1. Reduction of nitriles RCN with Raney nickel.

R	Yield of aldehyde (%) by the	
	Sodium hypophosphite method[13]	Formic acid method[15]
Ph	85	72
o-MeC_6H_4	10[a]	65–75
p-MeC_6H_4	80	—
o-ClC_6H_4	40[a]	70–83
p-ClC_6H_4	85	82
p-$MeOC_6H_4$	80	75–85
α-$C_{10}H_7$	15[a]	60–75
β-$C_{10}H_7$	80[a]	55
n-C_7H_{15}	20[a]	35–40
PhCH═CH	20[a]	64

[a] Isolated as DNP derivative.

The hydrogenation according to Backeberg and Staskun[13,15] has been used for the preparation of sensitive unsaturated aldehydes such as vitamin A aldehyde from the corresponding nitrile[17]. It was also found that excellent yields in the aromatic series were obtained if equal weights of the nitrile and Raney nickel are refluxed in 75 % aqueous formic acid[18].

The preparation of aldehydes such as pyridine aldehydes, benzaldehyde and *o*-substituted benzaldehydes via a catalytic hydrogenation of nitriles with reduced nickel catalyst in media of controlled pH (2–5), at which the intermediate imines undergo fast hydrolysis, has also been reported[19].

Since the main problem in reduction of nitriles to aldehydes is to prevent further reduction of the aldimine intermediate, the use of trapping reagents for the intermediate may be advantageous.

The partial hydrogenation over Raney nickel (at 1–70 atmospheres) in the presence of carbonyl reagents such as hydrazine, phenylhydrazine, semicarbazide and *N,N'*-diphenylethylenediamine leads to an equilibrium (e.g. equation 8):

$$RCH{=}NH + H_2NNHPh \rightleftharpoons RCH{=}NNHPh + NH_3 \qquad (8)$$

The aldehyde derivatives formed in this reaction, i.e. hydrazone, azine, phenylhydrazone, semicarbazone and diphenyltetrahydroimidazole, do not undergo further hydrogenation under these reaction conditions. Subsequent liberation of the aldehydes from their derivatives may present some difficulties, but up to 90% yields are attainable. This method was used as long ago as 1923 by Rupe and coworkers[20], but not much attention was paid to it at that time.

Pietra and Trinchera[21] prepared aldehydes by hydrogenation of nitriles in neat or ethanolic hydrazine hydrate in the presence of a hydrogenation catalyst. The products—the corresponding hydrazone and azine—are then hydrolysed to the aldehydes (equation 9). This

$$PhC{\equiv}N \xrightarrow[\text{Raney Ni}/H_2]{N_2H_4\cdot H_2O} PhCH{=}N{-}N{=}CHPh + PhCH{=}NNH_2 \qquad (9)$$

reaction was successful for aromatic aldehydes, even in those cases where the Stephen method gave poor results. For example, *o*-tolualdehyde and α-naphthaldehyde are obtained in 78% and 69% yields, respectively, compared to 7–9% in the Stephen method.

Excess of hydrazine should be used to shift the equilibrium to the desired direction, but one should consider the fact that hydrazine decomposes under the influence of Raney nickel, and that the rate of decomposition is temperature dependent. This procedure was applied on a small scale only and was not found useful for aliphatic aldehydes[22]. No discussion of the reaction mechanism was given.

Other carbonyl derivatives that did not undergo further hydrogenation and thus enabled the partial reduction, are the semicarbazone and diphenyltetrahydroimidazole[23–25]. The hydrogenation of the nitriles was carried out either at atmospheric or at high pressure, with Raney nickel catalyst. Special attention was paid to the preparation of phenylacetaldehydes from substituted benzyl cyanides, since these aldehydes are starting materials for the preparation of alkaloids[25]. Thus, diphenylethylenediamine or semicarbazide were used to trap 3,5-dimethoxyphenylacetaldehyde as its tetrahydroimidazole or semicarbazone, as an initial step in the synthesis of yohimbane derivatives (equation 10):

$$\text{3,5-}(CH_3O)_2C_6H_3CH_2CN \xrightarrow[(PhNHCH_2)_2]{H_2/Ni} \text{3,5-}(CH_3O)_2C_6H_3CH_2CH\langle N(Ph)CH_2CH_2N(Ph)\rangle \xrightarrow[20^\circ]{pH\ 3} \text{3,5-}(CH_3O)_2C_6H_3CH_2CHO$$

$$\text{3,5-}(CH_3O)_2C_6H_3CH_2CHO + \text{tryptamine} \longrightarrow \text{1-(3,5-dimethoxybenzyl)tetrahydro-}\beta\text{-carboline}$$

While 60 % yields of the aldehyde were obtained with both derivatives, the tetrahydroimidazole derivative is preferable, since it was hydrolysed very easily in 20 % hydrochloric acid. Raney cobalt gave poor results under the same reaction conditions. This method is useful for the preparation of succinic and adipic dialdehydes (equation 11) as well as for indole and imidazole aldehydes[26,27].

$$NC(CH_2)_nCN \xrightarrow[H_2NNHCONH_2]{\text{Raney Ni}/H_2} H_2NCONHN{=}CH(CH_2)_nCH{=}NNHCONH_2 \xrightarrow{H_2O} OHC(CH_2)_nCHO \quad (n = 2, 3) \tag{11}$$

Rupe[20] hydrogenated benzonitrile and some other aromatic nitriles in alcohol in the presence of phenylhydrazine and reduced nickel catalyst and obtained mixtures of products[20,28]. For example, 30 % benzaldehyde (as its phenylhydrazone) was isolated from benzonitrile while the remaining products consisted of amines. The reaction under these conditions failed to yield phenylacetaldehyde from benzyl cyanide. This author suggested that the aldimine formed was in equilibrium with the phenylhydrazone.

The hydrogenation in the presence of phenylhydrazine and Raney nickel instead of the reduced nickel catalyst used before[20] was studied in detail[29,30]. The method and its scope were described for aromatic, aliphatic and heterocyclic nitriles[31]. When optimum reaction conditions were applied (1–70 atmospheres pressure, Raney nickel catalyst and ethanol as solvent) the phenylhsydrazone of the corresponding aldehydes were obtained in good yields, together with further hydrogenation by-products—the primary and secondary amines. The mechanism depicted in equations (12)–(14) was suggested[31], although equation (13) should be formulated as an equilibrium process.

$$RC{\equiv}N + H_2 \longrightarrow RCH{=}NH \quad (12)$$

$$RCH{=}NH + H_2NNHPh \longrightarrow RCH{=}NNHPh + NH_3 \quad (13)$$

$$RCH{=}NH + H_2 \longrightarrow RCH_2NH_2 \quad (14)$$

When nitriles are catalytically hydrogenated, the uptake of the first mole of hydrogen is faster than that of the second one[24] and Gaiffe[31] therefore assumed that the yield of the phenylhydrazone would depend on the amount of the phenylhydrazine present. Mixtures of nitriles and phenylhydrazine in variable ratios were hydrogenated under standard conditions and the yield of the aldehyde could be raised from 50% for a nitrile:phenylhydrazine ratio of 1:2 to 90% for a 1:4 ratio. This method was applicable for the preparation of many aldehydes including phenylacetaldehyde, α-naphthaldehyde and pyridine aldehydes. A survey on the addition of other carbonyl reagents, e.g. *p*-toluenesulphohydrazide, hydroxylamine and bisulphite showed that these reagents have no synthetic value[23–25].

2. Chemical reduction

The chemical reduction of nitriles was found useful for the preparation of aldehydes[31]. Reagents such as zinc, iron and chromium powders and acids or aluminium powder and base were used. A competitive reaction is the hydrolysis of the nitriles under these reaction conditions to the corresponding amides and/or carboxylic acid salts (equation 15):

$$RC{\equiv}N \xrightarrow[\text{or } Al/OH^-]{Zn/H^+} [RCH{=}NH] \longrightarrow RCHO$$

$$RC{\equiv}N \xrightarrow[\text{or } OH^-]{H^+} RCONH_2 + RCOOH \quad (15)$$

Here again the aldimine which is presumably formed is hydrolysed to the aldehyde. Since specific conditions, e.g. controlled pH, must be found for each case, this method is only of limited use.

It was predicted[32] and shown[33] that Grignard reagents would be capable of reducing as well as of adding to nitriles, and that the amount of the reduction product increased at the expense of the addition product with the increase of the branching of the Grignard reagent. The yields, however, were poor[32]. In an attempt to improve the reductive properties of this reagent, the reduction of nitriles was carried out with the combination Grignard reagent–ferric chloride[33]. For example, the reduction of trimethylacetonitrile with *t*-butylmagnesium chloride–ferric chloride mixture gave 20% yield of trimethylacetaldehyde (equation 16)[34]. Treatment of 3-ethyl-3-

$$CH_3C(CH_3)_2CN \xrightarrow[\text{ether, 28 h, 20\%}]{t\text{-BuMgCl, FeCl}_3} CH_3C(CH_3)_2CHO \qquad (16)^{34}$$

methylvaleronitrile with this combination gave the corresponding aldehyde in 31 and 51% yields with *t*-butylmagnesium chloride and isobutylmagnesium iodide, respectively. Capronitrile could not be reduced to the corresponding aldehyde by this method[34].

D. Hydride Reduction

1. Lithium aluminium hydride

Hydride reductions would be very useful for the synthesis of aldehydes from nitriles if selective reactions that stop at the aldimine stage were available. Lithium aluminium hydride usually reduces nitriles to primary amines[35], probably due to the ease of the reduction of the intermediate aldimines. This was observed also in the previous aldehyde syntheses.

The possibility to stop the reduction at the aldimine stage has been realized by: (*a*) variation of the reaction conditions; (*b*) the indirect synthesis and (*c*) the variation of the reducing hydride reagents.

It was claimed that the controlled reduction of aliphatic nitriles by lithium aluminium hydride at low temperatures provides a convenient synthetic route to aldehydes[36]. Unfortunately, the experimental details have not appeared and difficulty has been experienced with this synthesis[37]. However, the successful partial reductions of specific nitriles with lithium aluminium hydride under controlled conditions had been reported[38–41], as exemplified in equations (17)–(19).

$$\text{c-C}_3\text{H}_5\text{CN} \xrightarrow{\text{LiAlH}_4,\,-70°} \text{c-C}_3\text{H}_5\text{CHO}\ \ 48\% \qquad (17)^{38}$$

$$\text{CF}_3\text{CN} \xrightarrow{\text{LiAlH}_4,\,-70°} \text{CF}_3\text{CHO}\ \ 46\% \qquad (18)^{39}$$

$$\text{Ph}_2\text{C(CN)CH}_2\text{CH(Me)NMe}_2 \xrightarrow{\text{LiAlH}_4} \text{Ph}_2\text{C(CHO)CH}_2\text{CH(Me)NMe}_2 \qquad (19)^{40,41}$$

2. Indirect aldehyde synthesis

An indirect aldehyde synthesis via hydride reduction was developed by Claus and Morgenthau[42,43]. In their method, an ortho ester is formed by the addition of ethanolic hydrogen chloride to the nitrile, and this in turn is reduced by lithium aluminium hydride to the corresponding acetal. The latter can be hydrolysed to the aldehyde (equation 20):

$$\text{RC}\equiv\text{N} \xrightarrow[\text{EtOH}]{\text{HCl}} \text{RC(OEt)}{=}\text{NH·HCl} \xrightarrow{\text{EtOH}} \text{RC(OEt)}_3 \xrightarrow{\text{LiAlH}_4} \text{RCH(OEt)}_2 \xrightarrow{\text{H}^+} \text{RCHO} \qquad (20)$$

In another indirect synthesis, nitriles are reduced to aldehydes in high yields by pyrolysis of the β-hydroxy ester formed in the reaction of ethyl α-bromoisobutyrate and zinc, followed by borohydride reduction[44].

3. Variation of the hydride reagent

Several variations of the hydride reagent are known. Hesse and Schrödel used sodium triethoxyaluminohydride, prepared via the reaction of sodium hydride with aluminium ethoxide, to successfully reduce aromatic nitriles to the aldehydes, but this method was

TABLE 2. Formation of aromatic aldehydes in the sodium triethoxyaluminohydride reduction of nitriles.

Nitrile	Aldehyde	Yield (%)
Benzonitrile	Benzaldehyde	92
o-Toluonitrile	*o*-Tolualdehyde	88
α-Naphthonitrile	α-Naphthaldehyde	87
β-Naphthonitrile	β-Naphthaldehyde	76
9-Phenanthrene nitrile	9-Phenanthraldehyde	78
Terephthalonitrile	Terephthaldehyde	70
Pyridine-3-nitrile	Pyridine-3-aldehyde	83

unsatisfactory for aliphatic nitriles[45,46]. The reduction was carried out in tetrahydrofuran and the yields of hindered or unhindered aromatic aldehydes were similar (Table 2). The reaction mechanism can be formulated as follows (equation 21):

$$\text{RCN} + [\text{Al(OEt)}_3\text{H}]^-\text{Na}^+ \longrightarrow [\text{RCH}{=}\text{N}\cdots\text{Al(OEt)}_3]^-\text{Na}^+ \xrightarrow{\text{H}_2\text{O}} \text{RCHO} \qquad (21)$$

Hindered nitrile groups, e.g. in angular positions in steroids[47], are reduced by lithium aluminium hydride either to the aldehyde or to the corresponding amine, depending on the stereochemistry of the nitrile group, while lithium triethoxyaluminium hydride reduced the nitrile to the aldehyde regardless of the stereochemistry of the starting material[47] (equations 22 and 23):

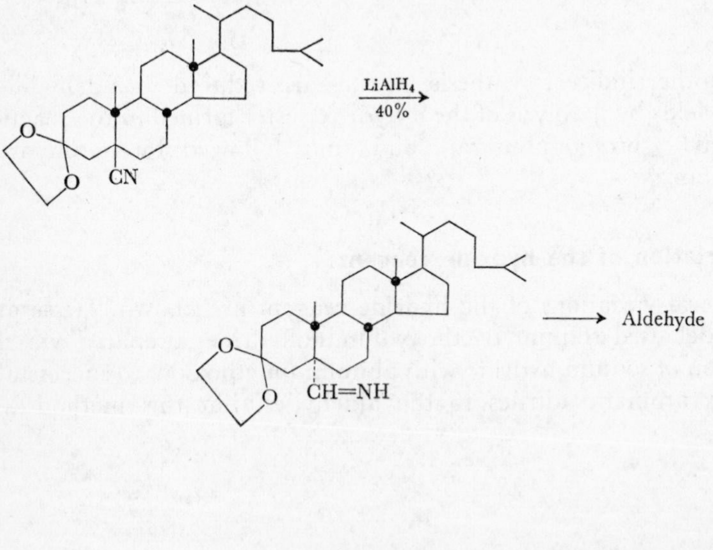

CN; $\xrightarrow[40\%]{\text{LiAlH}_4}$ CH=NH $\longrightarrow$ Aldehyde (22)

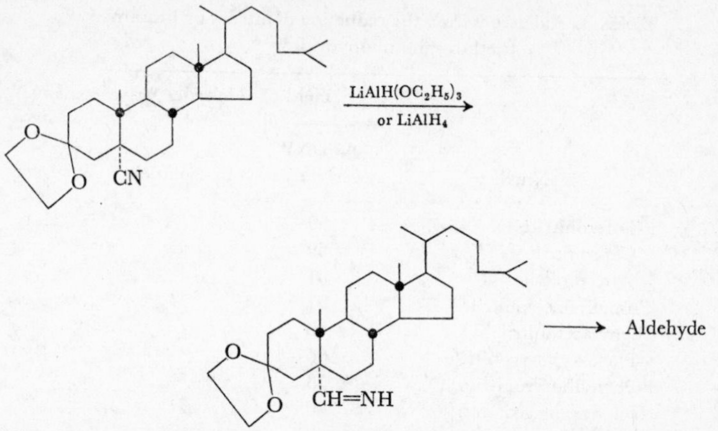

$\xrightarrow[\text{or } LiAlH_4]{LiAlH(OC_2H_5)_3}$

CN

CH=NH

$\longrightarrow$ Aldehyde (23)

Zakharkin and Khorlina used diisobutoxyaluminium hydride to reduce n-propyl cyanide to butyraldehyde in 93% yield[48].

An efficient reduction procedure which is applicable to both aliphatic and aromatic nitriles was developed by Brown and coworkers[49,50]. The hydride reagent was lithium triethoxyalumino-hydride, prepared by the reaction of 3 moles of ethanol or 1·5 moles of ethyl acetate with 1 mole of lithium aluminium hydride[51]. Aliphatic and aromatic nitriles can be converted to aldehydes by this procedure in 70 and 80–90% yields, respectively (Table 3). A systematic exploration of this reaction with the objective of developing a general synthetic procedure was carried out[50]. The relative utility of a number of alkoxy-substituted lithium aluminium hydrides as selective reducing agents for nitriles was investigated using butyronitrile as a model. Best yields were obtained with lithium triethoxy-, tris(n-propoxy)- and tris(n-butoxy)alumino-hydrides.

The first step of the reaction can be formulated as in equation 24:

$$RC{\equiv}N + HAl(OR)_3^- \longrightarrow R\overset{H}{\underset{|}{C}}{=}NAl(OR)_3 \qquad (24)$$

When aromatic nitriles are reacted with modified lithium aluminium hydride, the steric and electronic characteristics of the aryl group appear to stabilize the initial imine derivative, as judged by the isolation of excellent yields of aldehydes. Branching in the α-position of aliphatic nitriles evidently favours the first addition over the second (to yield an amine), resulting in an improved aldehyde yield. Presumably the rate of nucleophilic attack at the imine carbon is reduced as a

TABLE 3. Aldehydes from the reduction of nitriles by lithium triethoxyaluminohydride[49,50].

Nitrile	Yield of aldehyde (%) As DNP derivative	Yield of aldehyde (%) By isolation
n-Butyronitrile	69	
n-Capronitrile	69	55
Isobutyronitrile	81	
Trimethylacetonitrile	81	74
Phenylacetonitrile	0	
γ-Phenoxybutyronitrile	66	
γ-Phenylbutyronitrile	73	
Cyclohexanecarbonitrile	76	71
Cyclopropanecarbonitrile	69	
Cinnamonitrile	61	
Benzonitrile	96	76
o-Toluonitrile	87	
o-Chlorobenzonitrile	87	
p-Chlorobenzonitrile	92	84
p-Anisonitrile	81	
α-Naphthonitrile	80	
Nicotinonitrile	58	
Adiponitrile	60	

result of the increased inductive influence of the branched alkyl groups and their larger steric requirements. Similarly, it appears that an increase in the steric requirements of the substituents attached to aluminium also favours a single addition. This interpretation, however, does not account for the report that sodium triethoxyaluminohydride does not reduce aliphatic nitriles[45,46]. This indicates that the lithium ion is essential for the selective reduction. It was therefore suggested[50] that the lithium ion becomes coordinated to the nitrogen atom of the imine intermediate (equation 25) and assists in protecting the double bond from a second addition. Sodium

$$\begin{array}{l} \text{H}\quad \text{Li}^{+}\ \text{Solv} \\ \ | \\ \text{RC}{=}\ddot{\text{N}}\text{Al}^{-}(\text{OR})_3 \end{array} \tag{25}$$

will be much less prone to form a stable coordination linkage of this kind, and therefore sodium triethoxyaluminohydride cannot be considered a general reducing agent of nitriles to aldehydes, while lithium triethoxyaluminohydride is a reagent of general use for this reduction.

III. REDUCTION TO ALDIMINES

Although most of the reductions described in this chapter proceed via the aldimine stage, very few[47] intermediate aldimines are sufficiently unreactive in the usual reaction conditions to escape being either reduced to the amine or hydrolysed to the aldehyde (equation 26).

$$RC{\equiv}N + H_2 \longrightarrow RCH{=}NH \begin{cases} \xrightarrow{H_2O} RCHO \\ \xrightarrow{H_2} RCH_2NH_2 \end{cases} \tag{26}$$

Only recently a general method for the preparation of aromatic aldimines of the type ArCH=NH by reduction of nitriles was described[52]. In this modification of the Stephen aldehyde synthesis, the aldimine hydrochloride–stannic chloride complex is decomposed under anhydrous conditions. Several aldimines were prepared and characterized by hydrolysis to the corresponding aldehydes. The yields varied from poor to excellent (Table 4) and it was argued that

TABLE 4. Aldimines ArCH=NH from reduction of nitriles[52].

Ar	Yield (%)	Ar	Yield (%)
Ph	94	o-ClC_6H_4	22
$PhCH_2$	—	p-ClC_6H_4	72
$(Ph)_2CH$	—	p-$HO_2CC_6H_4$	27
o-$CH_3C_6H_4$	11	o-$Cl_3CC_6H_4$	44
m-$CH_3C_6H_4$	—	α-$C_{10}H_7$	49
p-$CH_3C_6H_4$	26	α-$C_{10}H_7CH_2$	23

the yields are more influenced by steric effects than by electronic ones. However, the electronic character of the substituents has an influence on the competing polymerization reaction of the aldimines, affecting the yields in this indirect way.

IV. REDUCTION TO AMINES

A. Introduction

Amines are the usual final reduction products of nitriles (equations 27–29). However, the reduction to primary amines is complicated by the equilibria (equation 28) which lead to secondary amines as by-products[20,28], by the von Brown mechanism (equations 28, 29)[53]. This difficulty may be overcome by performing the hydrogenation in the presence of noble metal catalysts in an acidic medium or in acetic anhydride. The primary amine formed (equation 27) is removed from these equilibria as its salt or substituted acetamide. For reductions over Raney nickel, where an acid cannot be used, an excess of ammonia is added to the reaction mixture, displacing the equilibrium (equation 28) to the left and thus avoiding secondary amine formation. The literature concerning reduction of nitriles to

$$RC{\equiv}N \xrightarrow{H_2} RCH{=}NH \xrightarrow{H_2} RCH_2NH_2 \tag{27}$$

$$RCH_2NH_2 + RCH{=}NH \rightleftharpoons RCH_2NHCHR(NH_2) \rightleftharpoons RCH_2N{=}CHR + NH_3 \tag{28}$$

$$RCH_2N{=}CHR \xrightarrow{H_2} RCH_2NHCH_2R \tag{29}$$

amines up to 1947 has been reviewed[1]. The synthetic aspects of the reaction have been summarized in several reviews[54,55].

B. Catalytic Hydrogenation

The catalytic hydrogenation of aliphatic and aromatic nitriles has been known for a long time to yield primary and secondary amines. When the reduction is carried out in a neutral solvent such as alcohol with platinum or palladium catalysts, useful yields of secondary amines may be obtained[56–59]. Rosenmund[60] has suggested carrying out the hydrogenation in acetic anhydride solvent, in order to trap the primary amines as their acetamide derivatives. Acetic anhydride was recommended as solvent for the hydrogenation with platinum oxide catalysts[61,62]. The yield of primary amine is also increased at the expense of side-products by using as solvent alcohol containing an equivalent of hydrochloric acid[63,64]. Raney nickel and Raney cobalt have been found very effective for reductions in alcoholic ammonia as solvent (Whitmore procedure)[65–68] or with an amine like methylamine or dimethylamine[69] under high pressure.

The catalytic hydrogenation of nitriles under mild conditions was reported by Freifelder[70]. The reaction is carried out at low pressure (2–3 atmospheres) with rhodium catalyst in ethanolic ammonia. For example, the hydrogenation of 3-indoleacetonitrile, using 10–20 % ratio of 5 % rhodium–alumina in 10 % ethanolic ammonia, was complete in two hours, giving a good yield of tryptamine (equation 30):

$$\text{indol-3-yl-}CH_2CN \xrightarrow[NH_3]{Rh/Al_2O_3} \text{indol-3-yl-}CH_2CH_2NH_2 \qquad (30)$$

The same result was obtained when strong base was used instead of ammonia. However, reduction in the absence of ammonia gave predominantly the secondary bis(3-indolyl)amine, contaminated with some tryptamine. The above procedure is advantageous since it requires mild conditions and the yields are good (Table 5). In

TABLE 5. Hydrogenation products of basic nitriles with rhodium catalyst[70].

$R(CH_2)_nNH_2$		$ROCH_2CH_2CH_2NH_2$	
R	Yield (%)	R	Yield (%)
$(CH_3)_2N$	69·0	CH_3	72·0
1-Piperidino	78·0	n-C_6H_{13}	90·5
$PhCH_2N(CH_3)$	67·2	Cyclopentyl	69·0
4-Methyl-1-piperazino	67·8	Hexahydrobenzyl	70·0
		Cycloheptyl	75·8

contrast to reduction with palladium catalyst, which is complicated by hydrogenolysis, the rhodium catalyst did not appear to cause any hydrogenolysis in the preparation of benzyl amines. At least five molar equivalents of ammonia are required to keep the formation of secondary amines at a low level.

Hydrogenation in acetic anhydride, in the presence of Raney nickel catalyst and a basic cocatalyst, is very effective for the reduction of a variety of nitriles to primary amines[71]. This seems to be the method of choice for catalytic hydrogenation to primary amines, since only moderate temperatures and low pressures are required, and high yields are obtained (see Table 6). With Raney nickel and sodium acetate as cocatalyst, hydrogenation of most nitriles is complete in 40–60 minutes at 50° and 3–4 atmospheres hydrogen pressure. Raney nickel-chromium appears to be as effective as Raney nickel, but Raney cobalt is somewhat less effective (Table 6). Sodium acetate is a very mild cocatalyst suitable for most purposes, but stronger bases such as sodium hydroxide aid in a more rapid reduction. Caution should be exercised when strong bases are used, since the very rapid reduction becomes vigorously exothermic.

TABLE 6. Hydrogenation of nitriles RCN in acetic anhydride over Raney catalysts[71].

R	Raney catalyst	Cocatalyst	Yield (%)
Ph	Ni	None	91
$PhCH_2$	Ni	NaOAc	97
$CH_3(CH_2)_{11}$	Ni	NaOAc	100
$CH_3(CH_2)_{11}$	Ni–Cr	NaOAc	89
$CH_3(CH_2)_{11}$	Co	NaOAc	60
$CH_2{=}CH$	Ni	NaOAc	92
$(CH_3)_2C(OH)$	Ni	NaOAc	40
$—(CH_2)_4—$	Ni	NaOAc	100
$—(CH_2)_4—$	Ni	NaOH	80
$—(CH_2)_4—$	Ni–Cr	NaOAc	77
$—(CH_2)_4—$	Co	NaOAc	25

Catalyst life is affected markedly by the nature of the cocatalyst. With sodium acetate, the activity of the Raney catalyst diminishes appreciably with each use and is too low to be of practical value after three or four cycles, whereas with strong base, recovered catalyst is as active as fresh and can be reused repeatedly. With some nitriles such as benzonitrile, no base is necessary, but in general its presence gives better yields and purer products. This general method was developed in order to overcome a crucial step in an elegant three-step synthesis of D,L-lysine from cyclohexanone[72–75] (equation 31). This step is the catalytic hydrogenation of both oxime and nitrile groups.

$$\text{cyclohexanone} \xrightarrow[HCl]{CH_3ONO} \text{HON=}\text{(2-oxocyclohexane-1,3-diylidene)}\text{=NOH} \xrightarrow[H_2SO_4]{Ac_2O} \text{AcON=}\text{(2-oxocyclohexane-1,3-diylidene)}\text{=NOAc} \xrightarrow[C_2H_5OH]{NaOC_2H_5}$$

$$NC(CH_2)_3C(=NOH)COOEt \xrightarrow{H_2} \text{D,L-}NH_2CH_2(CH_2)_3CH(NH_2\cdot HCl)COOEt \qquad (31)$$

The search for a suitable reducing system was carried out extensively in an effort to find a catalyst–solvent system which would permit hydrogenation of ethyl 5-cyano-2-oximinovalerate to lysine without the side reactions which apparently were responsible for the low yields obtained in earlier work. Although the combination of Raney nickel and acetic anhydride was claimed to be unsatisfactory in Adkin's classical work[56], it was shown recently that use of this system, especially in the presence of sodium acetate or other weak bases, enabled hydrogenation of oximes to acetylated amines in good yields[76]. The use of Raney nickel catalyst and acid anhydride solvent, which was not reported previously for the hydrogenation of nitriles, proved to be more effective in this reduction than the apparently very similar platinum–acetic anhydride system. Thus at only 3 atmospheres and 50° in the presence of sodium acetate, the hydrogenation was complete in about two hours, and lysine ester monohydrochloride was isolated in 92 % yield, while with a platinum catalyst after eight hours the yield was only 57 %. The use of strong bases such as potassium hydroxide or benzyltrimethylammonium hydroxide gave good yields even after fifteen minutes, and had the additional advantage that the catalyst could be reused many times. The platinum catalyst was completely inactive after a single experiment.

Attempts to synthesize 1,2-primary diamines by hydrogenation of α-aminonitriles (equation 32) were unsuccessful[77,78]. The difficulty in obtaining good yields is due to the small difference between the rates of hydrogenation and hydrogenolysis[79,80].

$$R^1R^2C(NH_2)CN \xrightarrow{H_2} R^1R^2C(NH_2)CH_2NH_2 \qquad (32)$$

Low-pressure reduction, in the presence of platinum oxide, of the aminonitrile base or the hydrochloride in alcoholic hydrogen chloride results, however, in good yields (Table 7) with minimum hydrogenolysis of the nitrile group[81]. Moreover, the use of the acid medium

TABLE 7. 1,2-Diamines by reduction of α-aminonitriles $R^1R^2C(NH_2)CN$[81].

R^1	R^2	Yield (%)
Et	H	76
$PhCH_2$	H	64
Me	Me	82
Et	Me	68
n-Pr	Me	75
n-Bu	Me	10
$PhCH_2$	Me	60
Et	Et	61
n-Bu	Et	65

eliminates the acylation and the subsequent required hydrolysis is avoided[79]. This method is not applicable to *N*-substituted-α-aminonitriles since extensive hydrogenolysis takes place.

Hydrogenation of nitriles in the presence of primary amines is a good method for the preparation of secondary amines (equation 33)[82]. This is an argument in favour of the von Brown mechanism[53] which explains the origin of secondary amines in the hydrogenation of nitriles (equations 28–29). This also seems to rule out

$$R^1CN + H_2 \longrightarrow R^1CH{=}NH \xrightarrow{R^2CH_2NH_2} R^1CH_2NHCHR^2(NH_2) \xrightarrow{H_2} R^1CH_2NHCH_2R^2 + NH_3 \quad (33)$$

the mechanism of Mignonac[83] which supports the hydrogenolysis of the Schiff base formed (equation 34).

$$R\text{-}C_6H_4\text{-}CH{=}NH \longrightarrow R\text{-}C_6H_4\text{-}CH(\text{-}N{=}CH\text{-}C_6H_4\text{-}R)_2 + NH_3 \xrightarrow{H_2} R\text{-}C_6H_4\text{-}CH_2N{=}CH\text{-}C_6H_4\text{-}R + R\text{-}C_6H_4\text{-}CH_2NH_2 \quad (34)$$

The reductions were carried out with a Raney nickel catalyst in ratio of 10–25 % to the nitrile. The results are summarized in Table 8.

The aliphatic secondary products formed are compatible with the von Brown mechanism, which would predict the formation of a

TABLE 8. Catalytic hydrogenation of nitriles R^1CN in presence of amines R^2NH_2[82].

R^1	R^2	Yield (%)
p-$CH_3OC_6H_4$	*p*-$OCH_3C_6H_4$	90
o-$CH_3OC_6H_4$	*o*-$OCH_3C_6H_4$	93
p-$CH_3C_6H_4$	*p*-$CH_3C_6H_4$	95
Ph	Ph	100

mixed secondary amine from the added primary amine. The authors[82] argue that the small effect that increasing primary amine concentration had on the preparation of the secondary amine formed indicates that the alkylation reaction occurs on the catalyst in competition with the hydrogenation reaction, which is favoured, and that the catalyst surface becomes saturated with respect to primary amine at an approximately 1:1 mole ratio of amine to nitrile. Aromatic nitriles give higher yields of the secondary product compared to those of aliphatic nitriles.

One or both nitrile groups of dinitriles can be reduced according to the reaction conditions. For example, adiponitrile was reduced to the corresponding mono- and diamine[84a] by using nickel–aluminium oxide, Raney nickel, nickel boride and other nickel catalysts.

Isophthalic dinitrile gave 79 % yield of the corresponding diamine with Raney cobalt in liquid ammonia under 120 atmospheres pressure[84b], but when the reaction was stopped after uptake of two moles of hydrogen, 80 % yield of 4-cyanobenzylamine was obtained[85].

Other hydrogenations with Raney nickel in the presence of ammonia or alcoholic ammonia were reported. *o*-Cyanoacetophenone and *o*-*t*-butylbenzonitrile are reduced in vapour phase over nickel copper–kieselguhr catalyst[86a]. Potassium hexacyanonickelate(I) $K_2[Ni(CN)_6]$ has been used recently to reduce benzonitrile to dibenzylamine[86b].

C. Chemical Reduction

I. Metal–hydrogen donor reductions

The reduction of nitriles with sodium and alcohol may lead in some cases to the formation of amines[87–89] but many difficulties arise[90] and this method cannot be considered as a general one. Similar metal–hydrogen donor combinations used in chemical reduction to amines were: sodium amalgam and water, chromous acetate in suspension in alcohol, tin and hydrochloric acid, zinc and hydrochloric acid and coppered magnesium[54]. None of these methods seem to have general application. It is interesting that 3-indoleacetonitrile and some of its derivatives are reduced with ethanolic hydrazine hydrate, in the presence of Raney alloy, to give only tryptamines[91]. Secondary amines were not detected in this reaction. With sodium and ethanol, nitriles are reduced to amines and hydrocarbons (see section V).

2. Hydride reduction

A general reduction method consists of using lithium aluminium hydride in ether[35,37]; it has been used for reduction of simple nitriles to amines as well as for the reduction of cyanohydrines to β-hydroxy primary amines. Typical results of this reaction (equation 35) taken from the very first hydride reduction study of nitriles[35] are summarized in Table 9. The absence of secondary amines in the

TABLE 9. Lithium aluminium hydride reduction of nitriles[35].

Nitrile	Amine	Yield (%)
Benzonitrile	Benzylamine	72
o-Toluonitrile	*o*-Xylylamine	88
Sebaconitrile	1,10-Decanediamine	40
Mandelonitrile	β-Hydroxy-β-phenylethylamine	48
Lauronitrile	Tridecylamine	90
Triphenylacetonitrile	—	—

product is a highly advantageous feature of this method.

$$2\,RC{\equiv}N + LiAlH_4 \xrightarrow{\text{ether}} (RCH_2N)_2LiAl \xrightarrow{H_2O} 2\,RCH_2NH_2 \quad (35)$$

Many nitriles are reduced satisfactorily with lithium aluminium hydride to primary amines and the reaction has been adequately reviewed[37,54]. The action of lithium aluminium hydride on nitriles was modified by Nystrom[92], who added to the reaction mixture a Lewis acid, aluminium chloride. When equimolar quantities of hydride and aluminium chloride are mixed, the first reaction which takes place is the formation of aluminium hydride and lithium chloride[93]. Next, this can be followed by a further reaction to form aluminium chlorohydride[94]. By using standard reducing solutions containing equimolar amounts of aluminium chloride and lithium aluminium hydride, nitriles containing active hydrogen are reduced completely without liberation of hydrogen during one hour. Reduction of other nitriles is also facilitated and the yields are improved by the use of this 'acidic' hydride. The results are compared in Table 10.

A comprehensive study of the reduction of nitriles in ether and tetrahydrofuran was carried out by using two different procedures[95]. In the direct addition the nitrile is added to the hydride solution, while in the reverse addition the hydride is added to the nitrile solution. Comparison of these two procedures was initiated, since it was previously reported[92,96] that the direct addition reductions of n-butyronitrile and of n-valeronitrile by lithium aluminium hydride in ether occurred with only limited hydrogen evolution, whereas much more gas was evolved when the identical reactions were performed in tetrahydrofuran. However, it was found[97] that in the reverse addition reductions of these nitriles in both ether and tetrahydrofuran, considerable amounts of hydrogen

TABLE 10. Reduction by $LiAlH_4$–$AlCl_3$ mixture[92].

Nitrile	Amine	$LiAlH_4$ Yield (%)	1:1 $LiAlH_4$–$AlCl_3$ Yield (%)
Diphenylacetonitrile	2,2-Diphenylethylamine	61	91
Phenylacetonitrile	2-Phenylethylamine	46	83
Sebaconitrile	1,10-Decanediamine	58	86
n-Valeronitrile	n-Pentylamine	63	75

were evolved. Both modes of addition were investigated with various types of nitriles[95], as well as with different molar ratios of hydride to nitrile. It can be seen (Table 11) that in the direct addition, high yields of primary amines are obtained when this ratio is one or higher, while little or no hydrogen is evolved. It was then concluded that at least one mole of hydride is necessary for optimum reduction. This is in line with the proposed sequence of reactions for the reduction of nitriles to primary amines (equations 36–39).

$$RCH_2C{\equiv}N + AlH_4^- \longrightarrow RCH_2CH{=}NAlH_3^- \quad (36)$$

$$RCH_2C{\equiv}N + RCH_2CH{=}NAlH_3^- \longrightarrow (RCH_2CH{=}N)_2AlH_2^- \quad (37)$$

$$(RCH_2CH{=}N)_2AlH_2^- + AlH_4^- \longrightarrow 2\ RCH_2CH_2NAlH_2^- \quad (38)$$

$$RCH_2CH_2NAlH_2^- + H_2O \longrightarrow RCH_2CH_2NH_2 \quad (39)$$

In the reverse addition, considerable amounts of hydrogen were evolved during the reduction of n-butyronitrile and n-capronitrile, and the yield of primary amine was much lower than in the analogous direct addition reductions. Significant amounts of gas were obtained only when the nitrile contained an α-hydrogen and the quantities of gas were dependent on the relative amounts of the hydride present. The other reduction products were mainly diamines. The following

TABLE 11. Lithium aluminium hydride reductions of nitriles RCN in ether and tetrahydrofuran[95].

R	Addition[a]	Molar ratio hydride: nitrile	Ether (at 34 °c) Mole % hydrogen evolved	Ether (at 34 °c) Primary amine, yield (%)	THF (at 30–35 °c) Mole % hydrogen evolved	THF (at 30–35 °c) Primary amine, yield (%)
n-Bu	DA	1·8	3	78·5	—	—
n-Bu	DA	1·1	8·8	77·0	45·8	48·7
n-Bu	DA	0·8	—	—	28·0	39·8
n-Bu	DA	0·5	8·2	39·7	14·0	29·9
Ph	DA	1·1	Trace	85·9	Trace	84·7
Ph	DA	0·78	Trace	58·3	—	—
$PhCH_2$	DA	1·1	61·4	41·0	—	—
n-C_5H_{11}	DA	1·1	—	—	36·4	36·0
n-Bu	RA	1·1	34·7	33·5	31·9	23·4
n-Bu	RA	0·8	18·6	39·0	28·0	19·5
n-Bu	RA	0·5	12·3	33·8	14·9	29·4
n-Bu	RA	0·28	14·3	0	—	—
n-C_5H_{11}	RA	1·1	32·5	33·6	25·3	26·9
n-C_5H_{11}	RA	0·5	15·0	24·4	—	—
o-$CH_3C_6H_4$	RA	1·1	Trace	83·6	—	—
o-$CH_3C_6H_4$	RA	0·5	Trace	20·6	—	—
Ph	RA	1·1	Trace	67·4	Trace	58·2

[a] DA: direct addition; RA: reverse addition.

mechanism (equation 40), proposing two competing reactions, accommodates these findings[95].

$$RCH_2{-}C{\equiv}N \xrightarrow[-H_2]{AlH_4^-} R\bar{C}H{-}C{\equiv}N{\cdot}AlH_3 \xrightarrow{RCH_2C{\equiv}N} \begin{array}{l} RCHCH_2NH_2 \\ | \\ RCHCH_2NH_2 \end{array} \quad (40)$$

$$RCH_2{-}C{\equiv}N \xrightarrow{AlH_4^-} RCH_2CH{=}NAlH_3^- \xrightarrow{AlH_4^-} RCH_2CH_2NAlH_2 \longrightarrow RCH_2CH_2NH_2$$

Similar results were obtained when both modes of reduction were performed in tetrahydrofuran.

No evidence is available that a carbon–lithium bond is present in the reaction intermediates, since none of the acid expected on carbonation could be isolated. Such carbonation of the intermediate from the reduction of n-capronitrile was carried out before hydrolysis, but only 7–19 % and 20–40 % yields of n-hexylamine were obtained in both the reverse and the direct addition, respectively, after hydrolysis.

While lithium triethoxyaluminohydride reduces nitriles to the corresponding aldehydes[49,50] (section II.D.3), lithium trimethoxyaluminohydride in tetrahydrofuran at 0 °C reduces capronitrile and benzonitrile to n-hexylamine and to benzylamine, respectively[98]. It is noteworthy that no hydrogen is evolved in the reduction of capronitrile. Such evolution is considerable with lithium aluminium hydride[95] and is believed to be responsible for the decreased yields observed in these reductions. On the other hand, lithium tris(*t*-butoxy)aluminohydride does not react with nitriles, either to evolve hydrogen or to transfer hydride for reduction[98]. It was shown[95] that only two of the four hydrides in lithium aluminium hydride are readily available for the reaction, and that the second addition does not occur intramolecularly, but involves a second molecule. It was also argued[50] that the reaction must largely proceed to the formation of a linear polymer (equation 41), and that the low availability of the remaining hydrides must be largely the result of physical factors.

$$n\,\mathrm{RC{\equiv}N} + n\,\mathrm{AlH_4^-} \longrightarrow \left(\begin{array}{c} \mathrm{H} \\ -\ddot{\mathrm{Al}}-\mathrm{N}- \\ \ddot{\mathrm{H}} \quad | \\ \mathrm{HCH} \\ | \\ \mathrm{R} \end{array} \right)_n \tag{41}$$

The reduction with lithium aluminium hydride is accompanied with condensation of the nitrile with the formation of 1,3-diamine. Reduction with lithium trimethoxyaluminohydride minimizes the effect of this side reaction and circumvents the difficulties arising from the formation of polymeric intermediates. A comprehensive discussion on the reductive properties of the various alkoxyaluminohydrides has been given by Brown and Shoaf[51].

A rather unexpected reaction was the sodium borohydride reduction of polyfluoroalkylnitriles to the corresponding amines[99,100] since regular nitriles are not reduced by this reagent (equation 42).

$$\mathrm{CF_3(CF_2)_{\mathit{n}}CN} \xrightarrow{\mathrm{NaBH_4/diglyme}} \mathrm{CF_3(CF_2)_{\mathit{n}}CH_2NH_2} \tag{42}$$

Addition of the nitrile to a slurry of sodium borohydride in diglyme gave good yields (Table 12) of amines. Apparently the amine was formed via a boron-containing complex, which was detected but not identified in gas chromatography of the reaction mixtures. At least two moles of sodium borohydride per mole of nitrile were employed, and the reduction was carried out for one to one-and-a-half hours at reflux temperature of diglyme, or at room temperature.

TABLE 12. Sodium borohydride reduction of perfluoronitriles in diglyme[99].

Nitrile	Amine	Reaction time (h)	Yield of amine (%)
$CF_3(CF_2)_2CN$	$CF_3(CF_2)_2CH_2NH_2$	1·5	46
$CF_3(CF_2)_6CN$	$CF_3(CF_2)_6CH_2NH_2$	1·0	73
$CF_2H(CF_2)_7CN$	$CF_2H(CF_2)_7CH_2NH_2$	1·5	61·8

In contrast to the inertness of regular nitriles towards sodium borohydride, nitriles are rapidly converted by diborane[101] (Table 13) and very slowly by disiamylborane[102,103] to the corresponding primary amines.

The high reactivity of the nitrile group is demonstrated by the selective reduction of aliphatic and aromatic oxonitriles to the oxoamines when only one equivalent of diborane is applied[104].

Brown argues[101,103] that the differences between the reduction ability of borohydride and diborane can be rationalized as follows: diborane is a Lewis acid which functions best as a reducing agent in attacking groups at positions of high electron density, whereas borohydride is a Lewis base which prefers to attack functional groups at positions of low electron density. Since the nitrile group is relatively insensitive to attack by nucleophilic reagents, its relative inertness toward attack by borohydride ion is not unexpected. At the same time, the nitrilic nitrogen atom is relatively basic, as shown by the formation of addition compounds of moderate stability with boron trifluoride. Presumably, the rapid reduction of nitriles by diborane involves an initial attack of the reagent at this relatively basic position. It should be mentioned, however, that nitriles undergo nucleophilic reaction at the carbon atom of the nitrile (see Chapter 6) and moreover that the reduction by lithium aluminium hydride should proceed by a nucleophilic mechanism.

TABLE 13. Reduction of nitriles by diborane[101].

Nitrile	Amine	Yield (%)
m-Nitrobenzonitrile	*m*-Nitrobenzylamine·HCl	88
Adiponitrile	1,6-Diaminohexane·2HCl	85
Benzonitrile	Benzylamine	83
Phenylacetonitrile	2-Phenylethylamine	84

The above explanation is therefore acceptable only if the borohydride is considered a weaker nucleophile than those known to attack cyano groups.

The behaviour of nitriles towards the different hydride reagents is summarized in Table 14.

In an indirect route to primary amines, the adduct of a nitrile and Grignard reagent is reduced by lithium aluminium hydride (equation 43)[105].

$$EtC{\equiv}N + PhMgBr \longrightarrow (Et)(Ph)C{=}NMgBr \xrightarrow{LiAlH_4} (Et)(Ph)CHNH_2 \quad (43)$$

V. REDUCTION TO HYDROCARBONS

Only one of the known reduction methods of nitriles to hydrocarbons is a real reduction to a methyl group with the preservation of the original carbon skeleton. A second method is making use of the nitrile group as an intermediate for the formation of aromatic hydrocarbon, but complete reduction is not involved. In other formal reductions of nitriles, a reductive cleavage (decyanation) takes place, with the formation of a hydrocarbon with a lower number of carbon atoms than the starting material.

A. Reduction of Nitrile to Methyl Group

Kindler and Luhrs carried out the catalytic reduction of nitriles to the corresponding methyl derivatives (equation 44)[106]. It is interesting that simple terpenes such as Δ^1-*p*-menthene, limonene or

$$RCN + 3\,H_2 \longrightarrow RCH_3 + NH_3 \quad (44)$$

α-phellandrene can be used as hydrogen donors, provided that the temperature is high enough (120–170°). The ability of the terpenes to undergo dehydrogenation and hydrogenation reactions under these conditions may make them the source of the active hydrogen required for the reduction of the nitriles. They are therefore carriers or mediators of hydrogen, being reduced during the reaction, while helping in the reduction of the nitrile function. Seven terpenes have been investigated in this respect[107], and comparison of their ability as well as the reaction conditions required was made by using the

TABLE 14. Behaviour of nitriles towards hydride reagents.

Reagent	$NaBH_4$	$NaBH_4 + LiCl$	$NaBH_4 + AlCl_3$	$NaBH_4 + BF_3$	B_2H_6	$BH(CHMeCHMe_2)_2$
Solvent	EtOH	Diglyme	Diglyme	Diglyme	THF	THF
Product	—[a,103]	—[a,103]	Amine[103]	Amine[103]	Amine[101,103]	Amine[b,103]
Reagent	$LiAHl(OBu\text{-}t)_3$	$LiAlH(OMe)_3$	$LiAlH(OEt)_3$	$NaAlH(OEt)_3$	$LiAlH(OPr\text{-}n)_3$	$LiAlH_4$
Solvent	THF	THF	THF	THF	THF	Ether
Product	—[a,103]	Amine[98]	Aldehyde[49,50]	Aldehyde[45,46]	Aldehyde[49,50]	Amine[35,36,92,95]

[a] No reaction.
[b] Slow reaction.

reduction of 2,4-dimethoxybenzyl cyanide to 2,4-dimethoxyethylbenzene as a model (equation 45)[106].

$$2,4\text{-}(CH_3O)_2C_6H_3CH_2CN \xrightarrow{3\,H_2} 2,4\text{-}(CH_3O)_2C_6H_3CH_2CH_3 \quad (45)$$

The most effective among the terpenes is Δ^1-*p*-menthene which is in turn reduced to *p*-menthane. This reaction can be used not only in nitrile reduction. Indeed, double bonds are saturated, and amines, halogen compounds and carbonates are also rapidly hydrogenated. The reduction of nitriles is therefore limited to nitriles which do not contain these reducible groups. The reaction which can be formulated as in equation (46) is very important for preparative and analytical purposes. The reaction conditions are

$$RCN \xrightarrow{2\,H} [RCH{=}NH \xrightarrow{2\,H} RCH_2NH_2] \xrightarrow{2\,H} RCH_3 + NH_3 \quad (46)$$

summarized in Table 15.

B. Reductive Decyanation and Ring Closure

Anthracene and anthracene derivatives can be formed from *o*-benzylbenzonitriles via a hydrogenation and ring closure[108]. This is really a combination of two reactions: (*a*) the partial reduction of nitriles to aldehydes by the method of Pietra and Trinchera[21,22] and (*b*) the Bradsher and Vingiello synthesis of anthracene derivatives from aldehydes[109] via a carbonium ion intermediate (equation 47).

$$o\text{-}ArCH_2C_6H_4CN \xrightarrow{N_2H_4\cdot H_2O,\ Ni} [o\text{-}ArCH_2C_6H_4CH{=}N{-}]_2 \xrightarrow{H^+/H_2O} [o\text{-}ArCH_2C_6H_4CHO \overset{H^+}{\rightleftharpoons} o\text{-}ArCH_2C_6H_4\overset{+}{C}H{-}OH] \longrightarrow \text{Polynuclear aromatic hydrocarbon} \quad (47)$$

This reaction was applied to a few 2-(arylmethyl)benzonitriles (Table 16). It is carried out by the partial hydrogenation with

TABLE 15. Reduction of nitriles to hydrocarbons.

Nitrile	Hydrocarbon product	Reaction time (min)	Yield
	A. *Aliphatic—Aromatic nitriles*[a]		
Phenylacetonitrile	Ethylbenzene	90	[b]
3,4-Dimethoxyphenylacetonitrile	3,4-Dimethoxyphenylethane	67	[b]
3,4-Dimethoxycinnamonitrile	1-(3,4-Dimethoxyphenyl)propane	70	[b]
α,β-Diphenylacrylonitrile	1,2-Diphenylpropane	100	[b]
α,β-Bis(4-methoxyphenyl) acrylonitrile	1,2-Bis(4-methoxyphenyl)propane	120	[b]
α-Phenyl-β-*p*-tolylacrylonitrile	2-Phenyl-1-*p*-tolylpropane	63	[b]
α-Phenyl-β-(3,4-methylendioxy-phenyl) acrylonitrile	2-Phenyl-1-(3,4-methylene-dioxyphenyl)propane	300	[b]
1,4-Diphenyl-1-cyanobutadiene	2,5-Diphenylpentane	210	[b]
2,5-Diphenylvaleronitrile	2,5-Diphenylpentane	180	[b]
	B. *Aromatic Nitriles*[c]		
Benzonitrile	Toluene	60–80	[d]
4-Methylbenzonitrile	*p*-Xylene	60	[d]
3,4-Dimethoxybenzonitrile	3,4-Dimethoxytoluene	40	[d]
3,4,5-Trimethoxybenzonitrile	3,4,5-Trimethoxytoluene	90	[d]
3,4-Methylenedioxy-5-methoxybenzonitrile	3,4-Methylenedioxy-5-methoxytoluene	50	[d]
2-Chlorobenzonitrile	Toluene	120	[d]
4-Chlorobenzonitrile	Toluene	120	[d]
1-Cyanonaphthalene	1-Methylnaphthalene	115	[d]
2-Cyanonaphthalene	2-Methylnaphthalene	60	[d]
9-Cyanophenanthrene	9-Methylphenanthrene	265	[d]
	C. *Cyano pyridines and quinolines*[a]		
2-Cyanopyridine	2-Methylpyridine	120	[e]
3-Cyanopyridine	3-Methylpyridine	150	[e]
4-Cyanopyridine	4-Methylpyridine	120	[e]
Pyridine-3-acetonitrile	3-Ethylpyridine	200	[e]
Pyridine-3-acrylonitrile	3-Propylpyridine	210	[e]
4-Cyanoquinoline	4-Methylquinoline	95	[e]
6-Methoxy-4-cyanoquinoline	6-Methoxy-4-methylquinoline	120	[e]

[a] 10 mmole nitrile, 70 ml Δ^1-*p*-menthene, 1 g of Pd/C(10%).
[b] 70–80%.
[c] 100 mmole nitrile, 120 ml Δ^1-*p*-menthene, 2 g of Pd/C(10%).
[d] 85–90%.
[e] 85–95% as picrates.

TABLE 16. Formation of anthracenes by reduction of 2-arylmethylbenzonitriles o-$NCC_6H_4CH_2Ar$[108].

Ar	Yield of azine (%)	Hydrocarbon	Yield of hydrocarbon (%)
Ph	90	Anthracene	95
2,3-$(CH_3)_2C_6H_3$	88	1,2-Dimethylanthracene	90
α-$C_{10}H_7$	92	Benz[a]anthracene	98
β-$C_{10}H_7$	91	Benz[b]anthracene	95

hydrazine hydrate and Raney nickel followed by the acid cyclization of the resulting azine, which is also isolated. It seems that this method of synthesis of polynuclear aromatic hydrocarbons has general application.

The reductive decyanation (or dehydrocyanation) to hydrocarbons is demonstrated in the reduction of dehydroabietonitrile with the solvated electrons formed either from sodium biphenyl ion radical or sodium in liquid ammonia to form high yields of $\Delta^{5,7,14(13)}$-abietatriene (equation 48)[110]. This is an application of a

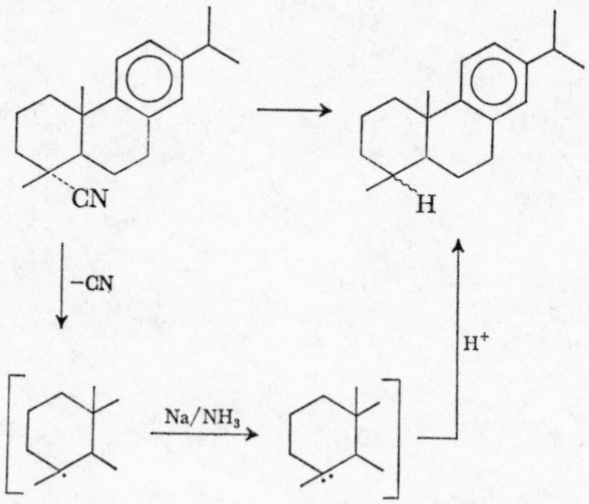

(48)

known procedure for reduction of organic halides[111,112] to the tertiary cyanides (pseudohalides). The reaction mechanism suggested[113,114] includes a stepwise two-electron transfer with initial cyanide loss to lead to the carbanion via the radical (equation 48). This reaction, which is useful only for reduction of tertiary nitriles, has also been applied to 1,1′-dicyanobicyclohexyl and tetramethylsuccinonitrile, both giving mixtures of the expected saturated and unsaturated hydrocarbons[110] (equation 49).

$$\text{1,1'-dicyano-bicyclohexyl} \xrightarrow[-33^\circ C]{Na/NH_3} \text{bicyclohexyl (95\%)} + \text{cyclohexylidenecyclohexane (5\%)}$$

$$(CH_3)_2C(CN)C(CN)(CH_3)_2 \xrightarrow{Li/EtNH_2} (CH_3)_2CH{-}CH(CH_3)_2 \ (88\%) + (CH_3)_2C{=}C(CH_3)_2 \ (11\%) \qquad (49)$$

The reduction of nitriles with sodium and alcohol may lead, in the case of some nitriles, to the formation principally of amines, but in a few cases, to the removal of the —CN group as sodium cyanide[87] (Table 17).

TABLE 17. The reduction of nitriles RCN with Na/EtOH[87].

R	Yield of NaCN (%)	Yield of RH (%)	Yield of RCH_2NH_2 (%)
n-Bu	16	—	76
Me_2CH	24	—	63
Me_3C	33	—	60
$(Ph)(PhO(CH_2)_3)CH$	91	89	5
$(C_5H_{11})(Me)(Me)C$	61	33	23
$(Bu)(Bu)(Pr)C$	10	7	54

VI. MISCELLANEOUS

A reaction which formally seems to be a transformation of a nitrile group to an amide group under hydrogenation conditions is actually an oxygen transfer followed by reduction[115]. *o*-Nitrobenzonitrile takes up three moles of hydrogen when it is hydrogenated over palladium or platinum with the formation of *o*-aminobenzamide in 90% yield (equation 50). If the hydrogenation is carried out in the presence

$$o\text{-}C_6H_4(CN)(NO_2) \xrightarrow{3\,H_2} o\text{-}C_6H_4(CONH_2)(NH_2) \tag{50}$$

of heavy water (^{18}O), no ^{18}O is incorporated in the product. Hence the amide oxygen does not come from the water formed during hydrogenation, but is apparently an oxygen atom originally belonging to the nitro group.

Aziridines can be formed by lithium aluminium hydride reduction of α-halonitriles[116] (equation 51). This is actually a ring closure to

$$RCH(Cl)CN \xrightarrow{LiAlH_4} \text{2-R-aziridine (N-H)} \tag{51}$$

aziridines by the intermediate *β*-haloamines in the reduction of the nitrile group to the amine.

Electroreduction of nitriles to amines and distribution of the products was reported[117,118] in the literature. The polarograms for unsaturated nitriles, e.g. acrylonitrile, were taken and the half-wave potentials were determined.

VII. REFERENCES

1. V. Migrdichian, *The Chemistry of Organic Cyanogen Compounds*, Reinhold, New York, 1947.
2. E. Mosettig, *Organic Reactions*, Vol. 8, John Wiley and Sons, New York, 1954, pp. 218–257.
3. J. Carnduff, *Quart. Rev.*, **20**, 169 (1966).
4. Houben-Weyl-Muller, *Methoden der Organischen Chemie*, 7/1, G. Thieme, Stuttgart, 1954, p. 299.
5. H. Stephen, *J. Chem. Soc.*, **1925,** 1874.
6. A. Sonn and E. Muller, *Ber.*, **52,** 1927 (1919).
7. H. Stephen, *J. Chem. Soc.*, **1930,** 2786.
8. T. Stephen and H. Stephen, *J. Chem. Soc.*, **1956,** 4695.
9. E. Lieber, *J. Am. Chem. Soc.*, **71,** 2862 (1949).
10. J. A. Knight and H. D. Zook, *J. Am. Chem. Soc.*, **74,** 4560 (1952).
11. P. S. Pyryalova and E. N. Zilberman, *Izv. Vysshikh Uchebn. Zavedenii, Khim. i Khim. Tekhnol.*, **8,** 82 (1965); *Chem. Abstr.*, **63,** 6913 (1965).
12. L. Turner, *J. Chem. Soc.*, **1956,** 1686.
13. O. G. Backeberg and B. Staskun, *J. Chem. Soc.* **1962,** 3961.
14. V. Guth, J. Leitich, W. Specht and F. Wessely, *Monatsh. Chem.*, **94,** 1262 (1963).
15. B. Staskun and O. G. Backeberg, *J. Chem. Soc.*, **1964,** 5880.

16. K. Tamaru, *Trans. Faraday Soc.*, **55**, 824 (1959).
17. N. O. Pastushak and A. V. Dombrovskii, *Zh. Org. Khim.*, **1**, 323 (1965); *Chem. Abstr.*, **62**, 9182 (1965).
18. T. van Es and B. Staskun, *J. Chem. Soc.*, **1965**, 5774.
19. G. Mignonac, *French Pat.* 1,186,960 (1957).
20. H. Rupe and E. Hodel, *Helv. Chim. Acta*, **6**, 865 (1923).
21. S. Pietra and C. Trinchera, *Gazz. Chim. Ital.*, **85**, 1705 (1955).
22. S. Pietra and C. Trinchera, *Gazz. Chim. Ital.*, **86**, 1045 (1956).
23. H. Plieninger, *Ger. Pat.* 957,029 (1957).
24. H. Plieninger and G. Werst, *Chem. Ber.*, **88**, 1956 (1955).
25. H. Plieninger and B. Kiefer, *Chem. Ber.*, **90**, 617 (1957).
26. H. Plieninger, *Ger. Pat.* 963,321 (1957).
27. H. Plieninger and G. Werst, *Angew. Chem.*, **67**, 156 (1955).
28. H. Rupe and W. Brentano, *Helv. Chim. Acta*, **19**, 588 (1936).
29. A. Gaiffe and R. Pollaud, *Compt. Rend.*, **252**, 1339 (1961).
30. A. Gaiffe and R. Pollaud, *Compt. Rend.*, **254**, 496 (1962).
31. A. Gaiffe, *Chim. Ind.*, **93**, 259 (1965).
32. C. R. Hauser and W. J. Humphlett, *J. Org. Chem.*, **15**, 359 (1950).
33. H. S. Mosher and W. T. Mooney, *J. Am. Chem. Soc.*, **73**, 3948 (1951).
34. N. Rabjohn and E. L. Crow, *J. Org. Chem.*, **28**, 2907 (1963).
35. R. F. Nystrom and W. G. Brown, *J. Am. Chem. Soc.*, **70**, 3738 (1948).
36. L. Friedman, *Abstr. 116. Nat. Meeting Amer. Chem. Soc. September 1949*, p. 5-M.
37. (a) N. C. Gaylord, *Reduction with Complex Metal Hydrides*, Interscience, New York, 1956;
(b) W. G. Brown, *Organic Reactions*, Vol. 6, John Wiley and Sons, New York, 1951.
38. L. I. Smith and E. R. Rogier, *J. Am. Chem. Soc.*, **73**, 4047 (1951).
39. A. L. Henne, R. L. Pelley and R. M. Alm, *J. Am. Chem. Soc.*, **72**, 3370 (1950).
40. T. D. Perrine and E. L. May, *J. Org. Chem.*, **19**, 775 (1954).
41. M. Yadnik and A. A. Larsen, *J. Am. Chem. Soc.*, **73**, 3534 (1951).
42. C. J. Claus and J. L. Morgenthau, Jr., *J. Am. Chem. Soc.*, **73**, 5005 (1951).
43. C. J. Claus and J. L. Morgenthau, Jr., *U.S. Pat.* 2,786,872 (1957).
44. H. Lapin and R. Malzieu, *Bull. Soc. Chim. France*, **1965**, 1864.
45. G. Hesse and R. Schrodel, *Ann. Chem.*, **607**, 24 (1957).
46. G. Hesse and R. Schrodel, *Angew. Chem.*, **68**, 438 (1956).
47. W. Nagata, S. Hirai, H. Itazaki and K. Takeda, *Ann. Chem.*, **641**, 196 (1961).
48. L. I. Zakharkin and I. M. Khorlina, *Dokl. Akad. Nauk SSSR*, **116**, 422 (1951); *Chem., Abstr.* **52**, 8040f (1958).
49. H. C. Brown and C. P. Garg, *Tetrahedron Letters*, **1959**, (3), 9.
50. H. C. Brown and C. P. Garg, *J. Am. Chem. Soc.*, **86**, 1085 (1964).
51. H. C. Brown and C. J. Shoaf, *J. Am. Chem. Soc.*, **86**, 1079 (1964).
52. T. L. Tolbert and B. Houston, *J. Org. Chem.*, **28**, 695 (1963).
53. J. von Brown, G. Blessing and F. Zobel, *Ber.* **56**, 1988 (1923).
54. (a) R· B. Wagner and H. D. Zook, *Synthetic Organic Chemistry*, John Wiley and Sons, New York, 1953;
(b) W. J. Hickinbottom, *Reactions of Organic Compounds*, Longmans-Green, London, 1950.
55. (a) H. O. House, *Modern Synthetic Methods*, W. A. Benjamin, New York, 1965;
(b) L. F. Fieser, *Reagents for Organic Synthesis*, John Wiley and Sons, New York, 1967.

56. H. Adkins, *Reaction of Hydrogen with Organic Compounds*, University of Wisconsin Press, Madison, Wisconsin, 1937.
57. C. F. Winas and H. Adkins, *J. Am. Chem. Soc.*, **54**, 306 (1932).
58. H. Adkins and H. I. Cramer, *J. Am. Chem. Soc.*, **52**, 4349 (1930).
59. W. Gulewitch, *Ber.*, **57**, 1645 (1924).
60. K. W. Rosenmund and E. Pfankuch, *Ber.*, **56**, 2258 (1923).
61. W. H. Carothers and G. A. Jones, *J. Am. Chem. Soc.*, **47**, 3051 (1925).
62. E. F. Degering and L. G. Boatright, *J. Am. Chem. Soc.*, **72**, 5137 (1950).
63. W. H. Hartung, *J. Am. Chem. Soc.*, **50**, 3370 (1928).
64. M. Danzig and H. P. Schultz, *J. Am. Chem. Soc.*, **74**, 1836 (1952).
65. F. C. Whitmore, *J. Am. Chem. Soc.*, **66**, 725 (1944).
66. E. J. Schwoegler and H. Adkins, *J. Am. Chem. Soc.*, **61**, 3499 (1929).
67. W. Reeve and W. M. Eareckson, *J. Am. Chem. Soc.*, **72**, 3299 (1950).
68. J. C. Robinson and H. R. Snyder, *Organic Syntheses*, Coll. Vol. 3, John Wiley and Sons, New York, 1955, p. 720.
69. B. S. Biggs and W. S. Bishop, *Ind. Eng. Chem.*, **38**, 1084 (1946).
70. M. Freifelder, *J. Am. Chem. Soc.*, **82**, 2386 (1960).
71. F. E. Gould, G. S. Johnson and A. F. Ferris, *J. Org. Chem.*, **25**, 1658 (1960).
72. A. F. Ferris, F. E. Gould, G. S. Johnson, H. K. Latourette and H. Stange, *Chem. Ind.*, **1959**, 996.
73. A. F. Ferris, G. S. Johnson, F. E. Gould and H. K. Latourette, *J. Org. Chem.*, **25**, 492 (1960).
74. A. F. Ferris, G. S. Johnson, F. E. Gould and H. Stange, *J. Org. Chem.*, **25**, 1302 (1960).
75. A. F. Ferris, F. E. Gould, G. S. Johnson and H. Stange, *J. Org. Chem.*, **26**, 2602 (1961).
76. M. Vignau, *Bull. Soc. Chim. France*, **1952**, 638.
77. W. L. Hawkins and B. J. Biggs, *J. Am. Chem. Soc.*, **71**, 2530 (1949).
78. J. Corse, J. T. Bryant and H. A. Shoule, *J. Am. Chem. Soc.*, **68**, 1907 (1946).
79. C. F. Winas and H. Adkins, *J. Am. Chem. Soc.*, **55**, 4172 (1933).
80. N. J. Leonard, G. W. Leubner and E. H. Burk, *J. Org. Chem.*, **25**, 982 (1960).
81. M. Freifelder and R. H. Hasbrouck, *J. Am. Chem. Soc.*, **82**, 696 (1960).
82. R. Juday and H. Adkins, *J. Am. Chem. Soc.*, **77**, 4559 (1955).
83. G. Mignonac, *Compt. Rend.*, **40**, 482 (1905).
84. (a) L. Kh. Freidlin and T. A. Sladkova, *Izv. Akad. Nauk SSSR, Odt. Khim. Nauk*, **1962**, 336; *Chem. Abstr.*, **57**, 11166 (1962); *Chem. Abstr.*, **59**, 6370 (1963);
(b) D. V. Sokolskii and F. Bizhanov, *Izv. Akad. Nauk Kaz. SSR, Ser. Khim.*, **1960**, 101; *Chem. Abstr.*, **57**, 9703, 14998 (1962).
85. L. D. Volkova and D. V. Sokolskii, *Izv. Akad. Nauk. Kaz. SSR. Ser. Khim.*, **1965**, 52.
86. (a) T. Hayashi, K. Watanabe and K. Hata, *Nippon Kagaku Zasshi*, **83**, 348 (1962); *Chem. Abstr.*, **59**, 3826 (1963);
(b) W. H. Dennis Jr., D. H. Rosenblatt, R. R. Richmond, G. A. Finseth and G. T. Davis, *Tetrahedron Letters*, **1968**, 1821.
87. L. A. Walter and S. M. McElvain, *J. Am. Chem. Soc.*, **56**, 1614 (1934).
88. M. S. Bloom, D. S. Breslow and C. R. Hauser, *J. Am. Chem. Soc.*, **67**, 539 (1945).
89. A. Ladenburg, *Ber.*, **18**, (1885); **19**, 780 (1886).
90. E. Bamberger, *Ber.*, **20**, 1703, 1711 (1887).
91. A. P. Terent'ev, M. N. Preobrazhenskaya and Ban-Lun Ge, *Khim. Nauka i Prom.*, **4**, 281 (1959); *Chem. Abstr.*, **53**, 21879 (1959).

92. R. F. Nystrom, *J. Am. Chem. Soc.*, **77**, 2544 (1955).
93. A. E. Finholt, A. C. Bond, Jr. and H. I. Schlesinger, *J. Am. Chem. Soc.*, **69**, 1199 (1947).
94. E. Wiberg and M. Schmidt, *Z. Naturforsch.*, **66**, 460 (1951).
95. L. M. Soffer and M. Katz, *J. Am. Chem. Soc.*, **78**, 1705 (1956).
96. L. M. Soffer and E. W. Parrotta, *J. Am. Chem. Soc.*, **76**, 3580 (1954).
97. H. E. Zaugg and B. W. Horron, *Anal. Chem.*, **20**, 1026 (1948).
98. H. C. Brown and P. M. Weissman, *J. Am. Chem. Soc.*, **87**, 5614 (1965).
99. S. E. Ellzey Jr., J. S. Wittman III and W. J. Connick Jr., *J. Org. Chem.*, **30**, 3945 (1965).
100. M. Sander, *Monatsh. Chem.*, **95**, 608 (1964).
101. H. C. Brown and B. C. Subba Rao, *J. Am. Chem. Soc.*, **82**, 681 (1960).
102. H. C. Brown and D. B. Bigley, *J. Am. Chem. Soc.*, **83**, 486 (1961).
103. H. C. Brown, *Hydroboration*, W. A. Benjamin, New York, 1962, Chap. 17.
104. B. C. Subba Rao and G. P. Thakor, *Current Sci.* (*India*), **32**, 404 (1963).
105. H. Pohland and H. R. Sullivan, *J. Am. Chem. Soc.*, **75**, 5898 (1953).
106. K. Kindler and K. Luhrs, *Ber.*, **99**, 227 (1966).
107. K. Kindler and K. Luhrs, *Ann. Chem.*, **685**, 36 (1965).
108. W. W. Zajac and R. H. Denk, *J. Org. Chem.*, **27**, 3716 (1962).
109. C. Bradsher and F. Vingiello, *J. Am. Chem. Soc.*, **71**, 1434 (1949).
110. P. G. Arapakos, *J. Am. Chem. Soc.*, **89**, 6794 (1967).
111. J. C. Bergmann and J. Slavik Jr., *Anal. Chem.*, **29**, 241 (1957).
112. L. M. Liggett, *Anal. Chem.*, **26**, 748 (1954).
113. J. F. Garst, D. W. Ayers and R. C. Lamb, *J. Am. Chem. Soc.*, **88**, 4260 (1966).
114. S. J. Cristol and R. V. Barbour, *J. Am. Chem. Soc.*, **88**, 4262 (1966).
115. H. Moll, H. Musso and H. Schroder, *Angew. Chem. Int. Ed.*, **2**, 212 (1963).
116. K. Ichimura and M. Ohta, *Bull. Chem. Soc. Japan*, **40**, 432 (1967).
117. A. P. Tamilov, L. V. Kaabak and S. L. Varshavskii, *Zh. Obsch. Khim.*, **33**, 2811 (1963); *Chem. Abstr.*, **60**, 1584 (1964).
118. I. G. Sevast'yanova and A. P. Tamilov, *Zh. Obsch. Khim.*, **33**, 21815 (1963); *Chem. Abstr.*, **60**, 1583 (1964).

CHAPTER 8

Additions to the cyano group to form heterocycles

A. I. MEYERS and J. C. SIRCAR

Louisiana State University in New Orleans, New Orleans, Louisiana, U.S.A.

I. INTRODUCTION 342
II. CYCLOADDITION REACTIONS (TYPE A) 344
A. General Considerations. 344
B. Three- and Four-Membered Heterocycles 345
C. Five-Membered Heterocycles 345
1. Heterocycles from nitriles and 1,3-dipoles of class 1 . 346
a. Nitrile oxides 346
b. Nitrile imines 348
c. Diazoalkanes 349
d. Nitrile ylids 350
e. Azides 350
2. Heterocycles from nitriles and 1,3-dipoles of class 2 . . 351
a. Azomethine imines 351
b. Azomethine ylids 353
3. Heterocycles from nitriles and 1,3-dipoles of class 3 . . 353
a. Ketocarbenes 353
b. Ketonitrenes 354
c. Thioketene. 356
D. Six-Membered Heterocycles 356
III. ELECTROPHILIC ADDITION TO THE CYANO GROUP FOLLOWED BY RING CLOSURE (TYPE B) 358
A. General Considerations. 358
B. Cyclization of Nitrilium Ions by Reaction with C=C Bonds . 359
1. Pyrroline derivatives 359
2. Pyridine derivatives 360
3. Isoquinoline derivatives 360
4. Seven- and higher-membered heterocycles 363
C. Cyclization of Nitrilium Ions by Reaction with C=O Bonds . 364
1. Oxazole derivatives 364
2. Oxazine derivatives 366

D. Cyclization of Nitrilium Ions by Reaction with OH Groups . 366
1. Oxazoles . 366
2. Oxazines . 367
3. Miscellaneous oxygen nucleophiles . 368
E. Cyclization of Nitrilium Ions by Reaction with SH Groups . 369
1. Thiazole derivatives . 369
2. Thiazine derivatives . 372
F. Cyclization of Nitrilium Ions by Reaction with Nitrogen-Containing Compounds . 373
1. Imidazole derivatives . 373
2. Quinazoline derivatives . 375
IV. INTRAMOLECULAR ELECTROPHILIC ADDITION TO THE CYANO GROUP (TYPE B_i) . 376
A. General Considerations . 376
B. Cyclization of Olefinic Nitriles . 377
C. Cyclization of Ketonitriles . 378
D. Cyclization of Dinitriles . 380
V. NUCLEOPHILIC ADDITION TO THE CYANO GROUP FOLLOWED BY RING CLOSURE (TYPE C) . 382
A. General Considerations . 382
B. Addition by Carbanionic Species Followed by Cyclization . 383
C. Addition by the Amino Group Followed by Cyclization . 387
D. Addition by Oxygen Nucleophiles Followed by Cyclization . 391
E. Addition by Sulphur Nucleophiles Followed by Cyclization . 393
VI. INTRAMOLECULAR NUCLEOPHILIC ADDITION TO THE CYANO GROUP (TYPE C_i) . 397
A. General Considerations . 397
B. Intramolecular Nucleophilic Addition by Carbon Nucleophiles . 397
C. Intramolecular Nucleophilic Addition by Nitrogen Nucleophiles 400
D. Intramolecular Nucleophilic Addition by Oxygen Nucleophiles . 405
E. Intramolecular Nucleophilic Addition by Sulphur Nucleophiles . 407
VII. POLYMERIZATION OF THE CYANO GROUP TO FORM HETEROCYCLES . 410
VIII. MISCELLANEOUS HETEROCYCLIC SYNTHESES FROM NITRILES . 411
IX. ACKNOWLEDGEMENTS . 414
X. REFERENCES . 414

I. INTRODUCTION

The utility of the cyano group in heterocyclic syntheses is probably surpassed only by the amino function. The high electron density present in the cyano group makes it an excellent candidate for a variety of electrophilic additions, whereas its strong dipole moment (3·5–3·6 D) allows for facile nucleophilic additions. If the attacking electrophile (E) also possesses a reactive nucleophilic centre (B) suitably situated in the molecule (i.e. a dipolar species), a ring closure can result giving rise to many varied heterocyclic systems. A simultaneous addition, or two of very rapid succession by either

$$R{-}C{\equiv}N + \overset{E}{\underset{B}{\big)}} \xrightarrow{\text{Type A}} R{-}C\overset{N-E}{\underset{B}{\big)}}$$

the electrophilic or nucleophilic terminus, without the formation of a stable intermediate, is now the well-known cycloaddition process (type A). Alternatively, the attack by an electrophile may occur as the first step of the reaction followed by a nucleophilic attack (type B) in the second step. Conversely, the nucleophile (B) could initially add to the carbon of the cyano group and, in a second step, the electrophile adds to the nitrogen atom (type C) which also would give rise to the heterocyclic system. These three types (A, B and C) of ring closures would incorporate the carbon and nitrogen atom of the cyano group and each type would depend upon the relative reactivity of the nucleophilic and electrophilic sites.

For example, type A would be expected to occur on the cyano group if both the nucleophilic and electrophilic sites were of relatively equal reactivity, thus causing the resulting cyclic transition state to be the important stage in the reaction. For the type B process, a strong electrophilic site (E) would result in the nitrilium salt intermediate **1** which now possesses a highly electrophilic carbon atom capable of attacking the poorly nucleophilic site, B. In the type C process, the presence of a strong nucleophile (B) would result in **2** which possesses a highly nucleophilic nitrogen atom able to coordinate with even the weakest electrophilic atoms. In the discussion to follow, examples of these three processes will be made evident.

A modification of types B and C is also very common when heterocyclic molecules are formed in the presence of the cyano group. This is seen when the nucleophile (B) or the electrophile (E) adding to the cyano group is already part of the nitrile-containing molecule.

$$R{-}C{\equiv}N \xrightarrow[\text{Type B}]{\widehat{B\;\;E}} \left[R{-}C{\equiv}\overset{+}{N}{-}E \;\cdots B \longleftrightarrow R{-}\overset{+}{C}{=}N{-}E \;\cdots B \right] \textbf{(1)} \longrightarrow R{-}C(=N{-}E){-}B \text{ (ring)}$$

$$R{-}C{\equiv}N \xrightarrow[\text{Type C}]{\widehat{B\;\;E}} R{-}C(=N^-){-}B \;\cdots E \;\textbf{(2)} \longrightarrow R{-}C(=N{-}E){-}B \text{ (ring)}$$

It is therefore possible to form a heterocycle (type B_i) if the electrophilic attack on nitrogen is intramolecular. Furthermore, intramolecular nucleophilic attack will also result in a heterocycle (type C_i). Although relatively few examples of type B_i are known, the literature abounds with examples of type C_i.

Type B_i

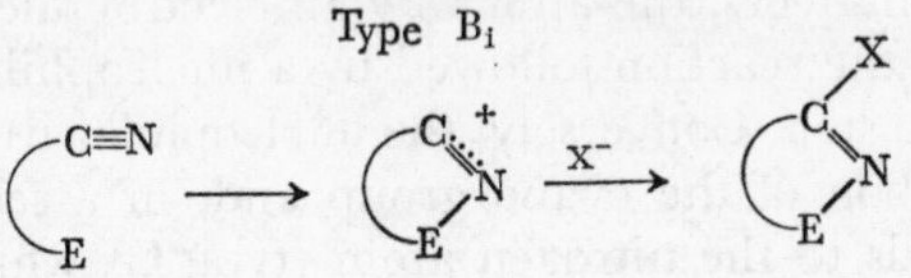

Type C_i

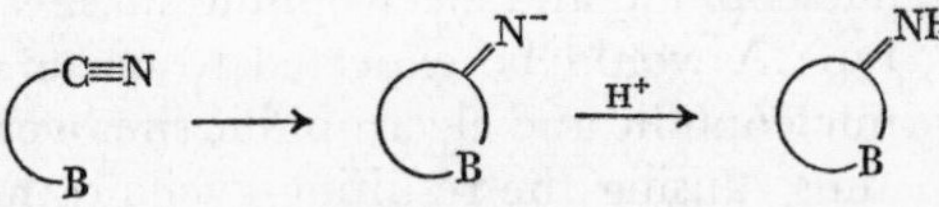

Many reviews dealing with heterocyclic syntheses from nitriles have appeared, but none have attempted to reduce the vast literature into the three basic mechanisms (A, B and C) and the two minor variations (B_i and C_i). Thus reactions initiated by nucleophilic attack on the cyano group (type C) have been compiled in the available compendia[1–4] on synthetic heterocyclic chemistry, whereas recent reviews[5–7] have been concerned with heterocyclic syntheses initiated by electrophilic reagents (type B). A report summarizing dipolar cycloadditions[8] has included nitriles as useful dipolarophiles (type A).

This chapter will describe representative examples of heterocyclic syntheses utilizing nitriles, with the intent to unify in the reader's mind the vast number of reported reactions into the basic mechanistic types mentioned. Although types B_i and C_i are mechanistically related to B and C, they will be treated as separate sections primarily because of the grossly different ring systems which result.

II. CYCLOADDITION REACTIONS (TYPE A)

A. General Considerations

In this reaction type, those cyclizations with nitriles are presented which involve (or appear to involve) concerted reactions. In some cases firm evidence for concerted cyclization is absent but nevertheless these processes do not clearly involve a two-step ring closure. A discussion on general considerations and definition of cycloaddition reactions has been published[12]. The heterocycles formed by this method will be presented according to ring size only and not by the number of heteroatoms in the molecule.

B. Three- and Four-Membered Heterocycles

Although it is theoretically possible for a carbene or nitrene[9] to add to the cyano group leading to azirines (**3**) or diazirines (**4**), respectively, no report has yet appeared describing the process. The successful production of these systems by this method would constitute the simplest cycloaddition process. Attempts to produce these compounds, which have been obtained by other routes[10,11], have resulted in five-membered rings via the 1,3-dipolar addition of ketocarbenes and ketonitrenes. These will be discussed later in the chapter.

Also absent from the literature are reports of cycloaddition of π-systems (alkenes, azines, carbonyl compounds) to the cyano group to form four-membered heterocycles.

$R^2—C≡N$ $\xrightarrow{R^1_2C:}$ (**3**) (R^1R^1C–R^2C=N ring)

$R^2—C≡N$ $\xrightarrow{R^1N:}$ (**4**) (R^1N–R^2C=N ring)

C. Five-Membered Heterocycles

The formation of five-membered heterocycles utilizing the cyano group is now a versatile process in view of the facile addition of 1,3-dipolar species[8,12]. Briefly, the 1,3-dipole **5** adds to the cyano group with the electrophilic terminus (E) adding to the nitrogen and the nucleophilic terminus (B) to the carbon to give the heterocycle **6**. The three atoms involved in the 1,3-dipole can be carbon, oxygen and nitrogen, with each serving as either the terminal and/or central atom. The 1,3-dipoles have been classified[8] as: (*a*) octet-stabilized with multiple linkages, i.e. nitrile oxides, nitrile imines, azides, etc., (*b*) octet-stabilized without multiple linkages, i.e. azomethines, ylids, azomethine imines, etc., and (*c*) species without octet-stabilization, i.e. ketocarbenes and ketonitrenes. The cycloadditions of nitriles with members of each class of 1,3-dipoles will be treated in the following sections.

$$R{-}C{\equiv}N + B{\overset{A}{\wedge}}E \longrightarrow R{-}C\langle N{=},\ E{-}A{-}B\rangle$$

(5) (6)

I. Heterocycles from nitriles and 1,3-dipoles of class I

a. Nitrile oxides. Nitrile oxides, isolated or generated *in situ*, condense with nitriles[13,14] to form 1,2,4-oxadiazoles (**7**). The ring closure may be carried out utilizing aliphatic or aromatic nitriles (Table 1). Simple aliphatic nitriles are not effective in this reaction,

$$\left.\begin{matrix} R^1{-}C{\equiv}\overset{+}{N}{-}\overset{-}{O} \\ \updownarrow \\ R^1{-}\overset{+}{C}{=}N{-}\overset{-}{O} \end{matrix}\right\} \xrightarrow{R^2CN} \text{(7)}$$

TABLE 1. Dipolar addition of nitrile oxides to nitriles to form 1,2,4-oxadiazoles (**7**)[13].

R^1	R^2	Percentage yield
Ph	Ph	69
Ph	3-Pyridyl	56
Ph	$ClCH_2$	89
Ph	CH_3OCH_2	31
Ph	$PhOCH_2$	62
Ph	PhCO	61
Ph	CH_3CO	68
Ph	CO_2Et	71
Ph	$CH_3C(CN)(OAc)$	64
Ph	CN	73
p-ClC_6H_4	Ph	~50
CO_2Et	Ph	Trace

whereas aliphatic nitriles containing an electron-attracting substituent condense smoothly. Aromatic nitriles containing an electron-attracting substituent likewise lead to good yields of the oxadiazoles. Thus the electrophilic character of the nitrile carbon is an important factor in the dipolar addition. This is borne out by a recent study[15] which employed boron trifluoride as a catalyst to enhance the electrophilicity of the cyano carbon through its Lewis acid complex (**8**)[16]. Moderate yields of the oxadiazoles were thus obtained (Table 2). Under neutral or alkaline conditions, ethyl

TABLE 2. Boron trifluoride catalysed additions of nitriles to benzonitrile oxides[15].

(**9**)

R	Percentage yield
CH_3	35
C_2H_5	40
n-C_3H_7	40
CH_2CO_2Et	14

cyanoacetate failed to give the 1,2,4-oxadiazole (**9**) but has been reported[17] to react with benzonitrile oxide to produce the isooxazole **10** under alkaline conditions. The use of boron trifluoride as catalyst resulted in **9a** in 14% yield. Therefore, employing a Lewis acid to

$$RC{\equiv}N + BF_3 \longrightarrow [RC{\equiv}\overset{+}{N} \longleftrightarrow R{-}\overset{+}{C}{=}\overset{-}{N}]BF_3 \xrightarrow{RCNO} \mathbf{9}$$
(**8**)

polarize the cyano group results in increased electrophilic character on carbon and nucleophilic character on nitrogen, forming the oxadiazole **9**. Both heterocyclic systems probably arise from the same initial zwitterion (**11**), but tautomeric shift of the methylene protons, readily feasible in neutral or alkaline media, produces **10** via the intermediate **11a** which is formed by a simple C—O bond rotation in **11**. The use of cyanotetrazoles (**12**) as a source of heterocyclic systems has been reported[18] to lead to tetrazyloxadiazoles (**13**). The nature and position of the R substituent on the tetrazole had little effect upon the cycloaddition, although substituents on the benzonitrile oxide exerted a profound effect. *ortho*-Nitro-substituted benzonitrile oxides[14,18] failed to give cycloaddition, whereas *para*-nitro-substituted ones enhanced the reaction over that of unsubstituted nitrile oxides. These results can be

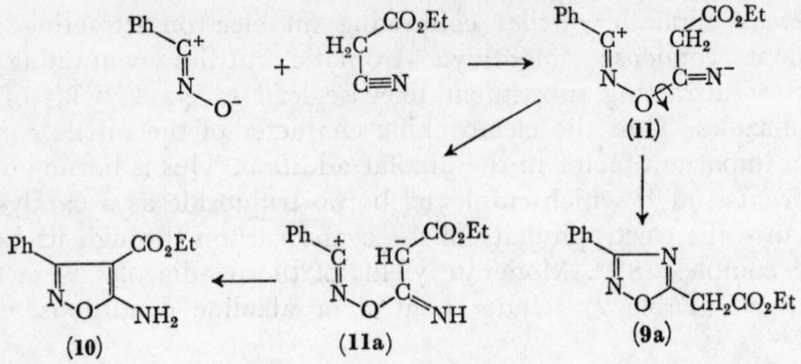

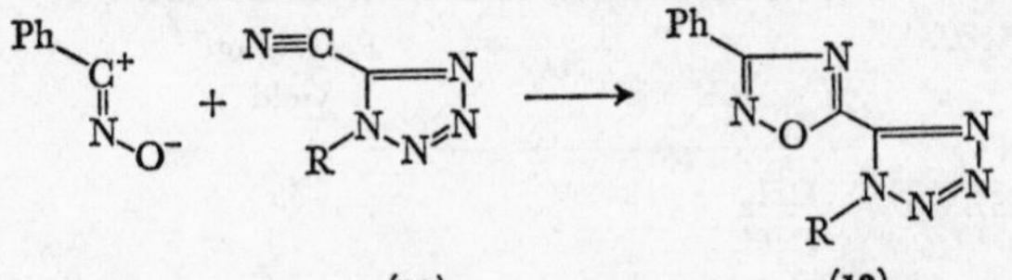

(12) (13)

explained if the nitro substituent is partially bonded to the nitrile oxide as in **14**, thereby reducing much of its electrophilic nature, whereas any electron-attracting substituent present on either the nitrile or the dipolar species would tend to increase its reactivity.

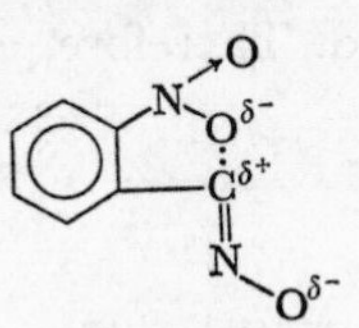

(14)

b. Nitrile imines. Dipolar addition of nitrile imines (**15**), prepared from tetrazoles[19] or hydrazidoyl halides[20], to nitriles leads to 1,2,4-triazoles (**16**). It is of interest that the contributing resonance structure **15c** does not manifest itself in the production of 1,2,3-triazoles (**17**). This is undoubtedly due to the higher energy associated with **15c** as compared with other resonance structures having electrophilic carbon. The formation of **16** is accomplished in good yields when aromatic nitriles are employed and in poor yields in the case of aliphatic nitriles. Once again, a substituent present in the nitrile which increases its electrophilic character tends to increase the conversion to **16.**

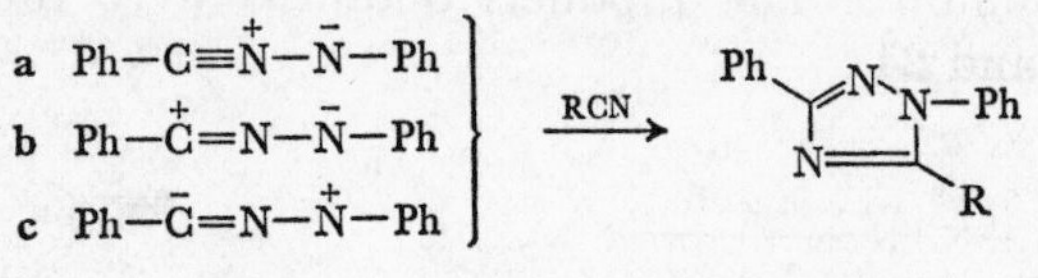

a $Ph-C\equiv\overset{+}{N}-\overset{-}{N}-Ph$
b $Ph-\overset{+}{C}=N-\overset{-}{N}-Ph$
c $Ph-\overset{-}{C}=N-\overset{+}{N}-Ph$

(15) $\xrightarrow{RCN}$ **(16)**

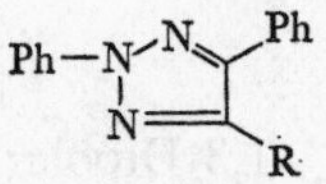

(17)

(R = aryl, alkyl)

c. Diazoalkanes. Although simple nitriles do not form cycloaddition products with diazoalkanes, electrophilic nitriles (**18**) do produce 1,2,3-triazoles (**19**)[21–23]. Of the several possible resonance structures for diazoalkanes, only that which contains the carbanion moiety appears in the products formed.

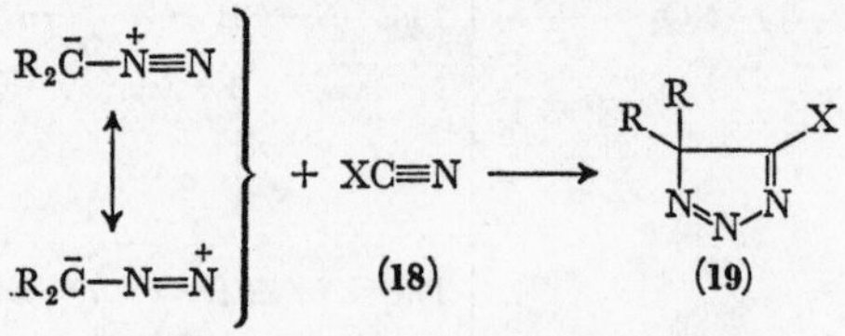

$R_2\overset{-}{C}-\overset{+}{N}\equiv N \leftrightarrow R_2\overset{-}{C}-N=\overset{+}{N}$ + $XC\equiv N$ **(18)** ⟶ **(19)**

(R = alkyl, aryl, X = Cl, CN, CO_2Et)

In the reaction of cyanogen bromide with diazomethane, three products are formed, which are trapped as their respective *N*-methyl derivatives **20–22** by reaction with excess diazomethane.

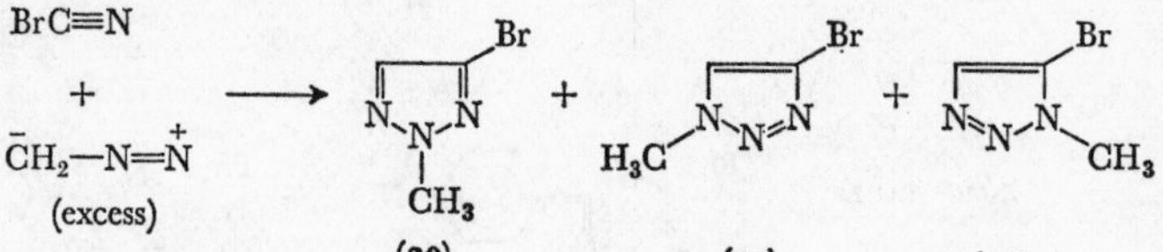

$BrC\equiv N$ + $\overset{-}{C}H_2-N=\overset{+}{N}$ (excess) ⟶ **(20)** + **(21)** + **(22)**

It is apparent that ordinary nitriles are not sufficiently electrophilic to react with the highly delocalized (poorly polarized) diazo function. However, if the electrophilic character of the cyano group is enhanced by a Lewis acid complex, it should as previously stated lead to successful ring closure. This point is amply exhibited by the cycloaddition of diazomethane to benzonitrile in the presence of triethylaluminium[24]. The products obtained were the *N*-methyltriazoles **23** and **24**.

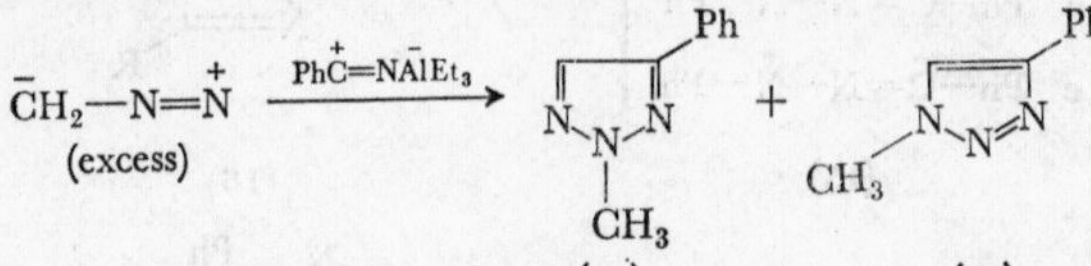

$$\overset{-}{C}H_2{-}N{=}\overset{+}{N} \text{ (excess)} \xrightarrow{Ph\overset{+}{C}{=}N\overset{-}{Al}Et_3} \textbf{(23)} + \textbf{(24)}$$

d. Nitrile ylids. 1,3-Dipolar cycloaddition to nitriles occurs when nitrile ylids (**26**), generated *in situ* from benzamidoyl chloride (**25**), are introduced to their solution[25,26]. The products **27** and **28** were obtained in 21 and 6% yield, respectively. It is surprising that **27** was obtained in higher yield than **28** in view of the fact that the former is derived from the highly energetic resonance form **26c**. However, this may be due to a kinetic effect which is the result of the highly electrophilic benzylic carbons in **26c**.

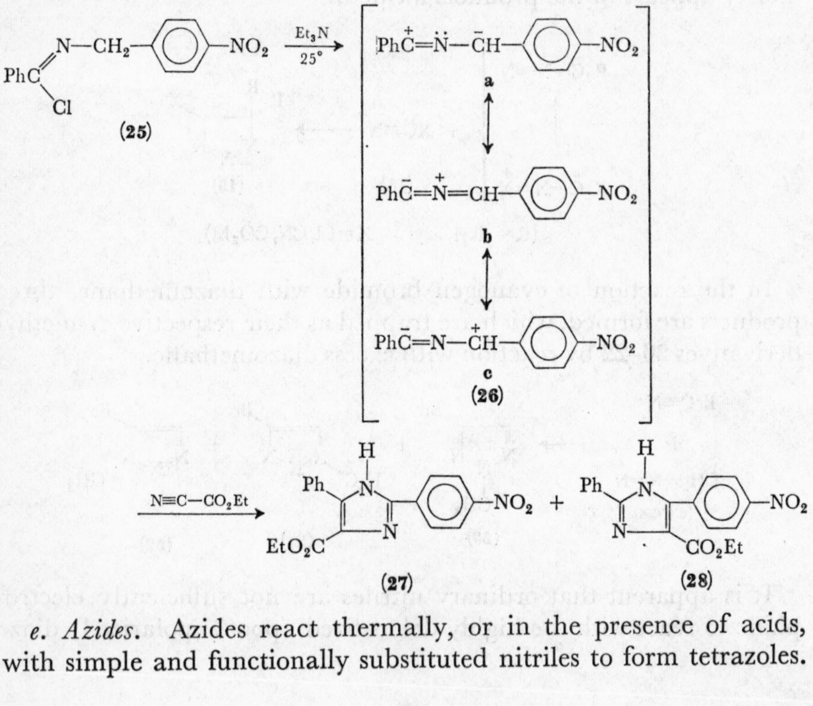

$$\textbf{(25)} \xrightarrow[25^\circ]{Et_3N} \textbf{(26)} \xrightarrow{N{\equiv}C{-}CO_2Et} \textbf{(27)} + \textbf{(28)}$$

e. Azides. Azides react thermally, or in the presence of acids, with simple and functionally substituted nitriles to form tetrazoles.

These cyclizations probably possess a concerted transition state[12], whereas tetrazole formation with azide ion proceeds initially through nucleophilic attack at the cyano carbon atom. This latter group of reactions will be treated in section V.C. Heterocycle formation from organic azides and nitriles produces a variety of interestingly substituted tetrazoles. For example, the thermal cyclization of 2-azido-2′-cyanobiphenyl (**29**) produces the tetrazolophenanthridine **30**[27]. Unactivated aliphatic nitriles are insufficiently polarized to react with the azido substituent, but chlorosulphonic acid catalyses the intramolecular cyclization of 4- and 5-azidoalkyl nitriles, leading to the bicyclic systems **31** and **32**[27,28]. The analogous

(**29**) $\xrightarrow{\Delta}$ (**30**)

intermolecular reaction was carried out at 150° with perfluoroalkyl nitriles[29] giving rise to perfluoroalkyl-substituted tetrazoles (**33**). The highly electrophilic character of fluoro-substituted nitriles was sufficient for cyclization, thus eliminating the need for acidic catalysts. It is somewhat surprising that the isomeric tetrazole **34** was not formed in this reaction.

$(CH_2)_n$(C≡N)(N_3) $\xrightarrow{ClSO_3H}$ (**31**, $n=3$) (**32**, $n=4$)

$R_fC{\equiv}N + RN_3 \longrightarrow$ (**33**) (**34**)

($R_f = CF_3, C_3F_7$, $R = Ph, n\text{-}C_8H_{17}$)

2. Heterocycles from nitriles and 1,3-dipoles of class 2

a. Azomethine imines. A few examples of the reactions of this recently prepared dipolar species with nitriles are known. The

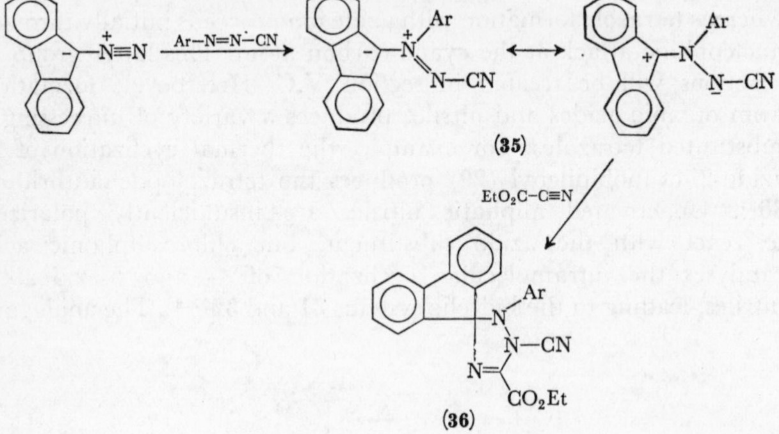

(35)

(36)

azomethine **35** prepared from diazofluorene and phenyldiazocyanide[30] condensed with ethyl cyanoformate to form the spirotriazole **36**[8]. A series of aromatic and electrophilic nitriles have been converted to fused triazoles (**38**) with the dipolar ion **37** (Table 3)[31].

TABLE 3. Triazoles **38** from nitriles.

R	Percentage yield of **38**
CO_2Et	83
Ph	55
o-$NO_2C_6H_4$	45
p-$NO_2C_6H_4$	69
o-NCC_6H_4	87

It is interesting to compare the efficiency of the reaction involving: (*a*) the *ortho*-nitro and (*b*) the *ortho*-cyano substituent in the nitrile undergoing cyclization. The yield of the latter is almost twice that of the former. This may be attributed to the usual *ortho* steric effect

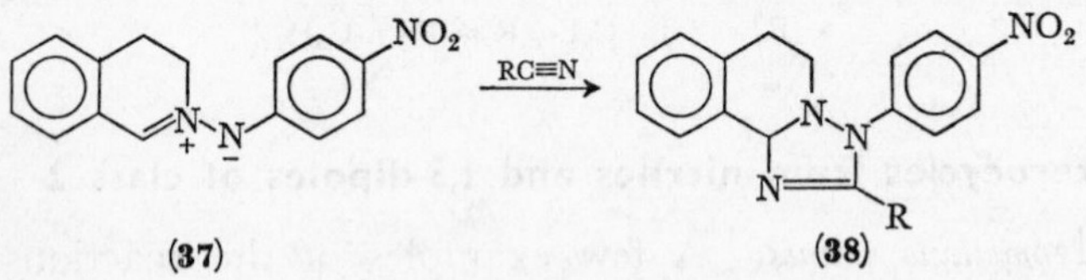

(37)

(38)

or, more convincingly, in the minds of the writers, to an intramolecular partial bonding between the nitro group and the electrophilic carbon of the cyano group as in **38a**, thus rendering the cyano group as a whole less electrophilic.

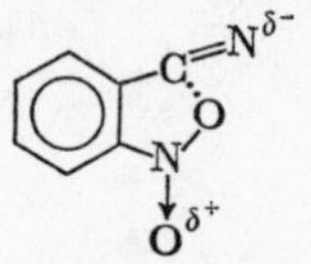

(**38a**)

b. Azomethine ylids. A recent report[32] described the condensation of the mesoionic oxazole **39**, whose canonical form **40** may be regarded as an azomethine ylid, with nitriles to form the bridged lactone **41**. Under the conditions of the reaction, carbon dioxide is lost due to the driving force toward the aromatic imidazole **42**.

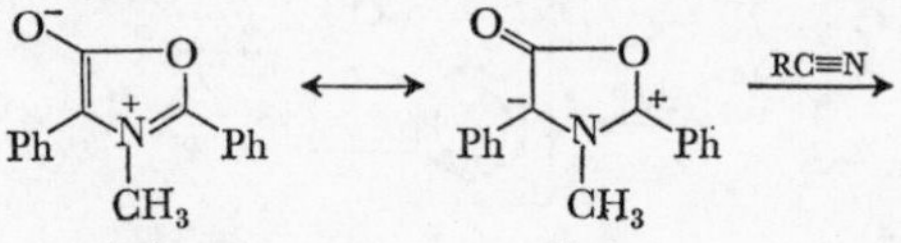

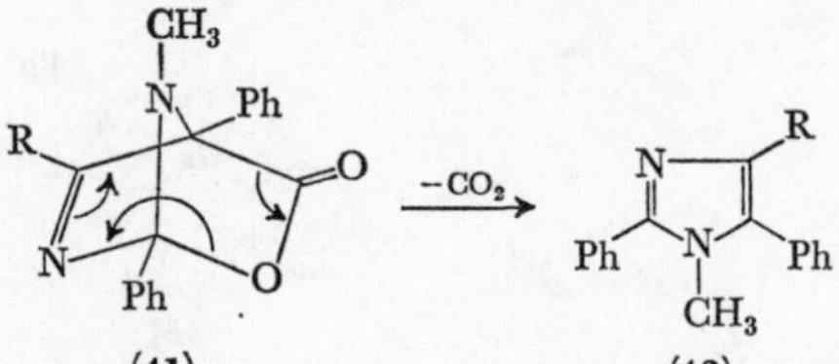

(**39**) (**40**)

(**41**) (**42**)

(R = CO_2Et, C(=O)Ph, 2, 6-$(CH_3)_2C_6H_3$)

3. Heterocycles from nitriles and 1,3-dipoles of class 3

a. Ketocarbenes. The thermal decomposition of a diazoacetophenone in the presence of benzonitrile has been reported[33] to produce 0·4% yield of the oxazole **43** which was undoubtedly formed from the ketocarbene before rearrangement. The yield was increased by the use of copper, presumably due to stabilization of the dipolar singlet species by complexation. A further example utilizing the dipolar ketocarbenes was depicted[8] when the cyclic diazoketone **44** was photochemically decomposed in the presence of benzonitrile to

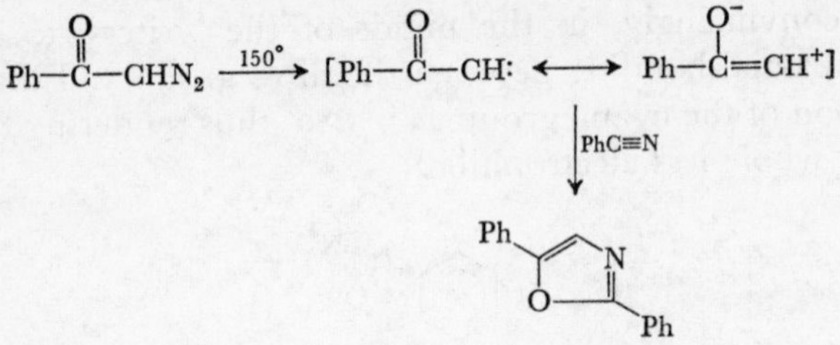

$$Ph{-}\overset{O}{\overset{\|}{C}}{-}CHN_2 \xrightarrow{150^\circ} [Ph{-}\overset{O}{\overset{\|}{C}}{-}CH: \longleftrightarrow Ph{-}\overset{O^-}{\overset{|}{C}}{=}CH^+] \xrightarrow{PhC{\equiv}N}$$

(43)

give the fused tricyclic oxazole **45** in 34% yield. These results clearly exhibit the delocalized and dipolar nature of ketocarbenes. An interesting case of a diazo compound which cannot undergo the competing rearrangement and therefore results in improved conversion to the oxazoles, is the diazoacetic ester **46** which was decomposed in the presence of nitriles to produce oxazoles (**47**)[34] (Table 4).

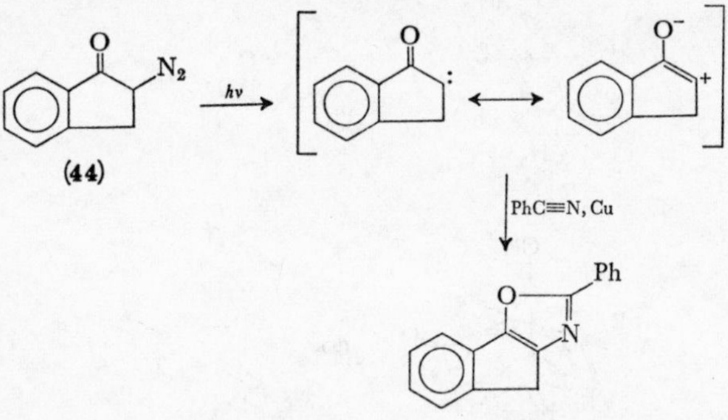

(44) $\xrightarrow{h\nu}$ [↔] $\xrightarrow{PhC{\equiv}N,\ Cu}$ **(45)**

The reaction was found to proceed poorly under photolytic conditions. Acetonitrile has also been reported to give the corresponding oxazole with the trifluoroacetylethoxycarbonyl carbenoid species[35].

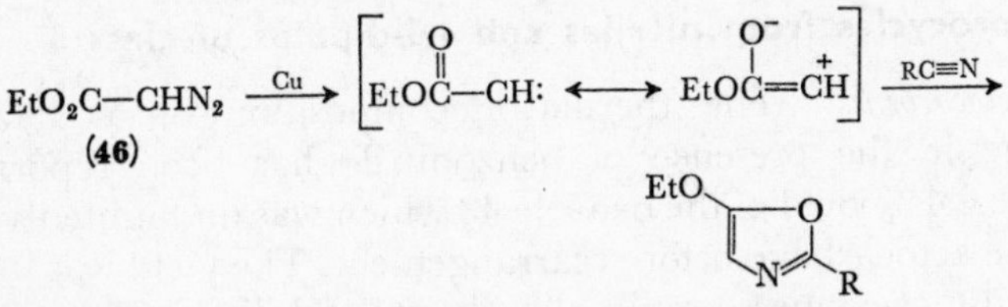

$$\underset{\mathbf{(46)}}{EtO_2C{-}CHN_2} \xrightarrow{Cu} \left[EtO\overset{O}{\overset{\|}{C}}{-}CH: \longleftrightarrow EtO\overset{O^-}{\overset{|}{C}}{=}\overset{+}{C}H\right] \xrightarrow{RC{\equiv}N}$$

(47)

b. Ketonitrenes. The cycloaddition of the nitrene **50** to a variety of nitriles was reported by Huisgen[37] and Lwowski[36]. Photolytic

TABLE 4. Oxazoles **47** from nitriles.

R	Percentage yield
Ph	42
CH_3	31
$PhCH_2$	11

decomposition of ethyl azidoformate (**48**) or the base-catalysed α-elimination of *N-p*-nitrobenzenesulphonoxy urethane (**49**) produced the nitrene in a nitrile solvent. The products were 1,3,4-oxadiazoles (**51**) (Table 5) with none of the 1,2,4-isomers (**52**)

TABLE 5. 1,3-Dipolar addition of ketonitrene to nitriles to form 1,3,4-oxadiazoles (**51**).

R	Percent yield	Reference
CH_3	55–60	36, 37
$(CH_3)_2CH$	64	36
Ph	16	37
$EtOCH_2CH_2$	73	36
CH_2CO_2Et	14	37
CO_2Et	1·7	37
$CH_2{=}CH$	14	36
$4\text{-}ClC_6H_4$	9·3	37
$4\text{-}O_2NC_6H_4$	4·5	37

formed. When acrylonitrile was used, the aziridine derivative was also obtained[36].

$$EtOC(=O){-}N_3 \ (\mathbf{48}) \xrightarrow{h\nu} \left[EtOC(=O){-}\ddot{N} \longleftrightarrow EtOC(O^-){=}N^+ \right] (\mathbf{50}) \xleftarrow{Et_3N} O_2N{-}C_6H_4{-}SO_2{-}ON(H){-}C(=O){-}OEt \ (\mathbf{49})$$

$$(\mathbf{50}) \xrightarrow{RC\equiv N} (\mathbf{51}) \quad \text{not} \quad (\mathbf{52})$$

(**48**) (**49**) (**50**) (**52**) (**51**)

c. Thioketene. This dipolar species (**53**) is much more stable than the corresponding forms of ketenes and will dimerize readily even in the presence of dipolarophiles. Reactions with nitriles do not occur, although highly dipolar nitrile imines do form cycloaddition products (**54**)[38].

(53) (54)

(Z=PhN, O)

D. Six-Membered Heterocycles

The cycloadditions of 1,4-dipolar species to nitriles (to form six-membered heterocyclic rings) are usually not as facile as those of their 1,3-dipolar analogues. Since 1,4-dipoles are not stabilized in the ground state by delocalization because of the extra atom present between the polar terminals, these substances must generate their dipolar character in the transition state. Therefore, many so-called 1,4-dipolar cycloadditions are stepwise rather than concerted and will be treated as examples of type B ring closures. There are several examples of Diels–Alder type concerted 1,4-cycloadditions which lead to heterocycles utilizing the nitrile as the dienophile. These will be discussed as examples of type A ring closures.

Recent reviews on diene syntheses[39,40] and Diels–Alder reactions[41,42] have included, to some extent, the behaviour of nitriles as dienophiles. Janz and his students[43–49] have extensively investigated the cycloaddition of dienes to nitriles. The latter function as dienophiles only at elevated temperatures (200–400°) in gas-phase cyclizations with the aid of acidic surfaces. Thus, the addition leading to the 3,6-dihydropyridines **55** has been successfully carried out, but **55** were never isolated since dehydrogenation to the pyridines **56** is so facile at these temperatures. On reaction with butadiene, acrylonitrile gives, in addition to the 4-cyanocyclohexene, a 2% yield of 2-vinylpyridine[50]. The heterocycle was formed only in the presence of acidic catalysts since in their absence the cyclohexene was the sole product. It was suggested that although Diels–Alder reactions in solution do not allow the cyano group to compete with the olefin as a dienophile, elevated temperatures and Lewis acid-induced polarization of the cyano group lower the energy of the

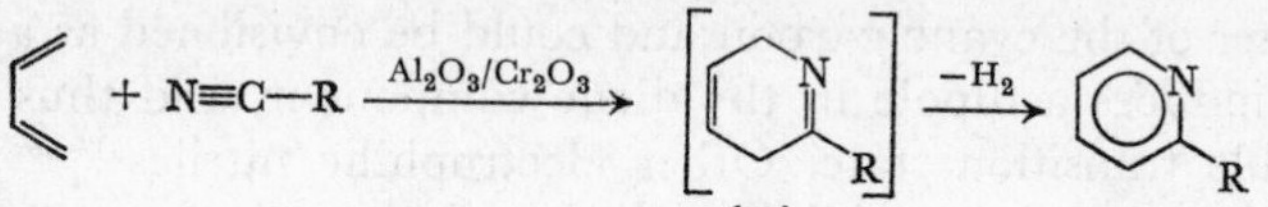

+ N≡C—R $\xrightarrow{Al_2O_3/Cr_2O_3}$ [(55)] $\xrightarrow{-H_2}$ (56)

(R = H, Me, Et, Ph)

transition state leading to the 2-vinylpyridine. The reaction of unsymmetrical dienes **57** with cyanogen give mixtures of both possible products **58** and **59**. The former product was reported[43]

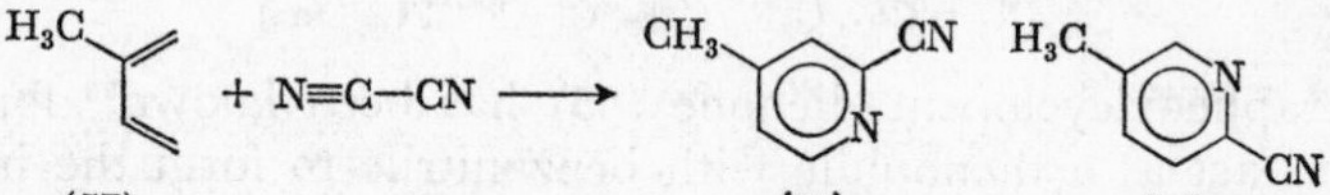

(57) + N≡C—CN ⟶ (58) + (59)

to predominate in the mixture and this is in accord with the slightly lower energy associated with the transition state leading to **58**. Consideration of the transition states **60** and **61** suggests that the former possesses electron deficiency at both the tertiary carbon and the terminal carbon, whereas the latter possesses its electron deficiency at a secondary and primary carbon. This would tend to cause a lowering of transition-state energy in **60**, resulting in a higher proportion of **58** formed. Support for this viewpoint is also found in the reaction of 1,3-pentadiene with benzonitrile at elevated temperatures[45]. Although the yield of **62** was low (18%) it was the only

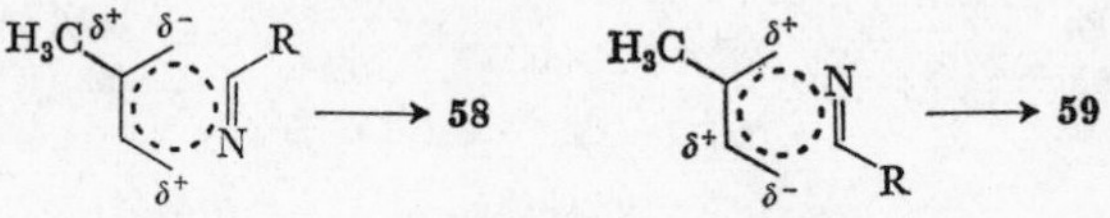

(60) ⟶ 58 (61) ⟶ 59

+ N≡CPh $\xrightarrow{-H_2}$ (62)

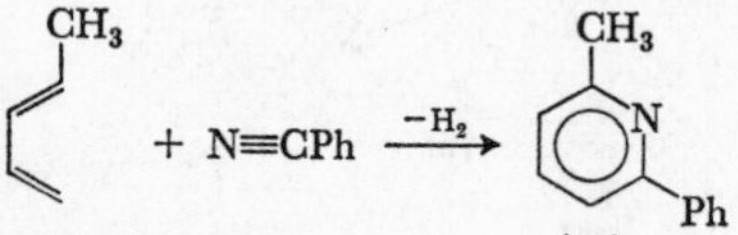

product obtained. It appears that the transition state leading to **62** is reached considerably more efficiently than that leading to the other isomer. The use of nitriles containing perhaloalkyl substituents (**63**) results, as expected, in high conversions to pyridines **64** under milder conditions[48,49]. This is due to the increased electrophilic character of the cyano carbon and could be envisioned as a process which induces a dipole in the diene component, and thus a more accessible transition state. Other electrophilic nitriles[51–55] such as α-ketonitriles, cyanopyridine and phenyl cyanoformate react with dienes to form pyridines at temperatures as low as 200°.

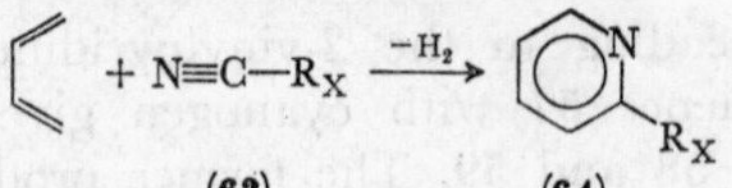

$$\text{diene} + N{\equiv}C{-}R_X \xrightarrow{-H_2} \text{2-}R_X\text{-pyridine}$$

(63) (64)

($R_X = CF_3, C_2F_5, C_3F_7, CF_2Cl, CFCl_2$, etc.)

Tetraphenylcyclopentadienone (**65**) has been known[56] for some time to act as a dienophile with benzonitrile to form the initially bridged adduct **66** which spontaneously loses carbon monoxide to form pentaphenylpyridine (**67**). Recently[55], phenyl cyanoformate was treated with **65** and the pyridine ester **68** was obtained. However, treatment of phenyl cyanoformate with the dimethyl ketal **69** afforded the pyridine dicarboxylic ester **71**, presumably due to the better leaving ability of chloride ion as compared with methoxyl in **70**.

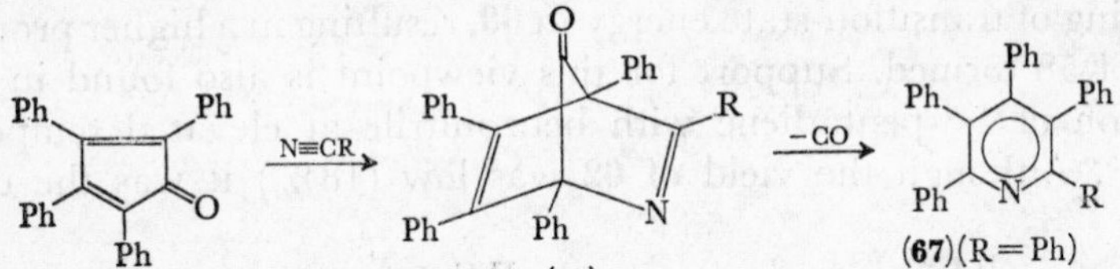

(65) —N≡CR→ (66) —−CO→ (67)(R = Ph) (68)($R = CO_2Ph$)

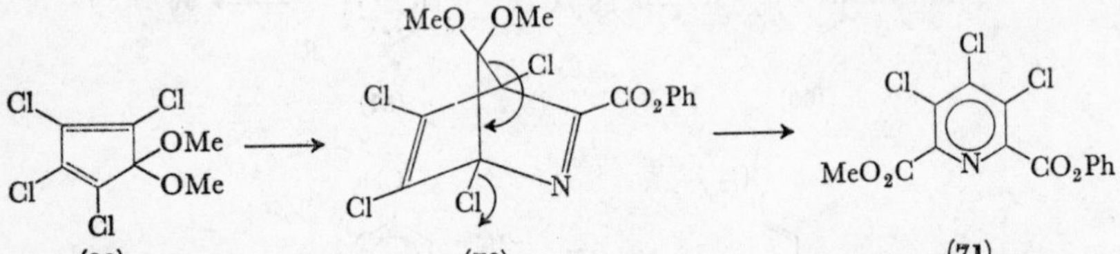

(69) → (70) → (71)

III. ELECTROPHILIC ADDITION TO THE CYANO GROUP FOLLOWED BY RING CLOSURE (TYPE B)

A. General Considerations

The observations by Meerwein[57] and Ritter[58] that the weakly basic nitrile group can accept a strong electrophile to form nitrilium salts which give amides (**72**) on hydrolysis provoked the idea that this behaviour would be useful for the synthesis of heterocycles. Thus if the attacking electrophile possesses a suitably situated nucleophilic site (**73**), ring closure will take place leading to a wide variety of heterocyclic systems (**74**). Since the nature of the nucleophilic site will determine the type of heterocyclic formed, this section

$$R^1CN \xrightarrow[H_2C{=}C(CH_3)_2,\ H_2SO_4]{Et_3O^+MX_4^-} [R^1{-}C{\equiv}\overset{+}{N}{-}R^2 \longleftrightarrow R^1{-}\overset{+}{C}{=}NR^2]\ X^-$$

$(R^2 = Et, t\text{-Bu};\ X^- = BF_4, AlCl_4, HSO_4)$

$$\downarrow H_2O$$

$$R^1C(=O){-}NHR^2$$

(**72**) $(R^2 = t\text{-Bu})$

$$R{-}\overset{+}{C}{=}N{-}R \ (\ddot{B}) \longrightarrow R{-}C(=N{-}R)(B)$$

(**73**) (**74**)

will be concerned with this aspect. Reviews have appeared[5–7] (see also Chapter 6) which deal in detail with generating the nitrilium ion, and therefore no special emphasis will be laid upon the latter except to illustrate their utility in initiation of the heterocyclic synthesis.

B. Cyclization of Nitrilium Ions by Reaction with C=C Bonds

1. Pyrroline derivatives

2,5-Dimethyl-2,4-hexadiene (**75**) and the corresponding glycol (**76**) condense with a variety of nitriles to form substituted 1-pyrrolines (**78**) by cyclization of the nitrilium ion **77** with the olefinic linkage[59]. The scope of this reaction has been reviewed[5,7]. There have been no other reports describing five-membered heterocycles utilizing this approach, although an attempt to induce 2,3-dimethylbutadiene to form the pyrrole **79** met with extensive polymerization and no recognizable products[59,60].

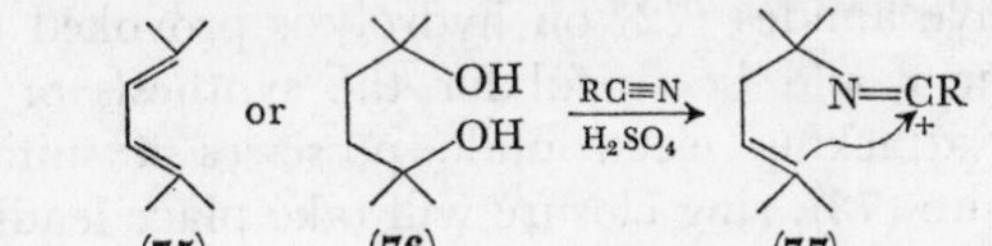

or

OH OH $\xrightarrow[H_2SO_4]{RC\equiv N}$ N=CR +

(75) (76) (77)

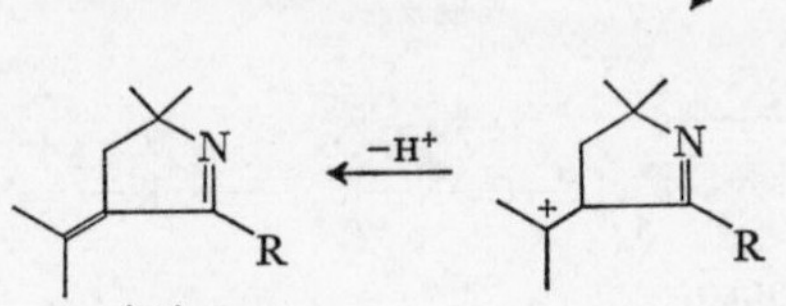

N $\xleftarrow{-H^+}$ N + R R

(78)

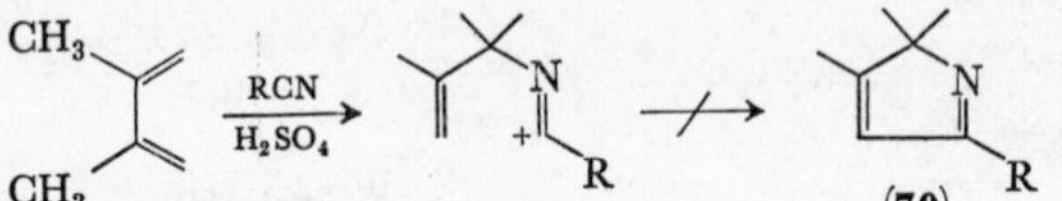

CH_3 CH_3 $\xrightarrow[H_2SO_4]{RCN}$ N + R ⟶/ N R (79)

2. Pyridine derivatives

In a similar fashion to that described above[59], the glycol **80** was transformed into a 5,6-dihydropyridine derivative (**81**) via attack of the olefin linkage (generated *in situ*) by the nitrilium ion **82**. An

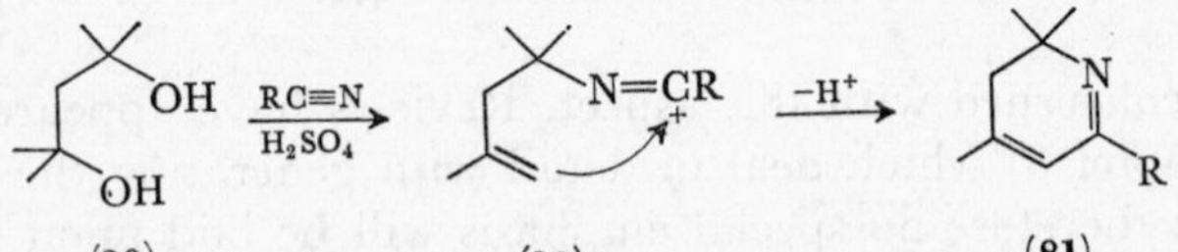

OH OH $\xrightarrow[H_2SO_4]{RC\equiv N}$ N=CR + $\xrightarrow{-H^+}$ N R

(80) (82) (81)

extension of this reaction was reported utilizing the glycol **83** which produced the pyridine system **84**. This and related reactions have been described in detail elsewhere[5,7].

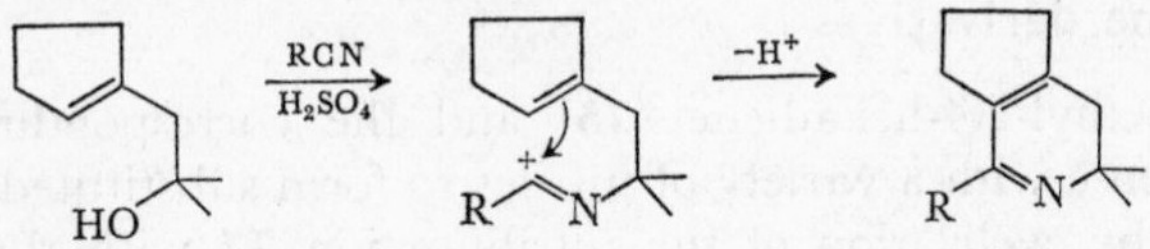

HO $\xrightarrow[H_2SO_4]{RCN}$ R + N $\xrightarrow{-H^+}$ R N

(83) (84)

3. Isoquinoline derivatives

Besides the use of an isolated double bond serving as a nucleophile, the aromatic nucleus is a useful substrate toward electrophilic attack.

Ritter and Murphy[61] showed that methyl eugenol (**85**) in sulphuric acid condenses with nitriles to form the nitrilium ions **86** which spontaneously cyclized to dihydroisoquinolines (**87**). Nitrilium ions (**88**), generated by using other Lewis acids, allowed Lora-Tamayo[62] to produce a variety of dihydroisoquinolines without the 3-methyl substituent. The advantage of the latter method lies in the fact that primary electrophilic carbons can be generated. This method has been employed[63] using ethyl cyanoacetate to prepare the 2-carbethoxymethyl derivative **89** (R = —CH_2CO_2Et). Employing stannous chloride as the Lewis acid has led to derivatives of **89**

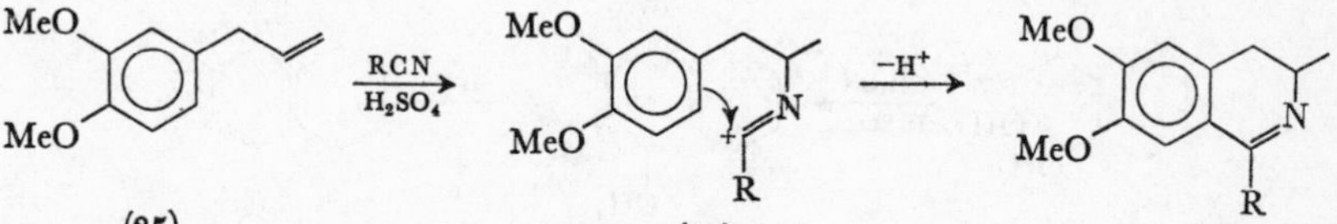

MeO MeO RCN H_2SO_4 MeO MeO N R −H⁺ MeO MeO N R

(**85**) (**86**) (**87**)

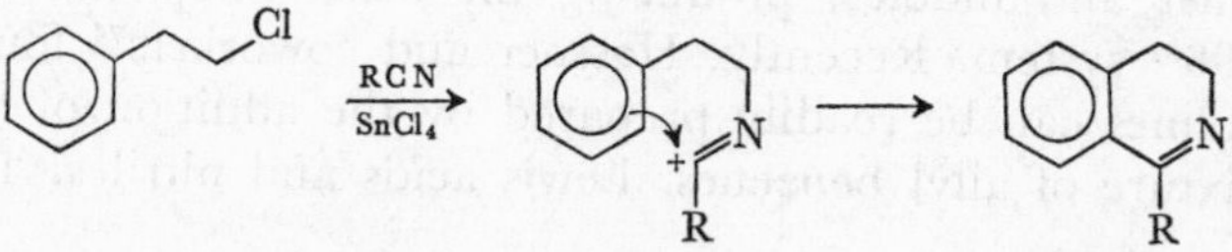

Cl RCN $SnCl_4$ N R N R

(**88**) (**89**)

containing longer carboxyalkyl side chains[64–66]. The use of ethyl cyanopropionate leads to puzzling results. When this nitrile is treated with 2-phenylethyl chloride in the presence of stannic chloride the product is reported to be **90**. Attempts to reproduce this experiment[67,68] resulted in a product which could not be identified; it was obtained as a deep-blue oil which resisted purification and characterization. The dihydroisoquinoline (**91**) could not be isolated presumably due to a facile intramolecular cyclization. It is conceivable that the product originally described as **90** is more accurately depicted as its tautomer (**92**), which would be expected to be somewhat unstable. The hexahydroisoquinoline **94** was formed by the cyclization of the intermediate nitrilium ion **93** in the presence of concentrated sulphuric acid[69]. Although a mixture of isomers formed, the isolated product was pure **94** due to isomerization to the cisoid system. Further examples involving nitrilium ion cyclization

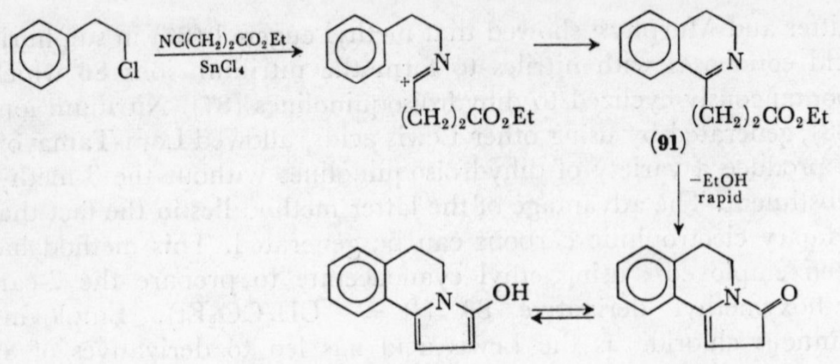

$NC(CH_2)_2CO_2Et$, $SnCl_4$ → (91) → −EtOH, rapid → (90) ⇌ (92)

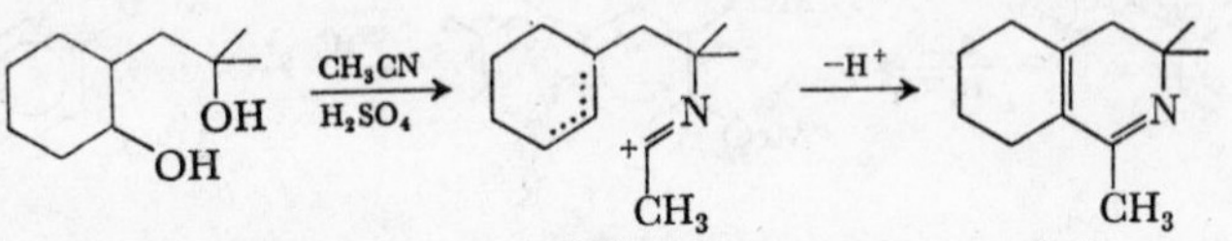

CH_3CN, H_2SO_4 → (93) → $-H^+$ → (94)

to unsaturated carbon bonds have been reported with 2-chloroethylthiophenes and indoles, producing the fused thiophene **95**[70] and indole **96**[62] system. Recently, Hassner and coworkers[71] found that isoquinolines can be readily prepared by the addition of halogens to a mixture of allyl benzenes, Lewis acids and nitriles. Thus, by

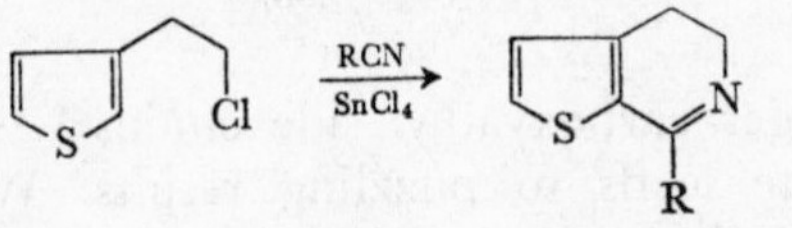

RCN, $SnCl_4$ → (95)

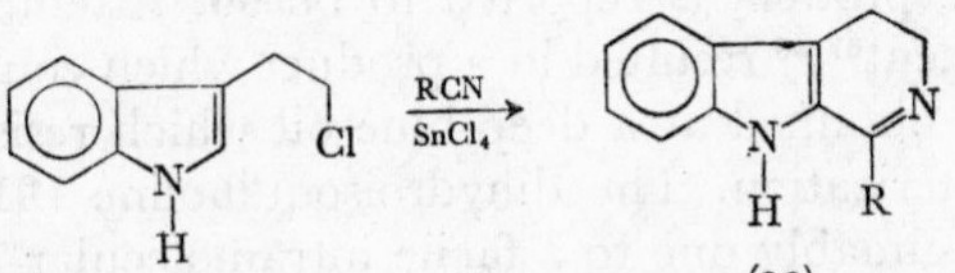

RCN, $SnCl_4$ → (96)

the addition of bromine or iodine to methallylbenzene the halonium salt **97** is produced. The latter is susceptible to attack by the nitrile to form the nitrilium salt **98**. The latter cyclizes spontaneously to the 2-halomethyldihydropyridine **99**. A variety of nitriles, which also serve as the solvent, have been employed. An interesting feature in this approach is that it allows the preparation of 3,3-disubstituted isoquinoline derivatives which cannot be obtained by the Bischler–Napieralski procedure. The fused aziridine derivative (**100**) was also obtained when **99** was reduced with sodium borohydride.

(R^1 = H, CH_3) —X_2 / Lewis acid→ (**97**) + [X-Lewis acid]$^-$

(**98**) —$-H^+$→ (**99**) —$NaBH_4$ / EtOH→ (**100**)

(R^2 = CH_3, Ph, $PhCH_2$)

A novel reaction involving nitrilium salts has been observed[72] when the diazonium fluoborate of 2-aminobiphenyl (**101**) is decomposed in acetonitrile solution. 6-Methylphenanthridine (**104**) was isolated in 70% yield. The reaction is believed to occur by virtue of the highly electrophilic cation **102** which reacts with the solvent to form the nitrilium ion **103**. The latter then cyclizes to **104**. Isolation of the *N*-arylacetamide (**105**) as a minor product lends support to the proposed mechanism.

4. Seven- and higher-membered heterocycles

There has appeared no report describing the cyclization of a nitrilium ion to a double bond leading to seven- or higher-membered systems.

C. Cyclization of Nitrilium Ions by Reaction with C=O Bonds

1. Oxazole derivatives

The oxygen atom of a carbonyl group may serve as a nucleophilic centre for a nitrilium ion. The earliest example of this type was reported by Japp and Murry[73], who found that heating benzoin with benzonitrile in concentrated sulphuric acid produced the oxazole **107**. Much later, Lora-Tamayo and coworkers[62,74] prepared the same oxazole utilizing stannic chloride and desyl chloride (**108**). The oxazole obtained from benzoin may be formed through the nitrilium ion **106** which cyclizes to the hydroxyl group and then suffers dehydration. A third pathway which may be considered, namely via the intermediate **109**, is not likely in view of the unstable species which results by generating a cation adjacent to a carbonyl group. The mechanism of the stannic chloride process (**108** → **107**) appears to involve the cyclization of the chloroamide and not the intermediate cation derived from **109**. It is to be recalled that primary halides (phenylethyl chloride) react smoothly with nitriles in the presence of stannic chloride and other Lewis acids. On the other hand, only tertiary, benzylic and on occasion secondary alcohols condense with nitriles in the presence of sulphuric acid or other Lewis acids. Although it is believed that both the alcohol and the halide reactions proceed via nitrilium ions and that the subsequent cyclizations are also comparable in scope, *the process leading to the nitrilium ion is different*. When a nitrilium ion (**110**) is formed from an alcohol (or olefin), carbonium ion production may be the rate-determining step, whereas reaction with the nitrile is a rapid

$$\text{PhCH(OH)COPh} \xrightarrow[\text{PhCN}]{H^+} \text{(106)} \xrightarrow{-H_2O} \text{(107)}$$

(106) (107)

$$\text{PhCH(OH)COPh} \xrightarrow{H^+,\ \text{PhCN}} \text{Ph(COPh)CH–}\overset{+}{N}{=}\text{CPh (109)}$$

$$\text{PhCH(Cl)–C(=O)Ph (108)} \xrightarrow{\text{SnCl}_4,\ \text{PhCN}} \text{(107)}$$

(109) (108)

process. However, when the nitrilium ion is formed from an alkyl halide (primary appears to be most favourable), the nitrile–Lewis acid complex[57] is probably displaced nucleophilically in the slow step by the non-bonding electrons on halogen (**111**) accompanied by a simultaneous displacement of halogen by the nitrile. This would make nitrilium ions formed by this process subject to steric requirements similar to those observed in nucleophilic substitutions. If this view is correct, then the transition state **111** should be more difficult

$$\left.\begin{array}{l}R^1\text{—OH}\\ R^1(\text{C}{=}\text{C})\end{array}\right\} \xrightarrow[\text{slow}]{H^+} R^{1+} \xrightarrow[\text{fast}]{R^2C{\equiv}N} R^2\text{—N}{=}\overset{+}{C}\text{—}R^1 \quad \textbf{(110)}$$

$$R^2\text{—C}{\equiv}N{\rightarrow}SnCl_4 \xrightarrow{R^1X} R^2\text{—C}{\equiv}N{\rightarrow}SnCl_4 \cdots R^1\text{—X} \ \textbf{(111)} \xrightarrow{\text{slow}} \textbf{(110)}$$

to reach when R^1 is a bulky group. This is borne out by the reluctance of 2-chlorocyclohexanone to react with the nitrile–stannic chloride complex[74]. Furthermore, if R^1 is secondary or tertiary, then considerable carbonium ion character can develop in the transition state, that is X leaves to coordinate with the stannic chloride before the nitrile nitrogen attacks R^1. If R^1 develops such carbonium ion character then the presence of strong electronegative groups (i.e. C=O) would tend to destabilize the transition state, as in the case of 2-chlorocyclohexanone. Studies of a systematic nature are obviously required before any firm conclusions regarding this process can be reached.

Benzilic acid has been reported[75–77] to condense with benzonitrile in the presence of concentrated sulphuric acid or stannic chloride forming the triphenyloxazolidone **112**. Although the intermediate is presumed to be the cation **113**[75] the nitrilium ion **114** could conceivably arise by attack on the α-lactone **115**. To date, there are no known reports of α-lactone formation and this process could be of interest in further studies on these strained systems.

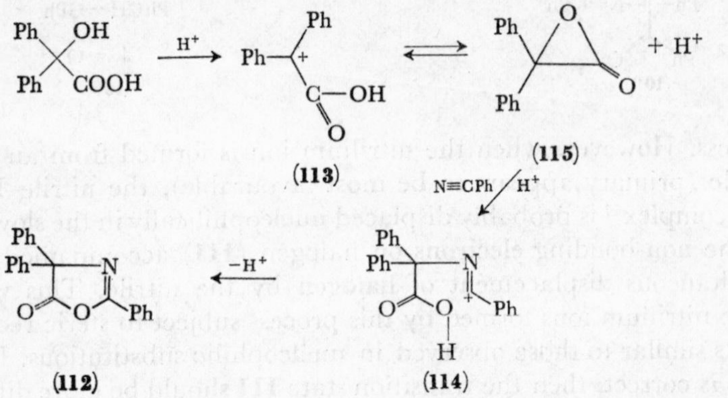

(113) (115) (112) (114)

2. Oxazine derivatives

β-Chloroketones have been treated with nitrile–stannic chloride complexes to produce 1,3-oxazines (**119**)[78]. The carbonyl group oxygen is acting as the nucleophile which coordinates with the nitrilium ion **118**. As mentioned in the previous section, the tertiary halide **116** probably forms the carbonium ion **117** by transfer of halide ion to the metal salt. Also produced in this reaction are the amides **120** resulting from hydrolytic cleavage of **119** or incomplete cyclization of **118**. An interesting variation to the above is the work of Ziegler and coworkers[79,80], who obtained 1,3-oxazine-4-ones (**121**) by treating nitriles with malonyl chloride derivatives.

D. Cyclization of Nitrilium Ions by Reaction with OH Groups

1. Oxazoles

An attempt to prepare the oxazoline **123** by condensation of pinacol with nitriles failed to produce the requisite nitrilium ion

(116) $\xrightarrow{SnCl_4 \leftarrow N{\equiv}CR}$ (117) $\xrightarrow{N{\equiv}CR}$ (118) $\xrightarrow{-H^+}$

(119) (120)

122. Apparently, the pinacol rearrangement proceeds faster than attack of the nitrile to the carbonium ion[60].

$\xrightarrow{RCN}$ $\xrightarrow{-H^+}$ (121)

$\xrightarrow[H_2SO_4]{N{\equiv}CR}$ (122) $\not\rightarrow$ (123)

2. Oxazines

Ritter and Tillmanns[81] were able to prepare dihydro-1,3-oxazines (**126**), in fair yield by treating 2-methyl-2,4-pentanediol (**124**) with several nitriles in cold sulphuric acid. It is interesting to note that the secondary hydroxyl group does not produce the carbonium ion (or the olefinic linkage) and can therefore function as a nucleophile toward the nitrilium ion **125**. A variety of nitriles have been employed[82], primarily to prepare tetrahydro-1,3-oxazines (**127**)

(124) $\xrightarrow[H_2SO_4]{N{\equiv}CR}$ (125) $\xrightarrow{-H^+}$ (126)

TABLE 6. Tetrahydro-1,3-oxazines (**127**)[83].

R	Percentage of **127**(D)
Et	86
CH_2CO_2Et	78
CH_2Ph	88
Ph	85
2-Pyridyl	72

(Table 6) which have been reported[83] to be useful precursors to aldehydes and their $C_{(1)}$ deuterated derivatives (**128**). The glycol **124** has been employed with dinitriles to form bisoxazinylalkenes (**129**)[84]. The details of this process have been discussed[5].

BH_4^- or BD_4^-, −30° → (**127**); H^+ → (**128**)

(**129**)

3. Miscellaneous oxygen nucleophiles

A novel heterocyclic synthesis has been described by Jones[85] while attempting to produce diphenylcyclopropenylidene (**132**). When the acid **130** was treated with the electrophile triphenylmethyl perchlorate, the resulting cyclopropenyl cation (**131**) reacted with the acetonitrile solvent to produce the oxazolone (**133**).

Another novel preparation of oxazines entails the use of haloalcohols (**134**) or haloolefins (**135**) in the presence of nitriles and sulphuric acid. For example, the nitrilium salt **136** when diluted with water is rapidly transformed into the chloroamide **137**, which spontaneously cyclizes to the oxazine **138**[86]. By employing γ-methallyl chloride and nitriles in acidic medium, the oxazolines (**139**) are similarly produced[87].

Epoxides have been employed under acidic conditions to produce β-hydroxynitrilium ions (**140, 141**) which can cyclize to oxazolines

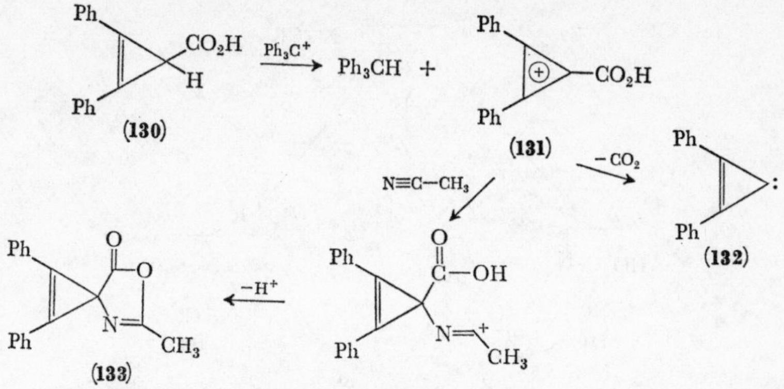

(130) (131) (132) (133)

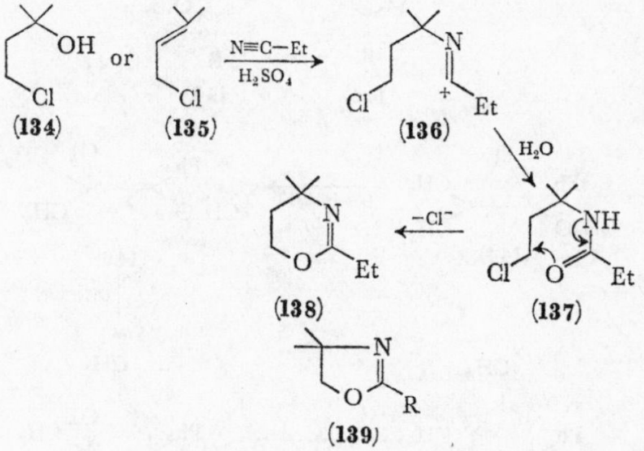

(134) (135) (136) (137) (138) (139)

(**142, 143**). Ring opening of the epoxide follows the expected course resulting in mixtures of products[88].

Epoxy ethers (**144**) have been used to form the oxazoline **145** by treatment with stannic chloride[89] or perchloric acid in acetonitrile solution[90]. In this case the more stable α-methoxyphenyl cation (**146**) is formed in preference to the tertiary carbonium ion.

E. Cyclization of Nitrilium Ions by Reaction with SH Groups

1. Thiazole derivatives

A thio-containing molecule capable of generating a carbonium ion will, as in previously described sections, form a nitrilium salt which will cyclize to the sulphur atom. β-Methallylmercaptan (**147**)

(140) (141) (142) (143) (144) (145) (146)

or its hydration derivative **148** condenses with nitriles in concentrated sulphuric acid to form the nitrilium ion **149** which proceeds to the thiazoline systems (**150**)[59,86]. The use of the mercapto alcohol gave improved yields of 2-thiazolines and it has been recently found[91] that the latter can be efficiently reduced to thiazolidines (**151**) with aqueous sodium borohydride.

A modification of the above synthesis was recently reported by Helkamp and coworkers[92,93] who claimed a new route to these compounds. The use of episulphides (**152**) to form the thiazolines **153** is, in effect, closely related to the scheme producing **150**. The novelty in this approach lies in the fact that stereospecific ring-opening results in the *trans*-thiazoline (**155**) from the *cis*-episulphide (**154**) and

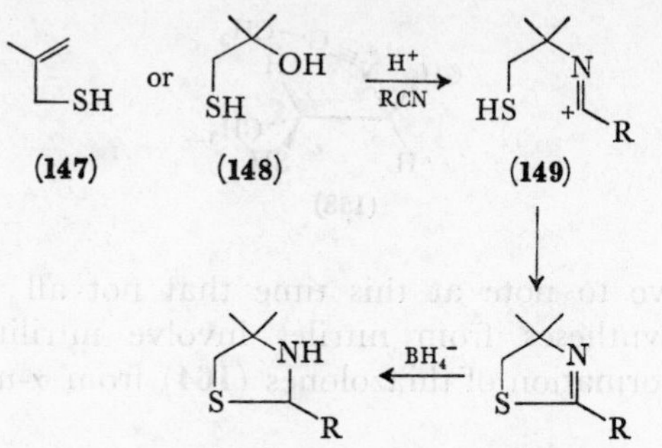

(147) (148) (149)

(151) (150)

(R = CH_3, Ph, CH_2CO_2Et, CH = CH_2, CH_2Ph)

the *cis*-thiazoline (**157**) from the *trans*-episulphide (**156**). The inversion of the episulphide configuration in the 2-thiazoline was depicted as proceeding through the intermediate **158.** It is somewhat surprising that these authors were not aware of the previous thiazoline synthesis from nitriles and mercaptoalkenes.

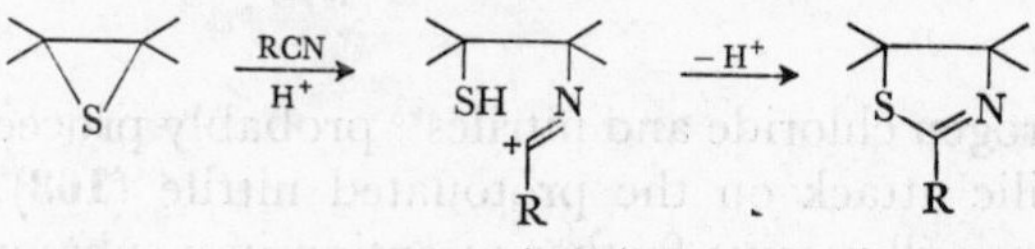

(152) (153)

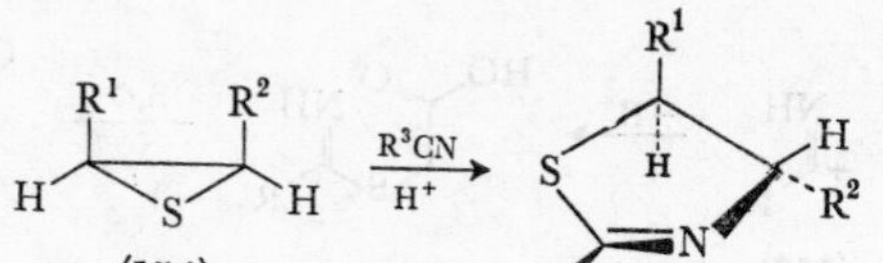

(154) (155)

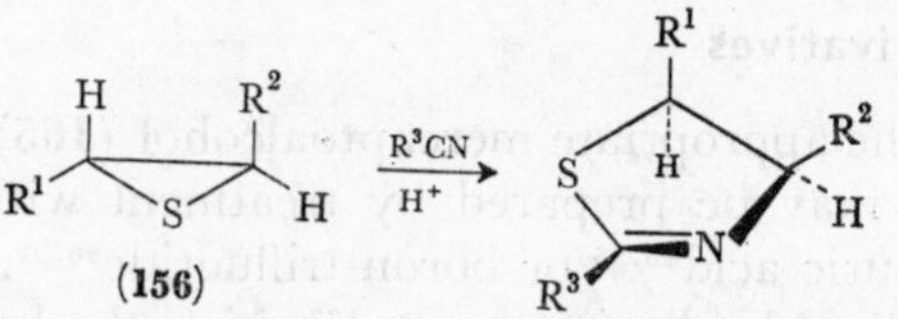

(156) (157)

The mercaptoketone **159** which may also form the electrophilic species **160** has been reported to form the thiazole derivative **161** when treated with hydrogen chloride in acetonitrile[94]. The alternative mechanism involving **162** cannot, however, be readily discarded.

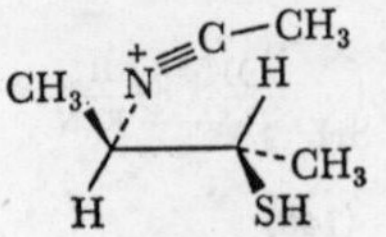

(158)

It is instructive to note at this time that not all acid-catalysed heterocyclic syntheses from nitriles involve nitrilium ions. For example, the formation of thiazolones (**164**) from α-mercaptoacetic

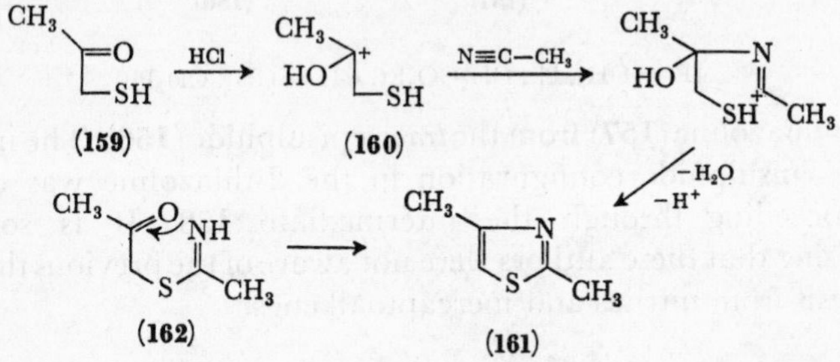

(159) (160) (161) (162)

acid, hydrogen chloride and nitriles[95] probably proceeds by an initial nucleophilic attack on the protonated nitrile (**163**). This type C mechanism will receive further attention in a subsequent section.

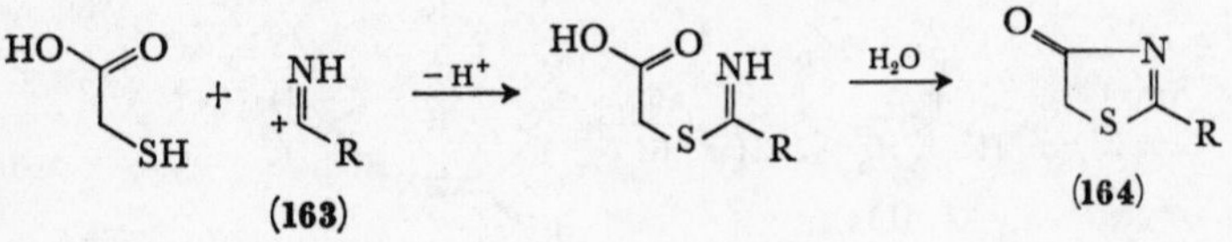

(163) (164)

2. Thiazine derivatives

By choosing the appropriate mercaptoalcohol (**165**), dihydro-1,3-thiazines (**167**) may be prepared by treatment with a variety of nitriles in sulphuric acid[86,95] or boron trifluoride[96,97]. The dihydrothiazines **167** have also been prepared[97] using the benzyl thioether of **165** ($R^2 = PhCH_2$) which ejects the benzyl group during the cyclization of the nitrilium ion **166.** Thioketones have produced, in low yield (10–15 %), thiazines containing an acylamino side chain (**171**)[98]. The reaction proceeds via the nitrilium salt **168** which cyclizes with concomitant removal of the benzyl group from the

CH_3 CH_3 OH SR^2 (**165**) $\xrightarrow[N{\equiv}C{-}R^1]{H^+}$ CH_3 CH_3 N $\overset{+}{S}R^2$ R^1 (**166**) $\xrightarrow{-R^{2+}}$ CH_3 CH_3 N S R^1 (**167**)

($R^1 = CH_3$, Ph, $PhCH_2$, CH_2CO_2Et, $CH{=}CH_2$, $R^2 = H$, $PhCH_2$)

CH_3 O S CH_2Ph $\xrightarrow[H_2SO_4]{N{\equiv}CR}$ CH_3 HO N R $\overset{+}{}$ S CH_2Ph (**168**) $\longrightarrow$ CH_3 HO N R S (**169**) + $PhCH_2^+$

(**169**) $\xrightarrow{-H_2O}$ CH_3 N R S (**170**)

(**169**) $\xrightarrow[N{\equiv}CR]{H^+}$ (**171**)

(**170**) $\xrightarrow[N{\equiv}CR]{H^+}$ CH_3 RCONH N R S (**171**)

(R = Me, Et)

4-hydroxythiazine **169**. The presence of excess nitrile allows a normal Ritter reaction to occur via the tertiary carbonium ion generated from **169** (or **170**).

F. Cyclization of Nitrilium Ions by Reaction with Nitrogen-Containing Compounds

1. Imidazole derivatives

The highly electrophilic aziridinium salt (**172**) attacks nitriles to give an intermediate nitrilium ion (**173**) which cyclizes with expulsion of the benzyl group, a behaviour similar to that observed in the thiazine series. The products **174** and **175** are obtained[99,100]. Aziridines have also been utilized to form imidazolinium salts (**176**) by treating them with perchloric acid in acetonitrile[90]. The site of ring opening appears to be specific as a result of the formation of the more stable (phenyl conjugated) carbonium ion. This reaction has been carried out[101,102] with aziridinium fluoborate, producing the imidazolinium fluoborates (**177**). It is reasonable to assume that

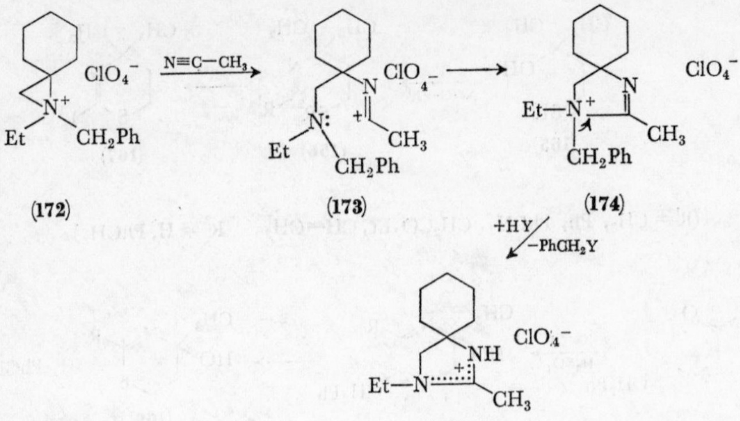

ClO_4^- $N{\equiv}C{-}CH_3$ ClO_4^- ClO_4^-

Et CH_2Ph Et CH_3 CH_2Ph Et—N+ CH_3 CH_2Ph

(172) (173) (174)

+HY, $-PhCH_2Y$

ClO_4^- NH Et—N CH_3

(175)

the very poorly nucleophilic nitrile is sufficient to open the protonated aziridine whereas the fluoborate anion is incapable of this chore.

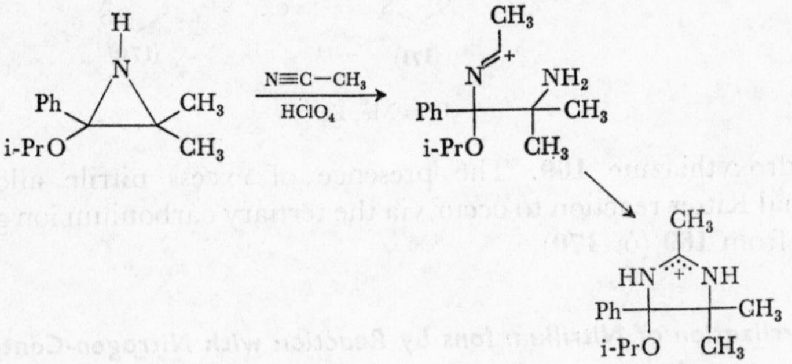

H N Ph CH_3 i-PrO CH_3 $N{\equiv}C{-}CH_3$ $HClO_4$ CH_3 N+ NH_2 Ph CH_3 i-PrO CH_3

CH_3 HN + NH Ph CH_3 i-PrO CH_3

(176)

H_2 N+ BF_4^- N≡CR R HN + NH BF_4^-

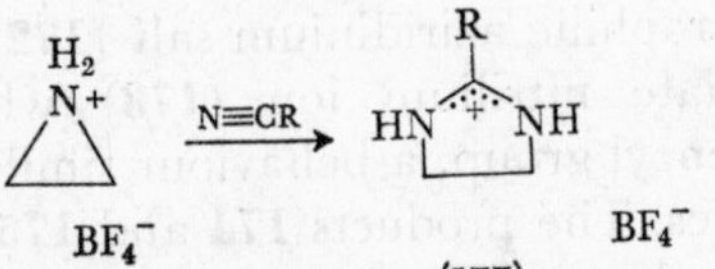

(177)

Azirines (**178**) also undergo ring expansion when treated with a solution of perchloric acid in acetonitrile[90]. The product obtained is the hydroxyimidazolinium perchlorate (**181**). The mode of ring cleavage in **178** was established by the use of acetonitrile-^{15}N which placed the label adjacent to the *gem*-dimethyl group.

Thus ring cleavage occurs, as expected, to give the nitrilium ion **179** followed by ring closure and addition of water to **180.** It is noteworthy to mention that all of the nitrilium ion cyclizations presented thus far may be classified as Ritter reactions.

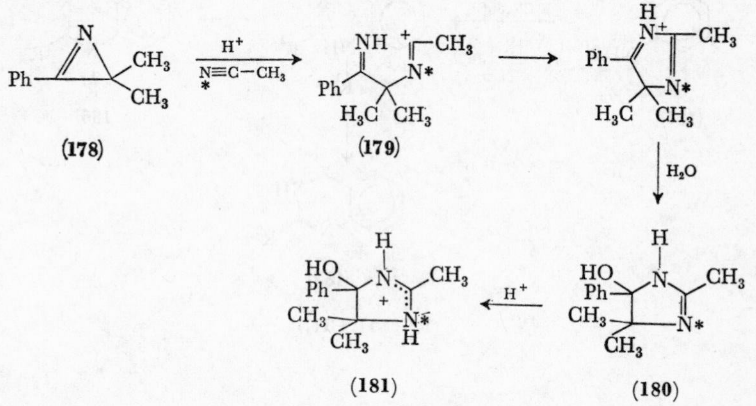

N, Ph, CH_3, CH_3 — H^+, $N{\equiv}C{-}CH_3$ → NH, +, CH_3, N*, Ph, H_3C, CH_3 → H, N^+, CH_3, Ph, N*, H_3C, CH_3

(178) (179)

H_2O

H, HO, Ph, N, CH_3, +, CH_3, N*, CH_3, H ← H^+ — H, HO, Ph, N, CH_3, CH_3, N*, CH_3

(181) (180)

2. Quinazoline derivatives

The nitrilium ion **183** formed by reaction of a nitrile and the *in situ*-generated aryl nitrilium ion **182** have been utilized extensively for 1,3-quinazoline (**184**) syntheses[103,104]. The aryl nitrilium ion can be considered as a powerful electrophile which serves the same purpose as carbonium ions in reactions with nitriles.

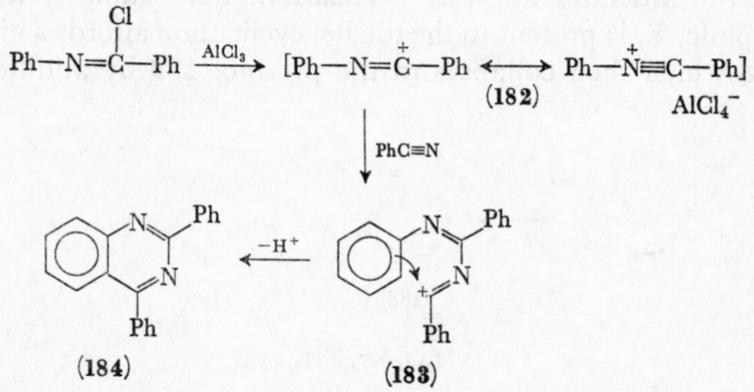

$$\text{Ph—N=C(Cl)—Ph} \xrightarrow{AlCl_3} [\text{Ph—N=}\overset{+}{\text{C}}\text{—Ph} \longleftrightarrow \text{Ph—}\overset{+}{\text{N}}\text{≡C—Ph}]\ AlCl_4^-$$

(182)

PhC≡N

N, Ph, N, Ph ← $-H^+$ — N, Ph, N, +, Ph

(184) (183)

Lora-Tamayo and his coworkers[105,106] have extended their studies on arylalkyl halides and nitriles in the presence of stannic chloride to the formation of quinazolines (**186**) by utilizing *o*-aminobenzyl chloride (**185**). The product obtained from the primary amino compound rearranges to the isomeric quinazoline **187**. Further examples of this reaction have already been discussed[5].

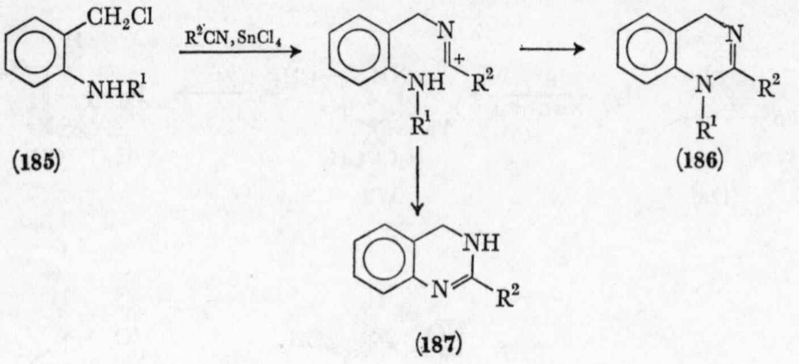

$R^2CN, SnCl_4$

(185) **(186)** **(187)**

$(R^1 = H, CH_3)$

IV. INTRAMOLECULAR ELECTROPHILIC ADDITION TO THE CYANO GROUP (TYPE B_i)

A. General Considerations

Although cyclizations of nitrilium salts, described in the previous section, can follow many varied paths in an intermolecular process, the major difference observed in an intramolecular process is the fate of the nitrilium ion *after* cyclization. For example, when an electrophile, E, is present in the nitrile, cyclization affords a nitrilium ion **188** which then collapses to the product **189** by addition of a

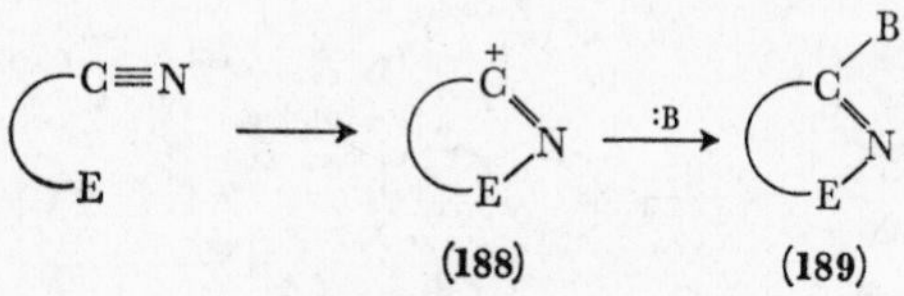

(188) **(189)**

$(B = H_2O, X^-, NH_3, \text{etc.})$

nucleophile, B. If the nucleophile is water or alcohol, lactams or iminoethers result whereas if the nucleophile is halide ion, iminohalides are formed. The electrophilic moiety can be any of a number of common species such as carbonium ion, oxocarbonium ion, nitrilium ion, etc.

B. Cyclization of Olefinic Nitriles

Only a few cases describing the intramolecular cyclization of a carbonium ion to the cyano group have been reported. 4-Pentenonitrile (**190**) produced[107], under typical Ritter conditions, 5-methylpyrrolidone (**192**). The cyclic nitrilium salt **191** can be considered as the intermediate. It is of interest that 2-cyclopentenone, derived from the ion **193**, did not form even though a similar cyclization process has been reported[108] to occur when the heterocyclic ring to

H_2SO_4 ; CH_3 ; C≡N ; (**190**) ; CH_3 N$^+$ (**191**) ; H^+ ; OH^- ; C=NH ; H_2O ; O ; (**193**) ; CH_3 N H O (**192**)

be formed contains less then seven atoms. For example, the unsaturated nitrile **194** produced the lactam **195** along with the unsaturated ketone **196** whereas **197** gave only the ketone **198** from the Hoesch cyclization. The report[109] that **199** under similar conditions gave **200** and the above-mentioned formation of **192** appear to be at variance with the cyclization mechanism, although the relative positions of the CN and olefinic linkages are the same. The factor which determines which reaction will occur (Ritter or Hoesch) must depend upon something other than the steric requirements in the reaction. Since the olefin linkage is considerably more basic than the nitrile, this would kinetically favour a carbonium ion formation followed by attack on the nitrile function (Ritter process). On the other hand, the strain produced in the transition state would favour some ring sizes over others, but since five-, six- and seven-membered rings are indeed reported, other factors are also operative. Additional studies are apparently required.

CH_2CH_2CN ... CH_2 (194) —H⁺→ ... H_3C CH_3 —H_2O→ ... NH, O, H_3C CH_3 (195)

... NH, CH_2, CH_3 —H_2O→ ... O, CH_3 (196)

Ph, Ph =CH$(CH_2)_3$CN (197) ⟶ O, Ph, Ph (198)

CH=CHR, N, CN (199) —H⁺→ R, N, N, + —H_2O→ R, NH, O (200)

C. Cyclization of Ketonitriles

The use of the protonated carbonyl group as an electrophile to condense with nitriles has been extended to intramolecular cyclizations. All the following may be considered as essentially proceeding through nitrilium ions. When δ-ketonitriles are exposed to acidic conditions, 2-pyridones (**201**) are obtained[110–117]. In one instance, when ketonitriles (**202**) are treated at room temperature with concentrated sulphuric acid, cyclization proceeds normally to **203** with concomitant aromatization[113] to the 2-pyridones (**204**). The aromatization results from the oxidation by the sulphuric acid solvent. A modification of an intramolecular electrophilic cyclization has recently been reported[118] involving the acid chloride **205** and leading to the isoquinoline derivative **206**. The cyclization of nitrile-substituted diketones leads, via nitrilium salts, to β-amino-α,β-unsaturated ketones (**207**)[119]. A further variation of this reaction

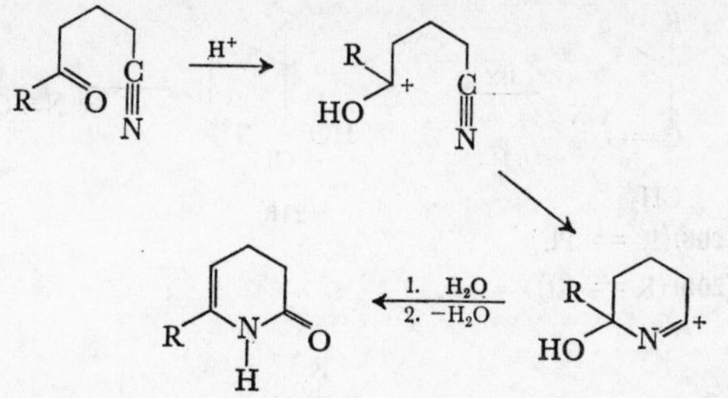

R O C≡N $\xrightarrow{H^+}$ R HO + C≡N

$\xleftarrow{1.\ H_2O,\ 2.\ -H_2O}$ R HO N+

R N O H

(201)

involves the cyclization of the keto dinitriles **208**[120] or **209**[121] to the bicyclic system **214** or **215**, respectively. This is an interesting example of both the Ritter and Hoesch reactions occurring in sequence. The initial step leading to **210** proceeds via the nitrilium ion **211** which, after loss of water, produces the olefin **212**. The Hoesch process follows by electrophilic attack on the vinyl group (**213**) which is converted to the bicyclic lactam upon hydrolysis.

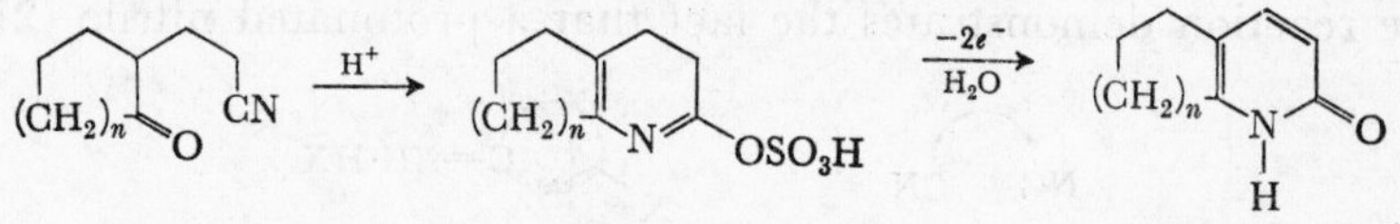

$(CH_2)_n$ O CN $\xrightarrow{H^+}$ $(CH_2)_n$ N OSO_3H $\xrightarrow[H_2O]{-2e^-}$ $(CH_2)_n$ N O H

(202) (203) (204)

($n = 1, 2, 3$)

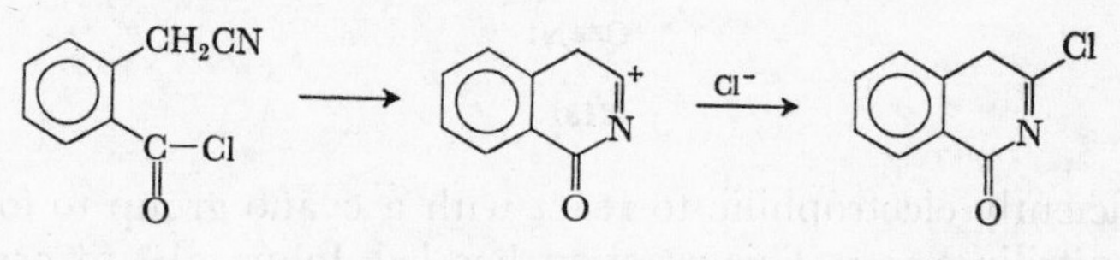

CH_2CN C–Cl O $\longrightarrow$ N+ O $\xrightarrow{Cl^-}$ Cl N O

(205) (206)

O O CH_3 CN $\xrightarrow{H^+}$ O CH_3 N O H

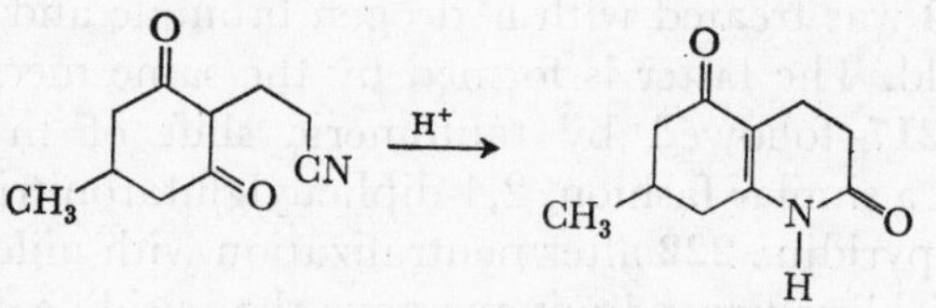

(207)

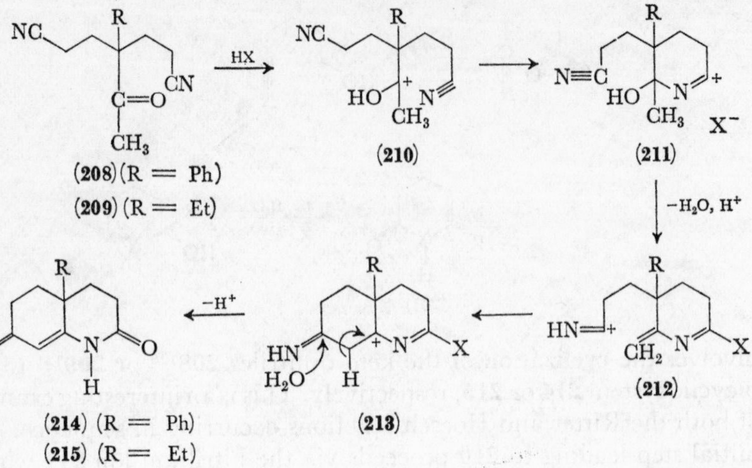

(208) (R = Ph)
(209) (R = Et)
HX
(210)
(211)
$-H_2O, H^+$
(212)
(213)
$-H^+$
(214) (R = Ph)
(215) (R = Et)

D. Cyclization of Dinitriles

The tendency of α,ω-dinitriles (**216**) to cyclize to the heterocycles **217** in the presence of hydrogen halides has been recently reviewed[5]. The reaction demonstrates the fact that a protonated nitrile (**218**)

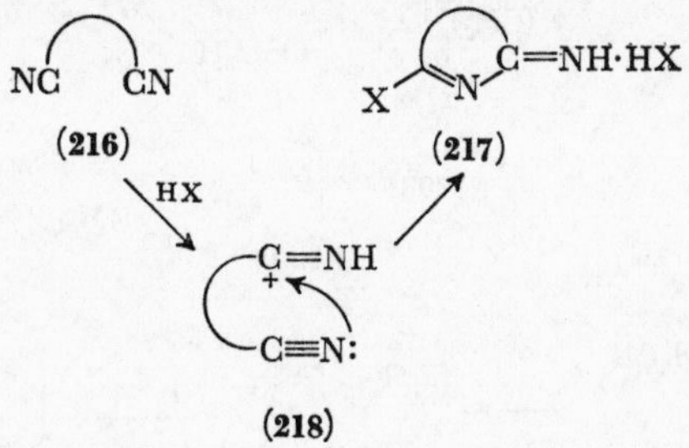

(216)
(217)
HX
(218)

is sufficiently electrophilic to react with a cyano group to form the usual nitrilium ion. This reaction has led Johnson[122] to prepare a host of interesting systems, whose structures have recently been confirmed by spectroscopic techniques. Recently[122], the unsaturated dinitrile **219** was treated with hydrogen bromide and produced **220** in good yield. The latter is formed by the same mechanism which results in **217** followed by tautomeric shift of the unsaturated linkages. In a similar fashion, 2,4-diphenylglutaronitrile (**221**) forms the dihydropyridine **222** after neutralization with mild base. Oxidation with dichlorodicyanoquinone gave the pyridine derivative **223**. A further extension of this versatile ring closure involving alicyclic

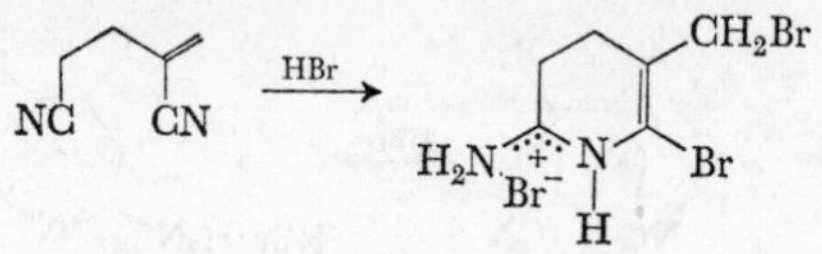

(219) (220)

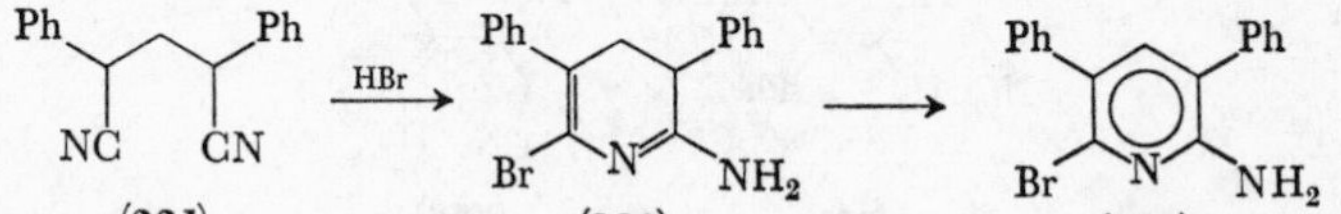

(221) (222) (223)

dinitriles (**224**) produced the azepine derivatives **225**[123] in good yield. The use of the unsymmetrical dinitrile **226** which has no tendency to isomerize resulted in a single azepine system (**227**). The alternative structure **228** was rejected on the ground of spectroscopic evidence. It therefore appears that the cyano group which

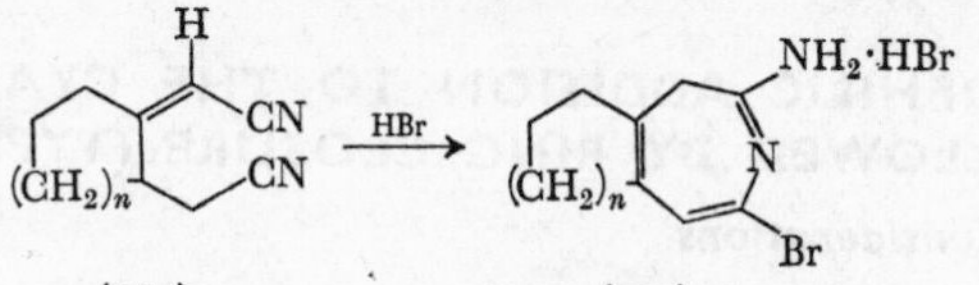

(224) (225)

($n = 1, 2, 3, 4$)

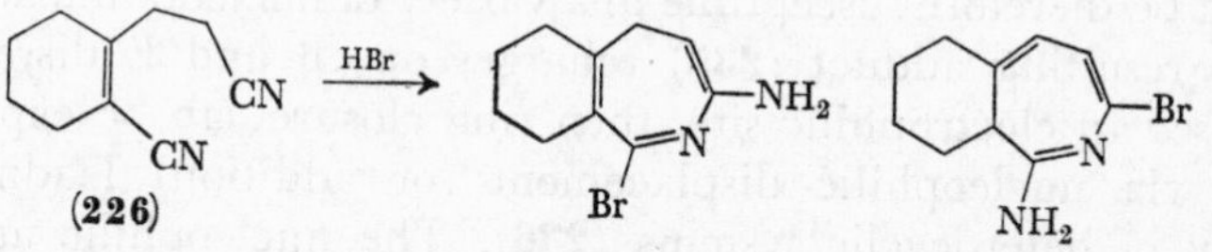

(226) (227) (228)

is protonated is that which is non-conjugated, and the nitrilium ion results from the less basic cyano group. This mode of cyclization has been observed by these authors in related systems. Monocyclic azepines (**230**) have been prepared[124] by treatment of *syn*-dicyanobutenes (**229**) with hydrogen bromide. However, the related *anti*-nitriles **231** and **232** failed to give cyclic products. This sets forth a limit to this cyclization process in which the cyano groups must be *syn* to each other or capable of facile isomerization. This latter fact was exemplified when the dinitrile **233a** underwent the expected cyclization as a result of its being in equilibrium with its isomer **233b**.

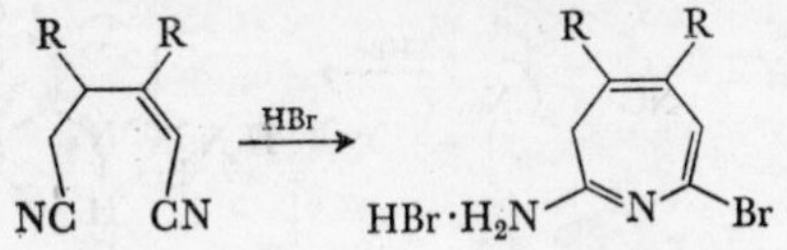

(229) (230)

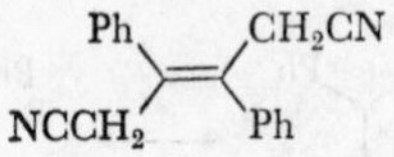

(231) 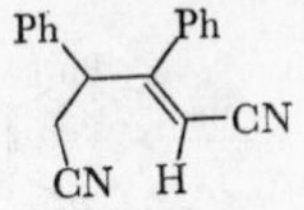 (232)

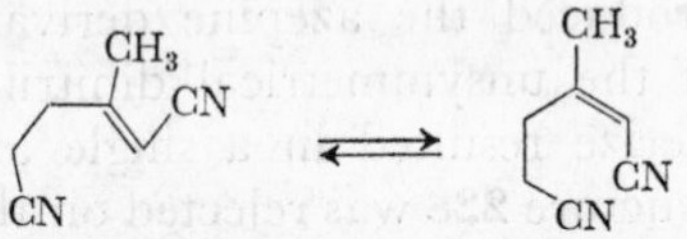

(233a) (233b)

V. NUCLEOPHILIC ADDITION TO THE CYANO GROUP FOLLOWED BY RING CLOSURE (TYPE C)

A. General Considerations

The strong dipole directed toward nitrogen in the cyano group results in an electron deficiency at the carbon atom (**234**) which should be therefore susceptible to a variety of nucleophilic additions. If the resulting adduct (**235**) (charges on :B and E disregarded) possesses an electrophilic site, then ring closure can be expected to occur via nucleophilic displacement (or addition) leading to a variety of heterocyclic systems (**236**). The nucleophilic atom can

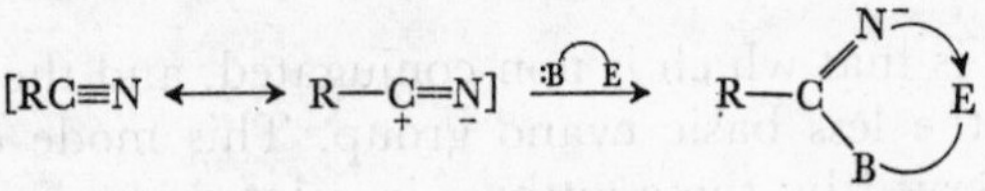

[RC≡N ⟷ R—C^{+}=N^{-}] (234) → (235) 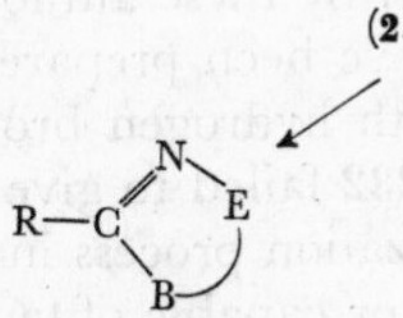 → (236)

(B: = C, O, N, S, E = C, N, S)

be carbon, nitrogen, oxygen, sulphur, etc., derived from a multitude of sources, whereas the electrophilic site (E) is almost always as electronegatively bound carbon (alkyl halides, carbonyl derivatives, nitriles). However, several cases are known, and will be discussed, in which nitrogen and sulphur act as electrophiles. It is clear from the above general scheme that many different types of heterocycles which involve anywhere from one to four (or five) heteroatoms are obtainable via this mechanism. The most efficient presentation of this process would seem to be one based upon the nature of the nucleophile adding to the cyano group rather than upon ring size or heterocyclic system produced. The literature contains a vast number of variations on this ring closure and this section will present only representative examples of the most interesting types.

B. *Addition by Carbanionic Species Followed by Cyclization*

The addition of the Reformatsky reagent **237** to nitriles is the only reported[125] example of a *β*-lactam (**240**) formation. The initial adduct (**238**) preferentially attacks a second molecule of the organozinc halide rather than proceed toward an intramolecular cyclization. The excessive substitution present in the bis-adduct (**239**) allows cyclization to the *β*-lactam derivative to compete effectively.

$$(CH_3)_2C(ZnBr)(CO_2R^1)\ \textbf{(237)} \xrightarrow{R^2C\equiv N} (CH_3)_2C(CO_2R^1)-C(R^2)=N^-\ \textbf{(238)} \xrightarrow{(CH_3)_2C(ZnBr)-C(=O)-OR^1}$$

$$\xrightarrow{-OR^{1-}} \textbf{(239)} \xrightarrow{H_2O} \textbf{(240)}$$

(239) **(240)**

(R^2 = 6-methoxy-2-naphthyl)

The use of Grignard reagents as precursors for carbon nucleophiles has resulted in the formation of 1-pyrrolines (**242**) and tetrahydropyridines (**243**) by treatment with ω-halonitriles[126–128]. Reaction of alkyl or aryl Grignard reagents with 4-chloro(or 4-bromo)butyronitriles produces the adduct **241** which then cyclizes via nucleophilic displacement of halide ion. In this manner, which is the most convenient method available, a wide variety of 2-substituted 1-pyrrolines and tetrahydropyridines may be obtained. An interesting and alternative mechanism for the formation of **242** can be considered in the light of the work of Cloke[129], who reported that cyclopropanecarbonitrile upon addition of phenylmagnesium

RMgX + (4-halobutyronitrile) ⟶ (**241**) ⟶ (**242**)

RMgX + (5-halovaleronitrile) ⟶ (**243**)

(R = Me, Et, Ph, $PhCH_2$, 2-naphthyl, 9-phenanthryl, 2-thienyl)

bromide produces the cyclopropyl imine **244** which is then thermally rearranged to the 1-pyrroline **245**. Attempts to reproduce this rearrangement have failed although the hydrochloride of **244** gave excellent conversions to the pyrroline in a thermal process[130,131a,b]. Examination of the experimental details leading to **242**[126] reveals that a two- and three-fold excess of the Grignard reagent was employed, which would suggest that an α-proton abstraction and subsequent cyclization to cyclopropanecarbonitrile may be occurring, followed by a second addition of the Grignard to give **246**. Since magnesium halide salts are present, these could function as Lewis acids to catalyse cyclization to **245**. The nature of the rearrangement and the role of the acid catalyst are still in question. Furthermore, detailed studies of the reactions leading to tetrahydropyridines (**243**) should reveal whether an initial cyclization to cyclobutanecarbonitrile (**247**) precedes their formation.

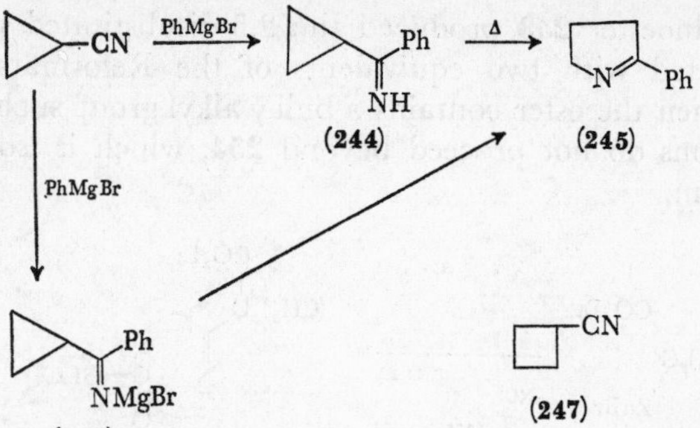

CN PhMgBr → Ph, NH (244) Δ → N, Ph (245)

PhMgBr ↓

Ph, NMgBr (246)

CN (247)

Acetylide salts have added to nitriles[132] and, depending upon the molar ratio of nitrile to acetylide, produce 2-substituted pyridines (**248**) or 2,4-disubstituted pyrimidines (**249**). This process required the utilization of elevated temperatures (~200°) and pressures (~15 atm). Excess nitrile, containing α-hydrogens, results in minor quantities of the aminopyridines **250** and **251**.

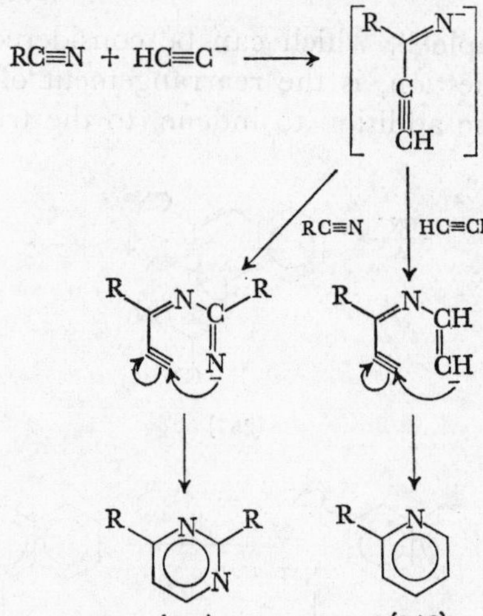

RC≡N + HC≡C⁻ ⟶ [R–C(=N⁻)–C≡CH]

RC≡N / HC≡CH

R, N, C, R, N / R, N, CH, CH

(249) (248)

(R = Me, Et, i-Bu, Ph)

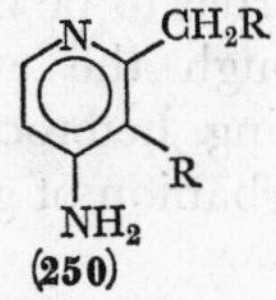

N, CH₂R, R, NH₂ (250) RCH₂, N, NH₂, R (251)

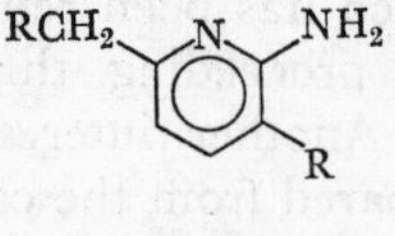

The cyanoester **253** produced the 2,5-disubstituted pyrrole **255** when treated with two equivalents of the Reformatsky reagent **252**[133]. When the ester contains a bulky alkyl group such as 2-butyl, the reactions do not proceed beyond **254**, which is isolated in its lactam form.

$(CH_3)_2C(CO_2Et)ZnBr$ (**252**) + $NC\text{–}CH_2CH_2\text{–}CO_2Et$ (**253**) → **254** → (**252**) → **255**

(**252**) (**253**) (**255**) (**254**)

A recent example[134], which can be considered as following the mechanism in question, is the rearrangement of **256**, obtained by tetracyanoethylene addition to indene, to the tricyclic system **258**.

(**256**) $\xrightarrow[\Delta]{CN^-}$ (**257**) → $\xrightarrow{H^+}$ → (**258**)

The process, which has been reported to be cyanide-ion catalysed, is postulated as proceeding through the anion of the valence tautomer (**257**). Another interesting heterocycle (**260**) has been described[135], prepared from the carbanion of guaiazulene (**259**) and two equivalents of benzonitrile. Although **260** was obtained only in 10% yield it represents the first example of the π-equivalent heteroanalogue of the non-benzenoid system, cyclopenta[e,f]heptalene (**261**).

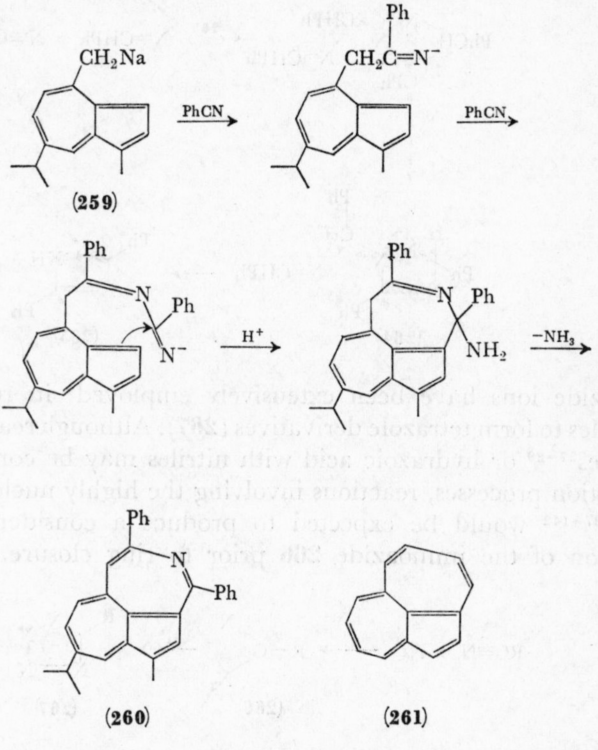

(**259**) $\xrightarrow{PhCN}$ $\xrightarrow{PhCN}$ $\xrightarrow{H^+}$ $\xrightarrow{-NH_3}$

(**260**) (**261**)

C. Addition by the Amino Group Followed by Cyclization

The amino nucleophile, originating from a variety of sources, has been utilized for additions to the cyano group. For example, the azine **262** when treated with *t*-butoxide ion has been reported to form the anion **263** which cleaves to benzonitrile and the imino nucleophile. Nucleophilic addition to another azine molecule and reaction with benzonitrile produce the cyclic system **264**. The latter subsequently fragments to the 1,2,4-triazole (**265**). The accompanying mechanism has been somewhat supported by the use of benzonitrile-^{14}CN which resulted in **265** containing approximately 30% of the radioactive carbon[136].

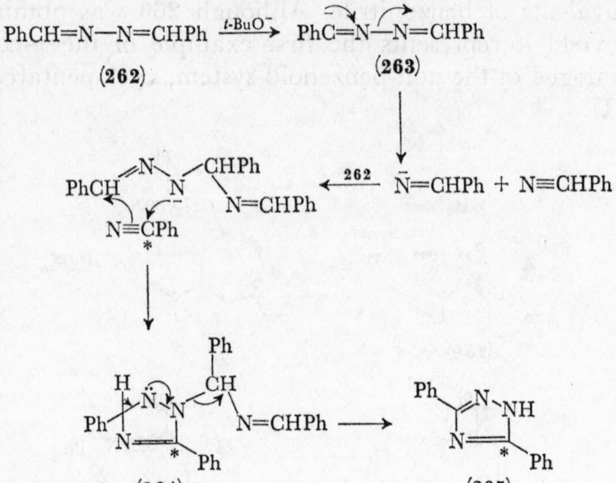

$$PhCH{=}N{-}N{=}CHPh \xrightarrow{t\text{-}BuO^-} Ph\bar{C}{=}N{-}N{=}CHPh$$

(262) (263)

$$\bar{N}{=}CHPh + N{\equiv}CHPh \xrightarrow{262} \ldots$$

(264) (265)

Azide ions have been extensively employed[4] in reactions with nitriles to form tetrazole derivatives (**267**). Although reactions of alkyl azides[27–29] or hydrazoic acid with nitriles may be concerted cycloaddition processes, reactions involving the highly nucleophilic azide ion[137–142] would be expected to produce a considerable concentration of the iminoazide **266** prior to ring closure. The ease of

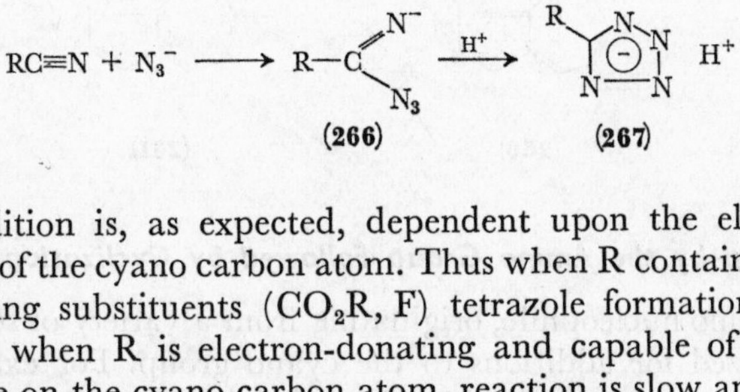

$$RC{\equiv}N + N_3^- \longrightarrow R{-}C({=}N^-)N_3 \xrightarrow{H^+} \ldots \; H^+$$

(266) (267)

azide addition is, as expected, dependent upon the electrophilic character of the cyano carbon atom. Thus when R contains electron-withdrawing substituents (CO_2R, F) tetrazole formation is facile. However, when R is electron-donating and capable of dispersing the charge on the cyano carbon atom, reaction is slow and requires Lewis acids to increase the electrophilicity[142–145].

That the use of Lewis acids increases the electrophilic nature of the cyano group is seen in the reaction of aniline with various nitriles[146] to form the amidinium salt **268.** Although the latter is a stable intermediate, it can be converted to the chloramine derivative **269** which cyclizes to the benzimidazole **270.** An unusual cyclization[147] is exhibited by the reaction of aryl hydrazines with two equivalents of nitrile, resulting in moderate conversions to

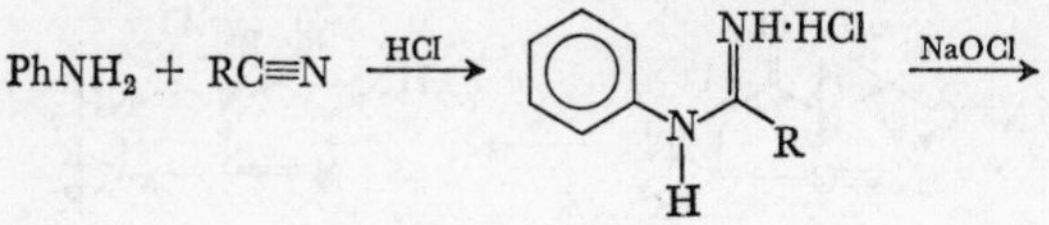

$$PhNH_2 + RC{\equiv}N \xrightarrow{HCl} \text{(268)} \xrightarrow{NaOCl}$$

(268)

$$\text{(269)} \xrightarrow{OH^-} \text{(270)}$$

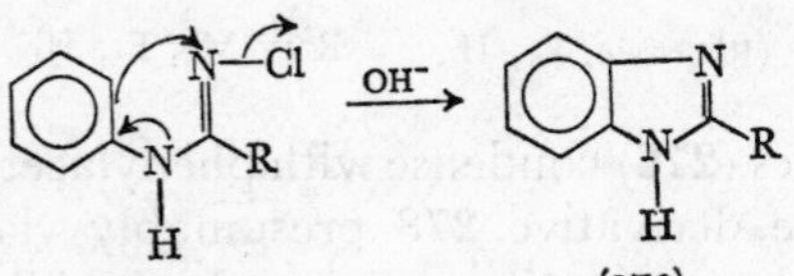

(269) **(270)**

(R = Et, Ph, 4-thiazyl)

triazoles (**273**). A variety of derivatives were prepared by this process (Table 7) which probably proceeds through the intermediates **271** and **272**. The amino group appears to have initiated

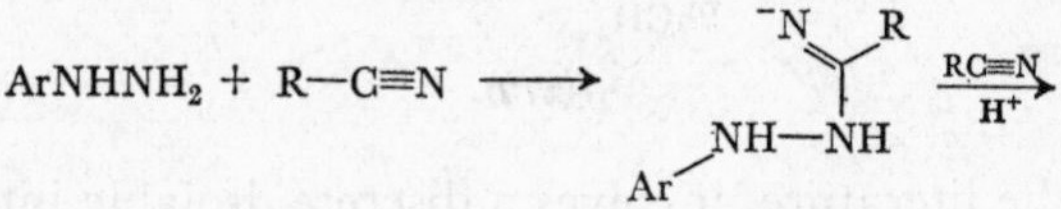

$$ArNHNH_2 + R{-}C{\equiv}N \longrightarrow \text{(271)} \xrightarrow[H^+]{RC{\equiv}N}$$

(271)

$$\text{(272)} \xrightarrow{-NH_3} \text{(273)}$$

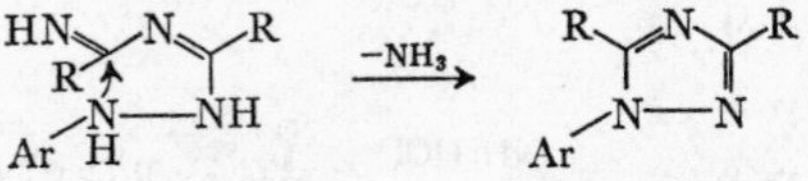

(272) **(273)**

the cyclization of an *N,N'*-disubstituted urea (**274**) to the bicyclic system **275**, which is formed by heating **274** above its melting point[148]. It is interesting that the weakly nucleophilic amido nitrogen was evidently sufficient to affect this tautomerization.

TABLE 7. *s*-Triazoles **273** from nitriles and hydrazines.

R	Ar	Yield of **273** (%)
Et	p-$O_2NC_6H_4$	32
$PhCH_2$	p-$O_2NC_6H_4$	65
$PhCH_2$	2,4-$(O_2N)_2C_6H_3$	73
$PhCH_2$	p-$CH_3C_6H_4SO_2$	63

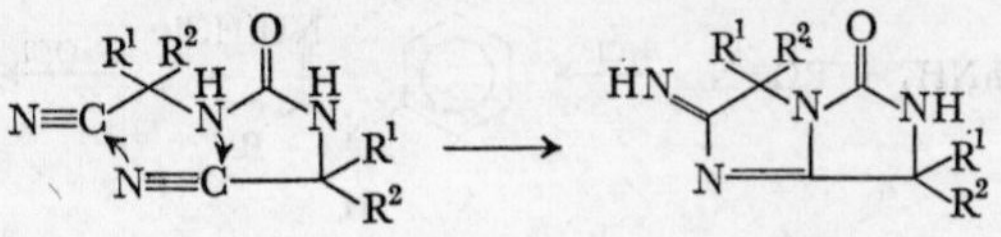

(274) (275)

(R^1 = Me, Et, H, R^2 = Me, Et, H)

Acyl guanidines (**276**) condense with phenylacetonitrile, producing the 1,3,5-triazine derivative **278** presumably via the intermediate **277**[149]. A related reaction[150], utilizing the amidine **280** in place of the guanidine results in the triazine **281**. This, like so many other

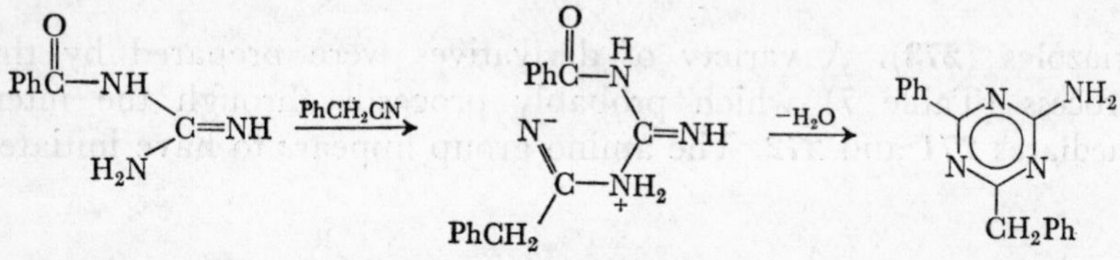

(276) (277) (278)

examples in the literature, involves a discrete, isolable intermediate, namely the imidate (**279**) which is then transformed into the heterocycle. The use of imidates in synthesis is very extensive and is outside the scope of this discussion. A review on this subject has recently appeared[151].

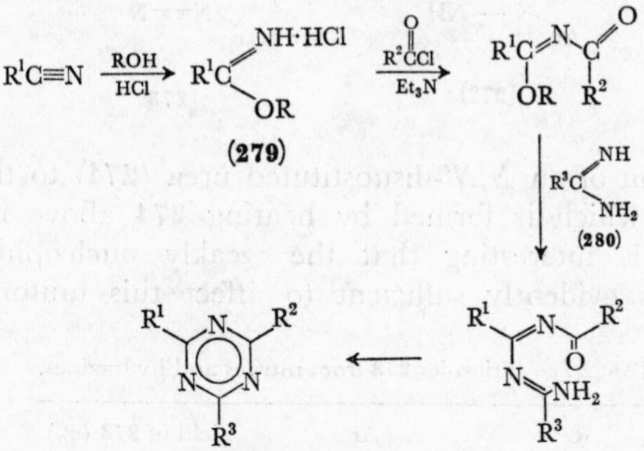

(279) (280) (281)

Another example of this type of ring closure is offered by Case, who treated nitriles with hydrazine to produce the adduct **282** which formed the triazine **283** upon treatment with α-diketones[152].

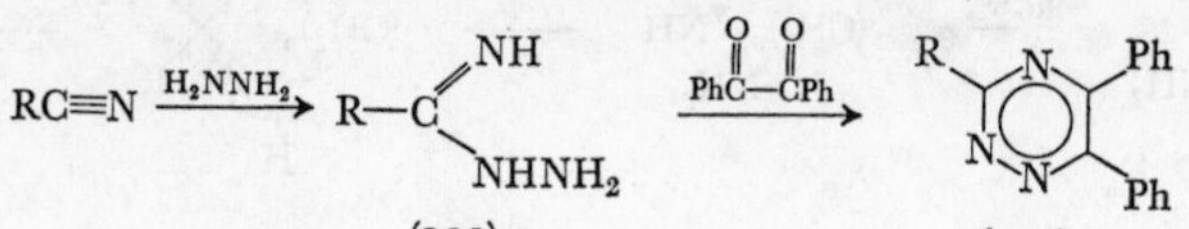

$$\mathrm{RC{\equiv}N} \xrightarrow{\mathrm{H_2NNH_2}} \mathrm{R{-}C({=}NH)NHNH_2} \xrightarrow{\mathrm{PhC(O){-}C(O)Ph}}$$

(282) (283)

A versatile method for the synthesis of tetrazines (**286**), recently reported by Yates[153], involved the condensation of two molecules of the imidate **284** with hydrazine hydrate to form the dihydro derivative **285**. The latter was aromatized with nitrous acid.

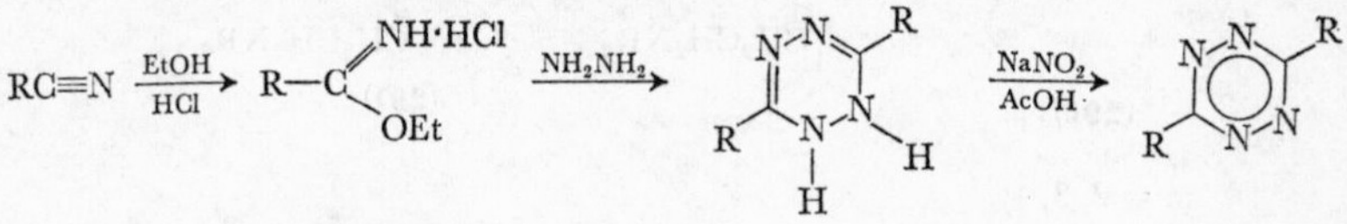

$$\mathrm{RC{\equiv}N} \xrightarrow[\mathrm{HCl}]{\mathrm{EtOH}} \mathrm{R{-}C({=}NH{\cdot}HCl)OEt} \xrightarrow{\mathrm{NH_2NH_2}} \xrightarrow[\mathrm{AcOH}]{\mathrm{NaNO_2}}$$

(284) (285) (286)

(R = Ph, $PhCH_2$, *p*-$MeOC_6H_4$)

A reaction which follows a slightly different course is the condensation of diamines (**287**) and nitriles to form imidazolines (**290**)[154]. Although attack by the amino nitrogen results in the adduct **288**, the nitrogen atom of the cyano group is probably expelled by a second intramolecular nucleophilic addition to form **289**. By use of homologues of ethylene diamine, ring systems up to seven members have been constructed. It is also possible to obtain[155] *N*-substituted imidazolines (**291**) as well as bis-imidazolinyl alkanes (**292**).

D. Addition by Oxygen Nucleophiles Followed by Cyclization

Since oxygen is a weaker nucleophile than nitrogen, cyclizations involving this species are found only under the influence of catalysts. For example, the highly electrophilic perfluoronitriles **293** react with 2-chloroethanol in the presence of a base to form the 2-perfluoroalkyloxazolines **294**[156,157].

The imidate **295** derived from substituted acetonitriles has been shown[158] to form the chloramine derivative **296** which when treated with strong base resulted in a novel α-amino acid synthesis. The process has been postulated to proceed through an unstable azirine intermediate (**297**).

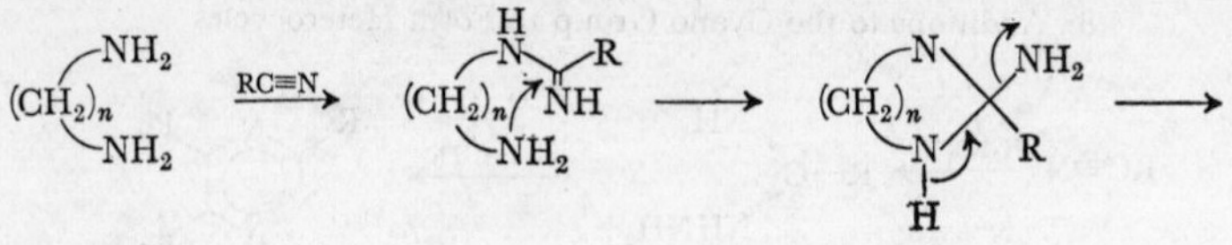

$(CH_2)_n(NH_2)_2$ $\xrightarrow{RC\equiv N}$ $(CH_2)_n$, NH, $\overset{H}{N}$—C(R)=NH, NH_2 $\longrightarrow$ $(CH_2)_n$, N, NH_2, R, $\underset{H}{N}$ $\longrightarrow$

(287) (n = 2, 3, 4) **(288)** **(289)**

$(CH_2)_n$, $\overset{H}{N}$, N, R

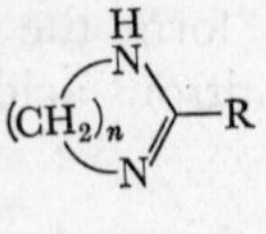

(290)

(n = 2, 3, 4)

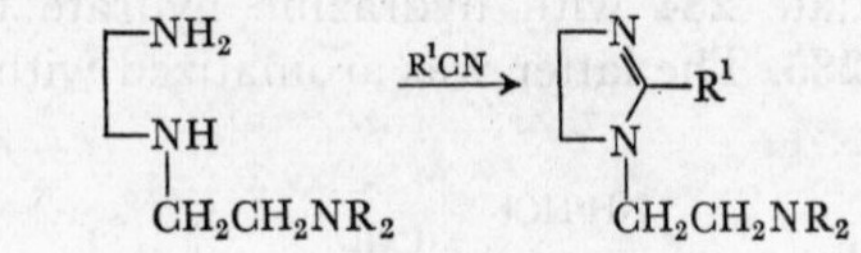

NH_2, NH, $CH_2CH_2NR_2$ $\xrightarrow{R^1CN}$ N, N, R^1, $CH_2CH_2NR_2$

(291)

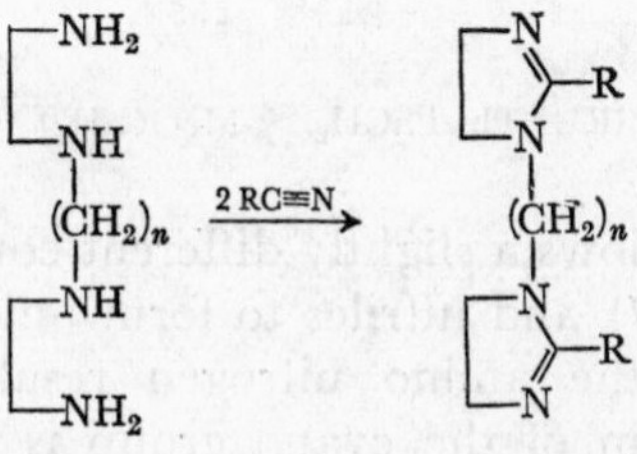

NH_2, NH, $(CH_2)_n$, NH, NH_2 $\xrightarrow{2\ RC\equiv N}$ N, N, R, $(CH_2)_n$, N, N, R

(292)

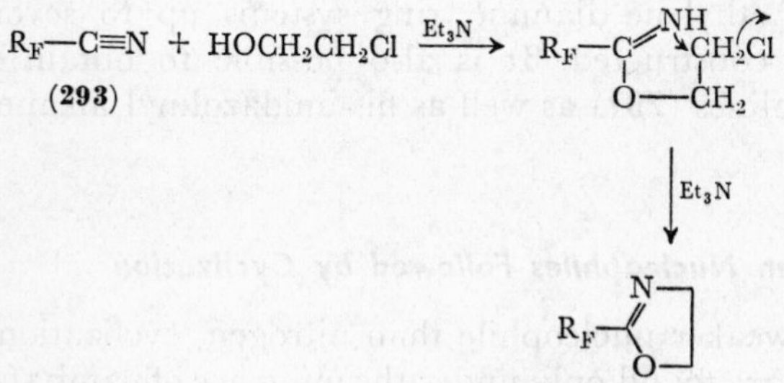

R_F—C≡N + $HOCH_2CH_2Cl$ $\xrightarrow{Et_3N}$ R_F—C(=NH)—O—CH_2—CH_2Cl

(293)

$\downarrow Et_3N$

R_F, N, O

(294)

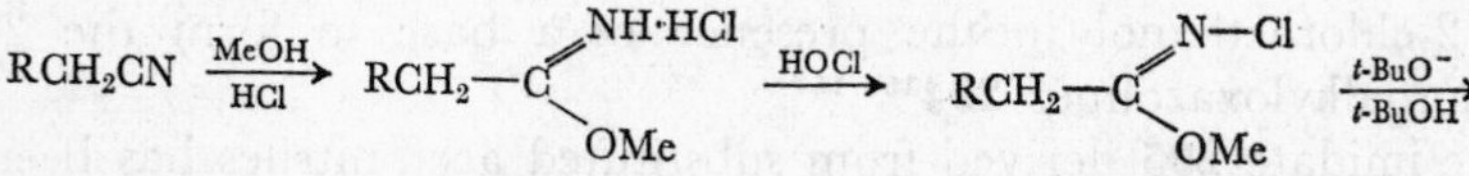

RCH_2CN $\xrightarrow[HCl]{MeOH}$ RCH_2—C(=NH·HCl)OMe $\xrightarrow{HOCl}$ RCH_2—C(=N—Cl)OMe $\xrightarrow[t\text{-}BuOH]{t\text{-}BuO^-}$

(295) **(296)**

R—$\bar{C}$H—C(=N—Cl)OMe $\longrightarrow$ [R—C, N, OMe] $\xrightarrow{H_3O^+}$ $RCH(NH_2)CO_2H$

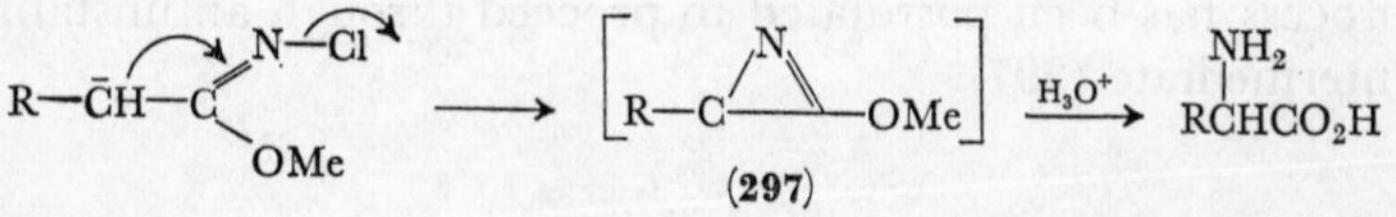

(297)

A recent report[159] described the novel behaviour of the dioxaphospholan **298** when warmed in acetonitrile solution for 20 hours. The oxazoline **300**, which is speculated to arise via the adduct **299**, was obtained in 70% yield.

$$\textbf{(298)} \xrightarrow{CH_3CN} \textbf{(299)} \longrightarrow \textbf{(300)} + OP(OEt)_3$$

E. Addition by Sulphur Nucleophile-Followed by Cyclization

The favourable nucleophilic properties of the sulphur atom coupled with the leaving group ability of sulphur nucleophiles result in some rather unique behaviour when applied to heterocyclic syntheses. An example of both properties is seen in the synthesis of isothiazoles (**304**)[160,161]. When the salt of dimercaptomethylenemalononitrile (**301**) is treated with sulphur in refluxing methanol, the existence of intermediates **302** and **303** has been envisioned. The former arises from nucleophilic attack by mercaptide ion on sulphur, whereas the latter involves a second nucleophilic attack on the nitrile with expulsion of the sulphur moiety by the nitrogen. Another example of this reaction involving the mononitrile **305** has been described[162], which presumably proceeds through the same path, leading to the isothiazole **306**.

A brief description of the thiadiazole **308** synthesis from nitriles utilizes elemental sulphur and a trialkylamine as catalyst at elevated temperatures[163]. Although little quantitative information is available on any of these nitrile–sulphur cyclizations, the formation of the product could be conceived as proceeding through the intermediate **307**.

$\bar{S}$ $\bar{S}$ NC CN (**301**) $\xrightarrow{S_x}$ $\bar{S}$ $S—S_x^-$ NC C≡N S_x^- (**302**) ⟶ $\bar{S}$ $S—S_x^-$ N NC $S—S_{x-1}$ (**303**)

$\downarrow -S_{2x-1}$

$\bar{S}$ S N NC S^- $\xrightarrow{2\,MeI}$ MeS S N NC SMe (**304**)

A reaction leading to 3-aminoisothiazoles (**310**)[164] is performed by treating a β-ketonitrile with morpholine and sulphur for several hours at 105°. This process may be considered as an example of initial nucleophilic attack by nitrogen on the enol thioether **309**.

$PhCH_2CN \xrightarrow[NaH]{CS_2}$ Ph S^- NC S^- (**305**) $\xrightarrow{S_x}$ HS Ph N S SH (**306**)

$2\ RC≡N \xrightarrow{Et_3N}$ R—C $\bar{N}$ R C N $^+NEt_3$

$\downarrow S_x$

R N R S—N (**308**) + Et_3N $\xleftarrow{-S_{x-1}}$ $^+NEt_3$ R—C N R N S_x (**307**)

(R = Ph, *p*-tolyl, 2-naphthyl)

A versatile synthesis[165] of pyrimidinethiones (**313**) from 2-aminonitriles (**311**) appears to include a nucleophilic addition of the sulphydryl ion to the cyano group of **312** followed by cyclization. The products are prepared directly from the crude ethoxymethyleneamino derivative (**312**) in good yields. The mechanism of this

$$\mathrm{Ar\overset{O}{\overset{\|}{C}}{-}CH_2CN \rightleftharpoons Ar\overset{OH}{\overset{|}{C}}{=}CHCN \xrightarrow{S_x^-}}$$ (309)

$\xrightarrow{-H^+,\ -S_{x-1}^-}$ (310)

(Ar = p-MeOC$_6$H$_4$, p-MeC$_6$H$_4$, p-ClC$_6$H$_4$)

cyclization is not firm and alternatives have been considered by the authors.

By treating the above aminonitriles with carbon disulphide in pyridine a simple and efficient preparation of pyrimidinedithiones (**316**) has been achieved[166,167]. The cyclization has been found to occur via the intramolecular nucleophilic attack of the sulphur on the nitrile **314**, leading to the thiazine **315** which exhibited the ring opening–ring closing sequence resulting in the product.

(311) $\xrightarrow{HC(OEt)_3}$ (312) $\xrightarrow{NaSH}$ (313)

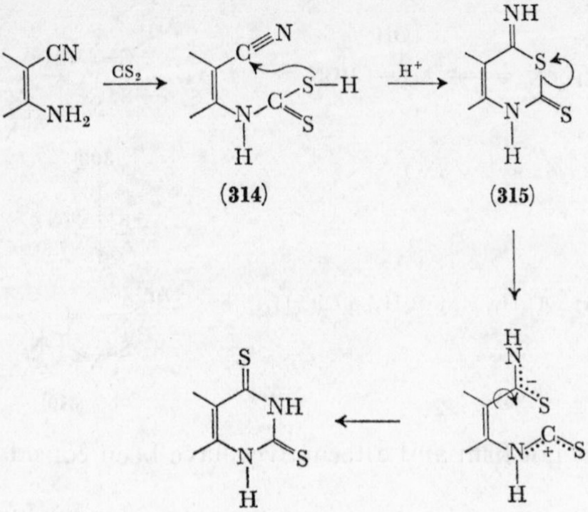

(314) (315) (316)

Taylor and his coworkers have prepared a variety of heterocyclic systems utilizing *o*-aminonitriles during the past fifteen years[165]. A synthesis of 2-thiazolines (**319**) from nitriles and mercaptoamines has been described by Kuhn and Drawert[168]. The reaction proceeds, as in the formation of imidazolines (**290**), via the adduct **317** followed by a second nucleophilic attack to form **318**, which produces the thiazoline after elimination of ammonia. In this manner, a variety of thiazolines were obtained including **320**, derived from dinitriles. This process is perhaps more in accord with the intramolecular cyclization (type C_i) to be discussed in the following section. However, since the intermediate **317** is only postulated and not actually observed, its description under the intermolecular cyclizations is preferred.

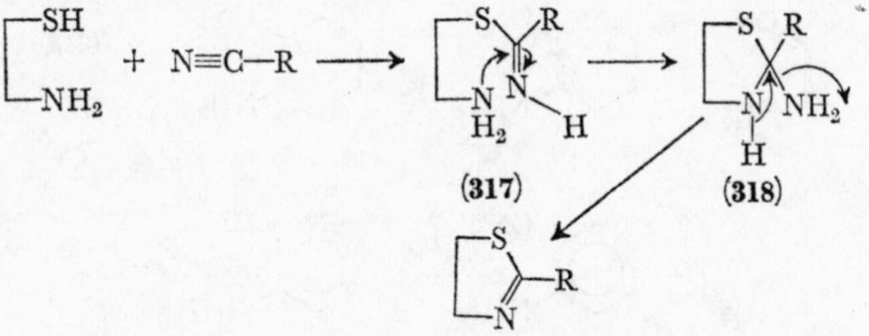

(317) (318) (319)

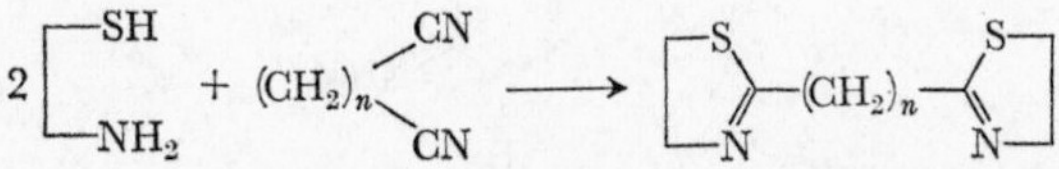

(320)

VI. INTRAMOLECULAR NUCLEOPHILIC ADDITION TO THE CYANO GROUP (TYPE C_i)

A. General Considerations

A simple extension of the cyclization process described in the preceding section involves a molecule **321**, which contains a nucleophile situated at an appropriate distance from the electrophilic site of the cyano group. Thus by nucleophilic addition to form **322,** followed by a proton acquisition, the ring closure will result in an amino-substituted heterocyclic **323**. A variety of nucleophiles may serve this function although a heterocycle will result only if A and/or B is a heteroatom.

(**321**) → (**322**) $\xrightarrow{H^+}$ (**323**)

(A = C, N, O, S, B = C, N, O, S)

B. Intramolecular Nucleophilic Addition by Carbon Nucleophiles

The most common example of this type of cyclization stems from the variety of Thorpe cyclizations in the literature[169]. For example, the base-catalysed cyclization of iminobis(propionitrile) (**324**) to the piperidine **326** proceeds through the carbanion **325**[170]. Acyl derivatives of 2-aminonitriles (**327**) undergo cyclization to the aminoquinolones **328** when treated with strong bases[171]. A similar reaction occurs when acyliminonitriles (**329**) are utilized producing the aminopyridones **330**. Many earlier examples of this type of ring closure have been reviewed[169].

The use of an olefin linkage derived from an enamine (**331**) has been reported[172,173] to produce pyrimidines (**332**), but due to the poor nucleophilic properties of the former, drastic conditions were necessary to affect cyclization. Reaction with 2-tetralone (**333**) produced the interesting system **335** presumably also through the intermediate enamine **334**. The driving force for these cyclizations may be attributed to the formation of the aromatic systems. Another example of this process is seen where the β-carbon atom of the

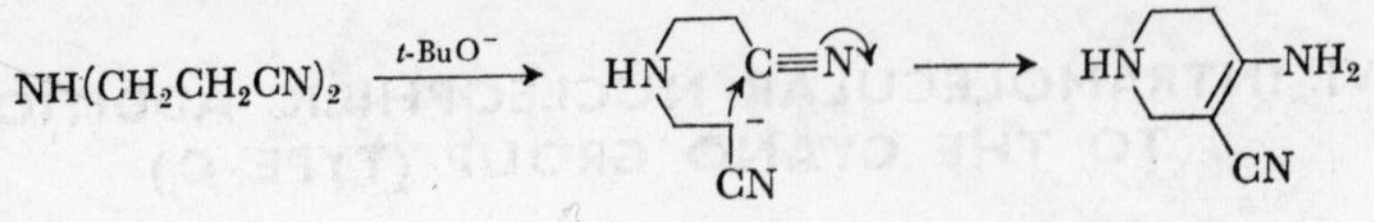

$NH(CH_2CH_2CN)_2$ $\xrightarrow{t\text{-BuO}^-}$ HN C≡N CN $\longrightarrow$ HN NH_2 CN

(324) (325) (326)

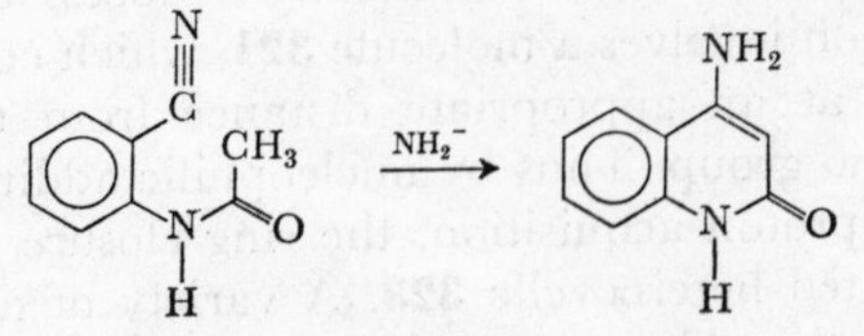

N C CH_3 N O H $\xrightarrow{NH_2^-}$ NH_2 N O H

(327) (328)

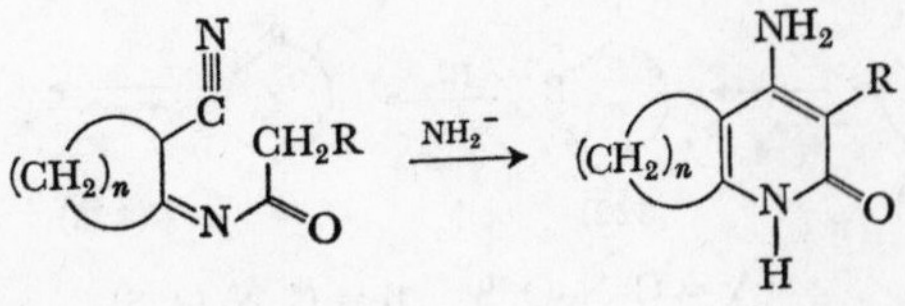

N C CH_2R $(CH_2)_n$ N O $\xrightarrow{NH_2^-}$ NH_2 R $(CH_2)_n$ N O H

(329) (330)

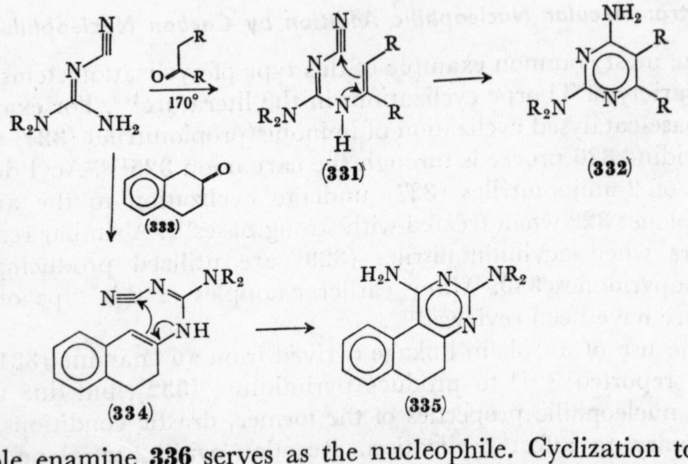

N N R_2N NH_2 $\xrightarrow[170^\circ]{}$ N N R R_2N N R H $\longrightarrow$ NH_2 R N R_2N N R

(331) (332)

(333)

N N NR_2 NH $\longrightarrow$ H_2N N NR_2 N

(334) (335)

simple enamine **336** serves as the nucleophile. Cyclization to **337** proceeds to a small extent (2%) at 200°, but employing zinc chloride as a Lewis acid increases the yield to 96%[174]. The latter fact attests to the need for a complexing agent to increase the electrophilic character of the cyano group.

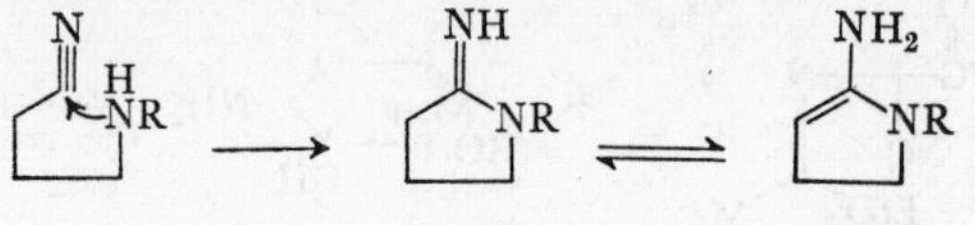

N H NR → NH NR ⇌ NH_2 NR

(346) (347)

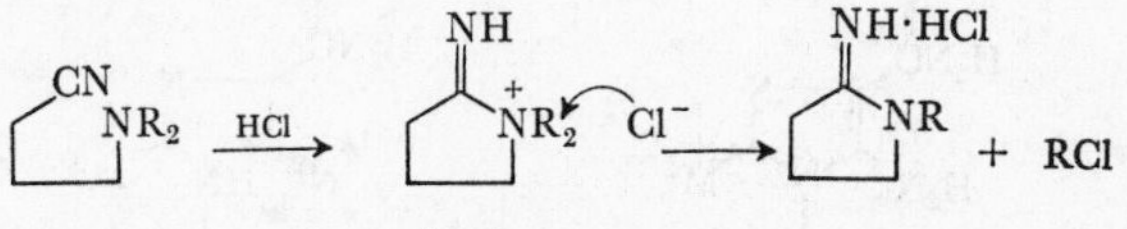

CN NR_2 —HCl→ NH $\overset{+}{N}R_2$ Cl^- → NH·HCl NR + RCl

(348)

involving nitrogen as the nucleophile is the cyclization of the hydrazine derivative **352** to **353**[184,185].

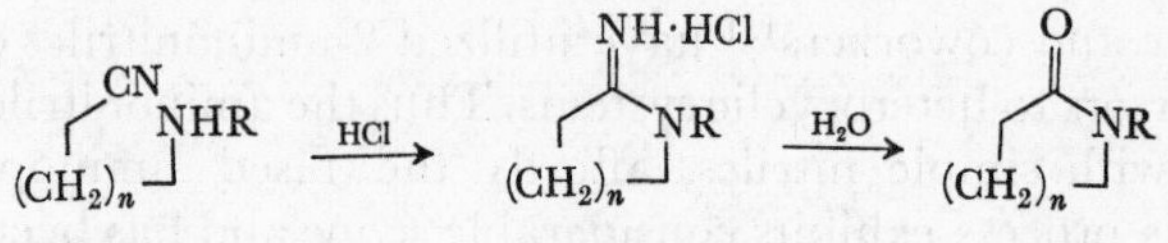

CN NHR $(CH_2)_n$ —HCl→ NH·HCl NR $(CH_2)_n$ —H_2O→ O NR $(CH_2)_n$

(349) (350) (351)

(n = 2, 3, 4)

N $CONH_2$ H_2N N NH_2 H —CS_2→ NH_2 OH N N HN N SH

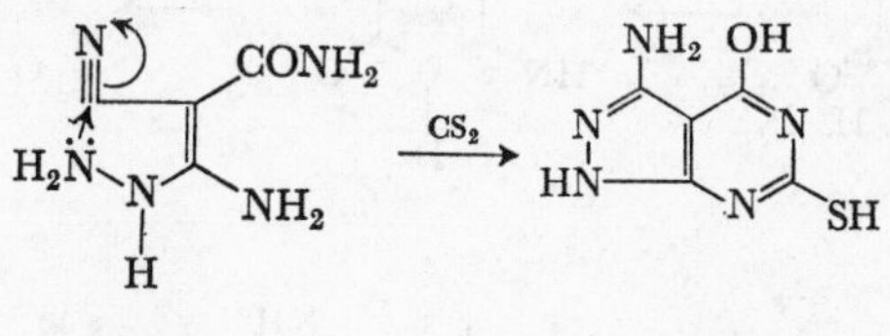

(352) (353)

A useful process leading to imidazole nucleosides (**356**)[186] is derived from the cyclization of the amidine derivative **355**, probably formed initially from D-xylopyranosylamine and the cyanoimidate **354**.

The amido function appears to possess a sufficiently nucleophilic nitrogen which allows cyclization to a cyano group[187]. Although oxygen is usually the more nucleophilic site in amides, the stability of the product **357** is probably responsible, in part, for the nitrogen acting as the nucleophile. Further examples where nitrogen nucleophiles are involved can be found in the cyclization of **358** to **359**[188] where previously it was thought that the product was **360**[189].

It is interesting that the thioamide **361**, exhibiting sulphur nucleophilicity, leads to the iminothionitrile **362** which cyclizes to the mesoionic system **363**[190], isolated as the acetyl derivatives **364**.

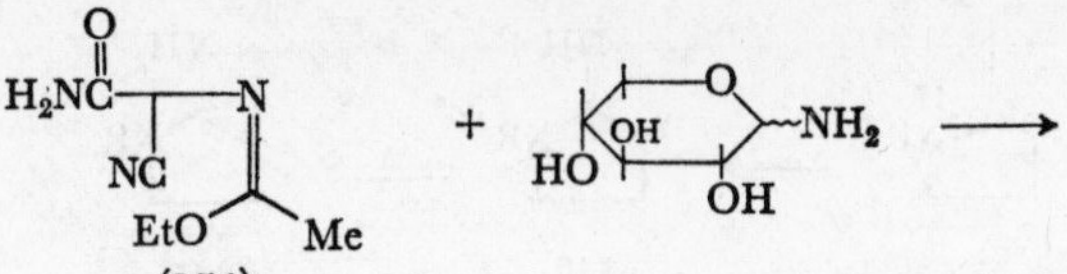

(354) (355) (356)

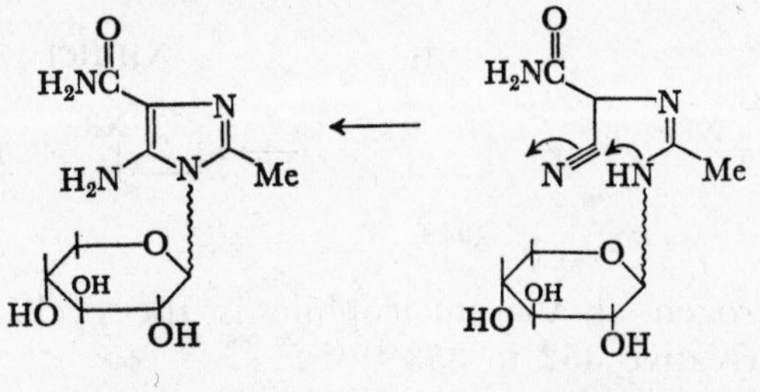

Taylor and coworkers[191] have utilized 2-aminonitriles extensively as precursors to heterocyclic systems. Thus the aminonitrile **365** when treated with simple nitriles, affords the fused aminopyrimidines **366**. This process exhibits considerable scope and has been discussed

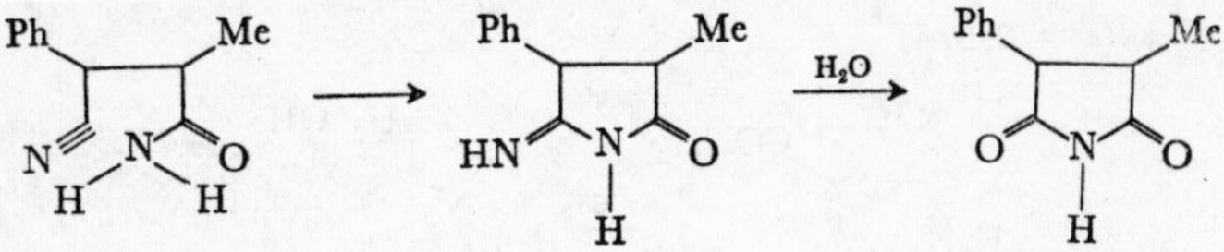

(357)

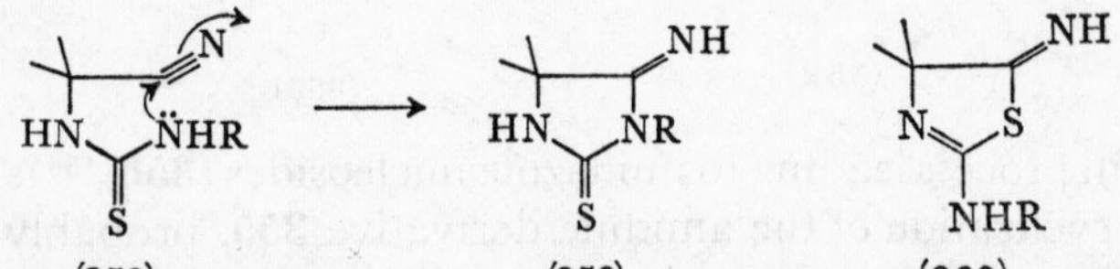

(358) (359) (360)

in detail[165]. Cinnamonitriles (**367**) have recently been employed[192] in a similar reaction which results in quinoline derivatives (**368**). Triazines **370** can be prepared from **369** by a thermal process[193].

An unusual double ring closure was recently reported[194] involving **371** which cyclized over basic alumina to give the tetracyclic system **372**. The weakly basic pyridine nitrogen can function as a nucleophile and add to the cyano group when heated at 100° in hydrochloric acid[195]. The product obtained is the azanthracene **373**.

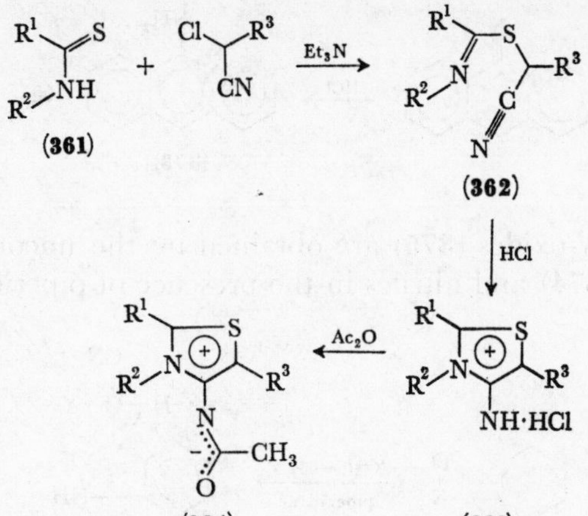

(361) (362)

(364) (363)

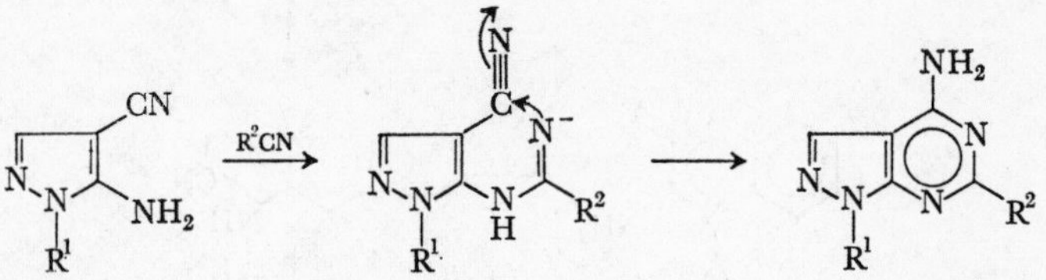

(365) (366)

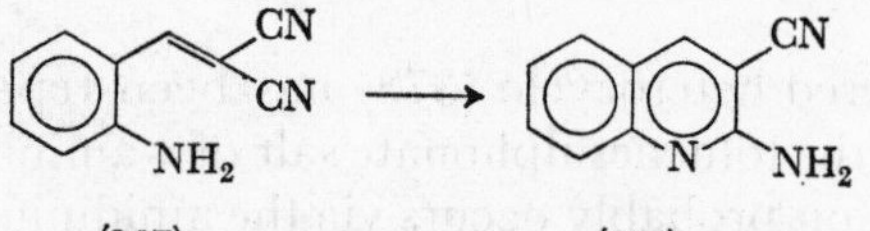

(367) (368)

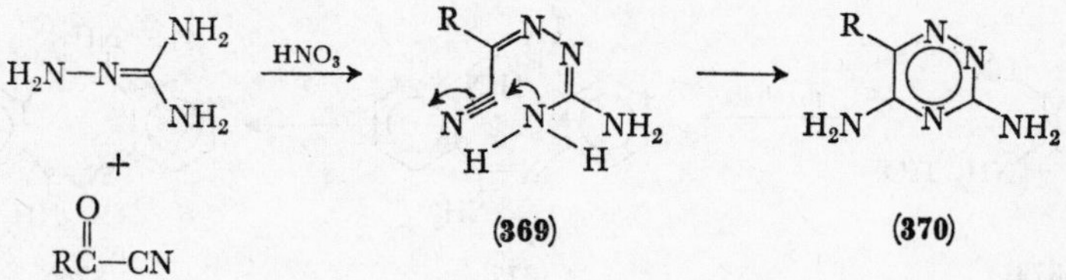

(369) (370)

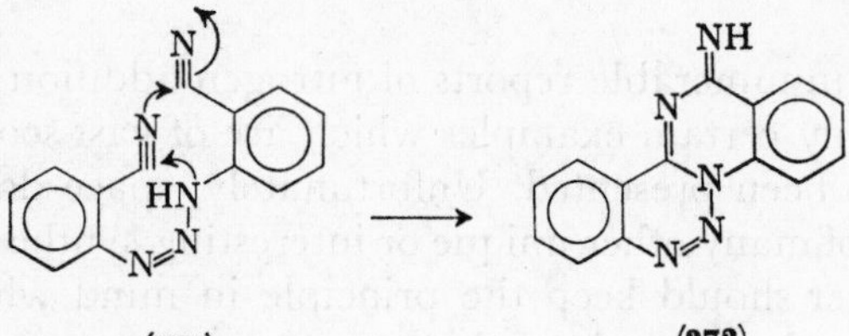

(371) (372)

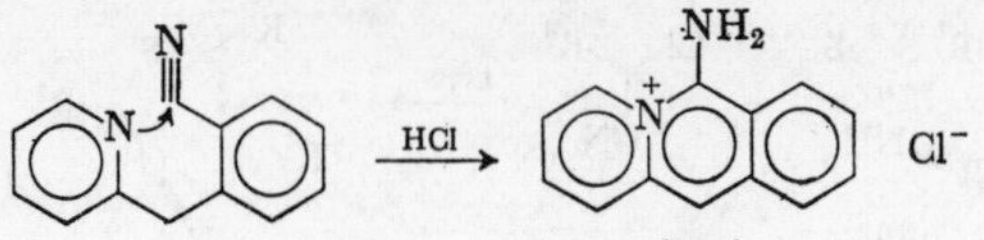

N N $\xrightarrow{\text{HCl}}$ NH_2 N^+ Cl^-

(**373**)

Quinoline *N*-oxides (**375**) are obtained by the uncommon route of anthranil (**374**) and nitriles in the presence of piperidines[196].

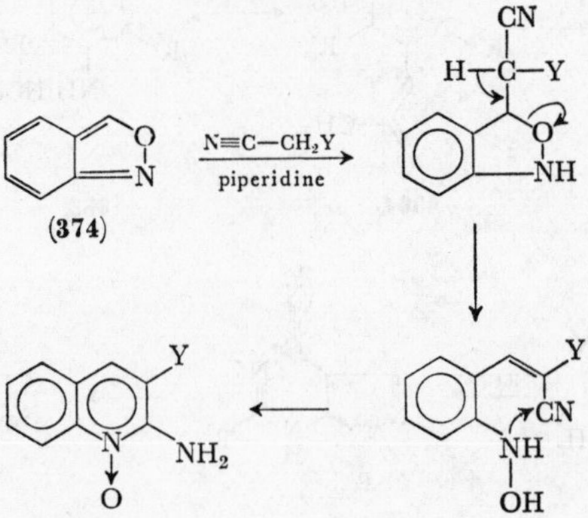

O N $\xrightarrow[\text{piperidine}]{N\equiv C-CH_2Y}$ CN H—C—Y O NH

(**374**)

Y NH2 N O ⟵ Y CN NH OH

(**375**)

An eight-membered heterocycle (**378**) has been reported[197] from the dimerization of the toluenesulphonate salt of *o*-aminobenzonitrile (**376**). The cyclization probably occurs via the amidinium salt (**377**).

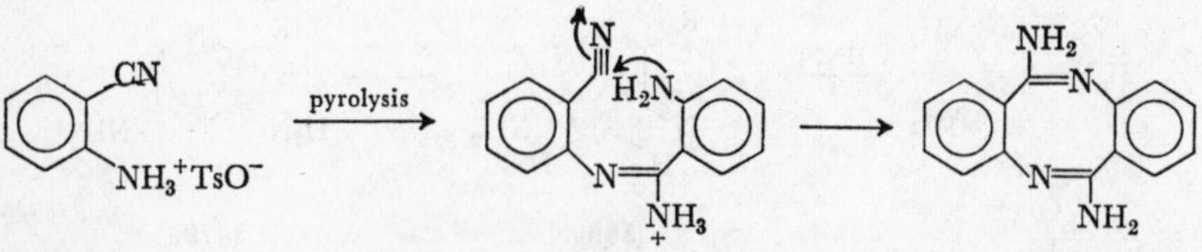

CN $NH_3^+TsO^-$ $\xrightarrow{\text{pyrolysis}}$ N H_2N N= $\overset{+}{N}H_3$ ⟶ NH_2 =N N= NH_2

(**376**) (**377**) (**378**)

There are innumerable reports of nitrogen addition to the cyano group and only certain examples which are of vast scope or special interest have been presented. Unfortunately, space does not allow descriptions of many other unique or interesting syntheses; nevertheless, the reader should keep the principle in mind when planning to synthesize a heterocycle of this type.

D. Intramolecular Nucleophilic Addition by Oxygen Nucleophiles

Both the hydroxyl group and the carbonyl group are suitable nucleophiles for addition to the cyano function. The latter is more frequently encountered.

The base-catalysed cyclization of the β-oximinonitrile **379** is an example of alkoxide addition which leads to the aminooxazole **380**[198]. If a heteroatom is absent from the hydroxynitrile, a variety

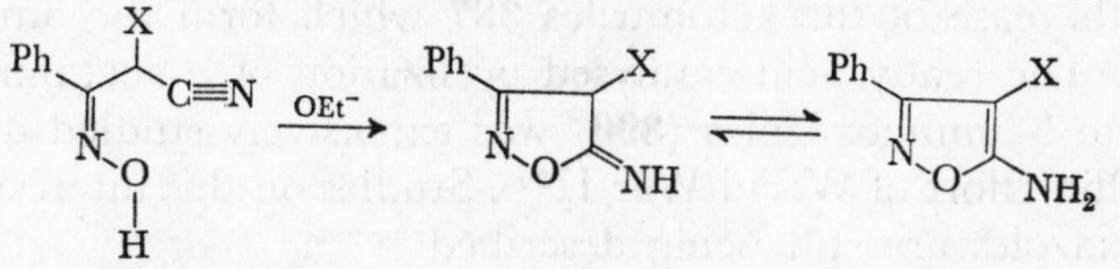

(379) (380)

($X = CO_2Et, CONH_2, PhCO$)

of lactones is formed by this mechanism. For example, the phenol derivative **381** and the hydroxybutyronitrile **383** cyclize readily to **382**[199] and **384**[200], respectively.

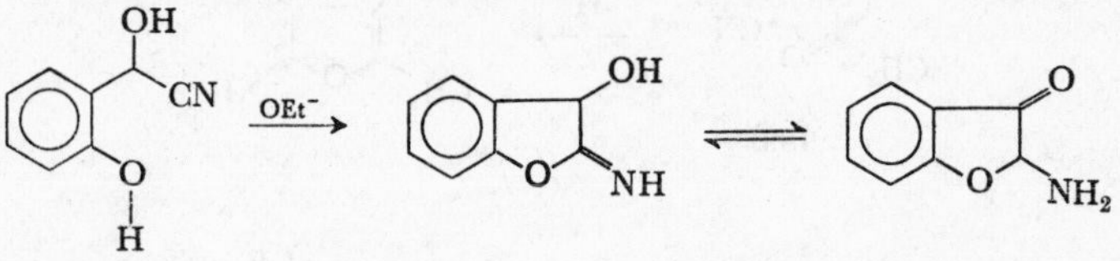

(381) (382)

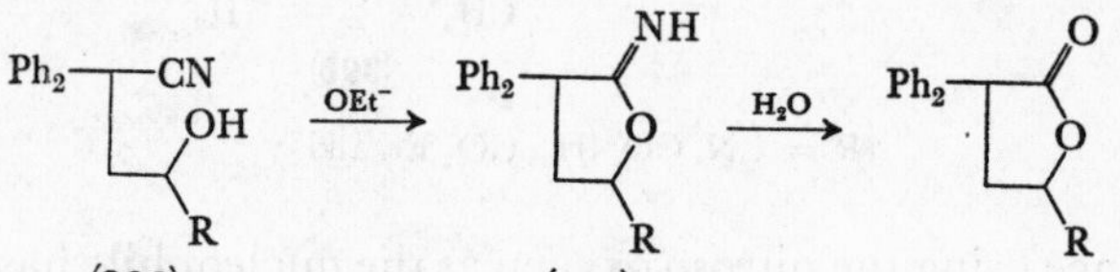

(383) (384)

It is of interest to note that ω-hydroxynitriles (**385**) capable of forming five- or six-membered rings do not show any nitrile sretching band in the infrared[201]. This supports the belief that the iminoethers **386** are present as ring–chain tautomers and that the equilibria lies heavily toward the cyclic structure. If the driving force for cyclization is so strong, it is surprising that so few examples

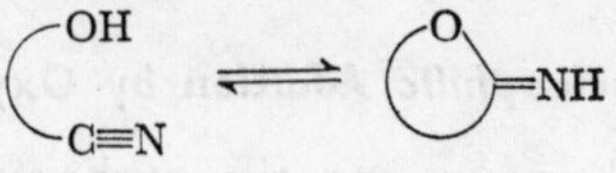

(385) ⇌ (386)

of heterocyclic syntheses of this type are encountered. One reason for this is probably that the reverse reaction is likewise facile, unless a subsequent process (as in **381–382**) occurs.

The carbonyl oxygen has been utilized extensively in cyclizations, as in the case of the ketonitriles **387** which form the aminofurans **388**[202]. The ready acid-catalysed cyclization of α-acetamidonitriles (**389**) to 5-aminooxazoles (**390**) was extensively studied during the penicillin effort of World War II[203]. Studies on this interesting entry into oxazoles are still being described[204,205].

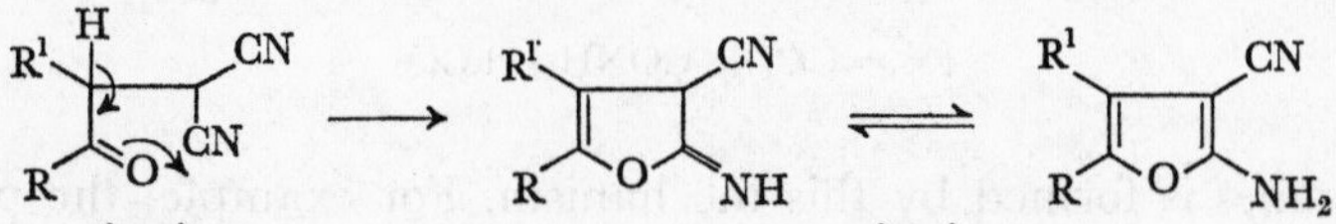

(387) → ⇌ (388)

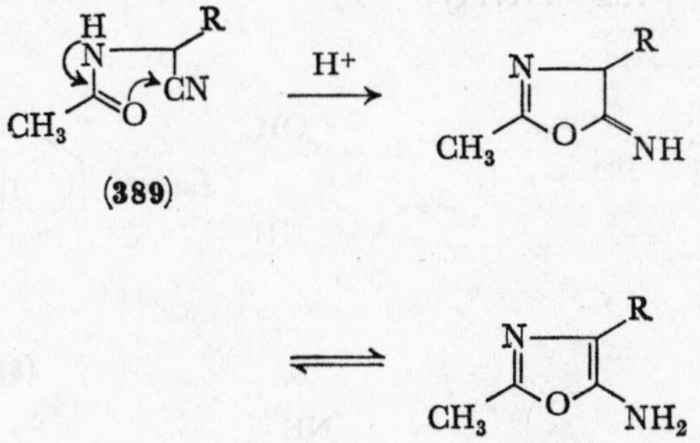

(389) $\xrightarrow{H^+}$ ⇌ (390)

(R = CN, $CONH_2$, CO_2Et, alkl)

An instance using the nitroso oxygen as the nucleophile has resulted in the conversion of the *N*-nitrosonitrile (**391**) to the sydnone imine **392**[206]. Once again, the cyclic product is stabilized so that reversal does not occur.

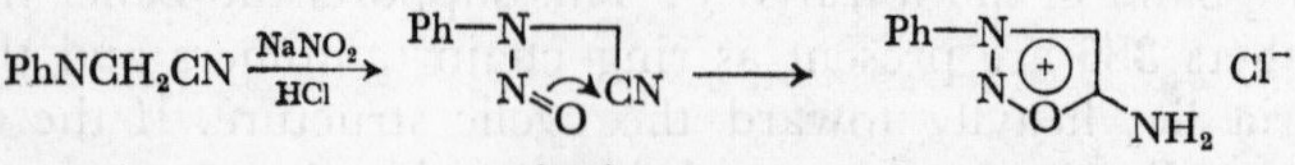

$PhNCH_2CN \xrightarrow[HCl]{NaNO_2}$ (391) → (392) Cl^-

E. Intramolecular Nucleophilic Addition by Sulphur Nucleophiles

As in the case of hydroxynitriles, mercaptonitriles will readily cyclize, usually with the aid of acids to increase the electrophilic nature of the nitrile. Stacy[207–209] has reported considerable data on the cyclization of the benzyl thioethers **393** which are dealkylated to form the thiol group **394** *in situ* which then proceeds to give the

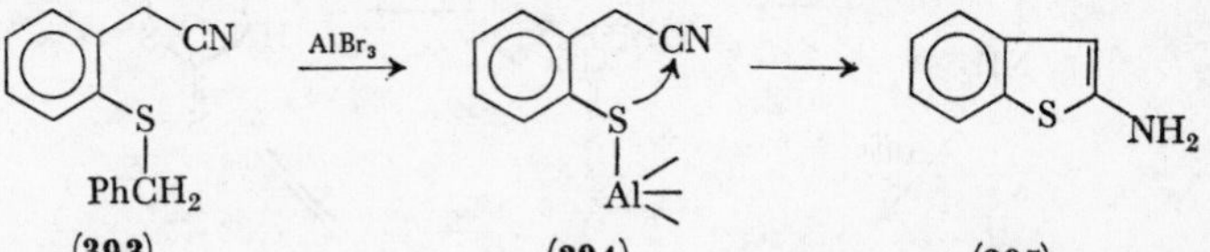

(393) (394) (395)

aminothiophene **395**. The evasive 2-aminothiophene[209] (**397**) has been prepared by Stacy and his students utilizing the *cis–trans* mixture of substituted crotononitriles (**396**).

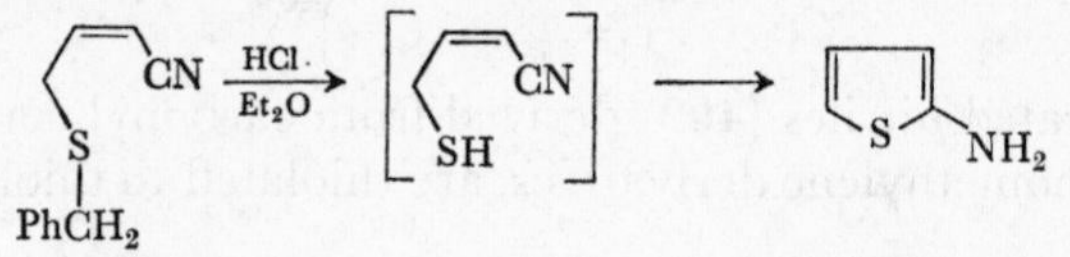

(396) (397)

Six-membered rings have been reported[210,211] by a mechanistically related process. The thiolactones **398** obtained from the thioesters can be opened and recyclized to the thiopyran or thiophene derivatives (**399**).

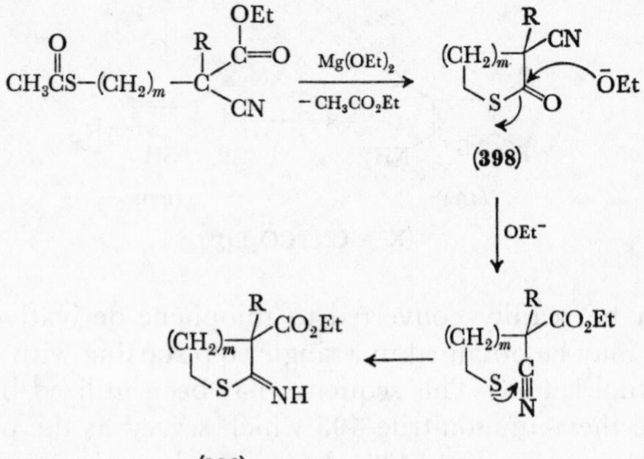

(398)

(399)

(R = H, CH_3, m = 1, 2)

A versatile synthesis of 5-amino-2-mercaptothiazoles (**400**) which proceeds by the addition of carbon disulphide to α-aminonitriles followed by nucleophile addition to the cyano group has been reported by Cook, Heilbron and their coworkers[212–217]. The use of carbon oxysulphide produced the corresponding 2-hydroxythiazole (**401**)[218].

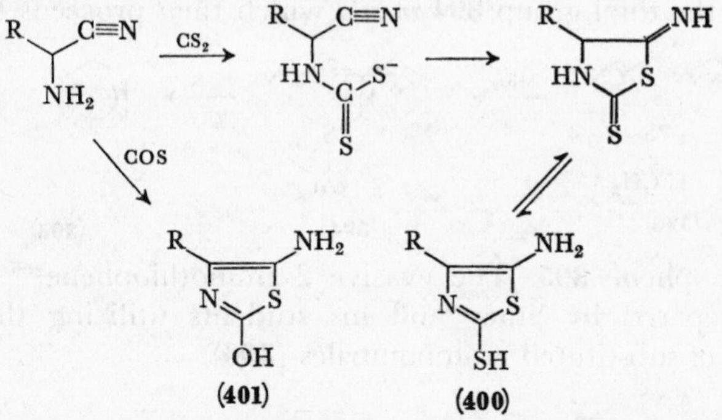

(401) (400)

Unsaturated nitriles (**402**) derived from carbonyl compounds and active cyanomethylene derivatives, are thiolated to thiolnitriles (**403**)

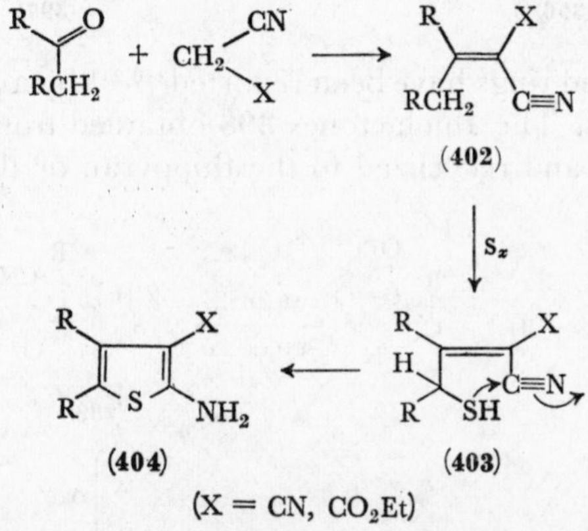

(402) (403) (404)

$(X = CN, CO_2Et)$

which can be readily converted to thiophene derivatives (**404**)[219]. The latter may be obtained in a single step starting with the ketone, or the α-thiol ketone. This sequence has been utilized by Taylor[220] to produce the *o*-aminonitrile **405** which served as the precursor to the thiophenopyrimidine **406**. Indanone has also served as the starting material for the formation of the indenothiophene **407**[221].

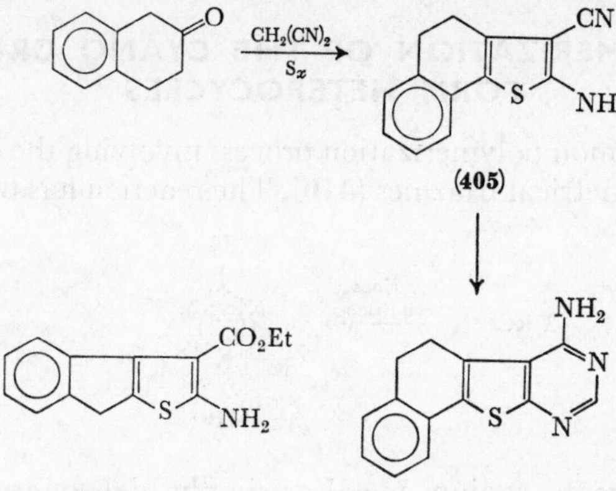

$CH_2(CN)_2$, S_x

(405)

(407)

(406)

Tetracyanoethylene has proved in recent years to be a valuable precursor for heterocyclic syntheses. A DuPont group reported[222,223] that tetracyanoethane obtained from tetracyanoethylene readily forms the thiophene **408** which, if desired, could be converted to the pyrrole **409** (R = H) upon treatment with alkali. Thus *o*-aminonitriles of the pyrrole and thiophene series became commonplace, and in view of the extensive work on these systems by Taylor[165], avenues were opened to a wide variety of previously inaccessible systems. It is appropriate to note here that the cyano group truly reached its present pinnacle of heterocyclic utility after these discoveries. The reaction of tetracyanoethylene with mercaptans (RSH)[223,224], in place of hydrogen sulphide, similarly reduced the olefinic linkage, producing only the pyrrole **409** (R = alkyl).

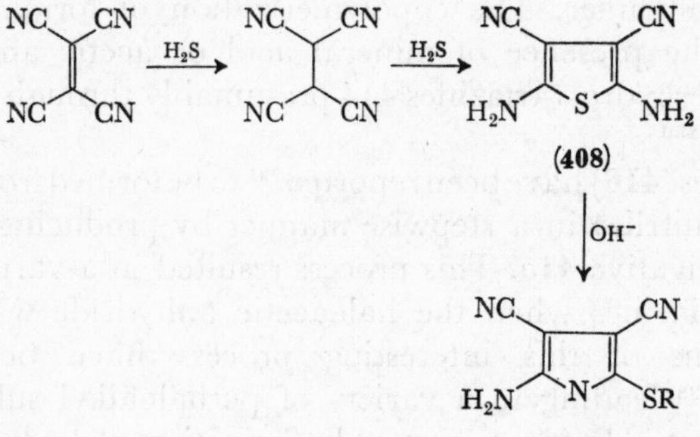

H_2S H_2S

(408)

OH^-

(409)

VII. POLYMERIZATION OF THE CYANO GROUP TO FORM HETEROCYCLES

The most common polymerization process involving the cyano group results in symmetrical triazines (**410**). The reaction has been initiated

$$3\ RC{\equiv}N \xrightarrow[\text{conditions}]{\text{various}} \text{(2,4,6-R}_3\text{-1,3,5-triazine)}$$

(**410**)

by acidic reagents, sodium metal or simply high temperatures and high pressures (Table 8). As expected, the more electrophilic the cyano carbon atom, the easier the cyclization occurs. Hence the use of acids[228] to increase the carbon electrophilicity increases the ease

TABLE 8. *s*-Triazines **410** from nitriles.

R	Conditions	Reference
Ph	Iron pentacarbonyl	225
Ph	High pressure	226
Ph	Sodium	227
CF_3	High pressure High temperature	228
CF_3	Hydrogen chloride	229

of cyclization. The trimerization leading to **410** may be formulated as a type B mechanism wherein the nitrilium ion **411** appears to be the initial intermediate followed by a second nitrilium ion **412** which precedes the trimer. The copolymerization of formaldehyde and nitriles in the presence of mineral acid or acetic anhydride produced the perhydro-*s*-triazines **414** presumably through the methylol adducts **413**[230].

s-Triazines (**416**) have been reported[231] to be formed from α-haloperfluoroalkyl nitriles in a stepwise manner by producing initially the amidine derivative **415**. This process resulted in a variety of mixed haloalkyl triazines when the haloacetic anhydride was employed. Modifications of this interesting process have been reported recently[232,233] leading to a variety of perhaloalkyl substituents on **416**. An extremely interesting polymerization of hydrogen cyanide

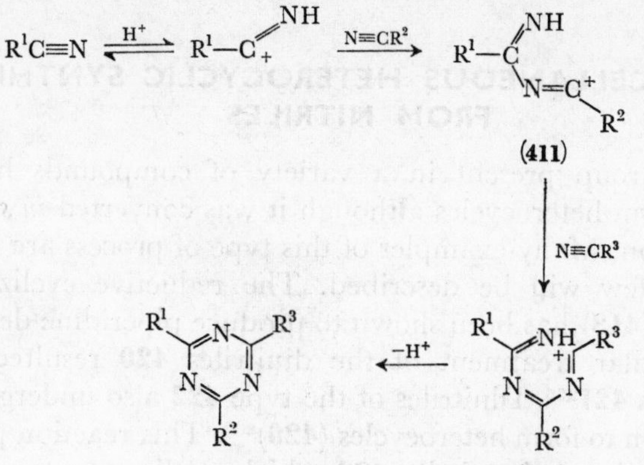

$$R^1C{\equiv}N \xrightleftharpoons{H^+} R^1{-}\overset{+}{C}{=}NH \xrightarrow{N{\equiv}CR^2} R^1{-}C({=}NH){-}N{=}\overset{+}{C}{-}R^2$$

(411)

$$\xrightarrow{N{\equiv}CR^3} \textbf{(412)} \xrightarrow{-H^+} \text{2-}R^1\text{-4-}R^3\text{-6-}R^2\text{-1,3,5-triazine}$$

$$RCN + CH_2O \xrightarrow{HX} [R{-}\overset{+}{C}{=}NCH_2OH] \longrightarrow \text{1,3,5-tris(acyl)hexahydro-1,3,5-triazine (O=CR, RCN(=O), NCR(=O))}$$

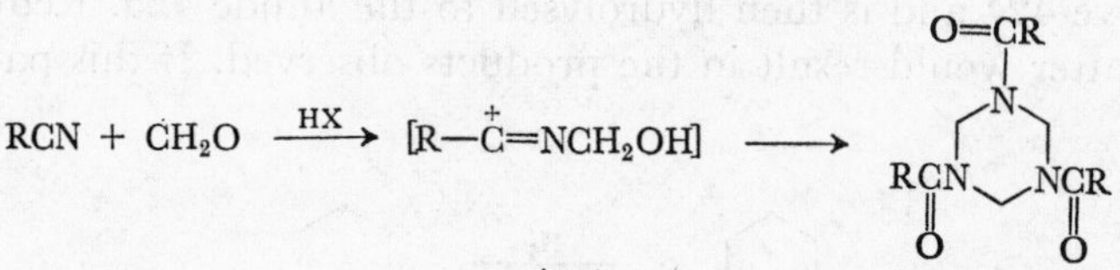

(413) **(414)**

has been shown to produce, in addition to other products, a 10–15% yield of the pentamer, adenine (**417**)[234]. This process has strong implications for studies of the origin of life if it can be reproduced under primitive Earth conditions.

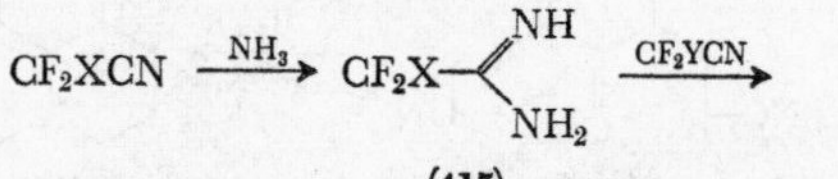

$$CF_2XCN \xrightarrow{NH_3} CF_2X{-}C({=}NH)NH_2 \xrightarrow{CF_2YCN}$$

(415)

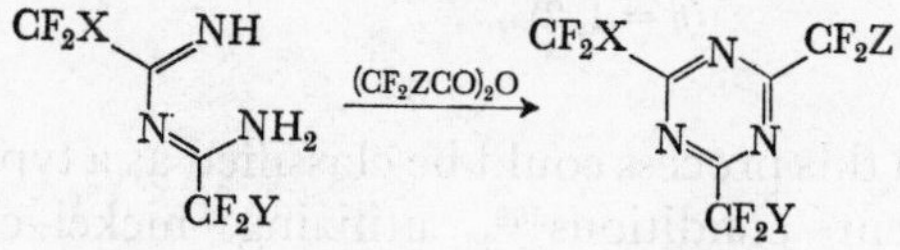

$$CF_2X{-}C({=}NH){-}N{=}C(NH_2){-}CF_2Y \xrightarrow{(CF_2ZCO)_2O} \text{2-}CF_2X\text{-4-}CF_2Z\text{-6-}CF_2Y\text{-1,3,5-triazine}$$

(416)

(X, Y, Z = F, Cl, Br I)

$$5\,HCN \longrightarrow \text{adenine}$$

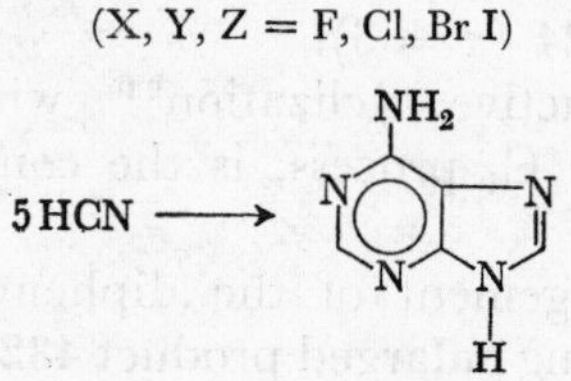

(417)

VIII. MISCELLANEOUS HETEROCYCLIC SYNTHESES FROM NITRILES

The cyano group present in a variety of compounds has been utilized to form heterocycles although it was converted *in situ* to an amino function. Many examples of this type of process are reported and only a few will be described. The reductive cyclization of δ-oxonitriles (**418**) has been shown to produce piperidine derivatives (**419**)[235]. Similar treatment of the dinitriles **420** resulted in the tricyclic bases **421**[236]. Dinitriles of the type **422** also undergo reductive cyclization to form heterocycles (**426**)[237]. This reaction probably proceeds via the aminonitrile **423** which cyclizes to the amidine derivative **424** and is then hydrolysed to the amide **425**. Reduction of the latter would result in the products observed. If this pathway

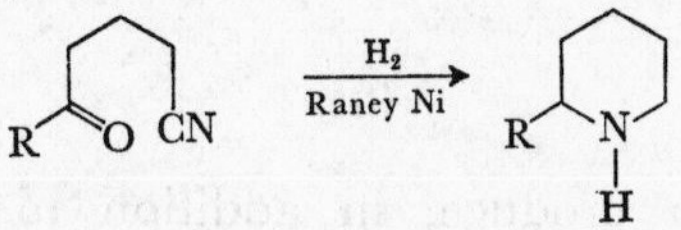

H_2 / Raney Ni

(418) **(419)**

H_2 / Raney Ni

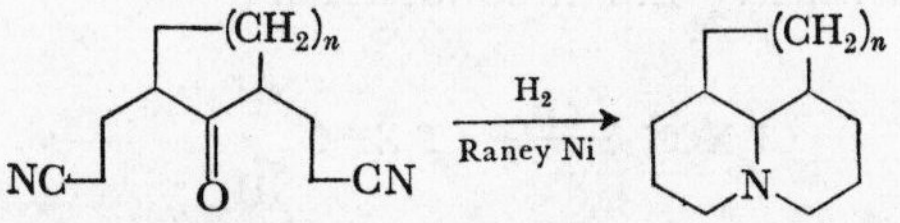

(420) **(421)**

($n = 1, 2$)

is correct, then this process could be classified as a type C_i mechanism. Under different conditions[238], utilizing nickel–cobalt catalysts, dinitriles can be cyclized to aminoazepines (**427**) presumably by direct reduction of **424** ($n = 3$).

An interesting reductive cyclization[239], which may be assumed to adhere to the type C_i process, is the conversion of **428** to the aminopyridine **429**.

A thermal rearrangement of the diphenylketene–diazocyanide adduct (**430**) to the ring enlarged product **432** represents an unusual process[240]. The eight-membered ring **431** has been postulated as

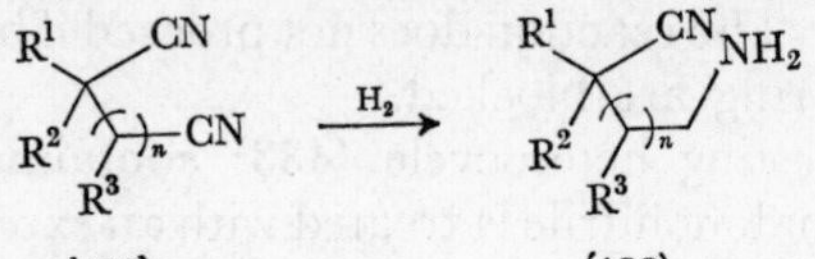

R^1 CN R^2 $(\ )_n$ CN R^3 $\xrightarrow{H_2}$ R^1 CN NH_2 R^2 $(\ )_n$ R^3

(422) ($n = 0, 1$) **(423)**

NH, NH, R^1, R^2, R^3 ⟶ O, NH, R^1, R^2, R^3 ⟶ NH, R^1, R^2, R^3

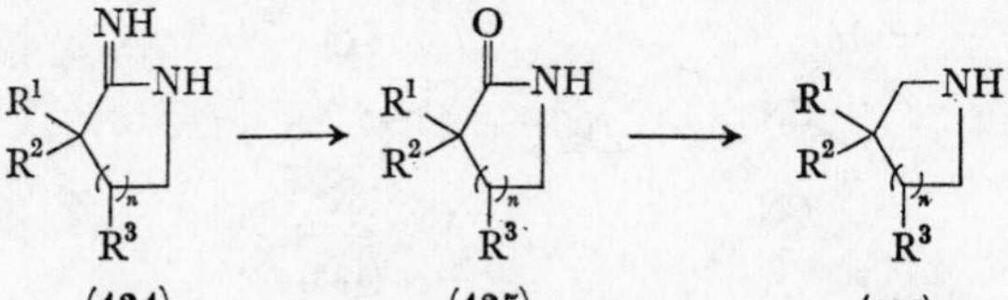

(424) **(425)** **(426)**

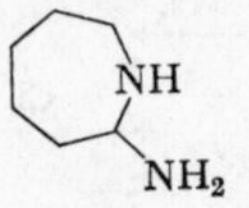

NH, NH_2

(427)

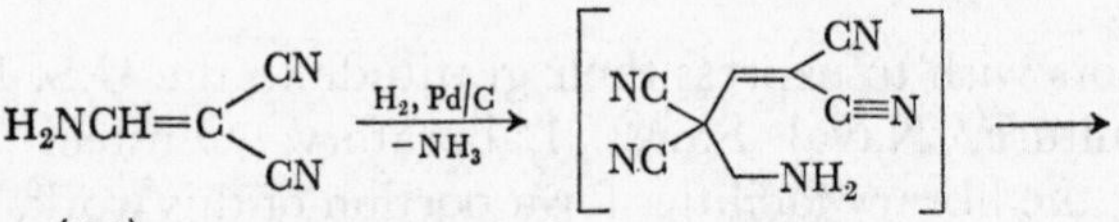

$H_2NCH{=}C(CN)_2$ $\xrightarrow[-NH_3]{H_2,\ Pd/C}$ [NC, NC, CN, C≡N, NH_2] ⟶

(428)

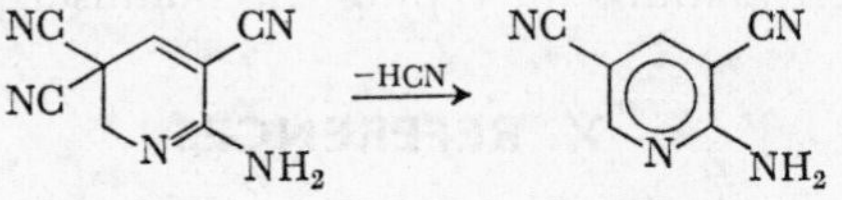

NC, NC, CN, N, NH_2 $\xrightarrow{-HCN}$ NC, CN, N, NH_2

(429)

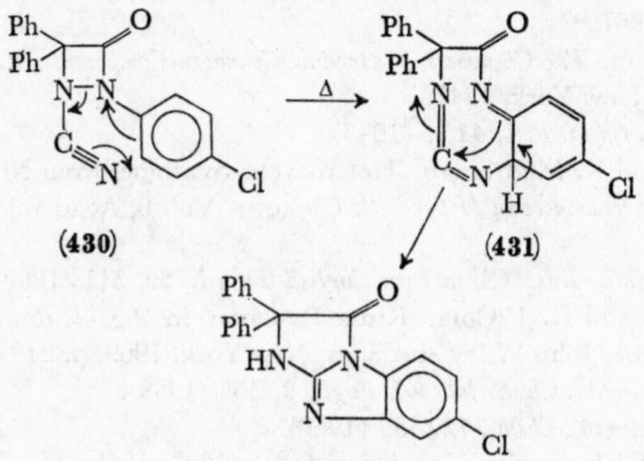

Ph, Ph, O, N, N, C, N, Cl $\xrightarrow{\Delta}$ Ph, Ph, O, N, N, C, N, H, Cl

(430) **(431)**

Ph, Ph, O, HN, N, N, Cl

(432)

the intermediate. The reaction does not proceed if both *ortho* positions on the aromatic ring are blocked.

Another interesting heterocycle (**433**) containing phosphorus is obtained when malononitrile is treated with an excess of a phosphorus halide[241]. However, when succinonitrile is employed the phosphorus atom is not incorporated in the ring and the result is **434**.

$CH_2(CN)_2 \xrightarrow{PCl_5}$ (**433**)

$NCCH_2CH_2CN \xrightarrow{PCl_5}$ (**434**)

IX. ACKNOWLEDGEMENTS

The authors wish to express their gratitude to the U.S. Department of Agriculture, Naval Stores Laboratory, Olustee, Florida for providing the library facilities for a portion of this work, to Professor R. C. Petterson, Loyola University for valuable comments and to Mrs. Linda D'Alessandro for typing the manuscript.

X. REFERENCES

1. A. Weissberger (Ed.), 21 Vols., *The Chemistry of Heterocyclic Compounds*, Interscience, New York, 1950–1965.
2. R. C. Elderfield (Ed.), 7 Vols., *Heterocyclic Compounds*, John Wiley and Sons, New York, 1950–1961.
3. V. Migrdichian, *The Chemistry of Organic Cyanogen Compounds*, Reinhold Publishing Corporation, New York, 1947.
4. F. R. Benson, *Chem. Rev.*, **41**, 1 (1947).
5. F. Johnson and R. Madronero, 'Heterocyclic Syntheses from Nitriles under Acidic Conditions' in *Advances in Heterocyclic Chemistry*, Vol. 6, Academic Press, New York, 1966, p. 95.
6. E. N. Zil'berman, *Russ. Chem. Rev. English Transl.*, **29**, 311 (1960).
7. L. I. Krimen and D. J. Cota, 'Ritter Reaction' in *Organic Reactions*, Vol. 17 (Ed. W. C. Douben), John Wiley and Sons, New York, 1969, p. 213.
8. R. Huisgen, *Angew. Chem. Int. Ed. Engl.*, **2**, 565 (1963).
9. R. Huisgen, *Angew. Chem.*, **72**, 359 (1960).
10. A. Hassner and F. Fowler, *Tetrahedron Letters*, **1967**, 1545.
11. W. H. Gram, *J. Org. Chem.*, **30**, 2108 (1965) and other references cited therein.

12. R. Huisgen, R. Grashy and J. Sauer in *The Chemistry of Alkenes* (Ed. S. Patai), Interscience, New York, 1964, p. 739. For a recent discussion concerning the mechanism of 1,3-dipolar additions see R. A. Firestone, *J. Org. Chem.*, **33**, 2285 (1968), and R. Huisgen, *J. Org. Chem.*, **33**, 2291 (1968).
13. R. Huisgen, W. Mack and E. Anneser, *Tetrahedron Letters*, **1961**, 587.
14. G. Leandri and M. Pallotti, *Ann. Chim. (Rome)*, **47**, 376 (1957).
15. S. Morrocchi, A. Ricca and L. Velo, *Tetrahedron Letters*, **1967**, 331.
16. A. Laubengayer and D. S. Sears, *J. Am. Chem. Soc.*, **67**, 164 (1945).
17. F. Tieman and H. Krueger, *Ber.*, **17**, 1685 (1884).
18. M. S. Chang and J. V. Lowe, *J. Org. Chem.*, **32**, 1577 (1967).
19. R. Huisgen, M. Seidel, J. Sauer, J. W. McFarland and G. Wallbillich, *J. Org. Chem.*, **24**, 892 (1959).
20. R. Huisgen, R. Grashey, M. Seidel, G. Wallbillich, H. Knupfer and R. Schmidt, *Ann. Chem.*, **653**, 105 (1962).
21. A. Peratoner and E. Azzarello, *Gazz. Chim. Ital.*, **38**, I, 76 (1908).
22. E. Oliveri-Mandala, *Gazz. Chim. Ital.*, **40**, I, 123 (1910).
23. C. Pedersen, *Acta. Chem. Scand.*, **13**, 888 (1959).
24. H. Hoberg, *Ann. Chem.*, **707**, 147 (1967).
25. R. Huisgen, S. Stangl, H. J. Sturm and H. Wagenhofer, *Angew. Chem. Int. Ed. Engl.*, **1**, 50 (1962).
26. R. Huisgen, *Helv. Chem. Acta.*, **50**, 2421 (1967).
27. P. A. S. Smith, J. M. Clegg and J. H. Hall, *J. Org. Chem.*, **23**, 524 (1958).
28. V. Kereszty and E. Wolf, *Ger. Pat.* 611,692; *Chem. Abstr.*, **29**, 5994 (1935).
29. W. R. Carpenter, *J. Org. Chem.*, **27**, 2085 (1962).
30. R. Huisgen, R. Fleischmann and A. Eckell, *Tetrahedron Letters*, **1960**, 1.
31. R. Grashey, H. Leitermann, R. Schmidt and K. Adelsberger, *Angew. Chem. Int. Ed. Engl.*, **1**, 406 (1962).
32. R. Huisgen, E. Funke, F. C. Schaefer, H. Gotthardt and E. Brunn, *Tetrahedron Letters*, **1967**, 1809.
33. R. Huisgen, H. Konig, G. Binsch and H. J. Sturm, *Angew. Chem.*, **73**, 368 (1961).
34. R. Huisgen, H. J. Sturm and G. Binsch, *Chem. Ber.*, **97**, 2864 (1964).
35. F. Weygand, H. Dworschak, K. Koch and S. Konstas, *Angew. Chem.*, **73**, 409 (1961); *Chem. Ber.*, **101**, 302 (1968).
36. W. Lwowski, A. Hartenstein, C. D. Vita and R. L. Smick, *Tetrahedron Letters*, **1964**, 2497.
37. R. Huisgen and H. Blaschke, *Ann. Chem.*, **686**, 145 (1965).
38. K. Dickore and R. Wegler, *Angew. Chem. Int. Ed. Engl.*, **5**, 970 (1966).
39. A. S. Onishchenko, *Diene Synthesis*, Israel Program for Scientific Translations, Jerusalem, 1964.
40. R. Huisgen, R. Grashey and J. Sauer in *The Chemistry of Alkenes* (Ed. S. Patai), Interscience, New York, 1964, p. 880.
41. S. B. Needleman and M. C. Chang Kuo, *Chem. Rev.*, **62**, 405 (1965).
42. J. Sauer, *Angew. Chem. Int. Ed. Engl.*, **5**, 211 (1966).
43. G. J. Janz, R. G. Ascah and A. G. Keenan, *Can. J. Res.*, **B-25**, 272, 283 (1947).
44. G. J. Janz and S. C. Wait, Jr., *J. Am. Chem. Soc.*, **76**, 6377 (1954).
45. G. J. Janz and W. J. McCulloch, *J. Am. Chem. Soc.*, **77**, 3014, 3143 (1955).
46. G. J. Janz, J. M. S. Jarrie and W. E. Fitzgerald, *J. Am. Chem. Soc.*, **78**, 978 (1956).
47. G. J. Janz and J. M. S. Jarrie, *J. Phys. Chem.*, **60**, 1430 (1956).

48. G. J. Janz and A. R. Monahan, *J. Org. Chem.*, **29,** 569 (1964).
49. G. J. Janz and M. A. DeCrescente, *J. Org. Chem.*, **23,** 765 (1958).
50. G. J. Janz and N. E. Duncan, *J. Am. Chem. Soc.*, **75,** 5389 (1953).
51. W. Polaczkowa and J. Wolinski, *Roczniki Chem.*, **26,** 407 (1952); *Chem. Abstr.*, **48,** 11359 (1954).
52. W. Polaczkowa, T. Jaworski and J. Wolinski, *Roczniki Chem.*, **27,** 468 (1953); *Chem. Abstr.*, **49,** 3181 (1955).
53. T. Jaworski and W. Polaczkowa, *Roczniki Chem.*, **34,** 887 (1960); *Chem. Abstr.*, **55,** 8407d (1961).
54. T. Jaworski, *Roczniki Chem.*, **35,** 1309 (1961); *Chem. Abstr.*, **57,** 588i (1962).
55. T. Jaworski and B. Korybut-Daszkiewicz, *Roczniki Chem.*, **41,** 1521 (1967).
56. W. Dilthey, *Ber.*, **68,** 1162 (1935).
57. H. Meerwein, P. Lasch, R. Mersch and J. Spille, *Chem. Ber.*, **89,** 209 (1956).
58. J. J. Ritter and P. P. Minieri, *J. Am. Chem. Soc.*, **70,** 4045, 4048 (1948).
59. A. I. Meyers and J. J. Ritter, *J. Org. Chem.*, **23,** 1918 (1958).
60. A. I. Meyers, *Ph.D. Dissertation,* New York University, 1958.
61. J. J. Ritter and F. X. Murphy, *J. Am. Chem. Soc.*, **74,** 763 (1952).
62. M. Lora-Tamayo, R. Madronero, D. Gracian and V. Gomez-Parra, *Tetrahedron Suppl.* **1966**(8), 305 and the references cited therein.
63. W. Sobotka, W. N. Beverung, G. G. Munoz, J. C. Sircar and A. I. Meyers, *J. Org. Chem.*, **30,** 3667 (1965).
64. S. G. Agbalyan, A. O. Nshanyan and L. A. Nersesyan, *Izv. Akad. Nauk Arm. SSR, Khim. Nauk,* **16(1),** 77 (1963).
65. S. G. Agbalyan, L. A. Nersesyan and A. O. Nashanyan, *Izv. Akad. Nauk Arm. SSR, Khim. Nauk,* **18(1),** 83 (1965); *Chem. Abstr.*, **63,** 6972g (1965).
66. (a) S. G. Agbalyan and L. A. Nersesyan, *Izv. Akad. Nauk Arm. SSR, Khim. Nauk*, **17,** 441 (1964); *Chem. Abstr.*, **62,** 521h (1965);
(b) S. G. Agbalyan, A. O. Nashanyan and L. A. Nersesyan, *Izv. Akad. Nauk. Arm. SSR, Khim. Nauk*, **15,** 399 (1962); *Chem. Abstr.*, **59,** 567 (1963).
67. A. I. Meyers and J. C. Sircar, unpublished results.
68. T. Kametani, R. Yanase and S. Kakano, *Yakugaku Kenkyu*, **37,** 23–31 (1966); *Chem. Abstr.*, **65,** 15320 (1966).
69. A. I. Meyers, B. J. Betrus, N. K. Ralhan and K. B. Rao, *J. Heterocyclic Chem.*, **1,** 13 (1964).
70. M. Lora-Tamayo, R. Madronero and M. G. Perez, *Chem. Ber.*, **95,** 2188 (1962).
71. A. Hassner, R. A. Arnold, R. Gault and A. Terada, *Tetrahedron Letters,* **1968,** 1241.
72. R. C. Petterson and T. G. Troendle, personal communication.
73. F. R. Japp and T. S. Murray, *J. Chem. Soc.*, **63,** 469 (1893).
74. M. Lora-Tamayo, R. Madronero and H. Leipprand, *Chem. Ber.*, **97,** 2230 (1964).
75. C. M. Welch and H. A. Smith, *J. Am. Chem. Soc.*, **75,** 1412 (1953).
76. C. W. Bird, *J. Org. Chem.*, **27,** 4091 (1962).
77. H. Hohenlohe-Ochringen, *Monatsh. Chem.*, **93,** 639 (1962).
78. M. Lora-Tamayo, R. Madronero, G. G. Munoz and H. Leipprand, *Chem. Ber.*, **97,** 2234 (1964).
79. E. Ziegler, G. Kleineberg and H. Meindl, *Monatsh. Chem.*, **94,** 544 (1963).
80. E. Ziegler, G. Kleineberg and H. Meindl, *Monatsh. Chem.*, **97,** 10 (1966). Similar reactions have also been reported by Elvidge and coworkers, *J. Chem. Soc.*, **1962,** 3553, 3638.
81. J. J. Ritter and E. J. Tillmanns, *J. Org. Chem.*, **22,** 839 (1957).

82. A. I. Meyers, *J. Org. Chem.*, **25,** 218 (1960).
83. A. I. Meyers and A. Nabeya, *Chem. Commun.*, **1967,** 1163.
84. A. I. Meyers, *J. Org. Chem.*, **25,** 145 (1960).
85. S. D. McGregor and W. M. Jones, *J. Am. Chem. Soc.*, **90,** 123 (1968).
86. A. I. Meyers, *J. Org. Chem.*, **25,** 1147 (1960).
87. S. Julia and C. Papantoniou, *Compt. Rend.*, **260,** 1440 (1965).
88. R. Oda, M. Okano, S. Tokiura and F. Misumi, *Bull. Chem. Soc. Japan*, **35,** 1219 (1962).
89. T. I. Temnikova and T. E. Zhesko, *Zh. Obshch. Khim.*, **33,** 3436 (1965).
90. N. J. Leonard and B. Zwanenburg, *J. Am. Chem. Soc.*, **89,** 4456 (1967).
91. A. I. Meyers and R. C. Bhattacharjee, unpublished results.
92. G. K. Helkamp, D. J. Pettit, J. R. Lowell, W. R. Mahey and R. G. Walcott, *J. Am. Chem. Soc.*, **88,** 1030 (1966).
93. J. R. Lowell and G. K. Helkamp, *J. Am. Chem. Soc.*, **88,** 768 (1966).
94. K. Miyatake and T. Yoshikawa, *Japan Pat.* 7926 (1956); *Chem. Abstr.*, **52,** 14698g (1958).
95. H. Behringer and D. Weber, *Ann. Chem.*, **682,** 196 (1965).
96. A. I. Meyers and J. M. Greene, *J. Org. Chem.*, **31,** 556 (1966).
97. D. S. Tarbell, D. A. Buckley, P. P. Brownlee, R. Thomas and J. S. Todd, *J. Org. Chem.*, **29,** 3314 (1964).
98. A. I. Meyers and P. Singh, unpublished results.
99. N. J. Leonard and L. E. Brady, *J. Org. Chem.*, **30,** 817 (1965).
100. N. J. Leonard, *Record Chem. Prog.*, **26,** 211 (1965).
101. E. Pfeil and U. Harder, *Angew. Chem. Int. Ed. Engl.*, **4,** 518 (1965).
102. N. J. Leonard, D. A. Durand and F. Uchimaru, *J. Org. Chem.*, **32,** 3607 (1967).
103. C. Grundmann, G. Weisse and S. Seide, *Ann. Chem.*, **577,** 77 (1952).
104. H. Meerwein, P. Laasch, R. Mersch and J. Nentwig, *Chem. Ber.*, **89,** 224 (1956); see also J. Leviesalles, *Ph.D. Thesis*, University of Paris, 1956.
105. M. Lora-Tamayo, R. Madronero and G. G. Munoz, *Chem. Ind.* (*London*), **1959,** 657.
106. M. Lora-Tamayo, R. Madronero and G. G. Munoz, *Chem. Ber.*, **94,** 208 (1961).
107. H. Schnell and J. Nentwig in *Methoden der Organische Chemie*, Vol. 11 (Eds. J. Houben and T. Weyl), Vol. 11, Part 2, Stuttgart Germany, 1958, p. 561.
108. R. T. Conley and R. J. Lange, *J. Org. Chem.*, **28,** 210 (1963), and earlier references cited therein.
109. J. M. Bobbitt and R. E. Doolittle, *J. Org. Chem.*, **29,** 2298 (1964).
110. E. P. Kohler and B. L. Souther, *J. Am. Chem. Soc.*, **44,** 2903 (1922).
111. C. R. Hauser and C. J. Eby, *J. Am. Chem. Soc.*, **79,** 728 (1957).
112. N. P. Shuskerina, A. V. Golovin and R. Y. Levina, *Zh. Obshch. Khim.*, **30,** 1762 (1960).
113. A. I. Meyers and G. G. Munoz, *J. Org. Chem.*, **29,** 1435 (1964).
114. J. J. Vill, T. R. Steadman and J. J. Godfrey, *J. Org. Chem.*, **29,** 2780 (1964).
115. A. B. Farbenfabriken, *Ger. Pat.* 1,092,919; *Chem. Abstr.*, **56,** 4626i (1962).
116. N. P. Shusherina, R. Y. Levina and K. Khua-min, *Zh. Obshch. Khim.*, **32,** 3599 (1962).
117. A. Vigier and J. Dreux, *Bull. Soc. Chim. France*, **10,** 2294 (1963).
118. G. Simchen, *Angew. Chem. Int. Ed. Engl.*, **5,** (7), 663 (1966).
119. H. Dugas, M. E. Hazenberg, Z. Valenta and K. Wiesner, *Tetrahedron Letters*, **1967,** 4931.
120. C. F. Koelsch and H. M. Walker, *J. Am. Chem. Soc.*, **72,** 346 (1950).

121. Y. Ban, I. Inove, M. Magai, T. Oishi, M. Terashima, O. Yonemitsu and Y. Kanaoka, *Tetrahedron Letters*, **1965**, (27) 2261.
122. L. G. Duquette and F. Johnson, *Tetrahedron*, **23**, 4517 (1967).
123. W. A. Nasutavicus and F. Johnson, *J. Org. Chem.*, **32**, 2367 (1967).
124. W. A. Nasutavicus, S. W. Tobey and F. Johnson, *J. Org. Chem.*, **32**, 3325 (1967).
125. A. Horeau, J. Jacques, H. B. Kagan and Y. Heng Hsuan, *Compt. Rend.*, **255**, 717 (1962).
126. P. J. A. Demoen and P. J. A. Janssen, *J. Am. Chem. Soc.*, **81**, 6281 (1959).
127. J. H. Burckhalter and J. H. Short, *J. Org. Chem.*, **23**, 1281 (1958).
128. P. M. Maginity and J. B. Cloke, *J. Am. Chem. Soc.*, **73**, 49 (1951).
129. J. B. Cloke, *J. Am. Chem. Soc.*, **51**, 1174 (1929); **67**, 2155 (1945).
130. R. V. Stevens and M. C. Ellis, *Tetrahedron Letters*, **1967**, 5185.
131. (a) R. V. Stevens, M. C. Ellis and M. P. Wentland, *J. Am. Chem. Soc.*, **90**, 5576, 5580 (1968);
(b) S. L. Keely, Jr., and F. C. Tahk, *J. Am. Chem. Soc.*, **90**, 5584 (1968).
132. T. L. Cairns, J. C. Sauer and W. K. Wilkinson, *J. Am. Chem. Soc.*, **74**, 3989 (1952).
133. H. Lapin and A. Horeau, *Chimia*, **15**, 551 (1961); *Chem. Abstr.*, **59**, 560d (1963).
134. C. F. Huebner, P. L. Strachan, E. M. Donoghue, N. Cahoon, L. Dorfman, R. Margerison and E. Wenkert, *J. Org. Chem.*, **32**, 1126 (1967).
135. L. L. Replogle, K. Katsumoto, I. C. Morrill and C. A. Minor, *J. Org. Chem.*, **33**, 823 (1968).
136. J. T. A. Boyle and M. F. Grandon, *Chem. Commun.*, **1967**, 1137.
137. A. Hantzoch and A. Vagt, *Ann. Chem.*, **314**, 339 (1901).
138. J. S. Mihina and R. M. Herbst, *J. Org. Chem.*, **15**, 1082 (1950).
139. R. M. Herbst and K. R. Wilson, *J. Org. Chem.*, **22**, 1142 (1957).
140. W. G. Finnegan, R. A. Henry and R. Lofquist, *J. Am. Chem. Soc.*, **80**, 3908 (1958).
141. R. Huisgen, J. Sauer, H. J. Sturm and J. H. Markgraf, *Chem. Ber.*, **93**, 2106 (1960).
142. H. Behringer and K. Kohl, *Chem. Ber.*, **89**, 2648 (1956).
143. W. P. Norris, *J. Org. Chem.*, **27**, 3248 (1962).
144. H. C. Brown and R. J. Kassal, *J. Org. Chem.*, **32**, 1871 (1967).
145. L. A. Lee, E. V. Carbtree, J. U. Lowe, M. J. Czlesla and R. Evans, *Tetrahedron Letters*, **1965**, 2885.
146. V. J. Grenda, R. E. Jones, G. Gal and M. Sletzinger, *J. Org. Chem.*, **30**, 259 (1965).
147. I. Yamase, N. Kuroki and K. Konishi, *Kogyo Kagaku Zasshi*, **67**, 102 (1964); *Chem. Abstr.*, **61**, 4515d (1964).
148. A. F. McKay, G. Y. Paris and D. L. Garmaise, *J. Am. Chem. Soc.*, **80**, 6276 (1958).
149. P. B. Russell, G. H. Hatchings, B. H. Chase and J. Walker, *J. Am. Chem. Soc.*, **74**, 5403 (1952).
150. H. Bader, *J. Org. Chem.*, **30**, 930 (1965).
151. W. Seeliger, E. Aufderhaar, W. Diepers, R. Feinauer, R. Nehring, W. Thier and H. Hellmann, *Angew. Chem. Int. Ed. Engl.*, **5**, 875 (1966).
152. F. H. Case, *J. Org. Chem.*, **30**, 931 (1965).
153. P. Yates and O. Meresz, *Tetrahedron Letters*, **1967**, 77.
154. P. Oxley and W. F. Short, *J. Chem. Soc.*, **1947**, 497.
155. A. Marxer, *J. Am. Chem. Soc.*, **79**, 467 (1957).
156. H. C. Brown and C. R. Wetzel, *J. Org. Chem.*, **30**, 3729 (1965).
157. T. Joyama, *U.S. Pat.* 2,846,439; *Chem. Abstr.*, **53**, 3085 (1959).
158. H. E. Baumgarten, J. E. Dirks, J. M. Petersen and R. L. Zey, *J. Org. Chem.*, **31**, 3708 (1966).

159. I. J. Borowitz, P. D. Readio and P. Rusek, *Chem. Commun.*, **1968**, 240.
160. W. R. Hatchard, *J. Org. Chem.*, **29**, 665 (1964).
161. R. Mayer and K. Gewald, *Angew. Chem. Int. Ed. Engl.*, **6**, 304 (1967).
162. M. Davis, S. Snowling and R. W. Winch, *J. Chem. Soc.* (*C*), **1967**, 124.
163. W. Mack, *Angew. Chem. Int. Ed. Engl.*, **6**, 1084 (1967).
164. A. Bruno and G. Purrello, *Gazz. Chim. Ital.*, **96**, 986 (1966).
165. E. C. Taylor, A. McKillop and S. Vromen, *Tetrahedron*, **23**, 885 (1967).
166. E. C. Taylor, A. McKillop and R. N. Warrener, *Tetrahedron*, **23**, 891 (1967).
167. A. Aviran and S. Vromer, *Chem. Ind.* (*London*), **1967**, 1452.
168. R. Kuhn and F. Drawert, *Chem. Ber.*, **88**, 55 (1955).
169. J. P. Schaefer and J. J. Bloomfield in *Organic Reactions*, Vol. 15 (Ed. A. C. Cope), John Wiley and Sons, New York, 1967.
170. D. Taub, C. H. Kuo and N. L. Wendler, *J. Chem. Soc.* (*C*), **1967**, 1558.
171. H. E. Schroeder and G. W. Rigby, *J. Am. Chem. Soc.*, **71**, 2205 (1949).
172. E. J. Modest, S. Chatterjee and H. K. Protopapa, *J. Org. Chem.*, **30**, 1837 (1965).
173. E. P. Burrows, A. Rosowsky and E. J. Modest, *J. Org. Chem.*, **32**, 4090 (1967).
174. J. A. Moore and L. D. Kornreich, *Tetrahedron Letters*, **1963**, 1277.
175. A. I. Meyers, J. C. Sircar and S. Singh, *J. Heterocyclic Chem.*, **4**, 461 (1967).
176. A. I. Meyers and J. C. Sircar, *J. Org. Chem.*, **32**, 1250 (1967).
177. A. I. Meyers, A. H. Reine, J. C. Sircar, K. B. Rao, S. Singh, H. Weidmann and M. Fitzpatrick, *J. Heterocyclic Chem.*, **5**, 151 (1968).
178. A. I. Meyers, A. C. Kovelesky and S. Singh, *Abstr. 155. Am. Chem. Soc. Meeting, Org. Chem.*, p. 49.
179. E. Walton, P. Ofner and R. H. Thorp, *J. Chem. Soc.*, **1949**, 648.
180. M. W. Gittos and W. Wilson, *J. Chem. Soc.*, **1955**, 2371.
181. W. Wilson, *J. Chem. Soc.*, **1955**, 3524.
182. R. Kwok and P. Pranc, *J. Org. Chem.*, **32**, 738 (1967).
183. F. F. Blicke, A. J. Zambito and R. E. Stenseth, *J. Org. Chem.*, **26**, 1826 (1961).
184. E. C. Taylor and K. S. Hartke, *J. Am. Chem. Soc.*, **81**, 2456 (1959).
185. J. Burkhardt and K. Hamann, *Chem. Ber.*, **100**, 2569 (1967).
186. L. H. Smith and P. Yates, *J. Am. Chem. Soc.*, **76**, 6080 (1954).
187. J. M. Eby and J. A. Moore, *J. Org. Chem.*, **32**, 1346 (1967).
188. A. F. McKay, C. Pondesva and M. E. Kreling, *J. Org. Chem.*, **27**, 2884 (1962).
189. A. H. Cook, J. D. Downer and I. M. Heilbron, *J. Chem. Soc.*, **1948**, 2028.
190. H. Chosho, K. Ichimura and M. Ohta, *Bull. Chem. Soc. Japan*, **37**, 1670 (1964).
191. E. C. Taylor and A. L. Borror, *J. Org. Chem.*, **26**, 4967 (1961); also see A. Rosowsky and E. J. Modest, *J. Org. Chem.*, **31**, 2607 (1966).
192. H. Junek, *Monatsh. Chem.*, **96**, 2046 (1965).
193. J. A. Settepani and A. B. Borkovec, *J. Heterocyclic Chem.*, **3**, 188 (1966).
194. M. F. G. Stevens, *J. Chem. Soc.* (*C*), **1967**, 1096.
195. C. K. Bradsher and J. P. Sherer, *J. Org. Chem.*, **32**, 733 (1967).
196. E. C. Taylor and J. Bartulin, *Tetrahedron Letters*, **1967**, 2337.
197. F. W. Cooper and M. W. Patridge, *J. Chem. Soc.*, **1954**, 3429.
198. A. Quilico and R. Fusco, *Rend. Ist. Lombardo Sci. Letters*, **69**, 439 (1936); *Chem. Abstr.*, **32**, 7454 (1938).
199. R. Howe, B. S. Rao and H. Heyneker, *J. Chem. Soc.* (*C*), **1967**, 2511.
200. N. R. Easton, J. Gardner and J. R. Stevens, *J. Am. Chem. Soc.*, **69**, 2941 (1947); J. Attenburrow, J. Elks, B. A. Hems and K. N. Speyer, *J. Chem. Soc.*, **1949**, 510.
201. H. Matsui, *Tetrahedron Letters*, **1966**, 1827.

202. T. I. Temnikova, Y. A. Sharanin and V. S. Karavan, *Zh. Org. Khim.*, **3**, 681 (1967); Engl. Transl.
203. J. W. Cornforth in *Chemistry of Penicillin* (Eds. H. T. Clarke, J. R. Johnson, and R. Robinson), Princeton University Press, 1949, p. 700.
204. J. P. Ferris and L. E. Orgel, *J. Am. Chem. Soc.*, **88**, 3829 (1966).
205. G. Kille and J. P. Fluery, *Bull. Soc. Chim. France*, **1967**, 4619.
206. G. S. Puranik and H. Suschitzky, *J. Chem. Soc.* (*C*), **1967**, 1006.
207. G. W. Stacy, F. W. Villaescusa and T. E. Wollner, *J. Org. Chem.*, **30**, 4074 (1965).
208. G. W. Stacy and T. E. Wollner, *J. Org. Chem.*, **32**, 3028 (1967).
209. G. W. Stacy and D. L. Eck, *Tetrahedron Letters*, **1967**, 5201.
210. F. Korte and F. F. Wiese, *Chem. Ber.*, **97**, 1963 (1964).
211. F. Korte and H. Wamhoff, *Chem. Ber.*, **97**, 1970 (1964).
212. A. H. Cook, I. Heilbron and A. L. Levy, *J. Chem. Soc.*, **1947**, 1594.
213. A. H. Cook, I. Heilbron and A. L. Levy, *J. Chem. Soc.*, **1947**, 1598.
214. A. H. Cook, I. Heilbron and A. L. Levy, *J. Chem. Soc.*, **1948**, 201.
215. H. C. Carrington, *J. Chem. Soc.*, **1948**, 1619.
216. A. H. Cook, I. M. Heilbron and E. S. Stern, *J. Chem. Soc.*, **1948**, 2031.
217. A. H. Cook and S. F. Cox, *J. Chem. Soc.*, **1949**, 2337.
218. A. H. Cook, I. Heilbron and G. D. Hunter, *J. Chem. Soc.*, **1949**, 1443; J. Parrod and L. V. Huyen, *Compt. Rend.*, **236**, 933 (1953).
219. K. Gewald, H. Bottcher and E. Schinke, *Chem. Ber.*, **99**, 94 (1966).
220. E. C. Taylor and J. C. Berger, *J. Org. Chem.*, **32**, 2376 (1967).
221. D. W. H. MacDowell and T. B. Patrick, *J. Org. Chem.*, **32**, 2441 (1967).
222. T. L. Cairns, R. A. Carboni, D. D. Coffman, V. A. Engelhardt, R. E. Heckert, E. L. Little, E. G. McGeer, B. C. McKusick, W. J. Middleton, R. M. Scribner, C. W. Scribner, C. W. Theobald and H. E. Winberg, *J. Am. Chem. Soc.*, **80**, 2775 (1958).
223. W. J. Middleton, V. A. Engelhardt and B. S. Fisher, *J. Am. Chem. Soc.*, **80**, 2822 (1958).
224. G. N. Sausen, V. A. Engelhardt and W. J. Middleton, *J. Am. Chem. Soc.*, **80**, 2815 (1958).
225. S. F. A. Kettle and L. E. Orgel, *Proc. Chem. Soc.*, **1959**, 307.
226. I. S. Bengelsdorf, *J. Am. Chem. Soc.*, **80**, 1442 (1958).
227. A. Lottermoser, *J. Prakt. Chem.*, **54**, 132 (1896); J. J. Ritter and R. D. Anderson, *J. Org. Chem.*, **24**, 208 (1959).
228. W. L. Reilly and H. C. Brown, *J. Org. Chem.*, **22**, 698 (1957).
229. C. Grundmann, *Chem. Ber.*, **97**, 3262 (1964).
230. W. D. Emmons, H. A. Rolewicz, W. N. Cannon and R. M. Ross, *J. Am. Chem. Soc.*, **74**, 5524 (1952).
231. G. A. Grindahl, W. X. Bajzer and O. R. Pierce, *J. Org. Chem.*, **32**, 603 (1967).
232. H. C. Brown, I. D. Shuman and J. Turnbull, *J. Org. Chem.*, **32**, 231 (1967).
233. J. A. Young and R. L. Dressler, *J. Org. Chem.*, **32**, 2237 (1967).
234. J. Oro and A. P. Kimball, *Arch. Biochem. Biophys.*, **94**, 217 (1961); H. Wakamatsu, Y. Yamada, T. Saito, I. Kumashiro and I. Takenishi, *J. Org. Chem.*, **31**, 2035 (1966); see R. E. Moser, A. R. Claggett and C. N. Matthews, *Tetrahedron Letters*, **1968**, 1599, 1605.
235. R. Longeray, A. Vigier and J. Dreux, *Compt. Rend.*, **253**, 1810 (1961).
236. L. Mandel, J. V. Piper and K. P. Singh, *J. Org. Chem.*, **28**, 3340 (1963); K. Schofield and R. J. Wells, *J. Chem. Soc.* (*C*), **1967**, 621.

237. F. Bergel, A. L. Morrison and H. Rinderknecht, *U.S. Pat.* 2,446,803; *Chem. Abstr.*, **43**, 695c (1949).
238. F. Bergel, A. L. Morrison and H. Rinderknecht, *U.S. Pat.* 2,446,804; *Chem. Abstr.*, **43**, 695f (1949); A. P. Terent'ev and V. G. Yashunskii, *Zh. Obshch. Khim.*, **24**, 291 (1954).
239. S. Trofimenko, *J. Org. Chem.*, **28**, 2755 (1963).
240. C. W. Bird, *J. Chem. Soc.*, **1964**, 5284.
241. V. I. Shevchenko and V. P. Kukhar, *Zh. Obshch. Khim.*, **36**, 735 (1966) and earlier references cited.

CHAPTER 9

Cyanocarbon and polycyano compounds

E. CIGANEK, W. J. LINN and O. W. WEBSTER

E. I. du Pont de Nemours and Co.
Wilmington, Delaware, U.S.A.

I. INTRODUCTION 427
II. CYANOGEN 428
A. Synthesis 428
B. Paracyanogen 428
C. Characteristic Reactions of Cyanogen 428
1. Reaction with amines and other nitrogen compounds . . 428
2. Reaction with alcohols and thiols 429
3. Reaction with activated methylene compounds 430
4. Reaction with Grignard reagents 430
5. Reaction with aromatic compounds 431
6. Cycloaddition reactions 432
7. Fluorination 432
8. Miscellaneous reactions 432
III. POLYCYANOALKANES 433
A. Polycyanomethanes 433
1. Tricyanomethane 433
a. Synthesis and structure 433
b. Salt formation 434
c. Additions to tricyanomethane 435
2. Halotricyanomethanes 437
3. Alkyl- and aryltricyanomethanes 437
4. Trifluoromethyldicyanomethane 438
B. Cyano-Substituted Ethanes 438
1. 1,1,2,2-Tetracyanoethane 439
2. Pentacyanoethane 441
3. Hexacyanoethane 441
C. Polycyanocyclopropanes 442
IV. POLYCYANOOLEFINS 445
A. Tetracyanoethylene 446
1. Synthesis 446
a. From malononitrile 446

b. From dichlorofumaronitrile 446
c. From tetracyano-1,4-dithiin 446
d. From 1,2-dicyano-1,2-di-*p*-tolylsulphonylethylene . . 447
e. From bis(acetoxyiminomethyl)methylenemalononitrile . 447
2. Physical properties 447
3. Additions to the double bond of tetracyanoethylene . . 449
a. 1,3-Dienes 449
b. Transannular addition to diene systems 453
c. Cycloaddition to form cyclobutanes 453
d. Strained hydrocarbons 456
e. Free radicals 457
f. Nucleophilic carbenes 457
g. Hydrogen 458
h. Hydrogen peroxide and ozonides 458
i. Chlorine 458
j. Diazomethane 458
k. Sulphurous acid 458
l. Cyanide ion 459
m. Ketones 459
n. Phenols and aromatic amines 459
o. Zero-valent platinum and palladium compounds . . 459
p. Triphenylphosphine 460
4. Additions to the nitrile groups of tetracyanoethylene . . 460
a. Azide 460
b. Fluorosulphonic acid 461
c. Water or methanol under extremely high pressure . . 461
d. Trifluoromethanesulphenyl chloride 462
e. Ethyl diazoacetate 462
f. Benzonitrile oxide 462
5. Replacement of cyano groups in tetracyanoethylene . . 463
a. Aromatic compounds 463
b. Water 466
c. Alcohols 467
d. Ammonia, hydrazines and amines 468
e. Phosphorus ylids 469
6. Fragmentation reactions of tetracyanoethylene . . . 469
a. Retrograde Michael reaction 469
b. Elimination of cyanogen 470
c. Reaction with isatogens 470
7. Tetracyanoethylene anion radical 471
8. Polymers from tetracyanoethylene 474
B. Tricyanoethylenes 474
1. Tricyanoethylene 474
2. Tricyanovinylalkanes and -arenes 475
3. 1-Chloro-1,2,2-tricyanoethylene (tricyanovinyl chloride) . 478
C. Dicyanoethylenes Containing Other Electronegative Substituents 481
1. Dicyanodihaloethylenes 481
a. 1,2-Dichloro-1,2-dicyanoethylene 481
b. 1,2-Dibromo- and 1,2-diiodo-1,2-dicyanoethylene . . 482

c. 1,2-Difluoro-1,2-dicyanoethylene 482
d. 1,1-Dichloro-2,2-dicyanoethylene 483
2. Dicyanobis(fluoroalkyl)ethylenes 484
3. 1,2-Dicyano-1,2-disulphonylethylenes 490
4. Diethyl 1,2-dicyanoethylene-1,2-dicarboxylate . . . 492
D. Dicyanoethylenes Containing Electron-Donating Substituents . 493
1. Dicyanoketene acetals 494
2. Dicyanoketene thioacetals 496
3. Dimercaptomaleonitrile and dimercaptofumaronitrile . . 499
4. Diaminomaleonitrile and diaminofumaronitrile . . . 506
E. Hexacyanobutadiene 509
V. Cyanoacetylenes 512
A. Synthesis of Cyanoacetylenes 513
B. Physical Properties of Cyanoacetylenes 515
C. Reactions of Cyanoacetylenes 517
1. Salt and complex formation 517
2. Addition reactions 518
3. Cycloaddition reactions 520
4. Polymerization 523
VI. Polycyanobenzenes and Related Compounds . . . 524
A. Tricyanobenzenes 524
B. Tetracyanobenzenes 527
C. Penta- and Hexacyanobenzene 530
VII. Tetracyanoquinodimethanes and Polycyanoquinones . 531
A. Tetracyanoquinodimethanes 531
1. Synthesis and physical properties 531
2. Reduction and anion-radical formation 533
3. Addition reactions of tetracyanoquinodimethanes . . 536
4. Displacement reactions 538
B. Polycyanoquinones 540
1. 2,3-Dicyano-*p*-benzoquinones 540
2. 2,5- and 2,6-dicyano-*p*-benzoquinone; tetracyano-*p*-benzoquinone 545
VIII. Azacyanocarbons 546
A. Dicyanamide 546
B. Tricyanamide 547
C. Cyanogen Azide 547
D. 2,2-Dicyanovinyl Azides 549
E. Azodinitrile 550
F. Cyanodiazo Compounds 550
IX. Polycyano-Substituted Heterocycles 554
A. Polycyanooxiranes 554
1. Synthesis and properties 554
2. Ring-opening reactions with cleavage of a C—O and C—C bond 557
a. Amines 557
b. Sulphides 559
c. Thiocarbonyl compounds 561
d. Miscellaneous nucleophiles 562

3. Ring-opening reactions with cleavage of the C—C bond . . 563
a. Addition to olefins 563
b. Addition to aromatic compounds 565
c. Addition to acetylenes 567
d. Addition to carbon–heteroatom bonds 568
4. Ring-opening reactions with cleavage of a C—O bond . . 568
a. Acid anhydrides and halides 568
b. Carbon–carbon unsaturation 568
5. Nucleophilic attack on oxygen 569
6. Attack on a cyano group 570
B. Tetracyanofuran 570
C. 2,3,4,5-Tetracyanopyrrole 571
D. 3,4,5-Tricyanopyrazole 572
E. 4,5-Dicyanotriazole 572
F. Polycyanopyridines 573
1. Tricyanopyridines 573
2. Tetra- and pentacyanopyridine 575
G. Tricyano-*s*-triazine 576
H. Tetracyano-1,4-dithiin 576
I. Tetracyanothiophene 579
X. Carbonyl Cyanide 580
A. Synthesis 580
B. Physical Properties 581
C. Reaction with Alcohols and Amines 582
D. Wittig-Type Reactions 582
E. Acylation of Aromatic Hydrocarbons 583
F. Reaction with Olefins 583
G. Perfluoroacyl Cyanides 588
XI. Cyanocarbon Anions 589
A. Cyanomethanides 590
1. Malononitrile salts 590
2. Tricyanomethanide 590
3. Other negatively-substituted malononitrile ions . . 590
B. Cyanoethanides 591
1. Pentacyanoethanide 591
2. Tetracyanoethanide 592
3. Tricyanoethanide 592
C. Cyanopropenides 593
1. Pentacyanopropenide 593
2. Tetracyanopropenides 594
a. 1,1,3,3-Tetracyanopropenides 594
b. 1,1,2,3-Tetracyanopropenides 595
D. Cyanobutenides 596
1. Hexacyanoisobutenediide 596
2. Tetracyanobutanediide 596
3. Hexacyanobutenediide 596
E. Cyanopentadienides 597
1. Cyanocyclopentadienides 597
a. Tetracyanocyclopentadienide 597

b. Diazotetracyanocyclopentadiene 602
2. Heptacyanopentadienide and other cyanopentadienides . 606
F. Miscellaneous Cyanocarbon Anions 607
G. Electronic Spectra of Cyanocarbon Anions 608
H. Cyanocarbon Acids 610
1. Acidities 610
2. The nature of the protonated species 613
XII. References 617

I. INTRODUCTION

This chapter deals with compounds whose unusual physical and chemical properties can be traced to the presence of a number of cyano groups within their molecules. Although some of these compounds have been known for a long time, their recognition as a distinct class of organic chemicals is rather recent. Polycyano compounds owe their characteristics to the fact that the cyano group uniquely combines strong electron-withdrawing power with small steric bulk. A cyano group, being linear and having a cylindrical π-cloud, will always be in conjugation with an attached π-system, even though steric crowding may exist. It is in this respect that it differs drastically from other strongly electronegative substituents such as the nitro or sulphonyl groups.

Examples of the unusual properties of polycyano compounds are that tetracyanoquinodimethane ion radical salts conduct electricity as well as graphite, that pentacyanocyclopentadiene is a stronger acid than perchloric acid, that tetracyanoquinone oxidizes water and that tetracyanoethylene oxide adds to olefins and even aromatic compounds with opening of the carbon–carbon bond. The effect of the dicyanomethylene group as a structural element has been compared to that of oxygen[1]. Thus, water and malononitrile have similar acidities, both *p*-benzoquinone and tetracyanoquinodimethane form ion radicals; 2,2-dicyanovinyl chlorides have reactivities similar to those of acid chlorides, and the mode of addition of tetracyanoethylene oxide to double bonds resembles that of ozone. This principle holds fairly well in many cases but breaks down in some, as for instance in the case of dicyanodiazomethane, which is much more reactive than nitrous oxide.

Although an attempt has been made to cover the field of cyanocarbon chemistry completely, some omissions are inevitable. The guiding principle for selecting compounds for inclusion in this chapter has been their unusual properties rather than the number of cyano groups present; thus, some compounds containing only one nitrile group are discussed, while many containing two are not. Tetracyanoethylene and related compounds have been reviewed briefly several times[2–7]. Reviews on malononitrile[8], and hydrogen cyanide[9], cyanide[9] and cyanogen chloride chemistry[9] are available. A recent review on cyanide chemistry is available from E. I. du Pont de Nemours and Co., Inc[10].

II. CYANOGEN

Cyanogen is the simplest cyanocarbon. As its chemistry was reviewed in 1959[11], only the most characteristic reactions are summarized here.

A. Synthesis

The best laboratory synthesis of cyanogen is the oxidation of sodium cyanide by copper sulphate[12]. The best commercial synthesis is probably the air oxidation of hydrogen cyanide catalysed by nitrogen oxides[13].

B. Paracyanogen

Paracyanogen is a dark brownish-black insoluble polymer produced as a by-product in many cyanogen reactions. It also forms when cyanogen is heated at 300–400° [14,15] or is irradiated[16,17]. The structure is unknown.

C. Characteristic Reactions of Cyanogen

I. Reaction with amines and other nitrogen compounds

Cyanogen reacts with anhydrous ammonia to form a black solid[18] of empirical formula $C_2H_3N_3$. At high temperatures in the presence of a dehydrogenation catalyst, cyanogen and ammonia form melamine and hydrogen cyanide[19].

Cyanogen adds two equivalents of primary amines and some secondary amines to give oxamidines[12,20–24] (equation 1). If the amino group is hindered or its basicity is low, the reaction stops

$$2\,RNH_2 + (CN)_2 \longrightarrow RHN{-}\underset{\displaystyle\|\atop\displaystyle HN}{C}{-}\underset{\displaystyle\|\atop\displaystyle NH}{C}{-}NHR \qquad (1)$$

after one amino group has added to give cyanoformamidine[22,25] (equation 2). Since the addition of the first amino group lowers the

$$R_2NH + (CN)_2 \longrightarrow R_2N{-}C({=}NH){-}CN \quad (2)$$

reactivity of the remaining cyano function, cyanoformamidines could probably be obtained from primary amines if one mole of amine were used and the temperature of the reaction were lowered. Ethylenediamine and 1,3-propanediamine give five- and six-membered ring compounds[26], respectively (equation 3). Aromatic

$$2\ H_2NCH_2CH_2NH_2 + (CN)_2 \longrightarrow \text{2,2'-bi(2-imidazoline)} + NH_3 \quad (3)$$

1,2-diamines cyclize in a different manner to give 2,3-diaminoquinoxalines[27] (equation 4). Triethylamine does not react with

$$o\text{-}C_6H_4(NH_2)_2 + (CN)_2 \longrightarrow \text{2,3-diaminoquinoxaline} \quad (4)$$

cyanogen[28] and therefore is an ideal catalyst for the addition of weakly acidic compounds to the triple bonds of cyanogen.

Hydrazine[29–31], semicarbazide[32] and hydroxylamine[33–35] react with cyanogen in a manner similar to amines:

$$2\ H_2NNH_2 + (CN)_2 \longrightarrow H_2NHNC({=}NH){-}C({=}NH)NHNH_2 \quad (5)$$

$$NH_2C({=}O)NHNH_2 + (CN)_2 \longrightarrow NH_2C({=}O)NHN{=}C(NH_2){-}CN \quad (6)$$

$$2\ NH_2OH + (CN)_2 \longrightarrow NH_2C({=}NOH){-}C({=}NOH)NH_2 \quad (7)$$

2. Reaction with alcohols and thiols

Cyanogen adds two moles of alcohol[36–38] or thiol[28] in the presence of a basic catalyst to give oxaldiimidates (**1**) or dithiooxaldiimidates (**2**). In some instances, cyanoformimidates are isolated.

$$\underset{(\mathbf{1})}{\mathrm{ROC(=NH)-C(=NH)OR}} \qquad \underset{(\mathbf{2})}{\mathrm{RSC(=NH)-C(=NH)SR}}$$

The base-catalysed reaction of water with cyanogen gives cyanide and cyanate[39]. Acid-catalysed hydrolysis produces oxamide[40,41] or under controlled conditions, cyanoformamide ($NCCONH_2$)[42,43].

Hydrogen sulphide adds to cyanogen to give either thiocyanoformamide, or dithiooxamide[44].

3. Reaction with activated methylene compounds

Compounds possessing an activated methylene group, such as diethyl malonate[45–47], ethyl acetoacetate[46,48,49], malononitrile[50] and nitroethane[50], add to one cyano group of cyanogen (equation 8). Under more forcing conditions, addition will sometimes take place on the second cyano group (equation 8a). The cyanoformimino and α,β-diimino structures assigned for these products are probably incorrect. They are more likely vinyl amines, which would have lower energy because of conjugation of the two activating groups with the amino group.

$$(CN)_2 + CH_2(CO_2Et)_2 \xrightarrow{\text{base}} NCC(=NH)-CH(CO_2Et)_2 \longrightarrow NCC(NH_2)=C(CO_2Et)_2 \quad (8)$$

$$(CN)_2 + 2\,CH_2(CO_2Et)_2 \longrightarrow (EtO_2C)_2C=C(NH_2)-C(NH_2)=C(CO_2Et)_2 \quad (8a)$$

Aminotricyanoethylene (**3**) which is prepared in this way, cannot be made by reaction of ammonia with tetracyanoethylene (section IV.A.5.d).

$$(CN)_2 + CH_2(CN)_2 \longrightarrow \underset{(\mathbf{3})}{(NC)_2C=C(NH_2)CN}$$

4. Reaction with Grignard reagents

The reaction of Grignard reagents with cyanogen via **4** and **5** can be directed to any of four different types of products; ketones[51], α-diketones[52], nitriles[53] or glycinonitriles[54,55], depending on conditions (equation 9).

$$\mathrm{RMgX + (CN)_2 \longrightarrow R{-}\overset{\displaystyle CN}{\overset{|}{C}}{=}NMgX \longrightarrow RCN + Mg(CN)X}$$
$$\mathbf{(4)}$$

$$\mathbf{4} + \mathrm{RMgX} \xrightarrow[-70°]{} \mathrm{R\overset{XMgN}{\overset{\|}{C}}{-}\overset{NMgX}{\overset{\|}{C}}R} \xrightarrow{\mathrm{HCl}} \mathrm{R\overset{O}{\overset{\|}{C}}{-}\overset{O}{\overset{\|}{C}}R} \tag{9}$$

$$\mathbf{4} + \mathrm{RMgX} \xrightarrow[70°]{} \mathrm{R_2\overset{N(MgX)_2}{\overset{|}{C}}CN}\ (\mathbf{5}) \xrightarrow[\mathrm{H_2O}]{\mathrm{NH_4Cl}} \mathrm{R_2\overset{NH_2}{\overset{|}{C}}CN}$$

$$\mathbf{5} + \mathrm{HCl} + \mathrm{H_2O} \longrightarrow \mathrm{R\overset{O}{\overset{\|}{C}}R}$$

Benzyl Grignard reagents react with cyanogen to give *o*-methylbenzonitriles rather than phenylacetonitriles[56–58].

5. Reaction with aromatic compounds

Aromatic compounds may yield benzils, glyoxylic acids or nitriles on treatment with a mixture of cyanogen, hydrogen chloride and aluminium chloride (equations 10 and 11)[51,59–62].

$$\text{1,3-}\mathrm{C_6H_4(OH)_2} + \mathrm{(CN)_2 + HCl + AlCl_3} \longrightarrow \mathrm{(2,4\text{-}(HO)_2C_6H_3)C(O){-}C(O)(C_6H_3(OH)_2\text{-}2,4)} + \mathrm{2,4\text{-}(HO)_2C_6H_3C(O){-}COOH} \tag{10}$$

$$\mathrm{C_6H_6 + (CN)_2 + HCl + AlCl_3 \longrightarrow C_6H_5CN} \tag{11}$$

6. Cycloaddition reactions

The uncatalysed vapour-phase reaction of cyanogen with butadienes at 500° gives 2-cyanopyridines in modest yields[63,64]:

$$\text{butadiene} + (CN)_2 \longrightarrow [\text{2-cyano-tetrahydropyridine}] \longrightarrow \text{2-cyanopyridine} + H_2 \qquad (12)$$

The reaction of cyanogen with hydrazoic acid can be controlled to give either **6** or **7**[65]. Diazomethane adds to cyanogen to give **8**[66].

(6) **(7)** **(8)**

7. Fluorination

In an unprecedented ring-closing oxidation, cyanogen is fluorinated by silver(II) fluoride at 115° to give 3,3,4,4-tetrafluoro-Δ^1-1,2-diazetine (**9**) in 90% yield[67] (equation 13). At 240° **9** cleaves to

$$(CN)_2 + AgF_2 \longrightarrow \underset{\mathbf{(9)}}{\text{N=N}\ /\ CF_2\text{—}CF_2} \qquad (13)$$

tetrafluoroethylene and nitrogen. Fluorination of cyanogen with cobaltic fluoride at 150° gives CF_4, C_2F_6 and $F_2NCF_2CF_2NF_2$ as major products[68]. Mercuric fluoride at 240° produces $CF_3N{=}CF_2$ and $Hg[N(CF_3)_2]_2$ [68]. Fluorination with elemental fluorine gives CF_3NF_2, $(CF_3)_2NF$, $(CF_3)_3N$, $NF_2CF_2CF_2NF_2$ and $CF_3N{=}NCF_3$ [69].

8. Miscellaneous reactions

Cyanogen and sulphur trioxide form an adduct to which structure **10** has been assigned (equation 14). The reaction is reversible[70].

$$NC\text{—}CN + 2\,SO_3 \rightleftharpoons \underset{\mathbf{(10)}}{^-O_3S\text{—}\overset{+}{N}{\equiv}C\text{—}C{\equiv}\overset{+}{N}\text{—}SO_3^-} \qquad (14)$$

Cyanogen reacts with hydrogen at 675° to give hydrogen cyanide in 98% yield. A kinetic study indicates a radical chain mechanism[71,72].

In the presence of a nickel chromium catalyst at 1500°, a small amount of succinonitrile is obtained from cyanogen and ethylene[73].

A cyanocarbon acid, assigned structure **11**[74], is formed in 75% yield on treatment of cyanogen with potassium cyanide in acetonitrile[75] (equation 15).

$$3\ (CN)_2 + KCN \longrightarrow \mathbf{(11)}\ K^+ \qquad (15)$$

(11)

Sulphur dichloride adds to cyanogen in the presence of a catalytic amount of tetraethylammonium chloride to give 3,4-dichloro-1,2,5-thiadiazole[76]:

$$(CN)_2 + SCl_2 \longrightarrow \text{3,4-dichloro-1,2,5-thiadiazole} \qquad (16)$$

Cyanogen reacts photochemically with hexacarbonylchromium, hexacarbonylmolybdenum and hexacarbonyltungsten to give the corresponding cyanogen bis(pentacarbonylmetal) compound[77]:

$$(CN)_2 + 2\ Cr(CO)_6 \longrightarrow (CO)_5CrNC{-}CNCr(CO)_5 \qquad (17)$$

III. POLYCYANOALKANES

A. Polycyanomethanes

1. Tricyanomethane

a. Synthesis and structure. Like most cyanocarbon acids, free tricyanomethane is unstable. The resonance-stabilized anion can be easily prepared and kept in the form of salts for which two methods of preparation have been described. The reaction of cyanogen chloride with malononitrile in the presence of two equivalents of sodium ethoxide gives sodium tricyanomethanide[78]:

$$NaCH(CN)_2 + ClCN \longrightarrow NaCl + HC(CN)_3 \xrightarrow{NaOEt} Na^+[C(CN)_3]^- \qquad (18)$$

The potassium salt is obtained in higher yield and purity by the treatment of a dihalomalononitrile with two equivalents of potassium cyanide[79]. The mechanism of the latter reaction probably involves an intermediate tricyanohalomethane which can be shown to react with cyanide ion to generate cyanogen halide and tricyanomethane anion:

$$Br_2C(CN)_2 + 2\ KCN \longrightarrow K^+[C(CN)_3]^- + KBr + BrCN \qquad (19)$$

Acidification of aqueous solutions of tricyanomethane in the presence of ether results in moist ether (aquoethereal) solutions of the free acid. By working rapidly on a small scale it is possible to isolate from the ether solution a colourless crystalline material with the composition of tricyanomethane[80]. The free acid can be sublimed in small amounts, but on standing at room temperature it decomposes to an orange-red, apparently polymeric, substance. A report that the free acid is stable and can be recrystallized from petroleum ether is apparently erroneous[81].

It was recognized early that tricyanomethane (**12**) can exist in a tautomeric form which was first called 'isocyanoform'[82]. The dicyanoketenimine structure (**13**) has been assigned to the solid on the basis of spectral evidence[80,83]. In aqueous solution, tricyano-

$$H{-}C(CN)_2{-}CN \rightleftharpoons (NC)_2C{=}C{=}NH \longleftrightarrow (NC)_2\overset{-}{C}{-}C{\equiv}\overset{+}{N}H$$

(12) **(13)**

methane is completely ionized[803] with the negative charge of the ion delocalized. A more complete discussion of the structure and position of protonation is reserved to the section on cyanocarbon anions (section XI.K).

b. Salt formation. Metal salts of tricyanomethane are readily prepared by reaction of a water-soluble metal salt with aqueous or alcoholic solutions of potassium or sodium tricyanomethanide[78,79,84–86]. Most of the salts are insoluble and crystallize with non-integral amounts of water of hydration. The insolubility is attributed to their polymeric nature proposed on the basis of spectral studies[84]. Spectral[87,88,813] and x-ray examination[85] of the potassium salt support a planar structure for the anion, although a slight deviation from planarity was found in the ammonium salt[823]. An extended Hückel calculation implies a significant electron delocalization as suggested above[84]. Furthermore, the charge density calculation indicates the nitrogens to be more negative than the central carbon atom. This is consistent with the conclusions concerning the position of ligand attachment in the organometallic tricyanomethanide complexes, **14** and **15**[91,814,815]. The structural assignments were based on the infrared spectra.

$$\underset{\textbf{(14)}}{\underset{(M = Sn, Pb)}{Ph_3MNCC(CN)_2}} \qquad \underset{\textbf{(15)}}{\underset{(M = Cr, Mo, W)}{K[M(CO)_5NCC(CN)_2]}}$$

Transition metal salts of tricyanomethane can be prepared in the presence of pyridine, resulting in complexes carrying 2, 3 or 4 molecules of pyridine as ligands. In contrast to the simple salts, these appear to be monomolecular and have some solubility in alcohol and pyridine[85].

Amines can form ammonium salts with tricyanomethane as well as addition products (see below)[79]. True salts can be distinguished easily by their infrared spectra, which show the characteristic tricyanomethyl anion bands including the strong nitrile at 2180 cm^{-1}. In addition to the ammonium salt, salts of *t*-butylamine, isopropylamine, triethylamine and pyridine have been prepared[79].

c. Additions to tricyanomethane. Hydrogen adds readily to moist ether solutions of tricyanomethane in the presence of acetic acid and palladium on carbon (equation 20)[83]. The major product is 3-amino-2-cyanoacrylonitrile (**16**), an expected reduction product of dicyanoketenimine. On the basis of spectral evidence and chemical transformations, the other product has been identified as 3-amino-2-cyanoacrolein (**17**). Tricyanomethanide anion is not hydrogenated

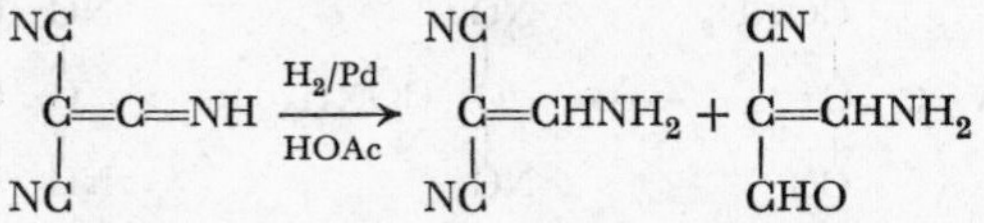

$$\underset{\textbf{(13)}}{(NC)_2C{=}C{=}NH} \xrightarrow[HOAc]{H_2/Pd} \underset{\textbf{(16)}}{(NC)_2C{=}CHNH_2} + \underset{\textbf{(17)}}{(NC)(OHC)C{=}CHNH_2} \qquad (20)$$

under the conditions where the ether solution of free cyanoform is readily reduced.

Hydrogen halides add readily to tricyanomethane to give 1-amino-1-halo-2,2-dicyanoethylenes (**18**)[86,811]. Infrared studies support the structure **18** rather than that of the tautomeric imino compound[812]. At room temperature **18** reacts with methanol or ethanol to form 1-amino-1-alkoxy-2,2-dicyanoethylenes (**19**), which are identical with the reaction products of ammonia with the corresponding dicyanoketene acetals (**20**) (equation 21)[89]. Alcohols add to tricyanomethane to form the 1-amino-1-alkoxy-2,2-dicyanoethylene directly[78,82].

$$(NC)_2C{=}C{=}NH \ (\mathbf{13}) \xrightarrow{ROH} (NC)_2C{=}C(NH_2)OR \ (\mathbf{19})$$

$$(NC)_2C{=}C(NH_2)X \ (\mathbf{18}) \ (X = \text{halogen}) \xrightarrow{ROH} (NC)_2C{=}C(NH_2)OR \ (\mathbf{19}) \qquad (21)$$

$$(NC)_2C{=}C(OR)_2 \ (\mathbf{20}) \xrightarrow{NH_3} (NC)_2C{=}C(NH_2)OR \ (\mathbf{19})$$

In spite of the rapidity with which tricyanomethane adds alcohols, water adds slowly giving carbamoyldicyanomethane (**21**) (equation 22)[79]. Like tricyanomethane, **21** is a strong acid, completely ionized

$$(NC)_2C{=}C{=}NH \ (\mathbf{13}) + H_2O \longrightarrow (NC)_2C{=}C(NH_2)OH \rightleftharpoons (NC)_2CH{-}C(=O)NH_2 \ (\mathbf{21}) \qquad (22)$$

in solution with the negative charge largely delocalized. Unlike

$$^{-}N{=}C{=}C(CONH_2)CN \longleftrightarrow (NC)_2\overset{-}{C}{-}CONH_2 \longleftrightarrow (NC)_2C{=}C(O^-)NH_2$$

tricyanomethane, free carbamoyldicyanomethane is relatively stable and can be isolated and kept as a crystalline solid that decomposes above 300°.

Reference has been made above to salts of tricyanomethane with primary and tertiary amines. The secondary amines piperidine and pyrrolidine, in addition to forming salts, react with aquoethereal tricyanomethane to give addition products analogous to those from alcohols and hydrogen halides (equation 23)[79]. The reaction of

$$(NC)_2C{=}C{=}NH + \text{piperidine (}C_5H_{10}NH\text{)} \longrightarrow (NC)_2C{=}C(NH_2)NC_5H_{10} \qquad (23)$$

aniline with 1-chloro-1-amino-2,2-dicyanoethylene results in removal of hydrogen chloride to give a mixture of anilinium hydrochloride and anilinium tricyanomethanide[88,812]. The latter on heating at 145–150° rearranges to 1-amino-1-phenylamino-2,2-dicyanoethylene, (**22**):

$$[(NC)_3C]^-C_6H_5NH_3^+ \xrightarrow{\Delta} \underset{(\mathbf{22})}{(NC)_2C{=}C(NH_2)NHC_6H_5} \quad (24)$$

2. Halotricyanomethanes

Bromotricyanomethane[92] and chlorotricyanomethane[79] have been described. The former was synthesized from silver tricyanomethanide and bromine (equation 25) whereas the latter was prepared from

$$[(NC)_3C]^-Ag^+ + Br_2 \longrightarrow AgBr + BrC(CN)_3 \quad (25)$$

the potassium salt. Both the bromo and chloro derivatives are extremely volatile compounds with pungent choking odours. As expected, the halogen is electropositive. For example, the reaction of chlorotricyanomethane with sodium cyanide gives cyanogen chloride and sodium tricyanomethanide.

Infrared and Raman spectra, as well as x-ray analysis, show bromotricyanomethane to be pyramidal[93,813].

3. Alkyl- and aryltricyanomethanes

Tricyanomethyl anion is not expected to be very nucleophilic, but the silver salt reacts with methyl iodide and benzyl iodide to give alkylated tricyanomethanes (**23**) (equation 25a)[79,819]. Yields

$$Ag^+[C(CN)_3]^- + RI \longrightarrow AgI + \underset{(\mathbf{23})}{RC(CN)_3} \quad (25a)$$

are poor and a better method of preparation is the reaction of cyanogen chloride with the anion of an appropriately substituted malononitrile (equation 26)[94]. Yields vary from about 40 to 90%.

$$R\bar{C}(CN)_2Na^+ + ClCN \longrightarrow RC(CN)_3 + NaCl \quad (26)$$

$$(R = Et, CH_2{=}CHCH_2, PhCH_2CH_2, PhC(CH_3)_2, PhCH_2)$$

A modification of the reaction has made possible the synthesis of aryl-substituted tricyanomethanes[94]. Arylacetonitriles, two moles of sodium hydride and two moles of cyanogen chloride give the aryltricyanomethanes without isolation of the intermediate arylmalononitrile (equation 27). A discussion of the electronic effects of the tricyanomethyl group will be found in Chapter 5.

$$ArCH_2CN \xrightarrow{NaH} H_2 + Ar\bar{C}HCN \xrightarrow{ClCN} ArCH(CN)_2$$

$$ArCH(CN)_2 \xrightarrow{Ar\bar{C}HCN} ArCH_2CN + Ar\bar{C}(CN)_2 \quad (ArCH_2CN \xrightarrow{NaH} Ar\bar{C}HCN)$$

$$Ar\bar{C}(CN)_2 \xrightarrow{ClCN} ArC(CN)_3 \qquad (27)$$

4. Trifluoromethyldicyanomethane

The trifluoromethyl group, like the cyano group, is strongly electron withdrawing, and therefore trifluoromethyldicyanomethane is closely related to tricyanomethane. When an attempt was made to synthesize 1,1-difluoro-2,2-dicyanoethylene by reaction of silver fluoride with the corresponding dichlorodicyanoethylene, the product was trifluoromethyldicyanomethane. Presumably the desired ethylene is an intermediate but adds silver fluoride (equation 28)[95]. The silver

$$Cl_2C{=}C(CN)_2 \xrightarrow{AgF} [F_2C{=}C(CN)_2] \xrightarrow{AgF} CF_3C(CN)_2Ag \qquad (28)$$

salt, but not the free acid, was isolated and characterized. Alkylation of the silver salt with benzyl bromide gives, in 42% yield, the substituted methane (equation 29).

$$CF_3C(CN)_2Ag + PhCH_2Br \longrightarrow AgBr + PhCH_2C(CN)_2CF_3 \qquad (29)$$

B. Cyano-Substituted Ethanes

This discussion is limited to the three known ethanes containing four or more cyano groups. All of these are derived from tetracyanoethylene and are legitimately classed as cyanocarbons, although hexacyanoethane is really the only known aliphatic saturated cyanocarbon.

I. 1,1,2,2-Tetracyanoethane[96]

Reduction of tetracyanoethylene can be accomplished by a number of reagents[97]. Although catalytic hydrogenation with a palladium catalyst has been demonstrated, chemical methods are more convenient and give higher yields. Hydrogen iodide or bromide and thiols are particularly convenient reagents. Mercapto-acetic acid has been recommended, because the acid and its disulphide are water soluble and easily separated from the product (equation 30).

$$(NC)_2C{=}C(CN)_2 + 2\,HSCH_2CO_2H \longrightarrow \underset{(\mathbf{24})}{(NC)_2CHCH(CN)_2} + (HO_2CCH_2S)_2 \quad (30)$$

The ease with which **24** can be formed is shown by the ability of tetracyanoethylene to dehydrogenate 1,4-cyclohexadiene and related dihydroaromatic compounds[98]. Both **24** and benzene are formed in essentially quantitative yield in 30 minutes under reflux in a mixture of benzene and dioxan or more slowly in dioxan at room temperature.

In turn, the tetracyanoethane can be dehydrogenated to tetra-cyanoethylene with sulphur or other conventional dehydrogenation reagents such as palladium on carbon, lead dioxide or cupric oxide.

Although **24** is volatile and can be sublimed, it is not particularly stable and turns brown or black on standing. As expected, the ethane is acidic and dissolves in aqueous sodium bicarbonate from which it reprecipitates on acidification. However, prolonged base treatment results in decomposition.

Most of the reported chemistry of **24** involves addition of acidic reagents to two *vicinal* cyano groups. A number of interesting heterocyclic compounds have been prepared in this manner[99]. In the presence of a base, hydrogen sulphide adds to **24** to give 2,5-diamino-3,4-dicyanothiophene (**25**) (equation 31). The reaction can

$$\underset{(\mathbf{24})}{(NC)_2CH{-}CH(CN)_2} + H_2S \longrightarrow \underset{(\mathbf{25})}{\text{2,5-(H}_2\text{N)}_2\text{-3,4-(NC)}_2\text{-thiophene}} \quad (31)$$

be carried out with tetracyanoethylene and sodium sulphide, but it can be shown that the ethane is an intermediate.

If a thiol is used instead of hydrogen sulphide, an analogous addition of two equivalents occurs to give a substituted butadiene (**26**) (equation 32). Hydrochloric acid can cause cyclization of **26** in

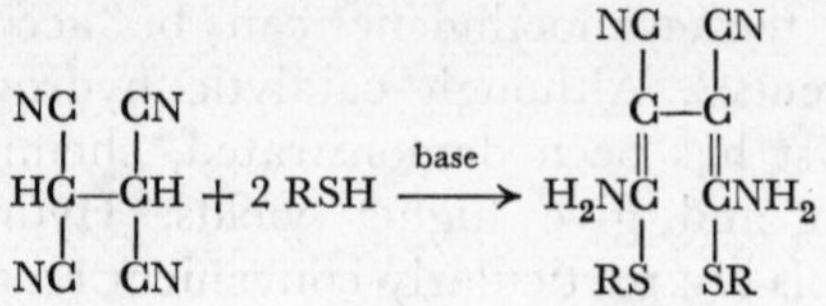

$$(NC)_2CH{-}CH(CN)_2 + 2\,RSH \xrightarrow{\text{base}} H_2N(RS)C{=}C(CN){-}C(CN){=}C(SR)NH_2 \qquad (32)$$

(**24**) (**26**)

one of two ways depending on the nature of the R group and to some extent on the concentration of the acid used. When R is phenyl, benzenethiol is eliminated and 2-amino-3,4-dicyano-5-phenylthiopyrrole (**27**) results regardless of acid strength (equation 33). In fact, no acid is required; the transformation will occur simply on heating. When R is methyl or ethyl, ammonia is eliminated with either dilute or concentrated acid. In this case a 2,5-dialkylthiopyrrole (**28**) is the product (equation 34). However, if R is hydroxyethyl, the course of the reaction is dependent on acid concentration.

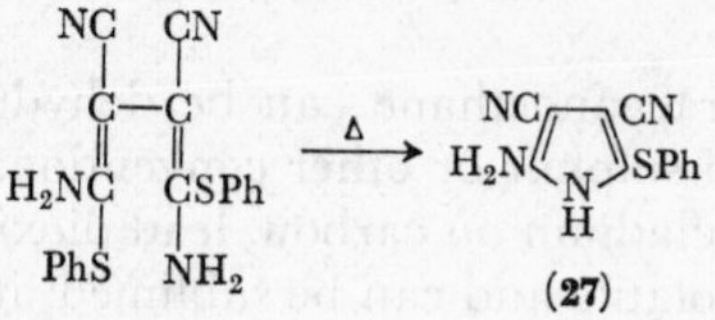

$$H_2N(PhS)C{=}C(CN){-}C(CN){=}C(SPh)NH_2 \xrightarrow{\Delta} \text{2-}H_2N\text{-3,4-(NC)}_2\text{-5-PhS-pyrrole (NH)} \qquad (33)$$

(**27**)

$$H_2N(RS)C{=}C(CN){-}C(CN){=}C(SR)NH_2 \xrightarrow{HCl} \text{2,5-(RS)}_2\text{-3,4-(NC)}_2\text{-pyrrole (NH)} \qquad (34)$$

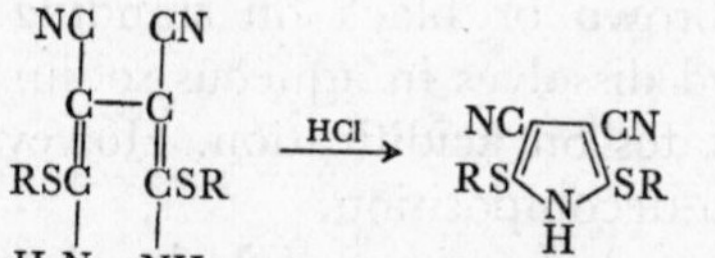

(**28**)

(R = Me, Et)

Dilute acid gives the product corresponding to **27** and cold concentrated hydrochloric acid gives **28** (R = CH_2CH_2OH). A study of the electronic factors governing the method of ring closure is needed to define more precisely this reaction.

Hydrogen bromide also converts **24** to a pyrrole (**29**) although the reported yield is poor (equation 35).

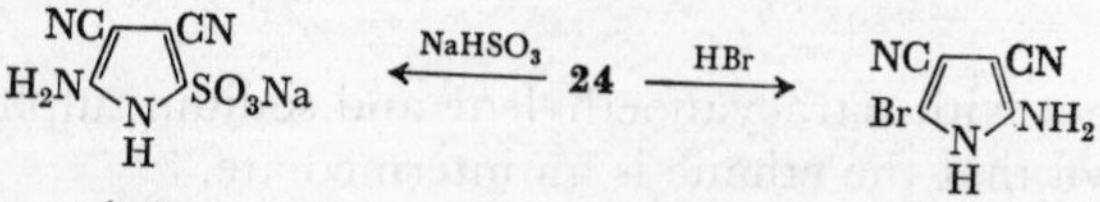

$$\text{2-}H_2N\text{-3,4-(NC)}_2\text{-5-}NaO_3S\text{-pyrrole (NH)} \xleftarrow{NaHSO_3} \mathbf{24} \xrightarrow{HBr} \text{2-Br-3,4-(NC)}_2\text{-5-}H_2N\text{-pyrrole (NH)} \qquad (35)$$

(**30**) (**29**)

Sodium bisulphite gives the sodium salt of the sulphonic acid **30** also in low yield (equation 35).

2. Pentacyanoethane

Equimolar amounts of sodium cyanide and tetracyanoethylene (**31**) mixed in acetonitrile give the corresponding anion radical in good yield (section IV.A.7). Use of only a slight excess of tetracyanoethylene results in a drastic decrease in the anion radical yield, and dilution of the acetonitrile solution with ether causes precipitation of the sodium salt of pentacyanoethane (**32**)[100]. If the salt is isolated and redissolved in acetonitrile, the anion radical is formed and this reaction is accelerated by excess cyanide.

These results are explained by the equilibrium of equation (36).

$$\underset{(\mathbf{31})}{(NC)_2C{=}C(CN)_2} + CN^- \rightleftharpoons \underset{(\mathbf{32})}{(NC)_3C\bar{C}(CN)_2} \xrightarrow{CN^-}$$

$$[(CN)_2] + \bar{C}(CN)_2\bar{C}(CN)_2 \xrightarrow{\mathbf{31}} (NC)_2\dot{C}C(CN)_2^- \quad (36)$$

Experimental verification of the equilibrium was obtained by the use of ^{14}C-labelled cyanide ion. After 3·5 hours at −30° the label had been incorporated into **32** to within 1% of the theoretical amount, assuming complete equilibration. A further confirmation was the incorporation of the label into originally inactive **32** when ^{14}C-labelled tetracyanoethylene was added to its solution in acetonitrile.

Treatment of an acetonitrile solution of sodium pentacyanoethanide with chlorine gives chloropentacyanoethane (**33**) in good yield (equation 37).

$$(NC)_3CC(CN)_2^-Na^+ + Cl_2 \longrightarrow NaCl + \underset{(\mathbf{33})}{(NC)_3CC(CN)_2Cl} \quad (37)$$

Acidification of sodium pentacyanoethanide with cold aqueous hydrochloric acid allows the isolation of crystalline pentacyanoethane[101]. The free acid is unstable and has been characterized only by its infrared spectrum.

3. Hexacyanoethane

Hexacyanoethane (**34**) is prepared from sodium pentacyanoethanide and cyanogen chloride[102] (equation 38). The percyanoalkane has been isolated only in low yield, but this is due, in part,

$$(NC)_3CC(CN)_2{}^-Na^+ + ClCN \longrightarrow NaCl + \underset{(\mathbf{34})}{(NC)_3CC(CN)_3} \quad (38)$$

to its instability. In solution or in the crystalline state at room temperature, **34** undergoes decomposition to tetracyanoethylene. The other product is presumably cyanogen, although this has never been isolated.

C. Polycyanocyclopropanes

Treatment of ethyl bromocyanoacetate with base[103,104] gives *trans*-triethyl 1,2,3-tricyanocyclopropane-1,2,3-tricarboxylate (**35**) (equation 39). The same product is realized starting with sodium

$$HC(CN)(Br)(CO_2Et) + KOAc \longrightarrow \text{(35)} \quad (39)$$

ethyl cyanoacetate and bromine[105]. Attempted synthesis of hexacyanocyclopropane by an analogous method using malononitrile results only in the pentacyanopropenide ion[106]. Hydrolysis and decarboxylation of **35** with potassium acetate in aqueous ethanol leads to a mixture of *cis*- and *trans*-1,2,3-tricyanocyclopropane in which the *cis* isomer predominates by a factor of two[103,104]. This ratio is the result of kinetic control, since *cis*-1,2,3-tricyanocyclopropane gives the *trans* isomer on treatment with potassium *t*-butoxide.

A large number of 1,1,2,2-tetracyanocyclopropanes are available by the Wideqvist reaction[107,108]. This involves the treatment of an aldehyde or ketone and bromomalononitrile with aqueous potassium iodide (equation 40). The crystalline cyclopropane derivative usually precipitates directly from the reaction mixture after a very short time under mild conditions. This reaction has been examined

$$CH_3COCH_3 + 2\,BrCH(CN)_2 \xrightarrow{KI} \text{3,3-}(CH_3)_2\text{-1,1,2,2-}(CN)_4\text{-cyclopropane} \quad (40)$$

extensively with regard to the generality and best conditions[109,110]. In the course of this work, a large number of tetracyanocyclopropanes substituted with both alkyl and aryl groups have been prepared. As a rule ketones react more slowly than aldehydes. The yields seem to depend on the extent of condensation of the carbonyl compounds with bromomalononitrile and in general parallel closely the activity of the aldehyde or ketone in other condensation reactions. Substituents which decrease the positive charge at the carbonyl carbon tend to diminish activity. Most aldehydes give cyclopropanes in high yield; poor yields are obtained with methyl ketones, and higher ketones such as 3-heptanone, dicyclopropyl ketone and benzophenone do not react at all. Exceptions to this rule are the smaller ring cyclic ketones, e.g. the spiro compound **36** is formed in 92 % yield from cyclohexanone[109]. As the ring size is increased above six, the yields drop markedly.

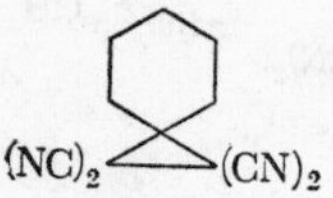

$(NC)_2$ $(CN)_2$

(36)

The following mechanism has been proposed for the reaction:

$$CH_3COCH_3 + BrCH(CN)_2 \longrightarrow (CH_3)_2C(OH)CBr(CN)_2 \xrightarrow{H^+}$$

$$(CH_3)_2C(^+OH_2)CBr(CN)_2 \xrightarrow{I^-} (CH_3)_2C{=}C(CN)_2 + IBr + H_2O \qquad (41)$$

$$(CH_3)_2C{=}C(CN)_2 \xrightarrow[-Br^-]{Br\bar{C}(CN)_2} \text{cyclopropane with } (CH_3)_2,\ (NC)_2,\ (CN)_2$$

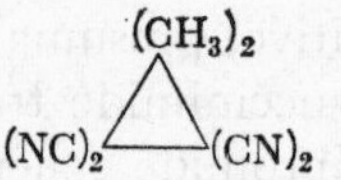

In support of this mechanism, a number of alkylidene malononitriles have been allowed to react with bromomalononitrile in the *absence* of iodide ion[111]. The tetracyanocyclopropanes form rapidly in high yield. This variant of the reaction makes possible the preparation in good yield of some cyclopropanes not accessible by the Wideqvist technique or, at the best, accessible in low yield. For example the cyclopropanes **37** can be prepared, whereas the Wideqvist reaction with dicyclopropyl ketone and 3-heptanone fails completely.

A closely related method has been used for the preparation of a few tetracyanocyclopropanes. Alkylidene bismalononitriles (**38**)

$$\text{(37)}\quad R^1R^2C\text{-}C(CN)_2\text{-}C(CN)_2 \text{ (cyclopropane)}$$

($R^1 = R^2$ = cyclopropyl;
(R^1 = Et, R^2 = n-butyl)

are available from aldehydes and malononitrile. These, on treatment with bromine, give the corresponding cyclopropane[112] (equation 42).

$$CH_3CH(CH(CN)_2)_2 \ \textbf{(38)} \xrightarrow[-HBr]{Br_2} CH_3\text{-}\triangle(CN)_2(CN)_2 \qquad (42)$$

The unsubstituted 1,1,2,2-tetracyanocyclopropane (**39**) has been prepared in 85% yield from aqueous formaldehyde and malononitrile by treatment with bromine water in the presence of a trace of β-alanine as a condensation catalyst[113]:

$$HCHO + 2\,CH_2(CN)_2 \xrightarrow[Br_2(H_2O)]{\beta\text{-alanine}} \triangle(CN)_2(CN)_2 \ \textbf{(39)} \qquad (43)$$

Two other preparations of **39** have been reported[113]. An unstable dibromo derivative, presumably the 1,3-dibromide (**41**) is obtained from *N*-bromosuccinimide treatment of **40**. Addition of potassium iodide to the dibromide generates **39** in 78% yield:

$$(NC)_2CHCH_2CH(CN)_2 \ \textbf{(40)} \xrightarrow{NBS} (NC)_2CBrCH_2CBr(CN)_2 \ \textbf{(41)} \xrightarrow{KI} \textbf{39} \qquad (44)$$

The cyclopropane **39** was also isolated in moderate yield from the reaction of tetracyanoethylene with diazomethane. No cyclopropane was found, however, in the products of the reaction of the cyanoolefin with ethyl diazoacetate. Instead, reaction apparently occurs at a nitrile group.

The chemistry of the tetracyanocyclopropanes has been little studied. Attempted bromination of **39** or condensations at the methylene group were unsuccessful. Acid hydrolysis does not give the cyclopropane-1,2-diacid, but instead leads to ring opening to give the itaconic acid derivative **42**[112]:

$$\text{R}\text{-cyclopropane-}(\text{CN})_2(\text{CN})_2 \xrightarrow[\text{H}^+]{\text{H}_2\text{O}} \text{RCH}{=}\text{C}(\text{COOH})\text{CH}_2\text{COOH} \quad \textbf{(42)} \tag{45}$$

The basic hydrolysis of the cyclopropanes is complex. Direct hydrolysis to the tetraacid does not occur, possibly because of the insolubility of some of the intermediate alkali metal salts. Reaction with aqueous alcoholic KOH affords the imide **43** as the first product (equation 46). A multistep process for conversion to the tetraester

$$(\text{CH}_3)_2\text{-cyclopropane-}(\text{CN})_2(\text{CN})_2 \xrightarrow[\text{MeOH, H}_2\text{O}]{2\text{N KOH}} (\text{CH}_3)_2\text{-cyclopropane}(\text{CO}_2\text{H})(\text{CONH}_2)\text{-imide (C=O, NH, C=O)} \quad \textbf{(43)} \tag{46}$$

corresponding to the original tetracyanocyclopropane has been worked out[114]. An examination of the n.m.r. spectra shows that cyclopropane ring hydrogen resonance ranges from τ 6·53 for the unsubstituted tetracyanocyclopropane to τ 4·77 for the derivative bearing a *m*-nitrophenyl group[110]. This demonstrates a considerable deshielding effect of the four cyano groups.

IV. POLYCYANOOLEFINS

The chemistry of polycyanoolefins is complex. Reagents can add to the double bond, or to the cyano groups, displace the cyano groups, cause fragmentation, be oxidized or reduced. The products are in many instances non-volatile and chromatography must be relied on heavily for separations. The cyanoolefins are readily attacked by nucleophiles but are surprisingly stable to acids. For example, tetracyanoethylene can be recovered from concentrated nitric acid solutions.

A. *Tetracyanoethylene*

I. Synthesis

a. From malononitrile. The oxidative coupling of malononitrile with sulphur monochloride was the first method by which tetracyanoethylene was prepared (equation 47)[115,116]. The use of chlorine

$$2\,CH_2(CN)_2 + 2\,S_2Cl_2 \longrightarrow (NC)_2C{=}C(CN)_2 + 4\,HCl + 4\,S \qquad (47)$$

at 450° in place of sulphur monochloride is cleaner[116]; however, the best procedure to conveniently make small amounts of tetracyanoethylene is to brominate the malononitrile and then debrominate with copper powder[116] (equation 48). The solvent used (benzene)

$$(CN)_2CBr_2 + Cu \longrightarrow (NC)_2C{=}C(CN)_2 \qquad (48)$$

is probably critical since tetracyanoethylene in acetonitrile reacts with copper (section IV.A.7.a). This debromination can be accomplished simply by pyrolysis of the dibromomalononitrile at about 500° [117]. Bromomalononitrile eliminates hydrogen bromide and produces tetracyanoethylene under the influence of diphenyl sulphide[118]:

$$HCBr(CN)_2 + Ph_2S \longrightarrow (NC)_2C{=}C(CN)_2 \qquad (49)$$

b. From dichlorofumaronitrile. Gaseous dichlorofumaronitrile and hydrogen cyanide react at 500° to produce tetracyanoethylene (equation 50)[119].

$$(NC)ClC{=}CCl(CN) + HCN \longrightarrow (NC)_2C{=}C(CN)_2 \qquad (50)$$

c. From tetracyano-1,4-dithiin. One of the most attractive starting materials for tetracyanoethylene from a commercial standpoint is tetracyano-1,4-dithiin, made from carbon disulphide, sodium cyanide and chlorine (section IX.H). This heterocycle reacts with cyanide ion to give tetracyanoethylene ion radical which in turn is oxidized to tetracyanoethylene with chlorine (equation 51)[120]. The oxidation

$$\text{(NC)}_2\text{C}_2\text{S}_2\text{C}_2\text{(CN)}_2 + \text{KCN} \longrightarrow \cdot\text{C(CN)}_2\text{—C}^-\text{(CN)}_2\ \text{K}^+ \xrightarrow{\text{Cl}_2} \text{(NC)}_2\text{C=C(CN)}_2 \qquad (51)$$

of tetracyanoethylene ion radical to tetracyanoethylene is discussed in section IV.A.7.c.

d. From 1,2-dicyano-1,2-di-p-tolylsulphonylethylene. Surprisingly, tetracyanoethylene forms directly from sodium cyanide and 1,2-dicyano-1,2-di-*p*-tolylsulphonylethylene without a subsequent oxidation step[121]. This reflects a stepwise addition–elimination process in which no cyanide is present when the tetracyanoethylene forms (equation 52) since tetracyanoethylene reacts with cyanide ion to give the ion radical (section IV.A.7.a).

$$\text{NC(}p\text{-MeC}_6\text{H}_4\text{O}_2\text{S)C=C(SO}_2\text{C}_6\text{H}_4\text{Me-}p\text{)CN} + \text{NaCN} \longrightarrow \text{(NC)}_2\text{C=C(SO}_2\text{C}_6\text{H}_4\text{Me-}p\text{)CN}$$

$$\xrightarrow{\text{NaCN}} {}^-\text{C(CN)}_2\text{—C(SO}_2\text{C}_6\text{H}_4\text{Me-}p\text{)(CN)—CN} \xrightarrow{60^\circ} \text{(NC)}_2\text{C=C(CN)}_2 + p\text{-MeC}_6\text{H}_4\text{SO}_2^- \qquad (52)$$

e. From bis(acetoxyiminomethyl)methylenemalononitrile. At 150 to 210° bis(acetoxyiminomethyl)methylenemalononitrile, which is obtained from 1,3-bis(acetoxyimino)-2-propanone and malononitrile, eliminates two molecules of acetic acid to give tetracyanoethylene (equation 53)[116].

$$\text{(AcON=CH)}_2\text{C=C(CN)}_2 \xrightarrow{210^\circ} \text{(NC)}_2\text{C=C(CN)}_2 + 2\ \text{HOAc} \qquad (53)$$

2. Physical properties

Tetracyanoethylene is a colourless crystalline solid, which is best purified by recrystallization from 1,2-dichloroethylene followed by sublimation. It is about as toxic as potassium cyanide[122] and thus should be handled with care. Table 1 gives some of the physical properties of tetracyanoethylene[6].

TABLE 1. Physical properties of tetracyanoethylene.

Property	Value
Melting point	198–200°
Boiling point	223°
Heat of formation	149 kcal/mole
Flame temperature	4000 °K
Density	1·318 g/cm^3
Index of refraction n_D^{25}	1·560
Dielectric constant	3·6
Specific heat	0·203 cal/g °C
Heat of sublimation	18·65 cal/mole

X-ray data show the carbon–carbon double bond distance to be 1·339 ± 0·008 Å[123], much shorter than the 1·37–1·38 Å predicted by HMO theory. As expected, the molecule is planar. The small dipole moment reported for tetracyanoethylene in dioxan (1·15 D) and in benzene (0·82 D) must be due to complex formation with the solvent (**44**)[124].

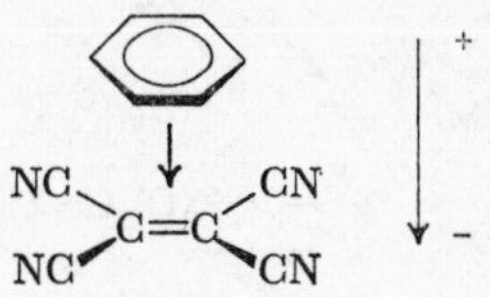

NC CN
C=C
NC CN
+ μ −

(**44**)

Tetracyanoethylene absorbs light at 267 and 277 mμ (log ε 4·13 and 4·08) in methylene chloride[125]. Its infrared and Raman spectra have been extensively studied[126–132] and comparison made with the other cyanoethylenes[133]. The stretching frequency of the double bond, 1525 cm^{-1}, indicates considerable resonance interaction with the cyano groups.

Based on the negative current between a filament and anode in a vacuum tube containing tetracyanoethylene, its electron affinity has been calculated to be 66·5 ± 1·4 kcal/mole[134]. Negative-ion mass spectrum studies confirm that tetracyanoethylene ion radical is the major ion[135]. This value is higher than those obtained by measurements of charge-transfer spectra (34, 51 and 64 kcal/mole)[136,137] possibly because the former is measured on gaseous tetracyanoethylene[134]. For a discussion of electron affinity determination from charge-transfer spectra see Chapter 10.

3. Additions to the double bond of tetracyanoethylene

a. 1,3-Dienes. By far the most studied reaction of tetracyanoethylene is its addition to dienes (Diels–Alder reaction). Tetracyanoethylene reacts with butadiene in a few minutes at 0° to give a quantitative yield of adduct (equation 54)[97]. The reaction proceeds

$$\text{CH}_2\!=\!\text{CH}\!-\!\text{CH}\!=\!\text{CH}_2 + (\text{CN})_2\text{C}\!=\!\text{C}(\text{CN})_2 \longrightarrow \text{4,4,5,5-tetracyanocyclohexene} \quad (54)$$

so well that it can be used for the quantitative determination of dienes[138]. A methylene chloride solution of the diene is heated with excess tetracyanoethylene and the solution is then back-titrated with a standard solution of cyclopentadiene. Pentamethylbenzene is used as the indicator. The break-up of the red pentamethylbenzene–tetracyanoethylene π-complex is taken as the end point. In a mechanistic study of the Diels–Alder reaction, tetracyanoethylene was shown to add to cyclopentadiene $6{\cdot}0 \times 10^3$ times faster than *N*-phenylmaleimide; $7{\cdot}7 \times 10^3$ faster than maleic anhydride; $5{\cdot}0 \times 10^5$ faster than maleonitrile; $1{\cdot}1 \times 10^6$ faster than dimethyl acetylenedicarboxylate; and $4{\cdot}6 \times 10^7$ faster than acrylonitrile[139,140].

Although tetracyanoethylene rapidly forms a π-complex with the diene, it has not been established that the complex is an intermediate to the adduct. The sensitivity to changes in the polarity of the solvent is low, therefore there is little charge separation in the transition state[818].

Because of their very high rate of reaction with dienes, tetracyanoethylene and dicyanomaleimide can be used to demonstrate that valence isomerization is occurring in cyclooctatetraene and similar systems (equation 55)[141]. When one gradually increases the concentration of the dienophile, a saturation effect is noted. Further increase in concentration does not increase the rate of adduct formation. From these data the rates of the forward and reverse reactions (k_1 and k_{-1}) and the equilibrium concentrations can be

$$\text{cyclooctatetraene} \underset{k_{-1}}{\overset{k_1}{\rightleftharpoons}} \text{bicyclo[4.2.0]octatriene} \xrightarrow{(\text{NC})_2\text{C}=\text{C}(\text{CN})_2} \text{adduct with } (\text{CN})_2,\ (\text{CN})_2 \quad (55)$$

calculated. Tetracyanoethylene reacts with cyclooctatetraene iron tricarbonyl to form an adduct[142] claimed to be **45**.

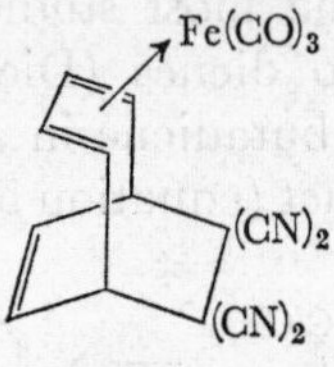

$Fe(CO)_3$ $(CN)_2$ $(CN)_2$

(45)

The reaction of cycloheptatriene with tetracyanoethylene gives an adduct containing a cyclopropyl ring (equation 56)[817]. The kinetics of the reaction were not studied, thus one is not certain whether the dienophile is attacking the cycloheptatriene directly or its valence isomer.

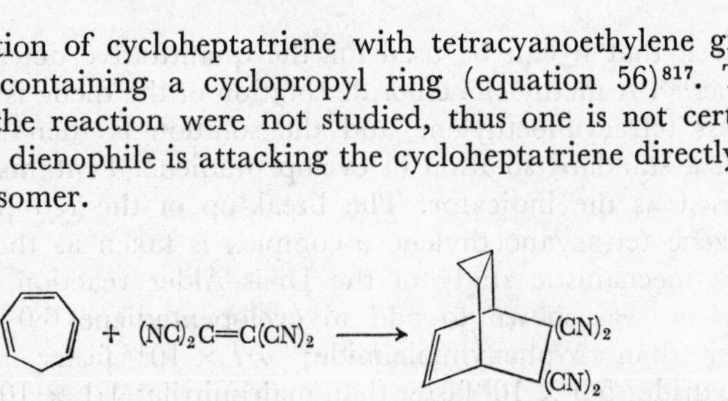

$$+ (NC)_2C{=}C(CN)_2 \longrightarrow (CN)_2\ (CN)_2 \qquad (56)$$

A similar structure containing a three-membered ring was assigned to the adduct from tetraphenylsesquifulvalene mainly on the basis of its ultraviolet spectrum (equation 57)[143]. Surprisingly,

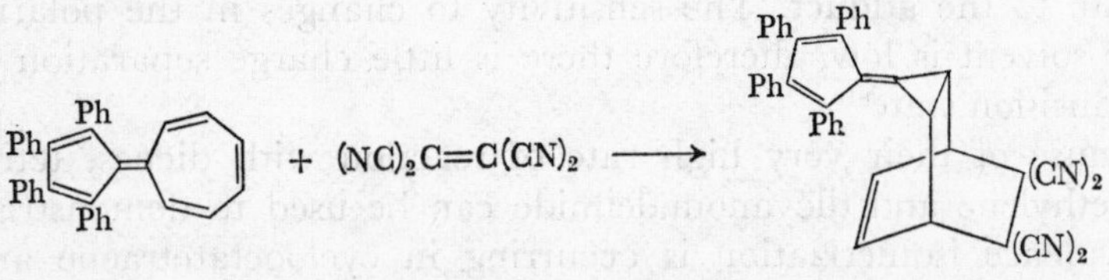

$$+ (NC)_2C{=}C(CN)_2 \longrightarrow (CN)_2\ (CN)_2 \qquad (57)$$

tetracyanoethylene adds to 8-benzyl-9,10-benzosesquifulvalene in a completely different manner, and again the structure was assigned mainly on the basis of the ultraviolet spectrum (equation 58)[143].

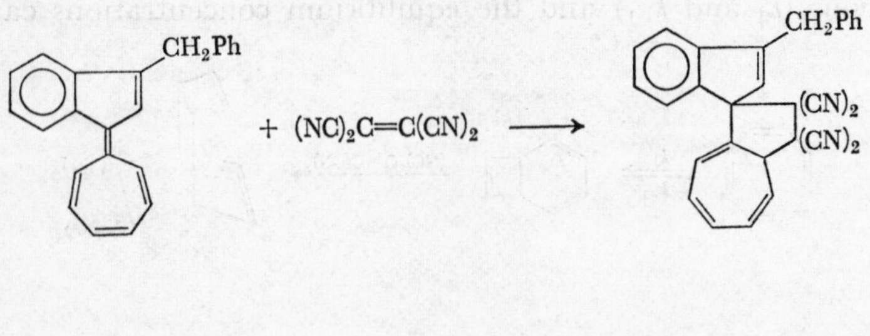

$$CH_2Ph \quad + (NC)_2C{=}C(CN)_2 \longrightarrow CH_2Ph\ (CN)_2\ (CN)_2 \qquad (58)$$

An adduct containing a three-membered ring was at first thought to result from the reaction of tetracyanoethylene with *N*-ethoxycarbonylazepine[144]. This was later shown to be a 1,4-addition product (equation 59)[145,146].

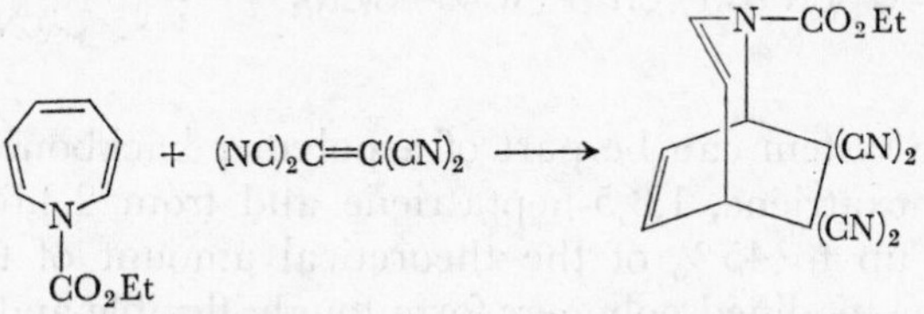

$$\text{N-}CO_2Et\text{-azepine} + (NC)_2C{=}C(CN)_2 \longrightarrow \text{adduct (N—}CO_2Et\text{, }(CN)_2\text{, }(CN)_2\text{)} \quad (59)$$

The adducts formed from tetracyanoethylene and 6-substituted fulvenes dissociate readily at room temperature (equation 60). As expected, the equilibrium is shifted to the left when R is electron withdrawing[147].

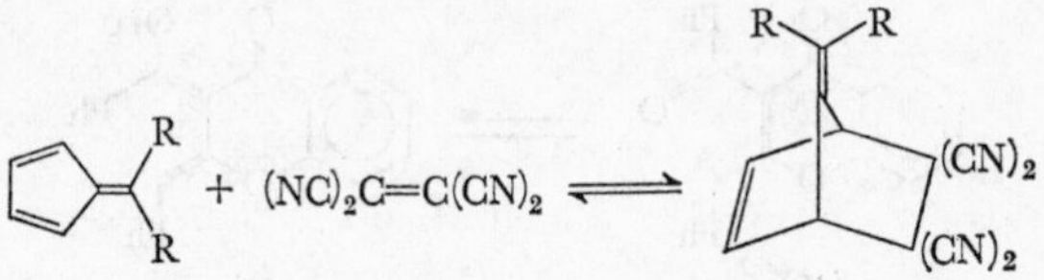

$$\text{fulvene (R, R)} + (NC)_2C{=}C(CN)_2 \rightleftharpoons \text{adduct (R, R; }(CN)_2\text{, }(CN)_2\text{)} \quad (60)$$

A similar case of reversal permits tetracyanoethylene to be used to store fulvalene (**46**). A crystalline diadduct from which fulvalene can be recovered is formed with this very unstable unsaturated compound[148].

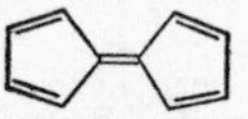

(46)

A thermal shift in the equilibrium constant for adduct formation is easily demonstrated for the system tetracyanoethylene/1,3-diphenylnaphtho[2,3-*c*]furan[149]. Solutions of this adduct **47** turn red (dissociation) when warmed and become colourless again when cooled (equation 61).

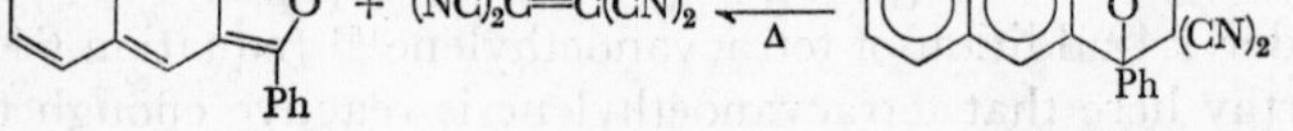

$$\text{1,3-diphenylnaphtho[2,3-}c\text{]furan (Ph, O, Ph)} + (NC)_2C{=}C(CN)_2 \underset{\Delta}{\rightleftharpoons} \text{adduct (Ph, O, Ph; }(CN)_2\text{, }(CN)_2\text{)} \quad (61)$$

(47)

1,4-Addition still takes place when part of the diene system is an allene (equation 62)[150].

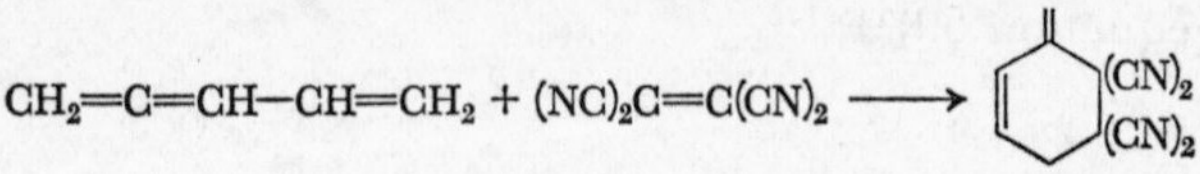

$$CH_2{=}C{=}CH{-}CH{=}CH_2 + (NC)_2C{=}C(CN)_2 \longrightarrow \quad (62)$$

(CN)$_2$
(CN)$_2$

The diene system can be part of a polymer backbone. Polymers from 1,3,5-hexatriene, 1,3,5-heptatriene and from 2,4,6-octatriene incorporate up to 45% of the theoretical amount of tetracyanoethylene. The modified polymers form tough, flexible and somewhat elastomeric films[151].

A striking thermo- and photochromic compound (**51**) forms on irradiation of a solution of 3-benzoyl-2-benzylchromone (**48**) and tetracyanoethylene in ethyl acetate (equation 63)[152,153]. Colourless

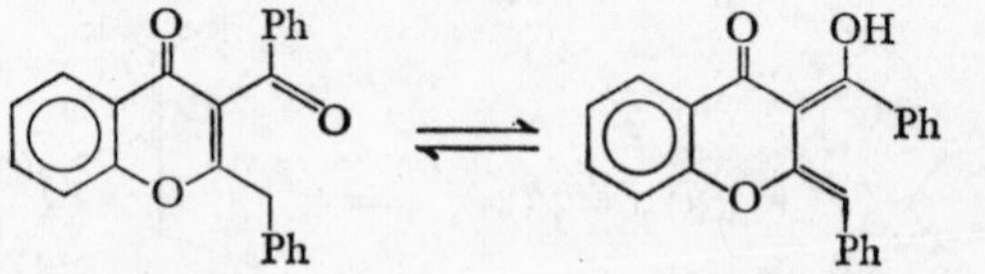

O Ph O Ph Ph
(48)

O OH Ph O Ph
(49)

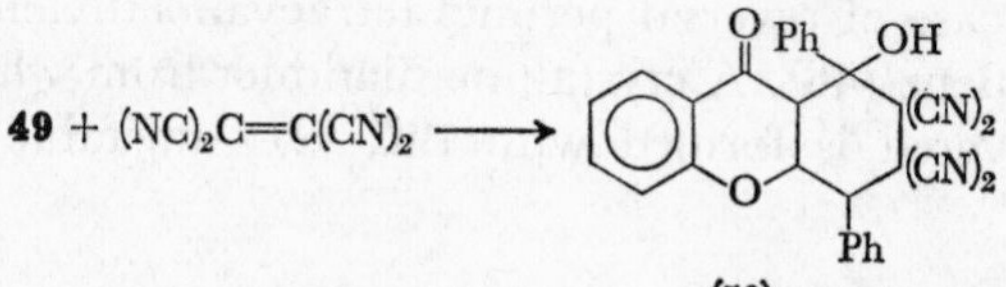

$$\mathbf{49} + (NC)_2C{=}C(CN)_2 \longrightarrow \quad (63)$$

O Ph OH (CN)$_2$ (CN)$_2$ O Ph
(50)

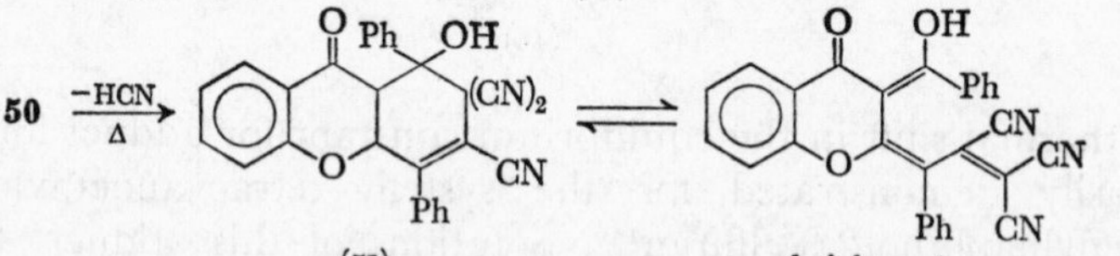

$$\mathbf{50} \xrightarrow[\Delta]{-HCN}$$

O Ph OH (CN)$_2$ O CN Ph
(51)

O OH Ph CN CN O Ph CN
bright red

solutions of **51** become bright red on heating or on irradiation for a few seconds with sunlight. On cooling in the dark, the original colourless solution forms.

The presence of a large R group in 2-vinylnaphthalenes (**52**) does not hinder 1,4-addition of tetracyanoethylene[154] (equation 64). It is noteworthy here that tetracyanoethylene is reactive enough to add to the naphthalene skeleton.

$$\text{(52)} + (NC)_2C{=}C(CN)_2 \longrightarrow \text{product} \quad (64)$$

b. Transannular addition to diene systems. If two isolated double bonds are forced into close proximity due to the conformational requirements of a molecule, they will react with tetracyanoethylene as if they were conjugated. Two noteable examples of this type of diene are 1,3,5,7-tetramethylenecyclooctane (**53**)[155] (equation 65) and norbornadiene (**54**)[156] (equation 66).

$$\text{(53)} + (NC)_2C{=}C(CN)_2 \longrightarrow \text{product} \quad (65)$$

$$\text{(54)} + (NC)_2C{=}C(CN)_2 \longrightarrow \text{product} \quad (66)$$

c. Cycloaddition to form cyclobutanes. Tetracyanoethylene will add to an isolated double bond to form a four-membered ring provided the bond is activated by steric strain or groups which supply electrons. Thus, vinyl ethers, vinyl sulphides, *N*-vinyl amides and *N*-vinyl sulphonamides all form tetracyanocyclobutanes with tetracyanoethylene[157] (equation 67). The vinyl group of *p*-methoxystyrene forms a four-membered ring with tetracyanoethylene, and styrene itself may react in this manner but at the temperature necessary for

$$MeO{-}CH{=}CH_2 + C(CN)_2{=}C(CN)_2 \longrightarrow \text{MeO-cyclobutane-}(CN)_2(CN)_2 \quad (67)$$

reaction (refluxing xylene) the isolated product is benzalmalononitrile (15% yield) (equation 68)[157].

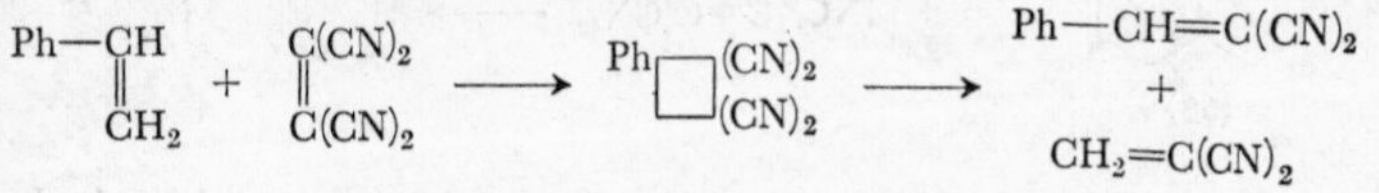

$$\mathrm{Ph{-}CH{=}CH_2} + \mathrm{(CN)_2C{=}C(CN)_2} \longrightarrow \text{Ph-cyclobutane-}(CN)_2(CN)_2 \longrightarrow \mathrm{Ph{-}CH{=}C(CN)_2} + \mathrm{CH_2{=}C(CN)_2} \qquad (68)$$

As in the case of diene addition, π-complex formation occurs. Unlike the diene addition though, the rate of four-membered ring formation is very sensitive to the polarity of the solvent. For example, the addition of *p*-methoxystyrene to tetracyanoethylene is over in one minute in nitromethane but is not complete in one month in carbon tetrachloride. This large solvent effect suggests that the reaction mechanism involves charge separation in the rate-determining step with a transition state or intermediate such as **55**. The reaction is first order in π-complex and roughly first order in *p*-methoxystyrene[158]. Activation parameters in ethyl acetate and cyclohexane show large negative entropies, −44 and −53 e.u. If **55** is an intermediate, it must be a tight ion pair, since there is little change in configuration during the reaction. In a similar reaction of *cis*- and of *trans*-fluoroalkyldicyanoethylene, a single different adduct is formed from each isomer (see section IV.C.2). Intermediate **55** could not be trapped with acetic acid, alcohol or phenyl isocyanate[158]. Some derivatives do, however, isomerize during the addition. *cis*-β-Methyl-*p*-methoxystyrene gives both *cis* and *trans* adducts[159].

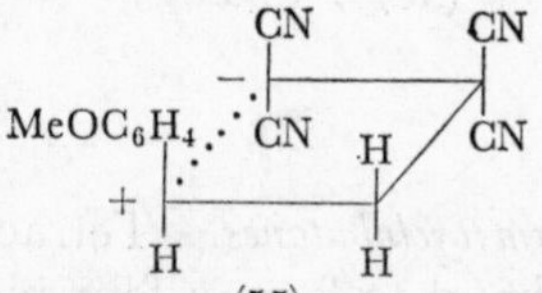

$$\mathrm{MeOC_6H_4}\overset{+}{\mathrm{C}}\mathrm{H}\cdots{}^{-}\mathrm{C(CN)_2{-}C(CN)_2{-}CH_2}$$

(55)

There is no correlation between the rate of these cycloaddition reactions and the tightness of the π-complex, therefore the π-complex is probably not an intermediate in the cycloaddition reactions[159]. The Hammett ρ value for cycloaddition to substituted styrenes is about $-7{\cdot}2$, a very high negative value indicative of a highly charged transition state[159].

An unusual activating group for cycloaddition is the azulene group in dimethylcycloheptatrienopentaene (equation 69)[160].

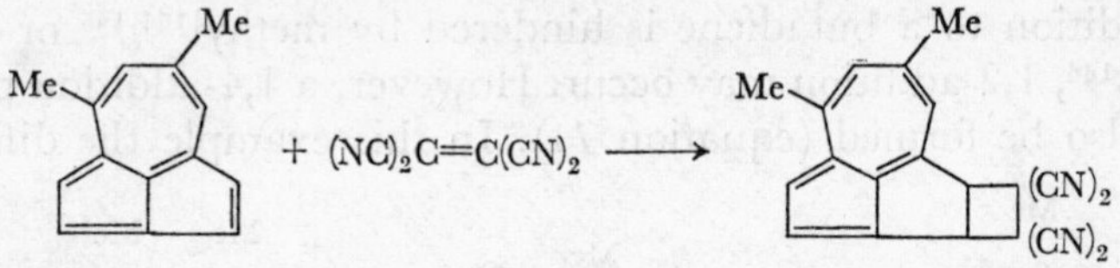

$$+ (NC)_2C{=}C(CN)_2 \longrightarrow \qquad (69)$$

An example of the addition of tetracyanoethylene to a strained double bond is its reaction with (1,4-dimethoxybenzo)bicyclo[2.2.1]-heptadiene[161]:

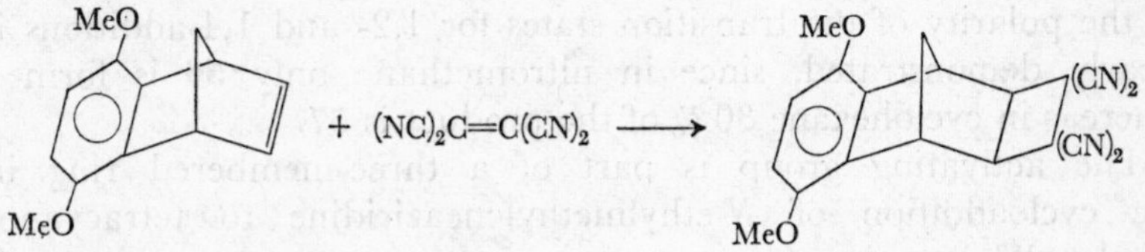

$$+ (NC)_2C{=}C(CN)_2 \longrightarrow \qquad (70)$$

When a 1,3-diene system possesses certain features which prevent or hinder 1,4-addition of tetracyanoethylene, 1,2-addition will take place. Thus, since 1,4-addition to diphenyldimethylenecyclobutene[156] or dimethyldimethylenecyclobutene[162] would give a cyclobutadiene structure, the reaction takes a different pathway and a 1,2-cycloadduct is formed:

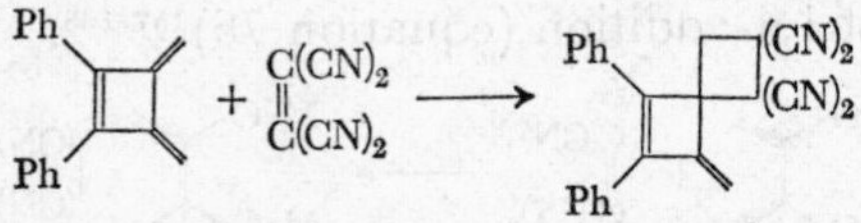

$$+ C(CN)_2{=}C(CN)_2 \longrightarrow \qquad (71)$$

1,4-Addition to the diene system in methylenecyclobutene[156,163] and in 3-methylenecyclohexene[156] would result in adducts with a double bond to a bridgehead, in violation of Bredt's rule. Thus, 1,2-addition results:

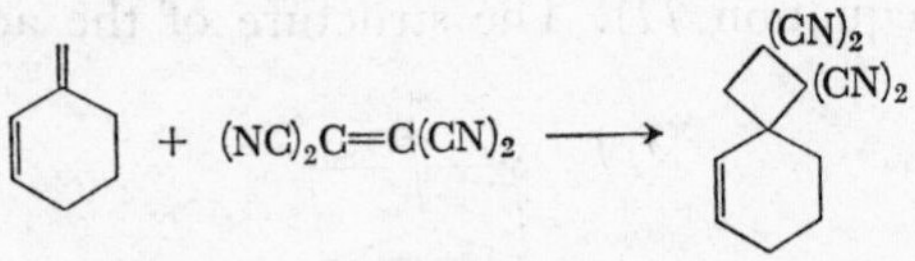

$$+ (NC)_2C{=}C(CN)_2 \longrightarrow \qquad (72)$$

If the exomethylene group is hindered by methyl groups, addition takes place on the *endo* double bond (equation 73)[161]. Similarly if

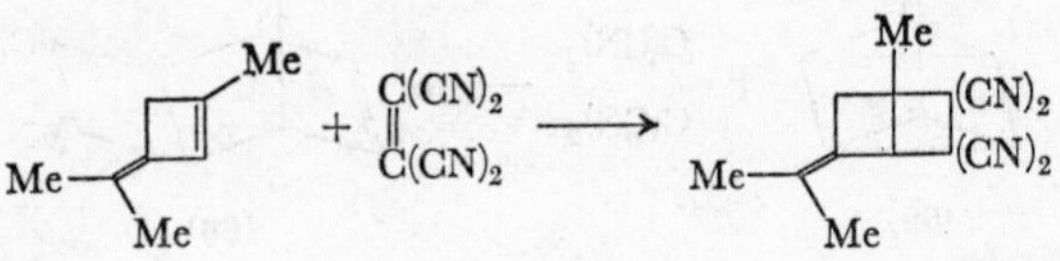

$$+ C(CN)_2{=}C(CN)_2 \longrightarrow \qquad (73)$$

1,4-addition to a butadiene is hindered by methyl[164,165] or phenyl groups[166], 1,2-addition may occur. However, a 1,4-addition product may also be formed (equation 74). In this example the difference

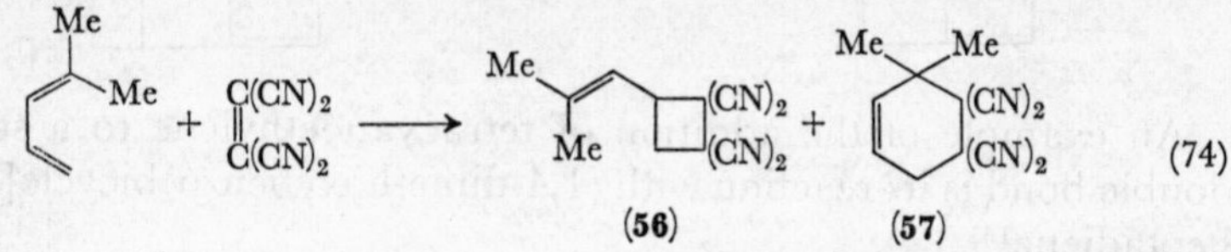

(74)

(56) (57)

in the polarity of the transition states for 1,2- and 1,4-additions is clearly demonstrated, since in nitromethane only **56** is formed whereas in cyclohexane 30% of the product is **57**.

The activating group is part of a three-membered ring in the cycloaddition of *N*-ethylmethyleneaziridine to tetracyanoethylene[167]:

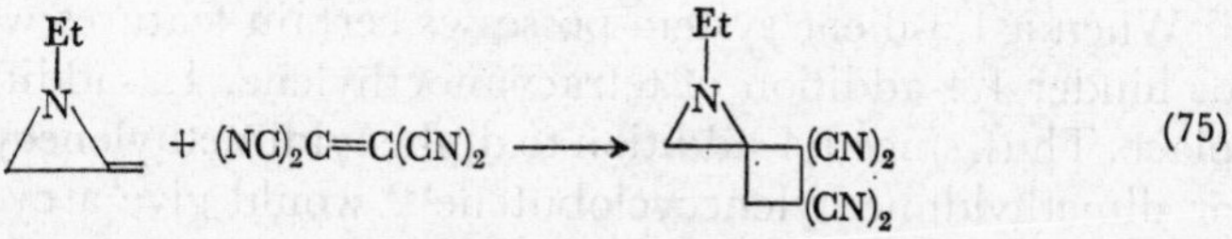

$$+ (NC)_2C{=}C(CN)_2 \longrightarrow \qquad (75)$$

The introduction of electron-donating groups on the 2- and 3-positions of a diene system does not cause 1,2-addition to take place at the expense of 1,4-addition (equation 76)[157,168].

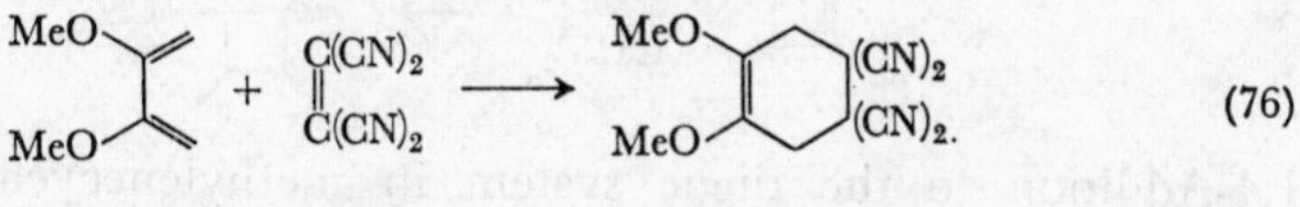

(76)

d. Strained hydrocarbons. If enough strain is present in a hydrocarbon, tetracyanoethylene may add to a carbon–carbon single bond. For example it adds to hexamethylprismane[170] (**58**) and to quadricyclane (**59**)[169] (equation 77). The structure of the adduct from **58**

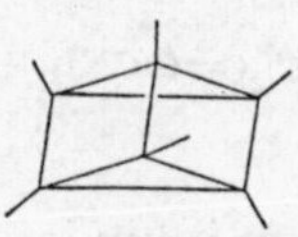

(58)

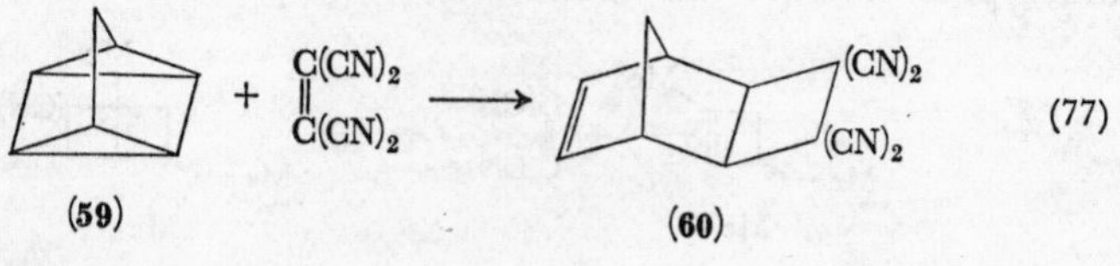

(77)

(59) (60)

was not given. The structure of the adduct **60** is different from that of the adduct of tetracyanoethylene with norbornadiene (equation 66), and thus **60** does not arise via isomerization of **59** to norbornadiene.

e. Free radicals. 2,5-Dimethyl-2,3,3,4,4,5-hexacyanohexane results from addition of α-cyanoisopropyl radical, generated from α,α′-azobis(isobutyronitrile), to tetracyanoethylene[97] (equation 78).

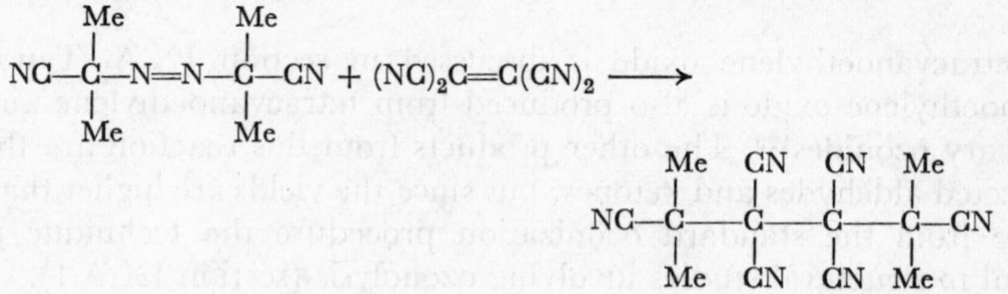

$$\mathrm{NC{-}C(Me)_2{-}N{=}N{-}C(Me)_2{-}CN} + \mathrm{(NC)_2C{=}C(CN)_2} \longrightarrow \mathrm{NC{-}C(Me)_2{-}C(CN)_2{-}C(CN)_2{-}C(Me)_2{-}CN} \quad (78)$$

Also, free-radical addition is by no doubt the mechanism by which tetracyanoethylene inhibits the polymerization of olefins[171].

A free-radical chain mechanism is postulated for the photoaddition of tetrahydrofuran to tetracyanoethylene[172]. Although two moles of tetracyanoethylene appear in the product, the n.m.r. spectrum indicates that it is an α-monosubstituted tetrahydrofuran. It is likely that the first step is addition of tetrahydrofuran to tetracyanoethylene (equation 79). The final product may be **61**.

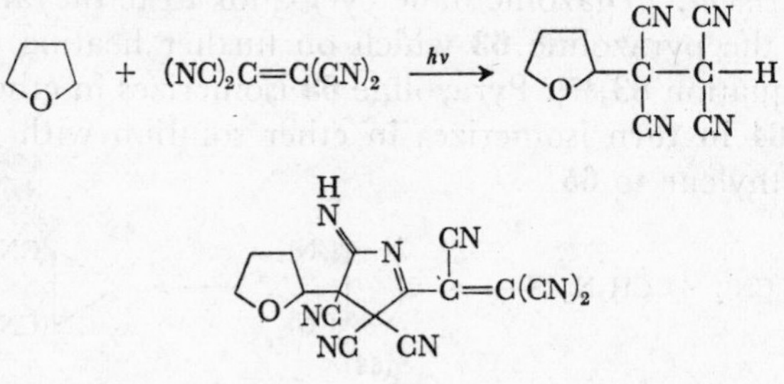

$$\text{tetrahydrofuran} + \mathrm{(NC)_2C{=}C(CN)_2} \xrightarrow{h\nu} \text{2-tetrahydrofuryl}{-}\mathrm{C(CN)_2{-}C(CN)_2{-}H} \quad (79)$$

(61)

f. Nucleophilic carbenes. The carbene formed from the dissociation of **62** adds to tetracyanoethylene to form a cyclopropane[173]:

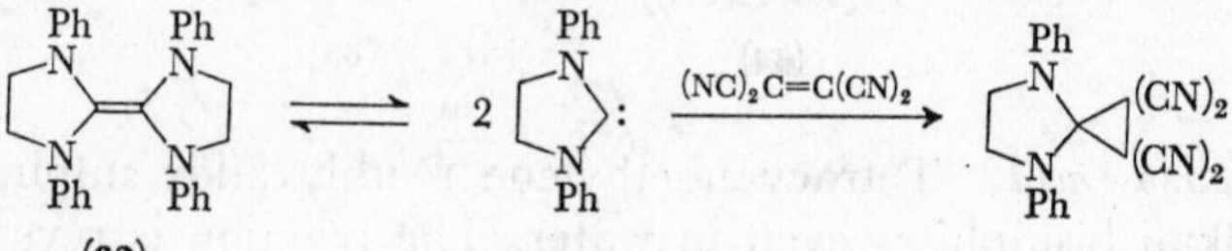

$$\mathbf{(62)} \rightleftharpoons 2\ \text{(1,3-diphenylimidazolidin-2-ylidene)} \xrightarrow{\mathrm{(NC)_2C{=}C(CN)_2}} \text{spiro-cyclopropane with } \mathrm{(CN)_2},\ \mathrm{(CN)_2} \quad (80)$$

g. Hydrogen. Tetracyanoethylene is readily reduced to tetracyanoethane (section III.B.1).

h. Hydrogen peroxide and ozonides. Hydrogen peroxide in acetonitrile epoxidizes tetracyanoethylene (equation 81)[174]. The chemistry

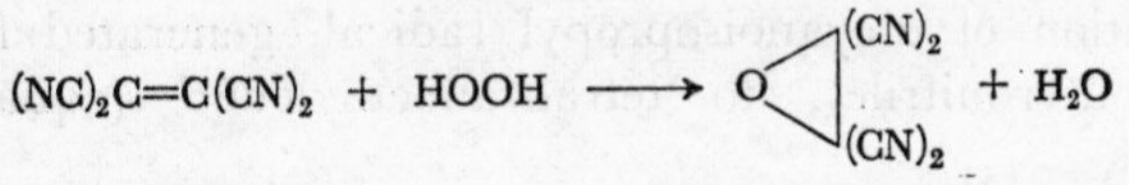

$$(NC)_2C{=}C(CN)_2 + HOOH \longrightarrow O\!\!\left\langle\begin{matrix}(CN)_2\\ \\(CN)_2\end{matrix}\right. + H_2O \qquad (81)$$

of tetracyanoethylene oxide is discussed in section IX.A. Tetracyanoethylene oxide is also produced from tetracyanoethylene and primary ozonides[175]. The other products from this reaction are the expected aldehydes and ketones, but since the yields are higher than those from the standard ozonization procedure the technique is useful in structural studies involving ozonolysis (section IX.A.1).

i. Chlorine. A reaction which clearly demonstrates the electrophilic character of the double bond in tetracyanoethylene is the addition of chlorine. This addition is catalysed by chloride ion. Evidently, a chlorocarbanion is an intermediate (equation 82)[176].

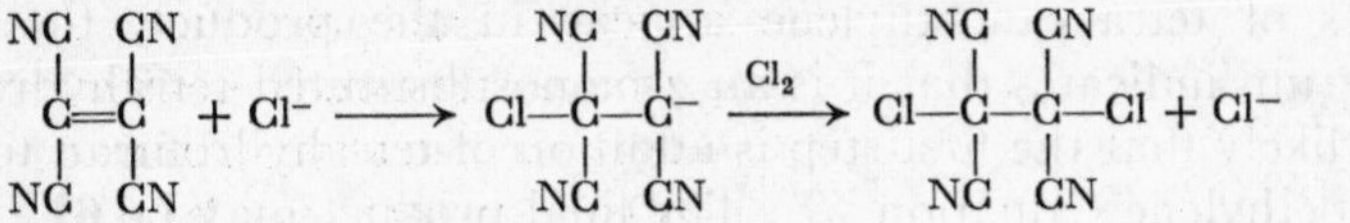

$$(NC)_2C{=}C(CN)_2 + Cl^- \longrightarrow Cl{-}C(CN)_2{-}C^-(CN)_2 \xrightarrow{Cl_2} Cl{-}C(CN)_2{-}C(CN)_2{-}Cl + Cl^- \qquad (82)$$

j. Diazomethane. Diazomethane cycloadds to tetracyanoethylene to produce the pyrazoline **63** which on further heating eliminates nitrogen (equation 83)[177]. Pyrazoline **63** isomerizes in ether solution to **64** and **64** in turn isomerizes in ether solution with a trace of tetracyanoethylene to **65**.

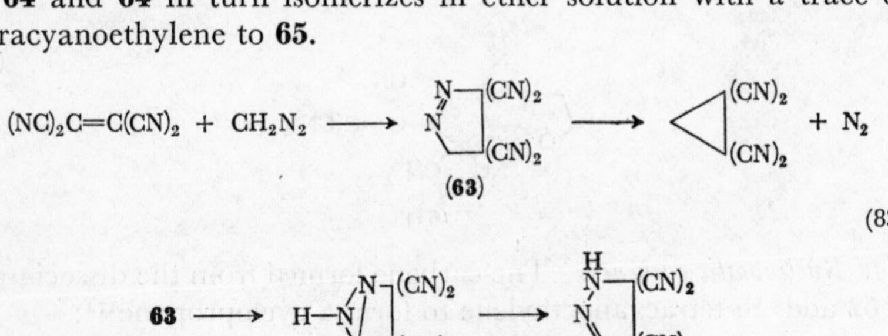

$$(NC)_2C{=}C(CN)_2 + CH_2N_2 \longrightarrow \mathbf{(63)} \longrightarrow \text{1,1,2,2-tetracyanocyclopropane} + N_2$$

$$\mathbf{63} \longrightarrow \mathbf{(64)} \longrightarrow \mathbf{(65)} \qquad (83)$$

k. Sulphurous acid. Tetracyanoethylene readily adds sulphurous acid or alkali bisulphites even in water. The reaction is reversible (equation 84)[97]. The adduct is a strong dibasic acid from which numerous salts have been made and characterized.

$$(NC)_2C{=}C(CN)_2 + H_2SO_3 \longrightarrow HC(CN)_2{-}C(CN)_2{-}SO_3H \qquad (84)$$

l. Cyanide ion. Alkali metal cyanides add to tetracyanoethylene in acetonitrile[100] (equation 84a). This reaction is discussed further in sections III.B.2 and IV.A.7.a.

$$(NC)_2C{=}C(CN)_2 + NaCN \longrightarrow (NC)_3C{-}C(CN)_2^{-}Na^{+} \qquad (84a)$$

m. Ketones. Ketones possessing an α-hydrogen readily add to tetracyanoethylene to give tetracyanopropyl ketones[97] (equation 85). The reaction is acid catalysed.

$$CH_3\overset{O}{\overset{\|}{C}}CH_3 + (NC)_2C{=}C(CN)_2 \longrightarrow HC(CN)_2{-}C(CN)_2{-}CH_2\overset{O}{\overset{\|}{C}}CH_3 \qquad (85)$$

n. Phenols and aromatic amines. The end-product from reaction of tetracyanoethylene with an electron-rich aromatic compound is a tricyanovinyl substituted aromatic derivative. In some cases (*N,N*-dimethylaniline[178], *N,N*-bis(β-chloroethyl)aniline[179], 4-hydroxycoumarins[180], *ortho*-substituted phenols[181], 2,6-dimethylaniline and 2,5-dimethoxyaniline[822]), however, an intermediate tetracyanoethyl compound can be isolated. This topic is discussed in more detail in section IV.A.5.a.

o. Zero-valent platinum and palladium compounds. Tetracyanoethylene displaces phenylacetylene from its bis(triphenylphosphine)platinum complex[182] (equation 86). An x-ray analysis[183] of **66** shows the angle

$$(Ph_3P)_2Pt(PhC{\equiv}CH) + \begin{matrix}C(CN)_2\\ \|\\ C(CN)_2\end{matrix} \longrightarrow (Ph_3P)_2Pt\langle C_{(1)}(CN)_2{-}C_{(2)}(CN)_2\rangle \qquad (86)$$

(66)

between the plane formed by $C_{(1)}$, $C_{(2)}$ and $C_{(1)}{-}CN$ of the tetracyanoethylene and the $C_{(1)}$, $C_{(2)}$, Pt plane is about 100°. The two platinum–carbon bond lengths are practically the same with the $C_{(1)}PtC_{(2)}$ bond angle 42°. The $C_{(1)}{-}C_{(2)}$ bond distance in the complex is 1·52 Å, close to the accepted value for C—C single bonds and is 0·21 Å longer than the C=C distance in uncomplexed tetracyanoethylene.

A similar complex forms in the reaction of tetrakis(triphenylphosphine)palladium with tetracyanoethylene[184]:

$$(Ph_3P)_4Pd + (NC)_2C{=}C(CN)_2 \longrightarrow (Ph_3P)_2Pd\begin{matrix} C(CN)_2 \\ | \\ C(CN)_2 \end{matrix} + 2\ Ph_3P \qquad (87)$$

p. Triphenylphosphine. Two molecules of tetracyanoethylene combine with one of triphenylphosphine to give octacyano-*P,P,P*-triphenylphosphacyclopentane (equation 88)[185].

$$2\ (NC)_2C{=}C(CN)_2 + Ph_3P \longrightarrow \text{(NC)}_2\text{C–C(CN)}_2,\ \text{(NC)}_2\text{C–C(CN)}_2 \text{ ring closed by } PPh_3 \qquad (88)$$

4. Additions to the nitrile groups of tetracyanoethylene

Only a few reagents have been observed to add to a nitrile group of tetracyanoethylene. This is surprising since each of the cyano groups of tetracyanoethylene is formally attached to a tricyanovinyl group, which is strongly electron-withdrawing. The cyano groups of tetracyanoethylene should thus be even more reactive than those in cyanogen. However, the carbon–carbon double bond of tetracyanoethylene is so much more reactive than the nitrile triple bonds that the former is no doubt attacked preferentially. Possibly a small amount of addition to nitrile groups does occur but the product is lost during the reaction work-up.

a. Azide. Approximately 65 % of the product from the reaction of sodium azide with tetracyanoethylene in acetonitrile results from nitrile addition and 25 % from carbon–carbon double bond addition[186]. Therefore, the azide ion, which is a strong nucleophile, may be adding indisciminately. The main reaction products are tricyanovinyltetrazole and bis(tricyanovinyl)amine isolated as tetramethylammonium salts (equation 89). Nitrogen but no cyanogen

$$(NC)_2C{=}C(CN)_2 + N_3^- \longrightarrow \text{NC–C(=C(CN)}_2\text{)–}CN_4^{-}\ \text{(tetrazolide)} + (NC)_2C{=}C(CN){-}\bar{N}{-}C(CN){=}C(CN)_2 \qquad (89)$$

was produced. However, as in many other cyanocarbon reactions where a cyanogen is needed to balance the products, it would be expected to end up as paracyanogen.

b. Fluorosulphonic acid. Dicyanomaleimide, a very reactive dienophile[141], is readily prepared by stirring tetracyanoethylene with fluorosulphonic acid and allowing the resulting imino salt **67** to hydrolyse in air (equation 90)[820]. The first step in this reaction is

$$(NC)_2C{=}C(CN)_2 + FSO_3H \longrightarrow \text{[NC, CN-substituted ring: HN=C–NH–CH(OSO}_2\text{F)]} \xrightarrow{FSO_3H}$$

$$\text{[NC, CN-substituted ring: } H_2\overset{+}{N}{=}C\text{–NH–C=O]}\ FSO_3^- \ (\mathbf{67}) + O(SO_2F)_2 \qquad (90)$$

$$\mathbf{67} + H_2O \longrightarrow \text{dicyanomaleimide (NC, CN; O=C–NH–C=O)}$$

most likely the protonation of one of the nitriles by fluorosulphonic acid.

c. Water or methanol under extremely high pressure. A brief communication claims that water hydrolyses one nitrile group of tetracyanoethylene at 5000–6700 atm and 110–150° to produce 2,3,3-tricyanoacrylamide (equation 91)[188]. Under similar conditions

$$(NC)_2C{=}C(CN)_2 + H_2O \longrightarrow (NC)_2C{=}C(CONH_2)(CN) \qquad (91)$$

methanol was reported to give *N*-methyl-2,3,3-tricyanoacrylamide (equation 92). This is a most unusual result, which is in marked

$$(NC)_2C{=}C(CN)_2 + MeOH \longrightarrow (NC)_2C{=}C(\overset{O}{\overset{\|}{C}}NHMe)(CN) \qquad (92)$$

contrast to that obtained under one atmosphere (see sections IV.A.5.b and IV.A.5.c).

d. Trifluoromethanesulphenyl chloride. Trifluoromethanesulphenyl chloride adds to a nitrile group of tetracyanoethylene at room temperature to form **68** (equation 93)[189]. The reaction, like the

$$(NC)_2C{=}C(CN)_2 + CF_3SCl \longrightarrow (NC)_2C{=}C(CN){-}C(Cl){=}NSCF_3 \quad \textbf{(68)} \qquad (93)$$

addition of chlorine to tetracyanoethylene (section IV.A.3.i), is catalysed by chloride ion. A mechanism involving addition of chloride ion to the double bond followed by attack of the sulphenyl chloride on the resulting anion is preferred (equation 94).

$$(NC)_2C{=}C(CN)_2 + Cl^- \longrightarrow Cl{-}C(CN)_2{-}C(CN){=}C{=}N^- \xrightarrow{CF_3SCl} Cl{-}C(CN)_2{-}C(CN){=}C{=}NSCF_3 + Cl^- \qquad (94)$$

$$\longrightarrow Cl{-}C(CN)_2{-}\overset{-}{C}(CN){-}C(Cl){=}NSCF_3 \xrightarrow{-Cl^-} \textbf{68}$$

Dienes add to **68** across the carbon–carbon double bond but electron-rich olefins add to the carbon–nitrogen double bond.

e. Ethyl diazoacetate. Although diazomethane adds to the double bond of tetracyanoethylene, ethyl diazoacetate adds to one of the cyano groups (equation 95)[113].

$$(NC)_2C{=}C(CN)_2 + \bar{N}{=}\overset{+}{N}{=}CH{-}CO_2Et \longrightarrow \text{1H-1,2,3-triazole bearing } C(CN){=}C(CN)_2 \text{ and } CO_2Et \qquad (95)$$

f. Benzonitrile oxide. It has been mentioned that benzonitrile oxide adds to a nitrile group of tetracyanoethylene[189] but no details were given.

5. Replacement of cyano groups in tetracyanoethylene

One or two cyano groups may be displaced from tetracyanoethylene by electron-rich reagents. The reaction probably takes place in two or more steps, addition of the reagent to the double bond being followed at some point by elimination of cyanide ion. Since cyanide ion itself reacts with tetracyanoethylene giving tetracyanoethylene ion radical[100], the detection of this species in the reaction mixture does not necessarily prove the abstraction of an electron from the attacking species.

a. Aromatic compounds. By far the most studied substitution reaction of tetracyanoethylene is the tricyanovinylation of aromatic systems. In general, any aromatic compound that will couple with benzenediazonium chloride will also undergo tricyanovinylation, for example: *N,N*-dimethylaniline (equation 96)[190], *o*-alkylphenols[181,191], pyrrole[192], indoles[192–194], phenanthrene[191], 2-methylfuran[195], azulenes[195,196], diazocyclopentadiene[197], phenylhydrazones[195] and cyclopentadienylidenetriphenylphosphorane[198]. The use of aluminium chloride to activate tetracyanoethylene makes possible the tricyanovinylation of benzene, toluene, naphthalene and fluorobenzene but not chlorobenzene[199].

$$C_6H_5NMe_2 + (NC)_2C{=}C(CN)_2 \longrightarrow p\text{-}Me_2NC_6H_4C(CN){=}C(CN)_2 \qquad (96)$$

The mechanism of tricyanovinylation of *N,N*-dimethylaniline has been extensively studied[178,200–205]. On the basis of the data now available the reaction probably proceeds through the intermediates shown in equations (97), (98) and (99).

A blue π-complex with an association constant of 15·0 (at 32·5° in chloroform) is first formed[200]. The conversion of the π-complex to **69** is first order in the π-complex and in the dimethylaniline[200]. The conversion of **69** to **71**, the slowest step, is first order in dimethylaniline as well as **69**[200]. The rate of formation of **69** is very sensitive to solvent polarity. The blue π-complex disappears quickly in acetonitrile but is stable for several days in carbon tetrachloride[200]. The further reaction of **69** is also sensitive to solvent polarity but less so.

$$\text{PhNMe}_2 + (NC)_2C{=}C(CN)_2 \underset{\text{fast}}{\rightleftharpoons} [\text{Me}_2\text{NC}_6\text{H}_5 \rightarrow (NC)_2C{=}C(CN)_2] \quad \text{blue } \pi\text{-complex} \tag{97}$$

$$\pi\text{-complex} \xrightarrow[\text{slow}]{\text{Me}_2\text{NPh}} \text{colourless } (\mathbf{69}) \longrightarrow p\text{-Me}_2\text{NC}_6\text{H}_4\text{C(CN)}_2\text{HC(CN)}_2 \; (\mathbf{70}) \text{ colourless} \tag{98}$$

$$\mathbf{69} \xrightarrow[\text{slow}]{\text{Me}_2\text{NPh}} p\text{-Me}_2\text{NC}_6\text{H}_4\text{C(CN)}{=}\text{C(CN)}_2 + \text{HCN} \quad (\mathbf{71}) \text{ red} \leftarrow \mathbf{70} \tag{99}$$

The tetracyanoethyl derivative **70** can be isolated if the reaction is conducted with equivalent amounts of tetracyanoethylene and *N,N*-dimethylaniline in dioxan[178]. Structure **69** was assigned to it. However, the absence of any appreciable nitrile absorption in its infrared spectrum would seem to rule out **69**. The two cyano groups conjugated with a full negative charge should have an intense nitrile absorption. On the other hand, tetracyanoethanes tend to have very weak nitrile absorption. Similar tetracyanoethyl products have been isolated from the reaction of tetracyanoethylene with *N,N*-bis(2-chloroethyl)aniline[179,206], 2,6-dimethylaniline[822] and with *ortho*-substituted phenols[191]. These tetracyanoethyl compounds readily eliminate hydrogen cyanide to form tricyanovinyl derivatives. The rate of this reaction, at least for 2,6-dimethyl-4-(1,1,2,2-tetracyanoethyl)aniline, shows no isotope effect[822]. Since the conversion of the colourless intermediate, formed in the reaction of tetracyanoethylene with *N,N*-dimethylaniline, to the tricyanovinyl derivative has an isotope effect, k_H/k_D, of 3·6[203], **70** is not the intermediate, but is merely a by-product formed by a proton shift when no excess base is present.

An ion-radical mechanism for the first step in the formation of **70** is doubtful, since *N,N*-diethylaniline reacts with tetracyanoethylene in solvents of moderate polarity to give large amounts of tetracyanoethylene ion radical, but little tricyanovinyl product[203]. This cannot be a steric problem, since *N,N*-bis(2-chloroethyl)-aniline[179,206] readily forms a tricyanovinyl derivative. In dimethylformamide, however, *N,N*-diethylaniline forms a tricyanovinyl derivative in good yield[190].

Phenols with large *ortho* groups such as 2,6-bis(*t*-butyl)phenol[181] undergo tricyanovinylation readily, whereas phenols with *meta*-alkyl substituents are unreactive.

The relative rates of tricyanovinylation of *N*-methylaniline and *N,N*-dimethylaniline have been examined[202,204]. The results are conflicting. One report states that the rate of formation of colourless intermediate is faster for *N*-methylaniline whereas the conversion of colourless intermediate to tricyanovinyl product is slower[202]. The other report states the opposite[204], but gives no quantitative data.

The 1,2-cycloaddition product **72** has been suggested as an intermediate in the tricyanovinylation of dimethylaniline[205]. This

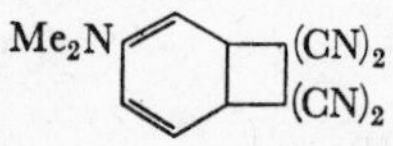

Me_2N — (benzocyclobutane with $(CN)_2$, $(CN)_2$)

(72)

idea is attractive, since it would unify the mechanism of tricyanovinylation of aromatic compounds with that of olefins in which a cyclobutane intermediate can be isolated (equation 100)[207]. The formation of an intermediate such as **72** from aromatic compounds, however, seems unlikely, since it would destroy the aromaticity of the system.

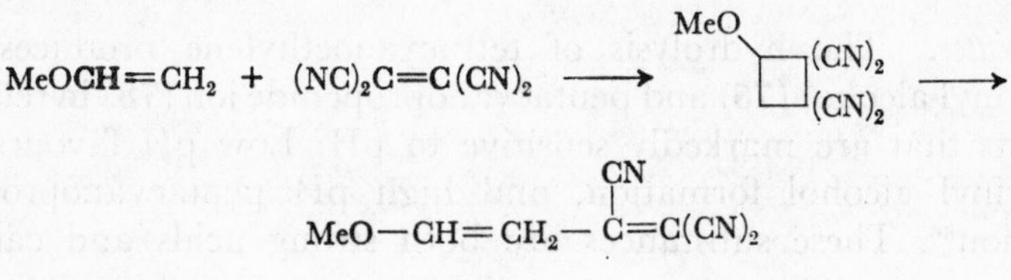

$$MeOCH{=}CH_2 + (NC)_2C{=}C(CN)_2 \longrightarrow \text{MeO-cyclobutane}(CN)_2(CN)_2 \longrightarrow MeO{-}CH{=}CH_2{-}\underset{|}{\overset{CN}{C}}{=}C(CN)_2 \quad (100)$$

$$\text{indole} + (NC)_2C{=}C(CN)_2 \longrightarrow \text{3-H, 3-}C(CN)_2C(CN)_2^-\text{-indoleninium} \longrightarrow \text{3-}[C(CN){=}C(CN)_2]\text{indole} \quad (101)$$

Kinetics parallel to those for *N*,*N*-dimethylaniline are observed for tricyanovinylation of indole[194]. Again a colourless intermediate forms (equation 101).

In the tricyanovinylation of dialkylanilines with tricyanovinyl chloride, the rate-determining step does not involve proton removal; the reaction is first order in both reagents, large alkyl groups on the nitrogen do not hinder the reaction, and no π-complex is observed at room temperature[201]. The mechanism suggested is given in equation (102).

$$C_6H_5NMe_2 + ClC(CN){=}C(CN)_2 \longrightarrow \underset{(\mathbf{73})}{\overset{+}{N}Me_2{=}C_6H_4(H)\text{-}C(Cl)(CN)\text{-}\bar{C}(CN)_2} \quad (102)$$

$$\mathbf{73} \longrightarrow \overset{+}{N}Me_2{=}C_6H_4(H)\text{-}C(NC){=}C(CN)_2 + Cl^- \longrightarrow p\text{-}Me_2NC_6H_4C(CN){=}C(CN)_2$$

b. Water. The hydrolysis of tetracyanoethylene produces tricyanovinyl alcohol (**75**) and pentacyanopropenide ion (**78**) in relative amounts that are markedly sensitive to pH. Low pH favours tricyanovinyl alcohol formation, and high pH pentacyanopropene formation[86]. These substances are both strong acids and can be characterized as their tetraalkylammonium salts. They form on exposure of tetracyanoethylene to moist air but can be readily removed by sublimation of the sample.

In acid solution, the initial attack is probably by water to form **74**, which loses hydrogen cyanide to give **75** (equation 103). At

$$HOH + (NC)_2C{=}C(CN)_2 \xrightarrow{H_2O} HC(CN)_2{-}C(CN)_2{-}OH \; (\mathbf{74}) \longrightarrow (NC)_2C{=}C(CN)OH \; (\mathbf{75}) \qquad (103)$$

high pH the initial attack is probably by hydroxide ion to form **76**. Since the supply of protons is low, **76** adds to another tetracyanoethylene to give **77**. This adduct fragments to pentacyanopropenide ion (**78**), carbonyl cyanide and cyanide ion (equation 104). Since

$$(NC)_2C{=}C(CN)_2 + OH^- \longrightarrow {}^-C(CN)_2{-}C(CN)_2{-}OH \; (\mathbf{76})$$

$$\mathbf{76} + (NC)_2C{=}C(CN)_2 \longrightarrow {}^-C(CN)_2{-}C(CN)_2{-}C(CN)_2{-}C(CN)_2{-}OH \; (\mathbf{77})$$

$$\mathbf{77} + OH^- \longrightarrow {}^-C(CN)_2{-}C(CN){=}C(CN)_2 \; (\mathbf{78}) + (NC)_2C{=}O + H_2O + CN^- \qquad (104)$$

malononitrile anion displaces cyanide ion from tetracyanoethylene[207] to give pentacyanopropenide ion, another possibility is that this ion is produced by the fragmentation of **76** and then adds to tetracyanoethylene.

c. Alcohols. The base-catalysed addition of alcohols to tetracyanoethylene proceeds stepwise. Either alkyltricyanovinyl ethers[208] or dicyanoketene acetals[89] can be isolated. If a molar amount of base is present, a trialkoxy salt is obtained[209] (equation 105).

$$\mathrm{MeOH} + \underset{\mathrm{NC}\quad\mathrm{CN}}{\overset{\mathrm{NC}\quad\mathrm{CN}}{\mathrm{C{=}C}}} \longrightarrow \underset{\mathrm{NC}\quad\mathrm{CN}}{\overset{\mathrm{NC}\quad\mathrm{OMe}}{\mathrm{C{=}C}}} \longrightarrow \underset{\mathrm{NC}\quad\mathrm{OMe}}{\overset{\mathrm{NC}\quad\mathrm{OMe}}{\mathrm{C{=}C}}} \longrightarrow \underset{\mathrm{NC}\quad\mathrm{OMe}}{\overset{\mathrm{NC}\quad\mathrm{OMe}}{{}^{-}\mathrm{C{-}C{-}OMe}}} \qquad (105)$$

d. Ammonia, hydrazines and amines. Tetracyanoethylene reacts with ammonia to give bis(tricyanovinyl)amine and similarly with hydrazine to give 1,2-bis(tricyanovinyl)hydrazine[86], isolated as their salts (equations 106 and 107).

$$2\,(\mathrm{NC})_2\mathrm{C{=}C(CN)_2} + \mathrm{NH_3} \longrightarrow (\mathrm{NC})_2\mathrm{C{=}}\overset{\mathrm{CN}}{\mathrm{C}}\mathrm{{-}\bar{N}{-}}\underset{\mathrm{CN}}{\mathrm{C}}\mathrm{{=}C(CN)_2} \qquad (106)$$

$$2\,(\mathrm{NC})_2\mathrm{C{=}C(CN)_2} + \mathrm{NH_2NH_2} \longrightarrow (\mathrm{NC})_2\mathrm{C{=}}\overset{\mathrm{CN}}{\mathrm{C}}\mathrm{{-}\bar{N}{-}\bar{N}{-}}\overset{\mathrm{CN}}{\mathrm{C}}\mathrm{{=}C(CN)_2} \qquad (107)$$

Tetracyanoethylene and other cyanoethylenes that contain replaceable groups react with monosubstituted hydrazines and hydrazides to give 5-amino-3,4-dicyanopyrazoles[210]:

$$(\mathrm{NC})_2\mathrm{C{=}C(CN)_2} + \mathrm{RNHNH_2} \longrightarrow \text{1-R-5-}\mathrm{H_2N}\text{-3,4-(NC)}_2\text{-pyrazole} \qquad (108)$$

Primary and secondary amines also give *N*-tricyanovinyl compounds when treated with tetracyanoethylene. If two equivalents of amine are used, a second cyano group is eliminated (equation 109)[211]. Aromatic primary amines are tricyanovinylated on the

$$\text{n-BuNH}_2 + (\mathrm{NC})_2\mathrm{C{=}C(CN)_2} \longrightarrow \text{n-BuNH}{-}\underset{\mathrm{CN}}{\mathrm{C}}\mathrm{{=}C(CN)_2}\quad \mathbf{(79)}$$

$$\mathbf{79} + \mathrm{Me_2NH} \longrightarrow \underset{\mathrm{Me_2N}}{\overset{\text{n-BuNH}}{\mathrm{C}}}\mathrm{{=}C(CN)_2} \qquad (109)$$

nitrogen atom unless hindered by alkyl groups. *N*-Alkyl and *N,N*-dialkyl aromatic amines are tricyanovinylated at the *para* position if it is open. The amino group on 2,4,6-trimethylaniline is tricyanovinylated, even though hindered, since the *para* position is blocked[822].

Tertiary aliphatic amines reduce tetracyanoethylene to tetracyanoethylene ion radical[100]. The triethylamine probably ends up as vinyldiethylamine as it does in other similar ion radical redox reactions[205] (equation 110).

$$(NC)_2C{=}C(CN)_2 + CH_3CH_2\ddot{N}Et_2 \longrightarrow (NC)_2\bar{C}{-}\dot{C}(CN)_2 + CH_3CH_2\dot{N}Et_2$$

$$(NC)_2\bar{C}{-}\overset{H}{C}(CN)_2 + CH_3CH{=}\overset{+}{N}Et_2 \longrightarrow (NC)_2\overset{H}{C}{-}\overset{H}{C}(CN)_2 + CH_2{=}CHNEt_2$$

$$(NC)_2\overset{H}{C}{-}\overset{H}{C}(CN)_2 + (NC)_2C{=}C(CN)_2 + 2\,Et_3N \longrightarrow 2\,Et_3\overset{+}{N}H\;(NC)_2\dot{C}{-}\bar{C}(CN)_2 \tag{110}$$

e. Phosphorus ylids. Triphenylphosphorus ylids react with tetracyanoethylene by substitution of one of the cyano groups[212,213] (equation 111). Substitution on cyclopentadienylidenetriphenylphosphorane occurs on the hindered 2-position in marked contrast to other aromatic tricyanovinylations which seem to avoid hindered sites (equation 112)[198].

$$Ph_3\overset{+}{P}{-}\underset{H}{\overset{-}{C}}{-}CN + (NC)_2C{=}C(CN)_2 \longrightarrow Ph_3\overset{+}{P}{-}\underset{CN}{\overset{-}{C}}{-}\overset{CN}{C}{=}C(CN)_2 \tag{111}$$

$$\text{(cyclopentadienide)}{-}\overset{+}{P}Ph_3 + (NC)_2C{=}C(CN)_2 \longrightarrow \text{2-}[\underset{CN}{C}{=}C(CN)_2]\text{(cyclopentadienide)}{-}\overset{+}{P}Ph_3 + HCN \tag{112}$$

6. Fragmentation reactions of tetracyanoethylene

a. Retrograde Michael reaction. 2H,3H-Thieno[3,2-b]pyrrol-3-one (**80**) and also 2H,3H-benzo[b]thiophene-3-one (**81**) react with tetracyanoethylene to give dicyanomethylene derivatives (equation 113)[214]. The tetracyanoethane **82** was not isolated, but malononitrile was. The acidity of the methylene compound required for this reaction must be critical. The very similar adducts of ketones with tetracyanoethylene (section IV.A.3.m) can be dissolved in sodium bicarbonate solution and recovered on acidification. Here the proton on the tetracyanoethyl group ionizes rather than the proton α to the keto function (equation 114).

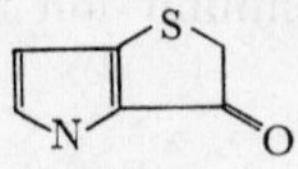

(80) 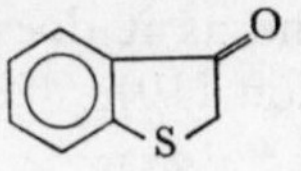 (81)

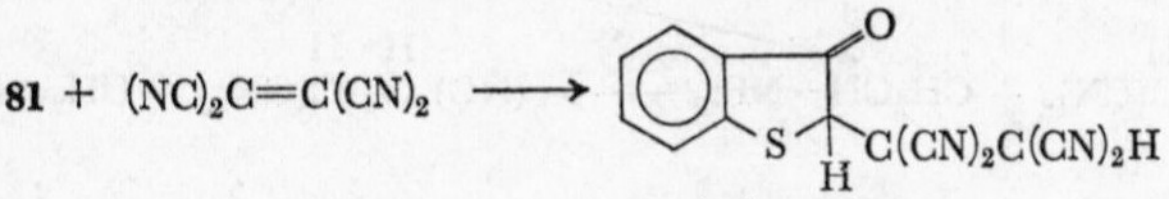

$$\mathbf{81} + (NC)_2C{=}C(CN)_2 \longrightarrow \mathbf{(82)}$$

$$\mathbf{82} \longrightarrow \text{CN CN; C—CH; CN CN} \longrightarrow \text{CN, C, CN} + {}^{-}CH(CN)_2 \quad (113)$$

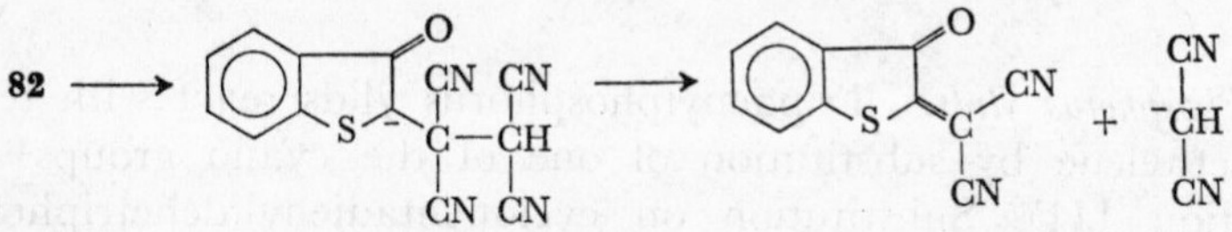

$$CH_3-\overset{O}{\overset{\|}{C}}-CH_2-C(CN)_2-CH(CN)_2 \xrightarrow{NaHCO_3} CH_3\overset{O}{\overset{\|}{C}}CH_2C(CN)_2-C^{-}(CN)_2 \quad (114)$$

b. Elimination of cyanogen. Tetracyanoethylene is cracked to cyanogen and dicyanoacetylene when passed through a tube packed with quartz chips at 800° (equation 115)[215].

$$(NC)_2C{=}C(CN)_2 \xrightarrow{800^\circ} NC-C{\equiv}C-CN + NC-CN \quad (115)$$

c. Reaction with isatogens. When equimolar quantities of 2-phenyl- or 2-methoxycarbonylisatogen and tetracyanoethylene are heated under reflux in xylene for 1·5–10 hours, 3,4-dihydro-2-phenyl- (**83**) or methoxycarbonyl-4-quinazolinone is produced in 30 to 40% yield (equation 116)[216]. These products are not formed when cyanide ion or ammonia are substituted for tetracyanoethylene, and thus it is assumed that tetracyanoethylene is the source of the extra nitrogen atom. The mechanism is not known.

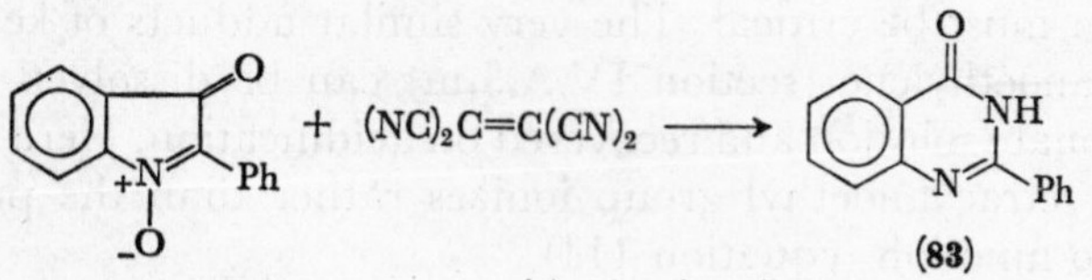

$$\text{(2-phenylisatogen)} + (NC)_2C{=}C(CN)_2 \longrightarrow \mathbf{(83)} \quad (116)$$

7. Tetracyanoethylene anion radical

Tetracyanoethylene reacts with a variety of reagents including metals, cyanides, iodides and tertiary amines to give tetracyanoethylene anion radical[100]. The reduction is reversible and occurs at $E_{\frac{1}{2}} + 0{\cdot}152$ v versus SCE in acetonitrile. A second wave occurs at $E_{\frac{1}{2}} - 0{\cdot}561$ v (equation 117). The sodium salt of the dianion **84**

$$(NC)_2C{=}C(CN)_2 \underset{}{\overset{e}{\rightleftharpoons}} (NC)_2\dot{C}{-}\bar{C}(CN)_2 \underset{}{\overset{e}{\rightleftharpoons}} (NC)_2\bar{C}{-}\bar{C}(CN)_2 \quad \textbf{(84)} \qquad (117)$$

is readily made from tetracyanoethane and sodium hydride. As expected, it transfers an electron to tetracyanoethylene and two equivalents of ion radical[100] result. In a variation of this method of synthesis, a mixture of tetracyanoethane and tetracyanoethylene is treated with a base to give two equivalents of ion radical. The reaction is reversible (equation 118). The reaction of tetracyanoethylene with sodium cyanide has been studied in detail[100]. A

$$(NC)_2CH{-}CH(CN)_2 + (NC)_2C{=}C(CN)_2 \underset{H^+}{\overset{OH^-}{\rightleftharpoons}} (NC)_2\dot{C}{-}\bar{C}(CN)_2 \qquad (118)$$

colourless isolable salt (**85**) is first formed. Tracer studies with $^{14}CN^-$ showed the first step to be reversible. The adduct reacts with cyanide to give the ion radical, most likely via tetracyanoethane dianion (equation 119). Cyanogen cannot be isolated from the

$$(NC)_2C{=}C(CN)_2 + CN^- \longrightarrow (NC)_2\bar{C}{-}C(CN)_3 \quad \textbf{(85)}$$

$$\textbf{85} + CN^- \longrightarrow (NC)_2\bar{C}{-}\bar{C}(CN)_2 + (CN)_2 \quad \textbf{(84)}$$

$$\textbf{84} + (NC)_2C{=}C(CN)_2 \longrightarrow 2\,(NC)_2\dot{C}{-}\bar{C}(CN)_2 \qquad (119)$$

reaction solution, but if solid sodium pentacyanoethanide (**85**) is heated it forms cyanogen and sodium tetracyanoethylene ion radical.

Since cyanide ion reacts with tetracyanoethylene to form an ion radical, the mere detection of ion radical by e.s.r. in a tetracyanoethylene reaction mixture does not necessarily mean that the tetracyanoethylene is abstracting an electron from the other reactant. Cyanide ion may in some way have been generated. We believe this is the case whenever the reagent involved is a poor reducing agent. For example, hydroxide[217], TiO_2, ZnO, MgO[218], pyridine[219] and dimethyl formamide[220] react with tetracyanoethylene to produce tetracyanoethylene ion radical (e.s.r.), whereas neither sodium bromide nor potassium thiocyanate, both fair reducing agents, transfers an electron to tetracyanoethylene[220] (u.v.).

Irradiation of tetracyanoethylene by γ-rays[221], x-rays[222], flash photolysis[223] or ultraviolet light[224] gives detectable amounts of its ion radical.

Electrolysis of 1,1,2,2-tetracyanocyclopropane and of tetracyanoethane produces tetracyanoethylene ion radical[220].

Nickel tetracarbonyl is stripped of its carbonyl groups by tetracyanoethylene and the nickel is oxidized to Ni^{2+} [100]. The grey area between π-complex formation and complete electron transfer or ion-radical formation is brought out by the reaction of tetracyanoethylene with metallocenes. The green crystalline solid from reaction of tetracyanoethylene with ferrocene is not ferricenium tetracyanoethylene ion radical as originally reported[100] but is a π-complex[225] (**86**). In polar solvents, however, this π-complex dissociates and a small amount of complete electron transfer occurs (equation 120).

$$\text{Fe}(C_5H_5)_2 + (NC)_2C{=}C(CN)_2 \rightleftharpoons (NC)_2C{=}C(CN)_2 \cdot \text{Fe}(C_5H_5)_2 \rightleftharpoons \text{ferricenium } C_6N_4^{\overline{\cdot}} \quad (120)$$

(**86**)

The equilibrium constant for formation of ferricenium tetracyanoethylenide in acetonitrile was determined[225] as $2{\cdot}5 \times 10^{-3}$, and in another study[226] as $8{\cdot}6 \times 10^{-4}$, by the intensity of the ion-radical absorption in the visible spectrum of its solution. Based on an oxidation–reduction potential of $+0{\cdot}30$ v for ferrocene in acetonitrile[227] and $+0{\cdot}15$ v for tetracyanoethylene in acetonitrile[100], the calculated equilibrium constant is $2{\cdot}9 \times 10^{-3}$. An equilibrium constant of

3×10^{-6} in acetone and 7×10^{-3} in dimethylformamide gives a measure of the solvent effect on electron transfer[226].

A similar solvent effect was noted in the interaction of tetracyanoethylene with tetramethyl-*p*-phenylenediamine[228]. In solvents of high dielectric constant, ion radicals were formed. In solvents of low dielectric constant only π-complexes result, whereas, in solvents of intermediate dielectric constant both were present in equilibrium with the starting materials. The Mössbauer spectrum of the tetracyanoethylene–ferrocene complex shows that the tetracyanoethylene is not bonded to the iron[229]. X-ray data confirm **86** as the structure of the complex[230].

Cobaltocene and bis(tetrahydroindenyl)iron both transfer an electron more or less completely to tetracyanoethylene in acetonitrile, but an equilibrium is set up in acetone and in dichloromethane[226].

Dibenzenechromium transfers an electron completely to tetracyanoethylene to give dibenzenechromium (I) tetracyanoethylene ion radical[231].

Solid tetracyanoethylene ion-radical salts vary in colour from brown to purple. Solutions are yellow with a characteristic cockscomb shaped absorption band centred at 435 mμ. The nitrile group absorbs strongly at 2210 and 2180 cm^{-1} in the infrared region[100]. The e.s.r. spectrum of tetracyanoethylene ion radical (g = 2·0026 ± 0·0002) shows all of the carbon-13 splitting (20 lines) arising from carbon-13 nuclei in natural abundance, as well as the nine lines arising from the four equivalent nitrogen atoms[220,232–236]. The hyperfine splitting constants[220] of **87** are $a^N = 1{\cdot}574 \pm 0{\cdot}005$ gauss, $a_1{}^C = 9{\cdot}541 \pm 0{\cdot}01$ gauss and $a_2{}^C = 2{\cdot}203 \pm 0{\cdot}01$ gauss. The e.s.r. spectrum has been studied in nematic liquid crystals[237]. The

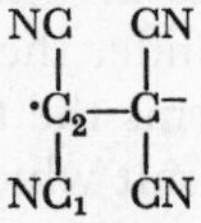

```
 NC    CN
  |    |
·C2 —  C⁻
  |    |
 NC1   CN
```

(87)

rate constant for electron exchange between tetracyanoethylene ion radical and added tetracyanoethylene is $2{\cdot}1 \times 10^8$ l/mol s [232]. The resistivity of tetracyanoethylene ion-radical salts ranges from 9×10^7 to $1{\cdot}6 \times 10^9$ ohm cm[100], too high to be classed as semiconductors. This is in marked contrast to the 1 ohm cm resistivity of some of the tetracyanoquinodimethane ion-radical salts (section VII.A.1).

Solid tetracyanoethylene ion-radical salts are relatively stable in air. In solution, however, they are very sensitive to oxygen and to a lesser degree to water. Either reagent causes formation of pentacyanopropenide ion and tricyanovinyl alcoholate[100]. On acidification, the ion radical disproportionates to tetracyanoethane and tetracyanoethylene. It is easily oxidized to tetracyanoethylene. On heating under reflux in dimethoxyethane, sodium tetracyanoethylenide slowly forms disodium 1,1,2,3,4,4-hexacyanobutenediide[238] (equation 121).

$$2\cdot\overset{NC}{\underset{NC}{C}}-\overset{CN}{\underset{CN}{C^-}} \longrightarrow {}^-\overset{NC}{\underset{NC}{C}}-\overset{CN}{C}=\underset{CN}{C}-\overset{CN}{\underset{CN}{C^-}} \qquad (121)$$

8. Polymers from tetracyanoethylene

Unlike acrylonitrile and vinylidene cyanide, tetracyanoethylene does not form a polymer with a carbon backbone. However, when heated to relatively high temperatures with metals, metal complexes, alcohols, phenols, amines or amides it forms dark-coloured polymeric materials with concurrent destruction of most of its nitrile functions[239–256]. These polymers are reported to be semiconductors and to have catalytic properties. A careful study has been made of this type of polymerization for a series of cyanocarbon acid salts[257]. These salts rearrange spontaneously when heated, with destruction of the cyano triple bonds. This rearrangement occurs with evolution of heat at temperatures which vary from one salt to another. The products of the rearrangement are black, refractory powders. Standard differential thermal analysis techniques have been used to determine the temperatures at which the exothermic rearrangement occurs. It is likely that under the conditions used to polymerize tetracyanoethylene, it first forms the tetracyanoethylene ion radical or another cyanocarbon acid salt which then undergoes the thermal rearrangement reaction.

B. Tricyanoethylenes

I. Tricyanoethylene

Tricyanoethylene (**88**) is prepared from ethyl cyanoacetate and glycolonitrile by the route outlined in equation (122)[258]. It is chemically similar to tetracyanoethylene, but reacts less vigorously. Thus,

$$NCCH_2OH + {}^{-}CH(CN)(CO_2Et) \longrightarrow NCCH_2CH(CN)(CO_2Et) \xrightarrow{NH_3} NCCH_2CH(CN)(CO_2NH_2)$$

$$NCCH_2CH(CN)(CO_2NH_2) \xrightarrow{POCl_3} NCCH_2CH(CN)_2 \xrightarrow{Br_2} NCCH_2CBr(CN)_2$$

$$NCCH_2CBr(CN)_2 + Et_3N \longrightarrow \underset{(88)}{NCCH{=}C(CN)_2} \qquad (122)$$

it forms a Diels–Alder adduct with anthracene, but at room temperature the reaction requires hours instead of minutes. The addition of chlorine to the double bond is catalysed by chloride ion. Tricyanoethylene adds to ketones and vinylates *N*,*N*-dimethylaniline. On the other hand, it is much more reactive toward water than is tetracyanoethylene, forming hydrogen cyanide and unidentified products[258].

2. Tricyanovinylalkanes and -arenes

Alkyl- and aryltricyanovinyl compounds can be prepared by three routes[259]. If the aryl compound is reactive enough, a tricyanovinyl derivative can be prepared by tricyanovinylation with tetracyanoethylene (section IV.A.5.a) or tricyanovinyl chloride (section IV.B.3).

A method starting with aldehydes first involves condensation with malononitrile, then addition of hydrogen cyanide and dehydrogenation:

$$C_6H_5CHO + CH_2(CN)_2 \longrightarrow C_6H_5CH{=}C(CN)_2 \xrightarrow{HCN} C_6H_5C(H)(CN){-}CH(CN)_2$$

$$C_6H_5C(H)(CN){-}CH(CN)_2 + \text{N-bromosuccinimide} \longrightarrow C_6H_5C(CN){=}C(CN)_2 + \text{succinimide (NH)} + HBr \qquad (123)$$

If the acyl cyanide is readily available the best route to tricyanovinyl compounds is condensation with malononitrile (equation 124).

$$\mathrm{Me_3C\overset{O}{\overset{\|}{C}}CN + CH_2(CN)_2 \longrightarrow Me_3C{-}\overset{NC}{\overset{|}{C}}{=}C(CN)_2 + H_2O} \qquad (124)$$

The reaction is catalysed by a salt of an amine with an organic or mineral acid in the presence of excess acid. Typical catalysts are piperidine–acetic acid, β-alanine–acetic acid and piperidine–sulphuric acid.

Tricyanovinylalkyl compounds fail to react with dienes at room temperatures but adducts are obtained in high yield at elevated temperatures.

Reaction of tricyanovinyl compounds with nucleophilic reagents generally results in replacement of the 1-cyano group. Basic hydrolysis results in 1-hydroxy-2,2-dicyanovinyl compounds of type **89** (equation 125). Similarly, alcohols produce 1-alkoxy and amines,

$$\mathrm{Ph\overset{NC}{\overset{|}{C}}{=}\underset{\underset{CN}{|}}{\overset{CN}{\overset{|}{C}}} + OH^- \longrightarrow Ph\overset{OH}{\overset{|}{C}}{=}\underset{\underset{CN}{|}}{\overset{CN}{\overset{|}{C}}}} \qquad (125)$$

(89)

1-amino derivatives[259–261].

The reaction of tricyanovinyl arenes with malononitrile anion provides a route to 2-substituted tetracyanopropene salts (equation 126).

$$\mathrm{Ph\overset{NC}{\overset{|}{C}}{=}\underset{\underset{CN}{|}}{\overset{CN}{\overset{|}{C}}} + {}^{-}CH(CN)_2 \longrightarrow \underset{\underset{NC}{|}}{\overset{NC}{\overset{|}{C}}}{=}\overset{Ph}{\overset{|}{C}}{-}\underset{\underset{CN}{|}}{\overset{CN}{\overset{|}{C^-}}} + HCN} \qquad (126)$$

The reaction of tricyanovinylbenzene with methylmagnesium iodide gives **90**, **91** and **92** in 50, 18 and 32% yields[262]. Thus the

$$\mathrm{Ph\overset{NC}{\overset{|}{C}}{=}\underset{\underset{CN}{|}}{\overset{CN}{\overset{|}{C}}} + MeMgI \longrightarrow \underset{(\mathbf{90})}{Ph\overset{Me}{\overset{|}{C}}{=}\underset{\underset{CN}{|}}{\overset{CN}{\overset{|}{C}}}} + \underset{(\mathbf{91})}{Ph\underset{\underset{Me}{|}}{\overset{Me}{\overset{|}{C}}}{-}\underset{\underset{CN}{|}}{\overset{CN}{\overset{|}{CH}}}} + \underset{(\mathbf{92})}{Ph{-}\overset{NC}{\overset{|}{C}}{=}\underset{\underset{Me}{|}}{\overset{Me}{\overset{|}{C}}}}}$$

relative reactivity in displacement of the 1- to the 2-cyano groups is 4:1. Tricyanovinyl compounds react readily with mercaptans to give substituted pyrroles:

$$\mathrm{PhC(CN)}=\mathrm{C(CN)_2} + \mathrm{RSH} \longrightarrow \mathrm{PhCH(CN)}-\mathrm{CH(CN)_2} \xrightarrow{\mathrm{RSH}} \text{2-amino-3-phenyl-4-cyano-5-(SR)pyrrole (Ph, CN, H}_2\text{N, SR, NH)} \qquad (127)$$

Diaminothiophenes are obtained when hydrogen sulphide is substituted for the mercaptan:

$$\mathrm{PhC(CN)}=\mathrm{C(CN)_2} + \mathrm{H_2S} \longrightarrow \text{2,5-diamino-3-phenyl-4-cyanothiophene (Ph, CN, H}_2\text{N, S, NH}_2) \qquad (128)$$

N,N-bis(*β*-chloroethyl)-*p*-tricyanovinylaniline reacts with 1,1-dialkylhydrazines to substitute the *α*-cyano group, but with 1-phenyl-1-alkylhydrazines with elimination of a malononitrile residue[261] (equations 129 and 130). Pyrimidines form on reaction of **93** with amidines[261] (equation 131). Sodium dialkylphosphites add to **93**[179] (equation 132).

$$\underset{\mathbf{(93)}}{p\text{-}\mathrm{(ClCH_2CH_2)_2NC_6H_4C(CN)}=\mathrm{C(CN)_2}} + \mathrm{Me_2NNH_2} \longrightarrow p\text{-}\mathrm{(ClCH_2CH_2)_2NC_6H_4C(NHNMe_2)}=\mathrm{C(CN)_2} \qquad (129)$$

$$\mathbf{93} + \mathrm{PhMeNNH_2} \longrightarrow p\text{-}\mathrm{(ClCH_2CH_2)_2NC_6H_4C(CN)}=\mathrm{N}-\mathrm{N(Me)}-\mathrm{Ph} \qquad (130)$$

$$\mathbf{93} + \mathrm{R}-\mathrm{C}(=\mathrm{NH})-\mathrm{NH_2} \longrightarrow p\text{-}\mathrm{(ClCH_2CH_2)_2NC_6H_4}\text{-(5-cyano-6-amino-2-R-pyrimidin-4-yl)} \qquad (131)$$

$$\mathbf{93} + \mathrm{(MeO)_2P(=O)^-\,Na^+} \longrightarrow p\text{-}\mathrm{(ClCH_2CH_2)_2NC_6H_4}-\mathrm{C(CN)(P(O)(OMe)_2)}-\mathrm{C^-(CN)_2\;Na^+} \qquad (132)$$

Tricyanovinylbenzene and 1,1-dicyanovinylbenzene react with nickel carbonyl to give 1:1 carbon monoxide-free complexes. These complexes give a strong e.s.r. signal[263].

Hydrolysis of tricyanovinylbenzene with concentrated hydrochloric acid gives **94** and **95**[259]. The electronic effects of a tricyanovinyl group is discussed in Chapter 5.

$$\underset{\textbf{(94)}}{\begin{array}{c} \mathrm{PhC{=}CCN} \\ \mathrm{HO_2C\quad CO_2H} \end{array}} \qquad \underset{\textbf{(95)}}{\begin{array}{c} \mathrm{PhC{=}CCN} \\ \mathrm{O{=}C{-}N(H){-}C{=}O} \end{array}}$$

3. 1-Chloro-1,2,2-tricyanoethylene (tricyanovinyl chloride)

Tricyanonovinyl chloride (**96**) was first prepared by addition of chlorine to tricyanoethylene followed by base-catalysed elimination of hydrogen chloride (equation 133)[208,264]. The chlorine addition,

$$\mathrm{NCCH{=}C(CN)_2 + Cl^- \longrightarrow [NCCHClC(CN)_2]^- \xrightarrow{Cl_2}}$$
$$\mathrm{Cl^- + NCCHClCCl(CN)_2 \xrightarrow[-HCl]{Et_3N} \underset{\textbf{(96)}}{ClC(CN){=}C(CN)_2}} \qquad (133)$$

like that to tetracyanoethylene, occurs at a reasonable rate only when catalysed by chloride ion and is apparently a nucleophilic addition. An alternate preparation from oxalyl chloride and the tetramethylammonium salt of tricyanovinyl alcohol gives lower yields (30–45 %), but is more convenient, because the starting material is readily available from tetracyanoethylene.

Tricyanovinyl chloride is a volatile, moderately stable crystalline solid that is similar to tetracyanoethylene, yet shows certain marked differences in reactivity.

Like tetracyanoethylene, tricyanovinyl chloride acts as a dienophile in the Diels–Alder reaction, although as might be expected, it adds more slowly[208]. Similarly, tricyanovinyl chloride is a weaker π-acid than tetracyanoethylene.

The tricyanovinylation of amines, alcohols and aromatic nuclei is the most studied reaction of tricyanovinyl chloride. It reacts faster than tetracyanoethylene and in some cases gives different products.

Addition of **96** to a solution of ethanol is a convenient route to ethyl tricyanovinyl ether (**97**) (equation 134). Although **97** can be

$$\underset{\textbf{(96)}}{\mathrm{(NC)_2C{=}C(CN)Cl}} + \mathrm{EtOH} \longrightarrow \mathrm{HCl} + \underset{\textbf{(97)}}{\mathrm{(NC)_2C{=}C(CN)OEt}} \qquad (134)$$

prepared directly from tetracyanoethylene, the reaction must be carefully controlled in the latter case to prevent replacement of two cyano groups. Hydrolysis of **96** can be controlled to give either tricyanovinyl alcohol (**98**) or 1,1,2,3,3-pentacyanopropene (**99**). An excess of water leads to a product containing mostly the alcohol, whereas the addition of a limited amount of water to a solution of the halide produces the propene **99** (equation 135)[208]. Reaction

$$(NC)_2C{=}C(Cl)CN\ (\mathbf{96}) \xrightarrow{\text{excess } H_2O} (NC)_2C{=}C(CN)OH\ (\mathbf{98})$$

$$(NC)_2C{=}C(Cl)CN\ (\mathbf{96}) \xrightarrow{H_2O} (NC)_2C{=}C(CN)CH(CN)_2\ (\mathbf{99}) \qquad (135)$$

between **96** and **98** has been shown to give **99** by an obscure mechanism and this is presumably the route for formation of pentacyanopropene when the hydrolysis is carried out in such a fashion that **96** is in excess.

When there is a choice between alkylation of an aromatic nucleus and an NH group, tricyanovinyl chloride substitutes on the nitrogen. Tricyanovinylation of *N*-methylaniline, for example, can be carried out either on the nitrogen or at the *para* position by using tricyanovinyl chloride (equation 136) or tetracyanoethylene (equation 137).

$$C_6H_5NHMe + (NC)_2C{=}C(Cl)CN \xrightarrow{-HCl} C_6H_5{-}N(Me){-}C(CN){=}C(CN)_2 \qquad (136)$$

$$C_6H_5NHMe + (NC)_2C{=}C(CN)_2 \xrightarrow{-HCN} (NC)_2C{=}C(CN){-}C_6H_4{-}NHMe \qquad (137)$$

This selectivity has been used to prepare different types of dyes by tricyanovinylation of hydrazones[195]. The phenylhydrazone of *p*-dimethylaminobenzaldehyde gives the 4-tricyanovinyl derivative **100** (λ_{max} 580 mμ) with tetracyanoethylene and the *N*-tricyanovinyl dye **101** (λ_{max} 414 mμ) with tricyanovinyl chloride.

Tricyanovinylation with **96** has also been used to advantage with

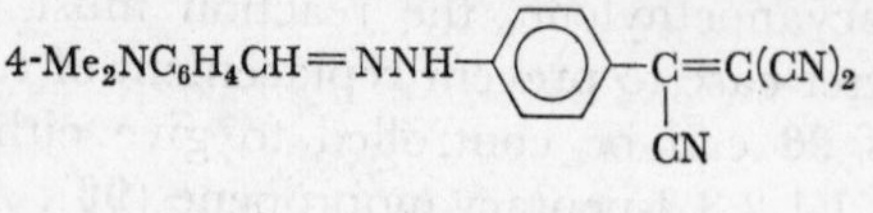

$$4\text{-}Me_2NC_6H_4CH{=}NNH\text{-}C_6H_4\text{-}C(CN){=}C(CN)_2 \qquad 4\text{-}Me_2NC_6H_4CH{=}NNPh\text{-}C(CN){=}C(CN)_2$$

(100) 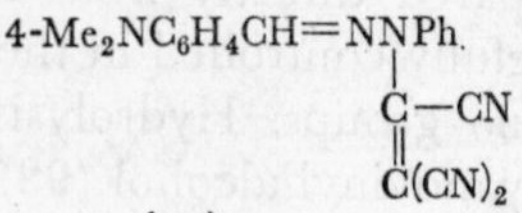 (101)

sterically hindered compounds[195]. Guaiazulene (**102**) and tetracyanoethylene form a π-complex at room temperature. The 4-methyl group hinders substitution at the nucleophilic 3-position, but under the same conditions, **96** rapidly reacts to form the tricyanovinyl derivative **103**.

Cyclopentadienylidenetriphenylphosphorane (**104**) forms a tricyanovinyl derivative (**105**) in quantitative yield with **96** in the *absence* of a catalyst, (equation 138), whereas the same reaction with tetracyanoethylene requires a large excess of triethylamine[198].

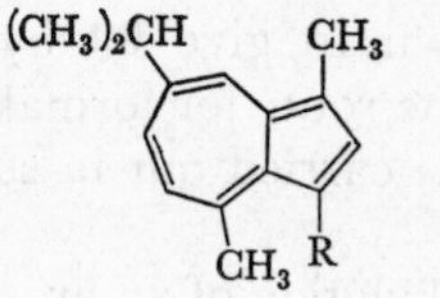

(**102**) (R = H)

(**103**) (R = $C(CN){=}C(CN)_2$)

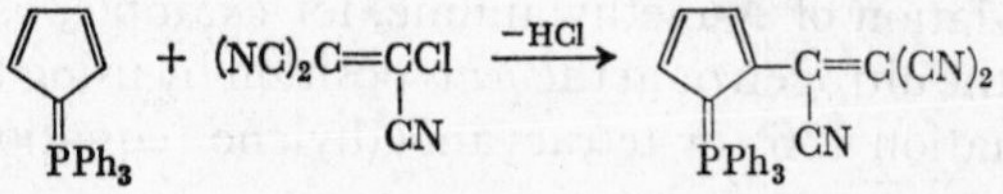

$$\textbf{(104)}\ (C_5H_4{=}PPh_3) + (NC)_2C{=}C(CN)Cl \xrightarrow{-HCl} \textbf{(105)}\ (Ph_3P{=}C_5H_3\text{-}C(CN){=}C(CN)_2) \qquad (138)$$

One of the most interesting reactions of tricyanovinyl chloride is the synthesis of 1,2,2-tricyano-1,3-butadienes from electron-rich olefins[207]. An example is the reaction with *p*-methoxystyrene at room temperature in tetrahydrofuran. Hydrogen chloride is evolved and the butadiene **106** is obtained as red crystals in 60% yield (equation 139).

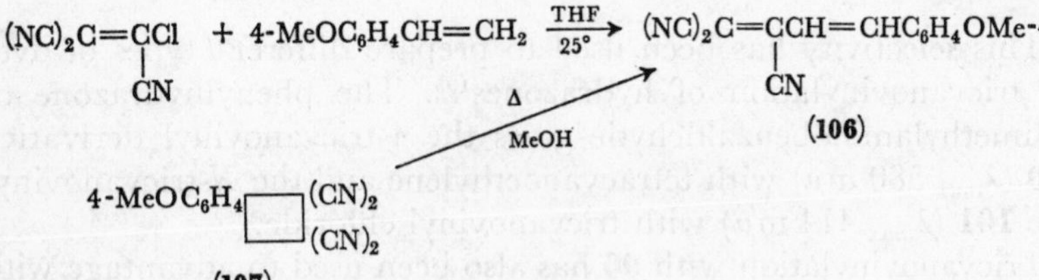

$$(NC)_2C{=}C(CN)Cl + 4\text{-}MeOC_6H_4CH{=}CH_2 \xrightarrow[25^\circ]{THF} (NC)_2C{=}C(CN)CH{=}CHC_6H_4OMe\text{-}4\ \textbf{(106)}$$

$$4\text{-}MeOC_6H_4\text{-cyclobutane-}(CN)_2(CN)_2\ \textbf{(107)} \xrightarrow[MeOH]{\Delta} \textbf{(106)}$$

Tetracyanoethylene reacts with this same styrene to give the cyclobutane **107** (section IV.A). When **107** is heated to reflux in methanol, the butadiene **106** is also formed. However, alcohol appears to be essential for the ring-cleavage reaction. Heating in other solvents, e.g. dioxan, acetonitrile or benzene, is ineffective. It is reasonable to assume that butadiene formation from tricyanovinyl chloride proceeds through intermediate formation of a cyclobutane, which readily loses hydrogen chloride and ring opens. In fact it is possible to isolate an unstable cyclobutane (**108**) from reaction with dihydropyran. Mild heating or merely solution in polar solvents at room temperature effects ring opening to give **109** (equation 140). Styrene does not give a cyclobutane with tetra-

dihydropyran + **96** ⟶ **(108)** [Cl, CN, CN, CN-substituted cyclobutane fused to tetrahydropyran] ⟶ **(109)** [dihydropyranyl–C(CN)=C(CN)$_2$] (140)

cyanoethylene, but does give the butadiene **110** in small yield when heated to reflux with **96** in acetonitrile or tetrahydrofuran (section IV.A).

$$PhCH{=}CHC(CN){=}C(CN)_2$$
(110)

Only in the case of dihydropyran has a cyclobutane been isolated from reaction with tricyanovinyl chloride. There is no evidence to support a cyclobutane intermediate in any of the other additions. The reaction with *p*-methoxystyrene was followed spectroscopically and the rate of appearance of butadiene found to approximately equal the rate of disappearance of the styrene. The relative rates in three solvents, acetonitrile, ether and cyclohexane, were found to be 200:80:1.

C. Dicyanoethylenes Containing Other Electronegative Substituents

1. Dicyanodihaloethylenes

a. 1,2-Dichloro-1,2-dicyanoethylene. Both isomers of 1,2-dichloro-1,2-dicyanoethylene form when succinonitrile is chlorinated in the presence of light[265,266] (equation 141). Small amounts of **111** and **112** have also been obtained along with other products from reaction of trichloroacetonitrile with copper powder[267]. The first fraction of

$$\mathrm{NCCH_2CH_2CN} \xrightarrow[-\mathrm{HCl}]{\mathrm{Cl_2}} \underset{\mathbf{(111)}}{\begin{matrix} \mathrm{NC} & \mathrm{Cl} \\ | & | \\ \mathrm{C} & = \mathrm{C} \\ | & | \\ \mathrm{Cl} & \mathrm{CN} \end{matrix}} + \underset{\mathbf{(112)}}{\begin{matrix} \mathrm{NC} & \mathrm{CN} \\ | & | \\ \mathrm{C} & = \mathrm{C} \\ | & | \\ \mathrm{Cl} & \mathrm{Cl} \end{matrix}} \qquad (141)$$

the distillate from the chlorination reaction readily solidifies and has been identified as the *trans* isomer **111**. Later fractions will crystallize on standing. After purification, both crops of solid have nearly the same melting point, but the latter (*cis* isomer **112**) can be distinguished on the basis of the infrared spectrum[268]. The less symmetrical *cis* isomer has a strong C=C stretching vibration at 1560 cm^{-1} that is almost completely absent in the spectrum of the *trans* isomer. As expected, **111** also has a considerably less complex infrared spectrum. The two isomers have been separated by gas chromatography and a slightly higher melting point is assigned to **111**[267].

b. 1,2-Dibromo- and 1,2-diiodo-1,2-dicyanoethylene. The *trans* isomers of these two ethylenes have been prepared by addition of the corresponding halogen to dicyanoacetylene. The *trans* configuration was assigned on the basis of spectral and x-ray examination[269].

c. 1,2-Difluoro-1,2-dicyanoethylene. Potassium fluoride in tetramethylene sulphone converts 1,2-dichloro-1,2-dicyanoethylene to a mixture of the *cis* (**113**) and *trans* (**114**) isomers of 1,2-difluoro-1,2-dicyanoethylene[270]. If the metathesis is carried out in the vapour phase at 250° over a bed of mixed fluorides, the difluoride portion of the product is reported to be only the *cis* isomer[271]. The major product of the vapour-phase reaction is the chlorofluorodicyanoethylene. The latter is reportedly a mixture of *cis* and *trans* isomers. A third method of preparation of 1,2-difluoro-1,2-dicyanoethylene, the pyrolysis of chlorofluoroacetonitrile at 700–900°, gives a mixture of the *cis* and *trans* isomers which can be separated by gas chromatography[272]. Each forms a Diels–Alder adduct with 1,3-cyclohexadiene (equations 142 and 143).

At room temperature, the fluorine atoms of **113** are readily replaced with morpholine[271]. The resulting olefin reportedly has the *cis* configuration. Hydrolysis of **113** with concentrated sulphuric acid gives the corresponding *cis* diamide, but dilute acid hydrolysis is claimed to give the *trans* diacid. This assignment is based on a rather poor comparison of very high melting points, and should be accepted with reservation pending a more thorough structural study. Heating 1,2-difluoro-1,2-dicyanoethylene with ethylene has

$$FFC{=}C(CN)CN\ (\mathbf{113}) + \text{cyclohexadiene} \longrightarrow \text{bicyclo adduct (F, F, CN, CN)} \qquad (142)$$

$$F(CN)C{=}C(F)CN\ (\mathbf{114}) + \text{cyclohexadiene} \longrightarrow \text{bicyclo adduct (CN, F, F, CN)} \qquad (143)$$

been reported to give α-fluoroacrylonitrile (equation 144)[273]. Presumably, the cyclobutane is an intermediate in this reaction.

$$FC(CN){=}CFCN + CH_2{=}CH_2 \longrightarrow \text{cyclobutane (NC, F, F, CN)} \xrightarrow{\Delta} 2\ CH_2{=}CFCN \qquad (144)$$

d. 1,1-Dichloro-2,2-dicyanoethylene. The only 1,1-dihalo-2,2-dicyanoethylene reported to date is the dichloro compound **115** prepared by the method of equation (145)[95,274]. The chlorination of

$$CH_2(CN)_2 \xrightarrow[HCOOEt]{KOEt} [OCH{=}C(CN)_2]^-K^+ \xrightarrow{PCl_5} ClCH{=}C(CN)_2 \xrightarrow{Cl_2}$$
$$Cl_2CHCCl(CN)_2 \xrightarrow{450\text{–}525°} Cl_2C{=}C(CN)_2\ (\mathbf{115}) \qquad (145)$$

dicyanoketene dimethylmercaptal also gives **115** in low yield[275].

Condensation of **115** with *N,N*-dimethylaniline occurs at room temperature (equation 146). Hydrazines add readily to **115** to give

$$C_6H_5NMe_2 + \mathbf{115} \xrightarrow{-HCl} p\text{-}Me_2NC_6H_4C(Cl){=}C(CN)_2 \qquad (146)$$

pyrazoles[210] (equation 147).

$$p\text{-}MeC_6H_4SO_2NHNH_2 + \mathbf{115} \longrightarrow \text{pyrazole (NC, Cl, } H_2N\text{, N-}SO_2C_6H_4Me\text{-}p) \qquad (147)$$

2. Dicyanobis(fluoroalkyl)ethylenes

Fluoroalkyl groups, like cyano groups, are strongly electron withdrawing but do not delocalize a negative charge to the same extent. Olefins containing a mixture of cyano and fluoroalkyl substituents behave in many reactions like tetracyanoethylene. In some instances, that require stabilization of dipolar forms in intermediates or transition states, these mixed olefins are more reactive.

The three possible isomers of dicyanobis(trifluoromethyl)ethylene have been synthesized. Condensation of hexafluoroacetone with malononitrile occurs readily in the presence of zinc chloride[276,277]. The resulting alcohol **116** is dehydrated with phosphorus pentoxide to 1,1-dicyano-2,2-bis(trifluoromethyl)ethylene (**117**) in an overall yield of 50% (equation 148). The same procedure starting with hexafluorocyclobutanone gives the ethylene **118**.

$$(CF_3)_2CO + CH_2(CN)_2 \longrightarrow \underset{\mathbf{(116)}}{(NC)_2HC{-}C(OH)(CF_3)_2} \xrightarrow{P_2O_5} \underset{\mathbf{(117)}}{(CF_3)_2C{=}C(CN)_2} \qquad (148)$$

$$\underset{\mathbf{(118)}}{\text{cyclo-}(CF_2)_3C{=}C(CN)_2}$$

Both isomers of 1,2-dicyano-1,2-bis(trifluoromethyl)ethylene are formed in the pyrolysis of the chlorosulphite of trifluoroacetaldehyde-cyanohydrin[278] (equation 149). The isomers have been separated

$$CF_3CHO + HCN \xrightarrow{C_5H_5N} CF_3CH(CN)OH \xrightarrow{SOCl_2} CF_3CH(CN)OSOCl \xrightarrow{450^\circ}$$

$$\underset{\mathbf{(119)}}{(CF_3)(NC)C{=}C(CF_3)(CN)} + \underset{\mathbf{(120)}}{(CF_3)(NC)C{=}C(CN)(CF_3)} + HCN + SO_2 \qquad (149)$$

by gas chromatography. The thermodynamically more stable *trans* isomer (**120**) is a colourless solid and may be obtained almost exclusively by treatment of the mixture with a mild base such as triethylamine. As expected, the cyanofluoroalkyl ethylenes form π-complexes with aromatic hydrocarbons. No quantitative data are available for complexes of **119** and **120**[278], but association constants have been determined for complexes of **117** and **118** with several π-bases[276]. These are somewhat weaker than the corresponding complexes of tetracyanoethylene.

Diels–Alder adducts of **119** and **120** have been reported only for 1,3-cyclohexadiene although reaction with dienes is undoubtedly general[279]. In each case only one isomer is formed. Even with the expected stereospecificity of the reaction, the *cis*-olefin **119** would be expected to give two products, **121** and **122**. Formation of one isomer may be an indication of the large steric requirements of the two trifluoromethyl groups, although this is not readily apparent from molecular models. The Alder rule favours **122** as the structure of the adduct. The unsymmetrical ethylenes **117** and **118** are both

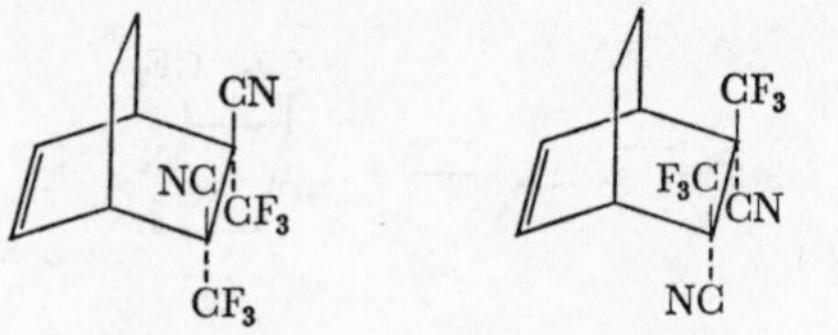

CN, NC, CF_3, CF_3 **(121)** CF_3, F_3C, CN, NC **(122)**

good dienophiles. In reactions of this type, **118** is more reactive than **117** and is about equivalent to tetracyanoethylene. This must be due in part to the relief in strain associated with changing the hybridization of the four-membered ring carbon from sp^2 to sp^3.

However, in the formation of the adduct **123** with bicycloheptadiene (equation 150), 1,1-dicyano-2,2-bis(trifluoromethyl)ethylene

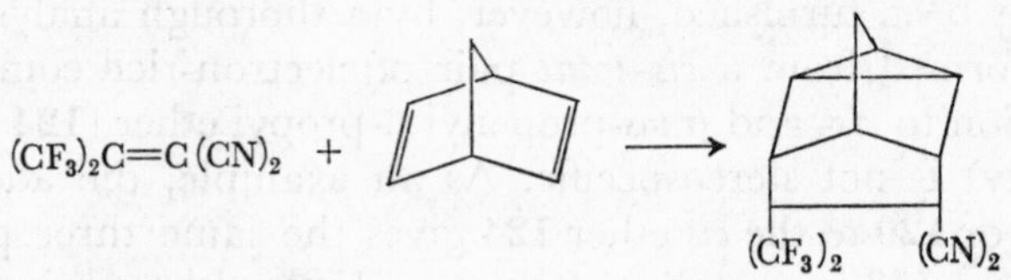

$$(CF_3)_2C{=}C(CN)_2 + \text{bicycloheptadiene} \longrightarrow (CF_3)_2 \ (CN)_2 \quad (150)$$

(117) **(123)**

reacts faster than does tetracyanoethylene. This has been attributed to the ready polarizability of **117**, which can better stabilize a negative charge on the carbon bearing the two cyano groups[276]. The argument assumes that the bicycloheptadiene addition involves more ionic character than does an ordinary Diels–Alder reaction.

The explanation is consistent with observations made in the additions of **117** to electron-rich olefins to form cyclobutanes (see below).

Cycloaddition of tetracyanoethylene to electron-rich olefins has been discussed (section IV.A). The mixed cyanofluoroalkylethylenes also undergo this reaction to form cyclobutanes in good yields. Use of the *cis* and *trans* isomers of 1,2-dicyano-1,2-bis(trifluoromethyl)-ethylene has made possible a detailed study of the stereochemistry of the reaction and allows certain conclusions concerning the mechanism of the reaction[279].

In general, the addition of **119** and **120** to ethylenes bearing one electron-donating substituent is slower than the corresponding reaction with tetracyanoethylene.

In each case two isomers are isolated, but all four cyclobutanes are different (equations 151 and 152). On the basis of these experiments one is tempted to generalize that the cycloaddition reaction

$$(CF_3)(NC)C{=}C(CF_3)(CN)\ (\mathbf{119}) + CH_2{=}CHX \longrightarrow \text{cyclobutane}\ (CF_3, CF_3, NC, NC, X) + \text{cyclobutane}\ (CF_3, CF_3, NC, NC, X) \tag{151}$$

$(X = EtO,\ t\text{-BuS},\ p\text{-}MeOC_6H_4)$

$$(CF_3)(NC)C{=}C(CN)(CF_3)\ (\mathbf{120}) + CH_2{=}CHX \longrightarrow \text{cyclobutane}\ (CF_3, CN, NC, F_3C, X) + \text{cyclobutane}\ (CF_3, CN, NC, F_3C, X) \tag{152}$$

is stereospecific with respect to the cyano olefin. A more revealing picture has been furnished, however, by a thorough analysis of the products formed from a *cis–trans* pair of electron-rich components. The addition to *cis*- and *trans*-propenyl n-propyl ether (**124** and **125**, respectively) is not stereospecific. As an example, the addition of either **119** or **120** to the *cis* ether **124** gives the same three products, **126**, **127** and **128** in each case (equation 153), although in different ratios. In this case, some of the products arise through loss of the stereochemistry of the cyanofluoroalkyl olefin. An analysis of the products from all the possible combinations showed that the stereochemistry is lost only in those cases involving the *cis* ether. It follows logically that the isomerization is due to steric crowding in the transition states and/or intermediates. In order for the rotation to

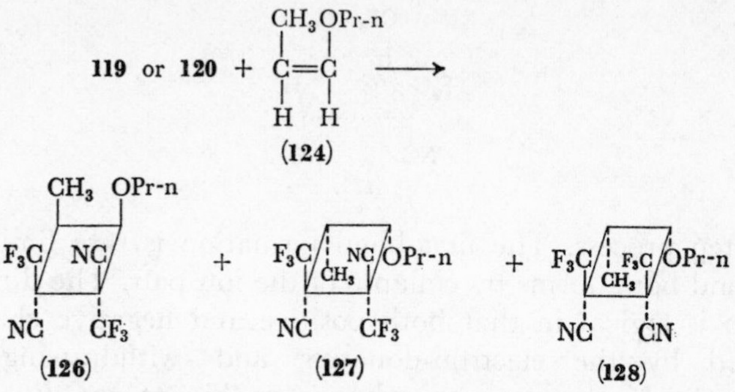

$$\mathbf{119} \text{ or } \mathbf{120} + \underset{\mathbf{(124)}}{\text{CH}_3(\text{H})\text{C}{=}\text{C}(\text{H})\text{OPr-n}} \longrightarrow \quad (153)$$

$$\mathbf{(126)} + \mathbf{(127)} + \mathbf{(128)}$$

occur, an intermediate must have a relatively long lifetime. Furthermore, solvent effects on the rate of reaction are large and suggest dipolar intermediates of the type **129** and **130** from the reaction of **120** and the *cis* ether. Both intermediates would be expected in the

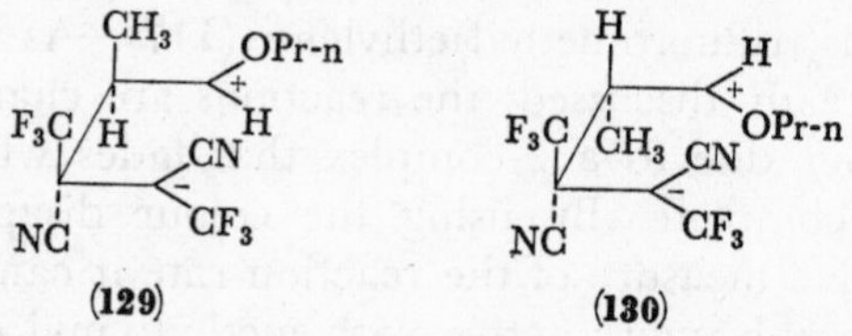

$$\mathbf{(129)} \qquad \mathbf{(130)}$$

reaction, but the rate of formation of **130** should be greater than the rate of formation of **129** where crowding of the methyl and trifluoromethyl groups in the transition state is severe. However, the major product of the reaction has *cis*-oriented trifluoromethyl groups. In **130** crowding in the dipolar part of the ion is severe, and rotation to relieve this strain apparently occurs. An alternate explanation, involving a reversal of the ions to starting materials after rotation, was ruled out by continuous monitoring of the reaction by n.m.r. It can be argued rather convincingly, on the basis of the products formed, that rotation occurs only around the bond at the carbanion portion of the intermediate. This same type of analysis can be applied to all of the possible *cis–trans* reaction pairs with a thoroughly consistent product analysis.

In addition to the large effect on the rate, the choice of solvent can also control, to some extent, the stereochemistry of the reaction. This too is compatible with an ionic intermediate wherein solvation would be expected to affect product formation.

The most logical explanation of this cycloaddition reaction is one involving an intermediate such as **131** in which bonding occurs as

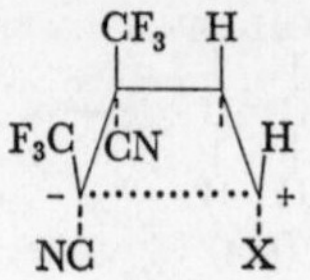

$$\begin{array}{c} CF_3 \quad H \\ F_3C\ /CN \quad \backslash H \\ -\cdots\cdots\cdots + \\ NC \qquad X \end{array}$$

(131)

a two-step process. The first bond formation is rate limiting and the second bond forms by collapse of the ion pair. The direction of addition is logical in that both positive and negative charges are stabilized by the electron-donating and -withdrawing groups, respectively. Unless steric crowding is significant, the dipole would be expected to collapse without a change of stereochemistry.

These observations cannot necessarily be extended to reactions of tetracyanoethylene. However, it is to be expected that these structurally similar molecules would react by an analogous mechanism.

Cyclobutane formation is also a characteristic reaction of 1,1-dicyano-2,2-bis(trifluoromethyl)ethylene (**117**). As in the case of the olefins already discussed, the reactions are characterized by a transient colour due to a π-complex that fades when the adduct formation is complete. By using the colour disappearance as a semiquantitative measure of the reaction rate it can be shown that **117** is considerably more active with methyl vinyl ether (equation 154) than is tetracyanoethylene. Furthermore, **117** will cycloadd to

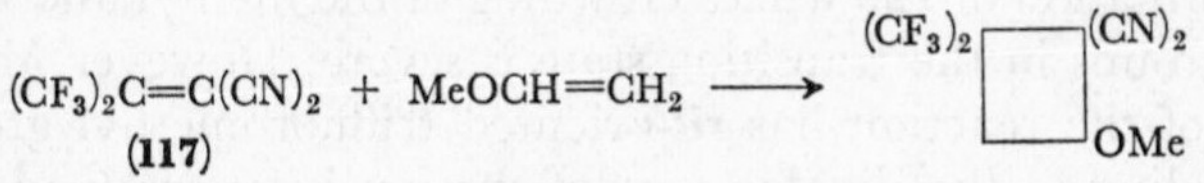

$$\underset{\textbf{(117)}}{(CF_3)_2C{=}C(CN)_2} + MeOCH{=}CH_2 \longrightarrow \text{cyclobutane: } (CF_3)_2,\ (CN)_2,\ OMe \qquad (154)$$

styrene at room temperature, whereas tetracyanoethylene and styrene do not react under these conditions. Stable cyclobutanes (**133**), have also been reported from reaction of **117** with ketene *S,N*-acetals (**132**) (equation 155)[280]. Tricarbonylcyclooctatetraene

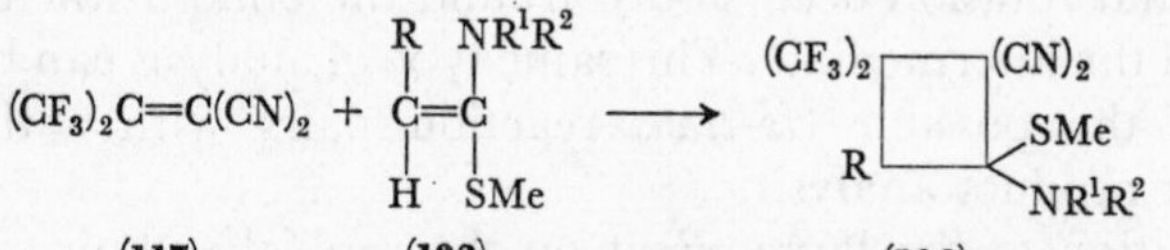

$$\underset{\textbf{(117)}}{(CF_3)_2C{=}C(CN)_2} + \underset{\textbf{(132)}}{R(H)C{=}C(NR^1R^2)(SMe)} \longrightarrow \underset{\textbf{(133)}}{\text{cyclobutane: } (CF_3)_2,\ (CN)_2,\ R,\ SMe,\ NR^1R^2} \qquad (155)$$

complexes of ruthenium and iron add to **117** to give cyclobutanes[281].

The greater reactivity of **117** compared with tetracyanoethylene can again be attributed to the polarizability as discussed above. It also supports the polar intermediate in the cyclobutane reaction.

Another example of the exceptional reactivity of **117** is the addition to simple olefins containing an allylic hydrogen atom (equation 156). Two products **134** and **135** are obtained from propylene. This

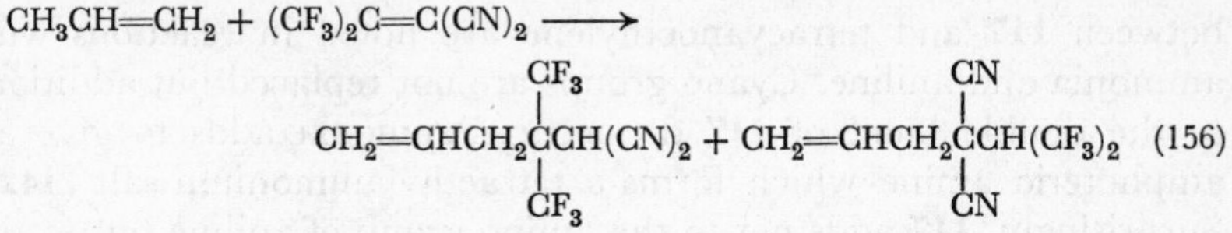

$$CH_3CH{=}CH_2 + (CF_3)_2C{=}C(CN)_2 \longrightarrow CH_2{=}CHCH_2C(CF_3)_2CH(CN)_2 + CH_2{=}CHCH_2C(CN)_2CH(CF_3)_2 \quad (156)$$

(134) **(135)**

'ene' reaction[282], which occurs in good yield at 150°, is reminiscent of the reaction of carbonyl cyanide with olefins (section X). A cyclic transition state (**136**) analogous to that postulated for the carbonyl cyanide reaction is logical. To explain the minor product **135** the polarization of **117** must be reversed. This is not expected but was rationalized with the assumption that **136** is sterically more crowded than the alternate **137**. Support for this argument is

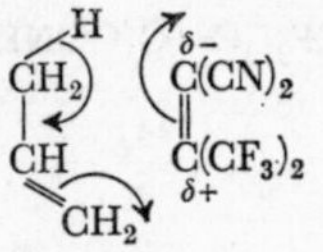

H, CH_2, CH, CH_2; $\delta-$ $C(CN)_2$ = $C(CF_3)_2$ $\delta+$

(136)

H, CH_2, CH, CH_2; $\delta-$ $C(CF_3)_2$ = $C(CN)_2$ $\delta+$

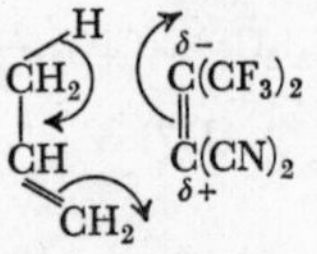

(137)

found in the reaction with 2,3-dimethyl-2-butene which gives only the olefin **138**. In this case a preference for the transition state **139**

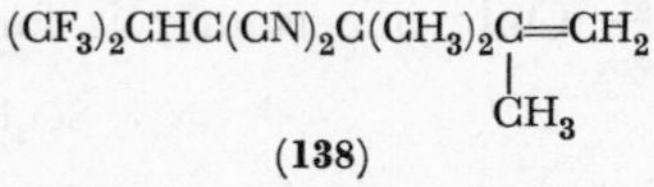

$$(CF_3)_2CHC(CN)_2C(CH_3)_2C(CH_3){=}CH_2$$

(138)

(versus **140**) is more readily apparent.

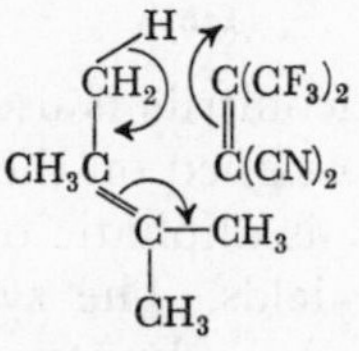

H, CH_2, CH_3C, $C{-}CH_3$, CH_3; $C(CF_3)_2$ = $C(CN)_2$

(139)

H, CH_2, CH_3C, $C{-}CH_3$, CH_3; $C(CN)_2$ = $C(CF_3)_2$

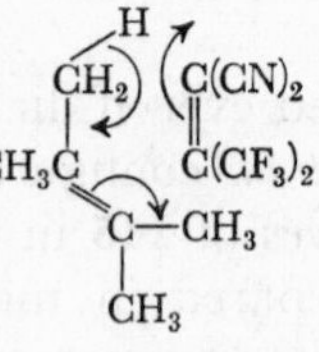

(140)

In contrast to the reaction between tetracyanoethylene and sodium cyanide (section III.B.2), anion-radical formation with 1,1-dicyano-2,2-bis(trifluoromethyl)ethylene evidently does not occur, but sodium cyanide adds to give **141** in good yield. Other major differences

$$NCC(CF_3)_2C(CN)_2Na$$
(141)

between **117** and tetracyanoethylene are noted in reactions with ammonia and aniline. Cyano groups are not replaced but additions to the double bond of **117** do occur. Ammonia adds to give an amphoteric amine which forms a tetraethylammonium salt (**142**). Surprisingly, **117** adds not to the amino group of aniline but to the *para* position to give **143.** To date this is the only reported example of this interesting reaction.

$$H_2NC(CF_3)_2C(CN)_2^-Et_4N^+ \qquad p\text{-}H_2NC_6H_4C(CF_3)_2CH(CN)_2$$
(142) **(143)**

Both water and alcohols add reversibly to the double bond of **117,** but in strong acid the two cyano groups are hydrolysed to give the corresponding diamide **144** (equation 157).

$$\underset{(\mathbf{117})}{(CF_3)_2C{=}C(CN)_2} \xrightarrow[H_2O]{H^+} \underset{(\mathbf{144})}{(CF_3)_2C{=}C(CONH_2)_2} \qquad (157)$$

3. 1,2-Dicyano-1,2-disulphonylethylenes

Many of the reactions of tetracyanoethylene are shown also by 1,2-dicyano-1,2-disulphonylethylenes (**145**). These negatively substituted ethylenes are prepared from dichlorofumaronitrile and a sulphinic acid salt[121]. Only one isomer is formed, and the same

$$\underset{NC\quad Cl}{\overset{Cl\quad CN}{C{=}C}} + 2\,NaSO_2R \longrightarrow \underset{\underset{(\mathbf{145})}{RO_2S\quad CN}}{\overset{NC\quad SO_2R}{C{=}C}} + 2\,NaCl \qquad (158)$$

isomer is formed even if dichloromaleonitrile is used as the starting material. The *trans* configuration is assigned on the basis of spectral data. Derivatives of **145** in which R is aliphatic or aromatic (**146**) have been prepared in moderate yields. The syntheses are best carried out at low temperatures in polar solvents.

Both Diels–Alder reactions and cycloaddition to electron-rich olefins occur readily as shown in equations (159) and (160). The stereochemistry of **147** and **148** was not defined, but the reactions are presumably stereospecific.

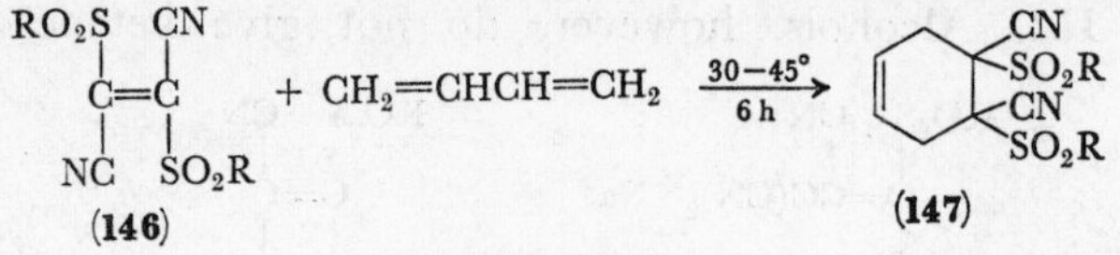

(159)

($R = p\text{-}MeC_6H_4$)

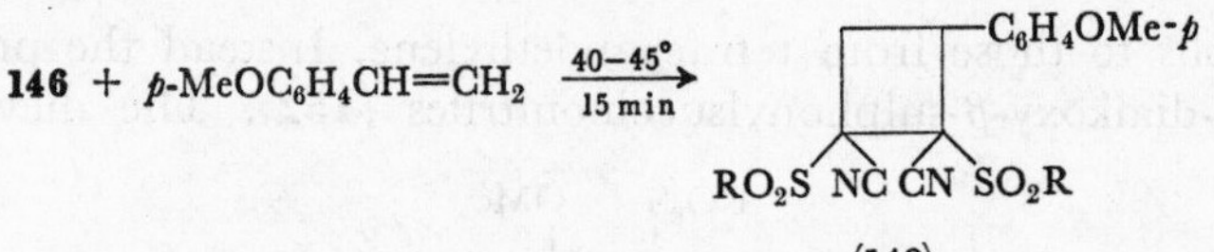

(160)

(148)

The reactions of **145** with amines are analogous to those of tetracyanoethylene except that the leaving group is sulphonyl rather than cyano. Primary aromatic amines give *N*-substituted derivatives, whereas secondary and tertiary amines are substituted in the 4-position. It is not necessary to start with the preformed dicyanodisulphonylethylene. The synthesis works as well using the amine, dichlorofumaronitrile and a sulphinate salt (equation 161). Based

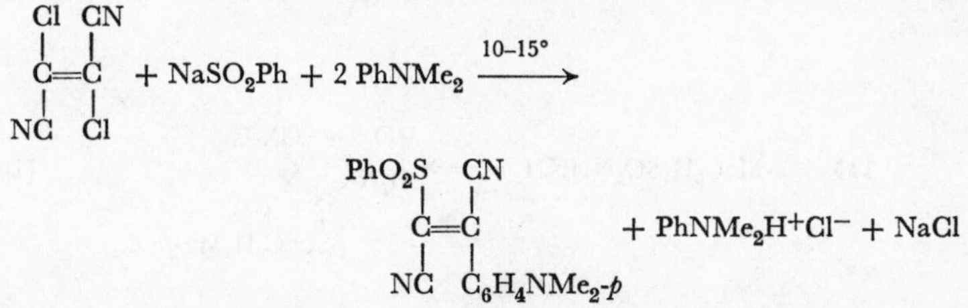

(161)

on the stoichiometry of the reaction, the actual intermediate is probably the sulphonylchloroethylene **149**. The dichloro compound

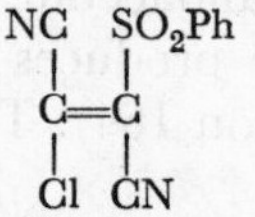

(149)

alone does not react with dimethylaniline other than to form a complex. A large number of dyes have been made using this technique[121]. They have properties similar to the tricyanovinyl dyes from tetracyanoethylene.

Dicyanodisulphonylethylenes will alkylate the sodium salt of malononitrile to form the tetracyanopropene anion **150**. Another reaction similar to that of tetracyanoethylene is hydrolysis to the alcoholate **151**. Alcohols, however, do not give ketene acetals

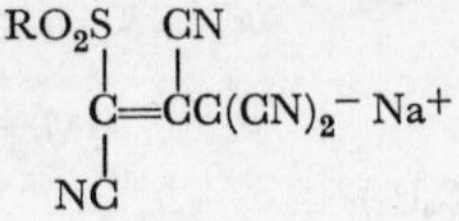

$$\begin{array}{c} RO_2S \quad CN \\ | \quad\;\; | \\ C{=}CC(CN)_2^- \; Na^+ \\ | \\ NC \end{array} \qquad \begin{array}{c} RO_2S \quad CN \\ | \quad\;\; | \\ C{=}C \\ | \quad\;\; | \\ NC \quad O^- \end{array}$$

(150) 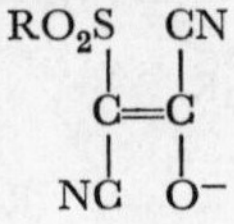 **(151)**

analogous to those from tetracyanoethylene. Instead the products are α,α-dialkoxy-β-sulphonylsuccinonitriles (**152**). The dicyanodi-

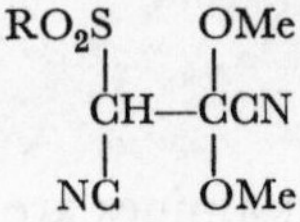

$$\begin{array}{c} RO_2S \quad OMe \\ | \qquad\; | \\ CH{-}CCN \\ | \qquad\; | \\ NC \quad OMe \end{array}$$

(152)

sulphonylethylenes are intermediates for heterocyclic syntheses as exemplified by the reactions with hydroxylamine and hydrazine derivatives (equations 162 and 163).

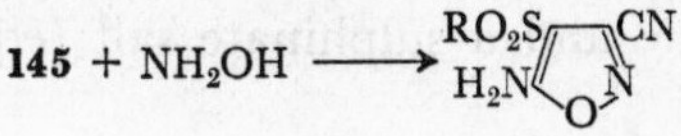

$$\mathbf{145} + NH_2OH \longrightarrow \text{(isoxazole: } RO_2S, CN, H_2N \text{ substituted)} \tag{162}$$

$$\mathbf{145} + p\text{-}MeC_6H_4SO_2NHNH_2 \longrightarrow \text{(pyrazole: } RO_2S, CN, H_2N, N\text{-}SO_2C_6H_4Me\text{-}p) \tag{163}$$

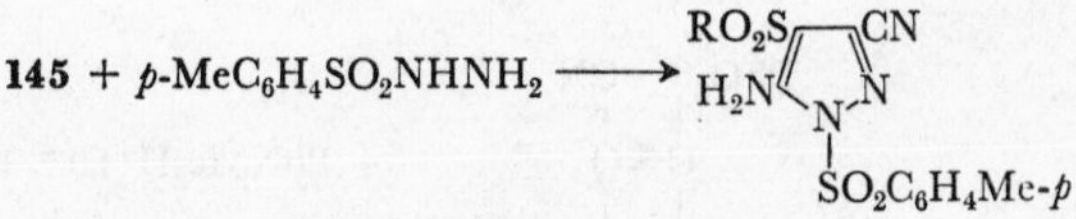

4. Diethyl 1,2-dicyanoethylene-1,2-dicarboxylate

The oxidation of ethyl cyanoacetate with sulphur monochloride[283] or with selenium dioxide[284] produces diethyl *trans*-1,2-dicyano-1,2-dicarboxylate (**153**) (equation 164). The *trans* structure was proved

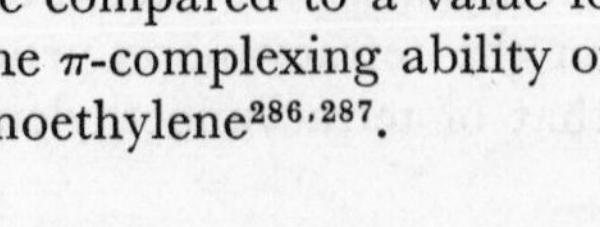

$$\begin{array}{c} CN \\ | \\ CH_2 \\ | \\ CO_2Et \end{array} \xrightarrow{S_2Cl_2} \begin{array}{c} NC \quad CO_2Et \\ | \qquad | \\ C{=}C \\ | \qquad | \\ EtO_2C \quad CN \end{array} \tag{164}$$

(153)

by infrared analysis[285] and chemical studies[286]. The electron affinity of **153** is 51 kcal/mole compared to a value for tetracyanoethylene of 65 kcal/mole[187]. The π-complexing ability of **153** is much weaker than that of tetracyanoethylene[286,287].

Diethyl 1,2-dicyanoethylene-1,2-dicarboxylate reacts with dimethylaniline[286] (equation 165) more slowly than does tetracyanoethylene. The reaction course with primary aromatic amines

$$\mathrm{PhNMe_2} + \underset{\mathrm{NC}\quad \mathrm{CO_2Et}}{\overset{\mathrm{EtO_2C}\quad \mathrm{CN}}{\mathrm{C{=}C}}} \longrightarrow \mathrm{Me_2N{-}C_6H_4{-}}\underset{\mathrm{NC}\quad \mathrm{CO_2Et}}{\overset{\mathrm{EtO_2C}\quad \mathrm{CN}}{\mathrm{C{-}CH}}} \longrightarrow \underset{p\text{-}\mathrm{Me_2NC_6H_4}\quad \mathrm{CO_2Et}}{\overset{\mathrm{EtO_2C}\quad \mathrm{CN}}{\mathrm{C{=}C}}} \qquad (165)$$

depends on reactant ratios[288]. Use of a molar excess of the olefin results in products in which ethyl formate has been eliminated (equation 166a), whereas a cyano group is lost if the amine is in excess (equation 166b). Although this may have mechanistic

$$\mathrm{ArNH_2} + \mathbf{153}\ \text{(excess)} \longrightarrow \mathrm{ArNH}\overset{\mathrm{CN}}{\mathrm{C}}{=}\underset{\mathrm{CN}}{\overset{\mathrm{CO_2Et}}{\mathrm{C}}} \qquad (166a)$$

$$\mathrm{ArNH_2}\ \text{(excess)} + \mathbf{153} \longrightarrow \mathrm{ArNH}\underset{\mathrm{EtO_2C}}{\mathrm{C}}{=}\underset{\mathrm{CN}}{\overset{\mathrm{CO_2Et}}{\mathrm{C}}} \qquad (166b)$$

significance, a more thorough investigation would be in order because of the low yields in most of the processes.

In order to effect the Diels–Alder addition of **153** to butadiene (equation 167), it is necessary to heat the reactants for a prolonged period in an autoclave at 180–190° [289].

$$\mathrm{CH_2{=}CHCH{=}CH_2} + \mathbf{153} \longrightarrow \text{cyclohexene-4,5-(CN)(CO}_2\text{Et)}_2\text{(CN)} \qquad (167)$$

D. Dicyanoethylenes Containing Electron-Donating Substituents

Our discussion of polycyano compounds has included olefins of the type **154** where X and/or Y are electronegative substituents. In this case, the properties of **154** parallel, to a certain extent, those of tetracyanoethylene. However, X and Y may be electron donating,

```
X  CN
|  |
C==C
|  |
Y  CN
```
(154)

leading to olefins stabilized by resonance contributing forms **154a**, **154b**, etc. Representative of this class of olefin dinitriles are the

```
                       N⁻
                       ‖
 NC  X+                C  X+
 |   ‖                 ‖  ‖
⁻C---C      <---->     C--C
 |   |                 |  |
 NC  Y                NC  Y
 (154a)               (154b)
```

dicyanoketene acetals derived from tetracyanoethylene. The corresponding thio compounds have not been prepared by a similar synthesis, but are available from malononitrile and carbon disulphide. Isomeric with the thioacetals are the derivatives of dimercaptomaleo- and fumaronitrile.

I. Dicyanoketene acetals

Synthesis of dicyanoketene acetals (**155**) has been discussed under the chemistry of tetracyanoethylene.

```
NC  OR
|   |
C===C
|   |
NC  OR
```
(155)

Although rapid addition of water or alcohols is characteristic of ordinary ketene acetals, the cyano-substituted acetals (**155**) are inert toward these reagents even in the presence of acid catalysts. In fact, the dicyanoketene acetals are surprisingly stable compounds especially when the acetal is formed from a glycol as in **156** (equation 168)[290].

$$(NC)_2C{=}C(CN)_2 + HOCH_2CH_2OH \longrightarrow (NC)_2C{=}C\langle O{-}CH_2CH_2{-}O\rangle \quad (168)$$

(156)

One of the acetal links is cleaved rapidly by anhydrous halogen halides. Simultaneous addition of a second mole of hydrogen halide occurs to one of the nitrile groups[89]. Two tautomeric forms, **157** and **158**, of the product are possible, with the latter evidently

$$\underset{\textbf{(157)}}{\mathrm{X{-}C({=}NH){-}C(CN){=}C(OR)(OH)}} \qquad \underset{\textbf{(158)}}{\mathrm{H_2N{-}C(X){=}C(CN){-}CO_2R}}$$

favoured on the basis of infrared spectral evidence[816]. The cyclic acetals also react rapidly in a similar way (equation 169). Attack

$$(NC)_2C{=}C\langle\mathrm{OCH_2CH_2O}\rangle + 2\ \mathrm{HCl} \longrightarrow \mathrm{H_2N{-}C(Cl){=}C(CN){-}C({=}O){-}OCH_2CH_2Cl} \qquad (169)$$

by nucleophilic reagents is especially facile. Sodium alkoxides add in the expected direction to give salts of 1,1-dicyano-2,2,2-trialkoxyethanes[209,291] (equation 170). With primary and secondary amines,

$$(NC)_2C{=}C(OR)_2 + NaOMe \longrightarrow (RO)_2MeOCC(CN)_2^{-}Na^{+} \qquad (170)$$

replacement of one of the alkoxy groups takes place under very mild conditions[89,292]. The critical variable is the ratio of reactants used. The stepwise nature of the reaction makes possible the preparation of diaminodicyanoethylenes with mixed amino groups (equation 171). These diaminodicyanoethylenes are stable, crystalline, neutral solids that bear little relationship to vinyl amines.

$$(NC)_2C{=}C(OEt)_2 \xrightarrow{CH_3NH_2} \underset{96\%}{(NC)_2C{=}C(OEt)(NHCH_3)} \xrightarrow{NH_3} \underset{84\%}{(NC)_2C{=}C(NH_2)(NHCH_3)} \qquad (171)$$

Reaction with tertiary amines also occurs readily, but well-defined products have been isolated only from the dicyanoketene cyclic acetals. An example is the zwitterion **159**, from triethylamine

$$\underset{\textbf{(159)}}{(NC)_2C{=}C(O^{-})(OCH_2CH_2\overset{+}{N}Et_3) \longleftrightarrow (NC)_2\overset{-}{C}{-}C({=}O){-}OCH_2CH_2\overset{+}{N}Et_3}$$

and dicyanoketene ethylene acetal, formed in nearly quantitative yield on mixing the two reagents in tetrahydrofuran. This inner salt formation is not limited to tertiary amines, but also takes place with dialkyl sulphides and certain thiocarbonyl compounds such as thiourea, thio amides and thiosemicarbazide[89,293].

Reaction of dicyanoketene acetals with compounds containing amino groups has been used to prepare heterocycles with hydrazine, hydroxylamine and amidines. Replacement of one alkoxy group and cyclization to a nitrile group gives a pyrazole (**160**) with hydrazine[294] (equation 172). The reaction occurs simply on mixing the two

$$(NC)_2C{=}C(OR)_2 + H_2NNH_2 \longrightarrow \text{H}_2\text{N-pyrazole(NC)(OR)-NH} \ (\mathbf{160}) + ROH \qquad (172)$$

reagents in water. A similar synthesis with hydroxylamine gives the isoxazole **161**. If an amidine is mixed with the acetals, a pyrimidine (**162**) results.

(**161**)

$$(NC)_2C{=}C(OR)_2 + R^1C({=}NH){-}NH_2 \longrightarrow \text{pyrimidine}(OR)(NC)(H_2N)(R^1) \ (\mathbf{162}) \qquad (173)$$

2. Dicyanoketene thioacetals

The first synthesis of a dicyanoketene thioacetal (**164**) was by condensation of carbon disulphide with an alkali metal salt of malononitrile, followed by alkylation[295] (equation 174). A number

$$CH_2(CN)_2 \xrightarrow[CS_2]{Na(NH_3)} \left[(NC)_2C{=}C\begin{matrix}S\\S\end{matrix}\right]^{2-} 2\,Na^+ \ (\mathbf{163}) \xrightarrow{CH_3I} (NC)_2C{=}C(SMe)_2 \ (\mathbf{164}) \qquad (174)$$

of investigators have made use of the easily prepared sodium salt **163**, as a synthetic intermediate, mainly for the preparation of a variety of heterocycles.

It is not necessary to preform the alkali metal salt of malononitrile; the reaction proceeds equally well when carbon disulphide is added to a solution of malononitrile and sodium or potassium hydroxide in alcohol. It is possible to use triethylamine as the base and isolate a crystalline triethylammonium salt.

It should be pointed out that the condensation with carbon disulphide is not limited to malononitrile. A condensation of the type discussed above has been demonstrated with a wide variety of active methylene compounds including dimethyl malonate, ethyl cyanoacetate, cyanoacetamide and benzyl cyanide. It is likely that many other extensions are possible. Although this discussion is limited to derivatives of dimercaptomethylenemalononitrile, it should be borne in mind that many of these syntheses can be extended to the related derivatives.

The disodium (or other alkali metal) salt can be alkylated to give a ketene thioacetal, although only a few of these have been made. In addition to the dimethyl and diethyl derivatives the cyclic thioacetals **165**[296] and **166**[297] are synthesized by alkylation with the corresponding alkyl halides.

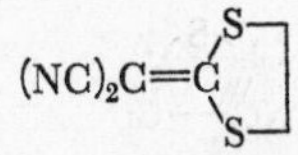

$(NC)_2C{=}C(S{-}CH_2CH_2CH_2{-}S)$ **(165)** $(NC)_2C{=}C(S{-}CH_2CH_2{-}S)$ **(166)**

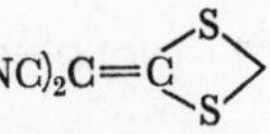

Like the corresponding oxygen analogues, the dicyanoketene thioacetals undergo reaction with amines with elimination of the thioalkyl groups, although slightly more severe conditions are necessary to introduce the second amino group[298] (equations 175 and 176).

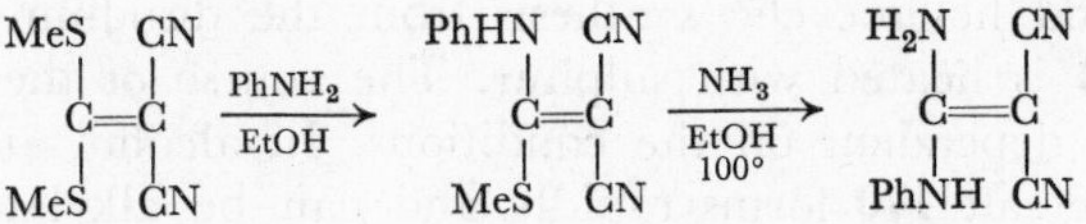

$$(MeS)_2C{=}C(CN)_2 \xrightarrow[EtOH]{PhNH_2} (PhHN)(MeS)C{=}C(CN)_2 \xrightarrow[EtOH,\ 100°]{NH_3} (H_2N)(PhNH)C{=}C(CN)_2 \quad (175)$$

$$(MeS)_2C{=}C(CN)_2 + H_2NCH_2CH_2NH_2 \longrightarrow \text{(imidazolidin-2-ylidene)}{=}C(CN)_2 \quad (176)$$

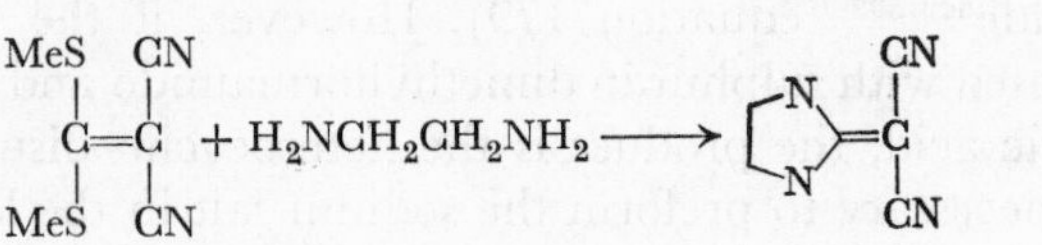

Reaction with hydrazine gives the substituted pyrazole (**167**) as observed with dicyanoketene dimethylacetal[299]. The condensation of the thioketal with an amidine gives a pyrimidine[294].

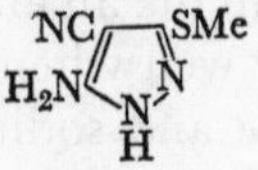

NC, SMe, H_2N, N, N, H

(167)

A thiophene synthesis via the thioacetal **168** has been worked out (equation 177)[300].

$$(NC)_2C{=}C(SNa)_2 + 2\,BrCH_2CO_2Et \longrightarrow$$

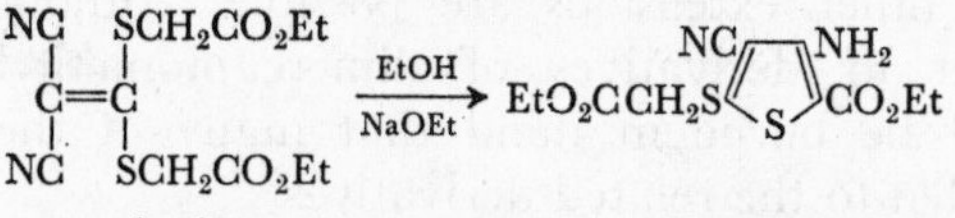

$$\underset{\textbf{(168)}}{(NC)_2C{=}C(SCH_2CO_2Et)_2} \xrightarrow[\text{NaOEt}]{\text{EtOH}} EtO_2CCH_2S\text{-thiophene}(NC)(NH_2)(CO_2Et) \qquad (177)$$

Some interesting heterocyclic syntheses have been developed starting from the disodium salt **163**. Chlorination in carbon tetrachloride at reflux gives 4-cyano-3,5-dichloroisothiazole (**169**) (equation 178)[301]. A sulphenyl chloride may be an intermediate in this reaction

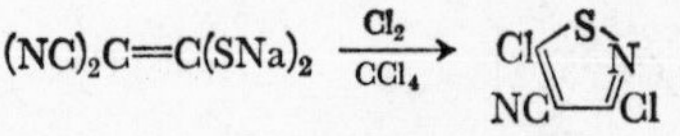

$$(NC)_2C{=}C(SNa)_2 \xrightarrow[CCl_4]{Cl_2} \underset{\textbf{(169)}}{\text{Cl, S, N, NC, Cl}} \qquad (178)$$

but this has not been demonstrated. In contrast to this reaction, chlorination of the dimethylthioacetal in acetic acid gives 1,1-dichloro-2,2-dicyanoethylene in low yield[275]. The major product of this latter reaction has not been identified, but may very well be a heterocycle.

A second heterocyclic synthesis from the disodium salt results when **163** is heated with sulphur. The course of the reaction is markedly dependent on the conditions. In alcohol at reflux, the isothiazole salt **170** forms rapidly and can be alkylated with an alkyl halide[302,303] (equation 179). However, if the disodium salt **163** is treated with sulphur in dimethylformamide and then acidified with acetic acid, the product is the heterocyclic disulphide **171**[304]. It is not necessary to preform the sodium salt in the latter reaction. Malononitrile, carbon disulphide and sulphur treated with a

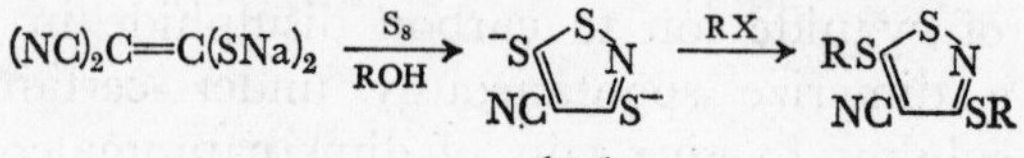

$$(NC)_2C{=}C(SNa)_2 \xrightarrow[ROH]{S_8} \text{(170)} \xrightarrow{RX} \text{RS, SR isothiazole} \quad (179)$$

(170)

catalytic amount of diethylamine in dimethylformamide also give **171** in high yield (equation 180). Treatment of **171** with 1 N sodium hydroxide results in reversal of the reaction with formation of **163**,

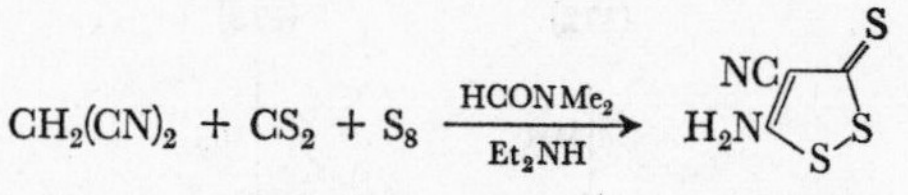

$$CH_2(CN)_2 + CS_2 + S_8 \xrightarrow[Et_2NH]{HCONMe_2} \text{(171)} \quad (180)$$

(171)

whereas 2 N potassium hydroxide causes isomerization to the isothiazole (equation 181). Rational mechanisms can be written for each of the two ring-closure reactions, but these do not adequately

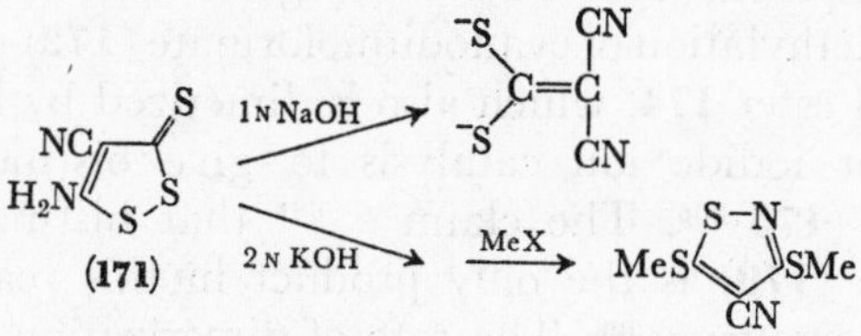

$$\text{(171)} \xrightarrow{1\,N\ NaOH} (^-S)_2C{=}C(CN)_2; \quad \text{(171)} \xrightarrow{2\,N\ KOH} \xrightarrow{MeX} \text{MeS, SMe, CN isothiazole} \quad (181)$$

explain the difference in the two solvent systems. Nucleophilic attack on sulphur is sufficiently documented and this is most likely the initial step in the formation of **171** (equation 182). Ring closure

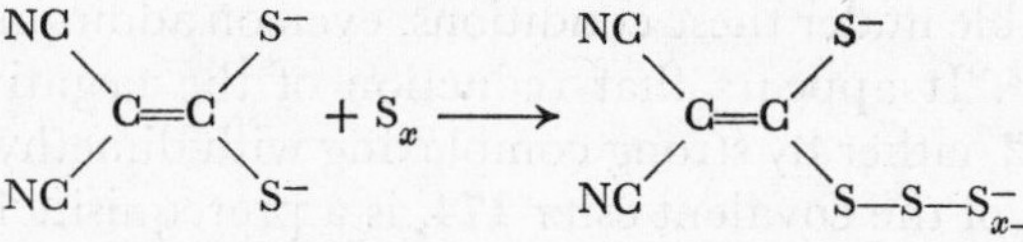

$$(NC)_2C{=}C(S^-)_2 + S_x \longrightarrow (NC)_2C{=}C(S^-)(S{-}S{-}S_{x-1}^-) \quad (182)$$

by attack of the disulphide on an adjacent nitrile with elimination of sulphur leads to **171**. It is possible that the isothiazole arises through the anion **170** as an intermediate, but the experiments to define the mechanism further have not been carried out.

3. Dimercaptomaleonitrile and dimercaptofumaronitrile

A wide field of sulphur-containing cyanocarbons was opened up by the observation that cyanodithioformates (**172**), readily obtained by addition of cyanide ion to carbon disulphide in dimethylformamide[305,306], dimerize spontaneously under certain conditions with loss of sulphur to give salts of dimercaptomaleonitrile (**173**; equation 183)[307–309]. Reaction of carbon disulphide with sodium

$$
\begin{array}{ccccc}
S{=}C{=}S + CN^- & \longrightarrow & (NC)({}^-S)C{=}S & \longrightarrow & (NC)({}^-S)C{=}C(S^-)(CN) + S \\
 & & \mathbf{(172)} & & \mathbf{(173)} \\
 & & \downarrow \text{MeI} & & \downarrow \text{MeI} \\
 & & (NC)(MeS)C{=}S & \xrightarrow{-S} & (NC)(MeS)C{=}C(SMe)(CN) \\
 & & \mathbf{(174)} & & \mathbf{(175)}
\end{array}
\qquad (183)
$$

cyanide in aqueous acetone at 50° gives the disodium salt **173** directly[310]. Methylation of cyanodithioformate (**172**) produces the unstable methyl ester **174**, which also is dimerized by heat, ultraviolet irradiation or iodide ion catalysis to give bis(methylmercapto)-maleonitrile (**175**)[306]. The claim[307–309] that bis(methylmercapto)-fumaronitrile (**179**) is the only product in this reaction has been shown to be erroneous[306]. The rate of dimerization of the salts **172** depends strongly on the counter ion. The sodium salt, which is obtained as a solvate with three molecules of dimethylformamide, dimerizes spontaneously in water or aprotic solvents, or on heating to 130° under vacuum, whereas the unsolvated tetraethylammonium salt **172** is stable under these conditions, even on addition of dimethylformamide[306]. It appears that reduction of the negative charge of the anion **172**, either by strong complexing with dimethylformamide, or formation of the covalent ester **174**, is a prerequisite for dimerization to occur. The mechanism of these dimerizations has not been elucidated; a reasonable suggestion, which also explains the fact that the *cis* isomers are the major products, has been made[306].

The *trans* isomers, disodium dimercaptofumaronitrile (**178**) and bis(methylmercapto)fumaronitrile (**179**), have been prepared by the sequence outlined in equation (184)[306]. Oxidation of the *cis* salt **173** with one-half equivalent of oxidizing agents such as bromine, iodine or thionyl chloride produces disodium *cis*, *cis*-bis(2-mercapto-1,2-dicyanovinyl)disulphide (**176**), which rearranges in dimethylformamide to the *trans* isomer (**177**). Reduction of **177** with potassium

(173) $\xrightarrow{Br_2}$ (176) $\xrightarrow{DMF}$ (177) $\xrightarrow{NaBH_4}$ (178) $\xrightarrow{MeX}$ (179)

borohydride gives the mixed sodium potassium salt **178**; the disodium salt **178** is obtained by benzoylation followed by cleavage with sodium methoxide, and the thioether **179** is produced by methylation[306].

Some physical properties of the isomeric salts and ethers are shown in Table 2. The structure assignments are supported by the dipole

TABLE 2. Physical properties of salts and ethers of dimercaptomaleonitrile and dimercaptofumaronitrile[306].

	(173)	(178)	(175)	(179)
M.p. (°C)	Decomposes	—	99	118–119
Dipole moment (D)	—	—	5·08	1·57
Infrared spectrum (C=C stretching frequency in cm^{-1})	1440	Missing	1495	Missing
Ultraviolet spectrum λ_{max}^{EtOH} (ε)	368 (11,400)	412 (10,100)	340 (12,900)	364 (10,300)
Oxidation potentials $E_{1/2}$ (V)	+0·08 +0·40	0·00 +0·43	—	—

moments and the ultraviolet and infrared spectra. The equilibrium constant between disodium dimercaptomaleonitrile (**173**) and its *trans* isomer **178** could not be measured accurately, but it was found qualitatively that the *cis* isomer **173** is more stable. The thioethers, bis(methylmercapto)maleonitrile (**175**) and bis(methylmercapto)-fumaronitrile (**179**), on the other hand, equilibrate cleanly at elevated temperatures in the presence of catalytic amounts of iodine. The extrapolated equilibrium constant $K_{cis/trans}$ at 25° is

3·23; the thermodynamic parameters, calculated for 204°, are ΔH −1·1 kcal/mole and ΔS −0·2 e.u.[306]. Irradiation of aqueous solutions of the salts **173** and **178** produces a photostationary state containing 73% of the *trans* isomer and 27% of the *cis* isomer at 55°. Curiously, the situation is completely reversed in the case of the thioethers, where irradiation of the *trans* isomer **179** causes complete rearrangement to the *cis* isomer **175**[306].

The obvious question as to why the *cis* salts **173** are more stable than their *trans* isomers has received detailed consideration. Two explanations have been advanced[306]: the first proposes that the unfavourable alignment of poles and dipoles in the *cis* salt **173** is imposed by the large energy gained on coordination to the counter ion. Another possibility is that there is some bonding between the two sulphur atoms resulting in the dianion **180**. In this representation, two electrons are in a $d\pi$-orbital extending over the two sulphur

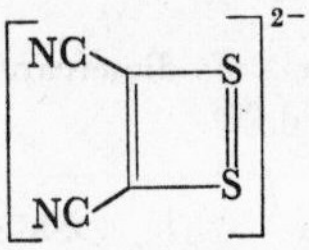

(180) 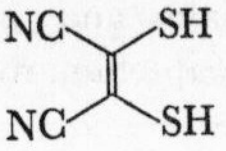 (181) 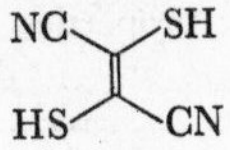 (182)

atoms, while four non-bonding electrons on sulphur and the π-electrons of the carbon–carbon double bond form an aromatic sextet. Repulsion between the negative charges on the two sulphur atoms is offset by delocalization of the $p\pi$-electrons into the cyano groups. Molecular-orbital calculations indicate the feasibility of such a representation, but a final decision will have to await determination of the crystal structure of the dianion.

The free acids, dimercaptomaleonitrile (**181**) and dimercaptofumaronitrile (**182**) have not been isolated. Acidification of the *cis* salt **173** gives a red, ether-soluble product which is too unstable to be characterized[309]. Dimercaptomaleonitrile dianion has been used as a ligand in a large number of transition metal complexes where it has been found to stabilize high oxidation states and unusual electronic configurations. Complexes reported include those of titanium[311–314], vanadium[311,315–318], chromium[317], molybdenum[314,317,319,320], tungsten[314,317,319], manganese[319,320], iron[319,321–328], cobalt[314,322–325,327–336], rhodium[322,334,337], nickel[322–325,327,328,330,331,334,337–342], palladium[307,308,322,323,325,328,330,331,336,341], platinum[322,323,325,330,331,336,341,342], copper[322,325,327,328,330,334,343] and gold[322,330,336,342,344].

A large number of derivatives of dimercaptomaleonitrile have been prepared (equation 185). Reaction with alkyl halides gives thio-

NC(SCOR)C=C(SCOR)CN **(190)** ← RCOX — **(173)**

NC(SR)C=C(SR)CN **(183)** ← RX — **(173)**

(184) ← $XCHRCH_2X$ — **(173)**

NC(SHgR)C=C(S^-)CN **(189)** ← RHgX — NC(S^-)C=C(S^-)CN **(173)** — CH_2I_2 → **(185)** (185)

(173) — $AsCl_3$ → **(188)**

(173) — $Cl_2C{=}X$ → **(186)**; **(187)** C=S=O

(**186a**, X = O; **b**, X = S;
c, X = NOH; **d**, X = NR)

ethers (**183**)[345,346], whereas 1,2-dihalo compounds and methylene iodide give cyclic products of type **184**[306,345,346] and **185**[307,328], respectively. Acylation yields diacyl derivatives (**190**)[306,347], reaction with phosgene and thiophosgene[346–350], phosgene oxime[350] and iminophosgenes[351] gives 4,5-dicyano-1,3-dithiole derivatives (**186**). The 1,3-dithiole ring in these compounds is stable to acid but opening to the dianion **173** occurs readily on treatment with base[349]. Pyrolysis of 4,5-dicyano-1,3-dithiolone (**186a**) gives dicyanoacetylene (section V)[215]. Oxidation of the thione **186b** furnishes the red *S*-oxide **187**[350]. The arsenic-containing heterocyclic compound **188** is obtained on treatment of the salt **173** with arsenic trichloride[328], while reaction with organomercurials gives compounds of type **189**[352].

Disodium dimercaptomaleonitrile is oxidized to tetracyano-1,4-dithiin (**196**; cf. section IX.H) by a wide variety of oxidizing agents. These include chlorine, bromine, iodine, triiodide ion, sulphur monochloride, sulphur dichloride, thionyl chloride, sulphuryl chloride, cyanogen chloride, methanesulphonyl chloride, tetracyanoethylene, ferricyanide and persulphate ions[328,353–355]. The course of the oxidation with thionyl chloride has been studied in detail; the proposed mechanism is shown in equation (186)[355]. The initial steps involve formation of 4,5-dicyano-1,2,3-trithiol-2-one (**191**), which can be isolated when thionyl chloride is used as solvent[353]. Attack of another molecule of the dianion **173** on **191**, followed by elimination of sulphur monoxide from the intermediate **192**, gives the disulphide **193**; oxidation stops at this stage if only one-half equivalent of oxidizing agent is used. In the presence of excess thionyl chloride, further attack on the dianion **193** results in formation of the intermediate **194**, which fragments into two molecules of 3,4-dicyano-1,2-dithiete (**195**); the latter finally

NC, S⁻ / NC, S⁻ (**173**) —$SOCl_2$→ NC, S, SO, NC, S (**191**) —173→

NC, S—S, CN / NC, S, SO⁻, ⁻S, CN (**192**) —(−SO)→ NC, S—S, CN / NC, S⁻ ⁻S, CN (**193**)

—$SOCl_2$, −Cl⁻→ NC, S—S, CN / NC, S, S, CN, SO, Cl (**194**) ⟶ 2 NC, S, S, NC (**195**) (186)

⟶ NC, S, CN, S, S, NC, S, CN —(−2S)→ NC, S, CN / NC, S, CN (**196**)

NC, S, S, NC (**195**) + RO ⟶ NC, S, OR / NC, S (**197**) (187)

dimerizes with the expulsion of sulphur to form tetracyano-1,4-dithiin (**196**). The dithiete **195** cannot be isolated but its intermediacy is made likely by formation of adducts of type **197** when the oxidation of **173**, **191** or **193** is carried out in the presence of electron-rich olefins such as vinyl ethers or eneamines (equation 187)[356]. Bis(trifluoromethyl)-1,2-dithiete (**198**)[357], the only example of this ring system known to date, undergoes similar addition reactions. The electronic structure of dithietes has been discussed in detail, and LCAO–MO calculations indicate that electron-withdrawing groups stabilize the dithiete over the open forms **199a** and **199b**; the possibility of a valence isomerization between **199a** and **195** has not been ruled out[356].

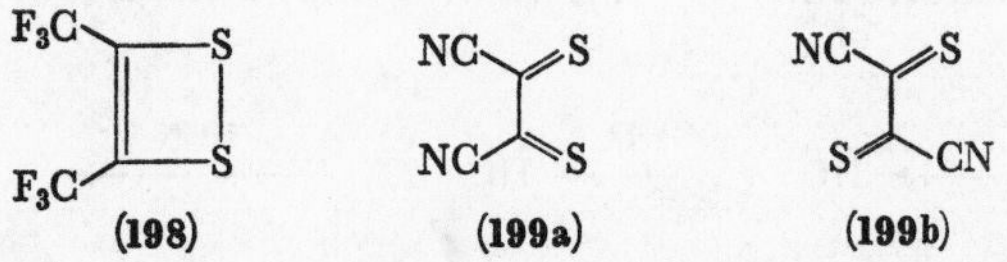

(**198**) (**199a**) (**199b**)

One-electron transfer reagents, such as tetracyanoethylene or ferricyanide ion, are believed to oxidize disodium dimercaptomaleonitrile by a different mechanism (equation 188)[356]. Tetracyanoethylene ion radical (**200**) and dianion (**193**) were detected

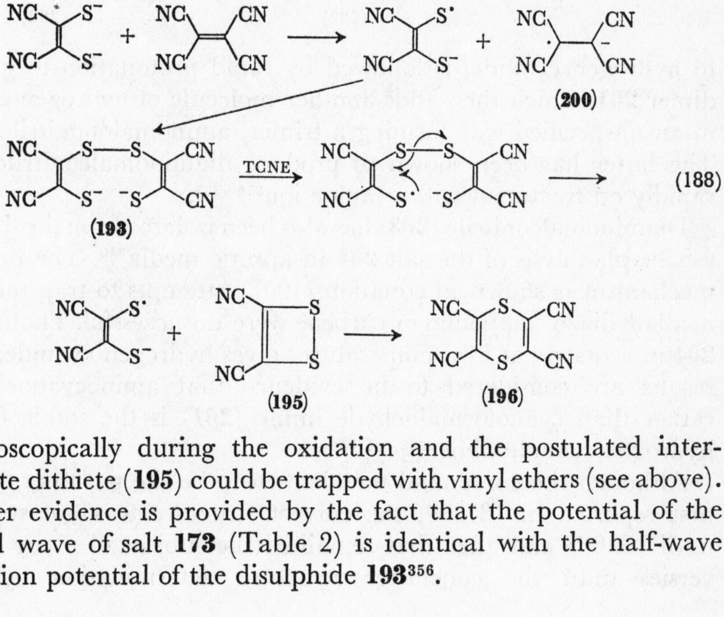

(188)

spectroscopically during the oxidation and the postulated intermediate dithiete (**195**) could be trapped with vinyl ethers (see above). Further evidence is provided by the fact that the potential of the second wave of salt **173** (Table 2) is identical with the half-wave oxidation potential of the disulphide **193**[356].

4. Diaminomaleonitrile and diaminofumaronitrile

Polymerization of hydrogen cyanide under the influence of a variety of bases such as alkali hydroxides[358,359], carbonates[359,360], cyanide ion[361–364], ammonia[359,360,362,363,365,366], tertiary amines[367], amine oxides[368], quaternary hydroxides[369] and alumina[370,371] or by irradiation with γ-rays[372], produces a tetramer, diaminomaleonitrile (**203**), in varying amounts in addition to adenine and brown polymers of higher molecular weight ('azulmic acid'). The mechanism of the tetramerization has not been elucidated with certainty. The rate of tetramer formation is first order in base[373] and quadratic in hydrogen cyanide concentration[373–375]. A proposed mechanism is shown in equation (189)[374,375]. The first step, addition of cyanide ion

$$NC^- + HCN \longrightarrow HC(CN){=}N^- \xrightarrow{H^+} HC(CN){=}NH\ \mathbf{(201)} \xrightarrow{HCN} \longrightarrow$$

$$(NC)_2CHNH_2\ \mathbf{(202)} \xrightarrow{HCN} \longrightarrow (NC)(H_2N)C{=}C(CN)(NH_2)\ \mathbf{(203)} \qquad (189)$$

to hydrogen cyanide, is followed by rapid protonation to give the dimer **201**, which then adds another molecule of hydrogen cyanide in an unspecified way forming a trimer, aminomalononitrile (**202**). The latter has been shown to produce diaminomaleonitrile (**203**) rapidly on treatment with cyanide ion[376–379].

Diaminomaleonitrile (**203**) has also been isolated from the thermolysis or photolysis of the salt **204** in aprotic media[380]. The proposed mechanism is shown in equation (190). Attempts to trap the intermediate diazo compound or carbene were unsuccessful. Photolysis of **204** in a matrix at low temperatures gives hydrogen cyanide. These results are considered to be evidence that aminocyanocarbene, rather than cyanoformaldehyde imine (**201**) is the stable form of hydrogen cyanide dimer[363,365,380].

Although the compound now known to be the tetramer **203** was first reported in 1873[381], the correct structure was not suggested until 1928[382] and after that remained the subject of many controversies until the conclusive structure determination by x-ray

$$RSO_2\bar{N}{-}N{=}C(CN)(NH_2) \longrightarrow RSO_2^- + [N_2{=}C(CN)(NH_2)]$$

(204)

$$\xrightarrow{-N_2} NC{-}\ddot{C}{-}NH_2 \qquad (190)$$

$$\xrightarrow{204} (NC)(H_2N)C{=}C(CN)(NH_2) \text{ (203)} \qquad \not\rightarrow NCCH{=}NH \text{ (201)}$$

techniques in 1960[383,384]. Other suggested structures included aminomalononitrile (**202**)[359,368,385,386], its hydrocyanide[387,388], aminoiminosuccinonitrile (**205**)[389–392] and cyanoformaldehyde imine (**201**)[393]. Diaminomaleonitrile is a colourless solid which decomposes

$$HN{=}C(CN){-}CH(CN)NH_2 \text{ (205)} \qquad (NC)(H_2N)C{=}C(NH_2)(CN) \text{ (206)}$$

at its melting point (184°). The ultraviolet spectrum in water shows a maximum at 295 mμ (log ε 4·08); the fact that the spectrum is very similar in 0·1 N hydrochloric acid (λ_{max} 290 mμ, log ε 4·06) indicates that no appreciable salt formation takes place in this medium[394]. An aqueous solution has a pH of 6·7[371]; the pK_a has not been determined accurately[394]. The observation of a double bond stretching band at 1620 cm^{-1} in the infrared as well as in the Raman spectrum supports the *cis* structure of the tetramer[395]. The dipole moment of 7·8 D (in dioxan) indicates considerable delocalization of the unshared electron on the amine nitrogen[394]. Polarographic reduction of the tetramer at pH 10 shows two waves at −1·24 and −1·58 V[396]. Other physical properties reported include the mass spectrum[395], n.m.r. spectrum[395] and solubility[367,388]. The x-ray diffraction study[383,384,397] shows that in the crystal, the molecule has no symmetry. The central double bond is twisted by 6°, and the two amino nitrogens are non-equivalent, one being planar, the other tetrahedral. Both amino groups are involved in intermolecular hydrogen bonding to the nitrile groups; one of the amino groups also acts as a hydrogen-bond acceptor. No explanation seems to have been advanced as to why the *cis* isomer, diaminomaleonitrile (**203**), is more stable than the *trans* isomer, diaminofumaronitrile (**206**). The occurrence of two signals in the ^1H n.m.r. spectrum of **203** has been interpreted as being due to a mixture of both isomers[395]; however, the ^{13}C n.m.r. spectrum of diaminomaleonitrile in dimethyl sulphoxide shows only two signals of equal intensities, indicating that only the *cis* isomer is present in detectable amounts[398].

Diaminofumaronitrile (**206**) has recently been prepared by irradiation of the *cis* isomer **203** with light of the wavelengths 295–335 mμ[825]. In the ultraviolet spectrum, **206** shows a maximum at 310 mμ (ε 8200). The occurrence of a C=C stretching band at 1618 cm^{-1} in the infrared spectrum is due to considerable deviation of the molecule from C_{2h} symmetry, as determined by an x-ray diffraction study. As in the *cis* isomer **203**, the two amino nitrogens are different, one being planar, the other tetrahedral. Diaminofumaronitrile (**206**) is very labile and reverts to the *cis* isomer **203** under the influence of acid, base, charcoal, light of wavelengths above 320 mμ[825] or on heating[375]. Tentative evidence for derivatives of **206** has been mentioned[366]. Further irradiation of **206** results in formation of 4-cyano-5-aminoimidazole (see below).

The diminished basicity of the amine groups in diaminomaleonitrile is evident from its chemical reactions. It forms only a monohydrochloride[389,399]; acetylation under normal conditions yields a monoacetyl derivative[389,399–401]; preparation of the diacetyl derivative requires more drastic conditions[389,401]. Reaction of the mono- or diacetyl derivative with acetic anhydride in dioxan gives a compound of proposed structure **207**[401,402]. Schiff bases of type **208** have been prepared from a number of aldehydes and ketones[385,389,391,400,401,403–405] whereas the action of α-diketones[364,382,389,401,406] and α-haloketones[406] results in the formation of pyrazines **209** (equation 191). Other heterocycles synthesized from this versatile intermediate include dicyanotriazole (**210**; section VIII.E)[389,401,403], 4,5-dicyanoimidazolin-2-one (**213**)[402,407], 4,5-di-

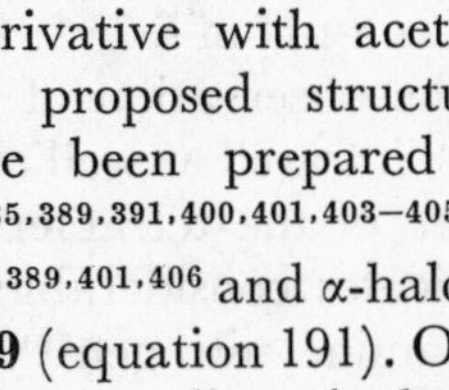

NC, NC, H, N, C, CH$_3$, N, OCOCH$_3$, COCH$_3$

(207)

NC, N=CR^{1}R^2, NC, NH$_2$

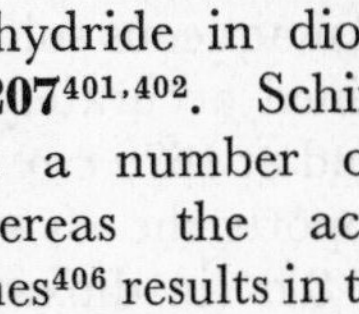

(208)

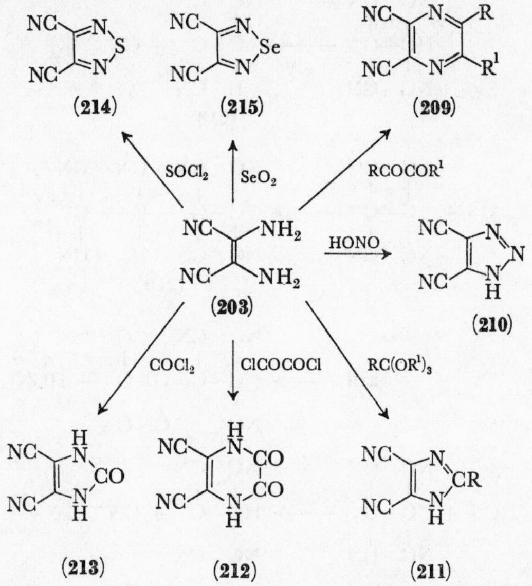

(191)

cyanoimidazoles (**211**)[366,408], 5,6-dicyanopyrazine-2,3-dione (**212**)[401] and 3,4-dicyano-1,2,5-thiadiazole (**214**) and its selenium analogue **215**[409]. Diaminomaleonitrile reduces Tillman's reagent and iodine

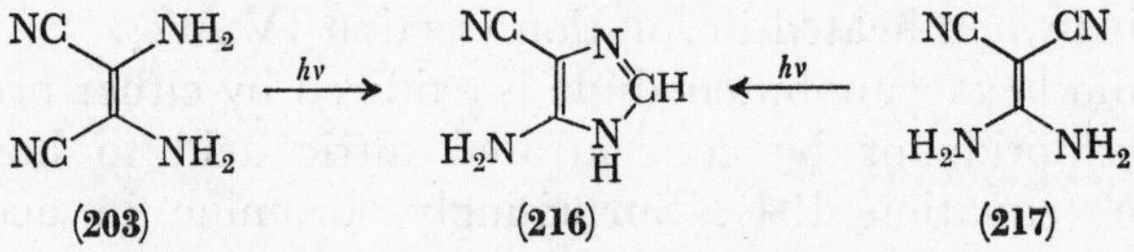

(192)

but these reactions have not been investigated in detail[410]. Heating aqueous solutions of diaminomaleonitrile to 100° results in the formation of peptide-like compounds in addition to glycine[411]. Irradiation of diaminomaleonitrile affects an interesting rearrangement to give 4-cyano-5-aminoimidazole (**216**; equation 192). 1,1-Diamino-2,2-dicyanoethylene (**217**) also gives **216** on irradiation[375,376]. The mechanism has not been elucidated, but isomerization of **203** to its *trans* isomer is almost certainly the initial step[375,825].

E. Hexacyanobutadiene

Disodium tetracyanoethanediide (section IV.A.7) readily eliminates cyanide ion and condenses to give hexacyanobutenediide[238]. The reaction is catalysed by the addition of either tetracyanoethane

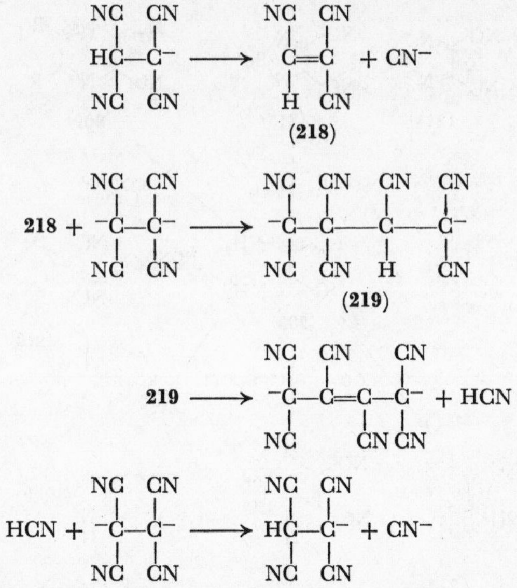

$$(NC)_2HC{-}C^-(CN)_2 \longrightarrow (NC)(H)C{=}C(CN)_2 + CN^-$$

(218)

$$\mathbf{218} + {}^-C(CN)_2{-}C^-(CN)_2 \longrightarrow {}^-C(CN)_2{-}C(CN)_2{-}C(CN)(H){-}C^-(CN)_2$$

(219)

$$\mathbf{219} \longrightarrow {}^-C(CN)_2{-}C(CN){=}C(CN){-}C^-(CN)_2 + HCN$$

$$HCN + {}^-C(CN)_2{-}C^-(CN)_2 \longrightarrow (NC)_2HC{-}C^-(CN)_2 + CN^- \quad (193)$$

or tricyanoethylene. Thus, a mechanism involving elimination of cyanide ion from the monosodium salt of tetracyanoethane is suggested (equation 193).

Tetracyanoethylene ion-radical salts also give hexacyanobutenediide when heated in solution (section IV.A.7).

Disodium hexacyanobutenediide is oxidized by either bromine in ethylene chloride or by concentrated nitric acid to hexacyanobutadiene (equation 194). Surprisingly, bromine in acetonitrile

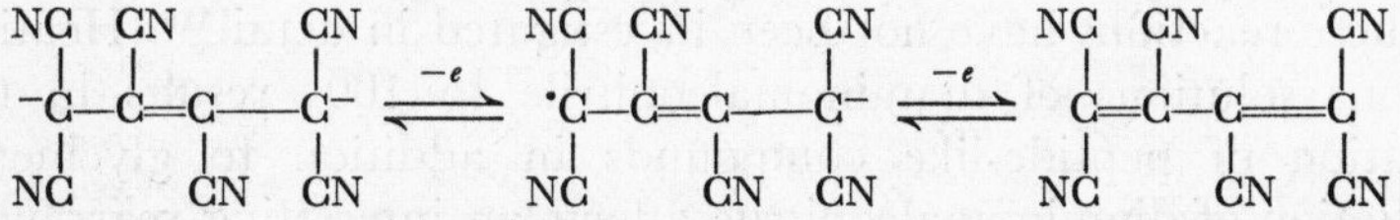

$${}^-C(CN)_2{-}C(CN){=}C(CN){-}C^-(CN)_2 \underset{}{\overset{-e}{\rightleftharpoons}} {}^{\bullet}C(CN)_2{-}C(CN){=}C(CN){-}C^-(CN)_2 \overset{-e}{\rightleftharpoons} (NC)_2C{=}C(CN){-}C(CN){=}C(CN)_2 \quad (194)$$

only removes one electron to give the ion radical. The polarographic oxidation of hexacyanobutenediide shows two reversible waves, ($E_{\frac{1}{2}}$ = 0·6 and 0·02 v versus SCE).

Hexacyanobutadiene is a colourless, crystalline solid which sublimes at 130° (0·3 mm) and melts at 253–255°. The infrared spectrum shows conjugated nitrile absorption at 2225 cm^{-1} and carbon–carbon double bond absorption at 1550 cm^{-1}. A planar structure is not likely because of the steric hindrance of its two tricyanovinyl groups, however, the ultraviolet spectrum $\lambda_{max}^{CH_3CN}$ 302 mμ (15,300) indicates considerable conjugation[238]. The electron affinity measured in a magnetron is 76·0 ± 2·4 kcal/mole[134] compared to 66·5 ± 1·4 for tetracyanoethylene.

Hexacyanobutadiene is a strong π-acid and forms charge-transfer complexes with aromatic compounds. Thus, its solutions in benzene are yellow-orange; in toluene, red; and in xylene, purple[238].

Hexacyanobutadiene is readily reduced to its ion radical by relatively mild reducing agents such as sodium bromide, and metallic mercury. The best way to obtain the ion radical, however, is by the transfer of one electron from the butenediide to the butadiene (equation 195)[238].

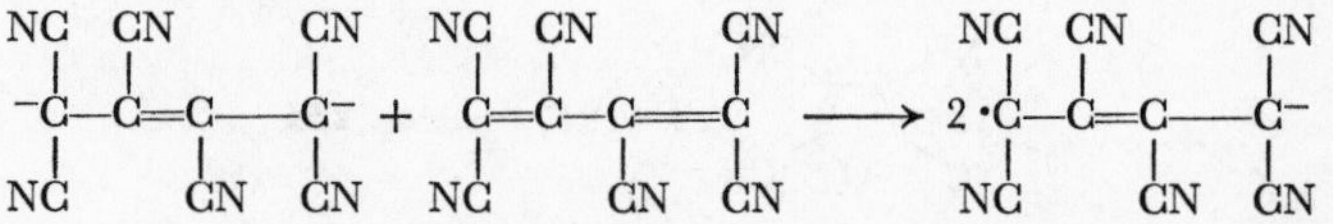

$$^{-}C(CN)_2{-}C(CN){=}C(CN){-}C^{-}(CN)_2 + (NC)_2C{=}C(CN){-}C(CN){=}C(CN)_2 \longrightarrow 2\,{\cdot}C(CN)_2{-}C(CN){=}C(CN){-}C^{-}(CN)_2 \quad (195)$$

The e.s.r. spectrum of the radical in 1,2-dimethoxyethane shows an overall pattern of 9 peaks with 1·16 gauss separation; each peak in turn is split five times into peaks of 0·22 gauss separation. The overall nine-line pattern is attributed to hyperfine interaction of the unpaired electron with the four nitrogen atoms of the terminal nitrile groups. The quintets are ascribed to the hyperfine interaction of the unpaired electron with the nitrogen atoms of the two central nitrile groups. The resistivity of a compaction of sodium hexacyanobutenide is in the semiconductor range at $4{\cdot}4 \times 10^7$ ohm cm. Sodium cyanide reduces hexacyanobutadiene to the butenediide[238].

Butadiene adds to hexacyanobutadiene across one of the tricyanovinyl groups to give a six-membered ring (equation 196). The

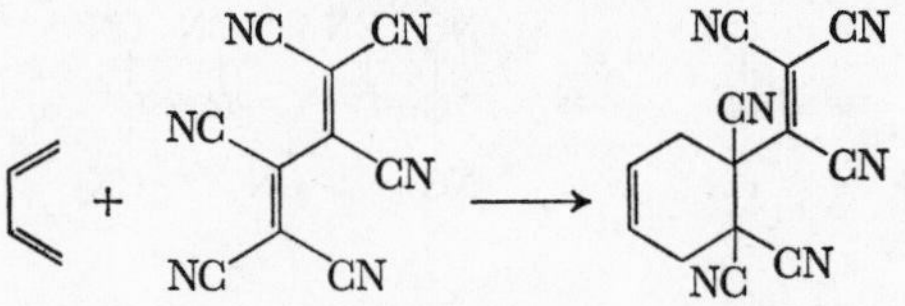

$$\text{butadiene} + (NC)_2C{=}C(CN){-}C(CN){=}C(CN)_2 \longrightarrow \mathbf{220} \quad (196)$$

(220)

remaining tricyanovinyl group in the adduct **220** reacts with dilute aqueous base, as do other tricyanovinyl compounds, to replace one of the cyano groups with OH (equation 197).

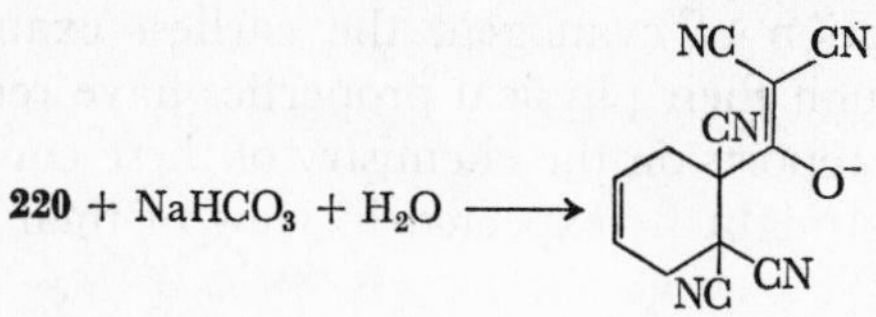

$$\mathbf{220} + NaHCO_3 + H_2O \longrightarrow \text{product } (O^{-}) \quad (197)$$

One of the terminal cyano groups is substituted by OH when hexacyanobutadiene is dissolved in aqueous sodium bicarbonate (equation 198). The same product (**221**) is produced by reaction of

$$(NC)_2C{=}C(CN){-}C(CN){=}C(CN)_2 + NaHCO_3 + H_2O \longrightarrow \underset{\mathbf{(221)}}{(NC)_2C{=}C(CN){-}C(CN){=}C(CN)O^-} \quad (198)$$

tetracyanofuran with cyanide ion (equation 199)[238]. If hexacyanobutadiene were reacting as two unconjugated tricyanovinyl groups

$$\text{2,3,4,5-tetracyanofuran} + CN^- \longrightarrow \mathbf{221} \quad (199)$$

as a consequence of non-planarity, the expected hydrolysis product would be 1,1,4,4-tetracyanobutadiene-2,3-diolate[238,412] (**222**). None of this compound is observed.

$$\underset{\mathbf{(222)}}{(NC)_2C{=}C(O^-){-}C(O^-){=}C(CN)_2}$$

On the other hand, products derived from substitution of cyano groups at both the 1- and 2-positions are obtained from the reaction of malononitrile anion with hexacyanobutadiene (equation 200)[238].

$$(NC)_2C{=}C(CN){-}C(CN){=}C(CN)_2 + HC^-(CN)_2 \longrightarrow (NC)_2C{=}C(CN){-}C(CN){=}C(CN){-}C^-(CN)_2 + \text{octacyanobenzene-type product} \quad (200)$$

V. CYANOACETYLENES

Cyanoacetylene, dicyanoacetylene and dicyanodiacetylene have been known for over fifty years. The two latter compounds are, with the exception of cyanogen, the earliest examples of cyanocarbons. Although their physical properties have received considerable attention, reports on the chemistry of these compounds are not as abundant as might be expected in view of their high reactivity.

A. Synthesis of Cyanoacetylenes

The original synthesis of cyanoacetylene (**223**), and probably still the most convenient, involves dehydration of the amide of propiolic acid with phosphorus pentoxide[413–416]:

$$HC{\equiv}CCOOH \longrightarrow HC{\equiv}CCONH_2 \longrightarrow \underset{(\mathbf{223})}{HC{\equiv}CCN} \qquad (201)$$

Other reported syntheses include the thermal decomposition of acrylonitrile[417], the gas-phase pyrolysis of mixtures of hydrogen cyanide and acetylene[417], acetonitrile and acetylene[418], acetonitrile and hydrogen cyanide[419], and acetylene and cyanogen[420], the dehydrogenation of propionitrile over magnesium oxide[421] and the dehydration of the oxime of propargyl aldehyde[422]. The three latter methods are claimed to be adaptable to the preparation of alkyl- and aryl-substituted cyanoacetylenes as well. Cyanoacetylene is also formed in the reaction of acetylene with active nitrogen[423], and by action of an electric discharge on mixtures of acetylene with cyanogen or hydrogen cyanide, and nitrogen and methane[424]. The latter observation, combined with the finding that cyanoacetylene can be converted to cytosine under mild conditions, has given rise to the proposal that cyanoacetylene is a precursor in the prebiotic formation of pyrimidines[425].

Chlorocyanoacetylene (**224**) is obtained by treatment of the lithium salt of cyanoacetylene with chlorine at $-70°$ (equation 202)[426], whereas bromo- and iodocyanoacetylene have been pre-

$$HC{\equiv}CCN \xrightarrow[-70°]{BuLi} LiC{\equiv}CCN \xrightarrow[-70°]{Cl_2} \underset{(\mathbf{224})}{ClC{\equiv}CCN} \qquad (202)$$

pared by the action of the corresponding potassium trihalide on cyanoacetylene[427–429].

Dicyanoacetylene (**225**) was originally prepared by dehydration of the diamide of acetylenedicarboxylic acid with phosphorus pentoxide (equation 203)[430,431]. Since acceptable yields (30–40%)

$$H_2NOCC{\equiv}CCONH_2 \xrightarrow{P_2O_5} \underset{(\mathbf{225})}{NCC{\equiv}CCN} \qquad (203)$$

can be attained only by working on a small scale, several unsuccessful attempts have been made to modify this synthesis[432,433]. Subsequently developed syntheses of dicyanoacetylene include the gas-phase pyrolysis of dichloromaleo- and -fumaronitrile[434], chloromaleo- and -fumaronitrile[434] and tetracyanoethylene[215]. Reaction of 3,4-dicyanofuran[435] with dimethyl acetylenedicarboxylate gives dicyanoacetylene by a sequence of Diels–Alder addition and retro-diene cleavage[436]:

$$\text{3,4-(NC)}_2\text{-furan} + \text{MeOOCC}{\equiv}\text{CCOOMe} \longrightarrow [\text{7-oxabicyclo adduct: NC, NC, COOMe, COOMe}] \longrightarrow \text{3,4-(MeOOC)}_2\text{-furan} + \underset{(225)}{\text{NCC}{\equiv}\text{CCN}} \quad (204)$$

Dicyanoacetylene, together with cyanogen and a C_6N_2 species (dicyanodiacetylene?) is also formed directly from the elements at 2500–2800° [437]. Probably the best synthesis of dicyanoacetylene currently known is the gas-phase pyrolysis of 4,5-dicyano-1,3-dithiol-2-one (**226**; equation 205), which is readily available in a

$$\underset{(226)}{\text{(NC)}_2\text{C}_2\text{S}_2\text{CO}} \xrightarrow[76\%]{600-800°} \text{NCC}{\equiv}\text{CCN} + \text{CO} + \text{COS} + \text{CS}_2 + \text{S} \quad (205)$$

two-step synthesis from sodium cyanide, carbon disulphide and phosgene (section IV.D.3)[215].

Attempts to prepare dicyanoacetylene by reaction of diiodoacetylene with silver cyanide[430], or treatment of the dilithium salt[433] or the bis-Grignard derivative[430,433] of acetylene with cyanogen halides were unsuccessful, probably due to its exceptional sensitivity toward nucleophiles.

The only synthesis of dicyanodiacetylene (**227**) reported to date is the oxidative coupling of the copper salt of cyanoacetylene[413,414,416,438–440]:

$$\text{NCC}{\equiv}\text{CCu} \xrightarrow{K_3Fe(CN)_6} \underset{(227)}{\text{NCC}{\equiv}\text{CC}{\equiv}\text{CCN}} \quad (206)$$

Other cyanopolyacetylenes reported are the enediyne **228** and the triyne **229**[441]; the antibiotic diatetryne nitrile **230** has been isolated from *clytocybe diatreta*[442].

MeCH=CHC≡CC≡CCN (**228**) MeC≡CC≡CC≡CCN (**229**)

HOOCCH=CHC≡CC≡CCN (**230**)

B. Physical Properties of Cyanoacetylenes

Selected physical properties of cyanoacetylenes, and references to others, are given in Table 3. All are linear, dicyanodiacetylene being the longest of these molecules whose structure is known with certainty[416]. In the solid state, the molecules of cyanoacetylene are oriented head-to-tail in chains to allow for intermolecular hydrogen bonding. The C≡N · · · H distance in the crystal is 3·27 Å, fairly long compared to the hydrogen bond in hydrogen cyanide (3·18 Å)[470]. The short liquid range of cyanoacetylene has been suggested as evidence that there is little change of molecular order on melting[449]. The average association is dimeric[448,449]; this relatively low degree of association, compared for instance with that of hydrogen cyanide, is primarily due to an enthalpy rather than an entropy effect[449]. Evidence for halogen–nitrogen interaction has also been obtained for the halocyanoacetylenes[472]. The various bond lengths in cyano- and dicyanoacetylene and some related molecules are shown in Table 4. They are surprisingly insensitive to changes in substitution. There is little change in the ionization potentials in the series cyanoacetylene, dicyanoacetylene and dicyanodiacetylene (Table 3); these potentials are identical to that of acetylene (11·4 ev) but 2 ev lower than that of cyanogen (13·6 ev). This has been taken as evidence that the electron is removed from the carbon–carbon triple bond of the cyanoacetylenes and that it comes from a comparatively localized orbital having little interaction with adjacent orbitals[446].

All cyanoacetylenes can be expected to have high positive heats of formation, although a determination of this parameter has been made only for dicyanoacetylene (ΔH° = 127·5 kcal/mole)[482]. The heat of combustion of dicyanoacetylene is almost 500 kcal/mole, and flame temperatures of mixtures of dicyanoacetylene and ozone are calculated to reach 6100°[475]. Neat dicyanoacetylene (and presumably the other compounds in this class as well) is thus potentially explosive, although in solution it has shown surprising thermal stability[215].

TABLE 3. Physical properties of cyanoacetylenes.

Property	HC≡CCN	NCC≡CCN	$NC(C{\equiv}C)_2CN$	ClC≡CCN	BrC≡CCN	IC≡CCN
Melting point (°C)	5[413]	20·5–21[430]	64·5–65·5[438]	42[426]	96–96·5[428]	152–152·5[429]
Boiling point (°C at ~760 mm)	42·5[413]	76–76·5[430]	154[a,439]	—	—	—
Density	0·8159[414]	1·0174[443]	—	—	—	—
Dipole moment (D)	3·6[444,445]	—	—	—	3·88[428]	4·59[429]
Ionization potential (ev)	11·6[446]	11·4[446]	11·4[446]	—	—	—
References to other physical properties of cyanoacetylenes[b]						
Infrared spectrum	441, 447–455	439, 456–458	416, 439	426, 459, 460	428, 459, 460	429, 459, 460
Raman spectrum	452	456, 451, 461	416, 462	—	—	—
Ultraviolet spectrum	441, 463	441, 464	441[c], 442[c]	426	428	429
Mass spectrum	446	446	446	—	—	—
Microwave spectrum	444, 445, 465–467	—	—	426	468	468
N.m.r. spectrum	469	—	—	—	—	—
Vapour pressure	—	439	439	—	—	—
X-ray structure determination	449, 470	471	—	—	—	472
Enthalpy, entropy of fusion, vaporization and sublimation	449	439, 449	439	—	—	—
Bond strengths	446, 473[d], 474[d]	446	—	—	—	—

[a] Extrapolated.
[b] The following additional data have been reported: flash point[431], flame temperatures[440,475] and magnetooptical parameters[476,477] of dicyanoacetylene, and molecular orbital calculations on cyanoacetylene[478,479] and dicyanoacetylene[478].
[c] Also includes data on related compounds.
[d] Calculated.

TABLE 4. Bond lengths in cyanoacetylenes.

Compound (dimension in Å)	References
H—C≡C—C≡N (1·06, 1·20, 1·38, 1·16)	444, 445, 470
NC—C≡C—C≡N (1·19, 1·37, 1·14)	471
NC—C≡N (1·38, 1·15)	480
HC≡C—C≡CH (1·19, 1·36)	481

C. Reactions of Cyanoacetylenes

I. Salt and complex formation

Cyanoacetylene has an acidic proton, and a number of salts have been prepared. The copper[414,416,438] and silver[414,415] salts are explosive in the dry state. Interaction of cyanoacetylene with titanium tetrachloride gives rise to a yellow powder of composition $TiCl_4 \cdot (NCC{\equiv}CH)_2$ [483]. Dicyanoacetylene readily forms a 1:1 complex with aluminium bromide[484], a finding that is surprising in view of the known electrophilicity of this molecule. The structure of the complex is not known, but by analogy with the many reported complexes of nitriles with Lewis acids[485] (Chapter 3) it is likely that the aluminium bromide is complexed with a cyano group. The complex is moisture-sensitive and reacts with aromatic hydrocarbons to produce Diels–Alder adducts of dicyanoacetylene (see below). Dicyanodiacetylene (**227**) is reported to give blue solutions with benzene; the colour has been ascribed to complex formation[416]. Since toluene, a better π-base than benzene, produces a blue solution only slowly[416], it is more likely that the colour formation is due to some other cause such as the presence of impurities. Solutions of dicyanoacetylene in aromatic hydrocarbons are colourless, and no change in the ultraviolet spectra due to charge-transfer transitions has been observed in these solutions[486]. This holds true even for such a potent π-base as hexamethylbenzene, which forms a strong, highly coloured complex with tetracyanoethylene[287]. Dicyanoacetylene adds to chloro(carbonyl)bis(triphenylphosphine)-iridium to give the complex chloro(carbonyl)bis(triphenylphosphine)-(dicyanoacetylene)iridium[628].

2. Addition reactions

Cyanoacetylenes are very susceptible to nucleophilic additions. Thus cyanoacetylene[425], its alkyl and aryl derivatives[487] and dicyanoacetylene[430,488–490] react at or below room temperature with ammonia and primary and secondary amines to give adducts such as **231** and **232**. The stereochemical aspect of this reaction has been

$$R_2NCH{=}CHCN \qquad NCCH{=}C(CN)(NR_2)$$

(231) **(232)**

investigated in the case of the piperidine addition to dicyanoacetylene, where only the product of *cis* addition could be detected[489], and in the addition of urea to cyanoacetylene, where both isomers were formed in unspecified amounts[425]. Preference for *cis* addition has also been observed in the reaction of methanol and isopropyl alcohol with dicyanoacetylene, although small amounts of the products of a *trans* addition were formed in this case (equation 207)[489]. *t*-Butyl alcohol gives an equal mixture of both isomers.

$$NCC{\equiv}CCN + ROH \longrightarrow (NC)(RO)C{=}C(CN)(H) + (NC)(RO)C{=}C(H)(CN) \tag{207}$$

The remarkable feature of the addition of alcohols to dicyanoacetylene is that it proceeds rapidly at room temperature in the absence of a catalyst. The corresponding addition to acetylenedicarboxylic ester requires heating to 150–200° [491]. One molecule of alcohol adds to cyanoacetylene in the presence of weak bases such as cyanide or carbonate ions, while a second addition occurs in the presence of alkoxide ion (equation 208)[415]. Both types of

$$HC{\equiv}CCN + ROH \xrightarrow{CN^-} ROCH{=}CHCN$$

$$HC{\equiv}CCN + ROH \xrightarrow{RO^-} (RO)_2CH{-}CH_2CN \tag{208}$$

adducts on acid hydrolysis give cyanoacetaldehyde[415]. The base-catalysed addition of phenols to cyanoacetylene is reported to give cyanovinyl aryl ethers of unspecified stereochemistry[492]. Cyanoacetaldehyde is formed on treatment of cyanoacetylene with aqueous sodium hydroxide at room temperature[425]. Other reported nucleophilic additions to cyanoacetylenes are: the addition of phosphate[493] and trialkyltin hydrides[494,495] to cyanoacetylene to give compounds **233** and **234**, respectively, and the addition of mercaptans to dicyanoacetylene[476] to give adducts **235** of unspecified stereochemistry.

$$NCCH{=}CHOPO_3^{2-} \quad H_2C{=}C(CN)(SnR_3) \quad NCCH{=}C(CN)(SR)$$

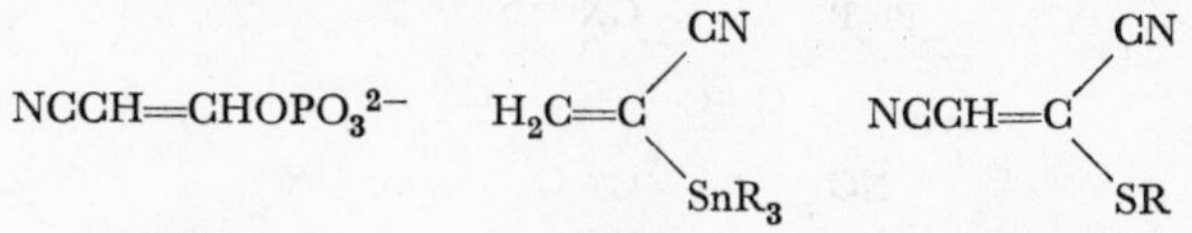

(233) **(234)** **(235)**

Whenever these addition reactions of cyanoacetylenes are compared to those of propiolic and acetylenedicarboxylic esters, it is found that the cyano compounds are considerably more reactive.

Somewhat surprisingly, dicyanoacetylene also reacts readily with hydrogen halides and with halogens. Thus hydrogen chloride, hydrogen bromide and hydrogen iodide add to dicyanoacetylene at room temperature to give halodicyanoethylenes of unspecified stereochemistry[430,496]. Addition of bromine to dicyanoacetylene also occurs at room temperature to give the *trans* adduct, dibromofumaronitrile, in 57% yield (equation 209)[269]. Iodine adds more

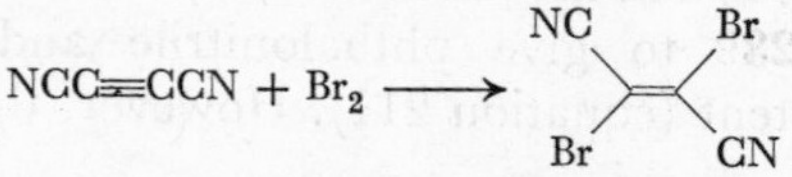

$$NCC{\equiv}CCN + Br_2 \longrightarrow (NC)(Br)C{=}C(Br)(CN) \qquad (209)$$

slowly than bromine, and ferrous iodide has to be used as an iodine carrier. The mechanism of this reaction has not been elucidated, but it may well proceed by a radical, rather than an ionic path.

Formal addition of hydrogen chloride and hydrogen cyanide to cyanoacetylene is the result of an apparent attempt to prepare dicyanoacetylene from hydrogen cyanide and cyanoacetylene (equation 210)[415]. Again, nothing is known about the mechanism

$$HC{\equiv}CCN + HCN \xrightarrow[\text{HCl, } O_2,\ 30°]{Cu_2Cl_2} NCClCHCH_2CN \qquad (210)$$

of this reaction. Carbon–metal bonds are formed in the reaction of dicyanoacetylene with pentacyanocobaltate ion[497] and mercuric chloride[490] to give products **236** and **237**, respectively. These

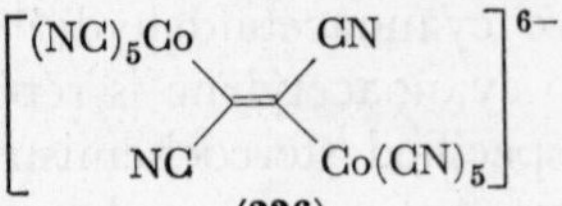

$$\left[\begin{matrix}(NC)_5Co & & CN \\ & C{=}C & \\ NC & & Co(CN)_5\end{matrix}\right]^{6-}$$

(236)

$Hg(C(CN){=}CClCN)_2$

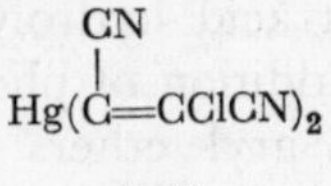

(237)

reactions also proceed with unactivated acetylenes[497–499]. Reaction of dicyanoacetylene with triphenylphosphine at room temperature gives a 3:2 adduct believed to be the diphosphorane **238**[824]; the previously[185] assigned tetracyanophosphole structure was shown to be wrong.

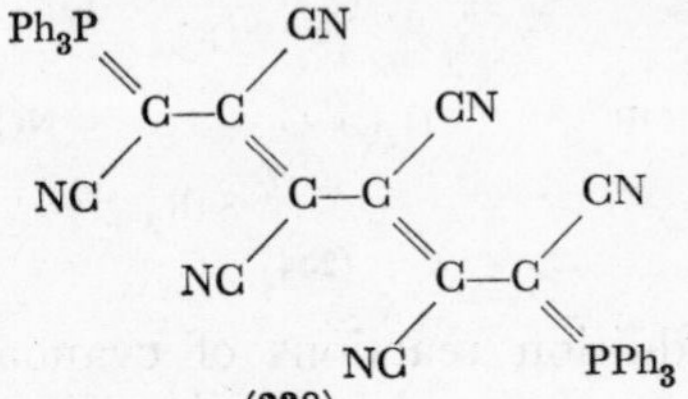

$Ph_3P{=}C(CN){-}C(CN){=}C(CN){-}C(CN){=}C(CN){-}C(CN){=}PPh_3$

(238)

3. Cycloaddition reactions

Dicyanoacetylene is one of the most powerful dienophiles known. It reacts with cyclopentadiene, 1,3-cyclohexadiene and butadiene at or below room temperature[432,490,500,501] and adds, albeit with decreasing ease, to anthracene[490,500,501], naphthalene[500,501] and even benzene itself[484]. The latter addition requires heating to 180° for extended periods and the yields are low, since retro-diene cleavage of the adduct **239** to give phthalonitrile and acetylene takes place to some extent (equation 211). However, the reaction can be

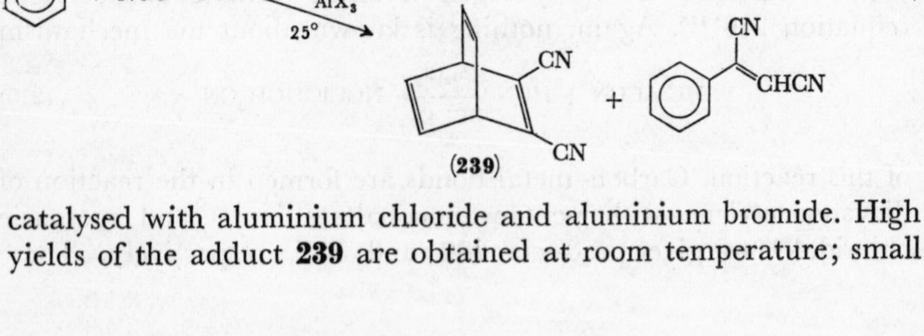

$$C_6H_6 + NCC{\equiv}CCN \xrightarrow[48\ h]{180^\circ} \textbf{(239)} + C_6H_4(CN)_2 + HC{\equiv}CH$$

$$C_6H_6 + NCC{\equiv}CCN \xrightarrow[25^\circ]{AlX_3} \textbf{(239)} + C_6H_5C(CN){=}CHCN \qquad (211)$$

catalysed with aluminium chloride and aluminium bromide. High yields of the adduct **239** are obtained at room temperature; small amounts of phenylmaleo- and phenylfumaronitrile, the products of a Friedel–Crafts addition of dicyanoacetylene to benzene, are also isolated[484]. The catalysed addition almost certainly proceeds by initial formation of an aluminium halide–dicyanoacetylene complex. Whether the subsequent addition to benzene is concerted, or proceeds through a dipolar intermediate, has not yet been determined[484]. Weaker Lewis acids, such as antimony pentachloride or boron trifluoride, are not effective as catalysts in this addition, nor could it be extended to other acetylenes, such as cyanoacetylene or acetylenedicarboxylic ester. By comparison, the reaction of tetracyanoethylene with aromatic hydrocarbons in the presence of aluminium chloride produces tricyanovinylbenzenes[199]; tetracyanoethylene does not react thermally with benzene at temperatures up to 300° [486].

The ultraviolet spectra of the adducts of dicyanoacetylene to aromatic hydrocarbons are of interest since they all exhibit a long wavelength absorption that has been ascribed to a charge-transfer transition, the dicyanoethylene bridge being the acceptor and the two remaining double bonds the donors[500,501]. This concept is supported by the fact that 2,3-dicyanobicyclo[2.2.2]oct-2-ene (**240**) shows only the maximum due to the maleonitrile chromophore, but no lower energy band. Furthermore, increased substitution of the donor bridges with methyl groups moves the charge-transfer transition to longer wavelengths as shown in Table 5 (only the longest wavelength maximum is given)[484,501]. That the maleonitrile grouping is a better acceptor than the corresponding diester is borne out by comparison of the ultraviolet spectra of the adducts **241** and **242** (Table 5)[501]. The adducts of dicyanoacetylene to 1,3-cyclohexadiene and anthracene have been reduced electrolytically to the corresponding anion radicals ($E_{\frac{1}{2}}$ $-1{\cdot}59$ and $-1{\cdot}4$ v, respectively, versus SCE in acetonitrile)[685].

Dicyanoacetylene reacts with styrenes by a sequence of Diels–Alder addition and ene reaction to give 1:2 adducts as illustrated in equation (212). The intermediate 1:1 adduct could not be isolated. Again, the reaction occurs under exceptionally mild conditions; other activated acetylenes, such as hexafluoro-2-butyne and acetylenedicarboxylic ester, also undergo this reaction, but higher temperatures are required[502]. Other examples of the high reactivity of dicyanoacetylene in the Diels–Alder reaction are its additions to furan[490] and furan derivatives[436,503], fulvenes[432,501], 1,4-diphenylnaphthlalene[810], cyclooctatetraene[490] and [2.2]paracyclophane[484].

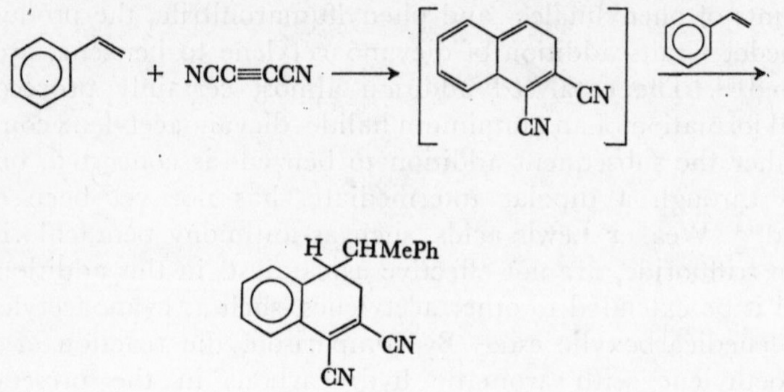

+ NCC≡CCN ⟶ [CN CN] ⟶ (212)

H CHMePh

CN

CN

Dicyanoacetylene reacts at room temperature with quadricyclane[169] and norbornadiene[490,501] to give adducts **243** and **244**, respectively. With bicyclo[2.1.0]pentane it undergoes both a cycloaddition and an ene reaction; the products are **245** and **246**[504]. Dicyanoacetylene acts as a 1,3-dipolarophile in reactions with diazomethane[490], ethyl diazoacetate[505] and benzonitrile oxide[490].

TABLE 5. Ultraviolet spectra of Diels–Alder adducts of dicyanoacetylene.

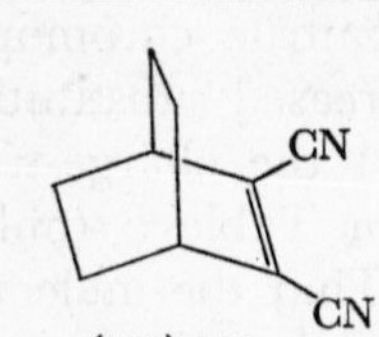

(**240**) 238 mμ	266 mμ	(**239**) 315 mμ
318 mμ	339 mμ	361 mμ
(**241**) 308 mμ	(**242**) 280 mμ	

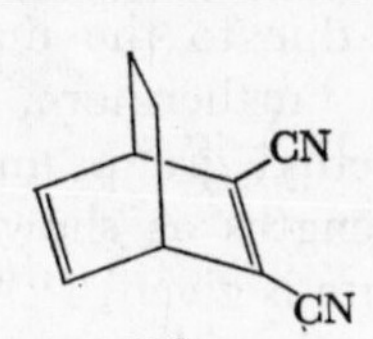

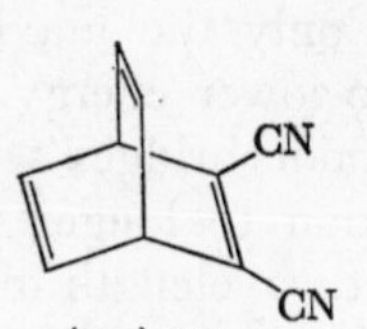

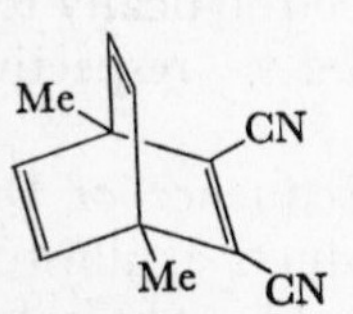

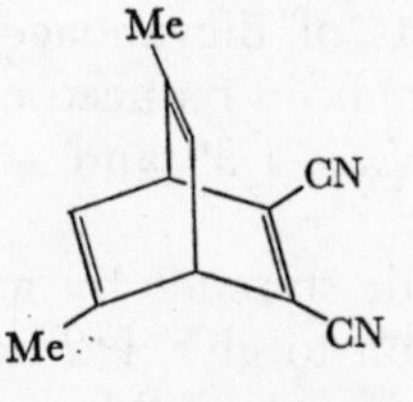

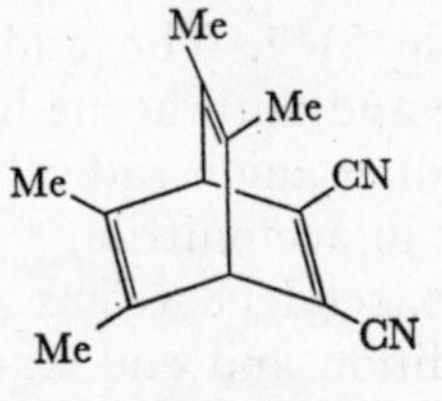

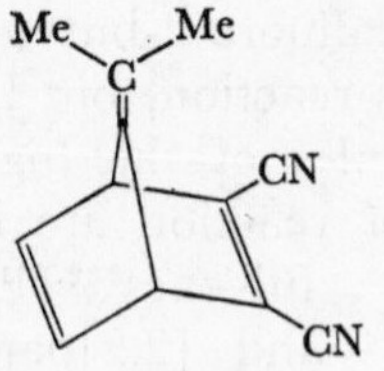

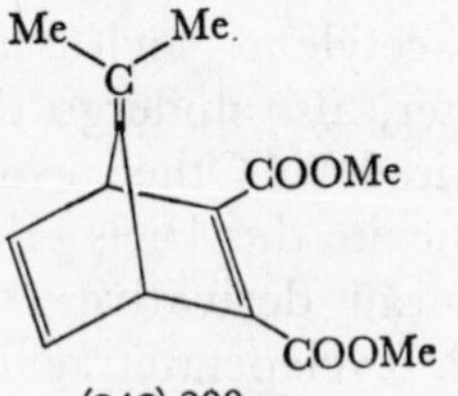

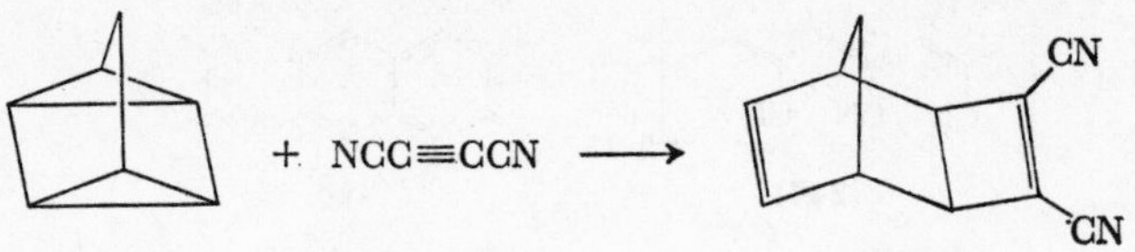

+ NCC≡CCN ⟶ (243)

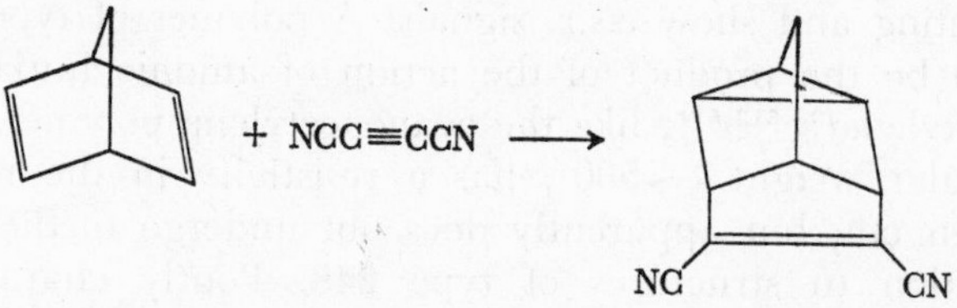

+ NCC≡CCN ⟶ (244)

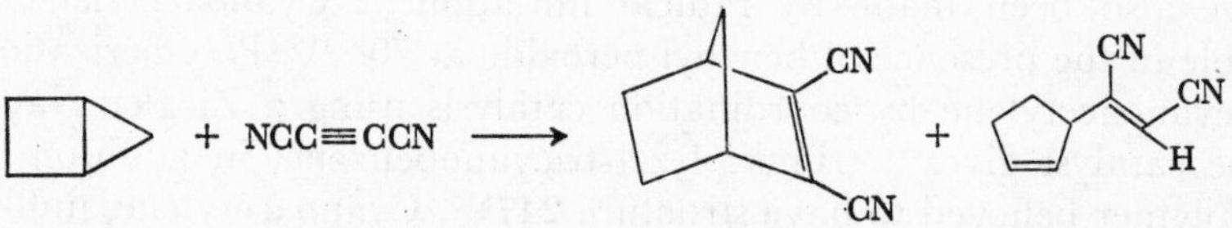

+ NCC≡CCN ⟶ (245) + (246)

In summary, dicyanoacetylene is a more active dienophile than any other activated acetylene including hexafluoro-2-butyne and acetylenedicarboxylic ester[490,500]. It has the added advantage over the latter of being thermally more stable, and excess dicyano-acetylene is more readily removed due to its much lower boiling point. Although no direct comparison has been made, it seems certain that dicyanoacetylene is also more reactive in most Diels–Alder reactions than maleic anhydride and tetracyanoethylene; in fact, it is probably only surpassed by benzyne, and some of the cyclic azo compounds such as 4-phenyl-1,2,4-triazoline-3,5-dione[506]. However, the temperature range within which the latter is stable is much smaller than that of dicyanoacetylene.

4. Polymerization

Cyanoacetylenes have attracted some attention as potential monomers for semiconducting polymers. Cyanoacetylene[507–510], methylcyanoacetylene[511] and phenylcyanoacetylene[507] have been polymerized with anionic initiators such as triethylamine, or sodium cyanide, to give black, low-molecular weight polymers claimed to have structure **247**; on heating, these are converted to thermostable polymers believed to be of type **248**, in analogy to the formation of black 'Orlon®' (registered trademark of E. I. duPont de Nemours

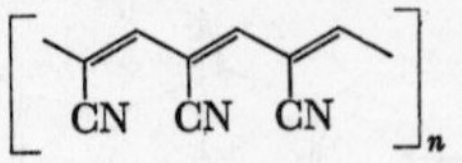

(247) 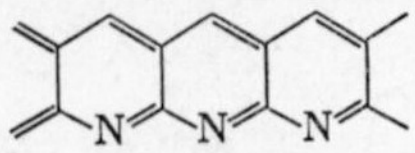 (248)

and Co.) by pyrolysis of polyacrylonitrile[512]. The polymers are semiconducting and show e.s.r. signals. A polymer of type **247** is believed to be the product of the action of anionic initiators on dicyanoacetylene[433,513,514]; like the cyanoacetylene polymer, it is of low molecular weight ($\sim$500), has a resistivity in the range of 10^9–10^{11} ohm cm, but apparently does not undergo further cyclopolymerization to structures of type **248**. Poorly characterized homopolymers of dicyanoacetylene and a copolymer with styrene have also been made by radical initiation[433]. Cyanoacetylene is stable in the presence of benzoyl peroxide at 70°[415]. Polymerization of cyanoacetylene by coordination catalysis using a Ziegler–Natta type catalyst gives a trimer, 1,2,4-tricyanobenzene, in addition to a polymer believed to have structure **247**[515]. Cyanoacetylene, unlike methyl propiolate, is not trimerized by bis(triphenylphosphine)-nickel dicarbonyl[516].

VI. POLYCYANOBENZENES AND RELATED COMPOUNDS

A. Tricyanobenzenes

Of the three isomeric tricyanobenzenes **249**, **250** and **251**, only the latter has been characterized and investigated to any extent.

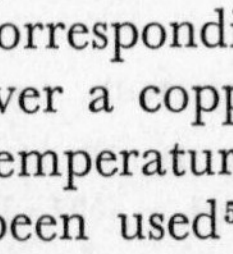

(249) 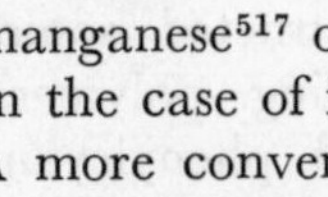 (250) 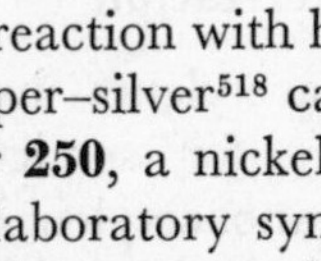 (251)

1,2,3-Tricyanobenzene (**249**) and the 1,2,4-isomer **250** are formed from the corresponding trichlorobenzenes on reaction with hydrogen cyanide over a copper–manganese[517] or copper–silver[518] catalyst at elevated temperatures. In the case of isomer **250**, a nickel catalyst has also been used[519]. A more convenient laboratory synthesis of 1,2,4-tricyanobenzene is the trimerization of cyanoacetylene with Ziegler–Natta catalysts at room temperature[515]. 1,3,5-Tricyanobenzene (**251**) has been prepared from trimesic acid via the acid chloride and amide[520,521], by reaction of the tris acid chloride of trimesic acid with phenylsulphonylphosphorimidic trichloride[522], by oxidation of mesitylene in the presence of ammonia over vanadium catalysts[523–526] and by the catalysed reaction of 1,3,5-trichlorobenzene with hydrogen cyanide[517,518]. Complexes of 1,3,5-tricyanobenzene have been studied to a limited extent[520,521,527,528], and it has been concluded that it is a poorer π-acceptor than 1,3,5-trinitrobenzene[520,527].

1,3,5-Tricyano-2,4,6-trimethylbenzene (**252**) has been prepared from 3,5-dicyano-2,4,6-trimethylaniline by a Sandmeyer reaction[529], from tribromomesitylene with copper cyanide in pyridine at 205° [530], and by dehydration of 3,5-dicyano-2,4,6-trimethylbenzamide[531]. The methyl groups in **252** are rendered acidic by the presence of three cyano groups; thus, condensation with *p*-nitrosodimethylaniline gives the black aldimine **253** (R = *p*-dimethylaminophenyl), which on hydrolysis yields the aldehyde **254**[532]. Similarly, reaction with ethyl nitrite and sodium ethoxide gives the oxime **253** (R = OH) as well as the dioxime **255**, both in the form of their sodium salts (equation 213). 2-Bromo-1,3,5-tricyanobenzene (**256a**), prepared

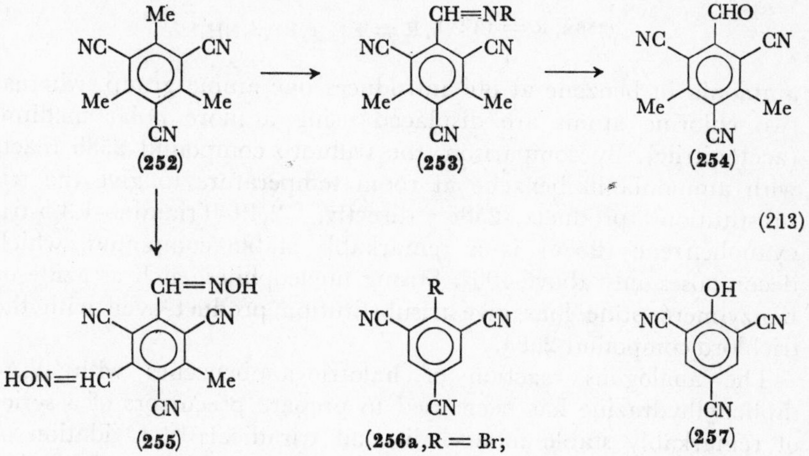

Me, NC, CN, Me, Me, CN (**252**) ⟶ CH=NR, NC, CN, Me, Me, CN (**253**) ⟶ CHO, NC, CN, Me, Me, CN (**254**) (213)

(**252**) ↓ CH=NOH, NC, CN, HON=HC, Me, CN (**255**)

R, NC, CN, CN (**256a**, R = Br; **b**, R = F; **c**, R = N(morpholino)O)

OH, NC, CN, CN (**257**)

by standard methods from bromomesitylene, can be converted to the fluoro derivative **256b** by halogen exchange with potassium or cesium fluoride, a reaction that is facilitated by the presence of the three cyano groups[533]. The halogen in these compounds is subject to ready nucleophilic displacement. Thus, whereas **256a** and **256b** react with morpholine at room temperature to give derivative **256c**, heating to 80° is required to effect this reaction in the corresponding dicyano compound[533]. Reaction of **256a** with aqueous potassium hydroxide at room temperature results in the instantaneous formation of the fluorescent tricyanophenolate ion. The corresponding free phenol **257** has a p*K* value of 1·0, similar to that of picric acid. Tricyanophenol (**257**) has also been prepared from 2-methoxytrimesic acid[534].

The trichloro, and especially the trifluoro derivatives of 1,3,5-tricyanobenzene, are even more susceptible to nucleophilic attack[535]. Reaction of 2,4,6-trichloro-1,3,5-tricyanobenzene (**258a**) with

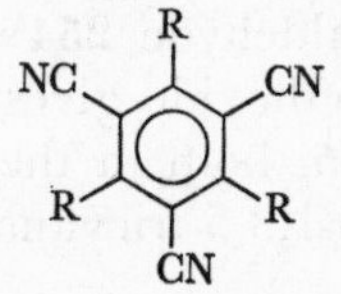

R
NC CN
R R
CN

(**258 a**, R = Cl ; **b**, R = F ; **c**, R = NH_2)

ammonia in benzene at 80° introduces one amine group, whereas two chlorine atoms are displaced using a more polar medium (acetonitrile). By comparison, the trifluoro compound **258b** reacts with ammonia in benzene at room temperature to give the trisubstitution product **258c** directly. 2,4,6-Triamino-1,3,5-tricyanobenzene (**258c**) is a remarkably stable compound, which decomposes only above 400°. Strong nucleophiles, such as azide or benzylmercaptide ions give trisubstitution product even with the trichloro compound **258a**.

The analogous reaction of halotricyanobenzenes with *N*,*N*-diphenylhydrazine has been used to prepare precursors of a series of remarkably stable mono-, di- and triradicals[536]. Oxidation of the monohydrazine derivatives **259a** and **259b** with lead dioxide produces the monoradicals **260** (equation 214); diradical **261** and triradical **262** are made analogously from the bis- and tris(diphenylhydrazine) derivatives, respectively. All are obtained in the form of stable black crystals having a metallic lustre. They are monomeric in solution. The monoradicals show five-line e.s.r. spectra, whereas only one broad line, due to interaction between the spins, is observed in the spectra of the di- and triradicals. The radicals are oxidizing

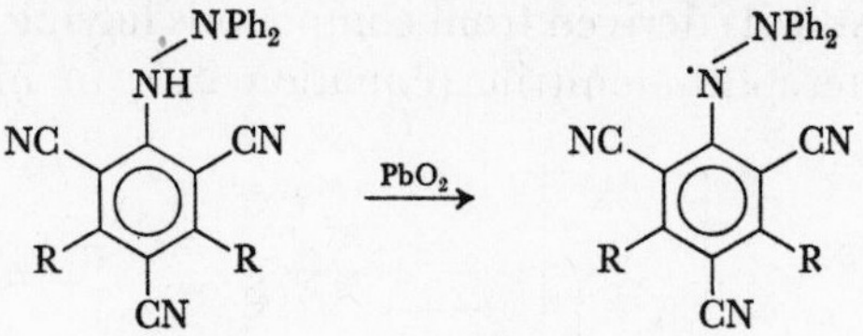

(259) $\xrightarrow{PbO_2}$ (260) (214)

(a, R = H; b, R = Cl)

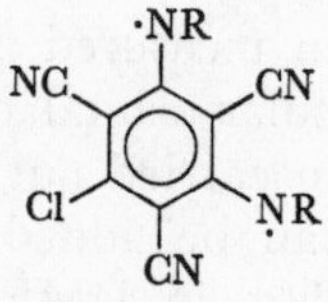

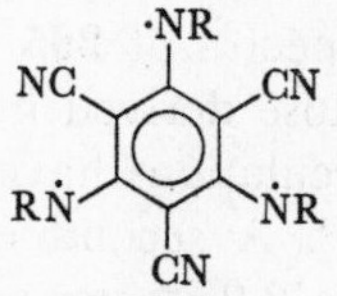

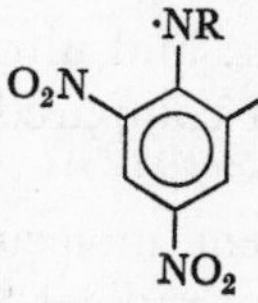

(261) (262) (263)

(R = NPh_2)

agents, for instance for dihydropyridines and hydroquinone, and are easily reconverted to the parent compounds by catalytic hydrogenation. It is of interest to compare the stabilities of these radicals with those of the corresponding nitro compounds. *N,N*-Diphenylpicrylhydrazyl (**263**) is comparable in stability to the monoradicals **260**, whereas the trinitrodiradical (corresponding to **261**) is very unstable. A triradical has not been obtained in the picryl series. This diminished stability of the picryl radicals has been explained as being a consequence of steric inhibition of resonance[536].

B. *Tetracyanobenzenes*

Of the three possible isomers, only 1,2,4,5-tetracyanobenzene (**264**) is known; the dimethyl derivative of one of the other isomers, 1,2,3,5-tetracyanobenzene, has been described[532]. 1,2,4,5-Tetracyanobenzene has been prepared via the tetraamide of the corresponding acid[537–539], by treatment of 7-oxa-2,3,5,6-tetracyanobicyclo[2.2.1]hepta-2,5-diene with triphenylphosphine[436], and from 3,6-diamino-1,2,4,5-tetracyanobenzene (see below)[540]. 1,2,4,5-Tetracyanobenzene forms an anion radical **265** on electrolytic reduction ($E_{\frac{1}{2}}$ −1·02 v versus $Ag/AgClO_4$ in DMF[541]; $E_{\frac{1}{2}}$ −0·66 v versus SCE in acetonitrile[134,542,543]) or chemical reduction with alkali metals[542,544]. This ion radical can also be generated by treatment of

264 with ion radicals derived from compounds having lower electron affinities, e.g. terephthalonitrile (equation 215) or nitrobenzene[545].

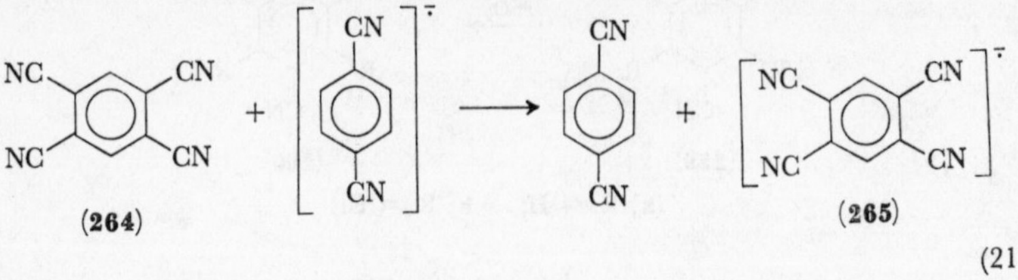

$$\mathbf{264} + [p\text{-}C_6H_4(CN)_2]^{\cdot-} \longrightarrow p\text{-}C_6H_4(CN)_2 + [1,2,4,5\text{-}C_6H_2(CN)_4]^{\cdot-}\ (\mathbf{265}) \tag{215}$$

The e.s.r. and ultraviolet spectra of **265** have been measured and the data compared with those derived from molecular-orbital calculations[236,542,544]. Such calculations have also been carried out on the parent molecule **264**[546]. A second electron can be added to the anion radical **265** ($E_{\frac{1}{2}}$ $-2{\cdot}07$ v versus $Ag/AgClO_4$ in DMF[541]; $E_{\frac{1}{2}}$ $-1{\cdot}63$ v versus SCE in acetonitrile[542]), but the nature of the product has not been determined[541].

A large number of complexes of 1,2,4,5-tetracyanobenzene with aromatic hydrocarbons and nitrogen bases has been prepared in solution[542,547–552]; the crystal structures of three such complexes have been determined by x-ray techniques[553–555]. Based on polarographic and spectral data, 1,2,4,5-tetracyanobenzene is a poorer π-acceptor than tetracyanoethylene[552] and about comparable in this respect to pyromellitic anhydride (**266**)[542].

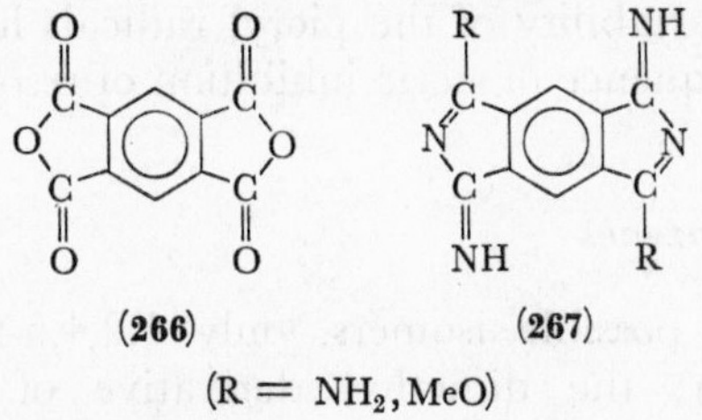

(**266**) (**267**)

($R = NH_2$, MeO)

Heating 1,2,4,5-tetracyanobenzene with cuprous chloride[537] or copper acetonylacetonate[556] gives black amorphous products, considered, mainly by analogy to the corresponding reaction of phthalonitrile, to have phthalocyanine-like structures. They are semiconducting[556] and have been used as cocatalysts for the polymerization of methyl methacrylate[557]. Reaction of 1,2,4,5-tetracyanobenzene with methanol, hydrogen sulphide or ammonia produces compounds considered to have structures **267**; pyrolysis of these gives black semiconducting linear or phthalocyanine-like polymers[256,558–561]. Heating 1,2,4,5-tetracyanobenzene in quinoline gives a black solid for which a triazine-type structure has been proposed[256]. A red copolymer stable to 600° is obtained from 1,2,4,5-tetracyanobenzene and diamines, such as *p*-phenylenediamine, in the presence of base[562]. Semiconducting polymers are also formed on pyrolysis of complexes of 1,2,4,5-tetracyanobenzene with naphthalene and anthracene[563].

The interesting orange-red 3,6-diamino-1,2,4,5-tetracyanobenzene (**268**) is formed in low yield on heating tetracyanoethane with potassium acetate in acetic acid[540]; the proposed mechanism, involving a double Thorpe condensation as the first step, is shown in equation (216). Other products of the reaction are tetracyanoethylene

$$(NC)_2CH{-}CH(CN)_2 + (NC)_2CH{-}CH(CN)_2 \longrightarrow \text{diamino(hexacyano)cyclohexadiene} \longrightarrow \text{diimine} \xrightarrow{(NC)_2CHCH(CN)_2} \underset{(\mathbf{268})}{C_6(NH_2)_2(CN)_4} + (NC)_2C{=}C(CN)_2 \qquad (216)$$

anion radical, potassium pentacyanopropenide, hydrogen cyanide and potassium heptacyanopentadienide. Reaction of 3,6-diamino-1,2,4,5-tetracyanobenzene (**268**) with nitrous acid and hydrochloric acid produces a mixture of the yellow monochloroamine **269** and the colourless dichloro compound **270** (equation 217). Reaction of the latter with diphenylhydrazine gives a monosubstitution product which on oxidation gives a stable radical analogous to the ones described in the tricyanobenzene series[532]. Treatment of 3,6-diamino-1,2,4,5-tetracyanobenzene with nitrosyl fluoroborate followed by hypophosphorous acid produces 1,2,4,5-tetracyanobenzene (**264**). Hydrolysis of the intermediate, believed to be the bis(diazonium)salt **271**, gives 3,6-dihydroxy-1,2,4,5-tetracyanobenzene (**272**). Solutions of **272** are yellow, showing blue fluorescence; the anion is red with yellow fluorescence[540].

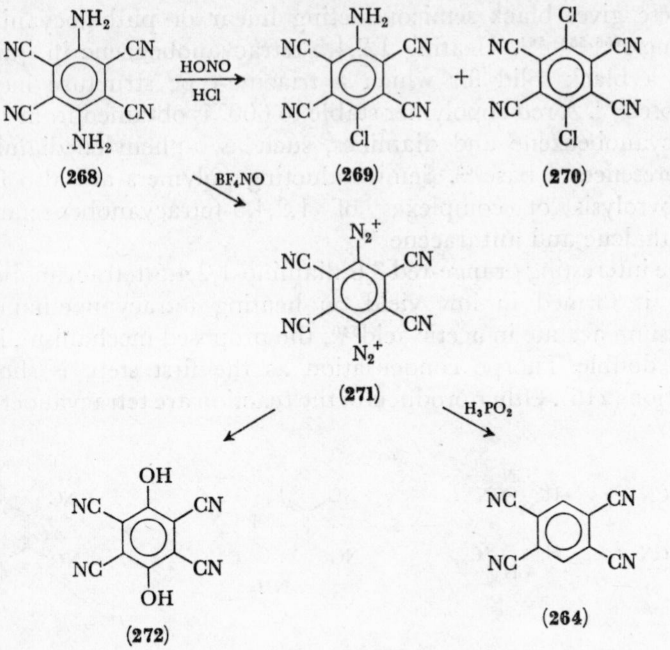

(268) $\xrightarrow[HCl]{HONO}$ (269) + (270)

(268) $\xrightarrow{BF_4NO}$ (271) $\xrightarrow{H_3PO_2}$ (264)

(271) → (272)

(217)

C. Penta- and Hexacyanobenzene

Although pentacyanobenzene itself has not been described, the methyl derivative **273** is formed on dehydration of the dioxime **255** (section VI.A) with thionyl chloride (equation 218)[532]. Methylpentacyanobenzene (**273**) is colourless and forms salts as a consequence of the considerable acidity of the methyl protons. Blue piperidine, pyridine and morpholine salts, as well as an unstable sodium salt, have been reported[532]. The acid character of the methyl protons is further demonstrated by the ready formation of the black Schiff base **274** (R = *p*-dimethylaminophenyl) with *p*-nitrosodimethylaniline and the oxime **274** (R = OH) with ethyl nitrite in dimethylformamide. Dehydration of the latter with thionyl chloride leads to the final member of this class, hexacyanobenzene (**275**; equation 218)[532]. Hexacyanobenzene is a colourless crystalline compound, which sublimes at 200°/10^{-3} mm and decomposes above 310° without melting. As expected for a highly symmetrical molecule, its infrared spectrum shows only four bands at 2260 (weak), 1430, 844 and 748 cm^{-1} (all strong). A structure determination by x-ray techniques shows that it has a planar benzene ring with Pa3

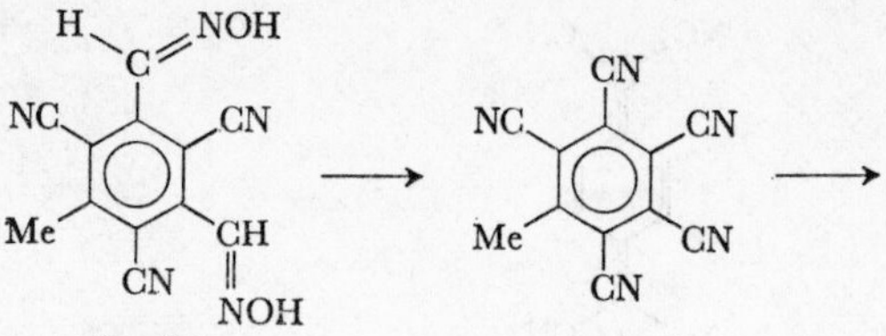

(255) → (273) → (274) → (275) (218)

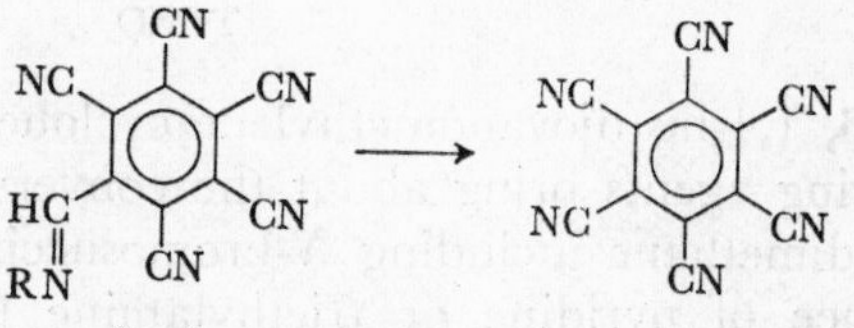

symmetry, a space group rare among organic compounds[564,565]. Hexacyanobenzene forms coloured complexes with aromatic hydrocarbons[532]; a black 1:2 complex with pyrene has been isolated[566]. The one-electron reduction potential of hexacyanobenzene has been measured (E_1 −0·60 v versus SCE) but the nature of the product has not been stated[134]. Its electron affinity has been calculated (E_0 = 58·6 ± 3·3 kcal/mole as compared to 50·8 ± 5·1 kcal/mole for 1,2,4,5-tetracyanobenzene)[134]. Hexacyanobenzene is readily attacked by water, alcohols and amines[532]; the course of the reactions has not yet been reported.

With the possible exception of a tricyanoperylene of unspecified structure[567], there appear to be no reports on condensed aromatic compounds containing more than two cyano groups.

VII. TETRACYANOQUINODIMETHANES AND POLYCYANOQUINONES

A. *Tetracyanoquinodimethanes*

The *Chemical Abstracts* name for **276**, 2,5-cyclohexadiene-$\Delta^{1,\alpha:4,\alpha'}$ dimalononitrile is cumbersome, and the simpler 7,7,8,8-tetracyanoquinodimethane (TCNQ) will be employed.

1. Synthesis and physical properties

Condensation of malononitrile with 1,4-cyclohexanedione in aqueous medium with β-alanine as a catalyst gives the intermediate

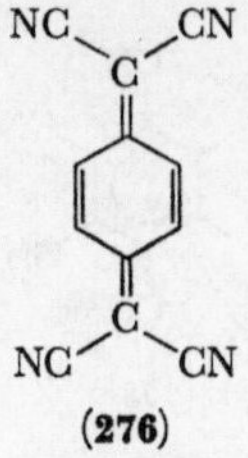

NC CN C (276) TCNQ

to **276**, 1,4-bis(dicyanomethylene)cyclohexane (**277**). A number of oxidizing agents bring about the conversion of **277** to tetracyanoquinodimethane including *N*-bromosuccinimide or bromine in the presence of pyridine or triethylamine (equation 219). Selenium

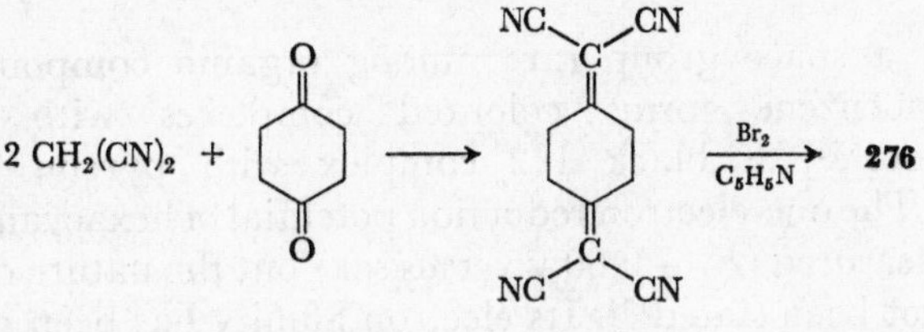

$$2\ CH_2(CN)_2 + \text{(cyclohexane-1,4-dione)} \longrightarrow \text{(277)} \xrightarrow[C_5H_5N]{Br_2} \mathbf{276} \qquad (219)$$

(277)

dioxide has also been used[90,568,569]. The same method of synthesis has been used to prepare tetracyanoquinodimethanes substituted with alkyl groups in the ring[570] (equation 220). Birch reduction of

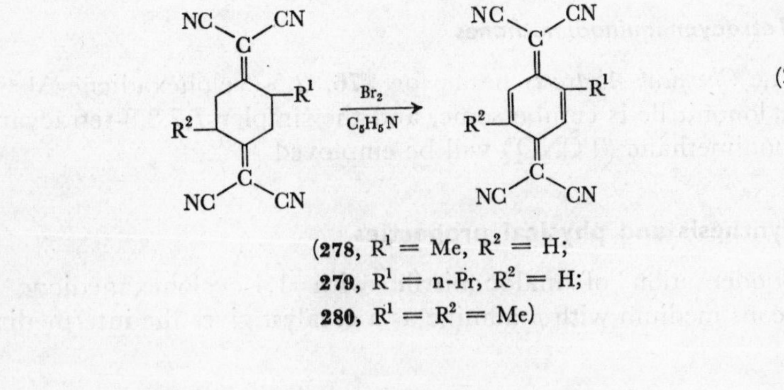

$$\xrightarrow{Na/NH_3} \xrightarrow{H^+} \xrightarrow{CH_2(CN)_2} \xrightarrow[C_5H_5N]{Br_2} \qquad (220)$$

(**278**, R^1 = Me, R^2 = H;
279, R^1 = n-Pr, R^2 = H;
280, R^1 = R^2 = Me)

the corresponding dimethoxybenzenes followed by acid hydrolysis is a convenient route to the substituted cyclohexanediones (**278–280**).

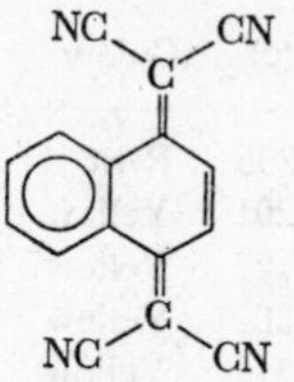

NC CN
C
C
NC CN

(281)

A benzo derivative of **276**, i.e. **281**, has been synthesized by an analogous method[571]. A different synthetic approach has been used for the preparation of 11,11,12,12-tetracyanonaphtho-2,6-quinodimethane (**282**) (TNAP) (equation 221).

Physical properties of all the tetracyanoquinodimethanes reported to date are summarized in Table 6.

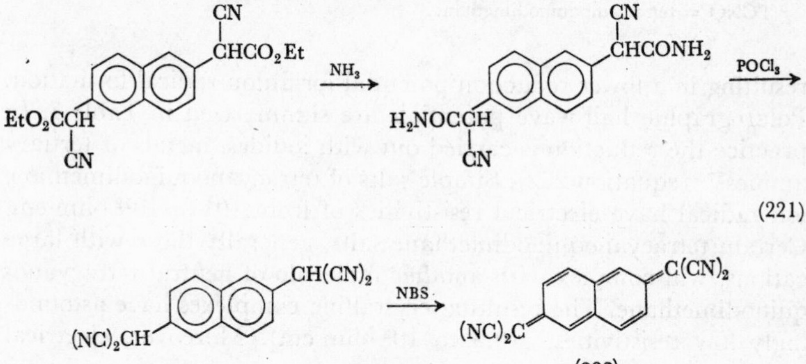

$$\mathrm{EtO_2CCH(CN)}\text{-}\mathrm{C_{10}H_6}\text{-}\mathrm{CH(CN)CO_2Et} \xrightarrow{NH_3} \mathrm{H_2NOCCH(CN)}\text{-}\mathrm{C_{10}H_6}\text{-}\mathrm{CH(CN)CONH_2} \xrightarrow{POCl_3}$$

$$\mathrm{(NC)_2CH}\text{-}\mathrm{C_{10}H_6}\text{-}\mathrm{CH(CN)_2} \xrightarrow{NBS} \mathrm{(NC)_2C}\text{=}\mathrm{C_{10}H_6}\text{=}\mathrm{C(CN)_2} \qquad (221)$$

(282)

TNAP

The heat of formation of tetracyanoquinodimethane is approximately 160 kcal/mole[91]. A complete crystal analysis has been carried out[572].

2. Reduction and anion-radical formation

Each of the tetracyanoquinodimethanes discussed above readily accepts an electron to form a stable anion radical. Tetracyanoquinodimethane is reduced about as easily as tetracyanoethylene. Substitution of the ring with an alkyl group increases the electron density

TABLE 6. Physical properties of tetracyanoquinodimethanes.

Compound	m.p. (°c)	Colour	λ_{max} (mμ) (ε)	C≡N absorption in the infrared region (cm^{-1})
TCNQ (**276**)[a]	293·5–296	Rust	395 (63,600)	2220
2-MeTCNQ (**278**)	200–201	Yellow	396 (~60,000)	
2-n-PrTCNQ (**279**)	125–127	Yellow	397 (~60,000)	
2,5-diMeTCNQ (**280**)	265–267	Yellow	403 (~60,000)	
BenzoTCNQ (**281**)	244–245	Yellow	288 (69,000)	2210
			392 (34,700)	
			409 (33,000)	
TNAP (**282**)	>420	Purple	472 (87,000)	2210
			258 (14,050)	
			248 (18,000)	
2-Benzhydryl TCNQ (**288**)	350–353	Yellow	361 (25,600)	2220
			258 (6,650)	
			221 (17,700)	

[a] TCNQ = tetracyanoquinodimethane.

resulting in a lower reduction potential for anion-radical formation. Polarographic half-wave potentials are summarized in Table 7. In practice the reduction is carried out with iodides, metals or tertiary amines[573] (equation 222). Simple salts of tetracyanoquinodimethane ion radical have electrical resistivities of from 10^4 to 10^{12} ohm cm. Certain tetracyanoquinodimethane salts, generally those with large cations, will complex with another molecule of neutral tetracyanoquinodimethane. The resulting crystalline complexes have astoundingly low resistivities, (10^{-2} to 10^3 ohm cm). Moreover, electrical

TABLE 7. Polarographic reduction of tetracyanoquinodimethanes[a].

Compound	1st half-wave potential (v)	2nd half-wave potential (v)
TCNQ (**276**)	+0·13	−0·28
2-MeTCNQ (**278**)	+0·12	−0·26
2-n-PrTCNQ (**279**)	+0·10	−0·31
2,5-diMeTCNQ (**280**)	+0·02	−0·28
TNAP (**282**)	+0·21	−0·17
Benzo TCNQ (**281**)	−0·09	−0·36
2-Benzhydryl TCNQ (**288**)	−0·28	−0·59

[a] In 0·1 M lithium perchlorate in acetonitrile versus SCE.

$$\text{(NC)}_2\text{C}{=}\text{C}_6\text{H}_4{=}\text{C(CN)}_2 + e \rightleftharpoons \text{(NC)}_2\dot{\text{C}}\text{—C}_6\text{H}_4\text{—}\bar{\text{C}}\text{(CN)}_2 \qquad (222)$$

measurements on single crystals have demonstrated marked anisotropy of resistivity. For example, the resistivity of triethylammonium tetracyanoquinodimethanide–tetracyanoquinodimethane is 0·25 ohm cm along the *a* axis, 20 ohm cm along the *b* axis and 1000 ohm cm along the *c* axis.

The e.s.r. spectrum of tetracyanoquinodimethane ion radical exhibits over forty lines. On the basis of the isotropic hyperfine contact interaction between the unpaired electron and the ^{1}H and ^{14}N nuclei, forty-five lines are expected[574]. The e.s.r. of the ion radical enriched with ^{13}C at the 1- and 4-positions has been studied[575].

The electronic absorption spectrum of tetracyanoquinodimethane ion radical exhibits major maxima at 420 and 842 mμ ($\varepsilon = 24{,}300$ and 43,000). There is also an electronic absorption band in the near infrared region at 6700–10,000 cm^{-1}.

In concentrated aqueous solutions, the green lithium tetracyanoquinodimethane assumes a vivid blue colour as a consequence of dimer formation. Dilution of these blue solutions results in reversion to the green monomeric ion radical[576]. As with other cyanocarbon ion radicals, treatment with strong mineral acids brings about disproportionation to dihydroquinodimethane and quinodimethane. A similar reaction occurs with tropylium iodide (equation 223)[573].

$$2\,\text{C}_7\text{H}_7^{+} + \cdot\text{C(CN)}_2\text{—C}_6\text{H}_4\text{—C(CN)}_2^{-} \longrightarrow \text{C}_7\text{H}_7\text{—C(CN)}_2\text{—C}_6\text{H}_4\text{—C(CN)}_2\text{—C}_7\text{H}_7 + \text{(NC)}_2\text{C}{=}\text{C}_6\text{H}_4{=}\text{C(CN)}_2 \qquad (223)$$

LCAO–MO crystal-field splitting calculations indicate that tetracyanoquinodimethane ion radical complexes will not display properties similar to conventional inorganic semiconductors[577].

In addition to these stable anion radical salts, tetracyanoquinodimethane forms π-complexes with a great many Lewis bases. Aromatic hydrocarbons, amines and polyhydric phenols all react to give, with few exceptions, 1:1 charge-transfer complexes.

Much of the research on the tetracyanoquinodimethanes has been concerned with their charge-transfer complexes and anion-radical salts[578]. These are covered in Chapter 10 and no further detail will be included here.

Two novel uses that have been made of the charge-transfer complexes of tetracyanoquinodimethane are in the preparation of electroconductive polymers[579] and as a reagent for metallic ions separated by thin-layer chromatography[580].

The two-electron reduction of tetracyanoquinodimethane corresponds to conversion to the anion of 1,4-bis(dicyanomethyl)-benzene (**283**). This reduction can be carried out with thiophenol, mercaptoacetic acid or hydrogen iodide (equation 224).

$$(NC)_2C{=}C_6H_4{=}C(CN)_2 \ (\mathbf{276}) \underset{}{\overset{HSCH_2CO_2H}{\rightleftharpoons}} (NC)_2CH{-}C_6H_4{-}CH(CN)_2 \ (\mathbf{283}) \quad (224)$$

3. Addition reactions of tetracyanoquinodimethanes

Additions to **276** occur in the 1- and 6-positions with aromatization of the ring[568]. Chlorine adds in the presence of a trace of chloride ion (equation 225) reminiscent of the addition of chlorine to tetracyanoethylene[208]. A sulphurous acid adduct forms when sulphur

$$\mathbf{276} + Cl_2 \longrightarrow (NC)_2ClC{-}C_6H_4{-}CCl(CN)_2 \quad (225)$$

dioxide is passed into an aqueous acetonitrile suspension of **276** (equation 226). *N,N*-Dimethylaniline also adds to give the adduct **284** (equation 227).

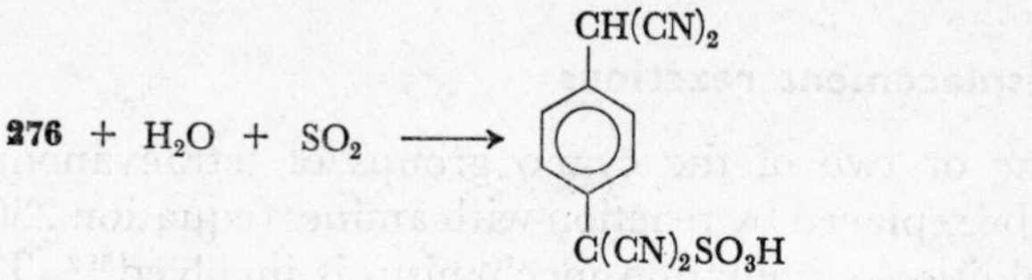

$$\mathbf{276} + H_2O + SO_2 \longrightarrow p\text{-}C_6H_4[CH(CN)_2][C(CN)_2SO_3H] \quad (226)$$

$$\mathbf{276} + PhNMe_2 \longrightarrow Me_2N\text{-}C_6H_4\text{-}C(CN)_2\text{-}C_6H_4\text{-}CH(CN)_2 \quad (227)$$

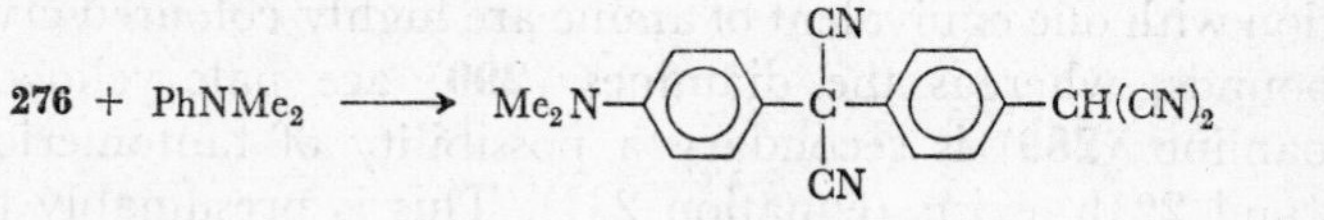

(284)

Free-radical attack occurs in a 1,6-manner. Decomposition of α,α′-azobis(isobutyronitrile) in the presence of **276** gives the adduct **285** in good yield. Another free-radical addition is observed when

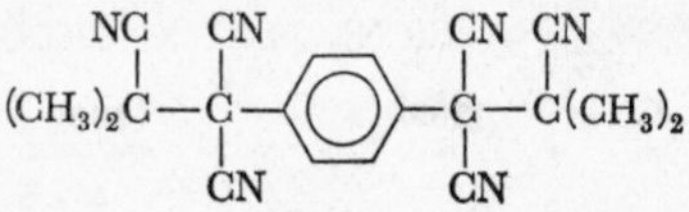

$$(CH_3)_2C(CN)\text{-}C(CN)_2\text{-}C_6H_4\text{-}C(CN)_2\text{-}C(CN)(CH_3)_2$$

(285)

tetracyanoquinodimethane is irradiated with a sunlamp in tetrahydrofuran solution[172]. The adduct **286** is formed in 55% yield along with a small amount of the reduction product **283** (equation 228). The only example of ring substitution is the reaction with

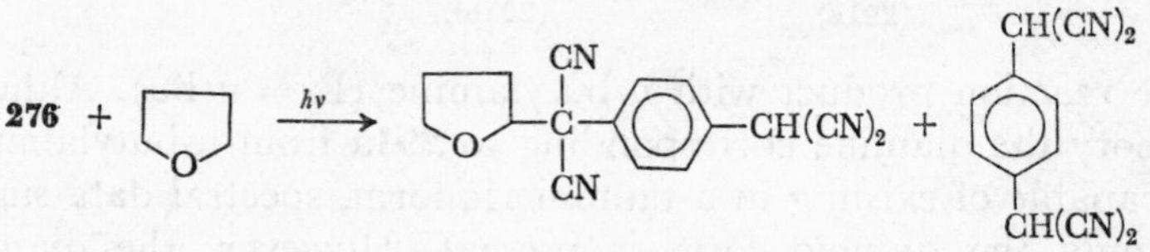

$$\mathbf{276} + \text{THF} \xrightarrow{h\nu} \text{(2-tetrahydrofuryl)}\text{-}C(CN)_2\text{-}C_6H_4\text{-}CH(CN)_2 + p\text{-}C_6H_4[CH(CN)_2]_2 \quad (228)$$

(286) **(283)**

diphenyldiazomethane at 60–80°[581]. Nitrogen is evolved and 2-benzhydryl-7,7,8,8-tetracyanoquinodimethane (**288**) is the product. The reaction probably involves attack on tetracyanoquinodimethane with intermediate formation of a pyrazoline **287** (equation 229).

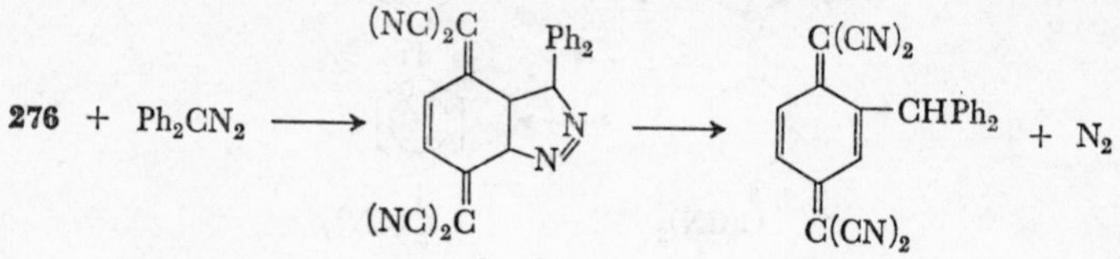

$$\mathbf{276} + Ph_2CN_2 \longrightarrow \mathbf{287} \longrightarrow \mathbf{288} + N_2 \quad (229)$$

(287) **(288)**

4. Displacement reactions

One or two of the cyano groups of tetracyanoquinodimethane may be replaced by reaction with amines (equation 230). Presumably an addition–elimination mechanism is involved[582]. The products of reaction with one equivalent of amine are highly coloured crystalline compounds, whereas the diamines (**290**) are pale yellow. If the monoamine (**289**) is secondary a possibility of tautomeric forms, **291a** and **291b**, exists (equation 231). This is presumably the case

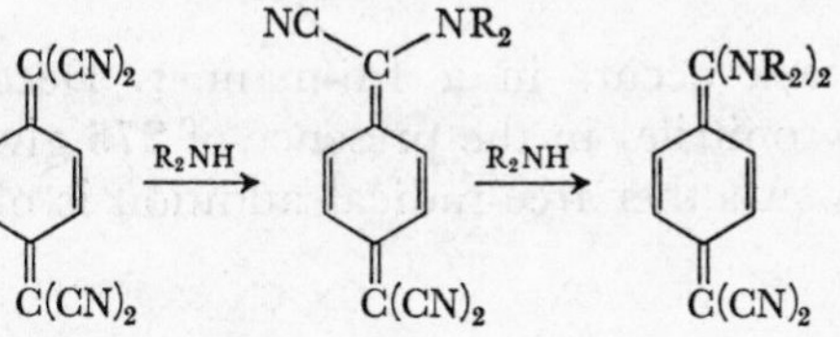

$C(CN)_2$ / $C(CN)_2$ $\xrightarrow{R_2NH}$ NC, NR_2 C / $C(CN)_2$ (**289**) $\xrightarrow{R_2NH}$ $C(NR_2)_2$ / $C(CN)_2$ (**290**) (230)

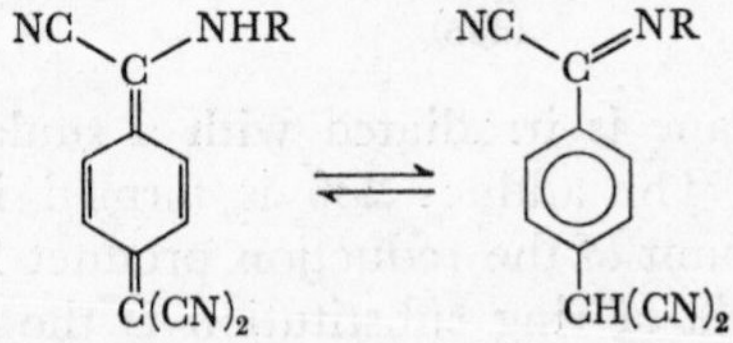

NC, NHR C / $C(CN)_2$ (**291a**) ⇌ NC, C=NR / $CH(CN)_2$ (**291b**) (231)

in the reaction product with n-butylamine (R = n-Bu). Although in theory the diamine corresponding to **291a** from n-butylamine is also capable of existing in a tautomeric form, spectral data suggest that only the quinoid form is present. However, the diamino-dicyanoquinodimethanes behave much like amidines. For example, they are monoacidic bases when titrated with acid. This behaviour, combined with infrared spectral evidence, indicates a strong contribution from forms such as **292**. The nitrile absorption is a strong

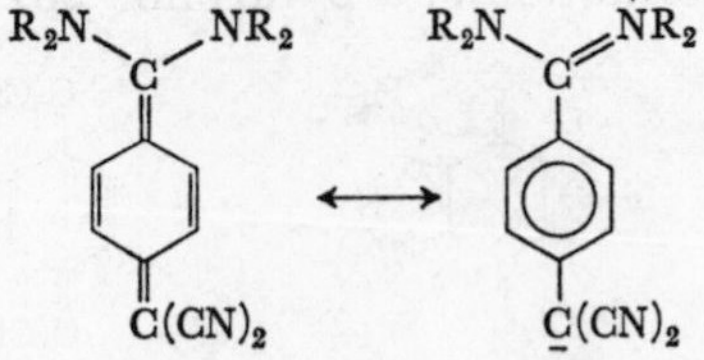

R_2N, NR_2 C / $C(CN)_2$ (**290**) ⟷ R_2N, $\overset{+}{N}R_2$ C / $\underset{-}{C}(CN)_2$ (**292**)

doublet at long wavelengths (2130 and 2175 cm^{-1}) characteristic of a monosubstituted malononitrile anion.

Use of a bifunctional derivative can give a cyclic substitution product. An example is the quinodimethane (**293**) from tetracyanoquinodimethane and monoethanolamine.

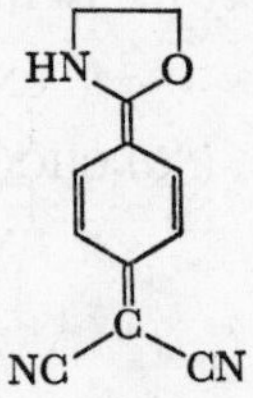

HN O

C

NC CN

(**293**)

Nitrite ion also displaces a cyano group from tetracyanoquinodimethane to give an ion believed to be the anion of α,α-dicyano-*p*-toluyl cyanide (**294**). The acyl cyanide has not been isolated but was alkylated with benzyl bromide to give, after hydrolysis, α-benzyl-α,α-dicyano-*p*-toluic acid (**295**) (equation 232). Related to

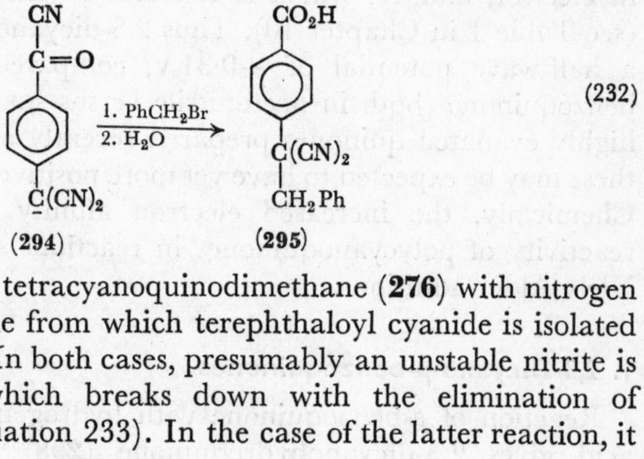

$$\underset{(\mathbf{294})}{NC{-}C(=O){-}C_6H_4{-}\bar{C}(CN)_2} \xrightarrow[2.\ H_2O]{1.\ PhCH_2Br} \underset{(\mathbf{295})}{HO_2C{-}C_6H_4{-}C(CN)_2{-}CH_2Ph} \qquad (232)$$

this is the reaction of tetracyanoquinodimethane (**276**) with nitrogen dioxide in acetonitrile from which terephthaloyl cyanide is isolated in up to 75% yield. In both cases, presumably an unstable nitrite is the intermediate, which breaks down with the elimination of nitrosyl cyanide (equation 233). In the case of the latter reaction, it

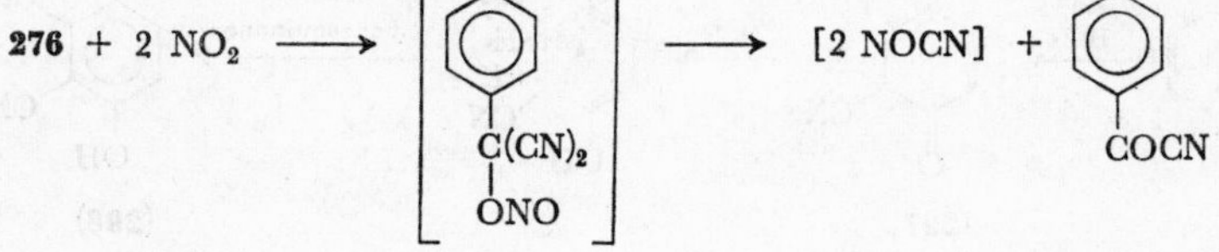

$$\mathbf{276} + 2\ NO_2 \longrightarrow \left[ONO{-}C(CN)_2{-}C_6H_4{-}C(CN)_2{-}ONO\right] \longrightarrow [2\ NOCN] + NCOC{-}C_6H_4{-}COCN \qquad (233)$$

is possible to isolate an unstable intermediate, which readily decomposes with the evolution of nitrogen oxides.

Malononitrile anion readily replaces one of the nitrile groups of tetracyanoquinodimethane to form the deep-blue *p*-tricyanovinylphenyldicyanomethide ion (**296**)[298]. Here again, addition, followed by elimination, seems the most likely mechanism (equation 234).

$$(NC)_2CH^- + \mathbf{276} \longrightarrow (NC)_2CHC(CN)_2\text{-}C_6H_4\text{-}\bar{C}(CN)_2 \xrightarrow{-CN^-}$$

$$(NC)_2CHC(CN){=}C_6H_4{=}C(CN)_2 \underset{+H^+}{\overset{-H^+}{\rightleftharpoons}} (NC)_2C{\cdots}C(CN){\cdots}C_6H_4{\cdots}C(CN)_2^- \quad (\mathbf{296}) \tag{234}$$

B. *Polycyanoquinones*

Introduction of cyano groups into a quinone results in an increase in electron affinity, which is reflected in the reduction potentials (see Table 1 in Chapter 10). Thus 2,3-dicyano-*p*-benzoquinone has a half-wave potential of +0·31 v, compared to −0·52 v for *p*-benzoquinone (both in acetonitrile versus SCE). Data on the more highly cyanated quinones prepared recently are not available, but these may be expected to have yet more positive reduction potentials. Chemically, the increased electron affinity results in enhanced reactivity of polycyanoquinones in reactions such as oxidations or Diels–Alder additions.

I. 2,3-Dicyano-*p*-benzoquinones

Reaction of *p*-benzoquinone with hydrogen cyanide in aqueous acid gives 2,3-dicyanohydroquinone (**298**)[583–586]. The proposed mechanism[600] is shown in equation (235). Monocyanoquinone, which

$$p\text{-benzoquinone} \xrightarrow{HCN} \underset{(\mathbf{297})}{\text{2-cyano-2H-quinol}} \xrightarrow{HCN} \text{2,3-dicyano-2,3-dihydrohydroquinone} \xrightarrow{\text{more } p\text{-benzoquinone}} \underset{(\mathbf{298})}{\text{2,3-dicyanohydroquinone}} \tag{235}$$

could arise by aromatization of the first intermediate (**297**) followed by oxidation, has been ruled out as an intermediate since it is too moisture-sensitive to survive the reaction conditions[587]. Oxidation of the hydroquinone **298** with nitrogen oxides gives the red 2,3-dicyano-*p*-benzoquinone (**299**)[584,586,588]. It forms complexes with donors such as polymethylbenzenes[528,589], azulene[590], hexamethylborazole[591] and phenothiazines[592,593]. The black hexamethylbenzene complex (m.p. >320°) shows a strong e.s.r. signal in the solid due to radical ions trapped in the lattice[589]. 2,3-Dicyano-*p*-benzoquinone is a strong oxidizing agent. Thus reaction with water gives the hydroquinone **298**; the nature of the oxidation product of water has not been stated[584]. It is somewhat more effective as a dehydrogenating agent than tetrachloro-*p*-benzoquinone, but much less so than its dichloro derivative (**303**; see below)[594]. Reduction with sodium amalgam is reported to produce 2,3-dicyanocyclohexane-1,4-dione (**300**; equation 236)[595]. The activity of **299** as a dienophile

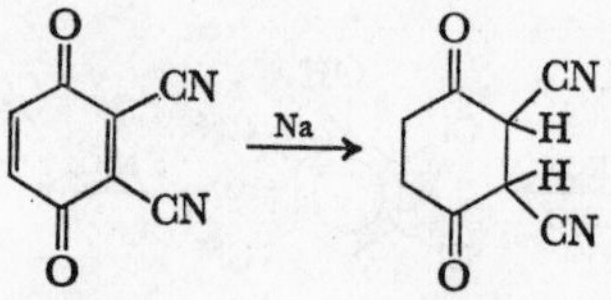

O, CN, CN, O —Na→ O, CN, H, H, CN, O (236)

(299) **(300)**

is illustrated by the ready reaction with dimethylenecyclobutane to give the adduct **301**[596]. The fact that addition occurs at the most electron-deficient double bond, the one carrying the two cyano groups, has been noted in other cases[597].

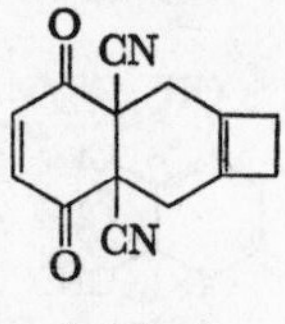

O, CN, CN, O

(301)

Addition of hydrogen chloride to 2,3-dicyano-*p*-benzoquinone, followed by oxidation, produces 5-chloro-2,3-dicyano-*p*-benzoquinone (**302**)[584,586,588], which by a similar sequence can be converted to 2,3-dichloro-5,6-dicyano-*p*-benzoquinone (**303**)[584,588], commonly known as 'DDQ' (equation 237). More direct syntheses of 2,3-dichloro-5,6-dicyano-*p*-benzoquinone from 2,3-dicyanohydroquinone (**298**)[598] and 2,3-dicyano-*p*-benzoquinone (**299**)[599] have

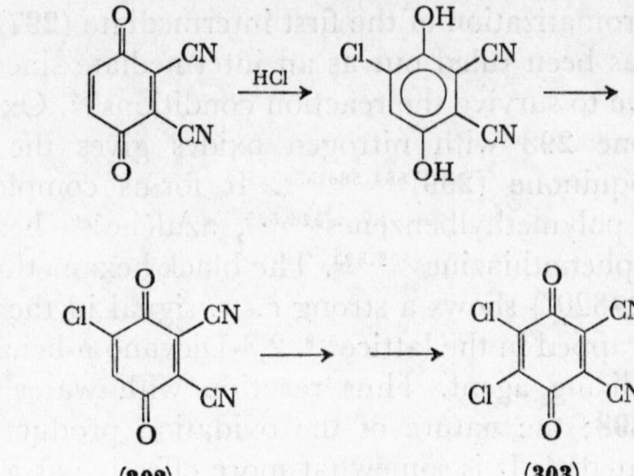

(237)

(302) (303)

been described; the compound is available commercially. Another direct route, which has also been used to prepare a number of other substituted 2,3-dicyano-*p*-benzoquinones (see below), involves reaction of perhalo-*p*-benzoquinone with potassium cyanide in aqueous

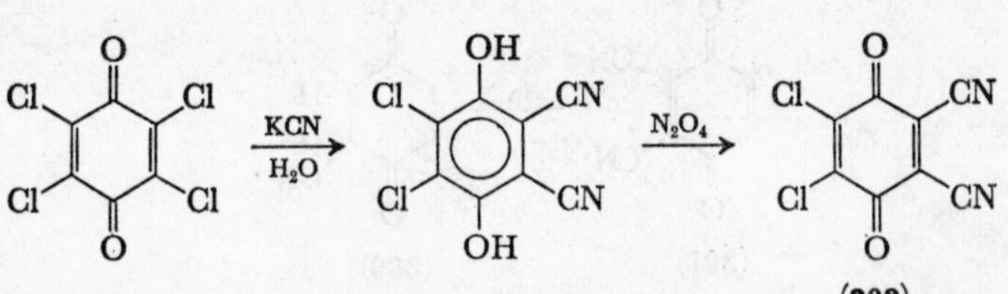

(238)

(303)

medium, followed by oxidation of the resulting hydroquinone, as illustrated in equation (238)[600]. The proposed[600] mechanism of the first step is shown in equation (239). The observations that the trichlorocyanoquinone **304**, synthesized independently[587], also gives

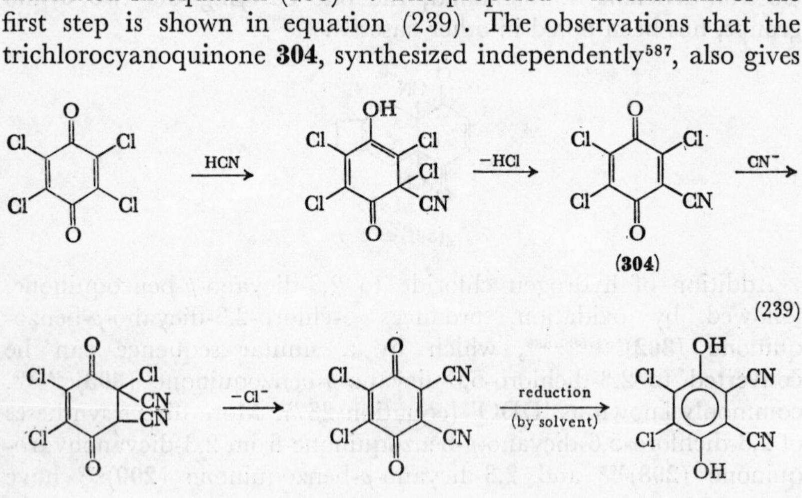

(304)

(239)

reduction (by solvent)

2,3-dichloro-5,6-dicyanohydroquinone under these conditions, and the fact that an oxidizable solvent is required for the success of the reaction, has been used as evidence for this mechanism.

2,3-Dichloro-5,6-dicyano-*p*-benzoquinone is a yellow crystalline solid. Its first half-wave potential (in acetonitrile versus SCE) is 0·51 v; a second electron is added at −0·30 v[601]. The ion radical **305** has been prepared by reduction of **303** with alkali iodides[602,603].

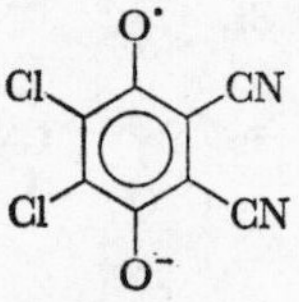

O˙ Cl CN Cl CN O⁻

(305)

Complete electron transfer is believed to occur also in some of the many complexes of **303** that have been studied. Interesting examples are the complexes with metallocenes such as ferrocene, cobaltocene and bis(tetrahydroindenyl)iron[226]. Cobaltocene (reduction potential E^0 −1·16 v) produces cobaltocinium ion radical salts even with weaker acceptors such as *p*-benzoquinone (E^0 −0·52 v), tetrachloro-*p*-benzoquinone (E^0 ≈0 v) and tetracyanoethylene (E^0 +0·15 v), whereas ferrocene (E^0 = +0·30 v) forms such salts only with strong acceptors such as **303** (E^0 +0·51 v). With the weaker acceptors, ferrocene forms π-complexes[226]. Other donor molecules studied include polymethylbenzenes[528,547,589,604,605], condensed aromatic hydrocarbons[547,604–608], aromatic ethers[552], *p*-phenylenediamine and its tetramethyl derivative[604,608–610], 1,6-diaminopyrene[611], 8-hydroxyquinoline and its chelates[550], dimethylalloxazine[612], phenothiazines[592,593,603,613,614], hexamethylborazole[591] and bis-(tropolono)silicone, -germanium and -boron chelates[615,616]. Polyvinylaromatics, such as polystryrene, form semiconducting complexes with **303**[617,618]. Anion radicals of **303** are believed to be the active species in the polymerization of *N*-vinylpyrrole[619], *N*-vinylindole[619] and *N*-vinylcarbazole[620,621], catalysed by **303**. In general, the latter is a stronger π-acid than tetracyanoethylene.

2,3-Dichloro-5,6-dicyano-*p*-benzoquinone has found wide application as a dehydrogenating agent, especially in the steroid field. This aspect of its chemistry has been reviewed in detail elsewhere[622–624]. Other reactions have received much less attention. Its reactivity in the Diels–Alder reaction has been compared to that of tetracyanoethylene[624], but a detailed study does not appear to have been made.

2,3-Dibromo-5,6-dicyano-*p*-benzoquinone (**307**) has been prepared by oxidation of the dibromohydroquinone **306**, which in turn is obtained by bromination of 2,3-dicyano-*p*-benzoquinone (**299**)[584], or 2,3-dicyanohydroquinone (**298**)[625]; the latter can be converted

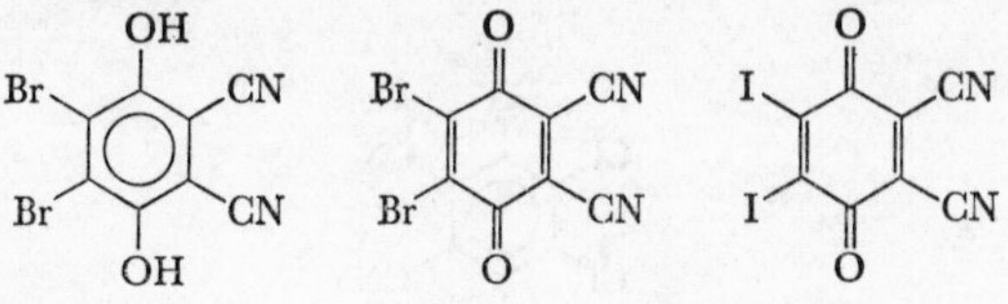

(**306**) (**307**) (**308**)

directly to **307** with *N*-bromosuccinimide[626]. The dibromoquinone **307** and its black-violet diiodo analogue **308** have also been prepared by treatment of the corresponding tetrahalo-*p*-benzoquinones with potassium cyanide, followed by oxidation (cf. equation 238). 2,3-Dicyano-5,6-difluoro-*p*-benzoquinone could not be prepared using this sequence, nor could the remaining halogen atoms in the 2,3-dicyano-5,6-dihalo-*p*-benzoquinone be replaced by further cyano groups in this way[600]. However, these reactions have been used to prepare 2,3-dicyano-1,4-naphthoquinone (**309**) and its nitro derivative **310**[600]. The latter forms a green-black 1 : 1 π-complex with pyrene. With anthracene it reversibly forms a π-complex which itself is in equilibrium with the colourless Diels–Alder adduct[600].

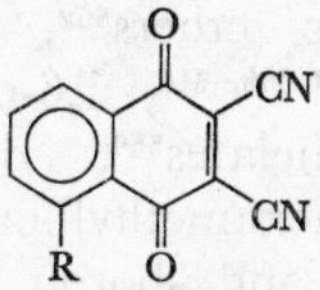

(**309**) (R = H)
(**310**) (R = NO_2)

2,3-Dicyano-5-phenylsulphonyl-*p*-benzoquinone (**311**) and 5-chloro-2,3-dicyano-6-phenylsulphonyl-*p*-benzoquinone (**312**) have been synthesized by the reactions outlined in equation (240)[627]. Quinone **311** has a first half-wave potential ($E_{\frac{1}{2}}$ +0·52 v versus SCE in acetonitrile) similar to that of 2,3-dichloro-5,6-dicyano-*p*-benzoquinone, while that of quinone **312** is considerably higher ($E_{\frac{1}{2}}$ +0·62 v). A polarographic study of a limited number of quinones showed the effect of substituents like chlorine, cyano or phenylsulphonyl to be additive[627]. Both **311** and its chloro derivative **312** are strong π-acceptors, forming coloured complexes with a

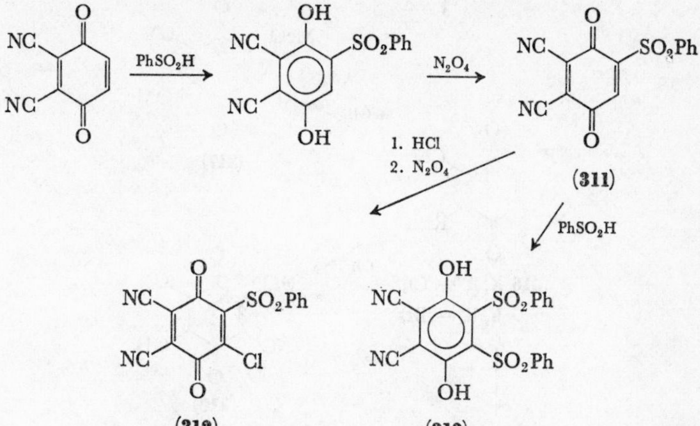

(240)

(311) (312) (313)

number of aromatic hydrocarbons[627]. In the Diels–Alder reactions of **311** and **312** with dimethylbutadiene, addition occurs at the double bond carrying the phenylsulphonyl group. Addition of phenylsulphinic acid to quinone **311** produces the quinol **313**, which, however, resists oxidation to the corresponding quinone[627].

2. 2,5- and 2,6-dicyano-*p*-benzoquinone; tetracyano-*p*-benzoquinone

2,5-Dicyano-*p*-benzoquinone (**314**) and the 2,6-isomer **315** have been prepared by conventional routes from 2,5-dihydroxyterephthalic acid and 1,3-dibromo-2,5-dimethoxybenzene, respectively. Both are quite unstable. They form π-complexes with pyrene[600].

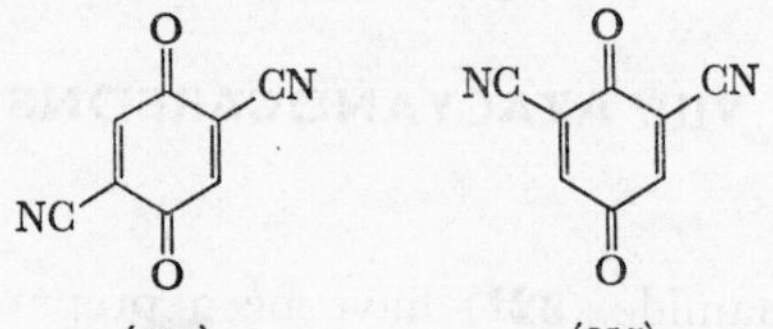

(314) (315)

2,5-Dichloro-3,6-dicyano-*p*-benzoquinone (**316a**) and its dibromo analogue (**316b**) have been synthesized from the corresponding 2,5-dihalo-3,6-dimethoxyterephthalic acids[600]. They react readily with methanol to give 2,5-dicyano-3,6-dimethoxy-*p*-benzoquinone (**317**) and with hydroxide ion to give the corresponding hydroquinone **318** ('cyanilic acid'; equation 241). Secondary amines react in a similar manner[629]. Reaction with cyanide ion in methanol produces

(316 a, R = Cl; b, R = Br) —MeOH→ (317); —OH^-→ (318); ↓ (319) → (320) (241)

the yellow tetracyanohydroquinone **319** (section VI.B) in 11% yield. This compound is remarkable for its ability to form π-complexes, a property not normally found in hydroquinones. Oxidation of **319** gives tetracyano-*p*-benzoquinone (**320**), a yellow compound that melts with decomposition at 205° [600,629]. It forms a blue 1:1 π-complex with pyrene. Chemical reactions of this interesting quinone have not been reported yet.

VIII. AZACYANOCARBONS

A. Dicyanamide

Salts of dicyanamide (**321**) have been prepared by reaction of sodamide with cyanogen bromide[630], and more conveniently by the action of cyanogen halides on salts of cyanamide (e.g. equation 242)[631–633]. The very unstable free acid **321** is obtained by treatment of the sodium or silver salts with hydrogen chloride and hydrogen sulphide, respectively[631,634]. It crystallizes with one mole of water[634] and is an acid whose strength approaches that of hydrogen chloride[631]. The dicyanamide ion is bent and has C_{2v} symmetry as derived from infrared spectral studies[635]. The ultraviolet spectra of **321** and some

$$BrCN + Na_2NCN \xrightarrow{NaBr} NaN(CN)_2 \xrightarrow{HCl} HN(CN)_2 \qquad (242)$$

(321)

of its derivatives have been studied[636]. Transition metal salts of dicyanamide form a series of complexes with dimethyl sulphoxide[637] and pyridines[85,638–640].

Alcohols[631], amines[631,641,642], hydrogen sulphide[643], hydrogen halides[644] and other reagents of this nature add to one or both nitrile groups of dicyanamide. Sodium dicyanamide trimerizes on heating to give the trisodium salt of tricyanomelamine (**322**; equation 243)[634,645–648]. The free tricyanomelamine could not be isolated but its strong acid character was proven by conductivity measurements[646].

$$3\ NaN(CN)_2 \longrightarrow \text{[triazine ring bearing three } \bar{N}CN \text{ groups]}\ 3\ Na^+ \qquad (243)$$

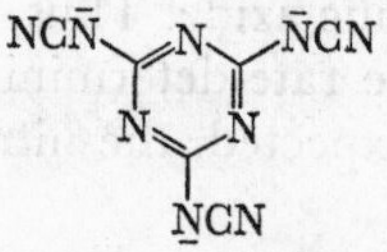

(322)

A number of aryl derivatives of dicyanamide have been prepared[630]. Reaction of sodium dicyanamide with methyl iodide is claimed to yield *N*-methyldicyanamide[631]; its structure assignment has been questioned[630]. The aryl derivatives are very prone to polymerization[630].

B. Tricyanamide

Tricyanamide (**323**) has not been reported to date. Attempts to prepare it by reaction of silver dicyanamide with excess cyanogen chloride (equation 244) were not successful[632]. Reported syntheses of **323** by pyrolysis of mercury thiocyanate[647] and mercury dicyanamide[634] are most probably in error.

$$AgN(CN)_2 + ClCN \longrightarrow N(CN)_3 \qquad (244)$$

(323)

C. Cyanogen Azide

Cyanogen azide (**324**) is prepared by reaction of cyanogen chloride with sodium azide in aprotic solvents (equation 245)[649]. Other cyanogen halides and metal azides may be used[650]. A compound prepared from cyanogen bromide and sodium azide in aqueous solution, claimed to be cyanogen azide[651], has been shown to be the dimer **325**[652]. Cyanogen azide is a colourless oil, which detonates with great violence on mechanical or thermal shock;

$$NaN_3 + NCCl \xrightarrow{-NaCl} N_3CN \quad \textbf{(324)} \tag{245}$$

$$(N_3)_2C{=}NCN \quad \textbf{(325)}$$

it may be handled with relative safety in solution. Photolysis[653–656,660–665] or thermolysis above 50° [657–659,666–669] produces cyanonitrene (see Chapter 11). Below 50°, cyanogen azide reacts as a highly electrophilic azide. Thus, it adds readily to electron-rich olefins to give, in the rate-determining step[649], unstable triazolines (equation 246). As expected, the nitrogen bearing the cyano group

$$N_3CN + Me_2C{=}CH_2 \longrightarrow \text{triazoline (1-NC, 5,5-Me}_2\text{, 4,4-H}_2) \longrightarrow NC{-}N^{-}{-}CMe_2{-}CH_2{-}N_2^{+} \ \textbf{(326)} \tag{246}$$

$$\textbf{(326)} \longrightarrow MeC(CH_2Me){=}NCN \ \textbf{(327)} \quad \text{and} \quad \text{2,2-Me}_2\text{-1-cyanoaziridine} \ \textbf{(328)}$$

becomes attached exclusively to the most highly substituted carbon atom of the olefin[670]. Decomposition of the intermediate triazoline, probably via a diazonium ylid of type **326**, gives, in addition to nitrogen, alkylidenecyanamides (e.g. **327**) and/or *N*-cyanoaziridines (e.g. **328**) as illustrated in equation (246)[670]. The ratio of these two products varies widely, depending on the type of olefin and solvent used. Formation of the alkylidenecyanamides involves hydrogen or alkyl migration; where possible, hydrogen migrates to the exclusion of alkyl groups. If two different alkyl groups may migrate, the ratio of products is determined by steric factors[670]. Cyclic olefins may undergo ring contraction (e.g. equation 247)[670]; this reaction has

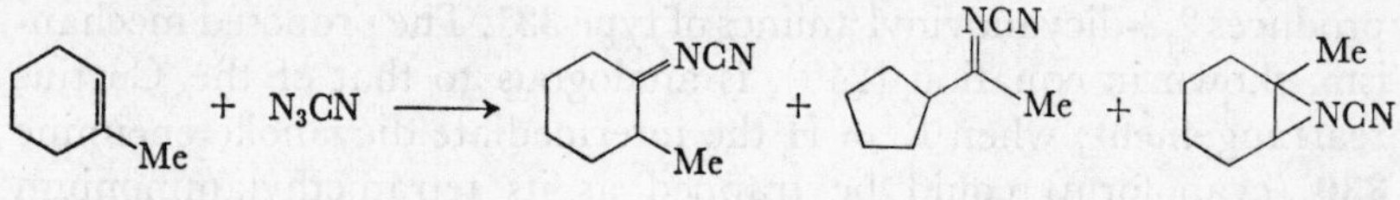

(247)

been used to prepare nor- and dinorsteroids[671]. Alkylidenecyanamides are high-boiling, unstable oils which are readily hydrolysed to the corresponding ketones by dilute acid, base or silver nitrate[649,670]. The reaction of cyanogen azide with olefins thus provides a general ketone synthesis under very mild conditions.

Reaction of cyanogen azide with acetylenes gives products that are equilibrium mixtures of triazoles and diazoimines as shown by spectral studies and chemical degradation (equation 248)[672].

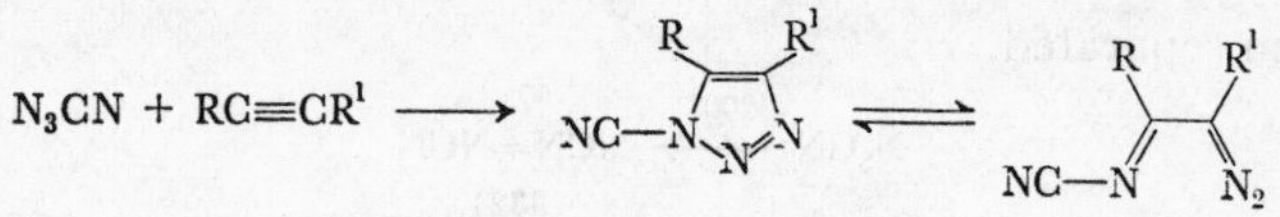

$$N_3CN + RC{\equiv}CR^1 \longrightarrow \text{NC—N(triazole)} \rightleftharpoons \text{NC—N=C(R)—C(R}^1\text{)=N}_2 \quad (248)$$

D. 2,2-Dicyanovinyl Azides

A series of 2,2-dicyanovinyl azides (**329**) has been prepared by the method outlined in equation (249)[673]. Thermolysis of these compounds in the presence of hydrogen halides, amines or alcohols

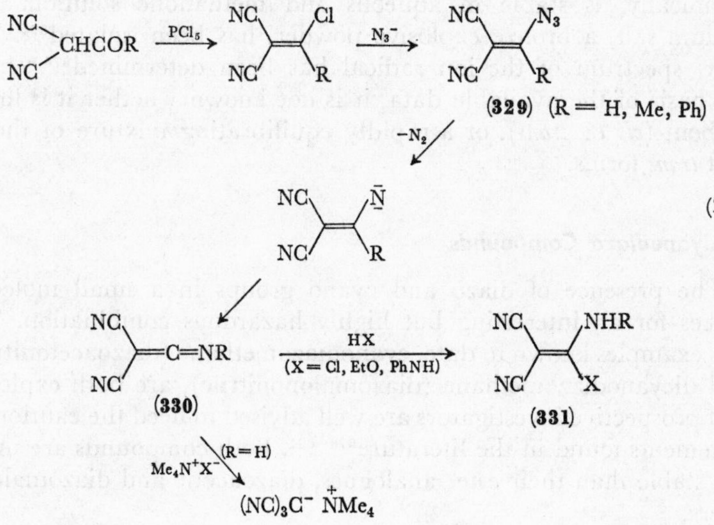

$$(NC)_2CHCOR \xrightarrow{PCl_5} (NC)_2C{=}C(Cl)R \xrightarrow{N_3^-} (NC)_2C{=}C(N_3)R$$

(**329**) (R = H, Me, Ph)

$$\xrightarrow{-N_2} (NC)_2C{=}C(\bar{N})R \quad (249)$$

$$(NC)_2C{=}C{=}NR \; (\mathbf{330}) \xrightarrow[(X = Cl,\ EtO,\ PhNH)]{HX} (NC)_2C{=}C(NHR)X \; (\mathbf{331})$$

$$(\mathbf{330}) \xrightarrow[Me_4N^+X^-]{(R=H)} (NC)_3C^- \overset{+}{N}Me_4$$

produces 2,2-dicyanovinyl amines of type **331**. The proposed mechanism, shown in equation (249), is analogous to that of the Curtius rearrangement; when R = H the intermediate dicyanoketeneimine **330** (cyanoform) could be trapped as its tetramethylammonium salt[673].

E. Azodinitrile

Gas-phase pyrolysis of cyanogen azide at 200° produces azodinitrile (**332**; equation 250), presumably by reaction of cyanonitrene with undecomposed azide[674]. Azodinitrile is an orange-red, volatile solid which detonates on thermal or mechanical shock. Vapour-pressure measurement and gas-chromatographic investigations indicate the presence of both the *cis* and the *trans* isomer; these have not been separated.

$$N_3CN \xrightarrow{200^\circ} \underset{(\mathbf{332})}{NCN{=}NCN} \qquad (250)$$

Azodinitrile is a potent dienophile; thus at room temperature it reacts instantaneously with cyclopentadiene or 2,3-dimethylbutadiene, and slowly even with anthracene. The one-electron reduction potential of azodinitrile ($E_{\frac{1}{2}}$ +0·40 v versus SCE) is considerably more positive than that of tetracyanoethylene (+0·15 v). The resulting ion radical, which can also be produced chemically, is stable in aqueous and methanolic solution. The sodium salt, a bronze explosive powder, has been isolated[674]. The e.s.r. spectrum of the ion radical has been determined[674,675]; on the basis of the available data, it is not known whether it is linear or bent (*cis* or *trans*), or a rapidly equilibrating mixture of the *cis* and *trans* forms.

F. Cyanodiazo Compounds

The presence of diazo and cyano groups in a small molecule makes for an interesting but highly hazardous combination. The two examples known to date, cyanodiazomethane (diazoacetonitrile) and dicyanodiazomethane (diazomalononitrile), are both explosive and prospective investigators are well advised to heed the cautionary statements found in the literature[676–681]. Both compounds are much less stable than their ester analogues, diazoacetic and diazomalonic esters.

Cyanodiazomethane (**333a**–**333c**) is prepared by diazotization of aminoacetonitrile[676–678,680]. It is a yellow oil of boiling point 35° (6 mm), which explodes violently on heating, mechanical shock,

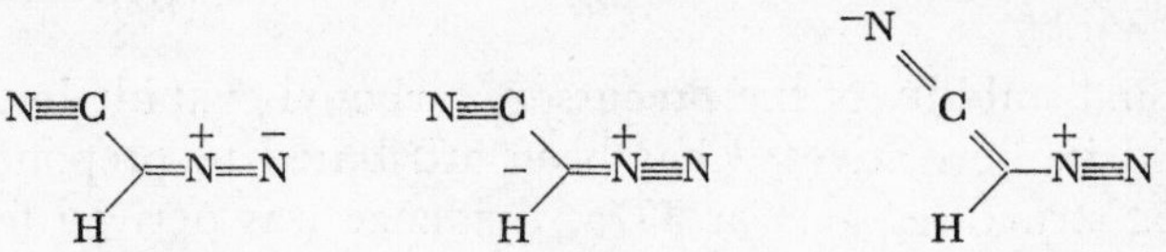

$$\underset{\textbf{(333a)}}{N{\equiv}C(H)C{=}\overset{+}{N}{=}\overset{-}{N}} \qquad \underset{\textbf{(333b)}}{N{\equiv}C(H)\overset{-}{C}{-}\overset{+}{N}{\equiv}N} \qquad \underset{\textbf{(333c)}}{{}^{-}N{=}C{=}C(H){-}\overset{+}{N}{\equiv}N}$$

friction[679] and contact with materials such as cuprous oxide[676]. Its isolation and especially its distillation should be avoided. The C=N and N=N bonds in cyanodiazomethane are shortened somewhat compared to those in diazomethane, whereas the C≡N bond is slightly longer than the corresponding bond in acetonitrile. This, together with the dipole moment of 3·45 D, has been taken as indicating considerable interaction between the diazo and cyano groups, i.e. as in **333c**[682]. Little is known about the chemistry of cyanodiazomethane. It forms an explosive mercury salt[676], and undergoes 1,3-dipolar additions with dimethyl acetylenedicarboxylate, but not with dicyanoacetylene[505]. Thermal[680], copper-catalysed[678] or photochemical[677,683,684] decomposition of cyanodiazomethane gives cyanocarbene (see Chapter 11). Two derivatives, trifluoromethylcyanodiazomethane (**334**) and phenylcyanodiazomethane (**335**) have been described and the intermediacy of two others, aminocyanodiazomethane (**336**; section IV.D.4)[380] and *p*-nitrophenylcyanodiazomethane[686], has been claimed. **334**

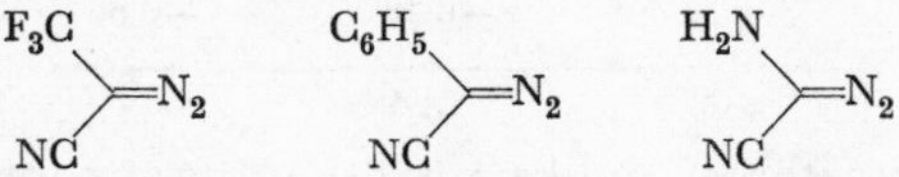

$$\underset{\textbf{(334)}}{(F_3C)(NC)C{=}N_2} \qquad \underset{\textbf{(335)}}{(C_6H_5)(NC)C{=}N_2} \qquad \underset{\textbf{(336)}}{(H_2N)(NC)C{=}N_2}$$

has been prepared by lead tetraacetate oxidation of its hydrazone[687]. It has been characterized only in the form of its triphenylphosphazine derivative and by conversion to cyanotrifluoromethylcarbene[687]. Phenylcyanodiazomethane, prepared by diazotization of phenylaminoacetonitrile[688,689], proved to be too unstable to be characterized other than by its infrared spectrum (bands at 2080 and 2220 cm^{-1})[689] and conversion to cyanophenylcarbene[688,689].

Dicyanodiazomethane (**338**) has been prepared from dibromomalononitrile via its hydrazone (**337**) as shown in equation (251)[681].

$$(NC)_2CBr_2 \xrightarrow{H_2NNH_2} (NC)_2C{=}NNH_2 \xrightarrow{Pb(OAc)_4} (NC)_2C{=}N_2 \qquad (251)$$

(**337**) (**338**)

The unusual stability of the precursor, carbonyl cyanide hydrazone (**337**), and its derivatives[681] has been attributed to preponderance of limiting structures such as **337a**. Evidence was derived from the n.m.r. and ultraviolet spectra and the high dipole moment (5·82 D) of **337**. Dicyanodiazomethane (**338**) is a yellow explosive solid, which melts with decomposition at about 75°. Its dipole moment (3·8 D), low vapour pressure and poor solubility in non-polar solvents all point to considerable contribution of the diazonium ylid structure **338a** to the ground state of dicyanodiazomethane. The two cyano

$$^{-}N{=}C{=}C(CN){-}N{=}\overset{+}{N}H_2 \qquad (N{\equiv}C)_2\overset{-}{C}{-}\overset{+}{N}{\equiv}N$$

(**337a**) (**338a**)

groups impart pronounced electrophilic character to dicyanodiazomethane. This is shown by comparison of its half-wave potentials with that of a number of other diazo compounds (Table 8)[690].

TABLE 8. Half-wave potentials of negatively substituted diazoalkanes in acetonitrile versus SCE, 0·1 M $LiClO_4$ added.

Compound	$(NC)_2CN_2$	$(PhSO_2)_2CN_2$	$(F_3C)_2CN_2$	Ph_2CN_2
$E_{\frac{1}{2}}$ (v)	−0·35	−0·48	−0·86	−1·55

Dicyanodiazomethane, unlike any other diazo compound studied so far, oxidizes primary and secondary alcohols at room temperature; in the process it is reduced to the hydrazone **337**. It is stable to acid, even 2 N sulphuric acid, but is readily attacked by nucleophiles such as triphenylphosphine and Grignard reagents (equation 252). Like other negatively substituted diazoalkanes[691], it undergoes a diazo coupling reaction, for instance with *N*,*N*-dimethylaniline (equation 252). Thermolysis[692] or photolysis[653,692] of dicyanodiazomethane gives dicyanocarbene (see Chapter 11).

Many negatively substituted diazo compounds have been prepared by reaction of *p*-toluenesulphonyl azide with active methylene

$$(NC)_2C{=}N_2 \xrightarrow{RMgBr} (NC)_2\bar{C}{-}N{=}N{-}R + \overset{+}{M}gBr \xrightarrow{H^+} (NC)_2C{=}N{-}NHR$$

$$(NC)_2C{=}N_2 \xrightarrow{PhNMe_2} (NC)_2\bar{C}{-}N{=}N{-}C_6H_4{=}\overset{+}{N}Me_2 \longrightarrow (NC)_2C{=}N{-}NH{-}C_6H_4{-}NMe_2 \quad (252)$$

compounds[686]. Application of this method to the preparation of dicyanodiazomethane results instead in the formation of the very stable nitrogen ylid **339** (equation 253); using aqueous base in place of a tertiary amine results in the formation of the rearrangement product **340**[686,693,694].

$$(NC)_2CH_2 + Me{-}C_6H_4{-}SO_2N_3 \xrightarrow{Et_3N} (NC)_2\bar{C}{-}N{=}N{-}\overset{+}{N}Et_3 \; (\mathbf{339})$$

$$\longrightarrow Me{-}C_6H_4{-}SO_2{-}NH{-}C{=}C(CN) \text{ (triazole ring: HN–N=N) } \; (\mathbf{340}) \quad (253)$$

3-Diazo-6-dicyanomethylene-1,4-cyclohexadiene (**341**), a vinylogue of dicyanodiazomethane (**338**), has been prepared from *p*-nitrophenylacetonitrile (equation 254)[695]. The zwitterionic structure **341a** is expected to make a large contribution to the resonance hybrid of **341**; this is borne out by its solubility characteristics. 3-Diazo-6-dicyanomethylene-1,4-cyclohexadiene is fairly unstable at room temperature. It readily forms a phosphazine derivative (**342**) which has a large dipole moment (12·01 D) as expected for a charge separated species of this type. The thermolysis and photolysis of **341** to give the corresponding carbene is described in Chapter 11.

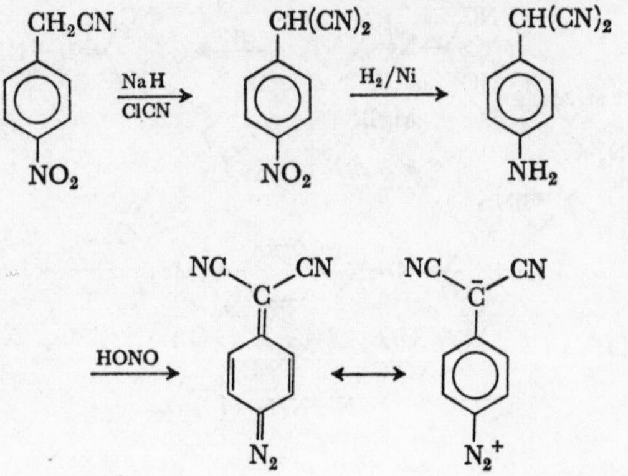

$$p\text{-}O_2NC_6H_4CH_2CN \xrightarrow[ClCN]{NaH} p\text{-}O_2NC_6H_4CH(CN)_2 \xrightarrow{H_2/Ni} p\text{-}H_2NC_6H_4CH(CN)_2 \tag{254}$$

$$\xrightarrow{HONO} (NC)_2C{=}C_6H_4{=}N_2 \longleftrightarrow (NC)_2\bar{C}\text{-}C_6H_4\text{-}N_2^+$$

(341) (341a)

The synthesis and chemistry of diazotetracyanocyclopentadiene are discussed in section XI.E.

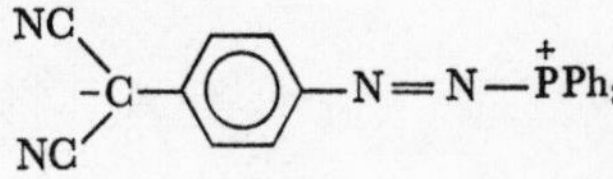

$$(NC)_2\bar{C}\text{-}C_6H_4\text{-}N{=}N{-}\overset{+}{P}Ph_3$$

(342)

IX. POLYCYANO-SUBSTITUTED HETEROCYCLES

A. Polycyanooxiranes

I. Synthesis and properties

Epoxidation with peracids commonly occurs by way of electrophilic attack on the olefin. Therefore, olefins bearing electronegative substituents are usually not susceptible to epoxidation with a peracid. The most widely used reagent for epoxidations of this type, has been alkaline hydrogen peroxide. Ordinarily it is successful for α,β-unsaturated esters and ketones, but the corresponding epoxyamide rather than epoxynitrile is formed in almost all instances from α,β-unsaturated nitriles. Although some of the epoxynitrile is often isolated, it is usually a minor product. In the simplest case, acrylonitrile, the epoxynitrile is accessible only by an indirect route[696].

It has been proposed that the intermediate in oxidations of α,β-unsaturated nitriles is a peroxyimidic acid **343** formed by addition of hydrogen peroxide to the carbon–nitrogen triple bond (equation 255)[697]. Evidence for this is found in the stoichiometry of the reaction under carefully controlled pH conditions. Only one equivalent of peroxide is required to convert the unsaturated nitrile to epoxyamide, whereas if the oxidation and hydrolysis were separate reactions, three equivalents would be required.

$$\mathrm{RCH{=}CHCN} \xrightarrow{H_2O_2} \underset{(\mathbf{343})}{\mathrm{RCH{=}CH\overset{\overset{NH}{\|}}{C}OOH}} \longrightarrow \mathrm{RCH\overset{O}{\frown}CHCONH_2} \quad (255)$$

An olefin with two cyano groups and one or two other electronegative substituents creates a special situation. The double bond becomes so electron-poor that it is susceptible to attack by hydrogen peroxide in the absence of any added base or with just a trace of mild base such as an amine[698]. Attack on the double bond, rather than on the nitrile group, is favoured by the increased resonance stabilization of the intermediate anion **344**. In the case of tetra-

$$(\mathrm{NC})_2\bar{\mathrm{C}}\text{—}\mathrm{C(CN)_2OOH} \longleftrightarrow {}^{-}\mathrm{N{=}C{=}C(CN)}\text{—}\mathrm{C(CN)_2OOH} \quad (\mathbf{344})$$

cyanoethylene no added base is required, and the epoxide **345** rapidly forms in good yield on mixing aqueous hydrogen peroxide and a solution of the olefin (equation 256)[174,699].

$$\mathrm{(NC)_2C{=}C(CN)_2 + H_2O_2} \longrightarrow \underset{(\mathbf{345})}{\mathrm{(NC)_2C\overset{O}{\frown}C(CN)_2}} + \mathrm{H_2O} \quad (256)$$

Synthesis in anhydrous systems is possible and has been carried out in ether solution with ethereal hydrogen peroxide[700]. A more convenient preparation in anhydrous medium uses *t*-butylhydroperoxide in benzene[700].

Not generally recognized is the fact that peracids can epoxidize a very electron-poor double bond by a *nucleophilic* attack. This has been demonstrated by the oxidation of tetracyanoethylene and 1,1-dicyano-2,2-bis(trifluoromethyl)ethylene (**346**) with *m*-chloroperbenzoic acid[701] and peracetic acid[702], respectively (equation 257).

No experimental work on the mechanism of formation of these

$$(NC)_2C{=}C(CF_3)_2 + CH_3CO_3H \longrightarrow (NC)_2C\overset{O}{\frown}C(CF_3)_2 + CH_3CO_2H \quad (257)$$

(346) **(347)**

epoxides has been reported. There can be little doubt that nucleophilic attack on the olefin is involved, but whether the attacking reagent is the peroxy acid or its corresponding anion is not determined. During the oxidation of tetracyanoethylene, the solution becomes dark amber, but the colour fades to a pale yellow when the theoretical amount of peroxide has been added. This implies the formation of an intermediate complex, which, however, may not be involved in the reaction mechanism.

Somewhat more complicated is the observation that tetracyanoethylene oxide forms when aqueous solutions of tetracyanoethylene anion radical are treated with acidic hydrogen peroxide[703]. This may involve intermediate formation of free olefin by acidification of the anion radical, followed by oxidation.

The question arises as to the degree of electronegative substitution on a double bond that is needed to allow oxidation with essentially neutral peroxide. No exhaustive study of this point has been made but all of the reported successful oxidations have been carried out with olefins that bear two cyano groups and at least one other electronegative substituent such as an ester or fluoroalkyl group. All of the olefins listed in Table 9 have been epoxidized.

TABLE 9. Epoxides of negatively substituted olefins.

Olefin	Method of preparation[a]	Melting point (°c)	Yield (%)	References
$(NC)_2C{=}C(CN)_2$	A, B, C	177–178	66–100	174, 700
$HC(CN){=}C(CN)_2$	A	76–77	31	174
$PhC(CN){=}C(CN)_2$	A	79–80	72	174
t-$BuC(CN){=}C(CN)_2$	A	44–45	76	174
$EtO_2CC(CN){=}C(CN)CO_2Et$	A	61–70[b]	93	174
$(CF_3)_2C{=}C(CN)_2$	D	46–47	77	702

[a] A = aqueous H_2O_2, B = ethereal H_2O_2, C = t-BuOOH in C_6H_6, D = peracid.
[b] Probably a mixture of *cis* and *trans* isomers.

The ease of epoxidation of tetracyanoethylene has been used to advantage in the ozonolysis of olefins[175,704,705]. Tetracyanoethylene itself is not attacked by ozone, but is converted to the epoxide in the presence of another olefin being ozonized. For example, 2,3-dimethyl-2-butene and tetracyanoethylene are oxidized directly to acetone and tetracyanoethylene oxide with ozone (equation 258).

$$(CH_3)_2C{=}C(CH_3)_2 + (NC)_2C{=}C(CN)_2 + O_3 \longrightarrow 2\,(CH_3)_2CO + \underset{\mathbf{(345)}}{(NC)_2C\overset{O}{\frown}C(CN)_2} \quad (258)$$

Normal ozonides, i.e. 1,2,4-trioxolanes, will not oxidize the cyano-olefin. Therefore it is most likely that reaction occurs with the primary ozonide (equation 259). This method can be

$$\text{primary ozonide } (R_2C\text{–}O\text{–}O\text{–}O\text{–}CR_2) + (NC)_2C{=}C(CN)_2 \longrightarrow 2\,R_2CO + \mathbf{345} \quad (259)$$

valuable in cases where normal ozonolysis results in the formation of side products, as in the oxidation of camphene[175]. Normally, ozonization in inert solvents leads to a complicated mixture of lactone, hydroxy acid and unsaturated acid. In the presence of tetracyanoethylene, camphenilone is obtained in 80% yield. The chief disadvantage of this procedure is the difficulty that may arise in separation of the product from the crystalline epoxide. A sufficiently volatile carbonyl compound may be removed by evaporation. This is rarely suitable and it may be necessary to resort to chromatography for separation, or to preferential conversion of the epoxide to water-soluble products, for example by reaction with halide ion.

2. Ring-opening reactions with cleavage of a C—O and C—C bond

a. Amines. Tetracyanoethylene oxide is cleaved by amines at room temperature or below into the elements of carbonyl cyanide and dicyanomethylene. With pyridine, for example, pyridinium dicyanomethylide (**348**) forms in better than 80% yield (equation 260)[699,700]. The carbonyl cyanide cannot be isolated, presumably due to reaction with excess pyridine. In the similar cleavage of

$$C_5H_5N + (NC)_2C\overset{O}{\frown}C(CN)_2 \longrightarrow \underset{\mathbf{(348)}}{C_5H_5\overset{+}{N}\text{–}\overset{-}{C}(CN)_2} + [CO(CN)_2] \quad (260)$$

1,1-dicyano-2,2-bis(trifluoromethyl)ethylene oxide (equation 261). The other cleavage product, i.e. hexafluoroacetone, was isolated[702].

$$\text{C}_5\text{H}_5\text{N} + (\text{NC})_2\overset{\text{O}}{\text{C}-\text{C}}(\text{CF}_3)_2 \longrightarrow (\text{CF}_3)_2\text{CO} + \mathbf{348} \qquad (261)$$

The reaction has not been studied with other cyanocarbon epoxides. Pyridine derivatives, such as 3- and 4-picoline, pyrazine and isoquinoline also give ylids, but the reaction fails if the pyridine contains an alkyl group in the 2-position as with 2-picoline or quinoline. The reasons for this are not clear. The dicyanomethylids of pyridine and substituted pyridines are extremely stable. An x-ray analysis of **348** has been carried out[706].

Decomposition of pyridinium dicyanomethylid at high temperature gives pyridine, but no other products have been isolated[700]. The cyano groups can be hydrolysed with base to give the corresponding amide (equation 262) whereas treatment with ethanol and hydrogen chloride converts only one cyano group to the ethyl ester (equation 263). The ability of **347** to exert 1,3-dipolar characteristics

$$\text{C}_5\text{H}_5\overset{+}{\text{N}}-\overset{-}{\text{C}}(\text{CN})_2 \xrightarrow[\text{H}_2\text{O}]{\text{NaOH}} \text{C}_5\text{H}_5\overset{+}{\text{N}}-\overset{-}{\text{C}}(\text{CONH}_2)_2 \qquad (262)$$

$$\text{C}_5\text{H}_5\overset{+}{\text{N}}-\overset{-}{\text{C}}(\text{CN})_2 \xrightarrow[\text{EtOH, H}_2\text{O}]{\text{HCl}} \text{C}_5\text{H}_5\overset{+}{\text{N}}-\overset{-}{\text{C}}(\text{CN})-\text{CO}_2\text{Et} \qquad (263)$$

has been exploited only in the addition of dimethyl acetylenedicarboxylate. In this case, hydrogen cyanide is eliminated to give the aromatic pyrrocoline derivative **349** (equation 264). The only reported example of the transfer of a dicyanomethylene group from pyridinium dicyanomethylid is that of the photolysis in benzene solution[707]. In addition to the (1,1-dicyanovinyl)pyrrole **350**, 7,7-dicyanonorcaradiene (**351**) was also isolated (equation 265).

The reactions of tetracyanoethylene oxide with simple aliphatic and aromatic amines have been less successful. Part of the problem may be caused by carbonyl cyanide generated in the reaction. More

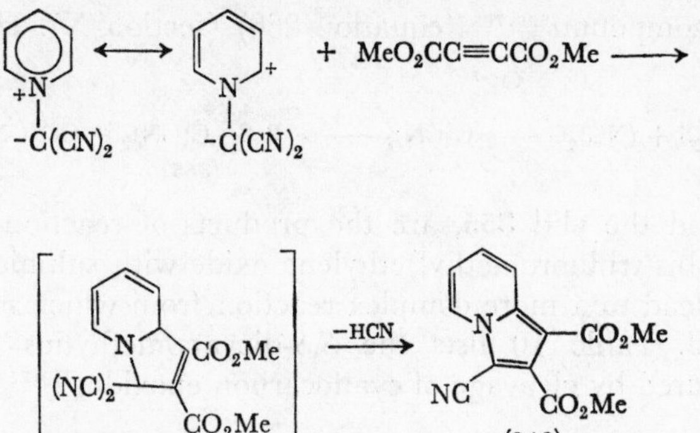

$$\text{Pyridinium-}C(CN)_2^- \longleftrightarrow \text{Pyridinium-}C(CN)_2^- + MeO_2CC{\equiv}CCO_2Me \longrightarrow$$

$$\left[\text{(NC)}_2\text{-dihydroindolizine-}CO_2Me,\ CO_2Me\right] \xrightarrow{-HCN} \text{NC-indolizine-}(CO_2Me)_2\ \textbf{(349)} \qquad (264)$$

success might be achieved with some of the other substituted cyanocarbon epoxides, but these reactions have not been reported.

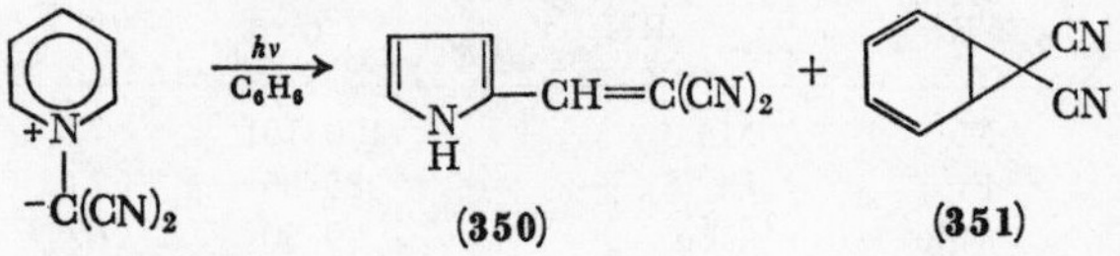

$$\underset{\textbf{(348)}}{\text{Pyridinium-}C(CN)_2^-} \xrightarrow[C_6H_6]{h\nu} \underset{\textbf{(350)}}{\text{2-pyrrolyl-}CH{=}C(CN)_2} + \underset{\textbf{(351)}}{\text{7,7-dicyanonorcaradiene}} \qquad (265)$$

Aniline and tetracyanoethylene oxide give *N*-cyanoformylaniline (**352**) in good yield and small amounts of the anil **353**[708]. The mechanism of formation of the latter is uncertain, although **352**

$$\underset{\textbf{(352)}}{PhNHCOCN} \qquad \underset{\textbf{(353)}}{PhN{=}C(CN)_2}$$

probably arises from reaction of the amine with carbonyl cyanide formed in the cleavage (section X).

N,*N*-dimethylaniline and tetracyanoethylene oxide give the same product (**354**) as does the amine and carbonyl cyanide[708].

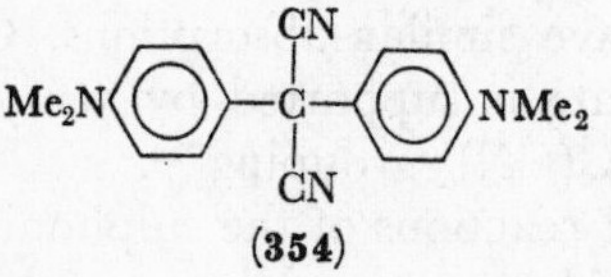

$$\underset{\textbf{(354)}}{Me_2N\text{-}C_6H_4\text{-}C(CN)_2\text{-}C_6H_4\text{-}NMe_2}$$

b. Sulphides. Reactions of cyanocarbon epoxides with dialkyl sulphides or arylalkyl sulphides parallel the reactions with pyridines. There is one important difference. In this case, the carbonyl cyanide resulting from the cleavage of tetracyanoethylene oxide can be isolated and in fact this provides a convenient synthesis of this carbonyl compound[699,709] (equation 266) (section X). Hexafluoro-

$$R_2S + (NC)_2\overset{\overset{O}{\diagup\;\diagdown}}{C\text{———}}C(CN)_2 \longrightarrow \underset{(\mathbf{355})}{R_2\overset{+}{S}\text{—}\overset{-}{C}(CN)_2} + CO(CN)_2 \qquad (266)$$

acetone and the ylid **355**, are the products of reaction of 1,1-dicyano-2,2-bis(trifluoromethyl)ethylene oxide with sulphides. Diaryl sulphides lead to a more complex reaction from which no ylid can be isolated. Table 10 lists the *S,S*-dicyanomethylids that have been prepared by cleavage of cyanocarbon epoxides[710,711].

TABLE 10. *S,S*-Sulphonium dicyanomethylids $R^1R^2\overset{+}{S}\text{—}\overset{-}{C}(CN)_2$ prepared from sulphides and cyanocarbon epoxides.

R^1	R^2	Melting point (°C)	Yield (%)
Me	Me	100–101	77
Et	Et	85–86	74
n-Bu	n-Bu	29–30	62
Me	n-$C_{12}H_{25}$	46–47	85
Me	Ph	77–78	70
Et	Ph	75–76	75
n-Bu	Ph	Oil	74
Me	2-Naphthyl	136–137	74
Me	*p*-$MeOC_6H_4$	92–93	67
Me	*p*-BrC_6H_4	124–125	45
Me	*p*-$CH_3SC_6H_4$	136–137	26

The sulphonium dicyanomethylids, like their pyridinium counterparts, are resonance stabilized and extremely stable. The contribution of ionic structures is indicated by the strong doublet nitrile absorption bands in the infrared at 2150 and 2180 cm^{-1}. Pyridinium dicyanomethylids have similar absorptions. Charge delocalization in the crystalline state is supported by x-ray studies showing the planarity of the —$SC(CN)_2$ grouping[712].

The only reported reactions of the sulphonium salts are cleavage by cold mineral acid to the original sulphide plus unidentified products and their thermal decomposition to the sulphide and tetracyanoethylene[710].

The first member of the series, dimethylsulphonium dicyanomethylid (**356**) has been prepared by two other methods (equations 267 and 268)[710,712,713].

$$(CH_3)_2SO + CH_2(CN)_2 \xrightarrow{SOCl_2} (CH_3)_2\overset{+}{S}-\overset{-}{C}(CN)_2\cdot 2HCl \xrightarrow{base} (CH_3)_2\overset{+}{S}-\overset{-}{C}(CN)_2 \; \textbf{(356)} \quad (267)$$

$$(CH_3)_2S + BrCH(CN)_2 \longrightarrow (CH_3)_2\overset{+}{S}-CH(CN)_2Br^- \xrightarrow{base} \textbf{356} \quad (268)$$

c. Thiocarbonyl compounds. Three different types of reaction products have been obtained from thiocarbonyl compounds and cyanocarbon epoxides. All three probably result from an initial nucleophilic attack and fragmentation analogous to that described above for pyridines and sulphides. In general, these reactions are cleaner with **347**, because the hexafluoroacetone produced does not lead to side reactions[714]. However, the reactions described below have been observed with both oxides[708,714]. The zwitterionic thiouronium salt **357** from thiourea is stable and is formed in good yield

$$(NH)_2C{=}S + (NC)_2C\overset{O}{\frown}C(CF_3)_2 \; \textbf{(347)} \longrightarrow (NH_2)_2C{=}S{=}C(CN)_2 \; \textbf{(357)} + (CF_3)_2CO \quad (269)$$

(equation 269). Similar salts are obtained from *N,N,N′,N′*-tetramethylthiourea and *N,N′*-ethylenethiourea. When **357** is heated in water it cyclizes to the thiazole **358** (equation 270).

$$(NH_2)_2C{=}S{=}C(CN)_2 \; \textbf{(357)} \longrightarrow \text{4-}H_2N\text{-5-NC-2-}NH_2\text{-thiazole} \; \textbf{(358)} \quad (270)$$

Presumably an ylid is also produced from thiobenzamide or from *S*-diphenylthiourea. However, in both cases cyclization occurs spontaneously (equations 271 and 272).

$$PhC(=S)-NH_2 + \textbf{347} \longrightarrow \left[PhC(NH_2){=}S{=}C(CN)_2\right] \longrightarrow \text{2-Ph-4-}NH_2\text{-5-CN-thiazole} \quad (271)$$

$$PhNHC(=S)NHPh + \textbf{347} \longrightarrow [(PhNH)_2C{=}S{=}C(CN)_2] \longrightarrow \text{4-}H_2N\text{-5-NC-3-Ph-2-(=NPh)-thiazoline} \quad (272)$$

The third type of reaction is more obscure. Overall, the thiocarbonyl group is eliminated as elemental sulphur and a dicyanoethylene is formed. Reaction with thioacetamide is representative

$$CH_3C(=S)NH_2 + \mathbf{347} \longrightarrow CH_3C(NH_2){=}C(CN)_2 + S + CF_3COCF_3 \qquad (273)$$

of this type (equation 273). Thiobenzophenone and tetracyanoethylene oxide behave similarly[708] (equation 274).

$$Ph_2C{=}S + (NC)_2C\overset{O}{\frown}C(CN)_2 \longrightarrow CO(CN)_2 + S + \underset{68\%}{(C_6H_5)_2C{=}C(CN)_2} \qquad (274)$$

It is logical to assume that in these reactions the first step is also formation of a thiouronium salt (**359**) which is not as stable as that from thiourea wherein the positive charge on carbon is stabilized by two amino groups. The charge could be neutralized by cyclization to an unstable thiirane **360** which would be expected to lose sulphur readily to give the observed olefin.

$$\underset{(\mathbf{359})}{(C_6H_5)_2\overset{+}{C}{-}S{-}\overset{-}{C}(CN)_2} \qquad \underset{(\mathbf{360})}{(C_6H_5)_2C\overset{S}{\frown}C(CN)_2}$$

d. Miscellaneous nucleophiles. At 80°, tetracyanoethylene oxide and benzalaniline give two products, neither in high yield (equation 275)[708]. Only one of these products qualifies the reaction for inclusion in this section on cleavage of both a C—C and a C—O bond. The Δ^4-oxazoline **361** presumably arises by nucleophilic ring opening followed by closure to the five-membered ring and loss of hydrogen cyanide. However, the second product **362** appears to be the result of addition of benzalaniline to the 1,3-dipole, **363** (equation 276). Confirmation of this is obtained by addition of a better dipolarophile, dimethyl acetylenedicarboxylate, to the reaction mixture. In this case neither **361** nor **362** is isolated, but the expected addition product **364** of the acetylene to **363** is formed (equation 277). The initial adduct readily loses hydrogen cyanide on heating to form the pyrrole **365**.

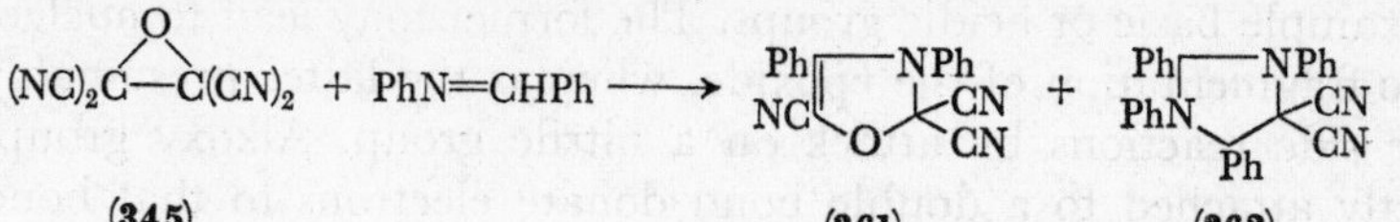

$$(NC)_2C\overset{O}{—}C(CN)_2 + PhN{=}CHPh \longrightarrow \textbf{(361)} + \textbf{(362)} \quad (275)$$

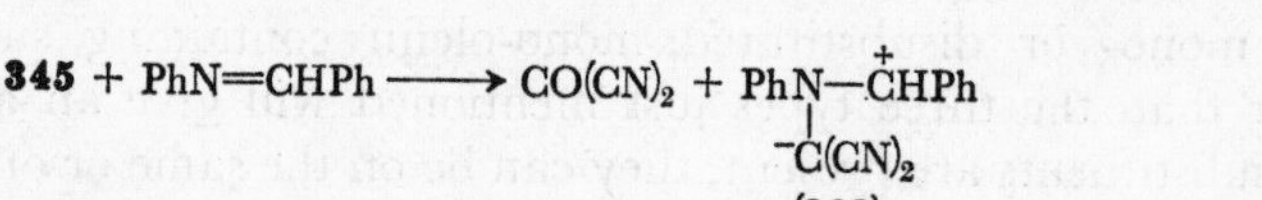

(345) (361) (362)

$$\textbf{345} + PhN{=}CHPh \longrightarrow CO(CN)_2 + Ph\overset{+}{N}{—}\overset{+}{C}HPh\ (^{-}C(CN)_2) \quad (276)$$

(363)

An even more intriguing reaction of tetracyanoethylene oxide occurs when it is heated with benzophenone azine in boiling

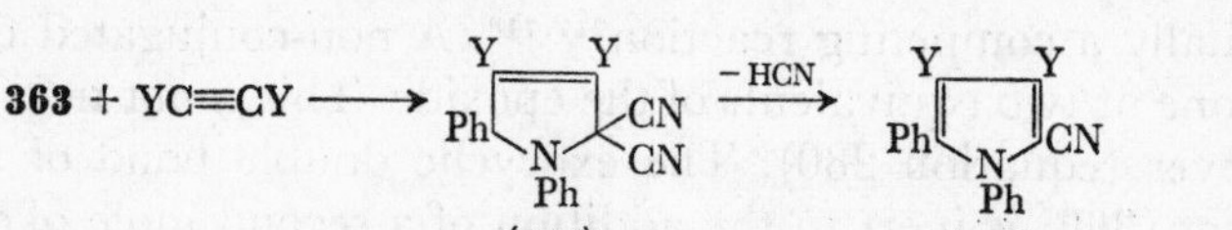

$$\textbf{363} + YC{\equiv}CY \longrightarrow \textbf{(364)} \xrightarrow{-HCN} \textbf{(365)} \quad (277)$$

(364) (365)

$(Y = CO_2Me)$

benzene. Nitrogen is evolved cleanly and 1,1-dicyano-2,2-diphenylethylene (**366**) and the epoxide of this olefin, **367**, are isolated in good yield[708] (equation 278). The mechanism of this reaction has not yet been elucidated.

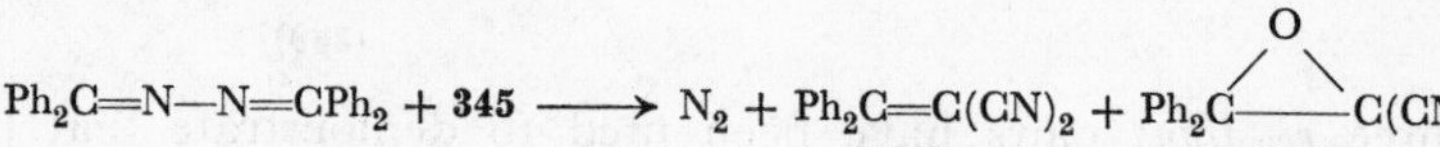

$$Ph_2C{=}N{—}N{=}CPh_2 + \textbf{345} \longrightarrow N_2 + Ph_2C{=}C(CN)_2 + Ph_2C\overset{O}{—}C(CN)_2 \quad (278)$$

(366) (367)

3. Ring-opening reactions with cleavage of the C—C bond

a. Addition to olefins. Tetracyanoethylene oxide undergoes a thermal addition to olefins with the formation of tetracyanotetrahydrofurans[715,716]. With ethylene, the adduct **368** is isolated in better than 80% yield (equation 279).

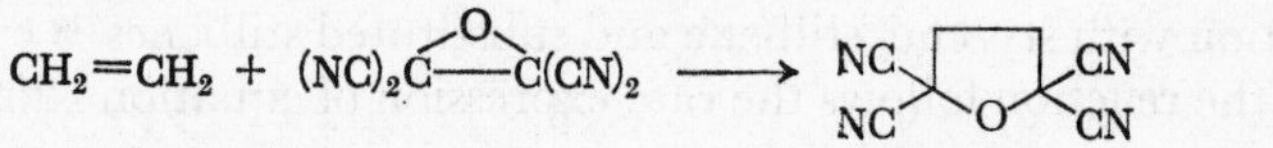

$$CH_2{=}CH_2 + (NC)_2C\overset{O}{—}C(CN)_2 \longrightarrow \textbf{(368)} \quad (279)$$

(368)

A large number of adducts of tetracyanoethylene oxide have been prepared, and it is possible to define rather precisely the structural requirements of the olefin. Certain substituents cannot be tolerated, for example basic or acidic groups. The former may lead to nucleophilic fragmentation of the epoxide, whereas the latter presumably cause side reactions by attack on a nitrile group. Alkoxy groups directly attached to a double bond donate electrons to that bond and lead to side reactions which will be discussed below. Almost any mono- or disubstituted mono-olefin containing substituents other than the three types just mentioned will give an adduct. If two substituents are present, they can be on the same or on different carbons. In general, electron-withdrawing groups decrease and electron-donating groups increase the reaction rate.

Addition of tetracyanoethylene oxide to conjugated dienes apparently occurs only in a 1,2-manner, but oxidation of the diene is usually a competing reaction[717,718]. A non-conjugated diene may add one or two equivalents of the epoxide. This is not true of allenes, however (equation 280). The exocyclic double bond of the allene adduct (**369**) is inert to the addition of a second mole of epoxide or apparently any other reagent. This is attributed to a very strong field effect exerted by the four cyano groups on the tetrahydrofuran ring.

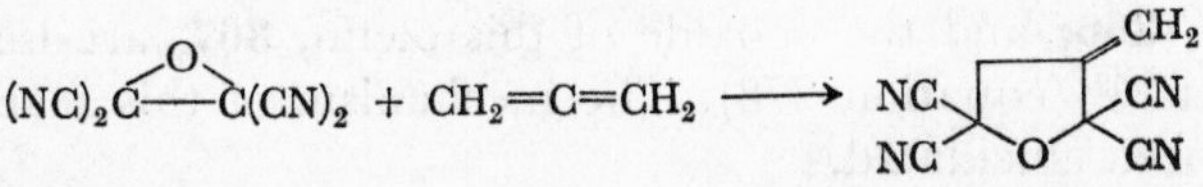

$$(NC)_2C(-O-)C(CN)_2 + CH_2{=}C{=}CH_2 \longrightarrow \textbf{369} \quad (280)$$

(**369**)

Three *cis–trans* pairs have been used to demonstrate that the addition of tetracyanoethylene oxide to olefins is stereospecific. In addition, a competition experiment with *cis*- and *trans*-1,2-dichloroethylene showed that the rate of addition to the *trans* olefin is larger by a factor of five (equation 281). A kinetic study of the olefin addition

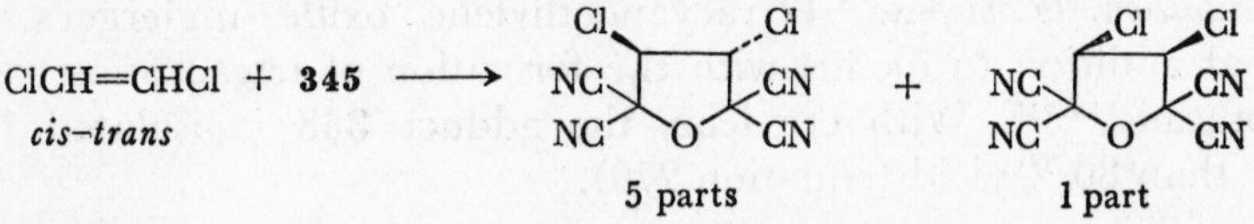

$$ClCH{=}CHCl + \textbf{345} \longrightarrow \text{(5 parts)} + \text{(1 part)} \quad (281)$$

cis–trans

5 parts 1 part

reaction with styrene, stilbene and substituted stilbenes[719] established that the reaction follows the rate expression of equation (282), where

$$\text{rate} = \frac{k_1k_2[\text{TCNEO}][\text{Olefin}]}{k_{-1} + k_2[\text{Olefin}]} \quad (282)$$

TCNEO = tetracyanoethylene oxide and the rate constants are for the reactions (283) and (284).

$$\text{TCNEO} \underset{k_{-1}}{\overset{k_1}{\rightleftharpoons}} \text{TCNEO*} \tag{283}$$

$$\text{TCNEO* + olefin} \xrightarrow{k_2} \text{product} \tag{284}$$

The activated species, TCNEO*, of equation (283) must be a ring-opened structure with a different conformation than the epoxide itself. It must approach planarity. Solvent and substituent effects suggest that there is very little charge separation in the transition state and that TCNEO* is a hybrid of zwitterionic and radical species, **370a–c**.

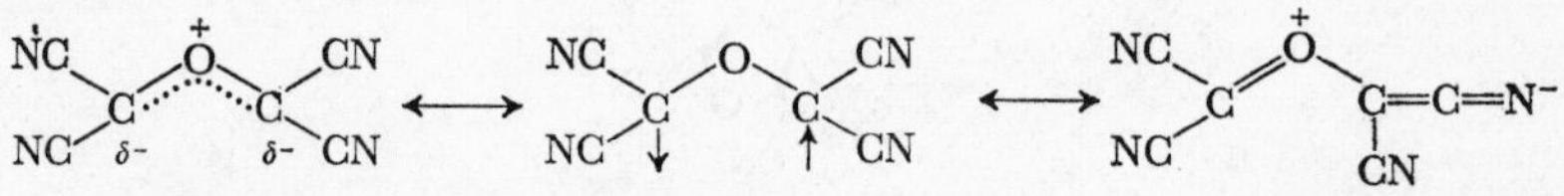

(370a) ⟷ **(370b)** ⟷ **(370c)**

The second step of the reaction, i.e. addition of TCNEO* to an olefin, has the characteristics of a concerted reaction. Criteria that have been established for processes of this type and that apply to the present reaction include the thermal nature of the reaction, insensitivity to solvent and structural changes, stereospecificity and faster reaction with the *trans* isomer of a *cis–trans* pair.

One other cyanocarbon epoxide has been reported to add to olefins. Phenyltricyanoethylene oxide gives the expected 1:1 product **371** with ethylene (equation 285).

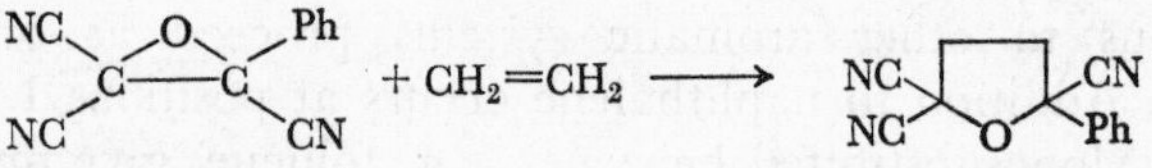

$$\text{(NC)}_2\text{C}\overset{\text{O}}{—}\text{C(Ph)(CN)} + \text{CH}_2{=}\text{CH}_2 \longrightarrow \textbf{(371)} \tag{285}$$

b. Addition to aromatic compounds. The reaction of tetracyano-ethylene oxide and olefins has no precedent in the chemistry of normal olefin epoxides. Still more unusual is the ability to add to aromatic hydrocarbons in the same manner[715]. For example the reaction with benzene at 135–150° gives the monoadduct **372** in addition to small amounts of a diadduct (equation 286). The latter

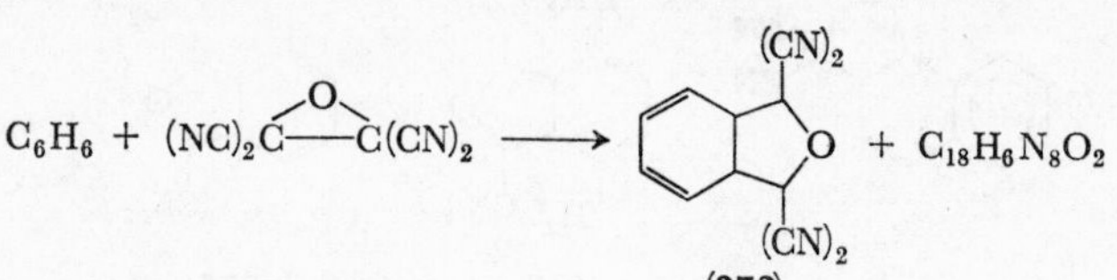

$$\text{C}_6\text{H}_6 + \text{(NC)}_2\text{C}\overset{\text{O}}{—}\text{C(CN)}_2 \longrightarrow \textbf{(372)} + \text{C}_{18}\text{H}_6\text{N}_8\text{O}_2 \tag{286}$$

can also be made starting with **372** and a second mole of the epoxide. The structure of the diadduct has not been determined with certainty but is most probably **373**. No examples of 1,4-addition of tetracyanoethylene oxide to a conjugated diene have been found[718]. (However, see section IX.4.b.) Steric considerations probably require the two heterocyclic rings of **373** to be *trans* to one another.

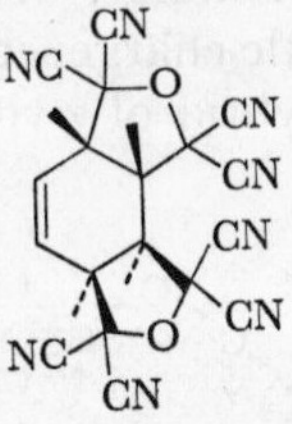

(373)

Although the double bonds of **372** are distant from the four cyano groups, there must be a definite field effect. The diene system undergoes the Diels–Alder reaction only with difficulty. Also surprising is the inability to aromatize the six-membered ring. Most diene systems are evidently too electron-rich to give clean 1:1 addition reactions with tetracyanoethylene oxide. This effect is evidently counterbalanced in **372** by the electron withdrawal caused by the cyano groups, and the addition of the second molecule of epoxide occurs smoothly.

Additions to other aromatic systems proceed as one might predict[715]. Addition to naphthalene occurs at positions 1 and 2 to give **374**. Monosubstituted benzenes, e.g. toluene, give mixtures of products as might be expected, however, it appears that the epoxide will not add at a position on an aromatic ring that is already substituted if another reaction path is available. Only one product, **375**, is isolated from reaction with *p*-xylene[715]. The report of an adduct, **376**, with durene demonstrates that addition to a substituted

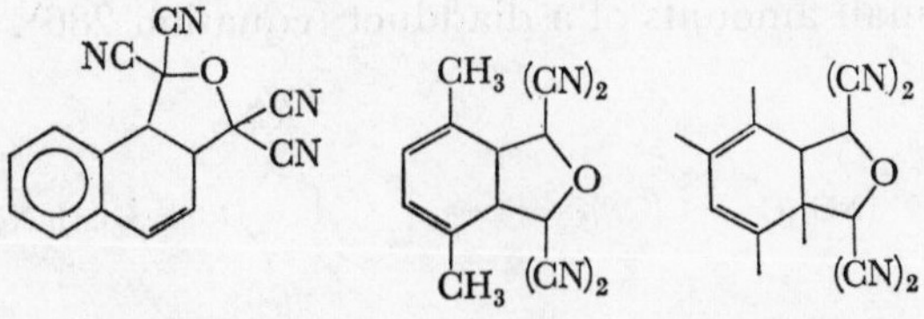

(374) **(375)** **(376)**

position can be forced[717,718]. The latter reaction is accompanied, however, by oxidation–reduction (see below).

c. Addition to acetylenes. Although tetracyanoethylene oxide will add to an acetylene in the same manner as it does to olefins[715], the reaction is considerably slower. A competition experiment between ethylene and acetylene showed the rate of formation of the dihydrofuran **377** to be about one-tenth as great as that of the ethylene product **368**.

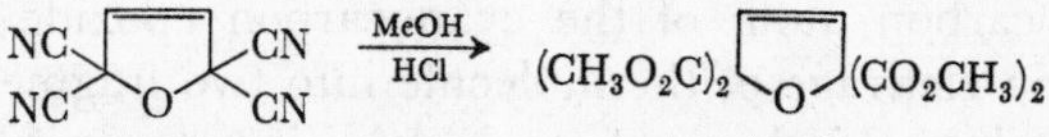

$$\text{(377)} \xrightarrow[\text{HCl}]{\text{MeOH}} (CH_3O_2C)_2\text{-dihydrofuran-}(CO_2CH_3)_2\ \text{(378)} \qquad (287)$$

The field effect of the four cyano groups shows up again in **377**. Catalytic reduction to the corresponding tetrahydrofuran is unsuccessful. Conversion of the cyano to ester groups occurs in good yield, however (equation 287), and the tetraester **378** can be easily reduced.

Another manifestation of the cyano group effect is the reaction with diynes. Only a monoadduct (**379**) of a conjugated diyne is formed even under forcing conditions with an excess of tetracyanoethylene oxide (equation 288). If two acetylenic bonds are not

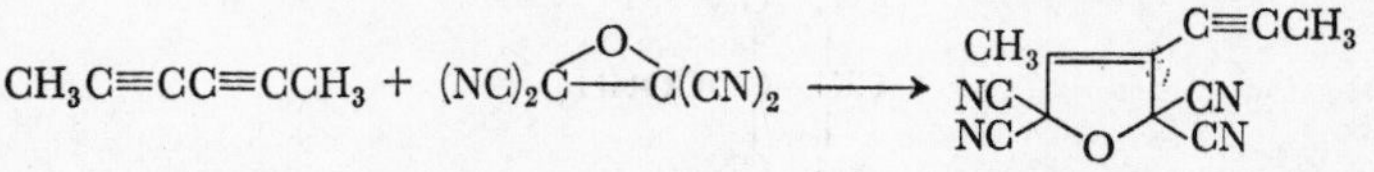

$$CH_3C{\equiv}CC{\equiv}CCH_3 + (NC)_2C\overset{O}{—}C(CN)_2 \longrightarrow \text{(379)} \qquad (288)$$

conjugated, either a monoadduct **380** or a diadduct **381** can be formed.

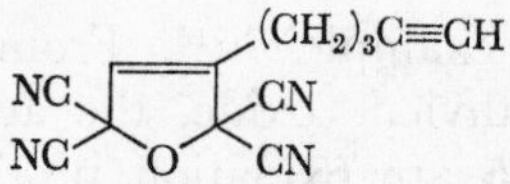

(380) 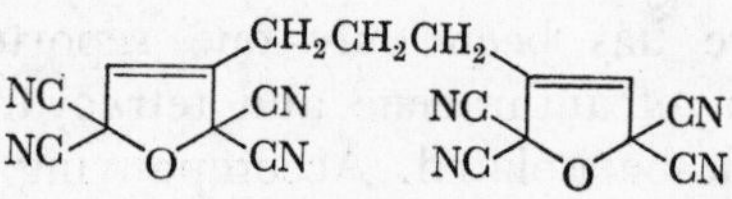 (381)

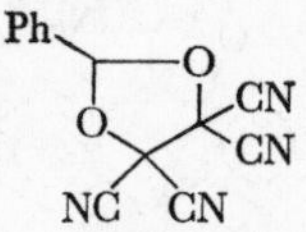

(382)

d. Addition to carbon–heteroatom bonds. At present there is only one example of the addition of a cyanocarbon epoxide to a carbon–heteroatom multiple bond[708]. Tetracyanoethylene oxide and benzaldehyde give a 1:1 adduct in 10% yield. Acid hydrolysis to benzaldehyde supports structure **382** for the product.

4. Ring-opening reactions with cleavage of a C—O bond

a. Acid anhydrides and halides. As was noted above, nucleophilic attack on a carbon atom of the cyanocarbon epoxide ring most often results in cleavage of the molecule into two fragments. There are reactions where this does not occur. Addition of a drop of pyridine to a solution of tetracyanoethylene oxide in acetic anhydride causes the precipitation of the diacetate of tetracyanoethylene glycol (**383**) (equation 289).

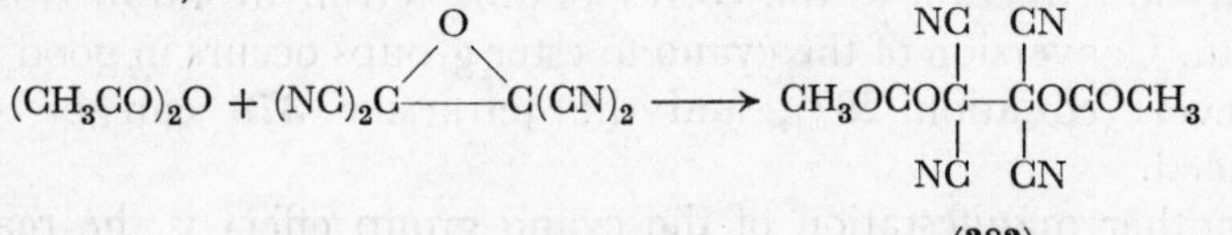

$$(CH_3CO)_2O + (NC)_2C\overset{O}{\frown}C(CN)_2 \longrightarrow CH_3OCOC(CN)_2{-}C(CN)_2OCOCH_3 \quad (289)$$

(**383**)

Acid halides behave similarly and 2-chlorotetracyanoethyl acetate (**384**) is synthesized in good yield by the same technique from acetyl chloride.

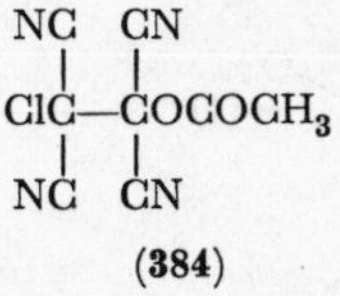

$$ClC(CN)_2{-}C(CN)_2OCOCH_3$$

(**384**)

b. Carbon–carbon unsaturation. Rupture of a carbon–oxygen bond of a cyanocarbon epoxide with 1:1 addition to a C=C bond system is a rare reaction.

There has been only one reported example[717,718]. From the reaction of anthracene and tetracyanoethylene oxide, the adduct **385** can be isolated. Accompanying **385** are oxidation products,

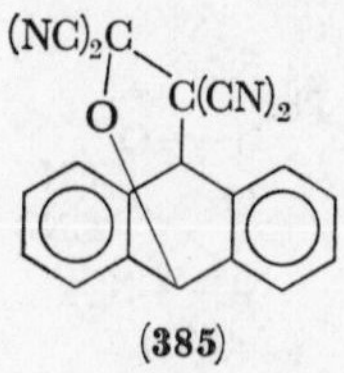

(**385**)

anthrone, anthraquinone and bianthrone, along with the tetracyanoethylene adduct of anthracene. All of the reaction products including **385** are undoubtedly the result of nucleophilic attack on oxygen rather than carbon (see below).

5. Nucleophilic attack on oxygen

It has already been mentioned that electron-rich olefins do not give simple adducts with tetracyanoethylene oxide. Instead, the reaction apparently proceeds by nucleophilic attack of the olefin on the oxygen atom of the epoxide ring. One of the examples of this is the reaction with 2,3-dimethyl-2-butene which occurs readily at temperatures below 100°. The products are 2,3-dimethyl-2-butene epoxide (**386**) and tetracyanoethylene. The overall result is a transepoxidation (equation 290). There are numerous other examples

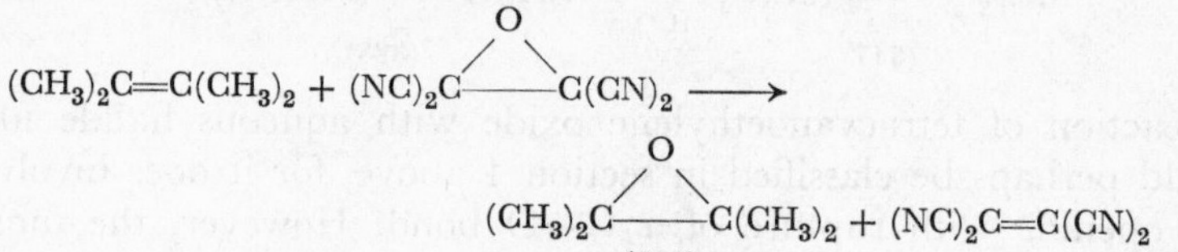

$$(CH_3)_2C{=}C(CH_3)_2 + (NC)_2C\overset{O}{\text{—}}C(CN)_2 \longrightarrow (CH_3)_2C\overset{O}{\text{—}}C(CH_3)_2 + (NC)_2C{=}C(CN)_2 \quad (290)$$

(386)

of oxidation–reduction occurring in these olefin reactions[718]. Many are not as clean-cut as the one described above, but all probably go by essentially the same mechanism. Nucleophilic attack of the olefin on the oxygen atom can produce an intermediate (**387**) which

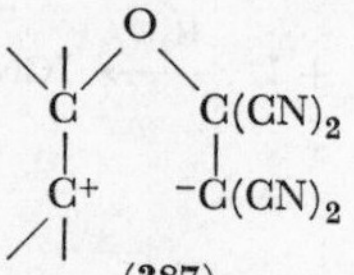

$$\begin{array}{ccc} & O & \\ >C & & C(CN)_2 \\ | & & | \\ C^+ & & {}^-C(CN)_2 \end{array}$$

(387)

collapses to tetracyanoethylene and an oxidation product of the olefin. Adducts of tetracyanoethylene are isolated when the olefin used can react with that cyanocarbon. This type of reaction has been observed with dihydropyran[715], cyclopentadiene, bicyclo-[2.2.1]heptadiene and cycloheptatriene[718].

By analogy with the reactions of amines and sulphides one might expect phosphines to cleave cyanocarbon epoxides to dicyanomethylids and carbonyl cyanide. Surprisingly, this is only a minor pathway in the reaction of triphenylphosphine and tetracyanoethylene oxide[717]. Instead, attack on oxygen occurs with generation of triphenylphosphine oxide and tetracyanoethylene (equation 291).

$$Ph_3P + (NC)_2C\overset{O}{\frown}C(CN)_2 \longrightarrow Ph_3PO + (NC)_2C{=}C(CN)_2 \qquad (291)$$

6. Attack on a cyano group

Very few transformations of the cyano groups of cyanocarbon epoxides can be carried out without disruption of the ring. Hydrolysis of 1,1-dicyano-2,2-bis(trifluoromethyl)ethylene oxide with dilute aqueous sodium hydroxide occurs with formation of the corresponding diamide (**388**) (equation 292). This reflects a difference in relative base stabilities of **347** and tetracyanoethylene oxide, which is completely destroyed by such treatment.

$$\underset{(\mathbf{347})}{(CF_3)_2C\overset{O}{\frown}C(CN)_2} \xrightarrow[H_2O]{NaOH} \underset{(\mathbf{388})}{(CF_3)_2C\overset{O}{\frown}C(CONH_2)_2} \qquad (292)$$

Reaction of tetracyanoethylene oxide with aqueous halide ion should perhaps be classified in section 4 above, for it does involve ring opening with rupture of a C—O bond. However, the most likely explanation of this reaction involves removal of cyanide ion by direct attack of halide ion (cf. reaction with halide ion in non-aqueous media, section 4). The products are the corresponding cyanogen halide and tricyanovinyl alcoholate ion (**389**) (equation 293).

$$(NC)_2C\overset{O}{\frown}C(CN)_2 + I^- \xrightarrow{H_2O} ICN + \underset{(\mathbf{389})}{(NC)_2C{=}\underset{CN}{\underset{|}{C}}{-}O^-} \qquad (293)$$

This reaction forms a basis for quantitative determination of tetracyanoethylene oxide. The cyanogen iodide reacts with excess iodide in the presence of acid to liberate iodine which can be titrated. Bromide ion attacks the epoxide similarly, but the reaction with chloride ion gives **389** in very poor yield.

B. *Tetracyanofuran*

Two methods of synthesis have been reported for tetracyanofuran (**392**)[435,720]. When tetraethyl furantetracarboxylate (**390**) is treated with aqueous ammonia in methanol the tetraamide (**391**) is formed, and this is dehydrated with $POCl_3$ (equation 294). A closely related synthesis uses the mixed cyanoester **393**, which is

$$\text{(390)}\ EtO_2C/CO_2Et\text{-furan} \longrightarrow \text{(391)}\ NH_2OC/CONH_2\text{-furan} \longrightarrow \text{(392)}\ NC/CN\text{-furan}\ 50\text{–}65\% \quad (294)$$

synthesized from succinonitrile and ethyl oxalate according to the scheme of equation (295). Tetracyanofuran is a moderately stable,

$$NCCH_2CH_2CN + 2\,(CO_2Et)_2 \xrightarrow[Et_2O]{NaH} EtO_2C\overset{ONa}{C}{=}C(CN)\text{—}C(CN){=}\overset{ONa}{C}CO_2Et\ 50\% \xrightarrow{SOCl_2}$$

$$\text{(393)} \xrightarrow{NH_3} NC,CN,NH_2OC,CONH_2\text{-furan} \xrightarrow[C_5H_5N]{POCl_3} \mathbf{392}\ 80\% \quad (295)$$

crystalline compound that turns brown on exposure to air and slowly evolves hydrogen cyanide.

The two alpha cyano groups of **392** appear to be the most reactive ones, as evidenced by reaction with ethanolic or aqueous hydrogen chloride (equation 296).

$$NC,CN,NH_2OC,CONH_2\text{-furan} \xrightarrow[HCl]{conc.} \mathbf{392} \xrightarrow[HCl]{EtOH} NC,CN,EtO_2C,CO_2Et\text{-furan} \quad (296)$$

C. 2,3,4,5-Tetracyanopyrrole

Salts of tetracyanopyrrole are obtained on treatment of tetracyano-1,4-dithiin with azide ion (section IX.H)[721]. The free pyrrole **394**, prepared from these salts using an ion-exchange resin (equation 297), is a tan solid, which can be sublimed at 200° (0·1 mm); it

$$\text{(394)}\ \text{tetracyanopyrrole (N–H)} \longleftarrow \text{tetracyanopyrrolide anion} \longrightarrow \text{(395)}\ \text{N-Me tetracyanopyrrole} \quad (297)$$

melts with decomposition at 203–212°. Tetracyanopyrrole is a strong acid whose pK_a in water is −2·71. It reacts readily with diazomethane to give the *N*-methyl derivative **395**; the latter may also be prepared under forcing conditions from the salts of tetracyanopyrrole with methyl iodide[721].

D. 3,4,5-Tricyanopyrazole

Tricyanopyrazole (**397**) has been prepared by two routes outlined in equation (298)[505]. Direct synthesis from diazoacetonitrile and

$$NCC{\equiv}CCN + N_2CHCOOEt \longrightarrow \text{(396a)} \xrightarrow{1.\ NH_3;\ 2.\ P_2O_5} \text{(397)}$$

$$MeOOCC{\equiv}CCOOMe + N_2CHCN \longrightarrow \text{(396b)} \xrightarrow{1.\ NH_3;\ 2.\ P_2O_5} \text{(397)} \quad (298)$$

dicyanoacetylene was unsuccessful[505]. No evidence has been presented that would favour the assignment of structures **396a** and **396b** to the intermediate adducts. Tricyanopyrazole is a strong acid as inferred from formation of stable salts. An *N*-methyl derivative has been prepared by methylation with diazomethane.

E. 4,5-Dicyanotriazole

4,5-Dicyanotriazole (**398**) is obtained on treatment of diaminomaleonitrile (section IV.D.4) with nitrous acid (equation 299)[389,401,403,722,723]. The claimed existence of two interconvertible

$$(NC)_2C{=}C(NH_2)_2 \xrightarrow{HONO} \text{(398)} \quad (299)$$

isomers[723,724] of dicyanotriazole has been disproved[725,726]. Dicyanotriazole is an acid whose strength is comparable to that of chloroacetic acid[726]. A series of salts has been prepared[722,724]; the silver salt is explosive. Methylation with diazomethane gives an *N*-methyl derivative[723].

F. Polycyanopyridines

I. Tricyanopyridines

Tricyanopyridine derivatives have been synthesized in a variety of ways. One method[727,728] starts from 3,5-dicyanopyridine, which on *N*-methylation to **399**, and treatment with potassium cyanide yields the tricyanodihydropyridine **400** (equation 300). Thermal

MeI → (**399**) → CN^- → (**400**) → 125° → (**401**) → N_2O_4 → (**402**) ⇌ $HClO_4$ / H_2O (**403**) ClO_4^- (300)

rearrangement of the latter leads to the isomer **401**. Both dihydropyridines **400** and **401** are surprisingly stable to acid. Oxidation of **401** with dinitrogen tetroxide produces 2-hydroxy-*N*-methyl-3,4,5-tricyano-1,2-dihydropyridine (**402**), which on treatment with perchloric acid is converted to *N*-methyl-3,4,5-tricyanopyridinium perchlorate (**403**). The latter is stable in the absence of moisture but reverts instantaneously to the dihydropyridine **402** in aqueous solution. Comparison of the equilibrium constant of the system **402** ⇌ **403** with that of the corresponding pair involving the dicyanopyridinium ion **399** shows that the latter is considerably more stable. The pK_{R^+} value for **402** is $-1{\cdot}0$, similar to that of 4,4′-dimethoxytriphenylcarbinol, whereas *N*-methyl-2-hydroxy-3,5-dicyano-1,2-dihydropyridine has a pK_{R^+} value of $+5{\cdot}5$, similar to that of tropyl alcohol. *N*-Methyl-3,4,5-tricyanopyridinium perchlorate (**403**) is a strong π-acid; its acceptor strength is comparable to that of hexacyanobenzene, but lower than that of tetracyanoethylene or 2,3-dicyanobenzoquinone.

Reaction of anhydrous hydrogen cyanide with **402** results in the formation of 2,3,4,5-tetracyano-*N*-methyl-1,2-dihydropyridine (**404**), which on oxidation yields the pyridone **405** instead of a tetracyanopyridinium salt or its hydration product (equation 301). Reduction

(**402**) $\xrightarrow{HCN}$ (**404**) $\xrightarrow{MeI}$ (**405**)

(**402**) $\xrightarrow{Na_2SO_2}$ (**406**) $\xrightarrow{O_2}$ (**407**) $\xleftarrow{Zn}$ (**403**) ClO_4^- (301)

of **402** gives a red compound considered to be *N*-methyl-3,4,5-tricyano-1,2-dihydropyridine (**406**). Air oxidation of **406** or treatment of **403** with zinc, produces solutions of a radical tentatively identified as *N*-methyl-3,4,5-tricyanopyridyl (**407**)[728].

The enhanced electron affinity of the tricyanopyridinium ion **403** is further demonstrated by the observation that it oxidizes tetramethyl-*p*-phenylenediamine to its cation radical ('Wurster's blue'), and fluorenol to fluorenone. Isolation of difluorenyl ether in the latter reaction indicates that **403** also acts as a Lewis acid. This was confirmed by the finding that it catalyses the Friedel–Crafts reaction of acetyl bromide with anisole[728].

A series of substituted 3,4,5-tricyanopyridines has been prepared by reaction of pentacyanopropenide anion with hydrogen halides (equation 302)[729]. The initial adducts **408** readily undergo addition–elimination reactions with nucleophiles such as alcohols and amines. Reaction of **408** with sodiomalononitrile followed by acidification

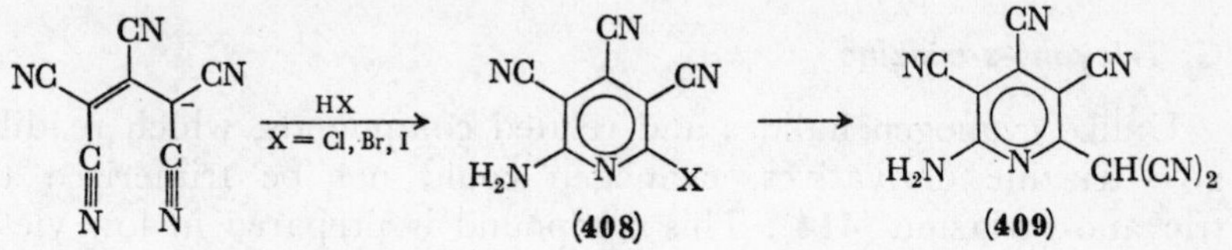

(302)

yields 2-amino-3,4,5-tricyano-6-dicyanomethylpyridine (**409**), a fairly strong acid (pK_a 2·3) of unusual thermal stability[729].

2,4,6-Tricyanopyridine *N*-oxide (**410**) has been prepared by reaction of the potassium salt of nitroacetonitrile with formaldehyde followed by treatment of the resulting 3-hydroxy-2-acinitropropionitrile with acid (equation 303)[730]. A mechanism for this unusual

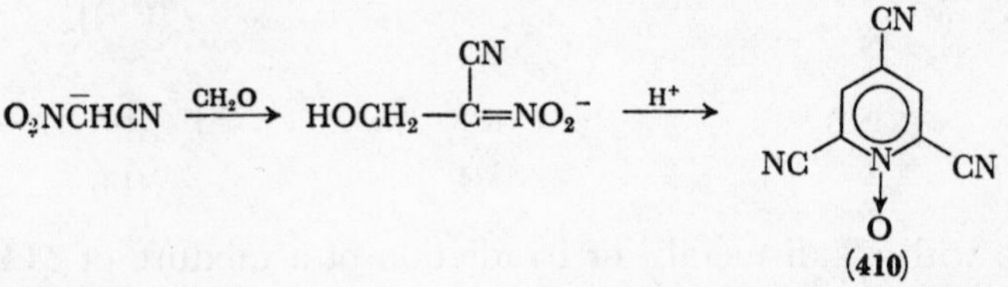

$$O_2N\bar{C}HCN \xrightarrow{CH_2O} HOCH_2{-}C(CN){=}NO_2^- \xrightarrow{H^+} \textbf{(410)} \quad (303)$$

reaction has not been advanced. The *N*-oxide **410** is a strong π-acid, which forms a red complex with *N,N*-dimethylaniline[730].

2. Tetra- and pentacyanopyridine

Although some physical properties of 2,3,5,6-tetracyanopyridine (**411**), 2,3,4,5-tetracyanopyridine (**412**) and pentacyanopyridine (**413**) have been published, details of the preparation of these

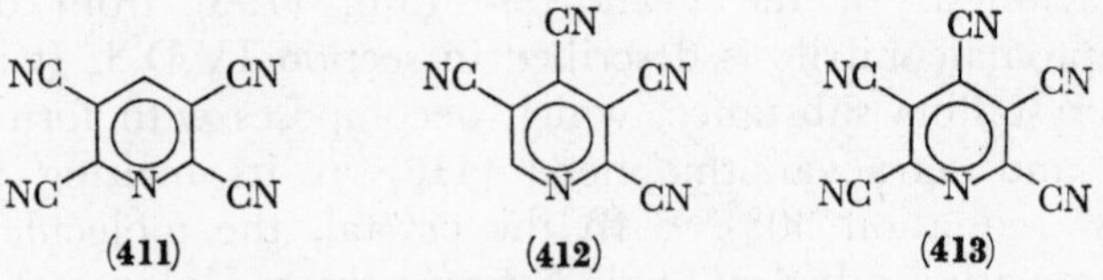

(**411**) (**412**) (**413**)

compounds so far have been reported only in a thesis[731]. The charge-transfer spectra are discussed in Chapter 10. The electron affinity of **412** has been measured[134], and the ion radicals of **412** and **413** have been prepared and their e.s.r. spectra studied[732].

G. Tricyano-s-triazine

Unlike cyanogen halides and related compounds, which readily give triazine derivatives, cyanogen could not be trimerized to tricyano-*s*-triazine (**414**). This compound is prepared in low yield by an indirect synthesis via cyanuric tricarboxylic acid and its amide[733]. Tricyano-*s*-triazine melts at 118–120° and is monomeric[734]. It is very prone to nucleophilic attack. Thus, reaction with alcohols leads ultimately to cyanuric esters (**415**; R = OR)[733]; the intermediate mono- (R^1, R^2 = CN; R^3 = OR)[520,733] and disubstitution products (R^1 = CN; R^2, R^3 = OR)[733] can be isolated (equation 304). Pyrolysis of **414** gives cyanogen[733,735]. Reduction of tricyano-*s*-

$$\text{2,4,6-}(H_2NOC)_3\text{-}s\text{-triazine} \xrightarrow{P_2O_5} \underset{(\mathbf{414})}{\text{2,4,6-}(NC)_3\text{-}s\text{-triazine}} \longrightarrow \underset{(\mathbf{415})}{\text{2-}R^1\text{-4-}R^2\text{-6-}R^3\text{-}s\text{-triazine}} \tag{304}$$

triazine with alkali metals, or irradiation of a mixture of **414** with hexamethylmelamine (**415**; R = NMe_2) results in the formation of a stable blue ion radical. Its e.s.r. spectrum is not that expected for a symmetrical species; the observed splitting has been explained in terms of a species permanently distorted due to the Jahn–Teller effect[734]. This rationalization has been questioned more recently and replaced by one involving an unspecified molecular rearrangement of the ion radical[732].

H. Tetracyano-1,4-dithiin

The synthesis of tetracyano-1,4-dithiin (**196**) from disodium dimercaptomaleonitrile is described in section IV.D.3. It is a pale yellow crystalline substance, which decomposes with formation of sulphur and tetracyanothiophene (**416**) at its melting point of 207–208° (equation 305)[353]. In the crystal, the molecule is in a boat form with a dihedral angle between the two planes of 124° [736]. This agrees with the dipole moment of 4·0 ± 0·5 D, measured in

$$\underset{(\mathbf{196})}{\text{tetracyano-1,4-dithiin}} \xrightarrow{-S} \underset{(\mathbf{416})}{\text{tetracyanothiophene}} \tag{305}$$

dioxan[353]. Polarographic reduction shows two waves at −0·24 and −0·72 v (versus SCE in 50 % aqueous acetic acid); addition of four and six electrons, respectively, is believed to take place[353]. Tetracyano-1,4-dithiin forms π-complexes, but it is a weaker acceptor than tetracyanoethylene[353,721].

The dithiin skeleton in **196** is stable to acid; hydrolysis of the nitrile groups is effected without attack on the ring[353]. Attempts to prepare sulphoxides or sulphones from the dithiin have been unsuccessful[721]. However, tetracyano-1,4-dithiin is very susceptible to nucleophilic attack at its double bond. The initial adduct either ring opens (see below) or undergoes an intramolecular addition. The latter course is followed in the conversion of tetracyano-1,4-dithiin to tetracyanothiophene catalysed by ethoxide[353] or fluoride ions[721]. The proposed mechanism is shown in equation (306).

(306)

The cases where the initial adduct of the nucleophile with tetracyano-1,4-dithiin undergoes ring opening are more numerous. The fate of the resulting anion depends on the nature of the nucleophile. It can either be attacked by a second molecule of the nucleophile, or undergo ring closure; in some cases, dianions derived from the initial ring-opened adduct can be isolated. Thus, reaction of the dithiin with sodiomalononitrile yields the dianion **417** (equation 307); similar products are obtained using acetate or alkylxanthates[721].

(307)

(417)

An example of attack of a second molecule of the nucleophile on the ring-opened anion is shown in equation (308)[721]. The dianion **418** undergoes homolytic cleavage to produce a mixture of dimercaptomaleonitrile anion radical and tetracyanoethylene anion radical. Oxidation of the mixture with chlorine yields tetracyano-1,4-dithiin and tetracyanoethylene[721]. In other cases, heterolytic cleavage of adducts of type **418** has been observed; addition of a second molecule of the nucleophile can also occur at the double bond not previously attacked[721].

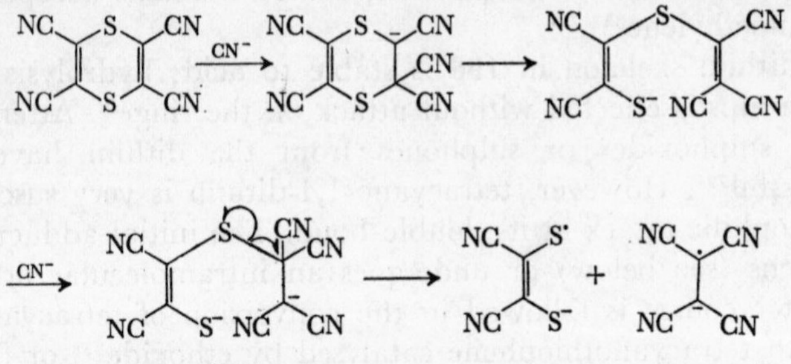

(308)

(418)

Finally, the formation of tricyano-1,4-dithiino[c]isothiazole (**419**)

$$S_8 + I^- \rightleftharpoons {}^-SS_7I$$

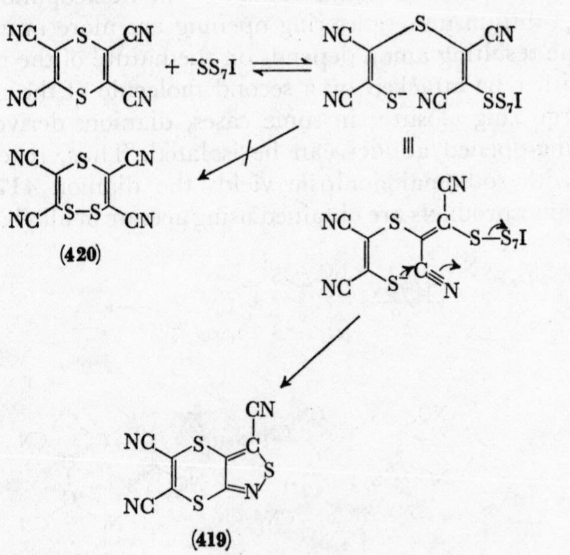

(309)

(420)

(419)

in the reaction of sulphur with tetracyano-1,4-dithiin illustrates a case where the ring-opened anion recloses. The reaction, which is catalysed by iodide or ethoxide ion, has been formulated as shown in equation (309)[353]. The product **419** is isomeric with tetracyano-1,2,5-trithiepin (**420**). This structure was ruled out, and the assigned structure confirmed by ^{13}C n.m.r. studies of a sample of **419** prepared from tetracyano-1,4-dithiin labelled at the nitrile groups[353]. **419** often occurs as a by-product in the formation of tetracyanodithiin by oxidation of disodium dimercaptomaleonitrile[353].

Tetracyano-1,4-dithiin reacts with azide ion to give tetracyanopyrrole, isolated in 63 % yield as the tetraethylammonium salt (equation 310). Since the yield is higher than 50 %, both halves of the dithiin

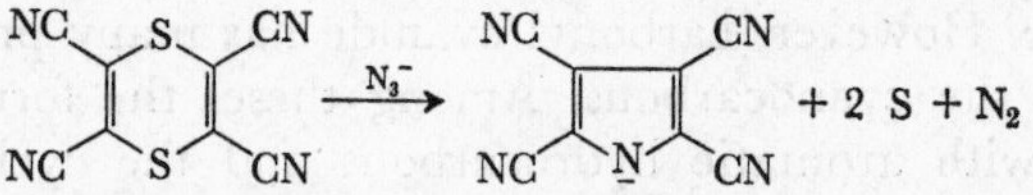

$$\text{tetracyano-1,4-dithiin} \xrightarrow{N_3^-} \text{tetracyanopyrrolide anion} + 2\ S + N_2 \qquad (310)$$

must end up in the product; tetracyanothiophene is not an intermediate since it does not react with azide ion under these conditions. Possible mechanisms for this remarkable transformation have been discussed[721].

I. Tetracyanothiophene

Tetracyanothiophene (**416**) is obtained in the thermal or catalysed decomposition of tetracyano-1,4-dithiin (section IX.H). Its physical and chemical properties are typically aromatic. Polarographic studies show the first wave of tetracyanothiophene to occur at more negative potential (−0·7 v compared to −0·24 v) than that of tetracyano-1,4-dithiin[353]. It is stable on heating to 900°. All four nitrile groups can be hydrolysed leading ultimately to the tetra-acid **421** (equation 311); various intermediate amides and imides have

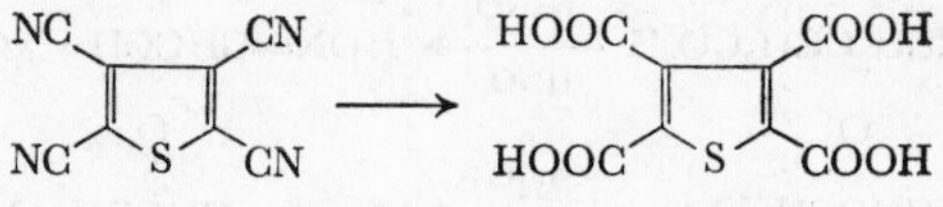

$$\underset{(416)}{\text{tetracyanothiophene}} \longrightarrow \underset{(421)}{\text{thiophenetetracarboxylic acid}} \qquad (311)$$

been isolated[353]. Tetracyanothiophene forms semiconducting phthalocyanine-like compounds on heating with copper acetonylacetonate[327,556]. Reaction with primary amines yields products of type **422**[737], whereas semiconducting polymers claimed to have **422** as structural units are formed when diamines are used in this reaction[327].

$$\begin{array}{c} \text{RN} \quad\quad \text{NHR} \\ \| \quad\quad\quad | \\ \text{C}-\text{C}\text{-}\text{C}\text{-}\text{C} \\ \text{N} \quad \text{S} \quad \text{N} \\ \text{C} \quad\quad\quad \text{C} \\ | \quad\quad\quad \| \\ \text{RHN} \quad\quad \text{NR} \end{array}$$

(422)

X. CARBONYL CYANIDE

Acyl cyanides exhibit the properties of both simple carbonyl compounds and acyl halides. A review of the chemistry of acyl cyanides has appeared[738] and no attempt is made to cover this subject here. However, carbonyl cyanide has many properties that relate it to the cyanocarbons. Among these, the formation of π-complexes with aromatic hydrocarbons and the cycloaddition to certain types of olefins might be mentioned. Reactions of this type, along with others, more than justify the inclusion of a section on carbonyl cyanide in this chapter. A short discussion of the closely related perfluoroacyl cyanides has also been included.

A. Synthesis

Early attempts to synthesize carbonyl cyanide by the action of ultraviolet light or a silent electric discharge on mixtures of cyanogen and carbon monoxide gave only yellow amorphous powders[739]. Ozonization of hydroxymethylenemalononitrile under various conditions was also unsuccessful[740].

The first successful preparation[741,742] of carbonyl cyanide is shown in equations (312)–(315). In spite of an improvement in the

$$HO_2CCH_2\overset{O}{\overset{\|}{C}}CH_2CO_2H \xrightarrow[HNO_3]{NaNO_2} HON{=}CH\overset{O}{\overset{\|}{C}}CH{=}NOH \tag{312}$$

$$HON{=}CH\overset{O}{\overset{\|}{C}}CH{=}NOH \xrightarrow{Ac_2O} AcON{=}CH\overset{O}{\overset{\|}{C}}CH{=}NOAc \tag{313}$$

$$AcON{=}CH\overset{O}{\overset{\|}{C}}CH{=}NOAc \xrightarrow[12\ mm]{60\text{–}130^\circ} AcON{=}CH\overset{O}{\overset{\|}{C}}CN \tag{314}$$

$$AcON{=}CH\overset{O}{\overset{\|}{C}}CN \xrightarrow[650\ mm]{150\text{–}180^\circ} NC\overset{O}{\overset{\|}{C}}CN \tag{315}$$

method of operation, which is reported to decrease the explosion hazard in reaction (314), the overall yield from acetonedicarboxylic acid is still only about 20%[743].

Nucleophilic cleavage of tetracyanoethylene oxide with a dialkyl sulphide (equation 316) gives carbonyl cyanide simply and in high

$$\text{n-Bu}_2\text{S} + (\text{NC})_2\text{C}\overset{\text{O}}{\text{—}}\text{C(CN)}_2 \longrightarrow \text{CO(CN)}_2 + \text{n-Bu}_2\text{S}{=}\text{C(CN)}_2 \quad (316)$$

yield but suffers from the disadvantage of using only half of the available cyano groups in the molecule[699,744].

A reaction which undoubtedly shows the most promise involves passing phosgene over a solid bed of silver cyanide at 275–300°[745]. The conversion to carbonyl cyanide is 40–50% based on the silver cyanide charged.

B. Physical Properties

Carbonyl cyanide is a nearly colourless liquid, which can be kept indefinitely in the absence of moisture. Physical properties are summarized in Table 11.

TABLE 11. Physical properties of carbonyl cyanide[746].

Property	Value
Boiling point (760 mm)	65·6°
Melting point	−37·9°
d_4^{20}	1·124 g/ml[741]
$n_D^{18.4}$	1·3547[746]
n_D^{20}	1·3919[741]
Dipole moment	1·35–1·5 D[747]
Heat of combustion	331·97 ± 0·64 kcal/mole[746]

Spectral characterization of carbonyl cyanide has been the subject of a number of studies[748–751]. The Raman spectrum in ether solution has been determined and the carbonyl and nitrile absorptions are found at 1720 and 2238 cm^{-1}, respectively[748]. The ultraviolet spectrum has two ranges, one from 2300 Å toward shorter wavelengths and the second from 3500 to 2570 Å. Irradiation with light in either wavelength region results in dissociation to carbon monoxide and cyanide radical which undergoes polymerization[752]. Pyrolysis of carbonyl cyanide at 750° results in dissociation to carbon monoxide and cyanogen[753].

As might be expected by analogy with other cyanocarbons, carbonyl cyanide readily forms charge-transfer complexes with aromatic hydrocarbons[754,755]. The colours formed with aromatic compounds[756,757] range from yellow to red.

C. Reaction with Alcohols and Amines

Addition of carbonyl cyanide to water results in almost explosive hydrolysis to carbon dioxide and hydrogen cyanide. A transient white solid, presumably cyanoformic acid, $NCCO_2H$, is observed when carbonyl cyanide is exposed to atmospheric moisture[741]. Esters of cyanoformic acid result when carbonyl cyanide is added dropwise to cold ether solutions of the appropriate alcohol[758].

Cyanoperformic acid has been suggested as a possible intermediate when carbonyl cyanide is treated with alkaline hydrogen peroxide[759]. The addition of 9,10-diphenylanthracene or rubrene to the reaction mixture produces a moderately strong chemiluminescence. This reaction is common to tetracyanoethylene, its epoxide and carbonyl cyanide, and it is possible that the reactions are all related by oxidation of the olefin to epoxide which is in turn cleaved to carbonyl cyanide.

Cyanoformyl amines have been prepared by reaction of carbonyl cyanide with the corresponding primary or secondary amine[760].

An especially interesting reaction with carbonyl cyanide is that with *N,N*-dimethylaniline. Two equivalents of amine react with the carbonyl group to give bis(4,4′-dimethylaminophenyl)dicyanomethane (**423**) (equation 317). The reaction occurs without added

$$2\ C_6H_5NMe_2 + (NC)_2CO \xrightarrow{-H_2O} Me_2N\text{-}C_6H_4\text{-}C(CN)_2\text{-}C_6H_4\text{-}NMe_2 \qquad (317)$$

(423)

catalyst in ether solution, but the yield is better when acetic acid is the solvent. An interesting comparison may be made with diethyl mesoxalate[761] and hexafluoroacetone[762]. These molecules condense with dimethylaniline to give only 1:1 products (equation 318).

D. Wittig-Type Reactions

A virtually untapped source of dicyanomethylene compounds is the reaction of Wittig reagents and carbonyl cyanide. The one

$$X_2C{=}O + C_6H_5NMe_2 \longrightarrow Me_2N{-}C_6H_4{-}C(X)_2{-}OH \qquad (318)$$

$$(X = CO_2Et,\ CF_3)$$

reported example (equation 319)[681] gives the product **424** under very mild conditions.

$$Ph_3P{=}N{-}N{=}CPh_2 + (CN)_2C{=}O \longrightarrow Ph_3PO + Ph_2C{=}N{-}N{=}C(CN)_2 \qquad (319)$$

(424)

E. Acylation of Aromatic Hydrocarbons

Benzene is acylated by carbonyl cyanide in the presence of aluminium chloride[763] (equation 320). This synthesis can be extended

$$C_6H_6 + CO(CN)_2 \xrightarrow[0^\circ]{AlCl_3} \underset{71\%}{C_6H_5COCN} + HCN \qquad (320)$$

to substituted benzenes. Only *para*-substituted derivatives were isolated from the reaction of monosubstituted benzenes. Strong *ortho–para* directing substituents in the benzene ring increase the yields of the aroyl cyanide, as expected. Cyanoacylation of naphthalene reportedly gives 2-naphthoyl cyanide, accompanied by a small amount of the 1-isomer[763].

Thiophene and pyrrole are likewise cyanoacylated, the latter in the absence of a catalyst[760]. The yield of 2-cyanoformylpyrrole is reported to be 56 %, whereas the corresponding 2-cyanoformylthiophene was obtained in only 22 % yield in the presence of aluminium chloride as catalyst.

F. Reaction with Olefins

Addition of carbonyl cyanide to 1,3-dienes to form dicyanodihydropyrans (equation 321) occurs readily under mild conditions. Only the adducts with butadiene and 2,3-dimethylbutadiene have been reported[764–766], but the reaction is undoubtedly general. Although there have been no reported additions to unsymmetrical

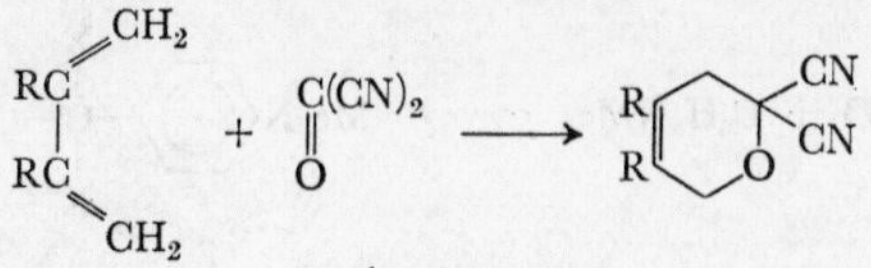

$$\text{RC}(=\text{CH}_2)\text{–C}(=\text{CH}_2)\text{R} + \text{O}{=}\text{C(CN)}_2 \longrightarrow \text{3,4-R}_2\text{-6,6-dicyano-5,6-dihydro-2}H\text{-pyran} \qquad (321)$$

(R = H, CH_3)

dienes, it is expected that the same type of effects occur as with fluorinated ketones[767]. The carbonyl group is highly polarized by the electron-withdrawing cyano groups and the carbonyl carbon atom should add to the most negative end of the diene.

Carbonyl cyanide undergoes the 'ene' reaction[282] with many olefins that contain an allylic hydrogen. Illustrative is the addition to α-methylstyrene (equation 322)[768]. The dicyanoalcohol **425**, as

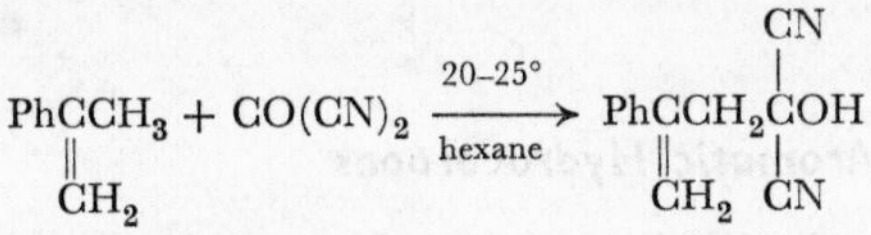

$$\text{PhC(=CH}_2\text{)CH}_3 + \text{CO(CN)}_2 \xrightarrow[\text{hexane}]{20\text{–}25^\circ} \text{PhC(=CH}_2\text{)CH}_2\text{C(CN)}_2\text{OH} \qquad (322)$$

(425)

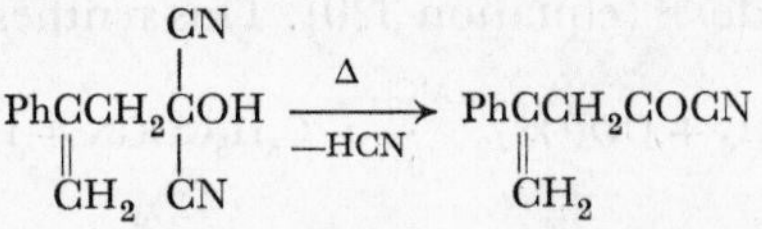

$$\text{PhC(=CH}_2\text{)CH}_2\text{C(CN)}_2\text{OH} \xrightarrow[-\text{HCN}]{\Delta} \text{PhC(=CH}_2\text{)CH}_2\text{COCN} \qquad (323)$$

(426)

$$\text{PhC(=CH}_2\text{)CH}_2\text{C(CN)}_2\text{OCOCN}$$

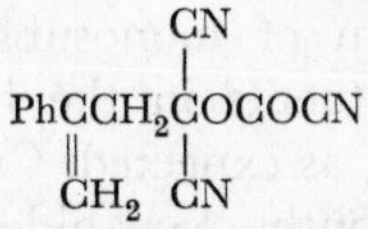

(427)

expected, is not stable and evolves hydrogen cyanide, when heated, to give the acyl cyanide **426** (equation 323). If two equivalents of carbonyl cyanide are used, the product is the cyanoformate **427**[769].

As the acyl cyanide **426** does not react with carbonyl cyanide, **427** probably arises by a simple cyanoacylation of **425**. This latter reaction has been demonstrated[769].

Most of the olefins that react in the manner of α-methylstyrene give only 1:2 adducts, i.e. the products are analogous to **427**. Another olefin that does form a 1:1 product is 2-methyl-2-butene. Here it becomes apparent that double-bond migration is involved in the reaction[770] (equation 324).

This bond migration is accommodated nicely by a concerted mechanism involving a six-membered ring transition state (equation

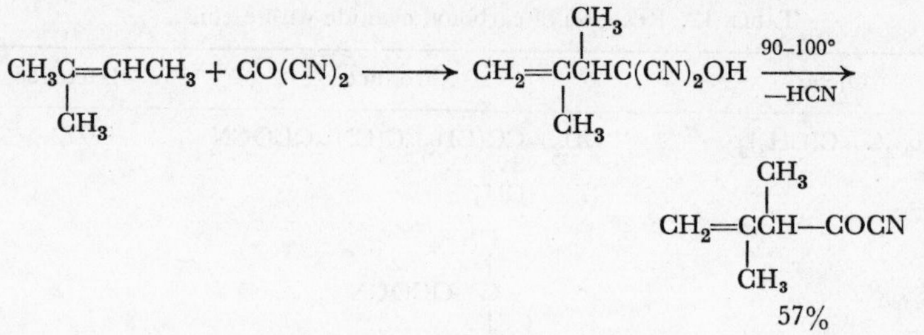

$$CH_3C(CH_3){=}CHCH_3 + CO(CN)_2 \longrightarrow CH_2{=}C(CH_3)C(CH_3)HC(CN)_2OH \xrightarrow[-HCN]{90\text{–}100^\circ} CH_2{=}C(CH_3)C(CH_3)H{-}COCN \quad (324)$$

57%

325). Absence of carbonium ion intermediates is indicated in the reaction with β-pinene (**428**)[771] (equation 326). The formation of the

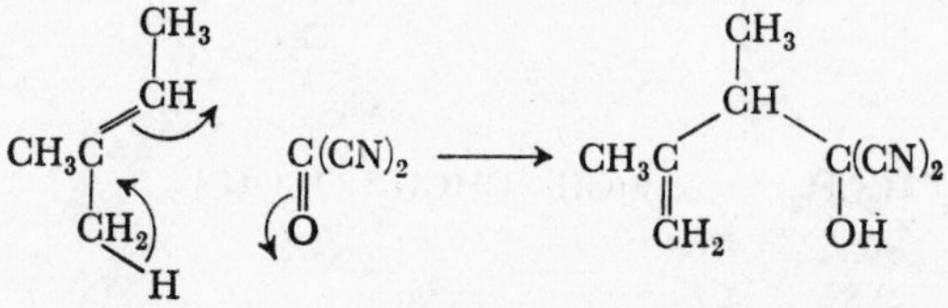

$$CH_3C(=CHCH_3)CH_2H + C(CN)_2{=}O \longrightarrow CH_3C(=CH_2)CH(CH_3)C(CN)_2OH \quad (325)$$

carbonium ion **429** should lead to skeletal rearrangement. Instead, the known carboxylic acid **430** is the end-product, as expected from a concerted cyclic mechanism as pictured above.

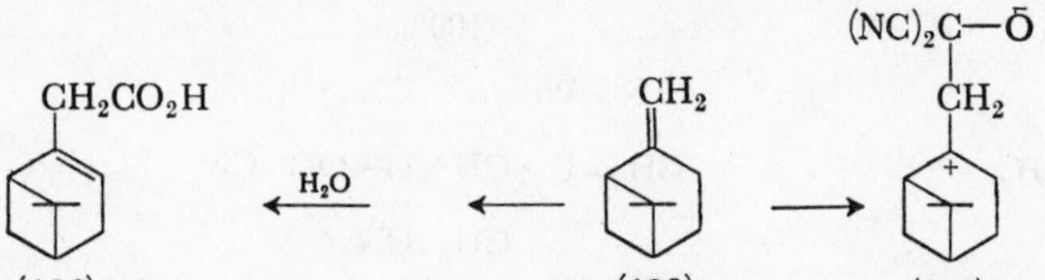

$$(\mathbf{430})\ CH_2CO_2H \xleftarrow{H_2O} \longleftarrow (\mathbf{428})\ CH_2 \longrightarrow (\mathbf{429})\ (NC)_2C(\bar{O}){-}CH_2{-}\overset{+}{} \quad (326)$$

Other reactions of this type are listed in Table 12. All of the products are formed at room temperature in an inert solvent such as hexane.

Not all olefins having allylic hydrogen atoms react with carbonyl cyanide. Examples of olefins that do not react include the substituted propenes **431**, **432** and **433**. In these olefins, the electron-donor

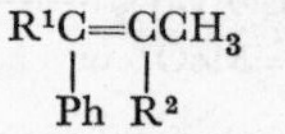

$$R^1C(Ph){=}C(R^2)CH_3$$

(**431**) ($R^1 = R^2 = H$)
(**432**) ($R^1 = H, R^2 = CH_3$)
(**433**) ($R^1 = Ph, R^2 = CH_3$)

properties of the aryl ring and the hyperconjugative effect of the alkyl group do not act together to increase the nucleophilicity of

TABLE 12. Reaction of carbonyl cyanide with olefins.

Olefin	Product	Reference
$(CH_3)_2C{=}C(CH_3)_2$	$CH_2{=}C(CH_3)C(CH_3)_2C(CN)_2OCOCN$	772
cyclohexene	2-cyclohexenyl–$C(CN)_2$–OCOCN	772
$PhCH_2CH{=}CH_2$	$PhCH{=}CHCH_2{-}C(CN)_2{-}OCOCN$	769
$CH_2{=}CHCH_2CH_2CH_3$	$EtCH{=}CHCH_2C(CN)_2OCOCN$ 85%	768
$CH_3CH{=}CHCH_2CH_3$	$CH_3CH{=}CH{-}CH(CH_3){-}C(CN)_2{-}OCOCN$ 100%	768
$PhC(CH_3){=}CHCH_3$ *cis* or *trans*	$CH_2{=}C(Ph){-}CH(CH_3){-}C(CN)_2{-}OCOCN$	773

either of the olefinic carbon atoms[774]. If there is an electron-releasing substituent in the *para* position of the ring, reactions of 1-arylpropenes and carbonyl cyanide (equation 327)[774] do take place. This reaction

$$p\text{-}XC_6H_4CH{=}CHCH_3 + CO(CN)_2 \xrightarrow{-HCN} p\text{-}XC_6H_4CH{=}C(CH_3)COCN \quad (327)$$
$$(X = MeO \text{ or } Me_2N)$$

differs from the 'ene' reactions discussed above in that double-bond migration does not occur. There is a relationship, however, to the addition of carbonyl cyanide to 1,1-diarylethylenes. In an inert solvent, at room temperature, carbonyl cyanide and 1,1-diphenylethylene (or 1,1-di-*p*-tolylethylene) cycloadd to form the oxetanes **434** and **435** (equation 328)[775].

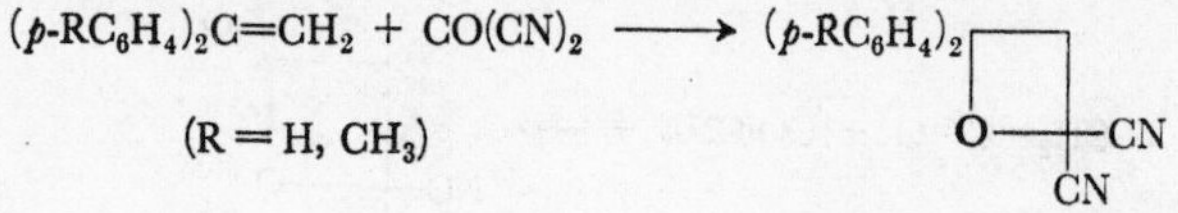

$$(p\text{-}RC_6H_4)_2C{=}CH_2 + CO(CN)_2 \longrightarrow (p\text{-}RC_6H_4)_2\overline{\underset{\displaystyle O\text{---}\underset{CN}{C}\text{---}CN}{\Box}} \quad (R = H, CH_3) \tag{328}$$

(**434**) (R = H) (58%)
(**435**) (R = CH_3) (66%)

Treatment of the oxetane **435** with glacial acetic acid at low temperature causes ring-opening to the hydroxyacetate **436**. The latter, on mild heating, eliminates hydrogen cyanide and acetic acid (equation 329)[776]. Simple styrenes that bear electron-donating groups in the *para* position (e.g. methyl or methoxyl) also undergo

$$\mathbf{435} \xrightarrow{HOAc} \underset{(\mathbf{436})}{(p\text{-}CH_3C_6H_4)_2\underset{OAc}{C}CH_2\overset{CN}{\underset{CN}{C}}OH} \xrightarrow[50\text{–}60°]{HOAc} \underset{(\mathbf{437})}{(p\text{-}CH_3C_6H_4)_2C{=}CHCOCN} \tag{329}$$

this reaction in acetic acid to give acyl cyanides analogous to **437**[777]. Styrene itself does not react with carbonyl cyanide under these conditions. However, in chloroacetic acid at 40–60° or in trichloroacetic acid at room temperature the adducts **438** and **439** have been isolated (equation 330). These reactions would appear to be

$$PhCH{=}CH_2 + CO(CN)_2 + RCO_2H \longrightarrow PhCH\underset{OCOR}{}CH_2\overset{CN}{\underset{CN}{C}}OH \tag{330}$$

(**438**) (R = $ClCH_2$—)
(**439**) (R = Cl_3C—)

dependent on acid strength and may involve protonation of carbonyl cyanide as in equation (331). Until further research is carried out it

$$PhCH{=}CH_2 + {}^{+}\overset{CN}{\underset{CN}{C}}\text{—}OH \longrightarrow Ph\overset{+}{C}HCH_2\overset{CN}{\underset{CN}{C}}OH \tag{331}$$

is not possible to distinguish between carbonium ion and oxetane intermediates in the reactions of carbonyl cyanide with styrenes and 1,1-diarylethylenes.

One additional example of a cycloaddition reaction is the addition of carbonyl cyanide to ketene giving the β-lactone **440**[778] (equation 332). This same β-lactone was prepared earlier from carbonyl

$$CH_2{=}C{=}O + CO(CN)_2 \longrightarrow \text{(440) 82\%} \qquad (332)$$

cyanide and acetic anhydride, but the correct structure was unrecognized[779].

G. Perfluoroacyl Cyanides

Although dimers of perfluoroacyl cyanides have been known for some time[780], the first successful synthesis of the monomers has only recently been accomplished[745]. The preparation is carried out by passing the perfluoroacetyl chloride through a tube of silver cyanide heated to 250–300° (equation 333).

$$CF_3COCl + AgCN \xrightarrow[\text{phase}]{\text{vapour}} \underset{\substack{\text{b.p. } 0° \\ \mathbf{(441)}}}{CF_3COCN} + AgCl \qquad (333)$$

Like carbonyl cyanide, **441** undergoes the Diels–Alder reaction at room temperature (equation 334). Condensation with isobutylene

$$\underset{\mathbf{(441)}}{CF_3COCN} + CH_2{=}C(CH_3){-}C(CH_3){=}CH_2 \longrightarrow \text{2-cyano-2-trifluoromethyl-4,5-dimethyl-3,6-dihydro-2H-pyran (73\%)} \qquad (334)$$

also occurs readily. The resulting adduct **442** readily loses hydrogen cyanide providing an interesting ketone synthesis (equation 335).

$$(CH_3)_2C{=}CH_2 + CF_3COCN \longrightarrow \underset{\mathbf{(442)}}{CH_2{=}C(CH_3)CH_2C(CF_3)(CN)OH} \longrightarrow CH_2{=}C(CH_3)CH_2C(O)CF_3 \qquad (335)$$

The lactone, **443**, results from addition to ketene, but cycloaddition to the electron-rich olefin, 4-methoxystyrene, gives a 2:1 adduct, **444**. The latter reaction is a departure from the analogies to carbonyl cyanide and suggests that further work on the chemistry of the perfluoroacyl cyanides will be rewarding. Another difference is dimer formation which has already been mentioned. Pyrolysis of **445**, the dimer of trifluoroacetyl cyanide, is a good route to bis-(trifluoromethyl)malononitrile (**446**; equation 336):

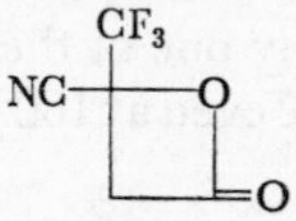

(**443**) 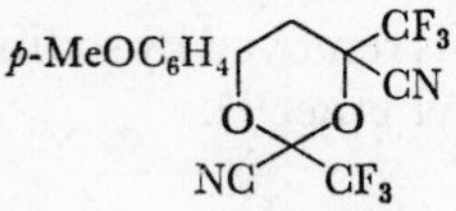 (**444**)

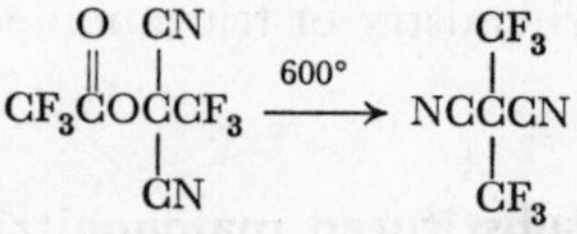

$$\mathrm{CF_3COC(CN)_2CF_3}\ (\mathbf{445}) \xrightarrow{600^\circ} \mathrm{NCC(CF_3)_2CN}\ (\mathbf{446}) \qquad (336)$$

XI. CYANOCARBON ANIONS

Cyanocarbon anions are resonance stabilized systems in which a negative charge is dispersed over all cyano groups in the structure. They are very stable and are thus the end-products of numerous cyanocarbon reactions. Because of this stability, some cyanocarbon anions are useful intermediates for the synthesis of neutral cyano-carbons. The stable anion or dianion is constructed and then it is oxidized to the reactive uncharged cyanocarbon.

Even though the anions possess a full negative charge many of them undergo reactions as if no charge were present. By changing the cation one can make the cyanocarbon anion soluble in either organic solvents or water. Usually the anions are isolated as tetra-alkylammonium salts since these can be handled like organic com-pounds, are not hygroscopic and have sharp melting points. For simplicity, the cation is not mentioned each time in this section. Most linear cyanocarbon anions are coloured and spectroscopy is relied on heavily for structural determinations. If an ion radical can be formed from the anion by either reduction or oxidation, the e.s.r. spectrum of this radical will often provide a conclusive proof of structure for the system.

Most cyanocarbon anions have been named by dropping the final 'e' from the name of the corresponding acid and adding 'ide'. If the acid forms a dianion, the 'e' is retained and diide is added.

A. Cyanomethanides

1. Malononitrile salts

Sodium malononitrile can be isolated by dilution of its alcohol solutions with ether. Its chemistry is similar to that of the salts of malonic esters and of β-keto esters[8,781]. Only one of the hydrogens of malononitrile is removed by sodium hydride even at 160° in diethylene glycol dimethyl ether[782].

2. Tricyanomethanide

The synthesis and chemistry of tricyanomethanide is discussed in section III.A.1.

3. Other negatively-substituted malononitrile ions

Salts of malononitrile substituted with most of the well-known electron-withdrawing groups have been prepared; these include, alkoxycarbonyl (alkyl dicyanoacetate)[783] (**447**); arylsulphonyl[783] (**448**), acyl[783] (**449**), cyanocarbonyl (**450**)[86], nitroso (**451**)[784], carbamoyl (**452**)[79], *p*-nitrophenyl (**453**)[785] and trimethoxymethyl (**454**) (section IV.A.5.c).

$^{-}C(CN)_2-C(=O)OMe$ (**447**) $^{-}C(CN)_2-SO_2Ph$ (**448**) $^{-}C(CN)_2-C(=O)CH_3$ (**449**)

$^{-}C(CN)_2-C(=O)CN \longleftrightarrow (CN)_2C{=}C(O^{-})CN$ (**450**)

$^{-}C(CN)_2-NO \longleftrightarrow (CN)_2C{=}N{-}O^{-}$ (**451**)

$^{-}C(CN)_2-C(=O)NH_2$ (**452**) $^{-}C(CN)_2-C_6H_4-NO_2$ (**453**)

$^{-}C(CN)_2-C(OMe)_3$ (**454**) $^{-}CBr(CN)_2$ (**455**)

When bromomalononitrile is added to excess methanolic potassium hydroxide at room temperature, potassium bromodicyanomethanide (**455**) is formed in solution[106]. This anion absorbs at 235 mμ (ε 15,000). Bromomalononitrile is recovered if the solution is quickly acidified. If the solution is allowed to stand, yellow pentacyanopropenide is formed. This suggests that the bromodicyanomethanide ion is eliminating bromide with formation of tetracyanoethylene. Tetracyanoethylene then reacts with the base to give pentacyanopropenide (section IV.A).

A number of dicyanomethane zwitterions (ylids) are known. The two cyano groups stabilize these ylids to such an extent that they do not have the reactivity commonly associated with ylids. Examples are: **456**[786], pyridinium dicyanomethylid (**457**) (section IX.A.2.a), triphenylphosphonium dicyanomethylid (**458**)[787] and dimethylsulphonium dicyanomethylid (**459**)[788]. In addition, the zwitterion form of dicyanodiazomethane (section VIII.F) no doubt contributes heavily to its structure (**460**).

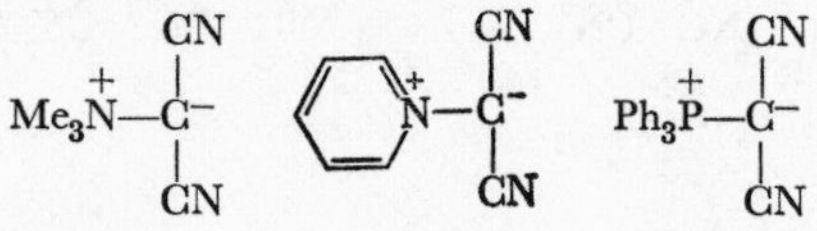

$$Me_3\overset{+}{N}-\overset{-}{C}(CN)_2 \quad (\mathbf{456}) \qquad C_5H_5\overset{+}{N}-\overset{-}{C}(CN)_2 \quad (\mathbf{457}) \qquad Ph_3\overset{+}{P}-\overset{-}{C}(CN)_2 \quad (\mathbf{458})$$

$$Me\overset{+}{S}-\overset{-}{C}(CN)_2 \quad (\mathbf{459}) \qquad N{\equiv}\overset{+}{N}-\overset{-}{C}(CN)_2 \longleftrightarrow \overset{-}{N}{=}\overset{+}{N}{=}C(CN)_2 \quad (\mathbf{460})$$

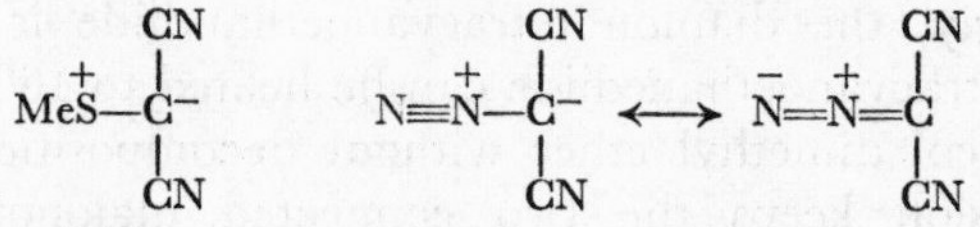

B. *Cyanoethanides*

Cyanoethanides are considerably less stable than cyanomethanides since only two cyano groups can participate in resonance. In addition, the β-cyano groups tend to be eliminated as cyanide.

I. Pentacyanoethanide

Sodium pentacyanoethanide results from addition of sodium cyanide to tetracyanoethylene (see section IV.A.3.1). It is an unstable, nearly white solid. Acidification gives pentacyanoethane (section III.B.2), chlorination, chloropentacyanoethane and cyanation with cyanogen chloride, hexacyanoethane (section III.B.3). Pentacyanoethanide is stable in solution in the presence of tetracyanoethylene; in its absence, it is converted to tetracyanoethylene ion radical (section IV.A.7).

2. Tetracyanoethanide

Tetracyanoethanides are probably the intermediates in all cyanosubstitution reactions of tetracyanoethylene (section IV.A.5). Since they readily eliminate cyanide, tetracyanoethanides are rarely isolated. This elimination of cyanide is demonstrated by the reaction of tetracyanoethanide with sodium malononitrile[238]. Cyanide is eliminated and the malononitrile anion adds to the resulting tricyanoethylene:

$$(NC)_2\overset{-}{C}-CH(CN)_2 \longrightarrow (NC)_2C{=}C(H)CN \xrightarrow{^{-}CH(CN)_2} (NC)_2\overset{-}{C}-C(H)(CN)-CH(CN)_2 \longrightarrow (NC)_2C{=}C(H)-\overset{-}{C}(CN)_2 \qquad (337)$$

Surprisingly, the dianion tetracyanoethandiide is quite stable. Disodium tetracyanoethanediide can be heated to 168° in refluxing ethylene glycol dimethyl ether without decomposition[238]. Electrostatic repulsion keeps the two connected malononitrile anions orthogonal, thus hindering β-elimination of a cyanide.

3. Tricyanoethanide

One synthetic route to tricyanovinyl compounds involves cyanide-catalysed addition of hydrogen cyanide to a 1,1-dicyanoethylene derivative. The intermediate here is no doubt a tricyanoethanide,

$$\text{2,2,4,4-}Me_4\text{-3-}[{=}C(CN)_2]\text{cyclobutanone} + CN^- \longrightarrow \text{2,2,4,4-}Me_4\text{-3-}(CN)\text{-3-}[\overset{-}{C}(CN)_2]\text{cyclobutanone} \qquad (338)$$

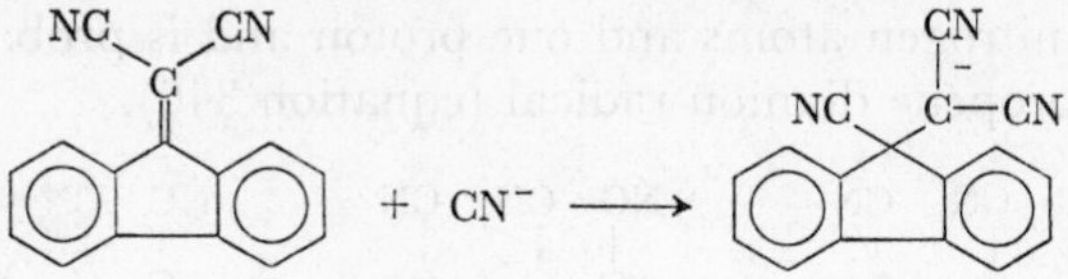

(339)

since in two examples such salts have indeed been isolated (equations 338 and 339)[789,790].

C. *Cyanopropenides*

I. Pentacyanopropenide

Pentacyanopropenide is a dicyanovinylogue of tricyanomethanide. It is made by the base-catalysed hydrolysis of tetracyanoethylene[86], by reaction of tetracyanoethylene with malononitrile or by treatment of bromomalononitrile[106] with basic reagents. Numerous salts have been made[86]. The acid is too unstable to be isolated. When an aqueous solution of the acid is made by ion exchange and the water evaporated completely, the residue turns black. Since the negative charge never resides on the central carbon atom, this position is positive with respect to the two external carbon atoms. Molecular-orbital calculations[791] indicate an effective charge distribution as shown in **461**. Conductivity studies show that pentacyanopropenide salts are not ion-paired in acetonitrile[791].

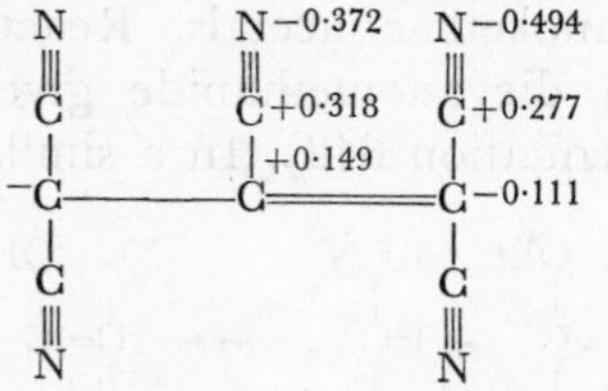

N−0·372 N−0·494
C+0·318 C+0·277
+0·149
C−0·111

(461)

Electrolysis of the yellow pentacyanopropenide anion at −2·0 v versus Ag–$AgClO_4$ in dimethylformamide produces the red pentacyanopropene dianion radical[729]. The e.s.r. spectrum shows 25 of the theoretical 27 lines. Further reduction at −2·25 to −2·3 v produces a pale yellow solution which has no e.s.r. spectrum. Electrolysis of this solution at −2·75 v gives another ion radical with an 18 line e.s.r. spectrum. This ion radical has a half-life of only about one minute. The spectrum would fit a radical with four equivalent nitrogen atoms and one proton and is probably 1,1,3,3-tetracyanopropene dianion radical (equation 340).

$$(NC)_2C{=}C(CN){-}C^-(CN)_2 \xrightarrow{e} {}^-C(CN)_2{-}\dot{C}(CN){-}C^-(CN)_2 \xrightarrow{e} {}^-C(CN)_2{-}\bar{C}(CN){-}C^-(CN)_2 \; (\mathbf{462})$$

$$\mathbf{462} \xrightarrow{-CN^-} {}^-C(CN)_2{-}\ddot{C}{-}C^-(CN)_2 \xrightarrow{H^+} (NC)_2C{=}CH{-}C^-(CN)_2 \xrightarrow{e} {}^-C(CN)_2{-}\dot{C}H{-}C^-(CN)_2 \qquad (340)$$

Pentacyanopropenide and 1,1,3,3-tetracyanopropenides ring close to pyridines on treatment with hydrogen halides (section IX.D).

2. Tetracyanopropenides

a. 1,1,3,3-Tetracyanopropenides. Sodium dicyanomethanide condenses with ethoxymethyldicyanoethylene or ethoxydicyanoethylene to give the corresponding tetracyanopropenide (equation 341)[792,793].

$$(NC)_2C{=}C(OEt)H + HC^-(CN)_2 \longrightarrow (NC)_2C{=}CH{-}C^-(CN)_2 \qquad (341)$$

A number of 2-substituted tetracyanopropenides can be made starting from dicyanoketene acetals. Reaction of dicyanoketene diethyl acetal with dicyanomethanide gives 2-ethoxytetracyanopropenide (**463**)[86] (equation 342). In a similar reaction, 2-methylthiotetracyanopropenide is prepared from dicyanoketene dimethyl thioacetal. Treatment of the ethoxy derivative with ammonia, or amines, gives 2-aminotetracyanopropenide (**464**) (equation 343).

$$(NC)_2C{=}C(OEt)_2 + HC^-(CN)_2 \longrightarrow (NC)_2C{=}C(OEt){-}C^-(CN)_2 \; (\mathbf{463}) \qquad (342)$$

$$\mathbf{463} + RNH_2 \longrightarrow (NC)_2C{=}C(NHR){-}C^-(CN)_2 \; (\mathbf{464}) \qquad (343)$$

Nitrosation of **464** does not give a stable diazonium inner salt. The 2-chloro- or 2-bromotetracyanopropenide is produced depending on which acid is used to generate the nitrous acid (equation 344)[86].

$$\mathbf{464} + HNO_2 \longrightarrow \left[(NC)_2C{=}C(N^+{\equiv}N){-}C^-(CN)_2 \right] \xrightarrow{HCl} (NC)_2C{=}C(Cl){-}C^-(CN)_2 \quad (344)$$

Reaction of tricyanovinyl compounds with dicyanomethanide gives the corresponding 2-substituted tetracyanopropenide (equation 345)[259].

$$Ph{-}C(CN){=}C(CN)_2 + HC^-(CN)_2 \longrightarrow (NC)_2C{=}C(Ph){-}C^-(CN)_2 \quad (345)$$

2-Trifluoromethyltetracyanopropenide is produced by replacement of the amino group of 1-amino-1-trifluoromethyl-2,2-dicyanoethylene with dicyanomethanide[794] (equation 346). Amino groups

$$(NC)_2C{=}C(NH_2)CF_3 + HC^-(CN)_2 \longrightarrow (NC)_2C{=}C(CF_3){-}C^-(CN)_2 \quad (346)$$

are not ordinarily replaced under the mild conditions employed in this reaction.

b. 1,1,2,3-Tetracyanopropenides. The reaction of 1,2-dicyano-1,2-bis(4-tolylsulphonyl)ethylene with sodium dicyanomethanide produces 1,1,2,3-tetracyano-3-(4-tolylsulphonyl)propenide (equation 346a)[121].

$$NC(4\text{-}MeC_6H_4SO_2)C{=}C(SO_2C_6H_4Me\text{-}4)CN + HC^-(CN)_2 \longrightarrow (NC)_2C{=}C(CN){-}C^-(CN)SO_2C_6H_4Me\text{-}4 \quad (346a)$$

A tetracyanopropenide containing a cyanocarbonyl group (pentacyanobutadienolate) is produced by hydrolysis of hexacyanobutadiene (section IV.E) or by oxidation of hexacyanobutadiendiide with nitrous acid (equation 347)[795].

$$\mathrm{^{-}C(CN)(NC)\!-\!C(CN)\!=\!C(CN)\!-\!C^{-}(CN)_2 + HNO_2 \longrightarrow (NC)_2C\!=\!C(CN)\!-\!C^{-}(CN)\!-\!C(=O)CN \longleftrightarrow (NC)_2C\!=\!C(CN)\!-\!C(CN)\!=\!C(CN)O^{-}} \quad (347)$$

D. *Cyanobutenides*

1. Hexacyanoisobutenediide

The reaction of dicyanoketene diethyl acetal with two moles of dicyanomethanide salt gives hexacyanoisobutenediide (equation 348)[86]. Salts of this dibasic acid are colourless with $\lambda_{max}^{H_2O}$ 335 mμ

$$\mathrm{(NC)_2C\!=\!C(OEt)_2 + 2\,HC^{-}(CN)_2 \longrightarrow (NC)_2C\!=\!C[C^{-}(CN)_2]_2} \quad (348)$$

(ε 32,700). X-ray studies show that the ion is propeller shaped with the arms tilted about 24° [796,797]. This tilt does not reduce the π-overlap appreciably, but does relieve intramolecular steric strain considerably.

2. Tetracyanobutanedionediide

The condensation of malononitrile with diethyl oxalate produces tetracyanobutane-2,3-dionediide in quantitative yield[412] (equation 349). The ion is not orange as originally reported but is colourless, $\lambda_{max}^{H_2O}$ 280 mμ (ε 13,400)[238].

$$\mathrm{EtOC(=O)\!-\!C(=O)OEt + 2\,HC^{-}(CN)_2 \longrightarrow {}^{-}C(CN)_2\!-\!C(=O)\!-\!C(=O)\!-\!C^{-}(CN)_2} \quad (349)$$

3. Hexacyanobutenediide

Heating disodium tetracyanoethanediide with a catalytic amount of acid produces hexacyanobutenediide in quantitative yield (section IV.E). The diide exists as a mixture of interconvertible

precipitation from 1,2-dimethoxyethane–acetonitrile with ether. *cis–trans* isomers (**465** and **466**), which are isolated by fractional

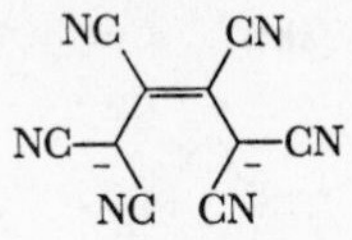

(465) 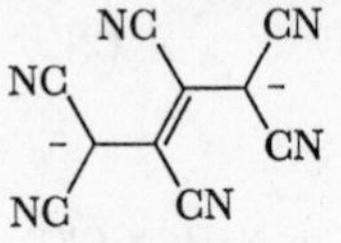 **(466)**

precipitation from 1,2-dimethoxyethane—acetonitrile with ether. The equilibrium mixture in 1,2-dimethoxyethane–acetonitrile contains approximately 25 % of the *cis* and 75 % of the *trans* form. As in the case of hexacyanoisobutenediide, the dicyanomethylene functions are probably twisted out of plane in both the *cis* and *trans* isomers.

E. *Cyanopentadienides*

1. Cyanocyclopentadienides

The cyanocarbon anion with the strongest conjugate acid is pentacyanocyclopentadienide. It can be heated in air to 400° without decomposition and it is not protonated by perchloric acid. This anion and all the other possible cyanocyclopentadienides are made by the stepwise cyanation of cyclopentadiene with cyanogen chloride[798]. The first three cyano groups are introduced in the presence of sodium hydride which serves to regenerate the cyclopentadiene anion. Introduction of the fourth and fifth cyano groups (to form **471** and **472**) requires aluminium chloride to increase the reactivity of the cyanogen chloride. The reaction can be stopped at the mono-, di- or tricyano stage merely by limiting the amount of cyanogen chloride (equation 350). The ratio of 1,3- (**467**) to 1,2-isomer (**468**) in the dicyano anion is 1:6, whereas that of the 1,2,4-(**469**) to 1,2,3-isomer (**470**) in the tricyano anion is 1:2.

Methyltricyano- and methyltetracyanocyclopentadienides are synthesized from methylcyclopentadiene in a similar fashion. Indene does not give 1,2,3-tricyanoindene. The product is the neutral 1,1,3-tricyanoindene (equation 351). Cyanation at the 2-position would require a high-energy *o*-quinodimethane-type transition state.

a. Tetracyanocyclopentadienide. Another route to the cyanocyclopentadienides is through protonation of hexacyanobutenediide[795] (**473**). 1-Amino-2,3,4,5,5-pentacyanocyclopentadiene (**474**) is produced by a ring closure. The reaction is reversible (equation 352).

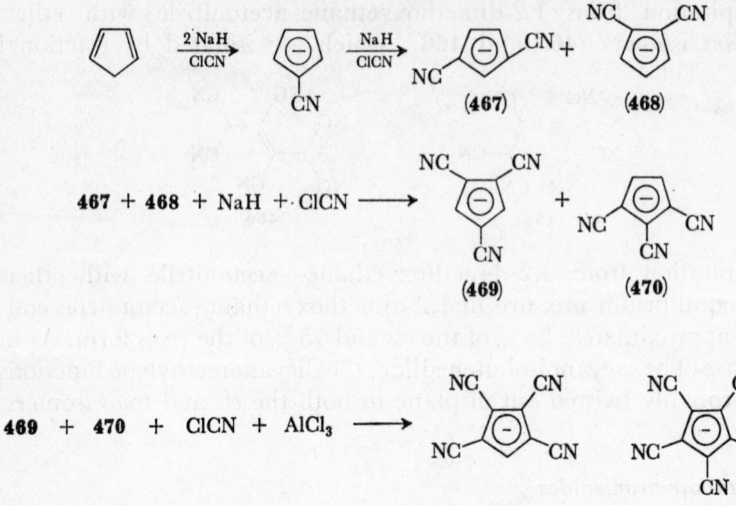

2 NaH / ClCN → (CN-cyclopentadienide) → NaH / ClCN → (467) + (468)

$$467 + 468 + NaH + ClCN \longrightarrow (469) + (470)$$

$$469 + 470 + ClCN + AlCl_3 \longrightarrow (471) \quad (472) \qquad (350)$$

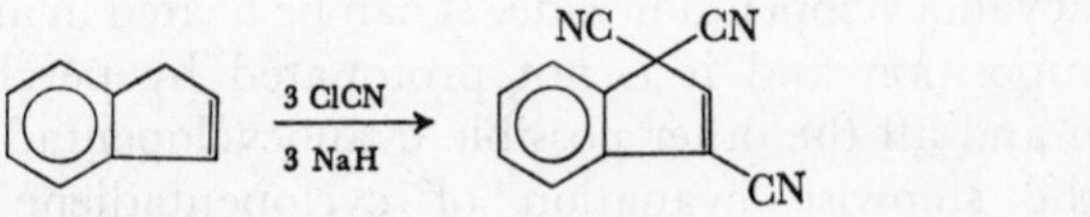

3 ClCN / 3 NaH (351)

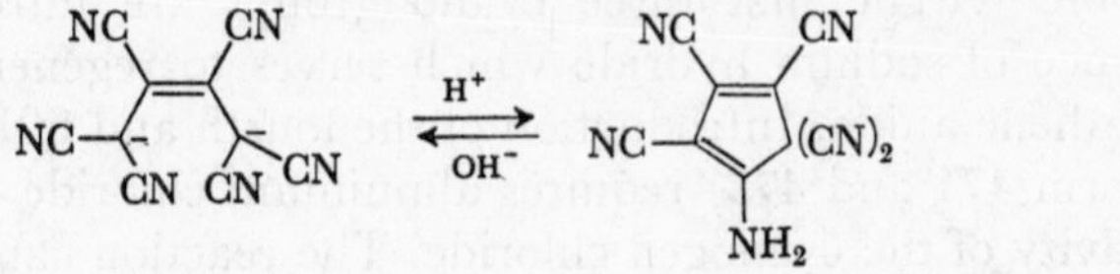

(473) ⇌ H⁺ / OH⁻ (474) (352)

This bright yellow cyanoamine readily loses one of its five cyano groups by hydrolysis and decarboxylation on treatment with concentrated hydrochloric acid (equation 353).

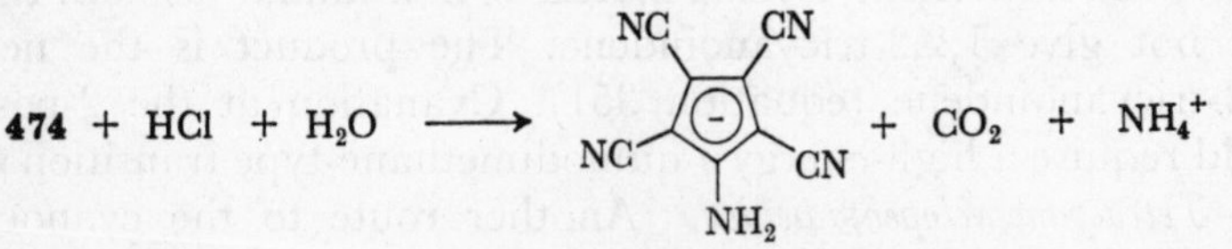

$$474 + HCl + H_2O \longrightarrow \text{(product)} + CO_2 + NH_4^+ \qquad (353)$$

Yet another route to the cyanocyclopentadienide system starts with cyclopentane-1,2,3,4-tetracarboxylic acid. This tetra-acid is converted via the tetra-amide **475** to the tetra-nitrile, which is then dehydrogenated with phosphorus pentachloride[799]:

$$\text{(HO}_2\text{C)}_4\text{-cyclopentane} \xrightarrow[\text{2. NH}_3]{\text{1. SOCl}_2} \text{(NH}_2\text{OC)}_4\text{-cyclopentane } \mathbf{(475)}$$

$$\mathbf{475} + \text{POCl}_3 \longrightarrow \text{(NC)}_4\text{-cyclopentane} \xrightarrow{\text{PCl}_5} \text{(NC)}_4\text{-cyclopentadienide}^{\ominus} \qquad (354)$$

2,3-Dicyano-1,4-diethoxycarbonylcyclopentadienides can be made from the diketo diester **476** by dehydration of the bis(cyanohydrin)[799]:

$$\mathbf{(476)} + 2\ \text{HCN} \longrightarrow \text{(HO)(NC)C–C(OH)(CN), EtO}_2\text{C, CO}_2\text{Et} \xrightarrow{\text{Ac}_2\text{O}} \text{(AcO)(NC)C–C(OAc)(CN), EtO}_2\text{C, CO}_2\text{Et } \mathbf{(477)} \qquad (355)$$

$$\mathbf{477} \xrightarrow{t\text{-BuO}^-} \text{2,3-(NC)}_2\text{-1,4-(EtO}_2\text{C)}_2\text{-cyclopentadienide}^{\ominus}$$

In general, the cyanocyclopentadienides undergo classical aromatic substitution reactions. The basicity of the cyclopentadienide is so low that strongly acidic conditions can be used. For example, cold, concentrated nitric acid nitrates tetracyanocyclopentadienide (**478**) in 15 minutes[799,800] (equation 356).

Acylation with acetic anhydride[800] in trifluoroacetic acid or with acetyl chloride–aluminium chloride[799] gives the acetyl derivative **479**. The keto group of **479** exhibits normal behaviour in that it is readily reduced with sodium borohydride. Dilute acids cause the

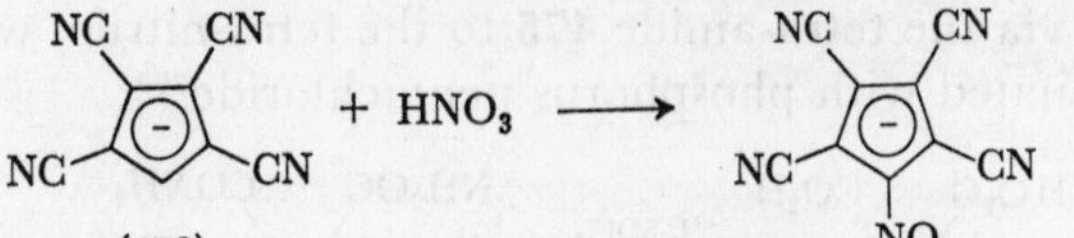

(478) + HNO_3 → (356)

resulting alcohol to form the ether **482**, possibly through the intermediacy of methyltetracyanofulvene. Treatment of the alcohol **480** with thionyl chloride in pyridine gives the pyridinium zwitterion **481**. When this substance is warmed in pyridine, vinyltetracyanocyclopentadienide is obtained, and this polymerizes readily at 135° to an acetonitrile-soluble polymer (equation 357)[800]. Formaldehyde

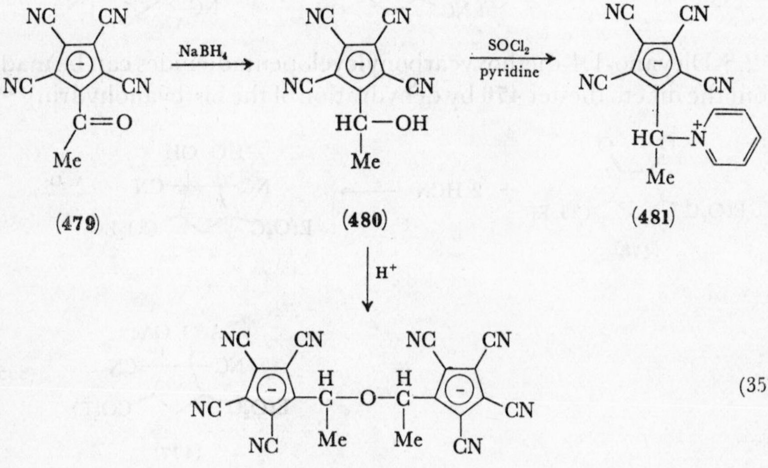

(479) —$NaBH_4$→ (480) —$SOCl_2$, pyridine→ (481)

(480) —H^+→ (482) (357)

481 + pyridine —Δ→

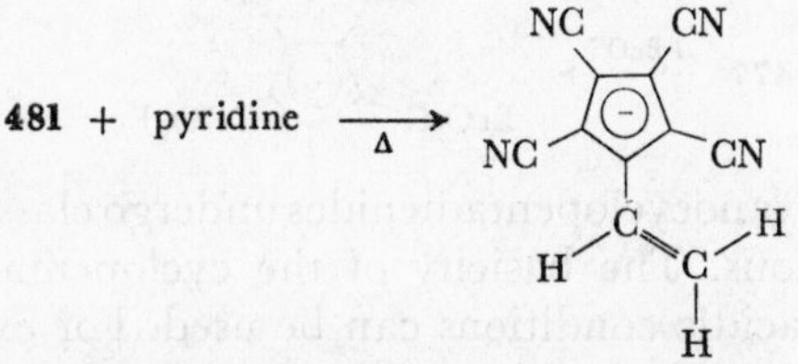

reacts with **478** to give a methylene compound (**482a**) rather than a hydroxymethyl derivative[800].

Chlorine[799,800] and bromine[800] but not iodine halogenate tetracyanocyclopentadienide at room temperature (equation 358). The pseudohalogen trifluoromethanesulphenyl chloride, however, requires a Friedel–Crafts catalyst (equation 359).

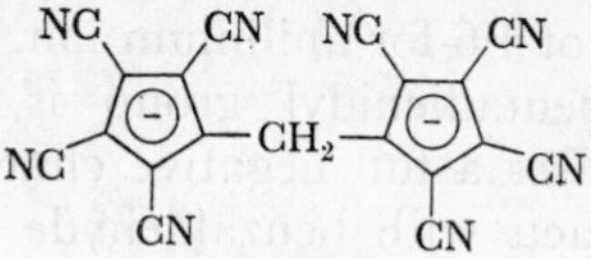

(482a)

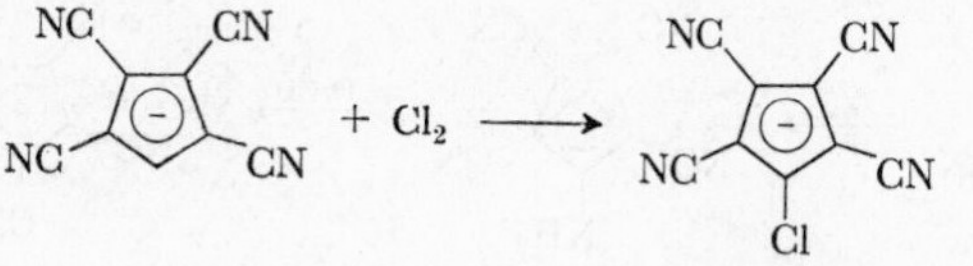

$$+ Cl_2 \longrightarrow \quad \textbf{(483)} \qquad (358)$$

$$+ CF_3SCl \xrightarrow{AlCl_3} \qquad (359)$$

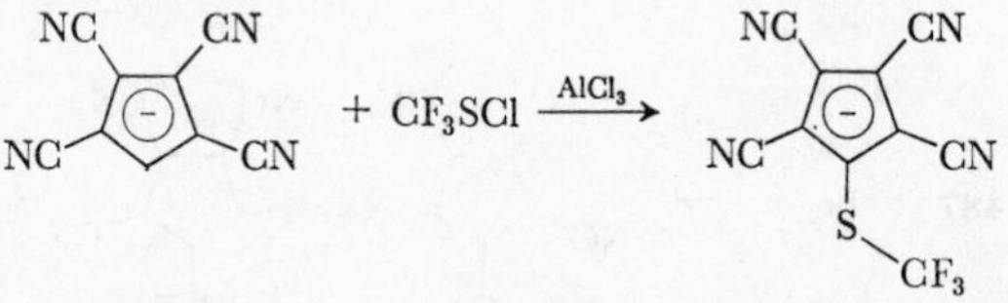

Chlorotetracyanocyclopentadienide (**483**) reacts with 1 mole of chlorine to produce 5,5-dichloro-1,2,3,4-tetracyanocyclopentadiene (**484**) and with two moles of chlorine to produce tetrachlorotetracyanocyclopentene (**485**). The chlorines of the dichloro compound are electropositive, and react with chloride ion to produce chlorine. The dichloro compound reacts with ethylene at one atmosphere and at room temperature to give the 7,7-dichloro-1,2,3,4-tetracyanonorbornene **486** (equation 360).

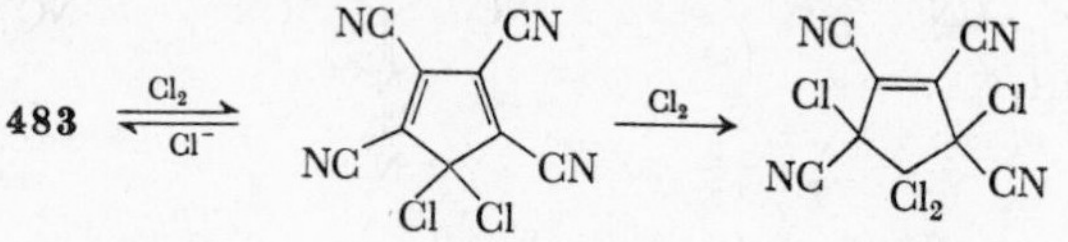

$$\textbf{483} \underset{Cl^-}{\overset{Cl_2}{\rightleftharpoons}} \textbf{(484)} \xrightarrow{Cl_2} \textbf{(485)}$$

$$\textbf{484} + CH_2{=}CH_2 \longrightarrow \textbf{(486)} \qquad (360)$$

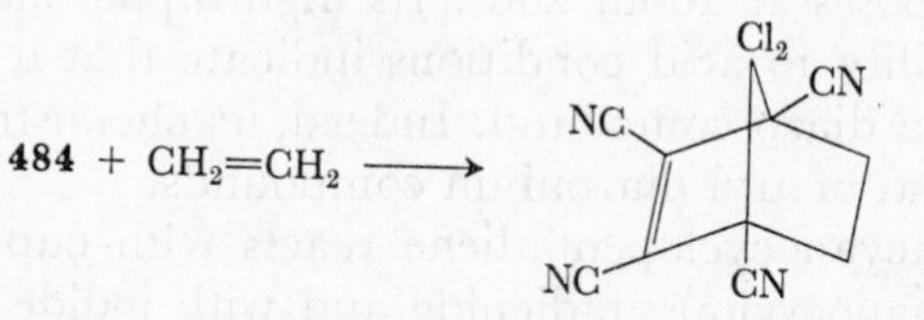

Nitrotetracyanocyclopentadienide is reduced by zinc and hydrochloric acid to aminotetracyanocyclopentadienide (**487**), also available from the cyclization of hexacyanobutenediide (equations 352 and 353). The zwitterion from this amine has a pK_a in water of 2·0

compared to a value of 4·6 for anilinium ion. These data show that the tetracyanocyclopentadienidyl group is electron-withdrawing even though it possesses a full negative charge. Aminotetracyanocyclopentadienide reacts with benzaldehyde to form the anil **488**, and with bromine to give the azo derivative **489** (equation 361).

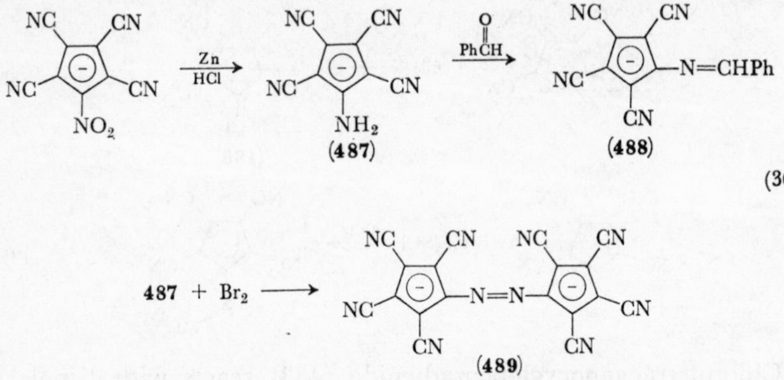

NC, CN, NC, CN, NO_2 $\xrightarrow[HCl]{Zn}$ NC, CN, NC, CN, NH_2 (**487**) $\xrightarrow{PhCHO}$ NC, CN, NC, CN, —N=CHPh (**488**)

(361)

487 + Br_2 ⟶ NC, CN, NC, CN, —N=N—, NC, CN, CN, CN (**489**)

b. Diazotetracyanocyclopentadiene. When aminotetracyanocyclopentadienide is diazotized under the conditions ordinarily used to prepare aromatic diazonium compounds, diazotetracyanocyclopentadiene (**490**) forms (equation 362)[795]. This stable, light-yellow

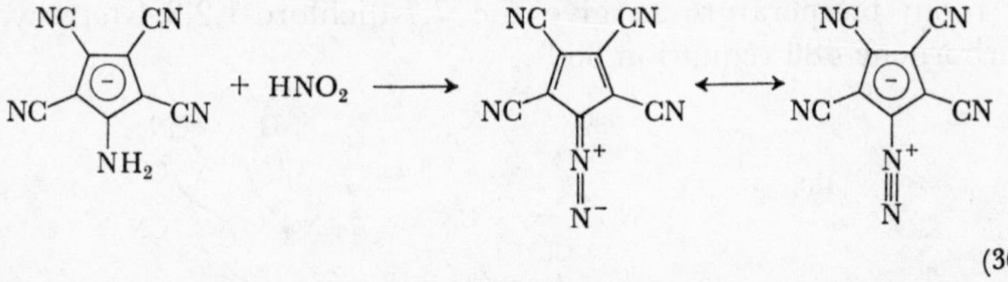

NC, CN, NC, CN, NH_2 + HNO_2 ⟶ NC, CN, NC, CN, N^+=N^- ⟷ NC, CN, NC, CN, N^+≡N

(362)

solid decomposes at about 200°. Its high dipole moment (11·44 D) and its stability to acid conditions indicate that it is a diazonium rather than a diazo compound. Indeed, its chemistry is remarkably similar to that of aryl diazonium compounds.

Diazotetracyanocyclopentadiene reacts with cuprous cyanide to give pentacyanocyclopentadienide and with iodide or bromide ion to give halotetracyanocyclopentadienides. Chloride ion requires copper as a catalyst (equations 363, 364 and 365). These and other substitutions proceed by a free-radical mechanism, since use of ethanol as the reaction solvent results in the formation of tetracyanocyclopentadienide ion by abstraction of an α-hydrogen from

(490) + CuCN ⟶ (363)

490 + NaI ⟶ (364)

490 + Cu + Cl^- ⟶ (365)

ethanol by tetracyanocyclopentadienidyl ion radical (**491**, equation 366).

The ion radical is generated irreversibly at about 0·23 v versus SCE in acetonitrile. It could not be detected, however, when the

$\xrightarrow{e}$ ⟶ (491) + N_2 (366)

491 + CH_3CH_2OH ⟶ + CH_3CHO

electrolysis was carried out in the cavity of an e.s.r. apparatus. The four cyano groups and the cyclopentadiene π-system provide no stabilization for the ion radical since the odd electron is in an orbital orthogonal to the system. Thus, this cyanocarbon ion radical is comparable to a phenyl radical rather than to a resonance-stabilized cyanocarbon ion radical. It is reactive enough to attack phenol or benzene but not nitrobenzene (equation 367). With phenol, a complex mixture of products (assumed to be the *ortho*, *meta* and

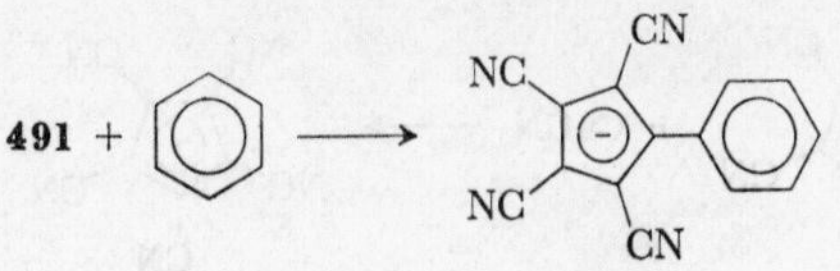

491 + [benzene] ⟶ [phenyltetracyanocyclopentadienide: NC, CN, NC, CN] (367)

para isomers) forms. The *para* isomer **492** was isolated and its pK_a in water determined as 9·66. Based on a pK_a for phenol of 9·92 and

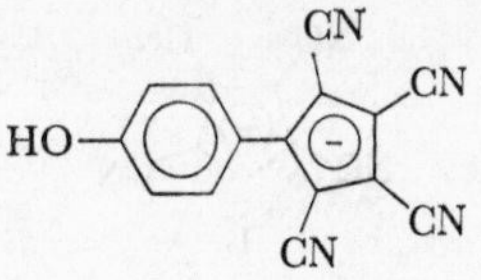

HO–[phenyl]–[tetracyanocyclopentadienide: CN, CN, CN, CN]

(**492**)

a reaction constant ρ of 2·23, the Hammett σ_p for the tetracyanocyclopentadienidyl group is +0·12 (slightly electron-withdrawing). This may reflect only an inductive effect since steric hindrance between the *ortho* hydrogens on the phenyl ring and the cyano groups on the cyclopentadienidyl ring must force the two groups out of conjugation. The electronic spectrum of phenyltetracyanocyclopentadienide, $\lambda_{max}^{CH_3CN}$ 300 mμ (ε 12,600) and 262 mμ (45,800) compared to $\lambda_{max}^{CH_3CN}$ 298 mμ (ε 13,500), 244 mμ (56,800) and 237 mμ (46,400) for tetracyanocyclopentadienide, points to a lack of conjugation between the phenyl and cyclopentadienidyl groups.

Carboxytetracyanocyclopentadienide, prepared by the copper-catalysed reaction of the diazonium compound with carbon monoxide and water (equation 368), has a pK_a value of 4·11 compared

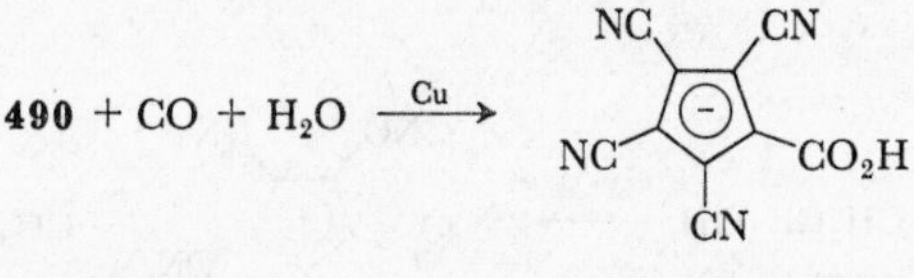

$$\mathbf{490} + CO + H_2O \xrightarrow{Cu}$$ [cyclopentadienide: NC, CN, NC, CO_2H, CN] (368)

to a value of 4·20 for benzoic acid. Here again the carboxy group may be forced out of plane by steric hindrance. The reaction of nitrite ion with **490** gives nitrotetracyanocyclopentadienide (equation 369). The thiol derivative **493** is made by reaction of **490** with xanthate ion followed by hydrolysis (equation 369a).

$$\mathbf{490} + NO_2^- \longrightarrow$$ [cyclopentadienide: NC, CN, NC, CN, NO_2] (369)

$$\mathbf{490} + {}^{-}SC(S)OEt \longrightarrow \text{tetracyanocyclopentadienide–}SC(S)OEt \xrightarrow[2.\ HCl]{1.\ KOH} \text{tetracyanocyclopentadienide–}SH\ (\mathbf{493}) \qquad (369a)$$

The reaction of diazotetracyanocyclopentadiene with sodium azide produces the colourless azidotetracyanocyclopentadienide ion. The thermal stability of this compound is lower than that of aryl azides. It decomposes on heating to 80° with formation of the bright orange azo derivative **489** (equation 370). The nitrene intermediate **494** could not be trapped with cyclohexene. The imine salt

$$\mathbf{490} + NaN_3 \longrightarrow \text{tetracyanocyclopentadienide–}N_3 \longrightarrow \mathbf{(494)} \longleftrightarrow \mathbf{(495)} \qquad (370)$$

$$\mathbf{494} \longrightarrow \text{(tetracyanocyclopentadienide)–}N{=}N\text{–(tetracyanocyclopentadienide)}\ \mathbf{(489)}$$

structure **495** may contribute heavily to the resonance hybrid, making it too stable to react with olefins.

The diazonium compound **490** couples with *N*,*N*-dimethylaniline and with 2-naphthol to give orange azo dyes (equation 371).

$$Me_2N\text{–}C_6H_5 + \mathbf{490} \longrightarrow Me_2N\text{–}C_6H_4\text{–}N{=}N\text{–(tetracyanocyclopentadienide)} \qquad (371)$$

As with other diazonium compounds, some reagents prefer to add to the terminal nitrogen. Triphenylphosphine gives the orange-red phosphazine **496**, sodium cyanide, the azocyanide **497**, and sodium bisulphite, the adduct **498**.

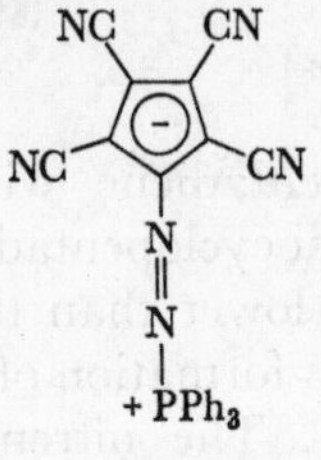

(496) 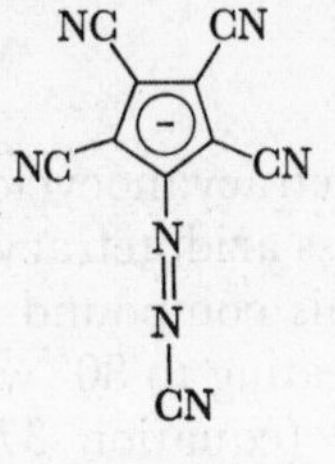 **(497)** 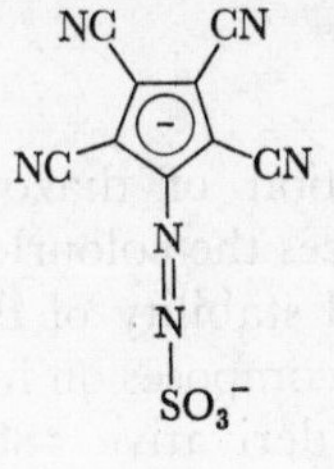 **(498)**

2. Heptacyanopentadienide and other cyanopentadienides

The action of malononitrile anion on hexacyanobutadiene gives the red heptacyanopentadienide **499**. 1,1,2,4,5,5-Hexacyanopentadienide (**500**) results on treatment of 4-methoxy-1,1,2-tricyanobutadiene with tricyanovinyl chloride[207].

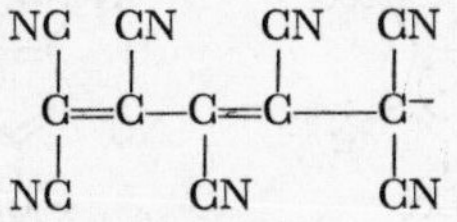

(499) 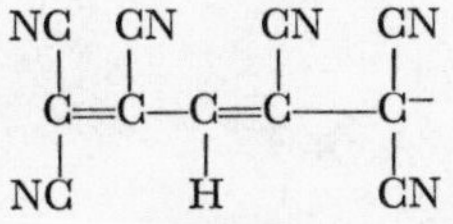 **(500)**

The corresponding anion with cyano groups only on the terminal carbons, 1,1,5,5-tetracyanopentadienide (**501**), is also known[801]. The chemical properties of these anions have not been studied.

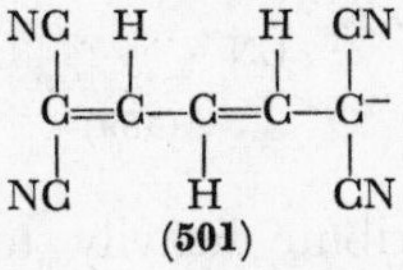

(501)

Introduction of nitrogen in the conjugated backbone of the pentadienide nucleus gives salts with properties similar to those of the all-carbon chains. The reaction of tetracyanoethylene with ammonia yields the red 1,1,2,4,5,5-hexacyano-3-azapentadienide **502**[86]. In a remarkable reaction, sodium cyanide and 1,2-dichlorohexafluorocyclopentene produce 3-fluoro-1,1,4,5,5-pentacyano-2-azapentadienide (**503**) in 30% yield (equation 372)[802]. The structure was confirmed by x-ray analysis.

```
NC  CN       CN
 |  |        |
 C=C—N=C——C⁻
 |       |   |
NC       CN  CN
       (502)
```

```
 Cl   Cl                        NC  CN       CN
  \==/                           |  |        |
F2/  \F2   + NaCN  ———→          C=C—C=N—C⁻          (372)
  \__/                           |    |      |
   F2                           NC    F      CN
                                     (503)
```

F. Miscellaneous Cyanocarbon Anions

1,1,2,6,7,7-Hexacyanoheptatrienide (**507**), a bright-blue anion, is formed when 4-methoxy-1,1,2-tricyanobutadiene (**504**) is refluxed in methanol. The mechanism is thought to involve addition of methoxide to the diene to form **505**, followed by condensation and

```
              CN  CN                              CN  CN
               |  |                      H  ⁻     |
MeOCH=CH—C==C   + MeO⁻ ——→  (MeO)2C—C—C==C
                  |                      H          |
                  CN                                CN
       (504)                            (505)
```

```
                 NC    HC(OMe)2 H           CN  CN
                  |     |       |       ⁻   |
504 + 505 ——→    C=C—C————C———C—C==C
                  |  |  H       |     |      |
                 NC  CN        OMe    H      CN
                               (506)
```

```
           NC  CN    H     CN  CN
            |  |     |     |   |
506 ——→     C=C—C=C—C=C——C⁻ + HC(OMe)3          (373)
            |     |     |      |
           NC     H     H      CN
                 (507)
```

elimination of methyl orthoformate[207] (equation 373). The corresponding blue anion with cyano groups on the ends of the chain (**508**) has been made, as well as the next higher vinylogue[801] **509**.

```
NC  H     H     H  CN       NC  H     H     H     H  CN
 |  |     |     |  |         |  |     |     |     |  |
 C=C—C=C—C=C—C⁻            C=C—C=C—C=C—C=C—C⁻
 |     |     |     |         |     |     |     |     |
NC     H     H     CN       NC     H     H     H     CN
      (508)                            (509)
```

Reaction of tetracyanoethylene with benzamidine gives the phenyldiazaheptatrienide **510**[86]. This ion is dark green ($\lambda_{max}^{acetone}$ 598 mμ).

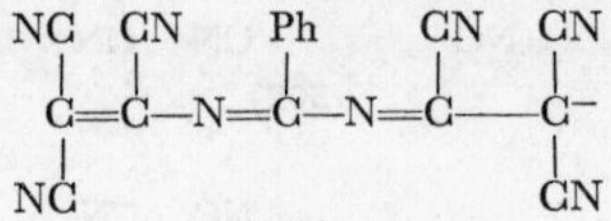

$$(NC)_2C{=}C(CN){-}N{=}C(Ph){-}N{=}C(CN){-}\bar{C}(CN)_2$$

(510)

The reaction of tetracyanoethylene with hydrazine gives 1,1,2,5,6,6-hexacyano-3,4-diazahexadienediide (**511**)[86]. As mentioned earlier

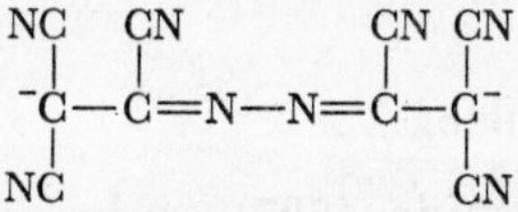

$$(NC)_2\bar{C}{-}C(CN){=}N{-}N{=}C(CN){-}\bar{C}(CN)_2$$

(511)

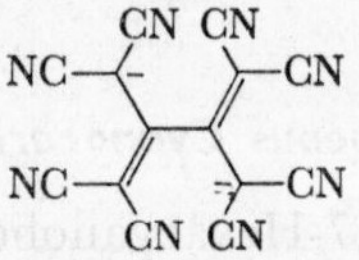

(512)

(section IV.E) one of the products from the reaction of dicyanomethanide with hexacyanobutadiene is 2,3-bis(dicyanomethyl)-1,1,4,4-tetracyanobutadienediide (**512**). Its electronic spectrum (λ_{max} 418 mμ compared to λ_{max} 412 mμ for pentacyanopropenide) indicates steric inhibition of conjugation between the two tetracyanopropenide groups.

As mentioned in section VII.A, several cyanocarbon anions are accessible from tetracyanoquinodimethane-type compounds; examples are: **513**[295], **514**[570] and **515**.

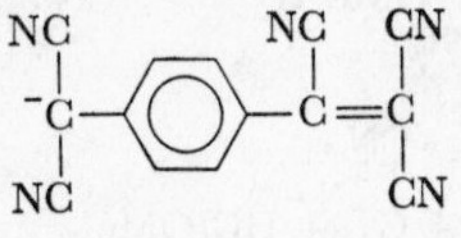

(513) 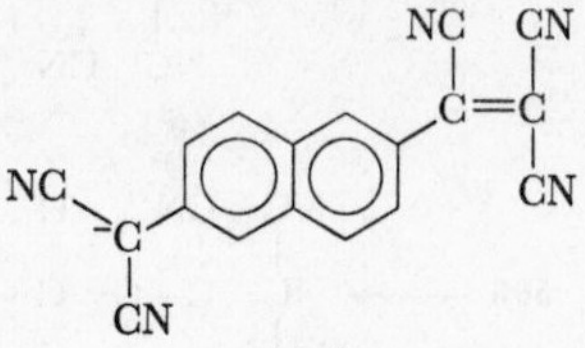 **(514)**

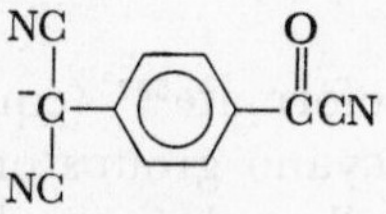

(515)

G. Electronic Spectra of Cyanocarbon Anions

In many cases, electronic spectral data must be relied on heavily for structural determination of cyanocarbon anions. The data in Tables 13 to 17 show the effect of chain length, substitution of

TABLE 13. Linear cyanocarbon anions with cyano groups at the ends of the chains.

Cyanocarbon anion	Position of longest wavelength maxima λ_{max} (mμ)	ε
$(NC)_2C{=}CH{-}C(CN)_2^-$	346	3,150
$(NC)_2C{=}CH{-}CH{=}CH{-}C(CN)_2^-$	440	8,300
$(NC)_2C{=}CH{-}(CH{=}CH)_2{-}C(CN)_2^-$	540	16,000
$(NC)_2C{=}CH{-}(CH{=}CH)_3{-}C(CN)_2^-$	632	—

TABLE 14. Cyanocarbon anions with cyano groups on all positions.

Anion	Position of longest wavelength maxima λ_{max} (mμ)	ε
$NC{-}C(CN)_2^-$	211	37,400
$(NC)_2C{=}C(CN){-}C(CN)_2^-$	412	22,100
$(NC)_2C{=}C(CN){-}C(CN){=}C(CN){-}C(CN)_2^-$	528	33,000
Pentacyanocyclopentadienide anion	291	10,250

TABLE 15. The effect of nitrogen in the chain of a cyanocarbon anion.

Anion	Position of longest wavelength maxima λ_{max} (mμ)	ε
$(NC)_2C{=}C(CN){-}N{=}C(CN){-}C^{-}(CN)_2$	464	45,300
$(NC)_2C{=}C(CN){-}C(F){=}N{-}C^{-}(CN)_2$	502	35,200
$(NC)_2C{=}C(CN){-}C(CN){=}C(CN){-}C^{-}(CN)_2$	528	33,000

nitrogen for carbon and substitution of cyano groups by hydrogens on the electronic spectra of selected cyanocarbon anions. The following generalizations can be made: the position of the lowest energy maximum shifts about 100 mμ to longer wavelengths for each added double bond in a linear cyanocarbon anion (Tables 13 and 14). Substitution of nitrogen for a C—CN group causes a large blue shift if the C—CN group is in an odd-numbered position, where it stabilizes the negative charge (Table 15). An increase in the number of cyano groups on a system usually causes a red shift (Table 16). A notable exception is the cyanocyclopentadienide series, in which the position of the longest wavelength maximum is fairly constant (Table 17).

Several linear cyanocarbon anions have very high extinction coefficients (Table 16).

H. Cyanocarbon Acids

1. Acidities

The reversible protonation of a number of cyanocarbon anions in sulphuric and perchloric acid solutions has been studied spectroscopically[803,804]. An indicator acidity function H_- has been set up for sulphuric acid solutions up to 12 molar (80 wt. %) based on equation (374)[805]. This H_- scale has roughly the same values

$$H^+ + A^- \rightleftharpoons HA$$

$$H_- \equiv pK_a + \log \frac{C_A}{C_{HA}} = -\log a_{H^+} - \log \frac{f_A}{f_{HA}} \tag{374}$$

as the familiar H_0 scale based on the protonation of neutral bases. The H_- function was used to determine the pK_a values of *p*-(tricyanovinyl)phenyldicyanomethane, methyl dicyanoacetate,

TABLE 16. The effect of the number of cyano groups on the electronic spectrum of an anion.

Anion	Position of longest wavelength maxima λ_{max} (mμ)	ε
$(NC)_2C{=}CH{-}C(CN)_2^-$	346	3,150
$(NC)_2C{=}C(CN){-}C(CN)_2^-$	412	22,100
$(NC)_2C{=}CH{-}CH{=}CH{-}C(CN)_2^-$	440	8,300
$(NC)_2C{=}C(CN){-}CH{=}C(CN){-}C(CN)_2^-$	538	82,000
$(NC)_2C{=}C(CN){-}C(CN){=}C(CN){-}C(CN)_2^-$	528	33,000
$(NC)_2C{=}C(CN){-}CH{=}CH{-}CH{=}C(CN){-}C(CN)_2^-$	635	165,000
$(NC)_2C{=}CH{-}CH{=}CH{-}CH{=}CH{-}C(CN)_2^-$	540	16,000

TABLE 17. Electronic spectra of cyanocyclopentadienides.

Cyclopentadienide	Solvent	Longest wavelength maxima λ_{max} (mμ)	ε
Monocyano-	H_2O	264	15,900
1,2-Dicyano-	CH_3CN	282	16,200
1,3-Dicyano-	CH_3CN	277	29,000
1,2,3-Tricyano-	CH_3CN	298	18,200
1,2,4-Tricyano-	CH_3CN	267	5,620
Tetracyano-	CH_3CN	298	14,500
Pentacyano-	CH_3CN	291	10,200

hexacyanoheptatriene, cyanoform, tricyanovinyl alcohol, bis-(tricyanovinyl)amine and hexacyanobutenediide (see Table 18).

Tetracyanopropenide, pentacyanopropenide and hexacyanoisobutenide are not protonated by 80% sulphuric acid. Above this concentration the nitrile groups begin to hydrolyse. The pK_a values of a number of cyanocyclopentadienes have been determined, the weakly acidic ones in water and the strongly acidic ones in acetonitrile[798]. The pK_a value for 1,2-dicyanocyclopentadiene is 9·1 units higher in acetonitrile than in water. Thus, to get a continuous scale for comparison with pK_a values determined in aqueous perchloric acid and sulphuric acid, 9·1 was subtracted from the pK_a value determined in acetonitrile. These corrected pK_a values are listed in Table 18, along with the pK_a values of various other cyanocarbon acids. In cases where protonation could occur on more than one position, such as in 1,3-dicyanocyclopentadienide (equation 375 for

NC–(cyclopentadienide)–CN + H⁺ ⇌ NC–(C₅H₄, H₂)–CN + NC–(C₅H₄, H₂)–CN (375)

example), the individual pK_a values were calculated from those observed for their mixture by using the expression[798]

$$pK_a^{\text{Acid 1}} = pK_a^{\text{obs}} - \log\left(1 + \frac{[\text{Acid 2}]}{[\text{Acid 1}]}\right)$$

The isomer ratios were determined by n.m.r. spectroscopy.

The striking strength of the cyanocarbon acids is due to resonance stabilization in the anion that is not possible in the protonated form.

2. The nature of the protonated species

The positions of protonation of several of the cyanocarbon anions have been studied spectroscopically[798,803]. Protonated tricyanomethanide is transparent down to 200 mμ which shows that the proton has destroyed conjugation and is attached to the central carbon[803]. When an aqueous–ethereal solution of tricyanomethane is partially evaporated, a white sublimable solid crystallizes[80]. This solid does not have the tricyanomethane structure **516** since its infrared spectrum, (ν_{max}^{nujol} 2500 (w), 2280 (w), 2200 (s), 1790 (w), 1250 (w), 1025 (s) and 823 cm^{-1}) contains unsaturated nitrile absorption. The product was therefore assigned a zwitterion structure (**518**)[80] (section II.A.1).

$$\mathrm{HC(CN)_3}\quad \mathbf{(516)} \qquad \mathrm{(NC)_2C{=}C{=}NH}\quad \mathbf{(517)} \longleftrightarrow \mathrm{(NC)_2\overset{-}{C}{-}C{\equiv}\overset{+}{N}H}\quad \mathbf{(518)}$$

Methoxycarbonyldicyanomethanide is protonated on the central carbon to give **519**, but tricyanovinyl alcoholate is protonated on oxygen as evidenced by the similarity of the spectra of the protonated form **520** and the anion.

$$\mathrm{HC(CN)_2{-}CO_2Me}\quad \mathbf{(519)} \qquad \mathrm{(NC)_2C{=}C(OH)CN}\quad \mathbf{(520)}$$

p-(Tricyanovinyl)phenyldicyanomethanide is protonated on the methyl carbon. The spectrum of the protonated form **521** ($\lambda_{max}^{H_2O}$

$$\mathrm{(NC)_2HC{-}C_6H_4{-}C(CN){=}C(CN)_2}\quad \mathbf{(521)}$$

332 mμ) is comparable to that of tricyanovinylbenzene (λ_{max}^{EtOH} 335 mμ) rather than tetracyanoquinodimethane (λ_{max} 395 mμ). This cyanocarbon acid can be isolated as a colourless covalent species. Solutions in water are blue as a consequence of complete ionization. The protonated form of hexacyanoheptatriene is probably that

TABLE 18. pK_a values of cyanocarbon acids.

Acid	pK_a	Reference
$CH_2(CN)_2$	11·19	806
H_2 / CN	9·78	798
HCN	9·21	807
$CHBr(CN)CO_2Et$	6	803
$PhC(CN)_2H$	5·80	95
$CH(CN)_2Br$	5	806
$HC(CN)_2C(CN)_2H$	3·9	97
$HC(CN)_2C(CN)_2SO_3^-$	2·6	97
NC / CN / H_2	2·40	798
p-$O_2NC_6H_4C(CN)_2H$	1·89	695
H_2 / NC / CN	1·40	798
NC / CN / H_2	0·81	798
NC / CN / H_2	0·81	798
p-[$(NC)_2C{=}C(CN)$]-$C_6H_4C(CN)_2H$	0.60	803
NC / NC / =CN / NC / CN / CN H	−2·5	803
CN / H—C—CO_2Me / CN	−2·78	803
NC CN CN CN / C=C—C=C—C—H / NC H CN	−3·55	803
$HC(CN)_3$	−5·13	803

(*Table continued*)

TABLE 18. (*continued*)

Acid	pK_a	Reference
$(NC)_2C{=}C(OH)CN$	−5·3	803
NC, CN, H, H, Me, CN (cyclopentadiene)	−5·7	798
$(NC)_2C{=}C(CN){-}N{=}C(CN){-}C(CN)_2H$	−6·07	803
NC, CN, H_2, CN (cyclopentadiene)	−6·1	798
H_2, NC, CN, CN (cyclopentadiene)	−7·2	798
$(NC)_2C{=}CH{-}C(CN)_2H$	−8	803
$(NC)_2C{=}C[CH(CN)_2]_2$	−8·5	803
NC, CN, NC, CN, H_2 (cyclopentadiene)	−9	798
NC, CN, NC, CN, H, CN (cyclopentadiene)	<−11	798
$R{-}C{\equiv}\overset{+}{N}H$	∼−12	808

shown in formula **522**, since the ultraviolet spectrum ($\lambda_{max}^{H_2O}$ 347 mμ) requires three double bonds in conjugation.

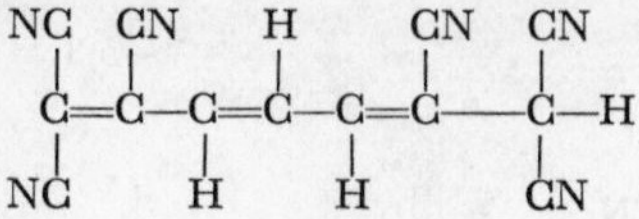

NC CN H CN CN
C=C—C=C—C=C—C—H
NC H H CN

(522)

Cyanocyclopentadienide is protonated exclusively to give 1-cyanocyclopentadiene[809] (**523**) rather than the 2-cyanocyclopentadiene as originally postulated[798].

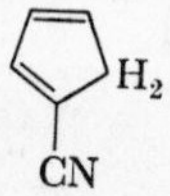

H2
CN
(523)

1,2-Dicyanocyclopentadienide is protonated to give equal amounts of isomers **524** and **525**. Statistically, the ratio should be 1:2. This

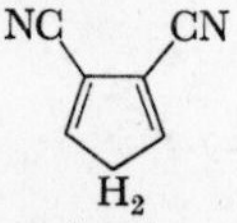

NC CN
H2
(524)

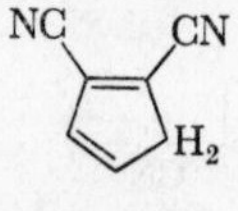

NC CN
H2
(525)

indicates that the methylene hydrogens in **524** are less acidic than those in **525**.

1,3-Dicyanocyclopentadienide is protonated to give **526** and **527** in a ratio of about 10:1. The cyclopentadiene (**526**) is favoured because it is fully conjugated.

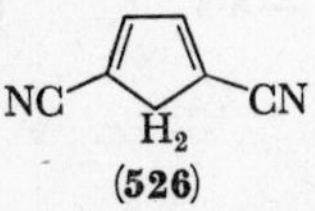

NC CN
H2
(526)

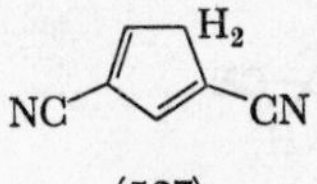

H2
NC CN
(527)

Protonation of 1,2,4- and 1,2,3-tricyanocyclopentadienides gives **528** and **529**, respectively.

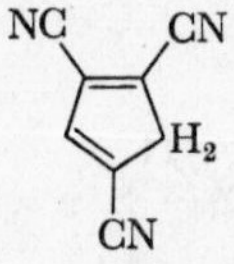

NC CN
H2
CN
(528)

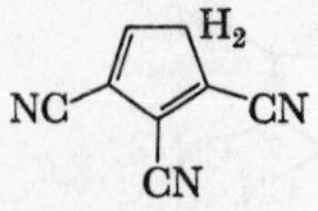

H2
NC CN
CN
(529)

Pentacyanocyclopentadiene cannot be protonated in acetonitrile. Indeed, if conditions strong enough to protonate this anion were used (SbF_5–FSO_3H, for example), the protonation would probably occur on one of the nitriles. Since simple nitriles are only partially protonated in 100 % sulphuric acid[808] with an estimated H_- of -11, their pK_a value is probably around -12.

XII. REFERENCES

1. K. Wallenfels, *Chimia (Aarau)*, **20**, 303 (1966).
2. D. N. Dhar, *Chem Revs.*, **67**, 611 (1967).
3. B. C. McKusick, *Trans. N.Y. Acad. Sci.*, [2] **27**, (7), 719 (1965).
4. B. C. McKusick and T. L. Cairns in *Encyclopedia of Chemical Technology*, Vol. 6 (Ed. Kirk-Othmer), 2nd ed., John Wiley and Sons, New York, 1965, pp. 625–633.
5. R. Winkler, *Chimia (Aarau)*, **16**, 360 (1962).
6. T. L. Cairns and B. C. McKusick, *Angew. Chem.*, **73**, 520 (1961).
7. B. C. McKusick and G. F. Biehn, *Chem. Eng. News*, **38**, 114 (1960).
8. F. Freeman, *Chem. Revs.*, **69**, 591 (1969).
9. V. Migrdichian, *The Chemistry of Organic Cyanogen Compounds*, Reinhold, New York, 1947.
10. A. O. Rogers, *Cyanides in Organic Reactions*, Electrochemicals Department, E. I. du Pont de Nemours and Co., Wilmington, Delaware, 1962.
11. T. K. Brotherton and J. W. Lynn, *Chem. Revs.*, **59**, 841 (1959).
12. H. M. Woodburn and L. N. Pino, *J. Org. Chem.*, **16**, 1389 (1951).
13. W. L. Fierce and W. J. Sandner, *U.S. Pat.* 3,020,126 (1962); *Chem. Abstr.*, **56**, 13805 (1962).
14. E. Briner and A. Wroczynski, *Compt. Rend.*, **151**, 314 (1910).
15. L. Troost and P. Hautefeuille, *Compt. Rend.*, **66**, 736, 798 (1868).
16. D. Berthelot and H. Gaudechoz, *Compt. Rend.*, **155**, 207 (1912).
17. J. A. Bladin, *Ber.*, **18**, 1545 (1885).
18. O. Jacobsen and A. Emmerling, *Ber.*, **4**, 949 (1871).
19. B. L. Williams, *U.S. Pat.* 2,959,588 (1960); *Chem. Abstr.*, **55**, 9441 (1961).
20. H. M. Woodburn and J. R. Fisher, *J. Org. Chem.*, **22**, 895 (1957).
21. H. M. Woodburn and E. L. Graminski, *J. Org. Chem.*, **23**, 819 (1958).
22. H. M. Woodburn, B. A. Morehead and W. H. Bonner, *J. Org. Chem.*, **14**, 555 (1949).
23. H. M. Woodburn, B. A. Morehead and C. M. Chih, *J. Org. Chem.*, **15**, 535 (1950).
24. H. M. Woodburn and B. G. Pautler, *J. Org. Chem.*, **19**, 863 (1954).
25. H. M. Woodburn and W. S. Zehrung III, *J. Org. Chem.*, **24**, 1184 (1959).
26. H. M. Woodburn and R. C. O'Gee, *J. Org. Chem.*, **17**, 1235 (1952).
27. O. Hinsberg and E. Schwantes, *Ber.*, **36**, 4040 (1903).
28. H. M. Woodburn and C. E. Sroog, *J. Org. Chem.*, **17**, 371 (1952).
29. L. O. Brockway, *Proc. Nat. Acad. Sci. U.S.*, **19**, 868 (1933).
30. T. Curtius and G. M. Dedicken, *J. Prakt. Chem.*, [2] **50**, 245 (1894).
31. G. M. Dedicken, *Thesis, Avhandl. Norske Videnskaps-Akad. Oslo, Mat.-Naturv. Kl*, (5) 1936; *Chem. Zentr.*, **1**, 86 (1937).
32. J. Thiele and K. Schleussner, *Ann. Chem.*, **295**, 161 (1897).
33. R. Chatterjee, *J. Indian Chem. Soc.*, **15**, 608 (1938).

34. E. Fischer, *Ber.*, **22**, 1931 (1889).
35. F. Tiemann, *Ber.*, **22**, 1936 (1889).
36. J. U. Nef, *Ann. Chem.*, **287**, 265, 310 (1895).
37. A. B. Whitehouse, *The Reaction of Cyanogen with Glycols and Monoethers of Ethylene Glycol*, Publication No. 23,479, University Microfilms, Ann Arbor, Michigan, 1957.
38. H. M. Woodburn, A. B. Whitehouse and B. G. Pautler, *J. Org. Chem.*, **24**, 210 (1959).
39. R. Naumann, *Z. Elektrochem.*, **16**, 773 (1910).
40. R. Schmitt and L. Glutz, *Ber.*, **1**, 66 (1868).
41. N. Beketoff, *Ber.*, **3**, 872 (1870).
42. R. P. Welcher, *U.S. Pat.* 2,804,470 (1958); *Chem. Abstr.*, **52**, 2894 (1958).
43. R. P. Welcher, *U.S. Pat.* 2,804,471 (1958); *Chem. Abstr.*, **52**, 2894 (1958).
44. G. Jander and H. Schmidt, *Chemiker Ztg.* (*Wien*), **46**, 49 (1943); L. J. Gay-Lussac, *Ann. Chim.*, [1] **95**, 195 (1815); R. Anschutz, *Ann. Chem.*, **254**, 262 (1889); J. Liebig and F. Wohler, *Ann. Physik.*, **24**, 167 (1832).
45. W. Traube, *Ber.*, **31**, 191 (1898).
46. W. Traube, *Ber.*, **31**, 2938 (1898).
47. W. Traube and C. Hoepner, *Ann. Chem.*, **332**, 118 (1904).
48. W. Traube, *Ann. Chem.*, **332**, 104 (1904).
49. W. Traube and M. Braumann, *Ann. Chem.*, **332**, 133 (1904).
50. H. M. Woodburn and T. J. Dolce, *J. Org. Chem.*, **25**, 452 (1960).
51. D. Vorlander, *Ber.*, **44**, 2455 (1911).
52. K. R. Lynn, *Australian J. Chem.*, **7**, 303 (1954).
53. V. Grignard, E. Bellet and Ch. Courtot, *Ann. Chim.* (*Paris*), **12**, 364 (1920).
54. L. B. Lathroum, *The Reaction of Cyanogen with Selected Grignard Reagents*, Publication No. 5121, University Microfilms, Ann Arbor, Michigan, 1953.
55. H. M. Woodburn and L. B. Lathroum, *J. Org. Chem.*, **19**, 285 (1954).
56. J. F. Eastham and V. F. Raaen, *Proc. Chem. Soc.*, **1958**, 149.
57. J. F. Eastham and D. Y. Cannon, *J. Org. Chem.*, **25**, 1504 (1960).
58. V. F. Raaen and J. F. Eastham, *J. Am. Chem. Soc.*, **82**, 1349 (1960).
59. P. Karrer and J. Ferla, *Helv. Chim. Acta.*, **4**, 203 (1921).
60. H. Knobloch and E. Schraufstätter, *Chem. Ber.*, **81**, 224 (1948).
61. E. Schraufstätter, *Chem. Ber.*, **81**, 235 (1948).
62. A. Desgrez, *Bull. Soc. Chim.*, [3] **13**, 736 (1895).
63. G. J. Janz, *J. Am. Chem. Soc.*, **74**, 4529 (1952).
64. G. J. Janz in *Organic Chemistry*, Vol. 8 (Ed. J. Hamer), Academic Press, New York, 1967, pp. 97–125.
65. E. Oliveri-Mandala and T. Passlacqua, *Gazz. Chim. Ital.*, **41**(2), 430 (1911).
66. A. Peratoner and E. Azzarello, *Rend. Accad. Lincei.* [5], **16** (2), 237, 318 (1907); *Gazz. Chim. Ital.*, [1] **38**, 76, 84, 88 (1907).
67. H. J. Emelus and G. L. Hurst, *J. Chem. Soc.*, **1962**, 3276.
68. H. J. Emelus and G. L. Hurst, *J. Chem. Soc.*, **1964**, 396.
69. P. Robson, V. C. R. McLoughlin, J. B. Hynes and L. A. Bigelow, *J. Am. Chem. Soc.*, **83**, 5010 (1961).
70. H. A. Lehmann, L. Riesil, K. Höhne and E. Maier, *Z. Anorg. Allgem. Chem.*, **310**, 298 (1961).
71. N. C. Robertson and R. N. Pease, *J. Am. Chem. Soc.*, **64**, 1880 (1942).
72. N. C. Robertson and R. N. Pease, *J. Chem. Phys.*, **10**, 490 (1942).
73. G. W. Ayers, *U.S. Pat.* 2,780,638 (1957); *Chem. Abstr.*, **51**, 13,906 (1957).
74. D. W. Wiley, E. P. Blanchard and O. W. Webster, unpublished results.
75. O. W. Webster, *U.S. Pat.* 3,093,653 (1963); *Chem. Abstr.*, **59**, 11,507 (1963).

76. R. D. Vest, *U.S. Pat.* 3,115,497 (1963); *Chem. Abstr.*, **60**, 5512 (1964).
77. J. F. Guttenberger, *Angew. Chem.*, **79**, 1071 (1967).
78. H. Schmidtmann, *Ber.*, **29**, 1168 (1896).
79. S. Trofimenko, E. L. Little, Jr. and H. F. Mower, *J. Org. Chem.*, **27**, 433 (1962).
80. S. Trofimenko, *J. Org. Chem.*, **28**, 217 (1963).
81. E. Cox and A. Fontaine, *Bull. Soc. Chim. France*, **1954**, 948.
82. A. Hantzsch and G. Osswald, *Ber.*, **32**, 641 (1899).
83. S. Trofimenko, *J. Org. Chem.*, **28**, 2755 (1963).
84. J. H. Enemark and R. H. Holm, *Inorg. Chem.*, **3**, 1516 (1964).
85. H. Köhler, *Z. Anorg. Allgem. Chem.*, **331**, 237 (1964).
86. W. J. Middleton, E. L. Little, D. D. Coffman and V. A. Engelhardt, *J. Am. Chem. Soc.*, **80**, 2795 (1958).
87. D. A. Long, R. A. G. Carrington and R. B. Gravenor, *Nature*, **196**, 371 (1962).
88. F. A. Miller, *Pure Appl. Chem.*, **7**, 125 (1963).
89. W. J. Middleton and V. A. Engelhardt, *J. Am. Chem. Soc.*, **80**, 2788 (1958); W. J. Middleton, *U.S. Pat.* 2,883,638 (1959).
90. D. S. Acker, R. J. Harder, W. R. Hertler, W. Mahler, L. R. Melby, R. E. Benson and W. E. Mochel, *J. Am. Chem. Soc.*, **82**, 6408 (1960).
91. R. H. Boyd, *J. Chem. Phys.*, **38**, 2529 (1963).
92. L. Birkenbach and K. Huttner, *Ber.*, **62B**, 153 (1929).
93. W. Fensch and G. Wagner, *Z. Physik. Chem.*, **B41**, 1 (1938).
94. J. K. Williams, E. L. Martin and W. A. Sheppard, *J. Org. Chem.*, **31**, 919 (1966).
95. A. D. Josey, C. L. Dickinson, K. C. Dewhirst and B. C. McKusick, *J. Org. Chem.*, **32**, 1941 (1967).
96. R. E. Heckert, *U.S. Pat.* 2,788,356, (1957); *Chem. Abstr.*, **51**, 12,959 (1957).
97. W. J. Middleton, R. E. Heckert, E. L. Little and C. G. Krespan, *J. Am. Chem. Soc.*, **80**, 2783 (1958).
98. D. T. Longone and G. L. Smith, *Tetrahedron Letters*, **1962**, 205.
99. W. J. Middleton, V. A. Engelhardt and B. S. Fisher, *J. Am. Chem. Soc.*, **80**, 2822 (1958).
100. O. W. Webster, W. Mahler and R. E. Benson, *J. Am. Chem. Soc.*, **84**, 3678 (1962).
101. O. W. Webster, *U.S. Pat.* 3,144,478 (1964); *Chem. Abstr.*, **61**, 10,595 (1964).
102. S. Trofimenko and B. C. McKusick, *J. Am. Chem. Soc.*, **84**, 3677 (1962).
103. G. W. Griffin and L. I. Peterson, *J. Org. Chem.*, **28**, 3219 (1963).
104. T. Sadeh and A. Berger, *Bull. Res. Council Israel Sect. A*, **7**, 98 (1958).
105. G. Errera and F. Perciabosco, *Ber.*, **33**, 2976 (1900).
106. J. P. Ferris and L. E. Orgel, *J. Org. Chem.*, **30**, 2365 (1965).
107. L. Ramberg and S. Wideqvist, *Arkiv Kemi*, **12A**, (22) (1937).
108. L. Ramberg and S. Wideqvist, *Arkiv. Kemi*, **14B**, (37) (1941).
109. S. Wideqvist, *Arkiv Kemi*, **20B**, (4) (1945).
110. H. Hart and F. Freeman, *J. Org. Chem.*, **28**, 1220 (1963).
111. H. Hart and Y. C. Kim, *J. Org. Chem.*, **31**, 2784 (1966).
112. R. P. Mariella and A. J. Roth III, *J. Org. Chem.*, **22**, 1130 (1957).
113. R. M. Scribner, G. N. Sausen and W. W. Prichard, *J. Org. Chem.*, **25**, 1440 (1960).
114. H. Hart and F. Freeman, *J. Am. Chem. Soc.*, **85**, 1161 (1963).
115. T. L. Cairns and E. A. Graef, *U.S. Pat.* 3,166,584 (1965).
116. T. L. Cairns, R. A. Carboni, D. D. Coffman, V. A. Engelhardt, R. E. Heckert, E. L. Little, E. G. McGeer, B. C. McKusick, W. J. Middleton, R. M. Scribner, C. W. Theobald and H. E. Winberg, *J. Am. Chem. Soc.*, **80**, 2775 (1958).
117. E. L. Martin, *U.S. Pat.* 3,076,836 (1963); *Chem. Abstr.*, **58**, 13,802 (1963).

118. J. E. Harris, *U.S. Pat.* 3,330,853 (1967); *Chem. Abstr.*, **67**, 99,664 (1967).
119. E. L. Martin, *U.S. Pat.* 3,118,929 (1964); *Chem. Abstr.*, **60**, 15,742 (1964).
120. R. D. Vest, *U.S. Pat.* 3,101,365 (1963); *Chem. Abstr.*, **60**, 4015 (1964).
121. E. L. Martin, *J. Am. Chem. Soc.*, **85**, 2449 (1963).
122. I. K. Panov, *Toksikol Novykh Prom. Khim Veshchesiv No.*, **7**, 180 (1965); *Chem. Abstr.*, **63**, 7550 (1965).
123. K. N. Trueblood, *AD 601478* pp. VI-4; *Chem. Abstr.*, **62**, 2312 (1965).
124. H. Huber and G. F. Wright, *Can. J. Chem.*, **42**, 1446 (1964).
125. C. E. Looney and J. R. Downing, *J. Am. Chem. Soc.*, **80**, 2840 (1958).
126. F. A. Miller, O. Sala, P. Devlin, J. Overend, E. Lippert, W. Lüder, H. Moser and J. Varchmin, *Spectrochim. Acta.*, **20**, 1233 (1964).
127. J. Prochorow and A. Tramer, *Bull. Acad. Polon. Sci., Ser. Sci., Math, Astron. Phys.*, **12**, 429 (1964).
128. T. Takenaka and S. Hayaski, *Bull. Chem. Soc. Japan*, **37**, 1216 (1964).
129. J. Halper, W. D. Classon and H. B. Gray, *Theoret. Chim. Acta*, **4**, 174 (1966).
130. P. Heim and F. Dörr, *Tetrahedron Letters*, **1964**, 3095.
131. P. Heim and F. Dörr, *Ber. Bunsenges Physik. Chem.*, **69**, 453 (1965).
132. D. A. Long and W. O. George, *Spectrochim. Acta*, **19**, 1717 (1963).
133. A. Rosenberg and J. P. Develin, *Spectrochim. Acta*, **21**, 1613 (1965).
134. A. L. Farragher and F. M. Page, *Trans. Faraday Soc.*, **63**, 2369 (1967).
135. J. T. Herron, H. M. Rosenstock and W. R. Shields, *Nature*, **206**, 611 (1965).
136. C. Briegleb, *Angew. Chem. Int. Ed.*, **3**, 617 (1964).
137. M. Batley and L. E. Lyons, *Nature*, **196**, 573 (1962).
138. M. Ozolins and G. H. Schenk, *Anal. Chem.*, **33**, 1035 (1961).
139. J. Sauer, H. Wiest and A. Mielert, *Chem. Ber.*, **97**, 3183 (1964).
140. J. Sauer, *Angew. Chem.*, **73**, 545 (1961).
141. R. Huisgen, F. Mietzsch, G. Boche and H. Seidl., *Chem. Soc. (London) Spec. Publ.*, **19**, 3 (1964); *Chem. Abstr.*, **63**, 13,043 (1965).
142. G. N. Schrauzer and S. Eichler, *Angew. Chem. Int. Ed.*, **1**, 454 (1962).
143. H. Prinzbach, *Angew. Chem.*, **73**, 169 (1961).
144. K. Hafner, *Angew. Chem. Int. Ed.*, **3**, 165 (1964).
145. J. H. van den Hende and A. S. Kende, *Chem. Commun.*, **1965**, 384.
146. R. A. Smith, J. E. Baldwin and I. C. Paul, *J. Chem. Soc. (B)*, **1967**, 112.
147. G. Kresze, S. Rau, G. Sabelus and H. Goetz, *Ann. Chem.*, **648**, 57 (1961).
148. W. V. E. Doering, *Abstr. 16. Nat. Org. Chem. Sympo., U.S.A.*, 1959, p. 25.
149. M. P. Cava, J. P. Van Meter, *J. Am. Chem. Soc.*, **84**, 2008 (1962).
150. E. R. H. Jones, H. H. Lee and M. C. Whiting, *J. Chem. Soc.*, **1960**, 341.
151. V. L. Bell, *J. Polymer Sci.*, **A2**, 5305 (1964).
152. K. R. Huffman, M. Loy, W. A. Henderson, Jr. and E. F. Ullman, *Tetrahedron Letters*, **1967**, 931.
153. E. F. Ullman, W. A. Henderson, Jr. and K. R. Huffman, *Tetrahedron Letters*, **1967**, 935.
154. L. H. Klemm, W. C. Solomon and A. J. Kohlik, *J. Org. Chem.*, **27**, 2777 (1962).
155. J. K. Williams and R. E. Benson, *J. Am. Chem. Soc.*, **84**, 1257 (1962).
156. A. T. Blomquist and Y. C. Meinwald, *J. Am. Chem. Soc.*, **81**, 667 (1959).
157. J. K. Williams, D. W. Wiley and B. C. McKusick, *J. Am. Chem. Soc.*, **84**, 2210 (1962).
158. D. W. Wiley, unpublished results.
159. P. D. Bartlett, unpublished results.

160. K. Hafner and J. Schneider, *Ann. Chem.*, **624**, 37 (1959).
161. R. C. Cookson, J. Dance and J. Hudec, *J. Chem. Soc.*, **1964**, 5416.
162. R. Criegee, *Angew. Chem. Int. Ed.*, **1**, 519 (1962).
163. J. K. Williams, *J. Am. Chem. Soc.*, **81**, 4013 (1959).
164. C. A. Stewart, Jr., *J. Org. Chem.*, **28**, 3320 (1963).
165. C. A. Stewart, Jr., *J. Am. Chem. Soc.*, **84**, 117 (1962).
166. J. J. Eisch and G. R. Husk, *J. Org. Chem.*, **31**, 589 (1966).
167. R. C. Cookson, B. Halton, I. D. R. Stevens and C. T. Watts, *J. Chem. Soc.* (*C*), **1967**, 928.
168. J. B. Miller, *J. Org. Chem.*, **25**, 1279 (1960).
169. D. M. Lemal and J. P. Lokensgard, *J. Am. Chem. Soc.*, **88**, 5934 (1966).
170. C. D. Smith, *J. Am. Chem. Soc.*, **88**, 4273 (1966).
171. M. L. Owens, *U.S. Pat.* 2,975,221 (1961); *Chem. Abstr.*, **55**, 17,096 (1961).
172. J. Diekmann and C. J. Pedersen, *J. Org. Chem.*, **28**, 2879 (1963).
173. H. W. Wanzlick and E. Schikora, *Chem. Ber.*, **94**, 2389 (1961).
174. W. J. Linn, O. W. Webster and R. E. Benson, *J. Am. Chem. Soc.*, **85**, 2032 (1963).
175. R. Criegee and P. Günther, *Chem. Ber.*, **96**, 1564 (1963).
176. C. L. Dickinson and B. C. McKusick, *J. Org. Chem.*, **29**, 3087 (1964).
177. J. Bastus and J. Castells, *Proc. Chem. Soc.*, **1962**, 216.
178. P. G. Farrell, J. Newton and R. F. M. White, *J. Chem. Soc.* (*B*), **1967**, 637.
179. W. Schulze, H. Willitzer and H. Fritzsche, *Chem. Ber.*, **100**, 2640 (1967).
180. H. Junek, *Monatsh. Chem.*, **96**, 1421 (1965).
181. B. Smith and U. Persmark, *Acta. Chem. Scand.*, **17**, 651 (1963).
182. W. H. Baddly and L. M. Venanzi, *Inorg. Chem.*, **5**, 33 (1966).
183. C. Panattoni, G. Bombieri, U. Belluco and W. H. Baddley, *J. Am. Chem. Soc.*, **90**, 798 (1968).
184. P. Fitton and J. E. McKeon, *Chem. Commun.*, **1968**, 4.
185. G. S. Reddy and C. D. Weis, *J. Org. Chem.*, **28**, 1822 (1963); see, however, G. Märkl, *Angew. Chem.*, **77**, 1109 (1965).
186. M. Brown and R. E. Benson, *J. Org. Chem.*, **31**, 3849 (1966).
187. M. Bately and L. E. Lyons, *Nature* **196**, 573 (1962).
188. M. Prince and J. Hornyak, *Chem. Commun.*, **1966**, 455.
189. H. D. Hartzler, *J. Org. Chem.*, **29**, 1194 (1964).
190. B. C. McKusick, R. E. Heckert, T. L. Cairns, D. D. Coffman and H. F. Mower, *J. Am. Chem. Soc.*, **80**, 2806 (1958).
191. R. E. Heckert, *U.S. Pat.* 2,762,833 (1956); *Chem. Abstr.*, **51**, 4422 (1957).
192. G. N. Sausen, V. A. Engelhardt and W. J. Middleton, *J. Am. Chem. Soc.*, **80**, 2815 (1958).
193. W. E. Noland, W. C. Kuryla and R. F. Lange, *J. Am. Chem. Soc.*, **81**, 6010 (1959).
194. R. Foster and P. Hanson, *Tetrahedron*, **21**, 255 (1965).
195. J. R. Roland and B. C. McKusick, *J. Am. Chem. Soc.*, **83**, 1652 (1961).
196. K. Hafner and K. Moritz, *Ann. Chem.*, **650**, 92 (1961).
197. D. J. Cram and R. D. Partos, *J. Am. Chem. Soc.*, **85**, 1273 (1963).
198. C. W. Rigby, E. Lord and C. D. Hall, *Chem. Commun.*, 714 (1967).
199. R. Henderson and W. A. Sheppard, *J. Org. Chem.*, **32**, 858 (1967).
200. Z. Rappoport, *J. Chem. Soc.*, **1963**, 4498.
201. Z. Rappoport, P. Greenzaid and A. Horowitz, *J. Chem. Soc.*, **1964**, 1334.
202. Z. Rappoport and A. Horowitz, *J. Chem. Soc.*, **1964**, 1348.
203. P. G. Farrell and J. Newton, *Tetrahedron Letters*, **1964**, 189.

204. N. S. Isaacs, *J. Chem. Soc.* (*B*), **1966**, 1053.
205. E. M. Kosower, *Progr. Phys. Org. Chem.*, **3**, 81 (1965).
206. F. D. Popp, *J. Org. Chem.*, **26**, 3019 (1961).
207. J. K. Williams, D. W. Wiley and B. C. McKusick, *J. Am. Chem. Soc.*, **84**, 2216 (1962).
208. C. L. Dickinson, D. W. Wiley and B. C. McKusick, *J. Am. Chem. Soc.*, **82**, 6132 (1960).
209. O. W. Webster, M. Brown and R. E. Benson, *J. Org. Chem.*, **30**, 3223 (1965).
210. C. L. Dickinson, J. K. Williams and B. C. McKusick, *J. Org. Chem.*, **29**, 1915 (1964).
211. B. C. McKusick, R. E. Heckert, T. L. Cairns, D. D. Coffman and H. F. Mower, *J. Am. Chem. Soc.*, **80**, 2806 (1958).
212. S. Trippett, *J. Chem. Soc.*, **1962**, 4733.
213. E. Zbiral, *Monatsh. Chem.*, **96**, 1967 (1965).
214. J. W. Van Dyke, Jr. and H. R. Snyder, *J. Org. Chem.*, **27**, 3888 (1962).
215. E. Ciganek and C. G. Krespan, *J. Org. Chem.*, **33**, 541 (1968).
216. W. E. Noland and D. A. Jones, *J. Org. Chem.*, **27**, 341 (1962).
217. G. V. Fomin, L. A. Blyumenfeld and B. I. Sukhorukov, *Dokl. Akad. Nauk. SSSR*, **157**, 1199 (1964); *Chem. Abstr.*, **61**, 12,824 (1964).
218. Y. D. Pimenov, V. E. Kholmogorov and A. N. Terenin, *Dokl. Akad. Nauk. SSSR*, **163**, 935 (1965); *Chem. Abstr.*, **63**, 14,081 (1965).
219. V. V. Pen'kovs'kii, *Teor. Eksp. Khim., Akad. Nauk. Ukr. SSR*, **2**, 282 (1966); *Chem. Abstr.*, **65**, 15,198 (1966).
220. P. H. Rieger, I. Bernal and G. K. Fraenkel, *J. Am. Chem. Soc.*, **83**, 3918 (1961).
221. M. R. Ronayne, J. P. Guarino and W. H. Hamill, *J. Am. Chem. Soc.*, **84**, 4230 (1962).
222. M. G. Ormerod and L. G. Stoodley, *Nature*, **195**, 262 (1962).
223. M. Sofue and S. Nagakura, *Bull. Chem. Soc. Japan*, **38**, 1048 (1965).
224. R. L. Ward, *J. Chem. Phys.*, **39**, 852 (1963).
225. M. Rosenblum, R. W. Fish and C. Bennett, *J. Am. Chem. Soc.*, **86**, 5166 (1964).
226. R. L. Brandon, J. H. Osiecki and A. Ottenberg, *J. Org. Chem.*, **31**, 1214 (1966).
227. J. Tirouflet, E. Laviron, R. Dabard and J. Komenda, *Bull. Soc. Chim. France*, **1963**, 857.
228. W. Liptay, G. Briegleb and K. Schindler, *Z. Elektrochem.*, **66**, 331 (1962).
229. R. L. Collins and R. Pettit, *J. Inorg. Nucl. Chem.*, **29**, 503 (1967).
230. E. Adman, M. Rosenblum, S. Sullivan and T. N. Margulis, *J. Am. Chem. Soc.*, **89**, 4540 (1967).
231. J. W. Fitch, III and J. J. Lagowski, *Inorg. Chem.*, **4**, 864 (1965).
232. W. D. Phillips, J. C. Rowell and S. I. Weissman, *J. Chem. Phys.*, **33**, 626 (1960).
233. J. A. Brivati, J. M. Gross, M. C. R. Symons and D. J. A. Tinling, *J. Chem. Soc.*, **1965**, 6504.
234. W. Müller-Warmuth, *Z. Naturforsch*, **19a**, 1309 (1964).
235. J. Gendell, J. H. Freed and G. K. Fraenkel, *J. Chem. Phys.*, **41**, 949 (1964).
236. P. H. Rieger and G. K. Fraenkel, *J. Chem. Phys.*, **37**, 2795 (1962).
237. A. Carrington and G. R. Luckhurst, *Mol. Phys.*, **8**, 401 (1964).
238. O. W. Webster, *J. Am. Chem. Soc.*, **86**, 2898 (1964).
239. A. A. Berlin and N. G. Matveeva, *Dokl. Akad. Nauk. SSSR*, **167**, 91 (1966); *Chem. Abstr.*, **64**, 17,719 (1966).
240. A. A. Berlin, A. I. Sherle, G. V. Belova and O. M. Boreev, *Vysokomolekul, Soedin.*, **7**, 88 (1965); *Chem. Abstr.*, **63**, 10,082 (1965).

241. A. A. Berlin, N. G. Matveeva, A. I. Sherle and N. D. Kostrova, *Vysokomol. Soedin.*, **4**, 860 (1962); *Chem. Abstr.*, **59**, 1766 (1963).
242. A. A. Berlin, L. I. Boguslavskii, R. K. Burshtein, N. G. Matveeva, A. I. Sherle and N. A. Schurmovskaya, *Dokl. Akad. Nauk. SSSR.*, **136**, 1127 (1961); *Chem. Abstr.*, **56**, 4198 (1962).
243. A. A. Berlin and N. G. Matveeva, *Dokl. Akad. Nauk. SSSR*, **140**, 368 (1961); *Chem. Abstr.*, **56**, 317 (1962).
244. A. A. Berlin, N. G. Matveeva and A. I. Sherle, *Izvest. Akad. Nauk. SSSR, Otdel. Khim. Nauk.*, **1959**, 2261; *Chem. Abstr.*, **54**, 10,854 (1960).
245. A. A. Berlin and N. G. Matveeva, *Acad. Sci. USSR Chem. Sect. Proc.*, **140**, 899 (1961).
246. L. I. Boguslovskii, A. I. Sherle and A. A. Berlin, *Zh. Fiz. Khim.*, **38**, 1118 (1964); *Chem. Abstr.*, **61**, 9007 (1964).
247. J. E. Katon, *U.S. Pat.* 3,229,469 (1966); *Chem. Abstr.*, **64**, 10,564 (1966).
248. S. D. Levina, K. P. Lobanova, A. A. Berlin and A. I. Sherle, *Dokl. Akad. Nauk. SSSR*, **145**, 602 (1962); *Chem. Abstr.*, **57**, 14,530 (1962).
249. T. Naraba, Y. Mizushima, H. Noake, A. Imamura, Y. Igarashi, Y. Torihashi and A. Nishioka, *J. Appl. Phys. (Japan)*, **4**, 977 (1965); *Chem. Abstr.*, **64**, 6776 (1966).
250. Y. Nose, N. Sera, M. Hatano and S. Kambara, *Kogyo Kogaku Zasshi*, **67**, 1600 (1964); *Chem. Abstr.*, **62**, 11,277 (1965).
251. S. Z. Roginskii, A. A. Berlin, L. N. Kutseva, R. M. Aseeva, L. G. Cherkashina, A. I. Sherle and N. G. Matveeva, *Dokl. Akad. Nauk. SSSR*, **148**, 118 (1963); *Chem. Abstr.*, **58**, 10,771 (1963).
252. S. Z. Roginskii, A. A. Berlin and M. M. Sokharov, *Katalitícheskie Reaktsii Zhidkoi Faze, Akad. Nauk. Kazakhsk. SSR Kazakhsk. Gos. Univ., Kazakhsk. Resp. Pravlenie, Mendeleevskogo Obshchestva. Tr. Vses. Konf., Alma Ata.*, **1962**, 334; *Chem. Abstr.*, **61**, 6440 (1964).
253. A. I. Sherle, Y. G. Aseev, E. L. Frankevich, A. A. Berlin and V. I. Kasatochkin, *Izv. Akad. Nauk. SSSR, Ser. Khim.*, **1964**, 1132; *Chem. Abstr.*, **61**, 6613 (1964).
254. M. Starke and I. Storbeck, *Ger. (East) Pat.* 36,825 (1965); *Chem. Abstr.*, **63**, 1899 (1965).
255. I. Storbeck and M. Starke, *Ber. Bunsenges. Physik. Chem.*, **69**, 343 (1965); *Chem. Abstr.*, **63**, 7124 (1965).
256. B. S. Wildi and J. E. Katon, *J. Polymer Sci.*, (A) **2**, 4709 (1964).
257. C. E. Looney and J. R. Downing, *J. Am. Chem. Soc.*, **80**, 2840 (1958).
258. C. L. Dickinson, D. W. Wiley and B. C. McKusick, *J. Am. Chem. Soc.*, **82**, 6132 (1960).
259. G. N. Sausen, V. A. Engelhardt and W. J. Middleton, *J. Am. Chem. Soc.*, **80**, 2815 (1958).
260. W. Schulze, H. Willitzer and H. Fritzsche, *Chem. Ber.*, **100**, 2640 (1967).
261. W. Schulze, H. Willitzer and H. Fritzsche, *Chem. Ber.*, **99**, 3492 (1966).
262. Y. Ohtsuka and M. Ohmori, *Bull. Chem. Soc. Japan*, **40**, 1734 (1967).
263. G. N. Schrauzer, S. Eichler and D. A. Brown, *Chem. Ber.*, **95**, 2755 (1962).
264. C. L. Dickinson, Jr., *U.S. Pat.* 2,942,022 (1960); *Chem. Abstr.*, **54**, 24,549 (1960).
265. O. W. Cass, *U.S. Pat.* 2,443,494 (1948); *Chem. Abstr.*, **42**, 7322 (1948).
266. N. R. Eldred and D. M. Young, *J. Am. Chem. Soc.*, **75**, 4338 (1953).
267. R. J. Wineman, R. M. Kliss and C. N. Matthews, *U.S. Pat.* 3,231,523 (1966); *Chem. Abstr.*, **64**, 11,345 (1966).
268. E. L. Martin, private communication.
269. E. Kloster-Jensen, *Acta. Chem. Scand.*, **17**, 1866 (1963).

270. H. J. Cenci, *Belg. Pat.* 618,526 (1962); *Chem. Abstr.*, **59**, 3773 (1963).
271. K. Wallenfels and F. Witzler, *Tetrahedron*, **23**, 1359 (1967).
272. S. Proskow, *U.S. Pat.* 3,121,734 (1964); *Chem. Abstr.*, **60**, 10,557 (1964).
273. H. J. Cenci, W. D. Niederhauser and P. L. DeBenneville, *Belg. Pat.* 636,521 (1964); *Chem. Abstr.*, **61**, 15,981 (1964).
274. A. E. Ardis, *U.S. Pat.* 2,774,783 (1956); *Chem. Abstr.*, **51**, 11,373 (1957).
275. R. Gompper and R. Kunz, *Chem. Ber.*, **99**, 2900 (1966).
276. W. J. Middleton, *J. Org. Chem.*, **30**, 1402 (1965).
277. W. J. Middleton, *U.S. Pat.* 3,162,674 (1964); *Chem. Abstr.*, **62**, 11,697 (1965).
278. S. Proskow, *U.S. Pat.* 3,133,115 (1964); *Chem. Abstr.*, **61**, 4230 (1964).
279. S. Proskow, H. E. Simmons and T. L. Cairns, *J. Am. Chem. Soc.*, **85**, 2341 (1963); **88**, 5254 (1966).
280. R. Gompper, W. Elser and H.-J. Muller, *Angew. Chem.*, *Int. Ed.*, **6**, 453 (1967).
281. M. Green and D. C. Wood, *Chem. Commun.*, **1967**, 1062.
282. W. R. Roth, *Chimia (Aarau)*, **20**, 229 (1966).
283. K. G. Naik, *J. Chem. Soc.*, **1921**, 1231.
284. D. G. I. Felton, *J. Chem. Soc.*, **1955**, 515.
285. D. G. I. Felton and S. F. D. Orr, *J. Chem. Soc.*, **1955**, 2170.
286. K. Kudo, *Bull. Chem. Soc. Japan*, **35**, 1490 (1962).
287. R. E. Merrifield and W. D. Phillips, *J. Am. Chem. Soc.*, **80**, 2778 (1958).
288. K. Kudo, *Bull. Chem. Soc. Japan*, **35**, 1730 (1962).
289. K. Kudo, *Bull. Chem. Soc. Japan*, **35**, 1842 (1962).
290. R. E. Heckert and W. J. Middleton, *U.S. Pat.* 2,980,698 (1961); *Chem. Abstr.*, **55**, 22,138 (1961).
291. O. W. Webster, *U.S. Pat.* 3,221,026 (1965); *Chem. Abstr.*, **64**, 4952 (1966).
292. W. J. Middleton, *U.S. Pat.* 2,883,368 (1959); *Chem. Abstr.*, **53**, 18,872 (1961).
293. V. A. Engelhardt and W. J. Middleton, *U.S. Pat.* 2,766,270 (1956); *Chem. Abstr.*, **51**, 11,373 (1957).
294. W. J. Middleton and V. A. Engelhardt, *J. Am. Chem. Soc.*, **80**, 2829 (1958).
295. H. D. Edwards and J. D. Kendall, *U.S. Pat.* 2,533,233 (1950); *Chem. Abstr.*, **45**, 2804 (1951).
296. R. Gompper and W. Töpfl, *Chem. Ber.*, **95**, 2861 (1962).
297. D. C. Dittmer, H. E. Simmons and R. D. Vest, *J. Org. Chem.*, **29**, 497 (1964).
298. J. K. Williams, *J. Am. Chem. Soc.*, **84**, 3478 (1962).
299. R. Gompper and W. Töpfl, *Chem. Ber.*, **95**, 2881 (1962).
300. R. Gompper, E. Kutter and W. Töpfl, *Ann. Chem.*, **659**, 90 (1962).
301. W. R. Hatchard, *J. Org. Chem.*, **29**, 660 (1964); *U.S. Pat.* 3,155,678 (1964); *Chem. Abstr.*, **62**, 2778 (1965).
302. E. Söderbäck, *Acta. Chem. Scand.*, **17**, 362 (1963).
303. W. R. Hatchard, *J. Org. Chem.*, **29**, 665 (1964); *U.S. Pat.* 3,230,229 (1966); *Chem. Abstr.*, **64**, 9733 (1966).
304. K. Gewald, *J. Prakt. Chem.*, **31**, 214 (1966).
305. G. Bähr and G. Schleitzer, *Chem. Ber.*, **88**, 1771 (1955).
306. H. E. Simmons, D. C. Blomstrom and R. D. Vest, *J. Am. Chem. Soc.*, **84**, 4756 (1962).
307. G. Bähr, *Angew. Chem.*, **68**, 525 (1956).
308. G. Bähr, G. Schleitzer and H. Bieling, *Chem. Techn.*, **8**, 597 (1956).
309. G. Bähr and G. Schleitzer, *Chem. Ber.*, **90**, 438 (1957).

310. E. Merck A. G. (by H. Hahn, G. Mohr and A. v. Schoor), *Ger. Pat.* 1,158,056 (1963); *Chem. Abstr.*, **60**, 5341 (1964).
311. R. J. H. Clark and W. Errington, *Inorg. Chem.*, **5**, 650 (1966).
312. M. A. Chaudhari and F. G. A. Stone, *J. Chem. Soc.* (*A*), **1966**, 838.
313. H. Köpf and M. Schmidt, *J. Organometal. Chem.*, **4**, 426 (1965).
314. J. Locke and J. A. McCleverty, *Inorg. Chem.*, **5**, 1157 (1966).
315. A. Davison, N. Edelstein, R. H. Holm and A. H. Maki, *Inorg. Chem.*, **4**, 55 (1965).
316. N. M. Atherton, J. Locke and J. A. McCleverty, *Chem. Ind.* (*London*), **1965**, 1300.
317. A. Davison, N. Edelstein, R. H. Holm and A. H. Maki, *J. Am. Chem. Soc.*, **86**, 2799 (1964).
318. E. I. Stiefel, Z. Dori and H. B. Gray, *J. Am. Chem. Soc.*, **89**, 3353 (1967).
319. M. Gerloch, S. F. A. Kettle, J. Locke and J. A. McCleverty, *Chem. Commun.*, **1966**, 29.
320. J. Locke and J. A. McCleverty, *Chem. Commun.*, **1965**, 102.
321. C. C. McDonald, W. D. Phillips and H. F. Mower, *J. Am. Chem. Soc.*, **87**, 3319 (1965).
322. R. Williams, E. Billig, J. H. Waters and H. B. Gray, *J. Am. Chem. Soc.*, **88**, 43 (1966).
323. J. F. Weiher, L. R. Melby and R. E. Benson, *J. Am. Chem. Soc.*, **86**, 4329 (1964).
324. E. Billig, H. B. Gray, S. I. Shupack, J. H. Waters and R. Williams, *Proc. Chem. Soc.*, **1964**, 110.
325. R. E. Benson, *U.S. Pat.* 3,255,195 (1966); *Chem. Abstr.*, **65**, 15,555 (1966).
326. W. C. Hamilton and I. Bernal, *Inorg. Chem.*, **6**, 2003 (1967).
327. G. Manecke and D. Wöhrle, *Angew. Chem.*, **79**, 1024 (1967).
328. G. Bähr, *Angew. Chem.*, **70**, 606 (1958); **73**, 628 (1961).
329. J. D. Forrester, A. Zalkin and D. H. Templeton, *Inorg. Chem.*, **3**, 1500 (1964).
330. A. Davison, N. Edelstein, R. H. Holm and A. H. Maki, *Inorg. Chem.*, **2**, 1227 (1963).
331. E. Billig, R. Williams, I. Bernal, J. H. Waters and H. B. Gray, *Inorg. Chem.*, **3**, 663 (1964).
332. A. Davison, N. Edelstein, R. H. Holm and A. H. Maki, *J. Am. Chem. Soc.*, **85**, 3049 (1963).
333. C. L. Langford, E. Billig, S. I. Shupack and H. B. Gray, *J. Am. Chem. Soc.*, **86**, 2958 (1964).
334. A. H. Maki, N. Edelstein, A. Davison and R. H. Holm, *J. Am. Chem. Soc.*, **86**, 4580 (1964).
335. A. L. Balch and R. H. Holm, *Chem. Commun.*, **1966**, 552.
336. S. I. Shupack, E. Billig, R. J. H. Clark, R. Williams and H. B. Gray, *J. Am. Chem. Soc.*, **86**, 4594 (1964).
337. E. Billig, S. I. Shupack, J. H. Waters, R. Williams and H. B. Gray, *J. Am. Chem. Soc.*, **86**, 926 (1964).
338. C. J. Fritchie, Jr., *Acta, Cryst.*, **20**, 107 (1966).
339. R. Eisenberg and J. A. Ibers, *Inorg. Chem.*, **4**, 605 (1965).
340. A. H. Maki, T. E. Berry, A. Davison, R. H. Holm and A. L. Balch, *J. Am. Chem. Soc.*, **88**, 1080 (1966).
341. A. Davison, N. Edelstein, R. H. Holm and A. H. Maki, *J. Am. Chem. Soc.*, **85**, 2029 (1963).
342. R. Eisenberg, J. A. Ibers, R. J. H. Clark and H. B. Gray, *J. Am. Chem. Soc.*, **86**, 113 (1964).

343. J. D. Forrester, A. Zalkin and D. H. Templeton, *Inorg. Chem.*, **3**, 1507 (1964).
344. J. H. Waters and H. B. Gray, *J. Am. Chem. Soc.*, **87**, 3534 (1965).
345. W. Wolf, E. Degener and S. Petersen, *Angew. Chem.*, **72**, 963 (1960).
346. E. Merck A. G. (by A. v. Schoor, E. Jacobi, S. Lust and H. Flemming), *Ger. Pat.* 1,060,655 (1959); *Chem. Abstr.*, **55**, 7748 (1961).
347. C. G. Krespan, *U.S. Pat.* 3,140,295 (1964); *Chem. Abstr.*, **61**, 8239 (1964).
348. Farbenfabriken Bayer A. G., *Brit. Pat.* 829,529 (1960); *Chem. Abstr.*, **54**, 19,497 (1960).
349. R. Mayer and B. Gebhardt, *Chem. Ber.*, **97**, 1298 (1964).
350. E. Klingsberg, *J. Am. Chem. Soc.*, **86**, 5290 (1964).
351. Farbenfabriken Bayer A. G. (by B. Anders and E. Kuehle), *Belg. Pat.* 632,578 (1963); *Chem. Abstr.*, **61**, 8321 (1964).
352. R. S. Waritz, *U.S. Pat.* 3,122,472 (1964); *Chem. Abstr.*, **60**, 14538 (1964).
353. H. E. Simmons, R. D. Vest, D. C. Blomstrom, J. R. Roland and T. L. Cairns, *J. Am. Chem. Soc.*, **84**, 4746 (1962).
354. E. J. Frazza and W. O. Fugate, *U.S. Pat.* 3,265,565 (1966); *Chem. Abstr.*, **65**, 15,391 (1966).
355. H. E. Simmons, D. C. Blomstrom and R. D. Vest, *J. Am. Chem. Soc.*, **84**, 4772 (1962).
356. H. E. Simmons, D. C. Blomstrom and R. D. Vest, *J. Am. Chem. Soc.*, **84**, 4782 (1962).
357. C. G. Krespan, *J. Am. Chem. Soc.*, **83**, 3434 (1961).
358. K. Onoda, *Nippon Nogeikagaku Kaishi*, **36**, 575 (1962); *Chem. Abstr.*, **61**, 14157 (1964).
359. R. Wippermann, *Ber.*, **7**, 767 (1874).
360. H. C. Adams and H. D. Green, *U.S. Pat.* 2,069,543 (1937); *Chem. Abstr.*, **31**, 1966 (1937).
361. G. S. Bohart, *U.S. Pat.* 1,464,802 (1923); *Chem. Abstr.*, **17**, 3259 (1923).
362. J. Oro and A. P. Kimball, *Arch. Biochem. Biophys.*, **96**, 293 (1962).
363. C. N. Matthews and R. E. Moser, *Nature*, **215**, 1230 (1967).
364. R. P. Linstead, E. G. Nobel and J. M. Wright, *J. Chem. Soc.*, **1937**, 911.
365. C. N. Matthews and R. E. Moser, *Proc. Natl. Acad. Sci. U.S.*, **56**, 1087 (1966).
366. Y. Yamada, I. Kumashiro and T. Takenishi, *J. Org. Chem.*, **33**, 642 (1968).
367. D. E. Carter, *U.S. Pat.* 2,813,423 (1957); *Chem. Abstr.*, **52**, 8187 (1958).
368. E. Bamberger and L. Rudolf, *Ber.*, **35**, 1082 (1902).
369. D. E. Carter, *U.S. Pat.* 2,722,540 (1955); *Chem. Abstr.*, **50**, 3499 (1956).
370. D. W. Woodward, *U.S. Pat.* 2,499,441 (1950); *Chem. Abstr.*, **44**, 5898 (1950).
371. H. Bredereck, G. Schmötzer and E. Oehler, *Ann. Chem.*, **600**, 81 (1956).
372. H. Ogura and M. Kondo, *Bull. Chem. Soc. Japan*, **40**, 2448 (1967).
373. H. H. Husteds, *Dissertation*, University of Marburg, 1957; see T. Völker, *Angew. Chem.*, **72**, 379 (1960).
374. R. A. Sanchez, J. P. Ferris and L. E. Orgel, *Science*, **153**, 72 (1966).
375. R. A. Sanchez, J. P. Ferris and L. E. Orgel, *J. Mol. Biol.*, **30**, 223 (1967).
376. J. P. Ferris and L. E. Orgel, *J. Am. Chem. Soc.*, **88**, 1074 (1966).
377. J. P. Ferris and L. E. Orgel, *J. Am. Chem. Soc.*, **87**, 4976 (1965).
378. J. P. Ferris and R. A. Sanchez, *Organic Syntheses*, in press.
379. J. P. Ferris and L. E. Orgel, *J. Am. Chem. Soc.*, **88**, 3829 (1966).
380. R. E. Moser, J. M. Fritsch, T. L. Westman, R. M. Kliss and C. N. Matthews, *J. Am. Chem. Soc.*, **89**, 5673 (1967).

381. O. Lange, *Ber.*, **6**, 99 (1873).
382. E. Gryszkiewicz-Trochimowski, *Rocz. Chem.*, **8**, 165 (1928); *Chem. Abstr.*, **22**, 4475 (1928).
383. B. R. Penfold and W. N. Lipscomb, *Tetrahedron Letters*, **1960**, 17.
384. B. R. Penfold and W. N. Lipscomb, *Acta Cryst.*, **14**, 589 (1961).
385. E. Gryszkiewicz-Trochimowski and A. Sementzova, *J. Russ. Phys. Chem. Soc.*, **55**, 547 (1924); *Chem. Abstr.*, **19**, 2810 (1925).
386. H. Lescoeur and A. Rigaut, *Compt. Rend.*, **89**, 310 (1879).
387. C. Bedel, *Compt. Rend.*, **176**, 168 (1923).
388. C. Bedel, *Bull. Soc. Chim. France*, **35**, 339 (1924).
389. L. E. Hinkel, G. O. Richards and O. Thomas, *J. Chem. Soc.*, **1937**, 1432.
390. L. E. Hinkel, *J. Chem. Soc.*, **1939**, 492.
391. L. E. Hinkel and T. I. Watkins, *J. Chem. Soc.*, **1940**, 1206.
392. W. Ruske, *Chem. Tech. (Berlin)*, **6**, 489 (1954).
393. T. Wadsten and S. Andersson, *Acta. Chem. Scand.*, **13**, 1069 (1959).
394. R. L. Webb, S. Frank and W. C. Schneider, *J. Am. Chem. Soc.*, **77**, 3491 (1955).
395. D. A. Long, W. O. George and A. E. Williams, *Proc. Chem. Soc.*, **1960**, 285.
396. K. Onoda, *Nippon Nogeikagaku Kaishi*, **36**, 167 (1962); *Chem. Abstr.*, **61**, 6916 (1964).
397. R. L. Sass and J. Donohue, *Acta. Cryst.*, **10**, 375 (1957).
398. J. D. Roberts and M. Jautelat, private communication.
399. K. Onoda, *Nippon Nogeikagaku Kaishi*, **36**, 255 (1962); *Chem. Abstr.*, **61**, 6917 (1964).
400. P. S. Robertson and J. Vaughan, *J. Am. Chem. Soc.*, **80**, 2691 (1958).
401. H. Bredereck and G. Schmotzer, *Ann. Chem.*, **600**, 95 (1956).
402. H. Bredereck, G. Schmotzer and H. J. Becher, *Ann. Chem.*, **600**, 87 (1956).
403. E. Gryszkiewicz-Trochimowski, *Rocz. Chem.*, **1**, 468 (1921); *Chem. Abstr.*, **17**, 1424 (1923).
404. M. P. Hartshorn and J. Vaughan, *Chem. Ind. (London)*, **1961**, 632.
405. W. Theilacker, *Naturwissenschaften*, **48**, 377 (1961).
406. W. Eckert and F. Quint, *U.S. Pat.* 2,200,689 (1940); *Chem. Abstr.*, **34**, 6455 (1940).
407. D. W. Woodward, *U.S. Pat.* 2,534,332 (1950); *Chem. Abstr.*, **45**, 5191 (1951).
408. D. W. Woodward, *U.S. Pat.* 2,534,331 (1950); *Chem. Abstr.*, **45**, 5191 (1951).
409. D. Shew, *Ph.D. Thesis*, Indiana University, 1959; *Dissertation Abstr.*, **20**, 1593 (1959).
410. H. v. Euler and H. Hasselquist, *Arkiv Kemi*, **11**, 407 (1957).
411. R. E. Moser, A. R. Claggett and C. N. Matthews, *Tetrahedron Letters*, **1968**, 1599.
412. R. Schenck and H. Finken, *Ann. Chem.*, **462**, 158 (1928).
413. C. Moureu and J. C. Bongrand, *Compt. Rend.*, **151**, 946 (1910).
414. C. Moureu and J. C. Bongrand, *Ann. Chim. (Paris)* [9], **14**, 47 (1920).
415. S. Murahashi, T. Takizawa, S. Kurioka and S. Maekawa, *Nippon Kagaku Zasshi*, **77**, 1689 (1956); *Chem. Abstr.*, **53**, 5163 (1959).
416. F. A. Miller and D. H. Lemmon, *Spectrochim. Acta. (A)*, **23**, 1415 (1967).
417. L. J. Krebaum, *J. Org. Chem.*, **31**, 4103 (1966).
418. L. J. Krebaum, *U.S. Pat.* 3,141,034 (1964); *Chem. Abstr.*, **61**, 8196 (1964).
419. L. J. Krebaum, *U.S. Pat.* 3,055,738 (1962); *Chem. Abstr.*, **58**, 2375 (1963).
420. J. L. Comp, *U.S. Pat.* 3,079,423 (1963); *Chem. Abstr.*, **59**, 2656 (1963).
421. I. G. Farbenindustrie A. G., *French Pat.* 790,262 (1935); *Chem. Abstr.*, **30**, 2991 (1936).
422. J. Happel, C. J. Marsel and A. A. Reidlinger, *U.S. Pat.* 3,006,948 (1958); *Chem. Abstr.*, 8574 (1962).

423. J. T. Herron, J. L. Franklin and P. Bradt, *Can. J. Chem.*, **37**, 579 (1959).
424. R. A. Sanchez, J. P. Ferris and L. E. Orgel, *Science*, **154**, 784 (1966).
425. J. P. Ferris, R. A. Sanchez and L. E. Orgel, *J. Mol. Biol.*, **38**, 121 (1968).
426. E. Kloster-Jensen, *Acta. Chem. Scand.*, **18**, 1629 (1964).
427. E. Kloster-Jensen, *Chem. Ind. (London)*, **1962**, 658.
428. E. Kloster-Jensen, *Acta. Chem. Scand.*, **17**, 1862 (1963).
429. E. Kloster-Jensen, *Acta. Chem. Scand.*, **17**, 1859 (1963).
430. C. Moureu and J. C. Bongrand, *Ann. Chim. (Paris)*, [9], **14**, 5 (1920).
431. C. Moureu and J. C. Bongrand, *Bull. Soc. Chim. France*, **1909**, 846.
432. A. T. Blomquist and E. C. Winslow, *J. Org. Chem.*, **10**, 149 (1945).
433. N. R. Byrd, *NASA Accession No. N 64-20601*, Report No. NASA-CR-56035; Report 166-F, 1964; *Chem. Abstr.*, **62**, 2831 (1965).
434. E. L. Martin, *U.S. Pat.* 3,070,622 (1962); *Chem. Abstr.*, **59**, 454 (1963).
435. C. D. Weis, *J. Org. Chem.*, **27**, 3514 (1962).
436. C. D. Weis, *J. Org. Chem.*, **27**, 3520 (1962).
437. P. D. Zavitsanos, *U.S. Pat.* 3,336,359 (1967).
438. F. J. Brockman, *Can. J. Chem.*, **33**, 507 (1955).
439. A. J. Saggiomo, *J. Org. Chem.*, **22**, 1171 (1957).
440. A. V. Grosse and C. S. Stokes, *U.S. Dep. Comm. Office Tech. Serv. P. B.*, Report 161,460 (1960); *Chem. Abstr.*, **56**, 9432 (1962).
441. F. Bohlmann and H. J. Mannhardt, *Chem. Ber.*, **89**, 2268 (1956).
442. M. Anchel, *Science*, **121**, 607 (1955).
443. H. Mommaerts, *Bull. Soc. Chim. Belg.*, **52**, 79 (1943).
444. J. K. Tyler and J. Sheridan, *Trans. Faraday Soc.*, **59**, 2661 (1963).
445. A. Westenberg and E. B. Wilson, Jr., *J. Am. Chem. Soc.*, **72**, 199 (1950).
446. V. H. Dibeler, R. M. Reese and J. L. Franklin, *J. Am. Chem. Soc.*, **83**, 1813 (1961).
447. G. C. Turrell, W. D. Jones and A. Maki, *J. Chem. Phys.*, **26**, 1544 (1957).
448. B. Wojtkowiak and G. Cornu, *Compt. Rend.*, **262C**, 305 (1966); B. Wojtkowiak and R. Queignec, *Compt. Rend.*, **262B**, 811 (1966).
449. W. Dannhauser and A. F. Flueckinger, *J. Chem. Phys.*, **38**, 69 (1963).
450. R. Queignec and B. Wojtkowiak, *Compt. Rend., Ser. A, B.*, **262B**, 486 (1966).
451. S. Murahashi, B. Ryutani and K. Hatada, *Bull. Chem. Soc., Japan*, **32**, 1001 (1959).
452. V. A. Job and G. W. King, *Can. J. Chem.*, **41**, 3132 (1963).
453. S. J. Cyvin and P. Klaeboe, *Acta. Chem. Scand.*, **19**, 697 (1965).
454. L. Lopez, J. F. Labarre, P. Castan and R. Mathis-Noël, *Compt. Rend.*, **259**, 3483 (1964).
455. R. West and C. S. Kraihanzel, *J. Am. Chem. Soc.*, **83**, 765 (1961).
456. F. A. Miller and R. B. Hannan, *J. Chem. Phys.*, **21**, 110 (1953).
457. F. A. Miller, R. B. Hannan, Jr. and L. R. Cousins, *J. Chem. Phys.*, **23**, 2127 (1955).
458. F. A. Miller, D. H. Lemmon and R. E. Witkowski, *Spectrochim. Acta.*, **21**, 1709 (1965).
459. S. J. Cyvin, E. Kloster-Jensen and P. Klaeboe, *Acta. Chem. Scand.*, **19**, 903 (1965).
460. P. Klaeboe, E. Kloster-Jensen and S. J. Cyvin, *Spectrochim. Acta. (A)*, **23**, 2733 (1967).
461. C. Nagarjan, E. R. Lippincott and J. M. Stutman, *Z. Naturforsch.*, **20A**, 786 (1965).
462. G. Nagarjan and J. R. Durig, *Bull. Soc. Roy. Sci. Liege*, **36**, 552 (1967).
463. V. A. Job and G. W. King, *J. Mol. Spectry.*, **19**, 155 (1966); **19**, 178 (1966).
464. F. A. Miller and R. B. Hannan, Jr., *Spectrochim. Acta.*, **12**, 321 (1958).
465. J. E. Boggs, C. M. Thompson and C. M. Crain, *J. Phys. Chem.*, **61**, 1625 (1957).

466. C. C. Costain, *J. Chem. Phys.*, **29**, 864 (1958).
467. A. H. Barrett, *Mem. Soc. Roy. Sci. Liege*, **7**, 197 (1962); *Chem. Abstr.*, **60**, 2444 (1964).
468. J. Sheridan, *U.S. Dep. Comm. Office Tech. Serv.*, *AD 417,348* (1963); *Chem. Abstr.*, **61**, 12,812 (1964).
469. A. A. Petrov, N. V. Elsakov and V. B. Lebedev, *Opt. i Spektroskopiya*, **16**, 1013 (1964); *Chem. Abstr.*, **61**, 10,205 (1964).
470. F. V. Shallcross and G. B. Carpenter, *Acta. Cryst.*, **11**, 490 (1958).
471. R. B. Hannan and R. L. Collin, *Acta. Cryst.*, **6**, 350 (1953).
472. B. Borgen, O. Hassel and C. Römming, *Acta. Chem. Scand.*, **16**, 2469 (1962).
473. A. V. Savitskii, *Dokl. Akad. Nauk. SSSR*, **87**, 631 (1952).
474. L. A. Errede, *J. Phys. Chem.*, **64**, 1031 (1960).
475. A. D. Kirshenbaum and A. V. Grosse, *J. Am. Chem. Soc.*, **78**, 2020 (1956).
476. A. Turpin and D. Voigt, *Compt. Rend.*, **256**, 1712 (1963).
477. F. Galais and J. F. Labarre, *J. Chim. Phys.*, **61**, 717 (1964).
478. J. B. Moffat, *Can. J. Chem.*, **42**, 1323 (1964).
479. E. Bayer and G. Hafelinger, *Chem. Ber.*, **99**, 1689 (1966).
480. A. Langseth and C. K. Moller, *Acta. Chem. Scand.*, **4**, 725 (1950).
481. L. Pauling, H. D. Springall and K. J. Palmer, *J. Am. Chem. Soc.*, **61**, 927 (1939).
482. G. T. Armstrong and S. Marantz, *J. Phys. Chem.*, **64**, 1776 (1960).
483. A. Misono and H. Noguchi, *Abstr, 19. Ann. Meeting Chem. Soc. Japan*, Yokohama, April, 1966, 4 T 104.
484. E. Ciganek, *Tetrahedron Letters*, **1967**, 3321.
485. N. N. Greenwood and K. Wade in *Friedel-Crafts and Related Reactions*, Vol. 1 (Ed George A. Olah), Interscience, New York, 1963, p. 569.
486. E. Ciganek, unpublished results.
487. C. Moureu and I. Lazennec, *Compt. Rend.*, **143**, 553 (1906); *Bull. Soc. Chim. France*, [3] **35**, 1179 (1906).
488. C. Moureu and J. C. Bongrand, *Compt. Rend.*, **1914**, 1092.
489. E. Winterfeldt, W. Krohn and H. Preuss, *Chem. Ber.*, **99**, 2572 (1966).
490. C. D. Weis, *J. Org. Chem.*, **28**, 74 (1963).
491. E. Winterfeldt and H. Preuss, *Chem. Ber.*, **99**, 450 (1966).
492. R. E. Miller and L. A. Miller, *U.S. Pat.* 3,128,300 (1964); *Chem. Abstr.*, **60**, 15,792 (1964).
493. J. P. Ferris, *Science*, **161**, 53 (1968).
494. A. J. Leusink, J. W. Marsman and H. A. Budding, *Rec. Trav. Chim.*, **84**, 689 (1965).
495. A. J. Leusink and J. W. Marsman, *Rec. Trav. Chim.*, **84**, 1123 (1965).
496. C. Moureu and J. C. Bongrand, *Compt. Rend.*, **170**, 1025 (1920).
497. M. E. Kimball, J. P. Martella and W. C. Kaska, *Inorg. Chem.*, **6**, 414 (1967).
498. W. P. Griffith and G. Wilkinson, *J. Chem. Soc.*, **1959**, 1629.
499. R. K. Freidlina, *Bull. Acad. Sci. URSS*, *Classe Sci. Chim.*, **14**, (1942); *Chem. Abstr.*, **37**, 3050 (1943).
500. R. C. Cookson and J. Dance, *Tetrahedron Letters*, **1962**, 879.
501. R. C. Cookson, J. Dance and M. Godfrey, *Tetrahedron*, **24**, 1529 (1968).
502. E. Ciganek, *J. Org. Chem.*, **34**, 1923 (1969).
503. C. D. Weis, *J. Org. Chem.*, **27**, 3693 (1962).
504. P. G. Gassman and K. T. Mansfield, *Chem. Commun.*, **1965**, 391; *J. Am. Chem. Soc.*, **90**, 1517, 1524 (1968).
505. C. D. Weis, *J. Org. Chem.*, **27**, 3695 (1962).
506. J. Sauer, *Angew. Chem.*, **78**, 233 (1966); **79**, 76 (1967).

507. S. A. Nazova, I. I. Patalakh and Ya. M. Paushkin, *Dokl. Akad. Nauk SSSR*, **153**, 144 (1963).
508. V. F. Belov and S. S. Oganesov, *Tr. Mosk. Inst. Neftekhim. i Gaz. Prom.*, (58), 136 (1965); *Chem. Abstr.*, **65**, 3989 (1966).
509. J. Manassen and J. Wallach, *J. Am. Chem. Soc.*, **87**, 2671 (1965).
510. B. J. MacNulty, *Polymer*, **7**, 275 (1966).
511. M. J. Beneš, J. Peška and O. Wichterle, *J. Polymer Sci.*, (*C*), **16**, 555 (1967).
512. M. A. Geiderikh, B. E. Davydov, B. A. Krentsel, I. M. Kustanovich, L. S. Polak, A. V. Topchiev and R. M. Voitenko, *J. Polymer Sci.*, **54**, 621 (1961).
513. M. J. Beneš, J. Peška and O. Wichterle, *Chem. Ind.* (*London*), **1962**, 562.
514. M. J. Beneš, J. Peška and O. Wichterle, *J. Polymer Sci.*, (*C*), **4**, 1377 (1963).
515. A. Misono, H. Noguchi and S. Noda, *J. Polymer Sci.*, (*B*) **4**, 985 (1966).
516. L. S. Meriwether, E. C. Colthup, G. W. Kennerly and R. N. Reusch, *J. Org. Chem.*, **26**, 5155 (1961).
517. R. Engelhardt and H. Arledter, *U.S. Pat.* 2,591,415 (1952); *Chem. Abstr.*, **47**, 613 (1953).
518. Farbenfabriken Bayer A. G. (by R. Engelhardt and H. Arledter), *Ger. Pat.* 842,045 (1952); *Chem. Abstr.*, **52**, 10,194 (1958).
519. A. V. Willett, Jr. and J. R. Pailthorp, *U.S. Pat.* 2,716,646 (1955); *Chem. Abstr.*, **50**, 1914 (1956).
520. A. S. Bailey, B. R. Henn and J. M. Langdon, *Tetrahedron*, **19**, 161 (1963).
521. G. M. Bennett and R. L. Wain, *J. Chem. Soc.*, **1936**, 1108.
522. A. V. Kirsanov and E. A. Abrazhanova, *Sbornik Statei Obshch. Khim.*, **2**, 865 (1953); *Chem. Abstr.*, **49**, 6920 (1955).
523. Badische Anilin- and Soda-Fabrik A. G. (by H. Kroeper, R. Platz, H. Nohe and R. Schanz), *Belg. Pat.* 632,702 (1963); *Chem. Abstr.*, **61**, 3032 (1964).
524. D. J. Hadley and B. Wood, *U.S. Pat.* 2,838,558 (1958); *Chem. Abstr.*, **52**, 17,191 (1958).
525. V. S. Kudinova, B. V. Suvorov and R. U. Umarova, *Teoriya i Prakt. Ionnogo Obmena, Akad. Nauk. Kas. SSR, Tr. Resp. Soveshch*, **1962**, 90; *Chem. Abstr.*, **61**, 13,233 (1964).
526. S. D. Mekhtiev and M. G. Guseinov, *Azerb. Neft. Khoz.*, **40**, 35 (1961); *Chem. Abstr.*, **55**, 27,204 (1961).
527. R. Foster and T. J. Thomson, *Trans. Faraday Soc.*, **59**, 2287 (1963).
528. R. Foster and C. A. Fyfe, *Trans. Faraday Soc.*, **62**, 1400 (1966).
529. F. W. Küster and A. Stallberg, *Ann. Chem.*, **278**, 207 (1894).
530. C. D. Weis, *J. Org. Chem.*, **27**, 2964 (1962).
531. J. Kuthan and J. Procházková, *Coll. Czech. Chem. Commun.*, **31**, 3832 (1966).
532. K. Wallenfels and K. Friedrich, *Tetrahedron Letters*, **1963**, 1223.
533. K. Wallenfels, F. Witzler and K. Friedrich, *Tetrahedron*, **23**, 1353 (1967).
534. K. Dimroth and K. J. Kraft, *Angew. Chem.*, **76**, 433 (1964).
535. K. Wallenfels, F. Witzler and K. Friedrich, *Tetrahedron*, **23**, 1845 (1967).
536. J. Bretschneider and K. Wallenfels, *Tetrahedron*, **24**, 1063 (1968).
537. A. Epstein and B. S. Wildi, *J. Chem. Phys.*, **32**, 324 (1960).
538. E. A. Lawton and D. D. McRitchie, *J. Org. Chem.*, **24**, 26 (1959).
539. M. T. Razomovskaya, A. I. Belyakova and G. I. Korel'sksya, *Metody Polucheniya Khim. Reaktivov i Preparatov*, **12**, 111 (1965); *Chem. Abstr.*, **65**, 3787 (1966).
540. O. W. Webster, M. Brown and R. E. Benson, *J. Org. Chem.*, **30**, 3250 (1965).
541. P. H. Rieger, I. Bernal, W. H. Reinmuth and G. K. Fraenkel, *J. Am. Chem. Soc.*, **85**, 683 (1963).

542. A. Zweig, J. E. Lehnsen, W. G. Hodgson and W. H. Jura, *J. Am. Chem. Soc.*, **85,** 3937 (1963).
543. M. E. Peover, *Trans. Faraday Soc.*, **58,** 2370 (1962).
544. A. Ishitani and S. Nagakura, *Theoret. Chim. Acta.*, **4,** 236 (1966).
545. A. Ishitani and S. Nagakura, *Bull. Chem. Soc. Japan*, **38,** 367 (1965).
546. H. E. Popkie and J. B. Moffat, *Can. J. Chem.*, **43,** 624 (1965).
547. M. E. Peover, *Trans. Faraday Soc.*, **60,** 417 (1964).
548. S. Iwata, J. Tanaka and S. Nagakura, *J. Am. Chem. Soc.*, **89,** 2813 (1967).
549. S. Iwata, J. Tanaka and S. Nagakura, *J. Am. Chem. Soc.*, **88,** 894 (1966); H. Hayashi, S. Nagakura and S. Iwata, *Mol. Phys.*, **13,** 489 (1967).
550. A. S. Bailey, R. J. P. Williams and J. D. Wright, *J. Chem. Soc.*, **1965,** 2579.
551. P. H. Emslie and R. Foster, *Rec. Trav. Chim.*, **84,** 255 (1965).
552. A. Zweig, J. E. Lehnsen and M. A. Murray, *J. Am. Chem. Soc.*, **85,** 3933 (1963).
553. S. Kumakura, F. Iwasaki and Y. Saito, *Bull. Chem. Soc. Japan*, **40,** 1826 (1967).
554. B. Kamenar, C. K. Prout and J. D. Wright, *J. Chem. Soc.* (*A*), **1966,** 661.
555. P. Murray-Rust and J. D. Wright, *J. Chem. Soc.* (*A*), **1968,** 247.
556. G. Manecke and D. Wöhrle, *Makromol. Chem.*, **102,** 1 (1967).
557. E. L. Kropa, *U.S. Pat.* 3,114,740 (1963).
558. B. S. Wildi, *U.S. Pat.* 3,157,687 (1964); *Chem. Abstr.*, **62,** 3522 (1965).
559. J. E. Katon and B. S. Wildi, *J. Chem. Phys.*, **40,** 2977 (1964).
560. J. E. Katon, *U.S. Pat.* 3,142,644 (1964); *Chem. Abstr.*, **61,** 10,169 (1964).
561. B. S. Wildi, *U.S. Patent* 3,046,323 (1962); *Chem. Abstr.*, **58,** 7495 (1963); 3,046,322 (1962); *Chem. Abstr.*, **58,** 2006 (1963); 3,086,001 (1963); *Chem. Abstr.*, **59,** 2279 (1963).
562. D. I. Packham and F. A. Rackley, *Chem. Ind.* (*London*), **1967,** 1254.
563. A. A. Berlin, V. P. Parini, E. L. Frankevich and L. G. Cherkashina, *Izv. Akad. Nauk SSSR, Ser. Khim.*, **1964,** 2108.
564. W. Littke and K. Wallenfels, *Tetrahedron Letters*, **1965,** 3365.
565. W. Littke, *Angew. Chem.*, **79,** 1002 (1967).
566. Farbwerke Hoechst A. G. (by K. Wallenfels and R. Friedrich), *Ger. Pat.* 1,183,900; *Chem. Abstr.*, **62,** 9074 (1965).
567. A. Zinke, A. Dadieu, K. Funke and A. Pongratz, *Monatsh. Chem.*, **50,** 77 (1928).
568. D. S. Acker and W. R. Hertler, *J. Am. Chem. Soc.*, **84,** 3370 (1962).
569. D. S. Acker and D. C. Blomstrom, *U.S. Pat.* 3,115,506 (1963); *Chem. Abstr.*, **60,** 14,647 (1964).
570. J. Diekmann, W. R. Hertler and R. E. Benson, *J. Org. Chem.*, **28,** 2719 (1963).
571. S. Chatterjee, *J. Chem. Soc.* (*B*), **1967,** 1170.
572. R. E. Long, R. A. Sparks and K. N. Trueblood, *Acta Cryst.*, **18,** 932 (1965).
573. L. R. Melby, R. J. Harder, W. R. Hertler, W. Mahler, R. E. Benson and W. E. Mochel, *J. Am. Chem. Soc.*, **84,** 3374 (1962).
574. P. H. H. Fischer and C. A. McDowell, *J. Am. Chem. Soc.*, **85,** 2694 (1963).
575. M. T. Jones and W. R. Hertler, *J. Am. Chem. Soc.*, **86,** 1881 (1964).
576. R. H. Boyd and W. D. Phillips, *J. Chem. Phys.*, **43,** 2927 (1965).
577. E. Menefee and Y. H. Pao, *J. Chem. Phys.*, **36,** 3472 (1962).
578. D. S. Acker and D. C. Blomstrom, *U.S. Pat.* 3,162,641 (1964); *Chem. Abstr.*, **63,** 549 (1965).
579. J. H. Lupinski and K. D. Kopple, *Science*, **146,** 1038 (1964).
580. L. F. Druding, *Anal. Chem.*, **35,** 1582 (1963).
581. H. D. Hartzler, *J. Org. Chem.*, **30,** 2456 (1965).

582. W. R. Hertler, H. D. Hartzler, D. S. Acker and R. E. Benson, *J. Am. Chem. Soc.*, **84**, 3387 (1962).
583. J. Thiele and J. Meisenheimer, *Ber.*, **33**, 675 (1900).
584. J. Thiele and F. Günther, *Ann. Chem.*, **349**, 45 (1906).
585. B. Helferich, *Ber.*, **54**, 155 (1921).
586. R. Geyer and H. Steinmetzer, *Wiss. Z. Tech. Hochsch. Chem. Leuna-Merseburg*, **2**, 423 (1959/60); *Chem. Abstr.*, **55**, 15,407 (1961).
587. K. Wallenfels, D. Hofmann and R. Kern, *Tetrahedron*, **21**, 2231 (1965).
588. A. G. Brook, *J. Chem. Soc.*, **1952**, 5040.
589. P. R. Hammond, *J. Chem. Soc.*, **1963**, 3113.
590. A. C. M. Finch, *J. Chem. Soc.*, **1964**, 2272.
591. R. Foster, *Nature*, **195**, 490 (1962).
592. R. Foster and P. Hanson, *Biochim. Biophys. Acta*, **112**, 482 (1966).
593. M. Ichikawa, M. Soma, T. Onishi and K. Tamaru, *J. Phys. Chem.*, **70**, 3020 (1966).
594. E. A. Braude, A. G. Brook and R. P. Linstead, *J. Chem. Soc.*, **1954**, 3569.
595. B. Helferich and H. G. Bodenbender, *Ber.*, **56B**, 1112 (1923).
596. H. D. Hartzler and R. E. Benson, *J. Org. Chem.*, **26**, 3507 (1961).
597. M. F. Ansell, G. C. Culling, B. W. Nash, D. A. Wilson and J. W. Lown, *Proc. Chem. Soc.*, **1960**, 405.
598. P. W. D. Mitchell, *Can. J. Chem.*, **41**, 550 (1963).
599. D. Walker and T. D. Waugh, *J. Org. Chem.*, **30**, 3240 (1965).
600. K. Wallenfels, G. Bachmann, D. Hofmann and R. Kern, *Tetrahedron*, **21**, 2239 (1965).
601. M. E. Peover, *J. Chem. Soc.*, **1962**, 4540.
602. R. Foster and T. J. Thomson, *Trans. Faraday Soc.*, **58**, 860 (1962).
603. Y. Matsunaga, *J. Chem. Phys.*, **41**, 1609 (1964).
604. M. E. Peover, *Trans. Faraday Soc.*, **58**, 1656 (1962).
605. R. D. Srivastava and G. Prasad, *Spectrochim. Acta*, **22**, 1869 (1966).
606. R. Foster and I. Horman, *J. Chem. Soc.* (*B*), **1966**, 171.
607. R. E. Rehwoldt and E. Boynton, *J. Chem. Educ.*, **42**, 648 (1965).
608. A. Ottenberg, R. L. Brandon and M. E. Browne, *Nature*, **201**, 1119 (1964).
609. R. Foster and T. J. Thomson, *Trans. Faraday Soc.*, **59**, 296 (1963).
610. K. M. C. Davis and M. C. R. Symons, *J. Chem. Soc.*, **1965**, 2079.
611. P. L. Kronick, H. Scott and M. M. Labes, *J. Chem. Phys.*, **40**, 890 (1964).
612. Y. Matsunaga, *Nature*, **211**, 182 (1966).
613. Y. Matsunaga, *J. Chem. Phys.*, **42**, 1982 (1965).
614. Y. Okamoto, S. Shah and Y. Matsunaga, *J. Chem. Phys.*, **43**, 1904 (1965).
615. E. L. Muetterties, *U.S. Pat.* 3,177,232 (1965); *Chem. Abstr.*, **63**, 14,422 (1965).
616. E. L. Muetterties, *U.S. Pat.* 3,177,240 (1965); *Chem. Abstr.*, **63**, 14,776 (1965).
617. National Research Development Corp. (by W. Slough), *Brit. Pat.* 1,009,361 (1965); *Chem. Abstr.*, **64**, 4423 (1966).
618. General Electric Co., *Neth. Appl.* 6,515,380 (1966); *Chem. Abstr.*, **65**, 13,062 (1966).
619. H. Nomori, M. Hatano and S. Kambara, *J. Polymer Sci.* (*B*), **4**, 623 (1966).
620. H. Scott, G. A. Miller and M. M. Labes, *Tetrahedron Letters*, **1963**, 1073.
621. British Oxygen Co. Ltd. (by L. P. Ellinger), *Brit. Pat.* 1,005,116 (1965); *Chem. Abstr.*, **63**, 18,294 (1965).
622. D. Walker and J. D. Hiebert, *Chem. Rev.*, **67**, 153 (1967).
623. P. J. Neustaedter in *Steroid Reactions*, (Ed. C. Djerassi), Holden-Day, San Francisco, California, 1963, p. 89; see also R. Owyang, p. 227.

624. L. M. Jackman, *Advan. Org. Chem.*, **2**, 329 (1960).
625. A. E. Hydorn, *U.S. Pat.* 3,035,050 (1962); *Chem. Abstr.*, **57**, 11,283 (1962).
626. Roussel-UCLAF, *French Pat.* 1,313,082 (1962); *Chem. Abstr.*, **59**, 2724 (1963).
627. R. M. Scribner, *J. Org. Chem.*, **31**, 3671 (1966).
628. W. H. Baddley and G. L. McClure, unpublished results cited in *J. Am. Chem. Soc.*, **90**, 3705 (1968).
629. Farbwerke Hoechst A. G., *French Pat.* 1,314,925 (1963); *Chem. Abstr.*, **59**, 2723 (1963).
630. J. Biechler, *Compt. Rend.*, **200**, 141 (1935).
631. W. Madelung and E. Kern, *Ann. Chem.*, **427**, 1 (1922).
632. C. Mauguin and L. J. Simon, *Compt. Rend.*, **170**, 998 (1920).
633. D. E. Nagy, *U.S. Pat.* 2,562,869 (1951); *Chem. Abstr.*, **45**, 10,521 (1951).
634. W. L. Burdick, *J. Am. Chem. Soc.*, **47**, 1485 (1925).
635. M. Kuhn and R. Mecke, *Chem. Ber.*, **94**, 3010 (1961).
636. M. Takimoto, *Nippon Kagaku Zasshi*, **85**, 159 (1964); *Chem. Abstr.*, **61**, 2937 (1964).
637. H. Köhler, *Z. Anorg. Allgem. Chem.*, **336**, 245 (1965).
638. H. Köhler and B. Seifert, *Z. Anorg. Allgem. Chem.*, **352**, 265 (1967).
639. H. Köhler and B. Seifert, *Z. Chem.*, **5**, 142 (1965).
640. H. Köhler and B. Seifert, *Z. Naturforsch.*, **22B**, 238 (1967) and references cited there.
641. G. Rembarz, H. Brandner and H. Finger, *J. Prakt. Chem.*, **26**, 314 (1964).
642. F. H. S. Curd, J. A. Hendry, T. S. Kenny, A. G. Murray and F. L. Rose, *J. Chem. Soc.*, **1948**, 1630.
643. N. H. Marsh and R. W. Hamilton, *U.S. Pat.* 2,557,984 (1951); *Chem. Abstr.*, **46**, 1586 (1952).
644. E. Allenstein, *Z. Anorg. Allgem. Chem.*, **322**, 265 (1963).
645. A. Bannow, *Ber.*, **13**, 2201 (1880).
646. W. Madelung and E. Kern, *Ann. Chem.*, **427**, 26 (1922).
647. E. C. Franklin, *J. Am. Chem. Soc.*, **44**, 486 (1922).
648. D. W. Kaiser and B. C. Redmon, *U.S. Pat.* 2,510,981 (1950); *Chem. Abstr.*, **44**, 9990 (1950).
649. F. D. Marsh and M. E. Hermes, *J. Am. Chem. Soc.*, **86**, 4506 (1964).
650. E. I. du Pont de Nemours and Co. (by F. D. Marsh), *Ger. Pat.* 1,181,182 (1964); *Chem. Abstr.*, **62**, 9010 (1965).
651. M. G. Darzens, *Compt. Rend.*, **154**, 1232 (1912).
652. C. V. Hart, *J. Am. Chem. Soc.*, **50**, 1922 (1928).
653. E. Wasserman, L. Barash and W. A. Yager, *J. Am. Chem. Soc.*, **87**, 2075 (1965).
654. D. E. Milligan, M. E. Jacox, J. J. Comeford and D. E. Mann, *J. Chem. Phys.*, **43**, 756 (1965).
655. D. E. Milligan, M. E. Jacox and A. M. Bass, *J. Chem. Phys.*, **43**, 3149 (1965).
656. G. J. Pontrelli and A. G. Anastassiou, *J. Chem. Phys.*, **42**, 3735 (1965).
657. A. G. Anastassiou, *J. Am. Chem. Soc.*, **89**, 3184 (1967).
658. A. G. Anastassiou and H. E. Simmons, *J. Am. Chem. Soc.*, **89**, 3177 (1967).
659. A. G. Anastassiou, *J. Am. Chem. Soc.*, **88**, 2322 (1966).
660. M. E. Jacox, D. E. Milligan, N. G. Moll and W. E. Thompson, *J. Chem. Phys.*, **43**, 3734 (1965).
661. L. J. Schoen, *J. Chem. Phys.*, **45**, 2773 (1966).
662. D. E. Milligan and M. E. Jacox, *J. Chem. Phys.*, **45**, 1387 (1966).
663. A. G. Anastassiou and J. N. Shepelavy, *J. Am. Chem. Soc.*, **90**, 492 (1968).
664. H. W. Kroto, *J. Chem. Phys.*, **44**, 831 (1966).

665. D. E. Milligan and M. E. Jacox, *J. Chem. Phys.*, **44**, 2850 (1966).
666. F. D. Marsh and H. E. Simmons, *J. Am. Chem. Soc.*, **87**, 3529 (1965).
667. A. G. Anastassiou, H. E. Simmons and F. D. Marsh, *J. Am. Chem. Soc.*, **87**, 2296 (1965).
668. A. G. Anastassiou, *J. Am. Chem. Soc.*, **87**, 5512 (1965).
669. R. M. Scribner, *Abstr. 155 Meeting Am. Chem. Soc., San Francisco, 1968*, p. P144.
670. M. E. Hermes and F. D. Marsh, *J. Org. Chem.*, in press.
671. R. M. Scribner, *Tetrahedron Letters*, **1967**, 4737.
672. M. E. Hermes and F. D. Marsh, *J. Am. Chem. Soc.*, **89**, 4760 (1967).
673. K. Friedrich, *Angew. Chem.*, **79**, 980 (1967).
674. F. D. Marsh and M. E. Hermes, *J. Am. Chem. Soc.*, **87**, 1819 (1965).
675. M. T. Jones, *J. Am. Chem. Soc.*, **88**, 227 (1966).
676. T. Curtius, *Ber.*, **31**, 2489 (1898).
677. M. J. S. Dewar and R. Pettit, *J. Chem. Soc.*, **1956**, 2026.
678. S. H. Harper and K. C. Sleep, *J. Sci. Food Agric.*, **6**, 116 (1955).
679. D. D. Phillips and W. C. Champion, *J. Am. Chem. Soc.*, **78**, 5452 (1956).
680. M. Lesbre and R. Buisson, *Bull. Soc. Chim., France*, **1957**, 1204.
681. E. Ciganek, *J. Org. Chem.*, **30**, 4198 (1965).
682. C. C. Costain and J. Yarwood, *J. Chem. Phys.*, **45**, 1961 (1966).
683. A. J. Mercer and D. N. Travis, *Can. J. Phys.*, **43**, 1795 (1965); **44**, 353 (1966).
684. R. A. Bernheim, R. J. Kempf, P. W. Humer and P. S. Skell, *J. Chem. Phys.*, **41**, 1156 (1964).
685. T. M. McKinney, *J. Am. Chem. Soc.*, **90**, 3879 (1968).
686. M. Regitz, *Angew. Chem.*, **79**, 786 (1967).
687. S. Proskow, unpublished; see E. Ciganek, *J. Am. Chem. Soc.*, **87**, 1149 (1965).
688. P. C. Petrellis, H. Dietrich, E. Meyer and G. W. Griffin, *J. Am. Chem. Soc.*, **89**, 1967 (1967).
689. R. Breslow and C. Yuan, *J. Am. Chem. Soc.*, **80**, 5991 (1958).
690. J. Diekmann, unpublished results.
691. M. Regitz and G. Heck, *Chem. Ber.*, **97**, 1482 (1964).
692. E. Ciganek, *J. Am. Chem. Soc.*, **88**, 1979 (1966); **89**, 1454 (1967).
693. J. P. Fleury, D. V. Assche and A. Bader, *Tetrahedron Letters*, **1965**, 1399.
694. J. P. Fleury and R. Mertz, *Bull. Soc. Chim. France*, **1967**, 237.
695. H. D. Hartzler, *J. Am. Chem. Soc.*, **86**, 2174 (1964).
696. G. B. Payne, *J. Am. Chem. Soc.*, **81**, 4901 (1959).
697. G. B. Payne and P. H. Williams, *J. Org. Chem.*, **26**, 651 (1961).
698. W. J. Linn, *U.S. Pat.* 3,238,228 (1966); *Chem. Abstr.*, **64**, 15,842 (1966).
699. W. J. Linn, O. W. Webster and R. E. Benson, *J. Am. Chem. Soc.*, **87**, 3651 (1965).
700. A. Rieche and P. Dietrich, *Chem. Ber.*, **96**, 3044 (1963).
701. W. J. Linn, unpublished observations.
702. W. J. Middleton, *J. Org. Chem.*, **31**, 3731 (1966).
703. O. W. Webster, *U.S. Pat.* 3,250,791 (1966); *Chem. Abstr.*, **65**, 2221 (1966).
704. S. Munavalli and G. Ourission, *Bull. Soc. Chim. France*, **1964**, 729.
705. H. Kwart and D. M. Hoffman, *J. Org. Chem.*, **31**, 419 (1966).
706. C. Bugg and R. L. Sass, *Acta Cryst.*, **18**, 591 (1965).
707. J. Streith and J.-M. Cassal, *Compt. Rend., C. Sci. Chim.*, **264**, 1307 (1967).
708. W. J. Linn and E. Ciganek, *J. Org. Chem.*, **34**, 2146 (1969).
709. W. J. Linn, *U.S. Pat.* 3,115,517 (1963); *Chem. Abstr.*, **60**, 7919 (1964).

710. W. J. Middleton, E. L. Buhle, J. G. McNally, Jr. and M. Zanger, *J. Org. Chem.*, **30**, 2384 (1965).
711. P. J. Graham, W. J. Linn and W. J. Middleton, *U.S. Pat.* 3,350,313 (1967).
712. A. F. Cook and J. G. Moffatt, *J. Am. Chem. Soc.*, **90**, 740 (1968).
713. H. Nozaki, D. Tunemoto, Z. Morita, K. Nakamura, K. Watanabe, M. Tokaku and K. Kondo, *Tetrahedron*, **27**, 4279 (1967).
714. W. J. Middleton, *J. Org. Chem.*, **31**, 3731 (1966).
715. W. J. Linn and R. E. Benson, *J. Am. Chem. Soc.*, **87**, 3657 (1965).
716. W. J. Linn, *U.S. Pat.* 3,317,567 (1967).
717. P. Brown and R. C. Cookson, *Proc. Chem. Soc.*, **1964**, 185.
718. P. Brown and R. C. Cookson, *Tetrahedron*, **24**, 2551 (1968).
719. W. J. Linn, *J. Am. Chem. Soc.*, **87**, 3665 (1965).
720. C. D. Weis, *U.S. Pat.* 3,060,198 (1962); *Chem. Abstr.*, **58**, 5638 (1963).
721. R. D. Vest and H. E. Simmons, in press; see H. E. Simmons, *U.S. Pat.* 3,221,024 (1965); *Chem. Abstr.*, **64**, 8359 (1966).
722. E. Gryszkiewicz-Trochimowski, *J. Russ. Phys. Chem. Soc.*, **55**, 548 (1924); *Chem. Abstr.*, **19**, 2810 (1925).
723. E. Gryszkiewicz-Trochimowski and L. Kotko, *J. Russ. Phys. Chem. Soc.*, **55**, 551 (1924).
724. J. A. Fialkoff, *Bull. Soc. Chim. France*, **41**, 1209 (1927).
725. J. A. Bilton and R. P. Linstead, *J. Chem. Soc.*, **1937**, 922.
726. E. G. Tailor, *Can. J. Res.*, **20B**, 161 (1942).
727. K. Wallenfels and W. Hanstein, *Angew. Chem.*, **77**, 861 (1965).
728. K. Wallenfels and W. Hanstein, *Ann. Chem.*, **709**, 151 (1967).
729. E. L. Little, Jr., W. J. Middleton, D. D. Coffman, V. A. Engelhardt and G. N. Sausen, *J. Am. Chem. Soc.*, **80**, 2832 (1958).
730. K.-D. Gundermann and H. U. Alles, *Angew. Chem.*, **78**, 906 (1966).
731. P. Neumann, *Dissertation*, University of Freiburg, 1967.
732. M. T. Jones, *J. Am. Chem. Soc.*, **88**, 5060 (1966).
733. E. Ott, *Ber.*, **52B**, 656 (1919).
734. A. Carrington, H. C. Longuet-Higgins and P. F. Todd, *Mol. Phys.*, **9**, 211 (1965).
735. I. B. Johns, E. A. McElhill and J. O. Smith, *J. Chem. Eng. Data*, **7**, 277 (1962).
736. W. A. Dollase, *J. Am. Chem. Soc.*, **87**, 979 (1965).
737. H. E. Simmons, Jr. (to E. I. du Pont de Nemours and Co.), *U.S. Pat.* 3,052,681 (1962); *Chem. Abstr.*, **58**, 531 (1963).
738. J. Thesing, D. Witzel and A. Brehm, *Angew. Chem.*, **68**, 425 (1956).
739. H. E. Williams, *Cyanogen Compounds*, 2nd ed., Edward Arnold, London, 1948, Chap. 1, p. 13.
740. O. Diels, H. Gärtner and R. Kaack, *Ber.*, **55B**, 3439 (1922).
741. R. Malachowski, L. Jurkiewicz and J. Wojtowicz, *Ber.*, **70B**, 1012 (1937).
742. R. Malachowski, *Ger. Pat.* 666,394 (1938); *Chem. Abstr.*, **33**, 2152 (1939).
743. O. Achmatowicz and M. Leplawy, *Roczniki Chem.*, **32**, 1375 (1958); *Chem. Abstr.*, **53**, 10,033 (1959).
744. W. J. Linn, *U.S. Pat.* 3,115,517 (1963); *Chem. Abstr.*, **60**, 7919 (1964).
745. S. Proskow, private communication.
746. O. Glemser and V. Häusser, *Z. Naturforsch.*, **3B**, 159 (1948).
747. M. Puchalik, *Acta Phys. Polon.*, **10**, 89 (1950); *Chem. Abstr.*, **44**, 10,407 (1950).
748. W. Kemula and A. Tramer, *Roczniki Chem.*, **27**, 522 (1953); *Chem. Abstr.*, **48**, 6255 (1954).

749. W. Kemula and K. L. Wierzchowski, *Roczniki Chem.*, **27**, 527 (1953); *Chem. Abstr.*, **48**, 6253 (1954).
750. A. Tramer and K. L. Wierzchowski, *Bull. Acad. Polon. Sci., Classe III*, **5**, 411 (1957); **5**, 417 (1957); *Chem. Abstr.*, **52**, 878 (1958).
751. J. F. Westerkamp, *Bol. Acad. Nac. Cienc.*, **42**, 191 (1961); *Chem. Abstr.*, **58**, 5170 (1963).
752. W. Kemula and K. Wierzchowski, *Roczniki Chem.*, **27**, 527 (1953); *Chem. Abstr.*, **48**, 6253 (1954).
753. W. J. Linn, unpublished observations.
754. A. Tramer, *Bull. Acad. Polon. Sci., Ser. Sci., Math., Astron. Phys.*, **12**, 669 (1964); *Chem. Abstr.*, **63**, 6471 (1965).
755. J. Prochorow and A. Tramer, *J. Chem. Phys.*, **44**, 4545 (1966).
756. R. Malachowski, *Roczniki Chem.*, **24**, 229 (1950); *Chem. Abstr.*, **47**, 8653 (1953).
757. R. Malachowski and L. Jurkiewicz, *Roczniki Chem.*, **24**, 88 (1950); *Chem. Abstr.*, **48**, 3914 (1954).
758. O. Achmatowicz, K. Belniak, C. Borecki and M. Leplawy, *Roczniki Chem.*, **39**, 1443 (1965); *Chem. Abstr.*, **64**, 17,457 (1966).
759. L. J. Bollyky, R. H. Whitman, R. A. Clarke and M. M. Rauhut, *J. Org. Chem.*, **32**, 1663 (1967).
760. R. Malachowski and J. Jankiewicz-Wasowska, *Roczniki Chem.*, **25**, 35 (1951); *Chem. Abstr.*, **49**, 10,483 (1953).
761. A. Guyot and E. Michel, *Compt. Rend.*, **148**, 229 (1909).
762. I. L. Knunyants, N. P. Gambaryan, C-X. Chen and E. M. Rokhlin, *Izv. Akad. Nauk. SSSR Old. Khim. Nauk.*, **1962**, 684; *Bull. Acad. Sci. USSR Div. Chem. Sci.* (Engl. Trans.), **1962**, 633.
763. O. Achmatowicz and O. Achmatowicz, Jr., *Roczniki Chem.*, **35**, 813 (1961); *Chem. Abstr.*, **56**, 7209 (1962).
764. O. Achmatowicz and A. Zamojski, *Bull. Acad. Polon. Sci. Classe III*, **5**, 927 (1957); *Chem. Abstr.*, **52**, 6333 (1958).
765. O. Achmatowicz and A. Zamojski, *Roczniki Chem.*, **35**, 799 (1961); *Chem. Abstr.*, **56**, 7257 (1962).
766. O. Achmatowicz and A. Zamojski, *Croat. Chem. Acta.*, **29**, 269 (1957).
767. W. J. Linn, *J. Org. Chem.* **29**, 3111 (1964).
768. O. Achmatowicz, M. Leplawy and A. Zamojski, *Bull. Acad. Polon. Sci. Classe III*, **3**, 539 (1955); *Chem. Abstr.*, **51**, 5013 (1957).
769. O. Achmatowicz, M. Leplawy and A. Zamojski, *Roczniki Chem.*, **30**, 215 (1956), *Chem. Abstr.*, **51**, 1087 (1957).
770. O. Achmatowicz and F. Werner-Zamojska, *Bull. Acad. Polon. Sci. Classe III*, **5**, 923 (1957); *Chem. Abstr.*, **52** 6333 (1958).
771. G. I. Birnbaum, *Chem. Ind.* (*London*), **1961** 1116.
772. R. Malachowski and L. Jurkiewicz, *Roczniki Chem.*, **24**, 88 (1950); *Chem. Abstr.*, **48**, 3914 (1954).
773. O. Achmatowicz and K. Belniak, *Roczniki Chem.*, **39**, 1685 (1965); *Chem. Abstr.*, **64**, 17,474 (1966).
774. O. Achmatowicz, O. Achmatowicz, Jr., K. Belniak and J. Wrobel, *Roczniki Chem.*, **35**, 783 (1961); *Chem. Abstr.*, **56**, 7209 (1962).
775. O. Achmatowicz and M. Leplawy, *Bull. Acad. Polon. Sci., Classe III*, **3**, 547 (1955); *Bull. Acad. Polon. Sci. Ser. Sci. Chim., Geol. Geograph.*, **6**, 409 (1958); *Roczniki Chem.*, **33**, 1349 (1959); *Chem. Abstr.*, **54**, 13,056 (1960).

776. O. Achmatowicz and A. Zwierzak, *Bull. Acad. Polon. Sci., Classe III*, **5**, 931 (1957); *Chem. Abstr.*, **52**, 6333 (1958).
777. O. Achmatowicz and A. Zwierzak, *Roczniki Chem.*, **35**, 507 (1961); *Chem. Abstr.*, **56**, 7208 (1962).
778. O. Achmatowicz and M. Leplawy, *Bull. Acad. Polon. Sci. Ser. Sci. Chim. Geol. Geograph*, **6**, 409 (1958); *Chem. Abstr.*, **53**, 3183 (1959).
779. R. Malachowski, *Roczniki Chem.*, **24**, 229 (1950); *Chem. Abstr.*, **47**, 8653 (1953).
780. R. H. Patton and J. H. Simons, *J. Am. Chem. Soc.*, **77**, 2016 (1955).
781. A. C. Cope, H. L. Homes and H. O. House, *Organic Reactions*, Vol. 9, John Wiley and Sons, New York, 1957, pp. 107–331.
782. O. W. Webster, unpublished results.
783. F. Arndt, H. Scholz and E. Frobel, *Ann. Chem.*, **521**, 95 (1935).
784. G. Ponzio, *Gazz. Chim. Ital.*, **61**, 561 (1931).
785. H. D. Hartzler, *J. Am. Chem. Soc.*, **86**, 2174 (1964).
786. Z. Arnold, *Coll. Czech. Chem. Commun.*, **26**, 1113 (1961); *Chem. Abstr.*, **55**, 18,584 (1961).
787. L. Horner and H. Oediger, *Chem. Ber.*, **91**, 437 (1958).
788. W. J. Middleton, E. L. Buhle, J. G. McNally, Jr. and M. Zanger, *J. Org. Chem.*, **30**, 2384 (1965).
789. E. A. LaLancette and R. E. Benson, *J. Am. Chem. Soc.*, **83**, 4867 (1961).
790. H. D. Hartzler, *J. Org. Chem.*, **31**, 2654 (1966).
791. R. H. Boyd, *J. Phys. Chem.*, **65**, 1834 (1961).
792. Y. Urushibara, *Bull. Chem. Soc. Japan*, **2**, 278 (1927); *Chem. Abstr.*, **22**, 579 (1928).
793. Y. Urushibara and M. Takelayashi, *Bull. Chem. Soc. Japan*, **11**, 557 (1936); *Chem. Abstr.*, **31**, 1769 (1937).
794. A. D. Josey, *J. Org. Chem.*, **29**, 707 (1964).
795. O. W. Webster, *J. Am. Chem. Soc.*, **88**, 4055 (1966).
796. K. N. Trueblood *AD601478* Avail OTS (1964); *Chem. Abstr.*, **62**, 2312 (1965).
797. D. A. Bekoe, P. K. Gantzel and K. N. Trueblood, *Acta Cryst.*, **22**, 657 (1967).
798. O. W. Webster, *J. Am. Chem. Soc.*, **88**, 3046 (1966).
799. R. C. Cookson and K. R. Friedrich, *J. Chem. Soc.* (*C*), **1966**, 1641.
800. O. W. Webster, *J. Org. Chem.*, **32**, 39 (1967).
801. M. Strell, W. B. Braunbruch, W. F. Fuhler and O. Huber, *Ann. Chem.*, **587**, 177 (1954).
802. W. R. Carpenter and G. J. Palenik, *J. Org. Chem.*, **32**, 1219 (1967).
803. R. H. Boyd, *J. Phys. Chem.*, **67**, 737 (1963).
804. R. H. Boyd and C. H. Wang, *J. Am. Chem. Soc.*, **87**, 430 (1965).
805. R. H. Boyd, *J. Am. Chem. Soc.*, **83**, 4288 (1961).
806. R. G. Pearson and R. L. Dillon, *J. Am. Chem. Soc.*, **75**, 2439 (1953).
807. J. J. Christensen, R. M. Izatt, J. D. Hale, R. T. Pack and G. D. Watt, *Inorg. Chem.*, **2**, 337 (1963).
808. M. Liler and Dj. Kosanovic, *J. Chem. Soc.*, **1958**, 1084.
809. R. C. Kerber and M. J. Chick, *J. Org. Chem.*, **32**, 1330 (1967).
810. C. Dufraisse, J. Rigaudy and M. Ricard, *Tetrahedron, Suppl.* **8**, (2), 491 (1966); J. Rigaudy and M. Ricard, **24**, 3241 (1968).
811. E. L. Little, Jr., *U.S. Pat.* 2,773,892 (1956); *Chem. Abstr.*, **51**, 8776 (1957).
812. E. Allenstein, *Chem. Ber.*, **96**, 3230 (1963).
813. F. A. Miller and W. K. Baer, *Spectrochim. Acta*, **19**, 73 (1963).
814. W. Beck, H. S. Smedal and H. Kohler, *Z. Anorg. Allgem. Chem.*, **354**, 69 (1967).

815. W. Beck, R. E. Nitzschmann and H. S. Smedal, *J. Organometal. Chem.*, **8**, 547 (1967).
816. E. Allenstein and P. Quis, *Chem. Ber.*, **96**, 1035 (1963).
817. N. W. Jordan and I. W. Elliott, *J. Org. Chem.*, **27**, 1445 (1962).
818. P. Brown and R. C. Cookson, *Tetrahedron*, **21**, 1977 (1965).
819. M. B. Frankel, A. B. Anister, E. R. Wilson, M. McCormick and M. McEachern, Jr. in 'Advanced Propellant Chemistry', *Advances in Chemistry Series 54*. (Ed. R .T. Holzmann), American Chemical Society, Washington, 1966, pp. 108–17.
820. E. G. Howard, Jr., *U.S. Pat.* 3,162,649 (1964); *Chem. Abstr.*, **62**, 11,783 (1965).
821. G. Briegleb, W. Liptay and R. Fick, *Z. Elektrochem.*, **66**, 859 (1962).
822. Z. Rappoport and E. Shohamy, (a) *Israel J. Chem.*, **6**, 865 (1968); (b) *J. Chem. Soc.* (*B*), **1969**, 77; (c) in press.
823. C. Bugg, R. Desiderato and R. L. Sass, *J. Am. Chem. Soc.*, **86**, 3157 (1964).
824. M. A. Shaw, J. C. Tebby, R. S. Ward and D. H. Williams, *J. Chem. Soc.* (*C*), **1968**, 1609.
825. Y. Yamada, N. Nagashima, A. Nakamura and I. Kumashiro, *Tetrahedron Letters*, **1968**, 4529.

CHAPTER **10**

Molecular complexes of polycyano compounds

L. Russell Melby

E. I. du Pont de Nemours and Co.
Wilmington, Delaware, U.S.A.

I.	Introduction	639
II.	Theoretical Aspects and Implications	640
III.	Acceptor Strengths of Polycyano π-Acids	643
	A. Molecular-Orbital Calculations	643
	B. Polarographic Reduction	644
	C. Electron Affinities of Polycyano Acceptors	647
IV.	Electronic Absorption Spectroscopy	649
	A. Absorption Maxima of Complexes	649
	B. Equilibrium Constants	652
	C. Thermodynamic Considerations	655
V.	Ion-Radical Formation	656
	A. General	656
	B. Anion Radicals in Solution	657
	C. Ion Radicals in the Solid State	659
VI.	Electrical Properties of Polycyano Complexes	662
VII.	Acknowledgements	667
VIII.	References	667

I. INTRODUCTION

Molecules having electron-deficient sites tend to form complexes or compounds with molecules having electron-rich sites and this tendency is defined in general terms as an acid–base interaction. When the interacting site in the electron-deficient component is an ethylenic or benzenoid π-molecular orbital which is depleted of electron density by conjugation with electronegative (electron-withdrawing) substituents, this component may be referred to as a π-acceptor or π-acid. Conversely, the electron-rich component may be a π-donor or π-base and a complex formed from such compounds is then referred to as a π-molecular complex. Since this review is mainly concerned with a particular type of π-acceptor, namely, cyano-substituted unsaturated compounds, we shall refer to the complexes as π-complexes irrespective of the nature of the donor component, be it a π- or a σ-donor.

The bonding or interaction between components in complexes of this type may be of such a nature that the complexes are not isolable, but are detectable only by, for example, spectral properties in solution. In other instances, union between components may be sufficiently strong that the complexes can be isolated in the solid state. Two excellent monographs on the general subject are available[1,2]. Interest in the subject was greatly stimulated by the discovery and investigation of the exceptional properties of the π-acids tetracyanoethylene[3,4] (**1**) and tetracyanoquinodimethane[5,6a] (**2**).

$$(NC)_2C{=}C(CN)_2$$

(**1**)

$$(NC)_2C{=}C_6H_4{=}C(CN)_2$$

(**2**)

Since that time an extensive literature dealing with molecular complexes of these and other polycyano compounds has accumulated. This review will deal almost exclusively with complexes derived from π-acids containing three or more cyano groups per molecule conjugated with ethylenic, benzenoid or quinonoid unsaturation; some attention will also be given to π-acids having only two cyano groups but with acidity augmented by other electronegative substituents such as halogen, nitro or carbonyl groups. This review will not discuss the involvement of such complexes as reaction intermediates, since this aspect was reviewed recently[6b].

II. THEORETICAL ASPECTS AND IMPLICATIONS

As expressed by Mulliken in his classical work on the theory of molecular complex formation[7], the interaction between an electron donor (D) and an electron acceptor (A) can result in electron transfer from donor to acceptor with the formation of a new species, the molecular complex, whose structure is described by combination of a no-bond form (D:A) and a dative or charge-transfer form ($D^{\dot{+}}A^{\dot{-}}$) (equation 1).

$$D + A \rightleftharpoons D{:}A \longleftrightarrow D^{\dot{+}}A^{\dot{-}} \tag{1}$$

Weiss[8] had earlier referred to the latter as a 'charge-transfer' salt. The electronic spectrum of such a complex contains a new band which is absent in the spectra of either component and which is referred to as a charge-transfer band since it is generated by energy uptake in transferring an electron from donor to acceptor. The energy of this electronic transition must thus relate to the ease with which the donor can provide electrons (determined by its ionization potential, I_p) and the avidity with which the acceptor receives electrons (determined by its electron affinity, E_A). It should be emphasized that a strong donor will have a small I_p and a strong acceptor will have a large E_A. Among the relationships expressing the energy of the transition is one due to McConnell and co-workers[9] (equation 2)

$$h\nu_{CT} = I_p - E_A - W \tag{2}$$

where ν_{CT} is the charge-transfer absorption maximum, h is Planck's constant and W is the dissociation energy of the charge-transfer excited state, the ground state dissociation term being negligibly small. Other relationships have been proposed, but this will suffice for the present discussion. The relative magnitudes of I_p and E_A may be such that the absorption band will appear in the visible spectrum accounting for the striking colours which often appear with complex formation.

In solution, the formation of a complex (C) according to equation (3) is an equilibrium process with an equilibrium constant K given by equation (4).

$$D + A \rightleftharpoons C \tag{3}$$

$$K = \frac{[C]}{[A][D]} = \frac{[C]}{\{[A]_0 - [C]\}\{[D]_0 - [C]\}} \tag{4}$$

where [A], [D] and [C] are the molar concentrations of free (uncomplexed) acceptor, the donor and the complex, respectively. $[A]_0$ and $[D]_0$ are the *total* concentrations (both free and complexed) of acceptor and donor. The intensity of an electronic absorption band resulting from complex formation will then be related to this equilibrium. Thus, following the method of Benesi and Hildebrand[10,11], when D is present in large excess, equation (4) is reduced to equation (5).

$$K = \frac{[C]}{\{[A]_0 - [C]\}\{D\}} \tag{5}$$

Substituting for [C] the value of its extinction coefficient ε at the wavelength of maximum absorption for the charge-transfer band by using the relationship $[C] = O.D./\varepsilon l$, where O.D. is the optical density at this wavelength and l is the cell length in cm, equation (6) is obtained.

$$y \equiv \frac{[A]_0 l}{O.D.} = \frac{1}{\varepsilon} + \frac{1}{K[D]\varepsilon} \tag{6}$$

Thus a straight line plot of y versus $1/[D]$ would provide ε as the reciprocal of the intercept and $K\varepsilon$ as the reciprocal of the slope with K in units l/mole*.

Clearly, K, ε and λ_{max} relate to various characteristics of donor and acceptor, to the stability of the derived complex and to the bonding forces involved. In a common non-interacting solvent, a common donor with a series of structurally related acceptors will give values of K and λ_{max} which increase in a more or less parallel fashion as the acceptor strength or donor capacity increases. However, the magnitudes of the extinction coefficients usually do not parallel K and λ_{max}^2 but discussion of this fact is beyond the scope of this review. Attention is drawn to the recent critique by Dewar and Thompson[12] concerning inferences from spectral measurements and their relationship to the bonding forces in complex formation. Nevertheless, these experimentally accessible properties are of great importance in the study of molecular complexes and a discussion of them is essential. However, caution must be exercised in comparison of acceptor strengths or complex stabilities based on electronic effects alone, since complex formation is very sensitive to a number of complicating factors such as steric effects and geometry of interacting species[4,6,13,14].

A recent review by Briegleb[15] extensively covers the general literature on the electron affinities of polycyano π-acids including polarographic reduction[16].

* Some equilibrium constants cited in the literature were determined using mole fractions and are thus dimensionless and differ from K by a factor dependent on the molar volume of the solvent. See References 2 and 4.

III. ACCEPTOR STRENGTHS OF POLYCYANO π-ACIDS

A. Molecular-Orbital Calculations

An electron acceptor, or π-acid, is a molecule whose π-electron system can undergo addition of an electron without increasing the energy of that system. A strong acceptor would thus be one in which the lowest unoccupied molecular orbital is a bonding orbital, or if not bonding, then at least only slightly antibonding. Using the LCAO–MO approximation, Phillips[17] has calculated the π-electron energy levels of tetracyanoethylene (TCNE) and tetracyanoquinodimethane (TCNQ) using Coulomb and resonance integral values suggested by Orgel and coworkers[18]. Both molecules were assumed to have D_{2h} symmetry, and the π-electron levels therefore fall into the A_u, B_{1u}, B_{2g} and B_{3g} symmetry species. The five lowest levels of TCNE are occupied by its ten π-electrons and the eight lowest levels of TCNQ by sixteen π-electrons. The results of the calculations are depicted in Figures 1 and 2 in which orbitals with energies greater than zero are bonding and those less than zero antibonding. For TCNE (Figure 1) it is seen that the lowest unoccupied orbital has an energy of $+0{\cdot}15\beta$ and this is actually a bonding orbital. For TCNQ (Figure 2) the orbital of concern has an energy of $-0{\cdot}23\beta$ and thus is antibonding, although only slightly so. The

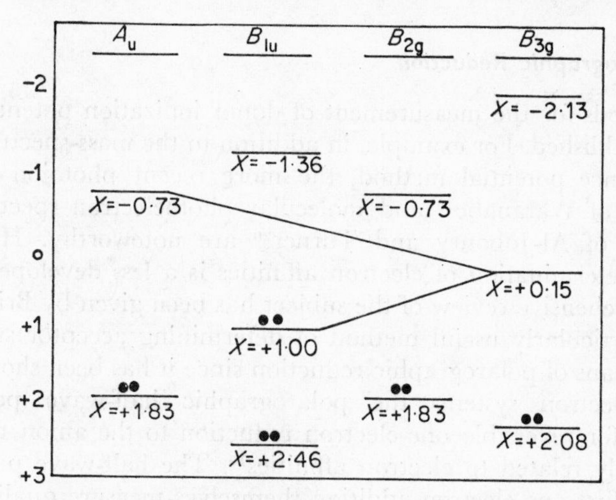

FIGURE 1. π-Electron energy level diagram for TCNE. The following parameters were used:
Coulomb integrals: $\alpha(\mathrm{C}) = \alpha$, $\alpha(\text{—N═}) = \alpha + \beta$
Resonance integrals: $\beta(\mathrm{CC}) = \beta$, $\beta(\mathrm{CN}) = 1{\cdot}2\beta$
Energies are expressed in units of β.

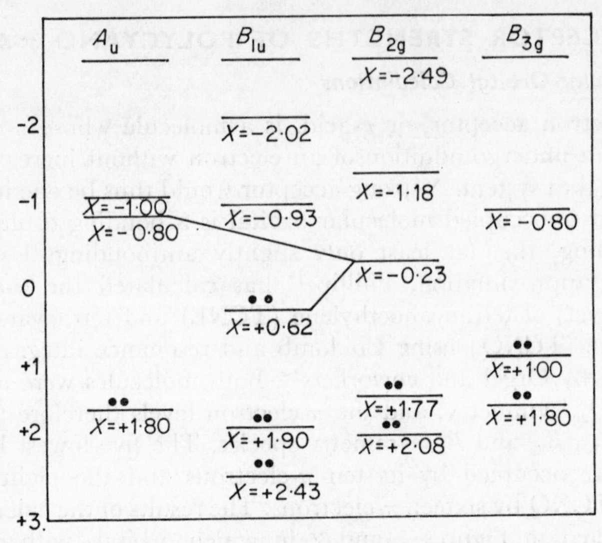

FIGURE 2. π-Electron energy level diagram for TCNQ. See legend of Figure 1.

results of the calculations are consistent with the acceptor properties of the two compounds and are qualitatively in the same order as determined by other means, i.e. with most donors TCNE is the stronger acceptor.

B. *Polarographic Reduction*

Methods for the measurement of donor ionization potentials are well established. For example, in addition to the mass-spectrometric appearance potential method, the more recent photoion current method of Watanabe[19] and molecular photoelectron spectroscopy method of Al-Joboury and Turner[20] are noteworthy. However, direct determination of electron affinities is a less developed area; a comprehensive review of the subject has been given by Briegleb[15]. One particularly useful method of determining acceptor strengths is by means of polarographic reduction since it has been shown that for π-electron systems the polarographic half-wave potential, ($E_{\frac{1}{2}\,\mathrm{red}}$) for reversible one-electron reduction to the anion radicals, is directly related to electron affinities[16]. The half-wave potentials for the first one-electron addition themselves measure qualitatively or semiquantitatively the acceptor strengths even without conversion to the related electron affinities[6a,16,21]. Qualitatively, the more positive the reversible $E_{\frac{1}{2}\,\mathrm{red}}$, the more facile the reduction and the greater the acceptor strength*. However, in comparing a series of compounds a common reference electrode must be used or corrections applied for differences in reference electrodes. In addition, the possible effects of changes in solvent and supporting electrolyte on $E_{\frac{1}{2}\,\mathrm{red}}$ values must be considered.

In Table 1 are listed the first polarographic half-wave reduction potentials for reversible one-electron reduction of a series of polycyano π-acids. For comparison purposes, values for the typical non-cyano π-acids *p*-benzoquinone and chloranil are also given.

As tabulated, the order of $E_{\frac{1}{2}\,\mathrm{red}}$ values (first column, not indented) determined in the same solvent with the same reference electrode and essentially identical electrolytes generally parallels acceptor strengths determined by other methods where such comparisons have been made[15]. The high positive reduction potential of hexacyanobutadiene is noteworthy and implies that it should be the strongest π-acid among those listed.

The ability of a given acceptor to be reversibly reduced in this manner does not, however, necessarily mean that it will form readily observable molecular complexes. For example, although 2-benzhydryl-7,7,8,8-tetracyanoquinodimethane (**3**) can be reduced to the anion radical[27], it does not form visible complexes such as those which unsubstituted tetracyanoquinodimethane will form. Presumably the steric effect of the bulky benzhydryl group lowers the association constant, *K*, to such a small value that the concentration of complex is below the limit of detection.

The reasons for the irreversibility of reduction of 2-chloro-5,6-dicyanoquinone and 9-dicyanomethylene-2,4,7-trinitrifluorene (**4**) are not clear, particularly since the fluorenone can be chemically converted to an anion radical which is stable in solution[13,26].

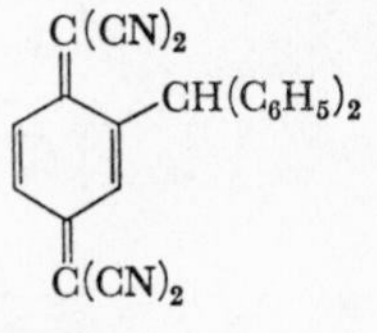

$C(CN)_2$ $CH(C_6H_5)_2$ $C(CN)_2$

(**3**)

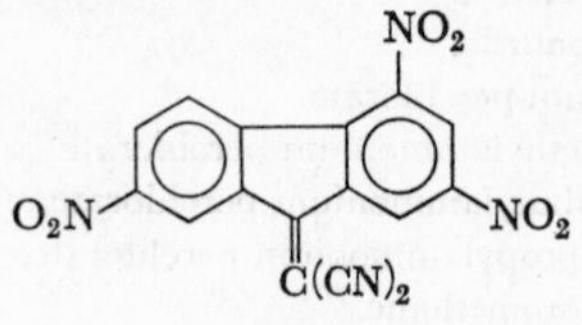

NO_2 O_2N NO_2 $C(CN)_2$

(**4**)

In addition to the first, usually reversible, reduction the acceptors listed in Table 1 also undergo a second, usually irreversible,

* Briegleb[15] suggests that negative values should be used for strong acceptors, but here we shall retain the electrochemical convention.

TABLE 1. First polarographic half-wave reduction potentials of polycyano acceptors[a] at 25°.

Acceptor	$E_{\frac{1}{2}}$ red (v)	Solvent	Supporting electrolyte	Electrode	Reference
Hexacyanobutadiene	+0·6	a	b	SCE[b]	22
2,3-Dichloro-5,6-	+0·5	a	b	SCE	23
dicyanoquinone	+0·51	a	c	SCE	16b
	+0·78	a	d	Ag/AgI	21
2-Chloro-5,6-dicyanoquinone	+0·41[d]	a	b	SCE	23
2,3-Dicyanoquinone	+0·31	a	b	SCE	23
TNAP[c]	+0·21	a	b	SCE	24
Tetracyanoethylene	+0·15	a	b	SCE	6
	+0·25	a	c	SCE	16b
	+0·54	a	d	Ag/AgI	21
7,7,8,8-Tetracyanoquino-	+0·13	a	b	SCE	6
dimethane	+0·57	a	d	Ag/AgI	21
2-Methyl-7,7,8,8-tetracyano-quinodimethane	+0·12	a	b	SCE	24
2,5-Dimethyl-7,7,8,8-tetra-cyanoquinodimethane	+0·02	a	b	SCE	24
2-Dicyanomethylene-1,3-indanedione	+0·02	a	e	SCE	25
Chloranil	+0.02	a	c	SCE	16b
	+0·35	f	d	Ag/AgI	21
9-Dicyanomethylene-2,4,7-trinitrofluorene	−0·08[d]	a	b	SCE	23, 26
2-Benzhydryl-7,7,8,8-tetra-cyanoquinodimethane	−0·28	a	b	SCE	27
Benzoquinone	−0·52	a	c	SCE	16b

[a] All reductions were carried out at 25° with a dropping mercury electrode. With two exceptions these reductions were reversible one-electron changes; for exceptions see superscript *d*.

[b] Aqueous saturated calomel.

[c] 11,11,12,12-Tetracyanonaphtha-2,6-quinodimethane.

[d] Not reversible.

a: acetonitrile.
b: lithium perchlorate.
c: tetraethylammonium perchlorate.
d: tetrabutylammonium perchlorate.
e: tetrapropylammonium perchlorate.
f: dichloromethane.

reduction at more negative potentials presumably corresponding to conversion to a dianion.

Briegleb has given a fairly detailed discussion directly relating reduction potentials to electron affinities, but examples specific to cyano compounds are not included[15]. Chaterjee[25] determined the electron affinity (E_A) of 2-dicyanomethylene-1,3-indanedione (**5**)

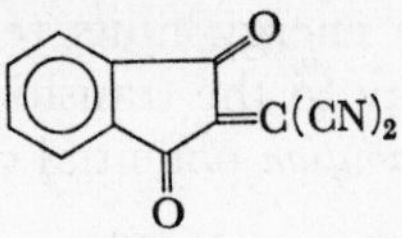

O
=C(CN)$_2$
O
(5)

by applying the relationship of equation (7)

$$-E_{\frac{1}{2}\,\mathrm{red}} = E_{\mathrm{A}} - \Delta F_{\mathrm{solv}} - 5{\cdot}07 \tag{7}$$

where $-E_{\frac{1}{2}\,\mathrm{red}}$ is the reversible polarographic reduction potential relative to the saturated calomel electrode and ΔF_{solv} is the solution energy of the anion; ΔF_{solv} was assigned an averaged value of $-3{\cdot}66$ ev for a number of acceptors having analogous structures. The value of E_{A} (1·38 ev) agreed well with the value obtained from charge-transfer spectra (1·33 ev) determined versus chloranil as a standard acceptor with an assigned $E_{\mathrm{A}} = 1{\cdot}37$ ev. This consistency is seen also in the $E_{\frac{1}{2}\,\mathrm{red}}$ values for **5** and chloranil listed in Table 1, which imply equivalent acceptor capacity for the two compounds.

C. Electron Affinities of Polycyano Acceptors

Using a series of structurally related condensed polynuclear aromatic hydrocarbons as reference donors, Briegleb estimated the electron affinities of a series of acceptors including a number of polycyano compounds[15], using, among others, the relationships in equations (8), (9) and (10). The first expression relates the electron affinity (E_{A} expressed in ev) to the energy difference between the highest occupied and lowest unoccupied orbital of the donor ($\bar{\nu}_0$), and the charge transfer transition energy ($\bar{\nu}_{\mathrm{CT}}$):

$$\bar{\nu}_{\mathrm{CT}} = 0{\cdot}701\bar{\nu}_0 + 0{\cdot}81 - E_{\mathrm{A}} \tag{8}$$

The second expression relates the relative acceptor capacities of one acceptor to another with the same donor to their electron affinities by the expression

$$(\bar{\nu}_{\mathrm{CT}})_{\mathrm{i}} - (\bar{\nu}_{\mathrm{CT}})_{\mathrm{k}} = E_{\mathrm{Ak}} - E_{\mathrm{Ai}} \tag{9}$$

where E_{Ai} is the electron affinity (in ev) of a reference acceptor, in this case chloranil which is assigned an $E_{\mathrm{A}} = 1{\cdot}37$ ev, E_{Ak} is the electron affinity of the new acceptor in question, and $(\bar{\nu}_{\mathrm{CT}})_{\mathrm{i}}$ and $(\bar{\nu}_{\mathrm{CT}})_{\mathrm{k}}$ are the charge-transfer transition energies of the reference acceptor (chloranil) and new acceptor, respectively, with the same donor(s) as determined from electronic absorption spectra and converted to appropriate energy units (ev). The third expression relates the electron affinity to the transition energy of the complex and the polarographic *oxidation* potential of the donor ($E_{\frac{1}{2}\,\mathrm{ox}}$):

$$E_{\mathrm{a}} = -\bar{\nu}_{\mathrm{CT}} + 1{\cdot}473E_{\frac{1}{2}\,\mathrm{ox}} + 1{\cdot}52 \qquad (10)$$

The results for several cyano compounds are summarized in Table 2 which includes for comparison the values for chloranil and benzoquinone.

TABLE 2. Electron affinities (E_{A}) of selected acceptors[15].

Acceptor	E_{A} (eV)		
	From Eq. (8)	From Eq. (9)	From Eq. (10)
2,3-Dichloro-5,6-dicyanoquinone	1·91	1·9	1·71
Tetracyanobenzoquinone	1·66	1·8	1·60
Tetracyanoethylene	1·53	2·2	1·5
2,3-Dicyanobenzoquinone		1·7	
7,7,8,8-Tetracyanoquinodimethane		1·7	
Chloranil	1·2	1·37	1·14
Benzoquinone	0·46	0·7	0·40

Excepting the value determined for tetracyanoethylene from equation (9), the orders of acceptor strength are self-consistent among the three methods. Also from equation (9), tetracyanoethylene and 2,3-dicyanobenzoquinone are out of order with respect to their relative strengths inferred from reduction potentials (Table 1) and other spectral data (see below). Among the quinones, acceptor strengths generally follow the order expected from the degree and type of substitution except for tetracyanobenzoquinone, which would be expected to be the most potent acceptor of all. The extraordinarily high value of 2·2 ev for tetracyanoethylene determined from equation (9) is suspect, and probably reflects errors imposed by the attempt to relate its π-electron system to the geometrically quite different cyclic π-system of the reference acceptor chloranil. These apparent anomalies serve to emphasize the difficulties in assigning absolute or even relative values to acceptor strengths.

More recently, Farragher and Page[28] have determined electron affinities of a number of polycyano compounds using the magnetron technique which presumably depends upon direct electron capture by the acceptors*. The electron affinities determined in this way were considerably higher than those shown in Table 2 (e.g. ~2·9 ev for both tetracyanoethylene and tetracyanoquinodimethane, 2·46 ev for chloranil and 1·38 ev for benzoquinone) but the order for any given structural type (substituted ethylene or benzene) showed a definite increase in electron affinity with increasing cyano substitution. Unfortunately, no quinones were included in this work, but hexacyanobutadiene was shown to have a very high electron affinity (~3·3 ev).

IV. ELECTRONIC ABSORPTION SPECTROSCOPY

A. Absorption Maxima of Complexes

It has been mentioned that extinction coefficients ε_{max} do not parallel in any reasonable way either the association constants K or λ_{max}. On the other hand, straightforward observations of K and λ_{max} can afford rapid and meaningful information concerning complex formation.

The relationship between the energy corresponding to the λ_{max}, the electron affinity of the acceptor and the ionization potential of the donor have already been discussed. Thus, the larger the E_A of the acceptor and the greater the electron donating capacity of the donor (small I_p) the lower will be the energy of the charge-transfer process responsible for light absorption, i.e. the longer the wavelength.

In Table 3 are listed the λ_{max} values for the first absorption band of a series of complexes of polycyano acceptors with several methylated benzene and condensed polynuclear hydrocarbon donors in dichloromethane. λ_{max} values for the chloranil and benzoquinone in CCl_4 are also included. It should be noted that direct comparison is justified only for data in identical solvents since it will be shown that the position of the λ_{max} for a given complex is affected by the solvent.

* In this technique the compound under study is introduced as a gas at low pressure (10^{-2}–10^{-4} mm Hg) into a chamber having a coaxial filament–grid–anode assembly mounted in a solenoidal coil. In the presence of a magnetic field, electrons from the heated filament are captured by the grid while heavier ions pass to the anode. The logarithm of the electron–ion current ratio is plotted against the reciprocal of the filament temperature and the slope of the plot gives the electron affinity at the mean of the temperature range investigated. This apparent electron affinity is then corrected to 0°K by an appropriate mathematical expression.

TABLE 3. λ_{max} for polycyano acceptors with hydrocarbon donors[a].

Acceptor	Benzene	Toluene	Mesitylene	Durene	Hexamethylbenzene	Naphthalene	Anthracene	Phenanthrene	Pyrene[b]	Reference
Hexacyanobutadiene									870 (605)	22
Tetracyanobenzoquinone									1129 (617)	29
2,3-Dichloro-5,6-dicyanobenzoquinone			518	595	629				(545)	17, 30
2-Chloro-5,6-dicyanobenzoquinone				565					773 (513)	17
2,3-Dicyanobenzoquinone			477	543	581				718 (478)	30
2,6-Dicyanobenzoquinone									703 (470)	29
Tetracyanoethylene	384	406	461	480	545	550		540	724	4
Chlorotricyanoethylene		334		410						31
Tricyanoethylene		317		365						31
Phenyltricyanoethylene				430	440				510	32
Tetracyanoquinodimethane				555	590				730	17
9-Dicyanomethylene-2,4,7-trinitrofluorene							662	500	650	13
2-Dicyanomethylene-1,3-indanedione					510	475	647	462	615	25
Hexacyanobenzene	343	368	425	429	505	480	624	466	637	33
Pentacyanobenzene			361	382	432	374	504		511	33
1,2,4,5-Tetracyanobenzene[c]						372			460	29
Pentacyanopyridine	338	358	410	418	488	469	603	455	613	33
2,3,5,6-Tetracyanopyridine				356	397		525		519	33
2,3,4,5-Tetracyanopyridine				367	418		480		487	33
3,4,5-Tricyanopyridinium perchlorate[c]					505				620	32, 34
Chloranil[d]	347			496	518	478	625	463	601	1
Benzoquinone[d]	284	311	344			367	457		448	1

[a] Dichloromethane solvent at 22–25° except where noted.
[b] The value in parentheses refer to the second λ_{max}.
[c] Acetonitrile solvent.
[d] Carbon tetrachloride solvent.

For a given class of donors, e.g. the methylated benzenes, the λ_{max} increases as the degree of methylation or increased π-donor propensity increases. This parallels the lowering of the ionization potentials from benzene to hexamethylbenzene[4]. Altering the class of donor to the condensed polynuclear hydrocarbons gives another set of λ_{max} values related to the changed π-orbital system of the donor, its size and geometry. Thus a given acceptor with naphthalene gives a λ_{max} considerably greater than with mesitylene even though naphthalene has the greater ionization potential (8·3 ev versus 8·14 ev[4]). Geometry effects in the donor are reflected in complexes with anthracene versus phenanthrene, the angular isomer phenanthrene producing a much lower λ_{max}. Among the most often-used polynuclear hydrocarbon donors, the highly symmetrical pyrene (**6**) gives complexes with very high λ_{max} values.

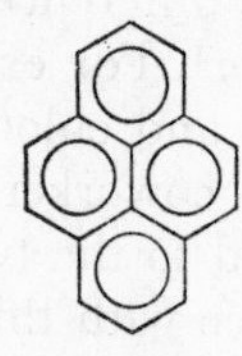

(**6**)

These general considerations are, of course, not unique to complexes derived from polycyano compounds. What is unique about polycyano compounds is their singularly high electron affinities shown in previous sections of this review and further reflected in the long wavelengths of the electronic absorption maxima of their complexes. In any given series, substituted quinones, ethylenes or pyridines, the higher the degree of cyano-substitution, the greater the acceptor strength, as might be expected from the increased depletion of electron density in the parent nucleus by the powerfully electron-withdrawing cyano groups. The cyanoquinone complexes in particular exhibit bands at very long wavelengths, λ_{max} with any given donor being considerably higher than that from the analogous acceptor chloranil. A comparison of pentacyano*benzene* and pentacyano*pyridine* is of interest, since in all cases the pyridine produces the longer wavelength absorption; this augmentation reveals the effect of the electronegative heterocyclic nitrogen atom assisting in electron depletion of the ring π-orbital. The ring nitrogen, however, is not as effective as the additional cyano substituent in hexacyanobenzene. The isomeric tetracyanopyridines show an interesting position effect, the symmetrical 2,3,5,6-isomer affording longer wavelength absorption with the polynuclear hydrocarbons and the less-symmetrical 2,3,4,5-isomer longer wavelength absorption with the methylated benzenes. Finally, the imposition of a permanent positive charge on the heterocyclic nitrogen of 3,4,5-tricyanopyridinium perchlorate is more effective in π-electron withdrawal than the additional two cyano groups in the neutral pentacyano compound. The very high λ_{max} of the hexacyanobutadiene–pyrene complex implies again a very high electron affinity of this acceptor as was also inferred from polarographic reduction data.

Many complexes exhibit two or more charge-transfer bands and the second band, of lower wavelength, is listed in Table 3 for pyrene complexes. The existence of two such bands has been attributed to the existence of two closely located occupied orbitals in the donor so that the energy difference between the absorption bands ($\Delta\nu_{CT}$) is about equal to the energy difference between the two highest occupied orbitals of the donor[1]. For example, in Table 4 the $\Delta\nu_{CT}$ values for tetracyanoethylene and chloranil are closely similar with any given donor. Iwata and coworkers[14] have suggested that this explanation cannot be applied to the two bands observed for tetracyanobenzene complexes, since with this acceptor $\Delta\nu_{CT}$ is relatively constant, irrespective of the donor. Furthermore, the observed energy difference for the tetracyanobenzene complexes coincides closely with the calculated energy difference between the two lowest vacant orbitals in the *acceptor* thus leading to a corollary explanation of multiple banding in some complexes. The relative constancy of $\Delta\nu_{CT}$ values for complexes of pyrene with several other acceptors (Table 4) supports the Briegleb explanation[1], but more data would be desirable.

The effect of solvent changes on λ_{max} for complexes has already been mentioned and Table 5 clearly shows such effects with polycyano complexes. Insufficient examples exist to allow detailed correlations to be made, but in general the λ_{max} differences are not large with a given complex among the closely related chlorinated hydrocarbons, although highly polar solvents can affect significant changes. Solubility considerations may of course dictate the solvent to be used, but it is desirable that measurements in several solvents be made in order to facilitate comparison with data already available.

B. Equilibrium Constants

The equilibrium constants (K) for 1:1 complex formation with a variety of polycyano acceptors are cited in Table 6. To the extent

Table 4. Frequency differences ($\Delta\nu_{CT}$ in cm^{-1}) between first and second charge-transfer bands.

Acceptor	Hexamethyl-benzene	Fluorene	Naphthalene	Pyrene	TMPD[a]	*p*-Dimethoxy-benzene	Reference
1,2,4,5-Tetracyanobenzene	6,400		6,300		6,400		14
Tetracyanoethylene		6,500	5,200	6,500	13,100	10,200	1
Chloranil			5,100	6,500		10,600	1
2,3-Dicyanobenzoquinone				6,900			29
2,5-Dicyanobenzoquinone				7,100			29
2,6-Dicyanobenzoquinone				8,400			35
2,3-Dicyano-5-nitro-1,4-naphthoquinone				6,200			35
Hexacyanobutadiene				5,000			22

[a] *N,N,N′,N′*-Tetramethyl-*p*-phenylenediamine.

Table 5. Solvent effects on charge-transfer absorption maxima.

Complex	Solvent	λ_{max} (mμ)	Reference
1,2,4,5-Tetracyanobenzene/hexamethylbenzene	$CHCl_3$	427	36
	CH_3OH	399	36
1,2,4,5-Tetracyanobenzene/naphthalene	$CHCl_3$	405	36
	CH_3CN	372	29
1,2,4,5-Tetracyanobenzene/pyrene	$CHCl_3$	495	37
	CH_3CN	460	29
	CH_3OH	449	36
2,3-Dicyanobenzoquinone/hexamethylbenzene	CCl_4	573	30
	CH_2Cl_2	581	30
2,3-Dichloro-5,6-dicyanobenzoquinone/hexamethylbenzene	CCl_4	615	30
	CH_2Cl_2	629	30
2,3-Dicyano-5-nitro-1,4-naphthoquinone/pyrene	CH_2Cl_2	760	35
	CH_3CN	725	35

that equilibrium constants measure the stability of the complexes, the order of increasing K for a given series of measurements reflects in general the acceptor capacities, as inferred from previous sections of this review. With respect to solvents we see here that marked changes in observed K's result from what might be thought to be relatively minor changes in solvent, so that in comparing K's it is mandatory that reference be made to the same solvent[2].

We have commented previously on the differences in magnitude and dimensions of K determined from molar concentration data versus mole fractions and some confusion on this point exists in the literature. For example, to compare the data with K's expressed in l/mole, Peover[21] has appropriately taken the dimensionless K's of Merrifield and Phillips[4] and divided by the molar volume of the solvent; this conversion was also made for the purposes of Table 6. In other instances those same data have been used incorrectly, without conversion, for direct comparison with K's expressed in l/mole (see Reference 2, p. 28).

In any case, the K values in Table 6 again reflect the greater stability of the polycyanoquinone complexes and thus the high π-acceptor capacity of the quinones. However, anomalies relative to previous criteria are evident. With respect to polarographic reduction and λ_{max} with pyrene, hexacyanobutadiene was inferred to be a much stronger acceptor than, for example, tetracyanoquinodimethane, and yet the K with pyrene ranks the butadiene considerably below the quinodimethane. A similar inconsistency in the

TABLE 6. Equilibrium constants K for complex formation[a].

Acceptor	Mesitylene	Durene	Hexamethylbenzene	Pyrene	Dimethylaniline	Reference
2,3-Dichloro-5,6-dicyanobenzoquinone	19·2[b]	72·3[b]	574[b] 97·5			30
2,3-Dicyanobenzoquinone	4·9[b]	12·3[b]	73·8[b] 13·4			30
2,3-Dicyano-5-nitro-1,4-naphthoquinone				12·0 5·2[c]		35
Tetracyanoethylene	1·1	3·5	16·7 35[d]	1·9	8·5[d,e]	4, 21, 38
Chlorotricyanoethylene		0·6				31
Tricyanoethylene		0·4				31
Phenyltricyanoethylene		0·6	0·8	0·4	0·4[c,f]	32
Tetracyanoquinodimethane		0·4	0·9	5·0		17
Hexacyanobutadiene				2·7		22
Chloranil			3.2 2·1[d]	1·5		21

[a] In units of l/mole; where the original literature cited dimensionless K's based on mole fractions, the value has been converted by dividing by the molar volume of the solvent. Temperature 22–25°, solvent dichloromethane except where noted.
[b] Carbon tetrachloride solvent.
[c] Acetonitrile.
[d] Chloroform.
[e] 48°.
[f] *N,N*-Dimethyl-*p*-toluidine.

relative order of tetracyanoethylene versus tetracyanoquinodimethane is seen using K or λ_{max} as criteria with the structurally different donors hexamethylbenzene and pyrene[6]. This again emphasizes the need for caution in assigning relative acceptor strengths without reference to the particular criteria and geometry or chemical constitution of interacting species.

C. Thermodynamic Considerations

Analysis of the temperature dependence of equilibrium constants affords estimations of heats of formation ($\Delta H°$), free energies of formation ($\Delta F°$) and entropy changes ($\Delta S°$) which accompany complex formation[1,2]. In general these are small compared to those encountered in covalent bond-forming reactions. We do not intend to consider these properties in detail, but by way of illustration the heats of formation of several cyano complexes and of chloranil are listed in Table 7. Since equilibrium constants, and thus $\Delta H°$, are

TABLE 7. Heats of formation ($-\Delta H$, kcal/mole).

Donor	2,3-Dicyano-benzoquinone	Tetracyano-ethylene	1,2,4,5-Tetra-cyanobenzene	Chlora-nil	Refer-ence
Benzene		2·3[a]	0·4[b]	−1·7[c]	1, 4, 14
Mesitylene	5·1[c]	4·5[a]			4, 30
Durene	6·9[c]	5·1[a]		4·4[c]	1, 4, 30
Hexamethylbenzene	7·8[c]	7·8[c]		5·4[c]	30, 40, 39
N,N-Dimethylaniline		6·9[d]		5·0[c]	38, 40
N,N,N′,N′-Tetramethyl-*p*-phenylenediamine			3·6[d]		36

[a] Solvent CH_2Cl_2.
[b] Solvent $C_2H_4Cl_2$.
[c] Solvent CCl_4.
[d] Solvent $CHCl_3$.

very sensitive to solvent changes some of the values in Table 7 are not directly comparable, but, within a self-consistent series of measurements, $\Delta H°$ should reflect increased strength of interaction in complex formation. This is seen to be the case for a given acceptor with a series of donors of increasing donor capacity. The values for 2,3-dicyanobenzoquinone and chloranil are directly comparable since they were determined in the same solvent, and here we see further confirmation of the greater acceptor strength of the cyano-quinone.

V. ION-RADICAL FORMATION

A. General

The possibility of complete electron transfer from donor to acceptor as depicted in equation (1) (tantamount to a one-electron oxidation–reduction reaction) implies the possibility of formation of cation radicals and anion radicals from donor and acceptor, respectively. Such ion-radical formation could occur either in solution or in the solid state and should be detectable by magnetic susceptibility or electron spin resonance (e.s.r.) measurements[41]. If there were no exchange interaction between the ion-radical pairs $D^{\dot{+}}A^{\dot{-}}$ then the individual components would exist in double electronic states such that the paramagnetic or e.s.r. signal intensity (I) would follow simple Curie-law temperature dependence (equation 11).

$$I \propto 1/T \tag{11}$$

If, however, there is extensive exchange interaction between pairs so as to give a singlet ground state and triplet excited state separated by the exchange energy J, then the intensity would follow equation (12)

$$I \propto 1/T[\exp(+J/kT) + 3] \tag{12}$$

At low temperatures the intensity would then be small or zero and would increase with temperature as the paramagnetic triplet state becomes thermally populated, reaching a maximum at a temperature of approximately J/k, and should then follow Curie-law dependence above that temperature. Conversely, a triplet ground state with excited singlet state should show greatest signal intensity at low temperature and decrease with increasing temperature more rapidly than Curie-law behaviour according to equation (13).

$$I \propto 1/T[\exp(-J/kT) + 3] \tag{13}$$

B. Anion Radicals in Solution

In early work on the reaction of tetracyanoethylene (TCNE) with N,N-dimethylaniline (**7**), it was observed that the reaction mixture first assumed a blue colour (λ_{max} 650 mμ) which slowly disappeared with the formation of a red colour characteristic of the final reaction product, the tricyanovinyl dye **8** (λ_{max} 520 mμ)[42]. The blue colour

$(NC)_2C{=}C(CN)_2$ + $C_6H_5N(CH_3)_2$ (**7**) ⟶ [intermediate] ⟶

$(CH_3)_2N{-}C_6H_4{-}C(CN){=}C(CN)_2$ (**8**)

of the intermediate was ascribed to the formation of a π-complex and Phillips[17] reinvestigated the intermediate by e.s.r. using N,N-dimethyl-p-toluidine to suppress the substitution reaction. The e.s.r.

spectrum of the solution in tetrahydrofuran exhibited a well-resolved nine-line pattern characteristic of the TCNE anion radical ($TCNE^{\cdot}$) superimposed on a broad resonance ascribed to donor cation radical. Subsequent detailed analysis of the nine-line spectrum showed the line intensity ratios to be very close to those expected from splitting by four equivalent ^{14}N (nuclear spin $I = 1$) nuclei[43]. A variety of other reducing agents such as iodide, cyanide and various metals will also serve to reduce TCNE to the anion radical providing a variety of stable anion radical salts[43-45]. Among the more interesting donors, the organometallic ferrocene (**9**) was found to interact with TCNE to give a complex which in solution exhibited a strong e.s.r. signal and was thus assigned the salt structure ferricenium $TCNE^{\cdot}$ (**10**) by Webster and coworkers, even though in the

Fe + TCNE ⟶ Fe^{+} $TCNE^{\cdot}$

(**9**) (**10**)

solid state the product was diamagnetic[44]. However, subsequent x-ray work by Adman and coworkers[46] and Mössbauer spectroscopy by Collins and Pettit[47] showed that in the solid state the compound is the molecular complex and not the charge-transfer salt. A variety of other metallocenes do, however, form salt-type structures with TCNE[48] (see below).

A number of other polycyano compounds have been converted to anion radicals either electrolytically[25,27,49,50] or chemically[6,13,22,26,44] and their solution e.s.r. spectra determined. Thus the anion radical of 7,7,8,8-tetracyanoquinodimethane ($TCNQ^{\cdot}$) in THF solution exhibits over forty lines compared with the forty-five lines expected on the basis of hyperfine interaction between the unpaired electron and the ^{1}H and ^{14}N nuclei[6]. Electrolytic reduction of 2-dicyano-methylene-1,3-indanedione (**5**) gave a twenty-five-line spectrum arising from splitting by two ^{14}N atoms and four equivalent protons[25]. The solution e.s.r. spectrum of hexacyanobutadiene anion radical (**11**) gave a forty-five line pattern consisting of nine major lines arising from splitting by the terminal ^{14}N nuclei each being further split into five lines by the internal ^{14}N nuclei[22].

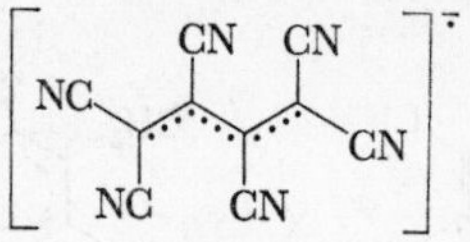

$$\left[\begin{array}{c} \text{NC} \quad \text{CN} \quad \text{CN} \\ \text{NC} \quad \text{CN} \quad \text{CN} \end{array} \right]^{\overline{\cdot}}$$

(11)

The e.s.r. spectrum of the $TCNQ^{\overline{\cdot}}$ ion radical in *aqueous* solution is of special interest. In dilute solution, the signal is strong as a result of absorption by the monomeric anion radical; in concentrated solution, the signal is very weak and this attenuation is ascribed to formation of radical dimers (**12**)[6,51]. The two forms have distinctly

$$2\,TCNQ^{\overline{\cdot}} \rightleftharpoons [TCNQ]_2^{\overline{\cdot}\,\overline{\cdot}}$$

(12)

different electronic absorption spectra and their designation as monomer and dimer was verified by equivalent conductance measurements[52]. The signal strength of the dimer solution increases with increasing temperature in a manner which suggests singlet–triplet behaviour as discussed above. This solution behaviour of the anion radical parallels the behaviour of the diaminodurene cation radical[6,53].

C. Ion Radicals in the Solid State

The formation of solid, often crystalline, complexes by reaction of appropriate acceptors and donors is a common phenomenon and the possibility of their existing as charge-transfer salts has long been recognized[8]. E.s.r. examination of such complexes and temperature dependence of signal intensities was first carried out by Bijl, Kainer and Rose-Innes working with complexes derived from substituted quinones and *p*-phenylenediamines[54]. In all cases, they found strict Curie-law dependence (equation 11) indicating that the ion radicals existed in a doublet electronic ground state. Subsequently, Singer and Kommandeur observed singlet–triplet behaviour (equation 12) in the perylene–iodine complex[55]. In examining the solid 1:1 chloranil–diaminodurene complex **13**, Chesnut and Phillips found that the e.s.r. behaviour could best be described by the singlet–triplet model, but with the solid containing as little as 0·04% of a doublet-state impurity[56]. Thus, the apparent Curie-law behaviour found by Bijl and coworkers was ascribed in part to the presence of dominating doublet state impurities.

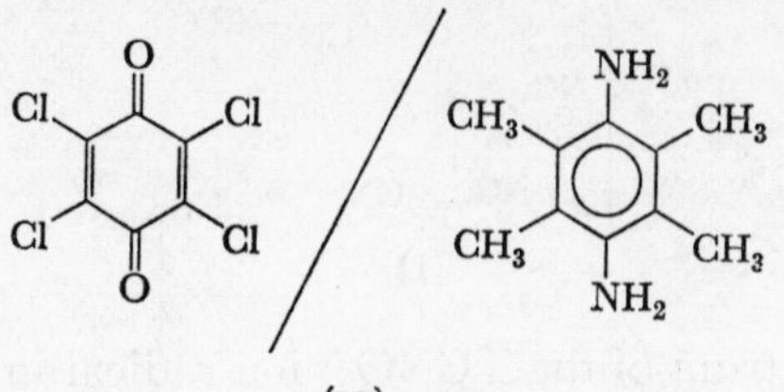

(**13**)

In any case, only a very small percentage of donor–acceptor molecules exist as paramagnetic species in complexes such as **13**; quantitative e.s.r. measurements showed only about 10^{21} unpaired electrons per mole of donor–acceptor pairs. Small spin concentrations are also found in a variety of polycyano complexes such as those listed in Table 8.

TABLE 8. Electron spin concentrations of solid polycyano complexes.

Acceptor[a]	Donor	Unpaired electrons per mole[b]	Reference
TCNE	3,3′-Dimethoxybenzidine	8×10^{19}	57
TCNQ	*p*-Phenylenediamine	4×10^{20}	6
	N,N-Dimethyl-*p*-phenylenediamine	4×10^{21}	6
	1,5-Diaminonaphthalene	2×10^{19}	6
2,3-Dichloro-5,6-	Dibenzo[c,d]phenothiazine	10^{20}	58
dicyanobenzoquinone		5×10^{20}[c]	58
	Durene	—[d]	30
	Pyrene	—[d]	30

[a] Acceptor: donor mole ratio 1:1 unless otherwise noted.
[b] E.s.r. absorption intensity measured versus diphenylpicrylhydrazyl as standard. Accuracy estimated to be ±25%.
[c] Acceptor: donor ratio 1:2.
[d] Resonance absorption observed, but not evaluated quantitatively.

Table 8 illustrates that irrespective of the nature of acceptor and donor the unpaired electron concentrations are very similar. The case of the 2,3-dichloro-5,6-dicyanoquinone(DDQ)–pyrene complex is of interest since the tetracyanoethylene analogue shows no e.s.r. absorption at room temperature[57]. This reflects the very high acceptor strength of the quinone, assuming that the observed resonance was not an impurity artifact. It would be valuable to have e.s.r. data on the solid hexacyanobutadiene–pyrene and –hexamethylbenzene complexes[22] since the butadiene has a very high acceptor strength; unfortunately such data are not available. In solution, these complexes have electronic absorption spectra showing the presence of the hexacyanobutadiene anion radical, but, as we have seen with the ferrocene–TCNE complex, extrapolation from solution data to the solid state can be misleading.

Stable anion radical salts represent the extreme of the charge-transfer phenomenon, and such salts are now known to be readily accessible from polycyano acceptors because of the high degree of resonance stabilization resulting from the multiplicity of conjugated or cross-conjugated cyano groups. For example, stable, solid salts of $TCNE^{\bar{\cdot}}$, $TCNQ^{\bar{\cdot}}$ and hexacyanobutadiene ($HCBD^{\bar{\cdot}}$) anion radicals can be isolated with appropriate precipitating cations[6,22,44,45]. Metallocenium salts of $TCNE^{\bar{\cdot}}$ and $DDQ^{\bar{\cdot}}$ have been prepared[48], as has the tropylium salt of $DDQ^{\bar{\cdot}}$ [57]. The derivatives of TCNQ constitute a unique series of compounds, since this acceptor will form typical π-molecular complexes with a variety of donors ranging in complexity from benzenoid hydrocarbons to planar metal chelates; it also forms simple anion radical salts of constitution $M^+TCNQ^{\bar{\cdot}}$ where the cation may be organic, inorganic or organo-inorganic; finally it forms a series of complex salts of types **14** and **15** all of which can be isolated in the solid state[6]. Type **14** constitutes,

$$M^+(TCNQ)^{\bar{\cdot}} \qquad M_2{}^+(TCNQ)(TCNQ^{\bar{\cdot}})_2$$

(**14**) (**15**)

in effect, a simple anion radical salt containing an extra molecule of neutral TCNQ; from the formulation one could infer that the electron is shared by two TCNQ moieties, but e.s.r. and x-ray studies[56,59] would suggest the molecular formulation $(M^+)_2(TCNQ)_4^{\bar{\cdot}\bar{\cdot}}$ to account for the pairwise spin correlation implied by the singlet–triplet e.s.r. behaviour. Type **15** is similar to **14** but the formulation indicates one mole of neutral TCNQ combined with two moles of simple salt; again singlet–triplet e.s.r. behaviour was observed with this type[56,59] and detailed x-ray analysis of the $(Cs^+)_2(TCNQ)_3^{\bar{\cdot}\bar{\cdot}}$ salt showed that the TCNQ moieties are packed face-to-face forming somewhat irregular columns of triads[59,60]. Other extensive x-ray[61–63] and e.s.r.[64,65] studies have been undertaken on TCNQ salts in relation to their unique properties as organic semiconductors.

VI. ELECTRICAL PROPERTIES OF POLYCYANO COMPLEXES

One of the most intriguing aspects of solid organic molecular complexes is that of electrical conductivity and their behaviour as semiconductors. Complexes of polycyano compounds have been of singular interest in this regard because some of them have been found to exhibit extremes of the properties related to semiconduction in organic compounds and have been a fruitful subject for elucidation of its theoretical aspects[66].

Conductivity in semiconductors is often referred to by the reciprocally related term 'volume electrical resistivity' (ρ), expressed in ohm cm. Resistivity is operationally determined by equation (14),

$$\rho = \frac{R \times A}{L} \tag{14}$$

where R is the electrical resistance (ohms), A is the cross-sectional area (cm^2) and L the length (cm) of the specimen being examined. Resistivity may further be defined by equation (15), where n is the

$$\rho = \frac{1}{n\mu e} \tag{15}$$

number of charge carriers, μ is their mobility and e the charge on each carrier. For insulators and semiconductors carrier production is a thermally activated process and so their resistivities show exponential temperature dependence according to equation (16),

$$\rho = \rho_0 \exp (E/kT) \tag{16}$$

where E is the activation energy, k is Boltzmann's constant and T is absolute temperature. A major distinction between insulators and semiconductors lies in the magnitude of their resistivities, which typically are of the order of 10^{14}–10^{22} ohm cm for insulators and 10^{-2}–10^{9} ohm cm for semiconductors. For intrinsic semiconductors the activation parameter is usually described by the energy gap, which is equivalent to twice the activation energy; for example, the typical intrinsic semiconductor, germanium, has a room-temperature resistivity of about 60 ohm cm and an energy gap of about 0·7 electron volts (ev), corresponding to an activation energy of 0·35 ev. Throughout this discussion the activation parameter will be expressed in terms of the activation energy. In contradistinction to semiconductors, metallic conductors exhibit room-temperature resistivities of the order of 10^{-6} ohm cm and their resistivities increase (conductivities decrease) with temperature.

Among organic derivatives, complexes of perylene with bromine or iodine have very low resistivities, in the range of 1–10 ohm cm, but such complexes are rather unstable[67]. The stable complexes of substituted quinones with *p*-phenylenediamines are highly conductive[68], e.g. the 1:1 chloranil–diaminodurene complex has a room-temperature resistivity of $\sim 3 \times 10^3$ ohm cm, with an activation energy for conduction of 0·25 ev[69] *. For comparison, the resistivities and conduction activation energies for π-molecular complexes derived from a number of polycyano π-acceptors and assorted π-donors are listed in Table 9. Clearly the resistivities run the gamut from values typical of insulators to values near the lower limit for semiconductors and there is no obvious correlation with either the structure of the acceptor or the donor. Values for the series of *p*-phenylenediamines with tetracyanoquinodimethane tend to suggest that within this group the greater the opportunity for coplanarity of the amino groups and the phenylene nucleus, the lower is the resistivity (i.e. the smaller the acceptor–donor separation, the greater the π-electron interaction); but even here the values for the *N*,*N*-dimethyl and *N*,*N*,*N′*,*N′*-tetramethyl derivatives are inconsistent. Of particular note are the values for the anomalous 1:2 and 1:3 complexes of the 2,3-dihalo-5,6-dicyanoquinones with dibenzophenothiazine which are the lowest reported for π-complexes. Matsunaga[58] has drawn attention to the analogy of these with respect to the highly conductive complex salts of tetracyanoquinodimethane (see below), but in the absence of further information on the state of ionization of the donor and acceptor components in the phenothiazine derivatives the analogy is at best a tenuous one; in the tetracyanoquinodimethane salts the 'donor' portion of the complex is in fact a stable cation while the anion portion is an anion radical complexed with its neutral counterpart.

In addition to monomolecular donors, a number of complexes of polycyano acceptors with polymeric donors have been investigated.

* This value for the resistivity was determined on a single crystal in the long needle-axis direction. When crystal size precludes single crystal measurements, resistivities are determined on mechanically compacted samples of polycrystalline material. Such compaction resistivities are higher than single crystal values by factors of about 10–100 because of electrical resistance conferred by imperfect interparticle contact.

TABLE 9. Volume electrical resistivities and conduction activation energies of solid polycyano π-complexes[a].

Acceptor	Donor	ρ^b (ohm cm)	E (ev)	Reference
Tetracyanoethylene	Naphthalene	3×10^{15}	1·24	70
	Hexamethylbenzene	4×10^{13}	0·58	70
	Ferrocene	3×10^{12}		48
		5×10^{9}		45
	Cobaltocene[c]	10^{12}	2·6	48
	Bis(tetrahydroindenyl)iron[c]	10^{9}		48
	Violanthrene	5×10^{8}		70
	3,3′-Dimethoxybenzidine	$4{\cdot}5 \times 10^{5}$		57
Hexacyanobutadiene	Hexamethylbenzene[d]	7×10^{10}		22
	Pyrene[d]	3×10^{9}		22
7,7,8,8-Tetracyano-quinodimethane	Pyrene	10^{12}		6
	Anthracene	10^{11}		6
	1,5-Diaminonaphthalene	10^{9}		6
	Diaminodurene	10^{9}		6
	N,N-Dimethyl-*p*-phenylenediamine	2×10^{9}		6
	N,N,N′,N′-Tetramethyl-*p*-phenylenediamine	10^{6}		6
	2-Methyl-*p*-phenylenediamine	3×10^{5}		6
	N,N′-Dimethyl-*p*-phenylenediamine	2×10^{4}		6
	p-Phenylenediamine	3×10^{3}	0·28	6
2,3-Dichloro-5,6-dicyanoquinone	Cobaltocene[c]	3×10^{13}	0·9	48
	Ferrocene[c]	3×10^{10}	0·7	48
	Tropylium[c]	5×10^{10}		57
	Perylene	10^{6}		71
	Dibenzo[c,d]phenothiazine	5×10^{3}		58
	Dibenzo[c,d]phenothiazine[d]	17	0·1	58
2,3-Dibromo-5,6-dicyanoquinone	Dibenzo[c,d]phenothiazine	10^{8}		58
	Dibenzo[c,d]phenothiazine[e]	240	0·13	58

[a] Acceptor: donor mole ratios are 1:1 unless otherwise noted.
[b] For polycrystalline compactions.
[c] These are ion radical salts, fully ionized in the solid state.
[d] Acceptor: donor mole ratio is 1:2.
[e] Acceptor: donor mole ratio is 1:3.

These included hydrocarbon polymers[72] and poly-*p*-dimethylamino-styrene[73]. The hydrocarbon polymers gave complexes with resistivities in the range 10^{13}–10^{14} ohm cm, not appreciably different from the uncomplexed donors. The polyamine gave complexes with resistivities from 10^{10} to 10^{15} ohm cm depending upon the nature of the acceptor.

Although the simple anion radical salts of $TCNQ^{\overline{\cdot}}$ are not molecular complexes they lead to a family of complex salts which can be regarded as derivatives of an anionic moiety consisting of the anion radical $TCNQ^{\overline{\cdot}}$ complexed with its neutral counterpart TCNQ. Such complexes are unprecedented and remain unique with respect to composition and, among organic compounds, to their electrical properties. Many of these $TCNQ^{\overline{\cdot}}$ derivatives are obtainable in macrocrystalline form such that resistivities are measurable in three major crystal directions; these measurements have revealed a marked anisotropy of resistivity with respect to crystal axes[6,74]. Anisotropy of resistivity has also been observed in complexes of the quinone–aromatic amine type but in these the resistivities vary only by a factor of about 10^2 relative to crystal direction[75], whereas with $TCNQ^{\overline{\cdot}}$ salts the values vary in some instances by as much as 10^4. The availability of these $TCNQ^{\overline{\cdot}}$ derivatives in macrocrystalline form has permitted considerably detailed study of properties related to the mechanism of conduction in organic solids[66]. It is not the purpose of this review to consider the physics of these conduction processes, but only to cite examples illustrating the magnitude of the values which have been observed and to make some qualitative observations with respect to composition and conductivity. In Table 10, are collected resistivity data and conduction activation energies for a number of $TCNQ^{\overline{\cdot}}$ salts, both simple and complex.

With organic cations it is generally true that the complex salt is much more highly conductive than the simple 1:1 salt, as illustrated by the triethylammonium derivatives. This is also true for a great variety of salts which are not listed in Table 10. The macrocrystalline forms of the triethylammonium and methyltriphenylphosphonium complex salts show very clearly the anisotropy of conduction in these salts, the resistivities varying by factors of about 10^3 and 10^4, respectively, with regard to the most conductive and least conductive crystal directions. In conjunction with x-ray data, it has been shown[74] that the most conductive direction is that perpendicular to the faces of the face-to-face stacked $TCNQ^{\overline{\cdot}}TCNQ$ moieties; thus these are inferred to be the predominant mediators of charge movement. The simple morpholinium salt illustrates the absence of anisotropy of conduction in this type of salt. Within the group of complex salts of the type $M^+(TCNQ)_2^{\overline{\cdot}}$, (where M^+ is a generalized cation) with aromatic cations, it is generally true that the nature of the cation does not greatly affect

TABLE 10. Resistivities of $TCNQ^{\overline{\cdot}}$ salts[a].

Salt	Resistivity[b] (ohm cm)	E^c (ev)
$Cs^+TCNQ^{\overline{\cdot}}$	3×10^4 [c]	0·15
$(Cs^+)_2(TCNQ)(TCNQ^{\overline{\cdot}})_2$	500; 6×10^4; 6×10^4; (9×10^4)[c]	0·36
$Cu^+TCNQ^{\overline{\cdot}}$	2×10^2 [c]	0·13
$Cu^{2+}(TCNQ^{\overline{\cdot}})_2$	2×10^2 [c]	0·12
Morpholinium $TCNQ^{\overline{\cdot}}$	6×10^8; 6×10^8; 6×10^8; (10^9)[c]	
$(Morpholinium)_2(TCNQ)(TCNQ^{\overline{\cdot}})_2$	5×10^3; (10^5)[c]	
N-Methylphenazinium $TCNQ^{\overline{\cdot}}$	7×10^{-3}; (0·5)[c]	
N-Methylphenazinium $(TCNQ)_2^{\overline{\cdot}}$	1·4 [c]	
N-Ethylphenazinium $TCNQ^{\overline{\cdot}}$	10^9; (10^{11})[c]	
N-Ethylphenazinium $(TCNQ)_2^{\overline{\cdot}}$	1·2 [c]	
Triethylammonium $TCNQ^{\overline{\cdot}}$	10^9 [c]	
Triethylammonium $(TCNQ)_2^{\overline{\cdot}}$	0·5; 40; 10^3; (20)[c]	0·13
Quinolinium $(TCNQ)_2^{\overline{\cdot}}$	10^{-2}	0·013
Methyltriphenylphosphonium $(TCNQ)_2^{\overline{\cdot}}$	60; 600; 10^5	0·25

[a] Values taken from References 6 and 76.
[b] Determined on single crystals except where noted; where three values are given they correspond to measurements perpendicular to principal crystal faces, the lowest value arbitrarily cited first. Where only one value is given, it corresponds to the long needle direction of the crystal.
[c] Determined on polycrystalline compaction.

the magnitude of the resistivity. It is also generally true that simple salts are much less conductive than the complex salts, but several notable exceptions to these generalizations are known. Thus, the simple *N*-methylphenazinium compound is the most highly conductive of the compounds listed, indeed it is the most highly conductive organic compound known. On the other hand, the simple *N*-ethylphenazinium salt has a low conductivity typical of usual simple salts; the presence of the ethyl versus methyl group must confer, among other things, a major difference in crystal structure in a manner that disrupts the conduction path[76]. X-ray data have shown that in the *N*-methyl compound the $TCNQ^{\overline{\cdot}}$ ions are stacked face-to-face in fairly regular columns; unfortunately x-ray data are not available on the ethyl homologue. However, the higher density of the methyl derivative suggests a more efficient molecular packing within its crystal.

The copper salts formulated as simple salts of cuprous and cupric ions further complicate attempts at generalization. The cuprous salt is quite highly conductive with respect to other simple salts.

The alleged cupric salt, which is very similar in electrical behaviour, could conceivably be formulated as the complex cuprous salt $Cu^{+}(TCNQ)_2^{\cdot-}$. However, other criteria such as magnetic susceptibility are not consistent with either formulation, so the existence of a complicated solid-state equilibrium involving Cu^{+}, Cu^{2+}, $TCNQ^{\cdot-}$ and neutral TCNQ has been proposed to account for the properties[6]. In any event, knowledge of the detailed electronic structure of many of these compounds remains vague and will require further study.

Finally, it should be noted that $TCNQ^{\cdot-}$ derivatives of polymeric quaternary ammonium salts have also been prepared and these also show increased conductivities resulting from the inclusion of neutral TCNQ, consistent with the behaviour of derivatives of monomeric cations[77,78].

In summary, these $TCNQ^{\cdot-}$ anion radical derivatives exemplify some of the structural requirements and attendant properties which may be sought or expected in the continuing search for practical organic semiconductors.

VII. ACKNOWLEDGEMENTS

I wish to express my thanks to Professor K. Wallenfels for providing considerable spectroscopic data and to Dr. E. A. Abrahamson and Miss Lucille E. Williams for some of the polarographic data. Helpful discussions with Drs. R. E. Merrifield, W. D. Phillips, P. E. Bierstedt and O. W. Webster are also gratefully acknowledged. Finally, my thanks to Kathleen Rowand for so ably typing the manuscript.

VIII. REFERENCES

1. G. Briegleb, *Electronen-Donator-Acceptor-Complexe*, Springer-Verlag, Berlin, 1961.
2. L. J. Andrews and R. M. Keefer, *Molecular Complexes in Organic Chemistry*, Holden-Day, San Francisco, 1964.
3. T. L. Cairns, R. A. Carboni, D. D. Coffman, V. A. Engelhardt, R. E. Heckert, E. L. Little, E. G. McGeer, B. C. McKusick, W. J. Middleton, R. M. Scribner, C. W. Theobald and H. E. Winberg, *J. Am. Chem. Soc.*, **80,** 2775 (1958).
4. R. E. Merrifield and W. D. Phillips, *J. Am. Chem. Soc.*, **80,** 2778 (1958).
5. D. S. Acker and W. R. Hertler, *J. Am. Chem. Soc.*, **84,** 3370 (1962).
6. (a) L. R. Melby, R. J. Harder, W. R. Hertler, W. Mahler, R. E. Benson and W. E. Mochel, *J. Am. Chem. Soc.*, **84,** 3374 (1962);
   (b) E. M. Kosower in *Progress in Physical Organic Chemistry*, Vol. 3, Interscience, New York, 1965, p. 81.
7. R. S. Mulliken, *J. Am. Chem. Soc.*, **74,** 811 (1952); *J. Phys. Chem.*, **56,** 801 (1952).
8. (a) J. J. Weiss, *Phil. Mag.*, **8,** 1169 (1963);
   (b) J. J. Weiss, *J. Chem. Soc.*, **1942,** 245.

9. H. McConnell, J. S. Ham and J. R. Platt, *J. Chem. Phys.*, **21**, 66 (1953).
10. H. A. Benesi and J. H. Hildebrand, *J. Am. Chem. Soc.*, **71**, 2703 (1949).
11. R. M. Keefer and L. J. Andrews, *J. Am. Chem. Soc.*, **72**, 4677 (1950).
12. M. J. S. Dewar and C. C. Thompson, Jr., *Tetrahedron*, **22**, *Suppl. No. 7*, 97 (1966).
13. T. K. Mukherjee and L. A. Levasseur, *J. Org. Chem.*, **30**, 644 (1965).
14. S. Iwata, J. Tanaka and S. Nagakura, *J. Am. Chem. Soc.*, **88**, 894 (1966).
15. G. Briegleb, *Angew. Chem. Intern. Ed. Engl.*, **3**, 617 (1964).
16. (a) A. Maccoll, *Nature*, **163**, 178 (1949);
(b) M. E. Peover, *Nature*, **191**, 702 (1961).
17. W. D. Phillips, unpublished work.
18. L. E. Orgel, T. L. Cottrell, W. Dick and L. E. Sutton, *Trans. Faraday Soc.*, **47**, 113 (1951).
19. K. Watanabe, *J. Chem. Phys.*, **26**, 542 (1957).
20. M. I. Al-Joboury and D. W. Turner, *J. Chem. Soc.*, **1964**, 4434.
21. M. E. Peover, *Trans. Faraday Soc.*, **60**, 417 (1964).
22. O. W. Webster, *J. Am. Chem. Soc.*, **86**, 2898 (1964).
23. E. A. Abrahamson and L. E. Williams, unpublished work.
24. J. Diekmann, W. R. Hertler and R. E. Benson, *J. Org. Chem.*, **28**, 2719 (1963).
25. S. Chaterjee, *Science*, **157**, 314 (1967).
26. H. D. Hartzler, *Abstracts, International Union of Pure and Applied Chemistry*, July, 1963.
27. H. D. Hartzler, *J. Org. Chem.*, **30**, 2456 (1965).
28. A. L. Farragher and F. M. Page, *Trans. Faraday Soc.*, **63**, 2369 (1967).
29. K. Wallenfels and G. Bachmann, private communication.
30. P. R. Hammond, *J. Chem. Soc.*, **1963**, 3113.
31. C. L. Dickinson, D. W. Wiley and B. C. McKusick, *J. Am. Chem. Soc.*, **82**, 6132 (1960).
32. W. A. Sheppard and R. M. Henderson, *J. Am. Chem. Soc.*, **89**, 4446 (1967).
33. K. Wallenfels and P. Neumann, private communication.
34. K. Wallenfels and W. Hanstein, private communication.
35. K. Wallenfels and R. Kern, private communication.
36. R. Foster and T. J. Thomson, *Trans. Faraday Soc.*, **59**, 2287 (1963).
37. A. S. Bailey, B. R. Henn and J. M. Langton, *Tetrahedron*, **19**, 161 (1963).
38. Z. Rappoport, *J. Chem. Soc.*, **1963**, 4498.
39. G. Briegleb, J. Czekalla and G. Reuss, *Z. Physik. Chem.*, **30**, 333 (1961).
40. G. Briegleb and J. Czekalla, *Z. Elektrochem.*, **58**, 249 (1954).
41. For a review of e.s.r. studies of ion radicals see B. J. McClelland, *Chem. Rev.*, **64**, 301 (1964) and A. Carrington, *Quart. Revs.*, **17**, 67 (1963).
42. B. C. McKusick, R. E. Heckert, T. L. Cairns, D. D. Coffman and H. F. Mower, *J. Am. Chem. Soc.*, **80**, 2806 (1958).
43. W. D. Phillips, J. C. Rowell and S. I. Weissman, *J. Chem. Phys.*, **33**, 626 (1960).
44. O. W. Webster, W. Mahler and R. E. Benson, *J. Am. Chem. Soc.*, **84**, 3678 (1962).
45. O. W. Webster, W. Mahler and R. E. Benson, *J. Org. Chem.*, **25**, 1470 (1960).
46. E. Adman, M. Rosenblum, S. Sullivan and T. N. Margulis, *J. Am. Chem. Soc.*, **89**, 4540 (1967).
47. R. L. Collins and R. Pettit, *J. Inorg. Nucl. Chem.*, **29**, 503 (1967).
48. R. L. Brandon, J. H. Osieki and A. Ottenberg, *J. Org. Chem.*, **31**, 1214 (1966).
49. P. H. Rieger, I. Bernal, W. H. Reinmuth and G. K. Fraenkel, *J. Am. Chem. Soc.*, **85**, 683 (1963).
50. P. H. H. Fischer and C. A. McDowell, *J. Am. Chem. Soc.*, **85**, 2694 (1963).

51. R. H. Boyd and W. D. Phillips, *J. Chem. Phys.*, **43**, 2927 (1965).
52. R. H. Boyd, *J. Chem. Phys.*, **38**, 2529 (1963).
53. L. Michaelis and S. Granick, *J. Am. Chem. Soc.*, **65**, 1747 (1943).
54. D. Bijl, H. Kainer and A. C. Rose-Innes, *J. Chem. Phys.*, **30**, 765 (1959).
55. L. S. Singer and J. Kommandeur, *Bull. Am. Phys. Soc.*, **4**, 421 (1959).
56. D. B. Chesnut and W. D. Phillips, *J. Chem. Phys.*, **35**, 1002 (1961).
57. R. J. Harder, L. R. Melby and W. D. Phillips, unpublished work.
58. Y. Matsunaga, *J. Chem. Phys.*, **42**, 1982 (1965).
59. D. B. Chesnut and P. Arthur, Jr., *J. Chem. Phys.*, **36**, 2969 (1962).
60. C. J. Fritchie, Jr., and P. Arthur, Jr., *Acta Cryst.*, **21**, 139 (1966).
61. A. W. Hanson, *Acta. Cryst.*, **19**, 610 (1965).
62. R. E. Long, R. A. Sparks and K. N. Trueblood, *Acta Cryst.*, **18**, 932 (1965).
63. C. J. Fritchie, Jr., *Acta Cryst.*, **20**, 892 (1966).
64. M. T. Jones, *J. Chem. Phys.*, **40**, 1837 (1964).
65. A. W. Merkl, R. C. Hughes, L. J. Berliner and H. M. McConnell, *J. Chem. Phys.*, **43**, 953 (1965).
66. For a comprehensive discussion of semiconduction in organic materials see F. Gutmann and L. E. Lyons, *Organic Semiconductors*, John Wiley and Sons, New York, 1967.
67. (a) J. Kommandeur and F. R. Hall, *Bull. Am. Phys. Soc., Series II*, **4**, 421 (1959); (b) H. Akamatu, H. Inokuchi and Y. Matsunaga, *Bull. Chem. Soc. Japan*, **29**, 213 (1956).
68. P. L. Kronick and M. M. Labes, *J. Chem. Phys.*, **35**, 2016 (1961).
69. R. G. Kepler, *J. Chem. Phys.*, **35**, 1002 (1961).
70. H. Kuroda, M. Kobayashi, M. Kinoshita and S. Takemoto, *J. Chem. Phys.*, **36**, 457 (1962).
71. A. Ottenberg, R. L. Brandon and M. E. Brown, *Nature*, **201**, 1119 (1964).
72. W. Slough, *Trans. Faraday Soc.*, **58**, 2310 (1962).
73. W. Klöpfer and H. Rabenhorst, *Angew. Chem. Intern. Ed. Engl.*, **6**, 268 (1967).
74. W. J. Siemons, P. E. Bierstedt and R. G. Kepler, *J. Chem. Phys.*, **39**, 3523 (1963).
75. P. L. Kronick and M. Labes in *Organic Semiconductors* (Eds. J. J. Brophy and J. W. Buttery), Macmillan, New York, 1962, p. 36.
76. L. R. Melby, *Can. J. Chem.*, **43**, 1448 (1965).
77. J. H. Lupinski and K. D. Kopple, Abstracts of the Chicago Meeting of the American Chemical Society, Division of Organic Coatings and Plastics Industry, Sept., 1964, pp. 72–76.
78. A. Mizoguchi, H. Moriga, T. Shimizu and Y. Amano, *Nat. Tech. Rep., Matsushita Elec. Ind. Co. Osaka, Japan*, **9**, (5) 407 (1963); *Chem. Abstr.*, **60**, 14626d (1964).

CHAPTER 11

Radicals with cyano groups

H. D. HARTZLER

E. I. du Pont de Nemours and Co.
Wilmington, Delaware, U.S.A.

I.	INTRODUCTION	671
II.	HYDROGEN ABSTRACTION FROM NITRILES	672
III.	FREE-RADICAL ADDITION OF NITRILES TO OLEFINS	674
	A. Cyanoesters	674
	B. Halonitriles	681
IV.	CYANOBENZYL RADICALS	682
V.	AZONITRILES	685
	A. Structure and Decomposition Rates	685
	B. Decomposition Products	688
	C. The Cage Effect	692
	D. Reactions of the 2-Cyano-2-propyl Radical	694
	1. Addition reactions	694
	2. Abstraction reactions	696
	3. Oxidations and reductions	697
VI.	RADICAL REARRANGEMENTS	698
VII.	RADICAL ADDITIONS TO ACRYLONITRILE AND OTHER CYANOOLEFINS	698
	A. Acrylonitrile and Vinylidene Cyanide	698
	1. Polymerizations	698
	2. Meerwein reaction	700
	3. Other ligand-transfer reactions	701
	B. Cyanobicyclobutanes	701
	C. Cycloadditions of α,β-Unsaturated Nitriles	702
	D. Additions to Cyanocarbons	703
VIII.	CYANOCARBENES	704
IX.	THE CYANO RADICAL	709
X.	REFERENCES	710

I. INTRODUCTION

The cyano substituent is quite inert to attack by free radicals. A wide variety of free-radical reactions are known where the cyano group is unaltered by the reaction. It does not follow, however, that the cyano group has no effect on the behaviour of free radicals. The following general characteristics may be given concerning the effects of the cyano group on free-radical reactions.

[1] As a substituent on a hydrocarbon the cyano group can activate or deactivate the hydrogen abstraction reaction. Its effect is very dependent upon the nature of the attacking radical. Abstraction reactions by acceptor (electrophilic) radicals such as halogen atoms are deactivated, but abstractions by donor (nucleophilic) radicals such as alkyl are promoted.

[2] As a substituent on an olefin, the cyano group greatly activates the olefin for radical attack. This effect is more pronounced for donor radicals, but acceptor radical addition is also promoted.

[3] α-Cyanoalkyl radicals are appreciably stabilized by electron delocalization.

$$R_2\dot{C}{-}C{\equiv}N \longleftrightarrow R_2C{=}C{=}\dot{N}$$

[4] While alkyl radicals are donor radicals, the electron-withdrawing cyano group makes cyanoalkyl radicals acceptor radicals.

II. HYDROGEN ABSTRACTION FROM NITRILES

Many studies have been made of polar effects in the hydrogen abstraction reaction by radicals. It is generally true that polar effects relating to the stability of the radical to be formed will be more important in reactions of high activation energy. This follows because the transition state will be closer to the (radical) products than in reactions of lower activation energy[1]. Even though adjacent cyano groups would stabilize a radical by electron delocalization, they could be deactivating in a hydrogen abstraction reaction by an acceptor radical, R·.

$$R\cdot + H{-}X \longrightarrow [R^{\delta-} \cdots H \cdots X^{\delta+}] \longrightarrow R{-}H + X\cdot$$

Activation energies for hydrogen abstraction from acetonitrile and propionitrile by trideuteriomethyl radicals are 10 and 8·5 kcal/mole, respectively[2]. The former value is only slightly smaller than the value of 10·4 kcal/mole for hydrogen abstraction from ethane.

The reactivities of the three types of hydrogen in butyronitrile have been determined in photochlorination[3] and in chlorination with *t*-butyl hypochlorite[4] and are given in Table 1. Deactivation

TABLE 1. Isomer distribution in chlorination of butyronitrile.

Reagent	Isomer distribution (%) α	β	γ
$Cl_2(h\nu)$[3]	0	69	31
t-C_4H_9OCl[4]	22·4	43·7	33·9

by the cyano group is greater for hydrogen abstraction by chlorine atoms than it is for *t*-butoxy radicals. The overall deactivation of the cyano group was shown by determining the relative reactivities of the three types of hydrogen of butyronitrile versus the primary hydrogen in 2,3-dimethylbutane in *t*-butylhypochlorite photochlorination. The values obtained were α 0·67, β 1·3 and γ 0·67 versus 1·00 for the primary hydrogen in 2,3-dimethylbutane[4]. Similar studies of isomer distribution in photochlorination have been done with propionitrile, isobutyronitrile and valeronitrile[3].

Bromination of nitriles with *N*-bromosuccinimide and benzoyl peroxide was shown to be a much more selective reaction[5]. Abstraction of primary hydrogen does not occur, so that good yields of α-bromonitriles were obtained from propionitrile and isobutyronitrile. Deactivation by the cyano group was observed, for valeronitrile gave 30% α- and 70% γ-bromovaleronitrile[5]. Hydrogen abstraction by bromine atoms is 15 kcal/mole less exothermic than hydrogen abstraction by chlorine atoms, so it would be expected that resonance stabilization of intermediate radicals would be more important in brominations than in chlorinations. The deactivation of hydrogen abstraction by adjacent cyano groups is not as great in bromination as in chlorination, for toluene underwent photochemical bromination with *N*-bromosuccinimide only twice as rapidly as phenylacetonitrile[6].

The effect of β-substituents in chlorinations with sulphuryl chloride initiated by azo-bis(isobutyronitrile) has been examined[7]. Pivalonitrile was only 0·17 times as reactive as *t*-butylbenzene. The cyano group was more deactivating than the benzoyl or chlorocarbonyl groups.

$$CH_3C(CH_3)_2\text{—}X + SO_2Cl_2 \longrightarrow ClCH_2C(CH_3)_2\text{—}X$$

$$(X = CN,\ C_6H_5,\ C_6H_5CO,\ ClCO)$$

Cyanoalkyl radicals have been generated from nitriles by reaction with hydroxyl radical. In aqueous solution, dimers were obtained[8]. Acetonitrile gave an 18% yield of succinonitrile, propionitrile gave a 60% yield of a dinitrile mixture and pivalonitrile gave a 52% yield of 2,5-dicyano-2,5-dimethylhexane. A claimed electrochemical formation of cyanomethyl radicals from acetonitrile[9] has been refuted[10].

III. FREE-RADICAL ADDITION OF NITRILES TO OLEFINS

A. Cyanoesters

Although an adjacent cyano group does not activate hydrogen for abstraction by donor radicals, many nitriles are reactive enough to undergo the free-radical chain olefin addition reaction. In most cases, hydrogen abstraction was accomplished by alkyl, acyloxy or alkoxy radicals. The following sequence of reactions is common to the olefin additions:

$$R^1_2CHCN + X\cdot \longrightarrow R^1_2C(CN)\cdot + HX \tag{1}$$

$$R^1_2C(CN)\cdot + R^2CH{=}CH_2 \longrightarrow \underset{(\mathbf{1})}{R^1_2C(CN){-}CH_2CH(R^2)\cdot} \tag{2}$$

$$\underset{(\mathbf{1})}{R^1_2C(CN){-}CH_2CH(R^2)\cdot} + R^1_2CHCN \longrightarrow \underset{(\mathbf{2})}{R^1_2C(CN){-}CH_2CH_2R^2} + R^1_2C(CN)\cdot \tag{3}$$

In order to obtain good yields of the 1:1 adduct (**2**), the rate of reaction (3) must be larger than the rate of reaction of (**1**) with more olefin which leads to telomer or polymer formation. To increase the yields of **2** large ratios of nitrile to olefin are used. In the *t*-butyl peroxide initiated addition of ethyl cyanoacetate to 1-octene, a molar ratio of nitrile to olefin of 4·5 to 1 gave a 31% yield of 1:1 adduct, while a ratio of 20 to 1 gave a 67% yield[11]. By the use of a very large excess of acetonitrile to 1-octene an 18% yield of 1-cyanononane was obtained[12] in the benzoyl peroxide initiated reaction.

Ethyl cyanoacetate has been shown to be twelve times more reactive than ethyl malonate in addition to 1-octene by product analysis in competition experiments[13]. At best the competition method measures relative rates of reactions exemplified by equation

(3). Product analysis does not measure the relative reactivities (equation 1) because of the other reactions which the intermediate radicals can undergo (disproportionation, dimerization, telomerization, etc.). In view of reaction complexities, such as reactions of the products with free radicals, the figures have probably only qualitative significance. Ethyl cyanomalonate has been added to 1-octene with bis(isopropyl) peroxydicarbonate as initiator[14] and has been shown to be about four times more reactive than ethyl malonate in addition to 1-octene[13].

Ethyl cyanoacetate underwent *trans* addition to 1-methylcyclohexene to give ethyl *cis*-2-methylcyclohexyl cyanoacetate[15]. Ethyl cyanoacetate was added to 1-hexyne to give 3·5% of ethyl hexylidenecyanoacetate[16]. The expected initial adduct (ethyl 2-cyano-3-octenoate) probably tautomerized to the observed product.

A variety of intramolecular cyclizations of unsaturated cyanoacetates have been accomplished by free-radical chain reactions. Ethyl 1-cyanocyclohexanecarboxylate (**4**) was prepared in 51% yield by the reaction of ethyl 2-cyano-6-heptenoate (**3**) with benzoyl peroxide in refluxing cyclohexane[17]. There are several important

CN, CH, CO_2Et (**3**) ⟶ CN, CO_2Et (**4**)

differences between the intramolecular and intermolecular addition reactions. The intramolecular cyclization is best run in very dilute solution in order to minimize intermolecular reactions. Doing this, however, decreases the rate of reaction of equation (3). This can be circumvented by running the reaction in a hydrogen donor solvent such as cyclohexane. With a hydrogen donor, chain lengths are small and larger amounts of initiator are required. The effect of dilution has been studied in the preparation of **4** from **3**[18]. The yield of **4** was increased from 35 to 70% by increasing the amount of cyclohexane diluent eightfold. A number of other cyclizations have been accomplished and are given in Table 2.

In view of the generality of low yields obtained in free-radical additions to internal double bonds[24a], the first three entries of Table 2 are of considerable interest. Evidently steric factors are less critical in intramolecular additions than in intermolecular additions. Example 7 indicates that the preference for a six-membered ring transition state rather than a seven-membered ring is of greater

TABLE 2. Cyclization of unsaturated nitriles.

	Reactant	Product	Yield (%)	Reference
1	CN, CH, CO_2Et	CN, CO_2Et + CH_3, CN, CO_2Et	49, 9	19
2	CH_3, CN, CH, CO_2Et	CH_3, CN, CO_2Et	90	19
3	Et, CN, CH, CO_2Et	Et, CN, CO_2Et	78	17
4	CH_3, CH_3, CN, CH, CO_2Et	$CH(CH_3)_2$, CN, CO_2Et	13·5	17
5	CN, CH, CO_2Et, CH_3	CN, CO_2Et, CH_3	22	17
6	CN, CH—CO_2Et, CH_3, CH_3	No cyclic isomer		17

7	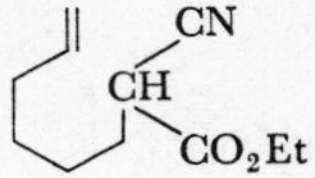 CN, CH, CO_2Et	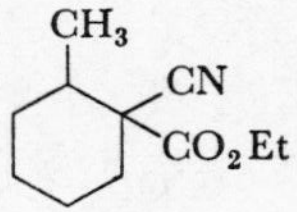 CH_3, CN, CO_2Et	39	17
8	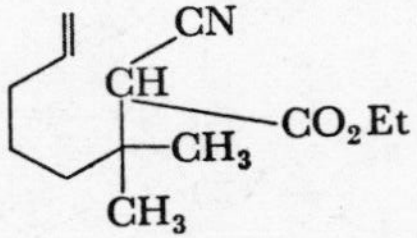 CN, CH, CO_2Et, CH_3, CH_3	No cyclic isomer		17
9	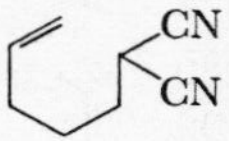 CN, CN	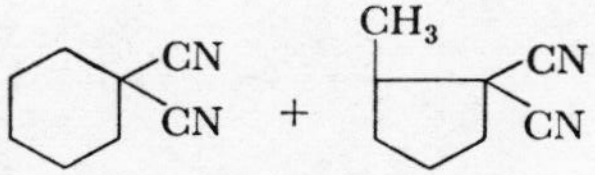 CN, CN + CH_3, CN, CN	56, 14	19
10	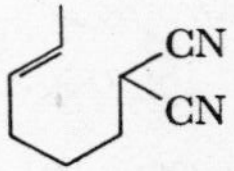 CN, CN	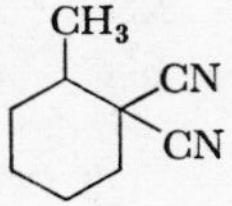 CH_3, CN, CN	50	19
11	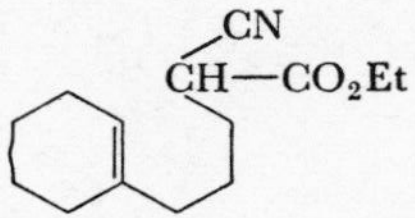 CN, CH—CO_2Et	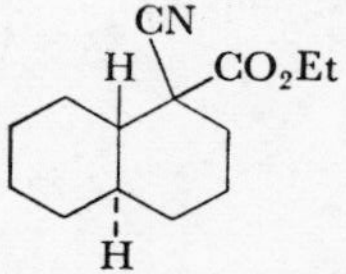 CN, H, CO_2Et, H	40	20
12	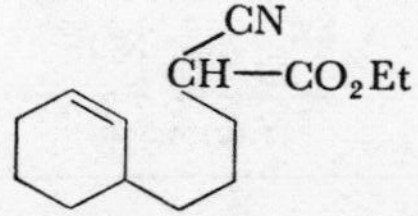 CN, CH—CO_2Et	No cyclic isomer		20

(*Table continued*)

TABLE 2 (*continued*)

	Reactant	Product	Yield (%)	Reference
13	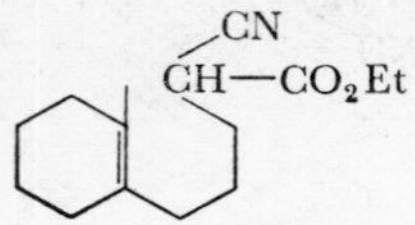 CN, CH—CO_2Et	No cyclic isomer		20
14	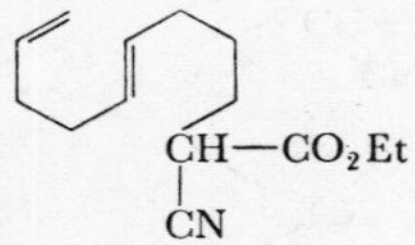 CH—CO_2Et, CN	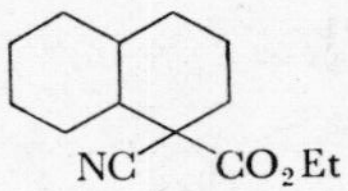 NC, CO_2Et (Mainly *trans*)	41	15
15	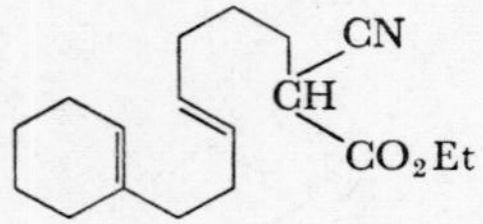 CN, CH, CO_2Et	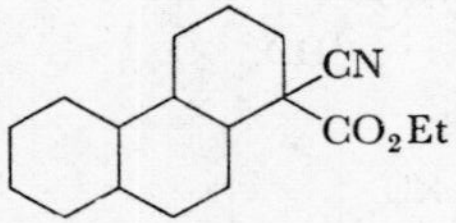 CN, CO_2Et	26	21
16	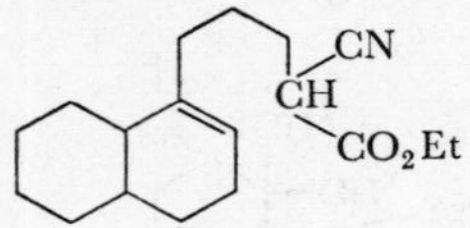 CN, CH, CO_2Et	No cyclic isomer		21
17	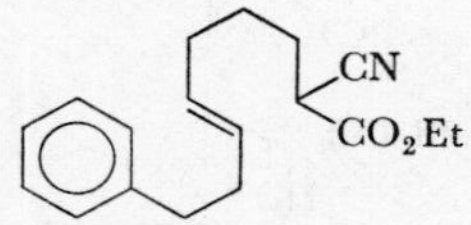 CN, CO_2Et	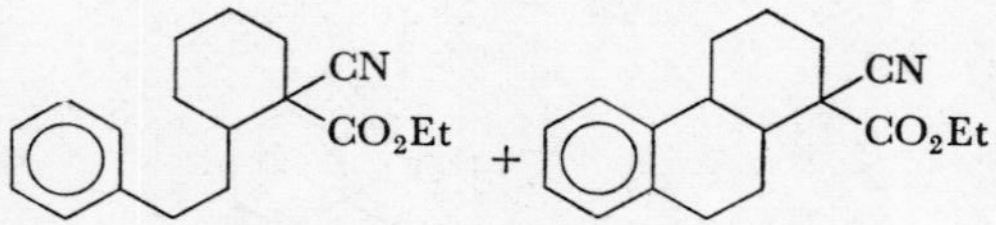 CN, CO_2Et + CN, CO_2Et	4, 42	22

18	$HC{\equiv}C-(CH_2)_3CH(CN)CO_2Et$	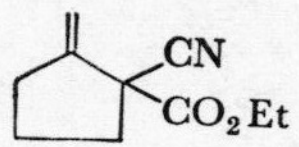 CN, CO_2Et	27	16
19	$CH_3C{\equiv}C-(CH_2)_3CH(CN)CO_2Et$	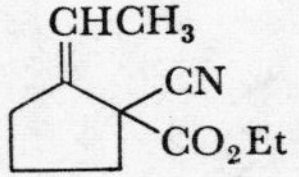 $CHCH_3$, CN, CO_2Et	63	16
20	$HC{\equiv}C(CH_2)_4CH(CN)CO_2Et$	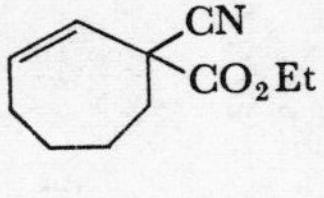 CN, CO_2Et	36	16
21	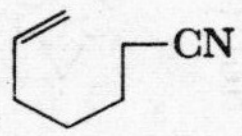 CN	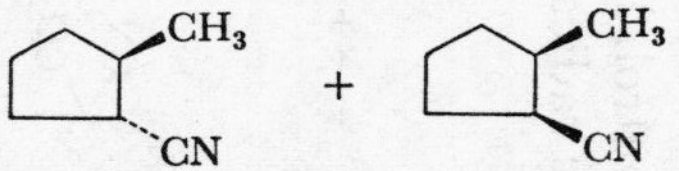 CH_3, CN + CH_3, CN	14, 18	19
22	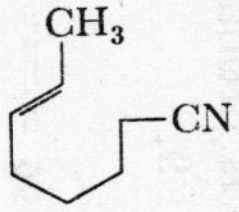 CH_3, CN	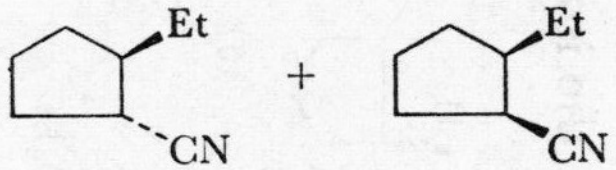 Et, CN + Et, CN	13, 6	19
23	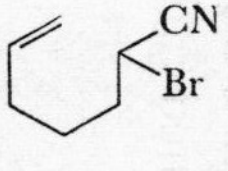 CN, Br	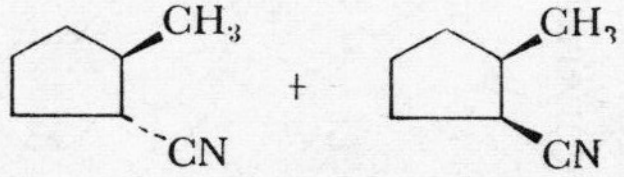 CH_3, CN + CH_3, CN	26, 12	23

importance than the energy differences between the primary and secondary radicals. The failure to observe cyclization in example 12 is probably the result of favourable geometry for abstraction of the allylic hydrogen. Polymerization takes precedence over cyclization in example 16. The cyclization to a seven-membered ring in example 20 occurs (in contrast to example 7) because of the rigid linearity of three of the carbon atoms.

A recent study has shown that the amounts of cyclization to five- and six-membered ring products in example 1 are very dependent upon temperature and upon solvent[24b]. The absence of six-membered ring products in examples 21, 22 and 23 suggests that more cyclization to a five-membered ring occurs when the initial radical is less stabilized. Initiation of the cyclization of example 23 was by bromine atom extraction rather than by hydrogen abstraction.

The products of example 1 have been shown to be those of thermodynamic rather than kinetic control. Decomposition of the peresters **5** and **7** generated the isomeric radicals **6** and **8** which gave nearly

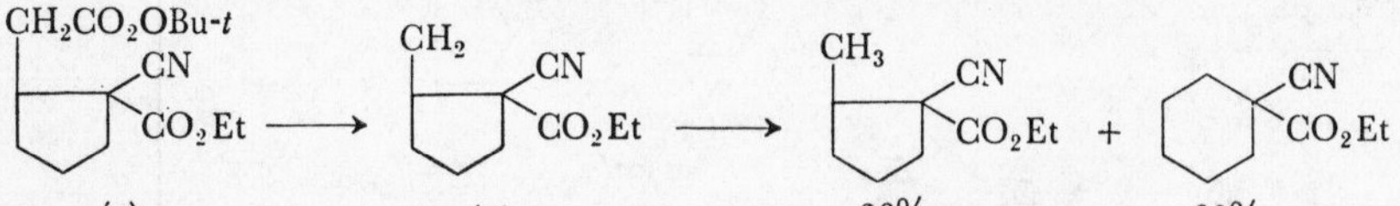

CH_2CO_2OBu-*t*, CN, CO_2Et (**5**) → CH_2, CN, CO_2Et (**6**) → CH_3, CN, CO_2Et (20%) + CN, CO_2Et (80%)

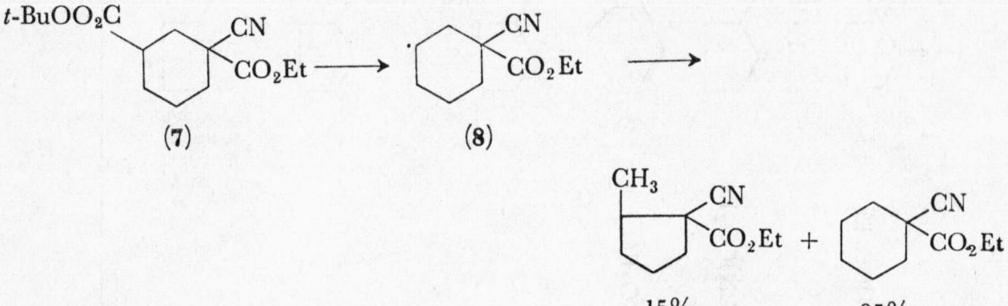

t-$BuOO_2C$, CN, CO_2Et (**7**) → CN, CO_2Et (**8**) → CH_3, CN, CO_2Et (15%) + CN, CO_2Et (85%)

identical product compositions[25]. The product composition was also that obtained in the cyclization of example 1. The interconversion of radicals **6** and **8** is probably not direct but rather by way of the ring-opened radical **9**.

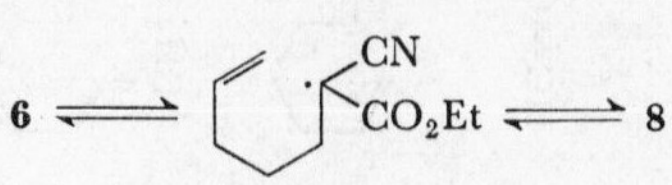

6 ⇌ CN, CO_2Et (**9**) ⇌ **8**

B. Halonitriles

A variety of bromonitriles have been added to olefins in peroxide induced reactions. These reactions proceeded by bromine atom rather than hydrogen atom abstraction.

$$R^1CHBr{-}CN \xrightarrow{X\cdot} XBr + R^1\underset{\displaystyle CN}{\underset{|}{C}}H\cdot$$

$$R^1\underset{\displaystyle CN}{\underset{|}{C}}H\cdot + R^2CH{=}CH_2 \longrightarrow R^1\underset{\displaystyle CN}{\underset{|}{C}}HCH_2\underset{\displaystyle R^2}{\underset{|}{C}}H\cdot$$

$$R^1\underset{\displaystyle CN}{\underset{|}{C}}HCH_2\underset{\displaystyle R^2}{\underset{|}{C}}H\cdot + R^1CHBrCN \longrightarrow R^1\underset{\displaystyle CN}{\underset{|}{C}}HCH_2CHBrR^2 + R^1\underset{\displaystyle CN}{\underset{|}{C}}H\cdot$$

Bromomalononitrile gave nearly quantitative yields of 1:1 adducts with olefins[26]. The reaction was shown to be accelerated by light, slowed by oxygen and inhibited by *t*-butylcatechol[26]. The products of the reaction, β-bromoalkylmalononitriles, were converted in high yields to dicyanocyclopropanes by reaction with triethylamine. The reaction of bromomalononitrile, olefins and triethylamine to give dicyanocyclopropanes had previously been thought to involve dicyanocarbene[27], but this interpretation is no longer tenable. The stereochemistry of addition of bromomalononitrile to cyclic olefins has been examined[28]. In most cases, slightly greater yields of *trans* than of *cis* adducts were obtained.

Dibromomalononitrile also has been added to olefins[29,30]. With an excess of dibromomalononitrile 1:1 adducts such as **10** were

$$CBr_2(CN)_2 + PhCH{=}CH_2 \longrightarrow \underset{(\mathbf{10})}{PhCHBr{-}CH_2CBr(CN)_2}$$

formed[29]. The products readily reacted with more olefin to give 1:2 adducts such as **11**[30]. The 1:2 adducts were hydrolysed by base

$$\mathbf{10} + PhCH{=}CH_2 \longrightarrow \underset{(\mathbf{11})}{PhCHBrCH_2C(CN)_2CH_2CHBrPh}$$

to spirolactones[30]. The radical nature of the olefin addition reaction was deduced by the structures of the products and by the initiation

$$\mathbf{11} + NaOH \longrightarrow \text{PhCH}\begin{matrix} \diagup \text{O—CO} \diagdown \\ \diagdown \text{CH}_2 \diagup \end{matrix}\text{C}\begin{matrix} \diagup \text{CH}_2 \diagdown \\ \diagdown \text{CO—O} \diagup \end{matrix}\text{CHPh}$$

of the reaction by peroxides, azonitriles or copper[30]. The formation of radicals by the reaction of dibromomalononitrile with copper indicates that the formation of tetracyanoethylene probably involves a radical dimerization rather than a carbene reaction as was suggested[31].

$$CBr_2(CN)_2 + Cu \longrightarrow (NC)_2CBr\cdot$$

$$(NC)_2CBr\cdot \longrightarrow [(NC)_2CBr{-}CBr(CN)_2] \longrightarrow (NC)_2C{=}C(CN)_2$$

The benzoyl peroxide induced addition of bromoacetonitrile to 1-octene has also been reported to give a 66% yield of the 1:1 adduct, 3-bromo-1-cyanononane[32]. The addition of dibromoacetonitrile to allyl acetate gave a 52% yield of 4-cyano-2,4-dibromobutyl acetate[32]. Trichloroacetonitrile gave only a trace of the 1:1 adduct with ethylene but a 66% yield of adduct with 1-octene[32]. The 1:1 adducts **12** and **13** were obtained from 2,2,3-trichloropropionitrile with 1-octene and allyl acetate, respectively[32].

$$CH_3(CH_2)_5CH(Cl)CH_2CCl(CH_2Cl)CN \quad \textbf{(12)} \qquad CH_3COOCH_2CH(Cl)CH_2CCl(CH_2Cl)CN \quad \textbf{(13)}$$

IV. CYANOBENZYL RADICALS

In 1889 it was observed that an acetic acid solution of tetraphenylsuccinonitrile (**14**) developed a red colour when heated and that the colour disappeared upon cooling[33]. It was not until 1925 that the phenomenon was attributed to the formation of cyanodiphenylmethyl radicals[34]. Tetra-*p*-anisylsuccinonitrile was shown to dissociate into radicals more readily[35]. Dissociation was confirmed by the formation of di-*p*-anisylacetonitrile upon reaction with phenylhydrazine[35].

A thermal and base-catalysed isomerization of tetraphenylsuccinonitrile to **15** has been observed[36]. Although **14** dissociated to

$$\underset{\textbf{(14)}}{Ph_2C(CN){-}C(CN)Ph_2} \longrightarrow \underset{\textbf{(15)}}{Ph_2C(CN){-}C_6H_4{-}CH(CN)Ph}$$

radicals in boiling naphthalene, there was no isomerization to **15** unless base was present[37]. The isomerization **14** → **15** can also be accomplished at temperatures as low as 50° in the presence of triethylamine[37]. No transient colour is seen at this temperature. The kinetics of the isomerization showed the reaction to be first order in

14 and nearly independent of triethylamine concentration. The probable mechanism is outlined below. The intermediate cyanodiphenylmethyl radicals have been trapped with NO_2 to give

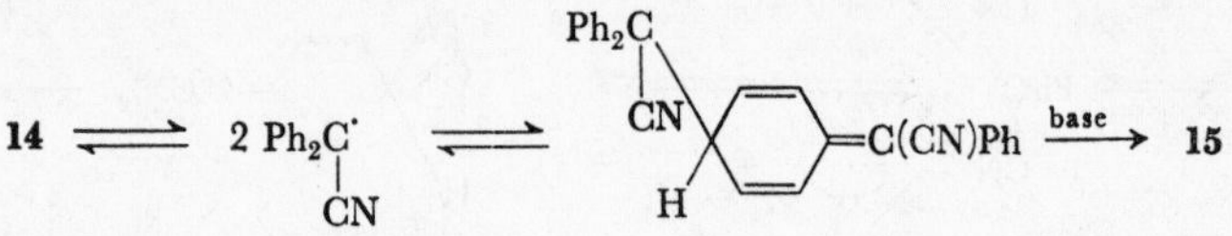

$$\mathbf{14} \rightleftharpoons 2\,Ph_2\dot{C}(CN) \rightleftharpoons \text{(Ph}_2\text{C(CN))(H)C}_6\text{H}_4\text{=C(CN)Ph} \xrightarrow{\text{base}} \mathbf{15}$$

cyanonitrodiphenylmethane (**16**)[38], with styrene to give 2,2,3,5,5-pentaphenyladiponitrile (**17**)[39] and with benzyl alcohol to give benzaldehyde and diphenylacetonitrile[40]. The polymerization of styrene was also initiated by **14** at 100° [39].

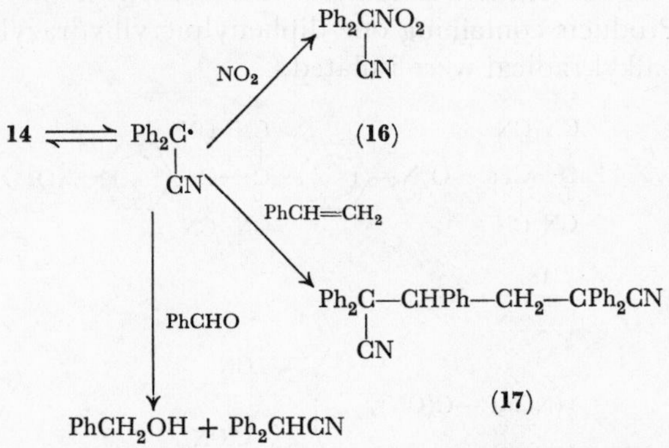

$$\mathbf{14} \rightleftharpoons Ph_2C^{\bullet}(CN) \xrightarrow{NO_2} Ph_2C(NO_2)CN\ (\mathbf{16})$$

$$Ph_2C^{\bullet}(CN) \xrightarrow{PhCH=CH_2} Ph_2C(CN)\text{—}CHPh\text{—}CH_2\text{—}CPh_2CN\ (\mathbf{17})$$

$$Ph_2C^{\bullet}(CN) \xrightarrow{PhCHO} PhCH_2OH + Ph_2CHCN$$

Similar reactions have been observed with *sym*-diphenyltetracyanoethane (**18**)[41]. Isomerization of **18** to **19** occurred at room

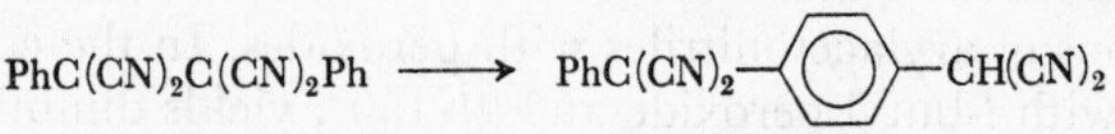

$$\underset{(\mathbf{18})}{PhC(CN)_2C(CN)_2Ph} \longrightarrow \underset{(\mathbf{19})}{PhC(CN)_2\text{—}C_6H_4\text{—}CH(CN)_2}$$

temperature and was unaffected by the presence of oxygen or styrene. The reaction of phenylmalononitrile with *t*-butyl peroxide also gave **19**[41]. Intermediate radicals in the isomerization **18** → **19** were trapped with dinitrogen tetroxide, tetrahydrofuran (hydrogen abstraction) and diphenylpicrylhydrazyl. The following reaction scheme was suggested. The nitrate ester is probably formed by oxidation of the initially formed nitrite. The isomerization of the intermediate cyclohexadiene to **19** did not require a basic catalyst.

Diphenylpicrylhydrazyl was found to be an efficient trap for polycyano radicals and was used to establish a qualitative ranking

$$18 \rightleftharpoons PhC^{\cdot}(CN)_2 \rightleftharpoons \text{Ph–C(CN)(CN)–}C_6H_5(H)\text{=C(CN)}_2 \longrightarrow 19$$

$PhC^{\cdot}(CN)_2 \xrightarrow{NO_2} PhC(CN)_2ONO_2$; $PhC^{\cdot}(CN)_2 \xrightarrow{\text{THF}} PhCH(CN)_2$; $PhCH(CN)_2 \xrightarrow{(t\text{-BuO})_2} PhC^{\cdot}(CN)_2$

of ease of dissociation of polycyanoethanes into radicals[41]. The rate of disappearance of diphenylpicrylhydrazyl was assumed to be indicative of the rate of homolytic dissociation of the substituted ethanes. Products containing one diphenylpicrylhydrazyl radical to one cyanoalkyl radical were isolated.

$$Ph_3C\text{—}CPh_3 \gg PhC(CN)_2\text{—}C(CN)_2Ph\ (18) \sim O_2N\text{—}C_6H_4\text{—}C(CN)_2\text{—}C(CN)_2\text{—}C_6H_4\text{—}NO_2 > Ph_2C(CN)\text{—}C(CN)Ph_2\ (14)$$

$$> (NC)_3C\text{—}C(CN)_3 > \text{9-cyano-9-(tricyanomethyl)fluorene } (C(CN)_3)$$

Dimerizations of other cyanobenzyl radicals have been observed in reactions of arylacetonitriles with peroxides. In the conversion of **20** to **21** with *t*-butyl peroxide at 140–150°, yields diminished as the

$$\underset{(20)}{PhCHRCN} \longrightarrow \underset{(21)}{PhCR(CN)\text{—}CR(CN)Ph}$$

size of the alkyl group increased[42]. This probably means that the steric requirements for the radical dimerization are greater than for disproportionation. The relative yields of *meso* and *dl* isomers of **21** did not vary in any way of apparent significance. Very high *meso*/*dl* ratios have been reported for the dehydrodimerizations of *p*-chlorophenylacetonitrile[43] and phenylacetonitrile[43,44]. The numbers obtained have only qualitative significance, for they were based upon product isolation.

The *meso* and *dl* isomers of 2,3-dimethyl-2,3-diphenylsuccinonitrile have been thermally interconverted at 125 to 175° [45]. At equilibrium a *dl* to *meso* ratio of 1·23 was obtained. This amounts to an energy difference of only 1·12 kcal/mole. The isomerization proceeded by way of radicals as was shown by the formation of 2-phenylpropionitrile when the reaction was run in the presence of benzenethiol. The intermediate radicals appeared to react only slowly with oxygen.

A dicyanobenzyl radical may have been an intermediate in the formation of **22** in the reaction of *N*,*N*-dimethylaniline and dicyanodiazomethane[46].

$$Me_2N{-}C_6H_4{-}C(CN)_2{-}C_6H_4{-}NMe_2$$

(22)

V. AZONITRILES

A. Structure and Decomposition Rates

Azonitriles are a class of compounds which have been extensively used as initiators of free-radical reactions. Since the discovery of the utility of azonitriles as polymerization initiators[47], decompositions of azonitriles have been widely studied and have led to fundamental understanding of free-radical processes. The azonitriles are easily prepared by oxidation of cyanoalkylhydrazines which are prepared from ketones, hydrazine and hydrogen cyanide[48]. The azonitriles

$$R_2CO + N_2H_4 + HCN \longrightarrow R_2C(CN)NHNHC(CN)R_2 \longrightarrow R_2C(CN){-}N{=}N{-}C(CN)R_2$$

serve as useful sources of radicals because they decompose at moderate temperatures and are not subject to induced decomposition as so many peroxides are. Azo-bis(isobutyronitrile) (**23**) has a half-life of 17 hours at 60° and 1·3 hours at 80°.

The energy of activation for decomposition of azo-bis(isobutyronitrile) was found to be approximately 31 kcal/mole[49]. The activation energy for azo-bis(isobutane) decomposition was 43 kcal/mole[50]. The difference in values suggests that bond breaking has proceeded to a considerable extent in the transition state, and that electron delocalization into the cyano group significantly stabilizes the transition state in azonitrile decomposition. The decrease of rate of decomposition of **23** with increasing pressure was explained by suggesting that in the transition state bond stretching had occurred to about 10 %[51]. There is no chemical or physical evidence to suggest that the carbon–nitrogen bonds are broken successively in thermal decompositions of azonitriles. A much higher activation energy than 31 kcal/mole for decomposition of **23** would be expected if radicals **24** and **25** were formed rather than nitrogen and **25**. In the photolysis of **23** in matrix at −196°, an electron spin resonance signal was seen which has been assigned to the radical **24**[52]. The spectrum consisted of five peaks separated by 65–70 gauss and in a

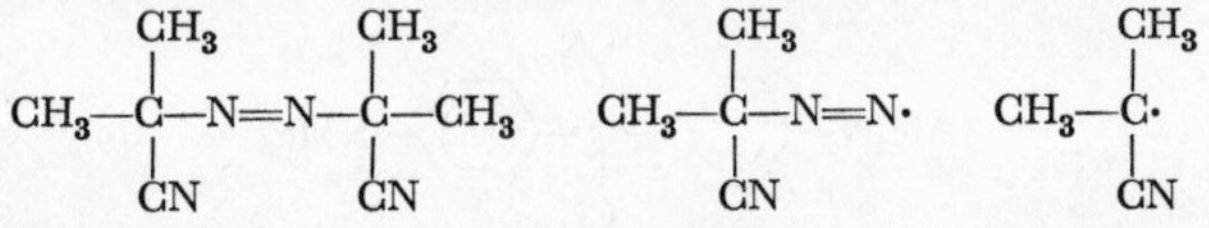

$$CH_3\text{—}\underset{\underset{CN}{|}}{\overset{\overset{CH_3}{|}}{C}}\text{—}N\text{=}N\text{—}\underset{\underset{CN}{|}}{\overset{\overset{CH_3}{|}}{C}}\text{—}CH_3 \quad (\mathbf{23}) \qquad CH_3\text{—}\underset{\underset{CN}{|}}{\overset{\overset{CH_3}{|}}{C}}\text{—}N\text{=}N\cdot \quad (\mathbf{24}) \qquad CH_3\text{—}\underset{\underset{CN}{|}}{\overset{\overset{CH_3}{|}}{C}}\cdot \quad (\mathbf{25})$$

1:2:2:2:1 intensity ratio. After a few days at −196° the spectrum changed to that of the 2-cyano-2-propyl radical (**25**). The electron spin resonance spectrum of **25** was obtained independently by photolysis in matrix at −196° of phenylazoisobutyronitrile[52].

The rate of decomposition of **23** was not greatly influenced by the nature of the solvent[53–55]. Maximum rate differences of 15 % have been observed. The decompositions were uniformly first order kinetically. This indicated that the azonitriles were not subject to radical induced decomposition. Catalysis of the thermal decomposition of **23** has been observed, however. Triethylaluminium and **23** form a complex as judged by changes in infrared spectra[56]. The complex has an activation energy for decomposition of only 12 kcal/mole as compared with 31 kcal/mole for **23** alone. Diethylaluminium chloride also formed a complex with **23** which also decomposed at greater rates than **23** alone. Both complexes served as good initiators for radical polymerizations at 40° [56]. Silver perchlorate also apparently formed a complex with **23**[57]. The rate of decomposition of the complex was four times that of **23**[57]. Copper, lithium and zinc salts did not affect the rate of decomposition of **23**[57].

Coordination of **23** with silver ion or triethylaluminium probably occurred at the nitrile nitrogen. The observed shift of 30–40 cm^{-1} of the nitrile absorption to lower frequency upon coordination is typical for such coordination. It will be interesting to learn whether the coordinating species plays any role in the subsequent reactions of the 2-cyano-2-propyl radical.

Many different azonitriles have been synthesized. Their rates of decomposition are given in Table 3. With few exceptions the spread of decomposition rates is quite small. The increased decomposition rates of the azonitriles with isobutyl and neopentyl substituents has been attributed to steric interactions between the groups destabilizing the ground state. The azonitriles with cyclopropyl and phenyl substituents also decomposed at higher rates. The cyclopropyl substituted compounds gave products which were different from those of the other azonitriles (see below). The benzyl–nitrogen bond in azo-bis(phenylpropionitrile) would be expected to be of lower

TABLE 3. Rates of decomposition of azonitriles $R^1R^2C(CN)N{=}NC(CN)R^1R^2$ at 80°.

$$R^1C(R^2)(CN)-N{=}N-C(R^2)(CN)R^1$$

R^1	R^2	10^4k (s^{-1})	Reference
Me	Me	1·7	54
Me	Et	0·9	54
Me	n-Pr	1·7	54
Me	i-Pr	1·0	54
Me	n-Bu	1·6	54
Me	i-Bu	7·1	54
Me	$(CH_2)_3CO_2H$	0·896	53
Me	$PhCH_2$—	1·16	58
Me	p-$ClC_6H_4CH_2$	0·88	58
Me	p-$O_2NC_6H_4CH_2$	1·00	58
Me	Cyclopropyl	33	59
Me	Cyclobutyl	1·51[a]	60
Me	Cyclopentyl	1·30[a]	60
Me	Cyclohexyl	2·27	60
Me	$(CH_3)_3CCH_2$	158[b]	61
Me	$(CH_3)_3CCH_2$	136[b]	61
Me	n-Am	1·63	61
i-Bu	i-Bu	49·5	61
i-Pr	i-Pr	1·25	61
Et	Et	0·84	62
Et	i-Pr	0·95	61
n-Pr	n-Pr	1·15	63
—$(CH_2)_3$—		0·00173	64
—$(CH_2)_4$—		0·726	64
—$(CH_2)_5$—		0·063	64
—$(CH_2)_6$—		12·2	64
—$(CH_2)_7$—		83·5	64
—$(CH_2)_9$—		18·42	64
—$CHCH_3(CH_2)_4$—		0·0743	64
Cyclopropyl	Cyclopropyl	347	65
Me	Ph	Very fast[c]	66

[a] Both stereoisomers decomposed at same rate.
[b] Stereoisomers.
[c] Compound decomposed in solution at room temperature.

bond energy than the other carbon–nitrogen bonds. Increased electron delocalization in the transition state for the incipient benzyl radical is to be expected. The variation in decomposition rates of the azo-bis(cyanocycloalkanes) with ring size parallels variation of solvolysis rates of 1-chloro-1-methylcycloalkanes[64]. Both types of reactions have transition states in which a tetrahedral carbon atom is converted to a trigonal intermediate, and rate differences were ascribed to ring strain energy differences.

B. Decomposition Products

The products of decomposition of azonitriles have been examined in almost all cases. Usually high yields of tetraalkylsuccinonitriles were obtained. Azo-bis(isobutyronitrile) has been studied most extensively. The ultimate products from **23** in refluxing toluene were nitrogen, tetramethylsuccinonitrile (84%), isobutyronitrile (3·5%) and 2,3,5-tricyano-2,3,5-trimethylhexane (**26**) (9%)[67]. Yields as high as 96% of tetramethylsuccinonitrile have been obtained[68a]. The isobutyronitrile and **26** resulted from radical disproportionation.

$$\mathbf{23} \longrightarrow (CH_3)_2\underset{\large CN}{\underset{|}{C}}\cdot \;\begin{cases} \nearrow (CH_3)_2\underset{\large CN}{\underset{|}{C}}-\underset{\large CN}{\underset{|}{C}}(CH_3)_2 \\ \searrow (CH_3)_2CHCN + CH_2{=}\overset{\large CH_3}{\overset{|}{C}}-CN \end{cases}$$

$$CH_2{=}\overset{\large CH_3}{\overset{|}{C}}-CN + (CH_3)_2\underset{\large CN}{\underset{|}{C}}\cdot \longrightarrow (CH_3)_2\underset{\large CN}{\underset{|}{C}}-CH_2-\underset{\large CN}{\underset{|}{\dot{C}}}-CH_3 \xrightarrow{(CH_3)_2\overset{CN}{\overset{|}{C}}\cdot} (CH_3)_2\underset{\large CN}{\underset{|}{C}}-CH_2-\overset{\large CH_3}{\underset{\large CN}{\overset{|}{\underset{|}{C}}}}-\overset{\large CH_3}{\underset{\large CN}{\overset{|}{\underset{|}{C}}}}-CH_3 \quad \mathbf{(26)}$$

The reaction is more complicated, however, and additional work has shown that the 2-cyano-2-propyl radicals also combine to give the ketenimine **27**[69,70]. Compound **27** was shown to be thermally unstable and to undergo thermal dissociation into radicals, so that

$$\mathbf{25} \rightleftharpoons (CH_3)_2C{=}C{=}N{-}\underset{\underset{CN}{|}}{C}(CH_3)_2 \xrightarrow[HCl]{H_2O} (CH_3)_2CHCONH{-}C(CH_3)_2CN$$

(27) (28)

the ultimate products were as given above. The amount of **27** formed was determined by decomposing **23** in refluxing toluene containing some aqueous hydrochloric acid. The ketenimine **27** underwent rapid irreversible hydration to the amide **28** and a 54·4% yield of **28** was obtained[71]. This was in close agreement with a value of 59% found for the photochemical yield of **27** from **23**[72]. Pure **27** has been isolated and its thermal decomposition studied[73]. It decomposed at a rate almost as great as **23**.

Similar products have been isolated from the azonitrile **29**[74]. It is very probable that ketenimines such as **30** are way stations in most

$$C_6H_{10}(CN){-}N{=}N{-}C_6H_{10}(CN)\ (\mathbf{29}) \longrightarrow C_6H_{10}^{\cdot}{-}CN \rightleftharpoons C_6H_{10}{=}C{=}N{-}C_6H_{10}(CN)\ (\mathbf{30})$$

$$\downarrow$$

$$(CN)C_6H_{10}{-}C_6H_{10}(CN)$$

thermal and photochemical decompositions of azonitriles. An interesting exception to this was the photolysis of **23** on silica gel suspended in benzene[75]. Nitrogen and tetramethylsuccinonitrile were obtained, but **27** was not. The quantum yield for disappearance of **23** was the same as without the silica gel. Evidently the radicals on the silica gel surface are not free to rotate to give the unsymmetrical coupling product as they can in solution.

Cyanoalkyl radicals were also generated by electrolysis[76a,76b] and by peroxydisulphate oxidation[76c] of cyanoacetates. The products were dialkylsuccinonitriles and amides formed by hydration of ketenimines[76a]. The major products were the amides rather than the

$$\underset{\underset{CN}{|}}{RCHCO_2H} \longrightarrow \underset{\underset{CN}{|}}{RCH\cdot} \longrightarrow \underset{\underset{CN}{|}}{RCH}{-}\underset{\underset{CN}{|}}{CHR} + RCH_2CONHCHRCN$$

succinonitriles. The electrolytic and the peroxydisulphate reactions complement the use of azonitriles as cyanoalkyl radical sources, for monosubstituted azonitriles normally do not decompose by radical pathways[76d].

The amounts of coupling versus disproportionation products varied with azonitrile structure. Compound **31** gave a 7% yield of the tetraalkylsuccinonitrile **32**[77]. The major products were those

$$t\text{-Bu}\!-\!\underset{\displaystyle CN}{\underset{|}{C}}(CH_3)\!-\!N\!=\!N\!-\!\underset{\displaystyle CN}{\underset{|}{C}}(CH_3)\!-\!\text{Bu-}t \longrightarrow \quad \textbf{(31)}$$

$$t\text{-Bu}\!-\!\underset{\displaystyle CN}{\underset{|}{C}}(CH_3)\underset{\displaystyle CN}{\underset{|}{C}}(CH_3)\!-\!\text{Bu-}t + t\text{-Bu}\!-\!\underset{\displaystyle CN}{\underset{|}{C}}HCH_3 + t\text{-Bu}\!-\!\underset{\displaystyle CN}{\underset{|}{C}}\!=\!CH_2 \quad \textbf{(32)}$$

of disproportionation. The symmetrical coupling reaction was more subject to steric hindrance than was the disproportionation reaction. Intermediate ketenimine formation was not investigated.

Thermal decomposition of **33** in benzene solution gave only a 19% yield of the succinonitrile **34**[78]. The major product was a polymer, and a 7·4% yield of the azine **35** was obtained. The decomposition of **33** in the solid state gave a 63% yield of **34** with

$$\triangleright\!-\!\underset{\displaystyle CN}{\underset{|}{\overset{\displaystyle CH_3}{\overset{|}{C}}}}\!-\!N\!=\!N\!-\!\underset{\displaystyle CN}{\underset{|}{\overset{\displaystyle CH_3}{\overset{|}{C}}}}\!-\!\triangleleft \longrightarrow \quad \textbf{(33)}$$

$$\triangleright\!-\!\underset{\displaystyle CN}{\underset{|}{\overset{\displaystyle CH_3}{\overset{|}{C}}}}\!-\!\underset{\displaystyle CN}{\underset{|}{\overset{\displaystyle CH_3}{\overset{|}{C}}}}\!-\!\triangleleft \;\textbf{(34)} + \text{(CH}_3\text{-cyclopentenyl)}\!-\!N\!=\!N\!-\!\text{(CH}_3\text{-cyclopentenyl)} \;\textbf{(35)} + \text{polymer}$$

a lowered yield of the polymer. A radical isomerization mechanism for the unusual formation of **35** was suggested[78].

$$\left[\triangleright\!-\!\overset{\displaystyle CH_3}{\overset{|}{C}}\cdot\!-\!C\!\equiv\!N \longleftrightarrow \triangleright\!-\!\overset{\displaystyle CH_3}{\overset{|}{C}}\!=\!C\!=\!N\cdot\right] \longrightarrow \text{(2-methylcyclopent-2-enylidene)}\!=\!N\cdot \longrightarrow \textbf{35}$$

An internal disproportionation reaction was found to be the major reaction path for **36**[79,80].

Many other indications that azonitrile decompositions are free-radical processes have been obtained. Oxygen was rapidly incorporated into azonitrile solutions undergoing decomposition[63]. Free-radical polymerizations of vinyl monomers including ethylene

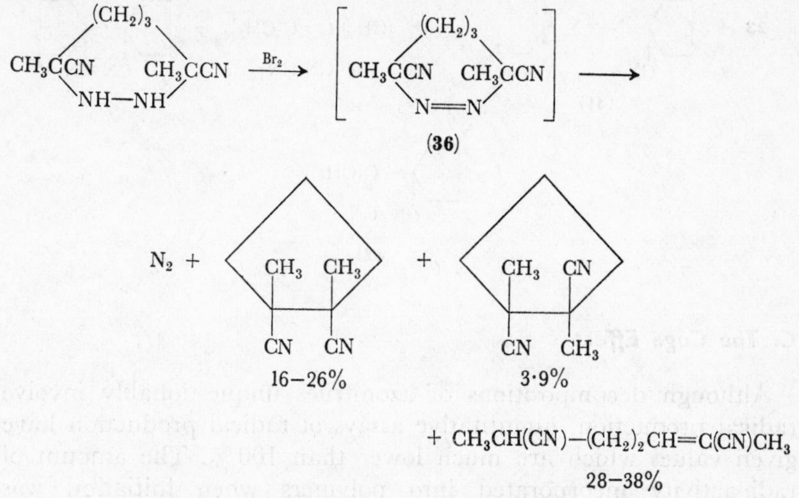

$$\underset{\text{NH—NH}}{\overset{(CH_2)_3}{CH_3CCN \quad CH_3CCN}} \xrightarrow{Br_2} \left[\underset{\text{N=N}}{\overset{(CH_2)_3}{CH_3CCN \quad CH_3CCN}} \right] \longrightarrow$$

(36)

$$N_2 + \text{(cyclopentane, 1,2-di-}CH_3\text{, 1,2-di-CN)} + \text{(cyclopentane, 1-}CH_3\text{, 1-CN, 2-CN, 2-}CH_3\text{)}$$

16–26% 3·9%

$$+ CH_3CH(CN)-(CH_2)_2CH{=}C(CN)CH_3$$

28–38%

have been initiated by decomposing azonitriles[81]. The rate of **23**-initiated polymerization of methyl methacrylate was proportional to the square root of the concentration of **23**[55]. Polymerization using ^{14}C-labelled **23** gave polymers with radioactivity incorporated into them[82]. *Meso* and racemic isomers of **37**[83] and **39**[84] have been separated and decomposed. Each isomer gave the same amounts of *meso* and racemic **38** and **39**, respectively. Similarly, decomposition of a mixture of **23** and **41** gave a mixture of succinonitriles including the crossed product **42**[83].

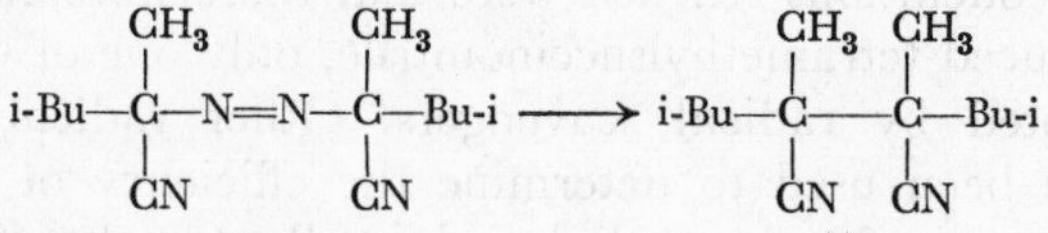

$$\text{i-Bu}-\underset{CN}{\overset{CH_3}{C}}-N{=}N-\underset{CN}{\overset{CH_3}{C}}-\text{Bu-i} \longrightarrow \text{i-Bu}-\underset{CN}{\overset{CH_3}{C}}-\underset{CN}{\overset{CH_3}{C}}-\text{Bu-i}$$

(37) (38)

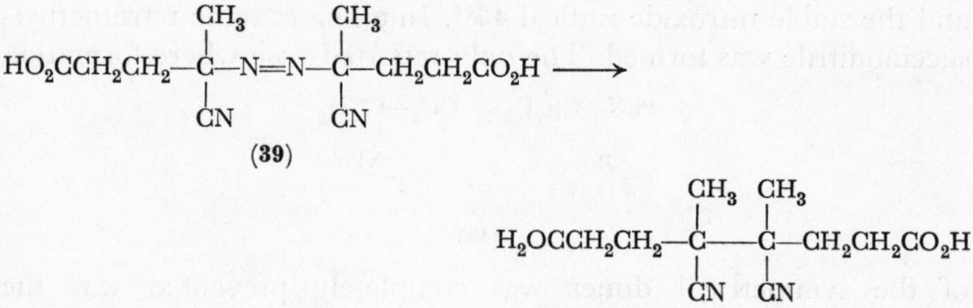

$$HO_2CCH_2CH_2-\underset{CN}{\overset{CH_3}{C}}-N{=}N-\underset{CN}{\overset{CH_3}{C}}-CH_2CH_2CO_2H \longrightarrow$$

(39)

$$H_2OCCH_2CH_2-\underset{CN}{\overset{CH_3}{C}}-\underset{CN}{\overset{CH_3}{C}}-CH_2CH_2CO_2H$$

(40)

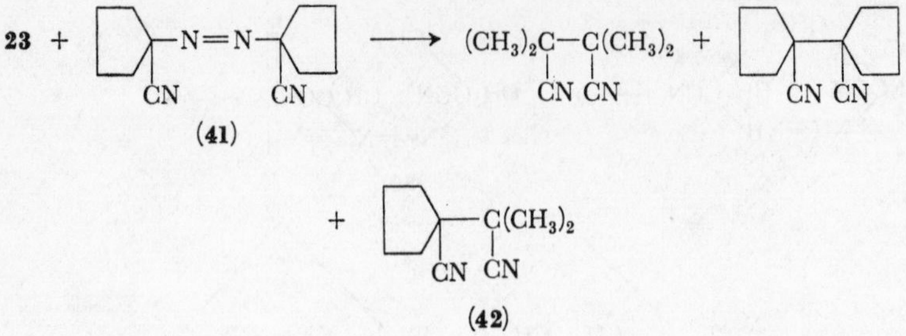

$$\mathbf{23} + \text{(41)} \longrightarrow (CH_3)_2\underset{CN}{C}-\underset{CN}{C}(CH_3)_2 + \text{(cyclopentyl)}\underset{CN}{C}-\underset{CN}{C}\text{(cyclopentyl)} + \mathbf{(42)}$$

(41): 1-cyanocyclopentyl–N=N–1-cyanocyclopentyl

(42): 1-cyanocyclopentyl–$\underset{CN}{C}(CH_3)_2$

C. The Cage Effect

Although decompositions of azonitriles unquestionably involve radical production, quantitative assays of radical production have given values which are much lower than 100%. The amount of radioactivity incorporated into polymers when initiation was caused by ^{14}C labelled **23** was only 52–63% with polymethyl methacrylate, 66–82% with polystyrene, 68–83% with polyvinyl-acetate and 70–77% with polyvinyl chloride[82]. A similar study showed 70% radioactivity in polystyrene with most of the remainder in tetramethylsuccinonitrile[85]. Addition of a variety of different radical trapping reagents to solutions of **23** decreased the yield but did not eliminate the formation of tetramethylsuccinonitrile. For example, the addition of increasing amounts of mercaptans to carbon tetra-chloride solutions of **23** cut its yield down to 20%[68a], but further increases in mercaptan concentration did not lower the yield any more. The conclusions reached were that there were two reactions which produced tetramethylsuccinonitrile, only one of which could be intercepted by radical scavengers. Other radical scavengers which have been used to determine the efficiency of free-radical production from **23** were diphenylpicrylhydrazyl[68a,86], iodine[68a], ferric chloride in dimethylformamide[87], triphenylmethyl radical[65] and the stable nitroxide radical **43**[88]. In all cases some tetramethyl-succinonitrile was formed. The only reported case where formation

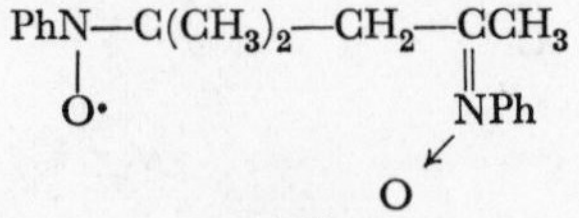

$$PhN(O\cdot)-C(CH_3)_2-CH_2-C(CH_3)=N(\rightarrow O)Ph$$

(43)

of the symmetrical dimer was completely prevented was the decomposition of **23** in liquid bromine[89].

The percentage efficiency of radical production from **23** (or the percentage of radicals that can be trapped by scavengers such as mercaptans, iodine, oxygen or diphenylpicrylhydrazyl) is greatly dependent upon the solvent. A variation between 46 % in carbon tetrachloride and 75 % in nitromethane has been observed[68a]. The efficiency of radical production varied more with solvent than did the rate of decomposition of **23** (at 62·5°, $k_{dec}(CH_3NO_2) = 13{\cdot}4 \times 10^{-6}\ s^{-1}$ and $k_{dec}(CCl_4) = 11{\cdot}8 \times 10^{-6}\ s^{-1}$).

It has been shown that the efficiency of radical production from the ketenimine **27** is lower than that of the azonitrile **23**[73,90,91], even though the two are producing the same radicals. The differences in efficiency of radical production were attributed to differences in spatial distribution at the moment of radical formation.

The portion of the decomposition reaction of **23** which cannot be intercepted by radical scavengers is thought to occur within a solvent cage. The cage reaction of a radical pair is their combination before diffusion into the solvent occurs. Recently, striking effects on cage combination of alkoxy radicals with changes in solvent viscosity have been observed[92]. The variation in radical efficiency with solvent in the decomposition of **23** can thus be understood if a diffusive process to leave the cage is required before scavenging occurs. At very high concentrations of scavenger where scavenger molecules have high probability of being nearest neighbours to the radicals where they are formed, the amount of cage radical combination was diminished[93].

The mechanism of thermal decomposition of azo-bis(isobutyronitrile) may be summarized by the following scheme, where RN_2R is the azonitrile, RR the symmetrical dimer tetramethylsuccinonitrile, RR′ the unsymmetrical dimer—ketenimine **27**, R· the 2-cyano-2-propyl radical **25**, S a scavenger and reactants in the cage are indicated by superscribed lines[93].

$$RN_2R \longrightarrow \overline{2\,R\cdot + N_2}$$

$$\overline{2\,R\cdot + N_2} \longrightarrow RR + N_2$$

$$\overline{2\,R\cdot + N_2} \longrightarrow RR' + N_2$$

$$\overline{2\,R\cdot + N_2} \longrightarrow 2\,R\cdot + N$$

$$R\cdot + S \longrightarrow RS$$

$$RR' \rightleftharpoons \overline{2\,R\cdot}$$

$$\overline{2\,R\cdot} \longrightarrow RR$$

$$\overline{2\,R\cdot} \longrightarrow 2\,R\cdot$$

$$\overline{2\,R\cdot + N_2} \longrightarrow \text{disproportionation products}$$

$$\overline{2\,R\cdot} \longrightarrow \text{disproportionation products}$$

While a scheme such as this accounts very adequately for the thermal decomposition of **23**, a complication arises in the photochemical decomposition. In principle, photodecomposition of either singlet or triplet azonitrile or ketenimine could give birth to a radical pair. A triplet radical pair might diffuse apart more readily than spin inversion occurs, and thus decrease the cage combination of radicals. While this does not appear to happen in the direct or sensitized photolysis of the azonitrile **29**, a decreased amount of cage recombination was found with the sensitized photolysis of the ketenimine **30**[94].

It should be mentioned that the cage effect in azonitrile decomposition has not been universally accepted[95].

D. Reactions of the 2-Cyano-2-propyl Radical

I. Addition reactions

The decomposition of **23** in *p*-xylene with oxygen gave *t*-butyl alcohol, hydrogen cyanide, cyanogen, acetone, *p*-tolualdehyde, tetramethylsuccinonitrile and the hydroperoxide **44**[96]. The hydroperoxide is thermally stable up to 120°, but is subject to induced

$$\mathbf{23} \longrightarrow (CH_3)_2\underset{\displaystyle CN}{\underset{|}{C}}\cdot \xrightarrow{O_2} (CH_3)_2\underset{\displaystyle CN}{\underset{|}{C}}{-}O{-}O\cdot \longrightarrow \underset{\displaystyle (\mathbf{44})}{(CH_3)_2\underset{\displaystyle CN}{\underset{|}{C}}{-}OOH}$$

decomposition. The peroxy radical precursor of **44** was trapped with 2,6-bis(*t*-butyl)-4-methylphenol[97,98]. This peroxy radical evidently

$$(CH_3)_2\underset{\displaystyle CN}{\underset{|}{C}}{-}O{-}O\cdot \; + \; \text{2,6-di-}t\text{-Bu-4-}CH_3\text{-phenol (OH, }t\text{-Bu, Bu-}t\text{, }CH_3\text{)} \longrightarrow \text{2,6-di-}t\text{-Bu-4-}CH_3\text{-4-}[O{-}O{-}C(CH_3)_2CN]\text{-cyclohexa-2,5-dienone}$$

reacted with diphenylpicrylhydrazyl to give diphenyl nitroxide radical and **45**[99]. This type of reaction probably accounts for the previously discovered inaccuracy of radical counting with diphenylpicrylhydrazyl when traces of oxygen are present[68a]. The reaction of the 2-cyano-2-propyl radical with sulphur gave the disulphide **45a** and none of the monosulphide[68b].

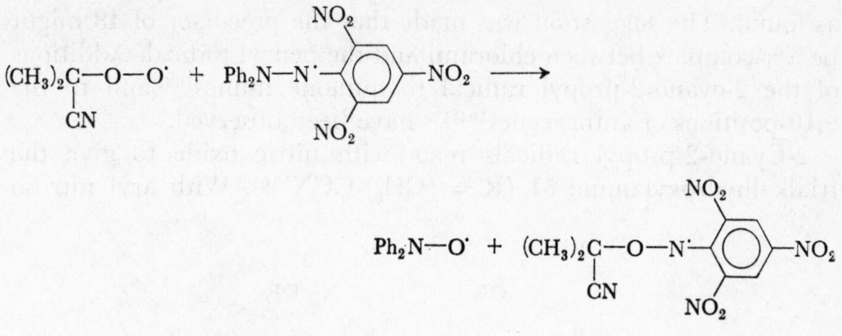

$$(CH_3)_2C(CN){-}O{-}O^\cdot + Ph_2N{-}N^\cdot{-}C_6H_2(NO_2)_3 \longrightarrow Ph_2N{-}O^\cdot + (CH_3)_2C(CN){-}O{-}N^\cdot{-}C_6H_2(NO_2)_3$$

(45)

$$(CH_3)_2C(CN){-}S{-}S{-}C(CN)(CH_3)_2$$

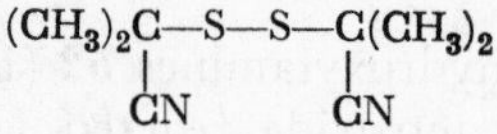

(45a)

Radicals from azo-bis(isobutyronitrile) have been added to benzoquinone to give the mono- (**46**) and bis(cyanoalkylethers) (**47**)[100]. Similar products have been obtained by additions to

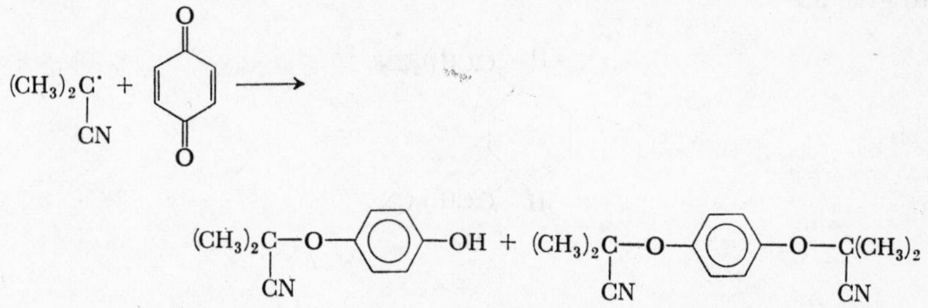

$$(CH_3)_2\dot{C}(CN) + O{=}C_6H_4{=}O \longrightarrow (CH_3)_2C(CN){-}O{-}C_6H_4{-}OH + (CH_3)_2C(CN){-}O{-}C_6H_4{-}O{-}C(CN)(CH_3)_2$$

(46) (47)

chloranil, 2,5-diacetoxybenzoquinone, toluquinone and phenanthraquinone[101]. An anomaly has been observed in the reactions of azonitriles with chloranil in refluxing toluene[102]. In addition to products analogous to **46** and **47**, the benzyl ether **48** was obtained in moderate yields. If the precursor of **48** was the phenoxy radical **49**, it did not seem reasonable that compound **50** would be absent,

$$PhCH_2O{-}C_6Cl_4{-}OH \qquad PhCH_2O{-}C_6Cl_4{-}O^\cdot \qquad PhCH_2O{-}C_6Cl_4{-}OCR_2(CN)$$

(48) (49) (50)

as found. The suggestion was made that the precursor of **48** might be a π-complex between chloranil and the benzyl radical. Additions of the 2-cyano-2-propyl radical to quinone imines[103] and to the 9,10-positions of anthracenes[104,105] have been observed.

2-Cyano-2-propyl radicals react with nitric oxide to give the trialkylhydroxylamine **51** (R = $(CH_3)_2CCN$)[106]. With aryl nitroso

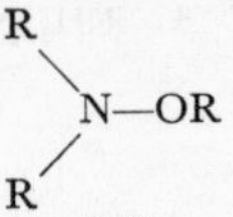

$$R_2N{-}OR \quad \textbf{(51)} \qquad Ar(R)N{-}OR \quad \textbf{(52)}$$

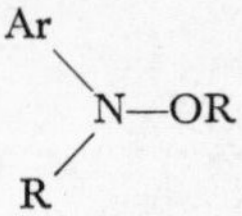

compounds, aryldialkylhydroxylamines **52** (R = $(CH_3)_2CCN$) were obtained[103]. Dinitrogen tetroxide reacted with azonitriles to give 2-nitronitriles[107]. 2-Cyano-2-propyl radicals underwent stepwise 1,3-addition to nitrones[108]. The intermediate nitroxide radicals have been observed in some cases[109]. Addition of 2-cyano-2-propyl radicals to furfural[110] and to 6,6-diphenylfulvene[111] have been reported. 2-Cyano-2-propyl radicals added 1,4 to cyclooctatetraene to give **53**[112].

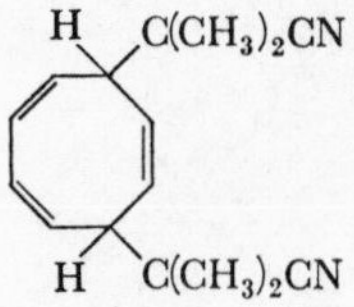

H C(CH$_3$)$_2$CN

H C(CH$_3$)$_2$CN

(53)

2. Abstraction reactions

Cyanoalkyl radicals from azonitriles abstracted hydrogen from thiols[113,114], cumene[73], tetralin[115] and fluorene[116]. Less reactive hydrogen donors such as toluene were not attacked, although it has been suggested that the 2-cyano-2-propyl radical can abstract hydrogen from methanol to give the hydroxymethyl radical[117]. Cyanoalkyl radicals reacted readily with iodine and bromine to give the 2-halonitriles[118]. 2-Halonitriles were also obtained from reactions of azonitriles with various positive halogen compounds[119]. Azonitriles were also useful as initiators in brominations with *N*-bromosuccinimide[120], and in the addition of hydrogen sulphide to olefins[121a].

3. Oxidations and reductions

Oxidation of cyanoalkyl radicals to carbonium ions would be expected to be difficult because of the electron-withdrawing properties of the cyano group. This was found to be the case, for attempted electrochemical oxidation of the 1-cyano-2,2-dimethylpropyl radical gave less than 1% of products derived from the carbonium ion[76b]. The approximate oxidation potential of cyanoalkyl radicals has been estimated to be slightly less than 0·15 v versus the saturated calomel electrode[84]. This value seems much too low, for the oxidation potential of the 2-propyl radical has been calculated to be +1·35 v versus the saturated calomel electrode[121b]. It appears probable that other more readily reactive intermediates were present in the experimental work. More recently the ionization potentials of 2-cyanoalkyl radicals have been determined by mass spectrometry[122]. The ionization potentials of the cyanoalkyl radicals were nearly 2 ev higher than the analogous alkyl radicals. Oxidation by ligand transfer can occur readily, however. The decomposition of azobis(isobutyronitrile) in aqueous dioxan with cupric chloride gave a 55% yield of α-chloroisobutyronitrile along with formation of

$$(CH_3)_2\underset{\displaystyle CN}{\underset{|}{C}}\cdot + CuCl_2 \longrightarrow (CH_3)_2CClCN + CuCl$$

cuprous chloride[123]. In acetonitrile yields as high as 84% of the α-chloroisobutyronitrile have been obtained[124]. Here the cupric chloride is as good or better a radical trap as oxygen. The cupric chloride oxidized the radical by a ligand-transfer process rather than by electron transfer. With cupric acetate in acetic acid or acetonitrile solution the decomposition of the azonitrile gave no α-acetoxyisobutyronitrile, but 42% tetramethylsuccinonitrile and 47% of the amide **54** were obtained (from reaction of the ketenimine). In aqueous or alcoholic solutions cupric acetate

$$(CH_3)_2CHCONH{-}C(CH_3)_2CONH_2$$

(54)

oxidized α-cyanoalkyl radicals to carbonium ions and products resulting from incorporation of solvent were found[125,126].

Reduction of the 2-cyano-2-propyl radical to the carbanion by tetraethyl-*p*-phenylenediamine has been observed[127].

VI. RADICAL REARRANGEMENTS

The effect of substituents on migrating aryl groups in free-radical rearrangements has been examined in several instances. The relative migratory rates in the *t*-butyl peroxide induced decarbonylations of *β*-aryl isovaleraldehydes varied over a factor of 50[128]. The *p*-cyanophenyl substituted aldehyde had the greatest rate which was

$$ArC(CH_3)_2CH_2CHO \longrightarrow ArC(CH_3)_3 + ArCH_2CH(CH_3)_2 + ArCH_2C(CH_3){=}CH_2 + ArCH{=}C(CH_3)_2$$

19 times faster than that of the phenyl derivative. A similar study of the thermal decomposition of *t*-butyl peresters of *β*-arylisovaleric acids showed the *p*-cyanophenyl group to migrate 35 times faster

$$\left[CH_3\overset{CH_3}{\underset{C_6H_4R}{C}}{-}CH_2^{\cdot} \longleftrightarrow CH_3\overset{CH_3}{\underset{C_6H_4R^{-}}{C}}{-}CH_2^{+} \longleftrightarrow CH_3\overset{CH_3}{\underset{C_6H_4R^{-}}{\overset{+}{C}}}{-}CH_2 \longleftrightarrow CH_3\overset{CH_3}{C}{-}CH_2 \text{ (spiro-bridged cyclohexadienyl, R)} \right]$$

than phenyl[129]. The data indicated that the ability of the aromatic substituent to accommodate negative charge stabilized the transition state for the rearrangement. Similar but smaller effects were observed in photorearrangements of arylcyclohexenones which occurred by triplet excited states[130,131].

VII. RADICAL ADDITIONS TO ACRYLONITRILE AND OTHER CYANOOLEFINS

A. Acrylonitrile and Vinylidene Cyanide

I. Polymerizations

Acrylonitrile is very susceptible to free-radical attack. Polymerization can occur at room temperature if inhibitors such as oxygen are absent. An extensive literature is available on homopolymerization and copolymerizations of acrylonitrile[132–136]. The polymerization of acrylonitrile is exothermic (ΔH (polymerization) = 17·3 kcal/mole)[137], and difficult to control in bulk polymerization. Acrylonitrile has frequently been polymerized in aqueous solution or in emulsion using redox initiation systems such as persulphate ion with ferrous ion and bisulphite or thiosulphate ions[138]. In these systems, the actual initiator is probably the sulphate anion radical $(SO_3^{\cdot -})$[139].

In general, acrylonitrile is less reactive to radicals than are styrene, butadiene and methyl vinyl ketone, and is more reactive than acrylates, vinyl chloride, ethylene, propylene, vinyl acetate and vinyl ethers. If an initiator can combine a nucleophilic attack with a mechanism for radical propagation, acrylonitrile polymerization becomes faster than that of styrene. Tris(trifluoroacetylacetonato)-manganese (III) (**55**) increased the rate of polymerization of acrylonitrile by nearly two powers of ten over th rate with tris(acetyl-acetonato)manganese (III)[140]. While **55** increased the rate of methyl methacrylate polymerization, it retarded the thermal polymerization of styrene. The following scheme was suggested for initiation:

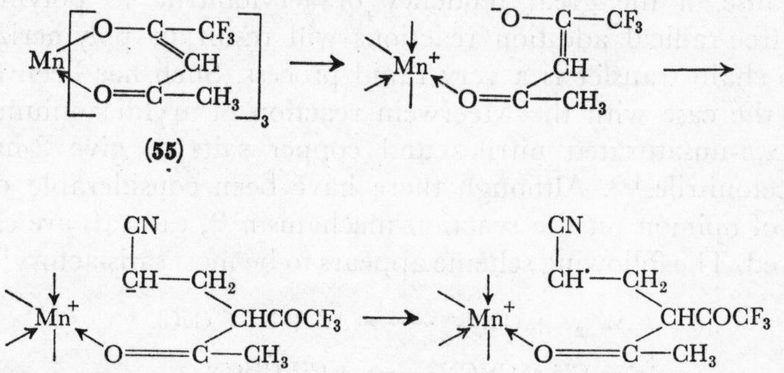

$$Mn\left[\begin{matrix} O-C-CF_3 \\ \quad CH \\ O=C-CH_3 \end{matrix}\right]_3 \longrightarrow \; >Mn^{+} \leftarrow \begin{matrix} ^{-}O-C-CF_3 \\ \quad \| \\ \quad CH \\ O=C-CH_3 \end{matrix} \longrightarrow$$

(55)

$$>Mn^{+} \leftarrow \begin{matrix} CN \\ | \\ CH-CH_2 \\ \quad CHCOCF_3 \\ O=C-CH_3 \end{matrix} \longrightarrow \; >Mn^{+} \leftarrow \begin{matrix} CN \\ | \\ CH^{\cdot}-CH_2 \\ \quad CHCOCF_3 \\ O=C-CH_3 \end{matrix}$$

The initial attack of **55** on acrylonitrile is thought to be nucleophilic rather than free radical, but the intermediate zwitterion can propagate homolytically.

The polymerization and copolymerization of vinylidene cyanide offers some interesting comparisons to acrylonitrile. Homopolymerization of vinylidene cyanide with free-radical initiation is very slow[141]. The slow rate of propagation can be ascribed to the unfavourable electrostatic interaction of a strong acceptor radical with an electron-poor double bond. Copolymerizations of vinylidene cyanide occur very readily and show a strong tendency for 1 : 1 alternation[142]. Vinylidene cyanide and vinyl acetate gave a 1 : 1 alternating copolymer over a wide range of monomer feed ratios. The tendency for 1 : 1 alternation was less pronounced with electron deficient comonomers such as maleic anhydride[143]. Copolymerization of styrene and vinylidene cyanide could only be prevented at room temperature with high concentrations of *t*-butylhydroquinone. By contrast, copolymers with acrylonitrile were much more random[143]. The composition of acrylonitrile–vinyl acetate copolymer was very dependent upon the monomer feed ratio used.

The differences in copolymerization behaviour may be explained by the much greater importance of polar effects in the copolymerizations with vinylidene cyanide. The dicyanoalkyl radical is a much stronger acceptor radical than the cyanoalkyl. The former radical will show a much greater tendency to add to electron-rich double bonds than will the latter. Radical addition to vinylidene cyanide will be much more rapid with donor radicals than with acceptor radicals.

2. Meerwein reaction

Because of the great tendency of acrylonitrile to polymerize, most free-radical addition reactions will result in polymerization unless chain transfer is a very rapid process. Such has been found to be the case with the Meerwein reaction of aryldiazonium salts with α,β-unsaturated nitriles and copper salts to give 2-halo-β-arylacetonitriles[144]. Although there have been considerable differences of opinion on the reaction mechanism[144], radicals are clearly involved. The following scheme appears to be most satisfactory[145,146]:

$$ArN_2^+ + CuCl_2^- \longrightarrow Ar^\bullet + N_2 + CuCl_2$$

$$Ar^\bullet + CH_2{=}CHCN \longrightarrow ArCH_2\dot{C}HCN$$

$$ArCH_2\dot{C}HCN + CuCl_2 \longrightarrow ArCH_2CHClCN + CuCl$$

Even though the reactions are frequently run with cupric halides, cuprous halides which are essential for the reaction are formed rapidly by reaction with the aqueous acetone medium[145,147]. The last ligand-transfer step must be very rapid, for polymerization of olefins such as styrene and acrylonitrile is not observed[144]. Phenyl radicals generated from nitrosoacetanilide, phenylazotriphenylmethane and benzoyl peroxide have also been added to acrylonitrile in the presence of cupric and ferric salts to accomplish the same overall reaction[146].

Similar reactions of diazonium salts with methacrylonitrile, cinnamonitrile and *p*-nitrophenylacrylonitrile have been described[144]. In the last two cases, the aryl radical added to give the β-cyano radical rather than the α-cyano radical because of the greater stabilities of the benzyl radicals which were formed. The products were α-arylcinnamonitriles. The yields of the Meerwein reaction are frequently high and the reaction is synthetically useful[144].

The Meerwein reaction has been used in competition experiments in order to determine relative reactivities of attack by aryl radical attack on various vinyl monomers[148].

3. Other ligand-transfer reactions

Additions of radicals from ketone peroxides to acrylonitrile followed by ligand-transfer reactions have been reported[149,150]. Ethyl radicals from methyl ethyl ketone peroxide with acrylonitrile and cupric chloride gave α-chlorovaleronitrile and with cupric bromide α-bromovaleronitrile. Other examples include those below.

$$\text{1-hydroxy-1-hydroperoxycyclopentane (HO, OOH)} + CH_2{=}CHCN \xrightarrow[HCl,\ H_2O]{FeCl_2,\ CuCl_2} HOOC(CH_2)_5CHClCN$$

$$\text{1-hydroxy-1-hydroperoxycyclohexane (HO, OOH)} + CH_2{=}CHCN \xrightarrow[HCl,\ H_2O]{FeO_2,\ CuCl_2} HOOC(CH_2)_6CHClCN$$

There have been many reported examples of the hydrodimerization of acrylonitrile to give adiponitrile. Suggestions have been made that the electrolytic hydrodimerization involved addition of an electron to acrylonitrile to give the anion radical, protonation to give the 2-cyanoethyl radical and dimerization[151]. A more generally accepted mechanism, however, is the addition of two electrons to give a dianion, addition of the dianion to acrylonitrile to give the 1,4-dicyano-1,4-butyl dianion and protonation[152].

B. Cyanobicyclobutanes

The photochemical addition of mercaptans to 1-cyano-3-methylbicyclobutane to give cyclobutanes has been observed[153]. The

$$CH_3\text{–(bicyclobutane)–}CN + RSH \xrightarrow{h\nu} CH_3\text{–(cyclobutane)}(CN)(SR) + RS(CH_3)\text{(cyclobutane)–}CN$$

preferred (4:1) attack at the carbon adjacent to the nitrile may be a reflection of the donor properties of mercapto radicals. It is also likely that the two intermediate radicals differ little in relative stability, for the cyanocyclobutyl radical **56** would not be greatly stabilized by electron delocalization because of the relative high energy of resonance structures involving an exocyclic double bond on the four-membered ring.

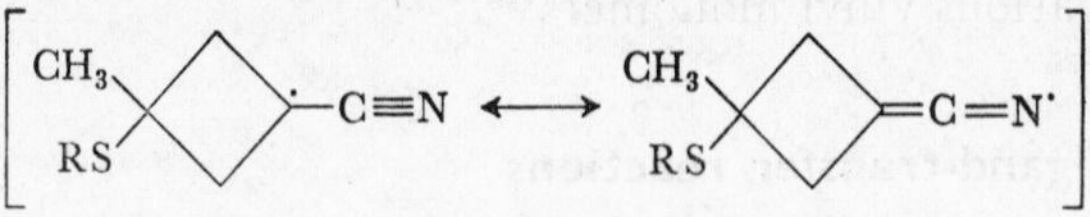

$$\left[\text{(CH}_3\text{)(RS)C}_4\text{H}_5\dot{}\text{—C}\equiv\text{N} \longleftrightarrow \text{(CH}_3\text{)(RS)C}_4\text{H}_5\text{=C=N}^{\cdot} \right]$$

(56)

C. Cycloadditions of α,β-Unsaturated Nitriles

When acrylonitrile was heated to 200–300° in the presence of polymerization inhibitors, dimerization to *cis*- and *trans*-1,2-dicyanocyclobutane occurred[154]. Thermal cycloadditions of this type are believed to involve 1,4-diradicals[155]. Ring closure of the 1,4-diradical is sufficiently rapid to prevent scavenging by the polymerization inhibitor. In addition to the two cyclobutanes, methacrylonitrile gave the ring-opened **57**[156]. The amount of this compound increased with increasing temperature and it probably resulted from ring

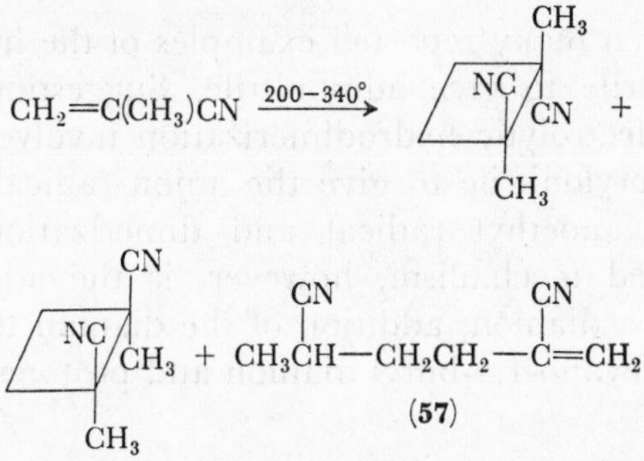

$$CH_2{=}C(CH_3)CN \xrightarrow{200-340°} \text{(cyclobutane with NC, CN, CH}_3\text{, CH}_3\text{)} +$$

$$\text{(cyclobutane with CN, NC, CH}_3\text{, CH}_3\text{)} + CH_3\overset{CN}{\overset{|}{C}}H{-}CH_2CH_2{-}\overset{CN}{\overset{|}{C}}{=}CH_2$$

(57)

opening of the cyclobutanes to give the diradical, which could undergo intramolecular disproportionation. Cyclodimerization of acrylonitrile has been accomplished photochemically[157,158]. That triplet sensitizers are required for the photodimerization suggests that the triplet state of acrylonitrile is formed and adds to acrylonitrile to form the same 1,4-diradical which is an intermediate in the thermal reaction.

Allenes and acrylonitrile underwent cycloaddition at 150–250° to give methylenecyclobutanes[159]. Similar additions occurred with methacrylonitrile and α-acetoxyacrylonitrile[159]. With dimethylallene

$$CH_2{=}C{=}CH_2 + CH_2{=}CRCN \longrightarrow \text{(3-methylenecyclobutane bearing R and CN)}$$

isomers were formed[159]. Acrylonitrile and tetrafluoroethylene gave an 84 % yield of cycloadduct[160]. Allyl cyanide also gave a cycloadduct with tetrafluoroethylene[160]. 1,1-Dichloro-2,2-difluoroethylene and

$$(CH_3)_2C{=}C{=}CH_2 + CH_2{=}CRCN \longrightarrow \text{(cyclobutane with R, CN and }{=}C(CH_3)_2\text{)} + \text{(cyclobutane with R, CN, }(CH_3)_2\text{ and }{=}CH_2\text{)}$$

acrylonitrile gave a similar cycloadduct[161]. Perfluoroacrylonitrile also cyclodimerized[162].

$$CF_2{=}CF_2 + CH_2{=}CHCN \xrightarrow{150°} \begin{array}{l} CF_2{-}CH_2 \\ |\quad\;\;\; | \\ CF_2{-}CHCN \end{array}$$

$$CF_2{=}CF_2 + CH_2{=}CHCH_2CN \longrightarrow \begin{array}{l} CF_2{-}CH_2 \\ |\quad\;\;\; | \\ CF_2{-}CH{-}CH_2CN \end{array}$$

Several α-mercaptoacrylonitriles have been prepared and been shown to cyclodimerize at room temperature[163–165].

$$\begin{array}{l} CH_2{=}CSR \\ \quad\;\;\;\; | \\ \quad\;\;\; CN \end{array} \longrightarrow \begin{array}{l} \qquad\;\; SR \\ \qquad\;\; | \\ CH_2{-}CCN \\ |\qquad\; | \\ CH_2{-}CCN \\ \qquad\;\; | \\ \qquad\;\; SR \end{array}$$

D. Additions to Cyanocarbons

A few examples of free-radical additions to tetracyanoethylene and tetracyanoquinodimethane have been described. Decomposition of azo-bis(isobutyronitrile) with tetracyanoethylene gave **58**[166], and with tetracyanoquinodimethane **59** was obtained[167]. Methyl radicals

$$(CH_3)_2C(CN){-}C(CN)_2{-}C(CN)_2{-}C(CN)(CH_3)_2 \quad \textbf{(58)}$$

$$(CH_3)_2C(CN){-}C(CN)_2{-}C_6H_4{-}C(CN)_2{-}C(CN)(CH_3)_2 \quad \textbf{(59)}$$

from the decomposition of *t*-butyl peroxide added to tetracyanoethylene to give 2,2,3,3-tetracyanobutane[41]. The preparation of terephthaloyl cyanide from tetracyanoquinodimethane and dinitrogen tetroxide probably involved addition of NO_2 radicals[168]. Photolysis of a tetrahydrofuran solution of tetracyanoquinodimethane resulted in formation of the adduct **60**[169]. The same product was

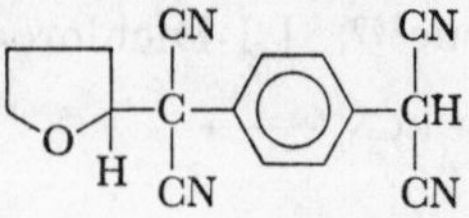

(60)

obtained by a *t*-butyl peroxide initiated addition. Under the photolytic conditions tetracyanoethylene and tetrahydrofuran gave a 2:1 adduct of unknown structure.

The addition of bicyclo[2.1.0]pentane to dicyanoacetylene occurred at room temperature, apparently by way of the diradical **61**[170]. The reaction was much faster than the analogous reaction with dicarbomethoxyacetylene.

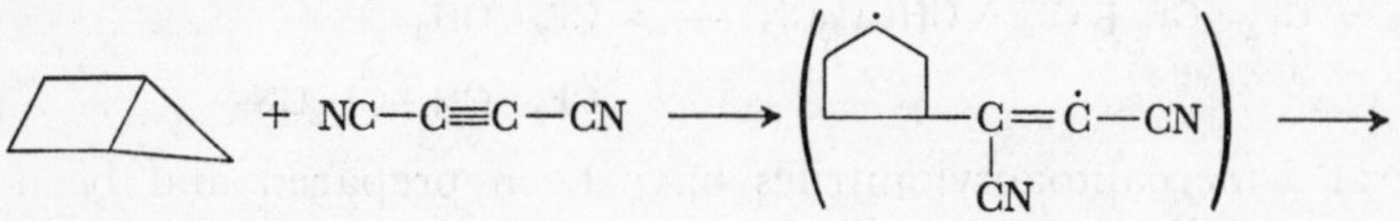

+ NC—C≡C—CN ⟶ (—C=Ċ—CN, CN) ⟶

(61)

—CN + —C=CHCN, CN

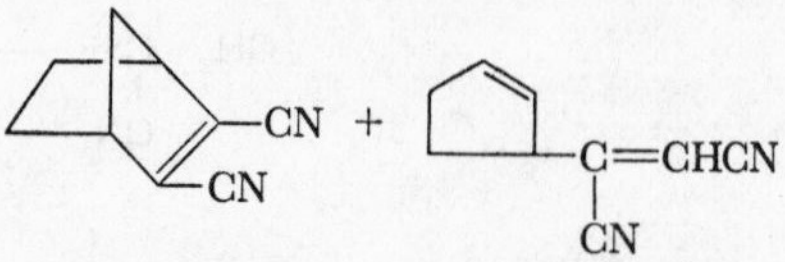

The electron-poor double bonds of polycyano olefins appear to react readily with donor radicals such as alkyl and alkoxyalkyl radicals, while reactions with acceptor radicals are not facilitated.

VIII. CYANOCARBENES

Very little work has been reported on reactions of cyanocarbene, probably because of the very explosive nature of diazoacetonitrile. Addition of the carbene to 2,5-dimethylhexa-2,4-diene to give *cis*- and *trans*-chrysanthemonitrile has been reported[171] as has the addition to benzene to give cyanonorcaradiene[172]. The latter reaction product is almost certainly cyanocycloheptatriene. Photolysis of diazoacetonitrile in matrix gave cyanocarbene in its ground triplet state[173,174]. Flash photolysis of diazoacetonitrile produced cyanocarbyne ($\cdot\ddot{C}$—CN) as well as cyanocarbene[175]. Analysis of the spectrum of cyanocarbyne indicated that its ground state is a doublet.

Dicyanocarbene has been more thoroughly investigated. Thermal or photochemical decomposition of dicyanodiazomethane generated the carbene which inserted into carbon–hydrogen bonds, added to olefins to give dicyanocyclopropanes, and to acetylenes to give dicyanocyclopropenes[176]. Dicyanocarbene does discriminate in its insertion reactions. The relative rates of insertion into primary, secondary and tertiary carbon–hydrogen bonds were 1:4·6:12. The addition to *cis*- and *trans*-2-butene was shown to be non-stereospecific. In neat olefin, less than 10% of the non-stereospecific adducts were formed. At high dilution (with cyclohexane) both olefins gave an identical mixture of 70% *trans*- and 30% *cis*-1,1-dicyano-2,3-dimethylcyclopropane[176]. The data suggested a ground triplet state, and this was confirmed by a study of the electron spin resonance of dicyanocarbene obtained by photolysis of the diazo compound in matrix[177].

Photolysis of dicyanodiazomethane in benzene gave dicyanonorcaradiene (**62**) in 82% yield[178,179]. The norcaradiene was stable

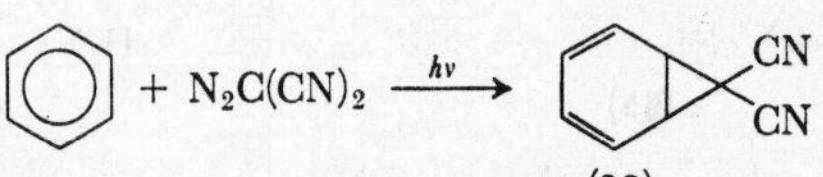

$$C_6H_6 + N_2C(CN)_2 \xrightarrow{h\nu} \text{(62)}$$

(62)

up to 160° at which temperature it rearranged to phenylmalononitrile[180]. Two isomeric norcaradienes were obtained in the reaction with *p*-xylene and three isomers were obtained from naphthalene[179]. Upon photolysis the dicyanonorcaradienes generated dicyanocarbene and their aromatic precursor[180]. The products obtained

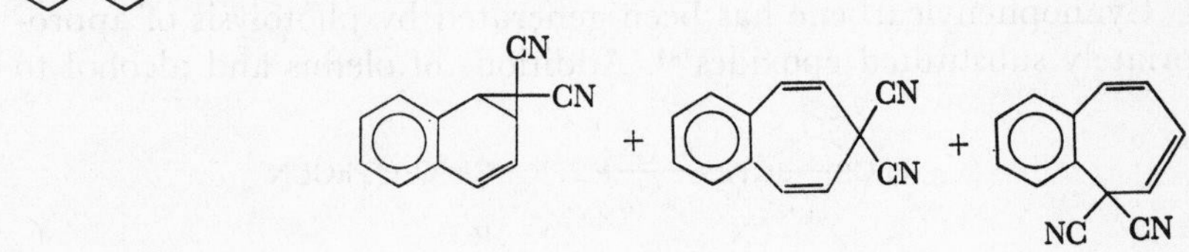

$$C_{10}H_8 + N_2C(CN)_2 \longrightarrow$$

CN, CN + CN, CN + NC, CN

from naphthalene underwent an interesting series of thermal 1,5-cyano shifts[180].

As discussed previously, dicyanocarbene is not an intermediate in the formation of dicyanocyclopropane, from bromomalononitrile, olefins and base, and probably not an intermediate in the formation of tetracyanoethylene from dibromomalononitrile and copper. Pyrolysis of dicyanodiazomethane at 220°, however, gave tetracyanoethylene, cyanogen and dicyanoacetylene[181], probably by way of dicyanocarbene. Thermal decomposition of tetracyanoethylene at 800° gave dicyanoacetylene and cyanogen[181].

The thermal decomposition of **63** in benzene gave a 77% yield of the equilibrium mixture **64** of the norcaradiene and cycloheptatriene[182]. Equilibration is rapid at room temperature.

$$CF_3CN_2CN + C_6H_6 \longrightarrow \text{7-CN-7-}CF_3\text{-norcaradiene} \rightleftharpoons \text{7-CN-7-}CF_3\text{-cycloheptatriene}$$

(63) **(64)**

Thermal or photochemical decomposition of the sodium salt of **65** and photolysis of the lithium salt of **65** gave hydrogen cyanide tetramer[183] (**66**). Aminocyanocarbene (**67**) was suggested as an

$$(NC)(H_2N)C{=}N{-}NHSO_2{-}C_6H_4{-}CH_3 \qquad (NC)(H_2N)C{=}C(CN)(NH_2)$$

(65) **(66)**

intermediate in the reactions. The intermediate could not be trapped with olefins. Photolysis of the lithium salt of **65** in 2-methyl-

$$[H_2N{-}\ddot{C}{-}C{\equiv}N \longleftrightarrow H_2\overset{+}{N}{=}\overset{-}{\ddot{C}}{-}C{\equiv}N \longleftrightarrow H_2\overset{+}{N}{=}C{=}C{=}\overset{-}{N}]$$

(67)

tetrahydrofuran glass at $-196°$ produced a yellow substance with absorption in the 332–450 mμ region. The spectrum was assigned to the carbene **67**, and from the absence of an electron spin resonance signal it was inferred that the carbene had a ground singlet state.

Cyanophenylcarbene has been generated by photolysis of appropriately substituted epoxides[184]. Additions of olefins and alcohol to

$$PhCR\overset{O}{\frown}C(CN)Ph \xrightarrow{h\nu} Ph{-}C(R){=}O + Ph\ddot{C}CN$$

the carbene were observed. Cyanophenylcarbene was suggested as an intermediate in the formation of *trans*-dicyanostilbene from the reaction of bromoiodostyryl azide (**68**) with zinc[185].

$$\underset{\textbf{(68)}}{\mathrm{PhC(I)\!=\!C(Br)N_3}} \xrightarrow{\mathrm{Zn}} [\mathrm{PhC{\equiv}C{-}N_3}] \longrightarrow [\mathrm{PhC{\equiv}C{-}\ddot{N}} \longleftrightarrow \mathrm{Ph\ddot{C}{-}C{\equiv}N}] \longrightarrow \mathrm{(NC)(Ph)C\!=\!C(Ph)(CN)}$$

Photolysis of the diazo compound **69** generated the intermediate **70** which reacted as an aryl radical in its arylation reactions with aromatics[186]. The intermediate **70** has been shown to be a ground-state triplet by observation of its electron spin resonance spectrum after irradiation of **69** in matrix[187].

$$\underset{\textbf{(69)}}{\mathrm{NC{-}C({=}C_6H_4{=}N_2){-}CN}} \xrightarrow{h\nu} \underset{\textbf{(70)}}{[\mathrm{NC{-}\dot{C}(C_6H_4\cdot){-}CN}]} \xrightarrow{\mathrm{ArH}} [\mathrm{NC{-}\dot{C}(C_6H_4{-}C_6H_5(H)R){-}CN}] \longrightarrow \mathrm{NC{-}CH(C_6H_4{-}C_6H_4R){-}CN}$$

Similar arylation reactions have been observed in the copper or cuprous chloride initiated decompositions of **71**[188].

$$\underset{\textbf{(71)}}{\mathrm{C_5(CN)_4N_2^+}^{\ominus}} \xrightarrow[\mathrm{Cu^I}]{\mathrm{Cu\ or}} [\mathrm{C_5(CN)_4\cdot}^{\ominus}] \xrightarrow{\mathrm{ArH}} \mathrm{C_5(CN)_4Ar}^{\ominus}$$

Cyanonitrene has been generated by thermal decomposition of cyanogen azide and found to react with saturated hydrocarbons to give alkyl cyanamides[189]. In diluted hydrocarbon solvent, the

$$\mathrm{N_3CN} \longrightarrow [\mathrm{\ddot{N}CN}] \xrightarrow{\mathrm{RH}} \mathrm{RNHCN}$$

relative rates of attack on tertiary, secondary and primary carbon–hydrogen bonds were 67:10:1[189]. The insertion into tertiary carbon–hydrogen bonds in undiluted hydrocarbon was highly stereospecific[189]. The stereospecificity of insertion was diminished as the reaction was diluted with an inert solvent, such as methylene chloride or ethyl acetate[190]. In methylene bromide solution, the reaction was stereorandom[190]. It was suggested that thermal decomposition of cyanogen azide initially produced singlet cyanonitrene

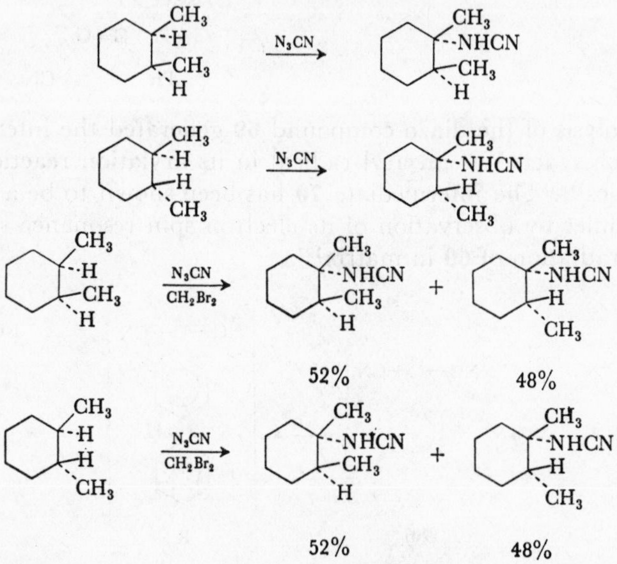

N_3CN

N_3CN

$\xrightarrow[CH_2Br_2]{N_3CN}$

52% 48%

$\xrightarrow[CH_2Br_2]{N_3CN}$

52% 48%

which could undergo collisional deactivation to the ground-state triplet cyanonitrene. The singlet nitrene was presumed to insert stereospecifically and the triplet nitrene non-stereospecifically (presumably by a hydrogen abstraction/radical recombination mechanism).

Photolysis of cyanogen azide also generated cyanonitrene[191]. The insertion of the photo-generated nitrene into *cis*-1,2-dimethylcyclohexane was found to be stereospecific, and it was concluded that the singlet nitrene was produced. The same stereochemical result was obtained over a range of wavelengths.

Cyanonitrene has been observed spectroscopically by flash photolysis of diazomethane[192] as well as cyanogen azide[193–196]. It was claimed that photolysis of cyanogen azide with wavelengths greater than 2750 Å gave triplet cyanonitrene by way of a primary photochemical process[196]. This interpretation would be in conflict with the stereospecificity of the carbon–hydrogen insertion of photogenerated cyanonitrene at these wavelengths[191]. The electron spin resonance spectrum of triplet cyanonitrene has been obtained[177].

Thermal decomposition of cyanogen azide in aromatic solvents produced *N*-cyanoazepines which were probably formed by way of the azanorcaradienes[197]. Thermal decomposition of cyanogen azide

$$C_6H_5X + N_3CN \xrightarrow{45-60^\circ} [\text{azanorcaradiene (NCN, X)}] \longrightarrow \text{N-cyanoazepine (X, N–CN)}$$

in cyclooctatetraene gave **72, 73** and **74,** the latter two being products

$$C_8H_8 + N_3CN \xrightarrow{78^\circ} \mathbf{(72)} + \mathbf{(73)} + \mathbf{(74)}$$

of the nitrene[198]. It was determined that triplet NCN gave primarily the 1,4-adduct **73** and singlet NCN primarily the 1,2-adduct **74**.

IX. THE CYANO RADICAL

The cyano radical has been extensively studied by physical chemists, but very little work has been done in organic systems. Photochemical and thermal decompositions of cyanogen and the cyanogen halides have been the principal methods of generation of the cyano radical. Two recent papers[199,200] provide entry into the physicochemical literature.

Photolysis of cyanogen chloride in hydrocarbons[201] or ethers[202] produced nitriles and hydrogen chloride by a free-radical chain reaction. The cyano radical is only formed in the initiation step, for the chlorine atom is the chain-carrying radical. If there is chlorine

$$\text{ClCN} \xrightarrow{h\nu} \text{Cl}\cdot + \text{CN}\cdot$$
$$\text{Cl}\cdot + \text{RH} \longrightarrow \text{HCl} + \text{R}\cdot$$
$$\text{R}\cdot + \text{ClCN} \longrightarrow \text{RCN} + \text{Cl}\cdot$$

present in the cyanogen chloride, the cyano radical is probably not involved in the reaction at all. The yields of nitriles (based upon the cyanogen chloride consumed) were quite high. With ethers it was necessary to perform the photolysis in the presence of a hydrogen chloride acceptor such as sodium bicarbonate. 2-Ethoxypropionitrile was formed in high yield from ethyl ether. The α-cyano ethers are the major isomers obtained from cyclic ethers. Homolytic cyanation of cyclohexane and heptane was obtained with cyanogen chloride and acetylcyclohexylsulphonyl peroxide[203]. Cyano radicals were probably not intermediates in this reaction.

Aromatic nitriles have been obtained by photolysis of cyanogen iodide[204,205]. The reaction appears to be a typical radical aromatic substitution reaction. It is possible that the iodine atoms oxidize the intermediate cyclohexadienyl radicals to the aromatic nitrile.

$$ICN \xrightarrow{h\nu} I^{\cdot} + CN^{\cdot}$$

$$R\text{—}C_6H_5 + CN^{\cdot} \longrightarrow R\text{—}C_6H_5(H)(CN)^{\cdot}$$

$$R\text{—}C_6H_5(H)(CN)^{\cdot} + I^{\cdot} \longrightarrow R\text{—}C_6H_4\text{—}CN + HI$$

Several suggestions have been made that the cyano radical is formed in thermal reactions of mercuric cyanide. At 100° cyanotrichlorosilane was formed from mercuric cyanide and disilicon hexachloride[206], and triphenylmethyl radical reacted with mercuric cyanide to give triphenylacetonitrile[207], but in neither reaction has there been evidence to substantiate these suggestions.

The formation of aromatic nitriles by electrochemical oxidation of aromatic compounds with sodium cyanide was originally suggested to be a homolytic reaction involving the cyano radical[208]. There now appears to be general agreement[205,209–211], however, that the reaction is polar.

X. REFERENCES

1. G. S. Hammond, *J. Am. Chem. Soc.*, **77**, 334 (1955).
2. M. H. J. Wijnen, *J. Chem. Phys.*, **22**, 1074 (1954).
3. A. Bruylants, M. Tits, C. Dieu and R. Gauthier, *Bull. Soc. Chim. Belg.*, **61**, 366 (1952).
4. C. Walling and B. B. Jacknow, *J. Am. Chem. Soc.*, **82**, 6113 (1960).
5. D. Couvreur and A. Bruylants, *J. Org. Chem.*, **18**, 501 (1953).
6. G. A. Russell and Y. R. Vinson, *J. Org. Chem.*, **31**, 1994 (1966).

7. R. VanHelden and E. C. Kooyman, *Rec. Trav. Chim.*, **73**, 269 (1954).
8. D. D. Coffman, E. L. Jenner and R. D. Lipscomb, *J. Am. Chem. Soc.*, **80**, 2864 (1958).
9. H. Schmidt and J. Noack, *Z. Anorg. Allgem. Chem.*, **296**, 262 (1958).
10. K. Koyama, T. Susuki and S. Tsutsumi, *Tetrahedron*, **23**, 2665 (1967).
11. J. C. Allen, J. I. G. Cadogan, B. W. Harris and D. H. Hey, *J. Chem. Soc.*, **1962**, 4468.
12. J. C. Allen, J. I. G. Cadogan and D. H. Hey, *J. Chem. Soc.*, **1965**, 1918.
13. J. I. G. Cadogan, D. H. Hey and J. T. Sharp, *J. Chem. Soc. (B)*, **1967**, 803.
14. J. I. G. Cadogan, D. H. Hey and J. T. Sharp, *J. Chem. Soc. (C)*, **1966**, 1743.
15. M. Julia, F. Le Goffic and L. Katz, *Bull. Soc. Chim. France*, **1964**, 1122.
16. M. Julia and C. James, *Compt. Rend.*, **255**, 959 (1962).
17. M. Julia, J. M. Surzur and L. Katz, *Bull. Soc. Chim. France*, **1964**, 1109.
18. J. I. G. Cadogan, D. M. Hey and S. H. Ong, *J. Chem. Soc.*, **1965**, 1932.
19. M. Julia and M. Maumy, *Bull. Soc. Chim. France*, **1966**, 434.
20. M. Julia, J. M. Surzur, L. Katz and F. LeGoffic, *Bull. Soc. Chim. France*, **1964**, 1116.
21. M. Julia and F. LeGoffic, *Bull. Soc. Chim. France*, **1964**, 1129.
22. M. Julia, J. C. Chottard and J. J. Basselier, *Bull. Soc. Chim. France*, **1966**, 3037.
23. M. Julia and P. Dostert, *Compt. Rend.*, **259**, 2872 (1964).
24. (a) C. Walling and E. S. Huyser in *Organic Reactions*, Vol. 13, John Wiley and Sons, New York, 1963, p. 91;
(b) M. Julia, *Free Radicals in Solution*, Butterworths, London, 1967, p. 174.
25. M. Julia, M. Maumy and L. Mion, *Bull. Soc. Chim. France*, **1967**, 2641.
26. P. Boldt, L. Schulz and J. Etzemüller, *Chem. Ber.*, **100**, 1281 (1967).
27. J. S. Swenson and D. J. Renaud, *J. Am. Chem. Soc.*, **87**, 1394 (1965).
28. P. Boldt and L. Schulz, *Tetrahedron Letters*, **1967**, 4351.
29. K. Torsell and K. Dahlqvist, *Acta. Chem. Scand.*, **16**, 346 (1962).
30. J. R. Roland, E. L. Little, Jr. and H. E. Winberg, *J. Org. Chem.*, **28**, 2809 (1963).
31. T. L. Cairns, R. A. Carboni, D. D. Coffman, V. A. Engelhardt, R. E. Heckert, E. L. Little, E. G. McGeer, B. C. McKusick, W. J. Middleton, R. M. Scribner, C. W. Theobald and H. E. Winberg, *J. Am. Chem. Soc.*, **80**, 2775 (1958).
32. E. C. Ladd, *U.S. Pat.* 2,615,915 (1952).
33. K. Auwers and V. Meyer, *Chem. Ber.*, **22**, 1227 (1889).
34. A. Löwenbein, *Chem. Ber.*, **58**, 601 (1925).
35. A. Löwenbein and R. F. Gagarin, *Chem. Ber.*, **58**, 2643 (1925).
36. G. Wittig and W. Hopf, *Chem. Ber.*, **65**, 760 (1932).
37. G. Wittig and H. Petri, *Ann. Chem.*, **513**, 26 (1934).
38. G. Wittig and U. Pockels, *Chem. Ber.*, **69**, 790 (1936).
39. G. V. Schulz and G. Wittig, *Naturwissenschaften*, **27**, 387 (1939).
40. A. Schönberg and A. Mustafa, *J. Am. Chem. Soc.*, **73**, 2401 (1951).
41. H. D. Hartzler, *J. Org. Chem.*, **31**, 2654 (1966).
42. R. L. Huang and K. T. Lee, *J. Chem. Soc.*, **1954**, 2570.
43. H. H. Huang and P. K. K. Lim, *J. Chem. Soc. (C)*, **1967**, 2432.
44. K. Schwetlick, J. Jentzsch, R. Karl and D. Wolter, *J. Prakt. Chem.*, [4] **25**, 95 (1964).
45. L. I. Peterson, *J. Am. Chem. Soc.*, **89**, 2677 (1967).
46. E. Ciganek, *J. Org. Chem.*, **30**, 4198 (1965).
47 M Hunt, *U.S. Pat.* 2,471,959 (1949).
48. J. Thiele and K. Heuser, *Ann. Chem.*, **290**, 1 (1896).

49. J. P. VanHook and A. V. Tobolsky, *J. Am. Chem. Soc.*, **80**, 779 (1958).
50. A. U. Blackham and N. L. Eatough, *J. Am. Chem. Soc.*, **84**, 2922 (1962).
51. A. H. Ewald, *Discussions Faraday Soc.*, **22**, 138 (1956).
52. P. B. Ayscough, B. R. Brooks and H. E. Evans, *J. Phys. Chem.*, **68**, 3889 (1964).
53. F. M. Lewis and M. S. Matheson, *J. Am. Chem. Soc.*, **71**, 747 (1949).
54. C. G. Overberger, M. T. O'Shaughnessy and H. Shalit, *J. Am. Chem. Soc.*, **71**, 2661 (1949).
55. L. M. Arnett, *J. Am. Chem. Soc.*, **74**, 2027 (1952).
56. T. Hirano, T. Miki and T. Tsuruta, *Makromol. Chem.*, **104**, 230 (1967).
57. C. H. Bamford, R. Denyer and J. Hobbs, *Polymer*, **8**, 493 (1967).
58. C. G. Overberger and H. Biletch, *J. Am. Chem. Soc.*, **73**, 4880 (1951).
59. C. G. Overberger and M. B. Berenbaum, *J. Am. Chem. Soc.*, **73**, 2618 (1951).
60. C. G. Overberger and A. Lebovits, *J. Am. Chem. Soc.*, **76**, 2722 (1954).
61. C. G. Overberger, W. F. Hale, M. B. Berenbaum and A. B. Finestone, *J. Am. Chem. Soc.*, **76**, 6185 (1954).
62. C. E. H. Bawn and S. F. Mellish, *Trans. Faraday Soc.*, **47**, 1216 (1951).
63. K. Ziegler, W. Deparade and W. Meye, *Ann. Chem.*, **567**, 141 (1950).
64. C. G. Overberger, H. Biletch, A. B. Finestone, J. Lilker and J. Herbert, *J. Am. Chem. Soc.*, **75**, 2078 (1953).
65. J. C. Martin, J. E. Schultz and J. W. Timberlake, *Tetrahedron Letters*, **1967**, 4629.
66. D. C. Pease and J. A. Roberston, *U.S. Pat.* 2,565,573 (1951).
67. A. F. Bickel and W. A. Waters, *Rec. Trav. Chim.*, **69**, 1490 (1950).
68. (a) G. S. Hammond, J. N. Sen and C. E. Boozer, *J. Am. Chem. Soc.*, **77**, 3244 (1955);
    (b) D. I. Rebyea, P. O. Tawney and A. R. Williams, *J. Org. Chem.*, **27**, 1078 (1962).
69. M. Talât-Erben and S. Bywater, *J. Am. Chem. Soc.*, **77**, 3710 (1955).
70. M. Talât-Erben and S. Bywater, *J. Am. Chem. Soc.*, **77**, 3712 (1955).
71. M. Talât-Erben and A. N. Isfendiyaroĝlu, *Can. J. Chem.*, **36**, 1156 (1958).
72. P. Smith and A. M. Rosenberg, *J. Am. Chem. Soc.*, **81**, 2037 (1959).
73. G. S. Hammond, O. D. Trapp, R. T. Keys and D. L. Neff, *J. Am. Chem. Soc.*, **81**, 4878 (1959).
74. H. P. Waits and G. S. Hammond, *J. Am. Chem. Soc.*, **86**, 1911 (1964).
75. P. A. Leermakers, L. D. Weis and H. T. Thomas, *J. Am. Chem. Soc.*, **87**, 4403 (1965).
76. (a) L. Eberson, *J. Org. Chem.*, **27**, 2329 (1962);
    (b) L. Eberson and S. Nilsson, *Acta Chem. Scand.*, **22**, 2453 (1968);
    (c) L. Eberson, S. Gränse and B. Olofsson, *Acta Chem. Scand.*, **22**, 2462 (1968);
    (d) L. Alderson, personal communication.
77. C. G. Overberger and M. B. Berenbaum, *J. Am. Chem. Soc.*, **74**, 3293 (1952).
78. C. G. Overberger, M. Tobkes and A. Zweig, *J. Org. Chem.*, **28**, 620 (1963).
79. C. G. Overberger, T. B. Gibb, Jr., S. Chibnik, P.-T. Huang and J. J. Monagle, *J. Am. Chem. Soc.*, **74**, 3290 (1952).
80. C. G. Overberger, P.-T. Huang and T. B. Gibb, Jr., *J. Am. Chem. Soc.*, **75**, 2082 (1953).
81. K. Ziegler, *Brennstoff-Chem.*, **30**, 181 (1949).
82. L. M. Arnett and J. H. Peterson, *J. Am. Chem. Soc.*, **74**, 2031 (1952).
83. C. G. Overberger and M. B. Berenbaum, *J. Am. Chem. Soc.*, **73**, 4883 (1951).
84. R. M. Haines and W. A. Waters, *J. Chem. Soc.*, **1955**, 4256.
85. J. C. Bevington, *Trans. Faraday Soc.*, **51**, 1392 (1955).

86. C. E. H. Bawn and D. Verdin, *Trans. Faraday Soc.*, **56**, 815 (1960).
87. J. Betts, F. S. Dainton and K. J. Ivin, *Trans. Faraday Soc.*, **58**, 1203 (1962).
88. J. C. Bevington and N. A. Ghanem, *J. Chem. Soc.*, **1956**, 3506.
89. O. D. Trapp and G. S. Hammond, *J. Am. Chem. Soc.*, **81**, 4876 (1959).
90. C.-H. S. Wu, G. S. Hammond and J. M. Wright, *J. Am. Chem. Soc.*, **82**, 5386 (1960).
91. G. S. Hammond, C.-H. S. Wu, O. D. Trapp, J. Warkentin and R. T. Keys, *J. Am. Chem. Soc.*, **82**, 5394 (1960).
92. R. Hiatt and T. G. Traylor, *J. Am. Chem. Soc.*, **87**, 3766 (1965).
93. H. P. Waits and G. S. Hammond, *J. Am. Chem. Soc.*, **86**, 1911 (1964).
94. J. R. Fox and G. S. Hammond, *J. Am. Chem. Soc.*, **86**, 4031 (1964).
95. C. H. Bamford, W. G. Barb, A. D. Jenkins and P. F. Onyon, *The Kinetics of Vinyl Polymerization by Radical Mechanisms*, Butterworth, London, 1958.
96. M. Talât-Erben and N. Önol, *Can. J. Chem.*, **38**, 1154 (1960).
97. A. F. Bickel and E. C. Kooyman, *J. Chem. Soc.*, **1953**, 3211.
98. C. E. Boozer, G. S. Hammond, C. E. Hamilton and J. N. Sen, *J. Am. Chem. Soc.*, **77**, 3233 (1955).
99. J. Osugi, M. Sato and M. Sasaki, *Nippon Kagaku Zasshi*, **85**, 307 (1964).
100. A. F. Bickel and W. A. Waters, *J. Chem. Soc.*, **1950**, 1764.
101. F. J. L. Aparicio and W. A. Waters, *J. Chem. Soc.*, **1952**, 4666.
102. G. S. Hammond and G. B. Lucas, *J. Am. Chem. Soc.*, **77**, 3249 (1955).
103. B. A. Gingras and W. A. Waters, *J. Chem. Soc.*, **1954**, 1920.
104. A. F. Bickel and E. C. Kooyman, *Rec. Trav. Chim.*, **71**, 1137 (1952).
105. J. W. Engelsma, E. Farenhorst and E. C. Kooyman, *Rec. Trav. Chim.*, **73**, 878 (1954).
106. B. Gingras and W. A. Waters, *Chem. Ind. (London)*, **1953**, 615.
107. J. F. Tilney-Bassett and W. A. Waters, *J. Chem. Soc.*, **1957**, 3129.
108. M. Iwamura and N. Inamoto, *Bull. Chem. Soc. Japan*, **40**, 702 (1967).
109. M. Iwamura and N. Inamoto, *Bull. Chem. Soc. Japan*, **40**, 703 (1967).
110. A. Nagasaka and R. Oda, *J. Chem. Soc. Japan, Ind. Chem. Sect.*, **56**, 42 (1953); *Chem. Abstr.*, **48**, 7597 (1954).
111. J. L. Kice and F. M. Parham, *J. Am. Chem. Soc.*, **80**, 3792 (1958).
112. J. L. Kice and T. S. Cantrell, *J. Am. Chem. Soc.*, **85**, 2298 (1963).
113. P. Bruin, A. F. Bickel and E. C. Kooyman, *Rec. Trav. Chim.*, **71**, 1115 (1952).
114. Y. Schaafsma, A. F. Bickel and E. C. Kooyman, *Rec. Trav. Chim.*, **76**, 180 (1957).
115. L. Horner and W. Naumann, *Ann. Chem.*, **587**, 81 (1954).
116. R. Huisgen, F. Jakob, W. Siegel and A. Cadus, *Ann. Chem.*, **590**, 1 (1954).
117. J. F. Bunnett and C. C. Wamser, *J. Am. Chem. Soc.*, **89**, 6712 (1967).
118. M. C. Ford and W. A. Waters, *J. Chem. Soc.*, **1951**, 1851.
119. M. C. Ford and W. A. Waters, *J. Chem. Soc.*, **1952**, 2240.
120. S. Pickholz and E. Roberts, *Brit. Pat.* 707,990 (1954).
121. (a) P. S. Pinkney, *U.S. Pat.* 2,551,813 (1951);
(b) L. Eberson, *Acta Chem. Scand.*, **17**, 2004 (1963).
122. R. F. Pottie and F. P. Lossing, *J. Am. Chem. Soc.*, **83**, 4737 (1961).
123. J. Kumamoto, H. E. DeLaMare and F. F. Rust, *J. Am. Chem. Soc.*, **82**, 1935 (1960).
124. J. K. Kochi and D. M. Mog, *J. Am. Chem. Soc.*, **87**, 522 (1965).
125. F. Minisci and R. Galli, *Chim. Ind. (Milan)*, **45**, 448 (1963).
126. F. Minisci, R. Galli and M. Cecere, *Chim. Ind. (Milan)*, **46**, 1064 (1964).

127. V. Franzen, *Chem. Ber.*, **88,** 1697 (1955).
128. C. Rüchardt and S. Eichler, *Chem. Ber.*, **95,** 1921 (1962).
129. C. Rüchardt and R. Hecht, *Chem. Ber.*, **98,** 2471 (1965).
130. H. E. Zimmerman, R. C. Hahn, H. Morrison and M. C. Wani, *J. Am. Chem. Soc.*, **87,** 1138 (1965).
131. H. E. Zimmerman, R. D. Rieke and J. R. Scheffer, *J. Am. Chem. Soc.*, **89,** 2033 (1967).
132. American Cyanamid Company, Petrochemicals Department, *The Chemistry of Acrylonitrile*, 2nd ed., New York, 1959.
133. F. R. Mayo and C. Walling, *Chem. Rev.*, **46,** 191 (1950).
134. C. Walling, *Free Radicals in Solution*, John Wiley and Sons, New York, 1957.
135. J. C. Bevington, *Radical Polymerization*, Academic Press, New York, 1961.
136. G. E. Ham (Ed.), *Copolymerization*, Interscience, New York, 1964.
137. L. K. J. Tong and W. O. Kenyon, *J. Am. Chem. Soc.*, **69,** 2245 (1947).
138. R. G. R. Bacon, *Trans. Faraday Soc.*, **42,** 140 (1946).
139. J. H. Merz and W. A. Waters, *Discussions Faraday Soc.*, **2,** 179 (1947).
140. C. H. Bamford and D. J. Lind, *Chem. Commun.*, **1966,** 792.
141. H. Gilbert, F. F. Miller, S. J. Averill, R. F. Schmidt, F. D. Stewart and H. L. Trumbull, *J. Am. Chem. Soc.*, **76,** 1074 (1954).
142. H. Gilbert, F. F. Miller, S. J. Averill, E. J. Carlson, V. L. Folt, H. J. Heller, F. D. Stewart, R. F. Schmidt and H. L. Trumbull, *J. Am. Chem. Soc.*, **78,** 1669 (1956).
143. B. S. Sprague, H. E. Greene, L. F. Reuter and R. D. Smith, *Angew. Chem. Int. Ed. Engl.*, **1,** 425 (1962).
144. C. S. Rondestvedt, Jr., *Organic Reactions*, Vol. 11, John Wiley and Sons, New York, 1960, p. 189.
145. S. C. Dickerman, K. Weiss and A. K. Ingberman, *J. Org. Chem.*, **21,** 380 (1956).
146. J. K. Kochi, *J. Am. Chem. Soc.*, **78,** 4815 (1956).
147. J. K. Kochi, *J. Am. Chem. Soc.*, **77,** 5274 (1955).
148. S. C. Dickerman, I. S. Megna and M. M. Skoultchi, *J. Am. Chem. Soc.*, **81,** 2270 (1959).
149. F. Minisci and U. Pallini, *Gazz. Chim. Ital.*, **91,** 1030 (1961).
150. J. K. Kochi and F. F. Rust, *J. Am. Chem. Soc.*, **84,** 3946 (1962).
151. I. L. Knunyants and N. P. Gambaryan, *Usp. Khim.*, **23,** 781 (1954).
152. M. M. Baizer, *J. Electrochem. Soc.*, **111,** 215 (1964).
153. E. P. Blanchard, Jr. and A. Cairncross, *J. Am. Chem. Soc.*, **88,** 487 (1966).
154. E. C. Coyner and W. S. Hillman, *J. Am. Chem. Soc.*, **71,** 324 (1949).
155. J. D. Roberts and C. M. Sharts, *Organic Reactions*, Vol. 12, John Wiley and Sons, New York, 1962, p. 1.
156. C. J. Albisetti, D. C. England, M. J. Hogsed and R. M. Joyce, *J. Am. Chem. Soc.*, **78,** 472 (1956).
157. S. Hosaka and S. Wakamatsu, *Tetrahedron Letters*, **1968,** 219.
158. J. Runge and R. Kache, *Brit. Pat.* 1,068,230 (1967).
159. H. N. Cripps, J. K. Williams and W. H. Sharkey, *J. Am. Chem. Soc.*, **81,** 2723 (1959).
160. D. D. Coffman, P. L. Barrick, R. D. Cramer and M. S. Raasch, *J. Am. Chem. Soc.*, **71,** 490 (1949).
161. G. N. B. Burch, Ph.D. Dissertation, Ohio State University, as quoted in reference 155.

162. J. D. LaZerte, D. A. Rausch, R. J. Koshar, J. D. Park, W. H. Pearlson and J. R. Lacher, *J. Am. Chem. Soc.*, **78**, 5639 (1956).
163. K.-D. Gundermann, *Chem. Ber.*, **88**, 1432 (1955).
164. K.-D. Gundermann and R. Thomas, *Chem. Ber.*, **89**, 1263 (1956).
165. K.-D. Gundermann and R. Huchting, *Chem. Ber.*, **92**, 415 (1959).
166. W. J. Middleton, R. E. Heckert, E. L. Little and C. G. Krespan, *J. Am. Chem. Soc.*, **80**, 2783 (1958).
167. D. S. Acker and W. R. Hertler, *J. Am. Chem. Soc.*, **84**, 3370 (1962).
168. W. R. Hertler, H. D. Hartzler, D. S. Acker and R. E. Benson, *J. Am. Chem. Soc.*, **84**, 3387 (1962).
169. J. Diekmann and C. J. Pedersen, *J. Org. Chem.*, **28**, 2879 (1963).
170. P. G. Gassman and K. T. Mansfield, *J. Am. Chem. Soc.*, **90**, 1517, 1524 (1968).
171. S. H. Harper and K. C. Sleep, *J. Sci. Food Agr.*, **6**, 116 (1955).
172. M. J. S. Dewar and R. Pettit, *J. Chem. Soc.*, **1956**, 2026.
173. R. A. Bernheim, R. J. Kempf, P. W. Humer and P. S. Skell, *J. Chem. Phys.*, **41**, 1156 (1964).
174. R. A. Bernheim, R. J. Kempf, J. V. Gramas and P. S. Skell, *J. Chem. Phys.*, **43**, 196 (1965).
175. A. J. Merer and D. N. Travis, *Can. J. Phys.*, **43**, 1795 (1965).
176. E. Ciganek, *J. Am. Chem. Soc.*, **88**, 1979 (1966).
177. E. Wasserman, L. Barash and W. A. Yager, *J. Am. Chem. Soc.*, **87**, 2075 (1965).
178. E. Ciganek, *J. Am. Chem. Soc.*, **87**, 652 (1965).
179. E. Ciganek, *J. Am. Chem. Soc.*, **89**, 1454 (1967).
180. E. Ciganek, *J. Am. Chem. Soc.*, **89**, 1458 (1967).
181. E. Ciganek and C. G. Krespan, *J. Org. Chem.*, **33**, 541 (1968).
182. E. Ciganek, *J. Am. Chem. Soc.*, **87**, 1149 (1965).
183. R. E. Moser, J. M. Fritsch, T. L. Westman, R. M. Kliss and C. N. Matthews, *J. Am. Chem. Soc.*, **89**, 5673 (1967).
184. P. C. Petrellis, H. Dietrich, E. Meyer and G. W. Griffin, *J. Am. Chem. Soc.*, **89**, 1967 (1967).
185. J. Boyer and R. Selvarajan, *Chem. Eng. News*, **45** (41), 52 (1967).
186. H. D. Hartzler, *J. Am. Chem. Soc.*, **86**, 2174 (1964).
187. E. Wasserman, private communication.
188. O. W. Webster, *J. Am. Chem. Soc.* **88**, 4055 (1966).
189. A. G. Anastassiou, H. E. Simmons and F. D. Marsh, *J. Am. Chem. Soc.*, **87**, 2296 (1965); A. G. Anastassiou and H. E. Simmons, *J. Am. Chem. Soc.*, **89**, 3177 (1967).
190. A. G. Anastassiou, *J. Am. Chem. Soc.*, **88**, 2322 (1966); **89**, 3184 (1967).
191. A. G. Anastassiou and J. N. Shepelavy, *J. Am. Chem. Soc.*, **90**, 492 (1968).
192. G. Herzberg and D. N. Travis, *Can. J. Phys.*, **42**, 1658 (1964).
193. G. J. Pontrelli and A. G. Anastassiou, *J. Chem. Phys.*, **42**, 3735 (1965).
194. D. E. Milligan, M. E. Jacox and A. M. Bass, *J. Chem. Phys.*, **43**, 3149 (1965).
195. H. W. Kroto, *J. Chem. Phys.*, **44**, 831 (1966).
196. L. J. Schoen, *J. Chem. Phys.*, **45**, 2773 (1966).
197. F. D. Marsh and H. E. Simmons, *J. Am. Chem. Soc.*, **87**, 3529 (1965).
198. A. G. Anastassiou, *J. Am. Chem. Soc.*, **87**, 5512 (1965); **90**, 1527 (1968).
199. J. C. Boden and B. A. Thrush, *Proc. Roy. Soc.* (*A*), **305**, 93 (1968).
200. J. C. Boden and B. A. Thrush, *Proc. Roy. Soc.* (*A*), **305**, 107 (1968).
201. E. Müller and H. Huber, *Chem. Ber.*, **96**, 670 (1963).
202. E. Müller and H. Huber, *Chem. Ber.*, **96**, 2319 (1963).

203. R. Graf, *Ann. Chem.*, **578**, 50 (1952).
204. N. Kharasch, W. Wolf, T. J. Erpelding, P. G. Naylor and L. Tokes, *Chem. Ind.* (*London*), **1962**, 1720; N. Kharasch and L. Göthlich, *Angew. Chem. Int. Ed. Engl.*, **1**, 459 (1962).
205. L. Eberson and S. Nilsson, *Discussions Faraday Soc.*, **45**, 242 (1968).
206. A. Kaczmarczyk and G. Urry, *J. Am. Chem. Soc.*, **81**, 4112 (1959).
207. C. W. Schmelpfenig, *J. Org. Chem.*, **26**, 4156 (1961).
208. K. Koyama, T. Susuki and S. Tsutsumi, *Tetrahedron Letters*, **1965**, 627.
209. K. Koyama, T. Susuki and S. Tsutsumi, *Bull. Chem. Soc. Japan*, **23**, 2675 (1966).
210. U. D. Parker and B. E. Burgert, *Tetrahedron Letters*, **1965**, 4065.
211. L. Eberson and K. Nyberg, *J. Am. Chem. Soc.*, **88**, 1686 (1966).

CHAPTER 12

The biological function and formation of the cyano group

J. P. Ferris

Rensselaer Polytechnic Institute, Troy, New York, U.S.A.

I. Introduction 718
II. Cyanogenic Glycosides in Plants 718
A. Occurrence 718
B. Biosynthesis 719
C. Hydrolysis 720
D. The Role of the Cyanogenic Glycosides in Asparagine Biosynthesis 721
III. Cyanogenic Compounds in Anthropods 722
A. Occurrence 722
B. Biosynthesis 723
IV. Lathyrogenic Cyano Compounds 723
A. Occurrence and Physiological Activity 723
B. Biosynthesis 725
V. Cyanide Formation by Molds and Microorganisms 727
A. Occurrence 727
B. Biosynthesis 728
1. Hydrogen cyanide formation 728
2. Amino acid synthesis from hydrogen cyanide 728
VI. Cyanopyridine Alkaloids 730
A. Occurrence 730
B. Biosynthesis 730
C. Hydrolysis 732
VII. Indoleacetonitrile 732
A. Occurrence 732
B. Biosynthesis 732
C. Hydrolysis 733
VIII. Detoxication of Nitriles and Cyanide 734
A. Cyanide Toxicity 734
B. Cyanide Detoxication 735

1. Conversion to thiocyanate 735
2. Conversion to 2-iminothiazolidine-4-carboxylic acid . . 736
C. Metabolism of Aliphatic Nitriles 736
D. Metabolism of Aromatic Nitriles 737
IX. Acknowledgements 738
X. References 739

I. INTRODUCTION

The literature on the biochemistry of the cyano group was almost non-existent five years ago. However, in the past few years a number of investigations have revealed the central role of nitriles in plant biochemistry. At the present state of our knowledge there exist no unifying principles by which one may classify the biological transformations of the cyano group. As a consequence in this review each type of cyano compound is discussed separately.

II. CYANOGENIC GLYCOSIDES IN PLANTS

A. Occurrence

HCN is released from a wide variety of plants when the stem or leaf is crushed. Cyanogensis, i.e. the formation of hydrogen cyanide, was one of the earliest approaches to chemical plant taxonomy and many plant species have been tested for hydrogen cyanide production[1–3]. Detailed isolation and structural investigations have shown that the hydrogen cyanide is usually bound as a cyanohydrin glycoside such as amygdalin (**1**)[4]. Some representative cyanogenic glycosides are listed in Table 1.

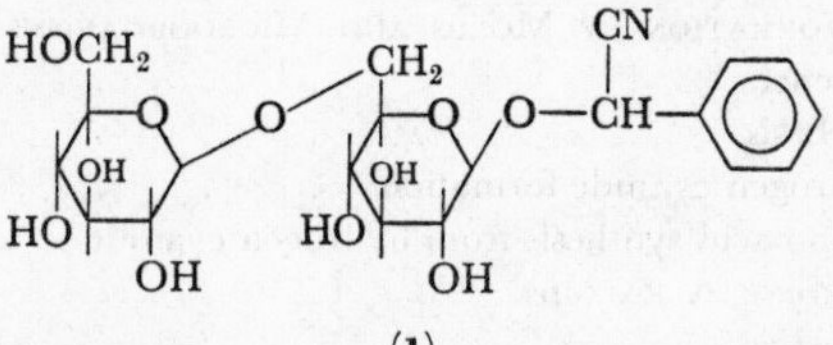

HOCH$_2$ CH$_2$ CN O O O O—CH— OH OH HO HO OH OH

(**1**)

It had been generally assumed that these compounds are end-products of the plant's metabolism, essentially biochemical refuse[3]. However, it has recently been demonstrated that one of these substances is rapidly metabolized with the resulting synthesis of asparagine[5]. Furthermore, the release of hydrogen cyanide from damaged plant cells may also serve as a defense against further attack as in the case of some anthropods (section III.A).

TABLE 1. Cyanogenic glycosides.

$$R^1-\overset{\overset{\large CN}{|}}{\underset{\underset{\large R^2}{|}}{C}}-OR^3$$

Compound	R^1	R^2	R^3
Linamarin	CH_3	CH_3	*d*-Glucose
Lotaustralin	C_2H_5	CH_3	*d*-Glucose
Prunlaurasin[a]	C_6H_5	H	*d*-Glucose
Sambunigrin[b]	C_6H_5	H	*d*-Glucose
Prulaurin[c]	C_6H_5	H	*d*-Glucose
Dhurrin	*p*-HOC_6H_4	H	*d*-Glucose
Vicianin	C_6H_5	H	Vicianose
d-*p*-Glucosyloxymandelonitrile	*p*-$C_6H_{12}O_5$—O—C_6H_4	H	H
Gynocardin	R^1R^2C = cyclic (HO, OH)		Glucose

[a] *d*-Form.
[b] *l*-Form.
[c] *d,l*-Form.

B. Biosynthesis

The pathway for the biosynthesis of these cyanogenic glycosides has been the subject of a number of recent investigations[6]. *A priori* one might postulate that these substances were formed in the plant from hydrogen cyanide and the appropriate aldehyde and sugar. Administration of $H^{14}CN$ gas and ^{14}C-acetone to the roots of *Linum usitatissimum* yielded no labelled linamarin[7]. However, administration of labelled acetone cyanohydrin to *L. usitatissium* resulted in labelled linamarin[8]. Further labelling experiments demonstrated that amino acids were the precursors of the aglycone portion of glycoside. For example, tyrosine (**2**) is the precursor of dhurrin (**3**). Double labelling experiments revealed that the cyano group is formed directly from tyrosine without cleavage of the C—N bond. Furthermore, other labelling studies showed that the intact amino acid, with the exception of the carboxyl group, is incorporated as a

$$HO-C_6H_4-CH_2CH(NH_2)-CO_2H \longrightarrow HO-C_6H_4-CH(CN)O-\text{glucose}$$

(**2**) (**3**)

unit in the cyanogenic glycoside[9–11]. Similar experiments have established the biosynthesis of linamarin (Table 1) from leucine[7], lotaustralin from isoleucine[7], prunasin from phenylalanine[12,13], *p*-glucosyloxymandelonitrile from tyrosine[5] and amygdalin from phenylalanine[14]. Gynocardin[15] is an apparent exception to this general biosynthetic pathway.

It was suggested several years ago that oximes may be the biosynthetic precursors of nitriles[16–18]. This hypothesis is supported by the recent observation that α-ketoisovaleric acid oxime and isobutyraldoxime (**5**) are efficiently incorporated into linamarin (**8**) by flax seedlings (*Linum usitatissimum*)[19]. Further investigation revealed that isobutyronitrile (**6**) and α-hydroxyisobutyronitrile (**7**) are also biosynthetic precursors of linamarin[8,20]. The observation of the direct conversion of valine (**4**) to the oxime **5** and the observation that isobutyraldehyde itself is not converted to linamarin provides support for the following biosynthetic pathway:

$$\underset{(\mathbf{4})}{(CH_3)_2CH\overset{NH_2}{\overset{|}{C}}HCO_2H} \longrightarrow \longrightarrow \longrightarrow \underset{(\mathbf{5})}{(CH_3)_2CH\overset{NOH}{\overset{\|}{C}}H} \longrightarrow$$

$$\underset{(\mathbf{6})}{(CH_3)_2\overset{CN}{\overset{|}{C}}H} \longrightarrow \underset{(\mathbf{7})}{(CH_3)_2\overset{CN}{\overset{|}{C}}OH} \longrightarrow \underset{(\mathbf{8})}{(CH_3)_2\overset{CN}{\overset{|}{C}}O\text{—glucose}}$$

The corresponding oxime and nitrile derivatives are also incorporated into prunasin by *Prunus laurocerasus*, a result which suggests the same biosynthetic pathway being operative for prunasin[8,20]. A glucosyl transferase has been isolated from *Linun usitatissium L.* which catalyses the formation of linamarin from UDP–glucose and acetone cyanohydrin[8].

No details are known concerning the enzyme system responsible for the conversion of the amino acids to the aglycones of the glycosides. It has been noted that a single set of enzymes probably catalyses the formation of both linamarin and lotaustralin[6]. The biosynthesis of these compounds is basically the same and the compounds occur together in the *Leguminosae*, *Linaceae* and *Compositae* families.

C. Hydrolysis

The hydrolysis of the cyanogenic glycosides to an aldehyde hydrogen cyanide and a sugar is catalysed by an enzyme system in the plant, referred to as emulsin. Presumably two steps are involved, the cleavage of the sugar moiety catalysed by a β-glycosidase[21] and the dissociation of the cyanohydrin to the aldehyde and cyanide catalysed by an oxynitrilase[22a]. The hydrolysis of amygdalin (**1**) proceeds in three stages and requires three different enzymes. The first stage is the formation of prunasin by the cleavage of one glucose unit. Enzymatic hydrolysis of prunasin cleaves the second glucose unit with the formation of the cyanohydrin of benzaldehyde. The third stage is the cleavage of the cyanohydrin to benzaldehyde and HCN[22b].

The oxynitrilase enzymes from *Sorghum vulgare* and *Prunus amygdalus* have been obtained in a high degree of purity[23,24a]. The substrates of the enzymes differ only slightly (*p*-hydroxymandelonitrile and mandelonitrile, respectively) yet the enzymes differ considerably. The molecular weights of the enzymes differ by a factor of two (being $1{\cdot}8 \times 10^5$ and $8{\cdot}2 \times 10^4$, respectively) and the *Prunus* enzyme contains a flavin prosthetic group which is not present in the *Sorghum* enzyme. The differences between these enzymes and the function of the flavin moiety is not understood.

The glucosidase enzyme linamarase in *Linum usitatissimum* has been purified 100-fold from the crude material[21]. This purified enzyme exhibits a moderate degree of aglycone specificity when tested with various cyanogenic glucosides. Two β-glucosidases have been isolated from *Sorghum vulgare*; one catalyses the hydrolysis of cyanogenic glucosides and the second aryl glucosides[24b].

D. The Role of the Cyanogenic Glycosides in Asparagine Biosynthesis

The role of the cyanogenic glycosides in the synthesis of asparagine has only been discovered recently. Administration of $H^{14}CN$ to plants producing cyanogenic glycosides results in the formation of labelled asparagine (section IV.B). It has now been observed that administration of tyrosine-2-^{14}C to *Nandina domestica* results in incorporation of 60 to 90 % of the radioactivity in the amide carbon atom of asparagine[5]. In a similar experiment, radioactivity from valine-2-^{14}C administered to *Lotus arabicus* also appeared in the amide carbonyl of asparagine[25]. These experiments suggest that the cyanogenic glycosides serve as a reservoir of hydrogen cyanide in the plant which is utilized in the synthesis of asparagine by the following route (shown at top of next page).

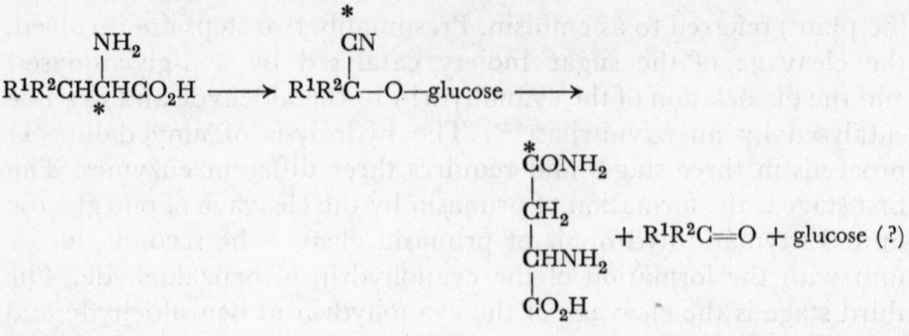

$$\underset{*}{R^1R^2CH\overset{NH_2}{\overset{|}{C}}HCO_2H} \longrightarrow R^1R^2\overset{\overset{*}{CN}}{\overset{|}{C}}\text{—O—glucose} \longrightarrow$$

$$\begin{array}{l} \overset{*}{C}ONH_2 \\ | \\ CH_2 \\ | \\ CHNH_2 \\ | \\ CO_2H \end{array} + R^1R^2C{=}O + \text{glucose (?)}$$

However, it has not been possible to establish this as the biosynthetic route to asparagine in the blue lupine (*Lupinus angustifolia*). Use of appropriately labelled phenylalanine, tyrosine, valine and isoleucine resulted in asparagine that was labelled in several positions[26].

Some of the plants which produce cyanogenic glycosides also respire trace amounts of hydrogen cyanide[27]. Apparently the cyanogenic glycosides are being degraded to hydrogen cyanide in the intact plant, a result which supports the hypothesis that these compounds have an important role in plant biochemistry.

III. CYANOGENIC COMPOUNDS IN ANTHROPODS

A. Occurrence

In millipedes (*Diplopoda*), hydrogen cyanide serves as a defensive agent[28–30]. Of the ten species which have been found to emit hydrogen cyanide, three have been found to emit benzaldehyde as well. This suggests that the hydrogen cyanide is stored as mandelonitrile (**9**) or some derivative of mandelonitrile such as the glucoside. In one instance, the glucoside of *p*-isopropylmandelonitrile (**10**) has been identified[31]. In a systematic study of the secretion of *Polydesmus collaris collaris*, the cyanogenic compound was identified as mandelonitrile benzoate (**11**)[28]. The source of hydrogen cyanide in *Alpheloria corrugata* has been positively identified as mandelonitrile (**9**). The

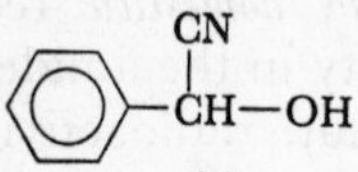

$$C_6H_5\text{—}\overset{CN}{\overset{|}{C}H}\text{—OH} \quad \textbf{(9)} \qquad (CH_3)_2CH\text{—}C_6H_4\text{—}\overset{CN}{\overset{|}{C}H}O\text{—glucose} \quad \textbf{(10)}$$

$$C_6H_5\text{—}\overset{CN}{\overset{|}{C}H}\text{—O—}\overset{O}{\overset{\|}{C}}\text{—}C_6H_5 \quad \textbf{(11)}$$

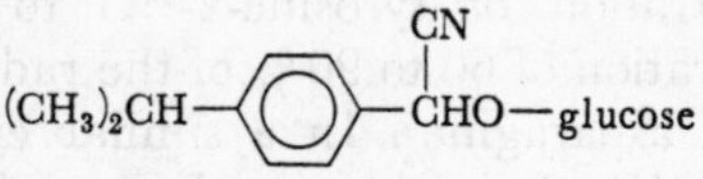

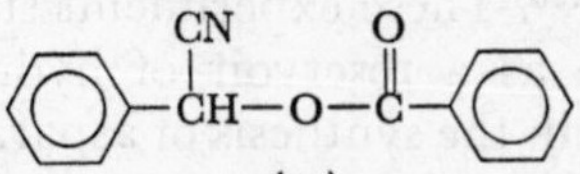

cyanogenic compound is stored in the inner compartment of a gland. The outer compartment contains the enzyme which catalyses the hydrolysis of mandelonitrile. When attacked, the mandelonitrile is released from the inner into the outer compartment where it is mixed with hydrolytic enzyme and the mixture discharged from the gland[32–35].

The larvae of the Australian leaf beetle (*Paropsis atomaria*) secretes a defensive fluid which contains hydrogen cyanide[36]. Other components of this fluid include benzaldehyde and glucose. These results suggest that the cyanide is stored by the larvae as the glucoside of mandelonitrile. In the same study, both hydrogen cyanide and benzaldehyde were tentatively identified in the defensive fluids of *Chryrophthasta variicollis* and *Chrysophtharta amoena.*

The crushed tissues, eggs, larvae and pupae of the moths *Zygaena filipendulae* and *Zygaena lonicerae* release hydrogen cyanide. It is not clear whether this factor is helpful in the defense since the moth also secretes a defensive fluid which does not contain hydrogen cyanide[37].

B. Biosynthesis

No investigations on the biosynthesis of these compounds have been undertaken. It would be especially interesting to know if cyanohydrins are formed by the same biosynthetic pathway in anthropods as in plants. The hydrogen cyanide is not derived from feeding on cyanogenic plants and must be synthesized by the anthropod[36].

IV. LATHYROGENIC CYANO COMPOUNDS

A. Occurrence and Physiological Activity

Consumption of lathyrus meal sometimes results in a nutritional disease, lathyrism, which is characterized by leg weakness and reflux irritability, or spastic paraplegia and occasionally death. Initially, *N*-(γ-L-glutamyl)-β-aminopropionitrile[38] was isolated as the active factor of *Lathyrus pusillus* and *Lathyrus odoratus*. Its physiological effects have received extensive study. This substance interferes with collagen formation and causes skeletal deformations and other teratogenic effects in a variety of test animals. This factor has been of considerable medical interest as a chemical agent whose effects can simulate Marfan's syndrome, a heritable collagen disease[39].

However, since the neurotoxic effects of lathyrus meal were not observed[38,40], this compound and substances of similar biological activity have been designated osteolathyrogens[41].

It was found that *Lathyrus sylvestris*, *Lathyrus latifolius* and *Vicia sativa* did produce neurotoxic symptoms in test animals. The neurotoxic effects associated with *L. sativus* consumption in humans have sometimes been attributed to contamination by *V. sativa*[42]. The active principles of *V. sativa* are β-cyanoalanine (**12**) and its *N*-(γ-glutamyl) derivative (**13**)[42–45] and the neurotoxic principle of *L. sylvestris* and *L. latifolius* is chiefly α,γ-diaminobutyric acid (**14**)[46]. Also present in the latter are much smaller amounts of β-*N*-oxalyl-α,γ-diaminobutyric acid and β-*N*-oxalyl-2,3-diaminopropionic acid[47], a factor isolated from *L. sativus*, a plant often associated with human lathyrism[48–50].

A series of cyano compounds with related structures have been tested for their neurotoxicity in rats. β-Cyano-L-alanine when fed at the 1% level causes hyperirritability, tremors, convulsions and death within 3–5 days. In chicks, less than 0·1% of it in the diet causes death with convulsions within 11 days. The D isomer is approximately one-third as active in the rat[42] and the homologous γ-cyano-L-α-aminobutyric acid produced toxic symptoms only at higher levels. Studies on other cyano compounds revealed that the osteolathyrogenic properties increased in the order bis(2-cyanoethyl)amine, β-aminopropionitrile (**15**), glycine nitrile, methyleneaminoacetonitrile. No osteolathyrogenic symptoms were observed with cyanide ion, acrylonitrile, acetonitrile, propionitrile, alanine nitrile, indoleacetonitrile, tris(2-cyanoethyl)amine, succinonitrile, cyanamide and cyanoguanidine[51,52].

The metabolism of β-aminopropionitrile by rats has been investigated[53,54]. Urine analyses showed that 40% of the β-aminopropionitrile is recovered unchanged, 20–30% is converted to cyanoacetic acid and 3% is converted to thiocyanate.

The osteolathyritic aminonitriles inhibit the germination of fungi and produce deformations in those fungi which do grow. This activity may be due only to the amino group, since simple amines cause the same biological response[55].

The osteolathyrogenic effect of glycine nitrile on rats can be moderated by administration of thyroxine, triiodothyronine and cortisol[56].

The neurotoxicity of a single dose of β-cyanoalanine on rats is reversed by injection of pyridoxal[57]. The nature of this protective effect is not known. β-Cyanoalanine is not an inhibitor of pyridoxal phosphate and produces few signs of vitamin B_6 deficiency[58,59]. Administration of β-cyanoalanine to rats does result in the excretion of large amounts of cystathionine, but pyridoxal does not reverse this effect. *In vitro* experiments demonstrated that L-β-cyanoalanine inhibits rat liver cystathionase and suggest that this is the reason for the formation of cystathionine *in vivo*[59]. It was observed that α-cyanoglycine, *N*-(γ-glutamyl)-β-cyanoalanine, L-γ-cyano-α-aminobutyric acid and malononitrile are also inhibitory. In studies using other compounds, it was found that the L-configuration (D-β-cyanoalanine was ineffective), and both the cyano and amino groups are probably essential structural features of the inhibition *in vitro*. Malononitrile is an obvious exception to this generalization[59].

The metabolism of β-cyanoalanine in rats was briefly investigated[58]. It can enter the brain. It is detoxified by the liver and possibly other tissues by conversion to the tripeptide *N*-(γ-glutamyl)-β-cyanoalanylglycine. This peptide is an analogue of glutathione, in which the cysteine residue is replaced by β-cyanoalanine[60]. However, this compound is probably hydrolysed by the kidney, since appreciable amounts of β-cyanoalanine are recovered in the urine. *N*-(γ-Glutamyl)-β-cyanoalanine is also excreted as β-cyanoalanine by rats.

Pseudomonas bacteria produce a nitrilase which converts β-cyanoalanine to aspartic acid without formation of asparagine. This enzyme differs from the enzyme produced by the same bacteria for the hydrolysis of ricinine (section VI.C)[61].

B. Biosynthesis

When $H^{14}CN$ is administered to *Lathyrus*, *Vicia* and many other species, asparagine-4-^{14}C (**16**) is the principal radioactive product in most instances[43,62,63]. Tracer studies have established that cyanide combines initially with serine or cysteine to form β-cyanoalanine (**12**) which in turn is hydrolysed to asparagine (**16**)[43,64–66]. Acetone extracts of *Sorghum vulgare*, *Vicia sativa*, *Lupinus augustifolia* and *Lotus tenius* catalyse the conversion of cysteine to β-cyanoalanine[67a]. A purified enzyme preparation has been isolated from blue lupine (*Lupinus angustifolia*) which utilizes only cysteine as a substrate[67a,b]. These results suggest that cysteine is the direct biosynthetic precursor of β-cyanoalanine.

N-(γ-Glutamyl)-β-cyanoalanine (**13**), and not asparagine, is produced by *Vicia sativa*[43a,45]. The efficiency with which volatile cyanide is incorporated *in vivo* into *N*-(γ-glutamyl)-β-cyanoalanine is noteworthy (30–60 % yields)[43,44]. *N*-(γ-Glutamyl)-β-cyanoalanine accumulates because *V. sativa* does not contain the nitrilase enzyme necessary for the hydrolysis of β-cyanoalanine to asparagine[74].

Labelled *N*-(γ-glutamyl)-β-aminopropionitrile (**17**) is also produced by feeding $H^{14}CN$ to *Lathyrus odoratus*[68]. This substance is probably formed by the conversion of β-cyanoalanine to β-aminopropionitrile (**15**) and then condensation of β-aminopropionitrile with glutamic acid. The synthesis does not proceed by the alternative route, β-cyanoalanine (**12**) $\rightarrow$ *N*-(γ-glutamyl)-β-cyanoalanine (**13**) $\rightarrow$ *N*-(γ-glutamyl)-β-aminopropionitrile (**17**)[69].

Labelled asparagine and *N*-substituted asparagines have been isolated from *Cucurbitaceae* seedlings fed with $H^{14}CN$. The *N*-substituted derivatives (**18**, R = alkyl) were formed by an exchange reaction between asparagine and the corresponding amine[70].

Finally the neurolathyrogenic factor, α,γ-diaminobutyric acid (**14**), has been reported to be labelled when $H^{14}CN$ was fed to *V. sativa* and *L. odoratus* seedlings[68]. Presumably the diaminobutyric acid was formed by reduction of β-cyanoalanine[46]. However, it has been reported by other workers studying *Lathyrus sylvestris*, which has a high content of α,γ-diaminobutyric acid, that only 0.27 % of the radioactivity of β-cyanoalanine-4-^{14}C was incorporated into α,γ-diaminobutyric acid while 33·5 % was incorporated into asparagine. The latter group found that homoserine and aspartic acid were the precursors of α,γ-diaminobutyric acid[71].

The conversion of amino acids to asparagine via the cyanogenic glycosides is described in section II.D. These biosynthetic reactions are summarized on the next page. The alpha carbon of the original amino acid is marked to assist in following the transformations.

It has been demonstrated that similar transformations of cyanide take place in insects. Administration of $H^{14}CN$ to *Sitophilus granarius* resulted in excretion of most of the radioactivity as a peptide. Hydrolysis of the peptide showed that most of the label was present in $C_{(1)}$ of aspartic acid. If the aspartic acid was formed via the asparagine pathway in plants, the label would have been in the 4-position. It is of interest that none of the mammalian detoxication products (section VIII) were observed in this study[72].

Extracts of *E. coli* cells will also catalyse the conversion of serine to β-cyanoalanine[73]. In contrast to similar studies on *Lotus tenius* extracts[66], ATP is required for this transformation. It is suggested that this represents a non-specific action of cysteine sulphydrase, an enzyme which converts serine to cysteine, and is without physiological significance. In support of this contention, it was observed

$$\underset{*}{RCH_2CH(NH_2)CO_2H} \longrightarrow R\overset{*}{C}HOR(CN) \longrightarrow H\overset{*}{C}N \xrightarrow[\text{cysteine}]{\text{serine}} \overset{*}{C}NCH_2CHNH_2CO_2H\ \mathbf{(12)}$$

$$\mathbf{(12)} \xrightarrow{?\ \mathbf{(14)}} \overset{*}{C}H_2NH_2CH_2CHNH_2CO_2H$$

$$\mathbf{(12)} \longrightarrow \overset{*}{C}NCH_2CH_2NH_2\ \mathbf{(15)} \longrightarrow \overset{*}{C}NCH_2CH_2NHCOCH_2CH_2CHNH_2CO_2H\ \mathbf{(17)}$$

$$\mathbf{(12)} \longrightarrow \overset{*}{C}NCH_2CHNHCOCH_2CH_2CH(NH_2)CO_2H\ \mathbf{(13)}$$

$$\mathbf{(12)} \longrightarrow \overset{*}{C}ONH_2CH_2CHNH_2CO_2H\ \mathbf{(16)} \longrightarrow \overset{*}{C}ONHRCH_2CHNH_2CO_2H\ \mathbf{(18)}$$

that the *Lotus tenius* enzyme has an affinity for cyanide that is 10^4 times as great as the *E. coli* enzyme[6].

V. CYANIDE FORMATION BY MOLDS AND MICROORGANISMS

A. Occurrence

A number of fungi and bacteria are known to be cyanogenic[74–76]. It is not clear whether the hydrogen cyanide is always combined in a cyanogenic compound or if in some instances free cyanide is present. Cyanogenic substances of unknown structures were isolated from the *Marasmus oreades* fungus and the snow mold fungus. The purpose of the cyanide may be defensive and/or as an intermediate in the synthesis of amino acids[77,78].

Diatretyne 2 (**19**), a nitrile antibiotic, has been isolated from the fungus *Clitocybe diatreta*[79]. The corresponding amide (diatretyne 1) is also present in the fungal extracts but exhibits no antibiotic activity.

$$\text{HOOCCH}{=}\text{CH}{-}\text{C}{\equiv}\text{C}{-}\text{C}{\equiv}\text{C}{-}\text{CN}$$
(19)

B. Biosynthesis

1. Hydrogen cyanide formation

In those instances studied, it appears that hydrogen cyanide is produced from glycine by a route that is similar to the conversion of the amino acids to the cyanogenic glycosides.

Cyanide formation from the bacterium *Chromobacterium violaceum* was enhanced by the addition of glycine or glycine ester[80]. A similar effect was noted in *Pseudomonas*[81]. In *Chromobacterium*, methionine and other methyl donors exhibited a synergistic effect in the hydrogen cyanide synthesis[74]. Studies with glycine labelled at $C_{(1)}$ and $C_{(2)}$ revealed that $C_{(1)}$ is eliminated as CO_2 and $C_{(2)}$ is incorporated into hydrogen cyanide. None of the biosynthetic intermediates between glycine and hydrogen cyanide have been isolated. However, the formation of cyanoformic acid has been postulated[80].

Glycine was also found to stimulate hydrogen cyanide synthesis from the snow mold fungus[82a]. Serine was less effective and other amino acids had no effect. Glycine esters were also stimulatory but *N*-methyl glycine derivatives had no effect. Preliminary evidence with labelled glycine suggests that $C_{(2)}$ is incorporated into the hydrogen cyanide.

Recently the synthesis of linamarin and lotaustralin by the snow mold fungus was reported. The biosynthesis of these cyanogenic glucosides proceeds from valine and isoleucine, respectively. In addition, there are enzymes present in the fungus which catalyse the hydrolysis of the glucosides to glucose, acetone and HCN. These data suggest that HCN biosynthesis proceeds by the same pathways in both fungi and higher plants[82b].

2. Amino acid synthesis from hydrogen cyanide

The synthesis of β-cyanoalanine and asparagine from glycine has been observed in *Chromobacterium violaceum*[83]. Labelling studies have established that all of the carbon atoms of β-cyanoalanine originate from the glycine. When $K^{14}CN$ and serine are incubated with non-proliferating cells, the resultant β-cyanoalanine is labelled on the nitrile carbon atom. These results suggest that glycine is converted to both cyanide and the cyanide acceptor in β-cyanoalanine synthesis. β-Cyanoalanine is hydrolysed to asparagine in the culture fluid. This pathway for asparagine biosynthesis in bacteria is almost identical with that observed in some plants (section IV.B).

Fungi synthesize alanine and glutamic acid via a Strecker synthesis. Presumably the overall biosynthetic pathway is glycine $\rightarrow$ HCN $\rightarrow$ alanine and glutamic acid. The individual steps in the sequence have been demonstrated but the overall process has not been verified experimentally.

α-Aminopropionitrile (alanine nitrile) and alanine are produced when cultures of a psychrophilic basidiomycete were fed $H^{14}CN$[84,85]. Early incorporation of label into alanine nitrile was observed. The amount of labelled alanine nitrile decreased as the amount of ^{14}C alanine increased. Double labelling with $H^{14}CN$ and $HC^{15}N$ revealed that the cyano group was incorporated intact into alanine nitrile. Only a small amount of the ^{15}N appeared in the amino group of the nitrile.

An extract of the fungus catalysed the addition of HCN and NH_3 to acetaldehyde, so presumably this is the biosynthetic pathway in the fungus. Cell free extracts will also hydrolyse alanine nitrile directly to alanine without formation of an amide intermediate.

Cyanide is also converted to glutamic acid (**22**) by the same psychrophilic basidiomycete fungus[78,84]. The rates of incorporation of $H^{14}CN$ into alanine and glutamic acid are independent of each other, suggesting the presence of two separate pathways for the biosynthesis of these compounds. A biosynthetic intermediate for the formation of glutamic acid, 4-amino-4-cyanobutyric acid (**21**), has been isolated. When $H^{14}CN$ was administered, the cyano group of the 4-amino-4-cyanobutryic acid was labelled and this in turn led to glutamic acid-1-^{14}C.

When $K^{13}C^{15}N$ ($^{15}N/^{13}C = 1$) and uniformly labelled ^{14}C-succinic semialdehyde (**20**) were administered to the fungus it was noted that the $^{15}N/^{13}C$ ratio was near unity in the **21** formed as would be expected by direct addition of cyanide. Degradation studies showed that all the ^{14}C was in carbon atoms other than the nitrile group, indicating that the main carbon skeleton is supplied by succinic semialdehyde. The synthesis of **21** and **22** from cyanide and succinic semialdehyde was also carried out in cell-free extracts.

$$\underset{(20)}{CHO-CH_2-CH_2-CO_2H} \xrightarrow[NH_3]{HCN} \underset{(21)}{CN-CHNH_2-CH_2-CH_2-CO_2H} \longrightarrow \underset{(22)}{CO_2H-CHNH_2-CH_2-CH_2-CO_2H}$$

VI. CYANOPYRIDINE ALKALOIDS

A. Occurrence

Two cyanopyridones have been isolated from plant sources, ricinine (**23**) from *Ricinus communis* L. (castor bean)[86] and nudiflorine (**24**) from *Trewia nudiflora* Linn[87,88].

OCH$_3$, CN, N, O, CH$_3$ **(23)**   NC, N, O, CH$_3$ **(24)**

B. Biosynthesis

The biosynthesis of ricinine (**23**) has been investigated in some detail. When labelled nicotinamide (**27**) and nicotinic acid (**26**) are fed to the castor bean, the label is incorporated into ricinine. Double-labelling experiments have shown that both 5-^{3}H, 6-^{3}H and ^{15}N-amide of nicotinamide are incorporated equally into ricinine. Furthermore, 5-^{3}H, 6-^{3}H and ^{14}C-carboxyl of nicotinic acid are uniformly incorporated as well. These experiments show that the nicotinamide skeleton is incorporated intact into ricinine and that the nitrile is formed directly by dehydration of the amide[89].

It was suggested that the first step in the conversion of nicotinamide to ricinine was formation of the nitrile **28**[90]. This proposal was based on the observations that an enzyme preparation from castor bean plants catalysed the oxidation of 1-methylnicotinonitrile to a mixture of the 2- and 4-pyridones, while the corresponding amide was not oxidized by this enzyme. However, recent experiments have shown that 1-methylnicotinonitrile is incorporated into ricinine only one-tenth as effectively as nicotinamide, a result which suggests that nitrile formation may take place at a later stage in the ricinine biosynthesis[91]. However, these data do not eliminate the possibility that nicotinonitrile is the next step in the biosynthesis and that this substance is converted to ricinine as a nucleotide derivative. The hypothesis of nucleotide intermediates receives some support from the observations that the nicotinic acid mononucleotide, nicotinic acid adenine dinucleotide and nicotinamide adenine dinucleotide are converted to ricinine as efficiently as nicotinamide and nicotinic acid[91]. Compounds (**25**), (**26**) and (**27**) are likely intermediates in the biosynthesis of ricinine. It is not certain whether the non-alkylated pyridine derivative is present as such, or whether the pyridine nucleotide derivative is the true biosynthetic intermediate.

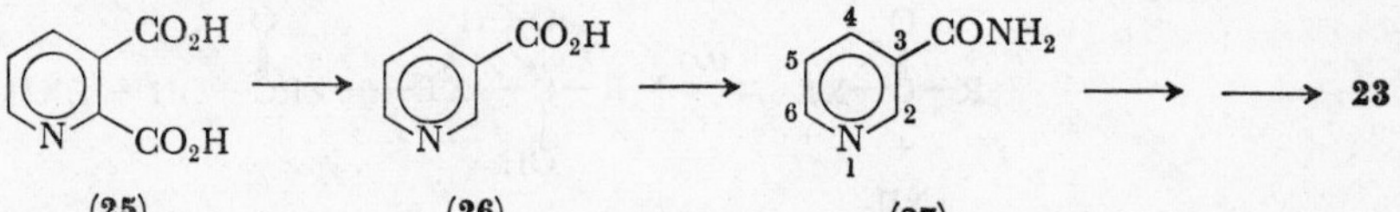

CO_2H / CO_2H (**25**) ⟶ CO_2H (**26**) ⟶ $CONH_2$ (**27**) ⟶ ⟶ **23**

High activity is observed in the cyano group of ricinine when $H^{14}CN$ is fed to *Ricinus communis*[92a]. Since it has been found that the amide group of asparagine (**16**) is also labelled when $H^{14}CN$ is fed to *R. Communis*, the biosynthesis of ricinine may proceed from cyanide via asparagine or *β*-cyanoalanine (**12**)[68]. The metabolic products of ricinine have recently been reported[92b].

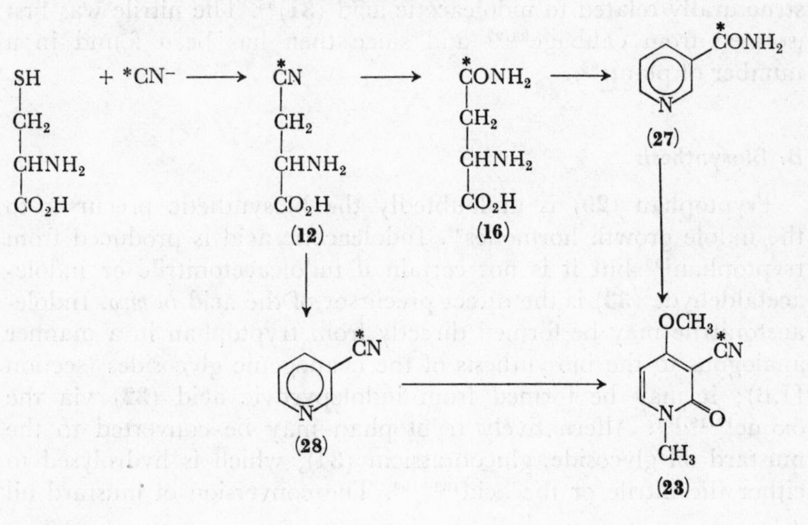

SH–CH_2–CHNH_2–CO_2H + $^*CN^-$ ⟶ $\overset{*}{C}N$–CH_2–CHNH_2–CO_2H (**12**) ⟶ $\overset{*}{C}ONH_2$–CH_2–CHNH_2–CO_2H (**16**) ⟶ $\overset{*}{C}ONH_2$ (**27**)

(**12**) ⟶ $\overset{*}{C}N$ (**28**) ⟶ OCH_3, $\overset{*}{C}N$, O, N–CH_3 (**23**)

(**27**) ⟶ (**23**)

C. *Hydrolysis*

Pseudomonas bacteria grown on ricinine as the sole carbon source produce an enzyme which catalyses the hydrolysis of the nitrile group of ricinine to the acid[61,93]. A low yield of the amide (9 %) is also obtained. This enzyme will also effect the hydrolysis of other 3-cyanopyridones. The mechanism postulated for ricinine nitrilase is similar to that first postulated for the related enzyme called nitrilase (section VII.B)[94].

$$\mathrm{RCN + EXH \rightleftharpoons R\overset{\displaystyle NH}{\overset{\|}{C}}{-}XE \xrightleftharpoons{H_2O} R{-}\overset{\displaystyle O{-}H}{\underset{\displaystyle XE}{C}}{-}NH_2 \longrightarrow}$$

$$\mathrm{R{-}\overset{\displaystyle O}{\overset{\|}{C}}{-}XE + NH_3 \xrightleftharpoons{H_2O} R{-}\overset{\displaystyle O{-}H}{\underset{\displaystyle OH}{C}}{-}XE \longrightarrow R\overset{\displaystyle O}{\overset{\|}{C}}{-}OH + EXH}$$

(enzyme = EXH, where X = S or O)

VII. INDOLEACETONITRILE

A. *Occurrence*

Indoleacetonitrile (**30**) is one of a family of plant hormones structurally related to indoleacetic acid (**31**)[95]. The nitrile was first isolated from cabbage[96,97] and since then has been found in a number of plants[98].

B. *Biosynthesis*

Tryptophan (**29**) is undoubtedly the biosynthetic precursor to the indole growth hormones[99]. Indoleacetic acid is produced from tryptophan[100] but it is not certain if indoleacetonitrile or indole-acetaldehyde (**33**) is the direct precursor of the acid *in vivo*. Indole-acetonitrile may be formed directly from tryptophan in a manner analogous to the biosynthesis of the cyanogenic glycosides (section II.B); it may be formed from indolepyruvic acid (**32**) via the oxime[17,18,101]. Alternatively tryptophan may be converted to the mustard oil glycoside, glucobrassicin (**34**), which is hydrolysed to either the nitrile or the acid[102–104]. The conversion of mustard oil glycosides to nitriles by the enzyme myrosinase is a general phenomenon. Allyl cyanide, benzyl cyanide, *p*-hydroylbenzyl cyanide and compounds **35** and **36** are examples of nitriles formed in this way[105–107].

There exists the possibility that indoleacetonitrile is an artefact[97,108].

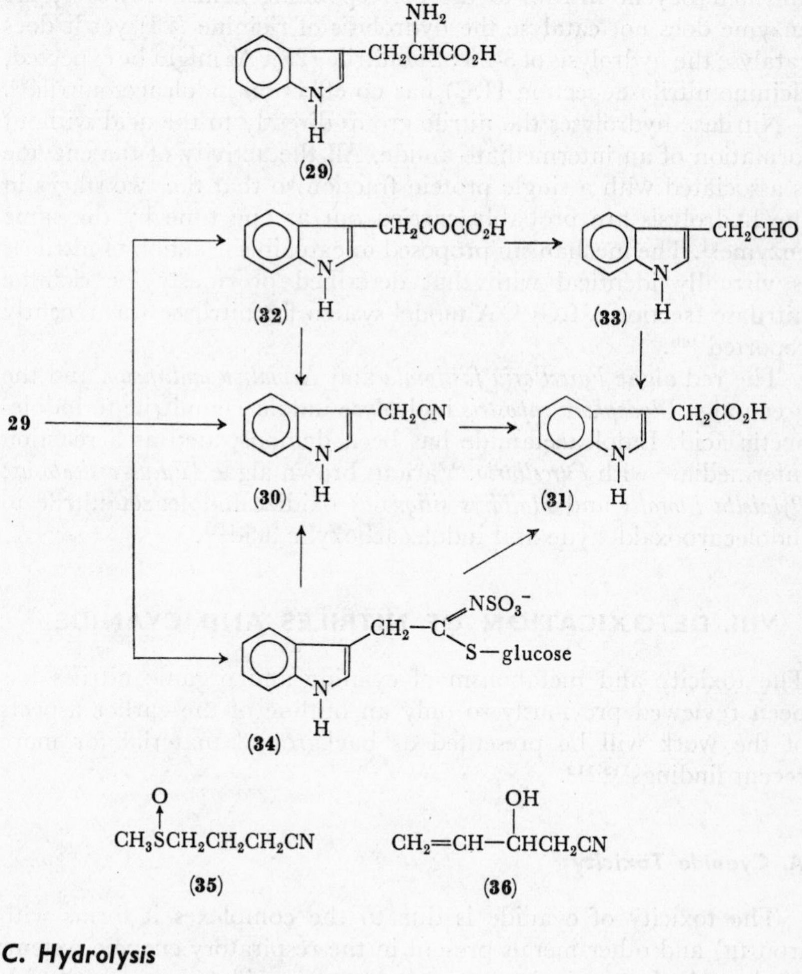

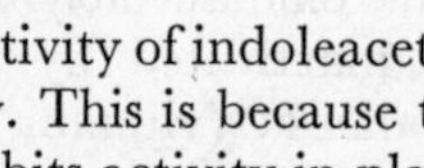

NH_2
CH_2CHCO_2H
N
H
(29)

CH_2COCO_2H → CH_2CHO

(32) **(33)**

29 → CH_2CN → CH_2CO_2H

(30) **(31)**

$CH_2-C(=NSO_3^-)-S-glucose$

(34)

$CH_3S(\rightarrow O)CH_2CH_2CH_2CN$

(35)

$CH_2{=}CH-CH(OH)CH_2CN$

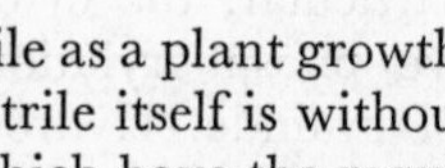

(36)

C. Hydrolysis

The activity of indoleacetonitrile as a plant growth hormone varies markedly. This is because the nitrile itself is without activity and it only exhibits activity in plants which have the requisite enzymes for its hydrolysis to indoleacetic acid[109,110a]. Of twenty-nine plant tested (representing twenty-one families) only in ten was indoleacetonitrile hydrolysed to the acid[110a]. The enzyme which catalyses the hydrolysis of indoleacetonitrile (nitrilase) has been isolated from barley and investigated in detail[94,110a]. It is not specific for indoleacetonitrile and will also catalyse the hydrolysis of aromatic, aliphatic and heterocyclic nitriles to the corresponding acids. However, the enzyme does not catalyse the hydrolysis of ricinine (**23**) yet it does catalyse the hydrolysis of 3-nicotinonitrile (**28**). As might be expected, ricinine nitrilase (section II.C) has no effect on indoleacetonitrile[93].

Nitrilase hydrolyses the nitrile group directly to the acid without formation of an intermediate amide. All the activity of the enzyme is associated with a single protein fraction so that the two stages in the hydrolysis are probably carried out at one time by the same enzyme[94]. The mechanism proposed to explain the action of nitrilase is virtually identical with that described previously for ricinine nitrilase (section VI.C)[94]. A model system for nitrilase was recently reported[110b].

The red algae *Furcellaria fastigiata* and *Nemalion multifidum* and the green alga *Cladophora rupostris* hydrolyse indoleacetonitrile to indoleacetic acid. Indoleacetamide has been demonstrated as a reaction intermediate with *Furcellaria*. Various brown algae (*Fucus vesiculosus*; *Pylaiella litoralis* and *Halidrys siliquosa*) oxidize indoleacetonitrile to indolecarboxaldehyde and indolecarboxylic acid[111].

VIII. DETOXICATION OF NITRILES AND CYANIDE

The toxicity and metabolism of cyanide and organic nitriles has been reviewed previously so only an outline of the earlier aspects of the work will be presented as background material for more recent findings[112,113].

A. Cyanide Toxicity

The toxicity of cyanide is due to the complexes it forms with iron(III) and other metals present in the respiratory enzyme systems of the cell. In particular, the cytochrome oxidase enzymes which catalyse oxidative phosphorylation are inactivated by very low concentrations of cyanide. Hydroxycobalamin (vitamin B_{12}) and methemoglobin (iron(III) haemoglobin) also form complexes with cyanide. These two substances bind cyanide as strongly as the cytochromes and have been used in the treatment of cyanide poisoning. Vitamin B_{12} is injected[114] and methemoglobin is produced by administration of amyl nitrite or nitrite ion. A rat liver enzyme converts cyanocobalamin to hydroxycobalamin[115]. The inhibition of growth of the T3 bacteriophage by cyanide may be a consequence of inhibition of oxidative phosphorylation in either the phage or the host bacterium. The cyanide inhibits the penetration of the phage into the bacterium[116]. Cyanide also produces chromosome aberrations in plants but the molecular basis of this effect is not known[117,118].

Carbonyl cyanide phenylhydrazones are more effective than cyanide in uncoupling oxidative phosphorylation in mitochondrial systems. Solutions 10^{-5} M in **37** and 10^{-8} M in **38** completely inhibit oxidative phosphorylation. The mechanism of action of these compounds is not known[119–121].

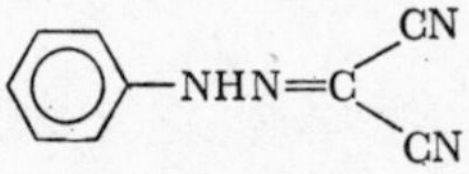

C_6H_5—NHN=C(CN)$_2$

(**37**)

F_3CO—C_6H_4—NHN=C(CN)$_2$

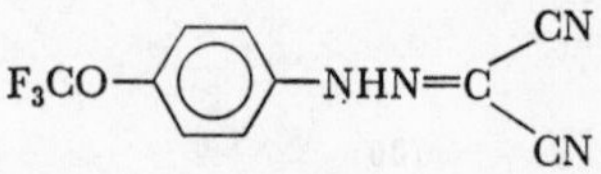

(**38**)

The toxicity of cyanide is probably due mainly to inhibition of oxidative phosphorylation. However, its physiological action may be due to other effects as well. Cyanide stimulates some enzymes by chelation (e.g. 6-phosphogluconate dehydrogenase[122]) by an anionic effect (kynurenine hydrosylase[123]) and by disulphide bond cleavage (inosinic acid dehydrogenase[124]). Undoubtedly cyanide has many other biological effects which may contribute to the observed toxicity.

B. *Cyanide Detoxication*

I. Conversion to thiocyanate

The main pathway for cyanide detoxication is conversion to thiocyanate. The liver enzyme rhodanese catalyses this reaction[125]. Rhodanese is a novel enzyme which when activated with sodium thiosulphate, binds a sulphur atom in a charge-transfer complex to the indole nucleus of tryptophan. Thiocyanate is formed by a nucleophilic attack of cyanide on the sulphur–rhodanese complex[126–130].

$$E + SSO_3^{2-} \longrightarrow ES + SO_3^{2-}$$

$$ES + CN^- \longrightarrow E + SCN^-$$

$$(E = \text{rhodanese})$$

2. Conversion to 2-iminothiazolidine-4-carboxylic acid (40)

Cyanide reacts directly with cystine (**39**) *in vivo* to form **40.** This is a secondary pathway of cyanide metabolism since it was demonstrated that 80 % of the cyanide is detoxified by formation of thiocyanate and 15 % by formation of **40**[131].

$$(\text{—SCH}_2\text{CH(NH}_2\text{)CO}_2\text{H})_2 + \text{HCN} \longrightarrow \text{H}_2\text{C—CHCO}_2\text{H (ring: S, NH, C=NH)} + \text{HSCH}_2\text{CH(NH}_2\text{)CO}_2\text{H}$$

(39) (40)

C. Metabolism of Aliphatic Nitriles

Most aliphatic nitriles are readily hydrolysed in the body to hydrogen cyanide. Compounds such as cyanogen, acetone cyanohydrin, benzyl cyanide, tetracyanoethylene[132], cyanogen halides[133] and diaminomaleonitrile[134] are also rapidly hydrolysed to hydrogen cyanide *in vivo* and cause the same symptoms as cyanide. Cyanide formation may involve more than a simple solvolysis of the cyano group, and in many instances the toxicity of organic nitriles is less than an equivalent amount of cyanide[112,113]. For example, cyanide is probably produced from cyanogen chloride by reaction with glutathione (**41**). Haemoglobin also reacts with cyanogen chloride to give an adduct which releases cyanide on treatment with glutathione[133].

Symptoms other than that of cyanide poisoning have been observed with some aliphatic nitriles. The lathyrogenic nitriles (section IV.A) are good examples. Acrylonitrile is a toxic compound, however; the toxic symptoms differ from those produced by cyanide itself. Presumably the molecule as a whole has a direct toxic effect[135].

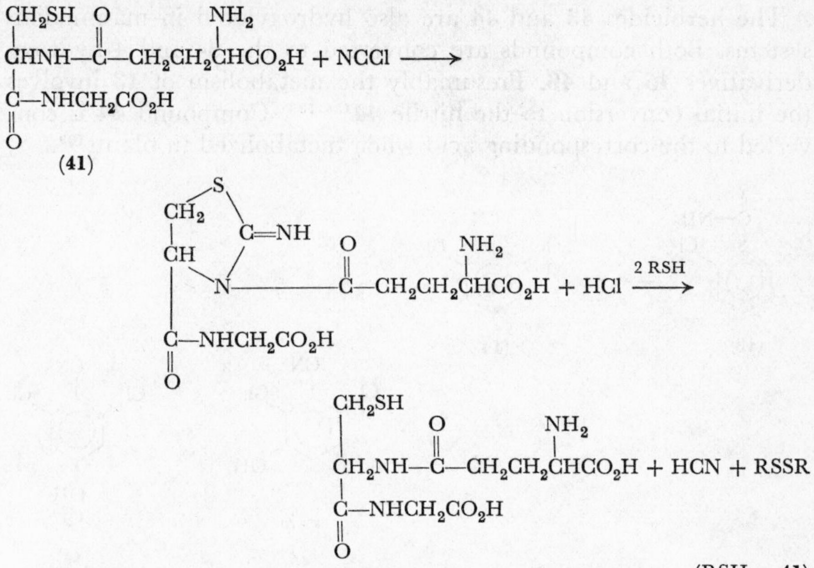

$$\begin{array}{l} CH_2SH \quad O \qquad NH_2 \\ | \qquad\quad \| \qquad\quad | \\ CHNH—C—CH_2CH_2CHCO_2H + NCCl \longrightarrow \\ | \\ C—NHCH_2CO_2H \\ \| \\ O \end{array}$$

(41)

$$\begin{array}{l} \text{(thiazolidine ring: } CH_2—S—C{=}NH,\ CH—N) \\ CH—N————C(=O)—CH_2CH_2CH(NH_2)CO_2H + HCl \xrightarrow{2\,RSH} \\ | \\ C(=O)—NHCH_2CO_2H \end{array}$$

$$\begin{array}{l} CH_2SH \\ | \qquad\qquad O \qquad\qquad NH_2 \\ | \qquad\qquad \| \qquad\qquad | \\ CH_2NH—C—CH_2CH_2CHCO_2H + HCN + RSSR \\ | \\ C—NHCH_2CO_2H \\ \| \\ O \end{array}$$

(RSH = **41**)

Not all aliphatic nitriles are toxic. Compounds with the general formula **42** are central nervous system stimulants with low toxicity[136].

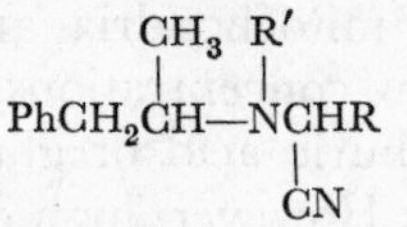

$$\begin{array}{c} CH_3\ R' \\ |\quad\ | \\ PhCH_2CH—NCHR \\ \qquad\quad | \\ \qquad\quad CN \end{array}$$

(42)

D. Metabolism of Aromatic Nitriles

The metabolism of aromatic nitriles has been studied in some detail after it was observed that substituted benzonitriles are herbicides[137]. These substances do not release hydrogen cyanide and hence exhibit a much lower level of toxicity than do aliphatic nitriles. The main detoxication process is oxidation to the cyanophenol in the liver and excretion as the glucuronic acid or the sulphuric acid ether.

Benzonitrile is oxidized and eliminated as *o*-, *m*- and *p*-cyanophenol and only a small portion of it is converted to benzoic acid. The major pathway of *m*- and *p*-tolunitrile metabolism is oxidation to the corresponding *m*- and *p*-cyanobenzoic acids[113]. *o*-Tolunitrile undergoes both oxidation of the methyl group and hydrolysis of the nitrile. Here nitrile hydrolysis is probably assisted by participation of hydroxymethyl and carboxyl groups in the *ortho* position.

The herbicides **43** and **44** are also hydroxylated in mammalian systems. Both compounds are converted to the 3- and 4-hydroxy derivatives **45** and **46**. Presumably the metabolism of **43** involves the initial conversion to the nitrile **44**[138,139]. Compound **44** is converted to the corresponding acid when metabolized in plants[139].

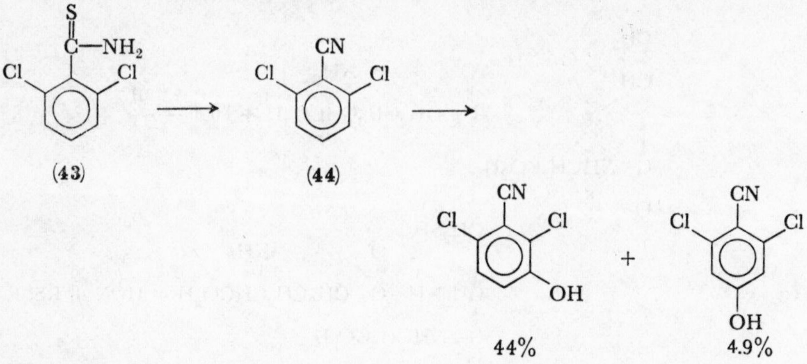

(43) → (44) → (45) 44% + (46) 4.9%

It was observed that injection of **44** results in damage to the liver[140]. The observation that **45** and **46** uncouple oxidative phosphorylation in rat liver mitochondria and yeast cell suspensions explains this result. In low concentrations, compound **44** is oxidized and combined with sulphuric acid or glucuronic acid in the liver and is readily eliminated. However, high concentrations of **44** result in the accumulation of high concentrations of **45** and **46**. These substances block oxidative phosphorylation and thereby prevent the ester formation which results in their elimination from the body.

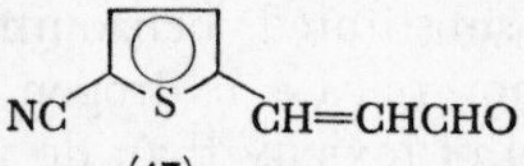

(47) 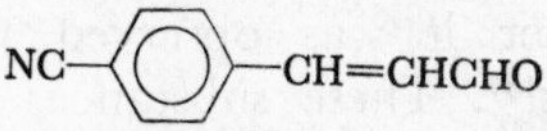 (48)

Some organic nitriles (**47**, **48**) have been reported to be potent antibacterial agents. The metabolic fate of these compounds is not known[141].

IX. ACKNOWLEDGEMENTS

I wish to thank E. Conn, H. Herbrandson, T. Eisner, R. Raffauf, C. Ressler and T. Robinson for reading and commenting on a preliminary draft of this chapter. This work was supported by a Career Development Award (GM 6380) from the National Institutes of Health.

X. REFERENCES

1. R. Hegenaur, *Chemotaxonomie der Pflanzen*, Vol. 1 (1962), Vol. 2 (1963), Vol. 3 (1964), Vol. 4 (1966), Birkhauser Verlag, Basel.
2. R. Darnley-Gibbs, *Chemical Plant Taxonomy*, (Ed. T. Swain), Academic Press, New York, 1963, pp. 58–88.
3. (a) R. E. Alston and B. Turner, *Biochemical Systematics*, Prentice–Hall, New York, 1963, pp. 181–190;
   (b) E. E. Conn, *J. Ag. Food Chem.*, **17,** 519 (1969).
4. W. Karrer, *Konstitution und Vorkommen der Organischen Pflanzenstoffe*, Birkhauser Verlag, Basel and Stuttgart, 1958, pp. 947–967.
5. Y. P. Abrol, E. E. Conn and J. R. Stoker, *Phytochemistry*, **5,** 1021 (1966).
6. L. Fowden, *Rev. Plant Physiol.*, **18,** 85 (1967).
7. G. W. Butler and E. E. Conn, *J. Biol. Chem.*, **239,** 1674 (1964).
8. K. Hahlbrock, B. A. Tapper, G. W. Butler and E. E. Conn, *Arch. Biochem. Biophys.*, **125,** 1013 (1968).
9. J. E. Gander, *J. Biol. Chem.*, **237,** 3229 (1962).
10. J. Koukol, P. Miljanich and E. E. Conn, *J. Biol. Chem.*, **237,** 3223 (1962).
11. E. G. Uribe and E. E. Conn, *J. Biol. Chem.*, **241,** 92 (1966).
12. J. Mentzer and J. Faure-Bonvin, *Compt. Rend.*, **253,** 1072 (1961).
13. S. Ben-Yehoshua and E. E. Conn, *Plant Physiol.*, **39,** 331 (1964).
14. Y. P. Abrol, *Indian J. Biochem.*, **4,** 54 (1967).
15. R. A. Coburn and L. Long, Jr., *J. Org. Chem.*, **31,** 4312 (1966).
16. W. N. Dannenburg and J. L. Liverman, *Plant Physiol.*, **32,** 263 (1957).
17. A. Ahmad and I. D. Spenser, *Can. J. Chem.*, **39,** 1340 (1961).
18. A. Ahmad and I. D. Spenser, *Can. J. Chem.*, **38,** 1625 (1960).
19. B. A. Tapper, E. E. Conn and G. W. Butler, *Arch. Biochem. Biophys.*, **119,** 593 (1967).
20. B. A. Tapper and G. W. Butler, *Arch. Biochem. Biophys.*, **120,** 719 (1967).
21. G. W. Butler, R. W. Bailey and L. D. Kennedy, *Phytochemistry*, **4,** 369 (1965).
22. (a) C. Bove and E. E. Conn, *J. Biol. Chem.*, **236,** 207 (1961);
    (b) D. R. Haisman and D. J. Knight, *Biochem. J.*, **103,** 528 (1967).
23. M. K. Seely, R. S. Criddle and E. E. Conn, *J. Biol. Chem.*, **241,** 4457 (1966).
24. (a) W. Becker and E. Pfeil, *Biochem. Z.*, **346,** 301 (1966);
    (b) C. H. Mao and L. Anderson, *Phytochemistry*, **6,** 473 (1967).
25. Y. P. Arbol and E. E. Conn, *Phytochemistry*, **5,** 237 (1966).
26. E. E. Conn, private communication.
27. Y. P. Abrol, *Indian J. Exp. Biol.*, **5,** 191 (1967).
28. G. Casnati, G. Nencini, A. Quilico, M. Pavan, A. Ricca and T. Salvalori, *Experientia*, **19,** 409 (1963).
29. T. Eisner, *Ann. Rev. Entomol.*, **7,** 107 (1962).
30. M. Jacobson, *Ann. Rev. Entomol.*, **11,** 403 (1966).
31. E. Sodi Pallares, *Arch. Biochem.*, **9,** 105 (1946).
32. T. Eisner, H. E. Eisner, J. J. Hurst, F. C. Kafatos and J. Meinwald, *Science*, **139,** 1218 (1963).
33. H. E. Eisner, T. Eisner and J. J. Hurst, *Chem. Ind.* (*London*), **1963,** 124.
34. H. E. Eisner, D. W. Alsop and T. Eisner, *Psyche*, **74,** 107 (1967).
35. T. Eisner and H. E. Eisner, *Natural History*, **74,** 30 (1965).
36. B. P. Moore, *J. Australian Ent. Soc.*, **6,** 36 (1967).

37. D. A. Jones, J. Parsons and M. Rothschild, *Nature*, **193**, 52 (1962).
38. E. D. Schilling and F. M. Strong, *J. Am. Chem. Soc.*, **77**, 2843 (1955).
39. V. A. McKusick, *Heritable Disorders of Connective Tissue*, 3rd. ed., C. V. Mosby Co., St. Louis, 1966, pp. 129–131.
40. G. F. McKay, J. J. Lalich and F. M. Strong, *Arch. Biochem. Biophys.*, **52**, 313 (1954).
41. L. Fowden, D. Lewis and H. Tristan, *Adv. Enzymol.*, **29**, 89 (1967).
42. C. Ressler, *J. Biol. Chem.*, **237**, 733 (1962) and references therein.
43. (a) C. Ressler, Y.-H. Giza and S. N. Nigam, *J. Am. Chem. Soc.*, **85**, 2874 (1963); (b) C. Ressler, S. N. Nigam and Y.-H. Giza, *J. Am. Chem. Soc.*, **91**, 2758 (1969); (c) C. Ressler, Y.-H. Giza and S. N. Nigam, *J. Am. Chem. Soc.*, **91**, 2766 (1969); (d) Y.-H. Giza and C. Ressler, *J. Lab. Compds.*, **5**, 142 (1969).
44. C. Ressler, *Federation Proc.*, **23**, 1350 (1964).
45. C. Ressler, S. N. Nigam, Y.-H. Giza and J. Nelson, *J. Am. Chem. Soc.*, **85**, 3311 (1963).
46. C. Ressler, P. A. Redstone and R. H. Erenberg, *Science*, **134**, 188 (1961).
47. E. A. Bell and J. P. O'Donovan, *Phytochemistry*, **5**, 1211 (1966).
48. S. L. N. Rao, P. A. Adiga and P. S. Sarma, *Biochemistry*, **3**, 432 (1964).
49. V. V. S. Murti, T. R. Sechadri and T. A. Venkitasubramanian, *Phytochemistry*, **3**, 73 (1964).
50. E. Jacob, A. J. Patel and C. V. Ramarkrishnan, *J. Neurochem.*, **14**, 1091 (1967).
51. I. V. Ponseti, S. Wawzonek, R. S. Shepard, T. C. Evans and G. Stearns, *Proc. Soc. Expt. Biol. Med.*, **92**, 366 (1956).
52. S. Wawzonek, I. V. Ponseti, R. S. Shepard and L. G. Wiedenmann, *Science*, **121**, 63 (1955).
53. J. J. Lalich, *Science*, **128**, 206 (1958).
54. A. Khogali, *Nature*, **214**, 920 (1967).
55. T. B. Norton and W. Dasler, *Proc. Soc., Expt. Biol. Med.*, **116**, 62 (1964).
56. I. V. Ponseti, *Proc. Soc. Expt. Biol. Med.*, **96**, 14 (1957) and references therein.
57. C. Ressler, J. Nelson and M. Pfeffer, *Nature*, **203**, 1286 (1964).
58. C. Ressler, J. Nelson and M. Pfeffer, *Biochem. Pharmacol.*, **16**, 2309 (1967).
59. M. Pfeffer and C. Ressler, *Biochem. Pharmacol.*, **16**, 2299 (1967).
60. K. Sadaoka, C. Lavinger, S. N. Nigam and C. Ressler, *Biochem. Biophys. Acta*, **156**, 128 (1968).
61. R. H. Hook and W. G. Robinson, *J. Biol. Chem.*, **239**, 4263 (1964).
62. S. Blumenthal-Goldschmidt, G. W. Butler and E. E. Conn, *Nature*, **197**, 718 (1963).
63. B. Tschiersch, *Flora*, **153**, 115 (1963).
64. H. G. Floss, L. Hadwiger and E. E. Conn, *Nature*, **208**, 1207 (1965).
65. S. N. Nigam and C. Ressler, *Biochem. Biophys. Acta*, **93**, 339 (1964).
66. H. G. Floss and E. E. Conn, *Nature*, **208**, 1207 (1965).
67. (a) S. G. Blumenthal, H. R. Hendrickson, Y. P. Abrol and E. E. Conn. *J. Biol. Chem.*, **243**, 5302 (1968).
(b) H. R. Hendrickson, *Federation Proc.*, **27**, 593 (1968).
(c) H. R. Hendrickson E. E. Conn, *J. Biol. Chem.*, **244**, 2632 (1969).
68. B. Tschiersch, *Phytochemistry*, **3**, 365 (1964).
69. L. Fowden and E. A. Bell, *Nature*, **206**, 110 (1965).
70. D. M. Frisch, P. M. Dunnell, H. Smith and L. Fowden, *Phytochemistry*, **6**, 921 (1967).
71. S. N. Nigam and C. Ressler, *Biochemistry*, **5**, 3426 (1966).
72. E. J. Bond, *Can. J. Biochem. Physiol.*, **39**, 1793 (1961).
73. P. M. Dunnill and L. Fowden, *Nature*, **208**, 1206 (1965).

74. R. Michaels and W. A. Corpe, *J. Bacteriol.*, **89,** 106 (1965).
75. E. W. B. Ward and G. D. Thorn, *Can. J. Bot.*, **44,** 95 (1966).
76. E. W. B. Ward, *Can. J. Bot.*, **42,** 319 (1964).
77. G. A. Strobel, *J. Biol. Chem.*, **241,** 2618 (1966).
78. G. A. Strobel, *J. Biol. Chem.*, **242,** 3265 (1967).
79. M. Anchel, *Science*, **121,** 607 (1955).
80. R. Michaels, L. V. Hankes and W. A. Corpe, *Arch. Biochem. Biophys.*, **111,** 121 (1965).
81. H. Lorck, *Physiol. Plantarum*, **1,** 142 (1948).
82. (a) E. W. B. Ward and G. D. Thorn, *Can. J. Bot.*, **43,** 997 (1965); (b) D. L. Stevens and G. A. Strobel, *J. Bact.*, **95,** 1094 (1968).
83. M. M. Brysk, W. A. Corpe and L. V. Hankes, *Bacteriol. Proc. Abstr.*, **P. 182,** 132 (1967). To be published in *Proc. Soc. Expt. Biol. Med.*
84. G. A. Strobel, *Can. J. Biochem.*, **42,** 1637 (1964).
85. G. A. Strobel, *J. Biol. Chem.*, **241,** 2618 (1966).
86. L. Marion in *The Alkaloids: Chemistry and Physiology*, Vol. 1 (Ed. R. H. E. Manske and H. G. Holmes), Academic Press, New York, 1950, pp. 206–209.
87. R. Mukherjee and A. Chatterjee, *Chem. Ind.* (*London*), **1964**, 1524.
88. R. Mukherjee and A. Chatterjee, *Tetrahedron*, **22,** 1461 (1966).
89. G. R. Waller and L. M. Henderson, *J. Biol. Chem.*, **236,** 1186 (1961).
90. T. Robinson, *Phytochemistry*, **4,** 67 (1965).
91. G. Waller, K. S. Yang, R. K. Gholson, L. A. Hadiniger and S. Chaykin, *J. Biol. Chem.*, **241,** 4411 (1966).
92. (a) U. Schiedt and G. Boeckh-Behrens, *Z. Physiol. Chem.*, **330,** 58 (1962); (b) G. R. Waller, L. Skursky and J. L.-C. Lee, *Abstracts 156th National Meeting Am. Chem. Soc.*, p. B–19.
93. W. G. Robinson and R. H. Hook, *J. Biol. Chem.*, **239,** 4257 (1964).
94. S. Mahadevan and K. U. Thimann, *Arch. Biochem. Biophys.*, **107,** 62 (1964).
95. K. V. Thimann, *Am. Sci.*, **42,** 589 (1954).
96. E. R. H. Jones, H. B. Henbest, G. F. Smith and J. A. Bentley, *Nature*, **169,** 485 (1952).
97. H. B. Henbest, E. R. H. Jones and G. F. Smith, *J. Chem. Soc.*, **1953,** 3796.
98. J. A. Bentley, *Ann. Rev. Plant Physiol.*, **9,** 47 (1958).
99. S. A. Gordon, *Encyclopedia of Plant Physiology*, Vol. 14, (Ed. W. Ruhland), Springer-Verlag, Berlin, 1960, pp. 620–646.
100. R. A. Khalifah, *Physiol. Plant.*, **20,** 355 (1967).
101. B. B. Stowe, *Fortschr. Chem. Org. Naturstoffe*, **17,** 248 (1959).
102. A. S. Anderson and R. M. Muir, *Physiol. Plant.*, **19,** 1038 (1966).
103. W. Zenk, *Colloq. Int. Centre Nat. Rech. Sci.* (*Paris*), **123,** 241 (1964).
104. R. Hegnauer, *Chemotaxonomie der Pflanzen*, Vol. 3, Birkhauser Verlag, Basel, (1964), pp. 587–605.
105. F. Challenger, *Aspects of the Organic Chemistry of Sulphur*, Academic Press, New York, 1959, pp. 115–161.
106. A. I. Virtanen, *Phytochemistry*, **4,** 207 (1965).
107. M. E. Daxenbichler, C. H. VanEtten and I. A. Wolff, *Chem. Commun.*, **1966,** 526.
108. H. Schraudolf and F. Bergmann, *Planta*, **67,** 75 (1965).
109. K. V. Thimann, *Arch. Biochem. Biophys.*, **44,** 242 (1953).
110. (a) K. U. Thimann and S. Mahadevan, *Arch. Biochem. Biophys.*, **105,** 133 (1964); (b) C. Zervos and E. H. Cordes, *J. Am. Chem. Soc.*, **90,** 6892 (1968).

111. U. Schiewer and E. Libbert, *Planta*, **66**, 377 (1965).
112. D. W. Fassett, *Industrial Hygiene and Toxicology*, (Eds. D. W. Fassett and D. D. Irish) Vol. 2, 2nd. rev. ed., Interscience, New York, 1963, pp. 1991–2036.
113. R. T. Williams, *Detoxication Mechanisms*, John Wiley and Sons, New York, 1959, pp. 390–409.
114. J. Delga, J. Mizoule, B. Veverko and R. Bon, *Ann. Pharm. Franc.*, **19**, 740 (1961).
115. L. Cima, C. Levorato and R. Mantovan, *J. Pharm. Pharmacol.*, **19**, 32 (1967).
116. H. Rieter, *Virology*, **21**, 636 (1963).
117. B. A. Kihlman, *J. Biophys. Biochem. Cytol.*, **3**, 363 (1957).
118. B. A. Kihlman, T. Merz and C. P. Swanson, *J. Biophys. Biochem. Cytol.*, **3**, 381 (1957).
119. P. G. Heytler and W. W. Prichard, *Biochem. Biophys. Res. Commun.*, **1**, 272 (1962).
120. B. Diehn and G. Tollin, *Arch. Biochem. Biophys.*, **121**, 169 (1967).
121. E. S. Bamberger, C. C. Black, C. A. Fewson and M. Gibbs, *Plant Physiol.*, **38**, 483 (1963).
122. B. L. Horecker and P. Z. Smyrniotis, *Methods in Enzymology*, (Eds. S.P. Colwick and N. O. Kaplan) Vol. 1, Academic Press, New York, 1955, p. 327.
123. O. Hayaishi, *Methods in Enzymology*, (Eds. S. P. Colwick and N. O. Kaplan) Vol. 5, Academic Press, New York, 1962, p. 807.
124. G. Weinbaum and R. J. Suhadolnik, *Biochem. Biophys. Acta*, **81**, 236 (1964).
125. B. Sorbo, *Acta. Chem. Scand.*, **16**, 2455 (1962) and previous references in this series.
126. J. Westley and T. Nakamoto, *J. Biol. Chem.*, **237**, 547 (1962).
127. B. Davidson and J. Westley, *J. Biol. Chem.*, **240**, 4463 (1965).
128. R. Mintel and J. Westley, *J. Biol. Chem.*, **241**, 3381 (1966).
129. R. Mintel and J. Westley, *J. Biol. Chem.*, **241**, 3386 (1966).
130. M. Volini, F. DeToma and J. Westley, *J. Biol. Chem.*, **242**, 5220 (1967).
131. J. L. Wood and S. L. Cooley, *J. Biol. Chem.*, **218**, 449 (1956).
132. E. I. du Pont Technical Information Document ES-3178.
133. W. N. Aldridge, *Biochem. J.*, **48**, 271 (1951).
134. C. Bedel, *J. Pharm. Chim.*, **30**, 189 (1924).
135. L. Magos, *Brit. J. Ind. Med.*, **19**, 283 (1962).
136. J. Klosa, *Ger. Pat.* 1,112,987, *Chem. Abstr.*, **57**, 3409 (1962).
137. H. Koopman and J. Doame, *Nature*, **186**, 89 (1960).
138. M. H. Griffiths, J. A. Moss, J. A. Rose and D. E. Hathway, *Biochem. J.*, **98**, 770 (1966).
139. J. G. Wit and H. van Genderen, *Biochem. J.*, **101**, 698 (1966).
140. J. G. Wit and H. van Genderen, *Biochem. J.*, **101**, 707 (1966).
141. A. Vecchi and G. Melone, *J. Org. Chem.*, **22**, 1636 (1957).

CHAPTER 13

Syntheses and uses of isotopically labelled cyanides

LOUIS PICHAT
Labelled Compounds Division
C.E.N.-Saclay, Gif-sur-Yvette (91), France

I. INTRODUCTION . . . 744
A. Unusual Starting Materials . . . 745
B. High Cost of the Isotope . . . 745
C. Necessity of Working with Microquantities . . . 745
II. ALKALI CYANIDES-^{14}C OR -^{13}C AND HYDROGEN CYANIDE-^{14}C OR -^{13}C 746
A. Reduction of Carbon Dioxide or of a Carbonate by an Alkali Metal in the Presence of Ammonia or an Ammonium Salt . . . 746
1. Carbon dioxide . . . 746
2. Barium carbonate . . . 747
B. Reduction of a Carbonate with Zinc Dust in the Presence of Ammonia . . . 747
1. Potassium carbonate . . . 747
2. Barium carbonate . . . 747
C. Reaction of Barium Carbonate with Azides . . . 748
1. Without addition of alkali metal . . . 748
2. With addition of alkali metal . . . 748
D. Direct Exchange of Labelled Carbonate with Unlabelled Potassium Cyanide . . . 749
E. Miscellaneous Methods . . . 749
III. CUPROUS CYANIDE-^{14}C . . . 751
IV. SODIUM CYANIDE-^{15}N . . . 751
V. METHODS OF PREPARATION OF ^{14}C- AND ^{13}C-NITRILES . . . 752
A. From Metallic Cyanides and Hydrogen Cyanide-^{14}C(-^{13}C) . . . 752
1. Additions of cyanides to unsaturated compounds . . . 752
2. Additions of hydrogen cyanide-^{14}C or -^{13}C to carbonyl compounds . . . 752
3. Addition of hydrogen cyanide-^{14}C or -^{13}C to carbonyl compounds in the presence of ammonia: the Strecker synthesis . 758

4. Addition of hydrogen cyanide-^{14}C or -^{13}C to carbonyl compounds in the presence of ammonium carbonate: Bücherer's modification of the Strecker synthesis 761
5. Condensation of alkali cyanides-^{14}C with 1,2-epoxides . . 762
6. Condensation of alkali cyanides-^{14}C with lactones . . 762
7. Reaction of metallic cyanides and halogen compounds . . 765
a. Saturated alkyl halides 765
b. Aralkyl halides 765
c. Dibromoalkanes 766
d. Haloalcohols and haloethers 766
e. Haloacids and haloesters 768
f. Unsaturated alkyl halides 768
g. Aryl halides and heterocyclic halides . . . 769
h. Acyl halides 771
i. Phthalimidoalkyl halides 771
8. Condensation of metallic cyanides and alkyl sulphates or alkyl *p*-toluenesulphonates (tosylates) . . . 772
9. Nucleophilic displacement of phthalimido group by alkali cyanides 773
10. Replacement of amino groups by cyanide in Mannich bases or their quaternary ammonium salts 773
11. Replacement of a diazonium group by cyanide: the Sandmeyer method 775
12. Reaction of alkali cyanide with thiocarbanilide . . 776
B. From ^{14}C- and ^{13}C-Carboxylic Acids and their Derivatives . . 776
1. Dehydration of amides 776
2. Catalytic dehydration of ^{14}C-carboxylic acids in the presence of ammonia 776
3. Exchange of ^{14}C-carboxylic acids with nitriles . . . 777
4. Reaction of cyanogen bromide with the sodium salts of ^{14}C-carboxylic acids 778
C. By Alkylation of Nitriles 779
D. Synthesis of α-Deuterated Alkyl Nitriles 779
VI. TRANSFORMATIONS AND USES OF LABELLED NITRILES . . . 779
A. Hydrolysis—Preparation of ^{14}C-Carboxylic Acids . . . 780
B. Reduction—Preparation of ^{14}C-Amines and ^{14}C-Aldehydes . . 780
C. Preparation of ^{14}C-Ketones by Action of Grignard Reagents . 782
D. Preparation of Amidines 783
E. Additions to the Cyano Group to Form ^{14}C-Heterocycles . . 783
F. Analytical Applications 789
VII. REFERENCES 789

I. INTRODUCTION

The synthesis of isotopically labelled compounds makes use of the general methods of classical organic chemistry. However, there are some interesting features due to various factors.

A. Unusual Starting Materials

Carbon-14, which is so often used nowadays in the labelling of organic compounds, is produced in nuclear reactors by neutron irradiation of aluminium nitride or beryllium nitride as targets according to the following nuclear reaction:

$$^{14}N(n, p)^{14}C$$

Carbon-14 is isolated from the target material, after oxidation, as barium carbonate. It is supplied in this form to the organic chemist who must himself build up the structures from $^{14}CO_2$ or barium carbonate. The fact that one is confined to this very unusual raw material is the most distinctive feature of the carbon-14 chemistry. The stable isotope carbon-13 is also provided as barium carbonate.

B. High Cost of the Isotope

Carbon-14 as barium carbonate remains a fairly expensive isotope, in spite of successive drops in price. The same is true of carbon-13. Most carbon-14 or carbon-13 syntheses pass through many stages from carbon dioxide. It is obvious that good yields at each step are desirable. The best experimental conditions have to be determined by a great number of 'blank experiments'. It is advisable to proceed then to a 'tracer run' involving only a few millicuries, and carry out all the reaction stages, so that the influence of trace impurities on the yield is not overlooked.

It is clear that in a given reaction involving both labelled and unlabelled reagents, an excess of the unlabelled reagent is preferably used. This means that very often the proportions of reagents commonly used in classical organic synthesis have to be reversed. It will be seen in this article that because of the cost, an excess of metallic radioactive cyanides cannot be employed. This has sometimes a deleterious effect on the yield and may increase the proportion of by-products.

C. Necessity of Working with Microquantities

Most frequently, labelled compounds are used in biological studies. Great dilution of the labelled compounds occurs in living organisms. The demand for high specific activities arises from the fact that many of the uses of labelled compounds are in the field of substances highly biologically active or whose normal physiological concentration is low. The lower limit at which five to ten stage organic syntheses with ^{14}C or ^{13}C cease to be practicable is about 1 mmole. If the specific activity required for the biological purpose is 40 mCi/mmole it can be attained only with a batch size of 40–80 mCi. Individual workers rarely or never require such quantities. Therefore the preparation of compounds at very high specific activities has become concentrated in a small number of laboratories specializing in this field and supplying a large number of users.

The purpose of this chapter is to present a review of the methods of synthesis and uses of isotopically labelled cyanides. A nitrile RCH_2CN or $ArCH_2CN$ can be labelled with ^{14}C, ^{13}C or ^{15}N on the cyano group and with carbon or deuterium on the chain. Most of the methods which have been devised serve to label the cyano group. Quite often the nitrile group is not the final stage; most frequently it is only an intermediate. Some examples of transformations of labelled nitriles will be given. Clearly the way through nitriles is one of the most efficient and versatile means of introducing isotopic carbon into a molecule.

II. ALKALI CYANIDES-^{14}C OR -^{13}C AND HYDROGEN CYANIDE-^{14}C OR -^{13}C

The methods of conversion of ^{14}C-carbonate into cyanide have already been reviewed[1,2]. The present author agrees with Catch's statement[2] that 'the multiplicity of published methods reflects the difficulty in reproducing some of them'. In our laboratory, we had opportunities in checking this statement experimentally.

A. Reduction of Carbon Dioxide or of a Carbonate by an Alkali Metal in the Presence of Ammonia or an Ammonium Salt

I. Carbon dioxide

The earliest method was that of Cramer and Kistiakowky[3] who heated carbon dioxide with ammonia and metallic potassium in a sealed tube at 525 °C for 10 minutes. This reaction was studied in greater detail by Lotfield[4] who recently gave a revised procedure[5]. According to Lotfield, heating 1 mmole carbon dioxide with 1 g potassium metal dispersed over the entire surface of a Pyrex glass tube and 2·2 mmoles NH_3 at 730–760 °C for 15 minutes gives potassium cyanide with a yield of 90–95 %. It seems that this method is a rather elaborate one, requiring a skillful and experienced operator. Bos[6] and Olynyk[7] have obtained less satisfactory yields by this technique.

2. Barium carbonate

Sixma[1] described in 1964 a simplification of the Cramer–Kistiakowsky process which apparently has since been in routine use in several laboratories. 1 mmole of barium carbonate, 25 to 50 mmole potassium metal and 2 mmole ammonium chloride are heated at 640 °C in a sealed tube (Supremax glass) for 1 hour. The chemical yield is about 94 % and a 5 % isotopic dilution occurs. A further publication by Isbell and Moyer[8] gives some more details and points out that a great pressure is built in the tube. Lambooy[9] modifies the technique slightly so that a 'Vycor' tube can be used. Silica reaction tubes are equally satisfactory[1].

B. Reduction of a Carbonate with Zinc Dust in the Presence of Ammonia

1. Potassium carbonate

McCarter[10] described in 1951 a method of the 'boat and tube' type which avoids sealed tubes. Cyanide is obtained in radioactive yields of 75 %, by heating potassium carbonate with zinc dust in a porcelain boat in a stream of ammonia which has previously passed over a plug of iron wool as a catalyst at 650 °C. Nystrom[11] and Lemmon[12] have carried out this procedure on 5 mmole batches of carbonate. A drawback of the method is the use of hygroscopic potassium carbonate which has to be prepared from barium carbonate.

2. Barium carbonate

Jeanes' procedure[13] avoids this preliminary conversion of barium carbonate into potassium carbonate. Sodium metal is added to the mixture of barium carbonate and zinc dust and then one proceeds just as in McCarter's method. Catch[2] reports that in his laboratory this procedure has not given yields better than 40 %. Pichat[14] has been more successful, but stresses that on a 1 mmole scale an important isotopic dilution takes place (chemical yield: 93–96 %, radioactive yield: only 60–75 %). This author abandoned Jeanes' procedure after he had made the observation that the yield strongly decreases when the reaction is performed on a 2–10 mmoles scale. The same observation is made when sodium is replaced by potassium. However, Okatova[15], carrying out the reaction on 18 mmoles $BaCO_3$ got a radioactive yield of 65–75 %.

Schuching[16] found that the yields varied from 38 to 100 %. The control of the temperature is indeed the governing factor for a high conversion yield of cyanide from carbonate. The yield is quantitative at 670–680 °c. She carried out the reaction in a stainless steel vessel and instead of gaseous ammonia, sodamide was used. Two runs performed on 0·75 mmole gave an 82–88 % yield of cyanide without isotopic dilution; this was isolated as the silver salt.

Musakin[17] reports that the reaction yield is improved by addition of finely powdered porcelain.

C. Reaction of Barium Carbonate with Azides

1. Without addition of alkali metal

Adamson[18] first described the preparation of labelled cyanide by simply heating barium carbonate (0·5 mmole) with sodium azide (15 mmoles) under nitrogen. Very soon some of the difficulties in employing this method were published. Explosions have frequently occurred. Most publications have been concerned with avoiding these by attention to such details as control of the reaction temperature[1,19]; portionwise addition of the reagents[20]; use of inert moderators[1,21]; crystal form of the carbonate and the intimate mixture of reagents[22] and preheating the azide to give sodium nitride which is then mixed with $BaCO_3$ and pyrolized for 15 minutes[23].

In conclusion, although this method may give good yields it is not satisfactory since it is not very reproducible.

2. With addition of alkali metal

Some of the above difficulties are overcome in Maimind's work[24]: potassium metal (15–50 mmole) is mixed with 1 mmole $BaCO_3$, and 3 mmoles KN_3. The mixture is heated at 300–400 °c for 5 min and for 2–3 min at 750–780 °c. The yield is 85–90 %. This has been confirmed by other workers[25]. Claus[22] uses sodium instead of potassium. Sixma[1] has published some minor modifications. Pichat[14] has stressed the necessity of a preliminary purification of potassium azide in order to get high yields (90–97 % chemical yield, 83–96 % radioactive yield). Sixma[1] and Rothstein and Claus[26] have noted isotopic dilutions of 7 and 11 %, respectively. In the author's laboratory, this process has been in use for some years, and better reproducibility was obtained by employing an automatic temperature programmed furnace.

D. *Direct Exchange of Labelled Carbonate with Unlabelled Potassium Cyanide*

The isotope exchange between $K^{12}CN$ and various ^{14}C carbonates, e.g. barium carbonate, has been studied by Andreeva and Kostikova[27]. At 800 °C the exchange between $K^{12}CN$ and $Ba^{14}CO_3$ is complete in about 2 h. The exchange is initiated when the KCN melts and continues between the liquid and solid phases. Separation of KCN from the reaction mixture is accomplished by extraction with liquid ammonia. Pichat[28] has simplified the method and noted that during the exchange some radioactive gas was evolved. It was burnt as carbon dioxide and recovered. The chemical purity was about 96 %. The specific activity of the KCN depends on the molecular proportions of $K^{12}CN$ and $Ba^{14}CO_3$. Usually it is 65–75 % of the specific activity of $Ba^{14}CO_3$. Pichat[28] has described an apparatus for the extraction of $K^{14}CN$ by liquid ammonia. Although this method is accompanied by some loss in specific activity it has been in use in the author's laboratory for many years and found to be most convenient.

E. *Miscellaneous Methods*

Abrams has prepared labelled hydrocyanic acid from elementary carbon[29] with yield of 60–70 %. However, the difficulty in making elementary carbon makes this method unattractive.

The preparation of sodium cyanide[30] by heating sodium formate with sodamide has been recorded. This method does not seem to have been much used, probably because it requires two more extra stages from barium carbonate.

The most recent contribution to this field is that of Vercier[31,32]. In his procedure, barium cyanamide is produced by passing NH_3 gas on $Ba^{14}CO_3$ and in the second stage the barium is displaced by heating at 330 °C for 20 min with divided sodium, to make sodium cyanamide, which is readily decomposed into sodium cyanide by heating at 800 °c for 10 min in the presence of iron as a catalyst.

$$BaCO_3 + 2\,NH_3 \longrightarrow BaCN_2 + 3\,H_2O$$

$$BaCN_2 + 2\,Na \xrightarrow{330\,°c} Na_2CN_2 + Ba$$

$$Na_2CN_2 \xrightarrow{800\,°c} NaCN + \tfrac{1}{2}\,N_2 + Na$$

According to Vercier[32] the addition of titanium dioxide to the mixture of barium cyanamide and divided sodium gives very reproducible results. The reaction can be performed on 1–20 mmole with radioactive yields of 88–93 % and chemical yields of 98–110 %.

Belleau and Heard[33] have devised an ingenious five-step method of preparing sodium cyanide from barium carbonate through the following sequence of reactions which does not require high temperature or pressure techniques.

$$Ph_3CNa \xrightarrow{^{14}CO_2} \underset{90\%}{Ph_3C^{14}CO_2H} \xrightarrow[2.\,NH_4OH]{1.\,SOCl_2} \underset{95\%}{Ph_3C^{14}CONH_2} \xrightarrow{P_2O_5}$$

$$\underset{97\%}{Ph_3C^{14}CN} \xrightarrow{Na,\,EtOH} Ph_3CH + \underset{90\%}{Na^{14}CN}$$

Triphenylacetonitrile undergoes hydrogenolysis to triphenylmethane and cyanide ion rather than reduction to triphenylethylamine. Overall yields of 68–72 % from barium carbonate were realized.

Vaughan and McCane[34] have reported a method based on the same principle through the following scheme:

$$PhCH_2MgCl \xrightarrow{^{14}CO_2} \underset{89\%}{PhCH_2{}^{14}CO_2H} \xrightarrow[\substack{silica\ gel\\490\,°c}]{NH_3} \underset{92\%}{PhCH_2{}^{14}CN} \xrightarrow{Na,\,EtOH}$$

$$PhCH_3 + \underset{90\%}{Na^{14}CN}$$

The chief advantage of this procedure over that of Belleau and Heard is the one-step conversion of the acid to the nitrile. However, high temperature is required. In spite of their ingenuity, none of these two methods has found practical use.

With the exception of the direct exchange[27,28] method, in which potassium cyanide is directly extracted by liquid ammonia, all the other methods require that hydrogen cyanide be purified by steam distillation from dilute sulphuric acid or phosphoric acid. Alternately, silver cyanide may be precipitated, and then reconverted to hydrogen cyanide by careful distillation from dilute sulphuric acid.

Aqueous solutions of sodium or potassium cyanides cannot be evaporated[2], even at low temperature under vacuum without loss of hydrogen cyanide as a result of the equilibrium:

$$KCN + H_2O \rightleftharpoons KOH + HCN$$

which can be displaced towards the left side by addition of a fairly high concentration of potassium hydroxide. The residual mixture of cyanide and hydroxide must be thoroughly dehydrated under high vacuum, otherwise the presence of moisture would cause hydrolysis to formate during storage. This excess of potassium or sodium hydroxide is sometimes the reason of poor yields in replacement reactions, or in reactions with lactones[35]. The excess of alkali can be removed by extraction of the cyanides by liquid ammonia. Hydrogen cyanide-^{14}C can also be neutralized by a solution of sodium methoxide in methanol[24].

III. CUPROUS CYANIDE-^{14}C

The preparation of cuprous cyanide-^{14}C from alkali cyanides has been described on several occasions[34,36,37,186] by adaptations of the

$$Na^{14}CN \xrightarrow[NaHSO_3]{CuSO_4} Cu_2(^{14}CN)_2$$

method of Barber[38]. The reduction of cupric sulphate is achieved at the expense of bisulphite instead of cyanide, one half of which would otherwise be lost as ^{14}C-cyanogen. On a 5 mmole scale the yield is usually 80%. Another method[39], consisting of shaking equimolar amounts of solid cuprous chloride and sodium cyanide in aqueous solution, seems to be used much less frequently.

Anhydrous hydrogen cyanide-^{14}C is sometimes necessary in the synthesis of labelled compounds. A detailed description of its preparation has been given by Urban[42].

IV. SODIUM CYANIDE-^{15}N

Sodium cyanide-^{15}N has been prepared[40] with a 97–100% yield by heating potassium phthalimide-^{15}N with sodium metal for 20 minutes at 700 °c in a small steel bomb.

V. METHODS OF PREPARATION OF ^{14}C- AND ^{13}C-NITRILES

Since there is generally no exchange between nitriles and metallic cyanides[41], the labelled cyano group has to be introduced by synthesis. Most of the methods of classical organic chemistry are used, with suitable modifications. They can be grouped under two main categories. Most frequently alkali or cuprous cyanides-^{14}C are employed to introduce the nitrile group. Much less frequently a ^{14}C-carboxyl group is transformed into a ^{14}C-cyano group.

A. From Metallic Cyanides and Hydrogen Cyanide-^{14}C (-^{13}C)

Nearly all of the known methods of nitrile preparation have been adapted to the use of ^{14}C-cyanides and, much less frequently, anhydrous hydrogen cyanide.

1. Additions of cyanides to unsaturated compounds

This group of methods has not been used frequently. For instance, acrylonitrile-1-^{14}C has never been prepared by the direct addition of hydrogen cyanide-^{14}C to acetylene, although a laboratory procedure[43] might well be adapted to semimicro work.

However, Westöö[44] has used the addition of sodium cyanide to ethyl crotonate (**1**) to prepare sodium 3-(^{14}C-cyano)butyrate (**2**) in 67 % yield based on sodium cyanide:

$$\underset{(\mathbf{1})}{MeCH{=}CHCO_2Et} + Na^{14}CN + H_2O \longrightarrow \underset{(\mathbf{2})}{MeCH(^{14}CN)CH_2CO_2Na} + EtOH$$

In this case, the alkaline conditions during the addition reaction are sufficiently strong to cause hydrolysis of the ester group.

2. Additions of hydrogen cyanide-^{14}C or -^{13}C to carbonyl compounds

The addition of hydrogen cyanide to carbonyl compounds gives α-hydroxynitriles (cyanohydrins):

$$RCHO + H^{14}CN \rightleftharpoons RCH(OH)^{14}CN$$

The reaction is reversible, the extent of the cyanohydrin formation depends upon the structure of the carbonyl compound. The formation of aliphatic and alicyclic cyanohydrins is favoured, the yields of alkyl aryl ketones cyanohydrins are lower and diaryl ketones do not react at all[45,46]. The reaction is base catalysed, e.g. by potassium cyanide, or potassium carbonate. The cyanohydrins are thermally unstable and decompose by heating to hydrogen cyanide and a carbonyl compound. However, some of them can be distilled *in vacuo* after stabilization with traces of sulphuric or phosphoric acids.

The usual procedures of cyanohydrin syntheses have been employed for ^{14}C-labelling. Some examples of ^{14}C-labelled compounds prepared through a cyanohydrin synthesis are given in Table 1. However, usually the reaction of the carbonyl compound with the cyanide is carried out with an excess of cyanide in order that the equilibrium will favour the cyanohydrin. This cannot be used in ^{14}C-labelling work, where an excess of aldehyde is commonly employed, since the radioactive cyanide is the costly reagent. It has been demonstrated that even under these conditions the reaction can be accomplished by the following routes:

[1] From a carbonyl compound and anhydrous hydrogen cyanide-^{14}C in the presence of an alkaline catalyst. This procedure can be exemplified by the preparation of acetone cyanohydrin-1-^{14}C (**3**) in 96% yield; it has been described in detail by Urban[42]:

$$CH_3COCH_3 + H^{14}CN \xrightarrow{K_2CO_3} (CH_3)_2C(OH)^{14}CN \quad (\mathbf{3})$$

Another example is the preparation of D,L-glyceric acid-1-^{14}C (**5**) from glycollaldehyde (**4**) and liquid hydrogen cyanide-^{14}C and subsequent hydrolysis of the cyanohydrin-^{14}C which is not isolated[47].

$$\underset{(\mathbf{4})}{HOCH_2CHO} + H^{14}CN \longrightarrow \left[HOCH_2CH(OH)^{14}CN\right] \longrightarrow \underset{(\mathbf{5})}{HOCH_2CH(OH)^{14}CO_2H}$$

The yield of 33% reported for the overall synthesis of calcium glycerate-1-^{14}C could not be reproduced by Ashworth[48]. This method is certainly in less use than those below since it requires a vacuum manifold.

[2] From a carbonyl compound and hydrogen cyanide-^{14}C generated in the reaction mixture by the action of an acid (sulphuric, acetic, hydrochloric or phosphoric acids).

This technique can be illustrated by the preparation of lactic acid-1-^{13}C by condensation of sodium cyanide-^{13}C with an excess of acetaldehyde[49]. Hydrogen cyanide-^{13}C being generated *in situ* by

TABLE 1. Some examples of ^{14}C-labelled compounds prepared through a cyanohydrin synthesis $RCHO + H^{14}CN \longrightarrow RCH(OH)^{14}CN$.

Compound	Formula	Starting carbonyl compound	References
Ascorbic acid-1-^{14}C	$HOCH_2CH(OH)CH\text{—}C(OH)\text{=}C(OH)\text{—}^{14}C\text{=}O$ (with —O— ring bridge)	$HOCH_2CH(OH)\text{—}CH(OH)COCHO$	72, 73
Citric acid-6-^{14}C	$HO_2CCH_2C(OH)(^{14}CO_2H)CH_2CO_2H$	$EtO_2CCH_2COCH_2CO_2Et$	74
1-Hydroxycyclopentane carbonitrile-^{14}C	HO ^{14}CN	O	75
2-Hydroxy-3-phenylsuccinic acid-1-^{14}C	$PhCH(CO_2H)\text{—}CH(OH)^{14}CO_2H$	$PhCH(CO_2H)CHO$	78

Intergerrinecic acid-[carboxyl-^{14}C]	MeCH=C(CO$_2$H)—CH$_2$CH(CH$_3$)—C(OH)(Me)^{14}CO$_2$H	MeCH=C(CO$_2$H)—CH$_2$CH(Me)COMe	79
Mandelic acid-1-^{14}C	PhCH(OH)^{14}CO$_2$H	PhCHO	115
2-Methyl-2-hydroxybutyric acid-1-^{14}C	EtC(Me)(OH)^{14}CO$_2$H	EtCOMe	71
2-Phenyllactic acid-1-^{14}C	PhC(Me$_3$)(OH)^{14}CO$_2$H	PhCOMe	76, 77
3,4,5-Trimethoxymandelic acid-[carboxyl-^{14}C]	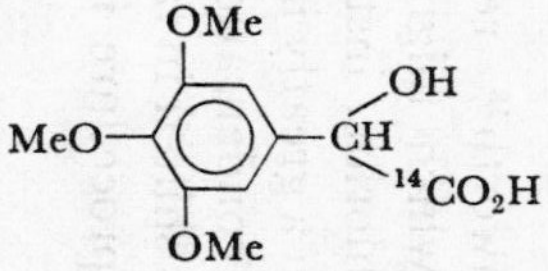 3,5-(MeO)$_2$-4-MeO-C$_6$H$_2$—CH(OH)^{14}CO$_2$H	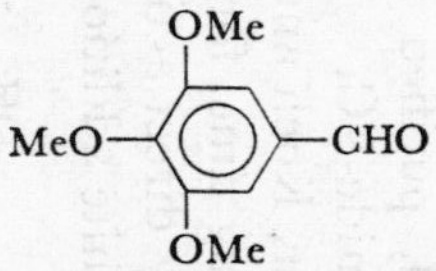 3,5-(MeO)$_2$-4-MeO-C$_6$H$_2$—CHO	80

sulphuric acid and the reaction mixture is then neutralized with sodium hydroxide solution.

$$CH_3CHO + H^{13}CN \longrightarrow \left[\begin{array}{c} CH_3\underset{\displaystyle OH}{\underset{|}{C}}H^{13}CN \end{array}\right] \longrightarrow CH_3\underset{\displaystyle OH}{\underset{|}{C}}H^{13}CO_2H$$

[3] Sometimes, the bisulphite addition product of the aldehyde is isolated, purified and then allowed to react directly with an alkali cyanide-^{14}C.

Schlesier, Koch and Büchner[50] have used this method, known for quite a long time, for the preparation of 2-hydroxynonanonitrile-1-^{14}C (**7**) by direct action of radioactive potassium cyanide-^{14}C upon the bisulphite addition product of octanal (**6**)

$$\underset{(\mathbf{6})}{CH_3(CH_2)_6\overset{\displaystyle OH}{\overset{|}{C}}HSO_3K} + K^{14}CN \longrightarrow \underset{(\mathbf{7})}{CH_3(CH_2)_6\overset{\displaystyle OH}{\overset{|}{C}}H^{14}CN} + K_2SO_3$$

[4] Ashworth[48] recently described a modified cyanohydrin synthesis which might be generally useful. It involves the use of benzoyl chloride, instead of the more usual hydrochloric acid, and it results in a greatly increased yield, suggesting that by benzoylating the initial condensation product the equilibrium of the cyanohydrin reaction is shifted in favour of the products. Ashworth employed this modified procedure in an improved preparation of glyceric acid-1-^{14}C (**10**).

$$\underset{(\mathbf{8})}{HOCH_2CHO} + Na^{14}CN \xrightarrow[NaOH]{PhCOCl} \underset{(\mathbf{9})}{PhCO_2CH_2\overset{\displaystyle ^{14}CN}{\overset{|}{C}}HOCOPh} \xrightarrow{50\%\ HCl} \underset{(\mathbf{10})}{HOCH_2\overset{\displaystyle OH}{\overset{|}{C}}H^{14}CO_2H}$$

The benzoylated cyanohydrin (**9**) is insoluble in the reaction mixture and is thus readily isolated. Ashworth suggests that this procedure may therefore be of general applicability in obtaining higher yields in cyanohydrin condensations.

[5] Reasoning that the cyanohydrin formation is a base-catalysed equilibrium, Kourim and Zikmund[93] could introduce ^{14}C into the cyanohydrin by exchange with H^{14}CN. They studied the exchange reaction:

$$CH_3OCH_2\overset{\displaystyle OH}{\overset{|}{C}}HCN + H^{14}CN \rightleftharpoons CH_3OCH_2\overset{\displaystyle OH}{\overset{|}{C}}H^{14}CN$$

under catalysis by *N*-ethylpiperidine or cyanide ions. In both cases, it was found that the exchange was complete at room temperature after 20 hours. This procedure does not seem to have found practical use, probably because of the isotopic dilution.

From the examples given above, one can already see that carbonyl compounds carrying a second functional group undergo the cyanohydrin reaction. The method is of a special importance in the synthesis of ^{14}C-sugars and is known as the Kiliani–Fischer synthesis of which it constitutes the initial step. This method of increasing the length of the carbon chain of sugars is particularly suitable for obtaining monosaccharides and disaccharides-1-^{14}C. It is illustrated here with the synthesis of glucose-1-^{14}C according to the scheme outlined below:

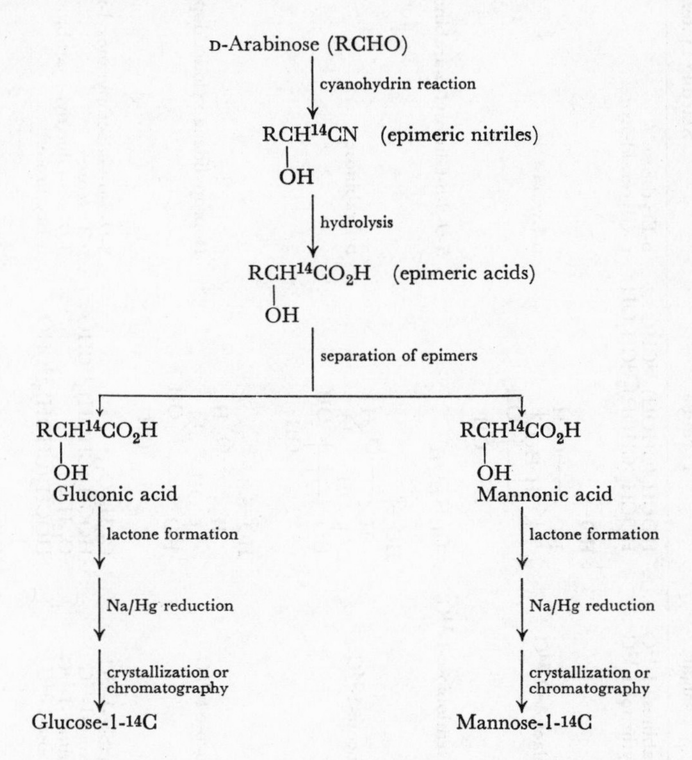

Isbell and coworkers[51] demonstrated that combination of an aldose with cyanide ion in stoichiometric proportions is nearly quantitative under a variety of conditions. This result was very important, since it allowed complete use of the ^{14}C-cyanide and a

TABLE 2. Some ^{14}C sugars prepared by the Kiliani–Fischer cyanohydrin synthesis $RCHO + H^{14}CN \rightarrow RCH(OH)^{14}CN$.

Sugar	Formula	Starting material	References
Arabinose-1-^{14}C	$HOCH_2(CHOH)_3{}^{14}CHO$	D-Erythrose	55
Erythritol-1-^{14}C	$HOCH_2(CHOH)_2{}^{14}CH_2OH$	D-Glyceraldehyde	53
Galactose-1-^{14}C	[structural formula]	D-Lyxose	66, 67
β-Gentiobiose-1-^{14}C	$C_{12}H_{22}O_{11}$	5-*O*-*β*-D-Glucosyl-D-arabinose	70
Glucose-1-^{14}C	[structural formula]	D-Arabinose	51, 59–61
Glucose-6-^{14}C	[structural formula]	Isopropylidene xylodialdopentofuranose	62–65
Lactose-1-^{14}C	$C_{12}H_{22}O_{11}$	3-*O*-*β*-D-Galactopyranosyl-D-arabinose	69
Lyxose-1-^{14}C	$HOCH_2(CHOH)_3{}^{14}CHO$	D-Threose	56, 57
Maltose-1-^{14}C	$C_{12}H_{22}O_{11}$	3-*O*-*α*-D-Glucosyl-D-arabinose	68
Ribose-1-^{14}C	$HOCH_2(CHOH)_3{}^{14}CHO$	D-Erythrose	55
Talose-1-^{14}C	[structural formula]	D-Lyxose	66

Kiliani–Fischer synthesis usually uses an excess of cyanide. Another problem concerns the relative amounts of the epimeric cyanohydrins that are formed, since these may vary from nearly equal amounts for one sugar to almost exclusively one isomer for another sugar. Isbell and coworkers[51,52] have shown that it is possible to control to a certain extent the proportions of the epimeric cyanohydrins by changing the experimental conditions.

The Kiliani–Fischer cyanohydrin synthesis has been successfully used on a semimicro scale for preparing a series of ^{14}C sugars, some of which are listed in Table 2. A short review on this subject has been published[54].

3. Addition of hydrogen cyanide-^{14}C or -^{13}C to carbonyl compounds in the presence of ammonia: the Strecker synthesis

Strecker[81] in 1850 inadvertently observed the formation of alanine instead of the expected lactic acid, from the interaction of acetaldehyde first with ammonia, then with hydrocyanic acid followed by acid hydrolysis. Since then, the reaction known as the Strecker synthesis has become a valuable method for the synthesis of α-amino acids. The α-aminonitriles usually formed in the initial step are not isolated but directly subjected to acid or alkaline hydrolysis.

$$R_2CO + NH_3 + {}^{14}CN^- \longrightarrow R_2\underset{\displaystyle NH_2}{\underset{|}{C}}{}^{14}CN \xrightarrow{H^+/H_2O} R_2\underset{\displaystyle NH_2}{\underset{|}{C}}{}^{14}CO_2H$$

Many modifications of the original procedure have been developed. The use of anhydrous hydrogen cyanide is not necessary, but has been employed by some authors. Indeed the most commonly employed method uses a mixture of alkali cyanide and ammonium chloride.

The formation of iminodinitrile (**11**) and of nitrilotrinitrile (**12**) is

$$\begin{array}{c} RCHCN \\ | \\ NH \\ | \\ RCHCN \\ (\mathbf{11}) \end{array} \qquad \begin{array}{c} RCHCN \\ | \\ N-CH(CN)R \\ | \\ RCHCN \\ (\mathbf{12}) \end{array}$$

ordinarily minimized by the use of an excess of cyanide. In radioactive work, once again this is not possible. We have found that the method of Holland[82] and Gaudry[83], which implies the inclusion of ammonia in the Strecker reaction, can be used with advantage for the preparation of α-amino acids-[carboxyl-^{14}C]. The addition of ammonia to the solution containing ammonium chloride, aldehyde and cyanide-^{14}C prevents the formation of iminodinitrile, even when a large aldehyde to cyanide ratio is used. The reaction mixture is left at room temperature for two days. Some results obtained in the author's laboratory are recorded in Table 3 and results from other laboratories in Table 4.

Kourim and Zikmund[93] have described an interesting modification of the Strecker's synthesis of serine-1-^{14}C (**13**):

$$CH_3OCH_2CH(OH)CN + H^{14}CN \xrightarrow{CN^-} [CH_3OCH_2CH(OH)^{14}CN] \xrightarrow{NH_3}$$

$$[CH_3OCH_2CH(NH_2)^{14}CN] \xrightarrow[2.\ C_5H_5N]{1.\ HBr} HOCH_2CH(NH_2)^{14}CO_2H \quad \mathbf{(13)}$$

Starting from the inactive cyanohydrin, the exchange with radioactive cyanide and the amination are carried out simultaneously in liquid ammonia. The obvious drawback of the method is the isotopic dilution.

Aminoacetonitrile-1-^{14}C has been prepared by Strecker's synthesis on formaldehyde[114]:

$$2\,CH_2O + Na^{14}CN + NH_4Cl \longrightarrow CH_2{=}NCH_2{}^{14}CN \xrightarrow[EtOH]{H_2SO_4} H_2NCH_2{}^{14}CN$$

TABLE 3. Some amino acids-1-^{14}C prepared by Strecker's synthesis as modified by Holland[82] and Gaudry[83] by saturation of the reaction mixture with ammonia.

Amino acid	Yields from K^{14}CN after chromatographic purifications (%)	References
D,L-Arginine-1-^{14}C	61	84
D,L-Glutamic acid-1-^{14}C	86	85
D,L-Methionine-1-^{14}C	61	86
D,L-Serine-1-^{14}C	30	85
D,L-Valine-1-^{14}C	40	85

TABLE 4. Other amino acids-1-^{14}C prepared by Strecker's synthesis from K^{14}CN or K^{13}CN.

Amino acid	Yields from K^{14}CN (%)	References
D,L-Alanine-1-^{14}C and 1-^{13}C	38, 48, 87	87–89
D,L-α-aminobutyric acid-1-^{14}C	67	89
2,2′-Dimethylcystine-1-^{14}C	38	94
D,L-Leucine-1-^{14}C	—	91
D,L-Phenylalanine-1-^{14}C	34	92
D,L-Serine-1-^{14}C	80	93
D,L-Valine-1-^{14}C	51	90

Although the overall yield of amino acids-1-^{14}C can be considered satisfactory, it has been found in the author's laboratory that a fairly large number of impurities are formed, either during the initial stage of the Strecker's synthesis or in the subsequent hydrolysis step. Column chromatography on ion-exchange resin, sometimes associated with preparative paper chromatography, have the necessary separation power for the complete removal of these impurities. Some impurities are also formed through parallel reactions when an impure aldehyde is subjected to the reaction. Lotfield[89] has found that the most effective way to ensure that the starting aldehydes are very pure is to begin with the purest available amino acids, which on reaction with ninhydrin yield pure aldehydes.

4. Addition of hydrogen cyanide-^{14}C or -^{13}C to carbonyl compounds in the presence of ammonium carbonate: Bücherer's modification of the Strecker synthesis

Bücherer and his collaborators[95] demonstrated in 1934 that 5-substituted hydantoins could be readily prepared in high yields by heating 1 mole of the aldehyde–sulphurous acid adduct with a 50% alcoholic solution of 2 moles potassium cyanide and 4 moles ammonium carbonate for 4–5 hours at 55 °C. The hydantoin crystallizes out of the reaction medium and is usually isolated.

$$RCH(OH)SO_3Na \xrightarrow{K^{14}CN} [RCH(OH)^{14}CN] \xrightarrow{(NH_4)_2CO_3} \text{hydantoin (14)}$$

(14)

Hydrolysis to α-amino acids is effected either in acid or basic medium. The reaction has been reviewed[96].

TABLE 5. Some amino acids-1-^{14}C prepared by the Bücherer method.

$$RCHO \rightarrow \underset{\displaystyle OH}{RCHSO_3Na} \xrightarrow[(NH_4)_2CO_3]{K^{14}CN} \left[\text{hydantoin (R, } ^{14}C{=}O\text{, HN, NH, C=O)} \right] \xrightarrow{\text{hydrolysis}} \underset{\displaystyle NH_2}{RCH^{14}CO_2H} \quad (\mathbf{15})$$

Amino acid-1-^{14}C	Aldehyde	Yield of hydantoin based on $KC^{14}N$ (%)	Yield of amino acid based on $KC^{14}N$ (%)	References
Allo isoleucine	3-Methylbutyraldehyde	84	82	89
3-(3,4-Dihydroxyphenyl)alanine	3,4-Dimethoxyphenylacetaldehyde	50	8·5	98
p-Fluorophenylalanine	*p*-Fluorophenylacetaldehyde	75	70	89
Isoleucine	3-Methylbutyraldehyde	88	85	89
Leucine	Isovaleraldehyde	92	86	89
Lysine	5-Benzamidovaleraldehyde	70	81	99
Lysine	5-Hydroxyvaleraldehyde	—	—	100
Lysine	Ethyl 5-formylvalerate	86	66	101
Norvaline	Butyraldehyde	82	78	89
Phenylalanine	Phenylacetaldehyde	80	70	89
Tyrosine	*p*-Methoxyphenylacetaldehyde	75–88	78–89	89, 97, 99
Valine	Isobutyraldehyde	86	81	89

To be applied to ^{14}C labelling, it was necessary either to use stoichiometric quantities of aldehyde and $K^{14}CN$ or an excess of aldehyde. According to Rothstein[101], a large excess of aldehyde does not improve the yield based on cyanide but complicates the isolation of pure hydantoin. The presence of an excess potassium hydroxide in the radioactive cyanide may sometimes lower the yields[97]. The Bücherer method has been extensively used by Lotfield[89] (see Table 5). However, this author used the Strecker synthesis for the preparation of alanine-1-^{14}C and α-aminobutyric acid-1-^{14}C because the hydantoins formed from acetaldehyde and propionaldehyde are too volatile. Some syntheses of amino acids-1-^{14}C (**15**) by the Bücherer method are collected in Table 5. Obviously the employed aldehyde should be of high purity[89,101].

5. Condensation of alkali cyanides-^{14}C and 1,2-epoxides

The preparation of 3-hydroxypropionitrile-1-^{14}C (**16**) has been accomplished by a number of authors[102–105] by reaction of alkali cyanides-^{14}C with ethylene oxide in an adaptation of Rambaud's procedure[106]:

$$\underset{\text{O}}{\text{CH}_2\text{—CH}_2} + {}^{14}\text{CN}^- + \text{H}^+ \longrightarrow \underset{(\mathbf{16})}{\text{HOCH}_2\text{CH}_2{}^{14}\text{CN}}$$

Yields of 90% based on cyanide were obtained for **16** which was not isolated and used without further purification in subsequent steps.

6. Condensation of alkali cyanides-^{14}C with lactones

The ring-opening of lactones to form salts of cyano acids by heating them with powdered alkali cyanides at 180 °C has been described by Wislicenus[107] as early as 1885. Pichat and coworkers have used this method as the initial step of several syntheses, for the introduction of carbon-14, without isolation of the cyano acid and found it well adapted[108].

$$\text{(lactone, } R^1, R^2\text{)} + K^{14}CN \longrightarrow N^{14}CCH_2CHR^2CHR^1COOK$$

(**17**, R^1 = H, R^2 = Me) (**18**, R^1 = H, R^2 = Me)
(**19**, R^1 = PhCONH, R^2 = H) (**20**, R^1 = PhCONH, R^2 = H)
(**21**, R^1 = R^2 = H) (**22**, R^1 = R^2 = H)

The possibilities of the method for isotopic work have been investigated first[108] with β-methylbutyrolactone (**17**). Rather drastic conditions were used: shaking the lactone without solvent, under nitrogen, in a sealed tube at 280°. Because of its inaccessibility, only a slight excess of the lactone **17** was used. Compound **18** was hydrolysed to β-methylglutaric acid-1-^{14}C.

Later α-benzamidobutyrolactone (**19**) was submitted[28] to the action of potassium cyanide-^{14}C in order to obtain the polyfunctional nitrile-^{14}C **20** which subsequently could be used in order to obtain successively: glutamic acid-5-^{14}C, ornithine-5-^{14}C and arginine-5-^{14}C. Pichat, Mizon and Herbert[28] studied the action of $K^{14}CN$ on the lactone **19** and found that the reaction could be conveniently carried out at reflux temperature in dimethylformamide or 'diethylcarbitol' solvents. A 100% excess of lactone over $K^{14}CN$ has only a slight influence upon the yield (about 58% from cyanide).

When the reaction was applied[109] to γ-butyrolactone (**21**) in order to obtain **22**, the intermediate for the synthesis of proline-[(5-ring)-^{14}C], the commercially available lactone, was used in large excess at its boiling point (190 °c) and played both the role of solvent and reagent. However, Muelder and Wass[110], probably unaware of Pichat's work[109], preferred another more time-consuming reaction scheme for the preparation of **22**.

The method has also been applied[111] with the δ-lactone **23**, for the preparation of α-aminoadipic acid-6-^{14}C (**24**) in 34% yield from $K^{14}CN$.

$$\textbf{(23)}\ \xrightarrow[\text{DMF}]{K^{14}CN} [N^{14}C(CH_2)_3CH(NHCOPh)COOK] \xrightarrow{\text{hydrolysis}} KO_2{}^{14}C(CH_2)_3CH(NH_2)CO_2H\ \textbf{(24)}$$

Much milder reaction conditions (20–45°) are used for the ring-opening of the very reactive β-propiolactone **25** with $K^{14}CN$.

$$\underset{\textbf{(25)}}{CH_2CH_2CO{-}O\ (\text{ring})} + K^{14}CN \longrightarrow \underset{\textbf{(26)}}{N^{14}CCH_2CH_2CO_2K}$$

By this method Frank[112] obtained β-cyanopropionic acid-4-^{14}C (**26**) in pure form by chromatography on Dowex-1. Pichat and coworkers[113] used this reaction as the initial step of a preparation of 4-aminobutyric acid-4-^{14}C and azetidin-2-carboxylic acid-(-4-ring-^{14}C) (**26a**).

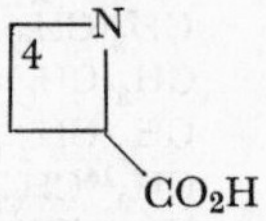

(**26a**)

7. Reaction of metallic cyanides and halogen compounds

The alkali cyanides react with alkyl halides to provide predominantly nitriles. The interaction of metallic cyanides-^{14}C and halogen compounds has been employed extensively for the introduction of carbon-14. Very often the intermediate nitriles have not been isolated or extensively purified but mainly hydrolysed to the corresponding carboxylic acid-1-^{14}C. On some occasions, labelled alkyl halide was employed with unlabelled metallic cyanide, so that nitriles labelled on the alkyl group were obtained. When a nitrile-1-^{14}C is required, an excess of alkyl halides has to be used, while excess cyanide is used with labelled alkyl halide.

The reaction is usually performed in an aqueous alcohol solvent[116]. Other solvent systems, such as 2-methoxyethanol[117] or polyethyleneglycol[118], have recently been suggested. However, dimethylsulphoxide[119,120], which permits the use of secondary alkyl chlorides in this type of reaction and presents advantages in conversion of alkyl bromides to nitriles, has been used more often than those in radioactive labelling.

a. Saturated alkyl halides. For simple saturated monofunctional alkyl halides the carbonation of the corresponding Grignard reagent with $^{14}CO_2$ is obviously preferred for the preparation of ^{14}C-carboxylic acids. However, in some circumstances a few acids-1-^{14}C have been prepared by the nitrile method. Some examples are collected in Table 6, which also contains some acids prepared via nitriles labelled on positions other than 1-^{14}C, and doubly labelled nitriles.

b. Aralkyl halides. Benzyl-type chlorides are converted to the corresponding nitriles more rapidly than allylic halides. Phenylacetonitrile-1-^{14}C [129,130] has been prepared by an adaptation of a

TABLE 6. Synthesis of acids and nitriles.

Acids and nitriles	RX	Cyanide	References
Acetic acid-1-^{13}C-2-^{14}C	$^{14}CH_3I$	$K^{13}CN$	121
Acetonitrile-2-^{13}C	$^{13}CH_3I$	KCN	128
Butyric acid-1-^{13}C	$CH_3CH_2CH_2I$	$K^{13}CN$	122
Butyric acid-1-^{14}C	$CH_3(CH_2)_2I$	$K^{14}CN$	126
Myristic acid-1-^{14}C	$CH_3(CH_2)_{12}Br$	$K^{14}CN$	127
Octanoic acid-1-^{13}C	$CH_3(CH_2)_6Br$	$K^{13}CN$	125
Propionic acid-2-^{14}C	$CH_3{}^{14}CH_2I$	—	124
Propionic acid-2-^{13}C-3-^{14}C	$^{14}CH_3{}^{13}CH_2I$	—	123
Propionic acid-3-^{13}C	$^{13}CH_3CH_2I$	—	125

standard procedure. More recently[132] ethylene glycol has been used as a solvent for this preparation. Ziegler and Shabica[133] also prepared benzyl cyanide-1-^{14}C by the standard procedure[131], but since their sample of labelled sodium cyanide contained sodium hydroxide, a compensatory excess of benzyl chloride was used to react with the sodium hydroxide, so that the active cyanide was conserved. Battersby has prepared 3-methoxy-4-benzyloxyphenylacetonitrile-1-^{14}C by the same method[134]. 3-Phenylpropionitrile-1-^{14}C obtained[145] by reaction of $K^{14}CN$ with 2-phenylethyl chloride in water–ethanol solvent has been an intermediate in the preparation of 3-phenylpropanoic acid-1-^{14}C.

c. Dibromoalkanes. Many ^{14}C-dinitriles have been prepared by action of $Na^{14}CN$ or $K^{14}CN$ on polymethylene dibromides. Table 7 lists some of the dibasic acids-[carboxyl-^{14}C] which have been prepared by hydrolysis of the corresponding dinitriles.

4-Chlorobutyronitrile-1-^{14}C has been prepared[146] from 1-bromo-3-chloropropane by taking advantage of the different reactivities of two dissimilar halogen atoms:

$$Cl(CH_2)_3Br + K^{14}CN \longrightarrow Cl(CH_2)_3{}^{14}CN + KBr$$

d. Haloalcohols and haloethers. Hydroxynitriles-1-^{14}C are obtained from haloalcohols as illustrated by the preparation of 3-hydroxypropionitrile-1-^{14}C from ethylene chlorohydrin[147]. 3-Hydroxypropionitrile-2,3-^{14}C is prepared analogously from 1-bromo-2-hydroxyethane-1,2-^{14}C[148]:

$$HO^{14}CH_2{}^{14}CH_2Br + KCN \longrightarrow HO^{14}CH_2{}^{14}CH_2CN + KBr$$

$$HOCH_2CH_2Cl + K^{14}CN \longrightarrow HOCH_2CH_2{}^{14}CN + KCl$$

Klenk and Pflüger[149] have prepared 7-hydroxyheptanonitrile-1-^{14}C as the primary stage of a nine-step synthesis of oleic acid-8-^{14}C.

Potassium cyanide-^{14}C and 3-methoxy-1-iodopropane refluxed in

TABLE 7. Dinitriles and dibasic acids-[carboxyl-^{14}C] prepared through polymethylene ^{14}C-nitriles.

Acids and dinitriles	$X(CH_2)_nX$	Metallic cyanides	References
Adiponitrile-1,6-^{14}C	$Br(CH_2)_4Br$	$K^{14}CN$	142
Azelaic acid-1,9-^{14}C	$Br(CH_2)_7Br$	$K^{14}CN$	137
Glutaric acid-1,5-^{14}C	$Br(CH_2)_3Br$	$K^{14}CN$	135, 136
Glutaronitrile-1,5-^{13}C	$Br(CH_2)_3Br$	$K^{13}CN$	144
Glutaronitrile-1,5-^{14}C	$Br(CH_2)_3Br$	$K^{14}CN$	141
Glutaronitrile-2-^{13}C	$BrCH_2CH_2{}^{13}CH_2Br$	KCN	144
Pimelic acid-1,7-^{14}C	$Br(CH_2)_5Br$	$K^{14}CN$	143
Sebacic acid-1,10-^{14}C	$Br(CH_2)_8Br$	$K^{14}CN$	138
Succinic acid-1,4-^{13}C	$BrCH_2CH_2Br$	$K^{13}CN$	140
Succinic acid-2,3-^{14}C	$Br^{14}CH_2{}^{14}CH_2Br$	KCN	141
Tetradecandioic acid-1,14-^{14}C	$Br(CH_2)_{12}Br$	$K^{14}CN$	139

methyl alcohol gave 4-methoxybutyronitrile-1-^{14}C in 50 % yield[150]

$$CH_3O(CH_2)_3I \xrightarrow[\text{MeOH}]{K^{14}CN} CH_3O(CH_2)_3{}^{14}CN$$

A novel synthesis via **28** of 2-deoxy-D-ribose-1-^{14}C, the carbohydrate component of deoxyribonucleic acid, has been described[171]. The initial step of this preparation is based on the smooth reaction of potassium cyanide-^{14}C in suspension in anhydrous dimethylformamide with the iodo compound **27**.

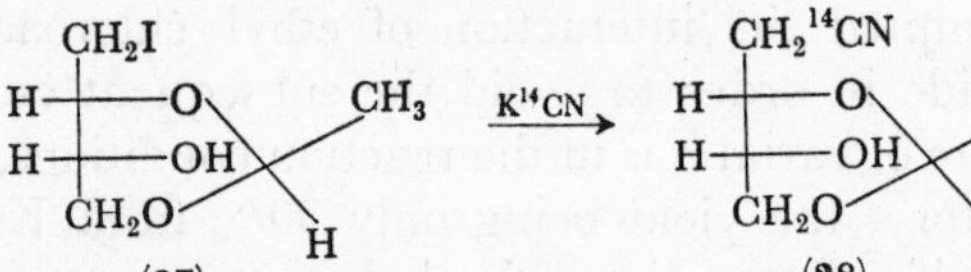

CH_2I, H—O, H—OH, CH_2O (acetonide with CH_3, H) **(27)** $\xrightarrow{K^{14}CN}$ $CH_2{}^{14}CN$, H—O, H—OH, CH_2O (acetonide with CH_3, H) **(28)**

Some bile acids given below labelled on the side chain have been obtained by Bergström, Rottenberg and Voltz[151] from the appropriate bromides, as exemplified by the preparation of lithocholic acid-24-^{14}C (**29**) through the corresponding nitrile in 80–90 % yield:

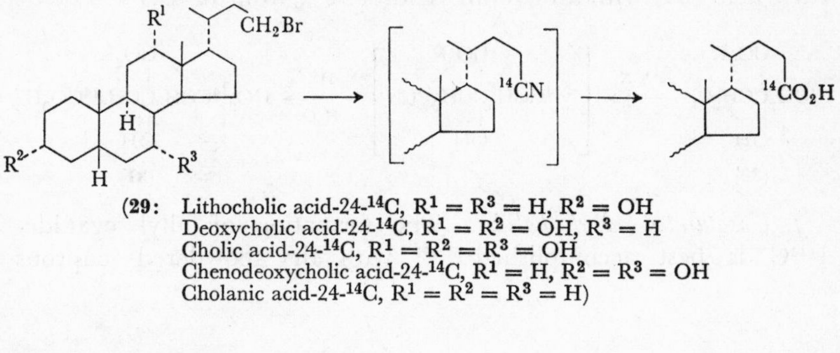

(steroid with R^1, R^2, R^3, side chain CH_2Br) $\longrightarrow$ [side chain ^{14}CN] $\longrightarrow$ side chain $^{14}CO_2H$

(**29**: Lithocholic acid-24-^{14}C, $R^1 = R^3 = H$, $R^2 = OH$
Deoxycholic acid-24-^{14}C, $R^1 = R^2 = OH$, $R^3 = H$
Cholic acid-24-^{14}C, $R^1 = R^2 = R^3 = OH$
Chenodeoxycholic acid-24-^{14}C, $R^1 = H$, $R^2 = R^3 = OH$
Cholanic acid-24-^{14}C, $R^1 = R^2 = R^3 = H$)

Hyodeoxycholic acid-24-^{14}C has also been prepared[152,153] by the ^{14}C-nitrile synthesis. However, the hydroxyl groups of the bromide were protected by acetylation. Muricholic acid-24-^{14}C has been obtained more recently[154].

An additional example is the synthesis of erythro-9,10-dihydroxy-stearic acid-1-^{14}C [170]:

$$CH_3(CH_2)_7CH(OH)CH(OH)(CH_2)_7Br \xrightarrow{K^{14}CN} \left[CH_3(CH_2)_7CH(OH)CH(OH)(CH_2)_7{}^{14}CN\right] \longrightarrow CH_3(CH_2)_7CH(OH)CH(OH)(CH_2)_7{}^{14}CO_2H$$

e. Haloacids and haloesters. Cyano acids are prepared by treating the aqueous solutions of sodium halocarboxylates with an alkali cyanide-^{14}C. This is illustrated by the preparation of cyano-^{14}C-acetic acid[155,156,223] which is an important synthetic intermediate either as the methyl ester[156] or as the amide. The dehydration of the latter leads to the very useful malononitrile-1,3-^{14}C [155]. Doubly-labelled cyanoacetic acid-3-^{14}C, ^{15}N has been obtained[157] through the use of $K^{14}C^{15}N$.

Korte[156] attempted the interaction of ethyl chloroacetate with potassium cyanide in order to avoid the subsequent esterification. However, in spite of variations in the reaction conditions, the results were unsatisfactory, the yield being only 50% from $K^{14}CN$, as a result of the self-condensation of ethyl cyanoacetate. However, ethyl 2-(cyano-^{14}C) propionate was obtained[158] by interaction of $K^{14}CN$ and ethyl 2-bromopropionate.

The reaction has also been applied[159] to potassium 2,2-bis(chloro-methyl)glycollate (**30**) as the first step in the preparation of 1,5-^{14}C-citric acid (**31**) with an overall yield of 40% from $K^{14}CN$

$$\underset{(\mathbf{30})}{ClCH_2C(CO_2K)(OH)CH_2Cl} \xrightarrow{K^{14}CN} \left[N^{14}CCH_2C(CO_2K)(OH)CH_2{}^{14}CN\right] \xrightarrow[H_2O]{H^+} \underset{(\mathbf{31})}{HO_2{}^{14}CCH_2C(CO_2K)(OH)CH_2{}^{14}CO_2H}$$

f. Unsaturated alkyl halides. The formation of allyl cyanide-1-^{14}C is best accomplished[160,161] with dry powdered cuprous cyanide-1-^{14}C. This avoids the rearrangement to crotononitrile which occurs when alkali cyanide is allowed to react with allyl bromide.

$$CH_2{=}CHCH_2Br \xrightarrow{Cu^{14}CN} CH_2{=}CHCH_2{}^{14}CN$$

Heating K^{14}CN with allyl bromide at 100 °C in a sealed tube, and hydrolysing and isomerizing with 50% H_2SO_4, served to prepare crotonic acid-1-^{14}C [169]:

$$CH_2{=}CHCH_2Br \xrightarrow{K^{14}CN} [CH_2{=}CHCH_2{}^{14}CN] \longrightarrow CH_3CH{=}CH^{14}CO_2H$$

Stoffel and Bierwirth[162] and Stoffel[163,164] treated various *cis*-polyene chlorides (**32**) with radioactive sodium cyanide in dimethylsulphoxide. The resulting polyene nitriles (**33**) were then smoothly esterified to the corresponding methyl esters in good yield:

$$\underset{(\mathbf{32})}{CH_3(CH_2)_4(CH{=}CHCH_2)_x(CH_2)_yCl} \xrightarrow{Na^{14}CN} \underset{(\mathbf{33})}{CH_3(CH_2)_4(CH{=}CHCH_2)_x(CH_2)_y{}^{14}CN}$$

$$\begin{array}{ll} x = 2, & y = 6 \\ x = 3, & y = 3 \\ x = 3, & y = 5 \\ x = 4, & y = 2 \end{array}$$

Würsch[165] prepared arachidonic acid-1-^{14}C from 5-hexynoic acid-1-^{14}C and K^{14}CN in dimethylsulphoxide.

$$HC{\equiv}C(CH_2)_3Cl \xrightarrow{K^{14}CN} [HC{\equiv}C(CH_2)_3{}^{14}CN] \longrightarrow HC{\equiv}C(CH_2)_3{}^{14}CO_2H$$

Previous attempts[166,167] to exchange the halogen atom in a polyacetylene chain by nitrile had failed when other solvents were used. 15-Hexadecenoic acid-1-^{14}C has also been prepared[168] via the ^{14}C-nitrile method.

g. Aryl halides and heterocyclic halides. The replacement of an aromatic halogen atom by the cyano group can be achieved by the action of anhydrous cuprous cyanide with or without solvent. In classical preparative organic chemistry, this has been frequently used.

$$2\,ArX + 2\,Cu^{14}CN \longrightarrow [Ar^{14}CN]_2CuX + CuX$$

$$[Ar^{14}CN]_2CuX \longrightarrow 2\,Ar^{14}CN + CuX$$

This method has been employed for the preparation of aryl cyanides-^{14}C. A typical example is the preparation[172] of veratronitrile-[nitrile-^{14}C] (**35**) by heating and stirring cuprous cyanide-^{14}C with a 10% excess of 4-iodoveratrole (**34**) at 250 °C.

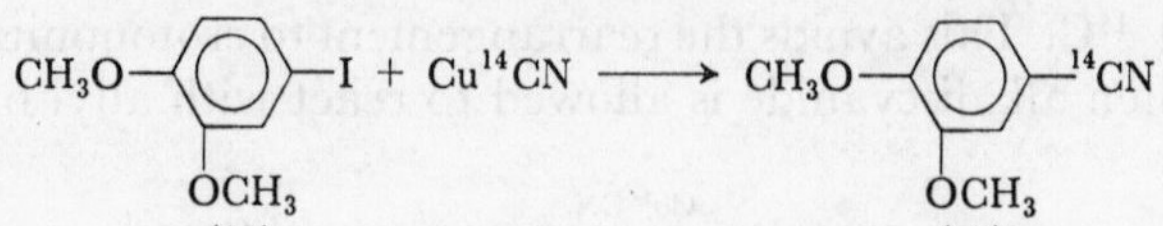

$$CH_3O-C_6H_3(OCH_3)-I + Cu^{14}CN \longrightarrow CH_3O-C_6H_3(OCH_3)-{}^{14}CN$$

(34) (35)

The preparation[173] of 3-methoxy-2-nitrobenzonitrile-[nitrile-^{14}C] (**37**), an intermediate in the preparation of 3-hydroxyanthranilic

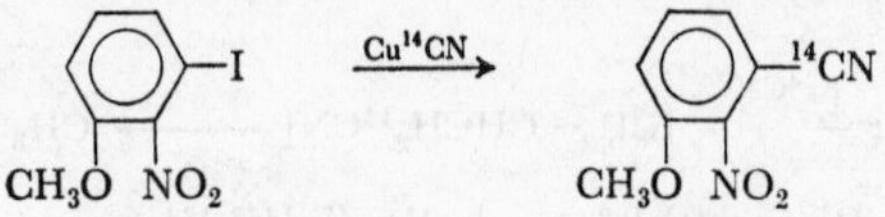

$$C_6H_3(OCH_3)(NO_2)-I \xrightarrow{Cu^{14}CN} C_6H_3(OCH_3)(NO_2)-{}^{14}CN$$

(36) (37)

acid from the rather reactive *o*-nitroaryl iodide (**36**) is another example where this reaction is performed without solvent. *o*-Nitrobenzonitrile-^{14}C has been prepared[186] in 80 % yield by heating 2 mmoles *o*-chloronitrobenzene and 1 mmole cuprous cyanide in a sealed tube at 160° for 2 h.

However, the displacement is most commonly effected in excellent yields (>85 %) by heating aryl halide with cuprous cyanide in the presence of pyridine or quinoline as a solvent.

The use of pyridine as promoter solvent can be exemplified by the preparation[174] of *trans*-4,4′-stilbenedicarbonitrile-$^{14}C_2$ (**38**).

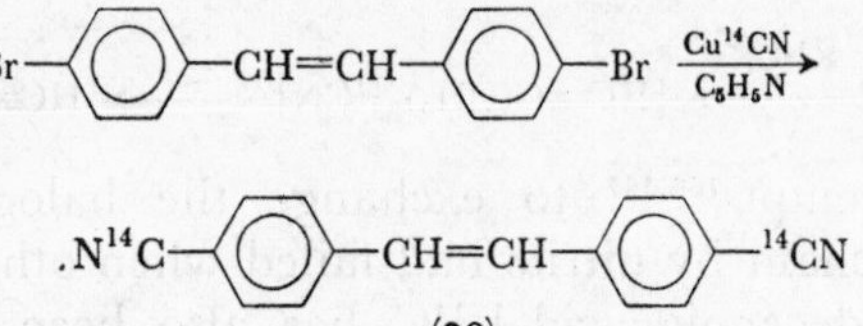

$$Br-C_6H_4-CH{=}CH-C_6H_4-Br \xrightarrow[C_5H_5N]{Cu^{14}CN}$$

$$N^{14}C-C_6H_4-CH{=}CH-C_6H_4-{}^{14}CN$$

(38)

More recently, Friedman and Shechter[175] demonstrated that the reaction of aryl bromides and activated aryl chlorides with cuprous cyanide occurs advantageously in refluxing dimethylformamide. Grisebach and Patschke[176] applied this method to the preparation of *p*-benzyloxybenzonitrile-[nitrile-^{14}C] (**40**) in 96 % yield:

$$PhCH_2O-C_6H_4-I \xrightarrow[DMF]{Cu^{14}CN} PhCH_2O-C_6H_4-{}^{14}CN$$

(39) (40)

The replacement of heterocyclic chlorine does not always require the use of cuprous cyanide. With reactive heterocyclic halides the replacement can be carried out successfully with potassium cyanide in DMSO at 130–140 °C as illustrated[177] by the preparation of 2-(cyano-^{14}C)-6-methoxybenzothiazole (**41**):

$$\text{2-chloro-6-methoxybenzothiazole} \xrightarrow{K^{14}CN} \text{2-(}^{14}\text{CN)-6-methoxybenzothiazole (41)}$$

h. Acyl halides. The conversion of aliphatic and aromatic acyl halides to acyl cyanides requires dry cuprous cyanide. Acyl bromides are preferred to acyl chlorides and sometimes are even essential, as in the case of the preparation of pyruvonitrile, where the reaction fails to occur with acetyl chloride. The method has been used very much. Typical examples are the preparations of pyruvonitrile-1-^{14}C as intermediate in the preparation of sodium pyruvate-1-^{14}C[178,179] and that of pyruvonitrile-2-^{14}C[180] and benzoylcyanide-1-^{14}C[181].

$$CH_3COBr \xrightarrow{Cu^{14}CN} CH_3CO^{14}CN \longrightarrow CH_3CO^{14}COONa$$

i. Phthalimidoalkyl halides. Sakami, Evans and Gurin[182] treated sodium cyanide-^{14}C with *N*-chloromethylphthalimide (**42**) in a methanol–dioxan mixture. The phthalimidoacetonitrile-1-^{14}C (**43**) formed was hydrolysed to glycine-1-^{14}C (**44**)

$$C_6H_4(CO)_2NCH_2Cl\ (\mathbf{42}) \xrightarrow{Na^{14}CN} C_6H_4(CO)_2NCH_2{}^{14}CN\ (\mathbf{43}) \xrightarrow{\text{hydrolysis}} NH_2CH_2{}^{14}COOH\ (\mathbf{44})$$

This method has been employed with some slight modifications by several authors[183,184]. Lotfield[89] considers it as the most consistently successful synthesis of glycine-1-^{14}C.

Fromm[185] generalized the procedure as outlined below:

$$C_6H_4(CO)_2NCH(R)COOAg\ (\mathbf{45}) \xrightarrow{Br_2} C_6H_4(CO)_2NCH(R)Br\ (\mathbf{46}) \xrightarrow{K^{14}CN} C_6H_4(CO)_2NCH(R)^{14}CN$$

$$(R = H, Me, Me_2CHCH_2)$$

The required phthalimidoalkyl bromides (**46**) are obtained by Hundsdiecker's degradation of the silver salts of *N*-phthaloylamino acids (**45**). It should be mentioned that the method cannot have the wide applicability of the Strecker or Bücherer methods discussed earlier, since it is restricted to amino acids which form phthaloyl derivatives.

Unexpected difficulties were encountered by Schilling and Strong[147] in the bromine replacement with cyanide in the next higher homologue (**47**)

$$C_6H_4(CO)_2NCH_2CH_2Br + K^{14}CN \not\longrightarrow C_6H_4(CO)_2NCH_2CH_2{}^{14}CN$$

(**47**) (**48**)

Only traces of the expected nitrile (**48**) could be isolated when 80 % ethanol was used as a solvent. A small yield (10 %) of this nitrile was nevertheless obtained when dimethyl formamide was tried as a solvent.

8. Condensation of metallic cyanides and alkyl sulphates or alkyl *p*-toluenesulphonates (tosylates)

It has been known for a long time that dimethyl and diethyl sulphate react readily with aqueous solutions of alkali cyanides forming respectively acetonitrile and propionitrile, which are contaminated with the corresponding isonitriles. The method has been used by Roberts[187] for the preparation of propionitrile-1-^{14}C in a 44 % yield:

$$(CH_3CH_2O)_2SO_2 \xrightarrow{Na^{14}CN} CH_3CH_2{}^{14}CN$$

Alkyl *p*-toluenesulphonates (alkyl tosylates) and alkali cyanides react also to give nitriles. Ethylene glycol bis(*p*-toluenesulphonate) (**49**) and sodium cyanide-^{14}C in ethanol–dioxan–water solution gave succinonitrile-1,4-^{14}C (**50**) in 76 % yield[188]. The preparation

$$p\text{-}CH_3C_6H_4SO_2OCH_2CH_2OSO_2C_6H_4CH_3\text{-}p \ (\mathbf{49}) \xrightarrow{2\,Na^{14}CN} N^{14}CCH_2CH_2{}^{14}CN \ (\mathbf{50})$$

of 3-(*p*-methoxyphenyl) propionitrile-1-^{14}C might serve as a further example[189] of the use of this procedure.

9. Nucleophilic displacement of phthalimido group by alkali cyanides

Egyed and coworkers[190] have shown that in compound **51** the phthalimido group can be readily replaced by anions of weak acids, e.g. cyanide ions in ethanol, with the formation of amino acid nitrile derivatives (**52**). Hydrolysis of the latter produces the amino acid:

$$\text{PhthN-CH(R)NHCO}_2\text{Et}\ \textbf{(51)} \xrightarrow{K^{14}CN}$$

$$N^{14}C\text{-CH(R)NHCO}_2\text{Et}\ \textbf{(52)} \xrightarrow{\text{hydrolysis}} HO_2{}^{14}C\text{-CH(R)NH}_2$$

The carbamic acid derivative **51** was prepared through the following sequence of reactions:

$$\text{PhthN-CH(R)CO}_2\text{H} \xrightarrow[\text{3. toluene + EtOH, 120°C}]{\text{1. SOCl}_2\text{; 2. NaN}_3} \text{PhthN-CH(R)NHCO}_2\text{Et}\ \textbf{(51)}$$

10. Replacement of amino groups by cyanide in Mannich bases or their quaternary ammonium salts

Indole-3-acetonitriles (**54**) are obtained in good yields by treatment of gramines or their quaternary salts (**53**) with an aqueous solution of alkali cyanides. All reactions of this type described in the literature use the alkali cyanide in a large excess, varying from three to twelve moles of alkali cyanide per mole of gramine or its quaternary salt[191,192]. For the reaction with radioactive alkali cyanides, some modifications to this procedure are required.

Keglevic-Brovet and coworkers[193] investigated the reaction between 5-benzyloxygramine methosulphonate (**53a**) and sodium cyanide-^{14}C with attempt to reduce the amount of alkali cyanide. Normally, when a large excess of alkali cyanide is used the reaction mixture

$$\left[\text{(R)-indole-3-}CH_2\overset{+}{N}Me_3\right]\bar{O}SO_3CH_3 \xrightarrow{Na^{14}CN}$$

(**53a**, R = $PhCH_2O$
53b, R = H)

$$\text{(R)-indole-3-}CH_2{}^{14}CN$$

(**54a**, R = $PhCH_2O$
54b, R = H)

remains alkaline during the whole process, although the trimethylamine formed during the reaction escapes to the atmosphere. When stoichiometric quantities are used under the same conditions, the alkalinity of the reaction mixture decreases with the evaporation of trimethylamine. This results in a poor yield of impure nitrile **54a**. It was found that if the reaction of stoichiometric amounts of reagents is carried out in a sealed tube under the slight pressure of the trimethylamine evolved, and if the mixture is heated for a longer period than described in the literature, the nitrile **54a** is obtained in a 82 % yield.

Stowe[194] has described the preparation of indoleacetic acid-[^{14}C-carboxyl] through the ^{14}C-nitrile (**54b**) using the same reaction. In an attempt to reduce the amount of radioactive cyanide, 0·2M K_2HPO_4 was added in order to maintain a basic reaction mixture. Erratic yields were obtained (50–80 %) in 'blank experiments'. In radioactive runs at medium specific activity (16·9 mCi/mmole), yields dropped to 18–25 %.

More recently Pichat, Herbert and Fabignon[195] reacted potassium cyanide-^{14}C with the methiodide of ethyl dimethylaminomethylacetamidomalonate (**55**) in order to obtain ethyl 3-cyano-2-acetamidopropionate (**56**):

$$K^{14}CN + \left[Me_3\overset{+}{N}CH_2\underset{|}{\overset{NHCOMe}{C}}(CO_2Et)_2\right]I^- \longrightarrow N^{14}CCH_2\overset{NHCOMe}{CH}(CO_2Et)_2$$

(**55**) (**56**)

The latter is a useful intermediate for the synthesis of some ^{14}C-amino acids such as aspartic acid-4-^{14}C, 2,4-diaminobutyric acid-4-^{14}C and homoserine-4-^{14}C. Very poor yields were obtained when an excess of the methiodide **55** was employed at atmospheric pressure in a stream of nitrogen as described by Hellmann[196]. However, a 65% yield of nitrile **56** was obtained when the reaction was performed in a sealed tube with a slight excess of the methiodide. It is likely that here too the evolved trimethylamine maintains a basic media which decreases the decomposition.

11. Replacement of a diazonium group by cyanide: the Sandmeyer method

The replacement of aromatic amino groups by cyanide is often accomplished in classical preparative organic chemistry by the action of cuprous cyanide with the diazonium compound (the Sandmeyer reaction):

$$ArNH_2 \xrightarrow{HCl/NaNO_2} ArN_2^+Cl^- \xrightarrow{CuCN} ArCN$$

Usually an excess of cuprous cyanide is used. Pichat, Baret and Audinot[14] studied the preparation of *p*-hydroxybenzonitrile-^{14}C by the Sandmeyer reaction with the diazonium salt derived from *p*-aminophenol, and found that an excess of cuprous cyanide was not necessary. The diazonium salt must be first neutralized to avoid losses of hydrogen cyanide-^{14}C. The yield was higher when a dilute (0·1M) neutralized solution of diazonium salt was used. Somewhat better yields were obtained by substituting nickel cyanide for the usual cuprous cyanide[197]. Nickel cyanide-^{14}C should find wider use in the Sandmeyer reaction with radioactive substances since it is very conveniently prepared by simply adding K^{14}CN to a solution of nickel sulphate. Some ^{14}C-nitriles prepared by the Sandmeyer reaction are shown in Table 8.

TABLE 8. ^{14}C-Nitriles prepared by the Sandmeyer reaction.

^{14}C-Nitrile	Yield based on K^{14}CN (%)	References
5-Cyano-2,2,8-trimethyl-4-*H*-*m*-dioxino-[4,5-*C*]-pyridine	20	200
p-Hydroxybenzonitrile	34	14
o-Nitrobenzonitrile	72–82	198, 199
3-Methoxy-2-nitrobenzonitrile	65	198

12. Reaction of alkali cyanide with thiocarbanilide

Ehrensvärd[224] has indicated that (cyano-^{13}C)-*N*,*N*′-diphenylformamidine (**58**) is obtained in 90 % yield by reaction of potassium cyanide-^{14}C with *N*,*N*′-diphenylthiourea (**57**) in the presence of basic lead carbonate. Compound **58** is a precursor of glycine-2-^{14}C. This type of reaction has not found general use.

$$\underset{(\mathbf{57})}{S{=}C(NHPh)_2} + K^{14}CN \xrightarrow{Pb(CO_3)_2} \underset{(\mathbf{58})}{N^{14}C\underset{\underset{NPh}{\|}}{C}NHPh}$$

B. *From ^{14}C- and ^{13}C-Carboxylic Acids and their Derivatives*

The synthetic methods belonging to this group have been much less frequently used for the preparation of ^{14}C-nitriles.

1. Dehydration of amides

$$R^{14}CONH_2 \xrightarrow{-H_2O} R^{14}CN$$

The preparation of nitriles by the removal of water from amides can be performed in high yields by using a variety of dehydrating agents, the most commonly used being phosphorus pentoxide. This process has been applied to some carboxylic amides-^{14}C. Dehydration on a 6 mmole scale of nicotinic acid-^{14}C-amide[201] gave a yield of 49 % of 3-(cyano-^{14}C)pyridine. (Cyano-^{14}C)acetamide has been dehydrated[202] to malononitrile-1-^{14}C on a 30–40 mmole scale, in a reaction–sublimation apparatus, with yields of 60–70 %. Dehydration of acetamide-1-^{14}C was used[203] to obtain acetonitrile-1-^{14}C.

2. Catalytic dehydration of ^{14}C-carboxylic acids in the presence of ammonia

Lauric acid-1-^{14}C (**59**) has been converted[204] into lauronitrile-1-^{14}C (**60**) by passing its vapour over aluminium oxide in the presence of ammonia. A yield of 97 % of crude radioactive nitrile was obtained

$$\underset{(\mathbf{59})}{C_{11}H_{23}{}^{14}CO_2H} + NH_3 \xrightarrow[320\text{–}340°\,C]{Al_2O_3} \underset{(\mathbf{60})}{C_{11}H_{23}{}^{14}CN} + 2\,H_2O$$

3. Exchange of ^{14}C-carboxylic acids with nitriles

Noszkó and Ötvös[205] have studied the nitrile–carboxyl exchange reaction which can take place at 260 °C in a sealed tube according to the two possibilities below:

$$R^1CN + R^2\,{}^{14}CO_2H \longrightarrow R^2CN + R^1\,{}^{14}CO_2H$$

$$R^1CN + R^2\,{}^{14}CO_2H \longrightarrow R^2\,{}^{14}CN + R^1CO_2H$$

at 260 °C in a sealed tube.

Two systems were studied: benzoic acid carboxyl-^{14}C + acetonitrile and benzoic acid carboxyl-^{14}C + *p*-brombenzonitrile. In the first system, benzonitrile-[cyano-^{14}C] was obtained in 19 % yield while in the second one its yield was 80 %. The authors could establish that the reaction occurs without rupture of the R^1—C and the R^2—C bonds. The results were explained by the following mechanism involving a O → N acyl migration:

$$R^1CN + R^2\,{}^{14}COOH \rightleftharpoons R^1C(=NH)-O-{}^{14}C(=O)R^2$$

$$R^1C(=NH)-O-{}^{14}C(=O)R^2 \rightleftharpoons R^2\,{}^{14}C(=O)-NH-C(=O)R^1$$

$$R^2\,{}^{14}C(=O)-NH-C(=O)R^1 \rightleftharpoons R^1C(=O)-O-{}^{14}C(=NH)R^2$$

$$R^1C(=O)-O-{}^{14}C(=NH)R^2 \rightleftharpoons R^1COOH + R^2\,{}^{14}CN$$

However, this reaction, which has been known since the 19th century, has not found general use for the synthesis of radioactive nitriles.

4. Reaction of cyanogen bromide with the sodium salts of ^{14}C-carboxylic acids

The reaction of cyanogen chloride with the sodium salts of aliphatic and aromatic carboxylic acids, at 200–300 °c, has been proposed[206] as a method of preparation of nitriles.

$$RCOONa + ClCN \longrightarrow NaCl + CO_2 + RCN$$

Douglas and coworkers[207] made a tracer study of the reaction of cyanogen bromide with sodium benzoate. It was shown that the carbon of the carbon dioxide is derived almost entirely from the cyanogen bromide according to the following overall reaction:

$$BrCN + R^{14}COONa \longrightarrow R^{14}CN + CO_2 + NaBr$$

In a further study of this reaction with the sodium salts of aliphatic acids (acetic, propionic, butyric, valeric, stearic), Douglas[208] observed the formation of an α-acyl isocyanate as the primary product with the lower aliphatic acids at 250–300 °c. The latter undergoes subsequent pyrolytic decomposition to a nitrile and carbon dioxide. This can be summarized as follows:

$$BrCN + RCOONa \longrightarrow RCONCO + NaBr$$
$$RCONCO \longrightarrow RCN + CO_2$$

The observed interchange of carbon atoms between the cyano group of the cyanogen bromide and the carboxyl group was explained[208] by the following mechanism:

$$Br{-}C{\equiv}N{:} + R{-}{}^{14}C({=}O){-}O^- \longrightarrow [Br{-}C({=}N{:}){-}O \text{ ring with } {}^{14}C{-}O^-] \xrightarrow{-Br^-} O{=}C{=}N{-}{}^{14}C({=}O){-}R \longrightarrow$$

$$[{:}N{-}C{=}O,\ {}^{14}C{-}O \text{ four-membered ring, } {}^{14}C{-}R] \longrightarrow \ddot{N}{\equiv}{}^{14}CR + CO_2$$

This method of synthesis of nitriles has not been used as a practical means of preparation of radioactive nitriles.

C. By Alkylation of Nitriles

The alkylation of the sodium derivatives of nitriles, which are prepared by action of sodium amide on nitriles in an inert solvent, is a general method of preparation of substituted acetonitriles. This procedure as applied to the preparation of radioactive nitriles can be exemplified by the preparation of 2-ethyl-2-phenylacetonitrile[209]:

$$PhCH_2{}^{14}CN \xrightarrow[2.\ EtBr]{1.\ NaNH_2} PhCH(Et)^{14}CN + NaBr$$

D. Synthesis of α-Deuterated Alkyl Nitriles

Leitch[210] has shown that α-deuterated nitriles can be prepared by base-catalysed exchange with deuterium oxide. The rate of exchange is 40 times as fast as hydrolysis in the system acetonitrile–deuterium oxide of slight enrichment. Acetonitrile exchanges rapidly when refluxed for 12 hours with a suspension of calcium deuteroxide in heavy water. After five exchanges a product containing 93·4 mole % CD_3CN and 4·4 mole % CHD_2CN was obtained. Propionitrile and butyronitrile required a higher temperature (120 °c). Equilibration took place with the formation of the 2,2-*D* derivatives and no significant exchange takes place elsewhere in the molecule.

Acrylonitrile exchanged rapidly at reflux temperature with heavy water and calcium deuteroxide, but a fraction of the nitrile polymerized under these conditions. Only the α-hydrogen atom was replaced with deuterium[210–212]. Base-catalysed deuteration of *trans*-cinnamonitrile employing deuteroethanol and sodium ethoxide gives a mixture of α-deuterated *trans*- and *cis*-cinnamonitriles[213].

$$PhCH{=}CHCN \xrightarrow[EtONa]{EtOD} PhCH{=}CDCN$$

VI. TRANSFORMATIONS AND USES OF LABELLED NITRILES

In the introduction of this chapter, it has already been pointed out that many labelled nitriles have been only intermediates in the preparation of radioactive compounds. Only a brief survey is given here of the transformations of the nitrile group most often encountered in the literature.

A. Hydrolysis—Preparation of ^{14}C-Carboxylic Acids

In the sections above, it has often been mentioned that acid or alkaline hydrolysis of ^{14}C-nitriles has been used for the preparation of ^{14}C-carboxyl labelled acids when the carbonation of a Grignard reagent is not feasible.

Saponification of an intermediate nitrile without its isolation is important in the preparation of:

[1] amino acids-1-^{14}C

$$RCH(NH_2)^{14}CN \longrightarrow RCH(NH_2)^{14}CO_2H \qquad \text{(sections V.A.3 and V.A.4)}$$

[2] monosaccharides and α-hydroxy acids-1-^{14}C:

$$RCH(OH)^{14}CN \longrightarrow RCH(OH)^{14}CO_2H \qquad \text{(section V.A.2)}$$

B. Reduction—Preparation of ^{14}C-Amines and ^{14}C-Aldehydes

Nitriles add hydrogen with the formation of primary amines.

$$RCN \longrightarrow RCH_2NH_2$$

The reduction can be performed either by catalytic hydrogenation or by chemical methods.

Some typical catalytic hydrogenations which were used for preparing labelled nitriles are:

[1] The preparation[214] of methylamine-^{14}C by the catalytic hydrogenation of hydrocyanic acid-^{14}C in the presence of hydrochloric acid.

$$H^{14}CN \xrightarrow{H_2/Pt} {}^{14}CH_3NH_2 \quad 85\%$$

[2] The preparation[215] of β-alanine-β-^{14}C by reduction of 2-(cyano-^{14}C)acetic acid.

$$N^{14}CCH_2CO_2H \xrightarrow[\text{Raney Ni}]{H_2,\ 2\ \text{atm}} H_2N^{14}CH_2CO_2H$$

[3] The preparation[216] of a mixture of diethylamine-1-^{14}C, ethylamine-1-^{14}C and ammonia by hydrogenation of acetonitrile:

$$CH_3{}^{14}CN \xrightarrow[\text{Raney Ni}]{H_2} (CH_3{}^{14}CH_2)_2NH + CH_3{}^{14}CH_2NH_2 + NH_3$$

[4] The catalytic reduction of some ^{14}C-cyano acids or esters (**61**, **61a** and **61b**) as intermediates in the synthesis of amino acids[28,109,195]:

$$\underset{(\mathbf{61})}{N^{14}CCH_2CH_2CH(CO_2H)NHCOPh} \xrightarrow[\text{AcOH/HCl}]{H_2/Pt} NH_2{}^{14}CH_2CH_2CH_2CH(CO_2H)NHCOPh$$

$$\underset{(\mathbf{61a})}{N^{14}CCH_2CH_2CH_2CO_2H} \xrightarrow[\text{AcOH/HCl}]{H_2/Pt} NH_2{}^{14}CH_2CH_2CH_2CH_2CO_2H$$

$$\underset{(\mathbf{61b})}{N^{14}CCH_2CH(CO_2Et)NHCOCH_3} \xrightarrow[\text{AcOH/HCl}]{H_2/Pt} NH_2{}^{14}CH_2CH_2CH(CO_2Et)NHCOCH_3$$

[5] The catalytic hydrogenation of cyanohydrin (**62**) was used for the preparation[225] of 5-hydroxylysine-6-^{14}C (**63**):

$$\underset{(\mathbf{62})}{N^{14}CCH(OH)(CH_2)_2C(NHCOMe)(CO_2Et)_2} \xrightarrow{H_2/PtO_2/HCl} \underset{(\mathbf{63})}{NH_2{}^{14}CH_2CH(OH)(CH_2)_2CH(NH_2)CO_2H}$$

Chemical reduction is nowadays accomplished by action of lithium aluminium hydride as exemplified by the preparation[217] of propylamine-1-^{14}C or by the preparation[218] of 3-amino-1-propanol-1-^{14}C:

$$CH_3CH_2{}^{14}CN \xrightarrow{LiAlH_4} CH_3CH_2{}^{14}CH_2NH_2$$

$$NCCH_2{}^{14}CO_2Et \xrightarrow{LiAlH_4} \underset{55\%}{H_2NCH_2CH_2{}^{14}CH_2OH}$$

When the catalytic hydrogenation is carried out in the presence of semicarbazide acetate[229] the semicarbazone of the ^{14}C-aldehyde is obtained as exemplified by the preparation[176] of *p*-hydroxybenzaldehyde-[carbonyl-^{14}C] semicarbazide

$$BzO-C_6H_4-{}^{14}CN \xrightarrow[\text{semicarbazide}]{H_2/\text{Raney Ni}} BzO-C_6H_4-{}^{14}CH{=}N-NH-CONH_2$$

In the sugar series, one mole of α-aminonitrile (**64a**) takes up only one mole of hydrogen (half-hydrogenation)[219] with the formation of α-amino aldehydes (α-aminodeoxy sugars) as exemplified in the preparations[220] of galactosamine-1-^{14}C (**64b**) and glucosamine-1-^{14}C.

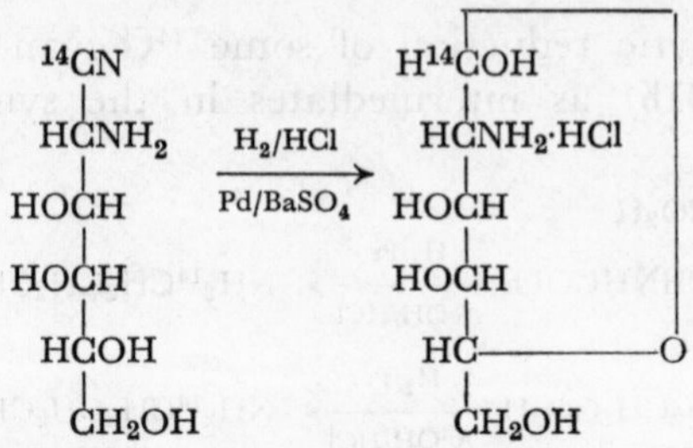

(**64a**) (**64b**)

Half-hydrogenation of the nitrile (**65**) is used for the preparation[171] of 2-desoxyribose-1-^{14}C (**66**).

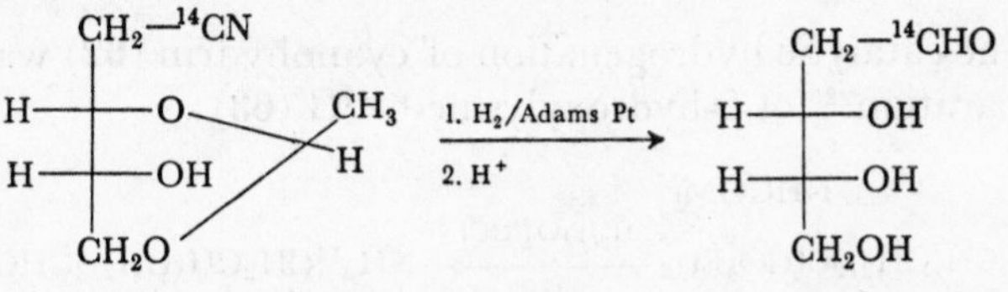

(**65**) (**66**)

C. *Preparation of ^{14}C-Ketones by Action of Grignard Reagents*

The addition of Grignard reagents to nitriles followed by hydrolysis with dilute acids gives ketones.

$$R^1CN + R^2MgX \longrightarrow R^1R^2C{=}NMgX \xrightarrow[H_2O]{H^+} R^1R^2C{=}O$$

The reaction has been used for the preparation of some labelled ketones as exemplified by:

[1] 3-hydroxy-2-butanone-1-^{14}C [221]

$$CH_3CH(OH)CN \xrightarrow[2.\ H_2O]{1.\ ^{14}CH_3MgI} CH_3CH(OH)CO^{14}CH_3$$

[2] 2-pentanone-1-^{14}C [222]

$$CH_3CH_2CH_2CN \xrightarrow[2.\ H_2O]{1.\ ^{14}CH_3MgI} CH_3CH_2CH_2CO^{14}CH_3$$

D. Preparation of Amidines

The addition of ammonia or amines to nitriles leads to amidines:

$$R^{14}CN \xrightarrow{NH_3} R^{14}C(NH_2){=}NH$$

Trans-4,4′-stilbenedicarboxamidine-$^{14}C_2$ (**68**) is prepared[174] by action of ammonium thiocyanate on the nitrile (**67**):

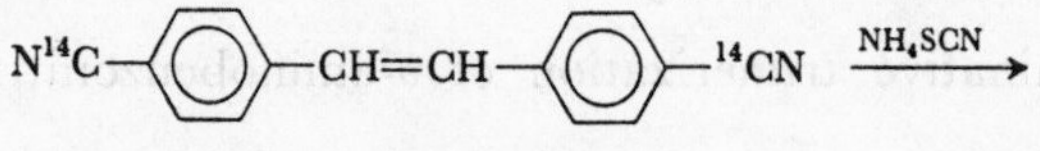
N^{14}C
CH=CH
^{14}CN
NH$_4$SCN

(**67**)

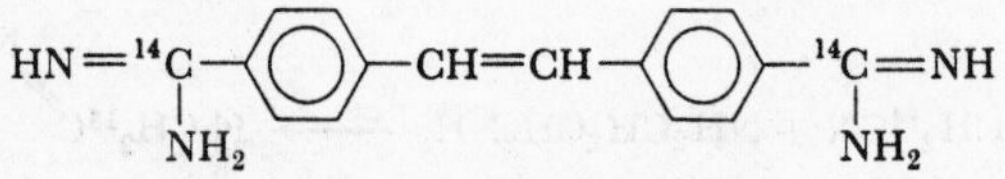
HN=^{14}C
NH$_2$
CH=CH
^{14}C=NH
NH$_2$

(**68**)

while 1,5-bis(*p*-amidino-$^{14}C_2$-phenoxy)pentane methanesulphonate was obtained[14] by action of ammonium methanesulphonate on the bis(imidate) derived from the dinitrile.

E. Additions to the Cyano Group to Form ^{14}C-Heterocycles

A few examples are given here to show how ^{14}C-nitriles had been used to make ring-labelled heterocycles (see Chapter 8)

[1] 4,5,6-Triaminopyrimidine-4,6-^{14}C[155]

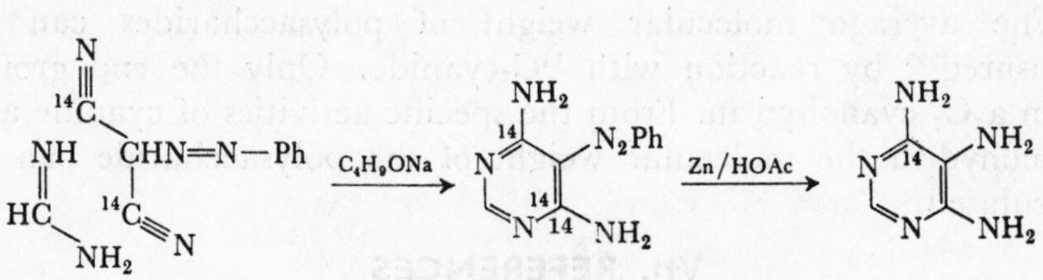
CHN=N—Ph
C$_4$H$_9$ONa
N$_2$Ph
Zn/HOAc

which serves then for the preparation of adenine-4,6-^{14}C.

[2] 5-(3-Pyridyl)tetrazole-^{14}C[201]

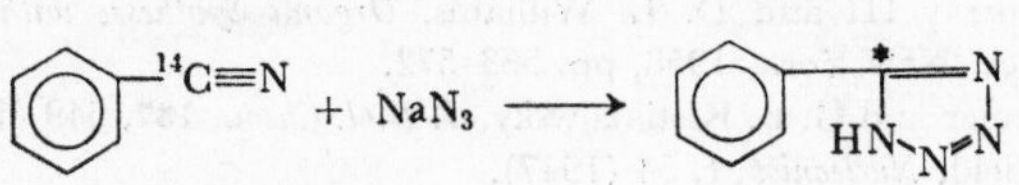
^{14}C≡N
+ NaN$_3$

[3] Tricycloquinazoline-^{14}C[186]

by the deaminative trimerization of *o*-aminobenzonitrile-[cyano-C^{14}].

[4] 2-Benzylimidazoline-2^{14} hyrochloride[133] from benzyl cyanide-^{14}*C*:

$$PhCH_2{}^{14}CN + NH_2CH_2CH_2NH_2 \longrightarrow PhCH_2{}^{14}C\begin{matrix} =N-CH_2 \\ \\ -NH-CH_2 \end{matrix}$$

F. Analytical Applications

The determination of reducing sugars by application of the cyanohydrin reaction could be made more sensitive by employing carbon-14 labelled cyanide[226,227]. Quantities of 0·04–0·2 mg of sugars can be determined with a standard deviation of $\pm 1{\cdot}4\%$. In this method, the sugar or polysaccharide is allowed to react in a buffered solution with ^{14}C-cyanide, previously standardized by reaction with D-glucose. The excess cyanide is volatilized as hydrogen cyanide, and the carbon-14 in the residue is determined by radioactivity measurement.

The average molecular weight of polysaccharides can be measured[228] by reaction with ^{14}C-cyanide. Only the end groups form a C_7 cyanohydrin. From the specific activities of cyanide and cyanohydrin the molecular weight of the polysaccharide can be calculated.

VII. REFERENCES

1. F. L. J. Sixma, H. Hendriks, K. Helle, U. Hollstein and R. Van Ling, *Rec. Trav. Chim.*, **73**, 161 (1964).
2. (a) J. R. Catch, *Carbon-14 Compounds*, Butterworths, London, 1961, p. 26;
   (b) A. Murray III and D. L. Williams, *Organic Syntheses with Isotopes*, Vol. 1, Interscience, New York, 1958, pp. 563–572.
3. R. D. Cramer and G. B. Kistiakowsky, *J. Biol. Chem.*, **137**, 549 (1941).
4. R. B. Lotfield, *Nucleonics*, **1**, 54 (1947).
5. R. B. Lotfield and E. A. Eigner, *Biochem. Biophys. Acta*, **130**, 449 (1966).

6. J. A. Bos, *Experientia*, **7**, 258 (1951).
7. P. Olynyk, D. B. Camp, A. M. Griffith, S. Woislowski and R. W. Helmkamp, *J. Org. Chem.*, **13**, 465 (1948).
8. J. D. Moyer and H. S. Isbell, *Anal. Chem.*, **29**, 393 (1957).
9. J. P. Lambooy, *Rec. Trav. Chim.*, **80**, 889 (1961).
10. J. A. McCarter, *J. Am. Chem. Soc.*, **73**, 483 (1951).
11. R. F. Nystrom and A. I. Promislow, *Univ. Microfilms Publ. 11,527; Chem. Abstr.*, **49**, 11334 (1955).
12. R. M. Lemmon, *U.C.R.L. 2026, Nucl. Sci. Abstr.*, **7**, 1656 (1953).
13. K. J. Jeanes, *Science*, **118**, 717 (1953).
14. L. Pichat, C. Baret and M. Audinot, *Bull. Soc. Chim. France*, **1956**, 151, 156.
15. G. P. Okatova, S. V. Tchinyakov and Y. M. Egorov, *Tr. Gos. Inst. Prikl. Khim.*, **45**, 45 (1960); *Chem. Abstr.*, **56**, 3100 (1962).
16. S. Von Schuching and T. Enns, *J. Am. Chem. Soc.*, **78**, 4255 (1956).
17. A. P. Musakin, T. M. Vladimirova, E. M. Inkova and V. A. Ossipov, *Radiokhimiya*, **1**, 734 (1959).
18. A. W. Adamson, *J. Am. Chem. Soc.*, **69**, 2564 (1947).
19. G. O. Henneberry and B. E. Baker, *Can. J. Chem.*, **28B**, 345 (1950).
20. A. G. MacDiarmid and N. F. Hall, *J. Am. Chem. Soc.*, **75**, 4850 (1953).
21. B. G. Van Den Bos and A. H. W. Aten, *Rec. Trav. Chim.*, **70**, 495 (1951).
22. C. J. Claus, D. C. Camp, J. L. Morgenthau, P. Olynyk and R. W. Helmkamp, *Abstr. 121. Meeting Am. Chem. Soc.*, **1952**, 51 k.
23. W. L. Carryck and A. Fry, *J. Am. Chem. Soc.*, **77**, 4381 (1955).
24. V. I. Maimind, B. V. Tokarev and M. M. Shemyakin, *Dokl. Akad. Nauk SSSR*, **81**, 195 (1951); *Chem. Abstr.*, **46**, 3889f (1952); *Zh. Obshch. Khim.*, **26**, 1962 (1956).
25. A. Stoll, J. Rutschmann, A. von Wartburg and J. Renz, *Helv. Chim. Acta*, **39**, 993 (1956).
26. M. Rothstein and J. C. Claus, *J. Am. Chem. Soc.*, **75**, 2981 (1953).
27. O. I. Andreeva and G. I. Kostikova, *Radioisotope Phys. Sci. Ind. Proc. Conf. Use, Copenhagen*, **3**, 111 (1962); *Chem. Abstr.*, **57**, 16084f (1962).
28. L. Pichat, J. Mizon and M. Herbert, *Bull. Soc. Chim France*, **1963**, 1787.
29. R. J. Abrams, *J. Am. Chem. Soc.*, **71**, 3835 (1949).
30. J. W. Spyker and A. Neish, *Can. J. Chem.*, **30**, 461 (1952).
31. P. Vercier, *Proceedings of the 2nd Int. Conf. on methods of preparing and Storing labelled Compounds*. Brussels, 1966, Publication Euratom 3746d-f-e, Brussels, 1968, p. 63.
32. P. Vercier, *J. Labelled Compounds*, **4**, 91 (1968); *Bull. Soc. Chim. France*, **1968**, 3915.
33. B. Belleau and R. D. Heard, *J. Am. Chem. Soc.*, **72**, 4268 (1950).
34. W. R. Vaughan and D. J. McCane, *J. Am. Chem. Soc.*, **76**, 2504 (1954).
35. L. Pichat, C. Baret, M. Audinot, M. Herbert and J. Lambin, *Radioisotope Conf. 1954*, Vol. 1, Butterworths, London, 1954, p. 245.
36. J. C. Reid and J. C. Weaver, *Cancer Res.*, **11**, 188 (1951).
37. W. G. Dauben, C. F. Hiskey and M. A. Muhs, *J. Am. Chem. Soc.*, **74**, 2082 (1952).
38. H. J. Barber, *J. Chem. Soc.*, **1943**, 79.
39. H. S. Anker, *J. Biol. Chem.*, **176**, 1337 (1948).
40. A. G. MacDiarmid and N. F. Hall *J. Am. Chem. Soc.*, **75**, 4850 (1953).
41. L. S. T'Sai, T. H. Chien, F. S. Chin and W. C. Hsu, *Chem. Abstr.*, **56**, 15363 (1962).
42. J. Urban, *Collection Czech. Chem. Commun.*, **24**, 4050 (1959).
43. P. Kurtz, *Methoden der Organischen Chemie*, Vol. 8 (Ed. Houben, Weyl), George Thieme Verlag, Stuttgart, 1952, p. 266.
44. G. Westöö, *Acta Chem. Scand.*, **11**, 204 (1957).

45. I. A. Ultee, *Rec. Trav. Chim.*, **28,** 1, 248 (1909).
46. M. Lapworth and R. H. F. Manske, *J. Chem. Soc.*, **1928**, 2533; **1930**, 1976.
47. H. Sallach, *J. Am. Chem. Soc.*, **74,** 2415 (1952).
48. J. M. Ashworth, *J. Chem. Soc.*, **1963**, 2563; *Biochem. Prep.*, **11,** 50 (1966).
49. W. Sakami, W. E. Evans and S. Gurin, *J. Am. Chem. Soc.*, **69,** 1110 (1947).
50. G. Schlesier, H. Koch and R. Büchner, *Kernenergie*, **6,** 65 (1963).
51. H. S. Isbell, J. V. Karabinos, H. L. Frush, N. B. Holt, A. Schwebel and T. T. Galkowski, *J. Res. Nat. Bur. St.*, **48,** 163 (1952).
52. H. S. Isbell, *U.S. Patent* 2,606, 918 (1952); *Chem. Abstr.*, **47,** 3873 (1953).
53. P. W. Kent and K. R. Wood, *J. Chem. Soc.*, **1963,** 2812.
54. M. Lorant, *Chem. Rundschau* (*Solothurn*), **17,** 19 (1964); *Chem. Abstr.*, **63,** 8465 (1965).
55. H. L. Frush and H. S. Isbell, *J. Res. Nat. Bur. Std.*, [A], **51,** 307 (1953).
56. H. S. Isbell, H. L. Frush and N. B. Holt, *J. Res. Nat. Bur. Std.*, [A], **53,** 325 (1954)
57. P. Kohn and B. L. Dmuchowski, *Biochim. Biophys. Acta*, **45,** 576 (1960).
58. A. C. Neish, *Can. J. Chem.*, **32,** 334 (1954).
59. D. E. Koshland and F. H. Westheimer, *J. Am. Chem. Soc.*, **72,** 3383 (1950).
60. H. R. V. Arnstein and D. Keglevic, *Biochem. J.*, **62,** 199 (1956).
61. U. Drehmann and K. Uhlig, *J. Prakt. Chem.*, **280,** 33 (1959).
62. F. Shafizadeh and M. L. Wolfrom, *J. Am. Chem. Soc.*, **77,** 2568 (1955).
63. J. C. Sowden, *J. Am. Chem. Soc.*, **74,** 4377 (1952).
64. S. Rosemann, *J. Am. Chem. Soc.*, **74,** 4467 (1952).
65. R. Schaffer and H. S. Isbell, *J. Res. Nat. Bur. Std.*, **56,** 191 (1956).
66. H. S. Isbell, H. L. Frush and N. B. Holt, *J. Res. Nat. Bur. Std.*, **53,** 217 (1954).
67. Y. J. Topper and D. Stetten, Jr., *J. Biol. Chem.*, **193,** 149 (1951).
68. H. S. Isbell and R. Schaffer, *J. Am. Chem. Soc.*, **78,** 1887 (1956).
69. H. L. Frush and H. S. Isbell, *J. Res. Nat. Bur. Std.*, **50,** 133 (1953).
70. R. Schaffer and H. S. Isbell, *J. Res. Nat. Bur. Std.*, **57,** 333 (1956).
71. M. Stiles and R. P. Mayer, *J. Am. Chem. Soc.*, **81,** 1497 (1959).
72. L. L. Salomon, J. J. Burns and C. G. King, *J. Am. Chem. Soc.*, **74,** 5161 (1952); *Science*, **111,** 257 (1950).
73. J. J. Burns and C. G. King, *Science*, **111,** 257 (1950).
74. A. P. Mussokin, T. M. Wladimorowa, J. N. Inkowa and W. A. Ossipow, *Radiokhimiya*, **1,** 734 (1959).
75. R. T. Arnold, AECU-575, cited reference 2b, p. 656.
76. W. A. Bonner and R. T. Rewick, *J. Am. Chem. Soc.*, **84,** 2334 (1962).
77. W. A. Bonner and T. W. Greenlee, *J. Am. Chem. Soc.*, **81,** 2122 (1959).
78. H. Banholzer and H. Schmid, *Helv. Chim. Acta*, **42,** 2584 (1959).
79. C. A. Hughes, C. G. Gordon-Gray, F. D. Schlosser and F. L. Warren, *J. Chem. Soc.*, **1965,** 2370.
80. A. R. Gennaro, N. Neff and G. V. Rossi, *J. Chem. Eng. Data*, **9,** 109 (1964); *Chem. Abstr.*, **60,** 13177 (1964).
81. A. Strecker, *Ann. Chem.*, **75,** 27 (1850).
82. D. O. Holland and J. H. C. Nayler, *J. Chem. Soc.*, **1952,** 3403.
83. R. Gaudry, *Can. J. Res.*, **24B,** 301 (1946).
84. L. Pichat, J. P. Guermont and P. N. Liem, *J. Labelled Compounds*, **4,** 251 (1968).
85. L. Pichat, M. Audinot and R. E. Wolff, unpublished results.
86. L. Pichat, J. P Guermont and D. Sharefkin, *C.E.A. Report 2989* (1966).
87. R. B. Lotfield, *Nucleonics*, **1**(3), 54 (1947).

88. S. Gurin, M. Delluva and D. W. Wilson, *J. Biol. Chem.*, **171,** 101 (1947).
89. R. B. Lotfield and E. A. Eigner, *Biochem Biophys. Acta*, **130,** 449 (1966).
90. R. E. Selff and B. M. Tolbert, in *Organic Syntheses with Isotopes*, Vol 1, New York, 1958, p. 189.
91. H. Borsook, C. L. Deasy, A. J. Haagen-Smit, G. Keighley and P. H. Lowy, *J. Biol. Chem.*, **184,** 529 (1950).
92. G. O. Henneberry, W. F. Oliver and B. E. Baker, *Can. J. Chem.*, **29,** 229 (1951).
93. P. Kourim and J. Zikmund, *Collection Czech. Chem. Commun.*, **26,** 717 (1961).
94. H. R. V. Arnstein, *Biochem. J.*, **68,** 333 (1958).
95. H. T. Bücherer and V. A. Lieb. *J. Pr.*, [2] **141,** 5 (1934).
96. E. Ware, *Chem. Rev.*, **46,** 403 (1950).
97. R. B. Lotfield, *J. Am. Chem. Soc.*, **72,** 2499 (1950).
98. G. R. Clemo, F. K. Duxbury and G. A. Swan, *J. Chem. Soc.*, **1952,** 3464.
99. V. I. Maimind, B. V. Tokarev and M. M. Shemyakin, *Dokl. Akad. Nauk SSSR*, **81,** 92 (1953); *Chem. Abstr.*, **48,** 10588 (1954); reference 2b, p. 275.
100. H. Borsook, C. L. Deasy, A. J. Haagen-Smit, G. Keighley and P. H. Lowry, *J. Biol. Chem.*, **184,** 529 (1950).
101. M. Rothstein, *J. Am. Chem. Soc.*, **79,** 2009 (1957).
102. H. Tiedmann, *Biochem. Z.*, **326,** 511 (1955).
103. H. Schmid and K. Schmid, *Helv. Chim. Acta*, **35,** 1879 (1952).
104. J. G. Burtle, J. P. Ryan and S. M. James, *Anal. Chem.*, **30,** 1640 (1958).
105. J. Falecki and R. Plejewski, *J. Prakt. Chem.*, **28,** 123 (1965).
106. E. Rambaud, *Bull. Soc. Chim. France*, **3**(5), 138 (1936).
107. W. Wislicenus, *Ber.*, **18,** 172 (1885).
108. L. Pichat, C. Baret and M. Audinot, *Bull. Soc. Chim. France*, **21,** 88 (1954).
109. L. Pichat, J. Mizon and M. Herbert, *Bull. Soc. Chim. France*, **1963,** 1792.
110. W. W. Muelder and M. V. Wass, *J. Agr. Food Chem.*, **15**(3), 508 (1967).
111. L. Pichat, G. Rochas and M. Herbert, *Bull. Soc. Chim. France*, **1965,** 1384.
112. L. Frank, *Anal. Biochem.*, **17,** 423 (1966).
113. L. Pichat, P. N. Liem and J. P. Guermont, *Bull. Soc. Chim. France*, **1968,** 4079.
114. I. V. Ponseti, S. Wawzonek, W. E. Franklin, R. E. Winnick and T. Winnick, *Proc. Soc. Exp. Biol. Med.*, **93,** 515 (1956).
115. C. J. Collins, *J. Am. Chem. Soc.*, **77,** 5517 (1955).
116. D. T. Mowry, *Chem. Revs.*, **42,** 189 (1948).
117. O. W. Cass, *Chem. Eng. News*, **32,** 2197 (1954).
118. A. Brändstrom, *Acta Chem. Scand.*, **10,** 1197 (1956).
119. R. A. Smiley and C. Arnold, *J. Org. Chem.*, **25,** 257 (1960).
120. L. Friedman and H. Shechter, *J. Org. Chem.*, **25,** 877 (1960).
121. V. Lorber, N. Lifson, H. G. Wood, W. Sakami and W. W. Shreeve, *J. Biol. Chem.*, **183,** 517 (1950).
122. I. Zabin and K. Bloch, *J. Biol. Chem.*, **192,** 261 (1951).
123. V. Lorber, N. Lifson, W. Sakami and H. G. Wood, *J. Biol. Chem.*, **183,** 531 (1950).
124. R. M. Roberts and S. G. Brandenberger, *Chem. Ind.* (*London*), **1957,** 227.
125. I. Siegel and V. Lorber, *J. Biol. Chem.*, **189,** 571 (1951).
126. R. M. Roberts, S. G. Brandenberger and S. G. Panayides, *J. Am. Chem. Soc.*, **80.** 2507 (1958).
127. H. S. Anker, *J. Biol. Chem.*, **194,** 177 (1952).
128. R. N. Renaud and L. C. Leitch, *Can. J. Chem.*, **49,** 2089 (1964).

129. J. D. Roberts and C. M. Regan, *J. Am. Chem. Soc.*, **75**, 2069 (1953).
130. D. Blackburn and G. Burghard, *J. Labelled Compounds*, **2**, 62 (1966).
131. R. Adams and A. F. Thal, *Organic Syntheses*, Coll. Vol. 1, John Wiley and Sons, New York, 1941, p. 107.
132. I. V. Wijngaarden and W. Soudiun, *J. Labelled Compounds*, **1**, 207 (1965).
133. J. B. Ziegler, A. C. Shabica, *J. Org. Chem.*, **24**, 1133 (1959).
134. A. R. Battersby, *J. Chem. Soc.*, **1964**, 3600.
135. M. Rothstein and L. L. Miller, *J. Biol. Chem.*, **199**, 199 (1952); **235**, 714 (1960).
136. G. Pajaro and A. Valvassori, *Gazz. Chim. Ital.*, **92**, 1446 (1962).
137. V. Prelog, H. H. Käge and G. H. White, *Helv. Chim. Acta*, **45**, 1658 (1952).
138. V. Prelog, H. J. Urech, A. A. Bothner-By and J. Würsch, *Helv. Chim. Acta*, **38**, 1095 (1955).
139. V. Prelog and S. Polyak, *Helv. Chim. Acta*, **40**, 816 (1957).
140. M. Kushner and S. Weinhouse, *J. Am. Chem. Soc.*, **71**, 3558 (1949).
141. S. A. Brown and A. C. Neish, *Can. J. Biochem. Physiol.*, **32**, 170 (1954).
142. F. Korte and G. Rechmeier, *Ann. Chem.*, **656**, 133 (1962).
143. L. H. Schmidt, M. Bubner and D. Dorr, *Kernenergie*, **6**, 411 (1963).
144. B. Bak and N. Clauson-Kaas, *Acta Chem. Scand.*, **12**, 995 (1958).
145. L. H. Slaugh, E. F. Magoon and V. P. Guinn, *J. Org. Chem.*, **28**, 2643 (1963).
146. H. Borsook, C. L. Deasy, A. J. Haagen-Smit, G. Keighley and P. H. Lowy, *J. Biol. Chem.*, **176**, 1383 (1948); reference 2a, p. 451.
147. E. D. Schilling and F. M. Strong, *J. Org. Chem.*, **22**, 349 (1957).
148. J. C. Westfahl and T. L. Gresham, *J. Org. Chem.*, **21**, 1145 (1950).
149. E. Klenk and M. Pflüger, *Z. Physiol. Chem.*, **336**, 20 (1964).
150. J. M. Walts, W. V. Kessler and J. E. Christian, *J. Pharm. Sci.*, **56**, 900 (1967).
151. S. Bergstrom, M. Rottenberg and J. Voltz, *Acta Chem. Scand.*, **7**, 481 (1953); reference 2a, p.1059.
152. J. T. Matschiner, *J. Biol. Chem.*, **225**, 803 (1957).
153. T. A. Mahowald and J. Matschiner, *J. Biol. Chem.*, **225**, 781 (1957).
154. G. D. Cherayil, S. L. Hsia, J. T. Matschiner, E. A. Doisyjr, W. H. Elliot, S. A. Trayer and E. A. Doisy, *J. Biol. Chem.*, **238**, 1973 (1963).
155. E. L. Bennett, *J. Am. Chem. Soc.*, **74**, 2420 (1952).
156. F. Korte and H. Barkemeyer, *Chem. Ber.*, **90**, 394 (1957).
157. D. Gross, J. W. Kurbatov and H. R. Schütte, *Z. Chem.*, **7**, 313 (1967).
158. J. H. Lukas and G. B. Gerber, *J. Labelled Compounds*, **1**, 231 (1965).
159. S. Rothchild and M. Fields, *J. Am. Chem. Soc.*, **74**, 2401 (1952).
160. S. Affrossman, D. Cormack and S. J. Thompson, *J. Chem. Soc.*, **1962**, 3217.
161. E. Rietz, *Organic Syntheses*, Coll. Vol. 3, John Wiley and Sons, New York, 1955, p. 851.
162. V. W. Stoffel and E. Bierwirth, *Angew. Chem.*, **74**, 904 (1962).
163. V. W. Stoffel, *Ann. Chem.*, **673**, 26 (1964).
164. V. W. Stoffel, *J. Am. Oil Chemist's Soc.*, **42**, 583 (1965).
165. J. Würsch, unpublished work quoted by J. M. Osbond in *Progress in the Chemistry of Fats and Other Lipids*, Vol. 9, Part. 1, Pergamon, London, 1966, p. 151.
166. W. J. Gensler and J. Bruno, *J. Org. Chem.*, **28**, 1254 (1963).
167. A. I. Rachlin, N. Wasyliw and M. W. Goldberg, *J. Org. Chem.*, **26**, 2688 (1961).
168. W. G. Knipprath and J. F. Mead, *Biochim. Biophys. Acta*, **116**, 198 (1966).
169. O. Simamura, N. Inamoto and T. Suehiro, *Bull. Chem. Soc. Japan*, **27**, 221 (1954); *Chem. Abstr.*, **49**, 7494 (1955).

170. S. Bergström, K. Pääbo and M. Rottenberg, *Acta Chem. Scand.*, **6,** 1127 (1952); reference 2b, p. 410.
171. R. J. Bayly and J. C. Turner, *J. Chem. Soc.*, **1966**, 704.
172. M. E. Gutzke, D. W. Fox, L. S. Cieresko and S. H. Wender, *J. Org. Chem.*, **22,** 1271 (1957).
173. L. V. Hankes, A. G. Fischer and R. J. Suhadolnik, *Biochem. Prep.*, **9,** 59 (1962).
174. E. L. Bennett, D. E. Pack, B. J. Krueckel and J. C. Weaver, *Cancer Res.*, **13,** 30 (1953); see reference 3, Vol. 1, p. 410.
175. L. Friedman and H. Shechter, *J. Org. Chem.*, **26,** 2522 (1961).
176. H. Grisbach and L. Patschke, *Chem. Ber.*, **95,** 2098 (1962).
177. E. H. White, H. Worther, G. F. Field and W. D. McElroy, *J. Org. Chem.*, **30,** 2344 (1965).
178. R. G. Gould, A. B. Hastings, C. B. Anfinsen, I. N. Rosenberg, A. K. Solomon and Y. J. Topper, *J. Biol. Chem.*, **177,** 727 (1949).
179. H. G. Wood, B. E. Christensen and H. S. Anker, *Nucleonics*, **7**(3), 60 (1950).
180. U. Drehmann and H. J. Born, *J. Prakt. Chem.*, **5,** 207 (1957).
181. K. Banholzer and M. Schmidt, *Helv. Chim. Acta*, **39,** 549 (1956).
182. W. Sakami, W. E. Evans and S. Gurin, *J. Am. Chem. Soc.*, **69,** 1110 (1947).
183. H. R. V. Arnstein, *Biochem. J.*, **48,** 27 (1951).
184. A. E. A. Mitta, A. M. Ferramola, H. A. Sancovich and M. Grinstein, *J. Labelled Compounds*, **3,** 20 (1967).
185. H. J. Fromm, *J. Org. Chem.*, **22,** 612 (1957).
186. H. G. Dean, R. J. Grout, M. W. Partridge and H. J. Vipond, *J. Chem. Soc.* (*C*), **1968,** 142.
187. J. D. Roberts and M. Halmann, *J. Am. Chem. Soc.*, **75,** 5759 (1953).
188. H. Müsso and D. Döpp, *Chem. Ber.*, **97,** 1147 (1964).
189. A. W. Fort and J. D. Roberts, *J. Am. Chem. Soc.*, **78,** 584 (1956).
190. J. Egyed, J. Meisel-Agoston and L. Ötvös, *Acta Chim. Acad. Sci. Hung.*, **38,** 123 (1963).
191. J. Thesing and F. Schülde, *Chem. Ber.*, **85,** 324 (1952).
192. H. R. Snyder and F. J. Pilgrim, *J. Am. Chem. Soc.*, **70,** 3770 (1948).
193. D. Keglevic-Brovet, S. Kveder and S. Iskrić, *Croat. Chem. Acta*, **29,** 351 (1957).
194. B. B. Stowe, *Anal. Biochem.*, **5,** 107 (1963).
195. L. Pichat, M. Herbert and Mlle C. Fabignon, *Proceedings of the 2nd Int. Conf. Methods Prep. Storing Labelled Compounds, Brussels 1966*, Publication Euratom, Abstr. No. 40, 1968.
196. H. Hellmann and E. Folz, *Chem. Ber.*, **88,** 1944 (1955).
197. H. Hodgson and F. Heyworth, *J. Chem. Soc.*, **1949,** 1131.
198. D. Munsche and H. R. Schütte, *Z. Chem.*, **3,** 230 (1963).
199. A. Butenandt and R. Beckmann, *Ann. Acad. Sci. Fennicae*, **60,** 275 (1955).
200. C. J. Argoudelis and F. A. Kummerow, *J. Org. Chem.*, **29,** 2663 (1964).
201. S. K. Figdor and M. S. Von Wittenau, *J. Med. Chem.*, **10,** 1158 (1967).
202. E. L. Bennett, *J. Am. Chem. Soc.*, **74,** 2420 (1952).
203. L. B. Dachkevitch, *J. Gen. Chem. USSR*, **27,** 2874 (1957).
204. H. J. Harwood and A. W. Ralston, *J. Org. Chem.*, **12,** 740 (1947).
205. L. H. Noszko and L. Ötvös, *Acta Chim Acad Sci. Hung.*, **25,** 123 (1960).
206. E. Zappi and O. Bouso, *Ann. Asoc. Quim. Argentina*, **35,** 137 (1947).
207. D. E. Douglas, J. Eccles and A. E. Almond, *Can. J. Chem.*, **31,** 1127 (1953).
208. D. E. Douglas and A. M. Burditt, *Can. J. Chem.*, **36,** 1256 (1958).
209. K. Bernhard, M. Just, J. P. Vuilleumier and G. Brubacher, *Helv. Chim. Acta*, **39,** 603 (1956).

210. L. C. Leitch, *Can. J. Chem.*, **35**, 345 (1957).
211. W. F. Cockburn and C. E. Hubley, *Appl. Spectr.*, **11**, 188 (1957).
212. R. Yamadera, H. Tadokoro and S. Murahashi, *J. Chem. Phys.*, **41**, 1233 (1964).
213. M. F. Zinn, T. M. Harris, D. G. Hill and C. R. Hauser, *J. Am. Chem. Soc.*, **85**, 71 (1963).
214. W. J. Skraba and A. R. Jones, *Science*, **117**, 252 (1953); R. D. Heard, J. R. Jamieson and S. Solomon, *J. Am. Chem. Soc.*, **73**, 4985 (1951).
215. P. Fritzon and L. Eldjarn, *Scand. Clin. Lab. Invest.*, **4**, 375 (1952); *Chem. Abstr.*, **47**, 9917 (1953).
216. A. Stoll, J. Rutschmann and A. Hofmann, *Helv. Chim. Acta*, **37**, 823 (1954).
217. J. D. Roberts and M. Halmann, *J. Am. Chem. Soc.*, **75**, 5759 (1953).
218. K. H. Schweer, *Chem. Ber.*, **95**, 1799 (1962).
219. R. Kuhn and W. Kirschenlohr, *Ann. Chem.*, **600**, 115 (1956).
220. R. Kuhn, H. J. Leppelmann and H. Fischer, *Ann. Chem.*, **620**, 15 (1959).
221. R. O. Brady, J. Rabinowitz, J. Van Baalen and S. Gurin, *J. Biol. Chem.*, **193**, 137 (1951).
222. E. Cerwonka, R. C. Anderson and E. V. Brown, *J. Am. Chem. Soc.*, **75**, 28 (1953).
223. K. Freudenberg and F. Bittner, *Chem. Ber.*, **86**, 155 (1953).
224. G. Ehrensvärd and R. Stjernholm, *Acta Chem. Scand.*, **3**, 971 (1949).
225. S. Lindstedt, *Acta Chem. Scand.*, **7**, 340 (1953).
226. J. D. Moyer and H. S. Isbell, *Anal. Chem.*, **29**, 1862 (1957).
227. J. D. Moyer and H. S. Isbell, *Anal. Chem.*, **30**, 1975 (1958).
228. H. S. Isbell, *Science*, **113**, 532 (1951).
229. H. Plieninger and A. Werst, *Chem. Ber.*, **88**, 1956 (1955).

CHAPTER 14

Nitrile oxides

Ch. Grundmann

Mellon Institute, Carnegie-Mellon University
Pittsburgh, Pa., U.S.A.

I. Introduction . . . 792
II. Preparation of Nitrile Oxides . . . 794
A. General . . . 794
B. Preparation of Fulminic Acid (Formonitrile Oxide) . . . 795
C. Preparation of Nitrile Oxides from Aldoximes . . . 800
1. Dehydrogenation of aldoximes . . . 800
2. Dehydrohalogenation of hydroximic acid halides . . . 802
D. Preparation of Nitrile Oxides from Primary Nitroparaffins . . . 804
1. Conversion to hydroximic acid chlorides . . . 804
2. Decomposition of nitrolic acids . . . 805
3. Dehydration of primary nitroparaffins . . . 806
E. Preparation of Nitrile Oxides from Fulminates . . . 807
F. Preparation of Nitrile Oxides *in situ* . . . 808
G. Nitrile Oxides with Functional Groups . . . 810
III. Physical Properties of Nitrile Oxides . . . 812
IV. Reactions of Nitrile Oxides . . . 814
A. General; Stability of Nitrile Oxides . . . 814
B. Rearrangement to Isocyanates . . . 815
C. Dimerization to Furoxans (1,2,5-Oxadiazole-2-Oxides) . . . 816
D. Polymerization . . . 819
1. Fulminic acid . . . 819
2. Other nitrile oxides . . . 820
E. Reduction to Nitriles . . . 822
F. Addition Reactions Leading to Open-Chain Structures . . . 822
1. Reactions with inorganic compounds . . . 822
a. Water (hydrolysis) . . . 823
b. Hydrogen halides . . . 824
c. Ammonia . . . 824
d. Azide ion . . . 824
e. Sulphide ion . . . 825
f. Thiocyanate ion . . . 826
g. Cyanide ion . . . 827

2. Reactions with organic compounds 828
a. Alcohols 828
b. Mercaptans 828
c. Organic acids 828
d. Amines and hydrazines 829
e. Grignard compounds 831
G. Addition Reactions Leading to Cyclic Structures (1,3-Dipolar Cycloadditions) 832
1. Reactions with olefins 833
2. Reactions with acetylenes 836
3. Reactions with carbonyl or thiocarbonyl compounds . . 837
4. Reactions with C=N compounds. 838
5. Reactions with nitriles 840
6. Reactions with systems containing N=N, N=S, N=B, C=P or N=P double bonds 841
H. Miscellaneous Reactions 843
V. References 846

I. INTRODUCTION

Nitrile oxides, RCNO, are organic compounds which contain the monovalent functional group —CNO bound directly to a carbon atom of the organic moiety of the molecule. Only the first member of this series, formonitrile oxide, H—CNO is an exception; accordingly its chemistry differs in some respects from that of its homologues.

In the nomenclature of nitrile oxides, the rules established for nitriles are generally applicable. Like the nitriles whose functional group, —CN, is often referred to as 'cyano-', (as derived from the parent compound hydrocyanic acid), the term 'fulmido-' is used sometimes for the functional group —CNO. Likewise, nitrile oxides are termed 'fulmides' in cases where one would use the term 'cyanide' in naming an organic nitrile.

The parent compound, formonitrile oxide or fulminic acid, has already been obtained as its mercury salt at the dawn of organic chemistry[1]. The problem of its structure, however, has occupied the most brilliant minds of chemistry for more than a century[2]. Although the work of Wieland and Quilico had produced much chemical evidence for the nitrile oxide structure, the generally accepted formula was that of carbonyl oxime, H—O—N=C, first proposed by Nef[3]. It was not before 1966 that W. Beck provided unequivocal evidence by infrared spectroscopy for the nitrile oxide formula H—C≡N→O[4].

Although the first higher homologue, benzonitrile oxide, PhCNO, was prepared by Werner in 1894[5], no detailed study of the chemistry of this new class of compounds was made until H. Wieland's classical work between 1907 and 1910. From 1946 onwards, Quilico and his associates have made especially important preparative contributions, mainly to cycloaddition reactions of nitrile oxides. It is interesting to note that both the above-mentioned scientists became interested in this field through their previous occupation with fulminic acid, a compound which at the time of their studies was not generally recognized as closely related to nitrile oxides. Finally, since 1960 much light has been shed on the reaction mechanisms by Huisgen's work on 1,3-dipolar cycloadditions.

Nitrile oxides are isomeric with cyanates (**1**) and isocyanates (**2**). While in **1** the organic group is connected to the oxygen and in **2** to the nitrogen atom, in the nitrile oxides it is the carbon atom which provides the link to the organic moiety. The older literature pre-

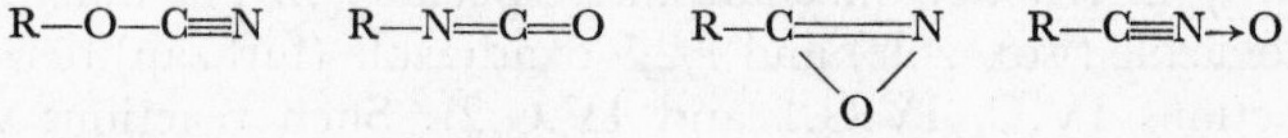

$$\underset{(\mathbf{1})}{\mathrm{R{-}O{-}C{\equiv}N}} \qquad \underset{(\mathbf{2})}{\mathrm{R{-}N{=}C{=}O}} \qquad \underset{(\mathbf{3})}{\mathrm{R{-}C{=}N\ (\text{bridged by }O)}} \qquad \underset{(\mathbf{4})}{\mathrm{R{-}C{\equiv}N{\rightarrow}O}}$$

ferred the cyclic structure **3**[5-8], but optical data ruled out **3** at a rather early stage in preference of **4**[9], which is the most generally accepted simplified expression of the structure of the nitrile oxides. Nevertheless, the structure formulation of a nitrile oxide as a resonance hybrid between the structures **5a–e** is a more accurate description[10,11]:

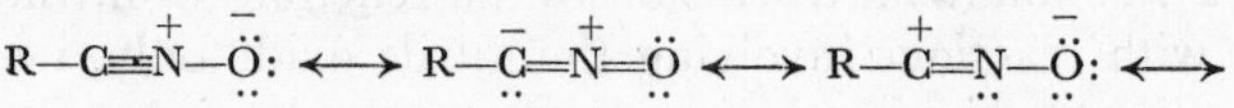

$$\underset{(\mathbf{5a})}{\mathrm{R{-}C{\equiv}\overset{+}{N}{-}\overset{-}{\ddot{O}}{:}}} \longleftrightarrow \underset{(\mathbf{5b})}{\mathrm{R{-}\overset{-}{\underset{..}{C}}{=}\overset{+}{N}{=}\underset{..}{\ddot{O}}}} \longleftrightarrow \underset{(\mathbf{5c})}{\mathrm{R{-}\overset{+}{C}{=}N{-}\overset{-}{\ddot{O}}{:}}} \longleftrightarrow$$

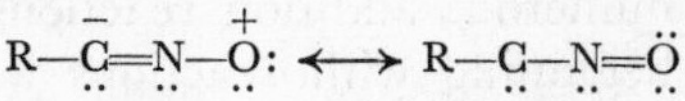

$$\underset{(\mathbf{5d})}{\mathrm{R{-}\overset{-}{\underset{..}{C}}{=}N{-}\overset{+}{\ddot{O}}{:}}} \longleftrightarrow \underset{(\mathbf{5e})}{\mathrm{R{-}\underset{..}{C}{-}N{=}\ddot{O}}}$$

Among the above mesomeric structures, the all-octet formulae **5a** and **5b** presumably represent the preferred electron distribution in the ground state, while the sextet formulae **5c** and **5d** express best most of the reactions of the nitrile oxides, especially the 1,3-dipolar additions.

As nitrogen-containing organic compounds, nitrile oxides are derivatives of hydroxylamine; the most important modes of preparation start with organic hydroxylamine derivatives. The complete hydrolysis of a nitrile oxide leads to hydroxylamine and the corresponding carboxylic acid (equation 1) in complete analogy to the corresponding hydrolysis of nitriles to ammonia and carboxylic acids (equation 2):

$$RCNO + 2\,H_2O \longrightarrow RCOOH + NH_2OH \quad (1)$$

$$RCN + 2\,H_2O \longrightarrow RCOOH + NH_3 \quad (2)$$

Otherwise, however, there are few analogies to the chemistry of nitriles. The direct oxidation of nitriles to nitrile oxides has not yet been achieved.

In the present chapter, the chemistry of nitrile oxides as defined in the first paragraph will be discussed. Accordingly, the chemistry of fulminato complexes of transition metals as well as the consideration of compounds which may contain a covalent bond between the —CNO group and an element other than carbon, e.g. PhSiCNO or PhHgCNO, is not included[12,13].

Nitrile oxides as transient intermediates have been postulated frequently in reaction mechanisms, especially in reactions leading to 1,2-oxazole (isoxazole) and 1,2,5-oxadiazole (furazan) derivatives (see sections IV.C, IV.G.1 and IV.G.2). Such reactions will be discussed in this chapter only in cases where unambiguous evidence exists for the occurrence of nitrile oxides.

The various methods for the synthesis of nitrile oxides are discussed first, including also procedures whereby the nitrile oxide is generated *in situ* in the presence of a suitable acceptor. In view of the instability of most nitrile oxides, such techniques have become of increasing importance. The physical properties of nitrile oxides will be reviewed next. There follows a discussion of the reactions of nitrile oxides, starting with reactions involving the nitrile oxide only, i.e. rearrangement, dimerization, polymerization and reduction. The numerous addition reactions of nitrile oxides are then presented, beginning with reactions with inorganic compounds and ending with the 1,3-dipolar cycloadditions to unsaturated systems leading to a wide variety of five-membered heterocyclic compounds, which are often rather inaccessible by other routes.

II. PREPARATION OF NITRILE OXIDES

A. General

All known methods for the synthesis of nitrile oxides start with organic systems already containing the C—N—O sequence of the nitrile oxide structure. There is no indication that any possible organic compound containing a C—N single, double or triple bond has ever been converted into a nitrile oxide. The fulminato radical, ·CNO, has been observed spectroscopically in mixtures of cyanogen and ozone subjected to flash photolysis[14,15], but it is not likely that such methods will ever be of practical value. Likewise, no example is known where an —N—O— compound has been introduced into an organic residue to generate directly a nitrile oxide.

The most important methods for the preparation of nitrile oxides start with aldoximes, from which by various techniques two hydrogen atoms are abstracted to form the nitrile oxide:

$$\mathrm{RCH{=}NOH} \xrightarrow{-2\,\mathrm{H}} \mathrm{RC{\equiv}N{\rightarrow}O}$$

Less widely-applied methods consist in the dehydration of primary nitroparaffins by various procedures:

$$\mathrm{RCH_2NO_2} \xrightarrow{-\mathrm{H_2O}} \mathrm{RC{\equiv}N{\rightarrow}O}$$

Contrary to the chemistry of nitriles, where the reaction of a metal cyanide with an organic halogen compound is a very important synthetic route, the analogous reaction, i.e. the reaction of a metal fulminate with an organic halide, has been realized only in one special case. Most attempts in this direction have usually resulted in a spontaneous rearrangement leading to the isomeric isocyanate:

$$\mathrm{R{-}Hal + MCNO} \longrightarrow \mathrm{[RCNO]} \longrightarrow \mathrm{RN{=}C{=}O}$$
$$(\mathrm{M} = \text{metal})$$

Table 1 includes only those nitrile oxides which have been prepared by the methods discussed in the following paragraphs and have been isolated in pure or approximately pure form.

B. Preparation of Fulminic Acid (Formonitrile Oxide)

Because of its unstability, fulminic acid (**9**) is always prepared first in form of its mercuric or silver salts. Although highly explosive, with the proper precautions both salts can be stored indefinitely at room temperature. The most convenient way to prepare mercuric fulminate is the reaction of ethanol with concentrated nitric acid in the presence of metallic mercury[19]. Equation (3) describes the mechanism first suggested by Wieland[2,58,59], and strongly supported by experimental evidence, since ethanol can be successfully replaced by acetaldehyde (**6**)[2], glyoxylic acid oxime (**7**)[59] or methylnitrolic acid (**8**). The silver salt is prepared analogously. Mercuric fulminate is

TABLE 1. Isolated nitrile oxides.

No.	Compound	Method of preparation[a]	Melting point (°C)	Melting point of the furoxan (°C)	Stability[b]	References
1	Formonitrile oxide (fulminic acid)	A, C, D	Polymerizes at −15°	—	40 min[c]	4, 16–20
2	Acetonitrile oxide	C, D, E	−5	106·5–107·5[d]	<1 min	21–24
3	2,2-Dimethylpropionitrile oxide	C	18[e]	67	2–3 days	24–26
4	2-Ethylbutyronitrile oxide	C	−33	−2	<1 min	24
5	Di-*t*-butylacetonitrile oxide	B	24–25[f]	—	Unlimited	27
6	Cyanogen bis(*N*-oxide)	C	Explodes at −45	—	5–6 h[i]	25, 28, 29
7	2,2,6-Trimethylcyclohexen-1-yl-1-fulmide	B	48–49[g]	—	Unlimited	27
8	2,2,6-Trimethylcyclohexyl fulmide	B	31[h]	—	Unlimited	27
9	Benzonitrile oxide	C, D, E	14–15	114–115	30–60 min	8, 22, 25, 30–32
10	2-Chlorobenzonitrile oxide	C	27–28	130–131	3–6 days	26
11	3-Chlorobenzonitrile oxide	C	42–43	96–97	50–60 min	26
12	4-Chlorobenzonitrile oxide	C	82–83	144–145	10 days	26, 33
13	2,6-Dichlorobenzonitrile oxide	B, C	86–87	199–200	30–35 days	26, 34
14	4-Bromobenzonitrile oxide	C	83–84	164–165	Not determined	26, 35
15	2-Nitrobenzonitrile oxide	C	76–77 (dec.)	199–200	1–2 days	26, 36
16	3-Nitrobenzonitrile oxide	C	82–83	183–186	20–25 days	26, 30, 33, 36
17	4-Nitrobenzonitrile oxide	C	95	205–206	>30 days	26, 36, 37
18	4-Methylbenzonitrile oxide	C	55–56	143–144	5–7 days	26
19	2,4,6-Trimethylbenzonitrile oxide	B, C	114	—	Unlimited	25, 34, 38–40
20	3,5-Dichloro-2,4,6-trimethylbenzonitrile oxide	C	138	—	Unlimited	41

21	2,3,5,6-Tetramethylbenzonitrile oxide	B, C	120	—	Unlimited	34, 38, 39
22	4-Methoxybenzonitrile oxide	C	69–70	113	7–10 days	42, 43
23	2,6-Dimethyl-4-methoxybenzonitrile oxide	B	66–68	—	Unlimited	44
24	2,4,6-Trimethoxybenzonitrile oxide	B	160–170 (dec.)	—	Unlimited	34, 39, 40
25	2,6-Dimethyl-4-dimethylamino-benzonitrile oxide	B	130–132 (dec.)	—	Unlimited	40, 45, 46
26	Triphenylacetonitrile oxide	F	153–154	—	Unlimited	47
27	Oximino phenylacetonitrile oxide	D	112–113	156	Not determined	48, 49
28	Oximino *p*-tolylacetonitrile oxide	D	112 (dec.)	—	Not determined	50
29	2-Methoxy-1-naphthonitrile oxide	B	101–103	—	Unlimited	40
30	2,6-Dimethoxy-1-naphthonitrile oxide	B	120–122	—	Unlimited	40
31	5-Bromo-2,6-dimethoxy-1-naphthonitrile oxide	B	192–194 (dec.)	—	Unlimited	40
32	2,7-Dimethoxy-1-naphthonitrile oxide	B	123–124	—	Unlimited	40
33	8-Bromo-2,7-dimethoxy-1-naphthonitrile oxide	B	122–124	—	Unlimited	40
34	Anthracene-9-nitrile oxide	B	127–128	—	Unlimited[j]	34
35	10-Methylanthracene-9-nitrile oxide	B	165	—	Unlimited[j]	40
36	*O*-Methylpodocarpinonitrile oxide	B[k]	132	—	Unlimited	51, 52
37	Isophthalobis(nitrile oxide)	C	92–94 (dec.)	—	Not determined	53
38	Terephthalobis(nitrile oxide)[l]	C	241–242 (dec.)	—	Not determined	54
			124–157 (dec.)	—	Not determined	55, 56
			~160 (dec.)	—	Not determined	57
39	2,4,6-Trimethylisophthalobis(nitrile oxide)	B	138–139 (dec.)	—	Unlimited	40
40	4-Dimethylamino-2,6-dimethyl-isophthalobis(nitrile oxide)	B	123–125 (dec.)	—	Unlimited	40, 45

(*Table continued*)

Table 1 (*continued*)

No.	Compound	Method of preparation[a]	Melting point (°C)	Melting point of the furoxan (°C)	Stability[b]	References
41	2,3,5,6-Tetramethylterephthalo-bis(nitrile oxide)	B	169–170 (dec.)	—	Unlimited	34, 39
42	5-Methyl-3-phenyl-1,2-oxazole-4-nitrile oxide	C	83	230	>30 days	26, 31
43	2,4,6-Trimethoxypyrimidine-5-nitrile oxide	B	134–136 (dec.)	—	Unlimited	40, 45
44	4-Chloro-2-dimethylaminopyrimidine-5-nitrile oxide	B	159–162 (dec.)	—	Unlimited	40
45	2-Dimethylamino-4-chloromethoxy-pyrimidine-5-nitrile oxide	B	154–155	—	Unlimited	40
46	4,6-Dimethyl-2-dimethylamino-pyrimidine-5-nitrile oxide	B	178–180 (dec.)	—	Unlimited	40, 45

[a] Method of preparation refers to: A. Special methods discussed in section II.B. B. Dehydrogenation of the aldoxime with hypobromite or *N*-bromosuccinimide. C. Dehydrohalogenation of the hydroximic acid chloride. D. Thermal decomposition of the nitrolic acid. E. Dehydration of the primary nitroparaffin with phenylisocyanate. F. Reaction of the organic halide with silver fulminate.

[b] Stability refers to the time at which the dimerization to the furoxan was found to be complete at 18 °C.

[c] Half-life in 0·4 N ethereal solution at 0°.

[d] B.p. (14 mm).

[e] B.p. 61° (15 mm).

[f] B.p. 55–56° (0·2 mm).

[g] B.p. (0·03 mm).

[h] B.p. (0·001 mm).

[i] In approximately 5% solution in methylene chloride at 0°.

[j] The compound is slowly decomposed by light and air.

[k] Dehydrogenation with $Pb(Ac)_4$.

[l] For a discussion of reasons for the wide difference of reported melting points. See p. 810.

$$C_2H_5OH \xrightarrow{HNO_3} \underset{(6)}{CH_3CHO} \xrightarrow{HNO_2} HON{=}CHCHO \xrightarrow{HNO_3}$$

$$\underset{(7)}{HON{=}CHCOOH} \xrightarrow{HNO_3} HON{=}C(NO_2)COOH \xrightarrow{-CO_2} \tag{3}$$

$$\underset{(8)}{HON{=}CHNO_2} \xrightarrow{-HNO_2} \underset{(9)}{HC{\equiv}N{\rightarrow}O} \xrightarrow{Hg^{2+}} Hg(CNO)_2$$

also obtained from malonic acid and a solution of mercury in dilute nitric acid[60], whereby presumably first mesoxalic acid oxime (**10**) is formed which then decarboxylates to glyoxylic acid oxime (**7**) (equation 4), from which the reaction proceeds according to equation (3). The mercuric salt of nitromethane decomposes directly into

$$HOOCCH_2COOH \xrightarrow{HNO_2} \underset{(10)}{HOOC(C{=}NOH)COOH} \xrightarrow{-CO_2} \underset{(7)}{HON{=}CHCOOH} \tag{4}$$

water and mercuric fulminate[61] (equation 5):

$$2\,CH_3NO_2 \xrightarrow{NaOH} 2\,CH_2{=}NO_2Na \xrightarrow{Hg^{2+}}$$

$$(CH_2NO_2)_2Hg \xrightarrow{-2\,H_2O} Hg(CNO)_2 \tag{5}$$

Another route to fulminic acid from nitromethane prepares first the methyl nitrolic acid (**8**) (equation 6) which then converts to fulminic acid by the route of equation (3). In this case, the acid is trapped as the silver salt[58,62]:

$$CH_3NO_2 \xrightarrow{OH^-} CH_2NO_2^- \xrightarrow{HNO_2} \underset{(8)}{HON{=}CHNO_2} \longrightarrow \underset{(9)}{HC{\equiv}N{\rightarrow}O} \tag{6}$$

Silver fulminate can also be obtained from formamide oxime (**11**) by heating with silver nitrate in nitric acid solution[63]:

$$\underset{(11)}{HON{=}CHNH_2} \xrightarrow{Ag^+,\ HNO_3} AgCNO + NH_4^+ \tag{7}$$

Under the same conditions, the potassium salt of aminomethyl nitrosolic acid (**12**) is likewise converted into silver fulminate[63]:

$$\underset{(12)}{H_2NC(NO){=}NOK} \xrightarrow{Ag^+,\ HNO_3} N_2 + H_2O + K^+ + AgCNO \tag{8}$$

Because of their complex nature (they actually contain the dibasic anion $C_2N_2O_2^{2-}$), the silver and the mercuric salts are not suited for many applications in organic chemistry, including the preparation of the free acid[64]. Therefore, they are often converted into the sodium fulminate by reaction with sodium amalgam in ethanol. Sodium fulminate is about as toxic as sodium cyanide[65]. Contrary to the alkali salts of some similarly unstable acids, e.g. hydrazoic acid, sodium fulminate is not storable and is also highly explosive[16]. More convenient, probably, is the conversion of mercuric fulminate into formhydroxamyl iodide (**13**) by means of hydroiodic acid and potassium iodide[18]:

$$Hg(CNO)_2 \xrightarrow{\text{HI, KI}} \underset{(\mathbf{13})}{2HIC{=}NOH} \tag{9}$$

Compound **13** is storable and non-explosive. Addition of triethylamine to an ethereal ice-cold solution of **13** liberates fulminic acid. Free fulminic acid can also be obtained by acidification of an aqueous solution of sodium fulminate with dilute sulphuric acid. To prevent rapid polymerization, the solution of the sodium fulminate should be added to the sulphuric acid, so that there is always an excess of mineral acid present. From such aqueous solutions fulminic acid can be extracted with ether[17]. The acid is volatile with ether vapours and has an odour similar to hydrocyanic acid, but much more aggressive[66a]. Recently, the free acid has been obtained in a crystalline state by fractional distillation of an aqueous solution in a high-vacuum system and freezing out the vapours of **9** at −78°. On warming to room temperature the crystals polymerize far below 0° [16]. At low pressure the vapour, however, is stable enough to determine the infrared spectrum. All reactions of fulminic acid have been carried out with dilute aqueous or ethereal solutions. At 0° the half-life time of an aqueous 0·4 N solution containing at least 0·2 N sulphuric acid is about 90 min[17] (for the stability of ethereal solutions see Table 1).

C. Preparation of Nitrile Oxides from Aldoximes

I. Dehydrogenation of aldoximes

Aromatic and heterocyclic oximate anions of aldoximes have been dehydrogenated to nitrile oxides in alkaline solution by potassium ferric cyanide or, preferably, hypohalites[34,39]:

$$Ar{-}CH{=}NO^- + OHal^- \longrightarrow Ar{-}C{\equiv}N{\rightarrow}O + OH^- + Hal^- \tag{10}$$

Alkaline sodium hydrobromite is the most preferable reagent since dehydrogenation is very fast occurring at temperatures around 0 °C, with good to excellent yields. The reaction with alkali hypoiodite is slower and the yields are much inferior. The reaction of sodium hypochlorite with aldoximes, which has been studied much earlier, yields a mixture of a minor amount of the desired nitrile oxide, together with a dimeric compound resulting from the abstraction of only one hydrogen atom from each molecule of aldoxime[66b]. These compounds have been reported in the earlier literature as 'aldoxime peroxides', but more recently the structures of an oxime anhydride *N*-oxide (**14**) or an aldazine bis(*N*-oxide) (**15**) have been proposed[67,68] (equation 11). Chemical reactions are compatible with both structures, but infrared spectra favour **14**:

$$2\,\text{ArCH}{=}\text{N}{-}\text{O}^- + \text{H}_2\text{O} + \text{OCl}^- \longrightarrow \text{ArCH}{=}\text{N}\underset{\downarrow\atop\text{O}}{\text{O}}\text{N}{=}\text{CHAr} \quad \textbf{(14)}$$

or

$$\text{ArCH}{=}\underset{\downarrow\atop\text{O}}{\text{N}}{-}\underset{\downarrow\atop\text{O}}{\text{N}}{=}\text{CHAr} + \text{Cl}^- + 2\,\text{OH}^- \quad \textbf{(15)} \qquad (11)$$

The different route taken by the oxidation in the case of hypochlorite can be attributed to the higher oxidation potential of this reagent. The method has been especially useful in cases where the nitrile oxides could not be prepared by the older route via the hydroximic acid chlorides because of side reactions during the chlorination step[25,34] (see next paragraph). However, the synthesis fails in cases where the aldoxime contains other functional groups which are alkali labile or unstable toward the oxidizing agent. Polyfunctional nitrile oxides are usually obtained in poor yields.

A milder and more selective dehydrogenation of aldoximes has been achieved with *N*-bromosuccinimide in the presence of alkali alkoxides or tertiary bases[40]. This modification allows the preparation, for example, of amino-substituted aromatic and heterocyclic nitrile oxides as well as of polyfunctional nitrile oxides, in satisfactory to very good yields; it is probably the most generally applicable procedure for the synthesis of this class of compounds. Dehydrogenation of aldoximes with lead tetraacetate has been claimed as a general method for the synthesis of nitrile oxides[51,52]. Only a few examples are known, however, and under apparently the same conditions other authors have obtained instead compounds of a different structure[68a].

2. Dehydrohalogenation of hydroximic acid halides

Although this route does not start directly from the aldoximes, the necessary precursors to the nitrile oxide, the hydroximic acid halides, are generally prepared from the corresponding aldoximes. For all practical purposes, only the hydroximic acid chlorides (**17**) are used (equation 12):

$$\mathrm{RCH{=}NOH} + \mathrm{Cl_2} \longrightarrow \underset{\mathbf{(16)}}{[\mathrm{RC(Cl)HNO}]} \longrightarrow \underset{\mathbf{(17)}}{\mathrm{RC(Cl){=}NOH}} + \mathrm{HCl} \qquad (12)$$

The halogenation of the aromatic aldoximes[68b], which is generally carried out either in an inert solvent, such as chloroform and carbon tetrachloride, or in acetic acid, water or hydrochloric acid, seems—at least in some cases—to proceed via the intermediate of a *geminal* chloronitroso compound **16**, as indicated by the transient blue-green colour of the reaction mixture. Aliphatic hydroximic acid chlorides are conveniently prepared in ether at −60 °C, via **16** (or its dimer) which rearranges to **17** within one hour[69]. Another route to hydroximic acid chlorides employs nitrosyl chloride as chlorinating agent[43]:

$$\mathrm{RCH{=}NOH} + 2\,\mathrm{NOCl} \longrightarrow \mathrm{RC(Cl){=}NOH} + 2\,\mathrm{NO} + \mathrm{HCl} \qquad (13)$$

Finally, hydroximic acid halides have been obtained by reaction of concentrated hydrochloric or hydrobromic acid on certain nitrolic acids[70–75], e.g.

$$\mathrm{RO_2CC(NO_2){=}NOH} + \mathrm{HHal} \longrightarrow \mathrm{RO_2CC(Hal){=}NOH} \qquad (14)$$

$$(\mathrm{R} = \mathrm{C_2H_5}, \mathrm{H};\ \mathrm{Hal} = \mathrm{Cl}, \mathrm{Br})$$

Nitrolic acids, however, can be converted in most cases directly into nitrile oxides (see section II.D.2), so this method offers little advantage except under special circumstances.

Hydroximic acid chlorides are stable indefinitely at room temperature[25] and are therefore conveniently storable precursors of the generally unstable nitrile oxides, which can be generated almost instantaneously from them when needed by dehydrochlorination with base:

$$\mathrm{R{-}C(Cl){=}NOH} \xrightarrow{\text{base}} \mathrm{R{-}C{\equiv}N{\rightarrow}O} + \mathrm{Base{\cdot}HCl} \qquad (15)$$

Any base, inorganic or organic, in either an aqueous or anhydrous medium may be used. Earlier investigators preferred aqueous sodium carbonate and by this method, the first nitrile oxide, benzonitrile oxide, was prepared[5]. More recently, the reaction of the hydroximic acid chloride, dissolved or suspended in an inert organic solvent, e.g. anhydrous ether, with one equivalent of a tertiary organic base, preferably triethylamine, has been recommended[33]. Using this technique at temperatures as low as −40 °C, even the most unstable aliphatic nitrile oxides have been successfully prepared[24,25]. Tertiary amines as dehydrohalogenating agents may, however, not be applicable in all cases, since they seem to form with some hydroximic acid chlorides rather stable addition compounds, which are apparently quaternary amidoximinium salts **18**[29,76,77]:

$$R^1\text{—}\underset{\displaystyle Cl}{\underset{|}{C}}\text{=NOH} + R_3N \longrightarrow \left[R^1\text{—}\underset{\displaystyle R_3N^+}{\underset{|}{C}}\text{=NOH}\right]Cl^- \qquad (16)$$

(18)

The main limitations of this synthesis lie in the chlorination step of the oldoximes. Unsaturated aldoximes add chlorine to the double bond, e.g. from 1-oximino-2-methylbutene-2 (**19**), 2,3-dichloro-2-methylbutanehydroximic acid chloride (**20**) and from 1-oximino-2, 2-dimethylpentene-4 (**21**), 4,5-dichloro-2,2-dimethylpentanehydroximic acid chloride (**22**) are obtained[78]:

$CH_3CH{=}C(CH_3)CH{=}NOH$ (**19**)

$CH_3CHClCCl(CH_3)CCl{=}NOH$ (**20**)

$H_2C{=}CHCH_2C(CH_3)_2CH{=}NOH$ (**21**)

$ClCH_2CHClCH_2C(CH_3)_2CCl{=}NOH$ (**22**)

Salicylaldoxime (**23**) and 2-methoxybenzaldoxime (**24**) cannot be transformed into the corresponding hydroximic acid chlorides, but products of additional chlorination in the benzene ring are obtained, i.e. 3,4-dichloro-2-hydroxybenzhydroximic acid chloride (**25**), or 5-chloro-2-methoxybenzhydroximic acid chloride (**26**)[78]. Likewise, the chlorination of the oximes of mesitylaldehyde (**27**) or 2,3,5,6-tetramethylbenzaldehyde leads to inseparable mixtures of the desired hydroximic acid chlorides and products of further chlorination of the aromatic ring and of the methyl groups[34]. By using three moles of chlorine, however, **27** is neatly converted into 3,5-dichloro-2,4,6-trimethylbenzhydroximic acid chloride[41] (**28**). Nonetheless, the hydroximic acid chlorides route is the most widely used and it

$$\text{(23, R = OH)} \rightarrow \text{(25)}; \quad \text{(24, R = OCH}_3\text{)} \rightarrow \text{(26)}$$

R–C$_6$H$_4$–CH=NOH (23, R = OH) (24, R = OCH_3) → HO–C$_6$H(Cl)$_2$–CCl=NOH (25); H_3CO–C$_6$H$_3$(Cl)–CCl=NOH (26)

H_3C–C$_6$H(CH$_3$)$_2$–CH=NOH (27) → H_3C–C$_6$(Cl)$_2$(CH$_3$)$_2$–CCl=NOH (28)

has been successfully applied to the preparation of aliphatic, aromatic and heterocyclic nitrile oxides.

D. Preparation of Nitrile Oxides from Primary Nitroparaffins

I. Conversion to hydroximic acid chlorides

The aci-salts of primary nitroparaffins, react with anhydrous hydrogen chloride to form hydroximic acid chlorides (**17**) (equation 17)[79], which are converted into nitrile oxides according to the preceding paragraph:

$$RCH_2NO_2 \xrightarrow{OH^-} RCH{=}N{\overset{\nearrow O}{\underset{\searrow O^-}{}}} \xrightarrow{2\,HCl} \underset{\text{(17)}}{RC(Cl){=}NOH} + H_2O + Cl^- \qquad (17)$$

$$(R = CH_3, Ph, PhCO, CO_2C_2H_5)$$

Compounds **17** are generally more easily accessible by the routes discussed above, and better methods to convert primary nitroparaffins into nitrile oxides are discussed below.

2. Decomposition of nitrolic acids

Nitrolic acids (**29**) are obtained by reaction of primary nitroparaffins with nitrous acid:

$$RCH_2NO_2 + HNO_2 \longrightarrow \underset{\underset{\textbf{(29)}}{NO_2}}{RC{=}NOH} + H_2O \tag{18}$$

All nitrolic acids easily lose the elements of nitrous acid with the formation of nitrile oxides. Sometimes this decomposition occurs spontaneously at room temperature; but generally, it is induced by gentle heating:

$$\underset{\underset{\textbf{(29)}}{NO_2}}{RC{=}NOH} \xrightarrow{\Delta} RC{\equiv}N{\rightarrow}O + HNO_2 \tag{19}$$

The applicability of this reaction has been only slightly investigated, but benzonitrile oxide (R = Ph) and oximino phenylacetonitrile oxide (**33**), were obtained in this manner[25,32].

33, which is obtained by reaction of phenylglyoxime (**30**) with dinitrogen tetroxide, presumably via the intermediate nitrolic acid **31** (equation 20), was first considered to be phenylfuroxan (**34**)[80]. A mild treatment with alkali isomerizes the compound to a material which was a 4-hydroxy-3-phenylfurazan (**35**), according to Wieland[81,82] or the nitrile oxide **33** according to Ponzio and coworkers[48,49,83–85].

The controversy has been settled in favour of structure **32** or **33**, since the infrared spectrum of the compound displays the characteristic frequencies of the $C{\equiv}N \rightarrow O$ group[86]. The properties of

$$\underset{\underset{\textbf{(30)}\ (\alpha\text{-form})}{NOH}}{PhCCH{=}NOH} \xrightarrow{N_2O_4} \Big[\underset{\underset{\textbf{(31)}}{NOH\quad NO_2}}{PhC{-}C{=}NOH}\Big] \xrightarrow{-HNO_2}$$

$$\underset{\underset{\textbf{(32)}}{HON}}{PhC{-}C{\equiv}N{\rightarrow}O} \xrightarrow{OH^-} \underset{\underset{\textbf{(33)}}{NOH}}{PhC{-}C{\equiv}N{\rightarrow}O} \tag{20}$$

Ph—C———CH (ring: C=N, CH=N→O, N—O—N) **(34)**

Ph—C———C—OH (ring: C=N, C=N, N—O—N) **(35)**

the primary reaction product, viz. the alleged phenylfuroxan **34**, suggest that it is a stereoisomer of **33**. The assignment of formulae **32** and **33** for these compounds is arbitrary since it is based only on the most probable configuration of α-phenylglyoxime. Homologues and their corresponding *O*-benzoyl derivatives have been obtained analogously[49,87].

3. Dehydration of primary nitroparaffins

A recent method achieves the dehydration reaction of primary nitroparaffins to the nitrile oxides in one step under rather mild conditions, using phenylisocyanate, or—less favourably—phosphorus oxychloride, as the dehydrating agent in the presence of catalytic amounts of triethylamine[21–23,88,89]. The original authors have suggested the following mechanism[21]:

$$RCH_2NO_2 + (C_2H_5)_3N \longrightarrow (C_2H_5)_3NH^+ + RCH{=}NO_2^- \qquad (21)$$

$$RCH{=}NO_2^- \xrightarrow{+PhN{=}C{=}O} R{-}CH{=}N(\rightarrow O){-}O{-}C(=O){-}\bar{N}Ph \xrightarrow{+H^+} RCH{=}N(\rightarrow O){-}O{-}C(=O){-}NHPh$$

$$RCH{=}N(\rightarrow O){-}O{-}C(=O){-}NHPh \longrightarrow RC{\equiv}N{\rightarrow}O + HOOCNHPh$$

The reaction proceeded well with nitroethane, 1-nitropropane and phenylnitromethane while it takes a slightly different course with nitromethane[90]. The fulminic acid (**9**), initially generated, reacted immediately with additional phenylisocyanate to give the *N*-phenylcarbamoylfulmide (**36**) which could be trapped by 1,3-dipolar cycloaddition to a suitable olefin present in the reaction mixture:

$$CH_3NO_2 \xrightarrow{-H_2O} \underset{(\mathbf{9})}{HC{\equiv}N{\rightarrow}O} \xrightarrow{PhNCO}$$

$$\underset{(\mathbf{36})}{PhNHCOC{\equiv}N{\rightarrow}O} \xrightarrow{CH_2{=}CHOC_2H_5} PhNHCOC{=}N{-}O{-}CH(OC_2H_5){-}CH_2 \ \text{(ring)} \qquad (22)$$

The original investigators did not attempt to isolate nitrile oxides thus prepared, but reacted them *in situ*, either by dimerization to furoxans or by 1,3-dipolar cycloaddition with olefins to isoxazoles. It is likely, however, that this method could also be used for isolation of the nitrile oxides. Since the lower aliphatic primary nitro compounds are more easily accessible than the corresponding hydroximic acid chlorides, there is no doubt about the usefulness of this method.

For the alleged preparation of monomeric carbethoxyfulmide (**44**) from ethyl bromoacetate and silver nitrite, which must involve the dehydration of the originally formed ethyl nitroacetate, reference is made to the literature[91–93].

E. Preparation of Nitrile Oxides from Fulminates

Early attempts to react metal salts of fulminic acid with alkyl or acyl halides were aimed at preparing esters of fulminic acid of structure **37**, a still unknown class of compounds. The only identifiable products obtained were the isomeric isocyanates **38** or products derived from them (equation 23)[3,77,94–97]:

$$\text{R—Hal} + \text{MCNO} \not\longrightarrow \underset{\mathbf{(37)}}{\text{RON}{\equiv}\text{C}} \quad (\text{R} = \text{alkyl or acyl}) \quad (\text{M} = \text{Ag, Hg, Na}) \qquad (23)$$

$$\downarrow$$

$$[\text{RC}{\equiv}\text{N}{\rightarrow}\text{O}] \longrightarrow \underset{\mathbf{(38)}}{\text{RN}{=}\text{C}{=}\text{O}}$$

The reaction of triphenylmethyl chloride (**39**) with silver fulminate, leading in good yield to triphenylacetonitrile oxide (**40**), is the only case in which a nitrile oxide has ever been isolated (equation 24) in this type of reaction[47]. If **39** was replaced by diphenylmethyl bromide (**41**), the only product which could be isolated was diphenylmethyl isocyanate (**42**) (equation 25). Picryl chloride which has a very reactive chlorine atom failed to react at all with silver fulminate[98].

$$\underset{\mathbf{(39)}}{\text{Ph}_3\text{CCl}} + \text{AgCNO} \longrightarrow \underset{\mathbf{(40)}}{\text{Ph}_3\text{CC}{\equiv}\text{N}{\rightarrow}\text{O}} \qquad (24)$$

$$\underset{\mathbf{(41)}}{\text{Ph}_2\text{CHBr}} + \text{AgCNO} \longrightarrow \underset{\mathbf{(42)}}{\text{Ph}_2\text{CHN}{=}\text{C}{=}\text{O}} \qquad (25)$$

Steric hindrance, which is undoubtedly present to a large degree in **40**, stabilizes nitrile oxides toward dimerization, but does not prevent rearrangement to isocyanates (see section IV.B). At present there is no explanation for the apparent stability of **40** under the reaction conditions, but its formation may indicate that nitrile oxides are also the primary reaction products in those reactions where earlier investigators isolated only isocyanates. It is possible that the metal halide formed catalyses the rearrangement of the nitrile oxide to the isocyanate, and it would be interesting to reinvestigate such reactions in the presence of a typical dipolarophile in order to trap the nitrile oxide *in situ*.

F. Preparation of Nitrile Oxides in situ

Since most nitrile oxides undergo irreversible changes very quickly at room temperature and above (see section IV.A), many of their chemical reactions have been investigated by generating the nitrile oxide from a stable precursor in the presence of a partner with whom it would react appreciably faster than with itself. In order to minimize the tendency toward autocondensation, in most cases dimerization to the furoxan (see section IV.C) or polymerization, it is advisable to generate the nitrile oxide very slowly, keeping its stationary concentration low. At the same time, the reaction partner (dipolarophile) should be present in as high a concentration as possible. This technique was first successfully applied by Huisgen[11,25,30,33].

Hydroximic acid chlorides are most frequently used to generate nitrile oxides *in situ*. The acid chloride is dissolved or suspended in ether together with the dipolarophile, and one mole of triethylamine, or another suitable tertiary amine, is added gradually at 0 to 20 °c. In cases where the dipolarophile itself might react with the base, ethereal solutions of the acid chloride and of triethylamine are added simultaneously in equivalent amounts to the solution of the dipolarophile, or the nitrile oxide is first generated at −20 °c and then the reaction partner is added while the reaction mixture warms slowly to room temperature.

The same technique has been applied to *in situ* reaction of nitrile oxides generated from primary nitroparaffins by dehydration with phenylisocyanate in the presence of catalytic amounts of triethylamine[22,23,89]. Here, a solution of the nitro compound and triethylamine in a suitable inert solvent, e.g. benzene, is added slowly to the solution of the dipolarophile and two moles of phenylisocyanate.

Hydroximic acid chlorides seem generally to be in equilibrium with the corresponding nitrile oxides (equation 26). The evidence

$$\underset{\displaystyle Cl}{RC}{=}NOH \rightleftharpoons RC{\equiv}N{\rightarrow}O + HCl \qquad (26)$$

for this is supplied by ultraviolet spectroscopy and polarographic studies of various hydroximic acid chlorides in aqueous solution[99–102a,102b]. If the dissociation takes place in an inert solvent, such as boiling toluene in the presence of a suitable scavenger for the nitrile oxide, the method provides a very elegant way of generating a nitrile oxide *in situ*, especially since its stationary concentration is presumably very low, the above equilibrium being far on the left side of equation (26) even at the temperature employed (110°)[103–107].

Thermal decomposition of nitrolic acids has also occasionally been used for the generation of unstable nitrile oxides *in situ*, e.g. pyruvic acid nitrile oxide (**43**) or carbethoxyfulmide (**44**)[72,108,109]. Such

$$\underset{(\mathbf{43})}{CH_3COC{\equiv}N{\rightarrow}O} \qquad \underset{(\mathbf{44})}{C_2H_5OOCC{\equiv}N{\rightarrow}O}$$

reactions are probably responsible for the production of a variety of heterocyclic compounds, mostly isoxazoles and furoxans from the action of concentrated nitric acid or higher oxides of nitrogen on unsaturated compounds. It is likely that nitrolic acids will form under such conditions and that the nitrile oxides resulting from them can either add to the unsaturated compound still present, or dimerize to the furoxans. An example is the formation of 3,3′-bis-(isoxazolyl) (**46**), from acetylene and a mixture of nitrogen monoxide and nitrogen dioxide in ethyl acetate at 60 °C under 10–15 atm pressure[110] (equation 27):

$$HC{\equiv}CH + 2\,NO + 2\,NO_2 \longrightarrow HON{=}\underset{O_2N}{C}{-}\underset{NO_2}{C}{=}NOH \xrightarrow{-2\,HNO_2}$$

$$\underset{(\mathbf{45})}{O{\leftarrow}N{\equiv}C{-}C{\equiv}N{\rightarrow}O} \xrightarrow{+2\,C_2H_2} \underset{(\mathbf{46})}{\text{3,3'-bis(isoxazolyl)}} \qquad (27)$$

Although the proposed reaction scheme involves the highly unstable cyanogen bis(*N*-oxide) (**45**) as an intermediate, yields of 60–70% of **46** can be obtained.

This type of reactions is outside the scope of this chapter; they are amply covered in several recent summaries[111–114].

G. Nitrile Oxides with Functional Groups

The wide variety of functional groups (including the nitrile oxide function itself) which react spontaneously with nitrile oxides (see section IV) considerably restricts the chances of obtaining nitrile oxides with functional groups. It has, however, been possible to prepare some difunctional nitrile oxides, such as **47**, **48**, **49** and **50**.

Terephthalic bis(nitrile oxide) (**47**) was synthesized via the corresponding hydroximic acid chloride; although earlier workers have claimed to have obtained polymers only[78], recent investigations have demonstrated that with proper precautions a product can be isolated that consists largely of the monomer[30,55–57,115]. The wide differences in melting points reported for **47** (Table 1) are attributed partly to the various states of crystallinity of the different preparations and partly to polymerization during slow heating. Since the molecular weight determinations of the purest samples are still unsatisfactory, these differences may also be caused by a varying amount of oligomers already present in freshly prepared **47**. No such difficulties seem to have been encountered in the preparation of **48**[53].

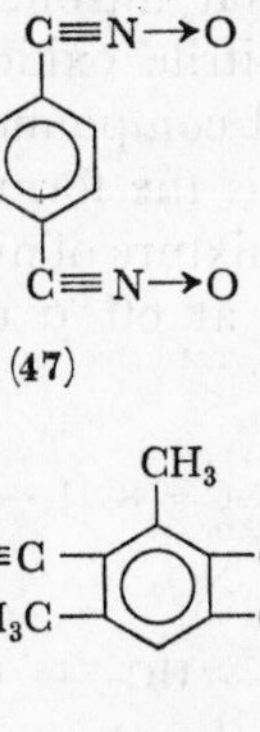

$C{\equiv}N{\rightarrow}O$ (para-substituted benzene) $C{\equiv}N{\rightarrow}O$

(47)

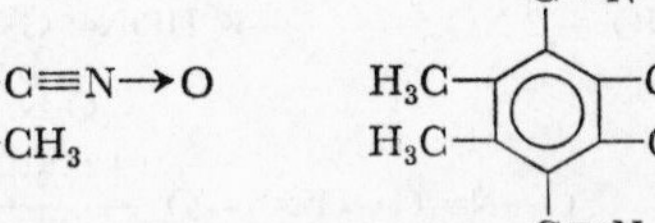

$O{\leftarrow}N{\equiv}C$–(meta-substituted benzene)–$C{\equiv}N{\rightarrow}O$

(48)

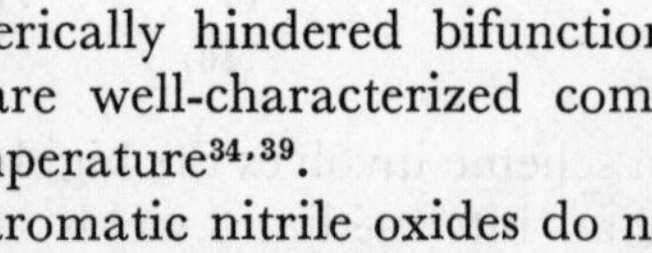

CH_3; $O{\leftarrow}N{\equiv}C$–, –$C{\equiv}N{\rightarrow}O$; H_3C–, –CH_3

(49)

$C{\equiv}N{\rightarrow}O$; H_3C–, –CH_3; H_3C–, –CH_3; $C{\equiv}N{\rightarrow}O$

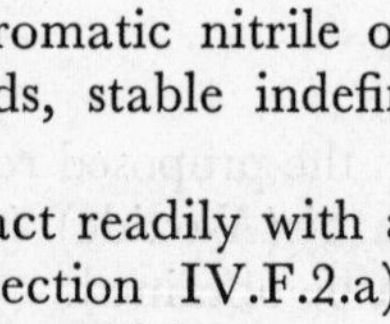

(50)

The sterically hindered bifunctional aromatic nitrile oxides **49** and **50** are well-characterized compounds, stable indefinitely at room temperature[34,39].

Since aromatic nitrile oxides do not react readily with alcoholic or phenolic groups without catalysts (section IV.F.2.a), nitrile oxides containing these functional groups could be prepared. The nitrile oxide **51** was obtained from the corresponding hydroximic acid chloride **25** (section II.C.2) but was isolated only as a dilute solution in CCl_4 and identified by its infrared spectrum and by subsequent dimerization to the furoxan[78]. An attempt to generate

C≡N→O, OH, Cl, Cl (**51**) C≡N→O, H_3C, CH_3, OH (**52**)

52 by dehydrogenation of the corresponding aldoxime with hypobromite failed because of the preferred attack of the oxidant on the ring[98]. Nitrile oxides in which the phenolic group is alkylated could be prepared as shown by several examples of methoxy-substituted nitrile oxides listed in Table 1. Other methoxylated aromatic nitrile oxides have been characterized only in solution by i.r. spectroscopy[78].

The stability of nitrile oxides with free carboxylic groups depends largely on the acidity of the carboxyl group, since the attack of the latter on the C≡N → O function starts with protonation to the conjugate acid. Thus, sterically hindered aromatic nitrile oxides can be recrystallized unchanged from hot acetic acid[116], while formic acid causes hydrolysis to the corresponding hydroxamic acid[56] (see sections IV.F.1.a and IV.F.2.c). This explains the formation of only polymeric products in the attempt to convert 3,5-dimethyl-4-formoximidophenoxyacetic acid (**53**) into the corresponding nitrile oxide by hypobromite, since the phenoxyacetic acids are generally strong acids[98]. An ester group is, however, compatible with the nitrile oxide function, and the synthesis of oximinophenylacetonitrile oxide (**33**) has already been discussed.

HC=NOH, H_3C, CH_3, OCH_2COOH (**53**) C≡N→O, H_3C, CH_3, $N(CH_3)_2$ (**54**) C≡N→O, H_3C, CH_3, C≡N→O, $N(CH_3)_2$ (**55**)

The ease with which nitrile oxides react with primary and secondary amines makes it unlikely that these functional groups will ever be found capable of existing together with a nitrile oxide group in a monomeric molecule, but tertiary amines seem to form stable adducts only in certain cases[29,76,77] (see section IV.F.2.d). Thus, the 2,6-dimethyl-4-dimethylaminobenzonitrile oxide (**54**) and the 2,6-dimethyl-4-dimethylaminoisophthalo bis(nitrile oxide) (**55**) could be prepared by oxidation of the corresponding aldoximes[40,45].

Attempts to prepare pyrrolo- or pyridinonitrile oxides have so far failed[98,117], but several pyrimidine nitrile oxides, (Table 1), were obtained by hypobromite or *N*-bromosuccinimide dehydrogenation of the corresponding aldoximes[40,45]. The isoxazolonitrile oxide (**56**) was prepared from the hydroximic acid chloride[26,31].

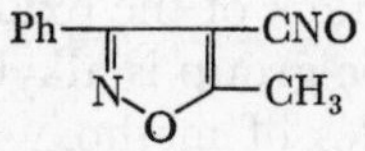

(**56**) 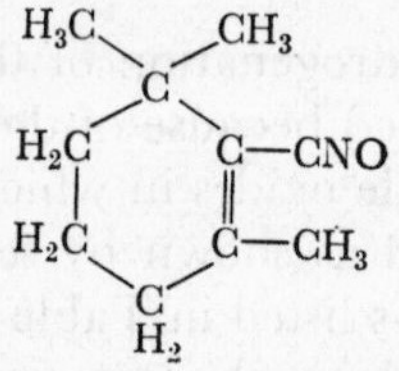 (**57**)

Nitrile oxides react generally with ethylenic double bonds (section IV.G.1), but the rate decreases sharply with increasing substitution[25]. Thus it was possible to obtain the sterically hindered, α,β-unsaturated nitrile oxide **57** which was indefinitely stable at room temperature[27].

III. PHYSICAL PROPERTIES OF NITRILE OXIDES

Nitrile oxides are energy-rich compounds; as derivatives of fulminic acid all low molecular-weight nitrile oxides and generally polynitrile oxides are to be considered as potentially *explosive*. Fulminic acid (**9**) and cyanogen bis(*N*-oxide) (**45**) explode already far below 0° (see Table 1) and even the polymers of **45** detonate violently on heating[28,29,118]. Terephthalic bis(nitrile oxide) (**47**) explodes at ~160° [119]. From the scattered data on the ultraviolet spectra of nitrile oxides, it seems that the CNO group has a similar influence as the CN group in conjugated and aromatic systems[86]. λ_{max} and log ε for cyanogen bis(*N*-oxide) (**45**) in n-hexane are 312 mμ (3·95), 288 mμ (4·0) and 256 mμ (3·90)[29].

More data are available on the infrared spectra of nitrile oxides, and due to the unstability of so many monomeric nitrile oxides, they have often been characterized by their infrared spectrum only. Aliphatic and aromatic nitrile oxides are characterized by two strong absorption bands at around 2330 cm^{-1} (C≡N stretching) and at around 1370 cm^{-1} (—N → O stretching). The former band is well suited for the identification of monomeric nitrile oxides. It is usually stronger and broader than the nitrile band which is often weak, but very narrow, and which appears in the corresponding nitriles in the same region but generally at ~70 cm^{-1} lower[25,38,78,86,120]. The two simplest nitrile oxides, fulminic acid (**9**) and cyanogen bis(*N*-oxide) (**45**) absorb at somewhat different wave numbers[4,29], their spectra are given in Table 2, together with that

TABLE 2. Infrared spectra of fulminic acid, cyanogen bis(*N*-oxide) and dinitrogen oxide.

Compound	Character of frequency (cm^{-1})			
	—C—H Stretching	—C≡N Stretching	—N=O Stretching	CN—O Deformation
H—C≡NO (gaseous)	3335	2190	1251	538
ON≡C—C≡NO (in CCl_4)	—	2190	1235	—[a]
N≡NO (gaseous)	—	2224[b]	1285	589[c]

[a] Could not be observed, because of adsorption of solvent.
[b] N≡N stretching.
[c] N≡NO deformation.

of dinitrogen oxide which is isoelectronic with **9** and shows a striking similarity with this compound[121,122]. The absence of bands around 3600 cm^{-1} (OH stretching) and 1400–1000 cm^{-1} (OH deformation) excludes definitely the old formula C⇇N—OH, at least for the gaseous fulminic acid.

The frequency shift of methanol upon hydrogen bonding to various types of *N*-oxides as proton acceptors, measures their ground-state basicity. The methanol $\Delta\nu$ are (in cm^{-1}), for trimethylamine oxide 486, pyridine *N*-oxide 256, bis(*t*-butyl)acetonitrile oxide 120, mesitonitrile oxide 78. This indicates the decreasing *s*-character in nitrogen from aliphatic amine oxides to aromatic nitrile oxides, together with a decreasing electronegativity of the nitrogen and a relatively low basicity of the oxygen. Since the electron density of oxygen is diminished by delocalization, structure **58** might represent a preferred resonance form of the ground state[123]. This is attributed to a redistribution of π-electrons from N towards C, resulting in inhibition of the expected increase in polarity for the nitrile oxide.

$$\overset{-}{\text{R—C}}\equiv\overset{+}{\text{N}}\text{—O}$$
(58)

Dipole moments of several aromatic nitrile oxides have been determined and found to be similar to those of the corresponding nitriles (~4·0–4·5 D)[124,125].

IV. REACTIONS OF NITRILE OXIDES

A. General; Stability of Nitrile Oxides

The nitrile oxide group is very similar in its electronic structure to the aliphatic diazo and to the azido groups. Consequently there are striking analogies in the chemical behaviour of these three classes of compounds[11,31,126]. On the other hand, there is a certain, although more superficial, resemblance between the analogous cumulated systems of nitrile oxides and those of ketenes and isocyanates. But in general nitrile oxides are far more prone to autocondensation than the latter compounds. All simple aliphatic and most aromatic nitrile oxides are permanently stable as monomers only at temperatures below $-70°$ [24,25,33,127]. Their most frequently observed mode of autocondensation is their dimerization to furoxans (1,2,5-oxadiazole-2-oxides). A notable exception is the autocondensation of fulminic acid which does not lead to a furoxan, and is discussed separately.

The electronic influence of substituents on the stability of aromatic nitrile oxides is not very pronounced and is apparently superimposed by other effects[128,129]. For instance, both electron-donor (methyl, methoxy) and -acceptor (halogen, nitro) substituents in the *para* position seem to stabilize nitrile oxides whereas the *ortho*-nitro group makes the nitrile oxide particularly unstable. Among the monochlorobenzonitrile oxides, however, the order of stability is *para* > *ortho* > *meta*.

The relative stability of cyanogen bis(*N*-oxide) (**45**) in dilute solution may be attributed to the higher number of mesomeric structures resulting from conjugation of the CNO groups, while cyanogen *N*-oxide (**59**), N≡C—C≡N→O, is as unstable as any of the lower aliphatic nitrile oxides[29,130]. The same effect might account for the unusual stability of terephthalo bis(nitrile oxide), while the stability of phenyloximinoacetonitrile oxide may be caused by the neighbouring oximino group, either by steric hindrance or by hydrogen bonding.

A very pronounced stabilization can be achieved by substituting aromatic or heterocyclic nitrile oxides in *o,o'*-positions with substituents of such spatial requirements that will block sterically the dimerization to furoxans without impairing the ability to react with other unhindered systems. Studies of Stuart–Briegleb models indicate that suitable groups are CH_3, C_2H_5, CH_3O or CH_3S, whereas Br, I, NO_2 or SO_2R are probably too large and F, Cl or OH too small for this purpose[34,39]. A higher degree of steric hindrance is obviously needed in the aliphatic series; *t*-butyl fulmide still dimerizes to the furoxan, although relatively slowly, but bis(*t*-butyl)-acetonitrile oxide is permanently stable at room temperature[27] (Table 1). This approach has led to the preparation of a number of nitrile oxides which are stable permanently as monomers and which are very useful in the study of reactions of nitrile oxides, especially those occurring at a slower rate than the dimerization[25,34,39,40,45,51,52,116,131].

Among the multitude of chemical reactions of nitrile oxides discussed in this paragraph, the overwhelming majority consists of addition reactions to the 1,3-dipolar mesomeric structures **5c** and **5d**, either of a nucleophile or of an unsaturated system. The most notable exceptions to this rule are the dimerization reaction to furoxans, which may involve the carbene structure **5e**, and the addition of strong inorganic acids in an aqueous medium, which proceeds through the protonation of **5a**.

Within the limitations of this chapter, it is impossible to cover completely the vast amount of recent literature. The following paragraphs are therefore mostly confined to some typical examples of the reaction under consideration. For more complete coverage, reference is made to the literature[25,111–114,132a,b].

B. Rearrangement to Isocyanates

On being heated in xylene solution to 110 °C, benzonitrile oxide (**60**) rearranges partly to phenylisocyanate (**61**), while most of the material dimerizes to diphenylfuroxan (**62**)[62,133]. When distilled under atmospheric pressure, the latter is also converted smoothly into **61**, presumably after initial depolymerization to the nitrile oxide (equation 28)[6,134,135].

Nitrile oxides which are sterically prevented from dimerization to furoxans by proper substitution in *o,o'*-positions, rearrange almost quantitatively to isocyanates[34,35]. At temperatures between 110 and

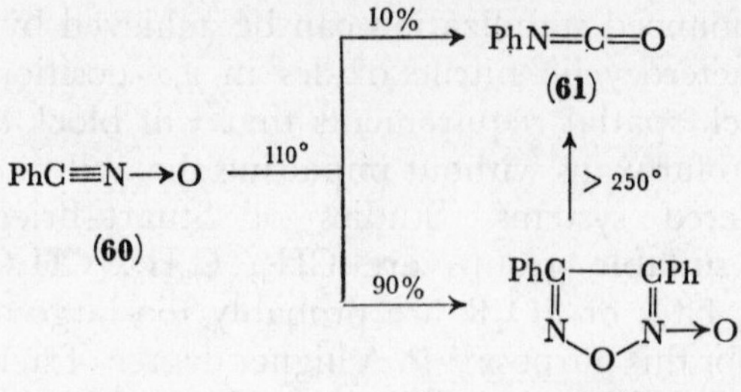

$$\text{PhC}{\equiv}\text{N}{\rightarrow}\text{O} \ (\mathbf{60}) \xrightarrow{110°} \begin{cases} \xrightarrow{10\%} \text{PhN}{=}\text{C}{=}\text{O} \ (\mathbf{61}) \\ \xrightarrow{90\%} \text{PhC}{-}\text{CPh (furoxan, } \mathbf{62}) \xrightarrow{>250°} \text{PhN}{=}\text{C}{=}\text{O} \ (\mathbf{61}) \end{cases} \tag{28}$$

140 °C, this reaction is complete in less than one hour. The isocyanates formed have mostly been identified as substituted diaryl ureas **63** following reaction with aniline. The mechanism of this reaction is unknown.

ArNHCONHPh
(**63**)

The polymeric nitrile oxides which are obtained from nitrolic acids by reaction with weak aqueous alkali (section IV.D.2) undergo the same thermolysis as the dimers, the furoxans. This rearrangement, however, seems not to proceed via initial depolymerization to the monomeric nitrile oxides. Isocyanates are obtained in moderate yields[62,136a,b]:

$$(\text{RCNO})_x \xrightarrow{\Delta} \longrightarrow \text{RN}{=}\text{C}{=}\text{O} \quad (\text{R} = \text{H, CH}_3\text{, Ph}) \tag{29}$$

The formation of isocyanates from alkyl and acyl halides with metal fulminates, presumably occurs via the initially formed nitrile oxides (section II.E).

C. Dimerization to Furoxans (1,2,5-Oxadiazole-2-Oxides)

This is the most frequently observed reaction of nitrile oxides. It occurs during the formation of nitrile oxides both in an acidic (from nitrolic acids) and in an alkaline (from hydroximic acid chlorides) environment and it is their normal reaction during storage under neutral conditions at room temperature. The dimerization rate is already very fast at 0 °C for the lower aliphatic nitrile oxides, the half-life of most aromatic nitrile oxides at room temperature being of the order of minutes to days. Bulky neighbouring substituents increase the stability, and the 2,2-dimethylpropionitrile oxide (*t*-butylfulmide) has a half-life of at least two orders of magnitude higher than the few known, not sterically hindered aliphatic nitrile oxides. Qualitative observations of the reaction rate are given in Table 1. Nitrile oxides with bulky enough substituents in the vicinity of the CNO group do not form furoxans at a rate comparable with that of the isomerization to the isocyanates. However, heating mesitonitrile oxide in a inert solvent for a day to 65–70°—conditions where the isomerization rate is still low—has produced a small yield of the extremely sterically hindered dimesitylfuroxan (**64**, R = 2,4,6-trimethylphenyl-), as shown by degradation and independent synthesis[137a].

The dimerization mechanism can be written as a 1,3-dipolar cycloaddition, where one molecule of the nitrile oxide is reacting in the mesomeric form (**5c**) and the other as the dipolarophile (**5b**) to give the furoxan **64**:

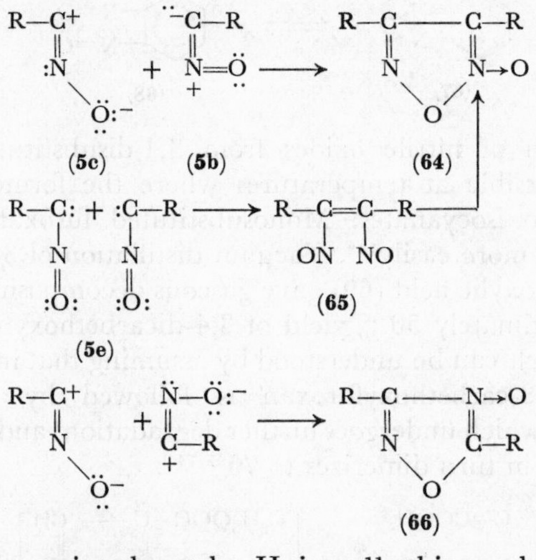

(5c) + (5b) ⟶ (64) (30)

(5e) + (5e) ⟶ R—C=C—R with ON NO (65) ⟶ (64) (31)

(5c) + (5c) ⟶ (66) (32)

However, as pointed out by Huisgen[11], this mechanism violates the principle of maximum gain in σ-bonding, invariably found valid in all of the many other types of 1,3-dipolar cycloadditions. To satisfy this principle, the dimerization would have to take the course of equation (32) leading to the isomeric 1,2,4-oxadiazole-4-oxides (**66**). Excluding structural elements common to **64** and **66** the value of $\Delta H°$ for **64** amounts to 151 kcal/mol (1 CC (81) + 2 NO (70)), while $\Delta H°$ for **66** is 186 kcal/mol (1 CN (68) + 1 NO (35) + 1 CO (83)). **66** are well known compounds[138], which have never been observed, even in traces, in the spontaneous dimerization of nitrile oxides. (For an apparent exception, see the formation of

3,5-diphenyl-1,2,4-oxadiazole 4-oxide (**66**, R = C_6H_5) from benzonitrile oxide (section IV.G.4)). It has, therefore, been suggested that the formation of the furoxan might occur in a two-step reaction via the dimerization of the mesomeric structure **5e** of the nitrile oxide with carbene character to a 1,2-dinitrosoethylene **65** (equation 31). **65** will then immediately stabilize itself by a simple regrouping of electrons as the furoxan **64**. Such a mechanism would naturally not interfere with the rule of maximal σ-bonding, which is only applicable to true 1,3-dipolar cycloadditions[137b,139]. This explanation is supported by recent n.m.r. studies of benzofuroxan (**67**) which indicate that it is in a mobile equilibrium with a modest amount of 1,2-dinitrosobenzene (**68**)[140–142]:

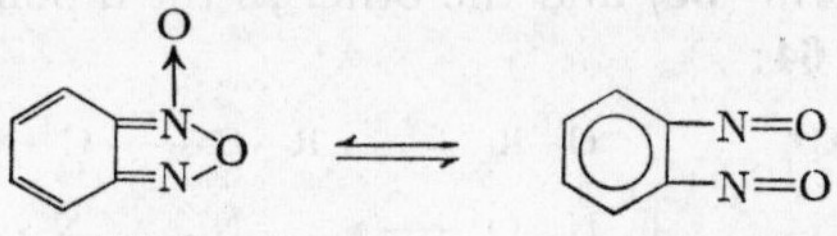

(67) ⇌ (68) [benzofuroxan ⇌ $C_6H_4(N{=}O)_2$]

A regeneration of nitrile oxides from 3,4-disubstituted furoxans seems only possible at temperatures where the former rearrange immediately to isocyanates. Monosubstituted furoxans, however, seem to cleave more easily[143]. Vacuum distillation of 3-carbethoxyfuroxan-4-carboxylic acid (**69**) gave gaseous decomposition products and an approximately 50% yield of 3,4-dicarbethoxyfuroxan (**70**), a reaction which can be understood by assuming that initial decarboxylation to 3-carbethoxyfuroxan is followed by cleavage to fulminic acid (which undergoes further degradation) and carbethoxyfulmide which in turn dimerizes to **70**[91,144]:

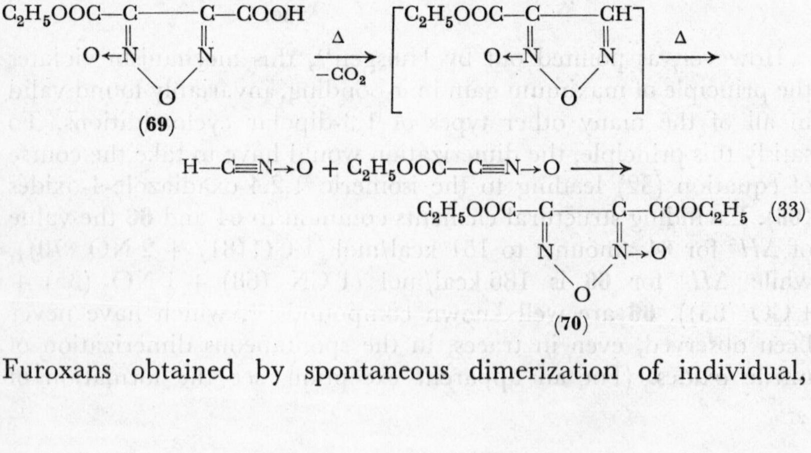

C_2H_5OOC—C——C—COOH (O←N, N, O ring) (**69**) $\xrightarrow[-CO_2]{\Delta}$ [C_2H_5OOC—C——CH (O←N, N, O ring)] $\xrightarrow{\Delta}$

$$H{-}C{\equiv}N{\rightarrow}O + C_2H_5OOC{-}C{\equiv}N{\rightarrow}O \longrightarrow \quad (33)$$

C_2H_5OOC—C——C—$COOC_2H_5$ (N, N→O, O ring) (**70**)

Furoxans obtained by spontaneous dimerization of individual, previously isolated nitrile oxides are listed in Table 1. A number of other routes by which furoxans have been prepared, may involve nitrile oxides as hypothetical transient intermediates[112,132a].

D. Polymerization

1. Fulminic acid

Ether solutions of fulminic acid (**9**) polymerize readily at room temperature. The major isolable compound is a trimer, metafulminuric acid, which has the structure of 4,5-dioximino-Δ^2-isoxazoline (**74**), while a minor portion consists of a tetramer, isocyanilic acid, which is furoxan dialdehyde dioxime (**73**)[2,20,66a,145]. The formation of the trimer and tetramer, however, is preceded by the formation of a very short-lived dimer whose existence was first recognized by kinetic measurements of the polymerization of **9**[17]. Much later, this intermediate was trapped when **9** was generated from its alkali salts at low temperature in the presence of acetylene or monosubstituted acetylenes. Besides the expected isoxazole **76**, a considerable amount of the corresponding isoxazole-3-aldoxime **75** was always obtained[146,147]. This indicates that the fulminic acid dimerizes at a rate comparable to that of the 1,3-dipolar cycloaddition and that the formed dimer reacts as oximinoacetonitrile oxide (**71**).

The kinetics of the subsequent further polymerization to **74** are, however, more compatible with a dimer of symmetrical structure. It is possible that **71** is in equilibrium with a modest amount of the tautomer **72**, dioximinoethylene. Metafulminuric acid (**74**) would result by a simple 1,3-dipolar cycloaddition of **9** to **72** while the familiar furoxan dimerization of **71** would yield isocyanilic acid (**73**) (equation 34) shown at top of next page.

Two more oligomers of **9** are known. Fulminuric acid, another trimer, is obtained by boiling mercuric fulminate with an aqueous solution of potassium chloride[148]. Its structure, nitrocyanoacetamide (**77**), was established long before the controversy over the parent compound could be settled[149–151], and did not help to solve the latter. The mechanism of the formation of **77** is still unknown, earlier explanations being unacceptable by the present knowledge of the chemistry of **9**[2]. It is not likely that either **73** or **74** are precursors of **77**. If nitrous acid is removed from methylnitrolic acid (**8**) in the presence of a mild base, e.g. sodium bicarbonate, fulminic

$$2\,HC{\equiv}N{\rightarrow}O \;(\mathbf{9}) \longrightarrow HON{=}CHC{\equiv}N{\rightarrow}O\;(\mathbf{71}) \rightleftharpoons HON{=}C{=}C{=}NOH\;(\mathbf{72})$$

$$(\mathbf{71}) \xrightarrow{\text{dimerization}} (\mathbf{73}); \quad (\mathbf{72}) \xrightarrow{+HCNO} (\mathbf{74}) \quad (34)$$

$$(\mathbf{71}) \xrightarrow{+RC\equiv CH} (\mathbf{75}); \quad (\mathbf{9}) \xrightarrow{+RC\equiv CH} (\mathbf{76})$$

acid is obtained not as the monomer (as under acidic conditions) but as a solid polymer. This highly explosive substance was originally thought to be a trimer ('trifulmin') having the structure **78**[62]. This

$NCCH(NO_2)CONH_2$ (**77**) (**78**)

structure is untenable in the light of our present knowledge of *s*-triazines, and the presumably correct structure (**79**, R = H) for trifulmin and higher homologues is discussed in the following section.

$$^{+}C{=}N{-}O{-}[{-}C{=}N{-}O{-}]_n{-}C{=}N{-}O^{-} \quad (\text{each C bearing } R)$$

(**79**)

$(R = H, CH_3, Ph, CO_2C_2H_5, CN)$

2. Other nitrile oxides

If nitrous acid is abstracted from nitrolic acids by means of mild aqueous alkalies, e.g. dilute sodium carbonate or ammonia, polymeric nitrile oxides which differ widely in their properties from the furoxans discussed previously are obtained[62,136a] (equation 35). A polymeric cyanofulmide which apparently belongs to the same class has been obtained by the spontaneous dehydrohalogenation of cyanoformhydroximic chloride in aqueous solution (equation 36)[70,130]:

$$\mathrm{RC}(=\mathrm{NOH})\mathrm{NO_2} \xrightarrow[-\mathrm{HNO_2}]{} [\mathrm{RCNO}]_n \qquad (35)$$

$$\mathrm{NCC}(=\mathrm{NOH})\mathrm{Cl} \xrightarrow[-\mathrm{HCl}]{} (\mathrm{NCCNO})_n \qquad (36)$$

It is unlikely that the formation of these polymers occurs via the monomeric nitrile oxides, since at least benzonitrile oxide is sufficiently stable to be detected under the conditions employed. The polymers are obtained as solids, some amorphous, some definitely crystalline and are characterized, contrary to the furoxans, by their high reactivity. The high sensitivity and the poor solubility of these products made a determination of their molecular weight impossible. The originally proposed structure of *s*-triazine tris(*N*-oxides) (**78**, R instead of H) had to be abandoned in view of our recent knowledge of the chemistry of *s*-triazines and the failure to obtain *s*-triazines from them by chemical reduction. In fact, the latter reaction yielded the corresponding nitriles exclusively. This and all other properties favour a dipolar chain structure **79**[152].

These polymers behave in many of their reactions like the parent monomeric nitrile oxides. Controlled thermal decomposition by heating in an inert solvent leads to the isomeric isocyanates, concentrated hydrochloric acid caused conversion to hydroximic acid chlorides or 1,2,4-oxadiazole-4-oxides (**66**) and amines formed the corresponding ureas via the (equation 37):

$$\mathbf{79} \begin{cases} \xrightarrow{\Delta} \mathrm{RN{=}C{=}O} \\ \xrightarrow{\mathrm{HCl}} \mathrm{R{-}C(=NOH)Cl} \\ \xrightarrow{\mathrm{H_2NR^1}} \mathrm{RNHCOHNR^1} \end{cases} \qquad (37)$$

A different type of polymeric nitrile oxide was obtained by the spontaneous or thermally induced polymerization of bifunctional nitrile oxides, e.g. cyanogen bis(*N*-oxide) (**45**), isophthalonitrile bis-(*N*-oxide) (**48**) or terephthalonitrile bis(N-oxide) (**47**). These polymers are undoubtedly polyfuroxans (equation 38)[28,29,54,55,57,78,115]:

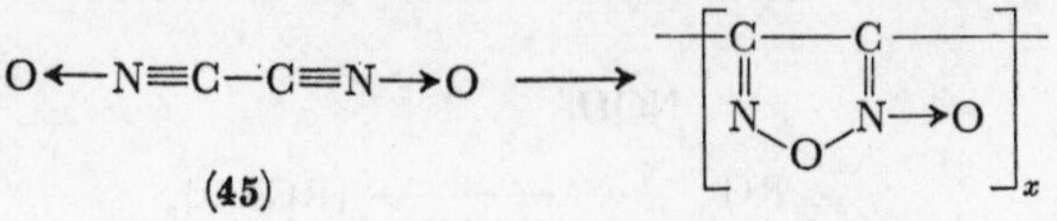

(**45**)

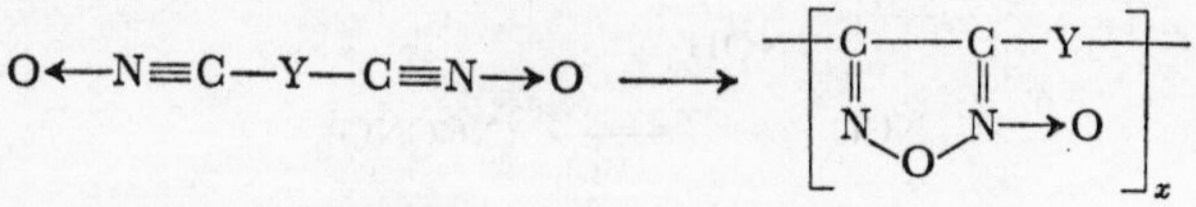

(38)

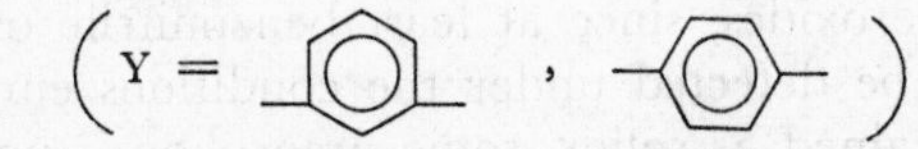

(**48**) (**47**)

E. Reduction to Nitriles

Benzonitrile oxide has been reduced to benzonitrile with acetic acid and zinc dust[8], or isocyanides[153,154], while triphenylacetonitrile oxide was converted to triphenylacetonitrile by means of tin and hydrochloric acid[47]. A more generally applicable method uses a trivalent phosphorus compound (equation 39)[131,155a]. The reaction

$$\text{R—C}\equiv\text{N}\rightarrow\text{O} + \text{PX}_3 \longrightarrow \text{R—C}\equiv\text{N} + \text{OPX}_3 \qquad (39)$$

(X = alkyl, aryl, alkoxyl)

is so specific that it reduces the very sensitive cyanogen bis(*N*-oxide) **45** quantitatively to cyanogen and can thus be employed for the determination of **45**[29,118]. While this reduction is of no preparative value, it can be sometimes helpful in establishing the structure of the nitrile oxides.

F. Addition Reactions Leading to Open-Chain Structures

I. Reactions with inorganic compounds

Nitrile oxides are inert towards chlorine, bromine and iodine[8,29,116]. An exception is fulminic acid which forms the dihalogenoformoxims, $Hal_2C{=}NOH$, as the primary products[155b].

a. Water (hydrolysis). The complete hydrolysis of nitrile oxides to hydroxylamine and carboxylic acids which is brought about by strong mineral acids or alkali has already been mentioned. Mesitonitrile oxide (**80**) was hydrolysed to the corresponding hydroxamic acid (**83**) by treatment with dilute sulphuric acid at 35 °C[116]. The reaction mechanism probably involves the protonation of **80** to the conjugate acid **81**, followed by addition of water to form **82**, which then deprotonates to **83**[91]:

$$\underset{(\mathbf{80})}{\mathrm{RC{\equiv}\overset{+}{N}{-}\overset{-}{O}}} \overset{H^+}{\rightleftharpoons} \underset{(\mathbf{81})}{\mathrm{RC{\equiv}\overset{+}{N}{-}OH}} \xrightarrow{H_2O} \underset{(\mathbf{82})}{\mathrm{RC(\overset{+}{O}H_2){=}NOH}} \rightleftharpoons \underset{(\mathbf{83})}{\mathrm{RC(OH){=}NOH}} + H^+ \qquad (40)$$

(R = 2,4,6-trimethylphenyl)

Terephthalonitrile bis(N-oxide) (**47**) was hydrolysed to the corresponding bis(hydroximic acid) (**84**) by anhydrous formic acid[56]. The formation of hydroxamic acids by the addition of hydroxide ion could not be achieved; the action of alkali on aromatic nitrile oxides under moderate conditions seems complicated and is not yet fully understood[98].

$$\mathrm{HON{=}C(OH){-}C_6H_4{-}C(OH){=}NOH}$$

(**84**)

Cyanogen bis(*N*-oxide) (**45**) reacts with NaOH to furnish yellow, very sensitive, water-soluble compounds; in absolute ethanolic solution an adduct with one mole each of NaOH and $NaOC_2H_5$ was obtained, to which the structure **85**, corresponding to a disodium salt of aci-nitroethylacethydroximic acid, has been tentatively ascribed[29]:

$$\mathrm{\overset{-}{O}{-}N{=}\overset{+}{C}{-}\overset{+}{C}{=}N{-}\overset{-}{O}} \longrightarrow \mathrm{\overset{-}{O}{-}N{=}\overset{+}{C}{-}\overset{-}{C}{=}\overset{+}{N}{=}O} \xrightarrow{+OH^-}$$

$$\mathrm{\overset{-}{O}{-}N{=}\overset{+}{C}{-}\overset{-}{C}{=}\overset{+}{N}(OH){-}\overset{-}{O}} \longrightarrow \mathrm{\overset{-}{O}{-}N{=}\overset{+}{C}{-}CH{=}N(\rightarrow O){-}\overset{-}{O}} \xrightarrow{EtO^-}$$

$$\left[\mathrm{\overset{-}{O}{-}N{=}C(OEt){-}CH{=}N(\rightarrow O){-}\overset{-}{O}}\right] 2\,Na^+ \qquad (41)$$

(**85**)

b. Hydrogen halides. Contrary to earlier statements[8], nitrile oxides easily add on hydrogen chloride, bromide or iodide to give hydroximic acid halides (equation 42). Aqueous hydrogen fluoride, however, does not react because of the less nucleophilic nature of the fluoride ion[24,28,29,34,39,47,50,87]. At higher temperatures, the reac-

$$RC{\equiv}N{\rightarrow}O + H{\cdot}Hal \rightleftharpoons \underset{\displaystyle Hal}{\underset{|}{RC}}{=}NOH \tag{42}$$

tion may be reversed, but the generated nitrile oxide will dimerize to the furoxan unless trapped by a suitable dipolarophile (sections II.F and IV.G).

c. Ammonia. In spite of the claim to the contrary by the original investigator[8], benzonitrile oxide reacts as easily as other nitrile oxides with ammonia to yield the corresponding amidoximes[29,45,49,116]:

$$PhC{\equiv}N{\rightarrow}O + NH_3 \longrightarrow \underset{\displaystyle NOH}{\underset{\|}{PhC}}{-}NH_2 \tag{43}$$

d. Azide ion. Aromatic nitrile oxides do not react easily with hydrazoic acid, but mesitonitrile oxide added azide ion to produce the azido oximate ion **86**, from which the free azido oxime could be isolated as a very unstable compound decomposing above room temperature into the nitrile, nitrogen and hyponitrous acid (equation 44)[116]. An analogous decomposition has been observed earlier with benzoximinoazide (**87**, Ar = Ph) prepared by a different route[156]. For a long time compounds of this type were considered to possess the isomeric cyclic 1-hydroxytetrazoles **88** structure, but recently the open-chain structure has been proven unequivocally[157]. The claim, however, that structures of type **87** are unusually stable azide derivatives cannot be generalized.

$$Ar\overset{+}{C}{=}N{-}O^- + N_3^- \longrightarrow \underset{\displaystyle N_3}{\underset{|}{ArC}}{=}N{-}O^- \xrightarrow{H^+} \underset{\displaystyle N_3}{\underset{|}{ArC}}{=}NOH$$

(**86**) (**87**)

$$\xrightarrow{\sim 50^\circ} ArCN + N_2 + (HON)_2 \qquad \text{(87)} \not\rightarrow \text{Ar—C(=N—N=N—)N—OH (1-hydroxytetrazole ring)} \tag{44}$$

(**88**)

(Ar = 2,4,6-trimethylphenyl)

e. Sulphide ion. The reaction of nitrile oxides with sulphide ion is dependent on the pH. At pH > 8, where the sulphide ion S^{-2} is the predominant species, the thiohydroximate ions (**89**) (from which the free acids (**90**) can be isolated) are formed in excellent yields. At pH < 8, **89** adds a second mole of nitrile oxide to form the diaroyl oximinosulphides **91** (equation 45)[116]. The symmetrical

$$\mathrm{Ar\overset{+}{C}{=}N{-}\overset{-}{O}} + S^{2-} \longrightarrow \underset{(\mathbf{89})}{\mathrm{ArC(S^{2-}){=}N{-}O^-}} \xrightarrow{H^+} \underset{(\mathbf{90})}{\mathrm{ArC(=S)NHOH}} \qquad (45)$$

$$(\mathbf{89}) \xrightarrow{+\mathrm{Ar\overset{+}{C}{=}N{-}\overset{-}{O}};\ +H^+} \underset{(\mathbf{91})}{\mathrm{ArC(=NOH){-}S{-}C(=NOH)Ar}}$$

structure of the diadduct was supported by the n.m.r. spectrum of **91** (Ar = 2,4,6-trimethylphenyl) and its diacetyl derivative, and, indirectly, by the synthesis of a diaroyl oximinosulphide in which the two Ar groups were different via two synthetic routes leading to the same product (equation 46).

$$2,6\text{-}Cl_2C_6H_3\text{-}C(=S)\text{-}NHOH + O{\leftarrow}N{\equiv}C\text{-}C_6H_2(CH_3)_3\text{-}2,4,6 \xrightarrow{Na_2S}$$

$$2,6\text{-}Cl_2C_6H_3\text{-}C(NOH){-}S{-}(HON)C\text{-}C_6H_2(CH_3)_3\text{-}2,4,6 \qquad (46)$$

$$2,6\text{-}Cl_2C_6H_3\text{-}C{\equiv}N{\rightarrow}O + HOHN{-}C(=S)\text{-}C_6H_2(CH_3)_3\text{-}2,4,6 \xrightarrow{Na_2S}$$

The thermal decomposition of **91** (Ar = 2,4,6-trimethylphenyl) at 200 °C gave an almost quantitative yield of 2,4,6-trimethylphenyl isothiocyanate and 42% of 1,3-bis(2′,4′,6′-trimethylphenyl)urea.

91 is an anhydride of a thiohydroximic acid. It is known that mixed anhydrides of hydroximic acids and carboxylic acids rearrange easily by acid migration to the mixed anhydrides of the hydroxamic acid and the carboxylic acid, e.g. PhC(NOH)OCOPh → PhCONHOCOPh [5]. An analogous rearrangement of **91**, followed by α-elimination of the thiohydroximyl group of **91a** leaves a nitrene intermediate which stabilizes as the isocyanate. The eliminated thiohydroxamic acid can simultaneously undergo an analogous Lossen rearrangement to the isothiocyanate while the water lost in this step converts the isocyanate in the known manner into the urea:

$$\underset{(\mathbf{91})}{\text{RC(NOH)—S—(HON)CR}} \longrightarrow \underset{(\mathbf{91a})}{\text{RC(NOH)—S—NHCOR}} \longrightarrow$$

$$\text{RC(NOH)SH} + [\ddot{\text{N}}\text{COR}] \longrightarrow \text{O=C=NR} \quad (47)$$

$$\text{RC(NOH)SH} \downarrow \text{RCSNHOH} \xrightarrow{-H_2O} [\text{RCS}\ddot{\text{N}}] \downarrow \text{RN=C=S}$$

$$\text{O=C=NR} \xrightarrow{H_2O} \text{RNHCONHR}$$

(R = 2,4,6-trimethylphenyl)

Cyanogen bis(*N*-oxide) (**45**) yielded both the oxalobis(thiohydroxamic) acid (**93**) and—by further addition of a second mole of **45**—the cyclic sulphide, tetraoximino-1,4-dithian (**92**)[29]:

$$\underset{(\mathbf{45})}{\text{O}\leftarrow\text{N}\equiv\text{C—C}\equiv\text{N}\rightarrow\text{O}} + 2\,\text{S}^{-2} \longrightarrow {}^{-}\text{O—N=}\underset{\text{S}^-}{\text{C}}\text{—}\underset{\text{S}^-}{\text{C}}\text{=N—O}^- \quad (48)$$

$$\xrightarrow{+\,\mathbf{45}} \underset{(\mathbf{92})}{\text{(HON=)}_2\text{C}_2\text{S}_2\text{C}_2\text{(=NOH)}_2} \qquad \xrightarrow{H^+} \underset{(\mathbf{93})}{\text{HOHN—C(=S)—C(=S)—NHOH}}$$

f. Thiocyanate ion. By analogy, the reaction of a nitrile oxide with thiocyanate ion would be expected to yield the thiocyanate oxime, RC(=NOH)SCN; however, in boiling methanol an almost quantitative yield of the isothiocyanate **96** was obtained together with an equivalent amount of cyanate ion, while the same reaction carried out at room temperature over a longer period of time yielded the methyl thiocarbamate **98**. These results were explained by assuming an initial nucleophilic attack on the nitrile oxide by the thiocyanate ion to give **94** followed by an electron shift to a 1,3,4-thiaoxazoline ion **95**. **95** then decomposes into cyanate ion and the nitrene intermediate **97** which stabilizes by rearrangement to the isothiocyanate **96**. Once formed, the isothiocyanates do not react with methanol under the prevailing conditions, and therefore it was assumed that the addition of the solvent which finally leads to **98** must occur during the decomposition or rearrangement of the intermediates **95** or **97** (equation 49). At 25 °C, the life of these intermediates is presumably long enough to allow this reaction to proceed to a major extent, while at 64 °C the decomposition is so fast that the isothiocyanate is the main product[116]. The proposed mechanism is consistent with the known reaction of aromatic nitrile oxides with thiocarbonyl compounds[30] (section IV.G.3).

$$R\overset{+}{C}{=}N{-}O^- + NCS^- \longrightarrow \left[RC(=N{-}O^-){-}S{-}C{\equiv}N \right] \textbf{(94)}$$

$$\longrightarrow \left[\text{R{-}C, S, N, C=}\ddot{N}^-\text{, O ring} \right] \textbf{(95)} \quad (49)$$

$$\xrightarrow{-OCN^-} \left[RC{=}S,\ :N: \right] \textbf{(97)}$$

$$RN{=}C{=}S\ \textbf{(96)} \longleftarrow \textbf{(97)} \xrightarrow{MeOH} RNHC(=S)OMe\ \textbf{(98)}$$

(R = 2,4,6-trimethylphenyl, 2,3,5,6-tetramethylphenyl)

g. Cyanide ion. Whereas hydrocyanic acid itself does not react with nitrile oxides[11,29,158], cyanide ion adds quickly to form the cyanooximate ion, from which the free α-oximinonitriles can be liberated[29,116,159]:

$$R\overset{+}{C}{=}N{-}O^- + CN^- \longrightarrow RC(CN){=}N{-}O^- \xrightarrow{H^+} RC({=}NOH)CN \qquad (50)$$

2. Reactions with organic compounds

a. Alcohols. Aromatic nitrile oxides are inert to aliphatic alcohols unless the addition reaction is promoted by the presence of the alkoxide ion (equation 51). The same reaction can be induced with dilute sulphuric acid as a catalyst (equation 52). Phenols, being more acidic, add without the aid of a catalyst[160]. The addition products are the alkylhydroximic acids (**99**), a class of compounds hitherto somewhat inaccessible[116]:

$$\mathrm{Ar\overset{+}{C}{=}N{-}O^-} \xrightarrow{\mathrm{RO^-}} \mathrm{ArC(OR){=}N{-}O^-} \xrightarrow{+\mathrm{H^+}} \mathrm{ArC(OR){=}NOH}\ (\mathbf{99}) \qquad (51)$$

$$\mathrm{ArC{\equiv}\overset{+}{N}{-}O^-} \underset{}{\overset{\mathrm{H^+}}{\rightleftharpoons}} \mathrm{ArC{\equiv}\overset{+}{N}OH} \xrightarrow{\mathrm{ROH}} \mathrm{ArC(\overset{+}{O}RH){=}NOH} \xrightarrow{-\mathrm{H^+}} \mathrm{ArC(OR){=}NOH} \qquad (52)$$

b. Mercaptans. As expected from their higher nucleophilicities, mercaptans add easily to aliphatic and aromatic nitrile oxides to give the alkylthiohydroximic acids (**100**) without the aid of a catalyst, although the presence of triethylamine is considered helpful[24,116]. While mechanisms have been postulated which would lead as well to the *cis* as to the *trans* forms of **100**, the *cis* structure according to equation (53) is supported by the use of this route for the synthesis of naturally occurring mustard oil glucosides, which are alkylthiohydroximic acid derivatives and have been shown by x-ray crystallography to possess the *cis* configuration[161–163].

$$\mathrm{RC{\equiv}\overset{+}{N}{-}O^-} + \mathrm{R^1{-}\ddot{S}{-}H} \longrightarrow \left[\mathrm{RC(\overset{+}{S}R^1H){=}N{-}O^-}\right] \longrightarrow \mathrm{(R)(R^1S)C{=}N{-}OH}\ (\mathbf{100}) \qquad (53)$$

c. Organic acids. Terephthalonitrile bis(*N*-oxide) reacted with formic acid without acylation to furnish the corresponding bis-(hydroxamic)acid (section IV.F), although sterically hindered stable aromatic nitrile oxides could be recrystallized unchanged from hot acetic acid[91]. In the presence of catalytic amounts of concentrated sulphuric acid, however, mesitonitrile oxide was converted to acetyl mesitohydroximic acid (**101**), presumably by an analogous mechanism to the acid-catalysed addition of alcohols discussed above. Structure **101** is supported by the intensely red ferric chloride reaction, indicative of an acylhydroximic acid with a free >NOH group[5]. With acetic anhydride, diacylation takes place already at room temperature to form the diacetylmesitohydroximic acid (**103**); in the presence of concentrated sulphuric acid, however, Lossen rearrangement occurred and the only identifiable product was *N*-acetylmesidine (**102**)[116]:

$$\mathrm{ArC{\equiv}N{\rightarrow}O \xrightarrow{AcOH,\ H_2SO_4} ArC(OOCCH_3){=}NOH\ (\mathbf{101})} \tag{54}$$

$$\mathrm{ArC{\equiv}N{\rightarrow}O \xrightarrow{Ac_2O} ArC(OOCCH_3){=}NOOCCH_3\ (\mathbf{103})}$$

$$\mathrm{ArC{\equiv}N{\rightarrow}O \xrightarrow{Ac_2O,\ H_2SO_4} ArNHCOCH_3\ (\mathbf{102})}$$

(Ar = 2,4,6-trimethylphenyl)

Aromatic nitrile oxides reacted with aliphatic or aromatic sulphonic and carboxylic acid chlorides in the presence of triethylamine to give arylhydroximoyl chloride esters[164]:

$$\mathrm{ArC{\equiv}N{\rightarrow}O + ClSO_2R^1 \xrightarrow{NEt_3} ArC(Cl){=}NOSO_2R^1}$$
$$\mathrm{ArC{\equiv}N{\rightarrow}O + ClOCR^2 \xrightarrow{NEt_3} ArC(Cl){=}NOCOR^2} \tag{55}$$

d. Amines and hydrazines. The reaction of nitrile oxides with amines which, according to equation (56), leads to substituted amidoximes, was first discovered by Ponzio with phenyloximinoacetonitrile oxide (**33**)[49], thus correcting contrary statements of the older literature[8]. The reaction is a general one for all types of nitrile

$$\mathrm{R^1C{\equiv}N{\rightarrow}O + HNR^2R^3 \longrightarrow R^1C(NR^2R^3){=}NOH} \tag{56}$$

oxides and for aliphatic and aromatic primary and secondary amines.

Bifunctional nitrile oxides react with bifunctional amines to yield polyamidoximes of varying degrees of polymerization; these, e.g. the polymer **104** from cyanogen bis(*N*-oxide) (**45**) and *p*-phenylene diamine[91], possess an interesting chelating affinity for various transition metals. **45** underwent ring closure with 1,2-diamines to yield 1,4-diazines; with *o*-phenylenediamine, 2,3-dioximino-1,2,3,4-tetrahydroquinoxaline (**105**), and with ethylenediamine, 2,3-dioximinopiperazine (**106**) were obtained[29].

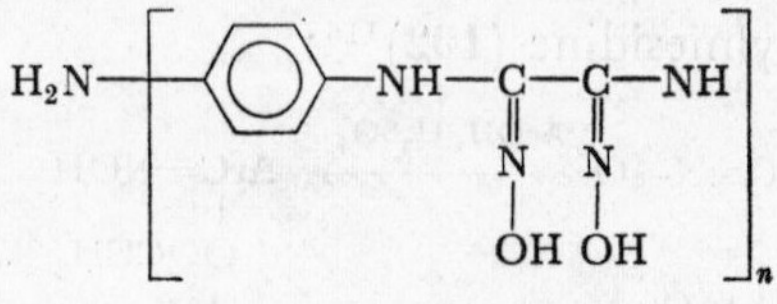

$$H_2N-\left[-C_6H_4-NH-\underset{\underset{OH}{|}}{\overset{}{\underset{N}{\overset{}{C}}}}-\underset{\underset{OH}{|}}{\underset{N}{C}}-NH-\right]_n-C_6H_4-NH_2$$

(**104**)

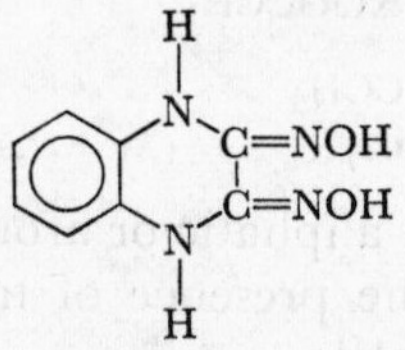

(**105**) 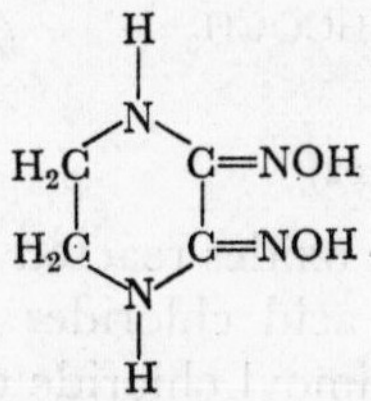 (**106**)

Tertiary amines add to nitrile oxides to furnish usually very labile compounds which are presumably inner salts of trisubstituted amidoximes **107**; these can be stabilized in some cases as halogenides of quaternary amidinium oximes (**108**) by the addition of hydrogen chloride[29,98]. The same compounds were occasionally obtained by the addition of a tertiary amine to a hydroximic acid chloride (equation 57)[76,77].

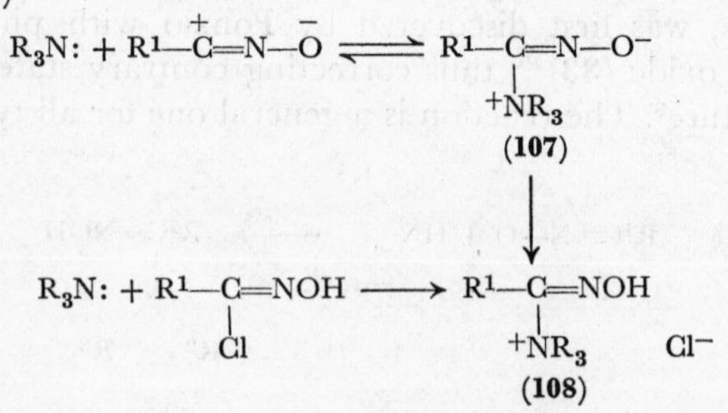

$$R_3N: + R^1-\overset{+}{C}{=}N-\overset{-}{O} \rightleftharpoons \underset{\underset{^+NR_3}{|}}{R^1-C{=}N-O^-} \quad (\mathbf{107})$$

$$\downarrow \qquad (57)$$

$$R_3N: + \underset{\underset{Cl}{|}}{R^1-C{=}NOH} \longrightarrow \underset{\underset{^+NR_3}{|}}{R^1-C{=}NOH}\quad Cl^- \quad (\mathbf{108})$$

As a powerful nucleophile, phenylhydrazine reacts readily with nitrile oxides, but the reaction course is complicated by a concomitant side reaction leading to the destruction of the hydrazine and formation of nitriles (which may react further with phenylhydrazine). Furthermore, the expected primary addition products, the hydrazido oximes (R—C(=NOH)NHNHPh), which have been synthesized by other routes, are particularly unstable compounds[156]. Only at −40 to −20 °C, cyanogen bis(*N*-oxide) (**45**) and phenylhydrazine yielded the expected adduct, oxalyl-bis(phenylhydrazide) dioxime (**109**), together with the product of its spontaneous dehydration, 3,4-diphenylhydrazino-1,2,5-oxadiazole (**111**). Even under

$$\mathbf{45} \xrightarrow{PhNHNH_2} PhNHNHC(=NOH)—C(=NOH)NHNHPh\ (\mathbf{109}) \xrightarrow{-H_2O} \text{3,4-(PhNHNH)}_2\text{-1,2,5-oxadiazole}\ (\mathbf{111})$$

$$\mathbf{45} \xrightarrow{PhNHNH_2} N{\equiv}C—C{\equiv}N \xrightarrow{PhNHNH_2} PhNHNHC(=NH)—C(=NH)NHNHPh\ (\mathbf{110}) \qquad (58)$$

these conditions, appreciable amounts of oxalyl bis(phenylhydrazide) diimide (**110**), resulting from the known reaction of cyanogen with phenylhydrazine were formed. At 0 °C, only **110** could be isolated (equation 58)[29]. Benzonitrile oxide yielded a small amount of benzohydroximyl phenyldiimide (phenylazobenzaldoxime) (**113**) after reaction with phenylhydrazine and subsequent oxidation with ferric chloride. This result is undoubtedly due to the initial formation of the hydrazide oxime intermediate (**112**)[116] (equation 59).

$$PhC{\equiv}N{\rightarrow}O \xrightarrow{PhNHNH_2} [PhC(=NOH)NHNHPh]\ (\mathbf{112}) \xrightarrow{-H_2} PhC(=NOH)N{=}NPh\ (\mathbf{113}) \qquad (59)$$

e. Grignard compounds. Benzonitrile oxide and methyl magnesium iodide added to form the iodomagnesium salt of acetophenone oxime, from which the oxime was liberated with acids (equation 60)[8]. Likewise, triphenylacetonitrile oxide and phenyl magnesium bromide yielded benzpinacolon oxime (**114**)[47].

$$\mathrm{Ph\overset{+}{C}{=}N{-}\overset{-}{O} + CH_3MgI \longrightarrow PhC(CH_3){=}N{-}O{-}MgI \xrightarrow{H^+} PhC(CH_3){=}NOH} \qquad (60)$$

$$\mathrm{Ph_3CC(Ph){=}NOH}$$

(114)

G. Addition Reactions Leading to Cyclic Structures (1,3-Dipolar Cycloadditions)

With the sole exception of the furoxan formation discussed in section IV.C, the reaction between a nitrile oxide and an unsaturated compound largely follows the scheme of equation (61), whereby the unsaturated dipolarophile always attacks with the more negatively charged end at the carbon atom of the nitrile oxide (the latter reacting in the 1,3-dipolar mesomeric structure **5c**). Reactions of this type are found already in the earlier literature, but without understanding of their mechanism[165]. Quilico first recognized the nitrile oxide as an intermediate[31,166], and Leandri postulated such a reaction as a 1,3-dipolar cycloaddition[167a], but not before Huisgen's extensive studies was the reaction mechanism fully understood and its wide applicability realized[10,11,139].

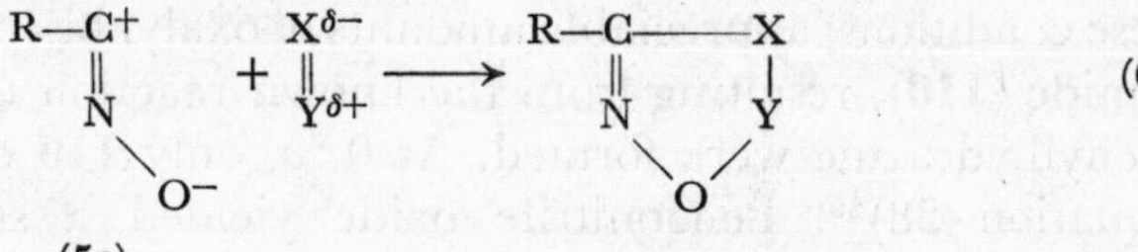

$$\mathrm{R{-}C^{+}{=}N{-}O^{-}}\ (\mathbf{5c}) + \mathrm{X^{\delta-}{=}Y^{\delta+}} \longrightarrow \mathrm{R{-}C{=}N{-}O{-}Y{-}X}\ (\text{ring}) \qquad (61)$$

Recently, a careful investigation of some 1,3-dipolar cycloaddition of nitrile oxides has demonstrated that the above scheme does not represent the reaction completely; to a minor extent (generally <1 to 10%), the inverse addition is also observed[18]. For example, in the addition of benzonitrile oxide and methyl acrylate, besides the 'normal' product (**115**), a minor amount of the 'wrong' adduct (**116**) was also obtained. Methyl β,β-dimethylacrylate and aliphatic as well as aromatic nitrile oxides yielded exclusively the 'wrong' adducts of type **116**[167b]. This serves to emphasize that no single charge distribution formula (**5a** to **5c**) should be considered to be solely responsible for the 1,3-dipolar cycloadditions and that steric factors are far more important in determining the reactivity of the dipolarophile than electronic influences.

$$\underset{\textbf{(115)}}{\underset{96\%}{\begin{array}{l}\text{Ph—C——CH}_2\\ \quad\ \|\qquad\ \ |\\ \quad\ \text{N}\quad\ \text{CH—COOCH}_3\\ \qquad\backslash\ \ /\\ \qquad\ \ \text{O}\end{array}}}\qquad\underset{\textbf{(116)}}{\underset{3.6\%}{\begin{array}{l}\text{Ph—C——CH—COOCH}_3\\ \quad\ \|\qquad\ \ |\\ \quad\ \text{N}\quad\ \text{CH}\\ \qquad\backslash\ \ /\\ \qquad\ \ \text{O}\end{array}}}$$

In cases where X and Y are elements other than carbon, the principle of maximum gain of σ-bond energy is also observed. Since the reaction is strictly stereospecific with stereoisomeric unsaturated compounds, e.g. maleic and fumaric or mesaconic and citraconic esters[139,168,169], it is assumed to occur as a true *cis* addition via a four-centre intermediate, where at no time the double-bond character of the dipolarophile is lost. As to be expected for this type of reactions which requires a high degree of order, large negative entropies of activation (∼ −22 to −40 cal/deg) are observed. Conjugation of the dipolarophilic double bond with a C=C or C=X (X = N, O, S) double bond enhances the reactivity to a marked degree. These general conclusions have been supported by kinetic measurements[139]. It has also been claimed that a carbon–carbon double bond with a length of more than 1·35 Å, corresponding to less than 80% double-bond character, will not undergo 1,3-dipolar cycloadditions with nitrile oxides[170–172].

I. Reactions with olefins

Olefins of all types react with nitrile oxides to furnish 4,5-dihydro-1,2-oxazoles (Δ^2-isoxazolines) (**117**):

$$\begin{array}{l}\text{R—C}\\ \quad\ \ |||\\ \quad\ \ \text{N}\\ \quad\ \ \downarrow\\ \quad\ \ \text{O}\end{array} + \begin{array}{l}\text{CR}^1\text{R}^2\\ \|\\ \text{CR}^3\text{R}^4\end{array} \longrightarrow \underset{\textbf{(117)}}{\begin{array}{l}\text{R—C——CR}^1\text{R}^2\\ \quad\ \|\qquad\ \ |\\ \quad\ \text{N}\quad\ \text{CR}^3\text{R}^4\\ \qquad\backslash\ \ /\\ \qquad\ \ \text{O}\end{array}} \tag{62}$$

Monosubstituted ethylenes react faster than disubstituted ones, tri- and tetrasubstituted ethylenes react so sluggishly that even the technique of generating slowly the nitrile oxide *in situ* does not lead to the formation of appreciable amounts of **117**[11]. With the stable mesitonitrile oxide, however, the addition of tetramethylethylene can be enforced; after refluxing for several days, an 18% yield of 4,4,5,5-tetramethyl-3-(2,4,6-trimethylphenyl)-Δ^2-isoxazoline (**117**, R = 2,4,6-trimethylphenyl, $R^1 = R^2 = R^3 = R^4 = CH_3$) has been obtained[25,137a]. Cyclohexene is an example of an especially unreactive olefin, while cyclopentene and even the unconjugated 1,4-cyclohexadiene react much faster under comparable conditions[29,137b]. Using the *in situ* technique, however, the isoxazolines **118** and **119** can be prepared from cyclohexene and the corresponding nitrile oxide[173a,b]. Exocyclic double bonds and double bonds in quinones add easily[174,175], while those which are part of true benzenoid or heterocyclic rings do not react with nitrile oxides under usual conditions[176]; here also, however, the reaction can be enforced with furans, pyrrole, indole and thiophene by generating the nitrile oxide *in situ*[177,178].

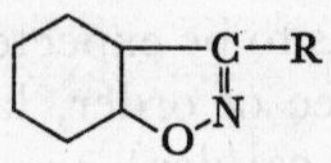

(**118**, R = C_6H_5)

(**119**, R = $C(CH_3)_3$)

Polyolefins will react, depending on the conditions with one or all double bonds[24,29,179], e.g. **120** or **121** were obtained from 1,3-butadiene. Difunctional nitrile oxides, such as **45** or **47**, yield with difunctional olefins 1:1 alternating copolymers containing Δ^2-isoxazoline rings[55,56,180]. The spirane **122** was obtained from allene and benzonitrile oxide[171,181]; other spiro-systems have been prepared from alkylidene and arylidene cycloalkenes[182].

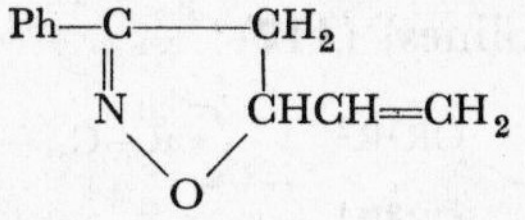

(**120**)

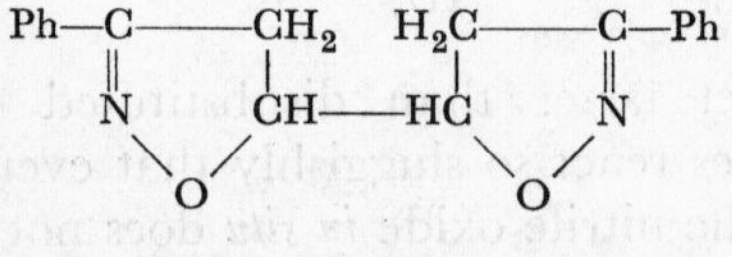

(**121**) 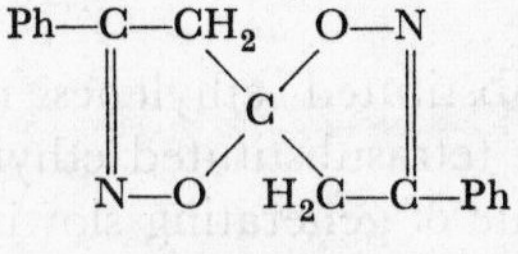 (**122**)

Vinyl ethers and esters react normally, the initially formed 5-alkoxy or 5-acyl-Δ^2-isoxazolines being easily aromatized to isoxazoles either thermally or by acids with the loss of alcohol or acid (e.g. equation 63)[22,90,183,184]:

```
PhC——CH2          H+      PhC——CH
 ‖    |         ——→        ‖    ‖     + ROH        (63)
 N    CH—OR                N    CH
  \  /                      \  /
   O                         O
```

(R = alkyl or acyl)

β-Chlorovinyl aryl ketones form 5-aroyl-4-chloro-4,5-dihydro-1,2-oxazoles which are dehydrohalogenated by alkali to 5-aroylisoxazoles[185,186] (e.g. equation 64):

```
PhC+        CHCl           PhC——CHCl        –HCl
 ‖      +   ‖       ——→     ‖    |         ——→
 N          CH—COPh         N    CH—COPh
  \                          \  /
   O–                         O

                                    PhC——CH
                                     ‖    ‖            (64)
                                     N    C—COPh
                                      \  /
                                       O
```

Even enolates react in the same manner[31,187]; however, dehydration occurs here spontaneously and the corresponding isoxazoles are the only products which can be isolated (equation 65):

```
RC+      HCR1          RC——CHR1        H+      RC——CR1
 ‖    +   ‖     ——→     ‖    |        ——→       ‖    ‖
 N        C—O–          N    C—O–     –H2O      N    CR2
  \       |              \  / |                  \  /          (65)
   O–     R2              O   R2                  O
```

(R = Ph; R^1 = CH_3CO, $CO_2C_2H_5$; R^2 = CH_3, Ph, $CO_2C_2H_5$)

An analogous reaction has been observed with carbanions stabilized by nitrile groups[31], e.g. ethyl cyanoacetate. The nitrile oxide adds at the C=C double bond, giving rise to 5-aminoisoxazoles (equation 66):

```
                       +
                       C—Ph              C2H5OOCC——CPh
C2H5OOCCH–             ‖       +H+               ‖    ‖
     |           +     N      ——→            H2NC    N       (66)
    N≡C               /                          \  /
                    –O                            O
```

Enamines react also in a similar fashion[167,188–191a,b]. Diphenylketene adds to aliphatic nitrile oxides with the formation of 3-alkyl-4,4-diphenylisoxazolin-5-ones (equation 67)[24,192], whereas ketene acetals and dithioacetals yield 5-alkoxy- or 5-alkylmercaptoisoxazoles via the intermediate **123** (equation 68)[193,194].

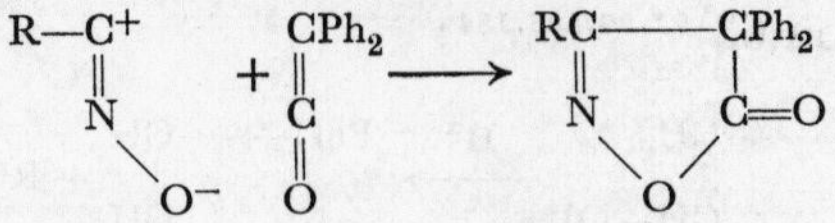

$$\text{R—C}^{+}{=}\text{N—O}^{-} + \text{Ph}_2\text{C}{=}\text{C}{=}\text{O} \longrightarrow \text{RC}{=}\text{N—O—C(=O)—CPh}_2 \text{ (ring)} \qquad (67)$$

$$\text{Ph—}\overset{+}{\text{C}}{=}\text{N—O}^{-} + \text{CH}_2{=}\text{C(XC}_2\text{H}_5)_2 \longrightarrow \left[\text{PhC}{=}\text{N—O—C(XC}_2\text{H}_5)_2\text{—CH}_2 \text{ (ring)}\right] \xrightarrow{-\text{HXC}_2\text{H}_5} \text{Ph—C}{=}\text{N—O—C(XC}_2\text{H}_5){=}\text{CH (ring)} \qquad (68)$$

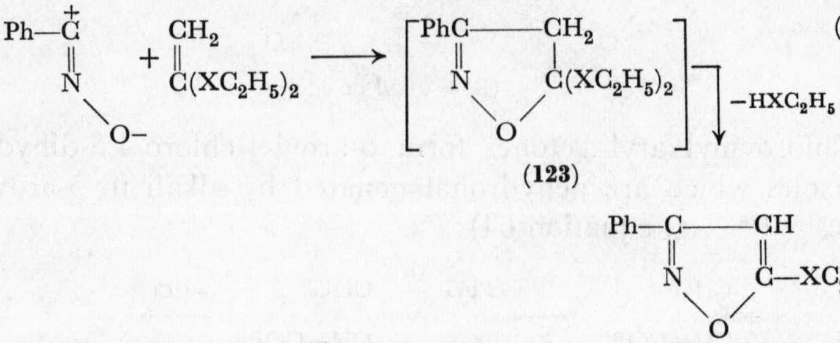

(123)

(X = O or S)

Apart from the exceptions mentioned above, the reaction of olefins with nitrile oxides is of such universal nature that benzonitrile oxide has been suggested as a suitable reagent for the identification of olefins[126].

2. Reactions with acetylenes

The addition of acetylenic compounds to nitrile oxides leads directly to isoxazoles (equation 69) and is one of the more important synthetic routes to this heterocycle[111,113,114]:

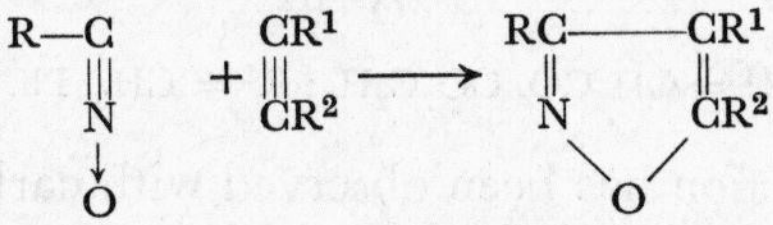

$$\text{R—C}{\equiv}\text{N}{\rightarrow}\text{O} + \text{CR}^1{\equiv}\text{CR}^2 \longrightarrow \text{RC}{=}\text{N—O—CR}^2{=}\text{CR}^1 \text{ (ring)} \qquad (69)$$

Acetylenes generally add more slowly than the corresponding olefins. Comparison of the reactivity of benzonitrile oxide with ethyl acrylate and ethyl propiolate and with styrene and phenylacetylene showed that the nitrile oxide reacted 5·7 and 9·2 times faster, respectively, with the ethylene derivative[139]. In the addition of vinylacetylene to benzonitrile oxide, the addition to the double bond is preferred, but the enynes $HOCH_2CH{=}CHC{\equiv}CH$ and $CH_3OOCCH{=}CHC{\equiv}CH$ are reported to add preferably at the triple bond[195], which seems to indicate that steric effects play a large role[196], an opinion further supported by the low reactivity of butynediol and acetylenedicarboxylic acid toward benzonitrile oxide[31,197].

Alkoxyacetylenes or dialkylaminoacetylenes react normally to produce the 5-alkoxy- and 5-dialkylaminoisoxazoles, respectively. For example, **124** and **125** have been obtained from methoxyacetylene and benzonitrile oxide, and from phenyldiethylaminoacetylene and terephthalonitrile bis(*N*-oxide)[198–200], respectively.

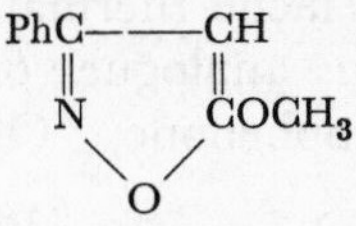

PhC——CH, N, $COCH_3$, O

(124)

PhC——C, $(C_2H_5)_2N—C$, N, O, C——CPh, N, $C—N(C_2H_5)_2$, O

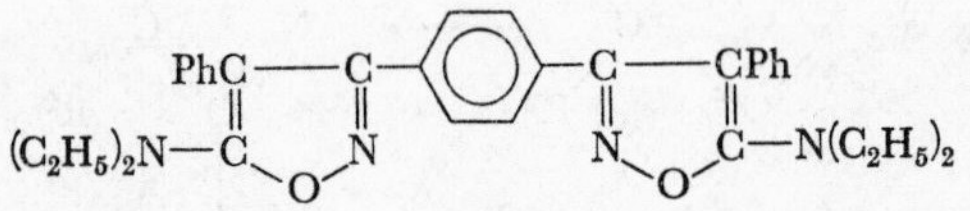

(125)

3. Reactions with carbonyl or thiocarbonyl compounds

Aromatic nitrile oxides add to the C=O double bond of aldehydes and ketones (but not of carboxylic acids, esters or amides), provided it is activated by an adjacent electron-withdrawing group, to form 1,4,2-dioxazoles (**126**) (equation 70), a class of heterocycles otherwise almost inaccessible[11,33]. As cyclic acetals, these compounds are readily hydrolysed by mineral acids to the original carbonyl compound and the corresponding hydroxamic acid. Suitable addends

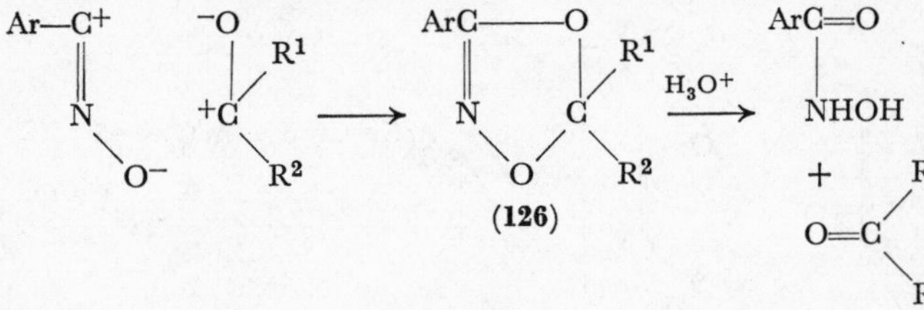

$$\text{Ar—C}^+{=}\text{N—O}^- + {}^-\text{O—}{}^+\text{C}(R^1)(R^2) \longrightarrow \text{ArC—O—C}(R^1)(R^2)\text{—O—N}{=} \ \textbf{(126)} \xrightarrow{H_3O^+} \text{ArC=O(NHOH)} + \text{O=C}(R^1)(R^2) \quad (70)$$

are, for example, chloral, ethyl pyruvate, 2,3-butanedione, diethyl mesoxalate, benzaldehyde, furfural and pyridine-2-aldehyde. Unactivated aliphatic aldehydes and ketones will react, however, in the presence of boron trifluoride etherate as a catalyst[201]. Enolizable 1,3-dicarbonyl compounds and quinones add to the ethylenic bond and not to C=O (section IV.F.1). However, *o*-quinones, which do not contain ethylenic bonds like phenanthrene-9,10-quinone or chrysene-5,6-quinone, reacted normally, but only at one of the C=O bonds[202].

The C=S double bond reacts with aromatic nitrile oxides not only in thioketones, but also if it is part of a thionocarboxylic acid ester or amide, with the formation of 1,4,2-oxathiazoles (equation 71). In this case, too, a similar activation of the thiocarbonyl group by electron-withdrawing groups is necessary. The 1,4,2-oxathiazoles are characterized by their facile thermal cleavage into the isothiocyanate and the oxygen analogue of the thiono compound employed[30,203]. Thiobenzophenone, O,O-diphenylthiocarbonate,

Ar—C⁺=N—O⁻ + ⁻S—C⁺(R¹)(R²) ⟶ ArC=N—O—C(R¹)(R²)—S (ring) —Δ→ [ArC≡S / :N:] + O=C(R¹)(R²); [ArC≡S / :N:] ⟶ ArN=C=S (71)

methyl dithio-α-naphthoate, diphenyl trithiocarbonate and thiobenzoic acid dimethylamide, have been employed in this reaction.

Carbon disulphide reacts with mesitonitrile oxide at both C—S bonds; the presumably formed intermediate (**127**) decomposes spontaneously into mesityl isothiocyanate and 3-mesityl-1,4,2-oxathiazoline-5-one (**128**) (equation 72)[204].

[spiro intermediate: ArC=N—O—C(—S—)—O—N=CAr—S] (**127**) ⟶ ArN=C=S + ArC(=N—O—C(=O)—S) (**128**) (72)

(Ar = 2,4,6-trimethylphenyl)

4. Reactions with C=N compounds

The carbon–nitrogen double bond of aldimines (Schiff bases) adds easily to aliphatic or aromatic nitrile oxides, apparently with much less structural prerequisitions than the C=O double bond (equation 73). In this reaction, 4,5-dihydro-1,2,4-oxadiazoles are formed[11,22,205]:

$$\text{R—C}^+(=\text{N—O}^-) + \text{R}^1\text{N}=\text{CHR}^2 \longrightarrow \text{RC}=\text{N—O—CHR}^2\text{—NR}^1 \text{ (ring)} \qquad (73)$$

Ethylidene butylamine, benzylidene aniline and furfurylidene methylamine gave excellent yields in this reaction. Aromatic *o*-quinone monoimides may react in a similar fashion[201].

If hydroximic acid chlorides were reacted with two moles of an alkyl imidate, 1,2,4-oxadiazoles (**129**) were obtained. Since the imidates are rather strong bases, it is likely that the first step of this synthesis consists in the dehydrohalogenation of the hydroximic acid chloride to the nitrile oxide which then undergoes a normal 1,3-dipolar cycloaddition to form a 5-alkoxy-4,5-dihydro-1,2,4-oxadiazole (**130**). This intermediate, however, loses alcohol spontaneoulsy to give **129**[91,104] (equation 74).

$$\text{R}^1\text{C(=NOH)Cl} \xrightarrow{+\text{R}^2\text{C(=NH)OR}^3} \text{R}^1\text{C}\equiv\text{N}\longrightarrow\text{O} + \text{R}^2\text{C(=}\overset{+}{\text{N}}\text{H}_2\text{)—OR}^3\cdot\text{Cl}^-$$

$$\text{R}^1\text{C}\equiv\text{N}\rightarrow\text{O} \xrightarrow{+\text{R}^2\text{C(=NH)—OR}^3} \left[\text{R}^1\text{C}=\text{N—O—C(R}^2\text{)(OR}^3\text{)—NH} \text{ (ring)}\right] \ (\mathbf{130}) \xrightarrow{-\text{R}^3\text{OH}} \text{R}^1\text{C}=\text{N—O—CR}^2=\text{N} \text{ (ring)} \ (\mathbf{129}) \qquad (74)$$

Nitrile oxides add to the C=N double bond of oximes in the presence of boron trifluoride etherate (equation 75)[206]. Hydroximic

$$\text{R}^1\text{—C}^+(=\text{N—O}^-) + \overset{\delta-}{\text{BF}_3}\text{—N(OH)=C}^{\delta+}\text{R}^2\text{R}^3 \longrightarrow \text{R}^1\text{C}=\text{N—O—C(R}^2\text{)(R}^3\text{)—NOH} \text{ (ring)} + \text{BF}_3 \qquad (75)$$

acid chlorides are apparently more active dipolarophiles. This explains an early observation that benzonitrile oxide is transformed into 3,5-diphenyl-1,2,4-oxadiazole-4-oxide in the presence of catalytic amounts of hydrogen chloride[8]. The nitrile oxide adds hydrogen chloride to give the benzhydroximic acid chloride which then undergoes a normal 1,3-dipolar cycloaddition with excess nitrile oxide to form the intermediate **131**. This aromatizes spontaneously to **132** with the loss of hydrogen chloride, thus regenerating the catalyst (equation 76)[91,139]. Aromatic carbodiimides

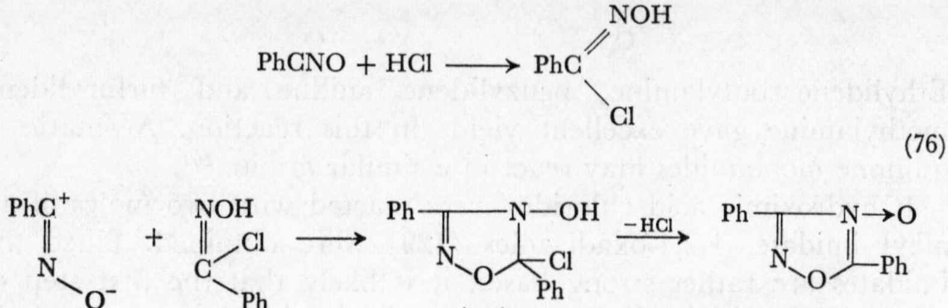

$$PhCNO + HCl \longrightarrow PhC(=NOH)Cl \tag{76}$$

$$PhC^{+}{=}N{-}O^{-} + (Ph)(Cl)C{=}NOH \longrightarrow \underset{(131)}{\text{Ph}{-}\text{C}{=}\text{N}{-}\text{O}{-}\text{C(Cl)(Ph)}{-}\text{N(OH)}} \xrightarrow{-HCl} \underset{(132)}{\text{Ph}{-}\text{C}{=}\text{N}{-}\text{O}{-}\text{C(Ph)}{=}\text{N}{\rightarrow}\text{O}}$$

reacted with aromatic nitrile oxides on both double bonds in the presence of boron trifluoride to form spiro adducts of type **133**[207]. Isocyanates, did not react under the same conditions[98].

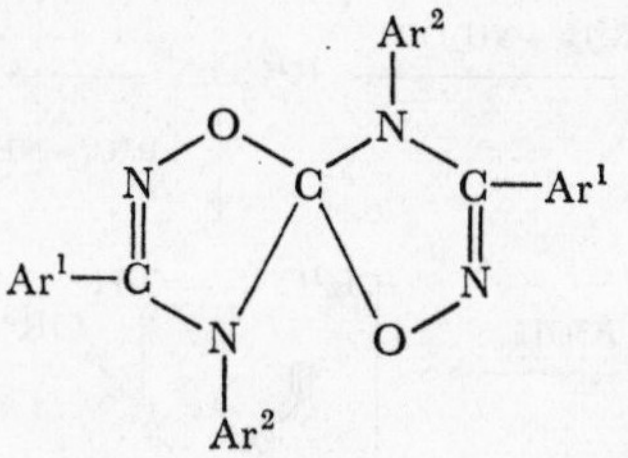

(133) — spiro compound: central C bearing two 1,2,4-oxadiazoline rings, each ring O–N=C(Ar¹)–N(Ar²)–C

(133)

5. Reactions with nitriles

In order to act as a dipolarophile, a nitrile must either be activated by an electron-withdrawing substituent or the reaction must be catalysed by boron trifluoride. Under these conditions, nitriles add to nitrile oxides with the formation of 1,2,4-oxadiazoles (equation 77), although the reaction is generally more sluggish than that of olefins and acetylenes[11,167a,201]. The reaction has been carried out

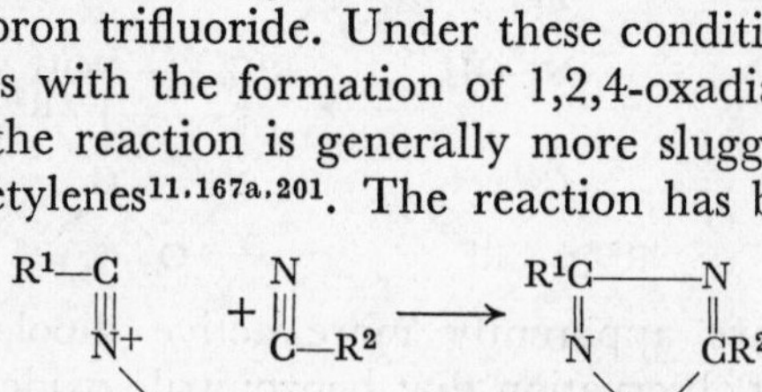

$$R^1{-}C{\equiv}N^{+}{-}O^{-} + N{\equiv}C{-}R^2 \longrightarrow R^1C{=}N{-}O{-}CR^2{=}N \ (\text{1,2,4-oxadiazole}) \tag{77}$$

mostly by generating the nitrile oxide *in situ* by one of the several techniques discussed earlier, which are especially advantageous in this case in view of the slow rate of the 1,3-dipolar cycloaddition[25,89,104,158,165]. Nitriles suitable for this reaction are acetyl cyanide, chloroacetonitrile, cyanogen, carbethoxycyanide and all aromatic nitriles. Difunctional nitriles and difunctional nitrile oxides yield 1,2,4-polyoxadiazoles[55,56]. Sterically unhindered esters of cyanic acid, ROCN, reacted with nitrile oxides to form 5-alkoxy- or 5-aryloxy-1,2,4-oxadiazoles (**134**)[208].

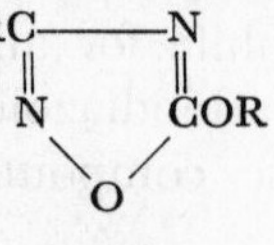

RC——N
‖ ‖
N COR
\ /
O

(**134**)

The synthesis and the chemistry of 1,2,4-oxadiazoles has been the subject of a recent exhaustive review[138].

6. Reactions with systems containing N=N, N=S, N=B, C=P or N=P double bonds

No true 1,3-dipolar cycloaddition reactions between a nitrile oxide and a N=O double-bond system are known. For the reaction of nitrile oxides with aromatic nitroso compounds, see section IV.H.

While the reaction of diazomethane with benzonitrile oxide is discussed in section IV.H, ethyl diazoacetate and diazoacetophenone were inert[209]. Azobenzene did not react with the stable mesitonitrile oxide, even on prolonged refluxing in ethanol[98], but dimethyl or diethyl azodicarboxylate reacted with aromatic nitrile oxides in a 1,3-dipolar cycloaddition at −15 °C to form initially 4-aryl-2,3-dicarbalkoxy-2,3-dihydro-1,2,3,5-oxatriazoles (**135**):

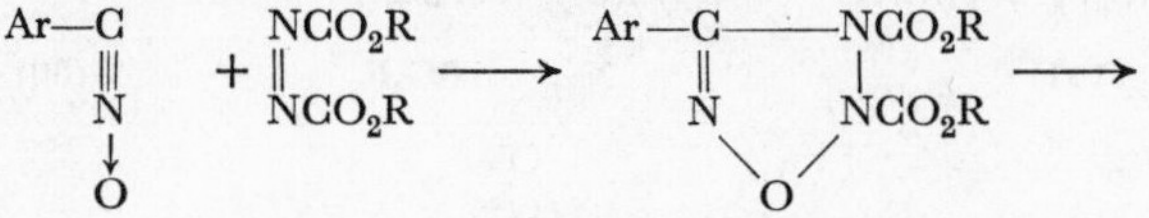

$$\text{Ar—C}{\equiv}\text{N}{\rightarrow}\text{O} + \text{RO}_2\text{CN}{=}\text{NCO}_2\text{R} \longrightarrow \text{Ar—C(=N—O—N(CO}_2\text{R)—N(CO}_2\text{R)—)} \longrightarrow$$

(**135**)

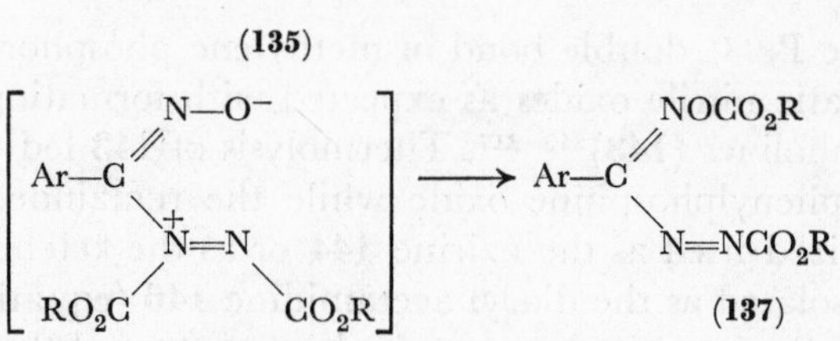

$$\left[\text{Ar—C}(=\text{N—O}^-)\text{—}\overset{+}{\text{N}}(\text{CO}_2\text{R}){=}\text{N—CO}_2\text{R}\right] \longrightarrow \text{Ar—C}(=\text{NOCO}_2\text{R})\text{—N}{=}\text{NCO}_2\text{R} \qquad (78)$$

(**136**) (**137**)

($R = CH_3, C_2H_5$)

These primary adducts **135** are, however, very unstable and rearrange very fast at room temperature, probably via the intermediate **136**, to the aroyl oxamidrazone derivative (**137**) (equation 78). In some cases, **135** is not isolable at all, but even at −20 °C the rearranged product **137** is immediately formed[210,211].

The N=S double bond in aliphatic or aromatic *N*-sulphinylamines is a good dipolarophile for aliphatic and aromatic nitrile oxides, forming 1,2,3,5-oxathiodiazole 2-oxides (**138**) (equation 79)[37,212]. On heating, those compounds of type **138** which are

$$R^1{-}C^+{=}N{-}O^- + \overset{\delta-}{N}(-R^2){=}\overset{\delta+}{S}{=}O \longrightarrow \underset{(138)}{R^1C{=}N{-}O{-}SO{-}NR^2} \xrightarrow{\Delta} \underset{(139)}{R^1N{=}C{=}NR^2} + SO_2 \qquad (79)$$

$$(138) \xrightarrow{\Delta} (140)\ \text{5,6-benzo-1,2,4-thiadiazine-1,1-dioxide } (N{=}CR^1,\ NH,\ SO_2)$$

derived from aromatic nitrile oxides decompose almost quantitatively to SO_2 and the carbodiimide **139**, whereas when R^1 is an aliphatic group and R^2 = phenyl, 5,6-benzo-1,2,4-thiadiazine-1,1-dioxides (**140**) are formed by rearrangement in yields up to 40%, together with **139**[37,213].

The *B*-pentafluorophenyl-*N*-arylborimides (**141**) react with benzonitrile oxide already at room temperature almost quantitatively to form 1,2,4,5-oxadiazaborolines (**142**)[214]:

$$\underset{(141)}{ArN{=}BC_6F_5} + PhCNO \longrightarrow \underset{(142)}{PhC{=}N{-}O{-}BC_6F_5{-}NAr} \qquad (80)$$

The P=C double bond of methylene phosphoranes reacted with aromatic nitrile oxides as expected with formation of 1,2,5-oxaazaphospholines (**143**)[215–217]. Thermolysis of **143** led to the elimination of triphenylphosphine oxide while the remainder of the molecule stabilized itself as the azirine **144** or as the ketene imine **145** which was isolated as the diaryl acetamidine **146** (equation 81).

In other cases, where one hydrogen atom of the methylene group of the phosphorane was substituted by a phenyl or carbomethoxy

$$\mathrm{RC{\equiv}\overset{+}{N}{-}\overset{-}{O} + H_2C{=}PPh_3 \longrightarrow (143) \xrightarrow{-OPPh_3} (144) \text{ or } H_2C{=}C{=}NR\ (145) \xrightarrow{PhNH_2} CH_3C({=}NR)NHPh\ (146)} \tag{81}$$

(R = Ph or 2,4,6-trimethylphenyl)

residue, the originally formed cycloadduct was unstable and underwent immediate decomposition to **145.** This reacted then with another molecule of the phosphorane to form **147** (equation 82).

$$\mathrm{R^1CH{=}C{=}NR^2\ (145) \xrightarrow{+R^1CH{=}PPh_3} R^1C({=}PPh_3){-}C({=}NR^2)CH_2R^1\ (147)} \tag{82}$$

(R^1 = Ph, CO_2CH_3; R^2 = Ph, 2,4,6-trimethylphenyl)

The analogous reaction with *N*-phenylimino triphenylphosphorane did not allow the isolation of the cycloaddition product **148** which, under the reaction conditions, decomposed into triphenylphosphine oxide and diphenylcarbodiimide[218] (equation 83).

$$\mathrm{Ph\overset{+}{C}{\equiv}N{-}\overset{-}{O} + PhN{=}PPh_3 \longrightarrow [148] \longrightarrow Ph_3PO + PhN{=}C{=}NPh} \tag{83}$$

H. Miscellaneous Reactions

In this section, a number of reactions of nitrile oxides are described which can neither be classified as nucleophilic additions leading to open-chain compounds nor as true 1,3-dipolarophilic cycloadditions, although in some cases, the reaction sequence may start with one of the above alternatives.

Nitrosobenzene reacts already at −20 °C with benzonitrile oxide. The initial nucleophilic attack on the nitrile oxide is followed by a cyclization of the insoluble intermediate **149** through abstraction of a proton from the *o*-position of the nitrosobenzene to give 1-hydroxy-2-benzimidazole-3-oxide (**151**).

p-Nitroso-*N*,*N*-dimethylaniline gives rise directly to the benzimidazole derivative **152** without permitting the isolation of any intermediate of type **149**. With *p*-nitrosophenol on the other hand, the intermediate **150** is stabilized as the quinone imide derivative **153** (equation 84)[219].

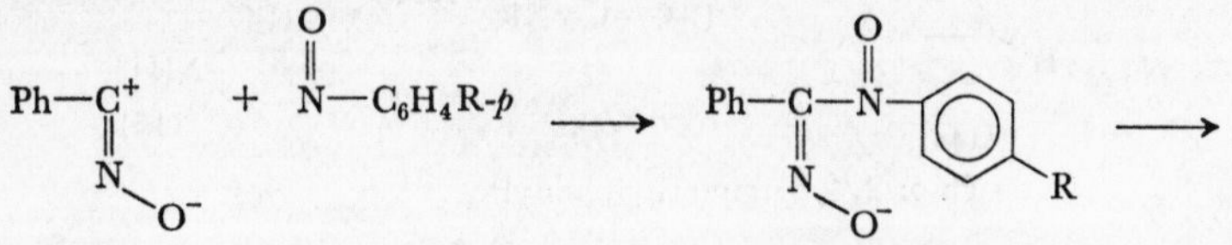

$$Ph{-}C^{+}{=}N{-}O^{-} + O{=}N{-}C_6H_4R\text{-}p \longrightarrow Ph{-}C(=N{-}O^{-}){-}N(=O){-}C_6H_4{-}R \longrightarrow \qquad (84)$$

(**149**, R = H)
(**150**, R = OH)

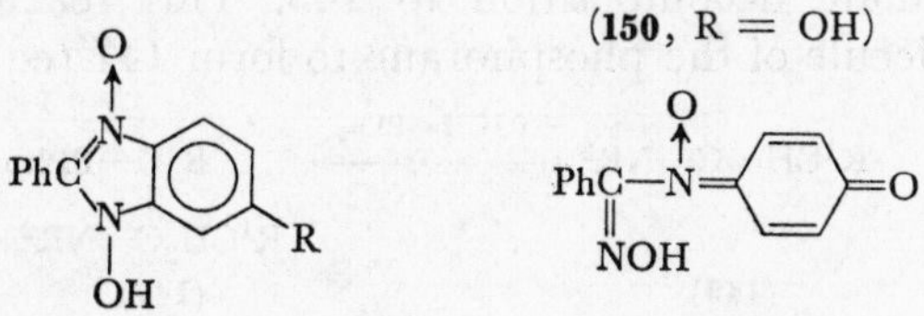

(**151**, R = H)
(**152**, R = $N(CH_3)_2$)

$PhC(=NOH){-}N(\rightarrow O){=}C_6H_4{=}O$

(**153**)

Two moles of diazomethane react with one mole of the nitrile oxide with the loss of nitrogen to form 1-nitroso-3-phenyl-2-pyrazoline (**154**). The following mechanism is offered[209]:

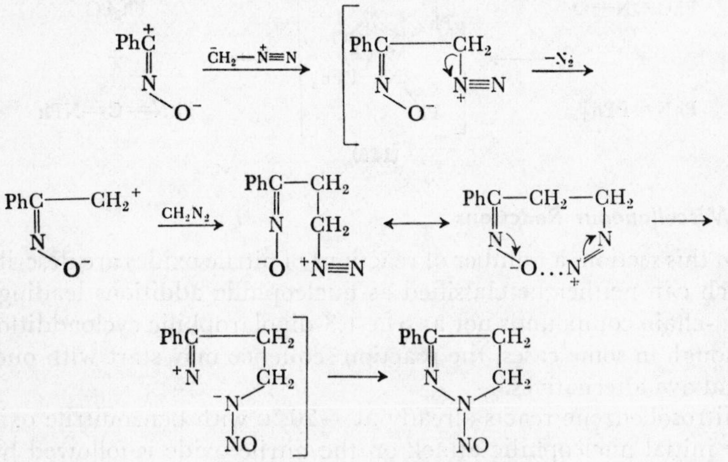

$$Ph\overset{+}{C}{=}N{-}O^{-} \xrightarrow{\bar{C}H_2{-}\overset{+}{N}{\equiv}N} \left[PhC(=N{-}O^{-}){-}CH_2{-}\overset{+}{N}{\equiv}N\right] \xrightarrow{-N_2}$$

$$PhC(=N{-}O^{-}){-}CH_2^{+} \xrightarrow{CH_2N_2} PhC(=N{-}O^{-}){-}CH_2{-}CH_2{-}\overset{+}{N}{\equiv}N \longleftrightarrow PhC{-}CH_2{-}CH_2 \cdots \longrightarrow$$

$$\left[PhC(=\overset{+}{N}){-}CH_2{-}CH_2{-}\bar{N}{-}NO\right] \longrightarrow \text{pyrazoline } PhC{=}N{-}N(NO){-}CH_2{-}CH_2 \qquad (85)$$

(**154**)

A similar reaction has been observed between benzonitrile oxide and dimethyloxosulphonium methylide (**155**) $\overset{-}{C}H_2\overset{+}{S}O(CH_3)_2$ [220–222]. The ylid **155** seems to be able to bring about two consecutive transfers of methylene to the nitrile oxide, probably through the zwitterionic intermediate **156**, from which by further transformation according to equation (86), the observed products, 3-phenyl-Δ^2-isoxazoline (**157**), phenyl vinyl ketoxime (**158**) and the benzhydroxamic ester of phenyl vinyl ketoxime (**159**) are derived. Further addition of benzonitrile to **158** leads oxide to the isoxazoline (**160**).

Ph—C⁺(=N—O⁻) —(+ 155)→ [PhC(=N—O⁻)CH₂S⁺(=O)(CH₃)₂] —(−DMSO, + 155)→ [PhC(=N—O⁻)(H)CH—CH₂—S⁺(=O)(CH₃)₂] (**156**) (86)

156 —(−DMSO)→ PhC=N—O—CH₂—CH₂ (ring) (**157**)

156 —(+ PhCNO)→ [PhC(=N—O—C(Ph)=N—O⁻)(H)CH—CH₂—S⁺(=O)(CH₃)₂] —(−DMSO, −H⁺)→ [PhC(=N—O—C(Ph)=N—O⁻)CH=CH₂] —(+ H⁺ (H₂O))→ PhC(=N—O—C(Ph)=NOH)—CH=CH₂ (**159**)

156 —(−DMSO, −H⁺)→ [PhC(=N—O⁻)—CH=CH₂] —(+H⁺ (H₂O))→ PhC(=NOH)—CH=CH₂ (**158**)

[PhC(=N—O⁻)—CH=CH₂] —(+ PhCNO, + H⁺ (H₂O))→ H₂C—CPh=N—O—CH(—C(Ph)=NOH) (ring) (**160**)

Dimethylsulphonium methylide, $\overset{-}{C}H_2\overset{+}{S}(CH_3)_2$, reacts analogously with benzonitrile oxide and compounds **157** and **158** are obtained as the major products.

V. REFERENCES

1. E. Howard, *Phil. Trans. Roy. Soc. London*, **1800**, 204.
2. H. Wieland, *Die Knallsäure* (Sammlung chemischer und chemisch-technischer Vorträge), Vol. 14, (Ed. F. B. Ahrens), F. Enke, Stuttgart, 1909, pp. 385–461.
3. J. U. Nef, *Ann. Chem.*, **280**, 291 (1893).
4. W. Beck and K. Feldl, *Angew. Chem.*, **78**, 746 (1966).
5. A. Werner and H. Buss, *Ber.*, **27**, 2193 (1894).
6. S. Gabriel and M. Koppe, *Ber.*, **19**, 1145 (1886).
7. L. Pauling and S. B. Hendricks, *J. Am. Chem. Soc.*, **48**, 641 (1926).
8. H. Wieland, *Ber.*, **40**, 1667 (1907).
9. K. V. Auwers, *Ber.*, **61**, 1041 (1928).
10. R. Huisgen, *Proc. Chem. Soc.*, **1961**, 577.
11. R. Huisgen, *Angew. Chem.*, **75**, 604 (1963).
12. W. Beck, *Angew. Chem.*, **75**, 872 (1963).
13. W. Beck, E. Schuirer and K. Feldl, *Angew. Chem.*, **77**, 722 (1965).
14. W. D. McGrath and T. Morrow, *Nature*, **203**, 619 (1964).
15. W. D. McGrath, *Nature*, **204**, 988 (1964).
16. W. Beck, private communication.
17. L. Birckenbach and K. Sennewald, *Ann. Chem.*, **512**, 45 (1934).
18. R. Huisgen and M. Christl, *Angew. Chem.*, **79**, 471 (1967).
19. P. Kurtz in *Methoden der Organischen Chemie*, Vol. 8 (Ed. E. Müller), 4th ed., Georg Thieme, Stuttgart, 1952, pp. 355–358.
20. H. Wieland, A. Baumann, C. Reisenegger, W. Scherer, J. Thiele, J. Will, H. Haussmann and W. Frank, *Ann. Chem.*, **444**, 7 (1925).
21. T. Hoshino and M. Mukaiyama, *Japan Pat.* 9855 (1959); *Chem. Abstr.*, **54**, 7738h (1960).
22. T. Mukaiyama and T. Hoshino, *J. Am. Chem. Soc.*, **82**, 5339 (1960).
23. P. Vita-Finzi and P. Grünanger, *Chim. Ind.* (*Milan*), **46**, 516 (1965).
24. G. Zinner and W. Günther, *Angew. Chem.*, **76**, 440 (1964).
25. C. Grundmann in *Methoden der Organischen Chemie*, Vol. 10/3 (Ed. E. Müller), 4th ed., Georg Thieme, Stuttgart, 1965, pp. 841–870.
26. G. Speroni, unpublished, cited in reference 127.
27. C. Grundmann and S. K. Datta, *J. Org. Chem.*, **34**, 2016 (1969).
28. C. Grundmann, *Angew. Chem.*, **75**, 450 (1963).
29. C. Grundmann, V. Mini, J. M. Dean and H.-D. Frommeld, *Ann. Chem.*, **687**, 191 (1965).
30. R. Huisgen, W. Mack and E. Anneser, *Angew. Chem.*, **73**, 656 (1961).
31. A. Quilico and G. Speroni, *Gazz. Chim. Ital.*, **76**, 146 (1946).
32. H. Wieland and L. Semper, *Ber.*, **39**, 2522 (1906).
33. R. Huisgen and W. Mack, *Tetrahedron Letters*, **1961**, 583.
34. C. Grundmann and J. M. Dean, *J. Org. Chem.*, **30**, 2809 (1965).
35. P. Grünanger, *Atti Accad. Naz. Lincei Rend. Classe Sci. Fis., Nat. Nat.*, [8] **24**, 163 (1958).

36. M. S. Chang and J. U. Lowe, *J. Org. Chem.*, **32**, 1577 (1967).
37. F. Eloy and R. Lenaers, *Bull. Soc. Chim. Belge*, **74**, 129 (1965).
38. S. Califano, R. Moccia, R. Scarpati and G. Speroni, *J. Chem. Phys.*, **26**, 1777 (1957).
39. C. Grundmann and J. M. Dean, *Angew. Chem.*, **76**, 682 (1964).
40. C. Grundmann and R. Richter, *J. Org. Chem.*, **33**, 476 (1968).
41. P. Beltrame, C. Veglio and M. Simonetta, *J. Chem. Soc. (B)*, **1967**, 867.
42. A. Dondoni, A. Mangini and S. Ghersetti, *Tetrahedron Letters*, **1966**, 2397.
43. H. Rheinboldt, M. Dewald, F. Jansen and O. Schmitz-Dumont, *Ann. Chem.*, **451**, 161 (1927).
44. C. Grundmann and K. Flory, unpublished.
45. C. Grundmann and R. Richter, *J. Org. Chem.*, **32**, 2308 (1967).
46. C. Grundmann and J. M. Dean, *Angew. Chem.*, **77**, 966 (1965).
47. H. Wieland and B. Rosenfeld, *Ann. Chem.*, **484**, 236 (1930).
48. G. Ponzio, *Gazz. Chim. Ital.*, **66**, 119 (1936).
49. G. Ponzio, *Gazz. Chim. Ital.*, **66**, 123 (1936).
50. G. Ponzio, *Gazz. Chim. Ital.*, **71**, 693 (1941).
51. G. Just and K. Dahl, *Tetrahedron Letters*, **1966**, 2441.
52. G. Just and W. Zehetner, *Tetrahedron Letters*, **1967**, 3389.
53. Y. Iwakura, M. Akiyama and S. Shiraishi, *Bull. Chem. Soc. Japan*, **38**, 335 (1965).
54. Y. Iwakura, M. Akiyama and K. Nagabuko, *Bull. Chem. Soc. Japan*, **37**, 767 (1964).
55. C. G. Overberger and S. Fujimoto, *Polymer Letters*, **3**, 735 (1965).
56. C. Overberger, private communication.
57. F. Eloy, *Bull. Soc. Chim. Belge*, **73**, 639 (1964).
58. H. Wieland, *Ber.*, **40**, 418 (1907).
59. H. Wieland, *Ber.*, **43**, 3362 (1910).
60. F. Angelico, *Atti Accad. Reale Lincei Rend. Classe Fis. Mat. Nat.*, **10**, 476 (1901).
61. J. U. Nef, *Ann. Chem.*, **280**, 263 (1893).
62. H. Wieland, *Ber.*, **42**, 803 (1909).
63. H. Wieland, *Ber.*, **40**, 821 (1907).
64. A. Quilico and L. Panizzi, *Gazz. Chim. Ital.*, **72**, 155 (1942).
65. L. Schischkoff, *Ann. Chem. Suppl.*, **1**, 109 (1861).
66. (a) H. Wieland and H. Hess, *Ber.*, **42**, 1347, 1352 (1909).
(b) G. Ponzio and G. Busti, *Gazz. Chim. Ital.*, **36II**, 338 (1906).
67. L. Horner, L. Hockenberger and W. Kirmse, *Chem. Ber.*, **94**, 260 (1961).
68. (a) H. Kropf and R. Lambeck, *Ann. Chem.*, **700**, 1, 18, (1966).
(b) H. Henecka and P. Kurtz in *Methoden der Organischen Chemie*, Vol. 8, (Ed. E. Müller), Georg Thieme, Stuttgart, 1952, pp. 690–691.
69. G. Casnati and A. Ricca, *Tetrahedron Letters*, **1967**, 327.
70. J. Houben and H. Kauffmann, *Ber.*, **46**, 2821 (1913).
71. J. Houben, *Ber.*, **46**, 4001 (1913).
72. M. Jovitschitsch, *Ber.*, **28**, 1213 (1895).
73. M. Jovitschitsch, *Ber.*, **35**, 151 (1902).
74. M. Jovitschitsch, *Ber.*, **39**, 784 (1906).
75. L. Semper and L. Lichtenstadt, *Ann. Chem.*, **400**, 302 (1913).
76. A. Quilico, G. Gaudiano and A. Ricca, *Gazz. Chim. Ital.*, **87**, 638 (1957).
77. H. Wieland and A. Höchtlen, *Ann. Chem.*, **505**, 237 (1933).
78. R. H. Wiley and B. J. Wakefield, *J. Org. Chem.*, **25**, 546 (1960).
79. W. Steinkopf and B. Jürgens, *J. Prakt. Chem.* [2], **84**, 686 (1911).
80. R. Scholl, *Ber.*, **23**, 3505 (1890).

81. H. Wieland, *Ann. Chem.*, **424**, 107 (1921).
82. H. Wieland and L. Semper, *Ann. Chem.*, **358**, 36 (1907).
83. G. Ponzio, *Gazz. Chim. Ital.*, **53**, 379 (1923).
84. L. Avogadro, *Gazz. Chim. Ital.*, **53**, 824 (1923).
85. G. Ponzio and G. Ruggiero, *Gazz. Chim. Ital.*, **56**, 733 (1926).
86. E. Borello and M. Columbo, *Ann. Chim. (Rome)*, **46**, 1158 (1956).
87. G. Ponzio, *Gazz. Chim. Ital.*, **61**, 561 (1931).
88. G. M. Bachman and L. E. Strom, *J. Org. Chem.*, **28**, 1150 (1963).
89. F. Eloy, *Bull. Soc. Chim. Belge*, **73**, 793 (1964).
90. R. Paul and S. Tchelitscheff, *Bull. Soc. Chim. France*, **1963**, 140.
91. C. Grundmann, *Fortschr. Chem. Forsch.*, **7**, 62 (1966).
92. R. Scholl, *Ber.*, **34**, 862 (1901).
93. R. Scholl and A. Schöfer, *Ber.*, **34**, 870 (1901).
94. G. Calmels, *Compt. Rend.*, **99**, 794 (1884).
95. A. F. Holleman, *Ber.*, **23**, 2998 (1890).
96. A. F. Holleman, *Rec. Trav. Chim.*, **10**, 65 (1891).
97. R. Scholl, *Ber.*, **23**, 3505 (1890).
98. C. Grundmann, unpublished results.
99. P. Souchay and J. Armand, *Compt. Rend.*, **256**, 4907 (1963).
100. J. Armand, *Bull. Soc. Chim. France*, **1966**, 882.
101. P. Souchay, J. Armand, J. P. Guetté and F. Valentini, *Compt. Rend.*, **262**, [C] 985, (1966).
102. (a) J. Armand, J. P. Guetté and F. Valentini, *Compt. Rend.*, **263**, [C] 1388 (1966); (b) J. Armand, P. Souchay and F. Valentini, *Bull. Soc. Chim. France*, **1968**, 4585.
103. M. Arbasino and P. Grünanger, *Ric. Sci.*, **34**, [IIA] 561 (1964).
104. F. Eloy and R. Lenaers, *Bull. Soc. Chim. Belge*, **72**, 719 (1963).
105. R. Lenaers and F. Eloy, *Helv. Chim. Acta*, **46**, 1067 (1963).
106. T. Sasaki and T. Yoshioka, *Bull. Chem. Soc. Japan*, **40**, 2604 (1967).
107. P. Vita-Finzi and M. Arbasino, *Ric. Sci.*, **35**, [IIA] 1484 (1965).
108. A. Quilico and M. Simonetta, *Gazz. Chim. Ital.*, **76**, 200 (1946).
109. A. Quilico and M. Simonetta, *Gazz. Chim. Ital.*, **77**, 586 (1947).
110. R. Cramer and W. R. McClellan, *J. Org. Chem.*, **26**, 2976 (1961).
111. R. A. Barnes in *Heterocyclic Compounds*, Vol. 5 (Ed. R. C. Elderfield), John Wiley and Sons, New York, 1957, pp. 452–483.
112. L. C. Behr in *The Chemistry of Heterocyclic Compounds*, Vol. 17 (Ed. A. Weissberger), Interscience, New York, 1962, pp. 283–320.
113. N. K. Kochetkov and S. D. Sokolov in *Advances in Heterocyclic Chemistry*, Vol. 2 (Ed. A. R. Katritzky), Academic Press, New York, 1963, pp. 365–422.
114. A. Quilico in *The Chemistry of Heterocyclic Compounds*, Vol. 17 (Ed. A. Weissberger), Interscience, New York, 1962, pp. 1–176.
115. J. M. Craven and R. Wehr, unpublished results, cited in reference 55.
116. C. Grundmann and H.-D. Frommeld, *J. Org. Chem.*, **31**, 157 (1966).
117. M. S. Chang and A. J. Matuszko, *J. Org. Chem.*, **28**, 2260 (1963).
118. C. Grundmann, *Chem. Ber.*, **97**, 575 (1964).
119. F. Eloy, private communication.
120. S. Califano, R. Scarpati and G. Speroni, *Atti Accad. Naz. Lincei Rend. Classe Sci. Fis. Mat. Nat.*, [8] **23**, 263 (1957).
121. R. P. Grosso and T. K. McCubbin, *J. Mol. Spectry.*, **13**, 240 (1964).
122. J. Pliva, *J. Mol. Spectry.*, **12**, 360 (1964).

123. P. V. R. Schleyer and L. Joris, private communication.
124. G. DelRe, *Atti Accad. Naz. Lincei Rend. Classe Fis. Mat. Nat.*, [8] **22,** 491 (1957).
125. G. Speroni, *Ric. Sci.*, **27,** 1195 (1957).
126. A. Quilico, *Nature,* **166,** 226 (1950).
127. Table II, p. 21 in reference 114.
128. P. Beltrame, A. Comoti and C. Veglio, *Chem. Commun.*, **1967,** 996.
129. A. Dondoni, A. Mangini and S. Ghersetti, *Tetrahedron Letters,* **1966,** 4789.
130. C. Grundmann and H.-D. Frommeld, *J. Org. Chem.*, **31,** 4235 (1966).
131. C. Grundmann and H.-D. Frommeld, *J. Org. Chem.*, **30,** 2077 (1965).
132. (a) J. H. Boyer in *Heterocyclic Compounds,* Vol. 7 (Ed. R. C. Elderfield), John Wiley and Sons, New York, 1961, pp. 462–507;
(b) C. Grundmann and P. Grünanger, *Nitrile Oxides,* Springer Verlag, Heidelberg, in press.
133. H. Wieland, *Ber.*, **42,** 4207 (1909).
134. K. Auwers and V. Meyer, *Ber.*, **21,** 784, 804 (1888).
135. K. Auwers and V. Meyer, *Ber.*, **22,** 705, 716 (1889).
136. (a) H. Wieland, *Ber.*, **42,** 816 (1909);
(b) C. Grundmann and G. F. Kite, unpublished results.
137. (a) C. Grundmann, H.-D. Frommeld, K. Flory and S. K. Datta, *J. Org. Chem.*, **33,** 1464 (1968);
(b) N. Barbulescu and P. Grünanger, *Gazz. Chim. Ital.*, **92,** 138 (1962).
138. F. Eloy, *Fortschr. Chem. Forsch.*, **4,** 807 (1965).
139. R. Huisgen, *Angew. Chem.*, **75,** 742 (1963).
140. P. Diehl, H. A. Christ and F. B. Mallory, *Helv. Chim. Acta,* **45,** 504 (1962).
141. A. R. Katritzky, S. Øksne and R. K. Harris, *Chem. Ind. (London)*, **1961,** 990.
142. F. B. Mallory and A. Cammarata, *J. Am. Chem. Soc.*, **88,** 61 (1966).
143. H. Wieland and E. Gmelin, *Ann. Chem.*, **367,** 52 (1909).
144. L. Bouveault and A. Bongert, *Bull. Soc. Chim. France,* [3] **27,** 1164 (1902).
145. F. Klages, *Naturwissenschaften,* **30,** 351 (1942).
146. A. Quilico and G. Stagno D'Alcontres, *Gazz. Chim. Ital.*, **79,** 654 (1949).
147. A. Quilico and G. Stagno D'Alcontres, *Gazz. Chim. Ital.*, **79,** 703 (1949).
148. J. v. Liebig, *Ann. Chem.*, **95,** 282 (1855).
149. M. Conrad and A. Schulze, *Ber.*, **42,** 735 (1909).
150. P. Seidel, *Ber.*, **25,** 431, 2756 (1892).
151. A. Steiner, *Ber.*, **9,** 779 (1876).
152. G. Speroni, *Sitzungsberichte d. 14. Internationalen Kongresses f. reine u. angewandte Chemie, Zürich 1955,* No. 52, p. 30.
153. R. A. Olofson and J. S. Michelman, *J. Am. Chem. Soc.*, **86,** 1863 (1964).
154. P. Vita-Finzi and M. Arbasino, *Tetrahedron Letters,* **1965,** 4645.
155. (a) P. Grünanger and M. R. Langella, *Atti Accad. Naz. Lincei Rend. Classe Fis. Mat. Nat.*, [8] **36,** 387 (1964);
(b) L. Birckenbach and K. Sennewald, *Ann. Chem.*, **489,** 18 (1931).
156. H. Wieland, *Ber.*, **42,** 4199 (1909).
157. F. Eloy, *J. Org. Chem.*, **26,** 952 (1961).
158. R. Huisgen, W. Mack and E. Anneser, *Tetrahedron Letters,* **1961,** 587.
159. G. Longo, *Gazz. Chim. Ital.*, **61,** 575 (1931).
160. N. E. Alexandrou and D. N. Nicolaides, *Chim. Chronika (Athens, Greece)*, **30A,** 49 (1965).
161. M. H. Benn, *Can. J. Chem.*, **42,** 2393 (1964).

162. M. H. Benn, *Can. J. Chem.*, **43,** 1 (1965).
163. J. Waser and J. W. Watson, *Nature,* **198**, 1297 (1963).
164. P. Rajagopalan and C. N. Talaty, *Tetrahedron Letters*, **1966**, 2101.
165. C. Weygand and E. Bauer, *Ann. Chem.*, **459**, 123 (1927).
166. A. Quilico and R. Fusco, *Gazz. Chim. Ital.*, **87**, 638 (1957).
167. (a) G. Leandri and M. Palotti, *Ann. Chim.* (*Rome*), **47**, 376 (1957); (b) M. Christl and R. Huisgen, *Tetrahedron Letters*, **1968**, 5209.
168. A. Quilico and P. Grünanger, *Gazz. Chim. Ital.*, **82**, 140 (1952).
169. A. Quilico, G. Stagno D'Alcontres and P. Grünanger, *Gazz. Chim. Ital.*, **80**, 479 (1950).
170. G. Lo Vecchio, *Gazz. Chim. Ital.*, **87**, 1413 (1957).
171. G. Lo Vecchio, G. Cum and G. Stagno D'Alcontres, *Tetrahedron Letters*, **1964**, 3495.
172. G. Lo Vecchio and P. Monforte, *Ann. Chim.* (*Rome*), **46**, 76 (1956).
173. (a) R. Huisgen and K. Bast, unpublished, cited in reference 25; (b) G. Zinner and H. Günther, *Chem. Ber.*, **98**, 1353 (1965).
174. N. Barbulescu, P. Grünanger, M. R. Langella and A. Quilico, *Tetrahedron Letters*, **1961**, 89.
175. A. Quilico and G. Stagno D'Alcontres, *Gazz. Chim. Ital.*, **80**, 140 (1950).
176. A. Grünanger and L. Grasso, *Gazz. Chim. Ital.*, **85**, 1271 (1955).
177. A. Corsico-Coda, P. Grünanger and G. Veronesi, *Tetrahedron Letters*, **1966**, 2911.
178. P. Grünanger, private communication.
179. A. Quilico, P. Grünanger and R. Mazzini, *Gazz. Chim. Ital.*, **82**, 349 (1952).
180. C. Grundmann, unpublished, cited in reference 29.
181. G. Stagno D'Alcontres and G. Lo Vecchio, *Gazz. Chim. Ital.*, **90**, 1239 (1960).
182. N. Barbulescu and A. Quilico, *Gazz. Chim. Ital.*, **91**, 326 (1961).
183. R. Paul and S. Tchelitcheff, *Bull. Soc. Chim. France,* **1962**, 2215.
184. G. Stagno D'Alcontres and P. Grünanger, *Gazz. Chim. Ital.*, **80**, 741 (1950).
185. P. Grünanger and S. Mangiapan, *Gazz. Chim. Ital.*, **88**, 149 (1958).
186. P. Vita-Finzi and M. Arbasino, *Ann. Chim.* (*Rome*), **54**, 1165 (1964).
187. S. Morrocchi, A. Ricca and L. Velo, *Chim. Ind.* (*Milan*), **49**, 168 (1967).
188. G. Bianchetti, D. Pocar and P. Dalla Croce, *Gazz. Chim. Ital.*, **93**, 1714 (1963).
189. G. Bianchetti, D. Pocar and P. Dalla Croce, *Gazz. Chim. Ital.*, **93**, 1726 (1963).
190. G. Bianchi and E. Frati, *Gazz. Chim. Ital.*, **96**, 559 (1966).
191. (a) G. Stork and J. E. McMurry, *J. Am. Chem. Soc.*, **89**, 5461 (1967); (b) M. E. Kuehne, S. J. Weaver and P. Franz, *J. Org. Chem.*, **29**, 1582 (1964).
192. R. Scarpati and P. Sorrentino, *Gazz. Chim. Ital.*, **89**, 1525 (1959).
193. R. Scarpati, C. Santacroce and D. Sica, *Gazz. Chim. Ital.*, **93**, 1706 (1963).
194. R. Scarpati and G. Speroni, *Gazz. Chim. Ital.*, **89**, 1511 (1959).
195. A. Quilico and P. Grünanger, *Rend. Ist. Lombardo Sci. Lettere A*, **88** ([3], **19**), 990 (1950); *Chem. Zentr.* **1959**, 8138.
196. A. Dondoni, A. Mangini and S. Ghersetti, *Tetrahedron Letters*, **1966**, 2397.
197. E. Mugnaini and P. Grünanger, *Atti Accad. Naz. Lincei Rend. Classe Sci. Fis. Mat. Nat.*, [8] **14**, 95 (1953).
198. P. Grünanger, *Atti Accad. Naz. Lincei Rend. Classe Sci. Fis. Mat. Nat.*, [8] **24**, 163 (1958).
199. P. Grünanger and M. R. Langella, *Gazz. Chim. Ital.*, **89**, 1784 (1959).
200. H. G. Viehe, R. Fuks and M. Reinstein, *Angew. Chem.*, **76**, 571 (1964).
201. S. Morrocchi, A. Ricca and L. Velo, *Tetrahedron Letters*, **1967**, 331.

202. W. L. Awad, S. M. Abdel Rahman Omran and M. Sobhy, *J. Org. Chem.*, **31**, 203, 331 (1966).
203. D. Noel and J. Vielle, *Bull. Soc. Chim. France*, **1967**, 2239.
204. W. D. Foye and J. M. Kauffmann, *J. Org. Chem.*, **31**, 2417 (1966).
205. F. Lauria, V. Vecchietti and G. Tosolino, *Gazz. Chim. Ital.*, **94**, 478 (1964).
206. S. Morrocchi and A. Ricca, *Chim. Ind.* (*Milan*), **49**, 629 (1967).
207. C. Grundmann and R. Richter, *Tetrahedron Letters*, **1968**, 963.
208. D. Martin, *Angew. Chem.*, **77**, 1033 (1965).
209. G. Lo Vecchio, M. Crisafulla and M. Aversa, *Tetrahedron Letters*, **1966**, 1909.
210. R. Huisgen, H. Blaschke and E. Brunn, *Tetrahedron Letters*, **1966**, 405.
211. P. Rajagopalan, *Tetrahedron Letters*, **1964**, 887.
212. P. Rajagopalan and H. U. Daeniker, *Angew. Chem.*, **75**, 91 (1963).
213. P. Rajagopalan and B. Advani, *J. Org. Chem.*, **30**, 3369 (1965).
214. P. Paetzold, *Angew. Chem.*, **79**, 583 (1967).
215. H. J. Bestmann and H. Kunstmann, *Angew. Chem.*, **78**, 1059 (1966).
216. R. Huisgen and J. Wulff, *Tetrahedron Letters*, **1967**, 917.
217. A. Umani-Ronchi, M. Acampora, G. Gaudiano and A. Selva, *Chim. Ind.* (*Milan*), **49**, 388 (1967).
218. R. Huisgen and J. Wulff, *Tetrahedron Letters*, **1967**, 921.
219. F. Minisci, R. Galli and A. Quilico, *Tetrahedron Letters*, **1963**, 785.
220. P. Bravo, G. Gaudiano and A. Umani-Ronchi, *Chim. Ind.* (*Milan*), **49**, 286 (1967).
221. P. Bravo, G. Gaudiano and A. Umani-Ronchi, *Gazz. Chim. Ital.*, **97**, 1664 (1967).
222. A. Umani-Ronchi, P. Bravo and G. Gaudiano, *Tetrahedron Letters*, **1967**, 3477.

CHAPTER 15

Isonitriles

PETER HOFFMANN, DIETER MARQUARDING, HELMUT KLIIMANN AND IVAR UGI

Bayer A.G., Leverkusen, W. Germany and University of Southern California, Los Angeles

I. GENERAL PROPERTIES 853
II. ISONITRILE SYNTHESES 854
A. Alkylation of Cyanides 855
B. Carbylamine Reaction 855
C. Dehydration of Formamides 856
D. Further Modes of Formation 857
E. Isocyano Groups Connected to Heteroatoms ('Heteroisonitriles'). 859
III. ISONITRILE REACTIONS 859
A. Cleavage of the C—N Single Bond 859
B. Simple α-Additions 860
C. α-Additions Followed by Secondary Reactions 862
1. Passerini reaction and related reactions . 863
2. Carbocations as α-addition partners . . 864
3. Four-component condensations . . . 865
D. Cycloadditions and Cyclizations . . . 869
IV. ISONITRILE COMPLEXES 873
V. BIOLOGICAL PROPERTIES AND TECHNOLOGICAL USES . . . 877
VI. REFERENCES 877

I. GENERAL PROPERTIES

Within organic chemistry isonitriles are a unique class of compounds, being the only stable organic compounds containing formally bivalent carbon. About a hundred years have passed since the discovery of the isonitriles in 1867 by Gautier[1] and Hofmann[2]. Basic problems of organic chemistry, such as the problem of isomerism[1–3] and the question whether carbon can also occur in bivalent form[4–8], were the reasons for the early investigations on isonitriles, but despite this and regardless of their versatile reactivity, their chemistry has been investigated very little.

The development of isonitrile chemistry has hardly been delayed by the unpleasant characteristic odour of volatile isonitriles. As isonitriles can be noticed in traces, the majority of their modes of formation was discovered by the occurrence of the isonitrile smell. The relatively small number of publications up to 1960 (see review articles[9–16]) is due primarily to the fact that convenient and generally applicable methods for the preparation of isonitriles were not available before 1958 (see section II).

In the early times of isonitrile chemistry, isonitriles were described as compounds of divalent carbon (**1**)[1,4]

$$\mathrm{R{-}N{=}C^{II}}$$

(1)

With increased knowledge of the physicochemical properties of isonitriles and with advances in understanding of the nature of the chemical bond, a semiionic formula, (**2**) written also as (**2a**), came into use[5–8].

$$\mathrm{R{-}\overset{+}{N}{\equiv}\overset{-}{C}} \quad = \quad \mathrm{R{-}N{\rightleftarrows}C}$$

(2) **(2a)**

The following physicochemical properties of the isonitriles have been investigated: i.r. spectra[17–22], Raman spectra[17–20], n.m.r. spectra[23,24], mass spectra[25], microwave spectra[26–28], dipole moments[29] and bond refractivity data[30], bond energies[31,32] and heats of formation[33], all of them being in favour of structure (**2**)[33a]. A particularly interesting property of isonitriles is their ability to accept protons to form hydrogen bonds[34–36].

II. ISONITRILE SYNTHESES

Except for the alkylation of cyanides, the formation of isonitriles always occurs by reactions in which a compound of tetravalent carbon loses two univalent groups or a bivalent group by α-eliminations or redox reactions (reactions 1 and 8). The tautomer (**4**) of hydrogen cyanide (**3**) already contains formally bivalent carbon

$$\mathrm{R{-}N{=}C} \begin{matrix} \diagup X \\ \diagdown Y \end{matrix} \longrightarrow \mathrm{R{-}N{\rightarrow}C} \qquad (1)$$

(reaction 2). The two classical isonitrile syntheses, the alkylation of

$$\underset{(3)}{\text{H—C}\equiv\text{N}} \rightleftharpoons \underset{(4)}{\text{H—N}\overrightarrow{=}\text{C}} \tag{2}$$

silver cyanide (reaction 3) and the reaction of primary amines with chloroform and alkali (reaction 4), remained until recently the only preparative access to isonitriles. Both methods are not generally applicable, and produce good yields only in exceptional cases. Isonitriles are now easily prepared by dehydrating *N*-monosubstituted formamides (**5**) (reaction 5)[15,21,37–40].

A. Alkylation of Cyanides

The alkylation of the free ambident cyanide ion by alkyl halides, alkyl sulphuric acid salts and dialkyl sulphates, predominantly yields nitriles; isonitriles are only formed in relatively low quantities. On the other hand, if the cyanide ion is complexed it is preferably alkylated at the nitrogen. Gautier[1] prepared isonitriles for the first time by combining silver cyanide and alkyl iodides to form isonitrile complexes, from which potassium cyanide liberates isonitriles (reaction 3).

$$\text{RI} \xrightarrow{\text{AgCN}} [(\text{RNC})\text{AgI}] \xrightarrow{\text{KCN}} \text{R—NC} \tag{3}$$

B. Carbylamine Reaction

The 'carbylamine reaction' [2] of primary amines with chloroform and strong bases, such as ethanolic potassium hydroxide, solid alkali metal hydroxides or potassium *t*-butylate, was recommended for qualitative detection of primary amines and was, for a long time, considered the most favourable method for the preparation of isonitriles. Nef[4] had already interpreted Hofmann's 'carbylamine' reaction in 1897 as an addition of dichlorocarbene to primary amines and subsequent *β*- and *α*-eliminations of hydrogen chloride (reaction 4). An analogous mechanism holds also for the formation

$$\text{R—NH}_2 \xrightarrow{\text{CHCl}_3,\ \text{KOH}} \text{R—}\overset{+}{\text{N}}\text{H}_2\text{—}\overset{-}{\text{C}}\text{Cl}_2 \longrightarrow \text{R—NC} \tag{4}$$

of isonitriles by the thermal decomposition of sodium trichloroacetate in the presence of aryl amines[41].

C. Dehydration of Formamides

Gautier[1] mentioned the idea that isonitriles, owing to their ease of hydrolysis, are to be considered derivatives of primary amines and formic acid. It should therefore be possible to prepare them by elimination of water from formates of primary amines (**5**) (reaction 5). His experiments in this respect were not successful.

$$\underset{(\mathbf{5})}{RNH_2 + HCOOH} \underset{+H_2O}{\overset{-H_2O}{\rightleftharpoons}} RNHCHO \underset{+H_2O}{\overset{-H_2O}{\rightleftharpoons}} R{-}NC \qquad (5)$$

In 1938, Wegler[42] took the idea up again and reacted *N*-monosubstituted formamides with thionyl chloride; among the reaction products he found traces of isonitriles. In connection with the elucidation of the constitution of xanthocillin (**6**), Hagedorn and Tönjes[37] for the first time carried out such an elimination of water from '*o,o*'-dimethylxanthocillin dihydrate' (**7**) in the presence of a base. The dehydrating agent was benzenesulphonyl chloride in

$$HO{-}C_6H_4{-}CH{=}C(CN){-}C(NC){=}CH{-}C_6H_4{-}OH \quad (\mathbf{6})$$

$$CH_3O{-}C_6H_4{-}CH{=}C(OHCHN){-}C(NHCHO){=}CH{-}C_6H_4{-}OCH_3 \quad (\mathbf{7})$$

$$\Big\downarrow C_6H_5SO_2Cl,\ \text{pyridine} \quad (-H_2O) \qquad (6)$$

$$CH_3O{-}C_6H_4{-}CH{=}C(CN){-}C(NC){=}CH{-}C_6H_4{-}OCH_3$$

pyridine (reaction 6). The antibiotic xanthocillin is the only isonitrile that has been found to occur in nature[43]. It is presumably formed from tyramine[44].

The dehydration of *N*-monosubstituted formamides (**5**) with acylating agents and bases takes place in two steps. A base-catalysed *O*-acylation is followed by α-elimination of a proton and an oxyacid anion (reaction 7).

$$\text{RNHCHO} \xrightarrow[\text{base}]{\text{Y-Cl}} \text{R—N═C}\begin{matrix}\text{H}\\ \\ \text{O—Y}\end{matrix} \xrightarrow{\text{base}} \text{R—NC} \qquad (7)$$

(5)

(e.g. Y = $ArSO_2$, $POCl_2$, COCl)

Good dehydrating agents are arylsulphonyl chlorides in pyridine[37,38] or quinoline[39], phosphorus oxychloride in combination with pyridine or potassium *t*-butylate[21] and, above all, phosgene in the presence of tertiary amines (trimethyl amine, triethyl amine, tri-n-butylamine, *N*-methylmorpholine, *N,N*-diethylaniline, pyridine and quinoline)[15,40]. The phosgene method is the method of choice, as it can be carried out conveniently and in good yield (70–95 %), is economic, and in most cases the product is stable towards moderate excess of the reagent.

D. Further Modes of Formation

The reduction of some readily accessible imino compounds of tetravalent carbon might be a useful supplement of the other main methods for preparing the isonitriles described above, although the reduction conditions required are normally drastic and the yields are moderate.

The reduction of isocyanates can be carried out successfully by heating them with triethyl phosphite[45] or cyclic phosphorus (III) amides[46]. Isothiocyanates are reduced by triethyl phosphine[47], by copper[48], by triphenyl tin hydride[49] or photolytically (reaction 8)[50].

$$\text{RN═C═Z} \xrightarrow{\text{reduction}} \text{R—NC} \qquad (8)$$

(Z = O, S)

Trifluoromethyl isocyanide, the first representative of perfluorinated aliphatic isocyanides, was prepared by reacting CF_3NHCF_2Br with magnesium in tetrahydrofuran (reaction 9)[51]. The *O*-tosyl

$$CF_3NHCF_2Br + Mg \longrightarrow CF_3NC + MgBrF + HF \qquad (9)$$

oximes of 3,5-disubstituted 4-hydroxybenzaldehydes (**8**) and of *p*-dimethylaminobenzaldehyde yield, by elimination of *p*-toluenesulphonic acid, mixtures of the corresponding nitriles (**9**) and isonitriles (**10**). The nitrile formation is a β-elimination of a proton and a tosylate anion, while the isonitrile originates from an abnormal Beckmann rearrangement of the *syn*-isomer (reaction 10)[52]. The

$$\text{HO-}\overset{R}{\underset{R}{C_6H_2}}\text{-CH=N-O-Ts} \; (8) \xrightarrow[-30^\circ C]{C_2H_5OH} \text{HO-}\overset{R}{\underset{R}{C_6H_2}}\text{-CN} \; (9) \; + \; \text{HO-}\overset{R}{\underset{R}{C_6H_2}}\text{-NC} \; (10) \qquad (10)$$

reaction of the *syn*-oximes of benzaldehyde and some of its derivatives (**11**) with methylketene diethyl acetal (**12**) (reaction 12) is

$$\text{R-}C_6H_4\text{-CH=NOH} \; (11) + (C_2H_5O)_2C{=}CH{-}CH_3 \; (12) \xrightarrow{BF_3/HgO} \text{R-}C_6H_4\text{-CH=NOC(CH}_2\text{CH}_3\text{)(OC}_2\text{H}_5\text{)}_2 \; (13)$$

$$13 \longrightarrow \text{R-}C_6H_4\text{-CN} + \text{R-}C_6H_4\text{-NC} + CH_3CH_2COOC_2H_5 + C_2H_5OH \qquad (11)$$

analogous[53]. In an attempt to subject γ-benzil monoxime (**14**) to Beckmann's reaction by treating it with benzenesulphonyl chloride and alkali, phenyl isocyanide, benzoate and benzenesulfonate ions were formed[54] from the oxime sulphonate (**15**) by fragmentation of the rearrangement product (**16**) (reaction 12). α-Bromo-*N*-*t*-butyl-

$$\underset{(14)}{\text{PhC(=N-OH)COPh}} \xrightarrow[OH^-]{PhSO_2Cl} \underset{(15)}{\text{PhC(=N-OSO}_2\text{Ph)-COPh}} \longrightarrow \underset{(16)}{\text{PhSO}_2\text{O-C(=NPh)-COPh}} \xrightarrow{OH^-} \text{Ph-NC} + PhCO_2^- + PhSO_3^- \qquad (12)$$

cyclohexane carboxamide (**17**) can be cyclized by potassium *t*-butylate to form an α-lactam (**18**). The α-lactam isomerizes on heating to form (**19**) which undergoes α-cycloelimination[55] giving *t*-butylisocyanide (reaction 13).

$$\text{(17)} \xrightarrow{(CH_3)_3COK} \text{(18)} \tag{13}$$

$$\mathbf{18} \xrightarrow{75°c} [\mathbf{(19)}] \longrightarrow \text{cyclohexanone} + (CH_3)_3CNC$$

E. *Isocyano Groups Connected to Heteroatoms ('Heteroisonitriles')*

In addition to the normal isonitriles 'heteroisonitriles', whose isonitrile group is attached to nitrogen, have been described recently. These compounds are obtained by dehydration of formyl hydrazine derivatives (**20** and **21**) (reactions 14 and 15)[56–58]. Isonitriles (**23**)

$$\underset{(\mathbf{20})}{R_2NNHCHO} \xrightarrow{-H_2O} R_2NNC \tag{14}$$

$$\underset{(\mathbf{21})}{R_2C{=}NNHCHO} \xrightarrow{-H_2O} R_2C{=}NNC \tag{15}$$

which are derived from silanes, germanes and stannanes are in equilibrium with the corresponding 'nitriles' ($R_3M^{IV}CN$) and can be prepared from metal cyanides (**22**) by 'alkylation' [59–63] (reaction 16).

$$\underset{(\mathbf{22})}{M^ICN} + XM^{IV}R_3 \longrightarrow \underset{(\mathbf{23})}{R_3M^{IV}NC} + M^IX \tag{16}$$

$$(M^I = K, Ag \quad M^{IV} = Si, Ge, Sn \quad X = Cl, Br)$$

III. ISONITRILE REACTIONS

A. *Cleavage of the C—N Single Bond*

When isonitriles are heated to more than 200 °c, they rearrange to form nitriles. This reaction has been investigated in great detail[48,64–68]. The rearrangement of isonitriles whose isonitrile group is attached to a chirality centre occurs in some of the cases with retention, but sometimes also with racemization. The rearrangement of **24** takes place, for example, at 250 °c with complete racemization[69] (reaction 17). When *t*-butyl isocyanide is treated with boron

$$\text{3,4-}(OCH_2O)C_6H_3-CH_2-C(CH_3)(NC)-CO_2CH_3 \;(\mathbf{24}) \longrightarrow \text{3,4-}(OCH_2O)C_6H_3-CH_2-C(CH_3)(CN)CO_2CH_3 \qquad (17)$$

trifluoride etherate, a 'dimer' (**25**) is obtained[70] (reaction 18). Sodium

$$2(CH_3)_3CNC \xrightarrow{BF_3} (CH_3)_3CC(CN){=}NC(CH_3)_3 \;(\mathbf{25}) \qquad (18)$$

in liquid ammonia reduces isonitriles to form cyanide and the corresponding hydrocarbon[71,72].

B. Simple α-Additions

The capability of isonitriles to undergo α-additions was demonstrated by the end of the last century on the basis of a few simple examples, such as the formation of isonitrile dichlorides (**26**)[4] and 2-oximidochlorides (**27**)[4,73] and served as the essential argument for the formal bivalency of the isonitrile carbon (reactions 19 and 20).

$$R{-}NC + Cl_2 \longrightarrow RN{=}CCl_2 \;(\mathbf{26}) \qquad (19)$$

$$R^1{-}NC + R^2COCl \longrightarrow R^1N{=}C(Cl){-}CO{-}R^2 \;(\mathbf{27}) \qquad (20)$$

Further examples of α-addition of electrophilic halides are the reactions of isonitriles with *N,N*-dialkylamido chlorides[74], trihalomethane sulphenyl chlorides[75] and *N*-chlorosulphonamides[76] in the presence of alcohols. Many compounds of the HX type are also capable of simple α-additions to isonitriles, for example hydrogen sulphide[77], mercaptans at extremely high pressure[78], thiolophosphoric acids[79] and hydroxylamine[80] (reaction 21). Alcohols[81] and

$$R{-}NC + HX \longrightarrow R{-}N{=}C(H)X \qquad (21)$$

amines[82] form adducts only in the presence of catalytically active transition metal compounds; amines can also be added to isonitriles in the form of their trialkylplumbyl derivatives[83]. Some amine hydrochlorides[84] also add to isonitriles.

Triphenylphosphonium fluoroborate (**28**) yields stable adducts (**29**)[85] with isonitriles (reaction 22). The reactions of isonitriles with

$$R{-}NC + H\overset{+}{P}Ph_3BF_4^- \longrightarrow R{-}N{=}C(H)(\overset{+}{P}Ph_3)\ BF_4^- \qquad (22)$$

(**28**) (**29**)

$(R = t\text{-}C_4H_9,\ l\text{-}C_6H_{11},\ 2,6\text{-}(CH_3)C_6H_3)$

pyrrole[86] (reaction 23) and hydroxypyrazolone[87,88] (reaction 24) derivatives also formally correspond to this general type of α-addition,

$$\text{2-}CH_3\text{-3-}C_2H_5OOC\text{-4-}CH_3\text{-pyrrole (N-H)} + PhNC \longrightarrow \text{2-}CH_3\text{-3-}C_2H_5OOC\text{-4-}CH_3\text{-5-}(CH{=}NPh)\text{-pyrrole (N-H)} \qquad (23)$$

$$\text{3-}CH_3\text{-5-}OH\text{-pyrazole (N-H)} + PhNC \longrightarrow \text{3-}CH_3\text{-4-}(CH{=}NPh)\text{-5-}OH\text{-pyrazole (N-H)} \qquad (24)$$

as well as the syntheses (reaction 25) of α-hydroxyarylglyoxal derivatives[89,90] and the reaction of isonitriles with benzene azocarboxylates[91] (reaction 26).

$$\text{HO-naphthalene-OH} + 2\,PhNC \longrightarrow \text{HO-naphthalene-OH bearing } C(=N{-}Ph){-}CH(=N{-}Ph) \qquad (25)$$

$$C_6H_5{-}N{=}N{-}CO{-}OH + CN{-}C_6H_{11} \longrightarrow C_6H_5{-}CH{=}N{-}C_6H_{11} + N_2 + CO_2 \qquad (26)$$

$$C_6H_{11}\text{—NC} + C_6H_5\text{—MgBr} \longrightarrow C_6H_{11}\text{—N=C(MgBr)Ph} \quad (27)$$

As was shown for the reaction of phenyl magnesium bromide with cyclohexyl isocyanide, Grignard compounds form primarily α-adducts (reaction 27) with isonitriles[92–94], which subsequently are converted into a number of secondary products. Isonitriles react with some radical generators by α-addition of radicals[95] (reaction 28).

$$C_6H_{11}\text{—NC} + \cdot\text{OOC—}C_6H_5 \longrightarrow C_6H_{11}\text{—N=COCO—}C_6H_5$$

$$C_6H_{11}\text{—N=COCO—}C_6H_5 \xrightarrow{H\cdot} C_6H_{11}\text{—N=CHOCO—}C_6H_5 \quad (28)$$

$$C_6H_{11}\text{—N=COCO—}C_6H_5 \xrightarrow{Ph\cdot} C_6H_{11}\text{—N=C(}C_6H_5\text{)OCO—}C_6H_5$$

Just as in the case of α-additions to isonitriles, the reaction of isonitriles with chalkogens takes place with transition of the formally bivalent isonitrile carbon into tetravalent carbon (reaction 29).

$$\text{R—NC} + [\text{Y}] \longrightarrow \text{R—N=C=Y} \quad (\text{Y} = \text{O, S, Se}) \quad (29)$$

The oxidations of isonitriles to form isocyanates by means of ozone[96], *t*-butyl hypochlorite[97], dimethyl sulphoxide[98,99] and nitrile oxides[100] are examples of this type of reaction. Sulphur[48] or sodium polysulphide[101] and selenium[102–105] easily react with isonitriles to form isothiocyanates and isoselenocyanates. The reactions of isonitriles with dichlorocarbene[106] and arenesulphonyl nitrene[107] also belong to this category of reactions.

The polymerization of isonitriles[108] can formally be considered as a poly-α-addition.

C. α-Additions Followed by Secondary Reactions

It is known that numerous further isonitrile reactions, whose end-products are not α-adducts, occur via primary α-additions; stable end-products are formed from unstable α-adducts by fast spontaneous secondary reactions.

The tetrazole synthesis of Oliveri-Mandalá and Alagna[109] (reaction 30), for example, is an α-addition of a proton and an azide ion to isonitriles, followed by secondary ring closure.

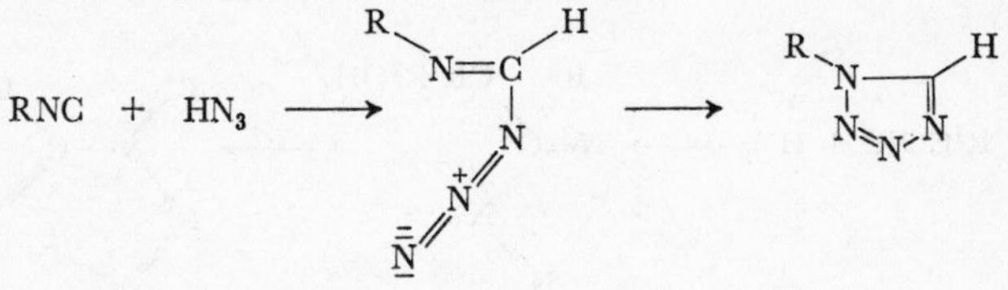

$$\text{RNC} + \text{HN}_3 \longrightarrow \text{R–N=C(H)–N=}\overset{+}{\text{N}}\text{=}\overset{-}{\text{N}} \longrightarrow \text{1-R-tetrazole (R–N, H on C; N=N–N ring)} \qquad (30)$$

I. Passerini reaction and related reactions

Passerini[110–115] found an elegant and generally applicable synthesis of α-acyloxycarboxamides (**30**) by reacting carboxylic acids and carbonyl compounds with isonitriles (reaction 31). A plausible

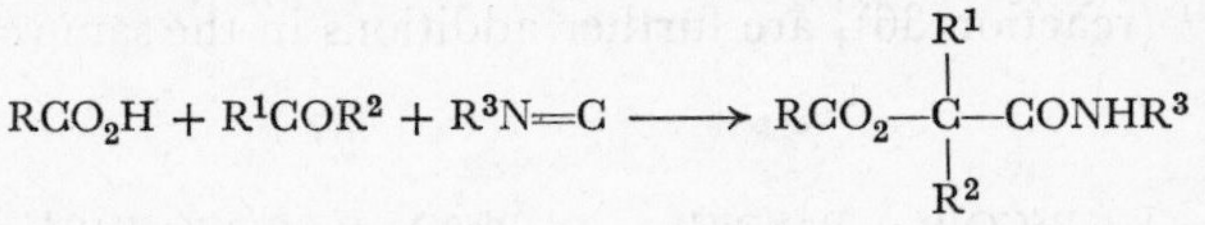

$$RCO_2H + R^1COR^2 + R^3N{=}C \longrightarrow RCO_2{-}C(R^1)(R^2){-}CONHR^3 \qquad (31)$$

(30)

reaction mechanism[116], which takes into account the ability of isonitriles to enter α-additions, is the α-addition of a hydrogen-bonded adduct of the carboxylic acid and carbonyl components, and a subsequent intramolecular acylation of the hydroxy group in the resulting α-adduct **31** (reaction 32). The Passerini reaction can be

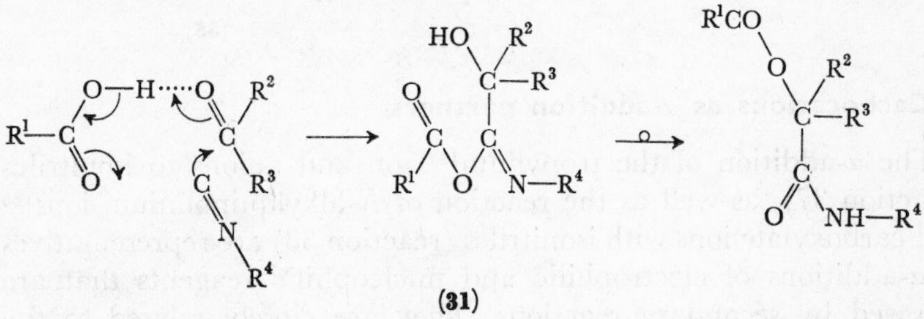

$$R^1{-}C(=O){-}O{-}H\cdots O{=}C(R^2)(R^3) + C{=}N{-}R^4 \longrightarrow R^1{-}C(=O){-}O{-}C(=N{-}R^4){-}C(OH)(R^2){-}R^3 \longrightarrow R^1CO{-}O{-}C(R^2)(R^3){-}C(=O){-}NH{-}R^4 \qquad (32)$$

(31)

used for the synthesis of depsipeptide (i.e. peptides in which some of the amide groups are replaced by ester groups) derivatives[117].

The synthesis of α-hydroxycarboxamides (**32**)[118,119] from isonitriles and carbonyl compounds in the presence of mineral acids (reaction 33) is closely related to the Passerini reaction, just as is the formation of tetrazole derivatives[116] from isonitriles, carbonyl compounds and hydrazoic acid (reaction 34).

$$R^3{-}NC + R^1R^2C{=}O + HX \longrightarrow \underset{(\mathbf{32})}{R^1R^2C(OH)CONHR^3} \qquad (33)$$

$$R^3{-}NC + R^1COR^2 + HN_3 \longrightarrow N{=}C(R^3)(CR^1R^2OH){-}N{=}N^+{=}N^- \longrightarrow \underset{(\mathbf{33})}{\text{1-}R^3\text{-5-}(CR^1R^2OH)\text{-tetrazole}} \qquad (34)$$

The preparation of α,γ-diketocarboxamides (**34**)[120] from isonitriles, ketenes and carboxylic acids (reaction 35), as well as the reaction of acyl isocyanates and carboxylic acids with isonitriles to yield **35**[121] (reaction 36), are further additions in the same category.

$$R^1R^2C{=}C{=}O + R^3CO_2H + R^4{-}NC \longrightarrow \underset{(\mathbf{34})}{R^3CO{-}C(R^1)(R^2){-}COCONHR^4} \qquad (35)$$

$$R^1CON{=}C{=}O + R^2CO_2H + R^3{-}NC \longrightarrow \underset{(\mathbf{35})}{R^1{-}CO{-}N(COR^2){-}CO{-}CO{-}NH{-}R^3} \qquad (36)$$

2. Carbocations as α-addition partners

The α-addition of the tropylium[122] ion and anions to isonitriles (reaction 37), as well as the reaction of *N*-alkylquinolinium ions[123] and carboxylate ions with isonitriles (reaction 38) are representatives of α-additions of electrophilic and nucleophilic reagents that are followed by secondary reactions. They are closely related to the four-component condensations with regard to the reaction mechanism. It appears advisable to mention these reactions here, since this would help to understand more easily the following discussion of four-component condensations.

RNC + [tropylium] ⟶ R—N≡C—(C₇H₆)H; NaOAc, H₂O ⟶ RNH—C(=O)—(C₇H₇) + NaI; NaN₃ ⟶ 1-R-5-(cycloheptatrienyl)tetrazole + NaN₃ (37)

R^1NC + [isoquinolinium $N^+—R^2$] + $R^3CO_2^-$ ⟶ $R^1N{=}C(O—C(=O)—R^3)$—(1,2-dihydroisoquinolin-1-yl, $N—R^2$) ⟶ $R^1NH—C(=O)$—(4-$C(=O)R^3$-1,2-dihydroisoquinoline, $N—R^2$) (38)

3. Four-component condensations*

Isonitriles (**40**) react with carbonyl compounds (**36**) (aldehydes or ketones), amines (**37**) and suitable acids (**38**) to form labile α-adducts (**41**), which are converted by spontaneous secondary reactions into stable end-products (reaction 39)[11–14,16,124,125,130]. The type of the secondary reaction depends mainly on the choice of the acid component.

Table 1 shows the various types of four-component condensation products (**42**) that arise from the various acid components (**38**) according to reaction 39.

* The terms 'α-aminoalkylation of isonitriles and acids' [11], or 'α-addition of immonium-ions and anions to isonitriles[6] followed by secondary reactions' [11], or 'Ugi reaction' [12,16,131,132,135] are also used for the 'four-component condensations', as well as the abbreviation 4CC.

$$R^2COR^3 + R^4NHR^5 + HX \rightleftharpoons R^2\overset{R^3}{C} \text{=====} \overset{R^4}{N}R^5X^- \quad (+)$$

(36) (37) (38) (39)

$$\mathbf{39} + R^1NC \xrightarrow{\alpha\text{-addition}} R^1N{=}\underset{X}{C}{-}\overset{R^2}{\underset{R^3}{C}}{-}\overset{R^4}{N}R^5 \xrightarrow[\text{reaction}]{\text{secondary}} \left[\begin{array}{c}\text{stable}\\ \text{end product}\end{array}\right] \qquad (39)$$

(40) (41) (42)

Ammonia, primary and secondary amines and hydrazine derivatives[130] can be used as amine components. Water, thiosulphuric acid, hydrogen selenide and hydrazoic acid are acid components which react with primary and secondary amines in an analogous manner, whereas hydrogen cyanate and thiocyanate (as well as carboxylic acids) are suitable acid components that react in different ways with primary and secondary amines. The α-adducts of these acids are acylating agents and they rearrange to form stable end-products, if the α-adduct contains an acylatable NH-group. If the reaction occurs in the absence of acylatable compounds, the combination of carboxylic acids and secondary amines leads to the formation of diacyl imides.

Hydrogen cyanide or hydrogen sulphide do not take part in four-component condensations as acid components, for these acids form stable α-amino alkylation products such as $CXR^2R^3{-}NR^4R^5$.

Four-component condensations are very easy to carry out. As a rule, the isonitrile component is added to the concentrated solution of the other three components while stirring and cooling (instead of the amines and the carbonyl compounds, it is also possible to use their condensation products, such as aminals, Schiff bases and enamines). The reaction product frequently crystallizes out. The yields of the four-component condensations are generally rather high (80–100%).

The formation of uniform products from four different reactants is accounted for by the fact that all possible side reactions are reversible, in contrast to the main reaction. As four different starting materials take part in the four-component condensation, the number of possible variations is exceptionally high.

The most rewarding preparative result of the four-component condensation will probably be the stereoselective peptide synthesis. By reacting an *N*-terminally protected α-amino acid (or a suitable

TABLE 1.

HX	Stable end-product	Note	Reference
HNCO	$R^1N{=}C$ ring: $C(R^2)(R^3)$ bonded to $R^1N{=}C$ and $N{-}R^4$; $R^1N{=}C{-}HN{-}C({=}O){-}N{-}R^4$	*a*	126
HSCN	$R^1N{=}C$ ring: $C(R^2)(R^3)$ bonded to $R^1N{=}C$ and $N{-}R^4$; $R^1N{=}C{-}HN{-}C({=}S){-}N{-}R^4$	*a*	127
R^6COOH	$R^1NHCO{-}C(R^2)(R^3){-}N(R^4){-}COR^6$	*a, b*	128, 129
R^6COOH	$R^6CO{-}N(R^1)CO{-}C(R^2)(R^3){-}NR^4R^5$	*b*	128, 131
H_2O	$R^1NHCO{-}C(R^2)(R^3){-}NR^4R^5$		11, 124
$H_2S_2O_3$	$R^1NHCS{-}C(R^2)(R^3){-}NR^4R^5$		11, 124
H_2Se	$R^1NHCSe{-}C(R^2)(R^3){-}NR^4R^5$		11, 124
HN_3	$R^1{-}N{-}C{-}C(R^2)(R^3){-}N(R^4)R^5$ (tetrazole ring: $R^1N{-}N{=}N{-}N{=}C$)		130–134

[a] $R^5 = H$; derivative of a primary amine.
[b] Primary amine components lead to the formation of α-acylamino carboxamides, whereas diacyl imides are obtained from secondary amine components in the absence of acylatable compounds; α-amino carboxamides[128,135] result in the presence of alcohols or amines.

peptide derivative) (**43**) with an optically active primary amine (**44**) that has a selectively-removable alkyl group (R^X)[129], an aldehyde (**45**) and a *C*-terminally protected α-amino acid (or a corresponding peptide derivative) with an isonitrile group replacing the amino group (**46**), it is possible to build up peptide derivatives (**47**). Two peptide linkages are formed by one single reaction step, and one new amino acid unit is built-up at the same time after removal of the R^X group to form **48**[11,14,136].

$$\underset{(43)}{\text{Peptide}_I\text{—COOH}} + \underset{(44)}{\underset{\displaystyle R^X}{\text{NH}_2}} + \underset{(45)}{\overset{\displaystyle R}{\text{CHO}}} + \underset{(46)}{\text{CN—Peptide}_{II}} \longrightarrow$$

$$\underset{(47)}{\text{Peptide}_I\text{—CON}\underset{\displaystyle R^X}{}\overset{\displaystyle R}{\text{CH}}\text{CONH—Peptide}_{II}} \longrightarrow$$

$$\underset{(48)}{\text{Peptide}_I\text{—CONH}\overset{\displaystyle R}{\text{CH}}\text{CONH—Peptide}_{II}} \qquad (40)$$

The synthesis of peptides by means of four-component condensation requires the newly-formed centre of chirality to be built-up highly stereoselectively, by using suitable chiral amine components. Model reactions have to be carried out as preliminary studies for stereospecific peptide syntheses.

The investigation of the concentration ratio of the diastereoisomeric reaction products of isobutyraldehyde-(S)-α-phenylethylimine with benzoic acid and *t*-butyl isocyanide as a function of varied reaction conditions yielded results from which it was possible to draw conclusions regarding the mechanism of the four-component condensations[136,137].

It is to be hoped that when the mechanism of the reaction is known, it may be possible to synthetize peptides by a stereoselective four-component condensation. This may well not only be superior to the classical syntheses regarding ease of operation and yields, but also with respect to building long chain peptides which so far could not be prepared for solubility reasons, since the four-component condensations allow a wider range of solvent choice than the classical acylation procedures[136].

D. Cycloadditions and Cyclizations

As isonitriles are capable of α-additions, they can undergo a large variety of different types of cycloadditions and ring-closure reactions. These will be described here in the order of the formal reaction pattern, disregarding the assumed reaction mechanism.

Isonitriles react with nitrosotrifluoromethane[138] and 1,1-dicyano-2,2-bis(trifluoromethyl)ethylene[139] by (1 + 1 + 2) cyclization (reaction 41). The formation of glyoxylic acid derivatives (**51**) from

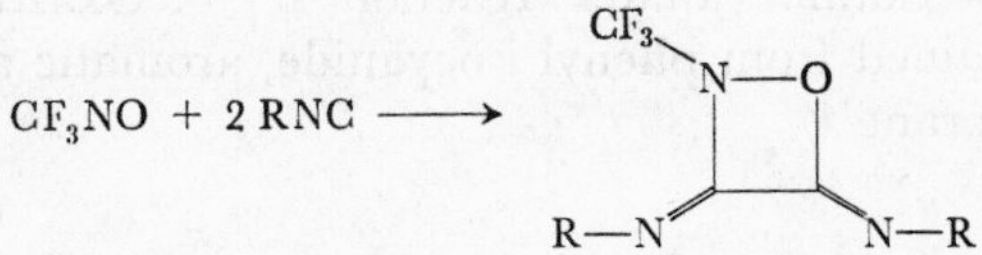

$$CF_3NO + 2\,RNC \longrightarrow \text{(49)} \qquad (41)$$

(49)

isonitriles and carbonyl compounds[140] in the presence of boron trifluoride takes place through an intermediate of the type **50**, which can be isolated[70] if the reaction is carried out under mild conditions (reaction 42). The reaction between α-naphthol and

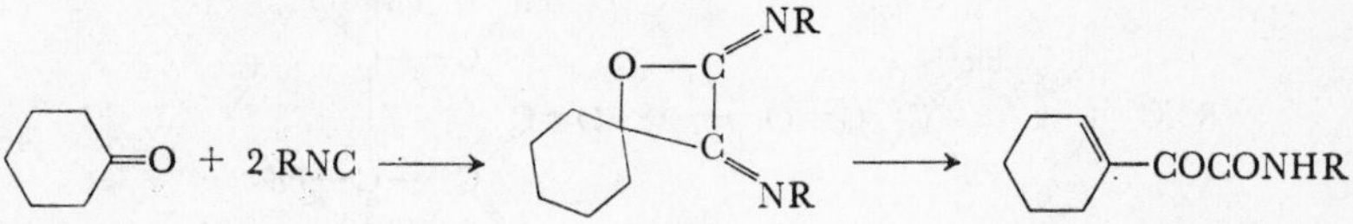

$$\text{cyclohexanone} + 2\,RNC \longrightarrow \text{(50)} \longrightarrow \text{(51)}\ \text{–COCONHR} \qquad (42)$$

(50) **(51)**

phenyl isocyanide[141] can be considered as a (1 + 1 + 3) ring closure (reaction 43). However, since a carbon atom seems to be lost, further investigation is necessary to clarify the reaction course.

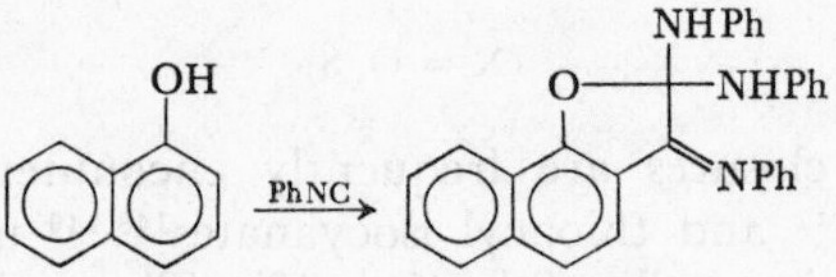

$$\text{α-naphthol (OH)} \xrightarrow{PhNC} \text{product with O–C(NHPh)}_2\text{–C(=NPh)} \qquad (43)$$

Cycloadditions and ring closures of the type (1 + 2 + 2) occur in many variations. Acetylenedicarboxylic esters (reaction 44)[142], fluoroketones (reaction 45)[139,143,144] and ketenes (reaction 46)[120], as well as ketenes in combination with chloral[143] form with isonitriles

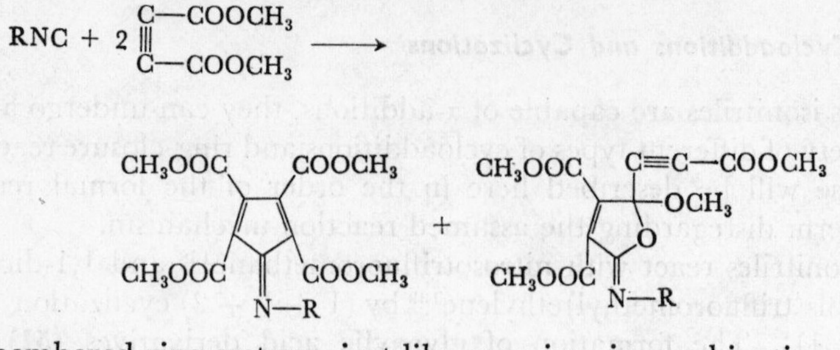

(44)

five-membered ring systems, just like enamines in combination with isocyanates or isothiocyanates (reaction 47)[145]. Oxazolidine derivatives are obtained from phenyl isocyanide, aromatic aldehydes and tri-n-butyl borane[146].

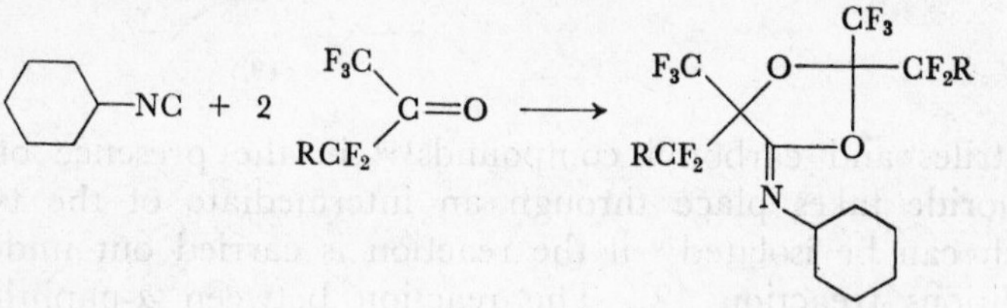

(45)

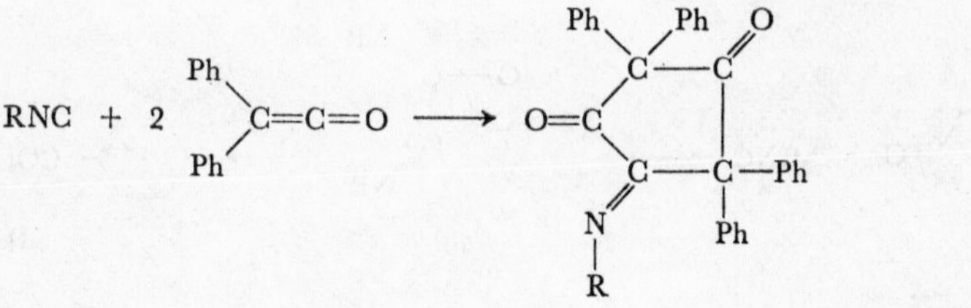

(46)

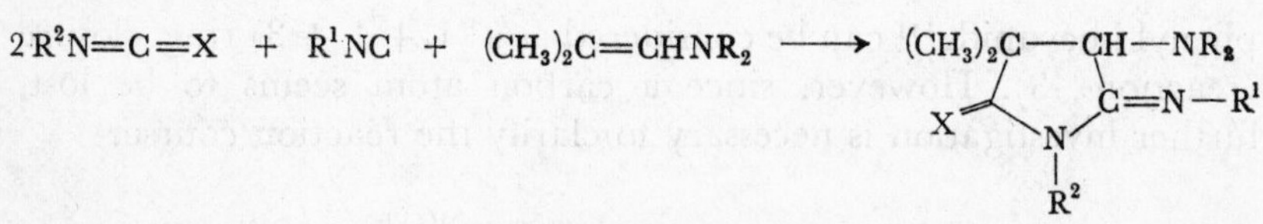

(X = O, S) (47)

(1 + 4) ring closures are frequently encountered in isonitrile chemistry. Acyl[147] and thioacyl isocyanates[148,149] undergo (1 + 4) cyclization with isonitriles (reaction 48). The union of trifluoroacetaldehyde *N*-acylimines (**52**) and isonitriles leads to cycloadducts (**53**)[144,150] that are intermediates for the synthesis of fluorinated peptides (reaction 49). The reaction of the zwitterion (**54**) that is formed by loss of nitrogen from benzenediazonium *o*-carboxylate,

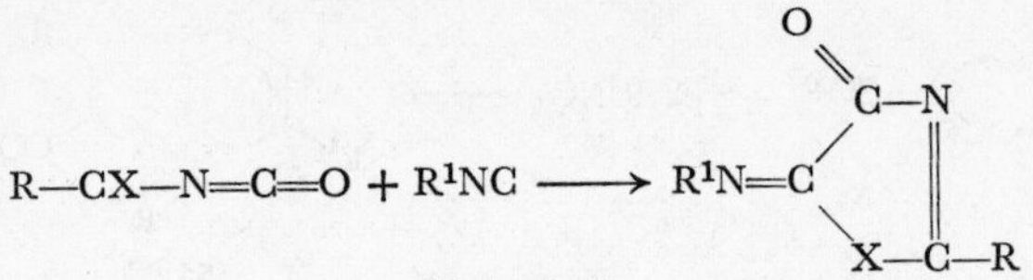

$$\text{R—CX—N=C=O} + \text{R}^1\text{NC} \longrightarrow \text{R}^1\text{N=C}\!\!\begin{array}{l}\text{C(=O)—N} \\ \text{X—C—R}\end{array} \qquad (48)$$

$$(\text{X} = \text{O, S})$$

with isonitriles (reaction 50)[151,152] and probably the reaction of α-nitroso-β-naphthol and 4-isocyanoazobenzene (reaction 51)[153] are

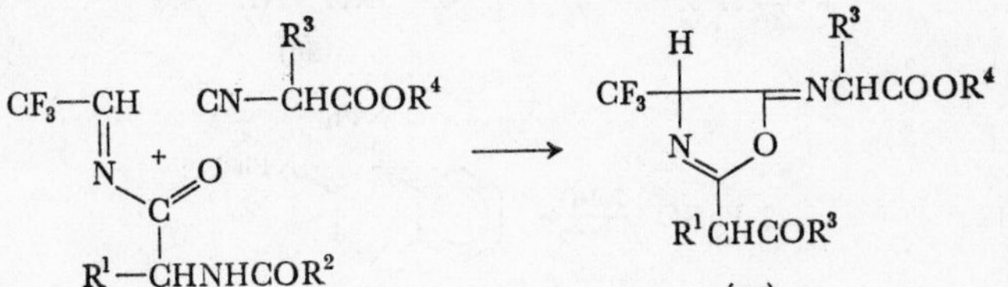

$$\text{CF}_3\text{—CH=N}^+\text{(O}^-\text{)—C—CHR}^1\text{NHCOR}^2 \; (\mathbf{52}) + \text{CN—CHR}^3\text{COOR}^4 \longrightarrow \mathbf{53} \qquad (49)$$

additional (1 + 4) cyclizations, although the details of the process are not known.

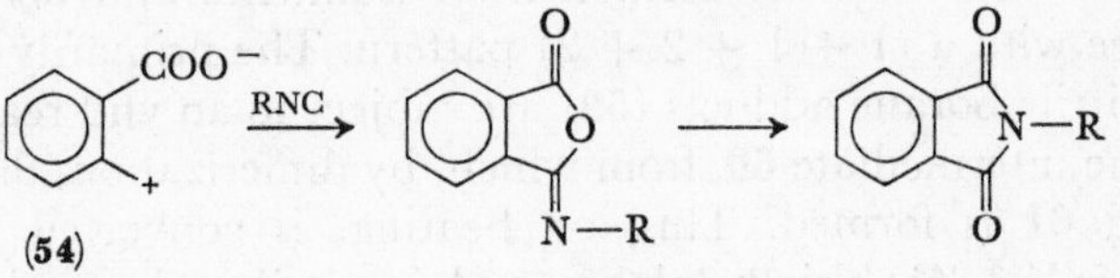

$$\mathbf{54} \xrightarrow{\text{RNC}} \longrightarrow \text{N—R} \qquad (50)$$

$$\alpha\text{-nitroso-}\beta\text{-naphthol} + \text{C}_6\text{H}_5\text{—N=N—C}_6\text{H}_4\text{—NC} \longrightarrow \text{NH—C(=N—C}_6\text{H}_4\text{—N=N—C}_6\text{H}_5\text{)—O} \qquad (51)$$

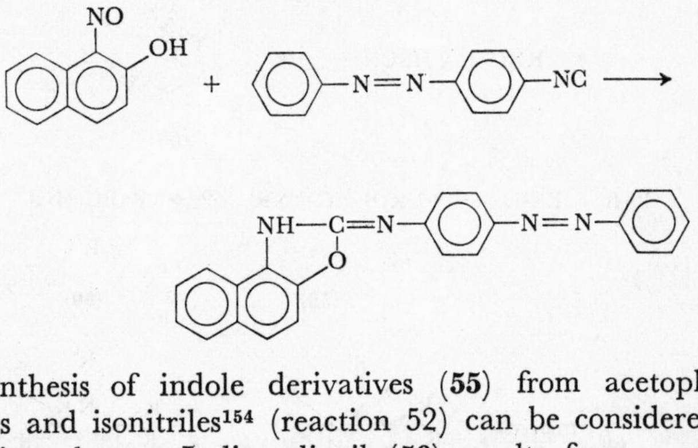

The synthesis of indole derivatives (**55**) from acetophenone derivatives and isonitriles[154] (reaction 52) can be considered as a (2 + 3) ring closure. Indigo dianil (**56**) results from a complex tetramerization (reaction 53)[155] of phenyl isocyanide. Compound **57** is formed from phenyl isocyanide in the presence of nitrosobenzene (reaction 54)[156]. Isonitriles and two moles of thiocyanic

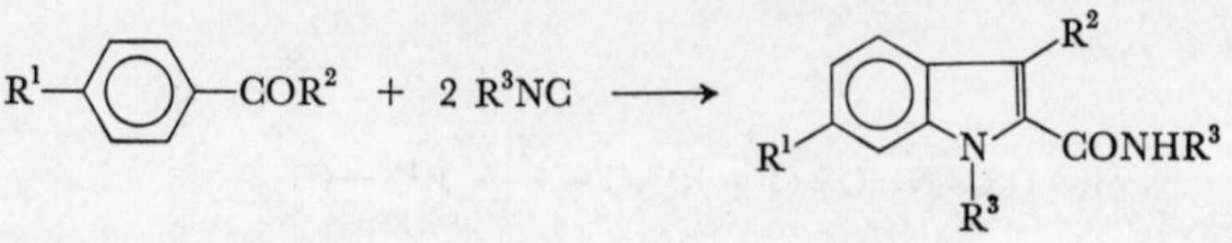

$$R^1\text{-}C_6H_4\text{-}COR^2 + 2\,R^3NC \longrightarrow \mathbf{(55)} \qquad (52)$$

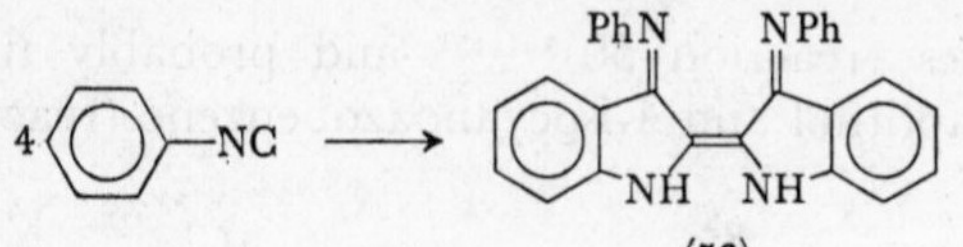

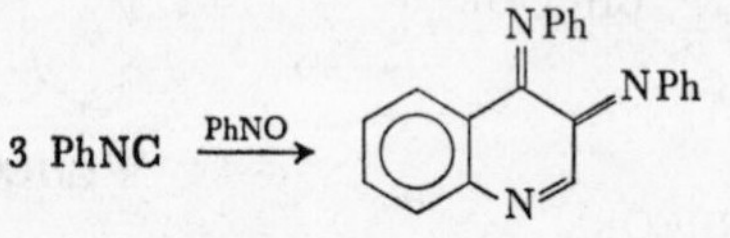

$$4\,C_6H_5NC \longrightarrow \mathbf{(56)} \qquad (53)$$

$$3\,PhNC \xrightarrow{PhNO} \mathbf{(57)} \qquad (54)$$

acid form by multistep (2 + 2 + 2) ring closure, 1-alkyl- or 1-aryl-5-azadithiouraciles (**58**) (reaction 55)[157,158].

Six-membered rings are formed from isonitriles and boranes in accordance with a (1 + 1 + 2 + 2) pattern. The primarily formed polar isonitrile borane adducts (**59**) are subject to an ylid rearrangement to the intermediate **60**, from which, by dimerization, the cyclic compound **61** is formed. This, on heating, is converted into **62** (reaction 56)[159–166]. Trialkylalanes and isonitriles also yield polar

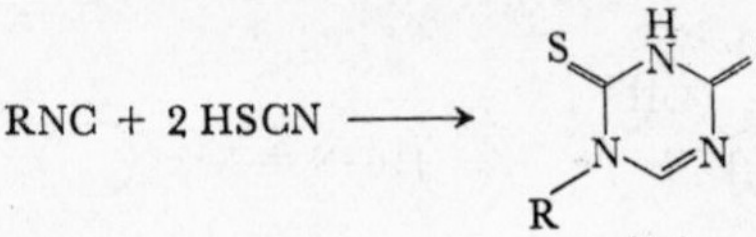

$$RNC + 2\,HSCN \longrightarrow \mathbf{(58)} \qquad (55)$$

$$R_3B + R^1NC \rightleftharpoons \underset{\mathbf{(59)}}{R_3\bar{B}\text{—}\overset{+}{C}\text{=}NR^1} \longrightarrow \underset{\mathbf{(60)}}{R_2BC(R)\text{=}NR^1}$$

$$\mathbf{60} \xrightarrow{\text{Dimerize}} \mathbf{(61)} \longrightarrow \mathbf{(62)} \qquad (56)$$

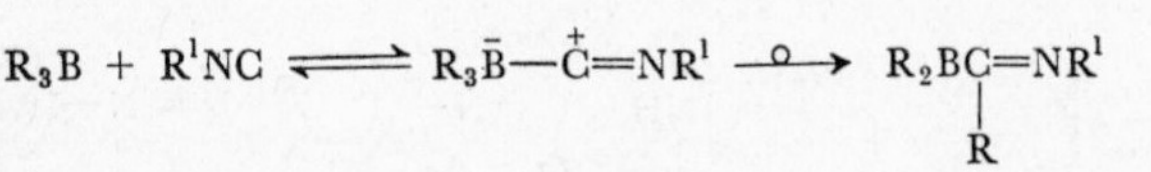

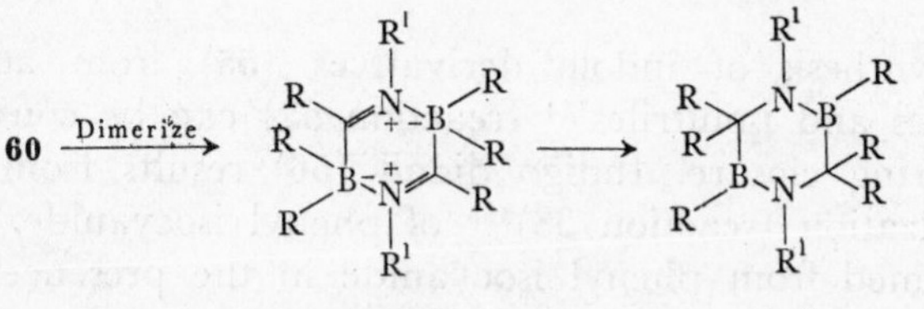

adducts, from which it has not been possible, however, to obtain cyclic dimers[167].

IV. ISONITRILE COMPLEXES

Numerous types of complexes are derived from the elements of the groups IB, IIB, VIB, VIIB and VIII as well as from the lanthanides, but no isonitrile complexes of the elements of the groups IA–VA are known.

Isonitrile complexes of almost all the valency stages of transition metals of the VI–VIII groups are known. These complexes are in

TABLE 2.

Complex	Reference	Complex	Reference
$CuI(CN—CH_3)$	169	$[Ag(CN—C_7H_7)_4]NO_3{\cdot}H_2O^a$	172
$CuCN(CN—C_2H_5)_2$	170	$AuCl(CN—Ph)$	174
$CuCN(CN—C_2H_5)_3$	171	$AuCl(CN—Ph)_4$	174
$[Cu(CN—Ph)_4]Cl{\cdot}6\,H_2O$	172	$AuCl_3(CN—C_7H_7)^a$	174
$AgCN(CN—CH_3)$	173	$ZnCl_2(CN—C_7H_7)_2{}^a$	175
$[Ag(CN—C_7H_7)_2]NO_3{}^a$	172	$CdCl_2(CN\text{-}C_7H_7)^a$	174

^{a} C_7H_7 = *p*-tolyl.

most cases derived from metal atoms and ions with free *d*-orbitals and belong to the 'inner' type. Compounds derived from lower valencies of the metals are the rule[168].

The isonitrile complexes of the elements of the groups IB and IIB are distinguished from the other ones by being derived from metal ions with filled outer orbitals of 18 electrons. The metals occur here in their usual valency, and the isonitrile complexes largely correspond to the complexes with other ligands. Some examples are given in Table 2. Very thermally stable isonitrile complexes (**63**) are obtained from tricyclopentadienyl derivatives of the lanthanides (**76**).

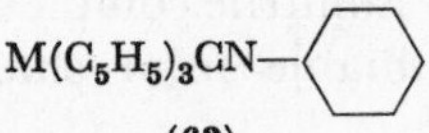

$$M(C_5H_5)_3CN\text{—}C_6H_{11}$$

(**63**)

(M = Yb, Ho, Tb)

Chromium, molybdenum and tungsten, the elements of group VIB form two classes of isonitrile complexes: complexes in which the central atom is zero-valent and complexes in which the central atom is tri- or tetravalent. The latter complexes are normally unstable, the stability rising with the atomic number of the central atom.

Isonitrile complexes of Cr, Mo and W are obtained as follows (reactions 57–62):

$$(C_2H_5)_4N^+[M(CO)_5X]^- + R{-}NC \longrightarrow M(CO)_5(CN{-}R) + M(CO)_4(CN{-}R)_2 + M(CO)_3(CN{-}R)_3 \quad (57)^{177}$$

(M = Cr, Mo, W X = Cl, I R = Ph, cyclohexyl)

$$3\,Cr^{2+} + 18\,Ph{-}NC \longrightarrow Cr(CN{-}Ph)_6 + 2[Cr(CN{-}Ph)_6]^{3+} \quad (58)^{178,179}$$

$$2\,MoCl_3 + 12\,Ph{-}NC + 3\,Mg \longrightarrow 2\,Mo(CN{-}Ph)_6 + 3\,MgCl_2 \quad (59)^{181}$$

$$K_4[Mo(CN)_8] \xrightarrow{(CH_3)_2SO_4} [Mo(CN)_4(H_2O)_2(CN{-}CH_3)_2]\cdot 4\,H_2O + K_2SO_4 \quad (60)^{180}$$

$$WCl_4 + 6\,Ph{-}NC + 2\,Mg \longrightarrow W(CN{-}Ph)_6 + 2\,MgCl_2 \quad (61)^{182}$$

$$Ag_4[W(CN)_8] + 4\,CH_3I \longrightarrow W(CN)_4(CN{-}CH_3)_4 + 4\,AgI \quad (62)^{183}$$

Isonitrile complexes of both manganese (I) as well as manganese (II) are known. The isonitrile manganese complexes were discovered relatively late, because the salts and the coordination compounds of bivalent and trivalent manganese do not react with isonitriles.

Sacco[184] found that anhydrous MnI_2 is an exception (reaction 63).

$$2\,MnI_2 + 12\,CH_3{-}NC \longrightarrow [Mn(CN{-}CH_3)_6]I + [Mn(CN{-}CH_3)_6]I_3 \quad (63)$$

64[185] is formed from cyclopentadienylmanganese tricarbonyl and cyclohexyl isocyanide on irradiation. The isonitrile complexes of

$$C_5H_5Mn(CO)_3 + C_6H_{11}{-}NC \xrightarrow{h\nu} C_5H_5Mn(CO)_2CN{-}C_6H_{11} \quad (64)$$

(**64**)

manganese(II) can be obtained from manganese(I) isonitrile complexes by oxidation[186].

Only a few rhenium (I) isonitrile complexes have been described[187]. An example is **65**, obtainable from chloropentacarbonylrhenium (I) and *p*-tolyl isocyanide.

$$[Re(CO)_4(CN{-}C_7H_7)_2]Cl$$

(**65**)

A large variety of isonitrile complexes are derived from different valence states of the elements of group VIII. The isonitrile complexes of iron have been investigated in great detail. In contrast to iron pentacarbonyl, which reacts with isonitriles only at temperatures above 70 °C[172] triiron dodecacarbonyl reacts quickly and easily with isonitriles at low temperature (reaction 65)[188]. Dinitrosyl iron dicar-

$$Fe_3(CO)_{12} + 3\ R\text{—}NC \longrightarrow 3\ Fe(CO)_4CN\text{—}R$$
$$Fe_3(CO)_{12} + 6\ R\text{—}NC \longrightarrow 3\ Fe(CO)_3(CN\text{—}R)_2 + 3\ CO \qquad (65)$$

bonyl[188] easily reacts with isonitriles to form complexes of the formula **66**, which are also accessible from isonitriles and $K_2[Fe_2(NO)_4S_2]$ or $NH_4[Fe(NO)_2(S_2O_3)]$[189]. The dinuclear complex **67** may formally

$$Fe(NO)_2(CN\text{—}R)_2$$
(66)

be considered as an iron (I) isonitrile complex[190]. Iron (II) complexes of the type **68** can be obtained by reacting iron pentacarbonyl

O, C, Fe, Fe, OC, CO, C, N, Ph

(67)

with iodine and isonitriles[188]. Only in exceptional cases, such as that

$$FeI_2(CO)_{4-x}(CN\text{—}R)_x \qquad (x = 1\text{–}3)$$
(68)

of reaction (66), is it possible to obtain iron (II) and iron (III)

$$FeCl_2 + 4\ CH_3\text{—}NC \longrightarrow FeCl_2(CN\text{—}CH_3)_4 \qquad (66)$$

isonitrile complexes from iron salts and isonitriles[191]. Most iron isonitrile complexes are obtained by alkylation of ferrocyanides[192–199]. The first example (reaction 67) was given by Freund in 1888[192]:

$$K_4[Fe(CN)_6] + 4\ C_2H_5I \longrightarrow Fe(CN)_2(CN\text{—}C_2H_5)_4 + 4\ KI \qquad (67)$$

As with the iron compounds, the cobalt isonitriles can be obtained by reacting isonitriles with cobalt nitrosocarbonyl[189] (reaction 68) or cobalt (II) salts (reaction 69)[191,200–204], or by alkylation of hexacyanocobaltates (III) (reaction 70)[193,205,206]. Cobalt (II) isonitrile complexes can be reduced to cobalt (I) isonitrile complexes (reaction 71)[207,208]. Cobalt (I) isonitrile complexes can also be obtained from

$$Co(NO)(CO)_3 + 2\ R{-}NC \longrightarrow Co(NO)(CO)(CN{-}R)_2 + 2\ CO \quad (68)$$

$$CoX_2 + nCN{-}R \longrightarrow CoX_2(CN{-}R)_n \quad (69)$$
$$(X = Cl, Br, I, SCN, ClO_4;\ n = 4, 5)$$

$$Ag_3[Co(CN)_6] + \xrightarrow{CH_3I} Co(CN)_3(CN{-}CH_3)_3 + 3\ AgI \quad (70)$$

$$[Co(CN{-}R)_5](ClO_4)_2 \xrightarrow{N_2H_4} [Co(CN{-}R)_5]ClO_4 \quad (71)$$

dicobalt octacarbonyl (reaction 72)[209–211]. The corresponding cobalt

$$Co_2(CO)_8 + 5\ R{-}NC \longrightarrow [Co(CN{-}R)_5]^+[Co(CO)_4]^- + 4\ CO \quad (72)$$

(III) compounds are obtained by oxidation of cobalt (II) isonitrile complexes[212].

Nickel (II) isonitrile complexes can also be obtained from isonitriles and nickel (II) salts[213], but the compounds obtained are not stable enough to be well characterized.

Nickel tetraisonitrile complexes are obtained from nickel tetracarbonyl and aryl isocyanides (reaction 73)[209,214,215]. Nickel tetra-

$$Ni(CO)_4 + 4\ Ar{-}NC \longrightarrow Ni(CN{-}Ar)_4 + 4\ CO \quad (73)$$

isonitrile complexes can also be obtained from nickel (II) compounds and isonitrile in the presence of reducing agents[216,217]. Alkyl isocyanides and nickel tetracarbonyl preferably form compounds of the type **69**[209]. Apart from just a few exceptions, the noble metal

$$Ni(CO)(CN{-}R)_3$$
(69)

complexes of the group VIII largely correspond to the complexes of iron, cobalt and nickel. Some examples to illustrate this are given below.

$RuX_2(CN{-}R)_4$[218]	X = Cl, Br, I, CN
$[Rh(CN{-}C_7H_7)_4]ClO_4$[219]	
$[Rh(CN{-}R)_2Ph_3Y_2]X$[220]	R = p-$CH_3OC_6H_4$, X = Cl, I, ClO_4, Y = P, As, Sb
$[Rh_2I_2(CN{-}Ph)_8]I_2$[221]	
$Pd(CN{-}c{-}C_6H_{11})_2$[222]	
$PdX_2(CN{-}Ph)_2$[223]	X = I
$PtX_2(CN{-}Ph)_2$[224]	X = Cl, Br, I, NO_2
$[Pt(CN{-}Ph)_4][PtCl_4]$[225]	
$[(CH_3{-}NC)_4Pt(NHNH_2)_2Pt(CN{-}CH_3)_4](ClO_4)_2 \cdot 2\ H_2O$[226]	

V. BIOLOGICAL PROPERTIES AND TECHNOLOGICAL USES

The biosynthesis of xanthocillin[43] shows that living organisms are capable of producing isonitriles. Only a few investigations on the biochemical properties have been carried out[227–229].

A relatively large number of isonitriles have a very strong acaricidal effect. This property, in combination with their insecticidal and fungicidal effects, as well as their generally low toxicity to warm-blooded animals, indicates possibilities of using isonitriles in agriculture[230–236].

VI. REFERENCES

1. A. Gautier, *Ann. Chem.*, **142,** 289 (1867); **146,** 119 (1868); **149,** 29, 155 (1869); **151,** 239 (1869); *Ann. Chim.* (4), **17,** 103, 203 (1869).
2. A. W. Hofmann, *Compt. Rend.*, **65,** 484 (1867); *Ann. Chem.*, **144,** 114 (1867); **146,** 107 (1868).
3. M. H. Guillemard, *Ann. Chim.* [8], **14,** 311 (1908).
4. I. U. Nef, *Ann. Chem.*, **270,** 267 (1892); **280,** 291 (1894); **287,** 265 (1895); **298,** 202, 368 (1897); **309,** 126 (1899).
5. H. Lindemann and L. Wiegrebe, *Ber.*, **63,** 1650 (1930).
6. D. L. Hammick, R. G. A. New, N. V. Sidgwick and L. E. Sutton, *J. Chem. Soc.*, **1930,** 1876.
7. N. V. Sidgwick, *Chem. Rev.*, **9,** 77 (1931).
8. R. G. A. New and L. E. Sutton, *J. Chem. Soc.*, **1932,** 1415.
9. V. Migrdichian, *The Chemistry of Organic Cyanogen Compounds*, Reinhold, New York, 1947, p. 393.
10. P. Kurtz, *Methoden der Organischen Chemie*, Vol. 8 (Ed. Houben-Weyl), Thieme, Stuttgart, 1952, p. 351.
11. I. Ugi, *Angew. Chem.*, **74,** 9 (1962); *Angew. Chem. Int. Ed. Engl.*, **1,** 8 (1962); *Neuere Präparative Methoden der Organischen Chemie*, Vol. 4, Verlag Chemie, Weinheim, Germany, 1966, p. 1.
12. K. Sjöberg, *Svensk Kem. Tidskr.*, **75,** 493 (1963).
13. R. Oda and T. Shono, *J. Soc. Org. Synt. Chem. Japan*, **22,** 695 (1964).
14. I. Ugi, *Akad. Wiss.* (*Göttingen*), Vandenhoeck and Rupprecht, Göttingen, 1965, p. 21.
15. I. Ugi, U. Fetzer, U. Eholzer, H. Knupfer and K. Offermann, *Angew. Chem.*, **77,** 492 (1965); *Angew. Chem. Int. Ed. Engl.*, **4,** 472 (1965).
16. N. P. Gambarjan, *Zh. Vses. Khim. Obshchestva im. D. I. Mendeleeva*, **12,** 65 (1967).
17. G. Herzberg, *Molecular Spectra and Molecular Structure*, Vol. 2, D. Van Nostrand Co., Princeton, N.Y., 1959, p. 332.
18. H. W. Thompson and R. L. Williams, *Trans. Faraday Soc.*, **48,** 502 (1952).
19. R. L. Williams, *J. Chem. Phys.*, **25,** 656 (1956).
20. M. G. K. Pillai and F. F. Cleveland, *J. Mol. Spectry.*, **5,** 212 (1960).
21. I. Ugi and R. Meyr, *Chem. Ber.*, **93,** 239 (1960).
22. R. G. Gillis and J. L. Occolowitz, *Spectrochim. Acta*, **19,** 873 (1963).

23. I. D. Kurtz, P. v. R. Schleyer and A. Allerhand, *J. Chem. Phys.*, **35**, 1533 (1961).
24. A. Loewenstein and Y. Margalit, *J. Phys. Chem.*, **69**, 4152 (1965).
25. R. G. Gillis and J. L. Occolowitz, *J. Org. Chem.*, **28**, 2924 (1963).
26. M. Dessler, H. Ring, R. Trambornlo and W. Gordy, *Phys. Rev.*, **79**, 54 (1950).
27. R. Trambornlo and W. Gordy, *Phys. Rev.*, **79**, 224 (1950).
28. B. Bok, L. Hansen-Nygaard and A. Rastrup-Andersen, *J. Mol. Spectry.*, **2**, 54 (1958).
29. C. P. Smyth, *Dielectric Behaviour and Structure*, McGraw-Hill, New York, 1955, p. 249, 283, 317, 321.
30. R. G. Gillis, *J. Org. Chem.*, **27**, 4103 (1962).
31. E. Pamain, *Proc. 11th Int. Congr. Pure Appl. Chem. London*, 241 (1947).
32. M. Sware and J. W. Taylor, *Trans. Faraday Soc.*, **47**, 1293 (1951).
33. (a) J. L. Franklin, *Ind. Eng. Chem.*, **41**, 1070 (1949);
    (b) I. Ugi, *Isonitrile Chemistry*, Academic Press, New York, in press.
34. P. v. R. Schleyer and A. Allerhand, *J. Am. Chem. Soc.*, **84**, 1322 (1962).
35. L. L. Ferstandig, *J. Am. Chem. Soc.*, **84**, 1323, 3553 (1962).
36. A. Allerhand and P. v. R. Schleyer, *J. Am. Chem. Soc.*, **85**, 866 (1963).
37. I. Hagedorn and H. Tönjes, *Pharmazie*, **12**, 567 (1957).
38. W. R. Hertler and E. J. Corey, *J. Org. Chem.*, **23**, 1221 (1958).
39. J. Casanova, R. E. Schuster and N. D. Werner, *J. Chem. Soc.*, **1963**, 4280.
40. I. Ugi, W. Betz, U. Fetzer and K. Offermann, *Chem. Ber.*, **94**, 2814 (1961).
41. A. P. Krapcho, *J. Org. Chem.*, **27**, 1089 (1962).
42. R. Wegler, private communication.
43. W. Rothe, *Pharmazie*, **5**, 190 (1950).
44. H. Achenbach and H. Grisebach, *Z. Naturforsch.*, **20B**, 137 (1965).
45. T. Mukaiyama, H. Nambu and M. Okamoto, *J. Org. Chem.*, **27**, 3651 (1962).
46. T. Mukaiyama and Y. Yokota, *Bull. Chem. Soc. Japan*, **38**, 858 (1965).
47. A. W. Hofmann, *Ber.*, **3**, 766 (1870).
48. W. Weith, *Ber.*, **6**, 210 (1873).
49. D. H. Lorenz and E. J. Becker, *J. Org. Chem.*, **28**, 1707 (1963).
50. U. Schmidt and K. H. Kabitzke, *Angew. Chem.*, **76**, 687 (1964); *Angew. Chem. Int. Ed. Engl.*, **3**, 641 (1964).
51. S. P. Makarov, M. A. Euglin, A. F. Bideiko and T. V. Nikolajeva, *Z. Obshch. Khim.*, **37**, 2781 (1967).
52. E. Müller and B. Narr, *Z. Naturforsch.*, **16B**, 845 (1961).
53. T. Mukaiyama, K. Tonooka and K. Inoue, *J. Org. Chem.*, **26**, 2202 (1961).
54. A. Werner and A. Piguet, *Ber.*, **37**, 4295 (1904).
55. J. C. Sheehan and I. Lengyel, *J. Am. Chem. Soc.*, **86**, 746, 1356 (1964).
56. H. Bredereck, B. Föhlisch and K. Walz, *Angew. Chem.*, **74**, 388 (1962).
57. H. Bredereck, B. Föhlisch and K. Walz, *Ann. Chem.*, **686**, 92 (1965); **688**, 93 (1965).
58. I. Hagedorn and U. Eholzer, *Angew. Chem.*, **74**, 499 (1962); *Angew. Chem. Int. Ed. Engl.*, **1**, 514 (1962).
59. T. A. Bither, W. H. Knoth, R. V. Lindsey and W. H. Sharkey, *J. Am. Chem. Soc.*, **80**, 4151 (1958).
60. J. J. McBride, *J. Org. Chem.*, **24**, 2029 (1959).
61. D. Seyferth and N. Kahlen, *J. Am. Chem. Soc.*, **82**, 1080 (1960); *J. Org. Chem.*, **25**, 809 (1960).
62. A. D. Craig, J. V. Urenovich and A. G. MacDiarmid, *J. Chem. Soc.*, **1962**, 548.
63. R. A. Cummins and P. Dunn, *Australian J. Chem.*, **17**, 411 (1964).
64. G. Kohlmaier and B. S. Rabinovitch, *J. Phys. Chem.*, **63**, 1793 (1959).

65. F. W. Schneider and B. S. Rabinovitch, *J. Am. Chem. Soc.*, **84**, 4215 (1962); **85**, 2365 (1963).
66. R. W. Horobin, N. R. Khan, J. McKenna and B. G. Hutley, *Tetrahedron Letters*, **1966**, 5087.
67. F. J. Fletcher, B. S. Rabinovitch, K. W. Watkins and D. J. Locker, *J. Phys. Chem.*, **70**, 2823 (1966).
68. J. Casanova, N. D. Werner and R. E. Schuster, *J. Org. Chem.*, **31**, 3473 (1966).
69. D. Marquarding, unpublished results.
70. H.-J. Kabbe, *Angew. Chem.*, **80**, 406 (1968); *Angew. Chem. Int. Ed. Engl.*, **7**, 389 (1968); *Chem. Ber.*, **102**, 1404 (1969).
71. I. Ugi and F. Bodesheim, *Chem. Ber.*, **94**, 1157 (1961).
72. W. Büchner and R. Dufaux, *Helv. Chim. Acta*, **49**, 1145 (1966).
73. I. Ugi and U. Fetzer, *Chem. Ber.*, **94**, 1116 (1961).
74. Y. Ito, M. Okano and R. Oda, *Tetrahedron*, **22**, 447 (1966).
75. E. Kühle (Farbenfabriken Bayer A.G.) *Ger. Pat.* 1,163,315 (1962, 1964).
76. W. Aumüller (Farbwerke Höchst) *Ger. Pat.* 1,187,234, (1963, 1965).
77. A. W. Hofmann, *Ber.*, **10**, 1095 (1877).
78. T. L. Cairns, A. W. Larchar and B. C. McKusick, *J. Org. Chem.*, **17**, 1497 (1952).
79. H. Malz and E. Kühle (Farbenfabriken Bayer A.G.) *Ger. Pat.* 1,158,962 (1961, 1963).
80. M. Passerini, *Gazz. Chim. Ital.*, **57**, 452 (1927).
81. T. Saegusa, Y. Ito, S. Kobayashi and K. Hirota, *Tetrahedron Letters*, **1967**, 521.
82. T. Saegusa, Y. Ito, S. Kobayashi, K. Hirota and H. Yoshioka, *Tetrahedron Letters*, **1966**, 6121.
83. W. P. Neumann and K. Kühlein, *Tetrahedron Letters*, **1966**, 3423.
84. J. V. Mitin, V. R. Glušenkova and G. P. Vlasov, *Zh. Obshch. Khim.*, **32**, 3867 (1962).
85. P. Hoffmann, unpublished results.
86. A. Treibs and A. Dietl, *Chem. Ber.*, **94**, 298 (1961).
87. M. Passerini and V. Casini, *Gazz. Chim. Ital.*, **67**, 332 (1937).
88. G. Losco, *Gazz. Chim. Ital.*, **67**, 553 (1937).
89. M. Passerini, *Gazz. Chim. Ital.*, **54**, 184, 633, 667 (1924); **56**, 365 (1926).
90. M. Passerini and A. Neri, *Gazz. Chim. Ital.*, **64**, 934 (1934).
91. A. Nagasaka and R. Oda, *J. Chem. Soc. Japan, Ind. Chem. Sect.*, **58**, 48 (1955).
92. F. Sachs and H. Loevy, *Ber.*, **37**, 874 (1904).
93. H. Gilman and L. C. Heckert, *Bull. Soc. Chim.*, **43**, 224 (1928).
94. I. Ugi and U. Fetzer, *Chem. Ber.*, **94**, 2239 (1961).
95. T. Shono, M. Kimura, Y. Ito, K. Nishida and R. Oda, *Bull. Chem. Soc. Japan*, **37**, 635 (1964).
96. H. Feuer, H. Rubinstein and A. T. Nielsen, *J. Org. Chem.*, **23**, 1107 (1958).
97. M. Okano, Y. Ito, T. Shono and R. Oda, *Bull. Chem. Soc. Japan*, **36**, 1314 (1963).
98. H. W. Johnson and P. H. Daughhetee, *J. Org. Chem.*, **29**, 246 (1964).
99. D. Martin and A. Weise, *Angew. Chem.*, **79**, 145 (1967); *Angew. Chem. Int. Ed. Engl.*, **6**, 168 (1967).
100. P. V. Finzi and M. Arbasino, *Tetrahedron Letters*, **1965**, 4645.
101. I. Ugi and C. Steinbrückner, unpublished results.
102. M. Lipp, F. Dallacker and I. Meier zu Köcker, *Monatsh. Chem.*, **90**, 41 (1959).
103. C. Collard-Charon and M. Renson, *Bull. Soc. Chim. Belg.*, **71**, 531 (1962).
104. E. Bulka and K. D. Ahlers, *Z. Chem.*, **3**, 348 (1963).
105. W. J. Franklin and R. L. Werner, *Tetrahedron Letters*, **1965**, 3003.

106. A. Halleux, *Angew. Chem.*, **76**, 889 (1964); *Angew. Chem. Intern. Ed. Engl.*, **3**, 752 (1964).
107. W. Aumüller, *Angew. Chem.*, **75**, 857 (1963); *Angew. Chem. Intern. Ed. Engl.*, **2**, 616 (1963).
108. F. Millich and R. Sinclair, *IUPAC Intern. Symp. Macromol. Chem., Brussels–Louvain; Chem. and Eng. News*, **45**, 30 (1967).
109. E. Oliveri-Mandala and B. Alagna, *Gazz. Chim. Ital.*, **40 II**, 441 (1910).
110. M. Passerini, *Gazz. Chim. Ital.*, **51 II**, 126, 181 (1921); **52 I**, 432 (1922); **53**, 331, 410 (1923); **54**, 529, 540 (1924); **55**, 721 (1925); **56**, 826 (1926); *Mem. Accad. Lincei* (**6**) **2**, 277 (1927).
111. M. Passerini and G. Ragni, *Gazz. Chim. Ital.*, **61**, 964 (1931).
112. R. H. Baker and A. H. Schlesinger, *J. Am. Chem. Soc.*, **67**, 1499 (1945).
113. R. H. Baker and L. E. Linn, *J. Am. Chem. Soc.*, **70**, 3721 (1948).
114. M. J. S. Dewar, *Electronic Theory of Organic Chemistry*, Clarendon Press, Oxford, 1949, p. 116.
115. H. Baker and D. Stanonis, *J. Am. Chem. Soc.*, **73**, 699 (1951).
116. I. Ugi and R. Meyr, *Chem. Ber.*, **94**, 2229 (1961).
117. U. Fetzer and I. Ugi, *Ann. Chem.*, **659**, 184 (1962).
118. I. Hagedorn, U. Eholzer and H. D. Winkelmann, *Angew. Chem.*, **76**, 583 (1964); *Angew. Chem. Int. Ed.*, **3**, 647 (1964).
119. I. Hagedorn and U. Eholzer, *Chem. Ber.*, **98**, 936 (1965).
120. I. Ugi and K. Rosendahl, *Chem. Ber.*, **94**, 2233 (1961).
121. R. Neidlein, *Z. Naturforsch.*, **19B**, 1159 (1964).
122. I. Ugi, W. Betz and K. Offermann, *Chem. Ber.*, **97**, 3008 (1964).
123. I. Ugi and E. Böttner, *Ann. Chem.*, **670**, 74 (1963).
124. I. Ugi and C. Steinbrückner, *Angew. Chem.*, **72**, 267 (1960).
125. C. Steinbrückner, *Ph.D. Thesis*, Munich University, 1961.
126. I. Ugi and K. Offermann, *Chem. Ber.*, **97**, 2276 (1964).
127. I. Ugi and F. K. Rosendahl, *Ann. Chem.*, **666**, 54 (1963).
128. I. Ugi and C. Steinbrückner, *Chem. Ber.*, **94**, 2802 (1961).
129. I. Ugi and K. Offermann, *Chem. Ber.*, **97**, 2996 (1964).
130. I. Ugi and F. Bodesheim, *Chem. Ber.*, **94**, 2797 (1961); *Ann. Chem.*, **666**, 61 (1963).
131. G. Opitz and W. Merz, *Ann. Chem.*, **652**, 158, 163 (1962).
132. G. Opitz, A. Griesinger and H. W. Schubert, *Ann. Chem.*, **665**, 91 (1963).
133. I. Ugi and C. Steinbrückner, *Chem. Ber.*, **94**, 734 (1961).
134. R. Neidlein, *Angew. Chem.*, **76**, 440 (1964); *Angew. Chem. Int. Ed.*, **3**, 382 (1964); *Arch. Pharm.*, **297**, 589 (1964).
135. J. W. McFarland, *J. Org. Chem.*, **28**, 2179 (1963).
136. I. Ugi, K. Offermann, H. Herlinger and D. Marquarding, *Ann. Chem.*, **709**, 1 (1967).
137. I. Ugi and G. Kaufhold, *Ann. Chem.*, **709**, 11 (1967).
138. S. P. Makarov, V. A. Španskij, V. A. Ginsburg, A. I. Sčekotichin, A. S. Filatov, L. L. Martynova, I. V. Pavlovskaja, A. F. Golovaneva and A. Ja. Jakubovič, *Dokl. Akad. Nauk. SSSR*, **142**, 596 (1962).
139. W. J. Middleton, *J. Org. Chem.*, **30**, 1402 (1965).
140. E. Müller and B. W. Zeeh, *Tetrahedron Letters*, **1965**, 3951.
141. M. Passerini, *Gazz. Chim. Ital.*, **55**, 555 (1925).
142. E. Winterfeldt, *Angew. Chem.*, **78**, 757 (1966); *Angew. Chem. Int. Ed.*, **5**, 741 (1966).
143. N. P. Gambarjan, E. M. Rokhlin, Yu. V. Zeifman, Chen Ching-Yun and I. L. Knunyants, *Angew. Chem.*, **78**, 1008 (1966); *Angew. Chem. Int. Ed.*, **5**, 947 (1966).

144. N. P. Gambarjan, E. M. Rochlin, Ju. V. Zeifman, L. A. Simonjan and I. L. Knunyants, *Dokl. Akad. Nauk SSSR*, **166**, 864 (1966).
145. K. Ley, U. Eholzer and R. Nast, *Angew. Chem.*, **77**, 544 (1965); *Angew. Chem. Int. Ed.*, **4**, 519 (1965).
146. G. Hesse, H. Witte and W. Gulden, *Angew. Chem.*, **77**, 591 (1965); *Angew. Chem. Int. Ed.*, **4**, 596 (1965).
147. R. Neidlein, *Angew. Chem.*, **76**, 500 (1964); *Angew. Chem. Int. Ed.*, **3**, 446 (1964); *Chem. Ber.*, **97**, 3476 (1964); *Arch. Pharm.*, **298**, 124 (1965).
148. J. Goerdeler and H. Schenk, *Chem. Ber.*, **98**, 3831 (1965).
149. H. Schenk, *Chem. Ber.*, **99**, 1258 (1966).
150. F. Weygand, W. Steglich, W. Oettmeier and A. Maierhofer, *Angew. Chem.*, **78**, 640 (1966); *Angew. Chem. Int. Ed.*, **5**, 600 (1966).
151. R. Knorr, *Chem. Ber.*, **98**, 4038 (1965).
152. S. Yaroslavsky, *Chem. Ind.*, **1965**, 765.
153. M. Passerini and N. Zita, *Gazz. Chim. Ital.*, **61**, 26 (1931).
154. B. Zeeh, *Tetrahedron Letters*, **1967**, 3881.
155. C. Grundmann, *Chem. Ber.*, **91**, 1380 (1958).
156. M. Passerini and T. Bonciani, *Gazz. Chim. Ital.*, **61**, 959 (1931).
157. I. Ugi and F. K. Rosendahl, *Ann. Chem.*, **670**, 80 (1963).
158. D. E. O'Brien, F. Baiochi and C. C. Cheng, *Biochemistry*, **2**, 1203 (1963).
159. G. Hesse and H. Witte, *Angew. Chem.*, **75**, 791 (1963); *Ann. Chem.*, **687**, 1 (1965).
160. J. Casanova and R. E. Schuster, *Tetrahedron Letters*, **1964**, 405.
161. S. Bresadola, G. Carraro, C. Pecile and A. Turco, *Tetrahedron Letters*, **1964**, 3185.
162. J. Casanova, H. R. Kiefer, D. Kuwada and A. H. Boulton, *Tetrahedron Letters*, **1965**, 703.
163. J. Tanaka and J. C. Carter, *Tetrahedron Letters*, **1965**, 329.
164. G. Hesse, H. Witte and G. Bittner, *Ann. Chem.*, **687**, 9 (1965).
165. S. Bresadola, F. Rossetto and G. Puosi, *Tetrahedron Letters*, **1965**, 4775; *Gazz. Chim. Ital.*, **96**, 1397 (1966).
166. G. Hesse, H. Witte and W. Guldon, *Tetrahedron Letters*, **1966**, 2707.
167. G. Hesse, H. Witte and P. Mischke, *Angew. Chem.*, **77**, 380 (1965).
168. L. Malatesta, *Progress in Inorganic Chemistry*, Vol. 1 (Ed. F. A. Cotton), Interscience, New York, 1959, p. 283.
169. E. G. J. Hartley, *J. Chem. Soc.*, **1928**, 780.
170. K. A. Hofmann and G. Bugge, *Ber.*, **40**, 1772 (1907).
171. L. Malatesta, *Gazz. Chim. Ital.*, **77**, 240 (1947).
172. F. Klages, K. Mönkemeyer and R. Heinle, *Chem. Ber.*, **83**, 501 (1950); **85**, 109 (1952).
173. H. Guillemard, *Ann. Chim. Phys.*, [8] **14**, 311 (1908).
174. A. Sacco and M. Freni, *Gazz. Chim. Ital.*, **86**, 195 (1956).
175. A. Sacco, *Gazz. Chim. Ital.*, **85**, 989 (1955).
176. E. O. Fischer and H. Fischer, *Angew. Chem.*, **77**, 261 (1965); *Angew. Chem. Int. Ed.*, **4**, 246 (1965).
177. H. D. Murdoch and R. Henzi, *J. Organometal. Chem.*, **5**, 166 (1966).
178. L. Malatesta, A. Sacco and S. Ghielmi, *Gazz. Chim. Ital.*, **82**, 516 (1952).
179. L. Malatesta and A. Sacco, *Atti Accad. Naz. Lincei. Rend. Classe Sci. Fis. Mat. Nat.*, (8) **12**, 308 (1952).
180. F. Hölzl and G. I. Xenakis, *Monatsh. Chem.*, **48**, 689 (1927).
181. L. Malatesta, A. Sacco and M. Gabaglio, *Gazz. Chim. Ital.*, **82**, 548 (1952).

182. L. Malatesta and A. Sacco, *Ann. Chim.* (Rome), **43**, 622 (1953).
183. F. Hölzl and N. Zymaris, *Monatsh. Chem.*, **51**, 1 (1929).
184. A. Sacco, *Gazz. Chim. Ital.*, **86**, 201 (1956).
185. E. O. Fischer and M. Herberhold, *Experimentia Suppl.*, **9**, 259 (1964).
186. L. Naldini, *Gazz. Chim. Ital.*, **90**, 871 (1960).
187. W. Hieber and L. Schuster, *Z. Anorg. Allgem. Chem.*, **287**, 214 (1956).
188. W. Hieber and D. v. Pigenot, *Chem. Ber.*, **89**, 193, 610 (1956).
189. L. Malatesta and A. Sacco, *Atti Accad. Naz. Lincei. Rend. Classe Sci. Fis. Mat. Nat.*, [8] **13**, 264 (1952); *Z. Anorg. Allgem. Chem.*, **274**, 341 (1953).
190. K. K. Joshi, O. S. Mills, P. L. Pauson, B. W. Shaw and W. H. Stubbs, *Chem. Commun.*, **1965**, 181.
191. K. A. Hofmann and G. Bugge, *Ber.*, **40**, 3759 (1907).
192. M. Freund, *Ber.*, **21**, 931 (1888).
193. E. G. J. Hartley, *J. Chem. Soc.*, **97**, 1066, 1725 (1910); **99**, 1549 (1911); **101**, 705 (1912); **103**, 1196 (1913); **105**, 521 (1914); **109**, 1296, 1302 (1916).
194. E. G. J. Hartley and H. M. Powell, *J. Chem. Soc.*, **1933**, 101.
195. F. Hölzl, W. Hauser and M. Eckmann, *Monatsh. Chem.*, **48**, 71 (1927).
196. J. Meyer, H. Domann and W. Müller, *Z. Anorg. Allgem. Chem.*, **230**, 336 (1937).
197. H. M. Powell and G. B. Stanger, *J. Chem. Soc.*, **1939**, 1105.
198. I. Ugi and C. Steinbrückner, *Angew. Chem.*, **71**, 386 (1959).
199. W. Z. Heldt, *J. Org. Chem.*, **26**, 3226 (1961); *J. Inorg. Nucl. Chem.*, **22**, 305 (1961).
200. L. Malatesta and L. Giuffré, *Atti Acad. Naz. Lincei Rend. Classe Sci. Fis. Mat. Nat.*, (8) **11**, 206 (1951).
201. L. Malatesta and A. Sacco, *Gazz. Chim. Ital.*, **83**, 499 (1953).
202. L. Malatesta, *Gazz. Chim. Ital.*, **83**, 958 (1953).
203. A. Sacco and M. Freni, *Angew. Chem.*, **70**, 599 (1958).
204. J. M. Pratt and P. R. Silvermann, *Chem. Commun.*, **1967**, 117.
205. C. E. Bolser and L. B. Richardson, *J. Am. Chem. Soc.*, **35**, 377 (1913).
206. F. Hölzl, T. Meier-Mohar and F. Viditz, *Monatsh. Chem.*, **53/54**, 237 (1929).
207. L. Malatesta and A. Sacco, *Z. Anorg. Allgem. Chem.*, **273**, 247 (1953); *Atti Accad. Naz. Lincei Rend. Classe Sci. Fis. Mat. Nat.*, (8) **15**, 94 (1953).
208. A. Sacco and M. Freni, *Gazz. Chim. Ital.*, **89**, 1800 (1959).
209. W. Hieber and E. Böckly, *Z. Anorg. Allgem. Chem.*, **262**, 344 (1950).
210. A. Sacco, *Gazz. Chim. Ital.*, **83**, 632 (1953).
211. W. Hieber and J. Sedlmeier, *Chem. Ber.*, **87**, 25 (1954).
212. A. Sacco, *Atti Accad. Naz. Lincei Rend. Classe Sci. Fis. Mat. Nat.*, (8) **15**, 82 (1953).
213. L. Malatesta, *Gazz. Chim. Ital.*, **77**, 240 (1947).
214. W. Hieber, *Z. Naturforsch.*, **5B**, 129 (1950).
215. F. Klages and K. Mönkemeyer, *Naturwissenschaften*, **37**, 210 (1950); *Chem. Ber.*, **83**, 501 (1950).
216. L. Malatesta and A. Sacco, *Atti Accad. Naz. Lincei Rend. Classe Sci. Fis. Mat. Nat.*, [8] **11**, 379 (1951).
217. W. Hieber and R. Brück, *Z. Anorg. Allgem. Chem.*, **269**, 28 (1952).
218. L. Malatesta, G. A. Padoa and A. Sonz, *Gazz. Chim. Ital.*, **85**, 1112 (1955).
219. L. Malatesta and L. Vallarino, *J. Chem. Soc.*, **1956**, 1867.
220. L. Vallarino, *Gazz. Chim. Ital.*, **89**, 1632 (1959).
221. L. Vallarino, *Int. Conf. Coord. Chem.*, London, April 5–11, 1959.
222. E. O. Fischer and H. Werner, *Chem. Ber.*, **95**, 703 (1962).
223. M. Angoletta, *Ann. Chim.* (*Rome*), **45**, 970 (1955).

224. L. Ramberg, *Ber.*, **40,** 2578 (1907).
225. L. Tschugajeff and P. Teearu, *Ber.*, **47,** 568 (1914).
226. L. Tschugajeff, M. Skanawy-Grigorjewa and A. Posnjak, *Z. Anorg. Allgem. Chem.*, **148,** 37 (1925).
227. J. Keilin, *Biochem. J.*, **45,** 440 (1949); *Nature,* **165,** 151 (1950).
228. Y. Imai and R. Sato, *Biochem. Biophys. Res. Commun.*, **25,** 80 (1966).
229. M. Kelly, J. R. Postgate and R. L. Richards, *Biochem. J.*, **102,** 1c (1967).
230. I. Ugi, U. Fetzer, G. Unterstenhöfer, W. Behrenz, P. E. Frohberger and H. Scheinpflug, (Farbenfabriken Bayer A.G.), *Ger. Pat.* 1,209,798 (1962/1966).
231. I. Ugi, U. Fetzer, F. Grewe, W. Behrenz, P. E. Frohberger, B. Blomeyer, H. Scheinpflug and G. Unterstenhöfer, (Farbenfabriken Bayer A.G.), *U.S. Pat.* 3,278,371 (1962/1966).
232. U. Fetzer, I. Ugi, G. Unterstenhöfer, W. Behrenz and P. E. Frohberger (Farbenfabriken Bayer A.G.), *Ger. Publ. Pat. Appl.* 1,211,853 (1962/1966).
233. U. Fetzer, I. Ugi and G. Unterstenhöfer (Farbenfabriken Bayer A.G.), *Ger. Pat.* 1,235,298 (1964/1967).
234. U. Fetzer, U. Eholzer, I. Ugi, I. Hammann and G. Unterstenhöfer (Farbenfabriken Bayer A.G.), *Ger. Pat.* 1,215,142 (1964/1966).
235. U. Fetzer, I. Ugi, H. Knupfer, J. A. Renner and F. Grewe (Farbenfabriken Bayer A.G.), *Brit. Pat.* 1,064,835 (1964/1967).
236. U. Eholzer, U. Fetzer, I. Ugi, I. Hammann and G. Unterstenhöfer (Farbenfabriken Bayer A.G.), *Ger. Pat.* 1,215,141 (1964/1966).

CHAPTER 16

Rearrangement reactions involving the cyano group

JOSEPH CASANOVA, JR.

California State College, Los Angeles, California, U.S.A.

I. INTRODUCTION	885
II. REARRANGEMENT REACTIONS DIRECTLY INVOLVING THE CYANO GROUP	886
A. Isonitrile–Nitrile Rearrangement	886
B. Electron-Deficient Rearrangements to the Cyano Group	892
C. Ketenimines and Ynamines	894
D. Cyanocarbenes	901
III. REARRANGEMENT REACTIONS OF THE CARBON SKELETON TO WHICH THE CYANO GROUP IS ATTACHED	906
IV. FRAGMENTATION REACTIONS INVOLVING THE CYANO GROUP	909
A. Rearrangement of Nitrilium Salts	912
1. The Ritter Reaction	913
2. Fragmentation of Ketoximes and their Derivatives	915
a. Simple ketoximes	916
b. α-Carbonyl- and α-hydroxyketoximes	925
3. Fragmentation of 1,3-Oxazet Rings, Oxazetones and Oxazoles	932
a. Oxazetes and oxazetones	932
b. Oxazoles	935
V. REARRANGEMENT REACTIONS INVOLVING INTERMEDIATE CYANO COMPOUNDS: THE VON RICHTER REACTION	936
VI. REFERENCES	940

I. INTRODUCTION

Of the many synthetically useful functional groups of organic chemistry, the cyano group ranks high in its utility. The ease with which it may be introduced by substitution, both at saturated and unsaturated carbon atoms, and its ready transformation into carbon–oxygen functional groups at both ends of the oxidation scale are no doubt responsible for this synthetic utility. Some time has elapsed since a comprehensive review has appeared on this topic[1]. It is the purpose of this chapter to examine the literature which deals with rearrangement reactions involving the cyano groups with particular attention given to: (*a*) rearrangement reactions in which the cyano group is a direct participant; (*b*) reactions involving the carbon skeleton to which the cyano group is attached; (*c*) reactions in which the cyano group appears in the starting material or product from a source other than direct substitution; (*d*) reactions for which the presence of the cyano group is considered essential but in which the group maintains its integrity during the course of reaction; and (*e*) those reactions in which the cyano group has been demonstrated or postulated to play an intermediate role.

It is intended to exclude from consideration direct substitution reactions involving the nucleophilic cyano group in aliphatic displacement reactions, addition reactions involving the cyano group such as cyanohydrin formation; aromatic nucleophilic substitutions such as the Sandmeyer reaction, the Rosenmund–von Braun reaction[1], the Houben–Fischer synthesis[2a] and the Hoesch synthesis[2b] or addition of cyano groups to carbon–carbon bonds, as in reactions such as the Reissert reaction[3]. Addition to or eliminations from a C—N group in which a cyano group appears in the starting material or product, wherein the C—N group does not suffer rearrangement, are also excluded, except as they might contribute to the understanding of other closely related rearrangement reactions. Examples of some reactions which are not included are the dehydration of aldoximes[4] or amides[5] to produce nitriles—the Wohl degradation[6] being a good example for the synthetic utility of the former—the hydration of nitriles to amides, ring-forming addition reactions such as the Büchner synthesis of hydantoins or addition of cyano groups to carbon–carbon bonds, as in the Reissert reaction[3]. These topics are the subject of other chapters of this volume, or have been treated in other reviews.

II. REARRANGEMENT REACTIONS DIRECTLY INVOLVING THE CYANO GROUP

A. Isonitrile–Nitrile Rearrangement

The thermal rearrangement of isonitriles to nitriles (equation 1) was recognized a century ago by Gautier[7a] and has been infrequently examined since that time[7b–h], but received more attention during

$$RN{\equiv}C \xrightarrow{\Delta} RC{\equiv}N \qquad (1)$$

the last decade. The stoichiometric simplicity of the transformation, the similarity of geometry of the starting material and product and the large free energy difference between them, which diminishes the necessity to consider the reverse reaction, suggested that this isomerization might be a useful vehicle for the study of unimolecular gas-phase reactions. The free energy difference between nitriles and isonitriles has not been calculated for many pairs of isomers. The most often quoted value is 15 kcal/mole for methyl isocyanide–acetonitrile[8a,b], in favour of the more stable nitrile. The Rossini measurement is based on differences in heats of combustion[8b]. Some recent work promises to yield more quantitative data[8c]. It seems probable that the reverse reaction may not be observable, although no attempts to verify this have been noted since the advent of modern chromatographic techniques. The advent of much improved syntheses for isocyanides[9], especially due to the considerable work of Ugi[9a] made these compounds readily available and has stimulated studies of the mechanism of the rearrangement. Rabinovitch[10] reported the first definitive study of the nature of the isomerization. He reported that *p*-tolyl isocyanide rearranged smoothly between 180–220° both in hydrocarbon solution and in the gas phase to give *p*-toluonitrile in high yields, free of the *meta* isomer (equation 2). The latter result suggests that the reaction does not lead to cross-products.

$$Me{-}C_6H_4{-}N{\equiv}C \longrightarrow Me{-}C_6H_4{-}C{\equiv}N \qquad (2)$$

The reactions were first order in isocyanide and twice as fast in the gas phase as in solution in mineral oil over the temperature range 180–220°. However, the similarity in rates is due to a balancing effect between the activating energy (E_a (solution) $= 36{\cdot}8$ kcal/mole, E_a (gas) $= 33{\cdot}8$ kcal/mole) and the preexponential factor (A (solution) $= 10^{13{\cdot}7}\,sec^{-1}$, A (gas) $= 10^{12{\cdot}7}\,sec^{-1}$). Rabinovitch has suggested a structure for the activated complex in which the development of the new C—C bond is nearly synchronous with breaking of the N—C bond. The work of other groups has served to reinforce this view, both experimentally[11] and theoretically[12]. Casanova and

Me—C₆H₄ (with N and C bridged, dotted bonds to ring carbon; N≡C)

coworkers demonstrated the stereochemical similarity between this rearrangement and other 'saturated' rearrangements such as the Hoffmann, Curtius, Schmidt, Lossen and Wolff rearrangements[13] by rearranging R-(+)-*s*-butylisonitrile (**1**) to R-(+)-2-methylbutyronitrile (**2**) with 87 % retention of optical activity (equation 3). These compounds were shown to be of the same absolute configuration by independent synthesis*. Two optically active isocyanides

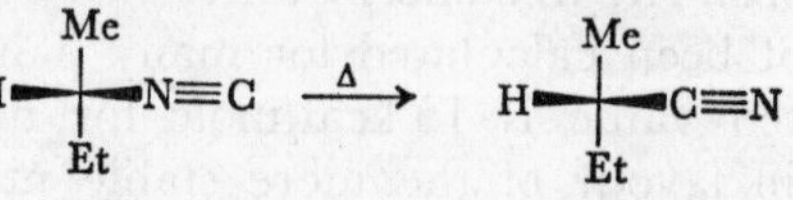

$$\text{H–C(Me)(Et)–N}\equiv\text{C}\ (\mathbf{1}) \xrightarrow{\Delta} \text{H–C(Me)(Et)–C}\equiv\text{N}\ (\mathbf{2}) \tag{3}$$

(**3** and **4**) of low optical purity had been previously reported[14] and **4** was thermally rearranged to an optically active nitrile. However, the asymmetry centre in this case was not located at the migrating

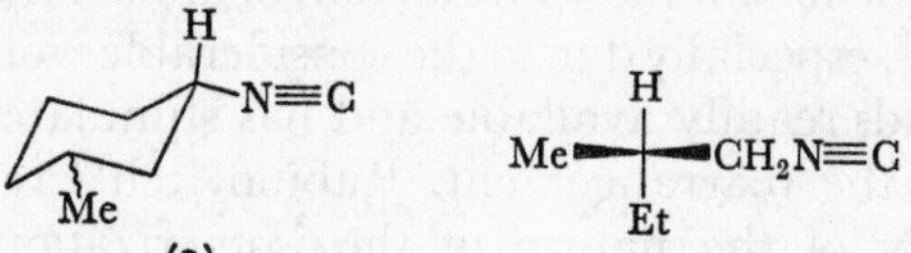

$$(\mathbf{3})\ \text{(cyclohexyl–N}\equiv\text{C, H, Me)} \qquad \text{Me–C(H)(Et)–CH}_2\text{N}\equiv\text{C}\ (\mathbf{4}) \tag{4}$$

carbon atom. Wright[15] has separately rearranged *cis*- and *trans*-2,2,4,4-tetramethyl-1,3-diisocyanide (**5a** and **5b**) stereospecifically

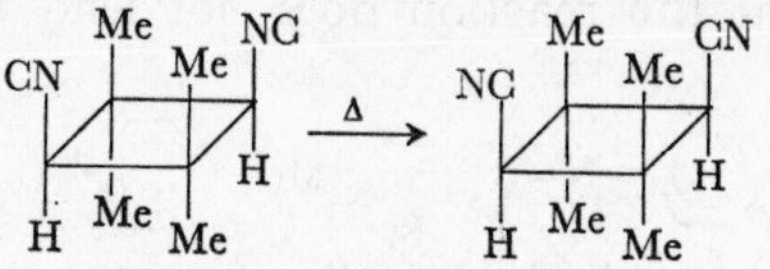

$$(\mathbf{5a}) \xrightarrow{\Delta} (\mathbf{6a}) \qquad (\mathbf{5b}) \xrightarrow{\Delta} (\mathbf{6b}) \tag{5}$$

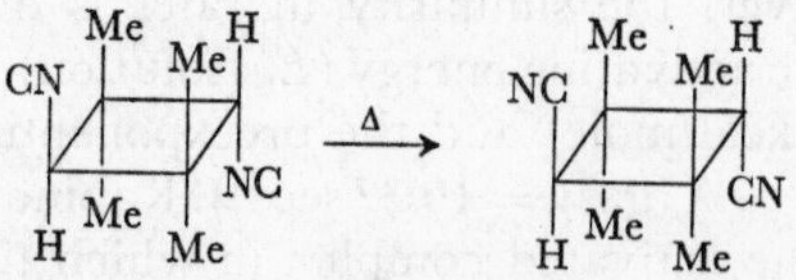

* *Note added in proof:* Yamada and coworkers (S. Yamada, K. Takoshima, T. Sato and S. Terashima, *Chem. Comm.*, **1969**, 811) have more recently provided additional evidence in support of these conclusions, reporting that *S*(−)-α-phenethyl isonitrile gave 57% retained configuration in the product. However, they found that carbethoxymethylbenzylcarbinyl isocyanide gave only 9% retention, and have proposed a radical rearrangement to account for racemization.

to high yields (85 % and 83 %, respectively) of *cis*- and *trans*-2,2,4,4-tetramethyl-1,3-dicyanide (**6a** and **6b**) (equations 4 and 5). Further evidence for the non-polar character of the transition state for the rearrangement was found in: (*a*) the failure of cyclobutylisonitrile (**7**) to undergo carbon skeleton rearrangement during its isomerization to the nitrile **8** (equation 6), (*b*) by a very small substituent effect in the rearrangement of *para*-substituted aryl isocyanides ($\rho \approx -0{\cdot}12$) in diglyme and (*c*) in alkyl isocyanides in which variation in rate of rearrangement from the slowest, *t*-butylisonitrile, to

$$\text{c-C}_4\text{H}_7\text{—N}{\equiv}\text{C} \xrightarrow{\Delta} \text{c-C}_4\text{H}_7\text{—C}{\equiv}\text{N} \qquad (6)$$

(**7**) (**8**)

the fastest, ethylisonitrile, was less than eight-fold[11]. Thermal decomposition of *N*-acyl-*N*-formylimides (**9**) at 400° produces nitriles[16] in fair yields by a reaction which may be formally likened to a reverse Passerini reaction followed by thermal isonitrile–nitrile rearrangement (equation 7). Formamidates from isonitriles have

$$\text{R—N(COR}^1\text{)(CHO)} \rightleftharpoons \text{R—N=C(OCOR}^1\text{)(H)} \longrightarrow \text{RN}{\equiv}\text{C} + \text{HOCOR}^1 \qquad (7)$$

(**9**)

$$\text{RN}{\equiv}\text{C} \xrightarrow{400°} \text{RC}{\equiv}\text{N}$$

been previously reported[17] although acyl formamides have not. The isonitrile–nitrile isomerization has also been observed to occur readily with neat samples in the presence of di-*t*-butyl peroxide as reported in the work of Shaw and Pritchard[18a,b]. These authors have proposed a radical-chain process (equations 8 and 9) to account for

$$\text{RN}{\equiv}\text{C} + \text{I}\cdot \longrightarrow \text{R}\cdot + \text{N}{\equiv}\text{CI} \quad \text{(initiation)} \qquad (8)$$

(I = initiator)

$$\text{RN}{\equiv}\text{C} + \text{R}\cdot \longrightarrow \text{R}\cdot + \text{N}{\equiv}\text{CR} \quad \text{(propagation)} \qquad (9)$$

(R = Me or Et)

their observations. A low activation energy, $E_a = 7{\cdot}8$ kcal/mole, and $A = 10^{12{\cdot}25}$ cc/mole sec were found for the reaction for which

$$\text{MeN}{\equiv}\text{C} + \text{Me}\cdot \longrightarrow \text{Me}\cdot + \text{N}{\equiv}\text{CMe}$$

the intermediate $\text{Me}\ddot{\text{N}}{=}\dot{\text{C}}\text{Me}$ had been proposed. Rabinovitch reported that in the absence of free-radical initiators the rate of thermal rearrangement was found to be independent of surface area[19a] for the gas-phase rearrangement. However, polymerization of both aryl and alkyl isonitriles by an unusual combination of concentrated sulphuric acid on powdered glass and free-radical sources, such as dibenzoyl peroxide or oxygen, have been reported[20a,b]. The formation of resins from phenyl isocyanide during rearrangement has been suppressed by the presence of amines[21], and suppression of polymerization by 'preconditioning' the glass vessel in which rearrangement is conducted by rinsing it with pyridine has been reported[11]. Claims by Wade[21] that contamination of isocyanides by weak protonic acids such as water or the primary amine from which the isocyanide was prepared, serve to effectively catalyse the rearrangement (equations 10–12) but not the polymerization of methyl and phenyl isocyanide, appear to be unfounded[10,11,19a] and have been disputed by Wright[15].

$$\mathrm{RN{\equiv}C + HA \longrightarrow RN{=}C(H)(A)} \qquad (10)$$

$$\mathrm{RN{=}C(H)(A) \longrightarrow HN{=}C(R)(A)} \qquad (11)$$

$$\mathrm{HN{=}C(R)(A) \longrightarrow HA + N{\equiv}CR} \qquad (12)$$

The nature of the intermediate or intermediates which are important in the isonitrile–nitrile reactions remains to be elucidated. Extended Hückel molecular orbital calculations for this rearrangement which have been reported by Hoffmann[12] support the activation energy results previously reported. They also suggest that substantial flattening of the migrating methyl group occurs in the transition state for rearrangement of methyl isocyanide, and that the methyl group is essentially equidistant between the seat and terminus of migration. Moreover, calculations for phenyl isocyanide suggest

[Structure: $\mathrm{H{-}C(H)(H)}$ bridging $\mathrm{N{\equiv}C}$ (transition state)]

that π-bridging in which the plane of the aromatic ring is perpendicular to the N—C bond axis in the transition state is energetically

preferred to σ-bridging, in which the plane of the aromatic ring contains the N—C bond axis. The former route utilizes primarily the 2*p*-orbital of carbon for overlap in the transition state and the latter utilizes primarily the sp^2-orbital of that atom. Hoffmann's calculations also suggest a substantial charge distribution in the transition state for rearrangement of both methyl and phenyl isocyanide.

High yields in the transformation of amines to isonitriles[9a] and the generally good yields (60%) in their stereospecific rearrangement into nitriles render the thermal rearrangement reaction of potential synthetic value (equation 13). However, the ready transformation

$$RNH_2 \longrightarrow RN{\equiv}C \longrightarrow RC{\equiv}N \tag{13}$$

of aryl isonitriles to highly coloured tetramers, shown to be the dianils of the appropriately substituted indigos[22] (**10**), and the propensity of many isonitriles to undergo polymerization[20], diminishes their utility in this regard.

NPh NH NH NPh

(10)

The careful and detailed analysis of the rearrangement of methyl-[19a,d] and methyl-d_3 [19b,c] isocyanide by Rabinovitch establishes this reaction as a model system for the study of unimolecular first order reactions in the gas phase. More recently, this same rearrangement has been employed as a vehicle for the study of the mechanism of deexcitation of methyl-t_1-isonitrile, produced by the substitution of energetic tritium atoms for hydrogen atoms[23] (equation 14).

$$\begin{aligned} T^* + MeN{\equiv}C &\longrightarrow H + CH_2TN{\equiv}\overset{*}{C} \\ CH_2TN{\equiv}C^* &\longrightarrow CH_2TC{\equiv}N \end{aligned} \tag{14}$$

In contrast to the dissimilar relative free energies of organic isonitriles and nitriles, it appears that trimethylsilyl isocyanide (**11**) is present in measurable quantity at equilibrium with trimethylsilyl cyanide (**12**)[24] (equation 15). Moreover, there has recently appeared

$$\underset{(\mathbf{11})}{Me_3SiN{\equiv}C} \longrightarrow \underset{(\mathbf{12})}{Me_3SiC{\equiv}N} \quad (15)$$

what may be the first case of the reverse reaction, namely, a nitrile to isonitrile transformation. Talat-Erbin[25] reported that under gamma irradiation such a transformation occurred. Although the cyanides of silver, copper and mercury undergo displacement reactions with alkyl halides to produce isocyanides[7a] (equation 16)

$$MC{\equiv}N + MeI \longrightarrow MI + MeN{\equiv}C \quad (16)$$
$$(M = Ag,\ Cu,\ HgMe)$$

it appears likely that such behaviour can be attributed to internal complexation[26,27] between the covalent cyanide and the organic halide. This reaction contrasts the normal substitution reaction observed with ionic cyanides.

B. Electron-Deficient Rearrangements to the Cyano Group

Although organic cyanides do not appear to form transition metal complexes as readily as do isocyanides[28], they do serve as Lewis bases for several group II and group III element reactions. Organometallic compounds in which these elements are the electrophilic species have been shown to react with nitriles, with metal to carbon alkyl or aryl migration. The Grignard reaction of nitriles illustrates this rearrangement[29a,b]. In this rearrangement a new carbon–carbon bond is formed, and the cyano group is destroyed (equation 17). Similar reactions have been described for

$$PhC{\equiv}N + PhMgBr \longrightarrow PhC{\equiv}\overset{+}{N}{-}\overset{-}{Mg}(Ph)(Br) \longrightarrow PhC(Ph){=}NMgBr \quad (17)$$

organoboranes[30], including decaborane[30d]. Lappert[30b] reported that at 150–160° the monomeric aldiminoborane (**13**) was formed from the reaction of *t*-butyl cyanide and tri-n-butylborane (equation 18)

$$t\text{-}BuC{\equiv}N + B(n\text{-}Bu)_3 \longrightarrow \underset{(\mathbf{13})}{t\text{-}BuCH{=}NB(n\text{-}Bu)_2} \quad (18)$$

and Wade[30c] observed rearrangement with dimerization to give **14** in the reaction between acetonitrile and trimethylborane (equation 19). In contrast to this result, it has been recently reported by Horn[31]

$$2\ MeC{\equiv}N + 2\ BMe_3 \longrightarrow \text{cyclic dimer } [Me(H)C{=}N{-}BMe_2]_2 \quad \textbf{(14)} \qquad (19)$$

that propionitrile reacts with tri-n-propylborane, when heated, to give over 90% yield of a dihydro-2,5-diborapyrazine (**15**) (equation 20). Hexamethyldialuminium reacts even more readily with alkyl

$$EtC{\equiv}N + (n\text{-}Pr)_3B \longrightarrow \text{cyclic } [Et\text{-}C{=}N(H){-}B(n\text{-}Pr)_2]_2 \quad \textbf{(15)} \qquad (20)$$

cyanides, in this instance giving simple metal to carbon alkyl migrations. Wade and coworkers[32] reported that pivalonitrile reacts readily with hexamethyldialuminium to give the adduct **16** which first rearranges to **17** and which then dimerizes to **18** (equation 21) at 150°. Pasynkiewitz[33a] reported a similar reaction of

$$t\text{-}BuC{\equiv}N + \tfrac{1}{2}\ Al_2Me_6 \longrightarrow [t\text{-}BuC{\equiv}\overset{+}{N}{-}\overset{-}{Al}Me_3]\ \textbf{(16)} \longrightarrow$$

$$\left[\begin{matrix} t\text{-}BuC{=}NAlMe_2 \\ | \\ Me \end{matrix}\right]\ \textbf{(17)} \longrightarrow \text{cyclic } [t\text{-}Bu(Me)C{=}N{-}AlMe_2]_2 \quad \textbf{(18)} \qquad (21)$$

benzyl cyanide (equation 22) but in this case the rearranged organo-aluminium compound **19** was not isolated. However, its formation can be readily inferred from the nature of the hydrolysis product **20**. This synthetic scheme for the preparation of ketones substantiates an earlier report[33b] that triethylaluminium and nitriles yield, following hydrolysis, ethyl ketones.

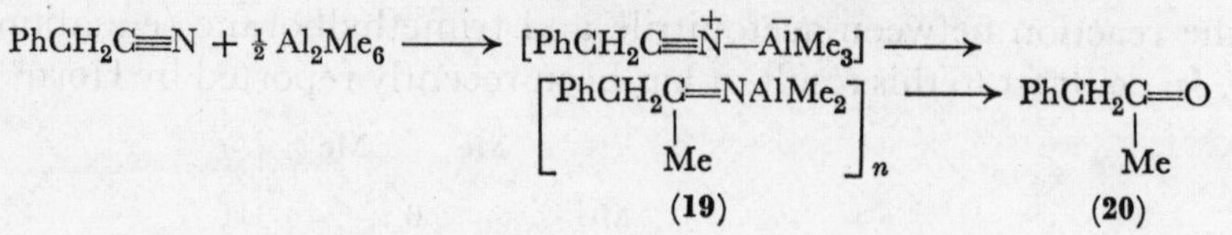

$$PhCH_2C{\equiv}N + \tfrac{1}{2}Al_2Me_6 \longrightarrow [PhCH_2C{\equiv}\overset{+}{N}{-}\overset{-}{Al}Me_3] \longrightarrow \left[\begin{array}{c}PhCH_2C{=}NAlMe_2\\ |\\ Me\end{array}\right]_n \longrightarrow \begin{array}{c}PhCH_2C{=}O\\ |\\ Me\end{array} \quad (22)$$

(19) (20)

A reaction was recently reported by Hooz and Linke[34] which involves alkyl rearrangement from boron to carbon in diazoacetonitrile (**21**) (equation 23). This reaction promises to be a very useful synthetic procedure for the high-yield (95%) preparation of nitriles and products derived therefrom.

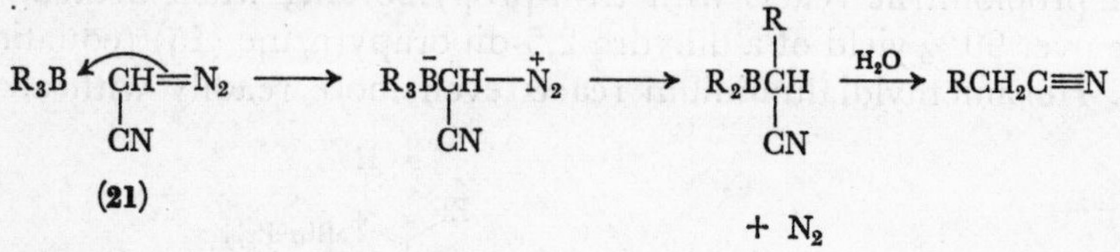

$$R_3B \curvearrowleft \underset{\underset{CN}{|}}{CH}{=}N_2 \longrightarrow R_3\overset{-}{B}\underset{\underset{CN}{|}}{CH}{-}\overset{+}{N}_2 \longrightarrow R_2B\overset{\overset{R}{|}}{\underset{\underset{CN}{|}}{CH}} + N_2 \xrightarrow{H_2O} RCH_2C{\equiv}N \quad (23)$$

(21)

C. Ketenimines and Ynamines

Cyano compounds which possess one or more hydrogen atoms attached to the α-carbon atom are potentially capable of tautomeric equilibrium between the nitrile and ketenimine (**22**) forms (equation 24). Indeed, in the case where two α-hydrogen atoms are present

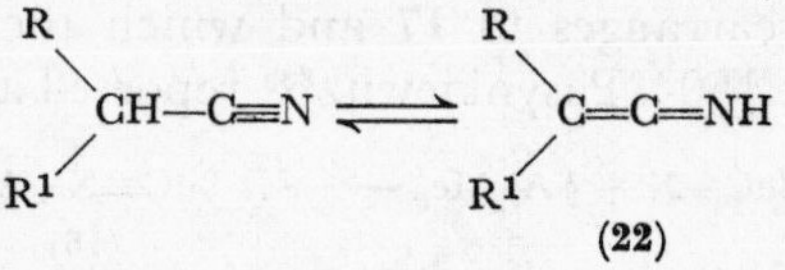

$$RR^1CH{-}C{\equiv}N \rightleftharpoons RR^1C{=}C{=}NH \quad (24)$$

(22)

both the ketenimine (**22a**) and the ynamine structure (**23**) may be written (equation 25). There is a paucity of experimental evidence

$$RCH_2{-}C{\equiv}N \rightleftharpoons \underset{(22a)}{RCH{=}C{=}NH} \rightleftharpoons \underset{(23)}{RC{\equiv}C{-}NH_2} \quad (25)$$

which bears on the properties and relative stability of these structures, although it is clear from carbanion alkylation reactions that the carbanion owes much of its stability to resonance structures such as:

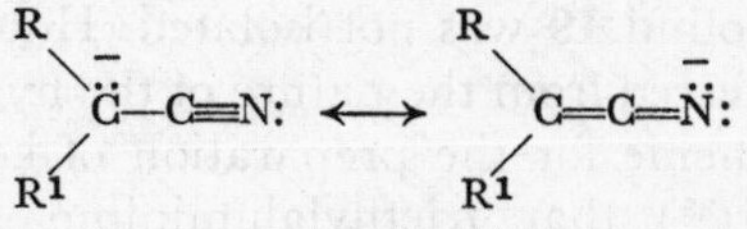

$$RR^1\overset{-}{C}{-}C{\equiv}N: \longleftrightarrow RR^1C{=}C{=}\overset{-}{N}:$$

Whereas the keteniminium anion form is important in stabilizing a charge in the anion, copious evidence suggests that the anion is much more nucleophilic at carbon than it is at nitrogen[35] since a preponderance of the products is found to be derived from *C*-alkylation rather than from *N*-alkylation (equation 26).

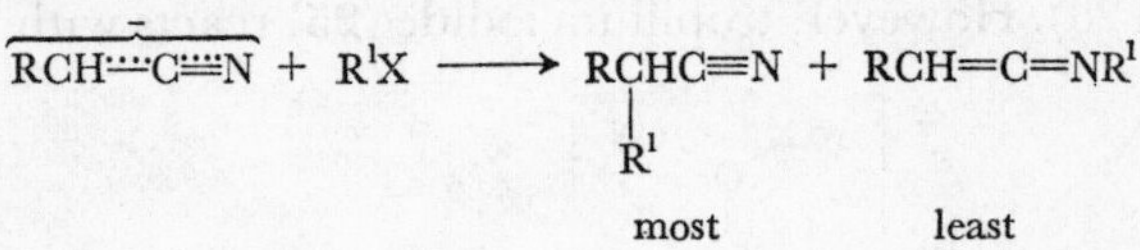

$$\overbrace{\text{RCH}\overset{-}{\cdots}\text{C}\equiv\text{N}} + \text{R}^1\text{X} \longrightarrow \underset{\text{most}}{\text{RCH(R}^1\text{)C}\equiv\text{N}} + \underset{\text{least}}{\text{RCH=C=NR}^1} \qquad (26)$$

One of the most extensively studied systems in which nitrile–ketenimine resonance can operate are the metal salts of 7,7,8,8-tetracyanoquinodimethane (TCNQ)[36a–f] (**24**) which exist as anion radical salts of the type $M^+TCNQ^{\overset{.}{-}}$ and as complex salts of the

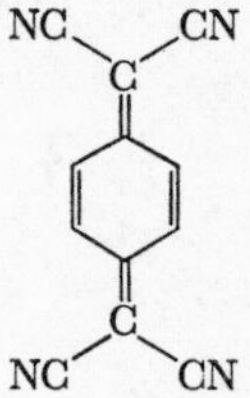

(**24**)

type $M^+TCNQ^{\overset{.}{-}}(TCNQ)$ in which M is a variety of metals or π-bases (see Chapter 9). This and other cyanocarbons are the subject of a recent review[37]. The anion radical $TCNQ^{\overset{.}{-}}$ may be represented as numerous resonance structures in which all four cyano groups become involved both in charge and odd-electron delocalization (equation 27). The infrared spectrum of simple salts

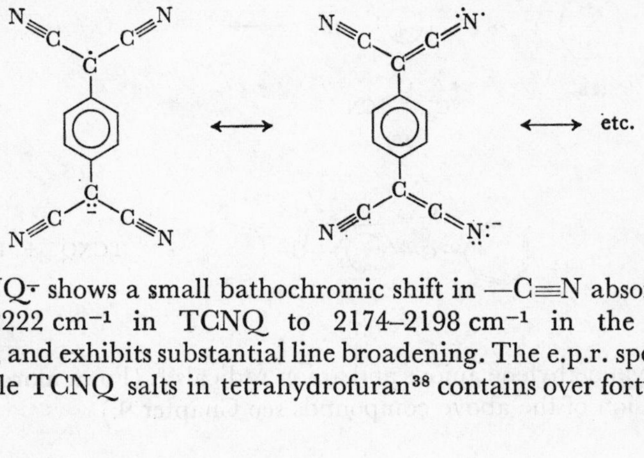

$$\longleftrightarrow \qquad \longleftrightarrow \text{ etc.} \qquad (27)$$

of $TCNQ^{\overset{.}{-}}$ shows a small bathochromic shift in —C≡N absorption, from 2222 cm^{-1} in TCNQ to 2174–2198 cm^{-1} in the anion radical, and exhibits substantial line broadening. The e.p.r. spectrum of simple TCNQ salts in tetrahydrofuran[38] contains over forty lines, with forty-five lines expected on the basis of isotropic hyperfine contact interaction between the ^{1}H and ^{14}N nuclei of the TCNQ radical. In chemical reactions, most simple salts of TCNQ react in a straightforward metathetical fashion, as in the reaction with acid (equation 28). However, tropilium iodide (**25**) reacts with $\overset{+}{Li}TCNQ^{\overline{\cdot}}$

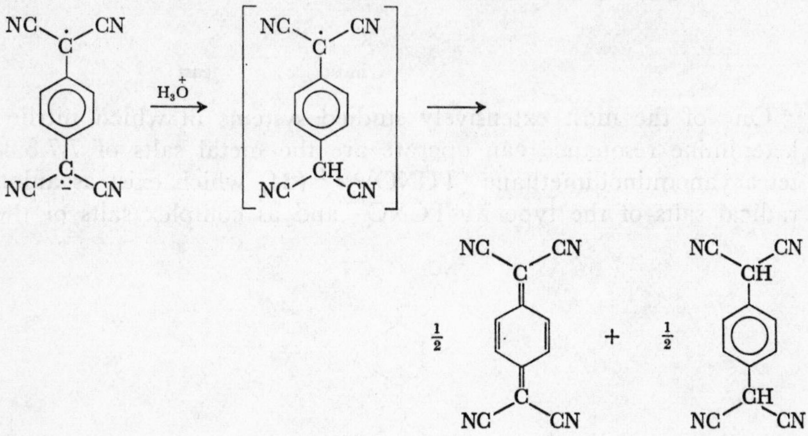

NC CN C· / benzene / C NC CN $\xrightarrow{H_3\overset{+}{O}}$ [NC CN C· / benzene / CH NC CN] $\longrightarrow$ $\frac{1}{2}$ (TCNQ) + $\frac{1}{2}$ (NC)₂CH–C₆H₄–CH(CN)₂ (28)

(**26**) to produce a high yield of α,α′-ditropyl-α,α,α′,α′-tetracyano-*p*-xylene (**27**) plus TCNQ (equation 29). It appears from these examples that although charge and spin delocalization are extensive, products are largely derived from reaction at carbon rather than at nitrogen. The same situation appears to prevail in the chemistry of

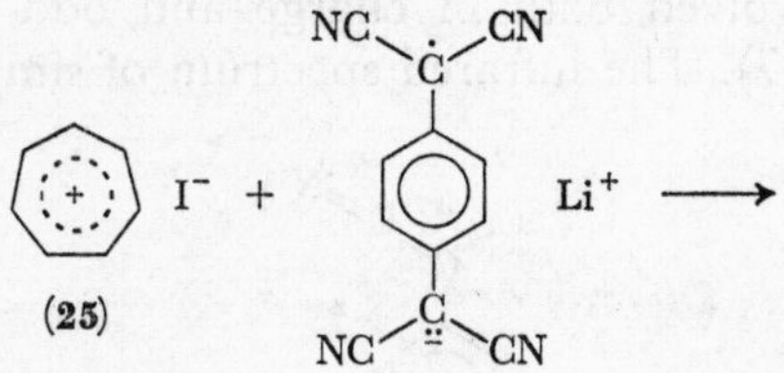

(**25**) I^- + (**26**) Li^+ $\longrightarrow$ (29)

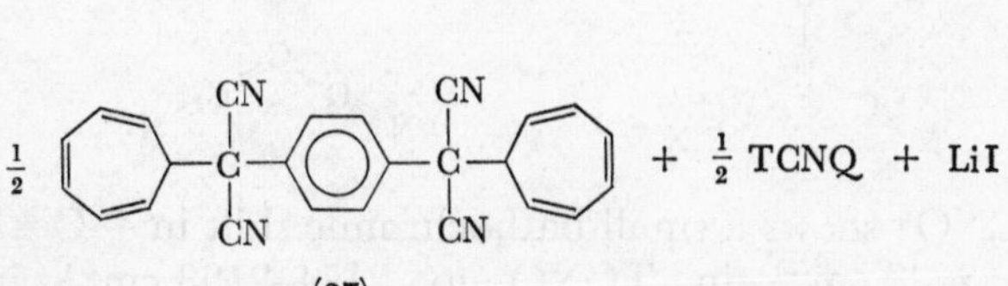

$\frac{1}{2}$ (**27**) + $\frac{1}{2}$ TCNQ + LiI

tetracyanoethylene anions and anion radicals[39]. (For a more detailed discussion of the above compounds see Chapter 9.)

A number of examples of stable ketenimines have been reported in which alkyl or aryl groups were substituted at nitrogen to preclude tautomerism to the cyano form. Stevens and coworkers reported the preparation[40a] and properties[40b] of *N*-substituted ketenimines (**28**).

$$Ar_2C{=}C{=}NR \quad RR^1C{=}C{=}NR^2$$
$$Ar_2C{=}C{=}NAr \quad RR^1C{=}C{=}NAr$$

(**28**)

(R, R^1, R^2 = alkyl or aryl)

These compounds, e.g. **30**, were prepared in good yields via the dehydrochlorination of the corresponding iminochloride (**29**) (equation 30). These workers found that these ketenimines with

$$R_2CH{-}C(Cl){=}NAr \xrightarrow{Et_3N} R_2C{=}C{=}NAr \qquad (30)$$

(**29**)

ammonia, or with primary or secondary aliphatic and aromatic amines to give amidines (**30**) in high yields [40b]. Two classical papers

$$R_2C{=}C{=}NAr + R^1NH_2 \longrightarrow R_2CH{-}C({=}NAr)NHR^1$$

(**30**)

of Hammond and coworkers, which demonstrated a cage effect in decomposition of 1,1′-azocyanocyclohexane[41a] (**31**) and α,α′-azoisobutyronitrile (AIBN)[41b] (**32**), serve to demonstrate that in the case of the radical at least, the product derived from attack by nitrogen (*N*-alkylation) may be observed as well. The thermal decomposition of

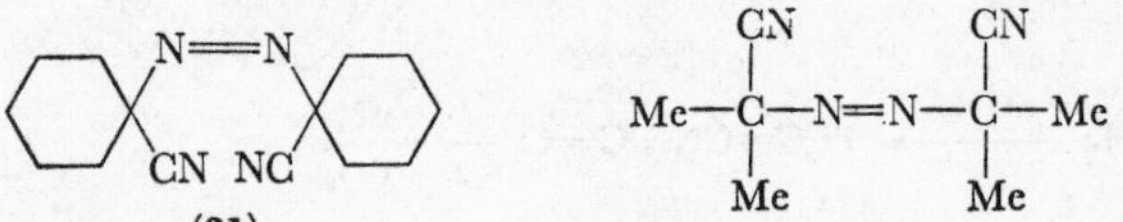

(**31**) (**32**) $Me{-}C(CN)(Me){-}N{=}N{-}C(CN)(Me){-}Me$

31 gave as one product the ketenimine (**33**) which could be isolated as a pure compound and which decomposed to 1,1′-dicyanodicyclohexane (**34**) at a rate comparable to the decomposition of the original azo compound (equation 31). A similar situation was encountered by

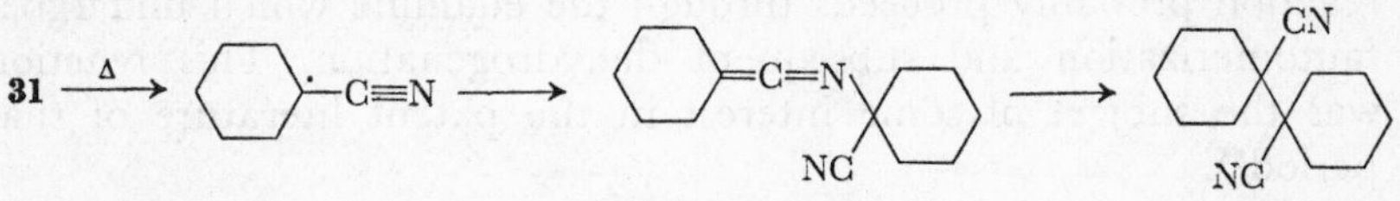

31 $\xrightarrow{\Delta}$ (cyclohexyl radical)$-C{\equiv}N$ ⟶ (**33**) ⟶ (**34**) (31)

these workers in studying the decomposition of AIBN. Here, ketenimine (**35**) was found to be a major product of the reaction. It also was found to decompose at a rate comparable to the rate of decomposition of AIBN itself to give 2,3-dicyano-2,3-dimethylbutane (**36**) (equation 32).

$$\underset{(\mathbf{32})}{\text{Me—C(CN)(Me)—N=N—C(CN)(Me)—Me}} \xrightarrow{\Delta} \text{Me}_2\dot{\text{C}}\text{—C}\equiv\text{N} + \text{N}_2 \longrightarrow \underset{(\mathbf{36})}{\text{Me—C(CN)(Me)—C(CN)(Me)—Me}} + \underset{(\mathbf{35})}{\text{Me}_2\text{C=C=N—C(CN)(Me)—Me}} \quad (32)$$

Isoelectronic with the ketenimine structure **33** is the ynamine structure **37** (equation 33), about which little has been reported. Hoffmann[42a,b], writing in the last century about the rearrangement

$$\underset{(\mathbf{37})}{\text{—C}\equiv\text{C—N}\langle} \rightleftharpoons \underset{(\mathbf{38})}{\rangle\text{C=C=N—}} \rightleftharpoons \text{—}\overset{|}{\underset{|}{\text{C}}}\text{—C}\equiv\text{N} \quad (33)$$

which bears his name, noted that phenylpropiolamide (**39**) underwent a degradation reaction with sodium hypobromite to give a good yield of benzyl cyanide (equation 34). It is likely that oxidation, rearrangement and decarboxylation produced aminophenylacetylene (**40**) as an intermediate which tautomerizes to benzyl cyanide (**41**).

$$\underset{(\mathbf{39})}{\text{PhC}\equiv\text{C}\overset{\text{O}}{\overset{\|}{\text{C}}}\text{NH}_2} \xrightarrow{\text{NaOBr}} \text{PhC}\equiv\text{CN=C=O} \xrightarrow[2.\ -\text{CO}_2]{1.\ \text{H}_2\text{O}} \underset{(\mathbf{40})}{\text{PhC}\equiv\text{C—NH}_2} \xrightarrow{\sim\text{H}} \underset{(\mathbf{41})}{\text{PhCH}_2\text{C}\equiv\text{N}} \quad (34)$$

Reports from the older literature[43a–c], that acetylene and ammonia can be converted to acetonitrile (equation 35) at elevated temperature, suggest the intervention of enamine in the reaction path. This reaction probably proceeds through the enamine which undergoes tautomerization and subsequent dehydrogenation. This reaction was the subject of some interest in the patent literature of that period[44].

$$\text{HC}\equiv\text{CH} + \text{NH}_3 \longrightarrow \text{H}_2\text{C=CH—NH}_2 \longrightarrow \text{CH}_3\text{CH=NH} \longrightarrow \text{CH}_3\text{CN} \quad (35)$$

A recent report has appeared[45] in which the behaviour of ynamines toward addition reactions has been noted. In contrast to primary ynamines, tertiary ynamines (**43**) are quite stable and many compounds with this structure have been reported. Bromination–dehydrobromination of enamines (**42**) at −80° gives good yields of ynamines[46] (equation 36). Indeed, tertiary ynamines have been

$$\underset{(\mathbf{42})}{RCH{=}CHNR_2^1} \xrightarrow[\substack{t\text{-BuOK, } -80^\circ \\ \text{tetrahydrofuran}}]{Br_2\text{, then}} \underset{(\mathbf{43})}{RC{\equiv}CNR_2^1} \qquad (36)$$

shown to be excellent dehydrating agents, producing amides from carboxylic acids and amines in nearly quantitative yields under very mild conditions[47a], a reaction which has led to the successful application of this reagent to peptide synthesis[47b,c].

It appears that the tautomeric equilibrium when hydrogen atoms are present as substituents, strongly favours the cyano compound, although Rappoport[48a] has recently found that strongly electron-withdrawing substitution on the α-carbon probably produces detectable concentrations of the ketenimine form. Thus in 1,1,2,2-tetracyanoethane (**44**) the enimine (**45**) is detectable (equation 37).

$$\underset{(\mathbf{44})}{(NC)_2CH{-}CH(CN)_2} \rightleftharpoons \underset{(\mathbf{45})}{(NC)_2CH{-}C(CN){=}C{=}NH} \qquad (37)$$

Fleury and Libis[48b] obtained high melting crystalline solid tautomers **47** from the treatment of acylmalononitriles (**46**) with acid followed by extraction into ether (equation 38).

$$RCOCl + CH_2(CN)_2 \xrightarrow{Et_3N} \underset{(\mathbf{46})}{RCOCH(CN)_2} \rightleftharpoons RC(OH){=}C(CN)_2$$

$$R(HO)C{=}C(CN)_2 \underset{}{\overset{H_2SO_4}{\rightleftharpoons}} \underset{(\mathbf{47})}{R(O{=})C{-}C(CN){=}C{=}NH} \qquad (38)$$

Trofimenko[49] has reported the preparation and properties of tricyanomethane (cyanoform) (**48**), which can be represented as tautomeric structures involving the ketenimine (**49**) (equation 39).

$$\underset{(48)}{HC(CN)_3} \rightleftharpoons \underset{(49)}{(NC)_2C{=}C{=}NH} \qquad (39)$$

In wet ether solution, secondary amines[49a], alcohols[49b] and hydrogen halides[49c], add to the ketenimino tautomer to give 1,1-dicyano-ethylene derivatives (**50**) (equation 40).

$$\underset{(49)}{(NC)_2C{=}C{=}NH} + HX \longrightarrow \underset{(50)}{(NC)_2C{=}C(NH_2)X} \qquad (40)$$

$$(X = OR, Cl, Br, NRR^1)$$

The reaction of *N,N*-disubstituted hydrazides with *α,β*-unsaturated aldehydes follows an unusual course which has been attributed to the intervention of an aminonitrile rearrangement. Ioffe and Zelenin[50] observed *β*-dialkylaminopropionitriles (**52**) as the major product of this reaction, a fact which they attribute to the the ring-opening of an intermediate pyrazolium ion (**51**) (equation 41).

$$CH_2{=}CHCHO + R_2NNH_2\cdot HX \longrightarrow CH_2{=}CHCH{=}NNR_2\cdot HX + H_2O$$

$$\longrightarrow \underset{(51)}{\text{[pyrazolinium, } \overset{+}{N}R_2\text{]}\ X^-} \xrightarrow{OH^-} \underset{(52)}{R_2NCH_2CH_2C{\equiv}N} \qquad (41)$$

These workers concluded that ring–chain transfer of the dialkyl-amino group intramolecularly was a more reasonable reaction path than a non-cyclic elimination–addition sequence.

In a related example of ambivalent tautomeric behaviour, Stacy and coworkers[51a] found that, while **53** displayed spectroscopic properties consistent with the ring-closed structure (**53a**), the chemical reactions with benzoyl chloride, alkaline hydrogen per-oxide or sodium borohydride–aluminium chloride resembled much more those of the open-chain tautomer (**53b**) (equation 42). Thus borohydride reduction of **53** gave the aminomercaptan (**54**) and dihydrothiophene (**55**), in spite of an n.m.r. spectrum and infrared

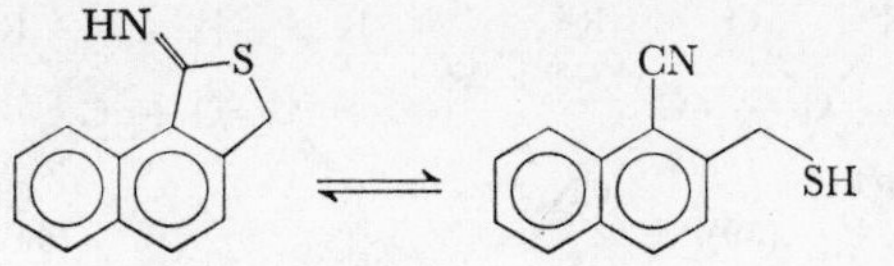

$$\mathbf{53a} \rightleftharpoons \mathbf{53b} \qquad (42)$$

frequency (3280 cm^{-1}) which suggest the predominance of the ring-closed tautomer **53a**[51b] (equation 43).

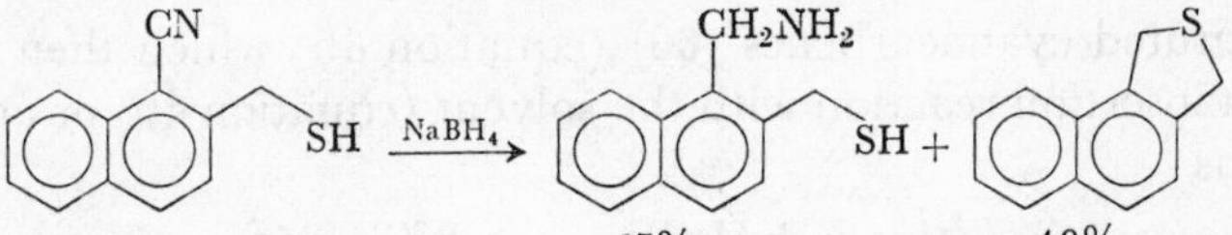

$$\mathbf{53b} \xrightarrow{NaBH_4} \mathbf{54}\ (17\%) + \mathbf{55}\ (48\%) \qquad (43)$$

D. Cyanocarbenes

Cyanocarbenes represent an interesting case of enimine structure, in that they can be represented in a dipolar resonance hybrid of the keteniminium form (**56a** and **56b**). A substantial effort has been devoted to the study of dicyanocarbene[36,52a–d] and aminocyanocarbene[53a,b] although the former was more intensively examined.

$$-\ddot{C}-C{\equiv}N: \longrightarrow -\overset{+}{C}{=}C{=}\overset{-}{N}: \qquad \mathbf{(56a)} \quad \mathbf{(56b)}$$

Ciganek had reported the preparation of the explosive dicyanodiazomethane[54] (**57**) from which dicyanocarbene (**58**) could be thermolytically generated (equation 44). This intermediate was found to add non-stereospecifically in high yields to olefins and, more

$$N{\equiv}C-\overset{N_2}{\overset{\|}{C}}-C{\equiv}N \longrightarrow N{\equiv}C-\ddot{C}-C{\equiv}N + N_2 \qquad (44)$$
$$\mathbf{(57)} \qquad\qquad \mathbf{(58)}$$

slowly to insert into tertiary, secondary and primary saturated C—H bonds, in that order of reactivity. Another very interesting method for the preparation of a cyanocarbene was reported by Griffin[55] and coworkers. 2-Cyanooxiranes (**59**), which are readily available from the epoxidation of substituted acrylonitriles, undergo photolytic cleavage in a number of different cases to give good yields

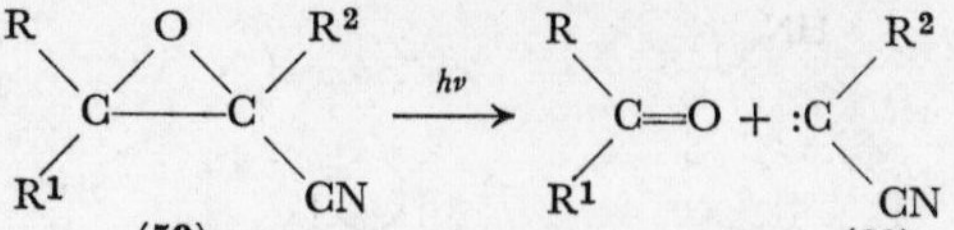

$$R(R^1)C\underset{O}{\text{—}}C(R^2)CN \xrightarrow{h\nu} R(R^1)C{=}O + :C(R^2)CN \qquad (45)$$

(59) (60)

R	R¹	R²
H	Ph	Ph
Ph	Ph	Ph
—$(CH_2)_5$—		Ph

of substituted cyanocarbenes (**60**) (equation 45) which then underwent an insertion reaction with the solvent (equation 46) or addition to olefins

$$R^2\ddot{C}\text{—CN} + MeOH \longrightarrow R^2CH(CN)OMe \qquad (46)$$

(60)

in unspecified yield and with only a slight stereoselectivity (equation 47).

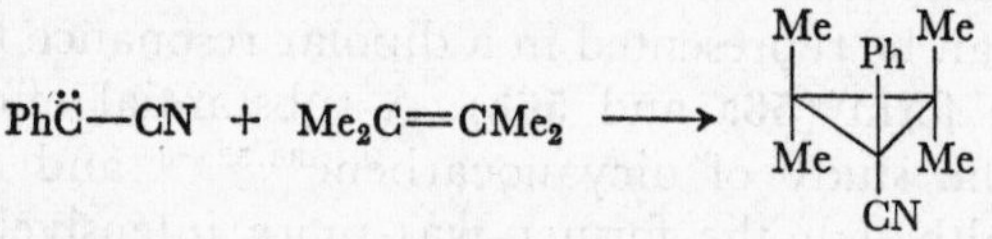

$$Ph\ddot{C}\text{—CN} + Me_2C{=}CMe_2 \longrightarrow \text{1-Ph-1-CN-2,2,3,3-tetramethylcyclopropane} \qquad (47)$$

(60) (R^2 = Ph)

Swenson[52b] prepared dicyanocarbene (**58**) by 1,1-dehydrobromination of bromomalononitrile (**61**) with triethylamine (equation 48).

$$BrCH(CN)_2 \xrightarrow[-8^\circ]{NEt_3} N{\equiv}C\text{—}\ddot{C}\text{—}C{\equiv}N + NEt_3{\cdot}HBr \qquad (48)$$

(61) (58)

The presence of a reactive, electron-deficient species was demonstrated by the formation of a 24% yield of 1,1-dicyanotetramethylcyclopropane (**62**) when the dehydrohalogenation was carried out in the presence of tetramethylethylene (equation 49).

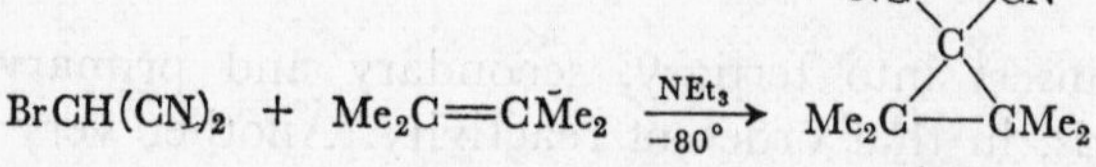

$$BrCH(CN)_2 + Me_2C{=}CMe_2 \xrightarrow[-80^\circ]{NEt_3} \text{(NC)}_2C<(CMe_2\text{—}CMe_2) \qquad (49)$$

(62)

Aminocyanocarbene (**64**) was reported by Moser and coworkers[53] to be formed during the photolysis of 1-cyanoformamide-*p*-toluenesulphonylhydrazone (**63**) in a 1-methyltetrahydrofuran glass at

$$\underset{\textbf{(63)}}{H_2N(NC)C{=}NNHSO_2\text{—}C_6H_4\text{—}CH_3} \xrightarrow[\text{2-Me-THF}]{-196°,\ h\nu} \underset{\textbf{(64)}}{H_2N\text{—}\ddot{C}\text{—}C{\equiv}N} \qquad (50)$$

−196° (equation 50). Several polar resonance structures can be drawn for this carbene (**65a–d**). It was concluded[53] that the carbene

$$\underset{\textbf{(65a)}}{H_2N\text{—}\ddot{C}\text{—}C{\equiv}N} \qquad \underset{\textbf{(65b)}}{H_2\overset{+}{N}{=}\overset{-}{\ddot{C}}\text{—}C{\equiv}N} \qquad \underset{\textbf{(65c)}}{H_2\overset{+}{N}{=}C{=}C{=}\overset{-}{\ddot{N}}:} \qquad \underset{\textbf{(65d)}}{H_2N\text{—}\overset{+}{C}{=}C{=}\overset{-}{\ddot{N}}:}$$

possesses singlet multiplicity in the ground state and that the polar charge-separated structures (i.e. **65b–65d**) make a considerable contribution to the structure. This was largely based on ultraviolet spectroscopic evidence.

Closely related to aminocyanocarbene is the nitrogen analogue, cyanonitrene (**67**), which has been recently reported[56a–e] from thermal decomposition of cyanogen azide (**66**) at 40–50° (equation

$$\underset{\textbf{(66)}}{N_3CN} \longrightarrow \underset{\textbf{(67)}}{NCN} + N_2 \qquad (51)$$

51). This unusual reagent was found to insert smoothly into the C—H bonds of saturated hydrocarbons, and displayed a selectivity for tertiary:secondary:primary C—H bonds of approximately 70:10:1. Thus with 2,3-dimethylbutane, nearly all of the product was derived from tertiary C—H insertion giving **68**, rather than from primary C—H insertion, which gives **69** (equation 52). Nitrene **67** also

$$\underset{\textbf{(67)}}{NCN} + (CH_3)_2CHCH(CH_3)_2 \xrightarrow{40\%} \underset{\substack{91{\cdot}6\% \\ \textbf{(68)}}}{(CH_3)_2CHC(CH_3)_2\text{—}NHCN} + \underset{\substack{8{\cdot}4\% \\ \textbf{(69)}}}{(CH_3)_2CHCH(CH_2NHCN)CH_3} \qquad (52)$$

reacted readily with the carbon–carbon double bond of norbornene to give a 1-cyanoaziridine derivative (**70**) (equation 53). From a

$$\text{norbornene} + \underset{\textbf{(67)}}{NCN} \longrightarrow \underset{\textbf{(70)}}{\text{norbornane-fused aziridine—NCN}} \qquad (53)$$

study of the variation in stereoselectivity of insertion of cyanonitrene into saturated C—H bonds, these same workers concluded that thermolysis of cyanogen azide at 41–53° produced singlet nitrene, which was readily converted to the ground state triplet form by collision deactivation. In contrast to this, Anastassiou has reported more recently[57] that when photolytically generated, with light between 2100–3000 Å wavelength, the cyanonitrene is of singlet multiplicity only, and that it inserts quite stereospecifically into tertiary C—H bonds. If methods can be devised for the safe handling of the very unstable cyanogen azide, this promises to become a very useful synthetic tool.

Aryl *ortho*-dinitrenes have been found to be an unusual source of unsaturated nitriles when they undergo ring scission. Nakagawa and Onoue[58a,b] found that *o*-phenylenediamine (**71**), when oxidized by lead tetraacetate or nickel (IV) oxide, cleaved smoothly to *cis,cis*-1,4-dicyanobutadiene (**72a**) (equation 54); Hall and Patterson[59] found

$$\text{(71)}\ C_6H_4(NH_2)_2 \xrightarrow[NiO_2]{Pb(OAc)_4\ \text{or}} \text{(72a)}\ NC\text{–}CH{=}CH\text{–}CH{=}CH\text{–}CN \quad (54)$$

later that a series of *ortho*-diazides (**73**) could be prepared and converted to the same type of product (**72**) (equation 55). The

$$\text{(73)}\ R\text{–}C_6H_3(N_3)_2 \longrightarrow \text{(72)}\ R\text{-substituted } NC\text{–}CH{=}CH\text{–}CH{=}CH\text{–}CN \quad (55)$$

	R
73a	H
b	3-Me
c	4-Me
d	4-OMe
e	4-Cl

latter authors postulated several processes which might lead to the dinitrile product, including direct decomposition of the dinitrene **73a** derived from the loss of nitrogen (equation 56). Hall demonstrated that the reaction uniquely requires an *ortho* relationship of the diazide by the failure of 1,4-diazidobenzene (**74**) to undergo a

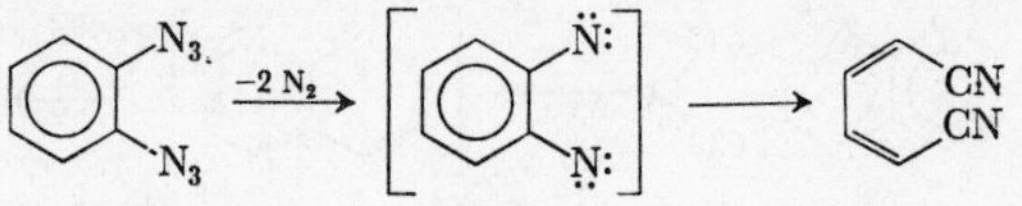

(56)

(73) (73a) (72a)

similar reaction (equation 57) and thus the direct decomposition noted above is less likely. Alternatively, the rearrangement might

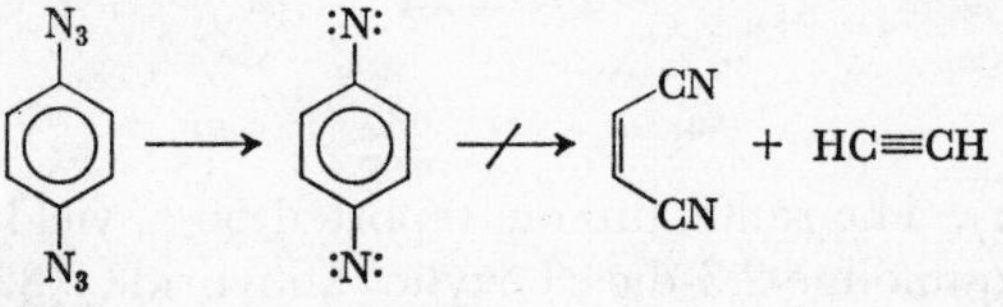

(57)

(74)

proceed by stepwise loss of nitrogen via a triazole (equation 58).

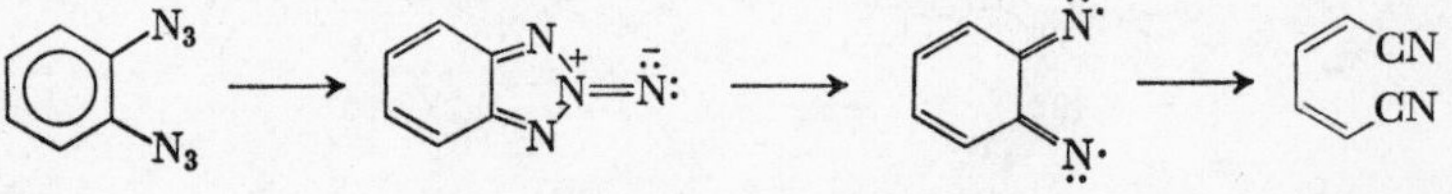

(58)

(71) (75) (73) (72a)

This contention is supported by observation that 2-aminobenzotriazole (**76**) could be oxidized by lead tetraacetate to *cis,cis*-1,4-dicyanobutadiene[60] (equation 59). Hall suggested that the Nakagawa

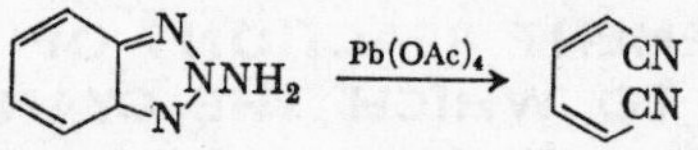

(59)

(76) (72a)

intermediate might be a metal-coordinated species such as **77** (equation 60). A more recent report by Cava[61] has noted a similar

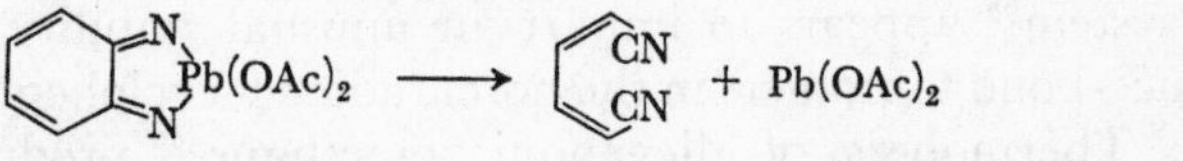

(60)

(77) (72a)

type of rearrangement reaction in the isocyano system. Quinoxaline-2,3-dicarboxylic anhydride (**78**) is reported to undergo a high-temperature gas phase pyrolysis to give 72 % of 1,2-dicyanobenzene. In this case, the proposed 'quinoxalyne' (**79**) intermediate decomposes to the diisocyanide (**80**), which is then presumed to undergo the isocyanide–cyanide rearrangement (see section II.A) to give **81**

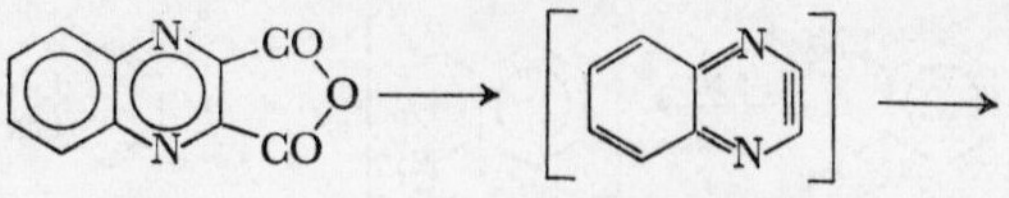

(78) → [(79)] →

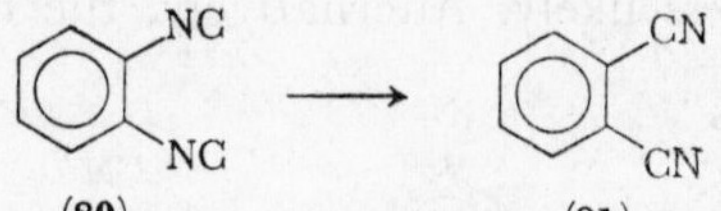

(80) → (81) (61)

(equation 61). The same authors reported 50% yield of the nitrile (**83**) from quinoline-2,3-dicarboxylic anhydride (**82**) under the same conditions, presumably by a similar process (equation 62).

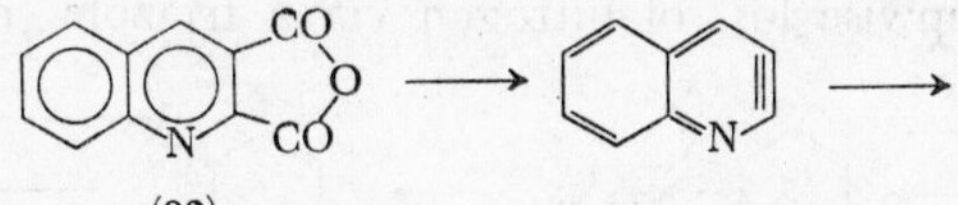

(82) → →

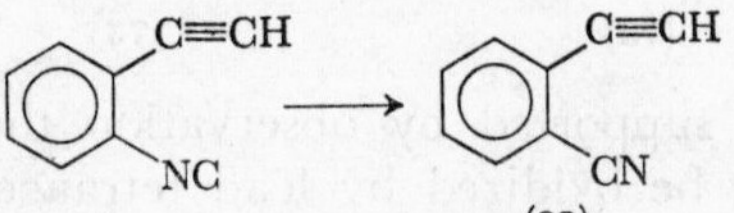

(83) (62)

III. REARRANGEMENT REACTIONS OF THE CARBON SKELETON TO WHICH THE CYANO GROUP IS ATTACHED

Several rearrangement reactions have been noted in which the carbon skeleton attached to cyano substituent groups has undergone rearrangement. The strongly electron-withdrawing dicyanomethylene system[62] appears to impart an unusual stability to one of the valence-bond tautomers in the norcaradiene–cycloheptatriene isomer pair. Thermolysis of dicyanodiazomethane[54] produced dicyanocarbene (**58**, see section II.D) which added in 60–80% yield to a variety of aromatic hydrocarbons. With benzene[52d], 7,7-dicyanonorcaradiene (**84**) was obtained on photolysis of the diazo compound

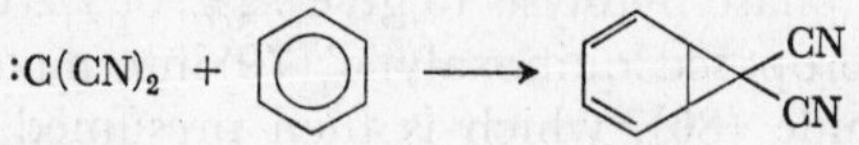

$$:C(CN)_2 + C_6H_6 \longrightarrow \quad (63)$$

(58) (84)

(equation 63). Thermolysis of **84** produced an interesting rearrangement to give phenylmalononitrile (**85**) and 3,7-dicyanocycloheptatriene (**86**)[63], the pathway to the benzene derivative being

(84) $\xrightarrow{\Delta}$ (85) + (86) (64)

slightly more favoured energetically (equation 64). With *p*-xylene, carbene **58** gave two isomeric norcaradienes (**87** and **88**) (equation 65). Thermal rearrangement of a 1:1 mixture of the dicyanodimethylnorcaradienes **87** and **88** occurred rapidly at 130° to give

:C(CN)$_2$ (58) + *p*-xylene $\longrightarrow$ (87) + (88) (65)

2,5-xylylmalononitrile (**89**) (equation 66). Similarly, dicyanocarbene added to naphthalene to give three isomeric compounds. Benzonorcaradiene (**90**) was obtained in 50% yield and the two dicyanobenzocycloheptatrienes (**91** and **92**) in combined yield of 12% (equation 67). The latter products appear to be formed as a result

(87) + (88) $\longrightarrow$ (89) (66)

:C(CN)$_2$ (58) + naphthalene $\longrightarrow$ (90) + (91) + (92) (67)

of the addition of dicyanocarbene to the 2,3- and 1,9-bonds of naphthalene followed by rearrangement of the carbon skeleton; 1,5-cyano shifts of the two dicyanobenzocycloheptatrienes **91** and **92** occur at a reasonable rate at elevated temperatures. Thus either **91** or **92** give 3,7-dicyano-1,2-benzocycloheptatriene (**93**) (equation 68), via (**90**), and through slower subsequent reactions **90** gave

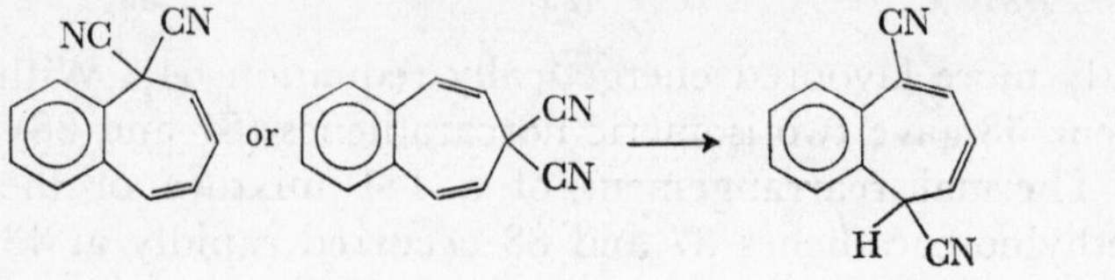

(**91**) or (**92**) → (**93**) (68)

1-naphthylmalononitrile (**94**) (equation 69), and **93** gave **96** (equation 70).

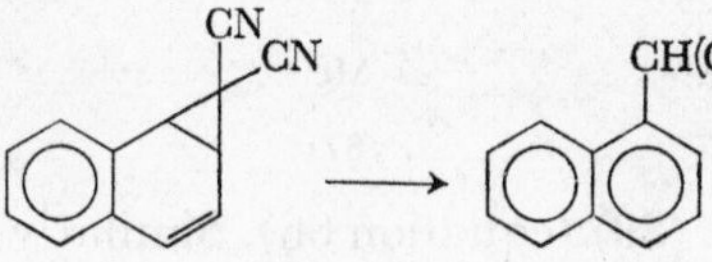

(**90**) → (**94**) (69)

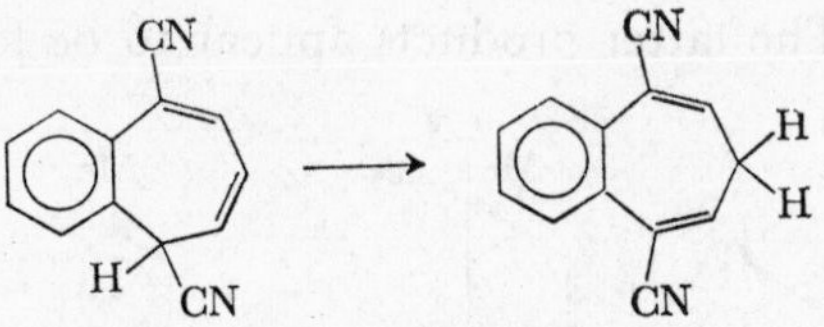

(**93**) → (**96**) (70)

It is noteworthy that while the presence of the *geminal* cyano groups in 7,7-dicyanonorcaradiene appears to lower the activation energy for the norcaradiene–toluene isomerization by approximately 10 kilocalories compared to the unsubstituted system[64a], (a fact attributed to stabilization of the intermediate diradical **97** by the cyano groups (equation 71)), it is not nearly as clear why the

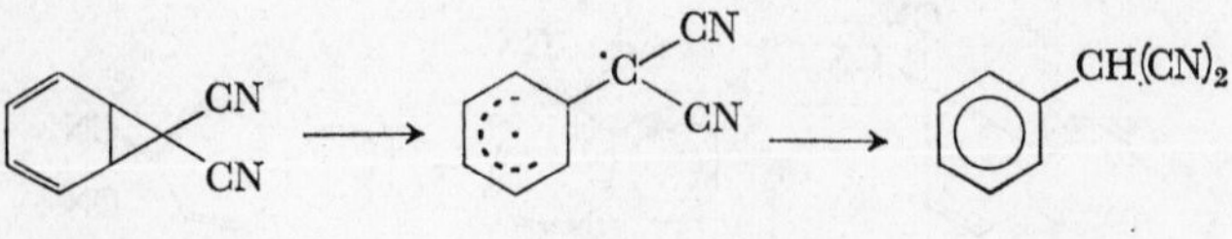

(**84**) → (**97**) → (**85**) (71)

norcaradiene form of these valence-bond tautomers are stable[64b]. The cycloheptatriene tautomer is the only one detected in a number of other systems, including 7-cyanobicycloheptatriene and 7,7-bis(trifluoromethyl)cycloheptatriene. The novel cyclopentadienyl anion (**100**) reported by Webster[65] to arise from the acid-catalysed ring closure of **98**, followed by rearrangement of **99** in strong acid (equation 72), is remarkable for its high acidity. Properties of this system are dealt with thoroughly in Chapter 9.

$$2\,Na^+(NC)_2\bar{C}C(CN){=}C(CN)\bar{C}(CN)_2 \xrightarrow{\text{dilute acid}} \mathbf{99} \xrightarrow[15^\circ]{\text{conc. HCl}} \mathbf{100} \qquad (72)$$

(**98**) (**99**) (**100**)

Tetracyano chromone derivative (**101**) is reported[66a,b] to undergo an unusual series of rearrangements of the carbon skeleton to give rise to an interesting photochromic system. Compound **101** was found to be thermally unstable and rearranges to imidate (**102**) upon standing in alcohol. Boiling alcohol converts **101** to **103**. Compound **102** is further rearranged to benzochromone (**104**) by extended reflux in xylene. The tricyano compound **102** undergoes a reversible photochemical ring–chain tautomerism with **105** in high quantum yield (equation 73). In the *O*-acetyl series (**106**), a ring-contracted product **108** is obtained, in addition to the normal ring-opened compound **107** (equation 74).

Isolated cases of rearrangements of the cyano bonded to nitrogen have been reported. Bird[67a] established that several 2-cyano-1-phenylpyrazole derivatives underwent isomerization which involved complex skeletal rearrangement when they were heated at 270°. Thus cyano pyrazole (**109**) gave **110** (equation 75). Similarly **111** gave **112** (equation 76). This author postulated, on the basis of earlier related work[67b,c], that the mechanism might involve a novel Cope-type of rearrangement involving electrophilic aromatic substitution with pyrazole ring opening (**113** → **114** → **115**) (equation 77).

IV. FRAGMENTATION REACTIONS INVOLVING THE CYANO GROUP

A variety of elimination and rearrangement reactions are known which lead to nitriles. One group of such reactions involves hydrogen

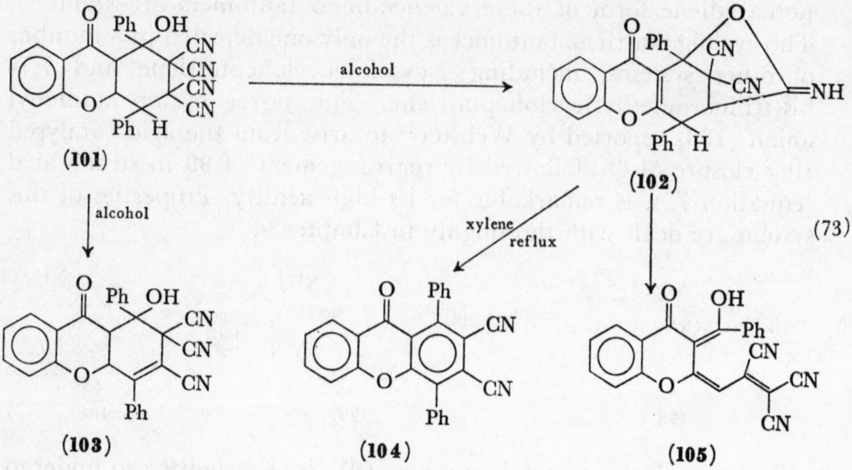

(101) alcohol → (102); (101) alcohol → (103); (102) xylene reflux → (104); (102) → (105)

(73)

(106) → (107) + (108)

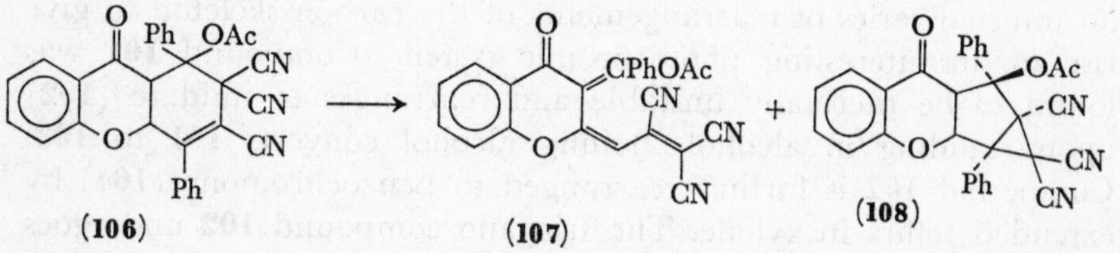

(74)

as the electrofuge and anions of strong acids as the nucleofuge. Such reactions as the dehydration of aldoximes (**116**) (equation 78) and derivatives, the Wohl degradation, and the 1,2-elimination reaction of chloramines[68] (**117**) (equation 79) or hydrazinium[4b,c,d] salts (**118**) (equation 80) are all examples of this type of reaction.

These reactions constitute elimination of small molecules from across the C—N bond, and do not involve rearrangement of the carbon skeleton. They will not be considered in this section. Reactions in which the cationoid species of an elimination scheme is carbon are generally referred to as fragmentation reactions. These reactions have been recently reviewed by the most active investigator of this area[69a–d]. Particular aspects of this reaction type, notably those reactions known as the 'abnormal' or 'second-order' Beckmann rearrangement have been the subject of earlier comprehensive review[4a–c,70]. These fragmentation reactions which involve the cyano group and result in rearrangement of the carbon skeleton and their related reactions are the topic of this section. It is the aim of this

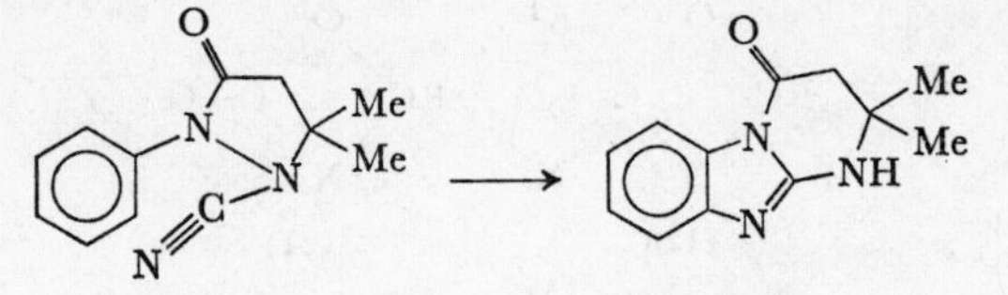

(109) → **(110)** (75)

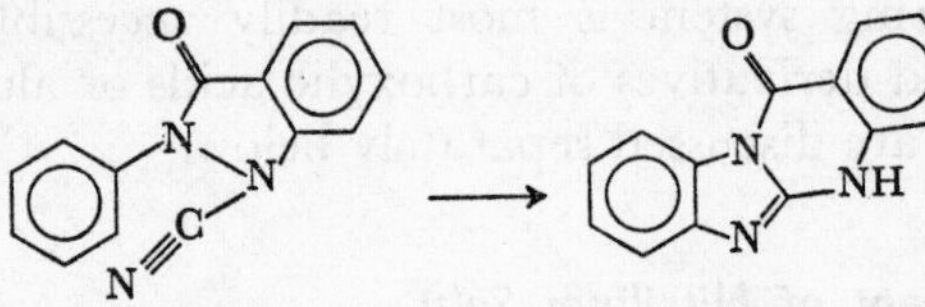

(111) → **(112)** (76)

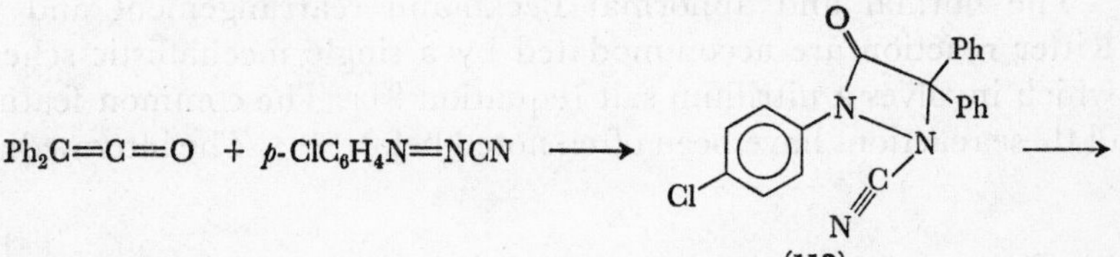

$Ph_2C{=}C{=}O + p\text{-}ClC_6H_4N{=}NCN \longrightarrow$ **(113)** $\longrightarrow$

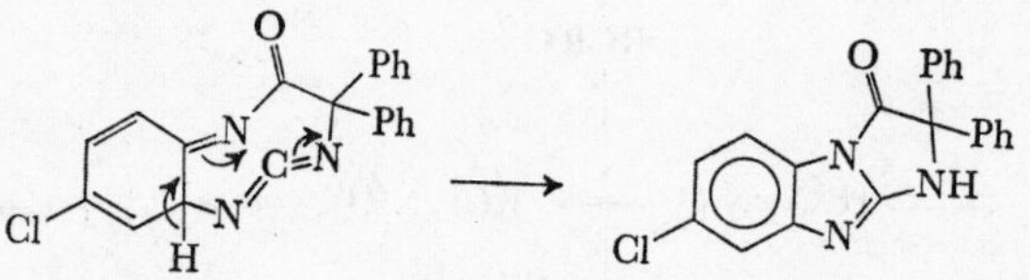

(114) → **(115)** (77)

$$\underset{\textbf{(116)}}{RCH{=}NOH} \xrightarrow{\text{acid}} RC{\equiv}N + H_2O \quad (78)$$

$$\underset{\textbf{(117)}}{RCH{=}NCl} \xrightarrow{\text{base}} RC{\equiv}N + Cl^- \quad (79)$$

$$\underset{\textbf{(118)}}{RCH{=}\overset{+}{N}NR_3{}^1} \xrightarrow{\text{base}} RC{\equiv}N + NR_3{}^1 \quad (80)$$

section to discuss those reactions which may be viewed as passing through a nitrilium ion (**119**) intermediate, or through an oxazete

$$\underset{\textbf{(119)}}{RC{\equiv}\overset{+}{N}R^1 \longleftrightarrow R\overset{+}{C}{=}\ddot{N}R^1}$$

ring (**120**) or an oxazetone (**121**). The nitrilium ion intermediate is attainable either from *N*-alkylation of the nitrile, a path generally referred to as the Ritter reaction[71a,b], or by rearrangement of the oxime

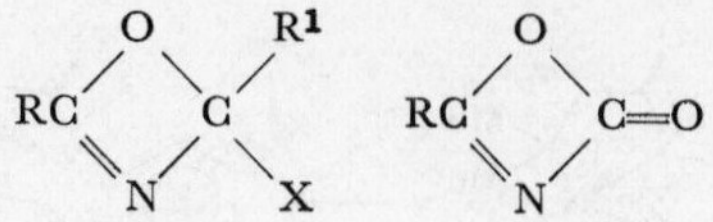

(120) (121)

or a derivative of the oxime in acidic or aprotic polar solvents[4a,69,70]. The oxazete ring system is most readily accessible from cyano compounds and derivatives of carboxylic acids or aldehydes. These reaction types are discussed separately below.

A. Rearrangement of Nitrilium Salts

The normal and abnormal Beckmann rearrangement and the Ritter reaction are accommodated by a single mechanistic scheme which involves a nitrilium salt (equation 81). The common features of these reactions have been often noted before[72a–c]. The intermediacy

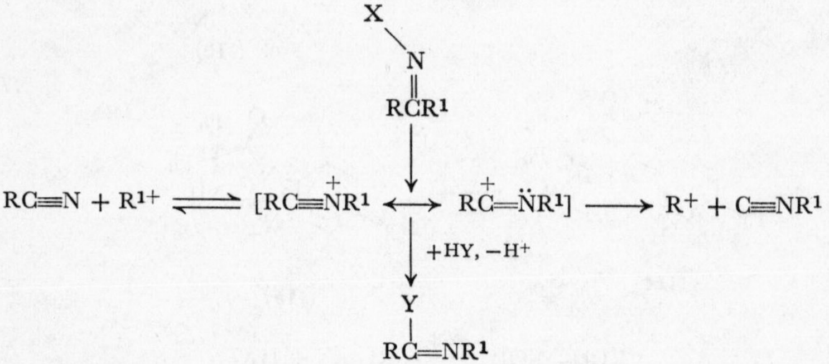

$$\mathrm{RC{\equiv}N + R^{1+} \rightleftharpoons [RC{\equiv}\overset{+}{N}R^1 \longleftrightarrow R\overset{+}{C}{=}\ddot{N}R^1] \longrightarrow R^+ + C{\equiv}NR^1} \quad (81)$$

(from X–N=CRR¹ above; with +HY, −H⁺ giving RC(Y)=NR¹ below)

of the nitrilium salt is usually inferred from the source of the alkyl group (alcohol or olefin plus acid) or the reaction conditions (usually strong acid). Occasionally intermediate nitrilium salts have been isolated[73a]. Indeed, a comprehensive study of nitrilium salts was recently reported[73b]. Olah has directly examined the p.m.r. spectra of nitrilium (**119a**) and alkyl nitrilium[73a] salts (**119b**) formed in 'super acid' (SO_2–FSO_3H–SbF_5), and concluded that both were linear in configuration. Olah has shown further that *N*-methyl

$$\mathrm{R{-}C{\equiv}\overset{+}{N}{-}H} \qquad \mathrm{R{-}C{\equiv}\overset{+}{N}{-}Me}$$

(**119a**) (**119b**)

acetonitrilium ion (**119c**) is directly accessible by the action of 'super acid' on acetone oxime (equation 82). The importance of a

$$\text{MeC(=NOH)Me} \longrightarrow \text{MeC(=}\overset{+}{\text{N}}\text{HOH)Me} \longrightarrow \text{MeC}{\equiv}\overset{+}{\text{N}}\text{Me} \quad \textbf{(119c)} \; (\text{R} = \text{Me, R}^1 = \text{Me}) \tag{82}$$

nitrilium salt in these reactions is further evidenced by the classical rate study[74] on the rearrangement of cyclic benzo-α-ketoxime picrates (**120**). Whereas nitrilium cation **121**, derived from the $n = 6$ ketoxime, is sufficiently non-linear to preclude substantial stabilization of the intermediate from the structure shown, cation **122** can

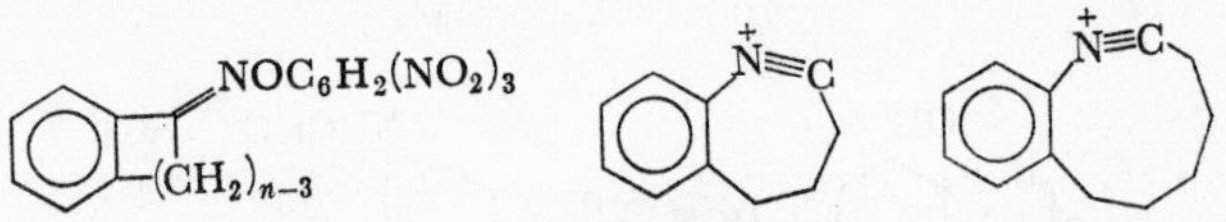

$NOC_6H_2(NO_2)_3$, $(CH_2)_{n-3}$ **(120)**; $\overset{+}{N}{\equiv}C$ from $n = 6$ **(121)**; $\overset{+}{N}{\equiv}C$ from $n = 8$ **(122)**

draw considerable stabilization from C≡N bond formation. Experimentally, **120** ($n = 6$) undergoes rearrangement very slowly relative to **120** ($n = 8$). Depending on the relative propensity of R and R^1 for formation of a stable carbonium ion, and on the nucleophilicity of the solvent, one or more of the products noted in the above scheme may be found.

I. The Ritter reaction

The reaction between carbonium ion sources and nitriles to give amides **124** (equation 83) may be referred to as the Ritter reaction[75a,b]. Carbonium ion sources have usually been tertiary or

$$\text{R}^{1+} + \text{N}{\equiv}\text{CR} \xrightarrow[\text{2. H}_2\text{O}]{\text{1. H}^+} \text{R}^1\text{NHC(=O)R} \quad \textbf{(124)} \tag{83}$$

benzylic alcohols, as in the reaction of benzyl alcohol with acrylonitrile, which gives a 60% yield of *N*-benzylacrylamide[76] (**125**)

$$\text{PhCH}_2\text{OH} + \text{CH}_2{=}\text{CHCN} \xrightarrow{\text{H}_2\text{SO}_4} \text{PhCH}_2\text{NHC(=O)CH}{=}\text{CH}_2 \quad \textbf{(125)} \tag{84}$$

(equation 84), or branched olefins[72c,77a,b] as illustrated by the reaction of isobutylene with actonitrile in aqueous sulphuric acid to give

N-*t*-butylacetamide (**126**) (equation 85). The generality of the

$$(CH_3)_2C{=}CH_2 + N{\equiv}CCH_3 \xrightarrow[H_2O]{H_2SO_4} (CH_3)_3CNH\overset{\overset{O}{\|}}{C}CH_3 \quad \textbf{(126)} \qquad (85)$$

reaction is demonstrated by the fact that large number of α- and β-substituted unsaturated acids, esters and hydroxy esters may serve as carbonium ion sources[77b] to give yields of amides that are generally above 50 %. Hydrocyanic acid has been employed as a source of the cyano group[78a,b] to give an *N*-substituted formamide (**127**) (equations

$$PhCH_2C(CH_3)_2OH + HCN \xrightarrow{H^+} PhCH_2C(CH_3)_2NHCHO \qquad (86)$$

$$\textbf{(128)}\ \text{(decalol, OH)} + HCN \xrightarrow[H_2O]{H^+} \textbf{(127)}\ \text{(decalyl NHCHO)} \qquad (87)$$

86 and 87). Unsymmetrical diamides have been observed from dinitriles[77a,b] (**128**) (equation 88) and the stabilizing influence of a β-halogen atom on the intermediate carbonium ion has been employed

$$(CH_3)_2C{=}CH_2 + NCCH_2CH_2CN \xrightarrow[H+]{H_2O} (CH_3)_3CNH\overset{\overset{O}{\|}}{C}CH_2CH_2\overset{\overset{O}{\|}}{C}NH_2 \quad \textbf{(128)} \qquad (88)$$

to produce a novel *N*-haloethylamide[79] (**129**) synthesis (equation 89).

$$\underset{\underset{CH_2X}{|}}{RR^1COH} + N{\equiv}CR^2 \xrightarrow{H_2SO_4} \underset{\underset{CH_2X}{|}}{RR^1CNH\overset{\overset{O}{\|}}{C}R^2} \quad (X = Cl, Br) \quad \textbf{(129)} \qquad (89)$$

In the case of an unsaturated nitrile providing both the cyano group and the carbonium ion, polymers have been reported[80a]. Thus acrylonitrile was reported to give insoluble polyalanine (**130**) in sulphuric–acetic acid mixtures (equation 90).

$$\frac{n+2}{n} CH_2{=}CHCN \xrightarrow[HOAc]{H_2SO_4} CH_2{=}CHCO(NHCHMeCO)_{\frac{n+2}{n}}CHMeCN \quad \textbf{(130)} \qquad (90)$$

Recently, Zilkha and coworkers[80b] have reported a soluble polymer in this same system using cold concentrated sulphuric acid.

Upon hydrolysis the polymer gives low yields of β-alanine. These workers postulate a trimeric sulphuric acid tetraester (**131**) as the intermediate before hydrolysis (equation 91).

$$\underset{\text{OSO}_2\text{OH}}{\text{CH}_2}\underset{\text{OSO}_2\text{OH}}{\text{CH}}{=}\text{CNHCH}_2\underset{\text{OSO}_2\text{OH}}{\text{CH}}{=}\text{CNHCH}_2\text{CH}_2\underset{\text{OSO}_2\text{OH}}{\text{C}}{=}\text{NH} \xrightarrow[\text{H}^+]{\text{H}_2\text{O}} 3\ \text{H}_2\text{NCH}_2\text{CH}_2\text{CO}_2\text{H} \qquad (91)$$

(**131**)

An interesting example of the 'electrochemical Ritter reaction' has been reported by Eberson[81]. Electrochemical oxidation of durene in acetonitrile leads to *N*-pentamethylbenzylacetamide (**132**), presumably via an intermediate carbonium ion and nitrilium salt (equation 92).

$$\text{MeC}{\equiv}\text{N} + \text{C}_6\text{Me}_6 \longrightarrow$$

$$\text{C}_6\text{Me}_5\text{CH}_2\overset{+}{\text{N}}{\equiv}\text{CMe} \longrightarrow \text{C}_6\text{Me}_5\text{CH}_2\text{NH}\overset{\text{O}}{\overset{\|}{\text{C}}}\text{Me} \qquad (92)$$

(**132**)

2. Fragmentation of ketoximes and their derivatives

The fragmentation of ketoximes and derivatives of ketoximes has been accomplished by a wide variety of reagents. In the case of ketoximes, most conceivable reagents which are capable of converting the oxime hydroxyl function to a good leaving group have been employed. Mineral acid, carboxylic acid chlorides, sulphonyl chlorides, thionyl chloride, phosphorous trihalides, phosphorous pentahalides and phosphorous pentoxide have been the most commonly used. Derivatives of oximes in which the oxime hydroxyl is already a good leaving group (i.e. tosylate) have been rearranged in a variety of polar aprotic solvents, pyridine being most commonly employed. The reaction has been widely studied and has been the subject of comprehensive mechanistic and stereochemical investigations.

a. Simple ketoximes. Perhaps the earliest description of an unsaturated nitrile among the products of oxime rearrangement is due to Wallach[82] who reported 4-pentenonitrile (**134**) as a major product in the boric oxide–alumina catalysed dehydration of cyclopentanone oxime (**133**) (equation 93). Subsequent reports by Lazier[83a] and

$$(133) \xrightarrow[200-250^\circ]{B_2O_3,\ Al_2O_3} H_2C{=}CHCH_2CH_2CN\ (134) + \text{2-piperidone} \tag{93}$$

Davydoff[83b] established the generality of the fragmentation reaction in simple six- (**135**) (equation 94) and seven-membered ring ketox-

$$(135) \xrightarrow[200-250^\circ]{B_2O_3,\ Al_2O_3} CH_2{=}CH(CH_2)_3CN\ (136) \tag{94}$$

imes (**137**) (equation 95) and in β-methylcyclohexanone oxime[84] (**139**) (equation 96), although the reported rearranged structure of the nitrile product in the latter case is unusual and merits scrutiny.

$$(137) \xrightarrow{P_2O_5} CH_2{=}CH(CH_2)_4CN\ (138) \tag{95}$$

$$(139) \xrightarrow{P_2O_5} CH_2{=}CHCH(Me)CH_2CH_2CN\ (140) \tag{96}$$

Grob has reported[85] the fragmentation of the oxime tosylate of menthone (**141**) in heated pyridine (equation 97). In this case the carbonium ion derived from fragmentation of the nitrilium salt would be secondary, and a trisubstituted double bond (**142**) results from elimination, enhancing the possibility of contribution from a second-order elimination pathway.

A number of simple bicyclic systems have been reported to produce unsaturated nitriles. β-Oximinoquinuclidine hydrochloride

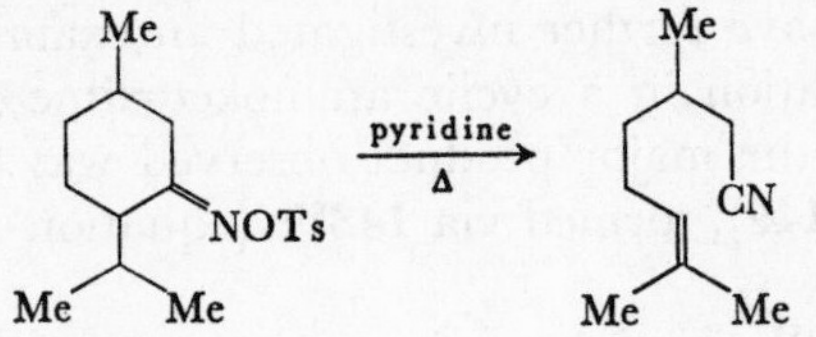

(141) →[pyridine, Δ] (142) (97)

(**143**) undergoes rearrangement with benzenesulphonyl chloride[86] to give 39 % of the nitrile which has lost one carbon and undergone

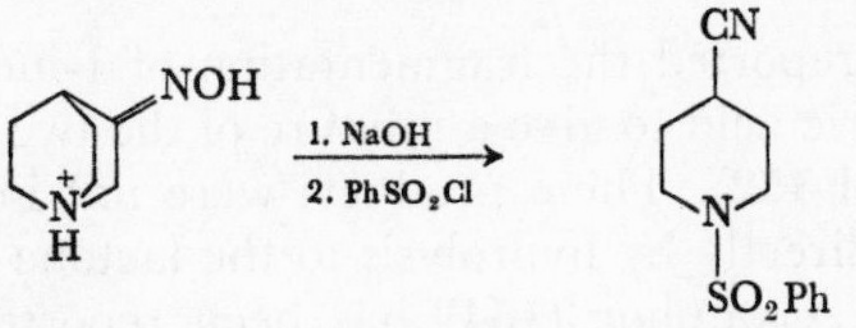

(143) →[1. NaOH; 2. $PhSO_2Cl$] (144) (98)

N-sulphonation (**144**) (equation 98). This fragmentation might proceed through an intermediate nitrogen ylid (**145**) (equation 99).

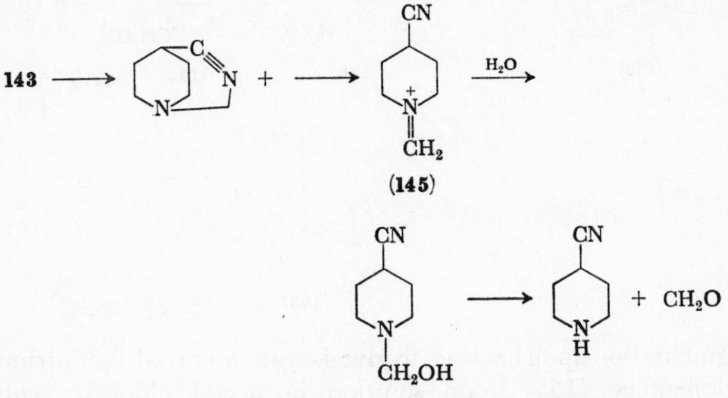

143 → ... + → (145) →[H_2O] ... → ... + CH_2O (99)

Nitrogen participation in the fragmentation of α-oximinoamines has been well documented in the recent work of Grob and Fischer[87a–e] and others, who have carried out fragmentation reactions of a series of α-oximinoamines (**146**) in ethanol at 80°[87f,g] (equation 100). Grob

$$RC(=NOH)CH_2NR_2 \ (\mathbf{146}) \xrightarrow[\text{EtOH}]{80^\circ} RC{\equiv}N + CH_2{=}\overset{+}{N}R_2 \quad (100)$$

$$CH_2{=}\overset{+}{N}R_2 \xrightarrow{H_2O} CH_2O + HNR_2$$

and coworkers have further investigated an example of α-oximino-amine fragmentation in a cyclic aminoketoxime (**147**), but under their conditions the major product observed was the normal Beckmann product (**149**), formed via **148**[87c] (equation 101).

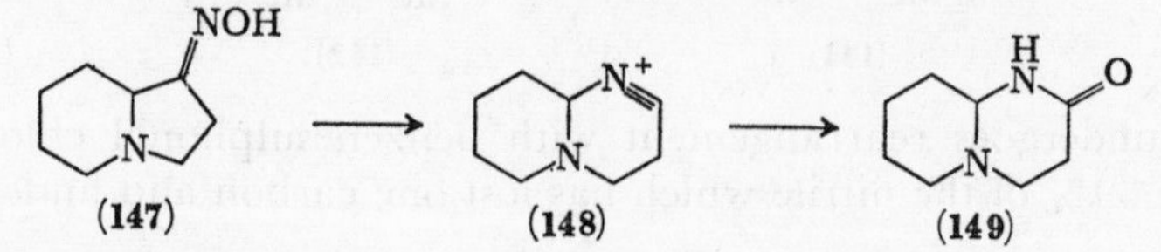

(**147**) ⟶ (**148**) ⟶ (**149**) (101)

Sauers[88] has reported the fragmentation of 1-methylnorcamphor (**150**) in sulphuric acid to give a mixture of the two expected cyano-olefins (**151** and **152**). These products were not isolated, however, but converted directly by hydrolysis to the lactone **153** in low yield (equation 102). Camphor (**154**) has been reported[89a] to undergo

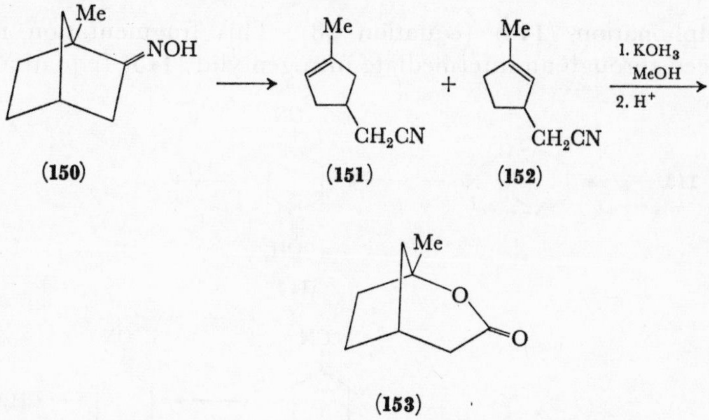

(**150**) ⟶ (**151**) + (**152**) $\xrightarrow[\text{2. } H^+]{\text{1. } KOH_3,\ MeOH}$ (**153**) (102)

fragmentation upon heating, to give 4-cyanomethyl-1,2,3,3-trimethyl-cyclopentene (**155**) upon solution in acetyl chloride (equation 103). However, it appears that this procedure leads to a complex mixture in low yield[89b]. In much earlier work, the fragmentation of

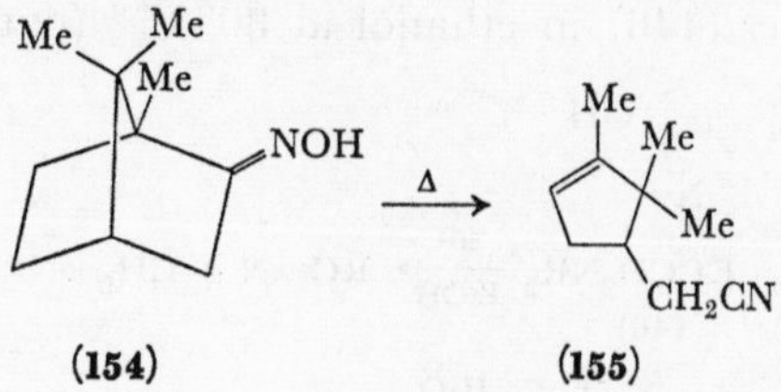

(**154**) $\xrightarrow{\Delta}$ (**155**) (103)

β-pericyclocamphenoneoxime (**156**) to **157** was reported[90a,b] (equation 104). In this case, *geminal* methyl groups preclude the possibility of two double bond isomers, which were possible in the case of 1-methylnorcamphor, reported by Sauers[88].

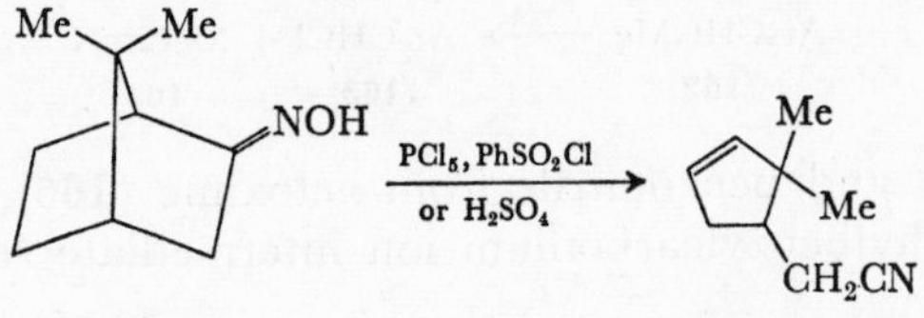

$$\textbf{(156)} \xrightarrow[\text{or } H_2SO_4]{PCl_5,\ PhSO_2Cl} \textbf{(157)} \qquad (104)$$

Hill and Conley[87g,91] have reported unsaturated cyclic nitriles (**159**) as the fragmentation products of spiroketoximes (**158**) by using thionyl chloride (equation 105). The normal Beckmann

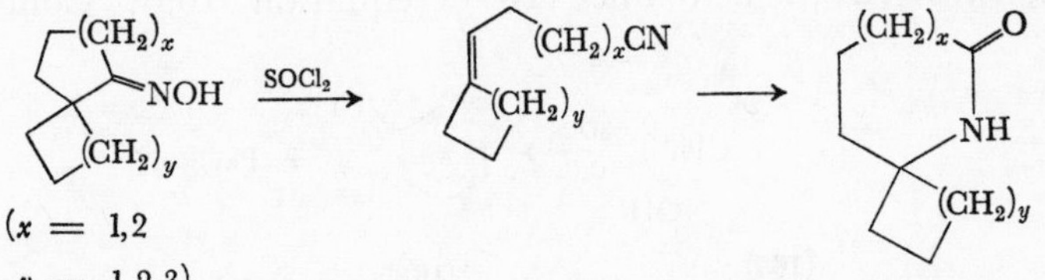

$$\textbf{(158)}\ (x = 1,2;\ y = 1,2,3) \xrightarrow{SOCl_2} \textbf{(159)} \longrightarrow \textbf{(160)} \qquad (105)$$

products (**160**) were also isolated in these cases. When polyphosphoric acid was used as the catalyst, an α,β-unsaturated decalone (**161**) was obtained by an unusual ring closure reaction (equation 106). These

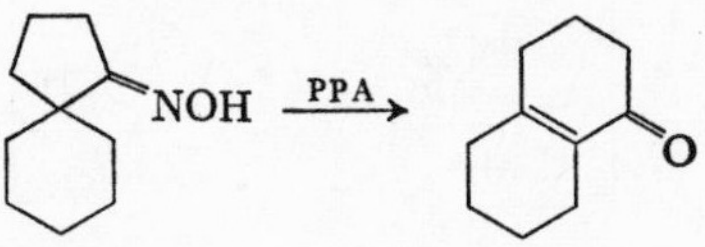

$$\textbf{(158)} \xrightarrow{PPA} \textbf{(161)} \qquad (106)$$

authors demonstrated that the decalone was a secondary product derived from the unsaturated nitrile. This phenomenon has been noted by other workers[72a,91].

Fragmentation of oximes and oxime derivatives, in which no heteroatom stabilization of the incipient carbonium ion is possible, appears to be the predominant or exclusive path of reaction for those compounds which have a fully substituted α-carbon atom. A good example of this is derived from the work of Hassner and Nash[92] on diarylmethyl ketoximes (**162**), which fragment readily in phosphorous pentachloride to give benzhydryl chlorides (**163**) and nitriles (**164**) (equation 107). Lyle and Lyle[93] obtained α-methylstyrene (**166**) and benzonitrile from ketoxime (**165**), which would give a dimethylbenzylcarbonium ion intermediate (equation 108).

$$\underset{(\mathbf{162})}{Ar_2CHC(=NOH)Me} \xrightarrow{PCl_5} \underset{(\mathbf{163})}{Ar_2CHCl} + \underset{(\mathbf{164})}{MeC{\equiv}N} \qquad (107)$$

$$\underset{(\mathbf{165})}{Me_2CPhC(=NOH)Ph} \xrightarrow[C_6H_6]{SOCl_2} PhC{\equiv}N + \underset{(\mathbf{166})}{CH_2{=}CMePh} \qquad (108)$$

Similarly these workers obtained 1-phenylcyclohexane (**168**) and benzonitrile from ketoxime (**167**) (equation 109). Conley and coworkers[72c,94a,b] have studied other α-trisubstituted oximes. 2,2-Dimethylcyclopentanone oxime (**169**) rearranges via the intermediate nitrile to 3-methylcyclohex-2-enone (**170**) in polyphosphoric acid (equation 110). In the case of 1,1-dimethyltetralone-2-oxime (**171**), Conley[94b] was able to isolate the normal (**172**) and the abnormal (**173**) product from the unsaturated nitrile (equation 111).

$$\underset{(\mathbf{167})}{\text{1-Ph-cyclohexyl-C(=NOH)Ph}} \xrightarrow[C_6H_6]{SOCl_2} \underset{(\mathbf{168})}{\text{1-Ph-cyclohexene}} + PhC{\equiv}N \qquad (109)$$

$$\underset{(\mathbf{169})}{\text{2,2-Me}_2\text{-cyclopentanone oxime}} \xrightarrow{PPA} \left[\text{CN, } \overset{+}{C}Me_2 \longrightarrow \text{CN, }C(Me){=}CH_2 \right] \longrightarrow \underset{(\mathbf{170})}{\text{3-Me-cyclohex-2-enone}} \qquad (110)$$

$$\underset{(\mathbf{171})}{\text{1,1-Me}_2\text{-tetralone-2-oxime}} \xrightarrow{PPA} \underset{(\mathbf{172})}{\text{lactam (NH, C=O, CMe}_2)} + \underset{(\mathbf{173})}{\text{benzocycloheptenone (Me, C=O)}} \qquad (111)$$

A study of the rate of decomposition of methyl ketoxime tosylates by Grob and coworkers[95] strongly suggests a common intermediate for both the normal Beckmann product and the fragmentation. Evidence has accumulated which suggests strongly that the nitrilium salt intermediates may undergo dissociation–recombination to produce intermolecular crossover products. Conley and coworkers found that rearrangement of a mixture of pinacolone oxime (**174**) and 2-methyl-2-phenylpropiophenone oxime (**175**) in polyphosphoric acid produced all four possible amides (**176–179**) derived from C to N migration of the most highly substituted groups[72c] (equation 112). These same workers carried out the fragmentation

$$\underset{\mathbf{(174)}}{Me_3C\overset{NOH}{\overset{\|}{C}}Me} + \underset{\mathbf{(175)}}{PhCMe_2\overset{NOH}{\overset{\|}{C}}Ph} \longrightarrow [Me_3C\overset{+}{N}{\equiv}CMe + PhCMe_2\overset{+}{N}{\equiv}CPh]$$

$$\rightleftharpoons [Me_3C^+ + N{\equiv}CMe + Ph\overset{+}{C}Me_2 + N{\equiv}CPh] \xrightarrow{H_2O} \left\{ \begin{array}{l} Me_3CNHCOMe\ \mathbf{(176)} \\ PhCMe_2NHCOMe\ \mathbf{(177)} \\ Me_3CNHCOPh\ \mathbf{(178)} \\ PhCMe_2NHCOPh\ \mathbf{(179)} \end{array} \right. \tag{112}$$

of the oxime of 9-acetyl-*cis*-decalin (**180**). They obtained predominantly *N*-(*trans*-9-decalyl)acetamide (**181**) using a large variety of dehydrating agents, and this was taken as compelling evidence for a fragmentation–recombination mechanism. Supporting evidence

$$\underset{\mathbf{(180)}}{\text{Me(C=NOH)-decalin}} \longrightarrow \text{9-decalyl cation} \xrightarrow{MeCN} \underset{\mathbf{(181)}}{\text{NHCOMe-decalin}} \tag{113}$$

$$\underset{\mathbf{(182)}}{\text{OH-decalin}} + CH_3CN \xrightarrow{H_2SO_4} \text{9-decalyl cation}$$

for this postulate was the fact that *cis*-β-decalol (**182**) and acetonitrile in sulphuric acid also gave the same product (equation 113). Evidence for the ionization of the oxygen leaving group to produce an intermediate nitrilium ion was recently presented[96]. A series of substituted acetophenone oximes (**183**) were rearranged in sulphuric acid containing 3·8 atom % excess ^{18}O. The substituted acetamide products had attained the ^{18}O concentration of the solvent, suggesting the existence of solvent separated nitrilium–bisulphate ion pairs (equation 114). It was also concluded that the ^{18}O isotopic labelling

$$\underset{\textbf{(183)}}{\mathrm{Ar\overset{\overset{NOH}{\|}}{C}Me}} \xrightarrow{H_2SO_4} \mathrm{ArN\overset{+}{\equiv}CMe} \xrightarrow{H_2{}^{18}O} \mathrm{ArNH\overset{\overset{^{18}O}{\|}}{C}Me} \qquad (114)$$

$$\underset{\textbf{(184)}}{\mathrm{Ar\overset{\overset{N^+}{\|}}{C}H\ OSO_2O^-}}$$

experiments preclude the formation of an unrearranged pre-equilibrium ion pair (**184**) of the type which has been postulated in earlier work[97a,b].

In what the authors[98] called an 'anomalous Ritter reaction', α-dialkylaminonitriles (**185**), which contained sterically large substituents either on nitrogen or on carbon in the Ritter reaction using *t*-butyl alcohol and sulphuric acid (equation 115) gave α-hydroxyamides (**187**) with loss of the amino substituent. It is possible that

$$\underset{\textbf{(185)}}{\mathrm{R\overset{\overset{R_2{}^1N}{|}}{C}HCN}} + t\text{-BuOH} \xrightarrow{H_2SO_4} \mathrm{R\overset{\overset{R_2{}^1N:}{|}}{C}H{-}C\overset{+}{\equiv}NBu\text{-}t} \longrightarrow$$

$$\underset{\textbf{(186)}}{\mathrm{RCH{-}C{=}NBu\text{-}}t\ (\text{bridged by } \mathrm{R_2{}^1N^+})} \xrightarrow[-H^+]{H_2O} \mathrm{R\overset{\overset{R_2{}^1N}{|}}{C}H{-}\underset{\underset{OH}{|}}{C}{=}NBu\text{-}t} \xrightarrow[H^+]{H_2O} \underset{\textbf{(187)}}{\mathrm{R\overset{\overset{HO}{|}}{C}H\overset{\overset{O}{\|}}{C}NHBu\text{-}t}} \qquad (115)$$

steric repulsion of the large R and R^1 groups contributes to the formation of an aziridinium ring intermediate (**186**) which undergoes substitution by water.

Besides assistance from neighbouring nitrogen noted here and earlier in this section[86,87], fragmentation of ketoximes, assistance by neighbouring ether oxygen, epoxy oxygen and carbonyl groups has been reported. Hill[99] noted that β-oximino ethers underwent fragmentation in the presence of phosphorous pentachloride, polyphosphoric acid or *p*-toluenesulphonyl chloride to give good yields of nitriles. Thus oxime **187a** fragmented to give, after cyclization and hydrolysis, γ-butyrolactone and benzophenone (equation

$$\text{(187a)} \rightleftharpoons [\,] \longrightarrow \text{NC(CH}_2)_3\text{OH} + Ph_2C{=}O \qquad (116)$$

116), and oxime (**188**) gave a dimethylacrylonitrile and acetone in good yield (equation 117). With an α-epoxyketoxime (**189**),

$$\text{(188)} \xrightarrow{\text{acid}} [\,] \longrightarrow \text{Me}_2\text{C}{=}\text{CHCN} + Me_2C{=}O \qquad (117)$$

fragmentation also occurred (equation 118). However, the products **190** and **191** were identified only tentatively and no yield was reported.

$$m\text{-}O_2NC_6H_4\text{-CH(O)CH-C(=NOH)Ph}\ \textbf{(189)} \xrightarrow{PCl_5} m\text{-}O_2NC_6H_4\text{-CH(OH)CHO}\ \textbf{(190)} + PhC{\equiv}N\ \textbf{(191)} \qquad (118)$$

A more recent study of tetralone oximes which possess a β-methylthio group (**192**), reported by Autrey and Scullard[100], showed that fragmentation occurred in these systems also giving **193** and **194** (equation 119).

The introduction of a new reagent for the Beckmann rearrangement into this system, 2-chloro-1,1,2-trifluorotriethylamine (**195**), is noteworthy and deserves comment. This reagent brings about rearrangement in excellent yields under especially mild conditions (70°, 18 min). Autrey and Scullard have noted the probable mode of the reaction by the amine (equation 120), and further observed that the intermediate *O*-acylated species (**196**) appeared to be formed

$$\text{(192)} \xrightarrow{Et_2NCF_2CHClF} \text{(193)} + \text{(194)} \quad (119)$$

MeO, SMe, HON (192); MeO, CN, SMe (193); MeO, CN, CHO (194)

rapidly. It is likely that this promising reagent will see further use in Beckmann and Lossen type rearrangements.

$$\underset{(195)}{Et_2NCF_2CHClF} \xrightarrow{-F^-} Et_2\overset{+}{N}{=}CFCHClF \xrightarrow{RC(=NOH)R^1} \underset{(196)}{RC(=NOC(=\overset{+}{N}Et_2)CHClF)R^1} \longrightarrow R\overset{+}{N}{\equiv}CR + Et_2NC(=O)CHClF \quad (120)$$

Grob has studied extensively the fragmentation of α-oximino ketones, especially in the decalyl system. When **197** was subjected to Beckmann conditions in ethanol, two normal Beckmann products **198** and **199** were obtained[101,102] (equation 121). However, when

$$\text{(197)} \xrightarrow[H^+]{EtOH} \text{(198)} + \text{(199)} \quad (121)$$

Me, O, TsON, H (197); Me, O, NH, O (198); Me, O, N, EtO (199)

the reaction was carried out in base, fragmentation involving alkoxide (or hydroxide) attack at carbonyl occurred, to give an unsaturated cyano ester (**201**) (equation 122). This fragmentation is especially interesting because it may be considered a 'seven-centred

$$\text{(197)} \underset{}{\overset{EtO^-}{\rightleftharpoons}} \text{(200)} \longrightarrow \text{(201)} + TsO^- \quad (122)$$

Me, O, TsON, H (197); Me, $\bar{O}$, OEt, TsON, H (200); Me, CO_2Et, CN (201)

fragmentation', and is nucleophile-induced at the carbonyl carbon atom. Grob found the reaction to be first order in both reactants. The kinetic order suggests that nucleophilic addition at the carbonyl group to form **200** is the initial step of the reaction. When such a base-induced fragmentation is carried out on an γ-oximinoketone which possesses an enolizable hydrogen (**202**), fragmentation which formally resembles an E1cB elimination process can occur[85], leading to an α,β-unsaturated ketonitrile (**203**) (equation 123).

TsON (**202**) $\xrightarrow{t\text{-BuO}^-\text{K}^+}$ TsON $\longrightarrow$ CN (**203**) (123)

b. α-Carbonyl- and α-hydroxyketoximes. The literature contains a large number of references to the fragmentation of ketoximes which contain an α-oxygen substituent, many of these references dating back to the beginning of this century. It was early ascertained[103a–h] that α-oximinoketones could undergo both a 'normal' and an 'abnormal' reaction, which can be rationalized in terms of *trans*-1,2-migration of the alkyl or acyl group, and hydration or fragmentation to give the more stable acylium ion. Thus benzil α-monoxime (**204**) in anhydrous medium gave benzonitrile and benzoic acid[103b] (equation 124) but the

$$\underset{(\mathbf{204})}{\mathrm{Ph\overset{O}{\overset{\|}{C}}\underset{\underset{NOH}{\|}}{C}Ph}} \xrightarrow{PhSO_2Cl} PhC{\equiv}N + PhCO_2H \qquad (124)$$

β-monoxime (**205**) under the same conditions, gave the isonitrile and benzoic acid (equation 125). The various modes of rearrangement

$$\underset{(\mathbf{205})}{\mathrm{Ph\overset{O}{\overset{\|}{C}}\underset{\underset{HON}{\|}}{C}Ph}} \xrightarrow{PhSO_2Cl} PhN{\equiv}C + PhCO_2H \qquad (125)$$

and fragmentation of α-oximinoketones was first put into a coherent framework by Ferris[104], who also clearly defined the terms 'normal', 'abnormal' and 'second-order' as they apply to the Beckmann rearrangement. A general scheme for the rearrangements is shown below in scheme (126). Many other examples of α-oximinoketone decomposition are recorded[105a–g], as well as for oximes with unsaturated

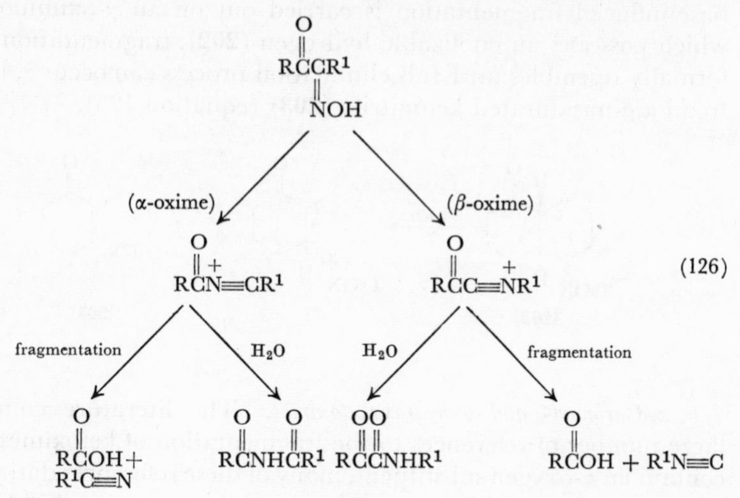

α-substituents, such as α-iminooximes[106]. α-Oximinoacids present an interesting case of α-carbonyloxime fragmentation. Dieckmann[107a] and Locquin[107b,c] reported early that α-oximinoacids undergo decarboxylation readily in acidic medium to give nitriles. A typical example of this, selected from six acyclic aliphatic systems studied by Locquin[107b], is α-oximinovaleric acid (**206**), which on treatment with strong sulphuric acid gives a good yield of butyronitrile (**207**) (equation 127). This reaction has also been the subject of a more

$$\underset{\mathbf{(206)}}{CH_3CH_2CH_2\overset{\overset{NOH}{\|}}{C}CO_2H} \xrightarrow{H_2SO_4} \underset{\mathbf{(207)}}{CH_3CH_2CH_2C{\equiv}N} + CO_2 \qquad (127)$$

recent investigation[97b]. A recent report on the preparation of nitriles in good yields from the reaction of α-oximinoesters with alkali in diethylene glycol[108] may be yet another example of Locquin's nitrile synthesis.

As in the case of the nitrilium salt studied by Hill and Conley[72c], fragmentation of acylnitrilium species appears to yield a cation, in this case an acylium ion. The same mode of decomposition accounts for the formation of a 2-cyanocinnamoyl chloride (**209**) from 1-nitroso-2-naphthol (**208**) when it is treated with benzenesulphonyl

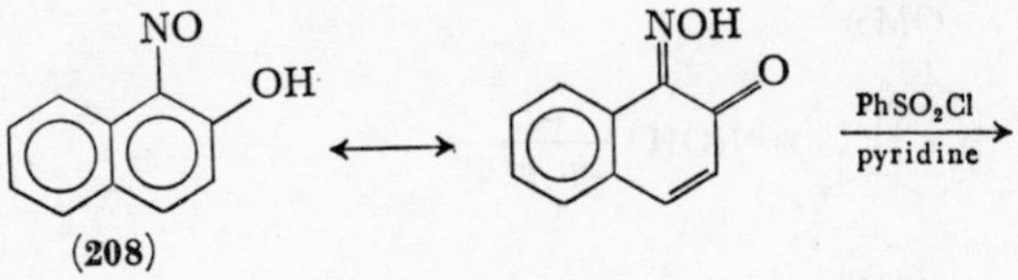

NO OH (208) ⟷ NOH O $\xrightarrow[\text{pyridine}]{PhSO_2Cl}$

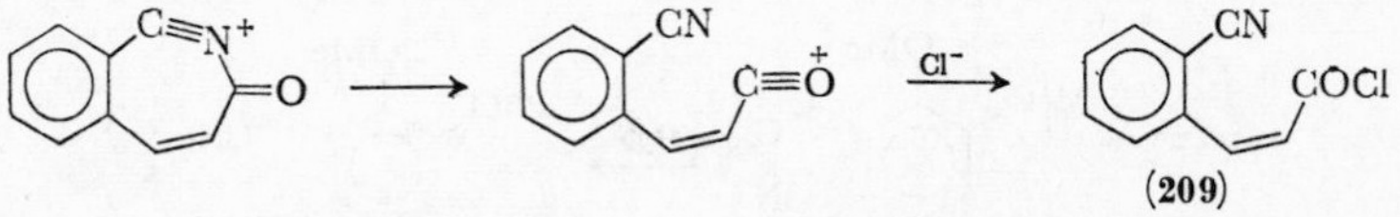

C≡N+ =O ⟶ CN $\overset{+}{C}{\equiv}O$ $\xrightarrow{Cl^-}$ CN COCl (209)

(128)

chloride and pyridine[103b,103d,109] (equation 128). Indeed, evidence for capture of an acylium species can be adduced from the report of Ferris and coworkers[110] on the base-induced fragmentation of acetylated α-oximinoketones (**210**) from which 50–80% yields of esters and nitriles were obtained (equation 129). These results can

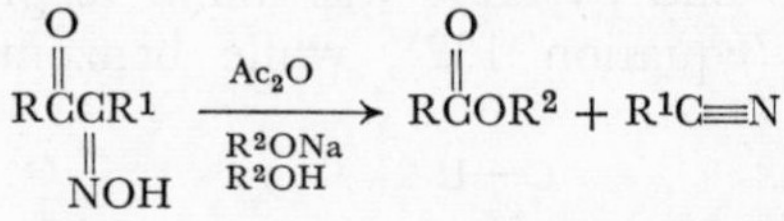

$$\underset{(\mathbf{210})}{RC(=O)C(=NOH)R^1} \xrightarrow[\substack{R^2ONa\\R^2OH}]{Ac_2O} RC(=O)OR^2 + R^1C{\equiv}N \qquad (129)$$

be most readily accommodated in terms of an intermediate acylnitrilium ion which fragments to the nitrile, and to an acylium ion, which in turn is captured by alcohol or alkoxide (equation 130).

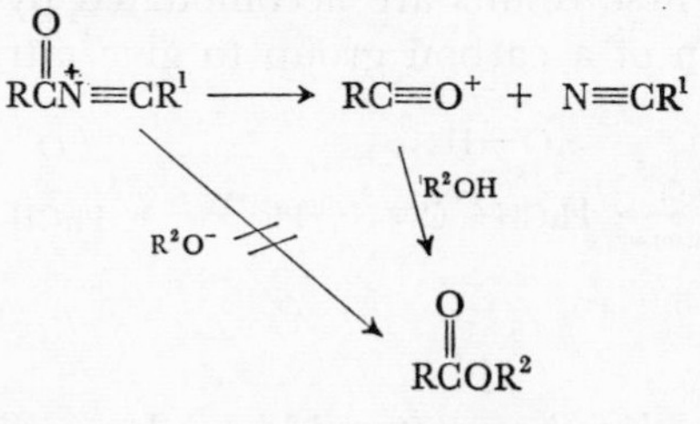

$$RC(=O)\overset{+}{N}{\equiv}CR^1 \longrightarrow RC{\equiv}O^+ + N{\equiv}CR^1 \qquad (130)$$

$$RC{\equiv}O^+ \xrightarrow{R^2OH} RC(=O)OR^2 \qquad RC(=O)\overset{+}{N}{\equiv}CR^1 \overset{R^2O^-}{\not\longrightarrow} RC(=O)OR^2$$

The results preclude S_N2 attack by alkoxide on the acylnitrilium salt, since the same esters were obtained when amines or other weak bases in alcohol instead of alkoxides, were employed. A cyclic α-oximinoketone indandione monoxime (**211**) in the presence of *p*-toluenesulphonyl chloride underwent fragmentation in basic aqueous solution to produce a 95% yield of the cyano acid (**212**)[111] (equation 131).

$$\text{(211)} \xrightarrow[\text{aq. NaOH}]{\text{TsCl}} \qquad (131)$$

$$\xrightarrow{H_2O} \text{(212) } 95\%$$

Many reports can be found in the literature regarding fragmentation of α-hydroxyketoximes, dating from the earliest work of Werner and Deutscheff[103c] in which benzoin α-oxime (**213**) in benzenesulphonyl chloride and pyridine was found to give benzaldehyde and benzonitrile (equation 132), while benzoin β-oxime (**214**),

$$\underset{\textbf{(213)}}{\text{PhCH(OH)}-\text{C(=NOH)Ph}} \xrightarrow[\text{pyridine}]{PhSO_2Cl} \text{PhC(OH)}-\overset{+}{N}\equiv\text{CPh} \longrightarrow \text{PhCHO} + \text{N}\equiv\text{CPh} \qquad (132)$$

under the same conditions, gave benzaldehyde and phenyl isocyanide (equation 133). These results are accomodated by *trans* carbon-to-nitrogen migration of a carbon group to give nitrilium salt which

$$\underset{\textbf{(214)}}{\text{PhCH(OH)C(=NOH)}-\text{Ph}} \xrightarrow[\text{pyridine}]{PhSO_2Cl} \text{PhCH(OH)}-\text{C}\equiv\overset{+}{N}-\text{Ph} \longrightarrow \text{PhCHO} + \text{C}\equiv\text{NPh} \qquad (133)$$

fragments so as to give the most stable products. The same authors reported that furoin β-oxime (**215**) also fragmented in a similar manner (equation 134). Buck and Ide[112] carried out a systematic

$$\underset{\textbf{(215)}}{\text{C}_4\text{H}_7\text{O}-\text{CH(OH)C(=NOH)}-\text{C}_4\text{H}_7\text{O}} \xrightarrow[\text{pyridine}]{PhSO_2Cl} \text{C}_4\text{H}_7\text{O}-\text{CHO} + \text{C}_4\text{H}_7\text{O}-\text{NC} \qquad (134)$$

investigation of the abnormal Beckmann rearrangement on seven mixed benzoins derived from a variety of substituted benzaldehydes using benzenesulphonyl chloride and sodium hydride. In six of these oximes, which possessed the α-configuration (**213a**), yields of 60–90% of the substituted benzaldehyde and benzonitrile were obtained (equation 135). In the only benzoin which possessed the

$$\underset{\substack{\| \\ \text{NOH} \\ (\mathbf{213a})}}{\overset{\substack{\text{OH} \\ |}}{\text{ArCHCAr}^1}} \xrightarrow[\text{aq. NaOH}]{\text{PhSO}_2\text{Cl}} \text{ArCHO} + \text{Ar}^1\text{C}{\equiv}\text{N} \tag{135}$$

β-configuration (**214a**), an 86% yield of the aldehyde was obtained, together with strong evidence for the presence of the isonitrile (equation 136).

$$\underset{\substack{\| \\ \text{HON} \\ (\mathbf{214a})}}{\overset{\substack{\text{OH} \\ |}}{\text{PhCHCAr}}} \xrightarrow[\text{aq. NaOH}]{\text{PhSO}_2\text{Cl}} \text{PhCHO} + [\text{ArNC}] \tag{136}$$

In the presence of benzenesulphonyl chloride, hydroxyketoxime **216** is reported to give benzophenone and benzonitrile[113a,b] (equation 137).

$$\underset{\substack{\| \\ \text{NOH} \\ (\mathbf{216})}}{\overset{\substack{\text{HO} \\ |}}{\text{Ph}_2\text{CCPh}}} \xrightarrow{\text{PhSO}_2\text{Cl}} \text{Ph}_2\text{C}{=}\text{O} + \text{PhC}{\equiv}\text{N} \tag{137}$$

An interesting variation on the fragmentation of α-hydroxyketoximes is encountered when cyanide ion is present during the fragmentation of the α- and β-oximes of benzoin. Unlike the earlier case[103c], no isonitrile was obtained from either oxime. This result has been rationalized in terms of aromatic nucleophilic displacement, with fragmentation as the mode of reaction for the nitrilium ion obtained from the β-oxime (equation 138). The present mechanism suggests at least a superficial resemblance between this reaction and the Sandmeyer reaction. In addition to the fragmentation of 1-nitroso-2-naphthol described earlier[103d], there is evidence that other

$$\overset{\substack{\text{O—H} \\ |}}{\text{PhCHC}}{\equiv}\overset{+}{\text{N}}\text{—C}_6\text{H}_5 \xrightarrow{\text{CN}^-} \overset{\substack{\text{O—H} \\ |}}{\text{PhCH}}\text{—C}{\equiv}\overset{+}{\text{N}}\text{—}(\text{C}_6\text{H}_5^-)\text{—C}{\equiv}\text{N} \longrightarrow \overset{\substack{\text{O} \\ \|}}{\text{PhCH}} + \text{CN}^- + \text{PhCN} \tag{138}$$

α-enolic ketoximes may be susceptible to cleavage. The indolenol (**217**) is cleaved by phosphorous pentachloride to give, after hydrolysis, a good yield of *N,N'*-(2-cyanophenyl) urea (**218**)[103c] (equation 139). One example of an α-oximinoketone with a β-sulphide link presents a situation which appears to be analogous

NOH, OH, N, (**217**), PCl_5, C≡N+, N, OH, $-H^+$ (139)

CN, NCO, H_2O, CN, CO, NH, 2, (**218**)

to that encountered earlier for β-amino and the β-ethers[99]. 2,3-Dihydro-2-oxo-3-oximinobenzothiophene (**219**), when treated with phosphorous pentachloride, yields 2-cyanobenzenesulphenyl chloride (**220**)[103d] (equation140). The similar behaviour[103d] of *N*-methyl-3-

NOH, O, S, (**219**), PCl_5, C≡N+, S, O (140)

CN, S=C=O, +, CN, SCl, (**220**)

oximinoisatin (**221**), very likely involves nitrogen lone-pair participation in the fragmentation step, to give *N*-(2-cyanophenyl)-*N*-methylcarbamoyl chloride (**222**) (equation 141). Although still quite uncertain as regards the mechanism, it is possible that the reduction of nitrile oxides by triarylphosphines and triphenylphosphite[114a,b] proceed through a nitrilium intermediate. Thus, benzonitrile oxide (**223**) reacts rapidly with triphenylphosphine (**224**) to give the nitrile and triphenylphosphine oxide (equation 142). It has since been

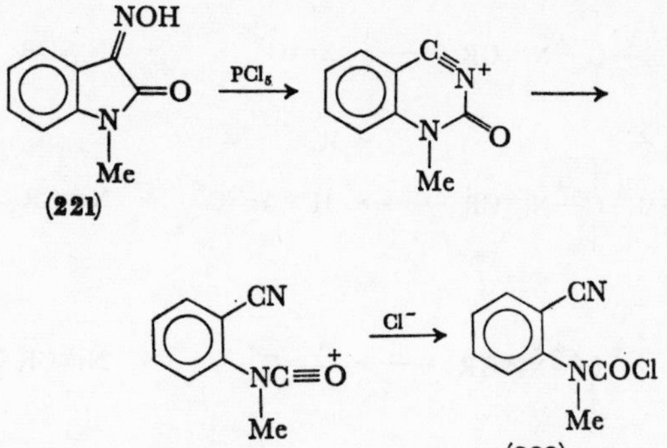

(221) (222) (141)

reported that even as mild an oxygen accepter as cyclohexyl iso-

$$\underset{(\mathbf{223})}{\mathrm{PhC{\equiv}\overset{+}{N}{-}\overset{-}{O}}} + \underset{(\mathbf{224})}{\mathrm{Ph_3P}} \longrightarrow \mathrm{PhC{\equiv}\overset{+}{N}{-}O{-}\overset{-}{P}Ph_3} \longrightarrow \mathrm{PhC{\equiv}N} + \mathrm{Ph_3P{=}O} \quad (142)$$

or

$$\begin{array}{c} \mathrm{N{-}O} \\ \| \quad | \\ \mathrm{PhC{-}PPh_3} \end{array}$$

nitrile (**225**) is sufficient to bring about the reduction[115] to give 30–60% yields under mild conditions (equation 143). In this case,

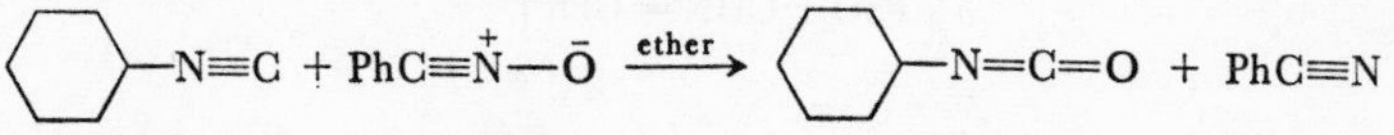

(225) (223) (143)

the intermediacy of a nitrilium species (**226**) (equation 144) remains an untested possibility.

$$\mathrm{RC{\equiv}\overset{+}{N}{-}\bar{O}} + \mathrm{C{\equiv}NR^1} \longrightarrow$$

$$\underset{(\mathbf{226})}{\mathrm{RC{\equiv}\overset{+}{N}{-}O{-}\bar{C}{=}\dot{N}R^1}} \longrightarrow \mathrm{RC{\equiv}N} + \mathrm{O{=}C{=}NR^1} \quad (144)$$

Hence, it appears generally true that fragmentation is the preferred route of decomposition of the nitrilium species whenever an atom in the β-position can stabilize the positive charge via the lone-pair electrons (equations 145–147) of a group V or group VI atom in that position. This concept has been elaborately generalized by Grob[69a] in a recent review.

Recent studies on the photolytic cleavage of azirines[116] to produce oxazoles have lead to the speculation[117] that a zwitterionic nitrilium

$$\ddot{X}-\overset{|}{\underset{|}{C}}-\overset{+}{N}\equiv CR \longrightarrow X{=}C\langle \quad + \quad N\equiv CR \tag{145}$$

$$H\ddot{Y}-\overset{|}{\underset{|}{C}}-\overset{+}{N}\equiv CR \longrightarrow H^{+}\ Y{=}C\langle \quad + \quad N\equiv CR \tag{146}$$

$$\ddot{Z}{=}\underset{|}{C}-\overset{+}{N}\equiv CR \longrightarrow \overset{+}{Z}{=}C\langle \quad + \quad N\equiv CR \tag{147}$$

species may be involved here. Azirine (**227**), when irradiated with 3130 Å light, gives oxazole (**228**) in nearly quantitative yield (equation 148). The formation of **228** cannot be quenched and is not influenced by sensitizers.

$$\text{PhCO-azirine(Ph)}\ (\mathbf{227}) \xrightarrow{h\nu} \left[\begin{array}{c} \text{Ph}\overset{O^-}{\overset{|}{C}}{=}\text{CH}\overset{+}{N}\equiv\text{CPh} \\ \updownarrow \\ \text{Ph}\overset{O}{\overset{\|}{C}}-\overset{-}{C}\text{H}\overset{+}{N}\equiv\text{CPh} \end{array}\right] \longrightarrow \text{2,5-diphenyloxazole}\ (\mathbf{228}) \tag{148}$$

3. Fragmentation of 1,3-oxazete rings, oxazetones and oxazoles

a. Oxazetes and oxazetones. Another group of reactions, which involve a carbon skeleton rearrangement and in which a cyano group appears in the product, display in common the feature that they may be written as if they passed through a 1,3-oxazete ring system. Only a few representatives of this group have been well studied. This may be due in part to the high temperatures which are normally required to achieve reaction. The reaction of carboxylate salts with cyanogen bromide (**229**) at 250–300° is one of the more closely studied representatives of this reaction. The reaction leads to nitrile and carbon dioxide by what appears to be a substitution (equation 149).

$$RCO_2^- + \underset{(\mathbf{229})}{BrC{\equiv}N} \longrightarrow RC{\equiv}N + CO_2 + Br^- \tag{149}$$

However, ^{14}C-labelling originally in the carboxylate carbon ultimately appears in the carbon atom of the cyano group and not in the carbon dioxide[118]. Moreover, retention of optical activity in the carbon grouping precluded an inversion mechanism[119]. These results strongly suggest the formation of an intermediate oxazetone ring (**231**) which undergoes reverse cycloaddition (equation 150). Douglas and coworkers were not able to isolate the intermediate acylisocyanate (**230**), but this is not surprising in view of

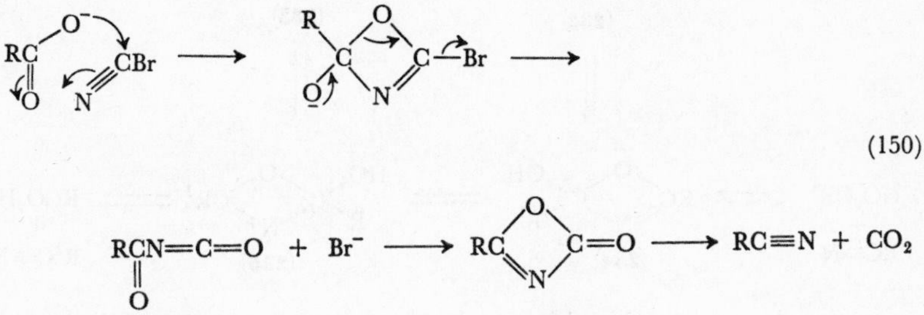

$$RCN{=}C{=}O \;(\mathbf{230}) + Br^- \longrightarrow \text{oxazetone } (\mathbf{231}) \longrightarrow RC{\equiv}N + CO_2 \quad (150)$$

the high temperature of the reaction. Durrell, Young and Dresdner[120] proposed a similar intermediate to account for the equilibration between a carboxylic acid–nitrile pair and imide (equation 151),

$$RCO_2H + R^1C{\equiv}N \rightleftharpoons RCONHCOR^1 \quad (151)$$

which can undergo bond scission to give a new acid–nitrile pair (equation 152). The reactions were base and acid induced, and

$$RCONHCOR^1 \rightleftharpoons RC{\equiv}N + R^1CO_2H \quad (152)$$

were also conducted thermally at 150–200°. These authors rationalized the exchange reaction in terms of an oxazete ring intermediate. The enolamides **232** and **233** may be in equilibrium with the oxazete rings **234** and **235** which may be equilibrating with each other. Ring opening of these compounds can produce either of the acid–nitrile pairs (equation 153). When R and R[1] are different the scheme shown here is further complicated by the symmetrical anhydride in the otherwise transparent reaction. This process is the subject of a patent[121], and a closely allied reaction between acid chlorides and acetonitrile in the presence of aluminium chloride has been attributed to Newman[122] (equation 154). Durrell and coworkers have rationalized the nitrile synthesis of Pomeroy and Craig[123] which appeared earlier. When *O*,*N*-bis(trifluoroacetyl) hydroxylamine (**236**) was treated with aldehydes in the presence of pyridine, nitriles

$$\begin{array}{c} RC(=O)NHC(=O)R^1 \\ \rightleftharpoons \quad \rightleftharpoons \\ RC(=O)-N=C(OH)-R^1 \ (\mathbf{232}) \qquad HO-C(R)=N-C(=O)R^1 \ (\mathbf{233}) \\ \updownarrows \qquad\qquad \updownarrows \\ HO_2CR^1 + RC{\equiv}N \rightleftharpoons \mathbf{234} \rightleftharpoons \mathbf{235} \rightleftharpoons RCO_2H + R^1C{\equiv}N \end{array} \tag{153}$$

were produced. Again, this can be described in terms of the oxazete ring intermediate (**237**) (equation 155). A little-studied method

$$RCOCl + CH_3C{\equiv}N \xrightarrow{AlCl_3} RC{\equiv}N + CH_3COCl \tag{154}$$

for the preparation of nitriles from carboxylic acids and inorganic thiocyanate (**238**), called the Letts synthesis[124a–e], may also involve

$$CF_3CONHOCOCF_3 \ (\mathbf{236}) \underset{}{\overset{\text{pyridine}}{\rightleftharpoons}} CF_3CO\ddot{N}OCOCF_3$$

$$CF_3C(=O)(O^-)\text{—}N(OCOCF_3)\text{—}CHR \longrightarrow \text{oxazetidine } CF_3C(O^-)\text{—}O\text{—}CHR\text{—}N(OCOCF_3) \longrightarrow \tag{155}$$

$$CF_3C(O^-)\text{—}O\text{—}CR{=}N \ (\mathbf{237}) + CF_3CO_2H \longrightarrow CF_3CO_2^- + RC{\equiv}N$$

the same kind of intermediate. A reasonable cyclic mechanism which involves an oxazete ring may be written for this reaction (equation 156). However, this possibility has not been examined.

$$RCO_2H + SCN^- \longrightarrow \underset{(238)}{} \text{R(HO)C}\langle\text{O–C(–S}^-\text{)=N}\rangle \longrightarrow \tag{156}$$

$$\text{R–C}\langle\text{O, N}\rangle\text{C(OH)(S}^-\text{)} \longrightarrow \text{R–C}\langle\text{O, N}\rangle\text{C(}^-\text{O)(SH)} \longrightarrow RC{\equiv}N + CO_2 + SH^-$$

Few examples of S_Ni-type displacement reactions involving the cyano group have been recorded, and it is likely that many reactions that appear to be substitutions are, in fact, more complex. Sheppard[125] has reported the pyrolysis of alkyl cyanoformates (**239**) in the gas phase at reduced pressure and at 200–800°, and has postulated a cyclic transition state (**240**) which involves unimolecular decomposition (equation 157).

$$\underset{(239)}{R_3CO\overset{O}{\overset{\|}{C}}CN} \longrightarrow \underset{(240)}{\left[R_3C\cdots O\cdots C{=}O,\ \cdots C{\equiv}N \right]} \longrightarrow R_3CCN + CO_2 \tag{157}$$

b. Oxazoles. Wasserman[126] has recently reported an unusual and especially promising new fragmentation reaction of oxazole derivatives which appears to be a general method for the transformation of acyloins (**241**) to ω-cyanoacids (**242**) via oxazoles (equation 158).

$$\underset{(241)}{(CH_2)_n\langle C{=}O,\ CH{-}OH\rangle} \longrightarrow \underset{(242)}{(CH_2)_n\langle CO_2H,\ CN\rangle} + CO \tag{158}$$

Acyloins are converted to oxazoles in high yield by acid condensation with formamide. Generation of singlet oxygen photochemically in the presence of the oxazole **243** leads to the ω-cyanoacid through 1,3-dipolar addition of the singlet oxygen to the oxazole, followed by fragmentation and loss of carbon monoxide (equation 159).

241 ⟶ $(CH_2)_n$ oxazole (243) $\xrightarrow{O_2^*}$ $(CH_2)_n$ peroxide ⟶ $(CH_2)_n$ C(=O)–O–CHO, C≡N $\xrightarrow{-CO}$ 242 (159)

V. REARRANGEMENT REACTIONS INVOLVING INTERMEDIATE CYANO COMPOUNDS: THE VON RICHTER REACTION

The von Richter reaction[127a–c] has been reviewed[128a] and characterized by Bunnett[128b] as a typical example of cine substitution—a term used to designate a reaction whose main feature is that the incoming group takes up a position other than the one vacated by the leaving group. In the von Richter reaction, an aromatic nitro compound, when treated with cyanide ion, gives a substituted benzoic acid, with the carboxyl group located *ortho* to the position which originally bore the nitro group (equation 160). For more than half a century

$$p\text{-}X\text{-}C_6H_4NO_2 \xrightarrow[H_2O]{CN^-} m\text{-}X\text{-}C_6H_4CO_2H \qquad (160)$$

it was believed that the rearranged nitrile **244** was an intermediate in the reaction, and that this compound subsequently hydrolysed to the acid. Doubt first arose as to the validity of this supposition

$m\text{-}X\text{-}C_6H_4CN$
(244)

when Bunnett and Rauhut[129] were unable to obtain 1-naphthoic acid (**246**) from 1-naphthyl cyanide (**245**) under conditions which converted 2-nitronaphthalene (**247**) to 1-naphthoic acid (equation 161). In the few years following that, a number of experiments

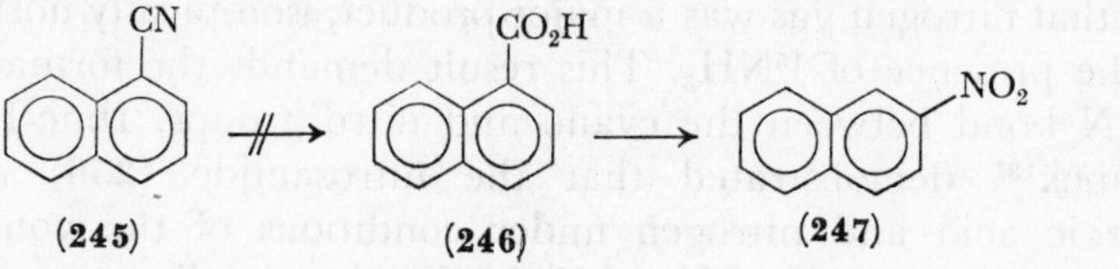

(245) (246) (247) (161)

were reported which led to the proposal of a new mechanism by Rosenblum[130] which accounts for all the experimental observations to date. The Rosenblum mechanism suggests two possible reaction paths (A and B of equation 162). Rosenblum reported that no ammonia or nitrite ion formed in the reaction as previously supposed,

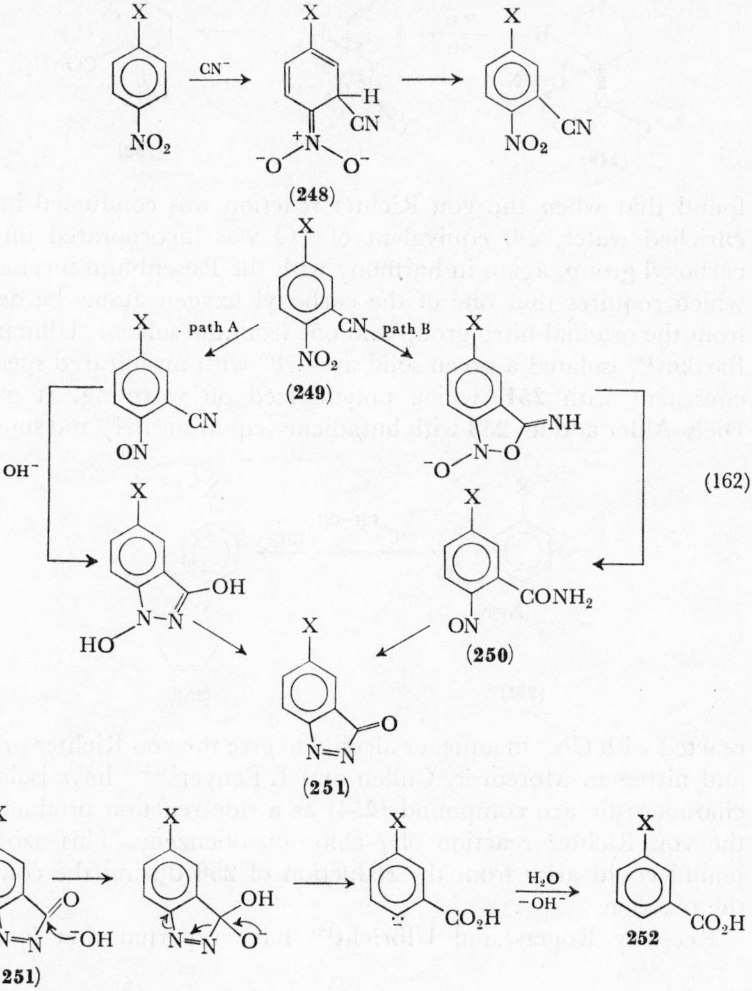

(248)

(249)

path A path B

(162)

(250)

(251)

(251) 252

and that nitrogen gas was a major product, isotopically normal even in the presence of $^{15}NH_3$. This result demands the formation of a N—N bond between the cyano and nitro groups. Ibne-Rasa and Koubek[131] demonstrated that the nitrosamide (**250**) *did* form benzoic acid and nitrogen under conditions of the von Richter reaction (aq. EtOH, CN^-, 140–160°), thus lending some support to the path B mechanism. Cullen and L'Écuyer[132] pointed out that the *ortho* nitrosamide was mechanistically accessible without proceeding through the dinegative ion (**248**) (equation 163). Samuel[133]

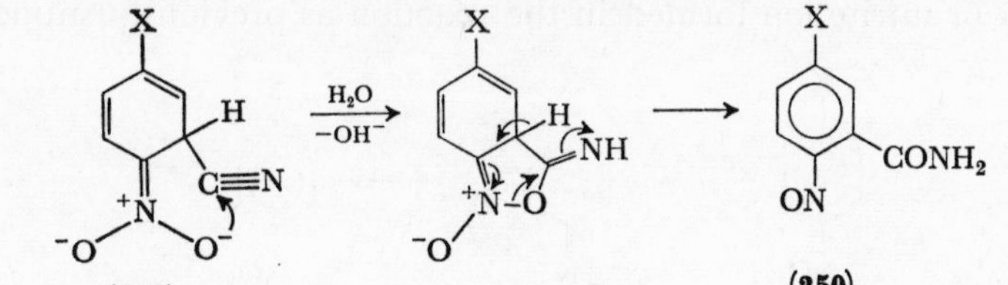

(163)

(**248**) (**250**)

found that when the von Richter reaction was conducted in ^{18}O-enriched water, 1·0 equivalent of ^{18}O was incorporated into the carboxyl group, again in harmony with the Rosenblum mechanism, which requires that one of the carboxyl oxygen atoms be derived from the original nitro group and one from the solvent. Ullman and Bartkus[134] isolated a green solid at −78° with an infrared spectrum consistent with **251**, which polymerized on warming. It gave a Diels–Alder adduct **253** with butadiene (equation 164) and smoothly

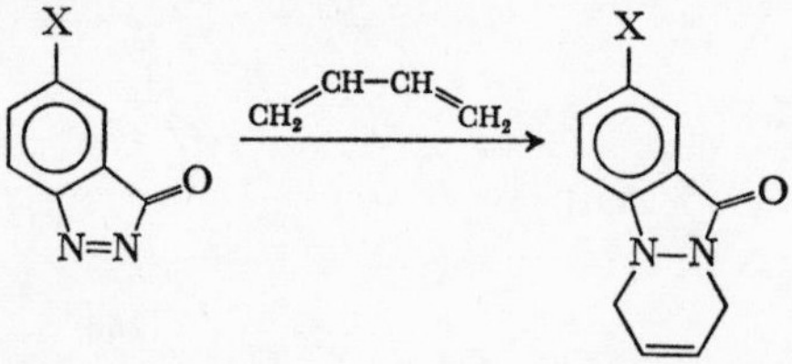

(164)

(**251**) (**253**)

reacted with CN^- in aqueous alcohol to give the von Richter product and nitrogen. Moreover, Cullen and L'Écuyer[135a–c] have isolated a characteristic azo compound (**254**) as a side reaction product from the von Richter reaction of *p*-chloronitrobenzene. This azo compound could arise from the reduction of **250** during the course of the reaction.

Recently Rogers and Ulbricht[136] have reported that five new

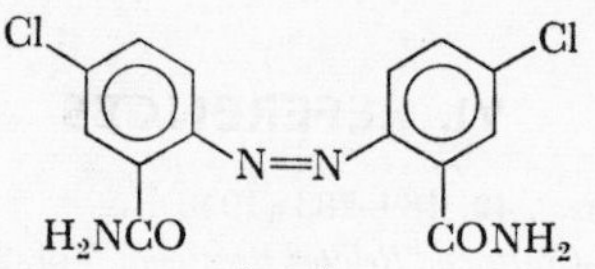

Cl Cl N=N H$_2$NCO CONH$_2$

(**254**)

types of products were obtained in a total yield of 40% when the von Richter reaction was carried out using dimethylsulphoxide as a solvent (equation 165). The rearrangement of *p*-chloronitrobenzene (**255**) in DMSO with potassium cyanide proceeded much more rapidly than did the rearrangement in alcohol, and gave five products. The structures of the products **258** and **260** were tentatively assigned. Products **259** and **260** were not found when the solvent was carefully dried. However, even products **256**, **257** and **258** are

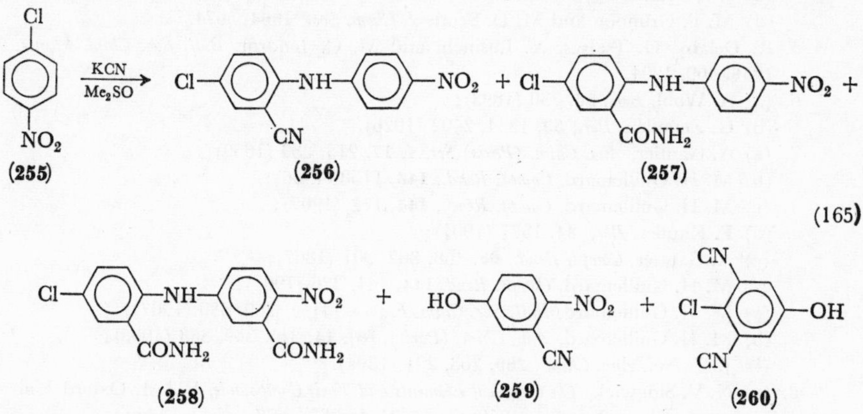

Cl NO$_2$ (**255**) $\xrightarrow[Me_2SO]{KCN}$ Cl—NH—NO$_2$ CN (**256**) + Cl—NH—NO$_2$ CONH$_2$ (**257**) +

(165)

Cl—NH—NO$_2$ CONH$_2$ CONH$_2$ (**258**) + HO—NO$_2$ CN (**259**) + Cl—OH CN CN (**260**)

difficult to rationalize in terms of the Rosenblum mechanism. It is not surprising that the normal von Richter product failed to be produced in a non-aqueous medium, since the decomposition of **251** requires water. In addition to this, however, the solvent obviously imparts some unusual features to the reactions which have still to be elucidated. Since **254** has been obtained as a product of the von Richter reaction, the possibility that **256**, **257** and/or **258** might arise by a semidine-type rearrangement of the hydrazo derivative of **253** must not be discounted.

Generally low yields and the hazards of employing potassium or sodium cyanide in an acidic medium have conspired to limit the utility of the von Richter reaction.

VI. REFERENCES

1. D. T. Mowry, *Chem. Rev.*, **42,** 189–283 (1948).
2. (a) W. Rusk in *Friedel-Crafts and Related Reactions*, Vol. 3 (Ed. G. Olah), Interscience, New York, 1964, p. 383;
   (b) P. E. Spoerri and A. S. DuBois in *Organic Reactions*, Vol. 5 (Ed. R. Adams), John Wiley and Sons, New York, 1949, Chap. 9, p. 387.
3. (a) A. Reissert, *Ber.*, **38,** 1603, 3415 (1905);
   (b) H. O. L. Fischer, *J. Am. Chem. Soc.*, **63,** 2021 (1941);
   (c) E. Mosettig in *Organic Reaction*, Vol. 8 (Ed. R. Adams), John Wiley and Sons, New York, 1954, Chap. 5, p. 218;
   (d) F. D. Popp, L. E. Katz, C. W. Klinowski and J. M. Wefer, *J. Org. Chem.*, **33,** 4447 (1968).
4. (a) L. G. Donaruma and W. Z. Heldt in *Organic Reactions*, Vol. 2 (Ed. R. Adams), John Wiley and Sons, New York, 1960, p. 41;
   (b) R. F. Smith and L. E. Walker, *J. Org. Chem.*, **27,** 4372 (1962);
   (c) I. I. Grandberg, *J. Gen. Chem. USSR*, **34,** 570 (1964);
   (d) M. F. Grundon and M. D. Scott, *J. Chem. Soc.*, **1964,** 5674.
5. R. Delaby, G. Tsalsas, X. Lusinchi and M. C. Jendrot, *Bull. Soc. Chim. France*, **1956,** 409, 1294.
6. (a) A. Wohl, *Ber.*, **26,** 730 (1893);
   (b) G. Zemplén, *Ber.*, **59,** 1254, 2402 (1926).
7. (a) A. Gautier, *Ann. Chim. (Paris) Ser. 4*, **17,** 215, 233 (1869);
   (b) M. H. Guillemard, *Compt. Rend.*, **143,** 1158 (1906);
   (c) M. H. Guillemard, *Compt. Rend.*, **144,** 112 (1907);
   (d) F. Kaufler, *Ber.*, **34,** 1577 (1901);
   (e) A. Gautier, *Compt. Rend.*, **65,** 468, 862, 901 (1867);
   (f) M. H. Guillemard, *Compt. Rend*, **144,** 141, 326 (1907);
   (g) M. H. Guillemard, *Bull. Soc. Chim. France*, [4], 1, 269, 530 (1907);
   (h) M. H. Guillemard, *Ann. Chim. (Paris)*, [8], **14,** 311, 349, 363 (1908);
   (i) J. U. Nef, *Ann. Chem.*, **280,** 263, 291 (1894).
8. (a) N. V. Sidgwick, *The Chemical Elements and Their Compounds*, Vol. 1, Oxford University Press, London, 1950, Group V, pp. 672–673;
   (b) F. D. Rossini, *Selected Values of Chemical Thermodynamic Properties*, National Bureau Standards Circular No. 500 (1952);
   (c) M. Hunt, J. A. Kerr and A. F. Trotman-Dickenson, *J. Chem. Soc.*, **1965,** 5074.
9. (a) I. Ugi, U. Fetzer, U. Eholzer, H. Knupfer and K. Offermann, *Angew. Chem. Int. Ed. Engl.*, **4,** 472 (1965);
   (b) W. R. Hertler and E. J. Corey, *J. Org. Chem.*, **23,** 1221 (1958);
   (c) J. Casanova, Jr., R. E. Schuster and N. D. Werner, *J. Chem. Soc.*, **1963,** 4280.
10. G. Kohlmaier and B. S. Rabinovitch, *J. Phys. Chem.*, **63,** 1793 (1959).
11. J. Casanova Jr., N. D. Werner and R. E. Schuster, *J. Org. Chem.*, **31,** 3473 (1966).
12. G. W. Van Dine and R. Hoffmann, *J. Am. Chem. Soc.*, **90,** 3227 (1968).
13. C. K. Ingold, *Structure and Mechanism in Organic Chemistry*, Cornell University Press, Ithaca, New York, 1953, Chap. 9.
14. H. Rupe and K. Glenz, *Ann. Chem.*, **436,** 184 (1924).
15. F. Lautenschlaeger and G. F. Wright, *Can. J. Chem.*, **41,** 863 (1963).
16. D. Hoy and E. J. Poziomek, *J. Org. Chem.*, **33,** 4050 (1968).

17. T. Saeguso, Y. Ito, S. Kobayashi, N. Takeda and K. Hirota, *Tetrahedron Letters*, **1967**, 1273.
18. (a) D. H. Shaw and H. O. Pritchard, *J. Phys. Chem.*, **70**, 1230 (1966);
    (b) D. H. Shaw and H. O. Pritchard, *Can. J. Chem.*, **45**, 2749 (1967).
19. (a) F. W. Schneider and B. S. Rabinovitch, *J. Am. Chem. Soc.*, **84**, 4215 (1962);
    (b) F. W. Schneider and B. S. Rabinovitch, *J. Am. Chem. Soc.*, **85**, 2365 (1963);
    (c) B. S. Rabinovitch, P. W. Gilderson and F. W. Schneider, *J. Am. Chem. Soc.*, **87**, 158 (1965);
    (d) F. J. Fletcher, B. S. Rabinovitch, K. W. Watkins and D. J. Locker, *J. Phys. Chem.*, **70**, 2823 (1966);
    (e) D. C. Tardy and B. S. Rabinovitch, *J. Chem. Phys.*, **48**, 1282 (1968).
20. (a) F. Millich and R. G. Sinclair, *J. Polymer Sci, A-1*, **6**, 1417 (1968) and references cited therein;
    (b) F. Millich, '12th Ann. Report on Research,' *Am. Chem. Soc., Pet. Res. Fund*, 1777-A1, August 31, 1967.
21. J. Wade, *J. Chem. Soc.*, **81**, 1596 (1902).
22. C. Grundmann, *Chem. Ber.*, **91**, 1380 (1958).
23. C. T. Ting and F. S. Rowland, *J. Phys. Chem.*, **72**, 763 (1968).
24. M. R. Booth and S. G. Frankiss, *Chem. Comm.*, **1968**, 1347.
25. M. Talat-Erbin, *AECCNAEM-28*, Atomic Energy Commission, U.S.A.; *Chem. Abstr.*, **65**, 6065f, (1966).
26. H. Bent, *Chem. Rev.*, **68**, 589 (1968).
27. (a) L. Pauling and S. E. Hendricks, *J. Am. Chem. Soc.*, **48**, 641 (1926);
    (b) F. Gallais, *Bull. Soc. Chim. France*, **12**, 657 (1945).
28. (a) F. A. Cotton and G. Wilkinson, *Advanced Inorganic Chemistry*, 2nd ed., John Wiley and Sons, New York, 1966, p. 744;
    (b) D. S. Matteson and R. A. Bailey, *J. Am. Chem. Soc.*, **89**, 6389 (1967);
    (c) J. M. Pratt and P. R. Silverman, *J. Chem. Soc. (A)*, **1967**, 1286;
    (d) H. Latka, *Z. Anorg. Allgem. Chem.*, **353**, 243 (1967).
29. (a) M. S. Kharasch and O. Reinmuth, *Grignard Reactions of Non-Metallic Substances*, Prentice-Hall, New York, 1954, Chap. 10;
    (b) P. L. Pickard and T. L. Tolbert in *Organic Synthesis*, Vol. 44 (Ed. W. E. Parham), John Wiley and Sons, New York, 1964, pp. 51–53.
30. (a) K. D. Williams Morse, *Diss. Abstr.*, **B28**, 2324-B (1967);
    (b) V. A. Dorokhov and M. Lappert, *Chem. Comm.*, **1968**, 250;
    (c) J. E. Lloyd and K. Wade, *J. Chem. Soc.*, **1964**, 1649;
    (d) D. E. Hyatt, D. A. Owen and L. J. Todd, *Inorg. Chem.*, **5**, 1749 (1966).
31. E. M. Horn, *U.S. Pat.* 3,338,962 (1967).
32. J. R. Jennings, J. E. Lloyd and K. Wade, *J. Chem. Soc.*, **1965**, 5083.
33. (a) S. Pasynkiewicz, K. Starowieski and Z. Rzepkowska, *Organometal. Chem.*, **10**, 527 (1967);
    (b) H. Reinheckel and D. Jahnke, *Tenside*, **2**, 249 (1965); *Chem. Abstr.*, **63**, 9805a (1965).
34. J. Hooz and S. Linke, *J. Am. Chem. Soc.*, **90**, 6891 (1968).
35. A. C. Cope, H. L. Holmes and H. O. House in *Organic Reactions*, Vol. 9 (Ed. R. Adams), John Wiley and Sons, New York, 1957, p. 107.
36. (a) D. S. Acker, R. J. Harder, W. R. Hertler, W. Mahler, L. R. Melby, R. E. Benson and W. E. Mochel, *J. Am. Chem. Soc.*, **82**, 6408 (1960);
    (b) R. G. Kepler, P. E. Bierstedt and R. E. Merrifield, *Phys. Rev. Letters*, **5**, 503 (1960);

(c) D. B. Chestnut, H. Foster and W. D. Phillips, *J. Chem. Phys.*, **34**, 684 (1961);
(d) L. R. Melby, R. J. Harder, W. R. Hertler, W. Mahler, R. E. Benson and W. E. Mochel, *J. Am. Chem. Soc.*, **84**, 3374 (1962);
(e) D. S. Acker and W. R. Hertler, *J. Am. Chem. Soc.*, **84**, 3370 (1962);
(f) W. R. Hertler, H. D. Hartzler, D. S. Acker and R. E. Benson, *J. Am. Chem. Soc.*, **84**, 3387 (1962).

37. E. Fischer, *Z. Chem.*, **8** (8), 281 (1968).
38. W. D. Phillips, J. C. Powell, H. Foster and D. B. Chestnut as cited in *J. Am. Chem. Soc.*, **84**, 3379 (1967).
39. D. N. Dhar, *Chem. Rev.*, **67**, 611 (1967).
40. (a) C. L. Stevens and J. C. French, *J. Am. Chem. Soc.*, **76**, 4398 (1954).
(b) C. L. Stevens, R. C. Freeman and K. Noll, *J. Org. Chem.*, **30**, 3718 (1965).
41. (a) C. S. Wu, G. S. Hammond and J. M. Wright, *J. Am. Chem. Soc.*, **82**, 5386 (1960);
(b) G. S. Hammond, C. S. Wu, O. D. Trapp, J. Warkentin and R. T. Keys, *J. Am. Chem. Soc.*, **82**, 5394 (1960).
42. (a) A. W. Hoffmann, *Ber.*, **17**, 1406, 1905 (1884);
(b) A. W. Hoffmann, *Ber.*, **19**, 1433, 1822 (1886).
43. (a) A. E. Chichibabin, *J. Russ. Phys. Chem. Soc.*, **47**, 709 (1915);
(b) A. E. Chichibabin, *J. Russ. Phys. Chem. Soc.*, **54**, 611 (1922);
(c) A. E. Chichibabin and P. A. Muschkin, *J. Prakt. Chem.*, **107**, 109 (1924).
44. O. Nicodemus, *Ger. Pat.* 547,518 (1929); L. Schlecht and H. Rotger, *U.S. Pat.* 1,936,995 (1933); B. K. Stuer and W. Grob, *Brit. Pat.* 109,983 (1916); B. K. Stuer and W. Grob, *U.S. Pat.* 1,421,743 (1922).
45. M. E. Kuehne and P. J. Sheeran, *J. Org. Chem.*, **33**, 4406 (1958).
46. Union Carbide Corp., *Neth. Appl. Pat.* 6,504,567 (1965).
47. (a) R. Buijle and H. G. Viehle, *Angew. Chem. Int. Ed. Engl.*, **3**, 582 (1964);
(b) F. Weygand, W. Koenig, R. Buijle and H. G. Viehle, *Chem. Ber.*, **98**, 3632 (1965);
(c) A. S. van Mourik, E. Harryvan and J. F. Arens, *Rec. Trav. Chim.*, **84**, 1344 (1965).
48. (a) Z. Rappoport, private communication;
(b) J.-P. Fleury and B. Libis, *Compt. Rend.*, **256**, 2419 (1963).
49. (a) S. Trofimenko, E. L. Little, Jr. and H. F. Mower, *J. Org. Chem.*, **27**, 433 (1962);
(b) H. Schmidtmann, *Ber.*, **29**, 1172 (1896);
(c) W. J. Middleton, E. L. Little, D. D. Coffman and V. A. Engelhardt, *J. Am. Chem. Soc.*, **80**, 2795 (1958).
50. B. V. Ioffe and K. N. Zelenin, *Zh. Obshch. Khim.*, **32**, 1708 (1962).
51. (a) G. W. Stacy, A. J. Papa and S. C. Ray, *J. Org. Chem.*, **26**, 4779 (1961);
(b) G. W. Stacy, A. J. Papa, F. W. Villaescusa and S. C. Ray, *J. Org. Chem.*, **29**, 607 (1964).
52. (a) E. Ciganek, *J. Am. Chem. Soc.*, **87**, 652 (1965);
(b) J. S. Swenson and D. J. Renaud, *J. Am. Chem. Soc.*, **87**, 1394 (1965);
(c) E. Ciganek, *J. Am. Chem. Soc.*, **88**, 1979 (1966);
(d) E. Ciganek, *J. Am. Chem. Soc.*, **89**, 1454 (1967).
53. (a) R. E. Moser, J. M. Fritsch, T. L. Westman, R. M. Kliss and C. N. Mathews, *154th Nat. Meeting, Am. Chem. Soc., San Francisco, Sept. 1967*, Paper S-110;
(b) R. E. Moser, J. M. Fritsch, T. L. Westman, R. M. Kliss and C. N. Mathews, *J. Am. Chem. Soc.*, **89**, 5673 (1967).
54. E. Ciganek, *J. Org. Chem.*, **30**, 4198 (1965).

55. P. C. Petrellis, H. Dietrich, E. Meyer and G. W. Griffin, *J. Am. Chem. Soc.*, **89**, 1967 (1967).
56. (a) F. D. Marsh and M. D. Hermes, *J. Am. Chem. Soc.*, **87**, 1819 (1965);
(b) A. G. Anastassiou, H. E. Simmons and F. D. Marsh, *J. Am. Chem. Soc.*, **87**, 2296 (1965);
(c) A. G. Anastassiou and H. E. Simmons, *J. Am. Chem. Soc.*, **89**, 3177 (1967);
(d) A. G. Anastassiou, *J. Am. Chem. Soc.*, **89**, 3185 (1967);
(e) A. G. Anastassiou, *J. Am. Chem. Soc.*, **90**, 1527 (1968).
57. A. G. Anastassiou and J. N. Shepelavy, *J. Am. Chem. Soc.*, **90**, 492 (1968).
58. (a) K. Nakagawa and H. Onoue, *Tetrahedron Letters*, **1965**, 1433;
(b) K. Nakagawa and H. Onoue, *Chem. Comm.*, **1965**, 396.
59. J. H. Hall and E. Patterson, *J. Am. Chem. Soc.*, **89**, 5856 (1967).
60. C. D. Campbell and C. W. Rees, *Chem. Comm.*, **1965**, 192.
61. M. P. Cava and L. Bravo, *Chem. Comm.*, **1968**, 1538.
62. W. A. Sheppard and R. M. Henderson, *J. Am. Chem. Soc.*, **89**, 4446 (1967).
63. E. Ciganek, *J. Am. Chem. Soc.*, **89**, 1458 (1967).
64. (a) K. N. Klump and J. P. Chesick, *J. Am. Chem. Soc.*, **85**, 130 (1963);
(b) J. A. Berson, *Acct. of Cml. Res.*, **1**, 152 (1968).
65. W. Webster, *J. Am. Chem. Soc.*, **87**, 1820 (1965).
66. (a) K. R. Huffman, M. Loy, W. A. Henderson Jr. and E. F. Ullman, *Tetrahedron Letters*, **1967**, 931;
(b) E. F. Ullman, W. A. Henderson and K. R. Huffman, *Tetrahedron Letters*, **1967**, 935.
67. (a) C. W. Bird, *Tetrahedron*, **21**, 2179 (1965);
(b) C. W. Bird, *Chem. Ind.* (*London*), **1963**, 1556;
(c) C. W. Bird, *J. Chem. Soc.*, **1964**, 5284.
68. C. R. Hauser, J. W. Le Maistre and A. E. Rainsford, *J. Am. Chem. Soc.*, **57**, 1056 (1935).
69. (a) C. A. Grob and P. W. Schiess, *Angew. Chem. Int. Ed. Engl.*, **6**, 1 (1967);
(b) C. A. Grob, *Gazz. Chim. Ital.*, **92**, 902 (1962);
(c) C. A. Grob, *Bull. Soc. Chim. France*, **1960**, 1360;
(d) C. A. Grob, 'Theoretical Organic Chemistry,' *The Kekulé Symposium*, Butterworths, London, 1959, p. 114.
70. (a) P. A. S. Smith in *Molecular Rearrangements* Vol. 1, (Ed. P. de Mayo), John Wiley and Sons, New York, 1963, pp. 483–507;
(b) C. A. Grob, *Experimentia*, **13**, 126 (1957);
(c) A. H. Blatt, *Chem. Rev.*, **12**, 215 (1933).
71. (a) J. J. Ritter and P. P. Minieri, *J. Am. Chem. Soc.*, **70**, 4045 (1948);
(b) J. J. Ritter and J. Kalish, *J. Am. Chem. Soc.*, **70**, 4048 (1948).
72. (a) R. K. Hill and O. T. Chortyk, *J. Am. Chem. Soc.*, **84**, 1064 (1962);
(b) R. T. Conley, *J. Org. Chem.*, **28**, 278 (1963);
(c) R. K. Hill, R. T. Conley and O. T. Chortyk, *J. Am. Chem. Soc.*, **87**, 5647 (1965).
73. (a) H. Meerwein, P. Laasch, R. Mersch and J. Spille, *Chem. Ber.*, **89**, 209 (1956);
(b) G. A. Olah and T. E. Kiovsky, *J. Am. Chem. Soc.*, **90**, 4666 (1968).
74. R. Huisgen, H. Witte and I. Ugi, *Chem. Ber.*, **90**, 1844 (1957).
75. (a) E. N. Zil'berman, *Russ. Chem. Rev.*, **29**, 331 (1960);
(b) E. M. Smolin, *J. Org. Chem.*, **20**, 295 (1955).
76. C. L. Parris, in *Organic Synthesis*, Vol. 42, (Ed. V. Boekelheide), John Wiley and Sons, New York (1962), p. 16.

77. (a) F. R. Benson and J. J. Ritter, *J. Am. Chem. Soc.*, **71**, 4128 (1949);
(b) L. W. Hartzel and J. J. Ritter, *J. Am. Chem. Soc.*, **71**, 4130 (1949).
78. (a) J. J. Ritter and J. Kalish in *Organic Synthesis*, Vol. 44, (Ed. W. E. Parham), John Wiley and Sons, New York, 1964, p. 44;
(b) H. Christol, R. Jacquier and M. Mousseron, *Bull. Soc. Chim. France*, **1957**, 1027.
79. R. M. Lusskin and J. J. Ritter, *J. Am. Chem. Soc.*, **72**, 5577 (1950).
80. (a) H. Plaut and J. J. Ritter, *J. Am. Chem. Soc.*, **73**, 4076 (1951);
(b) A. Zilkha, I. Barzilay, J. Naiman and B. Feit, *J. Org. Chem.*, **33**, 1686 (1968).
81. L. Eberson and K. Nyberg, *Tetrahedron Letters*, **1966**, 2389.
82. O. Wallach, *Ann. Chem.*, **309**, 1 (1889).
83. (a) W. A. Lazier and G. W. Rigby, *U.S. Pat.* 2,234,566 (1941);
(b) V. Davydov, *Chem. Tech.* (*Berlin*), **7**, 647 (1955).
84. O. Wallach, *Ann. Chem.*, **312**, 171 (1900).
85. W. Eisele, C. A. Grob and E. Renk, *Tetrahedron Letters*, **1963**, 75.
86. M. V. Rubtsov, E. E. Mikhlina, V. Ya Voreb'eva and A. D. Yanina, *Zh. Obsch. Khim.*, **34**, 2222 (1964).
87. (a) H. P. Fischer, C. A. Grob and E. Renk, *Helv. Chim. Acta*, **45**, 2539 (1962);
(b) H. P. Fischer and C. A. Grob, *Helv. Chim. Acta*, **46**, 936 (1963);
(c) C. A. Grob, H. P. Fischer, H. Link and E. Renk, *Helv. Chim. Acta*, **46**, 1190 (1963);
(d) H. P. Fischer, C. A. Grob and E. Renk, *Helv. Chim. Acta*, **42**, 872 (1959);
(e) H. P. Fischer and C. A. Grob, *Tetrahedron Letters*, **26**, 22 (1960);
(f) M. F. Bartlett, D. F. Dickel and W. I. Taylor, *J. Am. Chem. Soc.*, **80**, 126 (1958);
(g) R. K. Hill and R. T. Conley, *J. Am. Chem. Soc.*, **82**, 645 (1960).
88. R. R. Sauers and G. P. Ahearn, *J. Am. Chem. Soc.*, **83**, 2759 (1961).
89. (a) S. Yamaguchi, *Bull. Chem. Soc. Japan*, **1**, 35 (1926);
(b) J. Casanova, Jr., unpublished observation.
90. (a) W. Borsche and W. Sander, *Ber.*, **48**, 117 (1915);
(b) J. Bredt and W. Holz, *J. Prakt. Chem.*, [2] **95**, 1375 (1917).
91. R. K. Hill and R. T. Conley, *Chem. Ind.* (*London*), **1956**, 1314.
92. A. Hassner and E. G. Nash, *Tetrahedron Letters*, **1956**, 525.
93. R. E. Lyle and G. G. Lyle, *J. Org. Chem.*, **18**, 1058 (1953).
94. (a) R. T. Conley and B. E. Nowak, *J. Org. Chem.*, **27**, 3196 (1962);
(b) R. T. Conley and R. J. Lange, *J. Org. Chem.*, **28**, 210 (1963).
95. C. A. Grob, H. P. Fischer, W. Raudenbusch and J. Zergenyi, *Helv. Chem. Acta*, **47**, 1003 (1964).
96. B. J. Gregory, R. B. Moodie and K. Schofield, *Chem. Comm.*, **1968**, 1380.
97. (a) M. I. Vinnick and N. G. Zarakhani, *Russ. Chem. Rev.*, **36**, 62 (1967);
(b) A. Ahmad and I. D. Spenser, *Can. J. Chem.*, **39**, 1340 (1961).
98. D. Giraud-Clenet and J. Anatol, *Compt. Rend., Ser. C*, **262**, 224 (1966).
99. R. K. Hill, *J. Org. Chem.*, **27**, 29 (1962).
100. R. T. Autrey and P. W. Scullard, *J. Am. Chem. Soc.*, **90**, 4924 (1968).
101. W. Eisele, C. A. Grob, E. Renk and H. von Tschammer, *Helv. Chim. Acta*, **51**, 816 (1968).
102. C. A. Grob and H. von Tschammer, *Helv. Chim. Acta*, **51**, 1082 (1968).
103. (a) C. Bulow and H. Grotowsky, *Ber.*, **34**, 1479 (1901);
(b) A. Werner and A. Piguet, *Ber.*, **37**, 4295 (1904);
(c) A. Werner and T. Deutscheff, *Ber.*, **38**, 69 (1905);

(d) W. Borsche and W. Sander, *Ber.*, **47,** 2815 (1914);
(e) G. Bishop and O. L. Brady, *J. Chem. Soc.*, **121**, 2364 (1922);
(f) E. Beckmann and E. Bark, *J. Prakt. Chem.*, [2] **105,** 327 (1923);
(g) O. L. Brady and G. Bishop, *J. Chem. Soc.*, **1926**, 810;
(h) J. Meisenheimer, H. O. Kauffmann, U. von Kummer and J. Link, *Ann. Chem.*, **468,** 202 (1929).

104. A. F. Ferris, *J. Org. Chem.*, **25,** 12 (1960).
105. (a) R. T. Conley and F. A. Mikulski, *J. Org. Chem.*, **24,** 97 (1959);
(b) A. Ferris, *J. Org. Chem.*, **24,** 580 (1959);
(c) G. B. Backman and D. E. Welton, *J. Org. Chem.*, **12,** 221 (1947);
(d) W. Nagata and K. Takeda, *J. Pharm. Soc. Japan*, **72,** 1566 (1952);
(e) D. Murakami and N. Tokura, *Bull. Chem. Soc. Japan*, **31,** 1044 (1958);
(f) A. I. Green and B. Saville, *J. Chem. Soc.*, **1956,** 3887;
(g) S. Wawzonek and J. V. Hallum, *J. Org. Chem.*, **24,** 364 (1959).
106. E. C. Taylor, C. W. Jefford and C. C. Cheng, *J. Am. Chem. Soc.*, **83,** 1261 (1961).
107. (a) W. Dieckmann, *Ber.*, **33,** 579 (1900);
(b) R. Locquin, *Bull. Soc. Chim. France*, [3], **31,** 1068 (1904);
(c) L. Bouveault and R. Locquin, *Bull. Soc. Chim. France*, [3], **31,** 1142 (1904);
(d) E. E. Blaise and H. Gault, *Bull. Soc. Chim. France*, [4], **1,** 75 (1907).
108. J. O. Bledose Jr., *Diss. Abstr.*, **26,** 1913 (1965).
109. E. Beckmann, O. Liesche and E. Correns, *Ber.*, **56B,** 341 (1923).
110. A. F. Ferris, G. S. Johnson and F. E. Gould, *J. Org. Chem.*, **25,** 1813 (1960).
111. S. N. Chakravarti and M. Swaminathan, *J. Indian Chem. Soc.*, **11,** 101 (1934).
112. J. S. Buck and W. S. Ide, *J. Am. Chem. Soc.*, **53,** 1912 (1931).
113. (a) A. H. Blatt and R. P. Barnes, *J. Am. Chem. Soc.*, **56,** 1148 (1934);
(b) T. S. Oakwood and J. Tessieri, *112th Nat. Meeting Am. Chem. Soc., New York, Sept. 1947*, paper 71L.
114. (a) P. Grunanger and M. R. Langella, *Rend. Accad. Naz. Lincei*, **36** (8), 2077 (1964);
(b) D. P. Martin and A. Weise, *Chem. Ber.*, **99,** 976 (1966).
115. P. V. Finzi and M. Arbasino, *Tetrahedron Letters*, **1965,** 4645.
116. E. F. Ullman and B. Singh, *J. Am. Chem. Soc.*, **88,** 1844 (1966).
117. E. F. Ullman, *Acct. of Cml. Res.*, **1,** 353 (1968).
118. (a) D. E. Douglas, J. Eccles and A. E. Almond, *Can. J. Chem.*, **31,** 1127 (1953);
(b) D. E. Douglas and A. M. Burditt, *Can. J. Chem.*, **36,** 1256 (1958).
119. J. A. Barltrop, A. C. Day and B. D. Bigley, *J. Chem. Soc.*, **1961,** 3185.
120. W. S. Durrell, J. A. Young and R. D. Dresdner, *J. Org. Chem.*, **28,** 831 (1963).
121. D. J. Loder, *U.S. Pat.* 2,377,795 (1945).
122. M. S. Newman, as cited in reference 1 of this chapter, footnote 431a.
123. J. H. Pomeroy and C. A. Craig, *J. Am. Chem. Soc.*, **81,** 6340 (1959).
124. (a) E. A. Letts, *Ber.*, **5,** 669 (1872);
(b) R. Leuckart, *Ber.*, **18,** 2341 (1885);
(c) G. Kruss, *Ber.*, **17,** 1766 (1884);
(d) E. E. Reid, *Am. Chem. J.*, **43,** 162 (1910);
(e) G. D. Van Epps and E. E. Reid, *J. Am. Chem. Soc.*, **38,** 2120 (1916).
125. W. A. Sheppard, *J. Org. Chem.*, **27,** 3756 (1962).
126. H. H. Wasserman and E. Druckrey, *J. Am. Chem. Soc.*, **90,** 2440 (1968).
127. (a) V. von Richter, *Ber.*, **7,** 1145 (1874);
(b) V. von Richter, *Ber.*, **8,** 1418 (1875);
(c) M. M. Holleman, *Rec. Trav. Chim.*, **24,** 194 (1905).

128. (a) J. Sauer and R. Huisgen, *Angew. Chem.*, **72**, 294 (1960);
(b) J. F. Bunnett, *Quart. Rev.*, **12**, 1 (1958), and references cited therein.
129. J. F. Bunnett and M. M. Rauhut, *J. Org. Chem.*, **21**, 944 (1956).
130. M. Rosenblum, *J. Am. Chem. Soc.*, **82**, 3796 (1960).
131. K. M. Ibne-Rase and E. Koubek, *J. Org. Chem.*, **28**, 3240 (1963).
132. E. Cullen and P. L'Écuyer, *Can. J. Chem.*, **39**, 862 (1961).
133. D. Samuel, *J. Chem. Soc.*, **1960**, 1318.
134. E. F. Ullman and E. A. Bartkus, *Chem. Ind.* (*London*), **1962**, 93.
135. (a) E. Cullen and P. L'Écuyer, *Can. J. Chem.*, **39**, 144 (1961);
(b) E. Cullen and P. L'Écuyer, *Can. J. Chem.*, **39**, 155 (1961);
(c) E. Cullen and P. L'Écuyer, *Can. J. Chem.*, **39**, 382 (1961).
136. G. T. Rogers and T. L. V. Ulbricht, *Tetrahedron Letters*, **1968**, 1029.

Author Index

This author index is designed to enable the reader to locate an author's name and work with the aid of the reference numbers appearing in the text. The page numbers are printed in normal type in ascending numerical order, followed by the reference numbers in parentheses. The numbers in *italics* refer to the pages on which the references are actually listed.

Abraham, R. J. 150 (241), *164*
Abrahamson, E. A. 646 (23), *668*
Abramowitch, R. A. 104 (312), *121*
Abrams, R. J. 749 (29), *785*
Abrazhanova, E. A. 102 (295), *120*, 525 (522), *630*
Abrol, Y. P. 718 (5), 720 (5, 14), 721 (5), 722 (37), *739*
Acampora, M. 842 (217), *851*
Achenbach, H. 856 (44), *878*
Achmatowicz, O. 581 (743), 582 (758), 583 (763, 764, 765, 766), 584 (768, 769, 770), 586 (768, 769, 773, 774), 587 (776, 777, 778), *635*, *636*, *637*
Achmatowicz, O., Jr. 583 (763), 586 (774), *636*
Acker, D. S. 224 (41), *237*, 532 (90, 568, 569), 536 (568, 578), 538 (582), *619*, *631*, *632*, 640 (5), *667*, 703 (167), 704 (168), *715*, 895 (36a, 36e, 36f), 901 (36), *941*, *942*
Adams, D. M. 154 (298), *166*
Adams, H. C. 506 (360), *626*
Adams, R. 75 (54), 95 (241), *112*, *118*, 766 (131), *788*
Adams, R. M. 138 (123), *162*
Adamson, A. W. 748, *785*
Addison, C. C. 149 (232, 233), *164*
Adelsberger, K. 352 (31), *415*
Adiga, P. A. 724 (48), *740*
Adkins, H. 320 (56, 57, 58, 66), 322, 323 (79), 324 (79, 82), 325 (82), *339*
Adler, J. 93 (210), *117*
Adman, E. 473 (230), *622*, 658 (46), *668*
Adolph, H. G. 227 (51), *237*
Advani, B. 842 (213), *851*
Affrossman, S. 768 (160), *788*
Agbalyan, S. G. 361 (64, 65, 66), *416*
Aggarwal, R. C. 154 (301), *166*
Ahearn, G. P. 918 (88), *944*
Ahlers, K. D. 862 (104), *879*
Ahmad, A. 720 (17, 18), 732 (17, 18), *739*, 922 (97b), 926 (97b), *944*
Ahmad, I. 127 (19), 128 (19), 133 (57), *159*, *160*
Ahmed, A. 95 (243), *118*
Ahrens, F. B. 182 (51), *204*
Akamatu, H. 663 (67), *669*
Akanuma, K. 83 (119), *114*
Akhrem, A. A. 74 (39, 40), *111*
Akiyama, M. 797 (53, 54), 810 (53), 822 (54), *847*
Alagna, B. 863 (109), *880*
Albers, E. 74 (41), *111*
Albers, H. 74 (41), *111*
Albisetti, C. J. 702 (156), *714*
Albright, J. A. 94 (231), *118*
Alderson, L. 689 (76d), *712*
Aldridge, W. N. 736 (133), *742*
Alexandrou, N. E. 828 (160), *849*
Al-Joboury, M. I. 644 (20), *668*
Allen, B. B. 281 (161), *303*
Allen, C. F. 71 (18), *110*
Allen, C. F. H. 72 (22), 75 (55), *111*, *112*
Allen, J. A. van 75 (55), *112*
Allen, J. C. 674 (11, 12), *711*
Allendorff, O. 83 (127), *114*
Allenstein, E. 128 (28, 29), *159*, *160*, 243 (11, 12), 245 (11, 17), *298*,

435 (812), 437 (812), 495 (816), 547 (644), *633*, *637*, *638*
Allerhand, A. 137 (101), 139 (103, 128), 140 (103), 142 (103), 157, 158 (103), *161*, *162*, 196 (126), *206*, 854 (23, 34, 36), *878*
Allerhand, A. A. 137 (101), *161*
Alles, H. U. 575 (730), *635*
Allinger, N. L. 148 (224), *164*, 211 (8), *236*
Alm, R. M. 315 (39), *338*
Almange, J. P. 191 (99), *205*
Almond, A. E. 778 (207), *789*, 933 (118), *945*
Alsop, D. W. 723 (34), *739*
Alston, R. E. 718 (3), *739*
Amamiya, H. 83 (119), *114*
Amano, Y. 667 (78), *669*
Amaya, K. 150 (244), *165*
Ambrush, T. I. 196 (156), *206*
Amr El Sayed, M. F. 196 (151, 155, 161), *206*, *207*
Anan'eva, L. I. 180 (47), *204*
Anastassiou, A. G. 108 (348), *122*, 548 (656, 657, 658, 659, 663, 667, 668), *633*, *634*, 707 (189), 708 (189, 190, 191, 193), 709 (191, 198), *715*, 903 (56b, 56c, 56d, 56e), 904, *943*
Anatol, J. 292 (217), *304*, 922 (98), *944*
Anchel, M. 515 (442), 516 (442), *628*, 728 (79), *741*
Anders, B. 503 (351), *626*
Andersen, K. K. 148 (226), 149 (226), *164*, 217 (24), 219 (24), 220 (24), 229 (24), *236*
Anderson, A. S. 732 (102), *741*
Anderson, D. M. W. 105 (328), *121*, 283 (176), *303*
Anderson, J. D. 188 (79, 80), *204*
Anderson, L. 721 (24b), *739*
Anderson, R. C. 782 (222), *790*
Anderson, R. D. 284 (181), *303*, 410 (227), *420*
Andersson, S. 507 (393), *627*
Andreeva, O. I. 749, 750 (27), *785*
Andrejevic, V. 103 (303), *120*
Andrews, L. J. 144 (178), *163*, 640 (2), 642 (2, 11), 654 (2), 655 (2), *667*
Anfinsen, C. B. 771 (178), *789*
Angelico, F. 799 (60), *847*
Angoletta, M. 876 (223), *882*
Anister, A. B. 437 (819), *638*
Anker, H. S. 751 (39), 766 (127), 771 (179), *785*, *787*, *789*
Anker, R. M. 284, 285 (182), *303*
Anneser, E. 346 (13), *415*, 796 (30), 808 (30), 810 (30), 827 (30, 158), 838 (30), 840 (158), *846*, *849*
Anschutz, R. 430 (44), *618*
Ansell, M. F. 541 (597), *632*
Aparicio, F. J. L. 695 (101), *713*
Arad-Talmi, Y. 179 (39), *203*
Arapakos, P. G. 335 (110), 336 (110), *340*
Arbasino, M. 809 (103, 107), 822 (154), 835 (186), *848*, *849*, *850*, 862 (100), *879*, 931 (115), *945*
Arbol, Y. P. 721 (25), *739*
Archer, S. 87 (161), *115*
Ardis, A. E. 483 (274), *624*
Arens, J. F. 899 (47c), *942*
Argabright, P. A. 78 (79), *113*
Argoudelis, C. J. 775 (200), *789*
Arledter, H. 524 (517, 518), 525 (517, 518), *630*
Armand, J. 809 (99, 100, 101, 102a, 102b), *848*
Armstrong, G. T. 515 (482), *629*
Armstrong, M. D. 95 (243), *118*
Arndt, F. 88 (179), *116*, 590 (783), *637*
Arnett, E. M. 125, 126 (6), 141 (147), *159*, *162*
Arnett, L. M. 686 (55), 691 (55, 82), 692 (82), *712*
Arnold, C. 79 (82), *113*, 765 (119), *787*
Arnold, R. A. 362 (71), *416*
Arnold, R. T. 754 (75), *786*
Arnold, Z. 591 (786), *637*
Arnstein, H. R. V. 759 (60), 761 (94), 771 (183), *786*, *787*, *789*
Arthur, P., Jr. 661 (59, 60), *669*
Ascah, R. G. 356 (43), 357 (43), *415*
Aseev, Y. G. 474 (253), *623*
Aseeva, R. M. 474 (251), *623*
Ashworth, J. M. 753, 756, *786*
Ashworth, M. R. F. 177 (27), *203*
Assche, D. V. 553 (693), *634*
Aten, A. H. W. 748 (21), *785*
Atherton, N. M. 502 (316), *625*
Atkinson, E. F. J. 286 (188), *303*
Attenburrow, J. 405 (200), *419*
Audinot, M. 747 (14), 748 (14), 751 (35), 760 (85), 762 (108), 764 (108), 775, 783 (14), *785*, *786*, *787*

Aufderhaar, E. 390 (151), *418*
Augdahl, E. 135 (85), 144 (192), 145 (192), 146 (85), 157 (85), 158 (192), *161*, *163*, 196 (137), *206*
Augl, J. M. 255 (59, 61), 256 (62), *300*
Aumüller, W. 860 (76), 862 (107), *879*, *880*
Austin, T. E. 297 (240), *305*
Autrey, R. T. 923 (100), *944*
Auwers, K. 682 (33), *711*, 793 (9), 815 (134, 135), *846*, *849*
Auwers, K. v. 96 (247), *118*
Averill, S. J. 699 (141, 142), *714*
Aversa, M. 841 (209), 844 (209), *851*
Aviran, A. 395 (167), *419*
Avogadro, L. 805 (84), *848*
Awad, W. L. 837 (202), *851*
Ayers, D. W. 335 (113), *340*
Ayers, G. W. 433 (73), *618*
Ayscough, P. B. 686 (52), *712*
Azzarello, E. 349 (21), *415*, 432 (66), *618*

Bachman, G. M. 806 (88), *848*
Bachmann, G. 80 (100, 101), 86 (101), 99 (100), *113*, 540 (600), 542 (600), 544 (600), 545 (600), 546 (600), *632*, 650 (29), 653 (29), 654 (29), *668*
Backeberg, O. G. 309, 310 (13, 15), *337*
Backer, H. T. 221 (36), *237*, 270, (112), *301*
Backman, G. B. 926 (105c), *945*
Bacon, R. G. R. 699 (138), *714*
Baddiel, C. B. 133 (54), *160*
Baddley, W. H. 459 (183), 517 (628), *621*, *633*
Baddly, W. H. 459 (182), *621*
Bader, A. 553 (693), *634*
Bader, H. 390 (150), *418*
Baer, W. K. 434 (813), 437 (813), *637*
Bähr, G. 500 (305, 307, 308, 309), 502 (307, 308, 309, 328), 503 (307, 328), *624*, *625*
Bailey, A. S. 98 (264), *119*, 524 (520), 525 (520), 528 (550), 543 (550), 576 (520), *630*, *631*, 654 (37), *668*
Bailey, R. A. 892 (28), *941*
Bailey, R. W. 721 (21), *739*
Baiochi, F. 872 (158), *881*
Baizer, M. M. 110 (363), *122*, 188 (77–80), *204*, 701 (152), *714*
Bajzer, W. X. 410 (231), *420*
Bak, B. 62 (62), *65*, 767 (144), *788*
Baker, B. E. 748 (19), 761 (92), *785*, *787*
Baker, F. W. 218 (27), *236*
Baker, H. 863 (115), *880*
Baker, R. H. 863 (112, 113), *880*
Baker, W. 70 (12), 78 (73), *110*, *113*
Balasubramanian, C. 196 (144), *206*
Balch, A. L. 502 (335, 340), *625*
Baldauf, H-J. 264 (96), 268 (96, 109), *301*
Baldwin, J. E. 451 (146), *620*
Baldwin, S. 194, *205*
Baliah, V. 210 (6), *236*
Ballauf, A. 295 (230), *305*
Bamberger, E. 325 (90), *339*, 506 (368), 507 (368), *626*
Bamberger, E. S. 735 (121), *742*
Bamford, C. H. 686 (57), 694 (95), 699 (140), *712*, *713*, *714*
Ban, L. 273 (130), *302*
Ban, Y. 379 (121), *418*
Bandzaitis, A. A. 10, 19 (21), 39 (21), *64*
Banewicz, J. J. 144 (190), 145 (190), *163*
Banholzer, H. 754 (78), *786*
Banholzer, K. 771 (181), *786*, *789*
Bann, B. 275 (137), *302*
Bannow, A. 547 (645), *633*
Barash, L. 548 (653), 552 (653), *633* 705 (177), 709 (177), *715*
Barb, W. G. 694 (95), *713*
Barber, H. J. 88 (165), *116*, 751 (38), *785*
Barbour, R. V. 335 (114), *340*
Barbulescu, N. 834 (137b, 174, 182), *849*, *850*
Baret, C. 747 (14), 748 (14), 751 (35), 762 (108), 764 (108), 775, 783 (14), *785*, *787*
Bargain, M. 189 (85), *205*
Bark, E. 925 (103f), *945*
Barkemeyer, H. 768 (156), *788*
Barltrop, J. A. 933 (119), *945*
Barnard, D. 95 (240), *118*
Barnes, R. A. 809 (111), 815 (111), 836 (111), *848*
Barnes, R. P. 929 (113a), *945*
Barrett, A. H. 516 (467), *629*
Barrick, P. L. 293 (223a), *305*, 703 (160), *714*
Barrow, G. M. 138 (118), 139 (125), *162*
Bartkus, E. A. 938, *946*
Bartle, K. D. 202 (186), *207*

Bartlett, M. F. 917 (87f), 922 (87), 930 (87), *944*
Bartlett, P. D. 454 (159), *620*
Barton, D. H. R. 93 (208), *117*
Bartulin, J. 404 (196), *419*
Barzilay, I. 914 (80b), *944*
Basinski, J. E. 108 (351), *122*
Bass, A. M. 548 (655), *633*, 708 (194), *715*
Basselier, J. J. 678 (22), *711*
Bast, K. 834 (173a), *850*
Bastiansen, O. 211 (3), *235*
Bastus, J. 458 (177), *621*
Bately, M. 492 (187), *621*
Bateman, L. 95 (240), *118*
Batley, M. 448 (137), *620*
Battersby, A. R. 766 (134), *788*
Battershell, R. D. 98 (264), *119*
Bauer, E. 69 (5), *110*, 832 (165), 840 (165), *850*
Baumann, A. 796 (20), 819 (20), *846*
Baumgarten, H. E. 391 (158), *418*
Bawn, C. E. H. 687 (62), 692 (86), *712*, *713*
Bax, C. M. 150 (259), *165*
Baxmann, F. 286 (187), *303*
Bayer, E. 516 (479), *629*
Bayer, O. 296 (234), *305*
Bayliss, N. S. 144 (180), *163*, 196 (138), *206*
Bayly, R. J. 767 (171), 782 (171), *789*
Bazouin, A. 100 (282), *120*
Beattie, I. R. 135 (82), 150 (251), 151 (82), 153 (251, 295), 154 (299, 300), *161*, *165*, *166*
Becher, H. J. 508 (402), *627*
Beck, W. 435 (814, 815), *637*, *638*, 792, 794 (12, 13), 796 (4, 16), 800 (16), 813 (4), *846*
Becke, F. 103 (299), *120*
Becker, E. I. 150 (255), *165*, 277, 278 (149, 151, 152), 285 (150, 153), *302*
Becker, E. J. 857 (49), *878*
Becker, H. G. O. 94 (232), *118*
Becker, W. 721 (24a), *739*
Beckmann, E. 925 (103f), 927 (109), *945*
Beckmann, R. 775 (199), *789*
Bedel, C. 507 (387, 388), *627*, 736 (134), *742*
Bedford, G. R. 95 (237), *118*
Beech, W. F. 73 (30), *111*
Behr, L. C. 809 (112), 815 (112), 819 (112), *848*
Behrenz, W. 877 (230, 231, 232), *883*
Behringer, H. 372 (95), 388 (142), *417*, *418*
Beketoff, N. 430 (41), *618*
Bekoe, D. A. 596 (797), *637*
Bell, C. L. 138 (118), *162*
Bell, E. A. 724 (47), 726 (69), *740*
Bell, F. 105 (328), *121*, 283 (176), *303*
Bell, F. K. 194 (110), *205*
Bell, R. P. 227 (49), *237*
Bell, V. L. 452 (151), *620*
Bellamy, L. J. 138 (113, 114), 158 (324), *161*, *166*, 194, 196 (112), *205*, 232 (62), *237*, 242 (5), *298*
Belleau, B. 750 (33), *785*
Bellet, E. 431 (53), *618*
Belliet, E. 89 (181), *116*
Belluco, U. 459 (183), *621*
Belniak, K. 582 (758), 586 (773, 774), *636*
Belov, V. F. 523 (508), *630*
Belova, G. V. 474 (240), *622*
Beltrame, P. 796 (41), 803 (41), 814 (128), *847*, *849*
Belyakova, A. I. 527 (539), *630*
Benderly, A. A. 85 (147), *115*
Beneš, M. J. 289 (199), *304*, 523 (511), 524 (513, 514), *630*
Benesi, H. A. 144 (187), *163*, 642 (10), *668*
Bengelsdorf, I. S. 298 (244), *305*, 410 (226), *420*
Benn, M. H. 828 (161, 162), *849*, *850*
Bennett, C. 472 (225), *622*
Bennett, E. L. 768 (155), 770 (174), 776 (202), 783 (155, 174), *788*, *789*
Bennett, G. M. 97 (255), *119*, 524 (521), 525 (521), *630*
Benneville, P. L. de 75 (51), *112*, 258 (76, 77, 78), 270 (117), 272 (123), 275 (76, 139), *300*, *302*
Benson, F. R. 291 (211), 292 (211), *304*, 344 (4), 388 (4), *414*, 913 (77a), 914 (77a), *944*
Benson, R. E. 109 (354), *122*, 229 (53, 55), *237*, 441 (100), 453 (155), 458 (174), 459 (100), 460 (186), 463 (100), 467 (209), 468 (100), 471 (100), 472 (100), 473 (100), 474 (100), 495 (209), 502 (323, 325), 527 (540), 529 (540), 532 (90, 570), 534 (573), 535 (573), 538 (582), 541 (596), 555(174, 699), 556 (174), 557 (699), 560 (699), 563 (715), 565 (715), 566 (715), 567 (715), 569 (715), 581

(699), 593 (789), 608 (570), *619*, *620*, *621*, *622*, *625*, *630*, *631*, *632*, *634*, *635*, *637*, 640 (6a), 642 (6), 644 (6a), 646 (6, 24), 655 (6), 658 (6, 44, 45), 659 (6), 660 (6), 661 (6, 44, 45), 664 (6, 45), 665 (6), 666 (6), *667*, *668*, 704 (168), *715*, 895 (36a, 36d, 36f), 901 (36), *941*, *942*
Bent, H. 892 (26), *941*
Bentley, J. A. 732 (96, 98), *741*
Bentley, T. J. 93 (208), *117*
Ben-Yehoshua, S. 720 (13), *739*
Berbalk, H. 142 (167), *163*
Berenbaum, M. B. 75 (53), 76 (61), 110 (53), *112*, 687 (59, 61), 690 (77), 691 (83), *712*
Bergel, F. 412 (237, 238), *421*
Berger, A. 442 (104), *619*
Berger, G. 245 (22a), *299*
Berger, J. C. 408 (220), *420*
Bergmann, F. 733 (108), *741*
Bergmann, J. C. 335 (111), *340*
Bergström, S. 767, 768 (170), *788*, *789*
Berkeley, P. J. 139 (135, 142, 143), 140 (135, 142, 143), 156, *162*
Berlin, A. A. 474 (239, 240, 241, 242, 243, 244, 245, 246, 248, 251, 252, 253), 529 (563), *622*, *623*, *631*
Berliner, L. J. 661 (65), *669*
Bernal, I. 183 (62), 184 (62), 190 (62), *204*, 229 (56), *237*, 297 (241), *305*, 472 (220), 473 (220), 502 (326, 331), 527 (541), 528 (541), *622*, *625*, *630*, 658 (49), *668*
Bernhard, K. 779 (209), *789*
Bernheim, R. A. 551 (684), *634*, 705 (173, 174), *715*
Bernstein, H. J. 135 (66), *160*
Bernthsen, A. 271 (120), *302*
Berry, R. E. 73 (36), *111*
Berry, R. S. 193 (107), *205*
Berry, T. E. 502 (340), *625*
Berson, J. A. 909 (64b), *943*
Berthelot, D. 428 (16), *617*
Bertoluzza, A. 143 (175), *163*
Besnainou, S. 194, 196 (113), *205*
Bestmann, H. J. 842 (215), *851*
Bethe, H. 29 (35), *64*
Bethe, H. A. 40 (54), *65*
Betrus, B. J. 361 (69), *416*
Betts, B. E. 74 (40), *111*
Betts, J. 692 (87), *713*
Betz, W. 855 (40), 857 (40), 864 (122), *878*, *880*
Beugelmans, R. 198 (171), *207*
Beverung, W. N. 361 (63), *416*
Bevington, J. C. 692 (85, 88), 698 (135), *712*, *713*, *714*
Beyl, V. 243 (11), 245 (11), *298*
Beynon, J. H. 200 (173), 201 (179), *207*
Bezman, I. I. 102 (296), *120*
Bhattacharjee, R. C. 370 (91), *417*
Bianchetti, G. 835 (188, 189), *850*
Bianchi, G. 835 (190), *850*
Bianco, E. J. 99 (275), *120*
Bickel, A. F. 688 (67), 694 (97), 695 (100), 696 (104, 113, 114), *712*, *713*
Bideiko, A. F. 857 (51), *878*
Biechler, J. 546 (630), 547 (630), *633*
Biehn, G. F. 428 (7), *617*
Bielecke, J. 190 (92), *205*
Bieling, H. 500 (308), 502 (308), *624*
Bierl, B. A. 147 (221), *164*, 196 (145), *206*
Bierstedt, P. E. 665 (74), *669*, 895 (36b), 901 (36), *941*
Bierwirth, E. 769 (162), *788*
Bigelow, L. A. 297 (238, 239, 240), *305*, 432 (69), *618*
Biggs, A. I. 218 (29b), *236*
Biggs, B. J. 323 (77), *339*
Biggs, B. S. 101 (292), *120*, 320 (69), *339*
Bigley, B. D. 933 (119), *945*
Bigley, D. B. 330 (102), *340*
Bijl, D. 659 (54), *669*
Bikales, N. M. 277 (153), 285 (153), *302*
Biletch, H. 75 (53), 110 (53), *112*, 687 (58, 64), 688 (64), *712*
Billig, E. 502 (322, 324, 331, 333, 336, 337), *625*
Bilton, J. A. 573 (725), *635*
Biltz, H. 128 (25), *159*
Binas, H. 154 (296), *166*
Binsch, G. 353 (33), 354 (34), *415*
Bircumshaw, L. L. 289 (198), *304*
Bird, C. W. 366 (76), 412 (240), *416*, *421*, 909, *943*
Bird, W. L. 186, 187 (67), *204*
Birkenbach, L. 437 (92), *619*, 796 (17), 800 (17), 819 (17), 822 (155b), *846*, *849*
Birnbaum, G. I. 585 (771), *636*
Bishop, B. C. 297 (238, 239, 240), *305*
Bishop, G. 925 (103e, 103g), *945*
Bishop, W. S. 101 (292), *120*, 320 (69), *339*
Bither, T. A. 859 (59), *878*

Bittner, F. 768 (223), *790*
Bittner, G. 151 (267), *165*, 872 (164), *881*
Bizhanov, F. 325 (84b), *339*
Bjorvatten, T. 147 (213), *164*
Black, C. C. 735 (121), *742*
Black, R. D. 286 (186), *303*
Blackburn, G. 765 (130), *788*
Blackham, A. U. 685 (50), *712*
Blackwood, R. K. 77 (70), *112*
Bladé-Font, A. 286 (189), *303*
Bladin, J. A. 428 (17), *617*
Blaise, E. E. 282 (167), *303*, 927 (107d), *945*
Blanchard, E. P. 433 (74), *618*
Blanchard, E. P., Jr. 701 (153), *714*
Blaschke, H. 354 (37), 355 (37), *415*, 842 (210), *851*
Blatt, A. H. 910 (70c), 912 (70), 929 (113a), *943*, *945*
Blatter, H. M. 93 (221), *118*
Bledose, J. O., Jr. 926 (108), *945*
Bledsoe, J. O. 281 (163), *303*
Blessing, G. 320 (53), 324 (53), *338*
Blicke, F. F. 279 (158), *303*, 400 (183), *419*
Bloch, K. 766 (122), *787*
Blomeyer, B. 877 (231), *883*
Blomquist, A. T. 108 (352), *122*, 453 (156), 455 (156), 513 (432), 520 (432), 521 (432), *620*, *628*
Blomstrom, D. C. 217 (22), *236*, 500 (306), 501 (306), 502 (306), 503 (306, 353, 355), 504 (353, 355), 505 (356), 532 (569), 536 (578), 576 (353), 577 (353), 579 (353), *624*, *626*, *631*
Bloom, M. S. 325 (88), *339*
Bloomfield, J. J. 282 (171), 284 (171), *303*, 397 (169), *419*
Bloor, J. E. 193 (108), 194 (109), *205*
Bluestone, H. 98 (264), *119*
Blumenthal-Goldschmidt, S. 725 (62), *740*
Blyumenfeld, L. A. 472 (217), *622*
Boatright, L. G. 320 (62), *339*
Bobbitt, J. M. 377 (109), *417*
Bobrova, M. I. 186 (66), 189 (82), *204*, *205*
Boche, G. 449 (141), 461 (141), *620*
Bock, H. 135 (90), 155 (90), *161*, 256 (63), *330*
Böcker, E. 87 (156), *115*
Böckly, E. 876 (209), *882*
Boden, H. 83 (121), *114*
Boden, J. C. 709 (199, 200), *715*
Bodenbender, H. G. 541 (595), *632*
Bodesheim, F. 860 (71), 865 (130), 866 (130), 867 (130), *879*, *880*
Bodnarchuk, N. D. 252 (53), *299*
Boeckh-Behrens, G. 731 (92a), *741*
Boehme, W. R. 109 (360), *122*
Boggs, J. E. 516 (465), *628*
Boguslavskii, L. I. 474 (242, 246), *623*
Bohart, G. S. 506 (361), *626*
Bohlmann, F. 515 (441), 516 (441), *628*
Böhme, H. 104 (315), *121*
Boivin, J. L. 101 (286), *120*
Bok, B. 854 (28), *878*
Boldt, P. 107 (343), *122*, 681 (26, 28), *711*
Boleslawski, M. 152 (286), *166*
Bollyky, L. J. 582 (759), *636*
Bolser, C. E. 875 (205), *882*
Bombieri, G. 459 (183), *621*
Bon, R. 735 (114), *742*
Bonciani, T. 871 (156), *881*
Bond, A. C., Jr. 326 (93), *340*
Bond, E. J. 726 (72), *740*
Bongert, A. 818 (144), *849*
Bongrand, J. C. 97 (254), *119*, 513 (413, 414, 430, 431), 514 (413, 414, 430), 516 (413, 414, 430, 431), 517 (414), 518 (430, 488), 519 (430, 496), *627*, *628*, *629*
Bonino, G. B. 143 (175), *163*
Bonner, W. A. 755 (76, 77), *786*
Bonner, W. H. 428 (22), 429 (22), *617*
Bonnier, J. M. 192 (101, 102), 193 (102), *205*
Booth, M. R. 892 (24), *941*
Boozer, C. E. 688 (68a), 692 (68a), 693 (68a), 694 (68a, 98), *712*, *713*
Borecki, C. 582 (758), *636*
Boreev, O. M. 474 (240), *622*
Borello, E. 805 (86), 812 (86), 813 (86), *848*
Borgen, B. 147 (214), *164*, 515 (472), 516 (472), *629*
Borkovec, A. B. 402 (193), *419*
Born, H. J. 771 (180), *789*
Borowitz, I. J. 393 (159), *419*
Borror, A. L. 402 (191), *419*
Borsche, W. 87 (156), *115*, 919 (90a), 925 (103d), 927 (103d), 929 (103d), 930 (103d), *944*, *945*
Borsook, H. 761 (91), 763 (100), 766 (146), *787*, *788*
Bos, J. A. 747 (6), *785*

Bosshard, H. H. 99 (273), *119*
Bothner-By, A. A. 767 (138), *788*
Bottcher, H. 408 (219), *420*
Böttner, E. 864 (123), *880*
Boudet, R. 101 (289), *120*
Bouguerra, M. L. 188 (81), *205*
Boulton, A. H. 872 (162), *881*
Bouso, O. 778 (206), *789*
Bouveault, L. 818 (144), *849*, 926 (107c), *945*
Bove, C. 721 (22a), *739*
Bowie, J. H. 200 (177, 178), *207*
Boyd, R. H. 191 (100), *205*, 224 (40), 227 (40, 45), *237*, 434 (803), 435 (91), 533 (91), 535 (576), 593 (791), 610 (803, 804, 805), 613 (803), 614 (803), 615 (803), *619*, *631*, *637*, 659 (51, 52), *669*
Boyer, J. 707 (185), *715*
Boyer, J. H. 815 (132a), 819 (132a), *849*
Boyle, J. T. A. 387 (136), *418*
Boynton, E. 543 (607), *632*
Boys, E. S. 8 (13), *63*
Brackman, W. 94 (227), *118*
Bradbrook, E. F. 84 (133), *114*
Bradsher, C. 333 (109), *340*
Bradsher, C. K. 402 (195), *419*
Bradt, P. 513 (423), *628*
Brady, L. E. 373 (99), *417*
Brady, O. L. 93 (209), *117*, 925 (103e, 103g), *945*
Brady, R. O. 782 (221), *790*
Bramley, A. 144 (190), 145 (190), *163*
Brandenberger, S. G. 766 (124, 126), *787*
Brandner, H. 547 (641), *633*
Brandon, M. 144 (194), *163*
Brandon, R. L. 472 (226), 473 (226), 543 (226, 608), *622*, *632*, 658 (48), 661 (48), 664 (48, 71), *668*, *669*
Brändström, A. 78 (76), *113*
Brändstrom, A. 765 (118), *787*
Bratoz, S. 194, 196 (113), *205*
Braude, E. A. 282 (164), *303*, 541 (594), *632*
Braumann, M. 430 (49), *618*
Braun, A. 83 (128), 100 (279), *114*, *120*, 320, 324 (53), *338*
Braun, J. v. 82, 97 (260, 261), *114*, *119*
Braunbruch, W. B. 606 (801), 607 (801), *637*
Bravo, L. 905 (61), *943*
Bravo, P. 845 (220, 221, 222), *851*
Bray, P. J. 201 (180), *207*
Breckpot, R. 80 (95), *113*
Bredereck, H. 69 (8), *110*, 506 (371), 507 (371), 508 (401, 402), 509 (401), 572 (401), *626*, *627*, 859 (56, 57), *878*
Bredereck, K. 290 (206), *304*
Bredt, J. 70 (10, 14), *110*, 919 (90b), *944*
Brehm, A. 580 (738), *635*
Bremer, N. J. 135 (91), 155 (91), *161*
Brentano, W. 312 (28), 320 (28), *338*
Bresadola, S. 872 (161, 165), *881*
Breslow, D. S. 325 (88), *339*
Breslow, R. 258 (72), 261 (72), 263 (72), *300*, 551 (689), *634*
Bretschneider, J. 102 (297), *120*, 526 (536), 527 (536), *630*
Briegleb, C. 448 (136), *620*
Briegleb, G. 143, 144 (177), 146 (199), *163*, *164*, 473 (228), *622*, *638*, 640 (1), 642, 644, 645, 646, 647, 648 (15), 650 (1), 652, 653 (1), 655 (1), 656 (1), *667*, *668*, 656 (39, 40), *668*
Briner, E. 428 (14), *617*
Britton, D. 135 (87), *161*
Brivati, J. A. 473 (233), *622*
Brockhurst, P. 193 (107), *205*
Brockman, F. J. 514 (438), 516 (438), 517 (438), *628*
Brockway, L. O. 429 (29), *617*
Brook, A. G. 541 (588), *632*, 541 (588, 594), *632*
Brooks, B. R. 686 (52), *712*
Brotherton, T. K. 268 (108), 281 (108), 285 (184), *301*, *303*, 428 (11), *617*
Brown, D. A. 477 (263), *623*
Brown, E. V. 782 (222), *790*
Brown, F. W. 74 (44), 79 (44), *111*
Brown, H. C. 151 (265), *165*, 218 (26, 28, 29c, 29d), 230, (58), *236*, *237*, 268 (110), 270 (113), 275 (136), 298 (244), *301*, *302*, *305*, 317 (49, 50, 51), 318 (49, 50), 329 (49, 50, 51, 98), 330 (101, 102, 103), 332 (49, 50, 98, 101, 103), *338*, *340*, 388 (144), 391 (156), 410 (228, 232), *418*, *420*
Brown, I. 150 (243), 158 (323), *165*, *166*
Brown, J. von, 320, 324, *338*
Brown, L. C. 201 (181a), *207*

Brown, M. 460 (186), 467 (209), 495 (209), 527 (540), 529 (540), *621*, *622*, *630*
Brown, M. E. 664 (71), *669*
Brown, P. 449 (818), 564 (717, 718), 566 (718), 567 (717, 718), 568 (717, 718), 569 (717, 718), *635*, *638*
Brown, P. M. 99 (274), *120*
Brown, S. A. 767 (141), *788*
Brown, T. L. 135 (80), 157 (321), 158 (321), *161*, *166*, 196 (152), *206*, 242 (9), 250 (42), *298*, *299*
Brown, W. G. 314 (35, 37), 326 (35, 37), 332 (35), *338*
Browne, M. E. 543 (608), *632*
Browne, M. F. 92 (200), *117*
Brownell, R. M. 221 (35), *237*
Brownlee, P. P. 372 (97), *417*
Brubacher, G. 779 (209), *789*
Brück, R. 876 (217), *882*
Bruckenstein, S. 127 (18), *159*
Brueckner, K. A. 35, 37 (49), *65*
Bruin, P. 696 (113), *713*
Brune, H. A. 135 (81), *161*, 196 (132), *206*
Brunn, E. 353 (32), *415*, 842 (210), *851*
Bruno, A. 394 (164), *419*
Bruno, J. 769 (166), *788*
Bruson, H. A. 105 (320), *121*
Bruylants, A. 79 (91), *113*, 278 (156), 279 (156), *303*, 672 (3), 673 (3, 5), *710*
Bruylants, P. 80 (93), *113*, 278 (156), 279 (156), *303*
Bruzzese, T. 196 (164), *207*
Bryant, J. T. 323 (78), *339*
Brysk, M. M. 728 (83), *741*
Bubner, M. 767 (143), *788*
Bucherer, H. 74 (37), 75 (49), *111*, *112*
Bücherer, H. T. 761 (95), *787*
Büchner, R. 756 (50), *786*
Büchner, W. 860 (72), *879*
Buck, J. S. 92 (197), *117*, 928, (112) *945*
Buckingham, A. D. 148 (231), 150 (242), *164*, *165*
Buckles, R. E. 146 (201, 203), *164*
Buckley, D. 187 (73), *204*
Buckley, D. A. 372 (97), *417*
Budding, H. A. 519 (494), *629*
Budeikiewitz, H. 198 (171), *207*
Budesinsky, Z. 98 (270), *119*
Budge, A. H. 150 (246), *165*
Bugg, C. 434 (823), 558 (706), *634*, *638*
Bugge, G. 873 (170), 875 (191), *881*, *882*
Buhle, E. L. 560 (710), 591 (788), *635*, *637*
Buijle, R. 899 (47a, 47b), *942*
Buisson, R. 550 (680), 551 (680), *634*
Bulka, E. 862 (104), *879*
Bulow, C. 925 (103a), *944*
Bunnett, J. F. 87 (157, 158), *115*, 230 (59), *237*, 260 (83b), 285 (184), *301*, *303*, 696 (117), *713*, 936, *946*
Burawoy, A. 193 (107), *205*
Burch, G. N. B. 703 (161), *714*
Burckhalter, J. H. 384 (127), *418*
Burdick, W. L. 546 (634), 547 (634), *633*
Burditt, A. M. 778 (208), *789*, 933 (118), *945*
Burdon, J. 73 (34), *111*
Burgert, B. E. 91 (194), *117*, 710 (210), *716*
Burghard, G. 765 (130), *788*
Burk, E. H., 323 (80), *339*
Burkhardt, J. 401 (185), *419*
Burkhardt, W. 273 (130), *302*
Burns, J. J. 754 (72, 73), *786*
Burrows, E. P. 397 (173), *419*
Burshtein, R. K. 474 (242), *623*
Burtle, J. G. 762 (104), *787*
Burton, W. H. 103 (302), *120*
Buss, A. 255 (61), *300*
Buss, A. A. 256 (62), *300*
Buss, H. 793 (5), 803 (5), 826 (5), 829 (5), *846*
Busti, G. 801 (66b), *847*
Butenandt, A. 775 (119), *789*
Butler, D. W. 101 (287), *120*
Butler, G. W. 719 (7, 8), 720 (7, 8, 19, 20), 721 (21), 725 (62), *739*, *740*
Buu-Hoi, N. P. 94 (233), *118*
Buzzell, A. 135 (86), *161*, 216 (18), *236*
Byrd, N. R. 513 (433), 514 (433), 524 (433), *628*
Bystrov, D. 135 (76), *161*
Bywater, S. 298 (242), *305*, 688 (69, 70), *712*

Cabana, A. 196 (153), *206*
Cadogan, J. I. G. 674 (11, 12, 13), 675 (13, 14, 18), *711*

Cadus, A. 696 (116), *713*
Cahoon, N. 386 (134), *418*
Cairncross, A. 701 (153), *714*
Cairns, T. 148 (222), *164*
Cairns, T. L. 92 (202), 109 (362), *117*, *122*, 293 (223a), 298 (244), *305*, 385 (132), 409 (222), *418*, *420*, 428 (4, 6), 446 (115, 116), 447 (6, 116), 463 (190), 465 (190), 468 (211), 485 (279), 486 (279), 503 (353), 504 (353), 576 (353), 577 (353), 579 (353), *617*, *619*, *621*, *622*, *624*, *626*, 640 (3), 657 (42), *667*, *668*, 682 (31), *711*, 860 (78) *879*
Calas, R. 100 (282), *120*
Calcott, W. S. 68 (3), 69 (3), *110*
Caldow, G. L. 135 (92), *161*, 196 (140), *206*
Caldow, L. L. 137 (107), *161*
Califano, S. 796 (38), 797 (38), 813 (38, 120), *847*, *848*
Callen, J. E. 82 (113), *114*
Calmels, G. 807 (94), *848*
Calov, U. 153 (294), *166*
Calvin, M. 246 (23), *299*
Cammarata, A. 818 (142), *849*
Camp, D. B. 747 (7), *785*
Camp, D. C. 748 (22), *785*
Campbell, C. D. 905 (60), *943*
Cannon, D. Y. 431 (57), *618*
Cannon, G. W. 293 (221), *305*
Cannon, W. N. 410 (230), *420*
Canon, J. M. 150 (247), *165*
Cantrell, T. S. 696 (112), *713*
Carangelo, M. G. 187 (74), *204*
Carboni, R. A. 92 (202), *117*, 284 (178), *303*, 409, (222), *420*, 446 (116), 447 (116), *619*, 640 (3), *667*, 682 (31), *711*
Carbtree, E. V. 388 (145), *418*
Carlson, E. J. 699 (142), *714*
Carlson, K. D. 44 (56), 45 (56), *65*
Carlson, R. D. 97 (260), *119*
Carnduff, J. 308 (3), 309 (3), *337*
Carothers, W. H. 320 (61), *339*
Carpenter, F. H. 104 (306), *120*
Carpenter, G. B. 147 (210), *164*, 515 (470), 516 (470), 517 (470), *629*
Carpenter, W. 197 (169), 198, 200 (169), *207*
Carpenter, W. R. 351 (29), 388 (29), *415*, 606 (802), *637*
Carpino, L. A. 99 (276), *120*
Carraro, G. 872 (161), *881*
Carrington, A. 473 (237), 576 (734), *622*, *635*, 656 (41), *668*
Carrington, H. C. 408 (215), *420*
Carrington, R. A. G. 434 (87), *619*
Carryck, W. L. 748 (23), *785*
Carten, J. W. 150 (263), *165*, 216 (20a), 217 (20a), *236*
Carter, A. S. 68 (3), 69 (3), *110*
Carter, D. E. 506 (367, 369), 507 (367), *626*
Carter, J. C. 872 (163), *881*
Casanova, J. 855 (39), 857 (39), 859 (68), 872 (160, 162), *878*, *879*, *881*
Casanova, J. Jr. 887 (9, 11), 889 (11), 890 (11), 918 (89b), *940*, *944*
Case, F. H. 285 (183), *303*, 390 (152), *418*
Casini, V. 861 (87), *879*
Casnati, G. 722 (28), *739*, 802 (69), *847*
Cason, J. 282 (168), *303*
Cass, O. W. 481 (265), *623*, 765 (117), *787*
Cassal, J.-M. 558 (707), *634*
Cast, J. 293 (222), *305*
Castan, P. 516 (454), *628*
Castells, J. 458 (177), *621*
Castille, A. 193 (104), *205*
Catch, J. R. 746, 747, 751 (2), *784*
Cava, M. P. 451 (149), *620*, 905 (61), *943*
Cavell, E. A. S. 142 (149), *162*
Cecere, M. 697 (126), *713*
Cenci, H. J. 482 (270), 483 (273), *624*
Cerwonka, E. 782 (222), *790*
Chabrier, P. 276 (143), *302*
Chakravarti, S. N. 927 (111), *945*
Challenger, F. 733 (105), *741*
Chambers, R. D. 215 (16), *236*
Chambers, W. J. 297 (238), *305*
Champion, W. C. 550 (679), 551 (679), *634*
Chan, S. I. 133 (53), *160*
Chandler, L. B. 294 (228), *305*
Chang, M. S. 347 (18), 348 (18), *415*, 796 (36), 812 (117), *847*, *848*
Chantooni, M. K. 127 (18), *159*
Chase, B. H. 390 (149), *418*
Chatterjee, A. 730 (87, 88), *741*
Chatterjee, R. 429 (33), *617*
Chatterjee, S. 397 (172), *419*, 533 (571), *631*, 646, 650 (25), 658 (25), *668*
Chaudhari, M. A. 502 (312), *625*

Chaykin, S. 731 (91), *741*
Chen, C.-X. 582 (762), *636*
Cheng, C. C. 872 (158), *881*, 926 (106), *945*
Cherayil, G. D. 768 (154), *788*
Cherkashina, L. G. 474 (251), 529 (563), *623*, *631*
Cheronis, N. D. 168 (5), 169 (5), 170 (7), 171 (7), 172 (15), *203*
Chesick, J. P. 908 (64a), *943*
Chesnut, D. B. 659, 661 (56, 59), *669*
Chestnut, D. B. 895 (36c), 896 (38), 901 (36), *942*
Chiang, M-C. 268 (106), *301*
Chibnik, S. 75 (53), 110 (53), *112*, 690 (79), *712*
Chichibabin, A. E. 898 (43a, 43b, 43c), *942*
Chick, M. J. 616 (809), *637*
Chien, S. L. 98 (266), *119*
Chien, T. H. 752 (41), *785*
Chih, C. M. 428 (23), *617*
Child, W. C. 149 (238), *164*
Chin, F. S. 752 (41), *785*
Ching-Yun, Chen 869 (143), *880*
Chirikova, Z. P. 142 (157), *162*
Chizkov, O. S. 292 (218b), *305*
Chortyk, O. T. 912 (72a, 72c), 913 (72c), 919 (72a), 920 (72c), 921 (72c), 926 (72c), *943*
Chosho, H. 401 (190), *419*
Chottard, J. C. 678 (22), *711*
Christ, H. A. 818 (140), *849*
Christe, M. 832 (167b), 835 (167), *850*
Christensen, B. E. 771 (179), *789*
Christensen, J. J. 614 (807), *637*
Christenson, R. M. 291 (215), 296 (231), *304*, *305*
Christian, J. E. 767 (150), *788*
Christl, M. 796 (18), 800 (18), 832 (18), *846*
Christol, H. 914 (78b), *944*
Cieresko, L. S. 769 (172), *789*
Ciganek, E. 107 (346), 109 (356), *122*, 470 (215), 503 (215), 514 (215), 515 (215), 517 (484, 486), 520 (484), 521 (484, 486, 502), 550 (681), 551 (681, 687), 552 (682, 692), 559 (708), 561 (708), 562 (708), 563 (708), 568 (708), 583 (681), *622*, *629*, *634*, 685 (46), 705 (176, 178, 179, 180), 706 (180, 181, 182), *711*, *715*, 901 (52a, 52c, 52d, 54), 906 (52d, 54), 907 (63), *942*, *943*
Cima, L. 735 (115), *742*
Citron, J. D. 277 (150), 285 (150), *302*
Claggett, A. R. 411 (234), *420*, 509 (411), *627*
Claisen, L. 82 (110), 93 (215), 96 (247), *114*, *117*, *118*
Clark, J. H. 98 (265), *119*, 273 (128), *302*
Clark, R. J. H. 154 (297), *166*, 502 (311, 336, 342), *625*
Clark, T. 291 (213), 292 (212), 293 (212), 304
Clarke, H. T. 88 (164, 170), 92 (200), *116*, *117*
Clarke, R. A. 582 (759), *636*
Clarke, R. L. 97 (256), *119*
Classon, W. D. 448 (129), *620*
Claus, C. J. 315, *338*, 748 (22, 26), *785*
Claus, J. C. 749, *785*
Clauson-Kass, N. 767 (144), *788*
Claver, G. C. 187, *204*
Clegg, J. M. 351 (27), 388 (27), *415*
Clementi, E. 10, 11 (14, 15), 13, 19 (26, 27), 23 (24), 26, 29 (34), 30 (27), 31, 32 (38, 39), 33 (40, 41, 42, 43, 44, 45, 46, 47), 35 (50), 37 (47), 39, 40, 41 (50), 43 (50, 55), 44 (56), 45 (56), 46, 48 (50), 49 (60), 54 (61), 55 (60), 62 (62), *63*, *64*, *65*
Clemo, G. R. 763 (98), *787*
Cleveland, F. F. 854 (20), *877*
Cloke, J. B. 384 (128, 129), *418*
Coates, G. E. 142 (153), 153 (287), *162*, *166*
Coates, J. E. 142 (153), *162*
Cobb, A. W. 245 (22a), *299*
Cobb, R. L. 76 (66), *112*
Coburn, R. A. 720 (15), *739*
Coburn, W. C. 157 (319), *166*
Cockburn, W. F. 779 (221), *790*
Coelho, R. A. 180 (48), *204*
Coerver, H. J. 132 (44), 135 (44), 151 (44), *160*
Coetzee, J. F. 127 (14, 16, 20), 133 (61, 62), 134 (62, 63), *159*, *160*
Coever, H. J. 196 (131), *206*
Coffman, D. D. 92 (202), *117*, 233 (64), *237*, 284 (178), 297 (238), *303*, *305*, 409 (222), *420*, 434 (86), 435 (86), 446 (116), 447 (116),

463 (190), 465 (190), 466 (68), 468 (86, 211), 574 (729), 575 (729), 590 (86), 593 (86, 729), 594 (86), 595 (86), 596 (86), 606 (86), 608 (86), *619*, *621*, *622*, *635*, 640 (3), 657 (42), *667*, *668*, 674 (8), 682 (31), 703 (160), *711*, *714*, 899 (49), 900 (49c), *942*
Cohen, M. S. 256 (65), *300*
Colby, C. E. 246 (29), *299*
Cole, A. R. H. 196 (138), *206*
Cole, R. H. 142 (155), *162*
Coleman, G. H. 82 (113), *114*
Collander, R. 142 (163), *163*
Collard-Charon, C. 862 (103), *879*
Colligiani, A. 242 (2), *298*
Collin, R. L. 516 (471), 517 (471), *629*
Collins, C. J. 755 (115), *787*
Collins, R. L. 473 (229), *622*, 658, *668*
Colson, A. 249 (39), *299*
Colthup, E. C. 524 (516), *630*
Colthup, N. B. 242 (7), *298*
Colton, E. 249 (41), 258 (41), *299*
Columbo, M. 805 (86), 812 (86), 813 (86), *848*
Comeford, J. J. 548 (654), *633*
Comnick, W. J., Jr. 100 (284), *120*
Comoti, A. 814 (128), *849*
Comp, J. L. 513 (420), *627*
Condon, E. V. 6 (7), *63*
Conduit, C. P. 210 (7), *236*
Conley, R. T. 95 (235), *118*, 377 (108), *417*, 912 (72b, 72c), 913 (72c), 917 (87g), 919, 920, 921, 922 (87), 926 (72c, 105a), 930 (87), *943*, *944*, *945*
Conn, E. E. 718 (3, 5), 719 (7, 8), 720 (5, 7, 8, 10, 11, 13, 19), 721 (5, 22a, 23, 25), 722 (26), 725 (62, 64, 66), *739*, *740*
Connick, W. J., Jr. 329 (99), 330 (99), *340*
Conrad, M. 819 (149), *849*
Cook, A. F. 560 (712), *635*
Cook, A. H. 75 (55), *112*, 284, 285 (182), *303*, 401 (189), 408, *419*, *420*
Cook, D. J. 95 (240), *118*
Cook, G. B. 8 (13), *63*
Cooks, R. G. 200 (178), *207*
Cookson, R. C. 73 (31), 98 (31), *111*, 449 (818), 455 (161), 456 (167), 520 (500, 501), 521 (500, 501), 522 (501), 523 (500), 564 (717, 718), 566 (718), 567 (717, 718), 568 (717, 718), 569 (717, 718), 599 (799), 600 (799), *621*, *629*, *635*, *637*, *638*
Cooley, S. L. 736 (131), *742*
Cooper, F. W. 404 (197), *419*
Cope, A. C. 105 (319, 325, 326), *121*, 590 (781), *637*, 895 (35), *941*
Copley, M. J. 142 (161, 162), 148 (161, 162), *163*
Cordes, E. H. 77 (69), *112*, 275 (138), *302*, 734 (110b), *741*
Cordier, M. 105 (325), *121*
Corey, E. J. 855 (38), 857 (38), *878*, 887 (9), *940*
Cormack, D. 768 (160), *788*
Cormack, J. F. 87 (157), *115*
Cornell, E. F. 273 (124), *302*
Cornforth, J. W. 406 (203), *420*
Cornu, G. 515 (448), 516 (448), *628*
Corpe, W. A. 726 (74), 727 (74), 728 (74, 80, 83), *740*, *741*
Correns, E. 927 (109), *945*
Corse, J. 323 (78), *339*
Corsico-Coda, A. 834 (177), *850*
Corson, B. B. 97 (258), *119*
Costain, C. C. 516 (466), 551 (682), *629*, *634*
Cota, D. J. 344 (7), 359 (7), 360 (7), *414*
Cott, W. J. 79 (86), *113*
Cotter, R. J. 99 (277), *120*
Cotton, F. A. 135 (79), *161*, 892 (28), *941*
Cottrell, T. L. 643 (18), *668*
Couch, J. F. 258 (73), *300*
Coulson, C. A. 2 (5), *63*
Courtot, C. 89 (181), *116*, 431 (53), *618*
Cousins, L. R. 516 (457), *628*
Couvreur, D. 673 (5), *710*
Cowdrey, W. A. 88 (172), *116*
Cox, E. 434 (81), *619*
Cox, R. F. B. 73 (32), *111*
Cox, S. F. 75 (55), *112*, 408 (217), *420*
Coyner, E. C. 702 (154), *714*
Crafts, J. M. 90 (189), *116*
Craig, A. D. 859 (62), *878*
Craig, C. A. 93 (213), *117*, 933 (123), *945*
Crain, C. M. 516 (465), *628*
Cram, D. J. 463 (197), *621*
Cramer, F. 264 (96), 268 (96, 109), *301*
Cramer, H. I. 320 (58), *339*
Cramer, R. 809 (110), *848*

Cramer, R. D. 703 (160), *714*, 746 (3), *784*
Craven, J. M. 810 (115), 822 (115), *848*
Criddle, R. S. 721 (23), *739*
Criegee, R. 455 (162), 458 (175), 556 (175), 557 (175), *621*
Cripps, H. N. 702 (159), 703 (159), *714*
Crisafulla, M. 841 (209), 844 (209), *851*
Cristol, S. J. 335 (114), *340*
Crompton, T. R. 187 (73), *204*
Cross, A. D. 202 (182), *207*
Cross, L. H. 196 (117), *205*
Crow, E. L. 314 (34), *338*
Cullen, E. 938, *946*
Culling, G. C. 541 (597), *632*
Cum, G. 833 (171), 834 (171), *850*
Cummins, R. A. 859 (63), *878*
Cunliffe-Jones, D. 135 (92), *161*, 196 (140), *206*
Cunningham, G. P. 141 (148), 142 (148), *162*
Curd, F. H. S. 547 (642), *633*
Curran, C. 132 (44), 135 (44), 151 (44), *160*, 196 (131), *206*
Currie, D. J. 70 (15), *110*
Curtin, D. Y. 168 (2), 169 (2), 170 (10), 171 (12), 172 (12), *203*
Curtius, T. 429 (30), 550 (676), 551 (676), *617*, *634*
Cutler, J. A. 190 (95), *205*
Cutter, H. B. 170 (8), *203*
Cymerman, J. 271 (118), *302*
Cyvin, S. J. 516 (453, 459, 460), *628*
Czekalla, J. 146 (199), *164*, 656 (39, 40), *668*
Czlesla, M. J. 388 (145), *418*

Dabard, R. 472 (227), *622*
Dachkevitch, L. B. 776 (203), *789*
Dachlauer, K. 270 (112), *301*
Dadieu, A. 531 (567), *631*
Daeniker, H. U. 842 (212), *851*
Dahl, K. 797 (51), 801 (51), 815 (51), *847*
Dahlqvist, K. 107 (344), *122*, 681 (29), *711*
Dainton, F. S. 692 (87), *713*
D'Alcontres, G. Stagno. 819 (146, 147), 833 (169, 171), 834 (171, 175, 181), 835 (184), *849*, *850*
Dallacker, F. 862 (102), *879*
Dalla Croce, P. 835 (188, 189), *850*
Daly, L. H. 242 (7), *298*
Dance, J. 455 (161), 520 (500, 501), 521 (500, 501), 522 (501), 523 (500), *621*, *629*
D'Andrea, R. 176 (25), *203*
Danielisz, M. 106 (334), *121*
Dannenburg, W. N. 720 (16), *739*
Dannhauser, W. 148 (228, 229), 149 (229), *164*, 515 (449), 516 (449), *628*
Danyluk, S. S. 127 (21), 128 (21, 32, 34), *159*, *160*
Danzig, M. 320 (64), *339*
Darmstaedter, L. 84 (135), *115*
Darnley-Gibbs, R. 718 (2), *739*
Darzens, M. G. 548 (561), *633*
Das, G. 10, 19 (25), *64*
Dasler, W. 724 (55), *740*
Datta, S. K. 796 (27), 812 (27), 815 (27), 817 (137a), 818 (137a), 833 (137a), *846*, *849*
Dauben, W. G. 751 (37), *785*
Daughhetee, P. H. 862 (98), *879*
Daves, G. W. 187 (70), *204*
Davey, W. 71 (18), 74 (40), *110*, *111*
Davidson, B. 735 (127), *742*
Davidson, D. 100 (280), *120*, 246 (28), 247 (28), *299*
Davies, D. S. 88 (172), *116*
Davies, M. 149 (238), *164*
Davies, W. 95 (238), *118*
Davies, W. H. 74 (38), 95 (38), *111*
Davis, B. A. 104 (312), *121*
Davis, G. T. 148 (226), 149 (226), *164*, 217 (24), 219 (24), 220 (24), 229 (24), *236*, 325 (86b), *339*
Davis, K. M. C. 543 (610), *632*
Davis, M. 393 (162), *419*
Davis, M. A. 100 (283), *120*
Davison, A. 502 (315, 317, 330, 332, 334, 340, 341), *625*
Davydov, B. E. 524 (512), *630*
Davydov, V. 916 (836), *944*
Davydov, V. Y. 143 (173), *163*
Daxenbichler, M. E. 733 (107), *741*
Day, A. C. 933 (119), *945*
Dean, H. G. 751 (186), 770 (186), 784 (186), *789*
Dean, J. M. 796 (29, 34, 39), 797 (34, 39, 46), 798 (34, 39), 800 (34, 39) 801 (34), 803 (29, 34), 810 (34, 39), 812 (29), 813 (29), 814 (29, 34), 815 (34, 39), 822 (29), 823 (29), 824 (29, 34, 39), 826 (29), 827 (29), 830 (29), 831 (29), 834 (29), *846*, *847*

Deasy, C. L. 761 (91), 763 (100), 766 (146), *787*, *788*
DeBenneville, P. L. 483 (273), *624*
Decker, H. J. 79 (84), *113*
DeCrescente, M. A. 356 (49), 357 (49), *416*
Dedicken, G. M. 429 (30, 31), *617*
De Diesbach, H. 84 (129), *114*
Degener, E. 503 (345), *626*
Degering, E. F. 320 (62), *339*
Degrez, M. A. 90 (188), *116*
Deichert, W. G. 289, 290 (205), *304*
Delaby, R. 80 (94), 98 (267), *113*, *119*, 886 (5), *940*
De La Mare, H. E. 697 (123), *713*
Delga, J. 735 (114), *742*
Delluva, M. 761 (88), *787*
Delmau, J. 202 (187), *207*
Del Re, G. 814 (124), *849*
Deming, P. H. 263 (90), *301*
Demoen, P. J. A. 384 (126), *418*
Derik, R. H. 333 (108), 335 (108), *340*
Dennis, W. H., Jr. 325 (86b), *339*
Deno, N. C. 126 (10, 13), 127 (13), *159*
Denyer, R. 686 (57), *712*
Deparade, W. 687 (63), 690 (63), *712*
Desgrez, A. 431 (62), *618*
Desiderato, R. 434 (823), *638*
Deskin, W. A. 144 (188), 146 (188, 204), *163*, *164*
Dessler, M. 854 (26), *878*
Dessy, R. E. 277 (154), 278 (154), 283 (154), *302*
DeToma, F. 735 (130), *742*
Deulofeu, V. 93 (207), *117*
Deutscheff, T. 925 (103c), 928, 929 (103c), 930 (103c), *944*
Develin, J. P. 448 (133), *620*
Devine, J. 291 (212), 292 (212), 293 (212), *304*
Devlin, P. 448 (126), *620*
Dewael, A. 79 (90), 85 (140), *113*, *115*
Dewald, M. 797 (43), 802 (43), *847*
Dewar, M. J. S. 223 (38), 229 (48), *237*, 550 (677), 551 (677), *634*, 642, *668*, 704 (172), *715*, 863 (114), *880*
Dewhirst, K. C. 438 (95), 483 (95), 614 (95), *619*
Deyrup, A. J. 125 (4), 126 (8), *159*
Dhar, D. N. 428 (2), *617*, 897 (39), *942*
Dibeler, V. H. 200 (176), *207*, 515 (446), 516 (446), *628*
Dick, W. 643 (18), *668*
Dickel, D. F. 917 (87f), 922 (87), 930 (87), *944*
Dicker, D. W. 291 (212), 292 (212), 293 (212), *304*
Dickerman, S. C. 88 (172), *116*, 700 (145), 701 (148), *714*
Dickinson, C. L. 438 (95), 458 (176), 467 (208), 468 (210), 474 (258), 475 (258), 478 (208), 479 (208), 483 (95, 210), 536 (208), 614 (95), *619*, *621*, *622*, *623*, 650 (31), 655 (31), *668*
Dickinson, C. L., Jr. 478 (264), *623*
Dickore, K. 356 (38), *415*
Dieck, H. tom 135 (90), 155 (90), *161*, 256 (63), *300*
Dieckmann, W. 926 (107a), *945*
Diehl, H. 78 (72), *112*
Diehl, P. 818 (140), *849*
Diehn, B. 735 (120), *742*
Diekmann, H. 77 (68), *112*
Diekmann, J. 457 (172), 532 (570), 537 (172), 552 (690), 608 (570), *621*, *631*, *634*, 646 (24), *668*, 704 (169), *715*
Diels, O. 580 (740), *635*
Diepers, W. 390 (151), *418*
Dietl, A. 861 (86), *879*
Dietrich, H. 551 (688), *634*, 706 (184), *715*, 901 (55), *943*
Dietrich, P. 555 (700), 556 (700), 557 (700), 558 (700), *634*
Dieu, C. 672 (3), 673 (3), *710*
Dillon, R. L. 227 (50), *237*, 614 (806), *637*
Dilthey, W. 358 (56), *416*
Dimroth, K. 526 (534), *630*
Dinsmore, H. L. 197 (120), *205*
Dirks, J. E. 391 (158), *418*
Disselnkötter, H. 81 (104, 105), *114*
Dittmer, D. C. 497 (297), *624*
Djerassi, C. 197 (169), 198 (169, 171), 200 (169), *207*
Dmuchowski, B. L. 759 (57), *786*
Doame, J. 737 (137), *742*
Dodge, F D. 246 (29), *299*
Doerffel, K. 197 (123), *206*
Doering, W. V. E. 451 (148), *620*
Doisy, E. A. 768 (154), *788*
Doisy, E. A., Jr. 768 (154), *788*
Dolce, T. J. 430 (50), *618*
Doleschall, G. 247 (33), *299*
Dollase, W. A. 576 (736), *635*
Domann, H. 875 (196), *882*

Dombrovskii, A. V. 310 (17), *338*
Donaruma, L. G. 886 (4), 912 (4a), *940*
Dondoni, A. 797 (42), 814 (129), 836 (196), *847*, *849*, *850*
Donoghue, E. M. 386 (134), *418*
Donohue, J. 507 (397), *627*
Donovan, G. 256 (65), *300*
Doolittle, R. E. 377 (109), *417*
Döpp, D. 772 (188), *789*
Dorfman, L. 386 (134), *418*
Dori, Z. 502 (318), *625*
Dornfeld, C. A. 82 (113), *114*
Dornow, A. 105 (316, 317), *121*, 286 (187), *303*
Dorokhov, V. A. 892 (30b), *941*
Doroshenko, Y. E. 287 (191), *304*
Dorr, D. 767 (143), *788*
Dörr, F. 448 (130, 131), *620*
Dorrer, E. 291 (208), *304*
Dostert, P. 679 (23), *711*
Douglas, D. E. 778 (208), *789*, 933 (118), *945*
Downer, J. D. 401 (189), *419*
Downing, F. B. 68 (3), 69 (3), *110*
Downing, J. R. 448 (125), 474 (257), *620*, *623*
Doyle, F. P. 92 (205), *117*
Drago, R. S. 135 (83), 136 (98), 138 (98), 139 (126, 138), 140 (98, 138), 141, 157 (98), *161*, *162*, 196 (133), *206*
Drawert, F. 396 (168), *419*
Drehmann, U. 759 (61), 771 (180), *786*, *789*
Dresdner, R. D. 297 (238), *305*, 933 (120), *945*
Dressler, R. L. 410 (233), *420*
Dreux, J. 378 (117), 412 (235), *417*, *420*
Drinkard, W. C. 139 (133), 140 (133), *162*
Drozdov, V. A. 153 (289), *166*
Druckrey, E. 935 (126), *945*
Druding, L. F. 536 (580), *631*
Dryden, J. S. 149 (236, 237), *164*
DuBois, A. S. 886 (2b), *940*
Due, M. 266 (100), 275 (100), *301*
Duennebier, F. C. 249 (41), 258 (41), *299*
Dufaux, R. 860 (72), *879*
Dufey, P. 202 (187), *207*
Duff, V. B. 78 (80), *113*
Duffield, A. M. 197 (169), 198 (169), 200 (169), *207*
Dufraisse, C. 521 (810), *637*
Dugas, H. 378 (119), *417*
Dulova, V. 142 (160), *163*
Dumas, J. 97 (252), *119*
Duncan, J. L. 105 (328), *121*, 135 (68), *160*, 283 (176), *303*
Duncan, N. E. 135 (69), *160*, 356 (50), *416*
Dunn, P. 859 (63), *878*
Dunnell, P. M. 726 (70), *740*
Dunnill, P. M. 726 (73), *740*
Duplan, J. C. 202 (187), *207*
Duquette, L. G. 380 (122), *418*
Durand, D. A. 373 (102), *417*
Durig, J. R. 516 (462), *628*
Durrell, W. S. 297 (238), *305*, 933 (120), *945*
Duxbury, F. K. 763 (98), *787*
Dworschak, H. 354 (35), *415*
Dyer, E. 187 (69), *204*

Eareckson, W. M. 320 (67), *339*
Eastham, J. F. 431 (56, 57, 58), *618*
Easton, N. R. 405 (200), *419*
Eatough, N. L. 685 (50), *712*
Eberson, L. 91 (193b, 194), 110 (364), *117*, *122*, 293 (223b, 223c), *305*, 689, (76a, 76b, 76c), 697 (121b), 710 (205, 211), *712*, *713*, *716*, 915 (81), *944*
Eby, C. J. 262 (86), 273 (127), *301*, *302*, 378 (111), *417*
Eby, J. M. 401 (187), *419*
Eccles, J. 778 (207), *789*, 933 (118), *945*
Eck, D. L. 407 (209), *420*
Eckell, A. 352 (30), *415*
Eckert, W. 508 (406), *627*
Eckmann, M. 875 (195), *882*
Edelstein, H. 277 (152), 278 (152), *302*
Edelstein, N. 502 (315, 317, 330, 332, 334, 341), *625*
Edwards, H. D. 496 (295), 608 (295), *624*
Eglinton, G. 148 (222), 158 (323), *164*, *166*
Egorov, Y. M. 748 (15), *785*
Egyed, J. 773 (190), *789*
Eholzer, U. 854 (15), 855 (15), 857 (15), 859 (58), 863 (118, 119), 870 (145), 877 (234, 236), *877*, *878*, *880*, *881*, *883*, 887 (9a), 891 (9a), *940*

Ehrenson, S. 217 (25), *236*
Ehrensvärd, G. 776 (224), *790*
Eicher, T. 277 (148), *302*
Eichler, S. 450 (142), 477 (263), *620*, *623*, 698 (128), *714*
Eigner, E. A. 746 (5), 761 (89), 762 (89), 763 (89), 771 (89), *784*, *787*
Einhorn, A. 99 (274), *120*, 295 (229), *305*
Eisele, W. 916 (85), 924 (101), 925 (85), *944*
Eisenberg, R. 502 (339, 342), *625*
Eisner, H. E. 723 (32, 33, 34, 35), *739*
Eisner, T. 722 (29), 723 (32, 33, 34, 35), *739*
Eitner, P. 246 (24, 26), 251 (47), *299*
El-Arbady, A. M. 72 (28), *111*
Elderfield, R. C. 344 (2), 345 (2), *414*
Eldjarn, L. 780 (215), *790*
Eldred, N. R. 481 (266), *623*
Elkind, V. T. 270 (117), 275 (139), *302*
Elks, J. 405 (200), *419*
Ellinger, L. P. 543 (621), *632*
Elliot, W. H. 768 (154), *788*
Elliott, I. W. 450 (817), *638*
Ellis, M. C. 384 (130, 131), *418*
Ellzey, S. E., Jr. 100 (284), *120*, 329 (99), 330 (99), *340*
Eloy, F. 272 (122), *302*, 796 (37), 797 (57), 806 (89), 808 (89), 809 (104, 105), 810 (57), 812 (119), 817 (138), 822 (57), 824 (157), 839 (104), 840 (89, 104), 841 (138), 842 (37), *847*, *848*, *849*
Elsakov, N. V. 516 (469), *629*
El Sayed, M. F. Amr 196 (151, 155, 161), *206*, *207*
Elser, W. 488 (280), *624*
Elston, C. T. 262 (87), *301*
Elvidge 366 (80), *416*
Emeléus, H. J. 151 (270), *165*
Emelus, H. J. 432 (67, 68), *618*
Emerson, M. T. 139 (134), 140 (134), *162*
Emmerling, A. 428 (18), *617*
Emmons, W. D. 73 (36), *111*, 410 (230), *420*
Emsley, J. W. 234 (66), *237*
Emslie, P. H. 528 (551), *631*
Enemark, J. H. 152 (277), *165*, 434 (84), *619*
Engelhard, H. 107 (339), *121*
Engelhardt, R. 524 (517, 518), 525 (517, 518), *630*
Engelhardt, V. A. 92 (202), *117*, 233 (64), *237*, 409 (222, 223, 224), *420*, 434 (86), 435 (86, 89), 439 (99), 446 (11b), 447 (116), 463 (192), 466 (86), 467 (89), 468 (86), 475 (259), 476 (259), 478 (259), 495 (89), 496 (89, 293, 294), 498 (294), 574 (729), 575 (729), 590 (86), 593 (86, 729), 594 (86), 595 (86, 259), 596 (86), 606 (86), 608 (86), *619*, *621*, *623*, *624*, *635*, 640 (3), *667*, 682 (31), *711*, 899 (49), 900 (49c), *942*
Engelsma, J. W. 696 (105), *713*
England, D. C. 702 (156), *714*
Enns, T. 748 (16), *785*
Entrikin, J. B. 168 (5), 169 (5), 170 (7), 171 (7), 172 (15), *203*
Epley, T. D. 136 (98), 138 (98), 140, 141 (98), 157 (98), *161*
Epstein, A. 527 (537), 528 (537), *630*
Erenberg, R. H. 724 (46), 726 (46), *740*
Erickson, J. G. 86 (150), *115*, 270 (114), *302*
Erickson, R. E. 146 (203), *164*
Erlenmeyer, E. 74 (47), 86 (151), *111*, *115*, 271 (119), *302*
Ermakova, I. V. 297 (237), *305*
Ernst, M. L. 247 (35), *299*
Erpelding, T. J. 710 (204), *716*
Errede, L. A. 516 (474), 629
Errera, G. 442 (105), *619*
Errington, W. 502 (311), *625*
Es, T. van 93 (216), *117*, 310 (18), *338*
Etienne, A. 95 (237), *118*
Etzemüller, J. 107 (343), *122*, 681 (26), *711*
Euglin, M. A. 857 (51), *878*
Eugster, C. H. 96 (249), *119*
Euler, H. v. 509 (410), *627*
Evans, H. E. 686 (52), *712*
Evans, J. C. 132 (45), 135 (45, 66), *160*, 285 (185), 286 (185), *303*
Evans, R. 388 (145), *418*
Evans, T. C. 724 (51), *740*
Evans, W. E. 753 (49), 771, *786*, *789*
Ewald, A. H. 686 (51), *712*
Exner, L. J. 75 (51), *112*, 258 (76, 77, 78), 275 (76), *300*
Eyman, D. P. 139 (138), 140 (138), *162*

Fabignon, C. 774, 781 (195), *789*
Facer, G. H. J. 147 (219), *164*
Fairfull, A. E. S. 274 (132), *302*
Fairweather, R. 258 (72), 261 (72), 263 (72), *300*
Falecki, J. 762 (105), *787*
Farenhorst, E. 696 (105), *713*
Faris, B. F. 294 (228), *305*
Farona, M. F. 135 (91), 154 (302, 305), 155 (91), *161*, *166*
Farragher, A. L. 448 (134), 511 (134), 527 (134), 531 (134), 575 (134), *620*, 648 (28), *668*
Farrell, P. G. 459 (178), 463 (178, 203), 464 (178, 203), 465 (203), *621*
Fassett, D. W. 734 (112), 736 (112), *741*
Faure-Bonvin, J. 720 (12), *739*
Feely, W. E. 76 (67), *112*
Feeney, J. 234 (66), *237*
Feigl, F. 172 (17), 173 (18), *203*
Fein, M. M. 256 (65), *300*
Feinauer, R. 390 (151), *418*
Feit, B. 914 (80b), *944*
Feldl, K. 792 (4), 794 (13), 796 (4), 813 (4), *846*
Felton, D. G. I. 196 (167), *207*, 492 (284, 285), *624*
Fensch, W. 437 (93), *619*
Fentress, J. 80 (96), *113*
Ferguson, J. W. 78 (77), *113*
Ferla, J. 431 (59), *618*
Ferramola, A. M. 771 (184), *789*
Ferrier, W. 92 (205), *117*
Ferris, A. 926 (105b), *945*
Ferris, A. F. 95 (236), *118*, 321 (71), 322 (71, 72, 73, 74, 75), *339*, 925, 927, *945*
Ferris, J. 289 (204), *304*
Ferris, J. P. 289 (202), *304*, 406 (204), *420*, 442 (106), 506 (374, 375, 376, 377, 378, 379), 508 (375), 509 (375, 376), 513 (424, 425), 518 (425), 519 (425, 493), 591 (106), 593 (106), *619*, *626*, *628*, *629*
Ferstandig, L. L. 137 (105), 139 (129, 130), 140 (130), 158 (105), *161*, *162*, 246 (29), *299*, 854 (35), *878*
Fetzer, U. 854 (15), 855 (15, 40), 857 (15, 40), 860 (73), 862 (94), 863 (117), 877 (230, 231, 232, 233, 234, 235, 236), *877*, *878*, *879*, *880*, *883*, 887 (9a), 891 (9a), *940*
Feuer, H. 862 (96), *879*
Fewson, C. A. 735 (121), *742*
Fialkoff, J. A. 573 (724), *635*
Fick, R. (821), *638*
Fickling, M. M. 218 (29a), *236*
Field, G. F. 771 (177), *789*
Fields, M. 768 (159), *788*
Fierce, W. L. 428 (13), *617*
Fierens, P. J. C. 103 (300), *120*
Fieser, L. F. 320 (55), *338*
Fiesselmann, H. 92 (199), *117*
Figdor, S. K. 776 (201), 783 (201), *789*
Figeys H. 196 (166), *207*
Figeys, H. P. 196 (165), *207*
Figeys-Fauconnier, M. 196 (166), *207*
Filatov, A. S. 869 (138), *880*
Filimonov, W. 135 (76), *161*
Finch, A. C. M. 541 (590), *632*
Findlay, J. A. 93 (218), *117*
Finestone, A. B. 75 (53), 76 (61), 110 (53), *112*, 687 (61, 64), 688 (64), *712*
Finger, H. 547 (641), *633*
Finholt, A. E. 326 (93), *340*
Finken, H. 283 (175), *303*, 512 (412), 596 (412), *627*
Finnegan, W. G. 388 (140), *418*
Finseth, G. A. 325 (866), *339*
Finzi, P. V. 862 (100), *879*, 931 (115), *945*
Firestone, R. A. 345 (12), 351 (12), *415*
Fischer, A. 218 (29a), *236*
Fischer, A. G. 770 (173), *789*
Fischer, E. 429 (34), *618*, 895 (37), *942*
Fischer, E. O. 873 (176), 874 (185), 876 (222), *881*, *882*
Fischer, H. 96 (245), *118*, 782 (220), *790*, 873 (176), *881*
Fischer, H. E. 95 (242), *118*
Fischer, H. O. L. 886 (3), *940*
Fischer, H. P. 917, 918 (87c), 921 (95), 922 (87), 930 (87), *944*
Fischer, P. H. H. 535 (574), *631*, 658 (50), *668*
Fischer, W. 91 (192), *117*
Fish, R. W. 472 (225), *622*
Fisher, B. S. 409 (223), *420*, 439 (99), *619*
Fisher, J. R. 428 (20), *617*
Fitch, J. W., III 473 (231), *622*
Fitton, P. 460 (184), *621*
Fitzgerald, W. E. 135 (72), 156 (72), *160*, 356 (46), *415*

Fitzpatrick, M. 399 (177), *419*
Fleischmann, R. 352 (30), *415*
Flemming, H. 503 (346), *626*
Fletcher, F. J. 859 (67), *879*, 891 (19d), *941*
Flett, M. St. C. 138 (121), 142 (168), *162*, *163*, 196 (119, 147), *205*, *206*
Fleury, J. P. 553 (693, 694), *634*, 899, *942*
Florian, W. 106 (336), *121*
Flory, K. 797 (44), 817 (137a), 818 (137a), 833 (137a), *847*, *849*
Floss, H. G. 725 (64, 66), 726 (66), *740*
Flueckinger, A. F. 148 (228, 229), 149 (229), *164*, 515 (449), 516 (449), *628*
Fluery, J. P. 406 (205), *420*
Fock, V. 8 (10), *63*
Fock, W. 150 (243), *165*
Foffani, A. 196 (139), *206*
Föhlisch, B. 859 (56, 57), *878*
Folt, V. L. 699 (142), *714*
Foltz, E. 775 (196), *789*
Fomin, G. V. 472 (217), *622*
Fontaine, A. 434 (81), *619*
Ford, M. C. 696 (118, 119), *713*
Forman, E. O. 78 (80), *113*
Forrester, J. D. 502 (329, 343), *625*, *626*
Forster, M. O. 93 (208), *117*
Fort, A. W. 772 (189), *789*
Foster, H. 895 (36c), 896 (38), 901 (36), *942*
Foster, R. 463 (194), 466 (194), 525 (527, 528), 528 (551), 541 (528, 591, 592), 543 (528, 591, 592, 602, 606, 609), *621*, *630*, *631*, *632*, 654 (36), 656 (36), *668*
Fowden, L. 719 (6), 720 (6), 724 (41), 726 (69, 70, 73), 727 (6), *739*, *740*
Fowler, F. 345 (10), *414*
Fox, D. W. 769 (172), *789*
Fox, I. R. 148 (226), 149 (226), *164*, 217 (24), 219 (24), 220 (24), 229 (24), *236*
Fox, J. R. 694 (94), *713*
Foye, W. D. 838 (204), *851*
Fraenkel, G. K. 183 (62), 184 (62), 190 (62), *204*, 229 (56, 57), *237*, 297 (241), *305*, 472 (220), 473 (220, 235, 236), 527 (541), 528 (236, 541), *622*, *630*, 658 (49), *668*
Frainnet, E. 100 (282), *120*
Frank, L. 765 (112), *787*
Frank, R. L. 73 (36), *111*
Frank, S. 507 (394), *627*
Frank, W. 796 (20), 819 (20), *846*
Frankel, M. B. 437 (819), *638*
Frankevich, E. L. 474 (253), 529 (563), *622*, *631*
Frankiss, S. G. 892 (24), *941*
Frankland, E. 283 (174), 285 (185), 286 (185), *303*
Franklin, J. L. 146 (197), *163*, 200 (176), *207*, 513 (423), 515 (446), 516 (446), *628*, 854 (33a), *878*
Franklin, T. C. 189 (84), *205*
Franklin, W. E. 760 (114), *787*
Franklin, W. J. 862 (105), *879*
Franz, P. 835 (191b), *850*
Franzen, V. 72 (27), *111*, 697 (127), *714*
Fraser, R. R. 211 (9), *236*
Frati, E. 835 (190), *850*
Frazza, E. J. 503 (354), *626*
Freed, J. H. 473 (235), *622*
Freeman, F. 106 (330), *121*, 428 (8), 442 (110), 445 (110, 114), 590 (8), *617*, *619*
Freeman, J. P. 73 (36), *111*
Freeman, R. C. 897 (40b), *942*
Freidlin, L. Kh. 325 (84a), *339*
Freidlina, R. K. 420 (499), *629*
Freifelder, M. 320, 321 (70), 323 (81), *339*
French, J. C. 897 (40a), *942*
Freni, M. 873 (174), 875 (203, 208), *881*, *882*
Frenkel, J. 10 (18), *63*
Freudenberg, K. 768 (223), *790*
Freund, M. 875 (192), *882*
Freure, B. T. 79 (84), *113*
Fricke, R. 150 (256), *165*
Friedel, C. 90 (189), *116*
Friedländer, L. 75 (48), *111*
Friedman, L. 79 (82), 83 (120), *113*, *114*, 765 (120), 770, *787*, *789*, 314 (36), 332 (36), *338*
Friedman, L. B. 152 (277), *165*
Friedman, O. M. 94 (226), *118*
Friedrich, D. 191 (97), *205*
Friedrich, K. 83 (122), 84 (130), 92 (195, 206), 98 (262), 106 (329), *114*, *117*, *119*, *121*, 231 (61, 61a), *237*, 525 (532), 526 (533, 535), 527 (532), 529 (532), 530 (532), 531 (532), 549 (673), 550 (673), *630*, *634*
Friedrich, K. R. 73 (31), 98 (31), *111*, 599 (799), 600 (799), *637*

Friedrich, R. 531 (566), *631*
Frisch, D. M. 726 (70), *740*
Fritchie, C. J., Jr. 502 (338), *625*, 661 (60, 63), *669*
Fritsch, J. M. 506 (380), 551 (380), *626*, 706 (183), *715*, 901 (53a, 53b), 902 (53), 903 (53), *942*
Fritzon, P. 780 (215), *790*
Fritzsche, H. 136 (97), 137 (97), 138, 156 (97), *161*, 459 (179), 464 (179), 465 (179), 476 (260, 251), 477 (179, 261), *621*, *623*
Frobel, E. 88 (179), *116*, 590 (783), *637*
Frohberger, P. E. 877 (230, 231, 232), *883*
Fröman, A. 39 (52), *65*
Fromm, H. J. 771 (185), *789*
Frommeld, H.-D. 796 (29), 803 (29), 811 (116), 812 (29), 813 (29), 814 (29, 130), 815 (116, 131), 817 (137a), 818 (137a), 821 (130), 822 (29, 116, 131), 823 (29, 116), 824 (29, 116), 825 (116), 826 (29), 827 (29, 116), 828 (116), 829 (116), 830 (29), 831 (29, 116), 833 (137a), 834 (29), *846*, *848*, *849*
Frush, H. L. 757 (51), 758 (51), 759 (51, 55, 56, 66, 69), *786*
Fry, A. 748 (23), *785*
Fugate, W. O. 503 (354), *626*
Fuhler, W. F. 606 (801), 607 (801), *637*
Fujimoto, S. 797 (55), 810 (55), 822 (55), 834 (55), 841 (55), *847*
Fujita, D. K. 139 (127), *162*
Fujiyama, T. 135 (73, 74), 156 (73), *160*
Fuks, R. 837 (200), *850*
Fukunaga, T. 100 (283), *120*, 294 (224), *305*
Fung, B. M. 133 (53), *160*
Funke, E. 353 (32), *415*
Funke, K. 531 (567), *631*
Furishiro, R. 141 (146), *162*
Fusco, R. 405 (198), *419*, 832 (166), *850*
Fuson, R. C. 83 (117), *114*, 168 (2), 169 (2), 170 (10), 171 (12), 172 (12), *203*
Fyfe, C. A. 525 (528), 541 (528), 543 (528), *630*

Gabaglio, M. 874 (181), *881*
Gabriel, S. 793 (6), 815 (6), *846*
Gadzhiev, A. Z. 132 (48), *160*, 196 (129), *206*
Gagarin, R. F. 682 (35), *711*
Gagnon, P. E. 101 (286), *120*
Gaiffe, A. 312 (29, 30, 31), 313, *338*
Gal, G. 388 (146), *418*
Galais, F. 516 (477), *629*
Galat, A. 258 (74), *300*
Galata, L. A. 142 (157), *162*
Galkowski, T. T. 757 (51), 758 (51), 759 (51), *786*
Gallais, F. 892 (27), *941*
Galli, R. 697 (125, 126), *713*, 844 (219), *851*
Gallo, G. 196 (158), *206*
Gambarjan, N. P. 854 (16), 865 (16), 869 (143, 144), 870 (144), *877*, *880*, *881*
Gambaryan, N. P. 582 (762), *636*, 701 (151), *714*
Gambelli, G. 180 (42), *204*
Gander, J. E. 720 (9), *739*
Gantzel, P. K. 596 (797), *637*
Gardner, J. 405 (200), *419*
Garg, C. P. 317 (49, 50), 318 (49, 50), 329 (49, 50), 332 (49, 50), *338*
Garmaise, D. L. 389 (148), *418*
Garst, J. F. 335 (113), *340*
Gärtner, H. 580 (740), *635*
Gassman, P. G. 109 (355), *122*, 522 (504), *629*, 704 (170), *715*
Gasson, E. J. 258 (75), *300*
Gattermann, L. 88 (167), *116*
Gaudechoz, H. 428 (16), *617*
Gaudemaris, M. de 192 (101), *205*
Gaudiano, G. 803 (76), 812 (76), 830 (76), 842 (217), 845 (220, 221, 222), *847*, *851*
Gaudry, R. 760 (83), *786*
Gaugler, R. W. 126 (10), *159*
Gault, H. 927 (107d), *945*
Gault, R. 293 (223a), *305*, 362 (71), *416*
Gauthier, R. 672 (3), 673 (3), *710*
Gautier, A. 72 (25), *111*, 246 (27), *299*, 853, 854 (1), 855, 856 (1), *877*, 886 (7a, 7e), 892 (7a), *940*
Gaylord, N. C. 314 (37), 326 (37), *338*
Gay-Lussac, L. J. 430 (44), *618*
Ge, Ban-Lun 325 (91), *339*
Gebhardt, B. 503 (349), *626*
Geiderikh, M. A. 196 (156), *206*, 524 (512), *630*

Geller, S. 147 (211), *164*
Gendell, J. 473 (235), *622*
Genderen, H. van 738 (139, 140), *742*
Gennaro, A. R. 755 (80), *786*
Gensler, W. J. 769 (166), *788*
Gentil, V. 173 (18), *203*
Geodakyan, K. T. 143 (174), *163*, 196 (125), *206*
George, W. O. 448 (132), 507 (395), 508 (395), *620*, *627*
Gerber, G. B. 768 (158), *788*
Gerhardt, C. 97 (257), *119*
Gerloch, M. 502 (319), *625*
Gerö, M. 95 (242), *118*
Gerrard, W. 150 (264), 151 (264), *165*, 216 (17), *236*
Geske, D. H. 133 (60), *160*
Geuskens, G. 196 (166), *207*
Gewald, K. 393 (161), 408 (219), *419*, *420*, 498 (304), *624*
Geyer, R. 540 (586), 541 (586), *632*
Ghanem, N. A. 692 (88), *713*
Ghersetti, S. 797 (42), 814 (129), 836 (196), *847*, *849*, *850*
Ghielmi, S. 874 (178), *881*
Gholson, R. K. 731 (91), *741*
Giauque, W. F. 142 (154), *162*
Gibb, T. B., Jr. 75 (53), 110 (53), *112*, 690 (79, 80), *712*
Gibbs, M. 735 (121), *742*
Gibson, D. T. 148 (222), *164*
Gibson, G. K. J. 152 (282), *165*
Gilbert, E. E. 274 (133), *302*
Gilbert, H. 699 (141, 142), *714*
Gilbert, T. L. 10 (20), 39 (20), *64*
Gilderson, P. W. 891 (19c), *941*
Giles, C. H. 142 (151), *162*
Gillaspie, A. G. 93 (222), *118*
Gillis, R. G. 854 (22, 25, 30), *877*, *878*
Gillois, M. 221 (37), *237*
Gillois-Doucet, J. 109 (361), *122*
Gilman, H. 282 (165), *303*, 862 (93), *879*
Gilson, G. R. 194 (109), *205*
Gilson, T. 135 (82), 151 (82), *161*
Ginaine, M. Benedicta 187 (74), *204*
Gingras, B. 696 (106), *713*
Gingras, B. A. 696 (103), *713*
Ginsburg, V. A. 869 (138), *880*
Giraud-Clenet, D. 292 (217), *304*, 922 (98), *944*
Gittos, M. W. 400 (180), *419*
Giuffré, L. 875 (200), *882*
Giza, Y.-H. 724 (43, 45), 725 (43, 45), 726 (43), *740*
Glemser, O. 92 (202), *117*, 297 (238), *305*, 581 (746), *635*
Glenz, K. 888 (14), *940*
Glew, D. N. 144 (183), 146 (183), *163*
Gloede, J. 100 (278), *120*
Glušenkova, V. R. 861 (84), *879*
Glushkova, O. A. 195 (115), *205*
Glutz, L. 430 (40), *618*
Gmelin, E. 818 (143), *849*
Gnad, J. 103 (299), *120*
Goates, J. R. 150 (246), *165*
Godfrey, J. J. 378 (114), *417*
Godfrey, M. 520 (501), 521 (501), 522 (501), *629*
Godon, F. de 93 (219, 224), *118*
Goerdeler, J. 276 (142), *302*, 870 (148), *881*
Goese, M. A. 84 (136), *115*
Goetz, H. 451 (147), *620*
Goldberg, A. A. 106 (337), *121*
Goldberg, M. W. 769 (167), *788*
Goldstein, H. 99 (271), *119*
Goldstone, J. 40 (54), *65*
Golodova, K. 86 (152), *115*
Golovaneva, A. F. 869 (138), *880*
Golovin, A. V. 378 (112), *417*
Golton, W. C. 144 (189), 145 (189), 157 (189), 158 (189), *163*
Gombas, P. 31 (36), *64*
Gomez-Parra, V. 361 (62), 362 (62), 364 (62), *416*
Gompper, R. 483 (275), 488 (280), 497 (296), 498 (275, 299, 300), *624*
Gordon, J. E. 129 (39), 130 (41), *160*, 250 (44), *299*
Gordon, M. 81 (102), *114*
Gordon, S. A. 732 (99), *741*
Gordon-Gray, C. G. 755 (79), *786*
Gordy, W. 138 (115, 116), *162*, 854 (26, 27), *878*
Göthlich, L. 91 (193b), *117*, 710 (204), *716*
Gotthardt, H. 353 (32), *415*
Gould, C. W. 170 (11), *203*
Gould, E. S. 77 (70), 88 (173), *112*, *116*
Gould, F. E. 321 (71), 322 (71, 72, 73, 74, 75), *339*, 927 (110), *945*
Gould, R. G. 771 (178), *789*
Gracian, D. 361 (62), 362 (62), 364 (62), *416*
Gradsten, M. A. 295 (230), *305*
Graef, E. A. 446 (115), *619*

Graf, R. 102 (297), 103 (298), *120*, 710 (203), *716*
Graham, P. J. 293 (223a), *305*, 560 (711), *635*
Gram, W. H. 345 (11), *414*
Gramas, J. V. 705 (174), *715*
Graminski, E. L. 428 (21), *617*
Gramstad, T. 156 (317), *166*
Gränacher, C. 95 (242), *118*
Gränacher, I. 139 (144), *162*
Grandberg, I. I. 886 (4), 910 (4c), *940*
Grandon, M. F. 387 (136), *418*
Granick, S. 659 (53), *669*
Gränse, S. 689 (76c), *712*
Grashey, R. 348 (20), 352 (31), 356 (40), *415*
Grashy, R. 345 (12), 351 (12), *415*
Grasselli, J. G. 154 (302, 305), *166*, 255 (60), 256 (62), *300*
Grasso, L. 834 (176), *850*
Gravenor, R. B. 434 (87), *619*
Gray, H. B. 448 (129), 502 (318, 322, 324, 331, 333, 336, 337, 342, 344), *620*, *625*, *626*
Gray, L. S. 180 (41), *204*
Graybill, B. M. 152 (275), *165*
Greaves, P. M. 85 (141), *115*
Grebber, K. K. 293 (221), *305*
Grebenyuk, A. D. 242 (6), *298*
Green, A. I. 926 (105f), *945*
Green, H. D. 506 (360), *626*
Green, J. 256 (65), *300*
Green, J. H. S. 135 (71), *160*
Green, M. 488 (281), *624*
Greene, H. E. 699 (143), *714*
Greene, J. M. 372 (96), *417*
Greenlee, J. W. 755 (77), *786*
Greenwood, H. H. 229 (48), *237*
Greenwood, N. N. 150 (248, 249), *165*, 517 (485), *629*
Greenzaid, P. 463 (201), 466 (201), *621*
Gregory, B. J. 922 (96), *944*
Greinacher, E. 138 (109), 139 (109), *161*
Grenda, V. J. 388 (146), *418*
Gresham, T. L. 766 (148), *788*
Gresham, W. 291 (210), *304*
Grewe, F. 877 (231, 235), *883*
Grewe, R. 85 (144), *115*
Griesinger, A. 865 (132), 867 (132), *880*
Griffin, C. E. 81 (102), *114*
Griffin, G. W. 108 (351), *122*, 200 (172), *207*, 442 (103), 551 (688), *619*, *634*, 706 (184), *715*, 901 (55), *943*
Griffith, A. M. 747 (7), *785*
Griffith, N. E. 135 (75), *161*, 194, 196 (111), *205*
Griffith, W. P. 520 (498), *629*
Griffiths, M. H. 738 (138), *742*
Grigat, E. 266 (101), 275 (135), 276 (135, 144), 286 (101), *301*, *302*
Grigg, R. 200 (177), *207*
Grignard, V. 89 (181), *116*, 431 (53), *618*
Grigsby, W. 291 (210), *304*
Grill, W. 128 (27), 129 (27), 130 (27), *159*, 251 (45), *299*
Grindahl, G. A. 410 (231), *420*
Grinstein, M. 771 (184), *789*
Grisbach, H. 770, 781 (176), *789*, 856 (44), *878*
Grivas, J. C. 270 (112), *301*
Grob, C. A. 96 (245), *118*, 910 (69a, 69b, 69c, 69d, 70b), 912 (69, 70), 917, 918 (87c), 921, 922 (87, 87d), 924, 930 (87), 931, *943*, *944*
Grob, C. A. 916, 925 (85), *944*
Grob, W. 898 (44), *942*
Grolée, A. 74 (37), *111*
Gross, D. 768 (157), *788*
Gross, H. 100 (278), *120*
Gross, J. M. 473 (233), *622*
Gross, S. T. 170 (11), *203*
Grosse, A. V. 514 (440), 515 (475), 516 (440, 475), *628*, *629*
Grosso, R. P. 813 (121), *848*
Grotowsky, H. 925 (103a), *944*
Grout, R. J. 751 (186), 770 (186), 784 (186), *789*
Grünanger, A. 834 (176), *850*
Grünanger, P. 796 (23, 35), 806 (23), 808 (23), 809 (103), 815 (35, 132b), 822 (155a), 833 (168, 169), 834 (137b, 174, 177, 178, 179), 835 (184, 185), 836 (195, 197), 837 (198, 199), *846*, *848*, *849*, *850*
Grunanger, P. 930 (114a), *945*
Grundmann, C. 375 (103), 410 (229), *417*, *420*, 796 (25, 27, 28, 29, 34, 39, 40), 797 (34, 39, 40, 44, 45, 46), 798 (34, 39, 40, 45), 800 (34, 39), 801 (25, 34, 40), 802 (25), 803 (25, 29, 34), 805 (25), 807 (91, 98), 808 (25), 810 (34, 39, 40), 811 (98, 116), 812 (25, 27, 28, 29,

45, 98, 118), 813 (25, 29), 814 (25, 29, 34, 130), 815 (25, 27, 34, 39, 40, 45, 116, 131, 132b), 816 (136b), 817 (137a), 818 (91, 137a), 821 (130), 822 (28, 29, 116, 118, 131), 823 (29, 91, 98, 116), 824 (28, 29, 34, 39, 45, 116), 825 (116), 826 (29), 827 (29, 116), 828 (116), 829 (91, 116), 830 (29, 91, 98), 831 (29, 116), 833 (25, 137a), 834 (29, 180), 839 (91), 840 (25, 91, 98, 207), 841 (98), *846*, *847*, *848*, *849*, *850*, *851*, 871 (155), *881*, 891 (22), *941*
Grundnes, J. 135 (70), 156 (70), *160*
Grundon, M. F. 886 (4), 910 (4d), *940*
Grunfeld, M. 190 (93), *205*
Grunwald, E. 157 (319), *166*
Gryszkiewicz-Trochimowski, E. 97 (254), *119*, 506 (382), 507 (385), 508 (382, 385, 403), 572 (403, 722, 723), 573 (722, 723), *627*, *635*
Gryszkiewicz-Trochimowski, O. 97 (254), *119*
Gualberto, G. G. 196 (148), *206*
Guarino, J. P. 472 (221), *622*
Gudmundson, A. G. 95 (243), *118*
Guermont, J. P. 760 (84, 86), 765 (113), *786*, *787*
Guerra, A. 180 (45), *204*
Guetté, J. P. 809 (101, 102a), *848*
Guibé, L. 242 (2), *298*
Guillemard, H. 873 (173), *881*
Guillemard, M. H. 853 (3), 854 (3), *877*, 886 (7b, 7c, 7f, 7g, 7h), *940*
Guinn, V. P. 766 (145), *788*
Gulden, W. 870 (146), *881*
Guldon, W. 872 (166), *881*
Gulewitch, W. 320 (59), *339*
Gulewitsch, W. 75 (49), *112*
Gundermann, K.-D. 575 (730), *635*, 703 (163, 164, 165), *715*
Günther, F. 540 (584), 541 (584), 544 (584), *632*
Günther, H. 834 (173b), *850*
Günther, P. 458 (175), 556 (175), 557 (175), *621*
Günther, W. 796 (24), 803 (24), 814 (24), 824 (24), 828 (24), 834 (24), 835 (24), *846*
Gurin, S. 753(49), 761 (88), 771, 782 (221), *786*, *787*, *789*, *790*
Guseinov, M. G. 525 (526), *630*
Guth, V. 309 (14), *337*
Gutmann, F. 662 (66), 665 (66), *669*
Guttenberger, J. F. 433 (76), *619*
Gutzke, M. E. 769 (172), *789*
Guyot, A. 582 (761), *636*

Haagen-Smit, A. J. 761 (91), 763 (100), 766 (146), *787*, *788*
Hadiniger, L. A. 731 (91), *741*
Hadley, D. J. 258 (75), *300*, 525 (524), *630*
Hadwiger, L. 725 (64), *740*
Hadzi, D. 156 (316), *166*, 242 (8), *298*
Haeseler, H. 297 (238), *305*
Hafelinger, G. 516 (479), *629*
Häfliger, O. 218 (26), *236*
Hafner, K. 106 (334), 451 (144), 454 (160), 463 (196), *620*, *621*
Hagedorn, I. 855 (37), 856, 857 (37), 859 (58), 863 (118, 119), *878*, *880*
Hagenest, H. 88 (166), *116*
Hager, H. 103 (299), *120*
Haggart, C. 101 (286), *120*
Haggett, E. 87 (161), *115*
Hahlbrock, K. 719 (8), 720 (8), *739*
Hahn, H. 500 (310), *625*
Hahn, R. C. 698 (130), *714*
Haigh, P. J. 242 (2), *298*
Haines, R. M. 691 (84), 697 (84), *712*
Haisman, D. R. 721 (22b), *739*
Haken, P. ten 83 (123), *114*
Haldna, Ü. L. 125 (5), 127 (5), 156 (5), *159*
Hale, C. H. 186 (67), *204*
Hale, J. D. 614 (807), *637*
Hale, W. F. 76 (61), *112*, 687 (61), *712*
Hall, C. D. 463 (198), 469 (198), 480 (198), *621*
Hall, D. W. 78 (79), *113*
Hall, F. R. 663 (67), *669*
Hall, J. H. 104 (311), *121*, 351 (27), 388 (27), *415*, 904 (59), *943*
Hall, N. F. 748 (20), 751 (40), *785*
Hallam, H. E. 138 (113, 114), *161*
Haller, A. 88 (175, 177), *116*
Halleux, A. 862 (106), *880*
Hallum, J. V. 926 (105g), *945*
Halmann, M. 772 (187), 781 (217), *789*, *790*
Halper, J. 448 (129), *620*
Halton, B. 456 (167), *621*
Ham, G. E. 698 (136), *714*
Ham, J. S. 146 (198), *164*, 641 (9), *668*

Hamann, K. 401 (185), *419*
Hamelin, J. 202 (185), *207*
Hamill, W. H. 472 (221), *622*
Hamilton, C. E. 694 (98), *713*
Hamilton, R. W. 78 (74), *113*, 547 (643), *633*
Hamilton, W. C. 502 (326), *625*
Hammann, I. 877 (234, 236), *883*
Hammett, L. P. 125, 126 (8), *159*, 217 (23d), *236*, 257, *300*
Hammick, D. L. 853 (6), 854 (6), 865 (6), *877*
Hammond, G. S. 298 (242), *305*, 672 (1), 688 (68a), 689 (73, 74), 692 (68a, 89), 693 (68a, 73, 90, 91, 93), 694 (68a, 94, 98), 695 (102), 696 (73), *710*, *712*, *713*, 897 (41a, 41b), *942*
Hammond, P. R. 541 (589), 543 (589), *632*, 650 (30), 654 (30), 655 (30), 656 (30), 660 (30), *668*
Hamner, W. F. 187 (70), *204*
Hankes, L. V. 728 (80, 83), *741*, 770 (173), *789*
Hanley, W. E. 129 (36), *160*
Hanna, J. G. 173, 174 (20), *203*
Hanna, M. W. 139 (135, 142, 143), 140 (135, 142, 143), 156, *162*
Hannan, R. B. Jr. 516 (456, 457, 464, 471), 517 (471), *628*, *629*
Hansen, A. v. 251 (48), *299*
Hansen-Nygaard, L. 854 (28), *878*
Hanson, A. W. 661 (61), *669*
Hanson, P. 463 (194), 466 (194), 541 (592), 543 (592), *621*, *632*
Hanstein, W. 77 (68), *112*, 573 (727, 728), 574 (728), *635*, 650 (34), *668*
Hantzoch, A. 388 (137), *418*
Hantzsch, A. 88 (178), *116*, 126, 128 (26), *159*, 434 (82), 435 (82), *619*
Happel, J. 513 (422), *627*
Hara, K. 288 (196), *304*
Harder, R. J. 229 (55), *237*, 532 (90), 534 (573), 535 (573), *619*, *631*, 640 (6a), 642 (6), 644 (6a), 646 (6), 655 (6), 658 (6), 659 (6), 660 (6, 57), 661 (6, 57), 664 (6, 57), 665 (6), *667*, *669*, 895 (36a, 36d), 901 (36), *941*, *942*
Harder, U. 373 (101), *417*
Harper, S. H. 550 (678), 551 (678), *634*, 704 (171), *715*
Harris, B. W. 674 (11), *711*
Harris, J. E. 446 (118), *620*
Harris, R. K. 818 (141), *849*
Harris, T. M. 779 (213), *790*
Harrison, G. C. 78 (72), *112*
Harrison, I. T. 202 (182), *207*
Harryvan, E. 899 (47c), *942*
Hart, C. V. 548 (652), *633*
Hart, H. 106 (330), 107 (345), *121*, *122*, 442 (110), 443 (111), 445 (110, 114), *619*
Hartenstein, A. 354 (36), 355 (36), *415*
Hartke, K. S. 401 (184), *419*
Hartley, E. G. J. 873 (169), 875 (193, 194), *881*, *882*
Hartman, H. 33 (45), *64*
Hartmann, H. 10, 11 (14), *63*
Hartree, D. R. 10, 39 (19), *63*
Hartree, W. 10, 39 (19), *63*
Hartshorn, M. P. 508 (404), *627*
Hartung, W. H. 76 (63), *112*, 320 (63), *339*
Hartzel, L. W. 291 (211), 292 (211), *304*, 913 (77b), 914 (77b), *944*
Hartzler, H. D. 462 (189), *621*, 537 (581), 538 (582), 541 (596), 553 (695), 590 (785), 593 (790), 614 (695), *631*, *632*, *634*, *637*, 645 (26, 27), 646 (26, 27), 658 (26, 27), *668*, 683 (41), 684 (41), 704 (41, 168), 707 (186), *711*, *715*, 895 (36f), 901 (36), *942*
Harwood, H. J. 776 (204), *789*
Hasbrouck, R. H. 323 (81), *339*
Hassel, O. 147 (209, 214), *164*, 211 (3), *235*, 515 (472), 516 (472), *629*
Hasselquist, H. 509 (410), *627*
Hassner, A. 293 (223a), *305*, 345 (10), 362, *414*, *416*, 919 (92), *944*
Hastings, A. B. 771 (178), *789*
Hastings, S. H. 146 (197), *163*
Hata, K. 83 (119), *114*, 325 (86a), *339*
Hata, T. 93 (211), *117*
Hatada, K. 516 (451), *628*
Hatano, M. 474 (250), 543 (619), *623*, *632*
Hatchard, W. R. 393 (160), *419*, 498 (301, 303), *624*
Hatchings, G. H. 390 (149), *418*
Hathaway, B. J. 132 (49, 50), *160*
Hathway, D. E. 738 (138), *742*
Hatton, J. V. 150 (240), *164*
Haurie, M. 138 (112, 124), *161*, *162*
Hauser, C. R. 75 (55), 87 (160), 93 (222), *112*, *115*, *118*, 127 (17), 131 (17),

150 (253), *159*, *165*, 251 (58), 255 (58), 262 (85, 86), 273 (127), 278 (157), 280 (157), 283 (157, 172, 173), 285 (183), *300*, *301*, *302*, *303*, 314 (32), 325 (88), *338*, *339*, 378 (111), *417*, 779 (213), *790*, 910 (68), *943*
Haüser, H. 108 (350), *122*
Hauser, W. 875 (195), *882*
Häusser, V. 92 (202), *117*, 581 (746), *635*
Haussmann, H. 796 (20), 819 (20), *846*
Hautefeuille, P. 428 (15), *617*
Havens, H. R. 103 (302), *120*
Hawkins, W. 175 (22), *203*
Hawkins, W. L. 323 (77), *339*
Hawthorne, M. F. 151, 152 (275), *165*
Hay, J. N. 288 (197), *304*
Hayaishi, O. 735 (123), *742*
Hayashi, H. 528 (549), *631*
Hayashi, T. 83 (119), *114*, 325 (86a), *339*
Hayaski, S. 448 (128), *620*
Hayter, R. G. 153 (287), *166*
Hazenberg, M. E. 378 (119), *417*
Heard, R. D. 750, 780 (214), *785*, *790*
Hechenbleikner, I. 244 (14), 266 (14), 270 (112), *298*, *301*
Hecht, R. 698 (129), *714*
Heck, G. 552 (691), *634*
Heckert, L. C. 862 (93), *879*
Heckert, R. E. 92 (202), 109 (358), *117*, *122*, 409 (222), *420*, 439 (96, 97), 446 (116), 447 (116), 449 (97), 457 (97), 459 (97), 463 (190, 191), 464 (191), 465 (190), 468 (211), 494 (290), 614 (97), *619*, *621*, *622*, *624*, 640 (3), 657 (42), *667*, *668*, 682 (31), 703 (166), *711*, *715*
Hedrick, J. L. 134 (63), *160*
Heess, R. 196 (142), *206*
Hegenaur, R. 718 (1), *739*
Hegnauer, R. 732 (104), *741*
Heiart, R. B. 147 (210), *164*
Heilbron, I. 408, *420*
Heilbron, I. M. 401 (189), 408 (212–214, 216, 218), *419*, *420*
Heilmann, R. 192, 193 (102), *205*
Heim, P. 448 (130, 131), *620*
Heinle, R. 873 (172), 875 (172), *881*
Helberger, J. H. 83 (119), *114*
Held, A. 88 (177), *116*
Heldt, W. Z. 875 (199), *882*, 886 (4), 912 (4a), *940*
Helferich, B. 540 (585), 541 (595), *632*
Helkamp, G. K. 370 (92, 93), *417*
Helle, K. 746 (1), 747 (1), 748 (1), 749 (1), *784*
Heller, H. J. 699 (142), *714*
Hellman, H. 87 (159), *115*
Hellmann, H. 390 (151), *418*, 775, *789*
Helmkamp, R. W. 747 (7), 748 (22), *785*
Hems, B. A. 405 (200), *419*
Henbest, H. B. 85 (146), *115*, 732 (96, 97), 733 (97), *741*
Henbest, J. 75 (53), 110 (53), *112*
Hende, J. H. van den 451 (145), *620*
Henderson, L. M. 730 (89), *741*
Henderson, R. 463 (199), 521 (199), *621*
Henderson, R. M. 219 (30), 220 (30), 222 (30), 223 (30b), 229 (30), *236*, 650 (32), 655 (32), *668*, 906 (62), *943*
Henderson, W. A. 909 (66b), *943*
Henderson, W. A., Jr. 452 (152, 153), *620*, 909 (66a), *943*
Hendricks, S. B. 142 (166), *163*, 793 (7), *846*
Hendricks, S. E. 892 (27), *941*
Hendrickson, H. R. 725 (67), *740*
Hendriks, H. 746 (1), 747 (1), 748 (1), 749 (1), *784*
Hendry, C. M. 247 (32), *299*
Hendry, J. A. 547 (642), *633*
Henecka, H. 256 (67), 262 (67), 263 (67), *300*, 801 (68), 802 (68b), *847*
Henn, B. R. 98 (264), *119*, 524 (520), 525 (520), 576 (520), *630*, 654 (37), *668*
Henne, A. L. 315 (39), *338*
Henneberry, G. O. 748 (19), 761 (92), *785*, *787*
Henri, V. 190 (92), *205*
Henry, R. A. 388 (140), *418*
Henze, H. R. 78 (80), *113*, 281 (161), *303*
Henzi, R. 874 (177), *881*
Hepp, E. 294 (226), *305*
Herberhold, M. 874 (185), *882*
Herbert, J. 687 (64), 688 (64), *712*
Herbert, M. 749 (28), 750 (28), 751 (35), 764 (28, 109, 111), 774, 781 (28, 109, 195), *785*, *787*, *789*

Herbison-Evans, D. 242 (3), *298*
Herbst, R. M. 388 (138, 139), *418*
Herlinger, H. 868 (136), *880*
Hermes, M. D. 903 (56a), *943*
Hermes, M. E. 547 (649), 548 (649, 670), 549 (649, 670, 672), 550 (674), *633*, *634*
Herron, J. T. 448 (135), 513 (423), *620*, *628*
Hershey, J. W. B. 98 (269), *119*
Hertler, W. R. 224 (41), 229 (55), *237*, 532 (90, 568, 570), 534 (573), 535 (573, 575), 536 (568), 538 (582), 608 (570), *619*, *631*, *632*, 640 (5, 6a), 642 (6), 644 (6a), 646 (6, 24), 655 (6), 658 (6), 659 (6), 660 (6), 661 (6), 664 (6), 665 (6), 666 (6), *667*, *668*, 703 (167), 704 (168), *715*, 855 (38), 857 (38), *878*, 887 (9), 895 (36a, 36d, 36e, 36f), 901 (36), *940*, *941*, *942*
Hertz, E. 258 (76), 275 (76), *300*
Herzberg, G. 708 (192), *715*, 854 (17), *877*
Hess, H. 800 (66a), 819 (66a), *847*
Hesse, G. 151 (267, 268), *165*, 316 (45, 46), 318 (45, 46), 332 (45, 46), *338*, 870 (146), 872 (159, 164, 166), 873 (167), *881*
Heuser, K. 685 (48), *711*
Hey, D. H. 674 (11, 12, 13), 675 (14), *711*
Hey, D. M. 675 (18), *711*
Heyneker, H. 405 (199), *419*
Heytler, P. G. 735 (119), *742*
Heyworth, F. 88 (169), *116*, 775 (197), *789*
Hiatt, R. 693 (92), *713*
Hibbert, H. 95 (242), *118*
Hickinbottom, W. J. 320 (54), 325 (54), 326 (54), *338*
Hickman, H. M. 179 (35), *203*
Hieber, W. 874 (187), 875 (188), 876 (209, 211, 214, 217), *882*
Hiebert, J. D. 543 (622), *632*
Higginbotham, L. 70 (11), *110*
Hilbert, G. E. 142 (166), *163*
Hildebrand, J. H. 144 (183, 187), 146 (183), *163*, 642 (10), *668*
Hill, D. G. 779 (213), *790*
Hill, R. K. 96 (244), *118*, 912 (72a, 72c), 913 (72c), 917 (87g), 919 (72a, 87g, 91), 920 (72c), 921 (72c), 922 (87, 99), 926, 930 (87, 99), *943*, *944*
Hillman, W. S. 702 (154), *714*
Hinkel, L. E. 244 (15), 266 (15), *298*, 507 (389, 390, 391), 508 (389, 391), 572 (389), *627*
Hinsberg, O. 429 (27), *617*
Hirai, S. 71 (20), *110*, 316 (47), 319 (47), *338*
Hirano, T. 686 (56), *712*
Hiroi, I. 288 (197), *304*
Hirota, K. 860 (81), 861 (82), *879*, 889 (17), *941*
Hiskey, C. F. 751 (37), *785*
Hoard, J. L. 135 (86), *161*, 216 (18), *236*
Hobbs, J. 686 (57), *712*
Hoberg, H. 350 (24), *415*
Höchtlen, A. 803 (77), 807 (77), 812 (77), 830 (77), *847*
Hockenberger, L. 801 (67), *847*
Hodel, E. 311 (20), 312 (20), 320 (20), *338*
Hodgkin, J. E. 93 (221), *118*
Hodgson, H. 775 (197), *789*
Hodgson, H. H. 88 (169), *116*
Hodgson, W. G. 527 (542), 528 (542), *631*
Hodnett, E. M. 168 (5), 169 (5), 170 (7), 171 (7), 172 (15), *203*
Hoepner, C. 430 (47), *618*
Hoffenberg, D. S. 262 (85), *301*
Hoffman, D. M. 556 (705), *634*
Hoffmann, A. K. 293 (223d), *305*
Hoffmann, A. W. 898 (42a, 42b), *942*
Hoffmann, P. 861 (85), *879*
Hoffmann, R. 109 (354), *122*, 887 (12), 890 (12), *940*
Hoffmann, R. W. 108 (350), *122*
Hofmann, A. 781 (216), *790*
Hofmann, A. W. 853, 854 (2), 855 (2), 857 (47), 860 (77), *877*, *878*, *879*
Hofmann, D. 72 (23), 80 (100), 99 (100), *111*, *113*, 540 (600), 541 (587), 542 (587, 600), 544 (600), 545 (600), 546 (600), *632*
Hofmann, K. A. 873 (170), 875 (191), *881*, *882*
Hogsed, M. J. 702 (156), *714*
Hohenlohe-Ochringen, H. 366 (77), *416*
Höhne, K. 432 (70), *618*
Holah, D. G. 132 (49, 50), *160*
Holland, D. O. 92 (205), *117*, 760 (82), *786*
Hollander, C. S. 272 (123), *302*

Holleman, A. F. 807 (95, 96), *848*
Holleman, M. M. 936 (127c), *945*
Hollstein, U. 746 (1), 747 (1), 748 (1), 749 (1), *784*
Holm, A. 266 (100), 275 (100), *301*
Holm, R. H. 434 (84), 502 (315, 317, 330, 332, 334, 335, 340, 341), *619*, *625*
Holmes, H. L. 70 (15), 105 (319), 109 (357), *110*, *121*, *122*, 895 (35), *941*
Holmes, J. R. 139 (133), 140 (133), *162*
Holt, N. B. 757 (51), 758 (51), 759 (51, 56, 66), *786*
Holubek, J. 183 (58), *204*
Holz, W. 919 (90b), *944*
Hölzl, F. 874 (180, 183), 875 (195, 206), *881*, *882*
Holzmann, R. T. 151 (269), *165*
Homes, H. L. 590 (781), *637*
Honour, R. J. 147 (221), *164*, 196 (145), *206*
Hook, R. H. 725 (61), 732 (61, 93), 734 (93), *740*, *741*
Hooker, S. W. 133 (52), *160*
Hooz, J. 894 (34), *941*
Hopf, W. 682 (36), *711*
Horák, M. 137 (102), 158 (325), *161*, *166*
Horean, A. 282 (169), *303*
Horeau, A. 383 (125), 386 (133), *418*
Horecker, B. L. 735 (122), *742*
Horman, I. 543 (606), *632*
Horn, E. M. 893 (31), *941*
Horn, P. 69 (8), *110*
Horner, L. 107 (340), *121*, 591 (787), *637*, 696 (115), *713*, 801 (67), *847*
Hornyak, J. 461 (188), *621*
Horobin, R. W. 859 (66), *879*
Horowitz, A. 463 (201, 202), 465 (202), 466 (201), *621*
Horron, B. W. 327 (97), *340*
Hosaka, S. 702 (157), *714*
Hoshino, T. 796 (21, 22), 806 (21, 22), 808 (22), 835 (22), 839 (22), *846*
Houben, J. 91 (192), *117*, 802 (70, 71), 821 (70), *847*
Houben-Weyl-Muller 308 (4), 309 (4), *337*
House, H. O. 105 (319), *121*, 320 (55), *338*, 590 (781), *637*, 895 (35), *941*
Houston, B. 319 (52), *338*
Howard, B. B. 139 (134), 140 (134), 142 (152), *162*
Howard, E. 792 (1), *846*
Howard, E. G. 284 (178), *303*
Howard, E. G., Jr. 461 (820), *638*
Howe, R. 405 (199), *419*
Howell, W. C. 79 (86), *113*
Howells, H. P. 172 (14), *203*
Hoy, D. 889 (16), *940*
Hoyle, K. E. 105 (325, 326), *121*
Hsia, S. L. 768 (154), *788*
Hsu, W. C. 752 (41), *785*
Hsu, Y.-K. 293 (221), *305*
Hsuan, Y. Heng 383 (125), *418*
Huang, H. H. 684 (43), *711*
Huang, P.-T. 75 (53), 110 (53), *112*, 690 (79, 80), *712*
Huang, R. L. 684 (42), *711*
Huber, H. 91 (193a), *117*, 448 (124), *620*, 709 (201, 202), *715*
Huber, O. 606 (801), 607 (801), *637*
Huber, W. Z. 175, 176, 177 (24), *203*
Hubley, C. E. 779 (211), *790*
Huchting, R. 703 (165), *715*
Hudec, J. 455 (161), *621*
Huebner, C. F. 386 (134), *418*
Huffman, K. R. 452 (152, 153), *620*, 909 (66a, 66b), *943*
Hughes, C. A. 755 (79), *786*
Hughes, D. W. 152 (282), *165*
Hughes, R. C. 661 (65), *669*
Huisgen, R. 344 (8), 345 (8, 9, 12), 346 (8, 13), 348 (19, 20), 350 (25, 26), 351 (12), 352 (8, 30), 353 (8, 32, 33), 354 (34, 37), 355 (37), 356 (40), 388 (141), *414*, *415*, *418*, 449 (141), 461 (141), *620*, 696 (116), *713*, 793 (10, 11), 796 (18, 30, 33), 800 (18), 803 (33), 808 (11, 30, 33), 810 (30), 814 (11, 33), 817, 818 (139), 827 (11, 30, 158), 832 (10, 11, 18, 139), 833 (11, 139), 834 (173a), 836 (139), 837 (11, 33), 838 (30), 839 (11), 840 (11, 139, 158), 842 (210, 216), 843 (218), *846*, *849*, *850*, *851*, 913 (74), 936 (128a), *943*, *946*
Huisgren, R. 832 (167b), 835 (167), *850*
Hullin, R. P. 273 (129), *302*
Humber, L. G. 100 (283), *120*
Humer, P. W. 551 (684), *634*, 705 (173), *715*
Humphlett, W. J. 278 (157), 280 (157), 283 (157, 172, 173), *303*, 314 (32), *338*

Humphrey, R. E. 146 (204, 206), *164*
Hunt, J. H. 93 (217), *117*
Hunt, M. 685 (47), *711*, 887 (8c), *940*
Hunter, G. D. 408 (218), *420*
Huong, P. V. 137 (108), *161*
Hurd, C. D. 76 (62), *112*
Hurley, A. C. 8 (11), *63*
Hurowitz, M. J. 258 (77, 78), *300*
Hurst, G. L. 432 (67, 68), *618*
Hurst, J. J. 723 (32, 33), *739*
Husk, G. R. 456 (166), *621*
Husteds, H. H. 506 (373), *626*
Hustedt, H.-H. 72 (27), *111*
Hutley, B. G. 859 (66), *879*
Huttner, K. 437 (92), *619*
Huyen, L. V. 408 (218), *420*
Huyser, E. S. 675 (24a), *711*
Hyatt, D. E. 151 (274), 152 (278), *165*, 892 (30d), *941*
Hydorn, A. E. 544 (625), *633*
Hylleraas, E. 8, *63*
Hylleraas, E. A. 29 (35), *64*
Hynes, J. B. 297 (238, 239, 240), *305*, 432 (69), *618*

Ibers, J. A. 502 (339, 342), *625*
Ibne-Rase, K. M. 938 (131), *946*
Ichikawa, M. 541 (593), 543 (593), *632*
Ichimura, K. 337 (116), *340*, 401 (190), *419*
Ide, W. S. 92 (197), *117*, 928 (112), *945*
Iffiand, D. C. 77 (70), *112*
Igarashi, Y. 474 (249), *623*
Ikan, R. 88 (171), *116*
Imai, Y. 877 (228), *883*
Imamura, A. 474 (249), *623*
Inamoto, N. 696 (108, 109), *713*, 769 (169), *788*
Ingberman, A. K. 88 (172), *116*, 700 (145), *714*
Ingham, H. 286 (188), *303*
Ingold, C. K. 888 (13), *940*
Inkova, E. M. 748 (17), *785*
Inkowa, J. N. 754 (74), *786*
Inokuchi, H. 663 (67), *669*
Inoue, K. 93 (214), *117*, 858 (53), *878*
Inove, I. 379 (121), *418*
Inukai, T. 218 (29d), *236*
Ioffe, B. V. 94 (229, 230), *118*, 137 (106), *161*, 900 (50), *942*
Iogansen, A. V. 137 (106), *161*
Isaacs, N. S. 463 (204), 465 (204), *622*
Isbell, H. S. 747, 757, 758, 759 (51, 55, 56, 65, 66, 68, 69, 70), 784 (226, 227, 228), *785*, *786*, *790*
Isfendiyaroĝlu, A. N. 689 (71), *712*
Ishitani, A. 527 (544), 528 (544, 545), *631*
Iskrić, S. 773 (193), *789*
Israel, M. 256 (65), *300*
Israeli, Y. 105 (321), *121*
Issleib, K. 286 (186), *303*
Itazaki, H. 316 (47), 319 (47), *338*
Ito, Y. 860 (74, 81), 861 (82), 862 (95, 97), *879*, 889 (17), *941*
Ivcher, T. S. 127 (22), 128 (22), *159*
Ivin, K. J. 692 (87), *713*
Iwakura, Y. 797 (53, 54), 810 (53), 822 (54), *847*
Iwamoto, R. T. 131 (43), 133 (56), *160*
Iwamura, M. 696 (108, 109), *713*
Iwasaki, F. 528 (553), *631*
Iwata, S. 528 (548, 549), *631*, 642 (14), 652, 653 (14), 656 (14), *668*
Izatt, R. M. 614 (807), *637*

Jackman, L. M. 543 (624), 544 (624), *633*
Jacknow, B. B. 672 (4), 673 (4), *710*
Jackson, W. R. 85 (146), *115*
Jacob, E. 724 (50), *740*
Jacobi, E. 503 (346), *626*
Jacobsen, O. 428 (18), *617*
Jacobson, M. 722 (30), *739*
Jacox, M. E. 548 (654, 655, 660, 662, 665), *633*, *634*, 708 (194), *715*
Jacques, J. 282 (169), *303*, 383 (125), *418*
Jacquesy, J. C. 202 (183), *207*
Jacquesy, R. 202 (183), *207*
Jacquier, R. 914 (78b), *944*
Jahnke, D. 893 (33b), *941*
Jain, S. R. 135 (88, 89), 153 (292) 154 (88, 89, 308), 155 (309), *161*, *166*
Jakob, F. 696 (116), *713*
Jakoubková, M. 137 (102), *161*
Jakubovič, A. Ja. 869 (138), *880*
James, C. 675 (16), 679 (16), *711*
James, S. M. 762 (104), *787*
Jamieson, J. R. 780 (214), *790*
Janardhan, P. B. 186 (63, 64), *204*
Jander, G. 430 (44), *618*
Jankiewicz-Wasowska, J. 582 (760), 583 (760), *636*
Jańsen, F. 797 (43), 802 (43), *847*

Janssen, P. J. A. 384 (126), *418*
Janz, G. J. 127 (19, 21), 128 (19, 21, 32, 34), 131 (42), 133 (42, 54, 57), 135 (69, 72), 156 (72), *159*, *160*, 183 (59), *204*, 298 (243), *305*, 356 (43, 44, 45, 46, 47, 48, 49, 50), 357 (43, 45, 48, 49), *415*, *416*, 432 (63, 64), *618*
Japp, F. R. 364 (73), *416*
Jarrie, J. M. S. 356 (46, 47), *415*
Jarvie, J. M. S. 183 (59), *204*
Jautelat, M. 508 (398), *627*
Jaworski, T. 358 (52, 53, 54, 55), *416*
Jeanes, K. J. 747 (13), *785*
Jefford, C. W. 926 (106), *945*
Jendrot, M. C. 98 (267), *119*, 886 (5), *940*
Jenkins, A. D. 694 (95), *713*
Jennen, J. 80 (97), *113*
Jenner, E. L. 674 (8), *711*
Jennings, J. R. 153 (288), *166*, 282 (170), *303*, 893 (32), *941*
Jennings, R. 152 (284), 153 (284), *166*
Jenny, E. 96 (249), *119*
Jensen, F. R. 148 (225), *164*, 211 (8), *236*
Jensen, K. A. 266 (100), 275 (100), *301*
Jentsch, W. 270 (115), *302*
Jentzsch, J. 684 (44), *711*
Jerrard, H. G. 142 (149), *162*
Jesson, J. P. 135 (93), 147 (93, 217), *161*, *164*, 196 (146), *206*
Jewell, D. J. 147 (218), *164*
Job, V. A. 516 (452, 463), *628*
Jodlowsky, H. A. 98 (269), *119*
Joesten, M. D. 139 (126), *162*
Johannesen, R. B. 151 (265), *165*
Johns, I. B. 576 (735), *635*
Johnson, A. W. 74 (38), 95 (38), *111*
Johnson, F. 78 (73), 86 (153), *113*, *115*, 245 (18), *298*, 344 (5), 359 (5), 360 (5), 368 (5), 376 (5), 380 (5, 122), 381 (123, 124), *414*, *418*
Johnson, G. S. 321 (71), 322 (71, 72, 73, 74, 75), *339*, 927 (110), *945*
Johnson, H. W. 862 (98), *879*
Johnson, J. B. 174, 175 (21), *203*
Johnson, R. N. 270 (113), *301*
Johnson, W. J. 96 (248), *119*
Johnston, D. L. 154 (307), 156 (311 315), *166*, 196 (130), *206*
Johnston, K. 78 (74), *113*
Jones, A. R. 780 (214), *790*
Jones, D. A. 470 (216), *622*, 723 (37), *740*
Jones, E. R. H. 452 (150), *620*, 732 (96, 97), 733 (97), *741*
Jones, G. A. 320 (61), *339*
Jones, M. T. 297 (241), *305*, 535 (575), 550 (675), 575 (732), 576 (732), *631*, *634*, *635*, 661 (64), *669*
Jones, R. E. 388 (146), *418*
Jones, W. D. 516 (447), *628*
Jones, W. M. 368 (85), *417*
Jordan, N. W. 450 (817), *638*
Joris, L. 141 (147), *162*, 813 (123), *849*
Josey, A. D. 438 (95), 483 (95), 595 (794), 614 (95), *619*, *637*
Joshi, K. K. 875 (190), *882*
Josien, M.-L. 138 (110, 111, 112), 139 (111), *161*
Jovitschitsch, M. 802 (72, 73, 74), 809 (72), *847*
Joyama, T. 391 (157), *418*
Joyce, R. M. 702 (156), *714*
Juday, R. 324 (82), 325 (82), *339*
Judd, H. M. 93 (208), *117*
Julia, M. 675 (15, 16, 17), 676 (17, 19), 677 (17, 19, 20), 678 (15, 20, 21, 22), 679 (16, 19, 23), 680 (24b, 25), *711*
Julia, S. 291 (214), *304*, 368 (87), *417*
Julien, P. L. 95 (242), *118*
Jullien, J. 78 (71), *112*
Jumper, C. F. 139 (134), 140 (134), 142 (152), *162*
Junek, H. 284 (177), *303*, 402 (192), *419*, 459 (180), *621*
Jungmann, P. 90 (185), *116*
Jungreis, E. 173 (18), *203*
Jura, W. H. 527 (542), 528 (542), *631*
Jürgens, B. 804 (79), *847*
Jurjewa, L. P. 91 (191), *116*
Jurkiewicz, L. 92 (202), *117*, 580 (741), 581 (741), 582 (741, 757), 586 (772), *635*, *636*
Just, G. 797 (51, 52), 801 (51, 52), 815 (51, 52), *847*
Just, M. 779 (209), *789*
Justoni, R. 74 (42, 43), 79 (43), *111*

Kaabak, L. V. 181 (49), 189 (83, 86), 190 (88), *204*, *205*, 337 (117), *340*
Kaack, R. 580 (740), *635*
Kabanov, V. A. 287 (191), *304*
Kabbe, H.-J. 860 (70), 869 (70), *879*
Kabitzke, K. H. 857 (50), *878*

Kache, R. 702 (158), *714*
Kaczmarczyk, A. 710 (206), *716*
Kaesz, H. D. 255 (60), *300*
Kafatos, F. C. 723 (32), *739*
Kagan, H. B. 282 (169), *303*, 383 (125), *418*
Käge, H. H. 767 (137), *788*
Kahlen, N. 859 (61), *878*
Kainer, H. 659 (54), *669*
Kaiser, D. W. 245 (19), *298*, 547 (648), *633*
Kaiser, E. M. 150 (253), *165*
Kakano, S. 361 (68), *416*
Kalish, J. 291 (211), 292 (211), *304*, 911 (71b), 914 (78a), *943*, *944*
Kallen, J. 70 (10, 14), *110*
Kamada, H. 196 (143), *206*
Kamat, S. S. 289 (203), *304*
Kambara, S. 288 (192), *304*, 474 (250), 543 (619), *623*, *632*
Kamenar, B. 528 (554), *631*
Kamernitskii, A. V. 74 (39, 40), 81 (103), *111*, *114*
Kametani, T. 361 (68), *416*
Kamlet, M. L. 227 (51), *237*
Kanaoka, Y. 379 (121), *418*
Kao, C. H. 98 (266), *119*
Karabinos, J. V. 757 (51), 758 (51), 759 (51), *786*
Karasch, N. 91 (193b), *117*
Karavan, V. S. 406 (202), *420*
Kargin, V. A. 287 (191), *304*
Karl, R. 684 (44), *711*
Karrer, P. 90 (190), *116*, 431 (59), *618*
Karrer, W. 718 (4), *739*
Kasatochkin, V. I. 474 (253), *623*
Kaska, W. C. 519 (497), 520 (497), *629*
Kaslow, C. E. 95 (240), *118*
Kassal, R. J. 388 (144), *418*
Kato, T. 200 (174), *207*
Katon, J. E. 474 (247, 256), 529 (256, 559, 560), *623*, *631*
Katritzky, A. R. 150 (259), *165*, 818 (141), *849*
Katsumoto, K. 386 (135), *418*
Katz, J. J. 138 (123), *162*
Katz, L. 675 (15, 17), 676 (17), 677 (17, 20), 678 (15, 20), *711*
Katz, L. E. 886 (3), *940*
Katz, M. 326 (95), 327 (95), 328 (95), 329 (95), 332 (95), *340*
Katzman, S. M. 103 (302), *120*
Kauffmann, H. 802 (70), 821 (70), *847*
Kauffmann, H. O. 925 (103h), *945*
Kauffmann, J. M. 838 (204), *851*
Kauffmann, Th. 273 (130), *302*
Kaufhold, G. 868 (137), *880*
Kaufler, F. 886 (7d), *940*
Kawamura, F. 186 (65), *204*
Kay, R. L. 141 (148), 142 (148), *162*
Kazitsyna, L. A. 195 (115), *205*
Keana, J. 258 (72), 261 (72), 263 (72), *300*
Kecki, Z. 132 (47), *160*
Keefer, R. M. 144 (178), *163*, 640 (2), 642 (2, 11), 654 (2), 655 (2), *667*, *668*
Keely, S. L., Jr. 384 (131), *418*
Keenan, A. G. 356 (43), 357 (43), *415*
Keglevic, D. 759 (60), *786*
Keglevic-Brovet, D. 773 (193), *789*
Keighley, G. 761 (91), 763 (100), 766 (146), *787*, *788*
Keilin, J. 877 (227), *883*
Kelly, M. 877 (229), *883*
Kelly, W. 106 (337), *121*
Kemp, A. D. 101 (293), *120*
Kempf, R. J. 551 (684), *634*, 705 (173, 174), *715*
Kemula, W. 581 (748, 749, 752), *635*, *636*
Kendall, E. C. 78 (73), 79 (89), *113*
Kendall, J. D. 496 (295), 608 (295), *624*
Kende, A. S. 451 (145), *620*
Kennedy, L. D. 721 (21), *739*
Kenner, G. W. 246 (23), *299*
Kennerly, G. W. 524 (516), *630*
Kenny, T. S. 547 (642), *633*
Kent, P. W. 759 (53), *786*
Kenyon, W. O. 698 (137), *714*
Kepler, R. G. 663 (69), 665 (74), *669*, 895 (36b), 901 (36), *941*
Kerber, R. C. 616 (809), *637*
Kereszty, V. 351 (28), 388 (28), *415*
Kern, E. 546 (631), 547 (631, 646), *633*
Kern, R. 72 (23), 80 (100), 99 (100), *111*, *113*, 540 (600), 541 (587), 542 (587, 600), 544 (600), 545 (600), 546 (600), *632*, 653 (35), 654 (35), 655 (35), *668*
Kerr, J. A. 887 (8c), *940*
Kessler, W. V. 767 (150), *788*
Ketelaar, J. A. A. 147 (212), *164*
Kettle, S. F. A. 256 (66), *300*, 410 (225), *420*, 502 (319), *625*
Keys, R. T. 298 (242), *305*, 689 (73),

693 (73, 91), 696 (73), *712*, *713*, 897 (41b), *942*
Khalifah, R. A. 732 (100), *741*
Khan, N. R. 859 (66), *879*
Kharaisch, M. S. 276 (146), 277, 278 (146), 280 (146), *302*
Kharasch, M. S. 892 (29a), *941*
Kharasch, N. 91, (193b), *117*, 710 (204), *716*
Khogali, A. 724 (54), *740*
Kholmogorov, V. E. 472 (218), *622*
Khomenko, Kh. 196 (160), *207*
Khorana, H. G. 247 (33), *299*
Khorlin, A. Ya. 292 (218a, 218b), *304*, *305*
Khorlina, I. M. 317 (48), *338*
Khua-min, K. 378 (116), *417*
Kice, J. L. 696 (111, 112), *713*
Kiefer, B. 311 (25), 313 (25), *338*
Kiefer, H. R. 872 (162), *881*
Kieffer, F. 221 (37), *237*
Kihlman, B. A. 735 (117, 118), *742*
Kiliani, H. 72 (26), *111*
Kille, G. 406 (205), *420*
Kilpatrick, M. 127 (15), *159*
Kilpatrick, M. L. 127 (15), *159*
Kilpatrick, M. L. 259, 260 (81), 265 (81), *301*
Kim, Y. C. 107 (345), *122*, 443 (111), *619*
Kim, Young-Ki 10 (16), *63*
Kimball, A. P. 289 (203), *304*, 411 (234), *420*, 506 (362), *626*
Kimball, M. E. 519 (497), 520 (497), *629*
Kimball, R. K. 71 (18), *110*
Kimura, M. 862 (95), *879*
Kindler, K. 274 (131), 275, *302*, 331 (106, 107), 333 (106), *340*
Kine, G. F. 816 (136b), *849*
King, C. G. 754 (72, 73), *786*
King, G. W. 516 (452, 463), *628*
King, J. A. 93 (221), *118*
Kinoshita, M. 664 (70), *669*
Kinoshita, Y. 156 (313, 314), *166*
Kiovsky, T. E. 246 (22b), 250 (22b), *299*, 912 (73b), *943*
Kirby, R. H. 282 (165), *303*
Kirmse, W. 801 (67), *847*
Kirsanov, A. V. 97 (259), 101 (285), 102 (295), *119*, *120*, 252 (53), *299*, 525 (522), *630*
Kirschenlohr, W. 782 (219), *790*
Kirshenbaum, A. D. 515 (475), 516 (475), *629*
Kiselev, A. V. 143 (172, 173, 174), *163*, 196 (125), *206*
Kissinger, L. W. 76 (64), *112*
Kistiakowsky, G. B. 746 (3), *784*
Kitahara, Y. 200 (174), *207*
Kitson, R. E. 135 (75), *161*, 194, 196 (111), *205*
Kivelson, D. 139 (133), 140 (133), *162*
Klaboe, P. 135 (70, 85), 144 (184, 185, 186, 192), 145 (184, 185, 186, 192), 146 (85, 184, 185, 186, 208), 147 (215), 156 (70), 157 (85), 158 (184, 185, 186, 192), *160*, *161*, *163*, *164*, 196 (137), *206*
Klaeboe, P. 516 (453, 459, 460), *628*
Klages, F. 127 (17), 128 (27), 129, 130 (27), 131 (17), *159*, 251 (45, 58), 255 (58), *299*, *300*, 819 (145), *849*, 873 (172), 875 (172), 876 (215), *881*, *882*
Kleinberg, J. 131 (43), *160*
Kleineberg, G. 366 (79, 80), *416*
Klemm, L. H. 452 (154), *620*
Klenk, E. 766 (149), *788*
Klingensmith, G. B. 217 (25), *236*
Klingsberg, E. 94 (234), *118*, 503 (350), *626*
Klinowski, C. W. 886 (3), *940*
Klint, D. 49 (60), 55 (60), *65*
Kliss, R. M. 481 (267), 482 (267), 506 (380), 551 (380), *623*, *626*, 706 (183), *715*, 901 (53a, 53b), 902 (53), 903 (53), *942*
Klofutar, C. 156 (316), *166*
Kloosterziel, H. 221 (36), *237*
Klopfenstein, A. 95 (242), *118*
Klöpfer, W. 664 (73), *669*
Klosa, J. 736 (136), *742*
Kloster-Jensen, E. 147 (215), *164*, 482 (269), 513 (426, 427, 428, 429), 516 (426, 428, 429, 459, 460), 519 (269), *623*, *628*
Klump, K. N. 908 (64a), *943*
Knight, D. J. 721 (22b), *739*
Knight, J. A. 296 (232), *305*, 309 (10), *337*
Knipple, W. R. 255 (59), *300*
Knipprath, W. G. 769 (168), *788*
Knobloch, H. 431 (60), *618*
Knoll, F. J. 256 (62), *300*
Knorr, A. 265 (99), *301*
Knorr, R. 871 (151), *881*
Knoth, W. H. 859 (59), *878*
Knott, E. B. 87 (161), *115*

Knunyants, I. L. 582 (762), *636*, 701 (151), *714*, 869 (143, 144), 870 (144), *880*, *881*
Knupfer, H. 348 (20), *415*, 854 (15), 855 (15), 857 (15), 877 (235), *877*, *883*, 887 (9a), 891 (9a), *940*
Kobayashi, M. 664 (70), *669*
Kobayashi, S. 860 (81), 861 (82), *879*, 889 (17), *941*
Kobelt, M. 72 (28), *111*
Koch, H. 756 (50), *786*
Koch, K. 354 (35), *415*
Kochetkov, N. K. 292 (218a, 218b), *304*, *305*, 809 (113), 815 (113), 836 (113), *848*
Kochi, J. K. 697 (124), 700 (146, 147), 701 (150), *713*, *714*
Kochi, J. R. 88 (172), *116*
Koda, A. 94 (225), *118*
Koebner, A. 101 (293), *120*
Koechlin, W. 280 (160), *303*
Koelsch, C. F. 80 (98), 82 (115), 83 (115, 124), *113*, *114*, 379 (120), *417*
Koenig, W. 899 (47b), *942*
Koepp, H.-M. 133 (58), *160*
Kofman, L. S. 142 (157), *162*, 178 (34), *203*
Kohl, K. 388 (142), *418*
Kohler, E. P. 74 (44), 79 (44), *111*, 378 (110), *417*
Kohler, H. 435 (814), *637*
Köhler, H. 434 (85), 435 (85), 547 (85, 637, 638, 639, 640), *619*, *633*
Kohlik, A. J. 452 (154), *620*
Kohlmaier, G. 859 (64), *878*, 887 (10), 890 (10), *940*
Kohn, P. 759 (57), *786*
Kolbe 283 (174), *303*
Kolditz, L. 153 (291, 293, 294), *166*
Kolos, S. 5, *63*
Kolos, W. 10, *63*
Kolthoff, I. M. 127 (16, 18), 133 (61, 62), 134 (62), *159*, *160*
Komenda, J. 472 (227), *662*
Kommandeur, J. 659, 663 (67), *669*
Kondo, K. 560 (713), *635*
Kondo, M. 506 (372), *626*
Konig, H. 353 (33), *415*
König, H. 286 (190), *303*
Konishi, K. 388 (147), *418*
Konstas, S. 354 (35), *415*
Koopman, H. 737 (137), *742*
Koopmann, H. 92 (198), *117*
Kooyman, E. C. 673 (7), 694 (97), 696 (104, 105, 113, 114), *711*, *713*
Kopecky, J. 98 (270), *119*
Köpf, H. 502 (313), *625*
Koppe, M. 793 (6), 815 (6), *846*
Kopple, K. D. 536 (579), *631*, 667 (77), *669*
Korczynski, A. 88 (168), *116*
Korel'sksya, G. I. 527 (539), *630*
Korinek, G. J. 139 (141), 140 (141), 156 (141), *162*
Kornblum, N. 77 (70), *112*
Kornreich, L. D. 398 (174), *419*
Kornuta, P. P. 252 (53), *299*
Korte, F. 407 (210, 211), *420*, 767 (142), 768 (156), *788*
Korybut-Daszkiewicz, B. 358 (55), *416*
Kosanovic, Dj. 126 (12), *159*, 615 (808), 617 (808), *637*
Koshar, R. J. 703 (162), *715*
Koshland, D. E. 759 (59), *786*
Kosower, E. M. 228 (47), *237*, 463 (205), 465 (205), 469 (205), *622*, 640 (6b), 642 (6), 646 (6), 655 (6), 658 (6), 659 (6), 660 (6), 661 (6), 664 (6), 665 (6), 666 (6), *667*
Kostikova, G. I. 749, 750 (27), *785*
Kostrova, N. D. 474 (241), *623*
Kotani, M. 2 (4), *63*
Kotko, L. 572 (723), 573 (723), *635*
Kotova, Z. N. 142 (157), *162*
Koubek, E. 938 (131), *946*
Koukol, J. 720 (10), *739*
Kourim, P. 756, 760, 761 (93), *787*
Koutenko, O. F. 197 (122), *206*
Kovaleva, V. P. 287 (191), *304*
Kovelesky, A. C. 399 (178), *419*
Koyama, K. 91 (194), *117*, 674 (10), 710 (208, 209), *711*, *716*
Krafft, F. 251 (47, 48), *299*
Kraft, F. 97 (253), *119*
Kraft, K. J. 526 (534), *630*
Kraihanzel, C. S. 516 (455), *628*
Kramer, D. N. 93 (223), *118*
Kranz, J. 246 (25), *299*
Krapcho, A. P. 271 (121), *302*, 855 (41), *878*
Krapcho, J. 98 (265), *119*
Krasnomolova, L. P. 196 (141, 159), *206*, *207*
Krause, R. A. 132 (51), *160*
Krebaum, L. J. 513 (417, 418, 419), *627*
Kreisley, J. W. 83 (117), *114*
Kreling, M. E. 401 (188), *419*
Krentsel, B. A. 524 (512), *630*

Kreshov, A. P. 153 (289), *166*
Krespan, C. G. 109 (356, 358), *122*, 439 (97), 449 (97), 457 (97), 459 (97), 470 (215), 503 (215, 347), 505 (357), 514 (215), 515 (215), 614 (97), *619*, *622*, *626*, 703 (166), 706 (181), *715*
Kresze, G. 451 (147), *620*
Kretov, G. E. 248 (37), *299*
Krewson, C. F. 258 (73), *300*
Krieble, V. K. 249 (40, 41), 258 (41, 79), *299*, *300*
Krimen, L. I. 344 (7), 359 (7), 360 (7), *414*
Kroeper, H. 525 (523), *630*
Krohn, W. 518 (489), *629*
Kronick, P. L. 543 (611), *632*, 663 (68), 665 (75), *669*
Kropa, E. L. 528 (557), *631*
Kropf, H. 801 (68), *847*
Kroto, H. W. 548 (664), *633*, 708 (195), *715*
Krueckel, B. J. 770 (174), 783 (174), *789*
Krueger, H. 347 (17), *415*
Krueger, P. J. 142 (169, 170), *163*
Kruss, G. 934 (124c), *945*
Kubiak, S. 153 (289), *166*
Kubicek, R. 183 (57), *204*
Kubota, M. 135 (80), 154 (306, 307), 156 (306, 311, 315), 157 (321), 158 (321), *161*, *166*, 196 (130), *206*, 250 (42), *299*
Kudinova, V. S. 525 (525), *630*
Kudo, K. 492 (286), 493 (286, 288, 289), *624*
Kuehle, E. 503 (351), *626*
Kuehne, M. E. 89 (182), *116*, 835 191b), *850*, 898 (45), *942*
Kuffner, F. 280 (160), *303*
Kühlcke, I. 286 (187), *303*
Kühle, E. 860 (75, 79), *879*
Kühlein, K. 861 (83), *879*
Kuhlmann, D. 273 (130), *302*
Kuhn, M. 546 (635), *633*
Kuhn, R. 396 (168), *419*, 782 (219, 220), *790*
Kuhn, S. J. 254 (54), *300*
Kukhar, V. P. 252 (53), *299*, 414 (241), *421*
Kulikova, A. E. 128 (33), *160*, 245 (21), 266 (21), *299*
Kumakura, S. 528 (553), *631*
Kumamoto, J. 697 (123), *713*
Kumashiro, I. 411 (234), *420*, 506 (366), 508 (366, 825), 509 (366, 825), *626*, *638*
Kummer, U. von 925 (103h), *945*
Kummerow, F. A. 775 (200), *789*
Kunstmann, H. 842 (215), *851*
Kunz, R. 483 (275), 498 (275), *624*
Kuo, C. H. 397 (170), *419*
Kuo, M. C. Chang 356 (41), *415*
Kurbatov, J. W. 768 (157), *788*
Kurioka, S. 513 (415), 517 (415), 518 (415), 519 (415), 524 (415), *627*
Kurkchi, G. A. 137 (106), *161*
Kuroda, H. 664 (70), *669*
Kuroki, N. 388 (147), *418*
Kürschner, Chr. 153 (294), *166*
Kurtz, I. D. 854 (23), *878*
Kurtz, P. 68 (2), 69 (2, 4, 7), 70 (9, 13), 71 (13, 19), 72 (29), 73 (29), 74 (39, 45), 75 (56), 76 (57), 77 (2), 79 (91), 80 (96), 81 (104, 105, 107), 82 (111), 84 (134, 138, 139), 85 (142), 86 (7, 154), 87 (163), 88 (174), 90 (187), 91 (192), 97 (251), *110*, *111*, *112*, *113*, *114*, *115*, *116*, *117*, *119*, 752 (43), *785*, 795 (19), 796 (19), 801 (68), 802 (68b), *846*, *847*, 854 (10), *877*
Kuryla, W. C. 463 (193), *621*
Kurz, P. 256 (67), 262 (67), 263 (67), *300*
Kushner, M. 767 (140), *788*
Kushnikov, Ya. A. 196 (141), *206*
Kushnikov, Yu. A. 196 (159), *207*
Kustanovich, I. M. 524 (512), *630*
Küster, F. W. 525 (529), *630*
Kuthan, J. 525 (531), *630*
Kutseva, L. N. 474 (251), *623*
Kutter, E. 498 (300), *624*
Kuwada, D. 872 (162), *881*
Kuznetsov, B. G. 143 (173), *163*
Kveder, S. 773 (193), *789*
Kwart, H. 556 (705), *634*
Kwok, R. 400 (182), *419*

Laasch, P. 129 (38), 130 (38, 40), 159 (38), *160*, 250 (43), 251 (43), 252 (43, 50), 253 (43, 50), *299*, 375 (104), *417*, 912 (73a), *943*
Labarre, J. F. 516 (454, 477), *628*, *629*
Labes, M. 665 (75), *669*
Labes, M. M. 543 (611, 620), *632*, 663 (68), *669*
Labriola, E. Restelli de 93 (207), *117*
Lach, B. 92 (203), *117*

Lacher, J. R. 703 (162), *715*
Ladd, E. C. 682 (32), *711*
Ladenburg, A. 325 (89), *339*
La Lancette, E. A. 105 (324), *121*, 593 (789), *637*
Lalich, J. J. 724 (40, 53), *740*
Lamb, R. C. 335 (113), *340*
Lambeck, R. 801 (68), *847*
Lambert, J. D. 148 (230), 149 (230), *164*
Lambin, J. 751 (35), *785*
Lambooy, J. P. 747 (9), *785*
Lanaers, R. 796 (37), 842 (37), *847*
Lancaster, J. E. 248 (38), *299*
Landor, S. R. 85 (141), *115*
Lane, J. F. 80 (96), *113*
Lane, L. A. 85 (147), *115*
Lang, R. P. 144 (194), *163*
Langdon, J. M. 98 (264), *119*, 524 (520), 525 (520), 576 (520), *630*
Lange, O. 506 (381), *627*
Lange, R. F. 463 (193), *621*
Lange, R. J. 377 (108), *417*, 920 (94b), *944*
Langella, M. R. 822 (155a), 834 (174), 837 (199), *849*, *850*, 930 (114a), *945*
Langford, C. L. 502 (333), *625*
Langley, W. D. 75 (54), *112*
Langseth, A. 517 (480), *629*
Langton, J. M. 654 (37), *668*
Lapin, H. 315 (44), *338*, 386 (133), *418*
Lappert, M. 892 (306), *941*
Lappert, M. F. 150 (264), 151 (264), *165*, 216 (17), *236*
Lapworth, A. 70 (11, 12), 72 (26), 78 (73), *110*, *111*, *113*
Lapworth, M. 753 (46), *786*
Larchar, A. W. 298 (244), *305*, 860 (78), *879*
Larsen, A. A. 315 (41), *338*
Lasarov, S. 188 (76), *204*
Lasch, P. 358 (57), 365 (57), *416*
Lascombe, J. 138 (112), *161*
Laszlo, P. 140 (145), *162*
Lathroum, L. B. 431 (54, 55), *618*
Latif, N. 106 (335), *121*
Latka, H. 892 (28), *941*
Latourette, H. K. 322 (72, 73), *339*
Laubengayer, A. 347 (16), *415*
Laubengayer, A. W. 150 (260), 151 (260), 156 (260), *165*
Laurent, P. 142 (150), *162*
Lauria, F. 839 (205), *851*
Lautenschlaeger, F. 888 (15), 890 (15), *940*
Lavinger, C. 725 (60), *740*
Laviron, E. 183 (57), *204*, 472 (227), *622*
Lawesson, S.-O. 200 (177, 178), *207*
Laws, D. R. J. 85 (141), *115*
Lawton, E. A. 83 (126), 99 (126), *114*, 527 (538), *630*
Lazaris, A. Ya. 244 (16), 266 (103), *298*, *301*
Lazennec, I. 518 (487), *629*
LaZerte, J. D. 703 (162), *715*
Lazier, W. A. 916 (83a), *944*
Leandri, G. 193 (106), *205*, 210 (5), *235*, 346 (14), 348 (14), *415*, 832, 835 (167), 840 (167a), *850*
Lebedev, V. B. 516 (469), *629*
Lebovits, A. 75 (53), 110 (53), *112*, 687 (60), *712*
L'Écuyer, P. 938 (132, 135a, 135b), *946*
Lednicer, D. 87 (160), *115*
Lee, H. H. 452 (150), *620*
Lee, J. L.-C. 731 (92b), *741*
Lee, K. T. 684 (42), *711*
Lee, L. A. 388 (145), *418*
Leermakers, P. A. 689 (75), *712*
LeGoffic, F. 675 (15), 677 (20), 678 (15, 20, 21), *711*
Lehmann, H. A. 432 (70), *618*
Lehnsen, J. E. 527 (542), 528 (542, 552), 543 (552), *631*
Leichner, L. 96 (249), *119*
Leipprand, H. 364 (74), 365 (74), 366 (78), *416*
Leitch, L. C. 766 (128), 779 (210), *787*, *790*
Leitermann, H. 352 (31), *415*
Leitich, J. 309 (14), *337*
Lemaire, H. 126 (7), *159*
Le Maistre, J. W. 910 (68), *943*
Lemal, D. M. 456 (169), 522 (169), *621*
Lemanceau, B. 139 (132), 140 (132), *162*
Lembert, K. 247 (33), *299*
Lemmon, D. H. 513 (416), 514 (416), 515 (416), 516 (416, 458), 517 (416), *627*, *628*
Lemmon, R. M. 246 (23), *299*, 747 (12), *785*
Lenaers, R. 272 (122), *302*, 809 (104, 105), 839 (104), 840 (104), *848*
Lengyel, I. 858 (55), *878*
Lennard-Jones, J. E. 8 (11), *63*

Lenthen, P. M. 158 (322), *166*
Leonard, N. J. 323 (80), *339*, 369 (90), 373 (90, 99, 100, 102), 374 (90), *417*
Leplawy, M. 581 (743), 582 (758), 584 (768, 769), 586 (768, 769, 775), 587 (778), *635, 636, 637*
Leppelmann, H. J. 782 (220), *790*
Lesbre, M. 550 (680), 551 (680), *634*
Lescoeur, H. 507 (386), *627*
Leslie, S. W. 180 (41), *204*
Lespieau, R. 79 (92), *113*
Lester, G. R. 200 (173), *207*
Lettré, H. 90 (185), *116*
Letts, E. A. 934 (124a), *945*
Leubner, G. W. 323 (80), *339*
Leuckart, R. 934 (124b), *945*
Leusink, A. J. 519 (494, 495), *629*
Levasseur, L. A. 642 (13), 645 (13), 650 (13), 658 (13), *668*
Levesque, C. L. 258 (76), 275 (76), *300*
Levina, R. Y. 378 (112, 116), *417*
Levina, S. D. 474 (248), *623*
Levisalles, J. 202 (183), *207*, 375 (104), *417*
Levorato, C. 735 (115), *742*
Levy, A. L. 408 (212, 213, 214), *420*
Levy, L. A. 293 (223a), *305*
Levy, M. 179 (39), *203*
Lewis, D. 724 (41), *740*
Lewis, F. M. 686 (53), 687 (53), *712*
Lewis, I. C. 148 (226), 149 (226), *164*, 217 (24), 219 (24), 220 (24), 229 (24), *236*
Lewis, R. N. 78 (76), *113*
Ley, K. 870 (145), *881*
Libbert, E. 734 (111), *741*
Liberek, B. 98 (267), *119*
Libis, B. 899 (486), *942*
Lichtenstadt, L. 802 (75), *847*
Lichtenthaler, F. W. 268 (109), *301*
Lichtenwalter, M. 282 (165), *303*
Liddel, U. 142 (166), *163*
Lieb, V. A. 761 (95), *787*
Lieber, E. 309 (9), *337*
Liebig, J. v. 430 (44), *618*, 819 (148), *849*
Liem, P. N. 760 (84), 765 (113), *786*, *787*
Liesche, O. 927 (109), *945*
Lifson, N. 766 (121, 123), *787*
Liggett, L. M. 335 (112), *340*
Liler, M. 126 (12), *159*, 615 (808), 617 (808), *637*
Lilker, J. 75 (53), 110 (53), *112*, 687 (64), 688 (64), *712*
Lim, E. C. 193 (105), *205*
Lim, P. K. K. 684 (43), *711*
Lind, D. J. 699 (140), *714*
Lindemann, H. 853 (5), 854 (5), *877*
Linderberg, J. 39 (51), *64*
Lindquist, R. N. 77 (69), *112*
Lindqvist, I. 135 (78), *161*
Lindsey, R. V. 859 (59), *878*
Lindstedt, S. 781 (225), *790*
Line, W. E. 179 (35), *203*
Linetskii, V. A. 257 (71), *300*
Lingens, E. 87 (159), *115*
Link, H. 917 (87c), 918 (87c), 922 (87), 930 (87), *944*
Link, J. 925 (103h), *945*
Linke, S. 894 (34), *941*
Linn, L. E. 863 (113), *880*
Linn, W. J. 109 (354), *122*, 458 (174), 555 (174, 698, 699, 701) 556 (174), 557 (699), 559 (708), 560 (699, 709, 711), 561 (708), 562 (708), 563 (708, 715, 716), 564 (719), 565 (715), 566 (715), 567 (715), 568 (708), 569 (715), 581 (699, 744, 753), 584 (767), *621, 634, 635, 636*
Linstead, R. P. 84 (129, 133), *114*, 506 (364), 508 (364), 541 (594), 573 (725), *626, 632, 635*
Lipp, M. 862 (102), *879*
Lippert, E. 225 (43), *237*, 448 (126), *620*
Lippincott, E. R. 516 (461), *628*
Lippmaa, E. 202 (184), *207*
Lipschitz, A. 172 (16), *203*
Lipscomb, R. D. 674 (8), *711*
Lipscomb, W. N. 151, 152 (276, 277), *165*, 256 (64), *300*, 507 (383, 384), *627*
Liptay, W. 473 (228), (821), *622, 638*
Lisnyanskii, L. I. 142 (165), *163*
Littke, W. 211 (2), *235*, 531 (564, 565), *631*
Little, E. L. 92 (202), 109 (358), *117*, *122*, 233 (64), *237*, 409 (222), *420*, 434 (86), 435 (86), 439 (97), 446 (116), 447 (116), 449 (97), 457 (97), 459 (97), 466 (86), 468 (86), 590 (86), 593 (86), 594 (86), 595 (86), 596 (86), 606 (86), 608 (86), 614 (97), *619*, 640 (3), *667*, 682 (31), 703 (166), *711*, *715*, 899 (49), 900 (49c), *942*
Little, E. L., Jr. 433 (79), 434 (79), 435 (79, 811), 436 (79), 437 (79),

574 (729), 575 (729), 590 (79), 593 (729), *619*, *635*, *637*, 681 (30), 682 (30), *711*, 899 (49), 900 (49), *942*
Little, J. G. 172 (14), *203*
Little, L. H. 143 (171), *163*, 196 (138), *206*
Liverman, J. L. 720 (16), *739*
Ljubavin, N. 75 (49), *112*
Lloyd, J. E. 152 (280, 283, 284), 153 (283, 284), *165*, *166*, 282 (170), *303*, 892 (30c), 893 (32), *941*
Lo, L. Y.-S. 132 (45), 135 (45), *160*
Lobanova, K. P. 474 (248), *623*
Locke, J. 502 (314, 316, 319, 320), *625*
Locker, D. J. 859 (67), *879*, 891 (19d), *941*
Locquin, R. 926 (107b,c), *945*
Loder, D. J. 933 (121), *945*
Loevy, H. 862 (92), *879*
Loewenstein, A. 854 (24), *878*
Loewenstein, D. 139 (136), 140 (136), *162*
Lofquist, R. 388 (140), *418*
Logothetis, A. L. 200 (175), *207*
Logowski, J. J. 473 (231), *622*
Lohaus, G. 102 (297), *120*
Lokensgard, J. P. 456 (169), 522 (169), *621*
Lokshin, B. V. 195 (115), *205*
Long, D. A. 434 (87), 448 (132), 507 (395), 508 (395), *619*, *620*, *627*
Long, F. A. 126 (9), *159*, 259 (82a), *301*
Long, L., Jr. 720 (15), *739*
Long, R. E. 533 (572), *631*, 661 (62), *669*
Longeray, R. 412 (235), *420*
Longfellow, J. M. 76 (62), *112*
Longo, G. 827 (159), *849*
Longone, D. T. 439 (98), *619*
Longuet-Higgins, H. C. 576 (734), *635*
Looney, C. E. 448 (125), 474 (257), *620*, *623*
Lopatina, K. I. 292 (218a), *304*
Lopez, L. 516 (454), *628*
Lorant, M. 758 (54), *786*
Lora-Tamayo, M. 361, 362 (62, 70), 364, 365 (74), 366 (78), 375, *416*, *417*
Lorber, V. 766 (121, 123, 125), *787*
Lorck, H. 728 (81), *741*
Lord, E. 463 (198), 469 (198), 480 (198), *621*
Lorenz, D. H. 857 (49), *878*
Lorquett, J. C. 23 (32), *64*
Losco, G. 861 (88), *879*
Losse, G. 248 (36), *299*
Lossing, F. P. 697 (122), *713*
Lotfield, R. B. 746, 761 (87, 89), 762 (89, 97), 763 (89, 97), 771, *784*, *786*, *787*
Lottermoser, A. 284 (181), *303*, 410 (227), *420*
Lough, C. E. 70 (15), *110*
Lo Vecchio, G. 833 (170, 171, 172), 834 (171, 181), 841 (209), 844 (209), *850*, *851*
Löwdin, P.-O. 19, 28 (30), *64*
Lowe, J. L. 274 (132), *302*
Lowe, J. U. 388 (145), *418*, 796 (36), *847*
Lowe, J. V. 347 (18), 348 (18), *415*
Lowell, J. R. 370 (92, 93), *417*
Löwenbein, A. 682 (34, 35), *711*
Lown, J. W. 541 (597), *632*
Lowry, P. H. 763 (100), *787*
Lowy, P. H. 761 (91), 766 (146), *787*, *788*
Loy, M. 452 (152), *620*, 909 (66a), *943*
Lucas, G. B. 695 (102), *713*
Lucas, H. J. 126 (7), *159*
Luck, J. M. 104 (308), *121*
Lucken, E. A. C. 242 (2), *298*
Luckhurst, G. R. 473 (237), *622*
Lüder, W. 225 (43), *237*, 448 (126), *620*
Luehrs, D. C. 131 (43), *160*
Luhrs, K. 331 (106, 107), 333 (106), *340*
Lukas, J. H. 768 (158), *788*
Lukaszewski, H. 93 (221), *118*
Lüpfert, S. 105 (316, 317), *121*
Lupinski, J. H. 536 (579), *631*, 667 (77), *669*
Lurin, A. F. 288 (194), *304*
Lusinchi, X. 98 (267), *119*, 886 (5), *940*
Luskin, L. S. 75 (51), *112*
Lussan, C. 139 (132), 140 (132), *162*
Lusskin, R. M. 291 (211), 292 (211), *304*, 914 (79), *944*
Lust, S. 503 (346), *626*
Lustig, M. 297 (237), *305*
Lütje, H. 133 (53), *160*
Lüttke, W. 138 (109), 139 (109), *161*
Lwowski, W. 354, 355 (36), *415*
Lygin, V. I. 143 (174), *163*, 196 (125), *206*

Lyle, G. G. 920, *944*
Lyle, R. E. 920, *944*
Lynn, J. W. 268 (108), 281 (108), *301*, 428 (11), *617*
Lynn, K. R. 431 (52), *618*
Lyons, L. E. 448 (137), 492 (187), *620*, *621*, 662 (66), 665 (66), *669*

Ma, S. K. 35 (49), 37 (49), *65*
Maas Reddy, J. van der 256 (64), *300*
Maccoll, A. 642 (16), 644 (16), *668*
MacDiarmid, A. G. 748 (20), 751 (40), *785*, 859 (62), *878*
MacDowell, D. W. H. 408 (221), *420*
Mack, C. H. 100 (284), *120*
Mack, W. 346 (13), 393 (163), *415*, *419*, 796 (30, 33), 803 (33), 808 (30, 33), 810 (30), 814 (33), 827 (30, 158), 837 (33), 838 (30), 840 (158), *846*, *849*
MacNulty, B. J. 523 (510), *630*
Madelung, W. 546 (631), 547 (631, 646), *633*
Madronero, R. 245 (18), *298*, 344 (5), 359 (5), 360 (5), 361 (62), 362 (62, 70), 364 (62, 74), 365 (74), 366 (78), 368 (5), 375 (105, 106), 376 (5), 380 (5), *414*, *416*, *417*
Madsen, P. 200 (177, 178), *207*
Maekawa, S. 513 (415), 517 (415), 518 (415), 519 (415), 524 (415), *627*
Maercker, A. 107 (338), *121*
Magai, M. 379 (121), *418*
Magat, E. E. 294 (228), *305*
Maginity, P. M. 384 (128), *418*
Magoon, E. F. 766 (145), *788*
Magos, L. 736 (135), *742*
Maguire, J. A. 144 (190), 145 (190), *163*
Mahadevan, S. 732 (94), 734 (94, 110a), *741*
Mahey, W. R. 370 (92), *417*
Mahler, W. 229 (53, 55), *237*, 441 (100), 459 (100), 463 (100), 468 (100), 471 (100), 472 (100), 473 (100), 474 (100), 532 (90), 534 (573), 535 (573), *619*, *631*, 640 (6a), 642 (6), 644 (6a), 646 (6), 655 (6), 658 (6, 44, 45), 659 (6), 660 (6), 661 (6, 44, 45), 664 (6, 45), 665 (6), 666 (6), *667*, *668*, 895 (36a, 36d), 901 (36), *941*, *942*
Mahowald, T. A. 768 (153), *788*
Maienthal, M. 85 (147), 109 (360), *115*, *122*
Maier, E. 432 (70), *618*
Maierhofer, A. 870 (150), *881*
Mailhe, A. 93 (219, 224), *118*
Maimind, V. I. 748, 751 (24), 763 (99), *785*, *787*
Makarov, S. P. 297 (237), *305*, 857 (51), 869 (138), *878*, *880*
Makarova, G. G. 129 (37), *160*
Maki, A. 516 (447), *628*
Maki, A. H. 502 (315, 317, 330, 332, 334, 340, 341), *625*
Malachowski, R. 92 (202), *117*, 580 (741, 742), 581 (741), 582 (741, 756, 757, 760), 583 (760), 586 (772), 588 (779), *635*, *636*, *637*
Malatesta, L. 124 (1), *159*, 873 (168, 171), 874 (178, 179, 181, 182), 875 (189, 200, 201, 202, 207), 876 (213, 216, 218, 219), *881*, *882*
Mallory, F. B. 818 (140, 142), *849*
Malz, H. 860 (79), *879*
Malzieu, R. 315 (44), *338*
Manahan, S. E. 133 (56), *160*
Manasse, O. 93 (215), *117*
Manassen, J. 523 (509), *630*
Mandel, L. 412 (236), *420*
Mander, M. R. 147 (216), *164*, 196 (150), *206*
Manecke, G. 502 (327), 528 (556), 579 (327, 556), *625*, *631*
Mangiapan, S. 835 (185), *850*
Mangini, A. 797 (42), 814 (129), 836 (196), *847*, *849*, *850*
Mann, B. R. 218 (29a), *236*
Mann, D. E. 548 (654), *633*
Mannhardt, H. J. 515 (441), 516 (441), *628*
Manousek, O. 181, 182 (50), *204*
Mansfield, K. T. 109 (335), *122*, 522 (504), *629*, 704 (170), *715*
Manske, R. H. F. 753 (46), *786*
Mantovan, R. 735 (115), *742*
Manz, G. 82 (112), *114*
Mao, C. H. 721 (24b), *739*
Marantz, S. 515 (482), *629*
Marchand, A. P. 223 (38), *237*
Marcinkowsky, A. E. 133 (57), *160*
Margalit, A. 139 (136), 140 (136), *162*
Margalit, Y. 854 (24), *878*
Margerison, R. 386 (134), *418*
Margulis, T. N. 473 (230), *622*, 658 (46), *668*
Mari, K. 288 (193), *304*

Mariella, R. P. 444 (112), 445 (112), *619*
Marion, L. 730 (86), *741*
Markgraf, J. H. 388 (141), *418*
Markham, Clare M. 187 (74), *204*
Märkl, G. 460 (185), 520 (185), *621*
Markova, T. A. 178 (34), *203*
Marquarding, D. 860 (69), 868 (136), *879*, *880*
Marsel, C. J. 513 (422), *627*
Marsh, F. D. 108 (347), *122*, 547 (649, 650), 548 (649, 666, 667, 670), 549 (649, 670, 672), 550 (674), *633*, *634*, 707 (189), 708 (189), 709 (197), *715*, 903 (56a, 56b), *943*
Marsh, N. H. 547 (643), *633*
Marshall, A. S. 104 (311), *121*
Marsman, J. W. 519 (494, 495), *629*
Marson, H. 273 (128), *302*
Marson, H. W. 98 (265), *119*
Marstokk, K. M. 134 (64), 135 (64), 146 (64), 147 (64), *160*
Martella, J. P. 519 (497), 520 (497), *629*
Martin, D. 79 (83), 90 (186), 91 (186), *113*, *116*, 252 (49), 253 (51), 293 (220), *299*, *305*, 841 (208), *851*, 862 (99), *879*
Martin, D. P. 930 (114b), *945*
Martin, D. R. 150 (247), *165*
Martin, E. L. 222 (34), *236*, 437 (94), 446 (117, 119), 447 (121), 482 (268), 490 (121), 491 (121), 514 (434), 595 (121), *619*, *620*, *623*, *628*
Martin, J. C. 687 (65), 692 (65), *712*
Martin, R. B. 260 (82c), 261 (82c), *301*
Martin, R. L. 150 (248), *165*
Martin-Smith, M. 158 (323), *166*
Martynova, L. L. 869 (138), *880*
Marvel, C. S. 78 (72), *112*, 142 (161), 148 (161), *163*
Marvel, G. S. 142 (162), 148 (162), *163*
Marxer, A. 275 (138), *302*, 391 (155), *418*
Mashlan, F. D. 142 (156), *162*
Maslan, F. D. 177 (30), *203*
Maslankiewicz, A. 173 (19), *203*
Matheson, M. S. 686 (53), 687 (53), *712*
Mathews, C. N. 901 (53a, 53b), 902 (53), 903 (53), *942*
Mathieson, D. W. 234 (66), 235 (66a), *237*
Mathis-Noël, R. 516 (454), *628*
Matkovic, B. 128 (30), *160*
Mato, F. 142 (158), *163*
Matschiner, J. T. 768 (152, 153, 154), *788*
Matsen, F. A. 146 (197), *163*
Matsubara, I. 156 (311, 313, 314), *166*
Matsui, H. 405 (201), *419*
Matsunaga, Y. 543 (603, 612, 613, 614), *632*, 660 (58), 663 (58, 67), 664 (58), *669*
Matteson, D. S. 892 (28), *941*
Matthews, C. N. 411 (234), *420*, 481 (267), 482 (267), 506 (363, 365, 380), 509 (411), 551 (380), *623*, *626*, *627*, 706 (183), *715*
Matthews, W. H. 78 (80), *113*
Matukawa, T. 182 (55), *204*
Matuszko, A. J. 812 (117), *848*
Matveeva, A. N. 186 (66), *204*
Matveeva, N. G. 474 (239, 241, 242, 243, 244, 245, 251), *622*, *623*
Matveeva-Kudasheeva, A. N. 189 (82), *205*
Matwiyoff, N. A. 133 (52), *160*
Mauguin, C. 546 (632), 547 (632), *633*
Maumy, M. 676 (19), 677 (19), 679 (19), 680 (25), *711*
Mavel, G. 139 (131), 140 (131), *162*
May, E. L. 315 (40), *338*
Mayer, R. 393 (161), *419*, 503 (349), *626*
Mayer, R. P. 755 (71), *786*
Mayes, N. 256 (65), *300*
Mayo, E. C. 221 (35), *237*
Mayo, F. R. 698 (133), *714*
Mazzini, R. 834 (179), *850*
McBee, E. T. 150 (257), *165*
McBride, J. J. 859 (60), *878*
McCane, D. J. 750, 751 (34), *785*
McCarter, J. A. 747 (10), *785*
McClellan, A. L. 136, *161*, 214 (11), *236*
McClellan, W. R. 809 (110), *848*
McClelland, B. J. 656 (41), *668*
McCleverty, J. A. 502 (314, 316, 319, 320), *625*
McClure, G. L. 517 (628), *633*
McColm, E. M. 78 (72), *112*
McConnel, H. 146 (198), *164*
McConnell, H. 641 (9), *668*

McConnell, H. M. 661 (65), *669*
McCormick, M. 437 (819), *638*
McCubbin, T. K. 813 (121), *848*
McCulloch, W. J. 356 (45), 357 (45), *415*
McDaniel, D. H. 218 (26, 28), *236*
McDonald, C. C. 502 (321), *625*
McDowell, C. A. 535 (574), *631*, 658 (50), *668*
McEachern, M., Jr. 437 (819), *638*
McElhill, E. A. 215 (14), *236*, 576 (735), *635*
McElroy, W. D. 771 (177), *789*
McElvain, S. M. 84 (136), 97 (256), *115*, *119*, 325 (87), 336 (87), *339*
McEwen, W. E. 286 (189), *303*
McFarland, J. W. 348 (19), *415*, 865 (135), 867 (135), *880*
McGeer, E. G. 92 (202), *117*, 409 (222), *420*, 446 (116), *619*, 640 (3), *667*, 682 (31), *711*
McGhie, J. F. 93 (208), *117*
McGrath, W. D. 795 (14, 15), *846*
McGregor, S. D. 368 (85), *417*
McGregor, W. H. 104 (306), *120*
McGuire, D. K. 127, 134 (63), *159* *160*
McHugh, G. P. 93 (209), *117*
McKay, A. F. 389 (148), 401 (188), *418*, *419*
McKay, F. C. 87 (157), *115*
McKay, G. F. 724 (40), *740*
McKee, R. H. 266 (102), 281 (102), *301*
McKenna, J. 859 (66), *879*
McKenzie, B. 78 (73), 79 (89), *113*
McKeon, J. E. 460 (184), *621*
McKillop, A. 394 (165), 395 (166), 396 (165), 402 (165), 409 (165), *419*
McKinney, T. M. 521 (685), *634*
McKoy, V. 39 (53), *65*
McKusick, B. C. 92 (202), 108 (349), *117*, *122*, 298 (244), *305*, 409 (222), *420*, 428 (3, 4, 6, 7), 438 (95), 441 (102), 446 (116), 447 (6, 116), 453 (157), 454 (157), 456 (157), 458 (176), 463 (190, 195), 465 (190, 207), 467 (207, 208), 468 (210, 211), 474 (258), 475 (258), 478 (208), 479 (195, 208), 480 (195, 207), 483 (95, 210), 536 (208), 606 (207), 607 (207), 614 (95), *617*, *619*, *620*, *621*, *622*, *623*, 640 (3), 650 (31), 655 (31), 657 (42), *667*, *668*, 682 (31), *711*, 860 (78), *879*
McKusick, V. A. 723 (39), *740*
McLafferty, F. W. 197, 198 (168), *207*
McLaughlin, D. E. 156 (318), *166*
McLean, A. D. 33 (41, 42), 46, 48 (58), 49 (59), *64*, *65*
McLean, J. D. 258 (80), *300*
McLoughlin, V. C. R. 432 (69), *618*
McMillan, A. 95 (243), *118*
McMurry, J. E. 835 (191a), *850*
McNally, J. G., Jr. 560 (710), 591 (788), *635*, *637*
McQuillan, G. P. 154 (299), *166*
McRitchie, D. D. 83 (126), 99 (126), *114*, 527 (538), *630*
McWeeny, R. 229 (48), *237*
Mead, J. F. 769 (168), *788*
Meakins, R. J. 149 (236), *164*
Mecke, R. 138 (109), 139 (109), *161*, 546 (635), *633*
Meek, D. W. 139 (137), 140 (137), *162*
Meerwein, H. 106 (336), *121*, 129, 130 (38, 40), 159 (38), *160*, 250 (43), 251 (43), 252 (43, 50), 253 (43, 50), *299*, 358, 365 (57), 375 (104), *416*, *417*, 912 (73a), *943*
Megna, I. S. 701 (148), *714*
Mehta, M. D. 92 (205), *117*
Meier, J. 131 (42), 133 (42), *160*
Meier-Mohar, T. 875 (206), *882*
Meier zu Köcker, I. 862 (102), *879*
Meindl, H. 366 (79, 80), *416*
Meinwald, J. 723 (32), *739*
Meinwald, Y. C. 108 (352), *122*, 453 (156), 455 (156), *620*
Meisel-Agoston, J. 773 (190), *789*
Meisenheimer, J. 71 (17, 21), *110*, *111*, 540 (583), *632*, 925 (103h), *945*
Meissner, B. 177 (28), *203*
Mekhtiev, S. D. 525 (526), *630*
Melby, L. R. 229 (55), *237*, 502 (323), 532 (90), 534 (573), 535 (573), *619*, *625*, *631*, 640 (6a), 642 (6), 644 (6a), 646 (6), 655 (6), 658 (6), 659 (6), 660 (6, 57), 661 (6, 57), 664 (6, 57), 665 (6), 666 (6, 76), *667*, *669*, 895 (36a, 36d), 901 (36), *941*, *942*
Mellish, S. F. 687 (62), *712*
Melone, G. 738 (141), *742*
Melton, J. W. 78 (80), *113*
Menefee, E. 535 (577), *631*

Mentzer, J. 720 (12), *739*
Mercer, A. J. 551 (683), *634*
Merer, A. J. 705 (175), *715*
Meresz, O. 391 (153), *418*
Meriwether, L. S. 524 (516), *630*
Merkl, A. W. 661 (65), *669*
Merrifield, R. E. 492 (287), 517 (287), *624*, 640 (4), 642 (4), 650 (4), 651 (4), 654, 655 (4), 656 (4), *667*, 895 (36b), 901 (36), *941*
Mersch, R. 129 (38), 130 (38, 40), 159 (38), *160*, 250 (43), 251 (43), 252 (43, 50), 253 (43, 50), *299*, 358 (57), 365 (57), 375 (104), *416*, *417*, 912 (73a), *943*
Mertz, R. 553 (694), *634*
Merz, J. H. 699 (139), *714*
Merz, T. 735 (118), *742*
Merz, V. 84 (131), *114*
Merz, W. 865 (131), 867 (131), *880*
Metcalfe, L. D. 179 (40), 180 (41), *203*, *204*
Mettee, H. D. 142 (170), *163*
Mettler, C. 99 (274), *120*
Metzger, H. 286 (190), *303*
Meye, W. 687 (63), 690 (63), *712*
Meyer, D. D. 150 (257), *165*
Meyer, E. 551 (688), *634*, 706 (184), *715*, 901 (55), *943*
Meyer, E. von 283 (174), *303*
Meyer, J. 875 (196), *882*
Meyer, V. 682 (33), *711*, 815 (134, 135), *849*
Meyers, A. I. 291 (211), 292 (211, 216), *304*, 359 (59, 60), 360 (59), 361 (63, 67, 69), 367 (60, 82), 368 (83, 84, 86), 370 (59, 86, 91), 372 (86, 96, 98), 378 (113), 399 (175, 176, 177, 178), *416*, *417*, *419*
Meyr, R. 854 (21), 855 (21), 857 (21), 863 (116), 864 (116), *877*, *880*
Michael, A. 70 (13), 71 (13), *110*
Michaelis, A. 99 (271), *119*
Michaelis, L. 659 (53), *669*
Michaels, R. 726 (74), 727 (74), 728 (74, 80), *740*, *741*
Michel, E. 582 (761), *636*
Michelman, J. S. 822 (153), *849*
Middleton, W. J. 92 (202), 109 (358), *117*, *122*, 233 (64), *237*, 409 (222, 223, 224), *420*, 434 (86), 435 (86, 89), 439 (97, 99), 446 (116), 447 (116), 449 (97), 457 (97), 459 (97), 463 (192), 466 (86), 467 (89), 468 (86), 475 (259), 476 (259), 478 (259), 484 (276, 277), 485 (276), 494 (290), 495 (89, 292), 496 (89, 293, 294), 498 (294), 555 (702), 556 (702), 558 (702), 560 (710, 711), 561 (714), 574 (729), 575 (729), 590 (86), 591 (788), 593 (86, 729), 594 (86), 595 (86, 259), 596 (86), 606 (86), 608 (86), 614 (97), *619*, *621*, *623*, *624*, *634*, *635*, *637*, 640 (3), *667*, 682 (31), 703 (166), *711*, *715*, 869 (139), *880*, 899 (49), 900 (49c), *942*
Mielert, A. 449 (139), *620*
Mietzsch, F. 449 (141), 461 (141), *620*
Miftakhova, R. A. 264 (97), 265 (97), *301*
Mignonac, G. 88 (176), 89 (176), *116*, 310 (19), 324 (83), *338*, *339*
Migrdichian, V. 308 (1), 320 (1), *337*, 344 (3), *414*, 428 (9), *617*, 854 (9), *877*
Migridician, V. 281 (162), *303*
Mihailovic, M. L. 103 (303), *120*
Mihina, J. S. 388 (138), *418*
Mikhlina, E. E. 917 (86), 922 (86), *944*
Miki, T. 686 (56), *712*
Mikulski, F. A. 95 (235), *118*, 926 (105a), *945*
Miljanich, P. 720 (10), *739*
Mill, T. 73 (35), *111*
Miller, C. S. 102 (294), *120*
Miller, F. A. 434 (88, 813), 437 (88, 813), 448 (126), 513 (416), 514 (416), 515 (416), 516 (416, 456, 457, 458, 464), 517 (416), *619*, *620*, *627*, *628*, *637*
Miller, F. F. 699 (141, 142), *714*
Miller, G. A. 543 (620), *632*
Miller, J. 273 (129), *302*
Miller, J. B. 456 (168), *621*
Miller, J. M. 150 (261, 262), 151 (261), *165*
Miller, L. A. 519 (492), *629*
Miller, L. L. 293 (223d), *305*, 767 (135), *788*
Miller, R. E. 519 (492), *629*
Miller, S. A. 275 (137), *302*
Miller, W. v. 76 (59), *112*
Millich, F. 862 (108), *880*, 890 (20a, 20b), 891 (20), *941*
Milligan, D. E. 548 (654, 655, 660, 662, 665), *633*, *634*, 708 (194), *715*

Mills, J. F. 146 (201), *164*
Mills, O. S. 875 (190), *882*
Milun, A. J. 197 (121), *205*
Mini, V. 796 (29), 803 (29), 812 (29), 813 (29), 814 (29), 822 (29), 823 (29), 824 (29), 826 (29), 827 (29), 830 (29), 831 (29), 834 (29), *846*
Minieri, P. P. 291 (211), 292 (211), *304*, 358 (58), *416*, 911 (71a), *943*
Minisci, F. 697 (125, 126), 701 (149), *713*, *714*, 844 (219), *851*
Minnis, J. W. 271 (118), *302*
Minor, C. A. 386 (135), *418*
Mintel, R. 735 (128, 129), *742*
Mion, L. 680 (25), *711*
Mischke, P. 873 (167), *881*
Mishriky, N. 106 (335), *121*
Misono, A. 94 (225), *118*, 517 (483), 524 (515), *629*, *630*
Misumi, F. 369 (88), *417*
Mitchell, J., Jr. 175 (22), *203*
Mitchell, P. W. D. 541 (598), *632*
Mitin, J. V. 861 (84), *879*
Mitra, S. S. 137 (100), 138, 140 (100), *161*, 196 (127), *206*
Mitta, A. E. A. 771 (184), *789*
Miwa, T. 294 (224), *305*
Miyatake, K. 371 (94), *417*
Mizoguchi, A. 667 (78), *669*
Mizon, J. 749 (28), 750 (28), 764 (28, 109), 781 (28, 109), *785*, *787*
Mizoule, J. 735 (114), *742*
Mizuchima, Y. 474 (249), *623*
Moccia, R. 796 (38), 797 (38), 813 (38), *847*
Mochel, W. E. 229 (55), *237*, 532 (90), 534 (573), 535 (573), *619*, *631*, 640 (6a), 642 (6), 644 (6a), 646 (6), 655 (6), 658 (6), 659 (6), 660 (6), 661 (6), 664 (6), 665 (6), 666 (6), *667*, 895 (36a, 36d), 901 (36), *941*, *942*
Modest, E. J. 397 (172, 173), 402 (191), *419*
Moffat, J. 103 (302), *120*
Moffat, J. B. 62 (65), *65*, 516 (478), 528 (546), *629*, *631*
Moffatt, J. G. 560 (712), *635*
Moffit, W. 2 (3), *63*
Mog, D. M. 697 (124), *713*
Mohr, C. C. 139 (125), *162*
Mohr, G. 500 (310), *625*
Mole, B. R. 260 (83), *301*
Mole, T. 152 (281), *165*
Moll, H. 336 (115), *340*
Moll, N. G. 548 (660), *633*
Moller, C. K. 517 (480), *629*
Möller, F. 291 (213), *304*
Mommaerts, H. 516 (443), *628*
Momsenko, A. P. 248 (37), *299*
Monagel, J. J. 690 (79), *712*
Monagle, J. J. 75 (53), 110 (53), *112*
Monahan, A. R. 356 (48), 357 (48), *416*
Monforte, P. 833 (172), *850*
Mönkemeyer, K. 873 (172), 875 (172), 876 (215), *881*, *882*
Moodie, R. B. 922 (96), *944*
Mooney, W. T. 314 (33), *338*
Moore, B. P. 723 (36), *739*
Moore, C. W. 284 (179), *303*
Moore, J. A. 398 (174), 401 (187), *419*
Moravec, J. 158 (325), *166*
Moreau, J. 105 (325), *121*
Morehead, B. A. 428 (22, 23), 429 (22), *617*
Moreland, W. T., Jr. 218 (27), *236*
Moreu, C. 97 (254), *119*
Morgan, D. J. 265 (98), *301*
Morgenthau, J. L., Jr. 315 (42, 43), *338*, 748 (22), *785*
Mori, R. 99 (273), *119*
Moriga, H. 667 (78), *669*
Morita, Z. 560 (713), *635*
Moritz, K. 463 (196), *621*
Morrill, I. C. 386 (135), *418*
Morrison, A. L. 412 (237, 238), *421*
Morrison, H. 698 (130), *714*
Morrison, J. D. 146 (196), *163*
Morrissette, R. A. 179 (35), *203*
Morrocchi, S. 347 (15), *415*, 835 (187), 837 (201), 839 (201, 206), 840 (201), *850*, *851*
Morrow, T. 795 (14), *846*
Moser, H. 448 (126), *620*
Moser, R. E. 411 (234), *420*, 506 (363, 365, 380), 509 (411), 551 (380), *626*, *627*, 706 (183), *715*, 901 (53a, 53b), 902, 903 (53), *942*
Mosettig, E. 308 (2), 309 (2), *337*, 886 (3), *940*
Mosher, H. S. 314 (33), *338*
Mosher, W. A. 93 (223), *118*
Moskvitin, M. N. 180 (47), *204*
Moss, J. A. 738 (138), *742*
Moureu, C. 92 (204), *117*, 513 (413, 414, 430, 431), 514 (413, 414, 430), 516 (413, 414, 430, 431),

517 (414), 518 (430, 487, 488), 519 (430, 496), *627, 628, 629*
Mourik, A. S. van 899 (47c), *942*
Mousseron, M. 78 (71), 89 (184), *112, 116,* 914 (78b), *944*
Movravec, J. 137 (102), *161*
Mower, H. F. 433 (79), 434 (79), 435 (79), 436 (79), 437 (79), 463 (190), 465 (190), 468 (211), 502 (321), 590 (79), *619, 621, 622, 625,* 657 (42), *668,* 899 (49), 900 (49a), *942*
Mowry, D. T. 68 (1), 69 (1), 71 (19), 72 (29), 73 (29), 74 (45), 75 (56), 76 (57), 77 (1), 79 (88), 80 (96), 81 (107), 82 (111), 83 (125), 84 (134, 138), 85 (142, 148), 86 (150, 154), 87 (162), 88 (172, 173, 174), 90 (187), 91 (192), 92 (1, 196), 93 (214), 97 (251), 101 (290), *110, 111, 112, 113, 114, 115, 116, 117, 119, 120,* 294 (227), *305,* 765 (116), *787,* 886 (1), *940*
Moyer, J. D. 747, 784 (226, 227), *785, 790*
Mrozinski, W. 88 (168), *116*
Muelder, W. W. 764 (110), *787*
Muetterties, E. L. 543 (615, 616), *632*
Mugnaini, E. 180 (42), *204,* 836 (197), *850*
Mühlbauer, E. 276 (144), *302*
Muhs, M. A. 751 (37), *785*
Muir, R. M. 732 (102), *741*
Mukaiyama, M. 796 (21), 806 (21), *846*
Mukaiyama, T. 93 (211, 212, 214), 104 (313), *117, 121,* 796 (22), 806 (22), 808 (22), 835 (22), 839 (22), *846,* 857 (45, 46), 858 (53), *878*
Mukeiyama, T. 103 (301), *120*
Mukherjee, R. 730 (87, 88), *741*
Mukherjee, T. K. 642 (13), 645 (13), 650 (13), 658 (13), *668*
Muller, E. 308 (6), *337*
Müller, E. 91 (193a), 93 (208), *117,* 709 (201, 202), *715,* 857 (52), 869 (140), *878, 880*
Muller, H.-J. 488 (280), *624*
Müller, W. 875 (196), *882*
Müller-Warmuth, W. 473 (234), *622*
Mulliken, R. S. 2, 8, 60 (2), 20, 21, 54 (61), *63, 64, 65,* 136 (94), 143, 144 (176, 181), 146 (176), *161, 163,* 640, *667*
Munavalli, S. 556 (704), *634*
Munoz, G. G. 361 (63), 366 (78), 375 (105, 106), 378 (113), *416, 417*
Munsche, D. 775 (198), *789*
Murahashi, S. 513 (415), 516 (451), 517 (415), 518 (415), 519 (415), 524 (415), *627, 628,* 779 (212), *790*
Murakami, D. 926 (105e), *945*
Murakami, S. 141 (146), *162*
Murano, M. 202 (188), *207*
Murata, Y. 179 (38), *203*
Murdoch, H. D. 874 (177), *881*
Murphy, F. X. 361 (61), *416*
Murphy, M. 187 (74), *204*
Murphy, M. E. 187 (72), *204*
Murray, A., III 746 (2), 751 (2), *784*
Murray, A. G. 547 (642), *633*
Murray, F. E. 129, 141 (35), 149, *160,* 216 (19), *236*
Murray, J. P. 260 (83), *301*
Murray, M. A. 528 (552), 543 (552), *631*
Murray, T. S. 364 (73), *416*
Murray-Rust, P. 528 (555), *631*
Murti, V. V. S. 724 (49), *740*
Murty, T. S. S. R. 141 (147), *162*
Musakin, A. P. 748 (17), *785*
Musante, C. 96 (248), *119*
Muschkin, P. A. 898 (43c), *942*
Musso, H. 336 (115), *340*
Müsso, H. 772 (188), *789*
Mussokin, A. P. 754 (74), *786*
Mustafa, A. 683 (40), *711*
Mutz, G. 103 (299), *120*

Nabeya, A. 368 (83), *417*
Nagabuko, K. 797 (54), 822 (54), *847*
Nagakura, S. 136 (94), 138 (117), 144 (193), *161, 162, 163,* 472 (223), 527 (544), 528 (544, 545, 548, 549), *622, 631,* 642 (14), 652 (14), 653 (14), 656 (14), *668*
Nagarjan, C. 516 (461), *628*
Nagarjan, G. 516(462), *628*
Nagasaka, A. 696 (110), *713,* 861 (91), *879*
Nagashima, N. 508 (825), 509 (825), *638*
Nagata, W. 71 (20), *110,* 316 (47), 319 (47), *338,* 926 (105d), *945*
Nagy, D. E. 546 (633), *633*
Nagy, S. M. 92 (200), *117*
Naik, K. G. 492 (283), *624*
Naiman, J. 914 (806), *944*

Najzarek, Z. 173 (19), *203*
Nakagawa, I. 135 (67), *160*
Nakagawa, K. 104 (305, 309, 310), *120*, *121*, 904 (58a, 58b), *943*
Nakagura, S. 144 (182), *163*
Nakamoto, T. 735 (126), *742*
Nakamura, A. 508 (825), 509 (825), *638*
Nakamura, K. 560 (713), *635*
Naldini, L. 874 (186), *882*
Nambu, H. 104 (313), *121*, 857 (45), *878*
Naraba, T. 474 (249), *623*
Narr, B. 93 (208), *117*, 857 (52), *878*
Nasarov, J. N. 74 (39), *111*
Nash, B. W. 541 (597), *632*
Nash, C. P. 139 (127), *162*
Nash, E. G. 919 (92), *944*
Nashanyan, A. O. 361 (65, 66), *416*
Nasielski, J. 196 (165, 166), *207*
Nast, R. 870 (145), *881*
Nasutavicus, W. A. 381 (123, 124), *418*
Naumann, R. 430 (39), *618*
Naumann, W. 696 (115), *713*
Nayler, J. H. C. 92 (205), *117*, 760 (82), *786*
Naylor, P. G. 710 (204), *716*
Nazarov, J. N. 81 (103), *114*
Nazova, S. A. 523 (507), *630*
Needleman, S. B. 356 (41), *415*
Neelakantan, L. 76 (63), *112*
Nef, I. U. 853 (4), 854 (4), 855, 860 (4), *877*
Nef, J. U. 267, *301*, 429 (36), *618*, 792, 799 (61), 807 (3), *846*, *847*, 886 (7i), *940*
Neff, D. L. 689 (73), 693 (73), 696 (73), *712*
Neff, N. 755 (80), *786*
Negita, H. 201 (180), *207*
Nehring, R. 390 (151), *418*
Neidlein, R. 104 (315), *121*, 864 (121), 867 (134), 870 (147), *880*, *881*
Neilson, D. G. 264 (94), *301*
Neish, A. 749 (30), *785*
Neish, A. C. 759 (58), 767 (141), *786*, *788*
Nelson, E. R. 85 (147), *115*
Nelson, J. 724 (45, 57), 725 (45, 58), *740*
Nelson, J. P. 197 (121), *205*
Nencini, G. 722 (28), *739*
Nentwig, J. 130 (40), *160*, 252 (50), 253 (50), *299*, 375 (104), 377 (107), *416*, *417*
Neri, A. 861 (90), *879*
Nersesyan, L. A. 361 (64, 65, 66), *416*
Nesbet, R. K. 19, 40, 44 (56), 45 (56), *64*, *65*
Nesmeyanov, A. N. 91 (191), 104 (312), *116*, *121*, 129 (37), *160*
Neumann, P. 575 (731), *635*, 650 (33), *668*
Neumann, W. P. 861 (83), *879*
Neustaedter, P. J. 543 (623), *632*
New, R. G. A. 853 (6, 8), 854 (6, 8), 865 (6), *877*
Newallis, P. E. 274 (133), *302*
Newberry, G. 273 (126), *302*
Newman, M. S. 78 (75), 82 (116), 83 (121), 85 (145), 100 (283), *113*, *114*, *115*, *120*, 294 (224), *305*, 933, *945*
Newton, J. 459 (178), 463 (178, 203), 464 (178, 203), 465 (203), *621*
Nicholaeva, N. M. 180 (46), *204*
Nicholl, L. 96 (250), *119*
Nichols, J. 109 (360), *122*
Nicholson, A. J. C. 146 (196), *163*
Niclas, H.-J. 79 (83), *113*
Nicodemus, O. 898 (44), *942*
Nicolaides, D. N. 828 (160), *849*
Niederhauser, W. D. 483 (273), *624*
Nielsen, A. T. 862 (96), *879*
Nieuwland, J. A. 68 (3), 69, *110*
Nigam, S. N. 724 (43, 45), 725 (43, 45, 60, 65), 726 (43, 71), *740*
Nikolajeva, T. V. 857 (51), *878*
Nilson, S. 91 (193b, 194), 110 (364), *117*, *122*
Nilsson, S. 689 (76b), 697 (76b), 710 (205), *712*, *716*
Nishida, K. 862 (95), *879*
Nishioka, A. 474 (249), *623*
Nitzschmann, R. E. 435 (815), *638*
Noack, J. 674 (9), *711*
Noake, H. 474 (249), *623*
Nobel, E. G. 506 (364), 508 (364), *626*
Noda, S. 524 (515), *630*
Noel, D. 838 (203), *851*
Noguchi, H. 517 (483), 524 (515), *629*, *630*
Nohe, H. 525 (523), *630*
Nohira, H. 93 (212), *117*
Noland, W. E. 463 (193), 470 (216), *621*, *622*

Noll, C. I. 258 (79), *300*
Noll, K. 897 (40b), *942*
Nolte, E. 85 (144), *115*
Nomori, H. 543 (619), *632*
Nomura, T. 156 (312), *166*
Normant, H. 79 (85), *113*
Norris, W. P. 388 (143), *418*
Norton, T. B. 724 (55), *740*
Nose, Y. 474 (250), *623*
Noszkó, L. H. 777 (205), *789*
Novak, A. 138 (124), *162*
Novikov, S. S. 107 (339), *121*
Nowak, B. E. 920 (94a), *944*
Nozaki, H. 560 (713), *635*
Nshanyan, A. O. 361 (64), *416*
Nyberg, K. 293 (223b, 223c), *305*, 710 (211), *716*, 915 (81), *944*
Nystrom, R. F. 314 (35), 326 (35, 92), 327 (92), 332 (35, 92), *340*, 747, *785*

Oakwood, T. S. 81 (107), *114*, 929 (113b), *945*
Oblak, S. 156 (316), *166*
O'Brien, D. E. 872 (158), *881*
Occolowitz, J. L. 854 (22, 25), *877*, *878*
Ochiai, E. 84 (137), *115*
Oda, E. 288 (195), *304*
Oda, R. 369 (88), *417*, 696 (110), *713*, 854 (13), 860 (74), 861 (91), 862 (95, 97), 865 (13), *877*, *879*
Odaina, Y. 108 (353), *122*
O'Donovan, J. P. 724 (47), *740*
Oediger, H. 107 (340), *121*, 591 (787), *637*
Oehler, E. 506 (371), 507 (371), *626*
Oettmeier, W. 870 (150), *881*
Offermann, K. 854 (15), 855 (15, 14), 857 (15, 40), 864 (122), 867(126, 129), 868 (129, 136), *877*, *878*, *880*, 887, 891 (9a), *940*
Ofner, A. 95 (242), *118*
Ofner, P. 400 (179), *419*
Oganesov, S. S. 523 (508), *630*
Ogata, Y. 263 (91), *301*
O'Gee, R. C. 429 (26), *617*
Ogura, H. 506 (372), *626*
Ogura, K. 182 (52), *204*
Ohgo, Y. 280 (159), *303*
Ohlinger, H. 273 (125), *302*
Ohmori, M. 476 (262), *623*
Ohta, M. 182, 186 (53), *204*, 337 (116), *340*, 401 (190), *419*
Ohtsuka, Y. 476 (262), *623*
Oikawa, E. 288 (192), *304*
Oishi, T. 379 (121), *418*
Okamoto, M. 857 (45), *878*
Okamoto, Y. 218 (29c, 29d), *236*, 543 (614), *632*
Okano, M. 369 (88), *417*, 860 (74), 862 (97), *879*
Okatova, G. P. 748 (15), *785*
Øksne, S. 818 (141), *849*
Olah, G. 246 (22b), 250 (22b), *299*
Olah, G. A. 254 (54), *300*, 912 (73b), *943*
Oliver, W. F. 761 (92), *787*
Oliveri-Mandala, E. 349 (22), *415*, 432 (65), *618*, 863 (109), *880*
Olivier, S. C. J. 245 (22a), *299*
Olivson, A. 202 (184), *207*
Olofson, R. A. 822 (153), *849*
Olofsson, B. 689 (76c), *712*
Olynyk, P. 747, 748 (22), *785*
Omran, S. M. Abdel Rahman 837 (202), *851*
Ong, S. H. 675 (18), *711*
Onishchenko, A. S. 356 (39), *415*
Onishi, T. 541 (593), 543 (593), *632*
Onoda, K. 506 (358), 507 (396), 508 (399), *626*, *627*
Önol, N. 694 (96), *713*
Onoue, H. 104 (309, 310), *121*, 904 (58a, 58b), *943*
Onyon, P. F. 694 (95), *713*
Onyszchuk, M. 150 (261, 262), 151 (261), *165*
Opitz, G. 865 (131, 132), 867 (131, 132), *880*
Orgel, L. E. 256 (66), 289 (202, 204), *300*, *304*, 406 (204), 410 (225), *420*, 442 (106), 506 (374, 375, 376, 377, 379), 508 (375), 509 (375, 376), 513 (424, 425), 518 (425), 519 (425), 591 (106), 593 (106), *619*, *626*, *628*, 643 (18), *668*
Ormerod, M. G. 472 (222), *622*
Oro, J. 289 (203), *304*, 411 (234), *420*, 506 (362), *626*
Orr, S. F. D. 196 (167), *207*, 492 (285), *624*
Orwille-Thomas, W. S. 156 (310), *166*
Orye, R. V. 150 (245), *165*
Osa, T. 94 (225), *118*
Osbond, J. M. 769 (165), *788*
O'Shaughnessy, M. T. 75 (53), 110 (53), *112*, 686 (54), 687 (54), *712*

Osiecki, J. H. 472 (226), 473 (226), 543 (226), *622*
Osieki, J. H. 658 (48), 661 (48), 664 (48), *668*
Ossipov, V. A. 748 (17), *785*
Ossipow, W. A. 754 (74), *786*
Osswald, G. 88 (178), *116*, 434 (82), 435 (82), *619*
Osteryoung, R. A. 183 (59), *204*
Osugi, J. 694 (99), *713*
Otsuka, T. 85 (145), *115*
Ott, E. 576 (733), *635*
Ott, J. B. 150 (246), *165*
Ottenberg, A. 472 (226), 473 (226), 543 (226, 608), *622*, *632*, 658 (48), 661 (48), 664 (48, 71), *668*, *669*
Ötvös, L. 773 (190), 777 (205), *789*
Ourission, G. 556 (704), *634*
Overberger, C. G. 75 (53), 76 (61), 110 (53), *112*, 686 (54), 687 (54, 58, 59, 60, 61, 64), 688 (64), 690 (77, 78, 79, 80), 691 (83), *712*, 797 (55, 56), 810 (55, 56), 811 (56), 822 (55), 823 (56), 834 (55, 56), 841 (55, 56), *847*
Overend, J. 448 (126), *620*
Owen, D. A. 152 (278), *165*, 892 (30d), *941*
Owen, T. B. 135 (86), *161*, 216 (18), *236*
Owens, M. L. 457 (171), *621*
Owyang, R. 543 (623), *632*
Oxley, P. 101 (293), *120*, 271 (118), *302*, 391 (154), *418*
Ozolins, M. 449 (138), *620*

Pääbo, K 768 (170), *789*
Pack, D. E. 770 (174), 783 (174), *789*
Pack, R. T. 614 (807), *637*
Packer, J. 218 (29a), *236*
Packham, D. I. 529 (562), *631*
Padoa, G. A. 876 (218), *882*
Paetzold, P. 842 (214), *851*
Page, F. M. 448 (134), 511 (134), 527 (134), 531 (134), 575 (134), *620*, 648 (28), *668*
Pailthorp, J. R. 524 (519), *630*
Pajaro, G. 767 (136), *788*
Pala, G. 196 (164), *207*
Palenik, G. J. 606 (802), *637*
Palfray, L. 169 (6), *203*
Pallares, E. Sodi 722 (31), *739*
Pallini, U. 701 (149), *714*
Pallotti, M. 346 (14), 348 (14), *415*
Palm, V. A. 125 (5), 127 (5), 156 (5), *159*
Palmer, K. J. 517 (481), *629*
Palotti, M. 832 (167a), 835 (167), 840 (167a), *850*
Pamain, E. 854 (31), *878*
Panattoni, C. 459 (183), *621*
Panayides, S.G. 766 (126), *787*
Panella, J. P. 78 (73), 86 (153), *113*, *115*
Panizzi, L. 800 (64), *847*
Panov, I. K. 447 (122), *620*
Pantelei, T. I. 197 (122), *206*
Pao, Y. H. 535 (577), *631*
Papa, A. J. 900 (51a), 901 (51b), *942*
Papantoniou, C. 291 (214), *304*, 368 (87), *417*
Paquot, C. 101 (291), *120*
Parameswaram, K. N. 94 (226), *118*
Parham, F. M. 696 (111), *713*
Parimskii, A. I. 180 (44), *204*
Parini, V. P. 529 (563), *631*
Paris, G. Y. 389 (148), *418*
Parish, R. C. 218 (27), *236*
Park, J. D. 703 (162), *715*
Parker, V. D. 91 (194), *117*, 710 (210), *716*
Parris, C. L. 291 (215), 296 (231), *304*, *305*, 913 (76), *943*
Parrod, J. 408 (218), *420*
Parrotta, E. W. 327 (96), *340*
Parsons, J. 723 (37), *740*
Partos, R. D. 463 (197), *621*
Partridge, M. W. 95 (237), 101 (293), *118*, *120*, 404 (197), *419*, 751 (186), 770 (186), 784 (186), *789*
Passavant, S. C. 74 (47), *111*
Passerini, M. 93 (215), *117*, 860 (80), 861 (87, 89, 90), 863, 869 (141), 871 (153, 156), *879*, *880*, *881*
Passlacqua, T. 432 (65), *618*
Past, J. 202 (184), *207*
Pasternak, Y. 81 (106), *114*
Pastushak, N. O. 310 (17), *338*
Pasynkiewicz, S. 152 (285, 286), *166*, 893 (33a), *941*
Patai, S. 69 (16), 70 (16), 105 (321), 106 (331), *110*, *121*
Patalakh, I. I. 523 (507), *630*
Patel, A. J. 724 (50), *740*
Pater, R. 275 (136), *302*
Patrick, T. B. 408 (221), *420*
Patschke, L. 770, 781 (176), *789*
Patterson, E. 104 (311), *121*, 904 (59), *943*

Patterson, J. M. 105 (327), *121*
Pattison, F. L. M. 79 (86), *113*
Patton, R. H. 82 (109), *114*, 588 (780), *637*
Paul, I. C. 451 (146), *620*
Paul, M. A. 126 (9), *159*, 259 (82a), *301*
Paul, R. 806 (90), 835 (90, 183), *848*, *850*
Paulik, F. 277 (154), 278 (154), 283 (154), *302*
Pauling, L. 210 (1), *235*, 517 (481), *629*, 793 (7), *846*, 892 (27), *941*
Pauling, L. C. 2, 60 (1), *63*
Paushkin, Ya. M. 288 (194), *304*, 523 (507), *630*
Pauson, P. L. 875 (190), *882*
Pautler, B. G. 268 (107), *301*, 428 (24), 429 (38), *617*, *618*
Pavan, M. 722 (28), *739*
Pavlovskaja, I. V. 869 (138), *880*
Pawelzik, K. 268 (109), *301*
Payne, G. B. 263 (89, 90, 92, 93), *301*, 554 (696, 697), *634*
Peak, D. A. 274 (132), *302*
Pearlson, W. H. 703 (162), *715*
Pearson, R. G. 227 (50), *237*, 614 (806), *637*
Pease, D. C. 687 (66), *712*
Pease, R. N. 432 (71, 72), *618*
Pecile, C. 196 (139), *206*, 872 (161), *881*
Pedersen, C. 349 (23), *415*
Pedersen, C. J. 457 (172), 537 (172), *621*, 704 (169), *715*
Peiffer, G. 81 (106), *114*
Peiker, A. L. 258 (79), *300*
Pelley, R. L. 315 (39), *338*
Penfold, B. R. 507 (383, 384), *627*
Pen'kovs'kii, V. V. 472 (219), *622*
Peover, M. E. 190 (89), 191 (98), *205*, 527 (543), 528 (547), 543 (547, 601, 604), *631*, *632*, 642 (16), 644 (16, 21), 646 (16b, 21), 654, 655 (21), *668*
Peratoner, A. 349 (21), *415*, 432 (66), *618*
Perciabosco, F. 442 (105), *619*
Perelygin, I. S. 132 (46), 138 (122a), *160*, *162*, 196 (128), *206*
Perepletchikova, E. M. 127 (22), 128 (22), *159*
Perewalowa, E. G. 91 (191), *116*
Perez, M. G. 362 (70), *416*
Perrine, T. D. 315 (40), *338*
Persmark, U. 459 (181), 463 (181), 465 (181), *621*
Person, W. B. 144 (189), 145 (189), 146 (203, 204, 205, 206, 207), 157 (189, 320), 158 (189), *163*, *164*, *166*
Perveev, F. Y. 86 (152), *115*
Peška, J. 289 (199), *304*, 523 (511), 524 (513, 514), *630*
Peter, J. P. 202 (183), *207*
Peters, G. A. 267 (105), 268 (105), 269 (105, 111), 270 (112), *301*
Petersen, J. M. 391 (158), *418*
Petersen, S. 503 (345), *626*
Peterson, J. H. 691 (82), 692 (82), *712*
Peterson, L. I. 108 (351), *122*, 200 (172), *207*, 442 (103), *619*, 685 (45), *711*
Peterson, L. R. 197 (121), *205*
Peterson, S. W. 128 (30, 31), *160*, 243 (12), *298*
Petrellis, P. C. 551 (688), *634*, 706 (184), *715*, 901 (55), *943*
Petri, H. 682 (37), *711*
Petrov, A. A. 516 (469), *629*
Petrovich, P. I. 180 (47), *204*
Petterson, R. C. 363 (72), *416*
Pettit, D. J. 370 (92), *417*
Pettit, R. 473 (229), 550 (677), 551 (677), *622*, *634*, 658 (47), *668*, 704 (172), *715*
Pfankuch, E. 320 (60), *339*
Pfeffer, M. 724 (57), 725 (58, 59), *740*
Pfeil, E. 72 (27), 87 (163), *111*, *115*, *116*, 373 (101), *417*, 721 (24a), *739*
Pflüger, M. 766 (149), *788*
Phillips, D. D. 550 (679), 551 (679), *634*
Phillips, W. D. 229 (52), *237*, 473 (232), 492 (287), 502 (321), 517 (287), 535 (576), *622*, *624*, *625*, *631*, 640 (4), 642 (4), 643, 650 (4, 17), 651 (4), 654, 655 (4, 17), 656 (4), 657, 658 (43), 659 (51, 56), 660 (57), 661 (56, 57), 664 (57), *667*, *668*, *669*, 895 (36c), 896 (38), 901 (36), *942*
Pichat, L. 747, 748, 749, 750 (28), 751 (35), 760 (84, 85, 86), 762, 764 (28, 108, 109, 111), 765, 774, 775, 781 (28, 109, 195), 783 (14), *785*, *786*, *787*, *789*
Pickard, P. L. 276 (145), *302*, 892 (29b), *941*
Pickholz, S. 696 (120), *713*
Pierce, O. R. 150 (257), *165*, 410 (231), *420*
Piest, R. 75 (48), *111*

Pietra, F. 196 (139), *206*
Pietra, S. 311 (21, 22), 333, *338*
Pigenot, D. v. 875 (188), *882*
Piggott, H. A. 73 (30), 74 (38), 95 (38), *111*
Piguet, A. 858 (54), *878*, 925 (103b), 927 (103b), *944*
Pilgrim, F. J. 99 (275), *120*, 773 (192), *789*
Pillai, M. G. K. 854 (20), *877*
Pimenov, Y. D. 472 (218), *622*
Pimentel, G. C. 136 (96), *161*
Pinkney, P. S. 696 (121a), *713*
Pinner, A. 264, 269 (95), *301*
Pino, L. N. 428 (12), *617*
Piper, J. V. 412 (236), *420*
Pitocelli, A. R. 152 (275), *165*
Platonova, M. N. 187 (71), *204*
Platt, J. R. 146 (198), *164*, 641 (9), *668*
Plattner, P. 72 (28), *111*
Platz, R. 525 (523), *630*
Plaut, H. 92 (196), *117*, 291 (211), 292 (211), *304*, 914 (80a), *944*
Plejewski, R. 762 (105), *787*
Plieninger, H. 311 (23, 24, 25), 312 (26, 27), 313 (23, 24, 25), *338*, 781 (229), *790*
Pliva, J. 137 (102), 158 (325), *161*, *166*, 813 (122), *848*
Plöchl, J. 76 (59), *112*
Pocar, D. 835 (188, 189), *850*
Pockels, U. 683 (38), *711*
Pohland, H. 331 (105), *340*
Pokorny, J. 179 (36), *203*
Polaczkowa, W. 358 (51, 52, 53), *416*
Polak, L. S. 524 (512), *630*
Poláková, P. 137 (102), *161*
Pollack, M. W. 295 (230), *305*
Polland, R. 312 (29, 30), *338*
Polyak, S. 767 (139), *788*
Pomeroy, J. H. 93 (213), *117*, 933 (123), *945*
Pominov, I. S. 132 (48), *160*, 196 (129), *206*
Pommer, H. 107 (339), *121*
Pondesva, C. 401 (188), *419*
Pongratz, A. 531 (567), *631*
Ponseti, I. V. 724 (51, 52, 56), *740*, 760 (114), *787*
Pontrelli, G. J. 548 (656), *633*, 708 (193), *715*
Ponzio, G. 590 (784), *637*, 797 (48, 49, 50), 801 (66b), 805, 806 (49, 87), 824 (49, 50, 87), 829 (49), *847*, *848*
Pool, W. O. 178 (32, 33), *203*
Poole, H. G. 95 (238), *118*
Popkie, H. E. 62 (65), *65*, 528 (546), *631*
Pople, J. A. 8 (11), *63*
Popov, A. I. 133 (60), 144 (188, 189), 145 (189), 146 (188, 200, 204, 205, 206), 157 (189), 158 (189), *160*, *163*, *164*
Popov, E. M. 196 (157), *206*
Popp, F. D. 464 (206), 465 (206), *622*, 886 (3), *940*
Porrmann, H. 276 (142), *302*
Posnjak, A. 876 (226), *883*
Postgate, J. R. 877 (229), *883*
Postlethwaite, J. D. 132 (49), *160*
Pottie, R. F. 697 (122), *713*
Powell, H. M. 149 (234, 235), *164*, 875 (194, 197), *882*
Powell, J. C. 896 (38), *942*
Poziomek, E. J. 93 (223), *118*, 889 (16), *940*
Prajsnar, B. 173 (19), *203*
Pranc, P. 400 (182), *419*
Prasad, G. 543 (605), *632*
Pratt, A. L. 147 (220), 148 (220, 223), *164*, 216 (20b), *236*
Pratt, J. M. 875 (204), *882*, 892 (28), *941*
Prausnitz, J. M. 150 (239, 245), *164*, *165*, 191 (96), *205*
Preiss, H. 153 (293), *166*
Prelog, V. 72 (28), 74 (41), *111*, 767 (137, 138, 139), *788*
Preobrazhenskaya, M. N. 325 (91), *339*
Preuss, H. 518 (489, 491), *629*
Prey, V. 142 (167), *163*
Price, C. C. 85 (149), *115*
Price, E. 148 (226), 149 (226), *164*, 217 (24), 219 (24), 220 (24), 229 (24), *236*
Prichard, W. W. 444 (113), 462 (113), *619*, 735 (119), *742*
Priest, W. J. 104 (311), *121*
Prince, M. 461 (188), *621*
Prinzbach, H. 450 (143), *620*
Pritchard, H. O. 889 (189a, 189b), *941*
Pritchard, R. B. 70 (15), *110*
Prober, M. 294 (225), *305*
Procházková, J. 525 (531), *630*
Prochorow, J. 448 (127), 582 (755), *620*, *636*
Promislow, A. I. 747 (11), 785
Proskow, S. 109 (362), *122*, 482 (272), 484 (278), 485 (278, 279),

486 (279), 551 (687), 581 (745), 588 (745), *624*, *634*, *635*
Protopapa, H. K. 397 (172), *419*
Prout, C. K. 528 (554), *631*
Prout, F. S. 105 (321), *121*
Puchalik, M. 581 (747), *635*
Puosi, G. 872 (165), *881*
Puranik, G. S. 406 (206), *420*
Purcell, K. F. 135 (83, 84), 139 (140), 140 (140), *161*, *162*, 196 (133, 134, 135, 136), *206*, 217 (21), *236*
Purrello, G. 394 (164), *419*
Putiev, Yu. P. 197 (124), *206*
Pütter, R. 266 (101), 275 (135), 276 (135, 144), 286 (101), *301*, *302*
Pyryalova, P. S. 296 (235), *305*, 309 (11), *337*
Pyszora, H. 150 (264), 151 (264), *165*

Queignec, R. 515 (448), 516 (448, 450), *628*
Quilico, A. 96 (246), *118*, 405 (198), *419*, 722 (28), *739*, 796 (31), 798 (31), 800 (64), 803 (76), 809 (108, 109, 114), 812 (31, 76), 814 (31, 126, 127), 815 (114), 819 (146, 147), 830 (76), 832, 833 (168, 169), 834 (174, 175, 179, 182), 835 (31), 836 (31, 114, 126, 195), 844 (219), *846*, *847*, *848*, *849*, *850*, *851*
Quint, F. 508 (406), *627*
Quis, P. 128 (29), *160*, 495 (816), *638*

Raab, R. E. 148 (231), *164*
Raaen, V. F. 431 (56, 58), *618*
Raasch, M. S. 703 (160), *714*
Rabenhorst, H. 664 (73), *669*
Rabinovitch, B. S. 257 (68), 258 (80), *300*, 859 (64, 65, 67), *878*, *879*, 887, 890 (10, 19a), 891 (9a–9d), *940*, *941*
Rabinowitz, J. 782 (221), *790*
Rabjohn, N. 83 (117), *114*, 314 (34), *338*
Rachlin, A. I. 769 (167), *788*
Rackley, F. A. 529 (562), *631*
Rackow, S. 90 (186), 91 (186), *116*
Ragni, G. 863 (111), *880*
Raimondi, D. L. 33 (42, 44), *64*
Rainsford, A. E. 910 (68), *943*
Rajagopalan, P. 829 (164), 842 (211, 212, 213), *850*, *851*
Ralhan, N. K. 361 (69), *416*
Ralston, A. W. 178 (32), *203*, 776 (204), *789*
Ramakrishnan, V. 210 (6), *236*
Ramarkrishnan, C. V. 724 (50), *740*
Rambaud, E. 762, *787*
Rambaud, R. 86 (151), *115*
Rambeck, O. W. 88 (176), 89 (176), *116*
Ramberg, L. 442 (107, 108), *619*, 876 (224), *883*
Rao, B. S. 405 (199), *419*
Rao, K. B. 361 (69), 399 (177), *416*, *419*
Rao, S. L. N. 724 (48), *740*
Rapoport, E. 88 (171), *116*
Rappoport, Z. 69 (16), 70 (16), 106 (331), *110*, *121*, 225 (42), *237*, 459 (822), 463 (200, 201, 202), 464 (822); 465 (202), 466 (201), 468 (822), *621*, *638*, 655 (38), 656 (38), *668*, 899, *942*
Rastrup-Andersen, A. 854 (28), *878*
Ratzkin, H. 100 (281), *120*
Rau, S. 451 (147), *620*
Rauch, E. 273 (130), *302*
Raudenbusch, W. 921 (95), *944*
Rauhut, M. M. 87 (1958), *115*, 582 (759), *636*, 936 (129), *946*
Rausch, D. A. 703 (162), *715*
Ray, S. C. 900 (51a), 901 (51b), *942*
Razomovskaya, M. T. 527 (539), *630*
Reade, R. R. 88 (164, 170), *116*
Readio, P. D. 393 (159), *419*
Rebmann, A. 90 (190), *116*
Rebyea, D. I. 694 (68b), *712*
Rechmeier, G. 767 (142), *788*
Reddy, G. S. 460 (185), 520 (185), *621*
Redmon, B. C. 547 (648), *633*
Redstone, P. A. 724 (46), 726 (46), *740*
Reed, W. R. 102 (296), *120*
Rees, A. L. G. 144 (180), *163*
Rees, C. W. 905 (60), *943*
Reese, C. B. 246 (23), *299*
Reese, R. M. 200 (176), *207*, 515 (446), 516 (446), *628*
Reesor, J. B. 70 (15), *110*
Reeve, W. 320 (67), *339*
Regan, C. M. 765 (129), *788*
Regitz, M. 551 (686), 552 (691), 553 (446), 516 (686), *634*
Rehwoldt, R. E. 543 (607), *632*
Reid, E. E. 934 (124d, 124e), *945*
Reid, J. C. 751 (36), *785*

Reidlinger, A. A. 513 (422), *627*
Reilly, W. L. 270 (113), 298 (244), *301*, *305*, 410 (228), *420*
Reine, A. H. 399 (177), *419*
Reinhardt, W. P. 33 (44), *64*
Reinheckel, H. 893 (33b), *941*
Reinmuth, O. 276 (146), 277, 278 (146), 280 (146), *302*, 892 (29a), *941*
Reinmuth, W. H. 183 (62), 184 (62), 190 (62), *204*, 229 (56), *237*, 297 (241), *305*, 527 (541), 528 (541), *630*, 658 (49), *668*
Reinstein, M. 837 (200), *850*
Reisenegger, C. 796 (20), 819 (20), *846*
Reissert, A. 76 (66), *112*, 886 (3), *940*
Reith, J. E. 294 (228), *305*
Rembarz, G. 547 (641), *633*
Renard, S. 276 (143), *302*
Renaud, D. J. 107 (342), *122*, 681 (27), *711*, 901 (52b), 902 (52b), *942*
Renaud, R. N. 211 (9), *236*, 766 (128), *787*
Renier, E. 276 (143), *302*
Renk, E. 96 (245), *118*, 916 (85), 917 (87a, 87c, 87d), 918 (87c), 922 (87), 924 (101), 925 (85), 930 (87), *944*
Renner, J. A. 877 (235), *883*
Renson, M. 862 (103), *879*
Renz, J. 748 (25), *785*
Replogle, L. L. 386 (135), *418*
Ressler, C. 100 (281), *120*, 724 (42, 43, 44, 45, 46, 57), 725 (43, 45, 58, 59, 60, 65), 726 (43, 44, 46, 71), *740*
Restelli de Labriola, E. 93 (207), *117*
Retherford, E. D. 139 (127), *162*
Reusch, R. N. 524 (516), *630*
Reuss, G. 656 (39), *668*
Reutenauer, G. 101 (291), *120*
Reuter, L. F. 699 (143), *714*
Revira, S. 169 (6), *203*
Rewick, R. T. 755 (76), *786*
Reynolds, G. A. 104 (311), *121*, 283, 285 (183), *303*
Rheinboldt, H. 797 (43), 802 (43), *847*
Ricard, M. 521 (810), *637*
Ricca, A. 347 (15), *415*, 722 (28), *739*, 802 (69), 803 (76), 812 (76), 830 (76), 835 (187), 837 (201), 839 (201, 206), 840 (201), *847*, *850*, *851*
Richards, G. O. 507 (389), 508 (389), 572 (389), *627*
Richards, R. E. 242 (3), *298*
Richards, R. L. 877 (229), *883*
Richardson, L. B. 875 (205), *882*
Richmond, R. R. 325 (86b), *339*
Richter, M. M. v. 80 (99), *113*
Richter, R. 290 (206), *304*, 796 (40), 797 (40, 45), 798 (40, 45), 801 (40), 812 (40, 45), 815 (40, 45), 824 (45), 840 (207), *847*, *851*
Richter, V. v. 87, *115*
Richter, V. von 936, *945*
Rickborn, B. 211 (8), *236*
Rickborn, R. 148 (225), *164*
Rieche, A. 555 (700), 556 (700), 557 (700), 558 (700), *634*
Riedel, H. W. 106 (334), *121*
Rieger, P. H. 183, 184 (62), 190 (62), *204*, 229 (56, 57), *237*, 297 (241), *305*, 472 (220), 473 (220, 236), 527 (541), 528 (236, 541), *622*, *630*, 658 (49), *668*
Rieke, R. D. 698 (131), *714*
Riesil, L. 432 (70), *618*
Rieter, H. 735 (116), *742*
Rietz, E. 768 (161), *788*
Rigaudy, J. 521 (810), *637*
Rigaut, A. 507 (386), *627*
Rigby, C. W. 463 (198), 469 (198), 480 (198), *621*
Rigby, G. W. 397 (171), *419*, 916 (83a), *944*
Rigler, N. E. 78 (80), *113*
Rinderknecht, H. 412 (237, 238), *421*
Rinehart, K. L. 282 (168), *303*
Ring, H. 854 (28), *878*
Ringwald, E. L. 294 (227), *305*
Ritchey, W. M. 255 (60), 256 (62), *300*
Ritchie, C. D. 147, 148 (220, 223), *164*, 196 (145), *206*, 216 (20b), 217 (23), *236*
Ritter, J. J. 92 (196), *117*, 284 (181), 291, 292 (211), *303*, *304*, 358, 359 (51), 360 (59), 361, 367, 370 (59), 410 (227), *416*, *420*, 911, 913 (77a, 77b), 914 (77a, 77b, 78a, 79, 80a), *943*, *944*
Rivest, R. 135 (88, 89), 154 (88, 89, 308), 155 (309), *161*, *166*
Robba, M. 98 (264), *119*
Roberts, E. 696 (120), *713*
Roberts, G. A. H. 148 (230), 149 (230), *164*

Roberts, J. D. 215 (14), 218 (27), *236*, 508 (398), *627*, 702 (155), *714*, 765 (129), 772 (187, 189), 781 (217), *788*, *789*, *790*
Roberts, R. M. 766 (124, 126), *787*
Robertson, J. A. 687 (66), *712*
Robertson, N. C. 177 (30), *203*, 432 (71, 72), *618*
Robertson, P. S. 508 (400), *627*
Robinson, J. C. 320 (68), *339*
Robinson, R. A. 218 (29b), *236*
Robinson, T. 730 (90), *741*
Robinson, W. G. 725 (61), 732 (61, 93), 734 (93), *740*, *741*
Roblin, R. O. Jr. 98 (265), *119*
Robson, P. 432 (69), *618*
Robson, T. D. 101 (293), *120*
Rochas, G. 764 (111), *787*
Rochlin, E. M. 869 (144), 870 (144), *881*
Rodin, J. O. 73 (35), *111*
Roemer, J. J. 245 (19), *298*
Roger, R. 264 (94), *301*
Rogers, A. O. 428 (10), *617*
Rogers, G. T. 938 (136), *946*
Rogers, R. G. 85 (149), *115*
Rogier, E. R. 315 (38), *338*
Roginskii, S. Z. 474 (251, 252), *623*
Rokhlin, E. M. 582 (762), *636*, 869 (143), *880*
Rol, N. C. 198 (170), *207*
Roland, J. R. 463 (195), 479 (195), 480 (195), 503 (353), 504 (353), 576 (353), 577 (353), 579 (353), *621*, *626*, 681 (30), 682 (30), *711*
Rolewicz, H. A. 410 (230), *420*
Rolfe, A. C. 196 (117), *205*
Romanov, V. G. 138 (122a), *162*
Rømming, C. 147 (209, 214), *164*
Römming, C. 515 (472), 516 (472), *629*
Ronayne, M. R. 472 (221), *622*
Rondestvedt, C. S., Jr. 700 (144), 701 (144), *714*
Roothaan, C. C. J. 8, 32 (38), *63*, *64*
Rorig, K. 78 (74), *113*
Rorig, K. J. 105 (323), *121*
Rose, F. L. 547 (642), *633*
Rose, J. 144 (179), *163*
Rose, J. A. 738 (138), *742*
Rose, T. J. 142 (151), *162*
Rose-Innes, A. C. 659 (54), *669*
Rosemann, S. 759 (64), *786*
Rosenberg, A. 448 (133), *620*
Rosenberg, A. M. 689 (72), *712*
Rosenberg, I. N. 771 (178), *789*
Rosenblatt, D. H. 325 (86b), *339*
Rosenblum, M. 472 (225), 473 (230), *622*, 658 (46), *668*, 937 (130), *946*
Rosendahl, F. K. 867 (127), 872 (157), *880*, *881*
Rosendahl, K. 864 (120), 869 (120), *880*
Rosenfeld, B. 797 (47), 807 (47), 822 (47), 824 (47), 831 (47), *847*
Rosenmund, K. W. 82 (60), *114*, 320 (114), *339*
Rosenstock, H. M. 448 (135), *620*
Roshchupkin, V. P. 196 (157, 160), *206*, *207*
Rosowsky, A. 397 (173), 402 (191), *419*
Ross, B. L. 154 (305), *166*, 255 (60), 256 (62), *300*
Ross, R. M. 410 (230), *420*
Rossetto, F. 872 (165), *881*
Rossi, G. V. 755 (80), *786*
Rossini, F. D. 887 (8b), *940*
Rotger, H. 898 (44), *942*
Roth, A. J., III 444 (112), 445 (112), *619*
Roth, W. R. 489 (282), 584 (282), *624*
Rothchild, S. 768 (159), *788*
Rothe, W. 856 (43), 877 (43), *878*
Rothschild, M. 723 (37), *740*
Rothstein, M. 749, 762, 763 (101), 767 (135), *785*, *787*, *788*
Rottenberg, M. 767, 768 (170), *788*, *789*
Rouschias, G. 253 (52), *299*
Rowell, J. C. 229 (52), *237*, 473 (232), *622*, 658 (43), *668*
Rowland, F. S. 891 (23), *941*
Rowlinson, J. S. 148 (230), 149 (230), *164*
Rubinstein, H. 862 (96), *879*
Rubtsov, M. V. 917 (86), 922 (86), *944*
Rüchardt, C. 698 (128, 129), *714*
Rudler, H. 202 (183), *207*
Rudolf, L. 506 (368), 507 (368), *626*
Rudolph, W. 97 (261), *119*
Ruehrwein, R. A. 142 (154), *162*
Ruff, J. K. 296 (236), 297 (236, 238), *305*
Ruggiero, G. 805 (85), *848*
Ruhnau, R. 127 (17), 131 (17), *159*, 251 (58), 255 (58), *300*
Rule, L. 154 (299, 300), *166*

Rumanowski, E. J. 274 (133), *302*
Rumpf, P. 221 (37), *237*
Runge, J. 702 (158), *714*
Runner, M. E. 133 (59), *160*
Runti, C. 105 (323), *121*
Rupe, H. 311, 312, 320 (20, 28), *338*, 888 (14), *940*
Ruppol, E. 193 (104), *205*
Ruschhaupt, F. 150 (256), *165*
Rusek, P. 393 (159), *419*
Rusk, W. 886 (2a), *940*
Ruske, W. 289 (201), *304*, 507 (392), *627*
Ruske, W. A. 254 (55), *300*
Russell, A. 86 (155), *115*
Russell, G. A. 673 (6), *710*
Russell, P. B. 390 (149), *418*
Rust, F. F. 697 (123), 701 (150), *713*, *714*
Rutschmann, J. 748 (25), 781 (216), *785*, *790*
Ruzicka, L. 72 (28), *111*
Ryan, J. P. 762 (104), *787*
Rybakova, N. A. 245 (20), 254 (20, 56), 255 (57), *299*, *300*
Rybinskaya, M. I. 104 (312), *121*
Rysselberge, J. van 103 (300), *120*
Ryutani, B. 516 (451), *628*
Rzepkowska, Z. 893 (33a), *941*

Saalfeld, J.-Ch. 90 (185), *116*
Sabelus, G. 451 (147), *620*
Sacco, A. 873 (174, 175), 874 (178, 179, 181, 182, 184), 875 (189, 201, 203, 207, 208), 876 (210, 212, 216), *881*, *882*
Sachs, F. 862 (92), *879*
Sadaoka, K. 725 (60), *740*
Sadeh, T. 442 (104), *619*
Saegusa, T. 860 (81), 861 (82), *879*
Saeguso, T. 889 (17), *941*
Safiullina, N. R. 138 (122a), *162*
Sager, W. F. 217 (23), *236*
Saggiomo, A. J. 97 (254), *119*, 514 (439), 516 (439), *628*
Sah, H. 190 (94), *205*
Saint-Ruf, G. 94 (233), *118*
Saito, G. 288 (193), *304*
Saito, T. 411 (234), *420*
Saito, Y. 156 (312, 313, 314), *166*, 528 (553), *631*
Sakami, W. 766 (121, 123), 771 (182), *786*, *787*, *789*
Sala, O. 448 (126), *620*
Salez, C. 31 (37), *64*
Salisbury, L. F. 294 (228), *305*
Sallach, H. 753 (47), *786*
Salmon, O. N. 135 (86), *161*, 216 (18), *236*
Salmon-Legagneur, F. 247 (34b), *299*
Salomon, L. L. 754 (72), *786*
Salvalori, T. 722 (28), *739*
Salzberg, P. L. 80 (93), *113*
Samuel, D. 938 (133), *946*
Sanchez, M. 142 (158), *163*
Sanchez, R. 289 (204), *304*
Sanchez, R. A. 506 (374, 375, 378), 508 (375), 509 (375), 513 (424, 425), 518 (425), 519 (425), *626*, *628*
Sancovich, H. A. 771 (184), *789*
Sander, M. 329 (100), *340*
Sander, W. 919 (90a), 925 (103d), 927 (103d), 929 (103d), 930 (103d), *944*, *945*
Sandmeyer, T. 87 (162), *115*
Sandner, W. J. 428 (13), *617*
Sandorfy, C. 196 (153), *206*
Santacroce, C. 835 (193), *850*
Santavy, F. 183 (57), 190 (90, 91), *204*, *205*
Sarma, P. S. 724 (48), *740*
Sasaki, M. 694 (99), *713*
Sasaki, T. 809 (106), *848*
Sass, R. L. 434 (823), 507 (397), 558 (706), *627*, *634*, *638*
Sato, M. 694 (99), *713*
Sato, R. 877 (228), *883*
Sato, T. 280 (159), *303*
Sato, Y. 138 (117), *162*
Sauer, J. 345 (12), 348 (19), 351 (12), 356 (40, 42), 388 (141), *415*, *418*, 449 (139, 140), 523 (506), *620*, *629*, 936 (128a), *946*
Sauer, J. C. 385 (132), *418*
Sauers, C. K. 99 (277), *120*
Sauers, R. R. 918 (88), *944*
Saum, A. M. 148 (227), *164*
Saumagne, P. 138 (110, 111), 139 (111), *161*
Saunders, R. A. 200 (173), *207*
Sausen, G. N. 200 (175), *207*, 409 (224), *420*, 444 (113), 462 (113), 463 (192), 475 (259), 476 (259), 478 (259), 574 (729), 575 (729), 593 (729), 595 (259), *619*, *621*, *623*, *635*
Saville, B. 926 (105f), *945*
Savitskii, A. V. 516 (473), *629*
Sawa, S. 182, 186 (54), *204*
Sawaki, Y. 263 (91), *301*

Sayed, M. F. Amr El 196 (151, 155, 161), *206*, *207*
Scala, A. A. 277 (151, 153), 278 (151), 285 (153), *302*
Scarpati, R. 796 (38), 797 (38), 813 (38, 120), 835 (192, 193, 194), *847*, *848*, *850*
Sčekotichin, A. I. 869 (138), *880*
Schaafsma, Y. 696 (114), *713*
Schaefer, F. C. 247 (31), 248 (38), 249 (31), 267 (105), 268 (105), 269 (105, 111), 270 (112), 271 (31, 121), *299*, *301*, *302*, 353 (32), *415*
Schaefer, J. P. 282 (171), 284 (171), *303*, 397 (169), *419*
Schaefer, T. 150 (242), *165*
Schaeffer, R. 151 (273), *165*, 256 (64), *300*
Schaffer, R. 759 (65, 68, 70), *786*
Schanz, R. 525 (523), *630*
Scharpf, W. G. 109 (360), *122*
Schawlow, A. L. 147 (211), *164*
Scheffer, J. R. 698 (131), *714*
Scheiner, P. 109 (359), *122*
Scheinpflug, H. 877 (230, 231), *883*
Schenck, R. 512 (412), 596 (412), *627*
Schenk, G. H. 449 (138), *620*
Schenk, H. 870 (148, 149), *881*
Schenk, R. 283 (175), *303*
Scherer, W. 796 (20), 819 (20), *846*
Schiedt, U. 731 (92a), *741*
Schiemenz, G. P. 107 (339), *121*
Schiess, P. W. 910 (69a), 912 (69), 931 (69a), *943*
Schiewer, U. 734 (111), *741*
Schikora, E. 457 (173), *621*
Schiller, J. C. 146 (197), *163*
Schilling, E. D. 723 (38), 724 (38), *740*, 766 (147), 772, *788*
Schimple, A. 246 (23), *299*
Schindler, K. 473 (228), *622*
Schinke, E. 408 (219), *420*
Schipper, E. 109 (360), *122*
Schischkoff, L. 800 (65), *847*
Schlatter, E. 95 (242), *118*
Schlecht, L. 898 (44), *942*
Schleitzer, G. 500 (305, 308, 309), 502 (308, 309), *624*
Schlesier, G. 756 (50), *786*
Schlesinger, A. H. 863 (112), *880*
Schlesinger, H. I. 326 (93), *340*
Schleussner, K. 429 (32), *617*
Schleyer, P. R. 137 (101, 103), 139 (103, 128), 140 (103), 141 (147), 142 (103), 157, 158 (103), *161*, *162*
Schleyer, P. v. Rague 196 (126), *206*, 813 (123), *849*, 854 (23, 34, 36), *878*
Schlosser, F. D. 755 (79), *786*
Schmelpfenig, C. W. 710 (207), *716*
Schmid, H. 754 (78), 762 (103), *786*, *787*
Schmid, K. 762 (103), *787*
Schmid, M. 99 (273), *119*
Schmidt, A. 128 (28), *159*, 243 (11, 12), 245 (11, 17), *298*
Schmidt, B. M. 201 (181a), *207*
Schmidt, H. 430 (44), *618*, 674 (9), *711*
Schmidt, K. F. 93 (220), *118*
Schmidt, L. H. 767 (143), *788*
Schmidt, M. 326 (94), *340*, 502 (313), *625*, 771 (181), *789*
Schmidt, R. 348 (20), 352 (31), *415*
Schmidt, R. F. 699 (141, 142), *714*
Schmidt, R. R. 251 (46), *299*
Schmidt, U. 857 (50), *878*
Schmidt, W. 97 (254), *119*
Schmidtmann, H. 88 (178), *116*, 433 (78), 434 (78), 435 (78), *619*, 899 (49), 900 (49b), *942*
Schmiegel, K. U. 109 (359), *122*
Schmir, G. L. 247 (35), *299*
Schmitt, R. 430 (40), *618*
Schmitz-Dumont, O. 797 (43), 802 (43), *847*
Schmotzer, G. 508 (401, 402), 509 (401), 572 (401), *627*
Schmötzer, G. 506 (371), 507 (371), *626*
Schmulbach, C. D. 152 (279), *165*
Schneider, F. W. 859 (65), *879*, 890 (19a), 891 (19a, 19b, 19c), *941*
Schneider, H. 133 (55), *160*
Schneider, J. 454 (160), *621*
Schneider, W. C. 507 (394), *627*
Schneider, W. G. 129, 136 (95), 139 (141), 140 (141), 141 (35), 149, 150 (240, 242), 156 (95, 141), *160*, *161*, *162*, *164*, *165*, 216 (19), *236*
Schnell, H. 377 (107), *417*
Schoen, L. T. 548 (661), *633*, 708 (196), 709 (196), *715*
Schöfer, A. 807 (93), *848*
Schofield, K. 412 (236), *420*, 922 (96), *944*
Scholer, F. R. 151 (274), *165*
Scholl, R. 93 (210), *117*, 805 (80), 807 (92, 93, 97), *847*, *848*

Scholz, H. 88 (179), *116*, 590 (783), *637*
Schön, N. 106 (336), *121*
Schönberg, A. 683 (40), *711*
Schoor, A. v. 500 (310), 503 (346), *625*, *626*
Schraudolf, H. 733 (108), *741*
Schraufstätter, E. 431 (60, 61), *618*
Schrauzer, G. N. 450 (142), 477 (263), *620*, *623*
Schrecker, A. W. 95 (241), *118*
Schreiber, R. S. 293 (223a), *305*
Schrodel, R. 316 (45, 46), 318 (45, 46), 332 (45, 46), *338*
Schroder, H. 336 (115), *340*
Schröder, H. 297 (238), *305*
Schroeder, H. E. 397 (171), *419*
Schroll, G. 200 (177, 178), *207*
Schubert, H. W. 865 (132), 867 (132), *880*
Schuerch, C., Jr. 93 (220), *118*
Schuirer, E. 794 (13), *846*
Schülde, F. 773 (191), *789*
Schultz, H. P. 320 (64), *339*
Schultz, J. E. 687 (65), 692 (65), *712*
Schulz, G. V. 683 (39), *711*
Schulz, L. 107 (343), *122*, 681 (26, 28), *711*
Schulze, A. 819 (149), *849*
Schulze, S. R. 154 (306), 156 (306), *166*
Schulze, W. 459 (179), 464 (179), 465 (179), 476 (260, 261), 477 (179, 261), *621*, *623*
Schuman, P. D. 270 (113), *301*
Schupp, R. 177 (27), *203*
Schurmovskaya, N. A. 474 (242), *623*
Schurz, J. 190 (94), *205*
Schuster, L. 874 (187), *882*
Schuster, R. E. 855 (39), 857 (39), 859 (68), 872 (160), *878*, *879*, *881*, 887 (9c, 11), 889 (11), 890 (11), *940*
Schütte, H. R. 768 (157), 775 (198), *788*, *789*
Schwantes, E. 429 (27), *617*
Schwarz, H. 79 (91), 81 (104, 105), *113*, *114*
Schwarz, M. 109 (360), *122*
Schwebel, A. 757 (51), 758 (51), 759 (51), *786*
Schweer, K. H. 781 (218), *790*
Schwetlick, K. 684 (44), *711*
Schwoegler, E. J. 320 (66), *339*
Scott, H. 543 (611, 620), *632*
Scott, M. D. 886 (4), 910 (4d), *940*
Scott, R. W. 97 (258), *119*
Scribner, C. W. 409 (222), *420*
Scribner, R. M. 409 (222), *420*, 444 (113), 446 (116), 447 (116), 462 (113), 544 (627), 545 (627), 548 (669), 549 (671), *619*, *633*, *634*, 640 (3), *667*, 682 (31), *711*
Scullard, P. W. 923 (100), *944*
Searles, S. 125, 127 (3), 144 (195), 156 (3, 318), *159*, *163*, *166*
Sears, D. S. 150 (260), 151 (260), 156 (260), *165*, 347 (16), *415*
Sease, J. W. 168 (3), *203*
Sechadri, T. R. 724 (49), *740*
Sedlmeier, J. 876 (211), *882*
Seefelder, M. 270 (115, 116), *302*
Seelert, C. 286 (190), *303*
Seeliger, W. 390 (151), *418*
Seely, M. K. 721 (23), *739*
Seide, S. 375 (103), *417*
Seidel, M. 348 (19, 20), *415*
Seidel, P. 819 (150), *849*
Seidl, H. 449 (141), 461 (141), *620*
Seifert, B. 547 (638, 639, 640), *633*
Seitz, F. 26 (33), 27 (33), 28 (33), 29 (33), 43 (33), *64*
Selff, R. E. 761 (90), *787*
Selva, A. 842 (217), *851*
Selvarajan, R. 707 (185), *715*
Semenovskii, A. V. 81 (103), *114*
Sementzova, A. 507 (385), 508 (385), *627*
Semper, L. 796 (32), 802 (75), 805 (32, 82), *846*, *847*, *848*
Sen, J. N. 688 (68a), 692 (68a), 693 (68a), 694 (68a, 98), *712*, *713*
Sennewald, K. 796 (17), 800 (17), 819 (17), 822 (155b), *846*, *849*
Sensi, P. 196 (148, 158), *206*
Sera, N. 474 (250), *623*
Serebryakov, B. R. 180 (46), *204*, 257 (71), *300*
Serencha, N. M. 173, 174 (20), *203*
Settepani, J. A. 402 (193), *419*
Sevast'yanova, I. G. 183, 187 (60, 61, 75), 188, 189 (60, 61, 75), *204*, 337 (118), *340*
Seyferth, D. 859 (61), *878*
Shabica, A. C. 766, 784 (133), *788*
Shafizadeh, F. 759 (62), *786*
Shah, S. 543 (614), *632*
Shah-Malak, F. 144 (191), 145 (191), *163*

Shalit, H. 75 (53), 110 (53), *112*, 686 (54), 687 (54), *712*
Shallcross, F. V. 515 (470), 516 (470), 517 (470), *629*
Shancke, P. N. 44 (56), 45 (56), *65*
Sharanin, Y. A. 406 (202), *420*
Sharefkin, D. 760 (86), *786*
Sharkey, W. H. 702 (159), 703 (159), *714*, 859 (59), *878*
Sharp, J. T. 674 (13), 675 (13, 14), *711*
Sharts, C. M. 702 (155), *714*
Shaw, B. W. 875 (190), *882*
Shaw, D. H. 889 (18), *941*
Shaw, G. 101 (287), *120*
Shaw, K. N. F. 95 (243), *118*
Shaw, M. A. 520 (824), *638*
Shechter, H. 79 (82), 83 (120), *113*, *114*, 765 (120), 770, *787*, *789*
Sheehan, J. C. 858 (55), *878*
Sheeran, P. J. 898 (45), *942*
Sheikh, Y. M. 197 (169), 198 (169), 200 (169), *207*
Shelbey, W. E. 96 (248), *119*
Sheldon, J. C. 149 (232, 233), *164*
Sheline, R. K. 196 (161), *207*
Shelomov, I. K. 180 (44), *204*
Shemiakin, M. M. 748 (24), 751 (24), *785*
Shemyakin, M. M. 763 (99), *787*
Shepard, R. S. 724 (51, 52), *740*
Shapelavy, J. N. 548 (663), *633*, 708 (191), 709 (191), *715*, 904 (57), *943*
Shepherd, R. G. 98 (265), *119*
Sheppard, W. A. 215 (15), 219 (30, 32), 220 (30, 31), 222 (30, 31, 34), 223 (30b, 31), 224 (39), 229 (30), *236*, *237*, 437 (94), 463 (199), 521 (199), *619*, *621*, 650 (32), 655 (32), *668*, 906 (62), 935, *943*, *945*
Sheppard, W. S. 105 (318), *121*
Sherer, J. P. 402 (195), *419*
Sheridan, J. 516 (444, 468), 517 (444), *628*, *629*
Sherkhgeimer, G. A. 107 (339), *121*
Sherle, A. I. 474 (240, 241, 242, 244, 246, 248, 251, 253), *622*, *623*
Sherwood, L. T., Jr. 80 (96), *113*
Shevchenko, V. I. 252 (53), *299*, 414 (241), *421*
Shew, D. 509 (409), *627*
Shields, W. R. 448 (135), *620*
Shigemitsu, Y. 108 (353), *122*
Shillady, D. D. 194 (109), *205*
Shimanouchi, T. 135 (67, 73, 74), 156 (73), *160*
Shimizu, T. 667 (78), *669*
Shiner, R. L. 92 (200), *117*
Shiraishi, S. 797 (53), 810 (53), *847*
Shoaf, C. J. 317 (51), 329 (51), *338*
Shohamy, E. 225 (42), *237*, 459 (822), 464 (822), 468 (822), *638*
Shono, T. 854 (13), 862 (95, 97), 865 (13), *877*, *879*
Short, J. H. 384 (127), *418*
Short, W. A. 97 (255), *119*
Short, W. F. 101 (293), *120*, 271, 273 (129), *302*, 391 (154), *418*
Shortley, G. H. 6 (7), *63*
Shorygin, P. P. 196 (156, 160), *206*, *207*
Shotwell, O. L. 73 (36), *111*
Shoule, H. A. 323 (78), *339*
Shreeve, W. W. 766 (121), *787*
Shriner, R. L. 168 (2), 169 (2), 170 (10), 171 (12, 13), 172 (12), *203*, 282 (166), *303*
Shull, H. 39 (51), *65*
Shul'man, M. L. 264 (97), 265 (97), *301*
Shuman, I. D. 410 (232), *420*
Shupack, S. I. 502 (324, 333, 336, 337), *625*
Shusherina, N. P. 378 (112, 116), *417*
Shuskerina, N. P. 378 (112), *417*
Shvekhgeimer, G. A. 264 (97), 256 (97), *301*
Sica, D. 835 (193), *850*
Sidgwick, N. V. 853 (6, 7), 854 (6, 7), 865 (6), *877*, 887 (8a), *940*
Siebert, H. 99 (271), *119*
Siegel, I. 766 (125), *787*
Siegel, W. 696 (116), *713*
Siemons, W. J. 665 (74), *669*
Siggia, S. 170 (9), 173, 174 (20), 175, 176 (23, 25), 177 (23), *203*
Silverman, P. R. 892 (28), *941*
Silvermann, P. R. 875 (204), *882*
Silverstein, R. M. 73 (35), *111*
Simamura, O. 769 (169), *788*
Simchen, G. 69 (8), *110*, 378 (118), *417*
Simmonds, B. A. W. 142 (149), *162*
Simmons, H. E. 108 (347), 109 (362), *122*, 217 (22), *236*, 485 (279), 486 (279), 497 (297), 500 (306), 501 (306), 502 (306), 503 (306, 353, 355), 504 (353, 355), 505 (356), 548 (658, 666, 667), 571 (721),

572 (721), 576 (353), 577 (353, 721), 578 (721), 579 (353, 721, 737), *624*, *626*, *633*, *634*, *635*, 707 (189), 708 (189), 709 (197), *715*, 903 (56b, 56c), *943*
Simmons, H. E., Jr. 579 (737), *635*
Simmons, J. H. 82 (109), *114*
Simon, L. J. 546 (632), 547 (632), *633*
Simonetta, M. 796 (41), 803 (41), 809 (108, 109), *847*, *848*
Simonjan, L. A. 869 (144), 870 (144), *881*
Simons, J. H. 588 (780), *637*
Simpson, M. 72 (25), *111*
Simpson, P. G. 152 (276), *165*
Sinanoğlu, O. 39 (53), *65*
Sinclair, R. 862 (108), *880*
Sinclair, R. G. 890 (20a), 891 (20a), *941*
Sindellari, L. 105 (323), *121*
Singer, L. S. 659, *669*
Singh, B. 931 (116), *945*
Singh, K. P. 412 (236), *420*
Singh, P. 372 (98), *417*
Singh, P. P. 154 (301), *166*
Singh, S. 399 (175, 177, 178), *419*
Sircar, J. C. 361 (63, 67), 399 (175, 176, 177), *416*, *419*
Sixma, F. L. J. 746 (1), 747, 748 (1), 749, *784*
Sjöberg, K. 854 (12), 865 (12), *877*
Skanawy-Grigorjewa, M. 876 (226), *883*
Skell, P. S. 551 (684), *634*, 705 (173, 174), *715*
Skelly, N. E. 146 (200), *164*
Skinner, M. W. 196 (118), *205*
Skoultchi, M. M. 701 (148), *714*
Skovronek, H. 100 (280), *120*, 246 (28), 247 (28), *299*
Skraba, W. J. 780 (214), *790*
Skursky, L. 731 (92b), *741*
Skysanskii, V. A. 297 (237), *305*
Sladkova, T. A. 325 (84a), *339*
Slater, J. C. 33 (48), 37 (48), *64*
Slaugh, L. H. 766 (145), *788*
Slavik, J., Jr. 335 (111), *340*
Sleep, K. C. 550 (678), 551 (678), *634*, 704 (171), *715*
Sletzinger, M. 388 (146), *418*
Slough, W. 543 (617), *632*, 664 (72), *669*
Smedal, H. S. 435 (814, 815), *637*, *638*
Smellie, R. H. 249 (40), *299*
Smick, R. L. 354 (36), 355 (36), *415*
Smiley, R. A. 77 (70), 79 (82), *112*, *113*, 765 (119), *787*
Smirnov, S. K. 189 (87), *205*
Smirnov, Yu. D. 189 (87), *205*
Smit, P. J. 94 (227), *118*
Smith, B. 459 (181), 463 (181),465 (181), *621*
Smith, C. D. 456 (170), *621*
Smith, D. C. 197 (120), *205*
Smith, G. 109 (359), *122*
Smith, G. F. 732 (96, 97), 733 (97), *741*
Smith, G. L. 439 (98), *619*
Smith, H. 726 (70), *740*
Smith, H. A. 366 (75), *416*
Smith, J. A. S. 202 (186), *207*
Smith, J. O. 576 (735), *635*
Smith, L. H. 401 (186), *419*
Smith, L. I. 315 (38), *338*
Smith, P. 689 (72), *712*
Smith, P. A. S. 351 (27), 388 (27), *415*, 910 (70a), 912 (70), *943*
Smith, R. A. 451 (146), *620*
Smith, R. D. 699 (143), *714*
Smith, R. F. 94 (228, 231), *118*, 886 (4), 910 (4b), *940*
Smith, T. J. 73 (34), *111*
Smolin, E. M. 105 (322, 323), *121*, 293 (219), *305*, 913 (75b), *943*
Smyrniotis, P. Z. 735 (122), *742*
Smyth, C. P. 124 (2), 147 (2), *159*, 214 (10), *236*, 854 (29), *878*
Snaprud, S. I. 156 (317), *166*
Snowling, S. 393 (162), *419*
Snyder, H. R. 95 (241), *118*, 262 (87), *301*, 320 (68), *339*, 469 (214), *622*, 773 (192), *789*
Sobel, J. 273 (130), *302*
Sobhy, M. 837 (202), *851*
Sobotka, W. 361 (63), *416*
Sobue, H. 288 (195, 196), *304*
Socrates, G. 139 (139), 140 (139), *162*
Söderbäck, E. 498 (302), *624*
Sodi Pallares, E. 722 (31), *739*
Soffer, L. M. 326 (95), 327 (95, 96), 328 (95), 329 (95), 332 (95), *340*
Sofue, M. 472 (223), *622*
Sokharov, M. M. 474 (252), *623*
Sokolov, N. D. 138 (122b), *162*
Sokolov, S. D. 809 (113), 815 (113), 836 (113), *848*
Sokolskii, D. V. 325 (84b, 85), *339*

Solomon, A. K. 771 (178), *789*
Solomon, S. 780 (214), *790*
Solomon, W. C. 452 (154), *620*
Soloway, S. 172 (16), *203*
Soma, M. 541 (593), 543 (593), *632*
Sonn, A. 308 (6), *337*
Sonz, A. 876 (218), *882*
Sorbo, B. 735 (125), *742*
Sorrentino, P. 835 (192), *850*
Sothern, R. D. 189 (84), *205*
Souchay, P. 809 (99, 101, 102b), *848*
Soudiun, W. 766 (132), *788*
Soundararajan S. 62 (64), *65*, 153, (292), *166*, 214 (12), *236*, 241 (1), *298*
Souther, B. L. 378 (110), *417*
Souty, N. 139 (132), 140 (132), *162*
Sowden, J. C. 759 (63), *786*
Sowerby, D. B. 151 (266), *165*
Ŝpanskij, V. A. 869 (138), *880*
Sparks, R. A. 533 (572), *631*, 661 (62), *669*
Specht, W. 309 (14), *337*
Speed, J. A. 142 (149), *162*
Spenser, I. D. 95 (243), *118*, 720 (17, 18), 732 (17, 18), *739*, 922 (97b), 926 (97b), *944*
Speroni, G. 796 (26, 31, 38), 797 (38), 798 (26, 31), 812 (26, 31), 813 (38, 120), 814 (31, 125), 821 (152), 832 (31), 835 (31, 194), 836 (31), *846*, *847*, *848*, *849*, *850*
Speyer, K. N. 405 (200), *419*
Spiegel, A. 73 (33), *111*
Spiers, D. B. 99 (274), *120*
Spiess, G. 294 (226), *305*
Spillane, L. 186 (68), *204*
Spille, J. 129 (38), 130 (38), 159 (38), *160*, 250 (43), 251 (43), 252 (43), 253 (43), *299*, 358 (57), 365 (57), *416*, 912 (73a), *943*
Spinelli, D. 193 (106), *205*, 210 (5), *235*
Spitzer, W. C. 231 (60), *237*
Spoerri, P. E. 886 (2b), *940*
Sprague, B. S. 699 (143), *714*
Springall, H. D. 517 (481), *629*
Springer, C. S. 139 (137), 140 (137), *162*
Spyker, J. W. 749 (30), *785*
Srivastava, R. D. 543 (605), *632*
Sroog, C. E. 429 (28), *617*
Stacy, G. W. 407 (207, 208, 209), *420*, 900, 901 (51b), *942*
Stadnikoff, G. 75 (50), *112*
Staehelin, A. 95 (237), *118*
Stahl, C. R. 170 (9), 175, 176 (23), 177 (23), *203*
Stallberg, A. 525 (529), *630*
Stanford, C. S. 138 (115), *162*
Stange, H. 322 (72, 74, 75), *339*
Stanger, G. B. 875 (197), *882*
Stangl, S. 350 (25), *415*
Stanonis, D. 863 (115), *880*
Starke, M. 474 (254, 255), *623*
Starowieski, K. 893 (33a), *941*
Starowieyski, K. 152 (285, 286), *166*
Staskun, B. 309, 310 (13, 15, 18), *337*, *338*
Stauf, F. 88 (166), *116*
Stauffer, B. 97 (253), *119*
Steadman, T. R. 378 (114), *417*
Stearns, G. 724 (51), *740*
Steel, G. 196 (149), *206*, 242 (8), *298*
Steglich, W. 870 (150), *881*
Steinbrückner, C. 862 (101), 865 (124, 125), 867 (124, 128, 133), 875 (198), *879*, *880*, *882*
Steiner, A. 819 (151), *849*
Steinkopf, W. 804 (79), *847*
Steinman, G. 246 (23), *299*
Steinmetzer, H. 540 (586), 541 (586), *632*
Stenseth, R. E. 400 (183), *419*
Stephen, H. 308, 309 (7, 8), *337*
Stephen, T. 309 (8), *337*
Stephens, C. R. 99 (275), *120*
Stephenson, N. C. 253 (52), *299*
Stephenson, O. 88 (167), *116*
Sterk, H. 284 (177), *303*
Stern, E. S. 408 (216), *420*
Stetten, D., Jr. 759 (67), *786*
Stevens, C. L. 897 (40), *942*
Stevens, D. L. 728 (82b), *741*
Stevens, G. de 93 (221), *118*
Stevens, I. D. R. 456 (167), *621*
Stevens, J. R. 405 (200), *419*
Stevens, M. F. G. 402 (194), *419*
Stevens, R. V. 384 (130, 131), *418*
Stevens, T. E. 104 (304), *120*
Stevens, T. S. 293 (222), *305*
Stevenson, G. W. 104 (308), *121*
Stewart, A. R. P. 258 (80), *300*
Stewart, C. A., Jr. 456 (164, 165), *621*
Stewart, F. D. 699 (141, 142), *714*
Stiefel, E. I. 502 (318), *625*
Stiles, M. 755 (71), *786*
Stilz, W. 107 (339), *121*

Stjernholm, R. 776 (224), *790*
Stock, L. M. 218 (27), 230 (58), *236*, *237*
Stoddard, E. A. 142 (156), *162*
Stoffel, V. W. 769 (163, 164), *788*
Stojiljkovic, A. 103 (303), *120*
Stoker, J. R. 718 (5), 720 (5), 721 (5), *739*
Stokes, C. S. 514 (440), 516 (440), *628*
Stoll, A. 748 (25), 781 (216), *785*, *790*
Stone, F. G. A. 150 (250), *165*, 502 (312), *625*
Stoodley, L. G. 472 (222), *622*
Stopp, G. 106 (336), *121*
Storbeck, I. 474 (254, 255), *623*
Storfer, S. J. 150 (255), *165*, 277 (149), 278 (149), *302*
Stork, G. 835 (191a), *850*
Stormont, R. T. 73 (32), *111*
Stothers, J. B. 242 (4), *298*
Stowe, B. B. 732 (101), *741*, 774 (194), *789*
Strachan, P. L. 386 (134), *418*
Straeten, P. van der 79 (91), *113*
Strauss, S. F. 187 (69), *204*
Strecker, A. 74 (46), *111*, 758 (81), *786*
Strehlow, H. 133 (55, 58), *160*
Streith, J. 558 (707), *634*
Strell, M. 606 (801), 607 (801), *637*
Strel'nikova, N. D. 142 (164), *163*
Strizhakov, O. D. 244 (16), *298*
Strobel, G. A. 727 (77, 78), 728 (82b), 729 (78, 84, 85), *741*
Strom, L. E. 806 (88), *848*
Strømme, K. O. 134 (64), 135 (64), 146 (64), 147 (64), *160*
Strong, F. M. 723 (38), 724 (38, 40), *740*, 766 (147), 772 (147), *788*
Strong, J. S. 270 (117), 275 (139), *302*
'Strotskire, T. D. 10, 19 (21), 39 (21), *64*
Struck, E. 82 (114), *114*
Stubbs, W. H. 875 (190), *882*
Stuer, B. K. 898 (44), *942*
Sturgis, B. M. 95 (242), *118*
Sturm, H. J. 350 (25), 353 (33), 354 (34), 388 (141), *415*, *418*
Stutman, J. M. 516 (461), *628*
Subba Rao, B. C. 330 (101, 104), 332 (101), *340*
Suehiro, T. 769 (169), *788*
Suen, Y.-H. 282 (169), *303*
Suhadolnik, R. J. 735 (124), *742*, 770 (173), *789*
Sukhorukov, B. I. 472 (217), *622*
Sullivan, H. R. 331 (105), *340*
Sullivan, M. J. 259 (81), 260 (81), 265 (81), *301*
Sullivan, S. 473 (230), *622*, 658 (46), *668*
Sullivan, W. I. 283 (172), *303*
Sumrell, G. 150 (252), *165*
Sundararjan, R. 62, *65*
Supniewski, J. V. 80 (93), *113*
Surrey, A. R. 98 (263), *119*
Surzur, J. M. 675 (17), 676 (17), 677 (17, 20), 678 (20), *711*
Suschitzky, H. 406 (206), *420*
Susi, P. V. 78 (76), *113*
Susuki, T. 91 (194), *117*, 674 (10), 710 (208, 209), *711*, *716*
Sutcliffe, L. H. 234 (66), *237*
Sutton, L. E. 150 (259), *165*, 643 (18), *668*, 853 (6, 8), 854 (6, 8), 865 (6), *877*
Suvorov, B. V. 525 (525), *630*
Suzuki, S. 186 (65), *204*
Swain, C. G. 150 (254), *165*, 277, 282 (147), *302*
Swamer, F. W. 283 (172, 173), 285 (183), *303*
Swaminathan, M. 927 (111), *945*
Swan, G. A. 763 (98), *787*
Swanson, C. P. 735 (118), *742*
Swanson, J. S. 107 (342), *122*
Swenson, J. S. 681 (27), *711*, 901 (52b), 902 (52b), *942*
Swirles, B. 10, 39 (19), *63*
Swoboda, O. 103 (299), *120*
Symons, M. C. R. 473 (233), 543 (610), *622*, *632*
Szczepaniak, K. 136 (94), *161*
Szewczvk, A. 177 (26), *203*
Szkrybalo, W. 148 (224), *164*, 211 (8), *236*
Szwarc, M. 854 (32), *878*

Tabata, Y. 288 (195, 196, 197), *304*
Tadokoro, H. 779 (212), *790*
Taft, R. W. 148, 149 (226), 150 (263), *164*, *165*, 216 (20a), 217 (20a, 23, 24, 25), 219 (24), 220 (24), 229 (24), *236*
Taft, R. W., Jr. 217 (23), 219 (23a), *236*
Tagaki, I. S. 96 (248), *119*
Tahk, F. C. 384 (131), *418*
Tai, T-C. 268 (106), *301*

Tailor, E. G. 573 (726), *635*
Tait, M. J. 131 (42), 133 (42, 54), *160*
Takeda, K. 316 (47), 319 (47), *338*, 926 (105d), *945*
Takeda, N. 889 (17), *941*
Takei, H. 103 (301), *120*
Takelayashi, M. 594 (793), *637*
Takemoto, S. 664 (70), *669*
Takenaka, T. 448 (128), *620*
Takenishi, I. 411 (234), *420*
Takenishi, T. 179 (38), *203*, 506 (366), 508 (366), 509 (366), *626*
Takimoto, M. 547 (636), *633*
Takizawa, T. 513 (415), 517 (415), 518 (415), 519 (415), 524 (415), *627*
Talât-Erben, M. 298 (242), *305*, 688 (69, 70), 689 (71), 694 (96), *712*, *713*, 892 (25), *941*
Talaty, C. N. 829 (164), *850*
Talvik, A. J. 125 (5), 127 (5), 156 (5), *159*
Tamaru, K. 310 (16), *338*, 541 (593), 543 (593), *632*
Tamilov, A. P. 337 (117, 118), *340*
Tamres, M. 125, 127 (3), 144 (195), 156 (3, 318), *159*, *163*, *166*
Tanabe, K. 196 (143), *206*
Tanaka, J. 144 (193), *163*, 528 (548, 549), *631*, 642 (14), 652 (14), 653 (14), 656 (14), *668*, 872 (163), *881*
Tanaka, S. 196 (143), *206*
Taniyama, M. 288 (197), *304*
Tapper, B. A. 719 (8), 720 (8, 19, 20), *739*
Taramasso, M. 180 (45), *204*
Taras, M. 170 (8), *203*
Tarbell, D. S. 372 (97), *417*
Tardy, D. C. 891 (19e), *941*
Tarsio, R. J. 96 (250), *119*
Tashpulatov, Yu. T. 197 (124), *206*
Tate, D. P. 255 (59, 61), 256 (62), *300*
Tatlow, J. C. 73 (34), *111*
Taub, D. 397 (170), *419*
Taurins, A. 270 (112), *301*
Tawney, P. O. 694 (68b), *712*
Taylor, E. C. 276 (141), *302*, 394 (165), 395 (166), 396, 401 (184), 402 (165, 191), 404 (196), 408, 409, *419*, *420*, 926 (106), *945*
Taylor, F. M. 289 (198), *304*
Taylor, H. M. 75 (55), *112*
Taylor, J. W. 854 (32), *878*
Taylor, W. I. 917 (87f), 922 (87), 930 (87), *944*
Tchelitcheff, S. 835 (183), *850*
Tchelitscheff, S. 806 (90), 835 (90), *848*
Tchingakov, S. V. 748 (15), *785*
Teague, P. T. 97 (255), *119*
Tebbens, W. G. 86 (155), *115*
Tebby, J. C. 520 (824), *638*
Teearu, P. 876 (225), *883*
Telonsky, T. J. 78 (74), *113*
Temnikova, T. I. 369 (89), 406 (202), *417*, *420*
Templeton, D. H. 502 (329, 343), *625*, *626*
Teng, C. S. 93 (218), *117*
Terada, A. 362 (71), *416*
Terashima, M. 379 (121), *418*
Terekhina, I. P. 287 (191), *304*
Terenin, A. 135 (76), *161*
Terenin, A. N. 472 (218), *622*
Terent'ev, A. P. 325 (91), *339*, 412 (238), *421*
Terruzzi, M. 74 (42, 43), 79 (43), *111*
Tessieri, J. 929 (113b), *945*
Thakor, G. P. 330 (104), *340*
Thal, A. F. 766 (131), *788*
Theilacker, W. 508 (405), *627*
Theobald, C. W. 409 (222), *420*, 446 (116), 447 (116), *619*, 640 (3), *667*, 682 (31), *711*
Thesing, J. 580 (738), *635*, 773 (191), *789*
Thiele, J. 71 (17, 21), *110*, *111*, 429 (32), 540 (583, 584), 541 (584), 544 (584), *617*, *632*, 685 (48), *711*, 796 (20), 819 (20), *846*
Thier, W. 390 (151), *418*
Thimann, K. V. 732 (94, 95), 734 (94, 109, 110a), *741*
Thomas, B. 194, 196 (113), *205*
Thomas, H. T. 689 (75), *712*
Thomas, O. 507 (389), 508 (389), 572 (389), *627*
Thomas, R. 372 (97), *417*, 703 (164), *715*
Thomas, W. M. 262 (84), *301*
Thompson, A. R. 193 (107), *205*
Thompson, C. C., Jr. 229 (48), *237*, 642 (12), *668*
Thompson, C. M. 516 (465), *628*
Thompson, H. W. 135 (92, 93), 137 (104, 107), 147, 157, 158 (104), 159 (104), *161*, *164*, 196 (118,

140, 146, 149, 150), *205*, *206*, 242 (8), *298*, 854 (18), *877*
Thompson, S. J. 768 (160), *788*
Thompson, W. E. 548 (660), *633*
Thomson, T. J. 525 (527), 543 (602, 609), *630*, *632*, 654 (36), 656 (36), *668*
Thorn, G. D. 727 (75), 728 (82a), *740*, *741*
Thornton, J. D. 282 (168), *303*
Thorp, R. H. 400 (179), *419*
Thorpe, J. F. 284, 286 (188), *303*
Thrush, B. A. 709 (199, 200), *715*
Thurman, J. C. 97 (251), 99 (272), *119*
Tiedmann, H. 762 (102), *787*
Tieman, F. 347 (17), *415*
Tiemann, F. 75 (48), *111*, 429 (35), *618*
Tillett, J. G. 260 (83), *301*
Tillmanns, E. J. 367, *416*
Tillmans, E. 291 (211), 292 (211), *304*
Tilney-Bassett, J. F. 696 (107), *713*
Timberlake, J. W. 687 (65), 692 (65), *712*
Timpe, H. J. 94 (232), *118*
Ting, C. T. 891 (23), *941*
Tinling, D. J. A. 473 (233), *622*
Tirouflet, J. 472 (227), *622*
Tits, M. 672 (3), 673 (3), *710*
Tivey, D. J. 71 (18), *110*
Tobey, S. W. 381 (124), *418*
Tobin, M. C. 289, 290 (205), *304*
Tobkes, M. 690 (78), *712*
Tobolsky, A. V. 685 (49), *712*
Todd, A. 246 (23), *299*
Todd, J. S. 372 (97), *417*
Todd, L. J. 151 (274), 152 (278), *165*, 892 (30d), *941*
Todd, P. F. 576 (734), *635*
Tokaku, M. 560 (713), *635*
Tokarev, B. V. 748 (24), 751 (24), 763 (99), *785*, *787*
Tokes, L. 710 (204), *716*
Tokiura, S. 369 (88), *417*
Tokizawa, M. 103 (301), *120*
Tokumaru, K. 135 (73), 156 (73), *160*
Tokura, N. 926 (105e), *945*
Toland, W. G. 101 (288), *120*, 246 (29), 275 (140), *299*, *302*
Tolbert, B. M. 761 (90), *787*
Tolbert, T. L. 276 (145), *302*, 319 (52), *338*, 892 (29b), *941*
Tollin, G. 735 (120), *742*
Tomilov, A. P. 181 (49), 183, 187 (60, 61, 75), 188, 189 (60, 61, 75, 83, 86, 87), 190 (88), *204*, *205*
Tong, L. K. J. 698 (137), *714*
Tönjes, H. 855 (37), 856, 857 (37), *878*
Tonooka, K. 858 (53), *878*
Tonouka, K. 93 (214), *117*
Topchiev, A. V. 524 (512), *630*
Töpfl, W. 497 (296), 498 (299, 300), *624*
Topper, Y. J. 759 (67), 771 (178), *786*, *789*
Torihashi, Y. 474 (249), *623*
Torsell, K. 681 (29), *711*
Torssell, K. 107 (344), *122*
Tosolino, G. 839 (205), *851*
Trabert, C. 92 (201), *117*
Trambornlo, R. 854 (26, 27), *878*
Tramer, A. 136 (94), *161*, 448 (127), 581 (748, 750), 582 (754, 755), *620*, *635*, *636*
Trapp, O. D. 298 (242), *306*, 689 (73), 692 (89), 693 (73, 91), 696 (73), *712*, *713*, 897 (41b), *942*
Traube, W. 89 (183), *116*, 430 (45, 46, 47, 48, 49), *618*
Travis, D. N. 551 (683), *634*, 705 (175), 708 (192), *715*
Trayer, S. A. 768 (154), *788*
Traylor, T. G. 693 (92), *713*
Treharne, G. J. 244 (15), 266 (15), *298*
Treibs, A. 861 (86), *879*
Trifonov, A. 188 (76), *204*
Trinchera, C. 311 (21, 22), 333, *338*
Triollais, M. R. 76 (58), *112*
Tripett, S. 104 (314), *121*
Trippett, S. 469 (212), *622*
Tristan, H. 724 (41), *740*
Troendle, T. G. 363 (72), *416*
Trofimenko, S. 168 (3), *203*, 412 (239), *421*, 433 (79), 434 (79, 80, 83), 435 (79, 83), 436 (79), 437 (79), 441 (102), 590 (79), 613 (80), *619*, 899, 900 (49a), *942*
Tronov, B. V. 142 (164), *163*
Troost, L. 428 (15), *617*
Trotman-Dickenson, A. F. 887 (8c), *940*
Truce, W. E. 254 (54), *300*
Trueblood, K. N. 448 (123), 533 (572), 596 (796, 797), *620*, *631*, *637*, 661 (62), *669*

Trumbull, H. L. 699 (141, 142), *714*
T'Sai, L. S. 752 (41), *785*
Tsalsas, G. 886 (5), *940*
Tsao, E-P. 279 (158), *303*
Tsatsas, G. 98 (267), *119*
Tschammer, H. von 924 (101, 102), *944*
Tscherniac, J. 83 (128), 100 (279), *114*, *120*
Tschiersch, B. 725 (63), 726 (68), 731 (68), *740*
Tschugajeff, L. 876 (225, 226), *883*
Tsubomura, H. 136 (94), 137, 144 (194), *161*, *163*
Tsuji, T. 104 (305), *120*
Tsuruta, T. 686 (56), *712*
Tsutsumi, S. 91 (194), 108 (353), *117*, *122*, 674 (10), 710 (208, 209), *711*, *716*
Tullock, C. W. 297 (238), *305*
Tunemoto, D. 560 (713), *635*
Turco, A. 872 (161), *881*
Turnbull, J. 270 (113), *301*, 410 (232), *420*
Turner, B. 718 (3), *739*
Turner, D. W. 644 (20), *668*
Turner, J. C. 767 (171), 782 (171), *789*
Turner, L. 296 (233), *305*, 309 (12), 310 (12), *337*
Turner, T. A. 171 (13), *203*
Turpin, A. 516 (476), 519 (476), *629*
Turrel, G. 137 (108), *161*
Turrel, G. C. 129 (39), 130 (41), *160*
Turrell, G. C. 250 (44), *299*, 516 (447), *628*
Tyler, J. K. 516 (444), 517 (444), *628*

Uchimura, F. 373 (102), *417*
Ugi, I. 854 (11, 14, 15, 21, 33), 855 (15, 21, 40), 857 (15, 21, 40), 860 (71, 73), 862 (94, 101), 863 (116, 117), 864 (116, 120, 122, 123), 865 (11, 14, 124, 130), 866 (130), 867 (11, 124, 126, 127, 128, 129, 130, 133), 868 (11, 14, 129, 136, 137), 869 (120), 872 (157), 875 (198), 877 (230, 231, 232, 233, 234, 235, 236), *877*, *878*, *879*, *880*, *881*, *882*, *883*, 887, 891 (9a), 913 (74), *940*, *943*
Uhlig, K. 759 (61), *786*
Ulbricht, T. L. V. 938 (136), *946*
Ullman, E. F. 452 (152, 153), *620*, 909 (66a, 66b), 931 (116, 117), 938 (134), *943*, *945*, *946*
Ullrich, A. 190 (94), *205*
Ultee, I. A. 753 (45), *786*
Ultee, J. A. 73 (30), *111*
Umani-Ronchoi, A. 842 (217), 845 (220, 221, 222), *851*
Umarova, R. U. 525 (525), *630*
Underhill, A. E. 132 (50), *160*
Ungnade, H. E. 76 (64), *112*
Unterstenhöfer, G. 877 (230, 231, 232, 233, 234, 236), *883*
Urata, Y. 81 (102), *114*
Urban, J. 751, 753 (42), *785*
Urech, F. 72 (26), 73 (32), *111*
Urech, H. J. 767 (138), *788*
Urenovich, J. V. 859 (62), *878*
Uribe, E. G. 720 (11), *739*
Urry, G. 710 (206), *716*
Urushibara, Y. 594 (792, 793), *637*
Usanovich, M. 142 (160), *163*
Utley, J. H. P. 144 (191), 145 (191), *163*

Vagt, A. 388 (137), *418*
Valenta, Z. 378 (119), *417*
Valentine, R. S. 177 (29), *203*
Valentini, F. 809 (101, 102a, 102b), *848*
Vallance, D. C. M. 142 (151), *162*
Vallarino, L. 876 (219, 220, 221), *882*
Valvassori, A. 767 (136), *788*
Van Allan, J. A. 104 (311), *121*
Van Baalen, J. 782 (221), *790*
Van Den Bos, B. G. 748 (21), *785*
Vander Werf, C. A. 286 (189), *303*
Van Dine, G. W. 887 (12), 890 (12), *940*
Van Dyke, J. W., Jr. 469 (214), *622*
Van Epps, G. D. 934 (124e), *945*
Van Etten, C. H. 733 (107), *741*
Van Helden, R. 673 (7), *711*
Van Hook, J. P. 685 (49), *712*
Van Ling, R. 746 (1), 747 (1), 748 (1), 749 (1), *784*
Van Meter, J. P. 451 (149), *620*
Varchmin, J. 448 (126), *620*
Varshavskii, S. L. 181 (49), 189 (83, 86), *204*, *205*, 337 (117), *340*
Vasilescu, V. 179 (37), 180 (43), *203*, *204*
Vaughan, J. 508 (400, 404), *627*
Vaughan, W. R. 97 (260), 109 (359), *119*, *122*, 750, 751 (34), *785*

Vaughn, J. 218 (29a), *236*
Vecchi, A. 738 (141), *742*
Vecchietti, V. 839 (205), *851*
Veglio, C. 796 (41), 803 (41), 814 (128), *847*, *849*
Veibel, S. 168 (4), *203*
Veillard, A. 10, 13, 19 (26, 27), 26, 30 (27), 39, 40, 62 (63), *64*, *65*
Velo, L. 347 (15), *415*, 835 (187), 837 (201), 839 (201), 840 (201), *850*
Venanzi, L. M. 459 (182), *621*
Venkateswarlu, K. 196 (144), *206*
Venkateswarlu, P. 135 (65), *160*
Venkitasubramanian, T. A. 724 (49), *740*
Vercier, P. 749, 750 (31, 32), *785*
Verdin, D. 692 (86), *713*
Veronesi, G. 834 (177), *850*
Vest, R. D. 317 (22), *236*, 433 (76), 446 (120), 497 (297), 500 (306), 501 (306), 502 (306), 503 (306, 353, 355), 504, (353, 355), 505 (356), 571 (721), 572 (721), 576 (353), 577 (353, 721), 578 (721), 579 (353, 721), *619*, *620*, *624*, *626*, *635*
Veverko, B. 735 (114), *742*
Viditz, F. 875 (206), *882*
Vidulich, G. A. 141 (148), 142 (148), *162*
Viehe, H. G. 837 (200), *850*
Viehle, H. G. 899 (47a, 47b), *942*
Vielau, W. 88 (168), *116*
Vielle, J. 838 (203), *851*
Vierk, A.-L. 142 (159), *163*
Vigier, A. 378 (117), 412 (235), *417*, *420*
Vignau, M. 323 (7b), *339*
Vill, J. J. 378 (114), *417*
Villaescusa, F. W. 407 (207), *420*, 901 (51b), *942*
Vingiello, F. 333 (109), *340*
Vinnick, M. I. 922 (97a), *944*
Vinograd, L. K. 107 (341), *122*
Vinogradova, V. S. 178 (34), *203*
Vinson, Y. R. 673 (6), *710*
Vipond, H. J. 751 (186), 770 (186), 784 (186), *789*
Virtanen, A. I. 733 (106), *741*
Vita, C. D. 354 (36), *416*
Vita-Finzi, P. 796 (23), 806 (23), 808 (23), 809 (107), 822 (154), 835 (186), *846*, *848*, *849*, *850*
Vitanov, T. 188 (76), *204*
Vizbaraire, Ya. I. 10, 19 (21), 39 (21), *64*
Vladimirova, T. M. 748 (17), 757 (74), *785*, *786*
Vlasov, G. P. 861 (84), *879*
Voegeli, R. 99 (271), *119*
Vofsi, D. 179 (39), *203*
Vogel, A. I. 168 (1), 169 (1), *202*
Voigt, D. 516 (476), 519 (476), *629*
Voitenko, R. M. 524 (512), *630*
Volini, M. 735 (130), *742*
Volke, J. 183 (57, 58), *204*
Völker, T. 289 (200), *304*, 506 (373), *626*
Volkova, L. D. 325 (85), *339*
Voltz, J. 767 (151), *788*
Von Schuching, S. 748, *785*
Von Wittenau, M. S. 776 (201), 783 (201), *789*
Voreb'eva, V. Ya 917 (86), 922 (86), *944*
Vorlander, D. 90 (188), *116*, 430 (51), 431 (51), *618*
Vose, C. E. 97 (258), *119*
Vromen, S. 394 (165), 396 (165), 402 (165), 409 (165), *419*
Vromer, S. 395 (167), *419*
Vuilleumier, J. P. 779 (209), *789*
Vulf'son, N. S. 107 (341), *122*
Vyks, M. F. 142 (165), *163*

Wada, S. 138 (119, 120), *162*
Wade, J. 890 (30c), *941*
Wade, K. 150 (249), 151 (270), 152 (280, 283, 284), 153 (283, 284, 288), *165*, *166*, 282 (170), *303*, 517 (485), *629*, 892 (30c), 893 (32), *941*
Wadsten, T. 507 (393), *627*
Wagenhofer, H. 350 (25), *415*
Wagner, G. 437 (93), *619*
Wagner, R. B. 256 (67), 262 (67), 263 (67), *300*, 320 (54), 325 (54), 326 (54), *338*
Wahl, A. C. 8 (9), 10, 19 (25), *63*, *64*
Wain, R. L. 97 (255), *119*, 524 (521), 525 (521), *630*
Wait, S. C., Jr. 356 (44), *415*
Waits, H. P. 689 (74), 693 (93), *712*, *713*
Wakamatsu, H. 411 (234), *420*
Wakamatsu, S. 702 (157), *714*
Wakefield, B. J. 803 (78), 810 (78), 811 (78), 813 (78), 822 (78), *847*
Walcott, R. G. 370 (92), *417*
Walden, P. 85 (143), *115*

Walker, D. 541 (599), 543 (622), *632*
Walker, H. M. 379 (120), *417*
Walker, J. 390 (149), *418*
Walker, L. E. 94 (228), *118*, 886 (4), 910 (4b), *940*
Walkley, J. 144 (183), 146 (183), *163*
Wallace, E. 233 (63), *237*
Wallach, J. 523 (509), *630*
Wallach, O. 915, 916 (84), *944*
Wallbillich, G. 348 (19, 20), *415*
Wallenfels, K. 72 (23), 77 (68), 80 (100), 81 (101), 83 (122), 86 (101), 92 (195, 206), 98 (262), 99 (100), 102 (297), *111*, *112*, *113*, *114*, *117*, *119*, *120*, 191 (97), *205*, 211, 221 (33), 231 (61, 61a), *235*, *236*, *237*, 427 (1), 482 (271), 525 (532), 526 (533, 535, 536), 527 (532, 536), 529 (532), 530 (532), 531 (532, 564, 566), 540 (600), 541 (587), 542 (587, 600), 544 (600), 545 (600), 546 (600), 573 (727, 728), 574 (728), *617*, *624*, *630*, *631*, *632*, *635*, 650 (29, 33, 34), 653 (29, 35), 654 (29, 35), 655 (35), *668*
Waller, G. 731 (91), *741*
Waller, G. R. 730 (89), 731 (92b), *741*
Walley, D. W. 98 (269), *119*
Walling, C. 672 (4), 673 (4), 675 (24a), 698 (133, 134), *710*, *711*, *714*
Wallis, J. W. 150 (264), 151 (264), *165*, 216 (17), *236*
Wallwork, S. C. 149 (235), *164*
Walter, L. A. 325 (87), 336 (87), *339*
Walton, E. 400 (179), *419*
Walton, J. H. 245 (22a), *299*
Walton, R. A. 150 (258), 153, 154 (258, 303, 304), *165*, *166*, 244 (13), 250 (13), *298*
Walts, J. M. 767 (150), *788*
Walz, K. 859 (56, 57), *878*
Wamhoff, H. 407 (211), *420*
Wamser, C. C. 696 (117), *713*
Wang, C. H. 610 (804), *637*
Wang, F. E. 152 (276), *165*
Wani, M. C. 698 (130), *714*
Wanmaker, W. L. 270 (112), *301*
Wanzlick, H. W. 457 (173), *621*
Ward, E. W. B. 727 (75, 76), 728 (82a), *740*, *741*
Ward, M. L. 83 (117), *114*
Ward, R. L. 472 (224), *622*
Ward, R. S. 520 (824), *638*
Ware, E. 761 (96), *787*
Waring, A. M. 94 (231), *118*
Waritz, R. S. 503 (352), *626*
Warkentin, J. 298 (242), *305*, 693 (91), *713*, 897 (41b), *942*
Warren, F. L. 755 (79), *786*
Warrener, R. N. 395 (166), *419*
Wartburg, A. von 748 (25), *785*
Waser, J. 828 (163), *850*
Wass, M. V. 764 (110), *787*
Wasserman, E. 548 (653), 552 (653), *633*, 705 (177), 707 (187), 709 (177), *715*
Wasserman, H. H. 935 (126), *945*
Wasyliw, N. 769 (167), *788*
Watanabe, K. 83 (119), *114*, 325 (86a), *339*, 560 (713), *635*, 644 (19), *668*
Waters, J. H. 502 (322, 324, 331, 337, 344), *625*, *626*
Waters, K. L. 95 (239), *118*
Waters, W. A. 88 (167), *116*, 129 (36), *160*, 688 (67), 691 (84), 695 (100, 101), 696 (103, 106, 107, 118, 119), 697 (84), 699 (139), *712*, *713*, *714*
Watkins, K. W. 859 (67), *879*, 891 (19d), *941*
Watkins, T. I. 507 (391), 508 (391), *627*
Watson, J. W. 828 (163) *850*
Watt, G. D. 614 (807), *637*
Watts, C. T. 456 (167), *621*
Waugh, T. D. 541 (599), *632*
Wawzonek, S. 105 (322, 323), *121*, 133 (59), *160*, 724 (51, 52), *740*, 760 (114), *787*, 926 (105g), *945*
Weaver, J. C. 751 (36), 770 (174), 783 (174), *785*, *789*
Weaver, S. J. 835 (191b), *850*
Webb, R. L. 215 (14), *236*, 507 (394), *627*
Weber, D. 372 (95), *417*
Webster, M. 153 (290, 295), 154 (299), *166*
Webster, O. W. 89 (180), 90 (180), 109 (354), *116*, *122*, 225, 226 (44), 227 (46), 228, 229 (53, 54), 232 (46), 233 (44, 65), 234 (68, 69), *237*, 433 (74, 75), 441 (100, 101), 458 (174), 459 (100), 463 (100), 467 (209), 468 (100), 471 (100), 472 (100), 473 (100), 474 (100, 238), 495 (209, 291), 509 (238), 510 (238), 511 (238), 512 (238), 527 (540), 529 (540),

555 (174, 699), 556 (174, 703), 557 (699), 560 (699), 581 (699), 590 (782), 592 (238), 595 (795), 596 (238), 597 (795, 798), 599 (800), 600 (800), 602 (795), 612 (798), 613 (798), 614 (798), 615 (798), 616 (798), *618*, *619*, *621*, *622*, *624*, *630*, *634*, *637*, 646 (22), 650 (22), 653 (22), 655 (22), 658 (22, 44, 45), 661 (22, 44, 45), 664 (22, 45), *668*, 707 (188), *715*
Webster, W. 273 (126), *302*, 909 (65), *943*
Weddige, H. 248 (36), *299*
Wefer, J. M. 886 (3), *940*
Wegler, R. 295 (230), *305*, 356 (38), *415*, 856 (42), *878*
Wehr, R. 810 (115), 822, (115), *848*
Weid, E. von der 84 (129), *114*
Weidinger, H. 246 (25), *299*
Weidmann, H. 399 (177), *419*
Weiher, J. F. 502 (323), *625*
Weimer, R. F. 150 (239), *164*, 191 (96), *205*
Weinbaum, G. 735 (124), *742*
Weiner, N. 70 (13), 71 (13), *110*
Weinhouse, S. 767 (140), *788*
Weis, C. D. 83 (118), 98 (268), *114*, *119*, 460 (185), 514 (435, 436), 518 (490), 519 (490), 520 (185, 490), 521 (436, 490, 503), 522 (490, 505), 523 (490), 525 (530), 527 (436), 551 (505), 570 (435, 720), 572 (505), *621*, *628*, *629*, *630*, *635*
Weis, L. D. 689 (75), *712*
Weise, A. 79 (83), 90 (186), 91 (186), *113*, *116*, 293 (220), *305*, 862 (99), *879*, 930 (114b), *945*
Weisgerber, C. S. 81 (107), *114*
Weiss, J. J. 641, 659 (8, 8a), *667*
Weiss, K. 88 (172), *116*, 700 (145), *714*
Weissberger, A. 344 (1), 345 (1), *414*
Weisse, A. 252 (49), 253 (51), *299*
Weisse, G. 375 (103), *417*
Weissert, N. H. 180 (48), *204*
Weissman, P. M. 329 (98), 332 (98), *340*
Weissman, S. I. 229 (52), *237*, 473 (232), *622*, 658 (43), *668*
Weith, W. 857 (48), 859 (48), 862 (48), *878*
Welch, C. M. 366 (75), *416*
Welcher, R. P. 430 (42, 43), *618*
Welker, D. M. 104 (314), *121*
Wells, C. H. J. 146 (202), *164*
Wells, R. J. 412 (236), *420*
Welton, D. E. 926 (105c), *945*
Wender, S. H. 769 (172), *789*
Wendler, N. L. 397 (170), *419*
Wenkert, E. 386 (134), *418*
Wentland, M. P. 384 (131), *418*
Wentrup, C. 266 (100), 275 (100), *301*
Werner, A. 793, 803 (5), 826 (5), 829 (5), *846*, 858 (54), *878*, 925 (103b, 103c), 927 (103b), 928, 929 (103c), 930 (103c), *944*
Werner, H. 876 (222), *882*
Werner, N. D. 855 (39), 857 (39), 859 (68), *878*, *879*, 887 (9, 11), 889 (11), 890 (11), *940*
Werner, R. L. 862 (105), *879*
Werner-Zamojska, F. 584 (770), *636*
Werst, A. 781 (229), *790*
Werst, G. 311 (24), 312 (27), 313 (24), *338*
Wessely, F. 309 (14), *337*
West, R. 516 (455), *628*
Westenberg, A. 516 (445), 517 (445), *628*
Westerkamp, J. F. 581 (751), *636*
Westfahl, J. C. 766 (148), *788*
Westheimer, F. H. 759 (59), *786*
Westley, J. 735 (126, 127, 128, 129, 130), *742*
Westman, T. L. 506 (380), 551 (380), *626*, 706 (183), *715*, 901 (53a, 53b), 902 (53, 53a), 903 (53), *942*
Westöö, G. 752, *785*
Wetzel, C. R. 268 (110), 270 (113), *301*, 391 (156), *418*
Weygand, C. 832 (165), 840 (165), *850*
Weygand, F. 354 (35), *415*, 870 (150), *881*, 899 (47b), *942*
Whalley, M. 99 (274), *120*
Wheeler, O. H. 72 (28), *111*, 193 (103), *205*
Wheland, G. W. 215 (13), 221 (35), 231 (60), *236*, *237*
White, E. H. 771 (177), *789*
White, G. H. 767 (137), *788*
White, R. F. M. 459 (178), 463 (178), 464 (178), *621*
White, R. W. 79 (86), *113*
White, S. C. 137, 157, 158 (104), 159 (104), *161*
Whitehouse, A. B. 268 (107), *301*, 429 (37, 38), *618*

Whitehurst, D. H. 174, 175 (21), *203*
Whiting, M. C. 452 (150), *620*
Whitman, R. H. 582 (759), *636*
Whitmore, F. C. 320 (65), *339*
Whitney, A. G. 82 (115), 83 (115), *114*
Wiberg, E. 326 (94), *340*
Wiberg, K. B. 257, 262 (69, 88), *300*, *301*
Wiberley, S. E. 242 (7), *298*
Wichelhaus, H. 84 (135), *115*
Wichterle, O. 289 (199), *304*, 523 (511), 524 (513, 514), *630*
Wickenden, A. E. 132 (51), *160*
Wideqvist, S. 247 (34a), *299*, 442 (107, 108, 109), 443 (109), *619*
Wiedenmann, L. G. 724 (52), *740*
Wiegrebe, L. 853 (5), 854 (5), *877*
Wieland, H. 291 (208), *304*, 792 (2), 793 (8), 795, 796 (8, 20, 32), 797 (47), 799 (58, 62, 63), 800 (66a), 803 (77), 805 (32, 81, 82), 807 (47, 77), 812 (77), 815 (62, 133), 816 (62, 136a), 818 (143), 819 (2, 20, 66a), 820 (62), 821 (62, 136a), 822 (8, 47), 824 (8, 47, 156), 829 (8), 830 (77), 831 (8, 47, 156), 839 (8), *846*, *847*, *848*, *849*
Wiemann, J. 188 (81), *205*
Wierzchowski, K. L. 581 (749, 750, 752), *636*
Wiese, F. F. 407 (210), *420*
Wiesner, K. 378 (119), *417*
Wiest, H. 449 (139), *620*
Wiffin, D. H. 289 (198), *304*
Wigner, E. 26 (33), 27, 28, 29, 43 (33), *64*
Wijnen, M. H. J. 672 (2), *710*
Wijngaarden, I. V. 766 (132), *788*
Wild, H. 72 (28), *111*
Wildi, B. S. 474 (256), 527 (537), 528 (537), 529 (256, 558, 559, 561), *623*, *630*, *631*
Wiley, D. W. 108 (349), *122*, 433 (74), 453 (157), 454 (157, 158), 456 (157), 465 (207), 467 (207, 208), 474 (258), 475 (258), 478 (208), 479 (208), 480 (207), 536 (208), 606 (207), 607 (207), *618*, *620*, *622*, *623*, 650 (31), 655 (31), *668*
Wiley, R. H. 803 (78), 810 (78), 811 (78), 813 (78), 822 (78), *847*
Wilhelm, H. 75 (52), *112*
Wilhelm, M. 74 (41), *111*
Wilk, W. D. 139 (125), *162*
Wilkinson, G. 253 (52), *299*, 520 (498), *629*, 892 (28), *941*
Wilkinson, V. J. 148 (230), 149 (230), *164*
Wilkinson, W. K. 385 (132), *418*
Will, J. 796 (20), 819 (20), *846*
Willett, A. V., Jr. 524 (519), *630*
Williams, A. E. 200 (173), *207*, 507 (395), 508 (395), *627*
Williams, A. R. 694 (68b), *712*
Williams, B. L. 428 (19), *617*
Williams, D. 201 (181a), *207*
Williams, D. H. 198 (171), 200 (177, 178), *207*, 520 (824), *638*
Williams, D. L. 746 (2), 751 (2), *784*
Williams, H. E. 580 (739), *635*
Williams, J. K. 95 (241), 108 (349, 352), *118*, *122*, 222 (34), *236*, 437 (94), 453 (155, 157), 454 (157), 455 (163), 456 (157), 465 (207), 467 (207), 468 (210), 480 (207), 483 (210), 497 (298), 540 (298), 606 (207), 607 (207), *619*, *620*, *621*, *622*, *624*, 702 (159), 703 (159), *714*
Williams, J. M. 128 (30, 31), *160*, 243 (12), *298*
Williams, L. E. 646 (23), *668*
Williams, P. H. 263 (89, 90), *301*, 554 (697), *634*
Williams, R. 502 (322, 324, 331, 336, 337), *625*
Williams, R. J. P. 528 (550), 543 (550), *631*
Williams, R. L. 138 (113), 158 (324), *161*, *166*, 854 (18, 19), *877*
Williams, R. T. 734 (113), 736 (113), 737 (113), *742*
Williams Morse, K. D. 892 (30), *941*
Willitzer, H. 459 (179), 464 (179), 465 (179), 476 (260, 261), 477 (179, 261), *621*, *623*
Wilmshurst, J. K. 196 (154), *206*
Wilson, C. V. 72 (22), *111*
Wilson, D. A. 541 (597), *632*
Wilson, D. W. 761 (88), *787*
Wilson, E. B., Jr. 516 (445), 517 (445), *628*
Wilson, E. R. 437 (819), *638*
Wilson, K. R. 388 (139), *418*
Wilson, S. T. 139 (140), 140 (140), *162*
Wilson, W. 400 (180, 181), *419*
Winas, C. F. 320 (57), 323 (79), 324 (79), *339*

Winberg, H. E. 409 (222), *420*, 446 (116), 447 (116), *619*, 640 (3), *667*, 681 (30), 682 (30, 31), *711*
Winch, R. W. 393 (162), *419*
Wineman, R. J. 481 (267), 482 (267), *623*
Winkelmann, H. D. 863 (118), *880*
Winkler, C. A. 257 (68), 258 (80), *300*
Winkler, F. 72 (24), *111*
Winkler, R. 428 (5), *617*
Winnick, R. E. 760 (114), *787*
Winnick, T. 760 (114), *787*
Winslow, E. C. 513 (432), 520 (432), 521 (432), *628*
Winterfeldt, E. 518 (489, 491), *629*, 869 (142), *880*
Winternitz, F. 89 (184), *116*
Wippermann, R. 506 (359), 507 (359), *626*
Wise, R. M. 85 (145), *115*
Wislicenus, W. 762 (107), *787*
Wisotsky, M. J. 126 (10, 13), 127 (13), *159*
Wit, J. G. 738 (139, 140), *742*
Witanowski, J. 132 (47), *160*
Witanowski, M. 202 (181b), *207*, 242 (3), *298*
Witkowski, R. E. 516 (458), *628*
Witt, O. 84 (132), *114*
Witte, H. 151 (267, 268), *165*, 870 (146), 872 (159, 164, 166), 873 (167), *881*, 913 (74), *943*
Wittig, G. 278 (155), *303*, 682 (36, 37), 683 (38, 39), *711*
Wittman, J. S., III 329 (99), 330 (99), *340*
Witzel, D. 580 (738), *635*
Witzler, F. 92 (195), 98 (262), *117*, *119*, 231 (61, 61a), *237*, 482 (271), 526 (533, 535), *624*, *630*
Wladimorowa, T. M. 754 (74), *786*
Wohl, A. 886 (6), *940*
Wohler, F. 430 (44), *618*
Wöhrle, D. 502 (327), 528 (556), 579 (327, 556), *625*, *631*
Woislowski, S. 747 (7), *785*
Wojtkowiak, B. 515 (448), 516 (448, 450), *628*
Wojtowicz, J. 92 (202), *117*, 580 (741), 581 (741), 582 (741), *635*
Wolf, E. 351 (28), 388 (28), *415*
Wolf, W. 503 (345), *626*, 710 (204), *716*
Wolff, I. A. 733 (107), *741*
Wolff, R. E. 760 (85), *786*
Wolfrom, M. L. 759 (62), *786*
Wolinski, J. 358 (51, 52), *416*
Wollner, T. E. 407 (207, 208), *420*
Wolniewicz, L. 5 (6), 10 (17), *63*
Wolter, D. 684 (44), *711*
Wood, B. 525 (524), *630*
Wood, D. C. 488 (281), *624*
Wood, H. G. 766 (121, 123), 771 (179), *787*, *789*
Wood, J. L. 736 (131), *742*
Wood, K. R. 759 (53), *786*
Woodburn, H. M. 268 (107), 270 (113), *301*, 428 (12, 20, 21, 22, 23, 24), 429 (22, 25, 26, 28, 38), 430 (50), 431 (55), *617*, *618*
Woods, L. L. 106 (333), *121*
Woodward, D. W. 506 (370), 508 (407), 509 (408), *626*, *627*
Woolf, C. 73 (35), *111*
Worther, H. 771 (177), *789*
Wotiz, J. H. 78 (75), *113*
Wren, J. J. 158 (322), *166*
Wright, G. F. 448 (124), *620*, 888 (15), 890 (15), *940*
Wright, J. D. 528 (550, 554, 555), 543 (550), *631*
Wright, J. M. 506 (364), 508 (364), *626*, 693 (90), *713*, 897 (41a), *942*
Wrobel, J. 586 (774), *636*
Wroczynski, A. 428 (14), *617*
Wu, C.-H. S. 693 (90, 91), *713*
Wu, C. S. 298 (242), *305*, 897 (41a, 41b), *942*
Wulf, O. R. 142 (166), *163*
Wulff, J. 842 (216), 843 (218), *851*
Wunderling, H. 96 (247), *118*
Würsch, J. 767 (138), 769 (165), *788*
Wyn-Jones, E. 156 (310), *166*
Wystrach, V. P. 270 (112), *301*

Xenakis, G. I. 874 (180), *881*

Yada, H. 144 (193), *163*
Yadnik, M. 315 (41), *338*
Yager, W. A. 548 (653), 552 (653), *633*, 705 (177), 709 (177), *715*
Yagi, Y. 146 (205), *164*
Yagudaer, M. R. 242 (6), *298*
Yamada, Y. 411 (234), *420*, 506 (366), 508 (366, 825), 509 (366, 825), *626*, *638*
Yamadera, R. 202 (188), *207*, 779 (212), *790*
Yamaguchi, S. 918 (89a), *944*
Yamanaka, H. 84 (137), *115*

Yamase, I. 388 (147), *418*
Yanase, R. 361 (68), *416*
Yang, K. S. 731 (91), *741*
Yanina, A. D. 917 (86), 922 (86), *944*
Yaroslavsky, S. 871 (152), *881*
Yarwood, J. 146 (207), *164*, 551 (682), *634*
Yashunskii, V. G. 412 (238), *421*
Yasuda, H. 96 (248), *119*
Yates, P. 391, 401 (186), *418*, *419*
Yen, J. Y. 98 (266), *119*
Yokota, Y. 857 (46), *878*
Yokoyama, A. 96 (248), *119*
Yoncoskie, R. A. 272 (123), *302*
Yonemitsu, O. 379 (121), *418*
Yoshikawa, T. 371 (94), *417*
Yoshimine, M. 32 (38), 33 (42), 46, 48 (58), 49 (59), *64*, *65*
Yoshimura, J. 280 (159), *303*
Yoshioka, H. 861 (82), *879*
Yoshioka, M. 71 (20), *110*
Yoshioka, T. 809 (106), *848*
Young, D. M. 481 (266), *623*
Young, J. A. 297 (238), *305*, 410 (233), *420*, 933 (120), *945*
Yuan, C. 551 (689), *634*
Yukhnovski, I. N. 195 (116), 196 (162, 163), *205*, *207*
Yutzis, A. P. 10, 19, 39 (20, 21), *64*

Zabicky, J. Z. 72 (28), *111*
Zabin, I. 766 (122), *787*
Zahn, H. 75 (52), *112*
Zajac, W. W. 333 (108), 335 (108), *340*
Zakharenko, E. T. 287 (191), *304*
Zakharkin, L. I. 317 (48), *338*
Zalkin, A. 502 (329, 343), *625*, *626*
Zambito, A. J. 400 (183), *419*
Zamojski, A. 583 (764, 765, 766), 584 (768, 769), 586 (768, 769), *636*
Zanger, M. 560 (710), 591 (788), *635*, *637*
Zappi, E. 778 (206), *789*
Zarakhani, N. G. 922 (97a), *944*
Zaugg, H. E. 327 (97), *340*
Zavitsanos, P. D. 514 (437), *628*
Zazin, A. B. 287 (191), *304*
Zbiral, E. 469 (213), *622*
Zeeh, B. W. 869 (140), 871 (154), *880*, *881*
Zehetner, W. 797 (52), 801 (52), 815 (52), *847*
Zehrung, W. S., III 429 (25), *617*
Zeifman, Ju. V. 869 (144), 870 (144), *881*
Zeifman, Yu. V. 869 (143), *880*
Zeil, W. 135 (81), *161*, 196 (132), *206*
Zelenaya, Sh. A. 197 (122), *206*
Zelenin, K. N. 94 (229, 230), *118*, 900 (50), *942*
Zelinsky, N. 75 (50), *112*
Zeller, E. 90 (190), *116*
Zellhoefer, G. F. 142 (161, 162), 148 (161, 162), *163*
Zeman, I. 179 (36), *203*
Zemplén, G. 886 (6), *940*
Zenk, W. 732 (103), *741*
Zergenyi, J. 921 (95), *944*
Zervos, C. 275 (138), *302*, 734 (110b), *741*
Zey, R. L. 391 (158), *418*
Zhesko, T. E. 369 (89), *417*
Zhukova, E. L. 135 (77), 149 (77), *161*
Ziegler, E. 366 (79, 80), *416*
Ziegler, J. B. 766, 784 (133), *788*
Ziegler, K. 273 (125), 284 (171, 180), *302*, *303*, 687 (63), 690 (63), 691 (81), *712*
Ziegler, M. L. 256 (62), *300*
Zikmund, J. 756, 760, 761 (93), *787*
Zil'berman, E. N. 127 (22), 128 (22, 23, 24, 33a, 33b), 129 (24), *159*, *160*, 243, 244 (16), 245 (20, 21), 246 (30), 247 (30), 249 (10), 254 (10, 20, 56), 255 (57), 264, 266 (21, 103), 291, 295 (10), 296 (235), *298*, *299*, *300*, *301*, *304*, *305*, 309 (11), *337*, 344 (6), 359 (6), *414*, 913 (75a), *943*
Zilkha, A. 914 (80b), *944*
Zimmerman, H. E. 698 (130, 131), *714*
Zingales, F. 135 (79), *161*
Zinke, A. 531 (567), *631*
Zinn, M. F. 779 (213), *790*
Zinner, G. 796 (24), 803 (24), 814 (24), 824 (24), 828 (24), 834 (24, 173b), 835 (24), *846*, *850*
Zita, N. 871 (153), *881*
Zobel, F. 320 (53), 324 (53), *338*
Zollinger, H. 99 (273), *119*
Zolotov, Y. M. 101 (285), *120*
Zoltewicz, J. A. 276 (141), *302*
Zook, H. D. 256 (67), 262 (67), 263 (67), 296 (232), *300*, *305*, 309 (10), 320 (54), 325 (54), 326 (54), *337*, *338*

Zubov, V. P. 287 (191), *304*
Zuman, P. 179, 182 (50), 190 (90, 91) *204*, *205*
Zwanenburg, B. 369 (90), 373 (90), 374 (90), *417*
Zwartsenberg, J. W. 147 (212), *164*
Zweig, A. 527 (542), 528 (542, 552), 543 (552), *631*, 690 (78), *712*
Zwierzak, A. 587 (776, 777), *637*
Zymaris, N. 874 (183), *882*

Subject Index

Absorption bands, charge-transfer 191
Absorption spectra—*see also* Infrared and Ultraviolet spectra
effect of cyano substitution 231–233
of nitrile group 242
λ_{max} for polycyano complexes 649–652
Acceptor radical 700
σ- & π-Acceptors 144
Acetaldehyde 729
Acetals, formation 315
hydrolysis to aldehydes 315
Acetamides 320
labelled, dehydration of 776
Acetic acids 218
Acetic anhydride,
as hydrogenation solvent 320–322
Acetolysis, of hydrogen cyanide 249
Acetone cyanohydrin 736
in formation of linamarin 720
labelled, preparation of 750
reaction with *Linum usitatissimum* 719
use in preparation of nitriles 81
Acetone oxime, conversion to *N*-methyl acetonitrilium ion 912
Acetonitrile 55, 201, 234, 724
acidity 227
alkoxy 281
boron trihalide complexes 151, 216
bromine adduct 146
deuteration of 779
labelled, synthesis of 766, 776
n.m.r. 201
vacuum ultraviolet spectrum 190
9-Acetyl-*cis*-decalin 921
Acetylenedicarboxylic esters 869
Acetylenes, n.m.r. chemical shift data 234
reactions with nitrile oxides 836, 837
Acidity scale, *H* 191
Acids, α-aminoalkylation 865
Acrylonitrile 69, 188, 698, 724, 736
exchange with heavy water 779
hydrodimerization 109, 188
polarographic reduction 186
polymerization 698
substituted 282
epoxidation of 901
p-Acylbenzonitriles 181
Acyl isocyanates, reaction with isonitriles 864
Acyl malononitrile 590
Acylnitrilium species 926
α-Acyloxycarboxamides 863
Addition—*see also* Cycloadditions, Electrophilic Addition and Nucleophilic Addition
Michael-type 70
to amines, of dichlorocarbene 855
to nitrile oxides 822–843
of inorganic compounds 822–827
of organic compounds 828–832
to nitriles, of alkenes 674
of ammonia and amines 269–274, 387–391
of carbanionic species 276–286, 383–387
of free radicals 296–298
of oxygen nucleophiles 263, 264, 267, 268, 391–393
of sulphur nucleophiles 274–276, 393–396
to tricyanomethane 435–437
α-Addition, of immonium ions and anions to isonitriles 865
of radicals to isonitriles 862
simple, of isonitriles 860–862
Addition compounds, from nitriles and Grignard reagent, conversion to ketones 171
Adiponitrile, labelled, synthesis of 767
mass spectrum 200
reduction 325

Aglycones 720
AIBN—*see* α,α′-Azobis(isobutyronitrile)
Alanine, formation of 729
β-Alanine-β-^{14}C, preparation from 2-(cyano-^{14}C) acetic acid 780
D,L-Alanine, labelled 761
Alanine nitrile 724, 729
Alcohols, addition to cyano compounds 263, 264, 267, 268, 429, 430, 435, 467, 860, 861
Aldehydes
  aromatic, formation of 309, 311, 315, 316
  formation from amides 308
  formation from nitriles 308–318
    by catalytic hydrogenation 309
    by chemical reduction 313, 314
    by hydride reduction 314–318
    by the Stephen synthesis 308, 309
  labelled 780–782
  reaction with nitriles 294–296
  trialkylhydrazonium salts 94
Aldimines, hydrolysis to aldehydes 313
  intermediates in aldehyde synthesis 308, 314, 319
  preparation by modified Stephen aldehyde synthesis 319
Aldiminoborane 892
Aldoximes 910
  conversion to nitrile oxides 800–804
Alkaloids, cyanopyridine 730–732
Alkoxyaluminohydrides 329
Alkoxycarbonyl malononitrile 590
Alkoxyl group migration 201
Alkylation, of nitriles 283, 290, 779, 855
N-Alkylation, of nitriles 291–294
Alkyl group migrations 200
Alkylhydroximic acids 828
N,N′-Alkylidenebisamides 294, 296
Alkylidenecyanamides 548
Alkylidenemalononitriles, cleavage and exchange reactions 106
N-Alkylnitrilium ion 291
o-Alkylphenols, reaction with tetracyanoethylene 463
O-Alkylpseudoureas 265
N-Alkylquinolinium ions 864
Alkylthiohydroximic acids 828
Alkyltricyanovinyl ethers, formation from alcohols and tetracyanoethylene 467
Alkyltriphenyldihydrotriazines 284
Allene, addition to tetracyanoethylene 452
Allo isoleucine 763
Allyl cyanide 733
Allyl cyanide-1-^{14}C 768
Alpheloria corrugata 722
Aluminium chloride 326
  with $LiAlH_4$ in reduction of nitriles 326
Aluminium chlorohydride 326
Aluminium compounds, coordination complexes with 152, 153
Aluminium hydride 326
Amides
  conversion to aldehydes 308
  dehydration 96, 97, 776
  reactions with nitriles 295
  N-substituted 291
  formation from nitrilium salts 293
Amides-^{14}C, carboxylic 776
Amidines 269, 270, 273, 783
  conversion to imidylamidines 270
  reaction with dicyanoketene thioacetals 498
  N-substituted 271
  synthesis 283
Amidine salts 271
Amidinium salts 270–273
Amidomethylation, of aromatic compounds 296
Amidoximes 272, 824, 829
Amidrazones 273
Amines
  addition, to isonitriles 861
    to nitriles 269–274
  aromatic, addition to tetracyanoethylene 459
  determination 170
  formation from nitriles 170, 308, 319–331
    by catalytic hydrogenation 170, 312, 320–325
    by chemical reduction 170, 325
    by hydride reduction 170, 326–331
  β-hydroxy, formation by reduction of cyanohydrins 326
  in formation of mixed secondary amines 325
  in isonitrile condensation 866
  labelled 780–782
  primary 319, 327
    addition of dichlorocarbene 855
  reaction, with cyanogen 428, 429
    with dicyanoketene thioacetals 497
    with tetracyanoethylene 468, 469

Amines—*cont.*
secondary 320
formation by hydrogenation of nitriles 320, 324, 325
in isonitrile condensation 866
reaction with cyanogen 428, 429
tertiary aliphatic, in reduction of tetracyanoethylene 468
transformation to isonitriles 891
trapped as acetamide derivatives 320
trialkylplumbyl derivatives 861
Aminoacetonitrile-1-^{14}C 760
Amino acids, oxidation to nitriles 104
Amino acids-1-^{14}C 780
α-Aminoacids 758
α-Aminoadipic acid-6-^{14}C 764
1-Amino-1-alkoxy-2,2-dicyanoethylenes formation 435
α-Aminoalkylation, of acids 865
of isonitriles 865
o-Aminobenzamide, formation from *o*-nitrobenzonitrile 336, 337
4-Aminobenzonitrile 185
2-Aminobenzotriazole, oxidation by lead tetraacetate 905
4-Aminobutyric acid-4-^{14}C 765
D,L-α-Aminobutyric acid 1-^{14}C 761
1-Amino-1-chloro-2,2-dicyanoethylene 437
3-Amino-2-cyanoacrolein 435
3-Amino-2-cyanoacrylonitrile 435
4-Amino-4-cyanobutyric acid as intermediate in formation of glutamic acid 729
Aminocyanocarbene 506, 706, 901, 903
formation during photolysis 902
Aminocyanodiazomethane 551
Aminocyanodurene 215
α-Aminodeoxy sugars 782
5-Amino-3,4-dicyanopyrazoles, formation from hydrazines and tetracyanoethylene 468
Aminodihydropyridinium salts 399
Amino group, addition to cyano group 387–391
reaction with dicyanoketene acetals 496
1-Amino-1-halo-2,2-dicyanoethylenes 435
Aminoketoxime 918
Aminomalononitrile 506
5-Amino-2-mercaptothiazoles 408
Aminonitrile rearrangement 900
α-Aminonitriles 280
hydrogenation 323
infrared spectra of hydrochlorides 195
Aminonitrodurene 215
Aminooxazoles 405, 406
Aminophenylacetylene 898
1-Amino-1-phenylamino-2,2-dicyanoethylene 437
3-Amino-1-propanol-1-^{14}C 781
β-Aminopropionitrile 724
condensation with glutamic acid 726
conversion, to cyanoacetic acid 724
to thiocyanate 724
production, from β-cyanoalanine 726
from psychrophilic basidiomycete 729
Aminopyrimidines 287
Aminopyrrolines 400
Aminosulphonylbenzonitriles 181
Aminotetracyanocyclopentadienide 226
formation from nitrotetracyanocyclopentadienide 601
2-Aminothiophene 407
2-Amino-3,4,5-tricyano-6-dicyanomethylpyridine 575
Aminotricyanoethylene 430
β-Amino-α,β-unsaturated nitriles 194
Ammonia 320
addition to nitriles 269–274
in isonitrile condensation 866
reaction with tetracyanoethylene 468, 469
Amygdalin 718
biosynthesis from phenylalanine 720
hydrolysis of 721
Amyl nitrite 735
Analysis, of nitriles 168–201
by infrared absorption 194–197
by mass spectrometry 197–201
by n.m.r. 201, 202
by qualitative chemical methods 168–173
by quantitative chemical methods 173–177
by separation techniques 177–180
distillation 177, 178
extraction 178
gas chromatography 178–180
liquid chromatography 178
by ultraviolet absorption 190–194

Aniline 490
  p-cyano- 215
  p-nitro- 215
  reaction with 1-chloro-1-amino-2,2-dicyanoethylene 437
  p-(trifluoromethyl) 215
Anilinium ions 218
Anion radicals 297, 656–661
  of polycyano compounds 229, 657–659
  salts of 660
Anisotropy effect, of C≡N 140
Anthracene, as complexing agent 650
  formation from *o*-benzylbenzonitriles 333
Anthropods 722, 723
  defense against attack 718
Arabinose-1-^{14}C 759
Arachidonic acid-1-^{14}C 769
Arenesulphonyl nitrene 862
Arginine-5-^{14}C 764
D,L-Arginine-1-^{14}C 760
Aroyl oxamidrazones 842
4-Aryl-2,3-dicarboxy-2,3-dihydro-1,2,3,5-oxatriazoles 841
2-(Arylmethyl)benzonitriles 333
Aryl sulphonyl malononitrile 590
Ascorbic acid-1-^{14}C 754
Asparagine 718, 721, 725, 726, 731
  as intermediate in ricinine synthesis 731
  biosynthesis 721, 722, 729
  formation, by hydrolysis of β-cyanoalanine 725, 726, 729
    from glycine 728
    via cyanogenic glycosides 726
  in blue lupine 722
  incorporation of radioactivity 726
  isolation from *Cucurbitaceae* 726
  labelled 722, 725
Aspartic acid, formation from β-cyanoalanine 725
  precursor of α,γ-diaminobutyric acid 726
Aspartic acid-4-^{14}C 774
Association, dipole-dipole 147
  hetero- 147
  self- 125, 148, 149
Azacyanocarbons 546–554
5-Azadithiouraciles, 1-alkyl- 872
  1-aryl- 872
Azelaic acid 1,9-^{14}C 767
Azepine derivatives 381
Azetidin-2-carboxylic acid-[carboxyl-^{14}C] 765
Azides 350, 351
  conversion to tetrazoles 351, 460
  reaction, with barium carbonate 748, 749
    with tetracyanoethylene 460
Azido oximes 824
Azidotetracyanocyclopentadienide ion 605
Azines 311
Aziridines
  formation, by reduction of α-halonitriles 337
    by ring closure 337
Azirines 842, 931, 932
  conversion to oxazole 932
Azo-bis-(alkyl) nitriles, decomposition 110
α,α′-Azobis(isobutyronitrile) (AIBN) 298, 457, 897, 898
1,1′-Azocyanocyclohexane 897
Azodinitrile 550
Azomethine imines 351–353
Azomethine system 75
Azomethine ylids 353
Azonitriles 685–697
  cage effect 692–694
  catalysts of thermal decomposition 686
  decomposition of, products 688–692
    rates 685–688
Azulenes, reaction with tetracyanoethylene 463

Basicity, of cyano group 124–134, 157
Basicity constants 125–127
Beckmann rearrangement 910, 924
  'abnormal' 94, 857
  'second-order' 94
Benzaldehyde 310
  emission by millipedes 722
  formation from benzonitrile 312
  in secretion, of Australian leaf beetle larvae 723
    of *Chryrophthasta variicollis* 732
    of *Chrysophtharta amoena* 723
  o-substituted 310
Benzalmalononitriles 70
α-Benzamidobutyrolactone 764
Benzene, as complexing agent 650
  reaction with tetracyanoethylene 463
Benzene azocarboxylates 861
Benzenediazonium *o*-carboxylate 870

Benzhydro· 173
Benzhydry l chloride 920
2-Benzhydryl-7,7,8,8-tetracyano-quinodimethane 646
Benzil α-monoxime 925
Benzils 431
Benzimidazoles 844
Benzoic acids 218
Benzo-α-ketoxime picrates 913
Benzonitrile, fluoro-substituted 215, 216, 219
 metabolism 737
 n.m.r. parameters 201
 reduction, to benzaldehyde 312
  to benzylamine 182, 329
  to dibenzylamine 325
  to 2,3-dihydrobenzonitrile 183
 ultraviolet spectrum 210
Benzonitrile-[cyano-^{14}C] 777
Benzonitrile oxide 930
 reaction with tetracyanoethylene 462
Benzoquinone 646, 648, 650
5,6-Benzo-1,2,4-thiadiazine-1,1-dioxides 842
2H, 3H-Benzo[b]thiophene-3-one, reaction with tetracyanoethylene 469
3-Benzoyl-2-benzylchromone 452
N-Benzylacrylamide, formation 913
Benzylamine, formation from benzonitrile 329
o-Benzylbenzonitriles, conversion to anthracenes 333
8-Benzyl-9,10-benzosesquifulvalene, addition of tetracyanoethylene 450
Benzyl cyanides 312, 733, 736
 conversion to phenylacetaldehydes 311
 labelled, conversion to hererocycles 784
  synthesis of 766
Benzylidenephosphorane 286
2-Benzylimidazoline-2-^{14}C hydrochloride, preparation from benzyl cyanide-^{14}C 784
p-Benzyloxybenzonitrile-[nitrile-^{14}C] 770
5-Benzyloxygramine methosulphonate 773
B_{12} icosahedron 152
Bicycloheptadiene 485
[2,2,2]-Bicyclooctanecarboxylic acids 218
Bile acids 767
Bis (acetoxyiminomethyl)methylenemalononitrile 447
1,5-Bis(*p*-amidino-$^{14}C_2$-phenoxy)-pentane methanesulphonate 783
N,N-Bis (2-chloroethyl)aniline 459, 464
N,N-Bis(β-chloroethyl)-*p*-tricyanovinylaniline, reaction with 1,1-dialkylhydrazines 477
Bis (2-cyanoethyl)amine 724
1,4-Bis(dicyanomethyl)benzene 536
1,4-Bis (dicyanomethylene)cyclohexane 532
2,3-Bis(dicyanomethyl)-1,1,4,4-tetracyanobutadienediide 608
Bis(tetrahydroindenyl)iron 473
Bis(tricyanovinyl)amine 612
 formation from ammonia and tetracyanoethylene 468
O,N-Bis(trifluoroacetyl) hydroxylamine 933
7,7-Bis(trifluoromethyl)cycloheptatriene 909
Bis(triphenylphosphine)platinum complex 459
Blue lupine 722, 725
Blue shifts 144
Bond energies, of isonitriles 854
Bond energy analysis 23–27
π-Bonding, cyano group involvement 155
Bond refractivity data, of isonitriles 854
Boranes 872
Borohydride reduction—*see* Reduction
Boron halides, coordination complexes with 151, 152
Boron hydrides, coordination complexes with 151, 152
Bromination, of malononitrile 446
 of nitriles 673
Bromocyanoacetylene 513
α-Bromoisobutyric acid, ethyl ester 315
Bromomalononitrile 591
 acidity 227
2-Bromo-1,3,5-tricyanobenzene 525
Bromotricyanomethane 437
Brønsted-Lowry acid 136
Bücherer synthesis 761–763
Büchner synthesis 886
Buna-N rubber, nitrile content 197
Bunnett ω criterion of mechanism 260

Butadiene, Diels-Alder addition to 493
reaction with cyanogen 432
N-t-Butylacetamide, formation from isobutylene 913
o-t-Butylbenzonitrile 325
4-*t*-Butylcyclohexanecarbonitrile 211
cis and *trans* 216
t-Butyl hypochlorite, chlorination with 672
R-(+)-*sec*-Butylisonitrile 888
rearrangement to R-(+)-2-methylbutyronitrile 880
t-Butylisonitrile 889
Butyric acid-1-^{13}C 766
Butyric acid-1-^{14}-C 766
γ-Butyrolactone 764, 923
Butyronitrile, reduction with $LiAlH_4$ 327
with trialkoxyaluminohydrides 317
Cage effect 692–694
Calorimetric studies 140, 141
Capronitrile 314
reduction to n-hexylamine 329
Carbamoyldicyanomethane 436, 590
Carbanionic species, addition to nitriles 276–286, 383–387
in synthesis of nitriles 89
Carbenes 345
addition to tetracyanoethylene 457
Carbodiimides 247, 840, 842
Carbomethoxymalononitrile, acidity 227
Carbonate, labelled, exchange with potassium cyanide 749
Carbonates, reduction 746, 747
Carbon dioxide, reduction 746, 747
Carbon, divalent 854
Carbonium ions, electrophilic addition 290–296
Carbon nucleophiles 397–400
Carbon skeleton rearrangement 889
Carbonyl compounds, reactions with nitrile oxides 837, 838
Carbonyl cyanide 580–589
acylation of aromatic hydrocarbons 583
physical properties 581
reaction with alcohols and amines 582
reaction with olefins 583–588
synthesis 580, 581
Wittig-type reactions 582, 583
Carbonyl cyanide hydrazone 552
Carbonyl cyanide phenylhydrazones 735
α-Carbonylketoximes, fragmentation 925–932
Carbonyl reagents, presence in hydrogenation reactions 311
p-Carboxybenzonitrile 181
Carboxylic acids, in isonitrile condensation 866
reactions with nitriles 246–249
Carboxylic acids, ^{14}C-, 780
exchange with nitriles 777, 778
preparation 780
Carboxylic amides-^{14}C 776
Carboxytetracyanocyclopentadienide 226, 604
Carbylamine reaction 855
Castor bean 730
reaction with nicotinamide and nicotinic acid 730
Catalysts—*see* Magnesium iodide Magnesium perchlorate, Metal catalysts, Nickel-aluminum oxide, Nickel boride, Palladium, Platinum, Raney cobalt, Raney nickel, Raney nickel-chromium catalyst, Rhodium, Rhodium-alumina, and Sodium acetate
C=C bonds—*see* Olefins
C—C≡N bending modes 154
Charge-transfer 54
absorption bands 144, 146, 191, 641
complexes 134, 143–147, 228
Chelates 154, 155
Chemical shifts,
of ^{13}C 202
of methyl groups in cyanoethylated α-methylcyclohexanones 202
of ^{14}N 242
of stereoisomers of 2,4,6-tricyanoheptane 202
Chenodeoxycholic acid-24-^{14}C 767
Chloranil 646, 648, 650, 653, 655, 656
Chlorination, with *t*-butyl hypochlorite 672
with sulphuryl chloride 673
Chlorine, addition to tetracyanoethylene 458
in preparation of tetracyanoethylene 446
4-Chlorobutyronitrile-1-^{14}C 766
Chlorocyanoacetylene 513
5-Chloro-2,3-dicyano-*p*-benzoquinone 541, 646, 650

5-Chloro-2,3-dicyano-6-phenylsulphonyl-*p*-benzoquinone 544
Chlorodinitromethane, acidity 227
Chlorofumaronitrile, in synthesis of dicyanoacetylene 514
Chloromaleonitrile, in synthesis of dicyanoacetylene 514
N-Chloromethylphthalimide 771
p-Chloronitrobenzene, von Richter reaction 938
Chloropentacyanoethane 441
N-Chlorosulphonamides 860
Chlorosulphonyl isocyanate 102
Chlorotetracyanocyclopentadienide in formation of 5,5-dichloro-1,2,3,4-tetracyanocyclopentadiene 601
1-Chloro-1,2,2-tricyanoethylene 478–481, 650, 655
Chlorotricyanomethane 437
2-Chloro-1,1,2-trifluorotriethylamine 923
N-Chlorovinylphosphoroimidic trichlorides 253
Cholanic acid-24-^{14}C 767
5α-Cholestane, substituent effect on 19-methyl group 202
Cholic acid-24-^{14}C 767
Chromatographic methods of analysis, gas 178–180
  conditions for 179, 180
  liquid 178
Chromium, dibenzene 473
  hexacarbonyl-, reaction with cyanogen 433
Chromobacterium violaceum 728
  addition of glycine and glycine ester to 728
  synthesis of β-cyanoalanine and asparagine in 728
Chronopotentiometry 188
Chryrophthasta variicollis 723
Chrysophtharta amoena 723
Cinnamonitrile, *trans*- 779
  electrolysis of 189
  m-fluoro-, *cis* and *trans* 219
  p-fluoro-, *cis* and *trans* 219
Citric acid, 1,5-^{14}C- 768
Citric acid-6-^{14}C 754,
Cladophora rupostris 734
Clathrates 149
Clitocybe diatreta 728
C—N bond, cleavage of 859, 860
C═N compounds, reactions with nitrile oxides 838–840
C≡N bond, anisotropy effect 140
  stretching frequency 135, 194, 242
C═O bonds, cyclization reaction with nitrilium ions 364–366
Cobaltocene 473
Cocatalysts, basic 321
Codimerization, of nitriles 286
Collagen formation 723
Colorimetric methods, of nitrile determination 177
Complexes, charge-transfer 134, 143–147
  coordination 134
    conformations of dinitriles in 156
    of heavy metal salts 153–156
  dipole-dipole, dissociation constants 147
  donor-acceptor 144, 191
  molecular addition 216
  nitrile-metal halide 154
  of halogens 145
  of isonitriles 124, 873–876
  of transition metal 892
  solid 659
Compositae 720
Condensations, four-component 865–868
  of nitriles, leading to 1,3-diamines 329
Configuration interaction 8
Conformations, of dinitriles in coordination complexes 156
Conjugation effects 157, 158
Coordination complexes 134
  conformations of dinitriles in 156
  of heavy metal salts 153–156
Copolymerizations of vinylidene cyanide 699
Copper cyanide 769, 892
  labelled 751
Correlation energy 28
  molecular extra 43, 49
Cortisol 724
Coupling constants 201
  of stereoisomers of 2,4,6-tricyanoheptane 202
Crotonic acid-1-^{14}C 769
Crotononitrile, constant current reduction 189
  hydrodimerization 188
Cucurbitaceae 726
t-Cumyl chlorides 218
Cyanalkynes 283
Cyanamides 241, 248, 252, 724
  conversion to guanidine salts 271

Cyanates 241, 275
aryl 90, 276, 286, 293
under Friedel-Crafts conditions 91
Cyanide ion 46, 724
addition to tetracyanoethylene 459
coordination chemistry 124
p-Cyanocinnamaldehyde 738
Cyano group—*see also* Cyanides and Nitriles
addition of free radicals 296–298
basicity 124
^{13}C resonance 242
donor strength of 156–159
effect on acidity 227
electron distribution 211
electronic structure 46
fluorination 296, 297
involvement in π-bonding 155
molecular orbital calculations 216
n.m.r. shielding effects 233–235
reduction 307–337
shielding of steroidal angular protons by 202
size and shape 209
steric requirements 211
substituent effect 217–226
on i.r. and u.v. spectra 231–233
Cyanoguanidine 724
conversion to diamino-*s*-triazines 274
reaction with nitriles 274
Cyanohydrins 72, 766
-^{14}C 752, 784
effect of ring size on stability 72
equilibrium constants 190
exchange with H^{14}CN 756
reduction to β-hydroxy primary amines 326
synthesis 754, 755
modified 756
1-Cyano-2-hydroxy-3-butene 733
2-(Cyano-^{14}C)-6-methoxybenzothiazole 771
Cyanomethyl group 223
Cyanomethyl-2,2,3-trimethyl-3-cyclopentene 918
Cyanonitrene 107, 548, 707, 708, 903, 904
2-Cyanooxiranes, from epoxidation of substituted acrylonitriles 901
Cyanophenanthrenes 202
Cyanophenol 737
Cyanophenol—*cont.*
o-, *m*-, and *p*-, from benzonitrile 737
Cyanophenylcarbene 706
from phenylcyanodiazomethane 551
N-(2-Cyanophenyl)-*N*-methylcarbamoyl chloride 930
2-Cyano-1-phenylpyrazole 909
β-Cyanopropionic acid-4-^{14}C 765
Cyanopyridine alkaloids 730–732
2-Cyanopyridines 432
n.m.r. parameters 201
polarographic reduction 183
3-(Cyano-^{14}C)pyridine 776
4-Cyanopyridine, n.m.r. parameters 201
polarographic reduction 183
3-Cyanopyridones 732
Cyano radical 709
Cyanotriazoles 549
5-Cyano-2,2,8-trimethyl-4-*H*-*m*-dioxino-[4,5-*C*]-pyridine 775
β-Cyanovinyl group 225
Cyanides, inorganic—*see also* specific compounds; for Organic cyanides—*see* Nitriles
alkali metal, -^{13}C 746–751
-^{14}C 746–751
copper, -^{14}C 751, 769, 892
hydrogen, -^{13}C 746–751
-^{14}C 746–751
detoxication 735, 736
toxicity 734, 735
mercury 892
metal, reaction with organic halogen compounds 77
nickel 775
silver 892
sodium, -^{15}N 751
(Cyano-^{14}C)acetamide 776
Cyanoacetates, unsaturated, cyclizations of 675
Cyanoacetic acid 724
3-^{14}C, ^{15}N 768
reaction with dimethylcyanamide 246
2-(Cyano-^{14}C)acetic acid 768
conversion to β-alanine-β-^{14}C 780
o-Cyanoacetophenone 325
Cyanoacetylenes 512-524
addition 518
cycloaddition 520
halo-substituted 512, 513
physical properties 515–517
polymerization 523, 524

Cyanoacetylenes—*cont.*
salt and complex formation 517
synthesis of 513–515
Cyano acids 768, 935
Cyanoacrylic acids 202
β-Cyanoalanine 724, 725, 726, 728, 729
as intermediate in ricinine biosynthesis 731
causing cystathionine excretion 725
conversion, to β-aminopropionitrile 726
to aspartic acid 725
formation, from cysteine 725
from glycine 728
from serine 725, 726
hydrolysis to asparagine 725, 726, 729
D and L isomers 724, 725
labelled 726
reduction 726
Cyanoalkyl radicals, oxidation of 697
Cyanoallenes 81
L-γ-Cyano-α-aminobutyric acid 724, 725
4-Cyano-5-aminoimidazole 508, 509
p-Cyanoaniline 215
N-Cyanoaziridines 548, 903
m- and *p*-Cyanobenzenesulphonamide 181
m- and *p*-Cyanobenzoic acid, from *m*- and *p*-tolunitrile 737
4-Cyanobenzylamine 325
Cyanobenzyl radicals 682–685
7-Cyanobicycloheptatriene 909
Cyanocarbenes 551, 704–709, 901–906
enimine structure 901
keteniminium form 901
preparation 901
N-radical character 298
Cyanocarbons 895
Cyanocarbon acids 191, 610–617
Cyanocarbon anions 589–617
electronic spectra 608–610
Cyanocarbon substituents, electronic effects 219
Cyanocarbonyl malononitrile 590
Cyanocarbyne 705
Cyanocobalamin, conversion to hydroxycobalamin 735
Cyano complexes, heats of formation 656
Cyanocyclopentadienes 612, 616
Cyanocyclopentadienide 234, 616
n.m.r. 234
Cyanodiazo compounds 550–554
4-Cyano-3,5-dimethylphenol 221
(Cyano-^{13}C)-*N*,*N'*-diphenylformamidine 776
Cyanodithioformates 499
Cyanoethanides 591–593
Cyanofluoroalkylethylenes, mixed 486
Cyanoform 612, 899
acidity of 227
Cyanoformamide 430
1-Cyanoformamide-*p*-toluenesulphonylhydrazone, 902
in formation of aminocyanocarbene 902
Cyanoformamidine 429
Cyanoformic acid 728
alkyl esters 935
Cyanoformimidates 429
Cyanogen 268, 281, 428, 576, 736, 822
addition, of diazomethane 432
of sulphur dichloride 433
to sulphur trioxide 432
elimination from tetracyanoethylene 470
fluorination of 432
in preparation of nitriles 89
polymerization 289
reactions of 428–433
with alcohol 429, 430
with amines, primary and secondary 428
with ammonia 428
with aromatic compounds 431
with butadienes 432
with diethyl malonate 430
with ethyl acetoacetate 430
with ethylene 433
with Grignard reagents 282, 430
with hexacarbonyl/metal complexes 433
with hydrazoic acid 432
with hydrogen 432
with hydrogen sulphide 430
with malononitrile 430
with nitroethane 430
with potassium cyanide 433
with thiols 429, 430
with water 430
Cyanogen azide 547–549, 904
Cyanogen bromide 778, 779, 932
Cyanogen chloride 88, 428, 778
reaction, with enamines 89
with glutathione 736

Cyanogen chloride—*cont.*
with haemoglobin 736
with malononitrile 433
use in formation of pentacyano-cyclopentadienide 597
Cyanogen halides 241, 736
reaction with organometallic compounds 88
Cyanogenic glucosides 728
Cyanogenic glycosides 718–722, 728
biosynthesis 719
in asparagine formation 726
α-Cyanoglycine 725
Cyclization, of isonitriles 869–873
of nitrilium ions by reaction
with C=C bonds 359–364
with C=O bonds 364–366
with OH groups 366–369
with nitrogen containing compounds 373–376
with SH groups 369–373
of olefinic nitriles 377, 378
of unsaturated cyanoacetates 675
Cycloadditions
of 1,2-dicyano-1,2-disulphonyl-ethylenes 490, 491
of isonitriles 869–873
of nitrile oxides 832–843
of nitriles 298
α,β-unsaturated 108, 702, 703
of tetracyanoethylene 453–456
Type A 344–358
1,4-Cycloadditions 356
Cyclobutanes, formation, from mixed cyanofluoroalkylethylenes 486
from tetracyanoethylene 453–456
Cyclobutylisonitrile 889
Cycloheptatriene, addition of tetra-cyanoethylene 450
Cyclohexanone, conversion to D,L-lysine 322
Cyclohexylisonitrile 931
Cyclooctatetraene, addition of tetracyanoethylene 450
Cyclopentadienide 234
reactivity of 228
Cyclopentadienylidenetriphenyl-phosphorane, reaction with tetra-cyanoethylene 463, 469
Cyclopenta[e, f]heptane 387
Cyclopentane-1,2,3,4-tetracarboxylic acid, use in formation of cyano-cyclopentadienide system 598
Cyclopentanone oxime 916
Cyclopropanecarbonitrile 384
conversion to cyclopropyl imine 384
Cyclopropylimine, formation from cyclopropanecarbonitrile 384
rearrangement to 1-pyrroline 384
Cystathionase 725
Cystathionine 725
Cysteine 725
formation from serine 726
Cystine 736
Cytochrome oxidase enzymes 734

Decaborane 256, 892
N-(*trans*-9-Decalyl)acetamide 921
Decyanation 331
reductive 333–336
Dehydration, of amides 96, 97, 856, 857
of oximes 92
Dehydroabietonitrile, reduction to $\Delta^{5,7,14(13)}$-abietatriene 335
Dehydrogenation, of terpenes 331
Deoxycholic acid-24-^{14}C 767
2-Deoxy-D-ribose-1-^{14}C 767, 782
Depsipeptide 863
Detoxication 734–738
Dhurrin 719
N,*N*-Dialkylamido chlorides 860
α-Dialkylaminonitriles, 922
conversion to α-hydroxyamides 922
β-Dialkylaminopropionitriles 900
1,1-Dialkylhydrazines, reaction with *N*,*N*-bis(β-chloroethyl)-*p*-tricyano-vinylaniline 477
2,5-Dialkylthiopyrrole 440
1,2-Diamines, aromatic 429
primary 323
1,3-Diamines, formation by reduction and condensation of nitriles 329
α,γ-Diaminobutyric acid 724, 726
homoserine and aspartic acid pre-cursors 726
in *Lathyrus sylvestris* 726
2,4-Diaminobutyric acid-4-^{14}C 774
Diaminodicyanoethylenes 495, 509
2,5-Diamino-3,4-dicyanothiophene 439
Diaminofumaronitrile 506–509
Diaminomaleonitrile 289, 506–509, 736
Diaminopyridines 283
2,3-Diaminoquinoxalines 429
3,6-Diamino-1,2,4,5-tetracyano-benzene 529

Diaminothiophenes 477
Diamino-*s*-triazines, formation 274
Diaroyl oximinosulphides 825
Diaryloxathiadiazine dioxides 246
Diatretyne 1 and 2 728
1,4-Diazidobenzene 904
Diazoacetonitrile 894
Diazoalkanes 349, 350
Diazocyclopentadiene, reaction with tetracyanoethylene 463
3-Diazo-6-dicyanomethylene-1,4-cyclohexadiene 553
Diazoimines 549
Diazomethane 844
addition to cyanogen 432
addition to tetracyanoethylene 458
Diazotetracyanocyclopentadiene 602–606
Dibenzenechromium 473
Dibenzylamine, formation by reduction of benzonitrile 325
Diborane, in amine formation from nitriles 330
reduction ability 330
2,3-Dibromo-5,6-dicyano-*p*-benzoquinone 544
1,2-Dibromo-1,2-dicyanoethylene 482
2,3-Dibromo-5,6-dicyanoquinone 544, 664
Dibromomalononitrile, pyrolysis at 500° 446
2,6-Dichlorobenzonitrile 738
Dichlorocarbene 862
addition to primary amines 855
2,3-Dichloro-5,6-dicyano-*p*-benzoquinone (DDQ) 541, 646, 648, 650, 654, 655, 660, 664
DDQ-pyrene complex 660
2,5-Dichloro-3,6-dicyano-*p*-benzoquinone 545
1,1-Dichloro-2,2-dicyanoethylene 483
1,2-Dichloro-1,2-dicyanoethylene 481, 482
Dichlorofumaronitrile, in synthesis, of dicyanoacetylene 514
of tetracyanoethylene 446
Dichloromaleonitrile, in synthesis of dicyanoacetylene 514
5,5-Dichloro-1,2,3,4-tetracyanocyclopentadiene, formation from chlorotetracyanocyclopentadienide 601
conversion to 7,7-dichloro-1,2,3,4-tetracyanonorbornene 601
7,7-Dichloro-1,2,3,4-tetracyanonorbornene, formation from 5,5-dichloro-1,2,3,4-tetracyanocyclopentadiene 601
3,4-Dichloro-1,2,5-thiadiazole 433
2,6-Dichlorothiobenzamide 738
Dicyanamide 546, 547
Dicyanoacetylene 512
elimination from tetracyanoethylene 470
formation from 4,5-dicyano-1,3-dithiolone 503
1,2-Dicyanobenzene, formation of 905
3,7-Dicyano-1,2-benzocycloheptatriene 908
2,3-Dicyanobenzoquinone 648, 650, 653–656
2,3-Dicyano-*p*-benzoquinones 540–545
2,3-dicyano-5,6-difluoro-*p*-benzoquinone 544
2,3-dicyano-5-phenylsulphonyl-*p*-benzoquinone 544
2,5-Dicyano-*p*-benzoquinone 545, 546, 653
2,6-Dicyano-*p*-benzoquinone 545, 546, 650, 653
1,1′-Dicyanobicyclohexyl 335
Dicyano bis(fluoroalkyl)ethylenes 484–490
1,1-Dicyano-2,2-bis(trifluoromethyl)-ethylene 485, 869
formation of 484
oxidation of 555
cis,cis-1,4-Dicyanobutadiene 904
formation from 2-aminobenzotriazole 905
Dicyanocarbene 552, 705, 706, 901
amino-substituted 506, 706, 901, 903
from dicyanodiazomethane 901
phenyl-substituted 551, 706
preparation 902
3,7-Dicyanocycloheptatriene 907
1,2-Dicyanocyclopentadienide 234, 616
1,3-Dicyanocyclopentadienide 234, 612, 616
Dicyanodiacetylene 512
mass spectra 200
synthesis 514
Dicyanodiazomethane 550, 901
1,1′-Dicyanodicyclohexane 897
2,3-Dicyano-1,4-diethoxycarbonylcyclopentadienides 599

2,3-Dicyano-5,6-difluoro-*p*-benzoquinone 544
Dicyanodihaloethylenes 481–483
2,5-Dicyano-3,6-dimethoxybenzoquinone 86
2,3-Dicyano-2,3-dimethylbutane 898
1,2-Dicyano-1,2-disulphonylethylenes 490–492
  Diels-Alder reactions 490, 491
3,4-Dicyano-1,2-dithiete 217, 504
4,5-Dicyano-1,3-dithiole derivatives 503, 514
Dicyanodithiooxalyl cyanide 217
1,2-Dicyano-1,2-di-*p*-tolylsulphonylethylene,
  in synthesis of tetracyanoethylene 447
Dicyanoethylenes 493, 494
3,4-Dicyanofuran, in synthesis of dicyanoacetylene 514
β,β-Dicyanovinyl group 202
Dicyanoheptafulvene, mass spectrum of 200
2,3-Dicyanohydroquinone 71, 540
4,5-Dicyanoimidazoles 509
4,5-Dicyanoimidazolin-2-one 508
Dicyanoketene acetals 494–496
  formation from alcohols and tetracyanoethylene 467
  reaction with amino groups 496
Dicyanoketeneimine 550
Dicyanoketene thioacetals 496–499
  heterocyclic compounds from 498
  reaction with amidine 498
    with amines 497
    with hydrazine 498
Dicyanomaleimide, formation from tetracyanoethylene 461
Dicyanomethanide, reaction with hexacyanobutadiene 608
Dicyanomethylene group 427
Dicyanomethyl group 223
2-Dicyanomethylene-1,3-indanedione 646, 650, 658
9-Dicyanomethylene-2,4,7-trinitrofluorene 646, 650
2,3-Dicyano-1,4-naphthoquinone 544
2,3-Dicyano-5-nitro-1,4-naphthoquinone 653, 654, 655
7,7-Dicyanonorcaradiene 558
2,3-Dicyano-5-phenylsulphonyl-*p*-benzoquinone 544
5,6-Dicyanopyrazine-2,3-dione 509
3,5-Dicyanopyridine 573
β,β-Dicyanostyrene, *m*- and *p*-fluoro- 219
1,1-Dicyanotetramethylcyclopropane 902
3,4-Dicyano-1,2,5-thiadiazole 509
4,5-Dicyanotriazole 508, 572, 573
4,5-Dicyano-1,2,3-trithiol-2-one 504
2,2-Dicyanovinylamines 550
2,2-Dicyanovinylazides 549, 550
1,1-Dicyanovinylbenzene, reaction with nickel carbonyl 477
β,β-Dicyanovinyl-*m*-fluorobenzene 219
β,β-Dicyanovinyl-*p*-fluorobenzene 219
β,β-Dicyanovinyl group 224, 225
(1,1-Dicyanovinyl)pyrrole 558
Dielectric polarization 148
Diels-Alder reactions 298, 356
  of 1,2-dicyano-1,2-disulphonylethylenes 490, 491
  of diethyl *trans*-1,2-dicyano-1,2-dicarboxylate 493
  of α,β-unsaturated nitriles 109
1,3-Dienes, addition of tetracyanoethylene to 449–453
Diethylamine-1-^{14}C, ethylamine-1-^{14}C mixture, preparation 781
Diethyl azodicarboxylate 841
Diethyl 1,2-dicyanoethylene-1,2-dicarboxylate 492, 493
  reaction with dimethylaniline 493
Diethyl malonate, reaction with cyanogen 430
1,1-Difluoro-2,2-dicyanoethylene 438
1,2-Difluoro-1,2-dicyanoethylene 482, 483
Dihalogenoformoximes 822
Dihalomalononitrile, reaction with potassium cyanide 433
Dihydro-2,5-diborapyrazine 893
Dihydroisoquinolines 361
4,5-Dihydro-1,2,4-oxadiazoles 838
4,5-Dihydro-1,2-oxazoles 833
2,3-Dihydro-2-oxo-3-oximinobenzothiophene 930
3,4-Dihydro-2-phenyl-4-quinazolinone 470
Dihydropyran 481
3-(3,4-Dihydroxyphenyl)alanine 763
3,6-Dihydroxy-1,2,4,5-tetracyanobenzene 529
1,2-Diiodo-1,2-dicyanoethylene 482
Diisobutoxyaluminium hydride 317
Diisocyanide 905
α,γ-Diketocarboxamides 864

α-Diketones 431
Dimercaptofumaronitrile 499–505
Dimercaptomaleonitrile 499–505
anion radical 578
disodium derivative 576
Dimerization, of imidyl halides 245
of nitrile oxides, mechanism 817
to furoxans 816–819
of nitriles 282–287
2,5-Dimethoxyaniline, addition to tetracyanoethylene 459
p-Dimethoxybenzene, as complexing agent 653
(1,4-Dimethoxybenzo)bicycloheptadiene, addition of tetracyanoethylene 455
2,4-Dimethoxybenzyl cyanide, reduction to 2,4-dimethoxyethylbenzene 333
2,4-Dimethoxyethylbenzene, formation by reduction of 2,4-dimethoxybenzyl cyanide 333
3,5-Dimethoxyphenylacetaldehyde 311
β,β-Dimethylacrylonitrile 923
Dimethylamine 320
2,6-Dimethylaniline 459, 464
N,N-Dimethylaniline 459, 493
reaction with tetracyanoethylene 463
Dimethylcyanamide 244, 290
reaction with cyanoacetic acid 246
Dimethylcycloheptatrienopentaene 454
2,2′-Dimethylcystine-1-^{14}C 761
Dimethyldimethylenecyclobutene, addition of tetracyanoethylene 455
N,N-Dimethylformamide (DMF) 78, 770
2,5-Dimethyl-2,3,3,4,4,5-hexacyanohexane, formation 457
3,5-Dimethyl-4-nitrophenol 221
Dimethyloxosulphonium methylide 845
Dimethylsulphonium dicyanomethylid 591
Dimethylsulphonium methylide 286, 846
Dimethyl sulphoxide 78, 293, 765, 771
in preparation of arachidonic acid-1-^{14}C 769
2,6-Dimethyl-4-(1,1,2,2-tetracyanoethyl)aniline 464
2,6-Dimethyl-4-(1,1,2,2-tetracyanomethyl)aniline 225
2,5-Dimethyl-7,7,8,8-tetracyanoquinodimethane 646
1,1-Dimethyltetralone-2-oxime 920
Dinitrenes 904
decomposition 904
Dinitriles, conformations in coordination complexes 156
cyclization 380–382
o-Dinitrobenzene, u.v. spectrum 210
Dinitromethane, acidity 227
1,4,2-Dioxazoles 837
Diphenyldimethylenecyclobutene, addition of tetracyanoethylene to 455
N,N′-Diphenylethylenediamine 311
N-(Diphenylmethyl)amide derivatives 173
1,3-Diphenylnaphtho[2,3-*c*]furan 451
3,5-Diphenyl-1,2,4-oxadiazole-4-oxide 839
Diphenyltetrahydroimidazole 311
N,N′-Diphenylthiourea 776
Diplopoda 722
1,3-Dipolar addition, of ketocarbenes 345
of ketonitrenes 345
Dipolar aprotic solvents 78, 83
dimethylformamide (DMF) 78, 183, 770
dimethylsulphoxide (DMSO) 78, 293, 765, 771
hexamethylamide of phosphoric acid (HMPT) 79
Dipole-dipole,
complexes, dissociation constants 147
interaction 125, 147, 148
Dipole moments 49, 342
of cyanide ion 57
of nitrile oxides 814
of nitriles 156, 214–216
Dipole pairs 148
Disiamylborane, in amine formation from nitriles 330
Disodium dimercaptomaleonitrile, in synthesis of tetracyano-1,4-dithiin 576
Dissociation constants, of dipole-dipole complexes 147
Dithiooxaldiimidates 429
Dithiooxamide 430
α,α′-Ditropyl-α,α,α′,α′-tetracyano-*p*-xylene 896

Donor-acceptor complexes 144, 191
Donor radicals 672, 700
n-Donors 144
π-Donor 53, 144
σ-Donor 53
Durene, as complexing agent 650

E. coli 726, 727
Electrical conductivity 153
of polycyano complexes 662–667
Electrochemical reduction—*see* Reduction
Electron acceptor 143
Electron affinities 641, 644, 647–649
Electron donor 143
Electronic absorption spectroscopy, of cyanocarbon anions 608–610
of polycyano molecular complexes 649–656
Electronic effects, of cyanocarbon substituents 219
Electron population analysis 20
Electron spin resonance 656
Electrophilic addition, of carbonium ions 290–296
Type B 358–376
Type Bi, intramolecular 376–382
Electrophilic aromatic substitution 230
Electroreduction of nitriles 337
Emulsin 721
Enamines 898
conversion to ynamines 899
reaction with cyanogen chloride 89
'ene' reaction 489
Energy gap 662
Energy levels of tetracyanoethylene 643
of tetracyanoquinodimethane 643
Enimine of 1,1,2,2-tetracyanoethane 899
Enimine structure, in cyanocarbenes 901
Enolamides, equilibrium with oxazetidine 933
Enthalpies of formation $\Delta H°$—*see* Heats of formation
Enzymes, cystathionase 725
emulsin 721
β-glycosidase 721
inosinic acid dehydrogenase 735
kynurenine hydrosylase 735
linamarase 721
myrosinase 733
nitrilase 725, 732, 734
Enzymes—*cont.*
oxynitrilase 721
6-phosphogluconate dehydrogenase 735
rhodanese 735
ricinine nitrilase 732, 734
cis-Episulphides, conversion to *trans*-thiazolines 370
trans-Episulphides, conversion to *cis*-thiazolines 371
Epoxidation of substituted acrylonitriles 901
Epoxides 86
α-Epoxyketoxime 923
Equilibrium constants 641, 652–655
for cyanohydrin formation 190
Erythro-9,10-dihydroxystearic acid-1-^{13}C 768
Erythritol-1-^{14}C 759
Esters, of sulphuric and phosphoric acids, in synthesis of aliphatic nitriles 85
preparation from nitriles 269
N-Ethoxycarbonylazepine, reaction with tetracyanoethylene 451
Ethyl acetoacetate, reaction with cyanogen 430
Ethylamine-1-^{14}C, diethylamine-1-^{14}C mixture, preparation 781
Ethyl bromocyanoacetate, acidity of 227
conversion to 1,2,3-tricyanocyclopropane-1,2,3-tricarboxylate 442
Ethyl cyanate 275
Ethyl 3-cyano-2-acetaminopropionate 774
Ethyl cyanoacetate, acidity of 227
in preparation of tricyanoethylene 474
Ethyl 5-cyano-2-oximinovalerate, hydrogenation to lysine 322
Ethyl 2-(cyano-^{14}C) propionate 768
Ethyl diazoacetate, reaction with tetracyanoethylene 462
Ethyl dimethylaminomethylacetamidomalonate, methiodide of 774
Ethylenediamine 429
Ethylene, elimination in nitrile fragmentation 198
reaction with cyanogen 433
Ethylisonitrile 889
N-Ethylmethyleneaziridine 456
Ethyl radical, elimination in nitrile fragmentation 198

Exchange, of ^{14}C-carboxylic acids with nitriles 777, 778
of $H^{14}CN$ with cyanohydrin 756
of labelled carbonate with potassium cyanide 749
Excited state, wave function 143
Extinction coefficient 642

^{19}F, n.m.r., of β,β-dicyano and α,β,β-tricyanovinyl groups on fluorobenzenes 202
Fermi hole 29
Ferrocene 472
Flavin prosthetic group 721
Flax 720
Fluorene, as complexing agent 653
Fluorescein chloride, fusion with nitriles 172
Fluorination, of cyanogen by silver(II) fluoride 432
of cyano group 296, 297
cis- and *trans*-Fluoroalkyldicyanoethylene 454
Fluorobenzenes 217
reaction with tetracyanoethylene 463
m-Fluorobenzonitrile 219
p-Fluorobenzonitrile 215, 216, 219
Fluorocarbons 297
m- and *p*-Fluoro-α-cyanotoluene 219
N-Fluoroiminonitriles, mass spectra of 200
Fluoroketones 869
m-Fluoronitrobenzene 219
p-Fluoronitrobenzene 215, 219
p-Fluorophenylalanine 763
m- and *p*-Fluorophenylmalononitrile 219
m- and *p*-Fluorostyrene 219
Fluorosulphonic acid, reaction with tetracyanoethylene 461
m- and *p*-Fluorotoluene 219
2-Fluoro-1,3,5-tricyanobenzene 525
m- and *p*-Fluoro-α,α,α-tricyanotoluene 219
Fluoro(tricyanovinyl)benzenes 229
Formamides, dehydration of 856, 857
N-substituted 914
Formation contants, for halogen complexes 145
Formic acid, with Raney nickel in reduction of nitriles 310
Formimidyl chloride 245
Formonitrile oxide 792
preparation 795, 799, 800
N-Formylamides, *N*-alkyl, thermal decomposition 889
Fragmentation 909–936
7-centred 924, 925
of α-carbonylketoximes 925–932
of α-hydroxyketoximes 925–932
of β-keto ether oximes 96
of ketoximes 915–925
of 1,3-oxazetidine 932–936
of oxazetidinones 932–936
of oxazoles 932–936
of α-oximinoketones 924
of γ-oximinoketones 925
Free-radicals—*see also* Radicals 890
addition to cyano group 296–298
Freezing point diagrams 141
Friedel-Crafts-Karrer method 90
Fucus vesiculosus 734
Fulminic acid 792, 812
conversion to dihalogenoformoximes 822
mercury salt 795
metal salts, conversion to nitrile oxides 807, 808
polymerization 819–822
preparation 795, 799, 800
silver salt 799
sodium salt 800
Fulminuric acid 819
Fulvalene 451
Fungi 729
Clitocybe diatreta 728
for synthesis, of alanine 729
of glutamic acid 729
Marasmus oreades 727
psychrophilic basidiomycete 729
snow mold 727, 728
Furcellaria 734
fastigiata 734
Furoxans 809, 814, 818
formation by dimerization of nitrile oxides 816–819

Galactosamine-1-^{14}C 782
Galactose-1-^{14}C 759
Gallium compounds, coordination complexes with 152, 153
Gamma irradiation 892
Gattermann aldehyde synthesis 254 255, 291
3-Gentiobiose-1-^{14}C 759
Glucobrassicin 732
Glucosamine-1-^{14}C 782

Glucose, in secretion of Australian leaf beetle larvae 723
Glucose-1-^{14}C 757, 759
Glucose-6-^{14}C 759
β-Glucosidases, from *Sorghum vulgare* 721
Glucosides, cyanogenic 728
  of mustard oil 828
d-*p*-Glucosyloxymandelonitrile 719
p-Glucosyloxymandelonitrile, biosynthesis from tyrosine 720
Glutamic acid 729
  biosynthesis of 729
  condensation with β-aminopropionitrile 726
Glutamic acid-5-^{14}C 764
D,L-Glutamic acid 1-^{14}C 760
N-(γ-Glutamyl)-β-aminopropionitrile 726
  production from *Lathyrus odoratus* 726
N-(γ-L-Glutamyl)-β-aminopropionitrile 723
N-(γ-Glutamyl)-β-cyanoalanine 724, 725, 726
  production by *Vicia sativa* 725
N-(γ-Glutamyl)-β-cyanoalanylglycine 725
Glutaric acid 1,5-^{14}C 767
Glutaronitrile-1,5-^{13}C 767
Glutaronitrile-1,5-^{14}C 767
Glutaronitrile-2-^{13}C 767
Glutathione 725, 736
  replacement of cysteine residue by β-cyanoalanine 725
Glyceric acid-1-^{14}C 756
D,L-Glyceric acid-1-^{14}C 753
Glycido nitriles 74
Glycine 728
  conversion, to alanine and glutamic acid 729
    to β-cyanoalanine and asparagine 728
  effect, on *Chromobacterium violaceum* 728
    on *Pseudomonas* 728
    on snow mold fungus 728
Glycine-1-^{14}C 771
  by hydrolysis of phthalimidoacetonitrile-1-^{14}C 771
Glycine-2-^{14}C 776
Glycine ester, effect, on *Chromobacterium violaceum* 728
  on *Pseudomonas* 728
  on snow mold fungus 728
Glycine nitrile 724
Glycinonitriles 431
Glycolonitrile, in preparation of tricyanoethylene 474
β-Glycosidase 721
Glycosides, cyanogenic 718–722, 728
  biosynthesis 719
  in formation of asparagine 726
Glyoxylic acids 431
Gramines 773
Grignard reagents 285, 314, 331, 384
  reaction, with cyanogen 282, 430
    with isonitriles 862
    with nitriles 171, 276–283
Ground state wave function (ψ_N) 143
Guaiazulene 480
Guanidine salts, preparation from cyanamides 271
Gynocardin 719, 720

Haemoglobin, reaction with cyanogen chloride 736
Half-wave potentials 133
Halides, aliphatic, saturated, synthesis of nitriles from 78
  allylic, reaction with cyanides 79
Halidrys siliquosa 734
N-(α-Haloalkylidene)amidines 244
β-Haloamines, intermediates in reduction of nitriles 337
N-Haloethylamide 914
Halogen acids, interaction with nitriles 243–245
Halogen complexes, formation constants 145
  formation enthalpies 145
Halogen compounds, reaction with metal cyanides 77
Halonitriles 681, 682
α-Halonitriles, reduction to aziridines 337
Halonium salts 362
Halotricyanobenzenes, stable radicals from 526
1-Halo-2,4,6-tricyanobenzenes 231
Hammett correlation 257, 262, 278
Hammett H_0 scale 125
Hartree-Fock model 8
Heats of formation, $\Delta H°$ of di- and polycyano complexes 655, 656
  of isonitriles 854
  of nitrile complexes,
    containing halogens 145
    containing hydrogen chloride 138
    containing hydroxyl compounds 140, 141

Heptacyanopentadienide 606
1,3,5-Heptatriene 452
Heteroassociation 147
Heterocycles, synthesis from nitriles 411–414
Heteroisonitriles 859
Hexabromobenzene 211
1,1,2,4,5,5-Hexacyano-3-azapentadienide 606
Hexacyanobenzene 211, 530, 650
Hexacyanobutadiene 233, 509–512 646, 650, 653, 655, 664
anion radical 658
reaction with dicyanomethanide 608
1,1,2,3,4,4-Hexacyanobutenediide 474, 596, 597, 612
1,1,2,5,6,6-Hexacyano-3,4-diazahexadienediide, formation from tetracyanoethylene 608
Hexacyanoethane 438, 441, 442
Hexacyanoheptatriene 612, 613
1,1,2,6,7,7-Hexacyanoheptatrienide 607
Hexacyanoisobutenediide 596
Hexacyanoisobutenide 612
1,1,2,4,5,5-Hexacyanopentadienide 606
15-Hexadecenoic acid-1-^{14}C 769
Hexafluoroacetone, reaction with malononitrile 484
Hexamethylamide of phosphoric acid (HMPT) 79
Hexamethylbenzene, as complexing agent 650, 653
Hexamethyldialuminium, reaction with pivalonitrile 893
Hexamethylprismane, addition of tetracyanoethylene to 456
1,3,5-Hexatriene 452
n-Hexylamine, formation from capronitrile 329
n-Hexyl cyanide 197, 198
Hoesch reaction 172, 252, 254, 255, 264, 379
Houben-Fischer nitrile synthesis 91, 886
Hückel calculations, extended 890
Hundsdiecker degradation 772
Hybridization 22
Hydantoins 761
Hydration, of nitriles 256–263
catalysis by metal ions 261
Hydrazido oximes 831
Hydrazines 311, 429
Hydrazines—*cont.*
derivatives, in isonitrile condensation 866
reaction with dicyanoketene thioacetals 498
with tetracyanoethylene 468, 469
Hydrazoic acid, as acid component in isonitrile condensation 866
reaction with cyanogen 432
Hydrazones 311
formation 311
reaction with tetracyanoethylene 463
Hydride method, of nitrile determination 176
Hydride reagent 315–318
Hydride reduction—*see* Reduction
Hydrocarbons, formation by reduction of nitriles 325, 331–336
polynuclear, aromatic, synthesis 335
Hydrocyanic acid—*see* Hydrogen cyanide
Hydrodimerization 188
electrolytic, of acrylonitrile 109
of crotononitrile 188
Hydrogen abstraction, from nitriles 672, 673
Hydrogen, addition to tetracyanoethylene 458
reaction with cyanogen 432
Hydrogenation, 170, 175, 309–313, 320–325
for nitrile determination 176, 177
of α-aminonitriles 323
of nitriles 297, 308, 310, 313, 320–322, 324
of oximes 322, 323
of terpenes 331
partial 311
selective 309–318
Hydrogen bonding 134, 136–143
intermolecular 136–142
intramolecular 142
of isonitriles 854
of nitrile oxides 813
of nitriles 136–143
Hydrogen cyanate, in isonitrile condensation 866
Hydrogen cyanide 46, 245, 270, 291, 428, 827
acetolysis 249
-^{14}C 751
conversion to methylamine-^{14}C 780
polymerization 289

Hydrogen cyanide—*cont.*
reaction with dichlorofumaronitrile 446
Hydrogen donors 331
Hydrogen halides, addition to tricyanomethane 435
Hydrogen peroxide, addition, to nitriles 262, 263
to tetracyanoethylene 458
Hydrogen selenide, as acid component in isonitrile condensation 866
Hydrogen sulphide 860
reaction with cyanogen 430
with tricyanovinylbenzene 477
Hydrolysis, of aldimines to aldehydes 313
of 2-cyanopyridine 183
of nitrile oxides 793, 823
of nitriles 168, 169, 173–175, 256
in acid media 168
in alkaline media 168
of tetracyanoethylene 461, 462
of tricyanovinylbenzene 478
p-Hydroxyphenylcyclopentadienide 226
Hydroximic acid halides 802–804, 808, 824, 839
Hydroximic acids, alkyl 828
α-Hydroxyamides, formation from α-dialkylaminonitriles 922
α-Hydroxyarylglyoxal derivatives 861
p-Hydroxybenzaldehyde-[carbonyl-^{14}C] 781
1-Hydroxy-2-benzimidazole-3-oxide 844
p-Hydroxybenzonitrile 775
-^{14}C 775
p-Hydroxybenzyl cyanide 733
3-Hydroxy-2-butanone-1-^{14}C 782
α-Hydroxycarboxamides 863
Hydroxycobalamin 734
production from cyanocobalamin 735
4-Hydroxycoumarins, addition to tetracyanoethylene 459
1-Hydroxycyclopentane carbonitrile-^{14}C 754
3-Hydroxy-2,6-dichlorobenzonitrile 738
4-Hydroxy-2,6-dichlorobenzonitrile 738
1-Hydroxy-2,2-dicyanovinyl compounds 476
7-Hydroxyheptanonitrile-1-^{14}C 766
α-Hydroxyisobutyronitrile, biosynthetic precursor of linamarin 720
α-Hydroxyketoximes, fragmentation 925–932
Hydroxylamine 429, 860
Hydroxylamine hydrochloride, test for nitriles 172
Hydroxyl group, cyclization reaction with nitrilium ions 366–369
migration of 201
5-Hydroxylysine-6-^{14}C 781
p-Hydroxymandelonitrile 721
2-Hydroxy-2-methylbutyric acid-1-^{14}C 755
2-Hydroxy-*N*-methyl-3,4,5-tricyano-1,2-dihydropyridine 573
α-Hydroxynitriles—*see* Cyanohydrins
2-Hydroxynonanonitrile-1-^{14}C 756
2-Hydroxy-3-phenylsuccinic acid-1-^{14}C 754
3-Hydroxypropionitrile-1-^{14}C 762, 766
3-Hydroxypropionitrile-2,3-^{14}C 766
2-Hydroxythiazoles 408
Hyodeoxycholic acid-24-^{14}C 768

Imidates 263, 264
as activated nitriles 268, 269
formation from nitriles 267
Imidazole aldehydes, preparation 312
derivatives of 373–375
Imidazoles 353
Imidazolines 396
Imides 246
Imidyl halides 249, 254 260, 265, 266, 272, 293, 295
dimerization 245
formation, from nitriles and halogen acids 243–245
Imidyl sulphate 296
Iminoalkylmercaptoacetic acid hydrochloride, formation 171
Iminochloride salt 308
α-Iminooximes 926
Iminopyrrolidines 400
2-Iminothiazolidine-4-carboxylic acid 736
Immonium ions, α-addition to isonitriles 865
Indandione monoxime 927
Indigo dianil, formation by tetramerization of phenyl isocyanide 871
Indoleacetaldehyde 732
Indoleacetamide 734

Indoleacetic acid 732, 734
formation by hydrolysis of indoleacetonitrile 734
production from tryptophan 732
Indoleacetic acid-[carboxyl-^{14}C] 774
Indoleacetonitrile 724, 732–734
formation, from indolepyruvic acid 732
from tryptophan 732
hydrolysis 734
to indoleacetic acid 734
occurrence 732
oxidation to indolecarboxaldehyde and indolecarboxylic acid 734
3-Indoleacetonitriles 773
reduction to tryptamines 320, 325
Indolecarboxaldehyde 312, 734
Indolecarboxylic acid 734
Indolepyruvic acid, conversion to indoleacetonitrile 732
Indoles 871
reaction with tetracyanoethylene 463
Inductive effects 157
Infrared spectra
frequency ranges, for C≡N stretching vibration 194
of isonitrile addition compounds 151
of isonitriles 854
of nitrile charge-transfer complexes 146
of nitrile coordination complexes in far infrared 154
of nitrile oxides 812
of nitriles 194–197
frequency shifts in 134, 135, 137, 139
Inosinic acid dehydrogenase 735
Intergerrinecic acid-[carboxyl-^{14}C] 755
Iodocyanoacetylene 513
4-Iodoveratrole 769
Ion-exchange resins 81
Ionization potentials 146, 641
of nitriles 156
Ion radicals—*see* Anion radicals
Iron, bis(tetrahydroindenyl) 473
Isatogens, reaction with tetracyanoethylene 470
Isobutylene, conversion to *N-t*-butylacetamide 913
Isobutyraldehyde 720
Isobutyraldoxime 720
from valine 720
Isobutyronitrile, biosynthetic precursor of linamarin 720
Isocyanates 840
acyl, cyclization with isonitriles 870
by rearrangement of nitrile oxides 815, 816
reduction 857
thioacyl, cyclization with isonitriles 870
Isocyanides—*see* Isonitriles
Isocyanilic acid 819
Isohexyl cyanide 197
Isoimides 247
Isoleucine 722, 728, 763
conversion to lotaustralin 720
Isomerization 887
of tetraphenylsuccinonitrile 682
isonitrile-nitrile 889
Isonitrile dichlorides 860
Isonitrile-nitrile isomerization 889, 905
Isonitriles—*see also* specific compounds
α-additions, followed by secondary reactions 862–868
simple 860–862
with ring closure 869–873
α-aminoalkylation of 865
biological properties 877
C—N bond cleavage 859, 860
complexes 124, 873–876
hydrogen bonding 139, 854
oxidation of 862
physicochemical properties 854
polymerization 862
rearrangement 859
to nitriles 103
syntheses 854–859
by alkylation, of cyanides 855
by carbylamine reaction 855
by dehydration of formamides 856, 857
by α-eliminations 854
toxicity of 877
Isophthalic dinitrile, reduction in presence of Raney cobalt catalyst 325
p-Isopropylmandelonitrile 722
Isoquinoline derivatives 360–363
Isoselenocyanates, formation from isonitriles 862
Isothiazoles 393
Isothiocyanates 826, 838, 862
reduction 857
Isoxazoles 96, 809, 835, 836

Isoxazoles—*cont.*
as starting material for preparation of β-ketonitriles 96
Δ^2-Isoxazolines 833

Karl Fischer reagent 175
K band shifts, of aromatic compounds 193
Ketenes 869
Ketenimines 689, 693, 694, 842, 894–901
and nitrile forms 894
from 1,1′-azocyanocyclohexane 897
structure 894
Keteniminium form, in cyanocarbenes 901
Ketimines, formation form nitriles and Grignard reagents 276
Ketocarbenes 353, 354
1,3-dipolar addition 345
α-Ketoisovaleric acid oxime 720
Ketones, addition to tetracyanoethylene 459
alkyl trihydroxyphenyl, from Hoesch reaction 172
-^{14}C 782
formation from addition compound of nitriles and Grignard reagent 171
Ketonitrenes 354, 355
1,3-dipolar addition 345
Ketonitriles, cyclization 378–80
preparation, by treatment of enamines with cyanogen chloride 89
using isoxazoles as starting material 96
Ketoximes 915
derivatives 915
fragmentation 915–925
of α-carbonylketoximes 925–932
of α-hydroxyketoximes 925–932
Kjeldahl distillation 174
Kiliani-Fischer synthesis 757
Knoevenagel-Bucherer method 75
Kynurenine hydrosylase 735

α-Lactam 858
Lactic acid-1-^{13}C 753
Lactones, condensation with alkali cyanides-^{14}C 762–765
Lactose-1-^{14}C 759
Lathyrism 723
Lathyrogenic nitriles 723–727, 736
Lathyrus 725
latifolius 724
Lathyrus—*cont.*
odoratus 723, 726
pusillus 723
sativus 724
sylvestris 724, 726
Lathyrus meal 723
Lauric acid-1-^{14}C, conversion to lauronitrile-1-^{14}C 776
LCAO-MO approximation 8
Lead tetraacetate 103
for oxidation of 2-aminobenzotriazole 905
Leguminosae 720
Leucine 763
conversion to linamarin 720
D,L-Leucine-1-^{14}C 761
Lewis acid 326
Lewis bases 892
Limonene, use as hydrogen donor 331
Linaceae 720
Linamarase, in *Linum usitatissimum* 721
Linamarin 719, 720
biosynthesis from leucine 720
formation 719, 720, 728
precursors of 720
Linum usitatissimum 719, 720
linamarase in 721
reaction with labelled acetone cyanohydrin 719
Lithium aluminium hydride 170, 175, 314, 315, 317, 326, 329
alkoxy-substituted 317
for reduction, of adduct of nitrile and Grignard reagent 331
of cyanohydrins to β-hydroxy primary amines 326
of α-halonitriles 337
of nitriles 314, 326–329
aromatic 317
hindered 316
Lithium reagents 280, 282
alkyl 282
aryl 282
Lithium triethoxyaluminohydride 316, 317, 318
for reduction of nitriles to aldehydes 329
Lithium trimethoxyaluminohydride 329
for reduction, of benzonitrile to benzyl amine 329
of capronitrile to n-hexylamine 329
Lithium tris(n-butoxy)aluminohydride 317

Lithium tris(*t*-butoxy)aluminohydride 329
Lithium tris(n-propoxy)aluminohydride 317
Lithocholic acid-24-^{14}C 767
Lone pair orbital 124
Lossen rearrangement 924
Lotaustralin 719, 720
  biosynthesis from isoleucine 720
  synthesis, by snow mold fungus 728
Lotus arabius 721
Lotus tenius 726, 727
Lupinus angustifolia 722, 725
D,L-Lysine 763
  formation from cyclohexanone 322
    from ethyl 5-cyano-2-oximinovalerate 322
Lyxose-1-^{14}C 759

Magnesium iodide, as catalyst 399
Magnesium perchlorate, as catalyst 399
Malononitrile 283, 428, 725
  acidity 227
  bromination 446
  m- and *p*-fluorophenyl- 219
  in tetracyanoethylene synthesis 446
  labelled 768, 776
  reactions, with cyanogen 430
    with cyanogen chloride 433
    with hexafluoroacetone 484
    with tricyanovinyl arenes 476
Maltose-1-^{14}C 759
Mandelic acid-1-^{14}C 755
Mandelonitrile 721, 722, 723
Mannich bases 773–775
Mannose-1-^{14}C 757
Marasmus oreades fungus 727
Marfan's syndrome 723
Mass spectrometry 200
  in determination, of amines 170
    of nitriles 197–201
  of isonitriles 854
  of nitriles, aliphatic 197
    saturated 197
    aromatic 201
    α,β-unsaturated 200
  of tricyanocyclopropane, *cis* and *trans* isomers 200
  pressure-dependent M + 1 peaks 197
MC SCF LCAO-MO technique 9
Meerwein reaction 700, 701
Melamine 428
p-Menthane, formation by reduction of Δ^1-*p*-menthene 333
Δ^1-*p*-Menthene, reduction to *p*-menthane 333
  use as hydrogen donor 331
Menthone, oxime tosylate 916
Mercaptans 860
Mercaptoacetic acid 171
Mercuric cyanide 892
Mercuric fulminate 795
Mesitylene, as complexing agent 650
Mesomeric structures 793
Metafulminuric acid 819
Metal carbonyl complexes 255, 256
Metal catalysts, noble 320
Metal cyanides—*see* Cyanides, inorganic
Metal halides 250
  electrophilic, complexes 250–254
  nitrile complexes with 287
    far infrared spectra 154
Metallocenes 472
Metal reductions—*see* Reduction
Metal salts, coordination complexes of 153–156
  solubility in nitriles 131
N-Methylacetonitrilium ion 912
Methacrylonitrile, constant current reduction 189
Methanol, reaction with tetracyanoethylene 461, 462
Methemoglobin 734, 735
D,L-Methionine-1-^{14}C 760
3-Methoxy-4-benzyloxyphenylacetonitrile-1-^{14}C 766
4-Methoxybutyronitrile-1-^{14}C 767
Methoxycarbonyldicyanomethanide 613
2-Methoxycarbonylisatogen, reaction with tetracyanoethylene 470
Methoxycarbonyl-4-quinazolinone 470
3-Methoxy-2-nitrobenzonitrile 775
3-Methoxy-2-nitrobenzonitrile-[nitrile-^{14}C] 770
3-(*p*-Methoxyphenyl)propionitrile-1-^{14}C 772
p-Methoxystyrene 453
4-Methoxy-1,1,2-tricyanobutadiene 607
Methylamine 320
Methylamine-^{14}C 780
o-Methylbenzonitriles 431
  ultraviolet spectrum 210
β-Methylbutyrolactone 764
R-(+)-2-Methylbutyronitrile 888
Methyl cyanide—*see* Acetonitrile

Methyl 3-cyanopropyl sulphoxide 733
β-Methylcyclohexanone oxime 916
α-Methylcyclohexanones 202
3-Methylcyclohex-2-enone 920
Methyl dicyanoacetate 611
Methyleneaminoacetonitrile 724
Methylenecyclobutene, addition of tetracyanoethylene 455
3-Methylenecyclohexene, addition of tetracyanoethylene 455
Methylene phosphoranes 842
Methylfuran, reaction with tetracyanoethylene 463
β-Methylglutaric acid-1-^{14}C 764
N-Methylglycine 728
Methyl group, formation by nitrile group reduction 331–333
Methyl isocyanide 55
Methylisonitrile, rearrangement 891
1-Methylnicotinonitrile 730
  incorporation into ricinine 730
Methylnitrolic acid 819
1-Methylnorcamphor 918
N-Methyl-3-oximinoisatin 930
2-Methyl-2-phenylpropiophenone oxime 921
2-Methyl-7,7,8,8-tetracyanoquinodimethane 646
1-Methyltetrahydrofuran, in formation of aminocyanocarbene 902
N-Methyl-3,4,5-tricyano-1,2-dihydropyridine 574
N-Methyl-3,4,5-tricyanopyridinium perchlorate 573
Michael reaction, retrograde 469, 470
Michael-type addition 70
Microwave spectra, of isonitriles 854
Migration, of alkoxyl groups 201
  of alkyl groups 200
  of hydroxyl group 201
Millipedes 722
Molds 727–730
Molecular addition complexes of aliphatic nitriles 216
Molecular extra correlation energy 43, 49
Molecular orbitals 2
  calculations for cyano group 216
Molybdenum, hexacarbonyl-, reaction with cyanogen 433
Monocyanoethylenes, synthesis by Wittig reaction 107
Muricholic acid-24-^{14}C 768
Mustard oil glucosides 828
Mustard oil glycosides 732, 733
Myristic acid-1-^{14}C 766
Myrosinase 733

Nandina domestica 721
α-Naphthaldehyde 311, 313
Naphthalene, as complexing agent 650, 653
  reaction with tetracyanoethylene 463
1-Naphthoic acid 936
α-Naphthol 869
1-Naphthyl cyanide 309, 936
Nemalion multifidum 734
Nickel-aluminium oxide catalyst 325
Nickel boride catalyst 325
Nickel carbonyl 472, 477
Nickel catalyst, reduced 310, 312
  and phenylhydrazine for nitrile hydrogenation 312
Nickel cyanide 775
Nicotinamide 730, 731
  adenine dinucleotide 371
  conversion to ricinine 730, 731
Nicotinic acid 730, 731
  adenine dinucleotide 731
  -^{14}C-amide 776
  conversion to ricinine 731
  mononucleotide 731
Nictinonitrile 731
Nieuwland-type catalyst 69, 84
Nitrenes 345, 903
Nitrilase 725, 732, 734
  for hydrolysis of β-cyanoalanine 726
  from *Pseudomonas* bacteria 725
Nitrile and ketenimine forms 894
Nitrile group—*see* Cyano group
Nitrile imines 348, 349
  conversion to 1,2,4-triazoles 348
Nitrile oxides 346–348, 791–846
  addition reactions with inorganic compounds 822–827
  addition reactions with organic compounds, leading to cyclic structures 346, 832–843
    leading to open-chain structures 828–832
  dimerization to furoxans 816–819
  dipole moments 814
  hydrogen bonding 813
  hydrolysis 793, 823
  infrared spectra 812
  isolated 796–798
  nomenclature 792
  physical properties 812–814

Nitrile oxides—*cont.*
polymeric 816, 820, 821
preparation 794–812
from aldoximes 800–804
from fulminates 807, 808
from nitroparaffins 804–807
in situ 808, 809
rearrangement to isocyanates 815, 816
reduction to nitriles 822
stability 814, 815
ultraviolet spectra 812
with functional groups 810–812
Nitriles—*see also* Cyanides, Cyano group and specific compounds
addition to, cyclo- 344–358
electrophilic 290–296, 358–382
leading to heterocycles 342–414
nucleophilic 263–286, 382–409
of free radicals 296–298
basicity of 125–134, 157
biochemistry of 717–738
chemical analysis 168–177
qualitative 168–173
quantitative 173–177
complexes of, coordination 150–156, 250–256
weak 134–150
dipole moments 214–216
electrochemistry of 179–190
electronic structure in 46–63
formation, by elimination 92–103
by HCN addition 68–77
by substitution 77–91
from cyanogen 431
from fragments 105–110
from nitrile oxides 822
hydration of 256–263
infrared spectra 194–197
labelled, synthesis and use 752–784
mass spectra 197–201
nuclear magnetic resonance spectra 201, 202, 233, 234, 242
polymerization of 287–290, 410, 411
reaction of, with acids 243–249
rearrangement of 885–939
reduction of 186, 296, 297, 308–337
separation of 177–179
ultraviolet spectra 190–194, 210, 233, 234
unsaturated 109, 187, 192, 193, 200, 377, 378, 702, 703
Nitrile ylids 350
Nitrilium ions, 265, 291
Nitrilium ions—*cont.*
cyclization by reaction, with C=C bonds 359–364
with C=O bonds 364–366
with nitrogen-containing compounds 373–376
with OH groups 366–369
with SH groups 369–373
Nitrilium salts 127–134, 245, 246, 250, 253, 265, 291, 293, 358, 359, 912, 916
formed in 'super acid' 912
hydrolysis of 260
preparation, from nitriles 252
N-substituted 129, 130
unsubstituted 130, 131
Nitrite ion 735
p-Nitroaniline 215
o-Nitrobenzonitrile 775
conversion to *o*-aminobenzamide 336, 337
o-Nitrobenzonitrile-^{14}C 770
Nitroethane, reaction with cyanogen 430
Nitrogen nucleophiles 400–404
Nitrogen oxides 428
Nitrogen ylid 917
Nitro group, size and shape 210
Nitrolic acids 805, 809, 820
Nitromethane, acidity 227
2-Nitronaphthalene 936
Nitroparaffins, primary, conversion to nitrile oxides 804–807
p-Nitrophenylcyanodiazomethane 551
p-Nitrophenylmalononitrile 590
1-Nitropropane 93
Nitrosobenzene 843
Nitroso compounds, aromatic, reaction with nitrile oxides 841
p-Nitroso-*N*,*N*-dimethylaniline 844
Nitrosomalononitrile 590
1-Nitroso-2-naphthol 871, 929
1-Nitroso-3-phenyl-2-pyrazoline 844
Nitrosotrifluoromethane 869
Nitrotetracyanocyclopentadienide, reduction of 601
o-Nitrotoluene, ultraviolet spectrum 210
Norbornadiene 453
Norvaline 763
Nuclear magnetic resonance spectra 139, 140, 171
^{19}F 202
for determination of amines 170

Nuclear magnetic resonance spectra—*cont.*
  for determination of nitriles 201, 202
  of isonitriles 854
  of nitrile coordination complexes 151
  shielding and deshielding effects of cyano group 233–235
Nucleophiles, oxygen 368, 369
  addition by 391–393
  sulphur, addition by 393–396
Nucleophilic addition 263–276
  and ring closure (Type C) 382–396
  intramolecular (Type C_i) 397–409
Nucleophilic substitution 230
Nudiflorine, isolation from *Trewia nudiflora* Linn 730

Octacyano-*p*,*p*,*p*-triphenylphosphacyclopentane 460
Octanal 756
Octanoic acid-1-^{13}C 766
2,4,6-Octatriene 452
O—H stretching frequency 157
Olefins, reaction, with nitrile oxides 833–836
  with nitriles 674–682
  with nitrilium ions 359–364
Oleic acid-8-^{14}C 766
Organoaluminium compounds 152, 282
Organoboranes 892
Organometallic compounds, reaction with cyanogen halides 88
'Orlon®', black 523
Ornithine-5-^{14}C 764
Ortho esters, formation 315
  reduction by lithium aluminium hydride 315
Osteolathyrogens 724
1,2,5-Oxaazaphospholines 842
Oxadiazoles 246
1,2,5-Oxadiazole 2-oxides 814
  formation 816–819
1,2,4-Oxadiazoles 839, 840
  formation from nitrile oxides 346
  4-oxides 817
1,3,4-Oxadiazoles 355
Oxaldiimidates 429
β-*N*-Oxalyl-*α*,*γ*-diaminobutyric acid 724
β-*N*-Oxalyl-2,3-diaminopropionic acid 724
Oxamide 430
Oxamidines 428
1,4,2-Oxathiazoles 838
1,2,3,5-Oxathiodiazoles 2-oxides 842
Oxazete ring system 912
Oxazetidines 932–935
Oxazetidinones 911, 932–935
Oxazine derivatives 366
1,3-Oxazines 366
Oxazoles 353, 932–936
  conversion to *ω*-cyano acid 935
  derivatives of 364–366
  from acyloins 935
  from azirine 932
Oxazoline 393
Oxazolone 368
Oxidation, of amino acids to nitriles 104
  of cyanoalkyl radicals 697
  of isonitriles 862
Oximes 720—*see also* specific compounds
  addition of nitrile oxides 839
  dehydration 92
  hydrogenation to acetylated amines 322, 323
2-Oximidochlorides 860
α-Oximinoacids 926
α-Oximinoamines, fragmentation of 917, 918
β-Oximinoethers 922
α- and *γ*-Oximinoketones 924, 925, 927
α-Oximinonitriles 827
β-Oximinoquinuclidine 917
Oximinosulphides, diaroyl 825
Oxoamines 330
Oxonitriles 330
Oxyacids, complexes with nitriles 245, 246
Oxygen nucleophiles 368, 369, 391–393, 405, 406
Oxynitrilase enzymes 721
  from *Prunus amygdalus* and *Sorghum vulgare* 721
Ozonides, addition to tetracyanoethylene 458

Palladium catalyst 321
  in nitrile reduction 320
Paracyanogen 289, 428
Paropsis atomaria 723
Passerini reaction 863, 864, 889
Pentacyanobenzene 530, 650
Pentacyanocyclopentadiene 617

Pentacyanocyclopentadienide 597
cyanogen chloride in formation 597
salt 228
Pentacyanoethane 441
Pentacyanoethanide 591, 592
1,1,2,3,3-Pentacyanopropene, formation 479
Pentacyanopropenide 233, 466, 593, 594, 612
Pentacyanopyridine 575, 650
Pentafluorobenzonitrile 215
Pentafluoronitrobenzene 215
2-Pentanone-1-^{14}C 782
4-Pentenonitrile 916
Peptides, fluorinated 870
stereoselective 866
synthesis of 248
Perfluoroacyl cyanides 588, 589
Perfluoroalkylamines 297
Perhydro-*s*-triazines 410
β-Pericyclocamphenone oxime 919
Peroxyimidic acids 263
α-Phellandrene, use as hydrogen donor 331
Phenanthrene, as complexing agent 650
reaction with tetracyanoethylene 463
Phenazomaleonitrile 735
Phenols 218, 459
addition to C≡N bond 263, 264
ortho-substituted 459, 463
Phenylacetaldehydes 312, 313
preparation from benzyl cyanides 311
Phenylacetonitrile-1-^{14}C 765
Phenylalanine 722, 763
conversion, to amygdalin 720
to prunasin 720
D,L-Phenylalanine-1-^{14}C 761
Phenyl cyanate 252
Phenylcyanodiazomethane 551
Phenylhydrazine, as trapping reagent in hydrogenation 311–313
Phenylhydrazones 311–313
carbonyl cyanide phenylhydrazone 735
Phenyl isocyanide 928
π-bridging in transition state 891
tetramerization to indigo dianil 871
3-Phenyl-Δ^2-isoxazoline 845
2-Phenyllactic acid-1-^{14}C 755
Phenylmagnesium bromide 862
3-Phenylpropanoic acid-1-^{14}C 766
Phenylpropiolamide 898
3-Phenylpropionitrile-1-^{14}C 766
Phenylthioureas, substituted, formation 170
Phenyltricyanoethylene 650, 655
oxide 565
Phenyl vinyl ketoxime 845
Phloroglucinol, reaction with nitriles 172
Phosgene method 857
6-Phosphogluconate dehydrogenase 735
Phosphoric acid esters, use for synthesis of aliphatic nitriles 85
Phosphorus pentachloride, conversion to *N*-chlorovinylphosphoroimidic trichlorides 253
Phosphorus ylids, reaction with tetracyanoethylene 469
Phosphorylation reactions 246
Photochlorination 672
Phthalimide-^{15}N 751
Phthalimidoacetonitrile-1-^{14}C, formation and hydrolysis 771
Phthalimidoalkyl bromides, formation by Hundsdiecker degradation 772
Phthalimido group 773
Phthalonitrile 185
ultraviolet spectrum 210
Pimelic acid-1,7-^{14}C 767
Pinacolone oxime 921
Pinner cleavage 264, 268
Pinner synthesis 264, 268
Pivalonitrile, reaction with hexamethyldialuminium 893
pK_a value 126
of nitriles 617
Platinum catalysts, in nitrile reduction 320
Polarization, dielectric 148
Polarographic reduction—*see* Reduction
Polyacrylonitrile 288
Polycyanoaromatics 230
Polycyanocarbons 190, 226
electrochemical reduction 190
ion radicals 229
Polycyanocyclopentadienes, acidity 228
Polycyanoolefins, electrophilic 228
Polycyanooxiranes 554–570
Polydesmus collaris collaris 722
Polyene nitriles 769
Polyfluoroalkyl nitriles, reduction by sodium borohydride 329

Polymerization 410, 411, 698, 890, 891
linear, of nitriles 287–290
of cyanogen 289
of fulminic acid 819–822
of hydrogen cyanide 289
of isonitriles 862
of vinylidene cyanide 699
suppression of 890
Polymers, from tetracyanoethylene 474
Polysaccharides, reaction with ^{14}C-cyanide 784
Potassium cyanide, exchange with labelled carbonate 749
reaction, with cyanogen 433
with dihalomalononitrile 433
Prebiotic synthesis 289
Proline-[(5-ring)-^{14}C] 764
1,3-Propanediamine 429
Propargyl cyanides 81
β-Propiolactone 764
Propionic acid, labelled 766
Propionitrile 724
vacuum ultraviolet spectrum 190
Propionitrile-1-^{14}C 772
Propylamine-1-^{14}C 781
Propylene 489
Prulaurin 719
Prunasin 720
biosynthesis from phenylalanine 720
formation from amygdalin 721
Prunlaurasin 719
Prunus amygdalus 721
Prunus laurocerasus 720
Pseudomonas bacteria 725, 728, 732
Pseudoureas, *o*-alkyl 265
Psychrophilic basidiomycete fungus 729
Purines 289
Pylaiella litoralis 734
Pyrazolines 285
Pyrazolone 861
Pyrene, as complexing agent 650, 653
Pyridine 770
trimeric 283
Pyridine aldehydes 310, 313
Pyridinium dicyanomethylid 558, 591
2-Pyridones 378, 730
4-Pyridones 730
Pyridoxal 724, 725
phosphate 725
5-(3-Pyridyl)tetrazole-^{14}C 783
Pyrimidines 273, 289
2,4-disubstituted 385
Pyrimidines—*cont.*
trimers 283
Pyrimidinethiones 394
Pyrolysis 315
of nitriles in presence of sulphur 173
Pyrrole 359, 861
2,5-disubstituted 386
reaction with tetracyanoethylene 463
Pyrroline derivatives 359, 360
1-Pyrroline, formation from cyclopropylimine 384
Pyruvonitrile 771

Quadricyclane, addition of tetracyanoethylene 456
Quaternary ammonium salt 87
Quinazoline derivatives 375, 376
Quinazoline synthesis 252, 253
Quinoline *N*-oxides 404
Quinolinium ions, *N*-alkyl- 864
Quinoxalines, 2,3-diamino- 429
Quinoxaline-2,3-dicarboxylic anhydride 905
Quinoxalyne 905

Radical process 889
Radical rearrangements 698
Radicals, acceptor 672, 700
α-addition to isonitriles 862
cyano 709
cyanoalkyl, oxidation 697
cyanobenzyl 682–685
dimers 659
donor 672, 700
Radziszewski reaction 262
Raman spectra, of isonitriles 854
of nitrile–halogen complexes 146
of nitriles 135
Rambaud's procedure 762
Raney cobalt catalyst, for nitrile hydrogenation 320, 321, 322, 325
Raney nickel catalyst, for nitrile hydrogenation 309–312, 320, 321, 323–325
Raney nickel-chromium catalyst 321
Rearrangements, alkyl, in diazoacetonitrile 894
electron-deficient 892–894
isonitrile-nitrile 886–892
of isonitriles 859
of nitrile oxides to isocyanates 815, 816
Reduction, by hydrides 314–318, 326–331

Reduction—*cont.*
chemical 313, 314, 325–331
for nitrile determination 175–177
constant current, of unsaturated nitriles 189
electrochemical, of cyano compounds 179–190, 297
of carbonates 746–749
of cyanohydrins, to β-hydroxy primary amines 326
of isocyanates 857
of isothiocyanates 857
of nitrile oxides to nitriles 822
of nitriles 170, 296, 307–337
to aldehydes 308–318, 333
to aldimines 319
to amines 170, 179, 182, 186, 319–331, 337
to hydrocarbons 325, 331–336
of polyfluoroalkyl nitriles, with sodium borohydride 329
partial, of nitriles to aldehydes 333
with lithium aluminium hydride 314
polarographic 181, 644–647
of 2- and 4-cyanopyridine 183
of nitriles 186
in non-aqueous solvents 183
α,β-unsaturated 186, 187, 189
with metals 325
Reductive cleavage 331
Reductive decyanation 333–336
of dehydroabietonitrile to $\Delta^{5,7,14(13)}$-abietatriene 335
Reformatsky reaction 107, 282
Reformatsky reagent 383
Reissert compounds 76
Reissert reaction 886
Resistivity 662
Resonance effect 158
Rhodanese 735
Rhodium-alumina catalyst 320
Rhodium catalyst, for nitrile hydrogenation 320, 321
Ribose-1-^{14}C 759
Ricinine 725, 730, 731, 732, 734
biosynthesis 730, 731
isolation from *Ricinus communis* L. 730
Pseudomonas bacteria grown on 732
Ricinine nitrilase 732, 734
Ricinus communis L. 730, 731
Ring closure 333–336, 337
Ring closure—*cont.*
Type B, following electrophilic addition 358–376
Type C, following nucleophilic addition 382–396
Ring size, effect on stability of cyanohydrins 72
Ritter reaction 291, 293, 375, 911, 912, 913–915
anomalous 922
electrochemical 915
with α-dialkylaminonitriles 922
Rosenmund-von Braun synthesis 82, 886

Sambunigrin 719
Sandmeyer nitrile synthesis 87, 775, 886
SCF LCAO-MO technique, multiconfiguration 9
Schmidt reaction 93
Sebacic acid-1,10-^{14}C 767
Self-association 125, 148, 149
Semicarbazide 311, 429
acetate 781
Semicarbazones 311
Semiconductors 662
Separation techniques 177–180
chromatography, gas 178
liquid 178
distillation 177, 178
extraction 178
Serine 725, 726, 728
conversion, to β-cyanoalanine 726
to cysteine 726
effect on snow mold fungus 728
D,L-Serine-1-^{14}C 760, 761
Sesquihydrohalides 245
Shielding, by cyano groups 202, 233–235
Shifts, blue 144
$\Delta\nu_{O-H}$ 137, 157
Silver cyanide 892
Silver(II) fluoride 432
Silver fulminate 799
Sitophilus granarius 726
Six-membered rings, formation from isonitriles and boranes 872
Snow mold fungus 727, 728
Sodamide, reaction with nitriles 273, 274
Sodium acetate, as cocatalyst 321
Sodium and alcohol, as reducing system 170
for nitrile reduction 336

Sodium borohydride reduction, of nitriles to amines 329
of polyfluoroalkyl nitriles 329
Sodium cyanide 471, 489, 490
Sodium cyanide-^{15}N 751
Sodium 3-(^{14}C-cyano)butyrate 752
Sodium formate 749
Sodium fulminate 800
Sodium hypophosphite reduction, of nitriles 309, 310
Sodium malononitrile 590
Sodium tricyanomethanide, formation from cyanogen chloride and malononitrile 433
Sodium triethoxyaluminohydride 315, 316, 318
Solid complexes 659
Solvation of metal ions by nitriles 131–134
Solvents, dipolar, aprotic 78, 83
dimethylformamide (DMF) 78
dimethylsulphoxide (DMSO) 78
hexamethylamide of phosphoric acid (HMPT) 79
in nitrile synthesis from saturated aliphatic halides 78
Sonn-Müller synthesis of aldehydes from amides 308
Sorghum vulgare, enzyme 721
Spot tests 172, 173
Statistical model 31
Stephen aldehyde synthesis 296, 308, 309
modification 319
Stephen reagent, modified form 309
Stereochemistry 888
Steric effects 158, 159
hindrance 158
Steroidal angular protons, shielding by cyano groups 202
Steroids, containing groups in angular positions 316
trans-4,4′-Stilbenedicarbonitrile-$^{14}C_2$ 770
trans-4,4′-Stilbenedicarboxamidine-$^{14}C_2$ 783
Strecker synthesis 75, 729, 758–761
Stretching frequency of C≡N bond 135, 194
of O—H 157
Structural effects, on nitrile reactivity 268
Substituent effects 889
empirical rules 193
Substituent effects—*cont.*
on infrared and ultraviolet spectra 231–233
on 19-methyl group of 5α-cholestane 202
on ultraviolet absorption of unsaturated nitriles 193
Substituent parameters 217
Succinic acid-1,4-^{13}C 767
Succinic acid-2,3-^{14}C 767
Succinic aldehydes, preparation 312
Succinic semialdehyde 729
Succinonitrile 724
formation from cyanogen and ethylene 433
Succinonitrile-1,4-^{14}C 772
Sugars, α-aminodeoxy 782
^{14}C 758
reducing, determination by cyanohydrin reaction 784
Sulphates, alkyl 772
N-Sulphinylamines 842
Sulphonates, aryl 84
conversion to nitriles 84
Sulphonium dicyanomethylids 560
Sulphur dichloride, addition to cyanogen 433
Sulphur, in pyrolysis of nitriles 173
Sulphuric acid esters, use for synthesis of aliphatic nitriles 85
Sulphur monochloride, in preparation of tetracyanoethylene 446
Sulphur nucleophiles 407–409
addition by 393–396
Sulphurous acid, addition to tetracyanoethylene 458
Sulphur trioxide, addition to cyanogen 432
reaction with nitriles 246
Sulphuryl chloride, chlorinations with 673
Sulphydryl compounds, addition to nitriles 274–276
'Super acid' 912
Surface adsorption 189

Taft σ*-constants 157
Taft σ° inductive effect constants 278
Taft σ*-plot 158
Talose-1-^{14}C 759
T3 bacteriophage 735
Teratogenic effects 723
Terpenes 331
dehydrogenation of 331
hydrogenation of 331

1,2,3,5-Tetracyanobenzene 527
1,2,4,5-Tetracyanobenzene 527, 529, 650, 653, 654, 656
Tetracyano-*p*-benzoquinone 545, 546, 648, 650
Tetracyanobutanedionediide 596
1,2,3,4-Tetracyanocyclopentadienide 234, 597–602
electronic character 226
salt 228
1,1,2,2-Tetracyanocyclopropanes 444, 472
formation by the Widequist reaction 442
Tetracyano-1,4-dithiin 503, 505
conversion to tetracyanoethylene 446
conversion to tetracyanothiophene 577
synthesis from disodium dimercapto-maleonitrile 576
1,1,2,2-Tetracyanoethane 439, 472, 899
conversion to 2,5-diamino-3,4-dicyanothiophene 439
enimine 899
Tetracyanoethanide 592
Tetracyanoethylene (TCNE) 108, 211, 229, 428, 446–474, 505, 578 640, 646, 648, 650, 653, 655, 656 660, 664, 736, 897
additions to the C=C double bond 449–460
of alcohols 467
of aromatic amines 459
of carbenes 457
of chlorine 458
of cyanide ion 459
of diazomethane 458
of 1,3-dienes 449–453
of hydrogen 458
of hydrogen peroxide 458
of ketones 459
of ozonides 458
of phenols 459
of sulphurous acid 458, 459
of triphenylphosphine 460
cracking, to cyanogen and dicyano-acetylene 470
cycloaddition to cyclobutanes 453–456
energy levels 643
in formation of dicyanoacetylene 514
in formation of 1,1,2,5,6,6-hexa-cyano-3,4-diazahexadienediide 608
Tetracyanoethylene—*cont.*
oxidation of 555
physical properties 447, 448
polymers 474
reactions with azide 460, 461
with benzene 521
with benzonitrile oxide 462
with ethyl diazoacetate 462
with fluorosulphonic acid 461
with isatogens 470
with methanol 461, 462
with olefins 556
with ozone 556
with trifluoromethanesulphenyl chloride 462
with water 461, 462
replacement of cyano groups 463–469
by alcohols 467, 468
by amines 468, 469
by ammonia 468, 469
by aromatic compounds 463–466
by hydrazines 468, 469
by phosphorus ylids 469
by water 466, 467
synthesis 446–447
from bis(acetoxyiminomethyl)-methylene malononitrile 447
from dichlorofumaronitrile 446
from 1,2-dicyano-1,2-di-*p*-tolyl-sulphonyl ethylene 447
from malononitrile 446
from tetracyano-1,4-dithiin 446
Tetracyanoethylene anion radical 229, 468, 471–474, 578, 658
Tetracyanoethylene oxide (TCNEO) 108, 556
addition, to acetylenes 567
to aromatic compounds 565–567
to carbon-heteroatom bonds 568
to olefins 563–565
hydrolysis of 570
nucleophilic attack on oxygen of 569, 570
reactions, with acid anhydrides and halides 568
with amines 557–559
with carbon-carbon unsaturation 568, 569
with iodide ion 570
with miscellaneous nucleophiles 562, 563
with sulphides 559–561
with thiocarbonyl compounds 561, 562

Tetracyanofuran 570, 571
Tetracyanohydroquinone 546
2,3,4,5-Tetracyano-*N*-methyl-1,2-dihydropyridine 574
11,11,12,12-Tetracyanonaphtho-2,6-quinodimethane 533, 646
Tetracyanopropenide 612
1,1,2,3-Tetracyanopropenides 595, 596
1,1,3,3-Tetracyanopropenides 594, 595
2,3,4,5-Tetracyanopyridine 575, 650
2,3,5,6-Tetracyanopyridine 575, 650
2,3,4,5-Tetracyanopyrrole 571, 572, 579
Tetracyanoquinodimethanes 224, 531–540, 608, 613, 640, 650, 655
  addition reactions 536, 537
  alkyl derivatives 532
  displacement reactions 538–540
  energy levels 643
  irradiation of 537
  reduction and anion-radical formation 533–536
7,7,8,8-Tetracyanoquinodimethane (TCNQ) 646, 648, 660, 664, 895
  anion radical (TCNQ$\bar{\cdot}$) 658, 895
  salts of 666
Tetracyanothiophene 577, 579, 580
Tetradecanedioic acid-1,14-^{14}C 767
Tetrafluorobenzonitrile 215
3,3,4,4-Tetrafluoro-Δ^1-1,2-diazetine 432
Tetrafluoronitrobenzene 215
Tetrahydrofuran, addition to tetracyanoethylene 457
Tetrakis(triphenylphosphine) palladium 460
Tetralone oximes, possessing a β-methylthio group 923
Tetramerization of phenyl isocyanide 871
cis- and *trans*-2,2,4,4-Tetramethylcyclobutane-1,3-dicyanide 889
cis- and *trans*-2,2,4,4-Tetramethylcyclobutane-1,3-diisocyanide 888
1,3,5,7-Tetramethylenecyclooctane 453
N,N,N',N'-Tetramethyl-*p*-phenylenediamine 473, 653
Tetramethylsuccinonitrile 335
Tetraphenylsesquifulvalene, addition of tetracyanoethylene 450
Tetraphenylsuccinonitrile, isomerization 682
Tetrazines 391
Tetrazole derivatives 351, 864
  synthesis 863
Thermal decomposition of *N*-alkyl *N*-formylamides 889
Thermal rearrangement 890
Thermodynamic data, for BX_3-acetonitrile complexes 151
1,3,4-Thiaoxazoline 827
Thiazine derivatives 372, 373
Thiazole derivatives 369–372
2-Thiazolines 396
cis- and *trans*-Thiazolines, formation from episulphides 371
2H, 3H-Thieno-[3,2-b-pyrrol-3-one], reaction with tetracyanoethylene 469
N-Thioacylamidines 276
Thioamides 274
  conversion to nitriles 101
  reaction with nitriles 276
Thiocarbanilide 776
Thiocyanates 252, 735, 736, 866
Thiocyanic acid 871, 872
Thiocyanoformamide 430
Thioglycolic acid 171
Thiohydroximic acids 825
  alkyl 828
Thioimidates 276
Thioketene 356
Thiolophosphoric acids 860
Thiols, reaction with cyanogen 429, 430
Thiophenes, from dicyanoketene thioacetals 498
Thiosulphuric acid, as acid component in isonitrile condensation 866
Thiourea 275
Thorpe-Ziegler reaction 284
Thyroxine 724
o-Tolualdehyde 311
Toluene, as complexing agent 650
  reaction with tetracyanoethylene 463
p-Toluenesulphonates, alkyl 772
m-Toluonitrile, oxidation to *m*-cyanobenzoic acid 737
o-Toluonitrile 309, 737
p-Toluonitrile 887
  oxidation to *p*-cyanobenzoic acids 737

p-Tolyl isocyanide 887
Tosylates 772
Transition metal, complexes 892
compounds, as catalysts 861
Trewia nudiflora Linn 730
1,3,5-Triacylhexahydro-*s*-triazines 295
Trialkylalanes, reaction with isonitriles 872
Trialkoxy salt formation from alcohols and tetracyanoethylene 467
Trialkylplumbyl amines 861
4,5,6-Triaminopyrimidine-4,6-^{14}C 783
2,4,6-Triamino-1,3,5-tricyanobenzene 526
s-Triazines 246, 410
formation by acid-catalysed nitrile trimerization 269
1,3,5-Triazine derivatives 390
Triazoles 246
1,2,3-Triazoles 348
1,2,4-Triazoles, formation from nitrile imines 348
Tri-n-butyl borane 870
Trichloroacetonitrile 201, 268, 270
2,4,6-Trichloro-1,3,5-tricyanobenzene 526
Tricyanamide 547
2,3,3-Tricyanoacrylamide, formation from tetracyanoethylene 461, 462
Tricyanobenzenes 524–527
1,2,2-Tricyano-1,3-butadienes 480
1,2,3-Tricyanocyclopentadienides 234, 616
1,2,4-Tricyanocyclopentadienides 234, 616
cis- and *trans*-Tricyanocyclopropane 442
mass spectra 200
1,2,3-Tricyanocyclopropane-1,2,3-tricarboxylate, from ethyl bromocyanoacetate 442
Tricyano-1,4-dithiino[c]isothiazole 578
Tricyanoethanide 592, 593
Tricyanoethylene 474, 475, 650, 655
preparation from ethyl cyanoacetate and glycolonitrile 474
2,4,6-Tricyanoheptane, stereoisomers, chemical shifts 202
coupling constants 202
2,4,6-Tricyanomesitylene 83
Tricyanomethane 433–437, 613, 899
additions to 435–437
alkyl 437, 438
Tricyanomethane—*cont.*
aryl 437, 438
m- and *p*-fluorophenyl- 219
Tricyanomethyl group 223
Tricyanoperylene 531
Tricyanophenol 526
Tricyanophenolate ion 526
3,4,5-Tricyanopyrazole 572
3,4,5-Tricyanopyridine derivatives 574
2,4,6-Tricyanopyridine *N*-oxide 575
3,4,5-Tricyanopyridinium perchlorate 650
α,α,α-Tricyanotoluene, *m*-and *p*-fluoro- 219
Tricyano-*s*-triazine 576
1,3,5-Tricyano-2,4,6-trifluorobenzene 526
Tricyanovinyl alcohol 466, 479, 612
Tricyanovinyl alcoholate 613
Tricyanovinylalkanes 475–478
Tricyanovinylarenes 475–478
reaction with malononitrile 476
Tricyanovinylation 466
Tricyanovinylbenzene 613
hydrolysis 478
reaction with nickel carbonyl 477
Tricyanovinyl chloride 466, 480
Tricyanovinyl ethers, alkyl 467
Tricyanovinyl-*m*- and *p*-fluorobenzene 219
α,β,β-Tricyanovinyl group 202, 223, 224
p-(Tricyanovinyl)phenyldicyanomethane 611
p-(Tricyanovinyl)phenyldicyanomethanide 613
Tricycloquinazoline-^{14}C 784
Triethylaluminium, reaction with nitriles 893
Triethylamine 429
Trifluoroacetaldehyde *N*-acylimines 870
Trifluoroacetonitrile 268, 270
Trifluoromethanesulphenyl chloride, reaction with tetracyanoethylene 462
p-(Trifluoromethoxy)phenazomaleonitrile 735
p-(Trifluoromethyl)aniline 215
Trifluoromethylcyanodiazomethane 551
Trifluoromethyldicyanomethane 438
Trifluoromethyl group, size and shape 210

Trifulmin 820
Trihalomethane sulphenyl chlorides 860
Triiodothyronine 724
3,4,5-Trimethoxymandelic acid-[carboxyl-^{14}C] 755
Trimethoxymethylmalononitrile 590
2,4,6-Trimethylaniline, reaction with tetracyanoethylene 468
Trimethylsilyl cyanide 892
Trimethylsilyl isocyanide 892
Triphenylacetonitrile 750
Triphenylphosphine, addition to tetracyanoethylene 460
Triphenylphosphonium dicyanomethylid 591
Triphenylphosphonium fluoroborate 861
Tris(2-cyanoethyl)amine 724
Tropylium iodide 896
Tropylium ion 864
Tryptamine 321, 325
Tryptophan 732, 735
Tungsten, hexacarbonyl-, reaction with cyanogen 433
Tyrosine 719–722, 763

UDP-glucose, in formation of linamarin 720
Ugi reaction 865
Ultraviolet spectra 190–194, 232, 903
 in vacuum, of acetonitrile 190
  of propionitrile 190
 of *o*-cyanobenzenes 210
 of nitrile oxides 812
 of nitriles, saturated, aliphatic 190
  unsaturated, rules for substituent effects 193
 of *o*-nitrobenzenes 210
 study indicating donor—acceptor complexes 191

Valence bond approximation 2
n-Valeronitrile, reduction by lithium aluminium hydride 327
Valine 722, 728, 763
 conversion to isobutyraldoxime 720
D,L-Valine-1-^{14}C 760, 761
Veratronitrile-[nitrile-^{14}C] 769
Vicia 725
 sativa 724, 725, 726
Vicianin 719
Vinylacetonitrile, constant current reduction 189
Vinylacrylonitrile, polarographic wave 189
Vinyl amines 430
Vinylidene cyanide, polymerization 699
Vinylisocyanide 54
2-Vinylpyridine 356
Virial coefficient, second 148
Viscosity 148
Vitamin A aldehyde 310
Vitamin B_{12} 734, 735
von Braun reaction 97, 320, 324
Von Richter reaction 936–939
 of *p*-chloronitrobenzene 938

Wave function, for excited state 143
 of ground state 143
Whitmore procedure 320
Wideqvist reaction 442
Wittig reaction, for synthesis of monocyanoethylenes 107
Wohl degradation 886, 910

Xanthocillin 877
X-ray analysis 661
X-ray diffraction patterns, for determination of amines 170
 of salts 169
Xylose-1-^{14}C 759

Ynamines 894–901
 from enamines 899
 primary 898
 structure 894
 tertiary 899

Zinc chloride 398
Zucker–Hammett hypothesis 259
Zygaena filipendulae 723
Zygaena Ionicerae 723